Neuroanatomy through Clinical Cases

NEUROANATOMY
through
Clinical Cases
SECOND EDITION

HAL BLUMENFELD, M.D., Ph.D.
Yale University School of Medicine

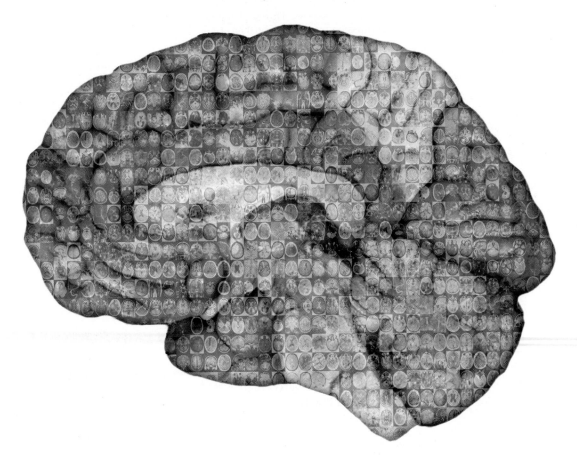

Sinauer Associates, Inc. Publishers
Sunderland, Massachusetts

Note

The author and the publisher have made every effort to provide clinical information in this book that is up-to-date and accurate at the time of publication. However, diagnostic and therapeutic methods evolve continuously based on new research and clinical experience. Because medical standards are constantly changing, and because of the possibility of human error, neither the author, nor the publisher, nor any other party who has been involved in the preparation or publication of this work warrants that the information contained herein is in every respect accurate or complete, and they disclaim all responsibility for any errors or omissions, or for clinical results obtained from use of the information contained in this work. Readers are strongly encouraged to consult other sources to confirm all clinical information when caring for patients, particularly in regard to medications, doses, and contraindications, which are subject to frequent changes and improvements, or when using new or infrequently used drugs.

The Cover

Base art for cover brain image modified with permission from the University of Washington Digital Anatomist Project.

Neuroanatomy through Clinical Cases, Second Edition

Copyright © 2010 by Sinauer Associates, Inc.

For information, address Sinauer Associates, P.O. Box 407, Sunderland, MA 01375 U.S.A.

www.sinauer.com

FAX: 413-549-1118

Email: publish@sinauer.com

Library of Congress Cataloging-in-Publication Data

Blumenfeld, Hal.

Neuroanatomy through clinical cases / Hal Blumenfeld. — 2nd ed.

    p. ; cm.

    Includes bibliographical references and indexes.

    ISBN 978-0-87893-613-7 (alk. paper)

    1. Neuroanatomy. 2. Neurologic examination—Case studies. I. Title.

    [DNLM: 1. Nervous System Diseases—diagnosis—Case Reports. 2. Nervous System—pathology—Case Reports. WL 141 B658n 2010]

    QM451.B68 2010

    616.8'0475—dc22

                  2010001339

Printed in U.S.A.

7 6 5

To Michelle
And I think to myself…what a wonderful world.

Brief Contents

Contents

CHAPTER 9 *Major Plexuses and Peripheral Nerves 357*

CHAPTER 10 *Cerebral Hemispheres and Vascular Supply 391*

CHAPTER *11* *Visual System* 459

CHAPTER *12* *Brainstem I: Surface Anatomy and Cranial Nerves* 493

CHAPTER *13* *Brainstem II: Eye Movements and Pupillary Control 565*

CHAPTER *14* *Brainstem III: Internal Structures and Vascular Supply 613*

CHAPTER *18* *Limbic System: Homeostasis, Olfaction, Memory, and Emotion* 819

CHAPTER *19* *Higher-Order Cerebral Function* 879

Preface

Neuroanatomy is a living, dynamic field that can bring both intellectual delight and aesthetic pleasure to students at all levels. However, by nature, it is also an exceedingly detailed subject, and herein lies the tragic pitfall of all too many neuroanatomy courses. Crushing amounts of memorization are often required of students of neuroanatomy, leaving them little time to step back and gain an appreciation of the structural and functional beauty of the nervous system and its relevance to clinical practice.

This book has a different point of view: instead of making the mastery of anatomical details the main goal and then searching for applications of this knowledge, actual clinical cases are used as both a teaching instrument and motivating force to encourage students to delve into further study of normal anatomy and function. Through this approach, structural details take on immediate relevance as they are being learned. In addition, each clinical case is an ideal way to integrate knowledge of disparate functional systems, since a single lesion may affect several different neural structures and pathways.

Over 100 clinical cases, accompanied by neuroradiological images, are presented in this text, and I am grateful to many neurologists, neurosurgeons, and neuroradiologists at the Columbia, Harvard, and Yale medical schools for helping me to amass enough material to present clinically relevant discussions of the entire nervous system. I have used this book's diagnostic method to teach neuroanatomy at these medical schools, and both students and faculty greeted the innovation enthusiastically. Through publication of *Neuroanatomy through Clinical Cases* I hope that students and faculty at many additional institutions will find this to be an enjoyable and effective way to learn neuroanatomy and its real-life applications.

Acknowledgments for First Edition

First and foremost, I must thank my wife Michelle, and our children Eva and Jesse, for their enthusiasm and support throughout the writing and publication of this book.

This project has spanned a number of years, and stints at several academic centers, so there is a formidable list of people who I must thank for their important contributions. This book was conceived while I was teaching neuroanatomy as an M.D., Ph.D. student at Columbia Medical School, where I was inspired by my teachers Eric Kandel, Jack Martin, and Steven Siegelbaum. They have remained invaluable sources of inspiration and advice ever since. I would also like to thank the following individuals who served as mentors, benefactors, or role models during my training as a neurologist and neuroscientist: Raymond D. Adams, Bernard Cohen, C. Miller Fisher, Jack Haimovic, Walter Koroshetz, Terry Krulwich, Elan Louis, Stephan Mayer, David McCormick, Thomas McMahon, Timothy Pedley, Pasko Rakic, Susan Spencer, Dennis Spencer, Stephen Waxman, Anne Young, and George Zubal. I would also like to offer special thanks to those who were my closest colleagues and friends during my neurology residency: Jang-Ho Cha, Mitchell Elkind, Martha Herbert, David Jacoby, Michael Lin, Guy Rordorff, Diana Rosas, and Gerald So.

The focus and main strength of this book is its clinical cases. Therefore, I am very grateful to the many colleagues who suggested the clinical cases used in this book: Robert Ackerman, Claudia Baldassano, Tracy Batchelor, Flint Beal, Carsten Bonneman, Lawrence Borges, Robert Brown, Jeffrey Bruce, Brad Buchbinder, Ferdinando Buonanno, William Butler, Steve Cannon, David Caplan, Robert Carter, Verne Caviness, Jang-Ho Cha, Paul Chapman, Chinfei Chen, Keith Chiappa, In Sup Choi, Andrew Cole, Douglas Cole, G. Rees Cosgrove, Steven Cramer, Didier Cros, Merit Cudkowicz, Kenneth Davis, Rajiv Desai, Elizabeth Dooling, Brad Duckrow, Mitchell Elkind, Emad Eskandar, Stephen Fink, Seth Finkelstein, Alice Flaherty, Robert Friedlander, David Frim, Zoher Ghogawala, Michael Goldrich, Jonathan Goldstein, R. Gilberto Gonzalez, Kimberly Goslin, Steven Greenberg, John Growdon, Andrea Halliday, E. Tessa Hedley-Whyte, Martha Herbert, Daniel Hoch, Fred Hochberg, J. Maurice Hourihane, Brad Hyman, Michael Irizarry, David Jacoby, William Johnson, Raymond Kelleher, Philip Kistler, Walter Koroshetz, Sandra Kostyk, Kalpathy Krishnamoorthy, James Lehrich, Simmons Lessell, Michael Lev, Susan Levy, Michael Lin, Elan Louis, David Louis, Jean Lud-Cadet, David Margolin, Richard Mattson, Stephan Mayer, James Miller, Shawn Murphy, Brad Navia, Steven Novella, Edward Novotny, Christopher Ogilvy, Robert Ojemann, Michael Panzara, Dante Pappano, Stephen Parker, Marie Pasinski, John Penney, Bruce Price, Peter Riskind, Guy Rordorff, Diana Rosas, Tally Sagie, Pamela Schaefer, Jeremy Schmahmann, Lee Schwamm, Michael Schwarzschild, Saad Shafqat, Barbara Shapiro, Aneesh Singhal, Michael Sisti, Gerald So, Robert Solomon, Marcio Sotero, Dennis Spencer, Susan Spencer, John Stakes, Marion Stein, Divya Subramanian-Khurana, Brooke Swearingen, Max Takeoka, Thomas Tatemichi, Fran Testa, James Thompson, Mark Tramo, Jean Paul Vonsattel, Shirley Wray, Anne Young, and Nicholas Zervas.

I am deeply indebted to the many individuals who provided critical reviews of one or more chapters, greatly enhancing the accuracy and clarity of the material in this book: Raymond D. Adams, Joshua Auerbach, William W. Blessing, Laura Blumenfeld, William Boonn, Lawrence Borges, Michelle Brody, Richard Bronen, Joshua Brumberg, Thomas N. Byrne, Mark Cabelin, Jang-Ho Cha, Jaehyuk Choi, Charles Conrad, Rees Cosgrove, Merit Cudkowicz, Mitchell Elkind, C. M. Fisher, David Frim, Darren R. Gitelman, Jonathan Goldstein, Gil Gonzalez, Charles Greer, Stephan Heckers, Tamas Horvath, Gregory Huth, Michael Irizarry, Joshua P. Klein, Igor Koralnick, John Krakauer, Matthew Kroh, Robert H. LaMotte, John Langfitt, Steven B. Leder, Elliot Lerner, Grant Liu, Andres Martin, John H. Martin, Ian McDonald, Lyle Mitzner, Hrachya Nersesyan, Andrew Norden, Robert Ojemann, Stephen Parker, Huned Patwa, Howard Pomeranz, Bruce Price, Anna Roe, David Ross, Jeremy Schmahmann, Mark Schwartz, Ted Schwartz, Michael Schwarzschild, Barbara Shapiro, Scott Small, Arien Smith, Adam Sorscher, Susan Spencer, Stephen M. Strittmatter, Larry Squire, Mircea Steriade, Ethan Taub, Timothy Vollmer, and Steven U. Walkley. I express my gratitude for their helpful suggestions, but accept full responsibility for any errors in this text.

Marty Wonsiewicz, John Dolan, Greg Huth, John Butler, and Amanda Suver were helpful in the early stages of editorial development of this book. Michael Schlosser and Tasha Tanhehco helped gather the references, and Jason Freeman and Susan Vanderhill helped obtain copyright permissions. Wendy Beck and BlackSheep Marketing designed and implemented the neuroexam.com website. The video segments for neuroexam.com and *The NeuroExam Video* were filmed by Douglas Forbush and Patrick Leone at Yale, and edited by Evan Jones of RBY Video. Milena Pavlova provided helpful suggestions, and played the role of the patient.

Finally, I thank the entire staff at Sinauer Associates for their tremendously helpful collaboration in all stages of producing this book. I have enjoyed working with, and am especially grateful to, Andrew D. Sinauer, Peter Farley, Kerry Falvey, Christopher Small, and Jefferson Johnson, but I extend my deep appreciation to all other members of the Sinauer staff as well. It is a pleasure to work with people who truly care about creating a fine book.

Additional Acknowledgments for Second Edition

My family again comes first in my acknowledgments, as they stood closest by me through the long process of revising and updating this book. I thank Michelle for her advice and support, and our children Eva, Jesse, and Lev for their enthusiasm and for always bringing a smile to my face. Also, none of this would have been possible without my parents who continue to be a source of inspiration. My sister, "the real writer in the family," and many other family members and close, lifelong friends complete the list of those most precious.

In addition to those listed in the Acknowledgments for the First Edition, I would also like to thank the following outstanding colleagues for their suggested cases or critical chapter reviews: Nazem Atassi, Joachim Baehring, Margaret Bia, William Blessing, Richard Bronen, Franklin Brown, Joshua Brumberg, Gordon Buchanan, Ketan Bulsara, Louis Caplan, Michael Carrithers, Jang-Ho Cha, Michael Crair, Merit Cudkowicz, Robin De Graaf, Daniel DiCapua, Mitchell Elkind, Carl Faingold, Susan Forster, Robert Fulbright, Karen Furie, Glenn Giesler, Darren Gitelman, Charles Greer, Stephen Grill, Noam Harel, Joshua Hasbani, Elizabeth Holt, Bahman Jabbari, Jason Klenoff, Igor Koralnick, Randy Kulesza, Robert LaMotte, Steven Leder, Ben Legesse, Robert Lesser, Albert Lo, Grant Lui, Steve Mackey, Andres Martin, Graeme Mason, Andrew Norden, Haakon Nygaard, Kyeong Han Park, Stephen Parker, Huned Patwa, Howard Pomeranz, Stephane Poulin, Sashank Prasad, Bruce Price, Diana Richardson, George Richerson, Anna Roe, David Russell, Robert Sachdev, Gerard Sanacora, Joseph Schindler, Michael Schwartz, Theodore Schwartz, Alan Segal, Nutan Sharma, Gordon Shepherd, Scott Small, Adam Sorscher, Joshua Steinerman, Daryl Story, Ethan Taub, Kenneth Vives, Darren Volpe, Jonathan Waitman, Howard Weiner, Norman Werdiger, Michael Westerveld, and Shirley Wray.

Medical students contributed in an important way to this edition by helping me find new cases and images. Wenya Linda Bi, Alexander Park, April Levin, Matthew Vestal, Kathryn Giblin, Alexandra Miller, Joshua Motelow, and Amy Forrestel spent many early morning hours reviewing case materials for this book. Dragonfly Media Group contributed to art revisions, Picture Mosaics created the cover mosaic, Jean Zimmer provided copy editing, and Nathan Danielson helped draft the concept for the cover design.

Once again, I am very grateful to the entire staff of Sinauer Associates for their outstanding attention to high-quality publishing, and for their collaboration in all aspects of producing this book. I have enjoyed working on the Second Edition with Sydney Carroll, Graig Donini, Joan Gemme, Christopher Small, Jason Dirks, Linda Vandendolder, Marie Scavotto, Dean Scudder, and Andrew D. Sinauer. Having worked on two editions with Sinauer, I have an ever deepening appreciation of the success of this group in producing excellent books.

How to Use This Book

The goal of this book is to provide a treatment of neuroanatomy that is comprehensive, yet enables students to focus on the most important "take-home messages" for each topic. This goal is motivated by the recognition that, while access to detailed information is often useful in mastering neuroanatomy, certain selected pieces of information carry the most clinical relevance, or are most important for exam review.

General Outline

The first four chapters of the book contain introductory material that will be especially useful to students who have little previous clinical background. Chapter 1 is an introduction to the standard format commonly used for presenting clinical cases, including an outline of the medical history, physical examination, neuroanatomical localization, and differential diagnosis. Chapter 2 is a brief overview of neuroanatomy which includes definitions and descriptions of basic structures that will be studied in greater detail in later chapters. Chapter 3 builds on this knowledge by describing the neurologic examination. It includes a summary of the structures and pathways tested in each part of the exam, which is essential for localizing the lesions presented in the clinical cases throughout the remainder of the book. Much of the material in this chapter is also covered on the neuroexam.com website described below, which provides video demonstrations for each part of the exam. For readers who are unfamiliar with neuroimaging techniques, Chapter 4 contains a concise introduction to CT, MRI, and other imaging methods. This chapter also includes a Neuroradiological Atlas showing normal CT, MRI, and angiographic images of the brain. Chapters 5–19 cover the major neuroanatomical systems and present relevant clinical cases.

Chapters 5–19

Chapters 5–19 have a common structure. An "Anatomical and Clinical Review" at the beginning of the chapters presents relevant neuroanatomical structures and pathways, and generously sized, carefully labeled color illustrations are used to vividly depict spatial relationships. The first part of each chapter also includes numbered sections called "Key Clinical Concept," or "KCC," which cover common disorders of the system being discussed.

CLINICAL CASES The second part of each chapter is a "Clinical Cases" section that describes patients seen by the author and colleagues, each presented in a numbered color box. Full-length cases include complete findings from the neurologic examination, while "Minicases" have a briefer format. Each case begins with a narrative of how the patient's symptoms developed and what deficits were found on neurologic examination. For example, one patient in Chapter 10 suddenly developed weakness in the right hand and lost the ability to speak. Another, in Chapter 14, experienced double vision and lapsed into a coma. Important symptoms and signs are indicated in boldface type. The reader is then challenged through a series of questions to deduce the neuroanatomical location of the patient's lesion and the eventual diagnosis.

A discussion follows each case, beginning with a summary of the key symptoms and signs. Answers to the questions are provided which refer to anatomical and clinical material presented in the first half of the chapter that is demonstrated by the case. Continual improvements in imaging technology have allowed us to make clear and detailed radiographs of the nervous system *in vivo*, and one of the most exciting features of the book is the inclusion of large-format, labeled CT, MRI, or other scans that show the lesion for each patient, and serve as a central tool for teaching neuroanatomy. These images reveal, with striking clarity, both the lesion's location and the anatomy of the system being studied. In addition, these radiographs help the reader develop skill in interpreting the kinds of diagnostic images employed on the wards. The neuroimaging studies for each case are provided in special boxes at least one page turn away from the case questions, so the answers to the questions are not "given away" by the imaging (see below).

The clinical course is also provided for each patient, and includes a discussion of how the patient was managed, and what outcome followed. Thus, by the end of each case, students learn the relevant material by application and diagnostic sleuthing rather than by rote memorization.

Special Features for Focused Study and Review

Since one of the goals of this book is to enable students to either read the material in depth, or to distill it down to the most clinically relevant points or to material most commonly covered on the national boards or other examinations, several special features have been included to expedite focused study and review:

- **Boldface type** is used rather differently than in most texts. In addition to identifying the text for all important topics and definitions, boldface is also used to facilitate rapid or focused reading.

- **Review Exercises** appear in the margins throughout the text, highlighting the most important anatomical concepts in each chapter, and providing practice exam questions.

- Helpful **mnemonics** are provided throughout the text, and these are flagged in the margins by a special icon (shown at right) showing a section of the hippocampus (a structure important in memory formation).

- A **Brief Anatomical Study Guide** appears at the end of each chapter, which summarizes the most important neuroanatomical material, and refers to the appropriate figures and tables needed for focused exam review.

- The **Neuroradiological Atlas** in Chapter 4 also provides a useful review of neuroanatomical structures in three-dimensional space, and can be used for reference and comparison to lesions seen in clinical cases.

REVIEW EXERCISE

MNEMONIC

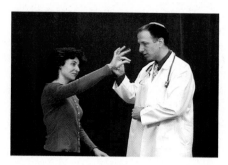

Rapid hand movements

- The **neuroexam.com** website includes much of the text from Chapter 3 describing the neurologic exam and its anatomical interpretation, and also features video demonstrations of each part of the exam that are cited in the text (e.g., "see **neuroexam.com Video 37**"). Selected video frames are also shown in the book margins, as shown at right, to illustrate relevant portions of the neurologic exam. Students or instructors who prefer to view a full-length video of the neurologic exam can purchase *The NeuroExam Video* from Sinauer Associates (www.sinauer.com).

- The **Key Clinical Concept** (**KCC**) sections provide a comprehensive introduction to clinical topics in neurology and neurosurgery, and enable an efficient review of these topics.

- Finally, the **Clinical Cases** can be used by themselves for study and review, since they consist of anatomical puzzles that reinforce the subject matter for each chapter in the most clinically relevant context. As noted above, the neuroimaging studies for each case are deliberately placed at least one page turn away from the case questions; the location of the images for each case are indicated by page numbers provided immediately after the images are cited in the text.

- The **Additional Cases** section at the end of each chapter, and the **Case Index** at the end of the book provide further cases relevant to the topics in each chapter.

Suggested Course Use

Neuroanatomy through Clinical Cases is intended primarily for first- or second-year medical students enrolled in a course in neuroanatomy or neuroscience, but it is a versatile text that could be used in many settings.

The topics covered in the book include all neuroanatomical material required for the medical school board examinations. Although fundamental concepts are emphasized, some advanced subject matter is also provided. Because the book includes chapters on peripheral nerves, students will also find this book useful in their general gross anatomy course in which peripheral nerves are usually covered. The Key Clinical Concept sections in this book also cover the major neurologic and neurosurgical disorders at a level appropriate for medical school pathophysiology courses, clinical rotations, and residents early in their training.

Students of other health professions, especially physical therapy, occupational therapy, nursing, dentistry, speech therapy, and neuropsychology will find this textbook useful as well, and it may also be of interest to graduate students of neuroscience. In addition to those learning neuroanatomy, it is hoped that the cases in this book will serve as a resource for advanced medical students in their clinical rotations, and residents in neurology, neurosurgery, and neuroradiology seeking examples of "typical" cases of neurologic disorders. Because each case is a real patient, the clinical cases in this book are, in effect, a collection of case reports that can serve as a useful resource, especially for teaching purposes and board review. It should be noted, however, that the cases presented here are highly selected for their teaching value and do not constitute an unbiased sampling of the kinds of cases found in clinical practice.

Here are some suggestions for using *Neuroanatomy through Clinical Cases* in various courses and curricula:

- For a comprehensive course in **medical school neuroanatomy**, students should read Chapters 2 and 5–18, with selected topics from Chapters 1, 3, 4 and 19. Reading assignments and large class lectures could focus on the Anatomical and Clinical Review sections at the beginning of each chapter. The clinical cases are most effectively discussed in small groups of students, where instructors can help students puzzle through the anatomical localization and diagnosis, and then discuss the neuroradiology and clinical outcome. An **Instructor's Resource Library** is available which contains material that will be useful for lectures, and **additional clinical cases** not found in the book that are ideal for use in small group teaching.

- For medical school courses covering neuroanatomy and other topics in **neuroscience**, additional readings from neuroscience texts such as *Neuroscience* by Purves et al. (2008, Sinauer Associates) or *Principles of Neural Science* by Kandel et al. (2000, McGraw-Hill) should be provided.

- For a comprehensive course in **clinical disorders of the nervous system**, students should read Chapters 3 and 4, and the Key Clinical Concept sections in Chapters 5–19. *The NeuroExam Video* should be viewed in class, and students referred to neuroexam.com for review. Clinical cases could then be presented in small groups, as described above.

- For a course focusing on **neuropsychological disorders** and anatomical correlations, students should read Chapters 2, 10, 18 and 19 and selected parts of Chapters 14 and 16.

- Finally, for a more **basic course in clinical neuroanatomy**, readings could be confined to selected topics in Chapters 2, 5–7, 10–16, and 18.

Neuroanatomy through Clinical Cases

CONTENTS

Chapter 1

Introduction to Clinical Case Presentations

Case presentations provide the framework for all communications about patient care. They lay down the basic information needed to formulate hypotheses about the location and nature of patients' problems. This information is then used to decide on further diagnostic tests or treatment measures. To diagnose and treat patients such as those described in this book, we must first learn how clinicians generally present a patient's medical history and the findings from their physical examination. In addition, we must learn how to formulate ideas about neurologic diagnosis and how the neurologic evaluation fits into the general context of patient assessment.

Introduction

NEUROANATOMY is one of the more clinically relevant courses taught in the first years of medical school. Principles learned in neuroanatomy are directly applicable to patient care, not just for the neurologist or neurosurgeon, but also for health care professionals in virtually every other field. However, medical students in their first years and other students of neuroanatomy are often unfamiliar with the basic principles of clinical case presentations used on the wards. Therefore, the first section of this chapter has been provided for the *nonclinician* or the *not-yet-clinician* as a brief orientation. Others may prefer to skip this section. The second section of this chapter discusses the neurologic differential diagnosis, a process through which several possible diagnoses are considered based on the available information. We will use this method when attempting to arrive at diagnoses in the cases throughout the remainder of the book.

Abbreviations will be avoided in the case presentations in this book whenever possible, although in reality they are used quite often on the wards. Therefore, some commonly used abbreviations will be introduced in this chapter.

The neurologic exam is only one part of the general physical exam. Nevertheless, the patient should always be treated as a whole and, in addition, much can be learned about neurologic illness from other parts of the physical exam. Therefore, in the final section of this chapter we will discuss the dynamic relationship between the general physical exam and the neurologic exam.

The General History and Physical Exam

While there are variations in personal styles, clinicians adhere to a fairly standard format when presenting cases so that all of the essential information can be succinctly communicated. Since this may be your first exposure to this format, we will first discuss the general structure of the history and physical examination that is used in all fields of medicine. Although the basic structure is always the same, the emphasis varies depending on the specialty. Therefore, in Chapter 3 we discuss the neurologic part of the physical exam in more detail. Note that case presentations in this book focus on the neurologic history and physical exam, although it is crucial to treat the patient as a whole and to never neglect symptoms and signs arising from other body systems. In addition, as described in the discussion that follows, certain features of the general physical exam often provide important information about neurologic illness.

One of the most daunting tasks confronting medical students as they first enter the wards is to master the art of case presentations. When a new patient is admitted to the hospital, it is the responsibility of the medical student and resident on call to obtain a good history and physical exam (H&P) and then to communicate this knowledge to the other members of the medical team. These skills are continually refined throughout a clinician's career as they see more patients.

The level of detail used in obtaining an H&P depends on both the setting and the patient. For example, the appropriate H&P when caring for an unfamiliar patient with multiple, active medical problems is much more detailed than the H&P for a familiar patient who is generally healthy and comes to the outpatient office with an injured finger. As a student's clinical skills develop, the H&P becomes a highly focused tool used both to investigate clinical problems of immediate concern and to screen for other potential problems that may be suspected on the basis of the overall clinical picture.

Remember that the whole point of the H&P is to *communicate.* The goal is to present the important points of the case to one's colleagues in the form of an interesting "story." They can then contribute to the patient's care through discussion of the case or by taking care of the patient in the middle of the night when the people who originally admitted the patient may be sound asleep at home. As one learns more clinical medicine, one gradually comes to know the difference between critical details not to be overlooked and irrelevant side issues that put listeners to sleep. This distinction is often surprisingly subtle, but it makes all the difference in effective case presentations.

The general format most commonly used for an H&P contains the following elements, which we will discuss in more detail in the sections that follow:

- Chief complaint, or why the patient was admitted
- History of the present illness
- Past medical history
- Review of systems
- Family history
- Social and environmental history
- Medications and allergies
- Physical exam
- Laboratory data
- Assessment and plan

Chief Complaint (CC)

This is a succinct statement that includes the patient's age, sex, and presenting problem. It may also include one or two very brief pieces of pertinent historical data.

> **Example:** "The patient is a 53-year-old man with a history of hypertension now presenting with crushing substernal chest pain of 1 hour's duration."

History of the Present Illness (HPI)

This is the complete history of the *current* medical problem that brought the patient to medical attention. It should include possible risk factors or other causes of the current illness as well as a detailed chronological description of all symptoms and prior care obtained for this problem. Pertinent negative information (symptoms or problems that are *not* present) helps exclude alternative diagnoses and is as important as pertinent positive information. Related medical problems can be mentioned as well; however, those that are not directly relevant to the present illness are usually covered instead in the section on past medical history (discussed in the next section).

> **Example:** "The patient has cardiac risk factors consisting of hypertension for 15 years and a family history of coronary artery disease. He does not smoke, nor does he have diabetes or elevated cholesterol. He has not had previous myocardial infarction. For the past 5 years he has had a stable pattern of chest pain on exertion, brought on by walking up two or more flights of stairs, lasting less than 5 minutes, not accompanied by other symptoms. The pain is relieved by rest and sublingual nitroglycerin. He has refused to undergo further cardiac workup, such as exercise stress testing, in the past. His hypertension is being treated with a beta-blocker. He denies symptoms of congestive heart failure and has no history of peripheral vascular or cerebrovascular disease. Today while sitting at his desk at work, he developed sudden 'crushing' substernal chest pain and pressure radiating to his neck, accompanied by tingling of the left arm, shortness of breath, sweating, and nausea without vomiting. The pain was not relieved by three

sublingual nitroglycerin tablets, and his coworkers called an ambulance to bring him to the emergency room, where he was afebrile with pulse 100, BP 140/90, and respiratory rate 20, and had an EKG with ST elevations, suggesting anterolateral myocardial ischemia. His pain was initially relieved by IV nitroglycerin and 2 mg of morphine, but then returned, lasting about 20 minutes with continued ST elevations, so he was started on the tissue plasminogen activator protocol for thrombolysis. He is now being admitted to the cardiac intensive care unit for further care, currently pain free."

Past Medical History (PMH)

Prior medical and surgical problems not directly related to the HPI are described here.

Example: "The patient has a history of a mildly enlarged prostate gland. He had a right inguinal hernia repair in 1978."

Review of Systems (ROS)

A brief, head-to-toe review of all medical systems—including head, eyes, ears, nose and throat, pulmonary, cardiac, gastrointestinal, genitourinary, OB/GYN, dermatologic, neurologic, psychiatric, musculoskeletal, hematological, oncologic, rheumatological, endocrine, infectious diseases, and so on—should be pursued with each patient to pick up problems or complaints missed in earlier parts of the history. If something comes up that is relevant to the HPI, it should be inserted in the HPI section, not buried in the ROS.

Example: "The patient has had mild upper respiratory symptoms for the past 4 days with nasal congestion but no cough, temperature, or sore throat."

Family History (FHx)

This section should list all immediate relatives and note familial illnesses such as diabetes, hypertension, asthma, heart disease, cancer, depression, and so on, especially those relating to the HPI. Family tree format is often a succinct and clear way to present this data.

Example: "Patient's mother died at 66 of myocardial infarction, had hypertension. Father had myocardial infarction at 55, had diabetes, died at 73 of stroke. Brother, 47 years old, healthy. Two children, healthy."

Social and Environmental History (SocHx/EnvHx)

This section should include the patient's occupation, family situation, travel history, sexual history (if not covered in ROS) and habits.

Example: "Electrical engineer. Married with two children. No recent travel. Denies ever smoking cigarettes or using drugs. Drinks 1–2 beers on Sundays."

Medications and Allergies

This section should list all medications currently being taken by the patient (including herbal or over-the-counter drugs), as well as any known general or drug allergies.

Example: "Atenolol 50 mg PO daily. Sublingual nitroglycerin as needed. No allergies. NKDA (no known drug allergies)."

Physical Exam

The examination generally proceeds from head to toe and includes the following sections:

- General appearance—for example, "A slightly obese man in no acute distress."

- Vital signs—temperature (T), pulse (P), blood pressure (BP), respiratory rate (R)

- HEENT (head, eyes, ears, nose, and throat)
- Neck
- Back and spine
- Lymph nodes
- Breasts
- Lungs
- Heart
- Abdomen
- Extremities
- Pulses
- Neurologic (see Chapter 3)
- Rectal
- Pelvic and genitalia
- Dermatologic

Laboratory Data

This comprises all diagnostic tests, including blood work, urine tests, electrocardiogram, and radiological tests (chest X-rays, CT scans, etc.).

Assessment and Plan

The **assessment** section usually begins with a one- or two-sentence **summary**, or **formulation**, that encapsulates the patient's main clinical features and most likely diagnosis. In more diagnostically uncertain cases, a brief discussion is added to the assessment, including a **differential diagnosis**—that is, a list of alternative possible diagnoses. With neurologic disorders, this discussion is often broken down into two sections: (1) localization and (2) differential diagnosis.

The **plan** section immediately follows the assessment and is usually broken down into a list of problems and proposed interventions and diagnostic procedures.

> **Example:** "This is a 53-year-old man with cardiac risk factors of hypertension and family history of coronary disease who presents with substernal chest pain and EKG changes suggestive of anterolateral wall myocardial infarction.
>
> 1. Coronary artery disease/hypertension: Will continue IV nitroglycerin and IV heparin after completion of tissue plasminogen activator protocol. Will resume beta-blocker, as patient has no evidence of congestive heart failure. Will check serial EKGs and cardiac enzymes to determine whether the patient has had a myocardial infarction.
> 2. Further cardiac workup: To include echocardiogram and an exercise stress test if cardiac enzymes are negative. If the patient develops further chest pain, he may require emergency cardiac catheterization."

Neurologic Differential Diagnosis

Reaching the correct diagnosis in patients with neurologic disorders sometimes presents a considerable challenge. As noted in the previous discussion, the assessment section of the H&P is therefore often broken down into several logical steps to facilitate this thought process. The first step is **localization** based on neuroanatomical clues gleaned from the H&P. This integration of anatomical and clinical knowledge will be the focus of this book. However, we will also briefly discuss the next step, the **neurologic differential diagnosis**.

FIGURE 1.1 **Arrowhead of Neurologic Differential Diagnosis** More acute diagnoses are often (but not always) at the top left point, and along the outer edges of the arrow. Examples and explanations of abbreviations are listed.

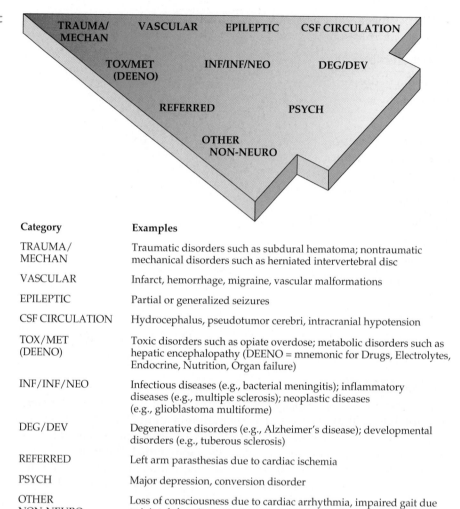

Category	Examples
TRAUMA/ MECHAN	Traumatic disorders such as subdural hematoma; nontraumatic mechanical disorders such as herniated intervertebral disc
VASCULAR	Infarct, hemorrhage, migraine, vascular malformations
EPILEPTIC	Partial or generalized seizures
CSF CIRCULATION	Hydrocephalus, pseudotumor cerebri, intracranial hypotension
TOX/MET (DEENO)	Toxic disorders such as opiate overdose; metabolic disorders such as hepatic encephalopathy (DEENO = mnemonic for Drugs, Electrolytes, Endocrine, Nutrition, Organ failure)
INF/INF/NEO	Infectious diseases (e.g., bacterial meningitis); inflammatory diseases (e.g., multiple sclerosis); neoplastic diseases (e.g., glioblastoma multiforme)
DEG/DEV	Degenerative disorders (e.g., Alzheimer's disease); developmental disorders (e.g., tuberous sclerosis)
REFERRED	Left arm parasthesias due to cardiac ischemia
PSYCH	Major depression, conversion disorder
OTHER NON-NEURO	Loss of consciousness due to cardiac arrhythmia, impaired gait due to joint deformity

MNEMONIC

When the diagnosis is uncertain and multiple possibilities must be considered, it is often helpful to have a mnemonic device handy, especially while being questioned on rounds by a more senior clinician. Such a mnemonic, the Arrowhead of Neurologic Differential Diagnosis, is shown in Figure 1.1. Disorders that tend to be more acute and require more immediate attention appear along the top and left leading edges of the arrowhead; disorders that are usually more chronic in nature appear on the inside. In visualizing and prioritizing one's clinical interventions, it can therefore be useful to move from the point on the left along the top row, and then along each subsequent row from left to right.

Relationship between the General Physical Exam and the Neurologic Exam

The neurologic exam is *part of the general physical exam*. Thus, although the neurologic exam is covered separately in Chapter 3, in reality the neurologic exam and the general physical exam should always be done and described as a single unit. The patient must be treated as a whole, with problems in different systems given priority depending on the situation. In addition, essential information about neurologic disease can be gleaned from all portions of the general physical exam. Some examples are given here (for explanations of unfamiliar terms refer to the Key Clinical Concepts sections throughout the rest of this book, or consult the Index):

- **General appearance**. How a person appears and behaves throughout the exam provides a wealth of information about his or her mental status and motor system.

- **Vital signs**. Hypertension, bradycardia, and other changes can be seen in elevated intracranial pressure. Exaggerated orthostatic changes (between reclining and upright positions) in heart rate and blood pressure can be seen in autonomic dysfunction and spinal cord injuries. Respiratory pattern provides important information about brainstem functioning. Elevated temperature suggests infection or inflammation, which may involve the nervous system.

- **HEENT**. Head shape can be a clue to congenital abnormalities, hydrocephalus, or tumors. Careful examination of the head, ears, and nose is essential in cranial trauma. Tongue abnormalities can suggest nutritional deficiencies, which may have neurologic manifestations. Oral thrush suggests immune dysfunction, which can predispose patients to a host of neurologic disorders. Palpation of the temporal and supraorbital arteries can give clues about vasculitis and collateral blood flow in cerebrovascular disease. A whooshing sound called a *bruit* can sometimes be heard with the stethoscope when intracranial vascular disease or arteriovenous malformations are present. Scalp tenderness may be present in migraine. The funduscopic exam is so relevant to neurologic disease that it is often included as part of the neurologic exam itself.

- **Neck**. Neck stiffness can be a sign of meningeal irritation. Cervical bruits can be heard with carotid artery disease. Thyroid abnormalities can cause mental status changes, eye movement disorders, and muscle weakness.

- **Back and spine**. Tenderness, misalignment, and curvature can give important information about possible fractures, metastases, osteomyelitis, and so on. Muscle stiffness and tenderness are diagnostically helpful in cases of back pain.

- **Lymph nodes**. Enlarged lymph nodes can be seen in neoplastic, infectious, and granulomatous disorders, which may involve the nervous system.

- **Breasts**. Breast cancer can metastasize to the nervous system or produce paraneoplastic disorders.

- **Lungs**. Unilateral decreased breath sounds with decreased movement of the diaphragm detected by percussion can be a sign of phrenic nerve dysfunction. An abnormal lung exam can also be associated with hypoxia and with infectious or neoplastic diseases, which may involve the nervous system.

- **Heart**. Clues to embolic sources can be provided by an irregular heartbeat with atrial fibrillation or by murmurs with valvular disease or endocarditis. Aortic stenosis can produce syncope, and severe heart failure results in cerebral hypoperfusion.

- **Abdomen**. Hepatomegaly may be palpated in Wilson's disease and other metabolic disorders that involve the nervous system. Abdominal aortic aneurysm, pancreatitis, and other abdominal pathology can produce back pain, which can occasionally be mistaken for disease of the spine.

- **Extremities**. Pain on straight-leg raising is a sign of nerve root compression. *Kernig's sign* (when examiner straightens the patient's knees with the hips flexed, patient has pain in the hamstrings) and *Brudzinski's sign* (when examiner flexes the patient's neck, patient flexes legs at the hips) are signs of meningeal irritation. Arthritis can be seen in autoimmune disorders, which often involve the nervous system. Clubbing or cyanosis suggests systemic illness, which may involve the nervous system. Leg edema can be seen in deep venous thrombosis, which is a common complication of neurologic disease, occurring with increased incidence in patients who rarely rise from bed.

- **Pulses**. Peripheral vascular disease suggests atherosclerosis, which may also involve intracranial vessels. Peripheral vascular disease produces symptoms such as pain, tingling, numbness, and even weakness, which can masquerade as neurologic disease.

- **Neurologic exam**. See Chapter 3.

- **Rectal**. Decreased rectal sphincter tone can signify pathology of the spinal cord or sacral nerve roots.

- **Pelvic and genitalia**. Gynecologic malignancies can be associated with paraneoplastic disorders involving the nervous system and, sometimes, with metastases. Testicular abnormalities can be seen in some neurodevelopmental disorders.

- **Dermatologic**. Several so-called neurocutaneous disorders have important dermatologic manifestations that can signal neurologic disease. These include neurofibromatosis, tuberous sclerosis, Sturge-Weber syndrome, and other disorders. A characteristic skin eruption is also seen in dermatomyositis. Local changes in skin texture, temperature, and color can be signs of chronic neurologic injury to the affected area. Other dermatologic abnormalities can, once again, signify systemic illnesses that may involve the nervous system.

Many more examples could be mentioned, but this list should at least illustrate that the neurologic exam and the general physical exam are inextricably linked.

Conclusions

In this chapter we have introduced the general format of the history and physical exam used in most medical settings. In the case presentations throughout the rest of the book, the same format will be used, with a focus on aspects of the H&P that reveal neurologic abnormalities. We have seen that much can be learned about neurologic disease from the *non-neurologic* portions of the general physical exam. Nevertheless, the *neurologic* portion of the general physical exam ultimately provides much of the information that enables us to localize lesions within the nervous system. Before we explore the neurologic exam in more detail, we must first lay down some foundations of basic neuroanatomy. This we will do in the next chapter, and we will then return to the neurologic exam in Chapter 3.

References

Bickley LS. 2008. *Bates' Guide to Physical Examination and History Taking*. 10th Ed. Lippincott Williams & Wilkins, Philadelphia.

Biller J (ed.). 2007. *The Interface of Neurology & Internal Medicine*. Lippincott Williams & Wilkins, Philadelphia.

Coulehan JL, Blok MR. 2005. *The Medical Interview: Mastering Skills for Clinical Practice*. 5th Ed. FA Davis, Philadelphia.

DeGowin RL, Brown DD, LeBlond RF. 2008. *DeGowin's Diagnostic Examination*. 9th Ed. McGraw-Hill, New York.

Gilman S. 2000. *Clinical Examination of the Nervous System*. McGraw-Hill, New York.

CONTENTS

Chapter 2

Neuroanatomy Overview and Basic Definitions

The nervous system is perhaps the most beautiful, elegant, and complex system in the body. Its interconnected networks perform processing that is simultaneously local and distributed, serial and parallel, hierarchical and global. Accordingly, structures of the nervous system can be described on multiple levels: in terms of macroscopic brain divisions; connecting pathways and cell groupings; individual brain cells; and, ultimately, receptors, neurotransmitters, and other signaling molecules.

In this chapter, we will learn about the nervous system's overall organization, and we will learn some basic terminology that will help us become oriented when we embark on a detailed study of the individual parts of the nervous system in subsequent chapters.

Basic Macroscopic Organization of the Nervous System

DECIDING TO STUDY NEUROANATOMY is somewhat like agreeing to paint a large mural that you will spend the rest of your life carefully improving and refining. To begin painting this mural we must first agree on the orientation of our subject matter—choose an up and a down, a forward and a backward. Then we will boldly sketch out the major features of the composition, paying particular attention to the relationships between different components and to how the composition works as a whole. The rough sketch provides a framework so that as we embark on painting various segments of the mural in ever-finer detail we never lose sight of the big picture, and we are therefore able to pass seamlessly from one area of the mural to another.

This chapter is devoted to the rough sketch. It would be a vain undertaking to attempt to learn the neuroanatomy of one system or region without some concept of how it relates both spatially and functionally to the whole nervous system. This is especially true when one uses clinical cases to learn neuroanatomy. Although each of the clinical cases in this book focuses on a particular neuroanatomical system, lesions almost invariably affect neighboring regions as well. These **neighborhood effects** are often critical in localizing neuroanatomical lesions. Therefore, in this chapter we will sketch out the main components of the nervous system and begin to describe some of the most important functions of each part.

After reading this chapter you should have some understanding of the nervous system as a whole. In addition, when you begin reading the clinical cases (Chapters 5–19), you should be able to localize lesions in certain general locations in the nervous system even though you have not yet studied those regions in detail. Finally, this chapter will provide the necessary background for understanding Chapter 3, in which we introduce the neurologic exam.

A caveat is in order before reading this chapter: This material is presented in the traditional style, without clinical cases. Therefore, do not become discouraged if you have trouble remembering all of the details. As you read the clinical cases in later chapters and refer back to this information to reach clinical diagnoses, the material will gradually be reinforced and solidified.

Main Parts of the Nervous System

The human nervous system can be divided into the **central nervous system (CNS)** and the **peripheral nervous system (PNS)**. The CNS includes the brain and spinal cord; the PNS is everything else (Figure 2.1; Table 2.1). During embryological development the CNS arises from a sheet of ectodermal cells that folds over to form the **neural tube**. The neural tube forms several swellings and outpouchings in the head that eventually develop into the brain, while the part of the neural tube running down the back of the embryo forms the spinal cord (Figure 2.2A,B). The fluid-filled cavities within the neural tube develop into the brain **ventricles**, which contain **cerebrospinal fluid (CSF)**.

The developing brain has three main divisions: the forebrain, or **prosencephalon**; the midbrain, or **mesencephalon**; and the hindbrain, or **rhombencephalon** (see Figure 2.2). The **forebrain** is the largest part of the nervous system in humans, and it is further subdivided into the telencephalon and diencephalon. The

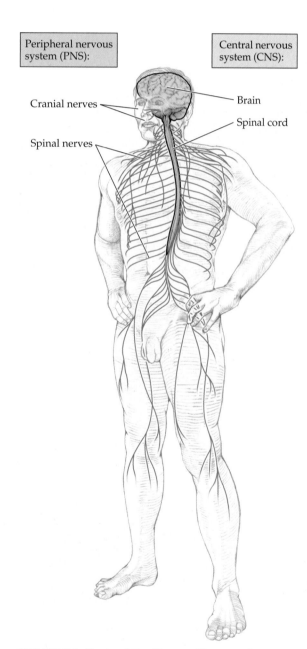

Peripheral nervous system (PNS):

Central nervous system (CNS):

Cranial nerves

Brain

Spinal cord

Spinal nerves

FIGURE 2.1 Parts of the Human Nervous System

telencephalon (meaning "end brain" in Greek) is made up of the cerebral hemispheres and includes structures discussed later in this chapter such as the cerebral cortex, white matter, and basal ganglia. The **diencephalon** is composed of the **thalamus**, **hypothalamus**, and associated structures. The **midbrain** is a relatively short and narrow region connecting the forebrain and hindbrain. The **hindbrain** is composed of the **pons** and **cerebellum** (**metencephalon**) together with the **medulla** (**myelencephalon**) (see Figure 2.2).

The midbrain, pons, and medulla together form a connection between the forebrain and the spinal cord. Since the forebrain sits on top of the **midbrain**, **pons**, and **medulla**, almost like a cauliflower on its stalk (see Figure 2.2C), these structures are often referred to as the **brainstem**.* The brainstem is the most evolutionarily ancient part of the human brain and is the part that most closely resembles the brains of fish and reptiles. It controls many of the most basic bodily functions necessary for survival, such as respiration, blood pressure, and heart rate.

*Some authors formerly defined the brainstem to include the cerebellum and diencephalon. In common clinical usage today, however, the term "brainstem" refers to the midbrain, pons, and medulla.

TABLE 2.1 Main Parts of the Human Nervous System
CENTRAL NERVOUS SYSTEM (CNS)
Brain
Spinal cord
PERIPHERAL NERVOUS SYSTEM (PNS)
Cranial nerves and ganglia
Spinal nerves and dorsal root ganglia
Sympathetic and parasympathetic nerves and ganglia
Enteric nervous system

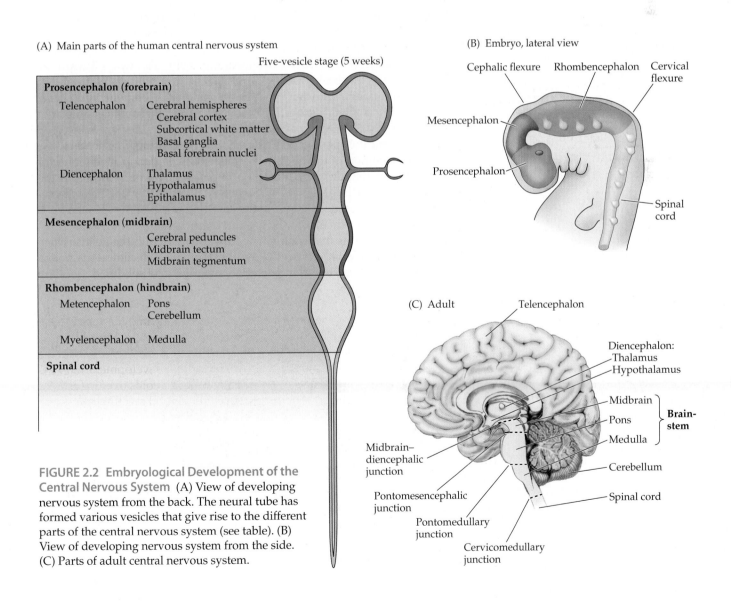

FIGURE 2.2 **Embryological Development of the Central Nervous System** (A) View of developing nervous system from the back. The neural tube has formed various vesicles that give rise to the different parts of the central nervous system (see table). (B) View of developing nervous system from the side. (C) Parts of adult central nervous system.

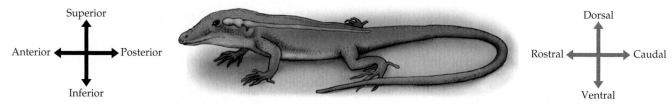

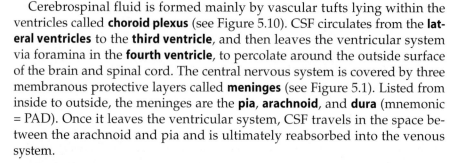

FIGURE 2.3 Orientation of the Central Nervous System in Reptiles The same terms apply above and below the midbrain.

MNEMONIC

Cerebrospinal fluid is formed mainly by vascular tufts lying within the ventricles called **choroid plexus** (see Figure 5.10). CSF circulates from the **lateral ventricles** to the **third ventricle**, and then leaves the ventricular system via foramina in the **fourth ventricle**, to percolate around the outside surface of the brain and spinal cord. The central nervous system is covered by three membranous protective layers called **meninges** (see Figure 5.1). Listed from inside to outside, the meninges are the **pia**, **arachnoid**, and **dura** (mnemonic = PAD). Once it leaves the ventricular system, CSF travels in the space between the arachnoid and pia and is ultimately reabsorbed into the venous system.

Orientation and Planes of Section

A variety of terms is used for different directions and planes of section in the nervous system. These terms are relatively simple in animals like fish and reptiles, in which the nervous system is linear in orientation (Figure 2.3). In these animals, **ventral** (from the Latin *venter*, meaning "belly") is always toward the earth, **dorsal** (Latin for "back," as in a shark's fin) is toward the sky, **rostral** (Latin for "beak"—think of a "rooster's rostrum") is toward the snout, and **caudal** (Latin for "tail") is toward the tail. However, since humans have an upright posture, the nervous system makes a bend of nearly 90° somewhere between the forebrain and the spinal cord (Figure 2.4). By definition, this bend is said to occur in the region of the midbrain–diencephalic junction. Therefore, for structures above the midbrain, the orientation of the nervous system is the same with respect to the ground as in reptiles. At the midbrain and below, however, there is a rotation of 90° since in the standing position the spinal cord is approximately perpendicular to the ground in humans.

Another set of terms that is often used for orientation in the nervous system remains constant with respect to the environment both above and below the midbrain. These are **anterior**, **posterior**, **superior**, and **inferior**. By looking

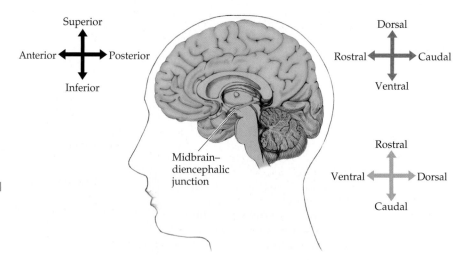

FIGURE 2.4 Orientation of the Central Nervous System in Humans The meaning of some terms (dorsal, ventral, rostral, caudal) changes at the midbrain–diencephalic junction.

at Figure 2.4 you should be able to confirm the definitions of the following terms as they apply to humans:

- Above the midbrain:

    Anterior = rostral

    Posterior = caudal

    Superior = dorsal

    Inferior = ventral

- Below the midbrain:

    Anterior = ventral

    Posterior = dorsal

    Superior = rostral

    Inferior = caudal

Thus, for example, above the midbrain the *anterior* commissure is *rostral*, the *posterior* commissure is *caudal*, the *superior* sagittal sinus is *dorsal*, and the *inferior* sagittal sinus is *ventral*. Meanwhile, below the midbrain the *anterior horn* of the spinal cord is *ventral*, the *posterior* horn is *dorsal*, the *superior* cerebellar peduncle is *rostral*, and the *inferior* cerebellar peduncle is *caudal*. In the midbrain itself, the same conventions are generally used as below the midbrain.

When the nervous system is studied pathologically or imaged radiologically, it is usually cut in one of three different orthogonal planes of section (Figure 2.5). **Horizontal** sections are parallel to the floor. Equivalent terms for horizontal sections in humans include **axial** or **transverse** sections, meaning sections perpendicular to the long axis of the person's body. The name **coronal** comes from the sectioning plane approximating that of a tiara-like crown. **Sagittal** sections are in the direction of an arrow shot, and this plane of section is best visualized by imagining the plane defined by a bow and arrow held by an archer (as in the star constellation Sagittarius). Sagittal sections passing through the midline are referred to as **midsagittal**, and when they are just off the midline they are referred to as **parasagittal** sections.

Note that the sagittal plane is orthogonal to the left–right axis, the coronal plane is orthogonal to the anterior–posterior axis, and the horizontal plane is orthogonal to the superior–inferior axis. When a plane of section lies somewhere between the three principal planes, it is referred to as oblique. The planes of section used for CT and MRI scans are approximately horizontal, coronal, or sagittal, with some slight adjustments (especially in the horizontal plane) often being necessary for technical reasons (see Chapter 4).

Basic Cellular and Neurochemical Organization of the Nervous System

Microscopically, the nervous system is composed of nerve cells, or **neurons**, and support cells called **glial cells** (or simply **glia**). Neurons are the basic units of signaling in the nervous system, although glial cells may contribute as well. Neuronal signaling is a complex phenomenon, presented here in a very simplified manner geared toward the clinical–anatomical discussions in this book (see the References at the end of this chapter to locate a more detailed treatment). A typical neuron has a **cell body** containing the nucleus, relatively short processes called **dendrites**, which receive most inputs to the cell, and long processes called **axons**, which carry most outputs (Figure 2.6). Most mammalian neurons are **multipolar**, meaning that they have several dendrites as well as several axons (see Figure 2.6A). Often, a single axon arising from the cell body will travel for a distance, and then one or several **axon collaterals** branch off the

(A) Horizontal plane

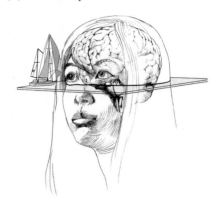

(B) Coronal plane

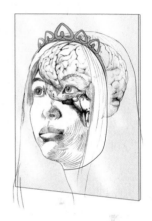

(C) Sagittal plane

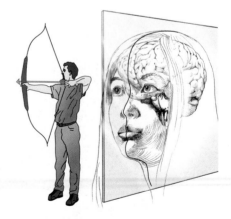

FIGURE 2.5 Anatomical Planes of Section (A) Horizontal (axial, transverse) plane. (B) Coronal plane. (C) Sagittal plane.

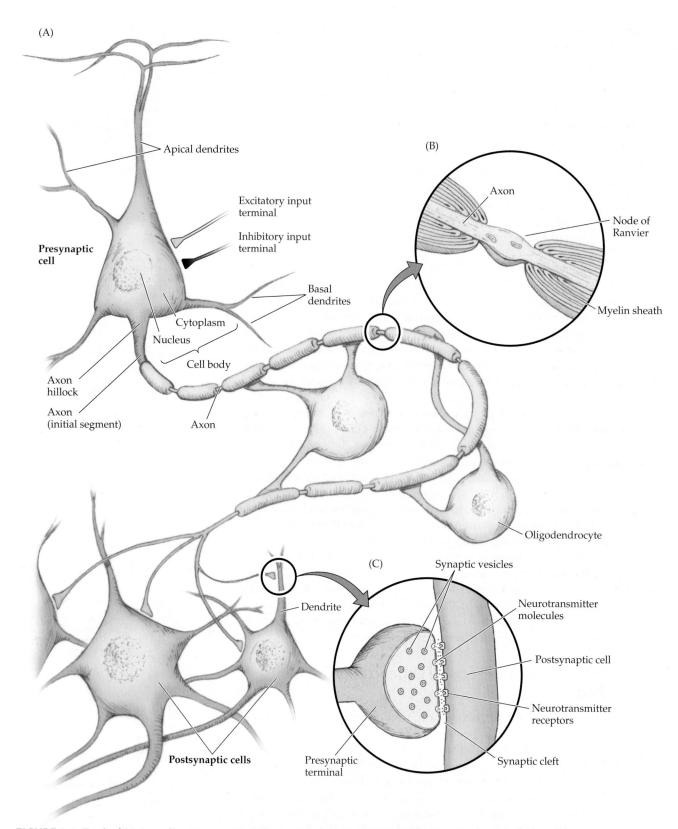

(A)

Apical dendrites

Excitatory input terminal

Inhibitory input terminal

Presynaptic cell

Basal dendrites

Cytoplasm

Nucleus

Cell body

Axon hillock

Axon (initial segment)

Axon

(B)

Axon

Node of Ranvier

Myelin sheath

Oligodendrocyte

(C)   Synaptic vesicles

Neurotransmitter molecules

Postsynaptic cell

Neurotransmitter receptors

Synaptic cleft

Dendrite

Presynaptic terminal

Postsynaptic cells

FIGURE 2.6 Typical Mammalian Neuron (A) The neuron mainly receives inputs on the dendrites and cell body and conveys signals via axonal electrical conduction to reach synapses, which send outputs other neurons. (B) Inset showing axonal myelin sheath and node of Ranvier. (C) Inset showing main elements of presynaptic and postsynaptic terminals.

main axon to reach different targets. Some neurons are **bipolar**, with a single dendrite and a single axon arising from the cell body. Bipolar cells are often sensory neurons, such as those involved in vision (see Figure 11.4) or olfaction (see Figure 18.5). Some bipolar neurons are called pseudo-unipolar, since their processes are initially fused and then split to produce two long axons. An example is dorsal root ganglion sensory neurons (see Figure 2.21). **Unipolar** neurons, in which both axons and dendrites arise from a single process coming off the cell body, occur mainly in invertebrates.

Communication between neurons takes place mainly at specialized regions called **synapses**. Classically, synapses carry information from the axon terminals of one neuron to the dendrites of the next neuron. However, there are also axo-axonic and dendro-dendritic synapses, and some forms of communication can even occur in reverse, traveling from dendrites back to axons. At **chemical synapses**, chemical **neurotransmitter** molecules, stored mainly in **synaptic vesicles**, are released from **presynaptic terminals** of the neuron (see Figure 2.6C). They then bind to **neurotransmitter receptors** on the **postsynaptic** neuron, giving rise to either excitation or inhibition of the postsynaptic neuron. In some cases, communication also takes place at **electrical synapses** where direct electrical coupling of neurons occurs through specialized junctions.

Neurons are electrically and chemically active. When excitatory synaptic inputs combine with endogenous transmembrane currents to sufficiently excite a neuron, a transient voltage change called an **action potential** occurs, lasting about 1 millisecond. Action potentials can travel rapidly throughout the length of a neuron, propagating at rates of up to about 60 meters per second along the cell membrane. Classically, action potentials travel from the dendritic end of the neuron along its axon to reach presynaptic terminals, where communication can occur with the next neuron (see Figure 2.6). Action potentials trigger release of neurotransmitter molecules from synaptic vesicles, allowing chemical communication with the postsynaptic cell (see Figure 2.6C).

Axons are often insulated by specialized glial cells that form a lipid **myelin sheath**, thereby speeding the rate of action potential conduction (see Figure 2.6B). The myelin-forming glial cells in the CNS are **oligodendrocytes**; in the PNS they are **Schwann cells**. Voltage-gated ion channels are concentrated in short, exposed segments of the axon called **nodes of Ranvier** (see Figure 2.6B). Conduction of action potentials from node to node occurs rapidly by a process called saltatory conduction.

Chemical neurotransmitters have two general types of functions. One is to mediate rapid communication between neurons through fast excitatory or inhibitory electrical events known as **excitatory postsynaptic potentials (EPSPs)** and **inhibitory postsynaptic potentials (IPSPs)**. Fast EPSPs and IPSPs occur on the timescale of tens of milliseconds and rapidly move the membrane voltage of the postsynaptic neuron between states more or less likely to fire an action potential. The postsynaptic neuron summates EPSPs and IPSPs arising from many presynaptic inputs. The second function of chemical neurotransmitters is **neuromodulation**, generally occurring over slower time scales. Neuromodulation includes a broad range of cellular mechanisms involving signaling cascades that regulate synaptic transmission, neuronal growth, and other functions. Neuromodulation can either facilitate or inhibit the subsequent signaling properties of the neuron.

Some of the more important and common neurotransmitters are summarized in Table 2.2. Note that neurotransmitters can be small molecules such as acetylcholine, amino acids like glutamate, or larger molecules such as peptides. Depending on the specific receptors present, neurotransmitters can mediate fast neurotransmission through EPSPs or IPSPs, or they may have

TABLE 2.2 Some Important Neurotransmitters

NAME	LOCATION OF CELL BODIES	MAIN PROJECTIONS	RECEPTOR SUBTYPES	MAIN ACTIONS
Glutamate	Entire CNS	Entire CNS	AMPA/kainate	Excitatory neurotransmission
			NMDA	Modulation of synaptic plasticity
			Metabotropic	Activation of second messenger systems
GABA	Entire CNS	Entire CNS	$GABA_A$, $GABA_B$	Inhibitory neurotransmission
		Retina	$GABA_C$	Inhibitory neurotransmission
Acetylcholine	Spinal cord anterior horns	Skeletal muscles	Nicotinic	Muscle contraction
	Autonomic preganglionic nuclei	Autonomic ganglia	Nicotinic	Autonomic functions
	Parasympathetic ganglia	Glands, smooth muscle, cardiac muscle	Muscarinic	Parasympathetic functions
	Basal forebrain: nucleus basalis, medial septal nucleus, and nucleus of diagonal band	Cerebral cortex	Muscarinic and nicotinic subtypes	Neuromodulation
	Pontomesencephalic region: pedunculopontine nucleus and laterodorsal tegmental nucleus	Thalamus, cerebellum, pons, and medulla	Muscarinic and nicotinic subtypes	Neuromodulation
Norepinephrine	Sympathetic ganglia	Smooth muscle, cardiac muscle	α and β subtypes	Sympathetic functions
	Pons: locus ceruleus and lateral tegmental area	Entire CNS	α_{1A-D}, α_{2A-D}, β_{1-3}	Neuromodulation
Dopamine	Midbrain: substantia nigra, pars compacta, and ventral tegmental area	Striatum, prefrontal cortex, limbic cortex, nucleus accumbens, amygdala	D_{1-5}	Neuromodulation
Serotonin	Midbrain and pons: raphe nuclei	Entire CNS	$5\text{-}HT_{1A-F}$, $5\text{-}HT_{2A-C}$, $5\text{-}HT_{3-7}$	Neuromodulation
Histamine	Hypothalamus: tuberomammillary nucleus; midbrain: reticular formation	Entire brain	H_{1-3}	Mainly excitatory neuromodulation
Glycine	Spinal cord; possibly also brainstem and retina	Spinal cord, brainstem, and retina	Glycine	Inhibitory neurotransmission[a]
Peptides	Entire CNS	Entire CNS	Numerous	Neuromodulation

[a]Glycine also has a modulatory role by binding to the NMDA receptor and increasing its response to glutamate.

facilitatory or inhibitory neuromodulatory effects on neuronal signaling. Some neurotransmitters have different actions at different synapses or even at the same synapse when a mixture of receptor types is present. In addition, more than one type of neurotransmitter molecule is often released, even at a single synapse.

In the CNS, the most common *excitatory* neurotransmitter is **glutamate**; the most common *inhibitory* neurotransmitter is **GABA** (gamma-aminobutyric acid). In the PNS, **acetylcholine** is the main transmitter at neuromuscular junctions, and both acetylcholine and **norepinephrine** are important in the autonomic nervous system (which we will discuss a little later). Aside from those listed in Table 2.2, numerous other neurotransmitters and neurotransmitter receptors have been described, and many more are yet to be discovered. Additional details of the functions of particular neurotransmitters will be provided in Chapters 6, 14, and 16.

CNS Gray Matter and White Matter; PNS Ganglia and Nerves

Areas of the CNS made up mainly of myelinated axons are called **white matter**. Areas made up mainly of cell bodies are called **gray matter**. Most of the local synaptic communication between neurons in the CNS occurs in the gray matter, while axons in the white matter transmit signals over greater distances. The surface of the cerebral hemispheres is covered by a unique mantle of gray matter called the **cerebral cortex**, which is far more developed in higher mammals than in other species. Beneath this lies the white matter, which conveys signals to and from the cortex (**Figure 2.7A**). Gray matter is also found in large clusters of cells called **nuclei** located deep within the cerebral hemispheres and brainstem. Examples include the **basal ganglia**, **thalamus**, and **cranial nerve nuclei** (see Figure 2.7A,B).

In the cerebral hemispheres the gray matter cortex is outside, while the white matter is inside. In the spinal cord the opposite is true: White matter pathways lie on the outside, while the gray matter is in the center (see Figure 2.7C). In the brainstem, gray matter and white matter regions are found both on the inside and on the outside, although most of the outside surface is white matter.

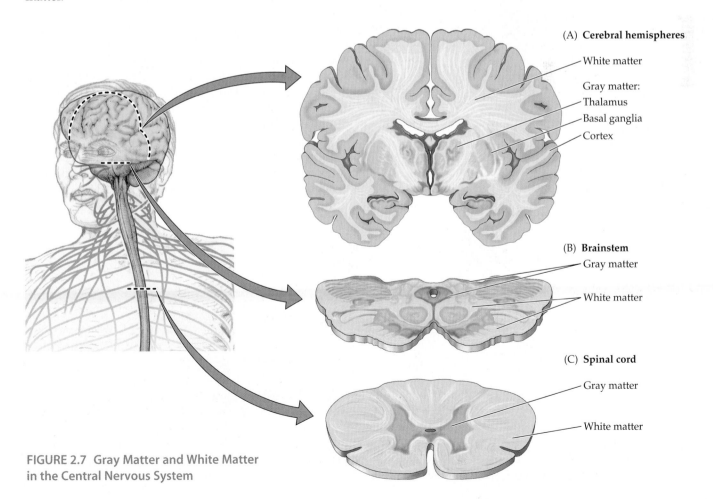

(A) **Cerebral hemispheres**
- White matter
- Gray matter:
- Thalamus
- Basal ganglia
- Cortex

(B) **Brainstem**
- Gray matter
- White matter

(C) **Spinal cord**
- Gray matter
- White matter

FIGURE 2.7 Gray Matter and White Matter in the Central Nervous System

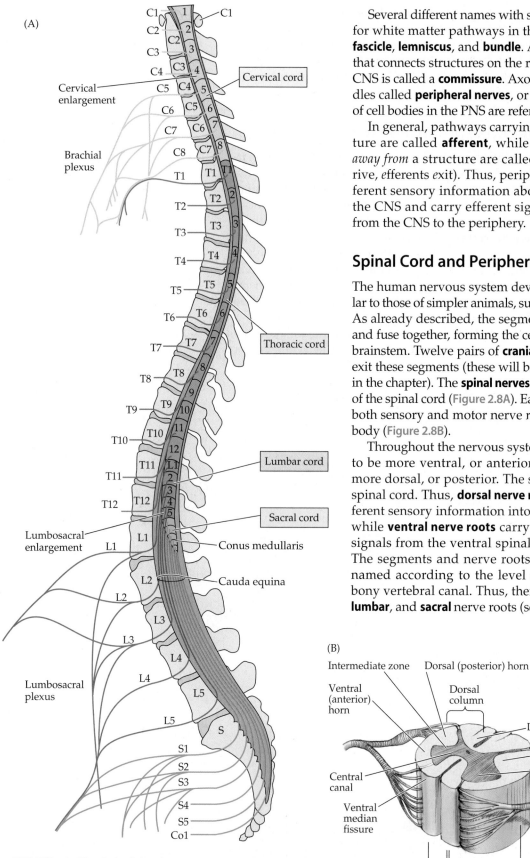

(A)

C1
C2
C3
C4
Cervical
enlargement
C5
C6
C7
C8
Brachial
plexus
T1
T2
T3
T4
T5
T6
T7
T8
T9
T10
T11
T12
Lumbosacral
enlargement
L1
L2
L3
L4
Lumbosacral
plexus
L5

C1
1
C2
2
3
C3
4
C4
5
C5
6
C6
7
C7
8
T1
1
T2
2
3
T3
4
T4
5
T5
6
T6
7
T7
8
T8
9
T9
10
T10
11
12
T11
L1
2
T12
3
4
5
L1
L2
L3
L4
L5
S
S1
S2
S3
S4
S5
Co1

C1

Cervical cord

Thoracic cord

Lumbar cord

Sacral cord

Conus medullaris

Cauda equina

Several different names with similar meaning are used for white matter pathways in the CNS, including **tract**, **fascicle**, **lemniscus**, and **bundle**. A white matter pathway that connects structures on the right and left sides of the CNS is called a **commissure**. Axons in the PNS form bundles called **peripheral nerves**, or simply nerves. Clusters of cell bodies in the PNS are referred to as **ganglia**.

In general, pathways carrying signals *toward* a structure are called **afferent**, while those carrying signals *away from* a structure are called **efferent** (afferents arrive, efferents exit). Thus, peripheral nerves convey afferent sensory information about the environment to the CNS and carry efferent signals for motor activity from the CNS to the periphery.

Spinal Cord and Peripheral Nervous System

The human nervous system develops in segments similar to those of simpler animals, such as segmented worms. As already described, the segments in the head expand and fuse together, forming the cerebral hemispheres and brainstem. Twelve pairs of **cranial nerves** (see Figure 2.1) exit these segments (these will be discussed further later in the chapter). The **spinal nerves** arise from the segments of the spinal cord (**Figure 2.8A**). Each segment gives rise to both sensory and motor nerve roots on each side of the body (**Figure 2.8B**).

Throughout the nervous system, motor systems tend to be more ventral, or anterior, and sensory systems more dorsal, or posterior. The same holds true for the spinal cord. Thus, **dorsal nerve roots** convey mainly afferent sensory information into the dorsal spinal cord, while **ventral nerve roots** carry mainly efferent motor signals from the ventral spinal cord to the periphery. The segments and nerve roots of the spinal cord are named according to the level at which they exit the bony vertebral canal. Thus, there are **cervical**, **thoracic**, **lumbar**, and **sacral** nerve roots (see Figure 2.8A).

(B)

Intermediate zone   Dorsal (posterior) horn

Ventral
(anterior)
horn

Dorsal
column

Dorsal median septum

White matter

Gray matter

Dorsal root
(sensory)

Central
canal

Spinal
nerve

Ventral
median
fissure

Ventral root
(motor)

Ventral
column

Lateral
column

FIGURE 2.8 The Spinal Cord (A) Cervical, thoracic, lumbar, and sacral spinal cord segments and nerves in relation to vertebral bones. (B) Dorsal sensory roots and ventral motor roots arise at each segment.

During development, the bony vertebral canal increases in length faster than the spinal cord. Therefore, the spinal cord ends at the level of the first or second lumbar vertebral bones (L1 or L2). Below this the spinal canal contains a collection of nerve roots known as the **cauda equina** (Latin for "horse's tail"), which continue down to their exit points. The sensory and motor nerve roots join together a short distance outside the spinal cord and form a mixed sensory and motor spinal nerve (see Figure 2.8B). Control of the arms and legs requires much more signal flow than does control of the chest and abdomen. Thus, the nerves controlling the extremities give rise to elaborate meshworks referred to as the **brachial plexus** for the arms and the **lumbosacral plexus** for the legs (see Figure 2.8A). In addition, the spinal cord contains a relatively increased amount of gray matter in these segments, causing the overall thickness of the cord to be greater. These regions of the cord are called the **cervical enlargement** and the **lumbosacral enlargement**, respectively.

In addition to the sensory and motor pathways already described, the PNS includes some specialized neurons that are involved in controlling such automatic functions as heart rate, peristalsis, sweating, and smooth muscle contraction in the walls of blood vessels, bronchi, sex organs, the pupils, and so on. These neurons are part of the **autonomic nervous system**. The autonomic nervous system has two major divisions (**Figure 2.9**): The **sympathetic**

FIGURE 2.9 Autonomic Nervous System The sympathetic division is shown on the left, and the parasympathetic division is shown on the right.

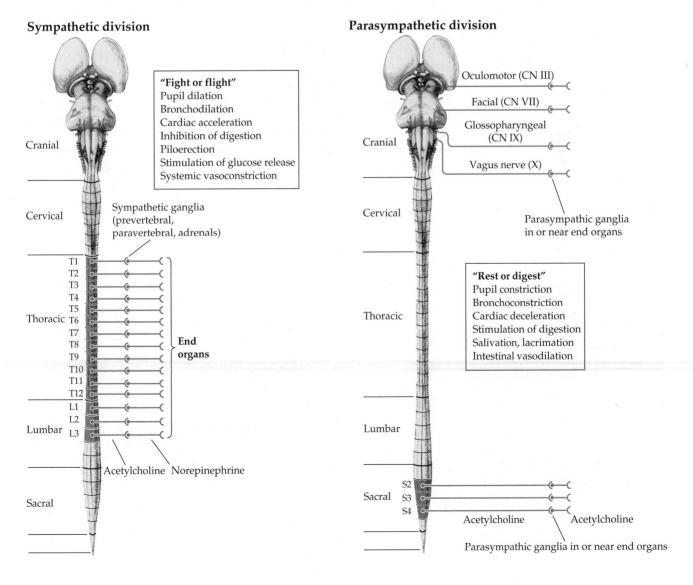

division arises from thoracic and lumbar spinal levels T1 to L3 (the **thoracolumbar** division). It releases the neurotransmitter **norepinephrine** onto end organs and is involved in such "fight or flight" functions as increased heart rate and blood pressure, bronchodilation, and increased pupil size. The **parasympathetic** division, in contrast, arises from the cranial nerves and from sacral spinal levels S2 to S4 (the **craniosacral** division). It releases **acetylcholine** onto end organs and is involved in more sedentary functions, such as increasing gastric secretions and peristalsis, slowing the heart rate, and decreasing pupil size. The sympathetic and parasympathetic pathways are controlled by higher centers in the hypothalamus and limbic system as well as by afferent sensory information from the periphery.

The enteric nervous system is considered a third autonomic division and consists of a neural plexus, lying within the walls of the gut, that is involved in controlling peristalsis and gastrointestinal secretions.

Cerebral Cortex: Basic Organization and Primary Sensory and Motor Areas

The cerebral cortex is not a smooth sheet, but rather has numerous infoldings or crevices called **sulci**. The bumps or ridges of cortex that rise up between the sulci are called **gyri**. Some sulci and gyri have particular names and functions, as we will learn shortly. The cerebral hemispheres have four major lobes: the frontal, temporal, parietal, and occipital (Figure 2.10).

Lobes of the Cerebral Hemispheres

The **frontal lobes** are, appropriately, in the front of the brain and extend back to the **central sulcus of Rolando**. The frontal lobes are separated inferiorly and laterally from the **temporal lobes** by an especially deep sulcus called the **Sylvian fissure**, or lateral fissure. (The term **fissure** is sometimes used to refer to deep sulci.) The **parietal lobes** are bounded anteriorly by the central sulcus but have no sharp demarcation from the temporal lobes or the **occipital lobes** when viewed from the lateral side of the brain (see Figure 2.10A). When viewed from

FIGURE 2.10 Cerebral Cortex: Frontal, Parietal, Temporal, and Occipital Lobes (A) Lateral view of left hemisphere. (B) Midsagittal view of right hemisphere.

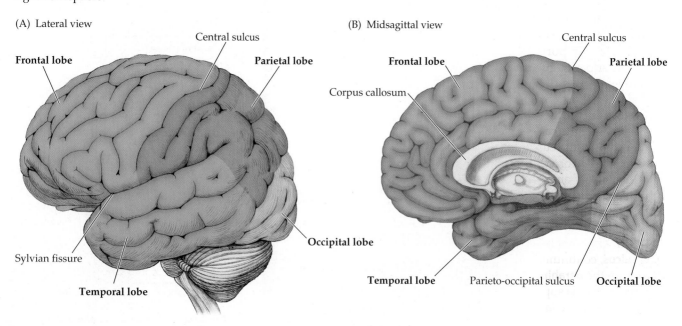

the medial aspect, the **parieto-occipital sulcus** can be seen more easily, separating the parietal from the occipital lobes (see Figure 2.10B).

In addition to these four major lobes, an additional region of cerebral cortex called the **insular cortex** lies buried within the depths of the Sylvian fissure. The insula is covered by a lip of frontal cortex anteriorly and parietal cortex posteriorly, called the **frontal operculum** and **parietal operculum**, respectively (*operculum* means "covering" or "lid" in Latin) (see Figure 2.24B). The limbic cortex (see Figure 2.25) was formerly referred to as the "limbic lobe," but this terminology is no longer generally used.

The two cerebral hemispheres are separated in the midline by the **interhemispheric fissure**, also known as the **sagittal** or **longitudinal fissure** (Figure 2.11D). A large, C-shaped band of white matter called the **corpus callosum** (meaning "hard body") connects both homologous and heterologous areas in the two hemispheres (see Figure 2.10B).

Surface Anatomy of the Cerebral Hemispheres in Detail

Although there is some variability, the sulci and gyri of the cerebral hemispheres form certain fairly consistent patterns. We will now briefly review the names of the major sulci, gyri, and other structures of the cerebral hemispheres (see Figure 2.11). Functions of these structures will be discussed in the next section and throughout the remainder of the book.

On the lateral surface (see Figure 2.11A), the frontal lobe is bounded posteriorly by the central sulcus, as already noted. The gyrus running in front of the central sulcus is called the **precentral gyrus**. The remainder of the lateral frontal surface is divided into the **superior, middle,** and **inferior frontal gyri** by the **superior** and **inferior frontal sulci**. Similarly, the lateral temporal lobe is divided into **superior, middle,** and **inferior temporal gyri** by the **superior** and **middle temporal sulci**. The most anterior portion of the parietal lobe is the **postcentral gyrus**, lying just behind the central sulcus. The **intraparietal sulcus** divides the **superior parietal lobule** from the **inferior parietal lobule**. The inferior parietal lobule consists of the **supramarginal gyrus** (surrounding the end of the Sylvian fissure) and the **angular gyrus** (surrounding the end of the superior temporal sulcus).

On the medial surface (see Figure 2.11B), the **corpus callosum** is clearly visible, consisting of the **rostrum, genu, body,** and **splenium**. The **cingulate gyrus** (cingulum means "girdle" or "belt") surrounds the corpus callosum, running from the paraterminal gyrus anteriorly to the isthmus posteriorly. The **cingulate sulcus** has a **marginal branch** running up to the superior surface that forms an important landmark, since the sulcus immediately in front of it, on the superior surface, is the central sulcus. The central sulcus does not usually extend onto the medial surface, but the region surrounding it is called the **paracentral lobule**. The portion of the medial occipital lobe below the **calcarine fissure** is called the **lingula** (meaning "little tongue"), while the portion above the calcarine fissure is called the **cuneus** (meaning "wedge"). Just in front of the cuneus, the medial parietal lobe is called the **precuneus**.

On the inferior surface (see Figure 2.11C), the **orbital frontal gyri** can be seen, which lay on top of the orbital ridges of the eye. More medially, the **olfactory sulcus** (containing the olfactory bulb) separates the orbital frontal gyri from the **gyrus rectus** (meaning "straight gyrus"). On the inferior surface of the temporal lobe, the **inferior temporal sulcus** separates the inferior temporal gyrus from the **occipitotemporal**, or **fusiform**, **gyri**. More medially, the **collateral sulcus**, continuing anteriorly as the **rhinal sulcus**, separates the fusiform gyri from the **parahippocampal gyrus**.

Finally, on the superior surface (see Figure 2.11D), many of the same landmarks seen on the lateral surface are again visible.

REVIEW EXERCISE

At this point you should become acquainted with the different lobes and with the major sulci and gyri of the cerebral hemispheres: Cover the labels in Figure 2.11A–D and attempt to name as many structures as possible. By the time you have completed this text, the functions of most of these sulci and gyri should be familiar.

(A)

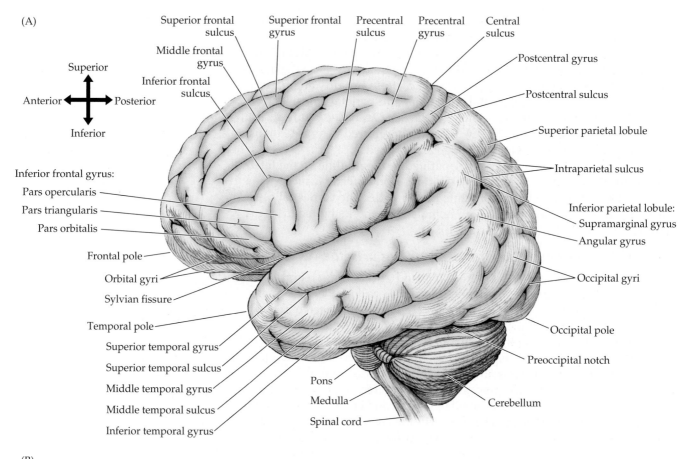

Superior
Anterior — Posterior
Inferior

Superior frontal sulcus
Superior frontal gyrus
Middle frontal gyrus
Inferior frontal sulcus
Precentral sulcus
Precentral gyrus
Central sulcus
Postcentral gyrus
Postcentral sulcus
Superior parietal lobule
Intraparietal sulcus
Inferior parietal lobule:
Supramarginal gyrus
Angular gyrus
Inferior frontal gyrus:
Pars opercularis
Pars triangularis
Pars orbitalis
Frontal pole
Orbital gyri
Sylvian fissure
Temporal pole
Superior temporal gyrus
Superior temporal sulcus
Middle temporal gyrus
Middle temporal sulcus
Inferior temporal gyrus
Occipital gyri
Occipital pole
Preoccipital notch
Pons
Medulla
Spinal cord
Cerebellum

(B)

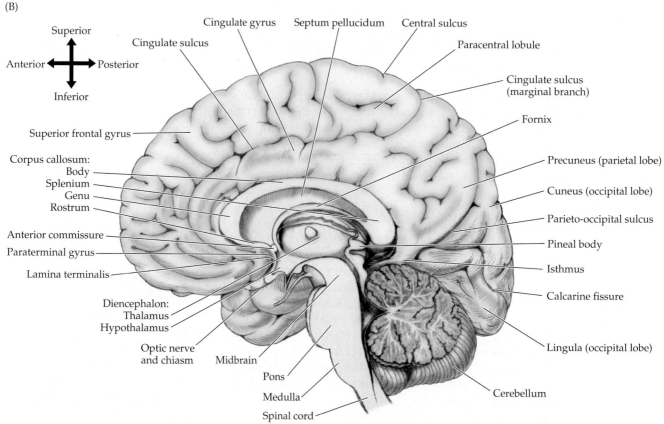

Superior
Anterior — Posterior
Inferior

Cingulate gyrus
Cingulate sulcus
Septum pellucidum
Central sulcus
Paracentral lobule
Cingulate sulcus (marginal branch)
Fornix
Superior frontal gyrus
Corpus callosum:
Body
Splenium
Genu
Rostrum
Anterior commissure
Paraterminal gyrus
Lamina terminalis
Diencephalon:
Thalamus
Hypothalamus
Optic nerve and chiasm
Midbrain
Pons
Medulla
Spinal cord
Precuneus (parietal lobe)
Cuneus (occipital lobe)
Parieto-occipital sulcus
Pineal body
Isthmus
Calcarine fissure
Lingula (occipital lobe)
Cerebellum

(C)

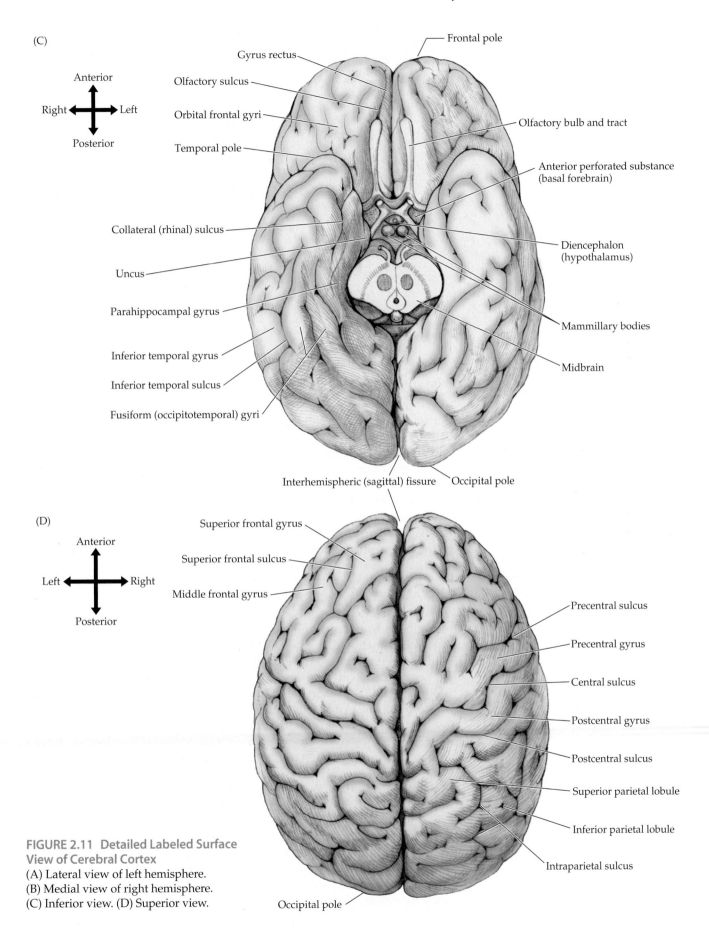

Anterior
Right ⟷ Left
Posterior

Frontal pole

Gyrus rectus

Olfactory sulcus

Orbital frontal gyri

Temporal pole

Olfactory bulb and tract

Anterior perforated substance (basal forebrain)

Collateral (rhinal) sulcus

Diencephalon (hypothalamus)

Uncus

Parahippocampal gyrus

Mammillary bodies

Inferior temporal gyrus

Inferior temporal sulcus

Midbrain

Fusiform (occipitotemporal) gyri

Interhemispheric (sagittal) fissure   Occipital pole

(D)

Anterior
Left ⟷ Right
Posterior

Superior frontal gyrus

Superior frontal sulcus

Middle frontal gyrus

Precentral sulcus

Precentral gyrus

Central sulcus

Postcentral gyrus

Postcentral sulcus

Superior parietal lobule

Inferior parietal lobule

Intraparietal sulcus

Occipital pole

FIGURE 2.11 Detailed Labeled Surface View of Cerebral Cortex
(A) Lateral view of left hemisphere.
(B) Medial view of right hemisphere.
(C) Inferior view. (D) Superior view.

FIGURE 2.12 Primary Sensory and Motor Cortical Areas

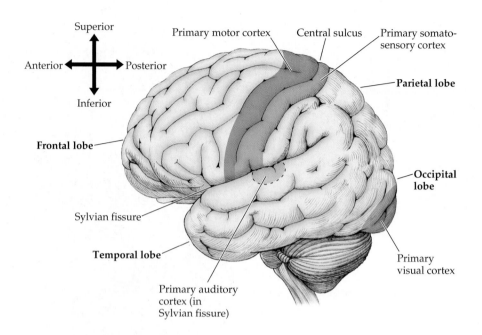

Primary Sensory and Motor Areas

The primary sensory and motor areas of the cortex are shown in Figure 2.12. The **primary motor cortex** lies in the **precentral gyrus** in the frontal lobe (see Figure 2.11A). This area controls movement of the opposite side of the body. The **primary somatosensory cortex** is in the **postcentral gyrus** in the parietal lobe and is involved in sensation for the opposite side of the body. Note that the precentral and postcentral gyri are separated by the **central sulcus** and that (as in the spinal cord) motor areas lie anterior to somatosensory areas. The **primary visual cortex** is in the occipital lobes along the banks of a deep sulcus called the **calcarine fissure** (see Figures 2.11B and 2.12). The **primary auditory cortex** is composed of the **transverse gyri of Heschl**, which are two fingerlike gyri that lie inside the Sylvian fissure on the superior surface of each temporal lobe (see Figures 2.12 and 2.24B). Higher-order sensory and motor information processing takes place in the association cortex, as will be discussed later in this chapter.

Sensory and motor pathways are usually **topographically organized**. This means that adjacent areas on the receptive (or motor) surface are mapped to adjacent fibers in white matter pathways and to adjacent regions of cortex. For example, in primary motor and primary somatosensory cortex, regions representing the hand are adjacent to regions representing the arm, and so on (Figure 2.13). These **somatotopic** maps on the cortex are sometimes called the motor or sensory **homunculus** ("little man"). Similarly, adjacent retinal areas are mapped in a **retinotopic** fashion onto the primary visual cortex, and adjacent regions of the cochlea, sensing different frequencies, have a **tonotopic** representation on the primary auditory cortex.

Interestingly, the primary somatosensory cortex and primary motor cortex represent sensation and movement, respectively, for the opposite side of the body. This relationship was first noted by physicians in ancient Greece, including Hippocrates, who observed that patients with head injuries had deficits affecting the side of the body opposite to the side of the injury. Knowledge of the levels at which the somatosensory and motor pathways cross over in the nervous system can be helpful for clinical neuroanatomical localization and will be discussed later in this chapter. The primary visual cortex represents visual inputs from the opposite *visual field.* Thus, the left

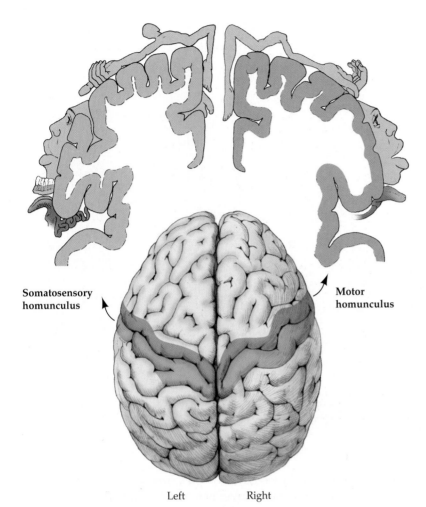

Somatosensory homunculus

Motor homunculus

Left   Right

half of the visual field *for each eye* is mapped to the right primary visual cortex (see Figure 11.15). Information reaching the primary auditory cortex is less lateralized and represents more of a mixture of inputs from both ears (the input from the opposite ear is slightly stronger, but this is usually not clinically detectable).

Cell Layers and Regional Classification of the Cerebral Cortex

The majority of the cerebral cortex is composed of **neocortex**, which has **six cell layers**, labeled I through VI, counting from the surface inward (Figure 2.14; Table 2.3). In a few regions associated with the limbic system, less than six layers are

TABLE 2.3 Cell Layers of the Neocortex

LAYER	NAME	ALTERNATIVE NAME	MAIN CONNECTIONS
I	Molecular layer		Dendrites and axons from other layers
II	Small pyramidal layer	External granular layer	Cortical–cortical connections
III	Medium pyramidal layer	External pyramidal layer	Cortical–cortical connections
IV	Granular layer	Internal granular layer	Receives inputs from thalamus
V	Large pyramidal layer	Internal pyramidal layer	Sends outputs to subcortical structures (other than thalamus)
VI	Polymorphic layer	Multiform layer	Sends outputs to thalamus

FIGURE 2.14 **Layers of the Neocortex**
(A–C from Parent, A. 1996. *Carpenter's Human Neuroanatomy*, 9th Ed. Williams & Wilkins, Baltimore.)

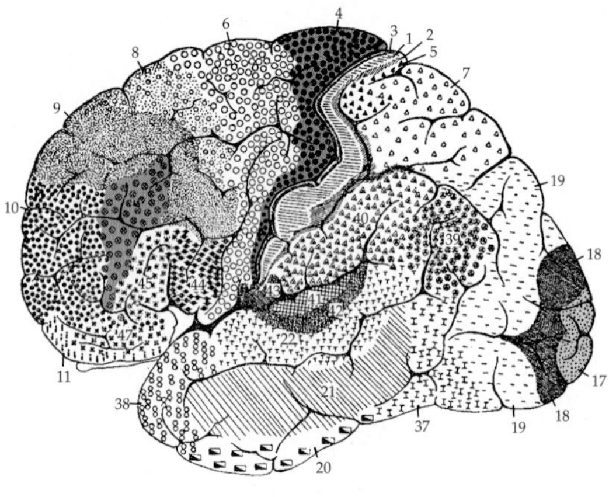

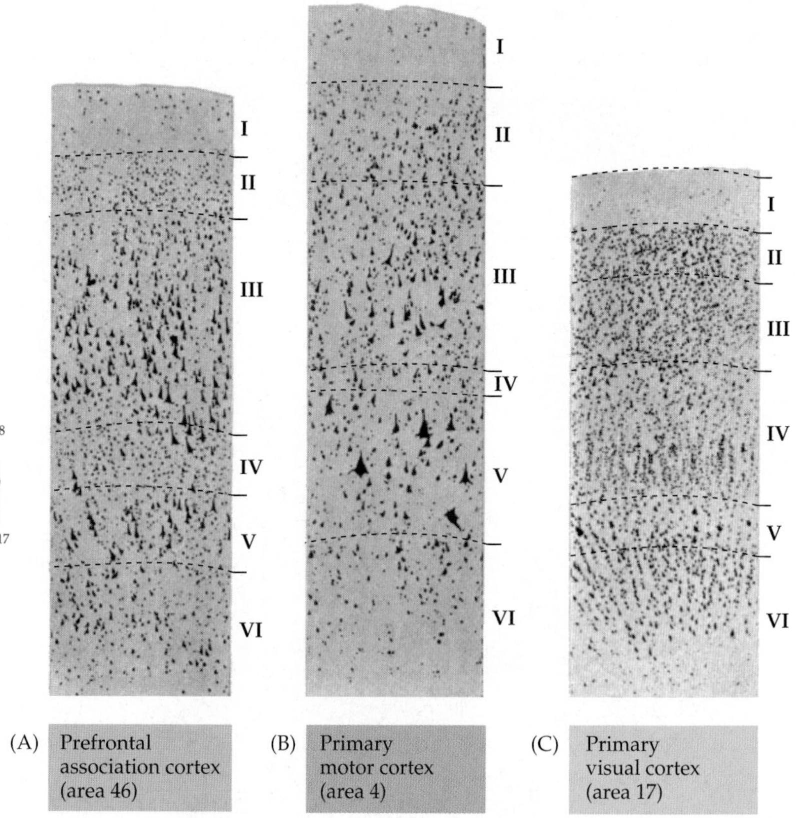

(A) Prefrontal association cortex (area 46)

(B) Primary motor cortex (area 4)

(C) Primary visual cortex (area 17)

present. Neocortical circuitry is quite complex, and we will describe only a few of the major connections of each layer here. **Layer I** contains mainly dendrites of neurons from deeper layers as well as axons. **Layers II** and **III** contain neurons that project mainly to other areas of cortex. **Layer IV** receives the majority of inputs from the thalamus. **Layer V** projects mostly to subcortical structures other than the thalamus, such as the brainstem, spinal cord, and basal ganglia. **Layer VI** projects primarily to the thalamus. In addition to these connections, numerous other circuits exist between and within the cortical layers that are beyond the scope of this discussion. The cortical layers I through VI have alternative names that will not be used here but are included in Table 2.3 for reference.

The relative thickness of the cell layers varies according to the main function of that area of cortex. For example, the primary motor cortex has large efferent projections to the brainstem and spinal cord, which control movement. It receives relatively little direct sensory information from thalamic relay centers. Therefore, in the primary motor cortex, layer V is thicker and has many more cell bodies than layer IV (see Figure 2.14B). The opposite holds for primary visual cortex, where layer IV contains many cell bodies and layer V is relatively cell poor (see Figure 2.14C). Association cortex has a cellular structure that is intermediate between these types (see Figure 2.14A).

A variety of classification schemes exist for different regions of the cerebral cortex based on microscopic appearance and function. The most widely known of these was published by Korbinian **Brodmann** in 1909. Based on his studies conducted with the microscope, Brodmann parceled the cortex into 52 **cytoarchitectonic areas**, each assigned a number corresponding to the order in which he prepared the slides (**Figure 2.15**; **Table 2.4**). It turns out that many of the areas identified by Brodmann correlate fairly well with various functional areas of the cortex, and therefore his nomenclature is still often used today.

TABLE 2.4 Brodmann's Cytoarchitectonic Areas

BRODMANN'S AREA	FUNCTIONAL AREA	LOCATION	FUNCTION
1, 2, 3	Primary somatosensory cortex	Postcentral gyrus	Touch
4	Primary motor cortex	Precentral gyrus	Voluntary movement control
5	Tertiary somatosensory cortex; posterior parietal association area	Superior parietal lobule	Stereognosis
6	Supplementary motor cortex; supplementary eye field; premotor cortex; frontal eye fields	Precentral gyrus and rostral adjacent cortex	Limb and eye movement planning
7	Posterior parietal association area	Superior parietal lobule	Visuomotor; perception
8	Frontal eye fields	Superior, middle frontal gyri, medial frontal lobe	Saccadic eye movements
9, 10, 11, 12[a]	Prefrontal association cortex; frontal eye fields	Superior, middle frontal gyri, medial frontal lobe	Thought, cognition, movement planning
17	Primary visual cortex	Banks of calcarine fissure	Vision
18	Secondary visual cortex	Medial and lateral occipital gyri	Vision; depth
19	Tertiary visual cortex, middle temporal visual area	Medial and lateral occipital gyri	Vision, color, motion, depth
20	Visual inferotemporal area	Inferior temporal gyrus	Form vision
21	Visual inferotemporal area	Middle temporal gyrus	Form vision
22	Higher-order auditory cortex	Superior temporal gyrus	Hearing, speech
23, 24, 25, 26, 27	Limbic association cortex	Cingulate gyrus, subcallosal area, retrosplenial area, and parahippocampal gyrus	Emotions
28	Primary olfactory cortex; limbic association cortex	Parahippocampal gyrus	Smell, emotions
29, 30, 31, 32, 33	Limbic association cortex	Cingulate gyrus and retrosplenial area	Emotions
34, 35, 36	Primary olfactory cortex; limbic association cortex	Parahippocampal gyrus	Smell, emotions
37	Parietal–temporal–occipital association cortex; middle temporal visual area	Middle and inferior temporal gyri at junction of temporal and occipital lobes	Perception, vision, reading, speech
38	Primary olfactory cortex; limbic association cortex	Temporal pole	Smell, emotions
39	Parietal–temporal–occipital association cortex	Inferior parietal lobule (angular gyrus)	Perception, vision, reading, speech
40	Parietal–temporal–occipital association cortex	Inferior parietal lobule (supramarginal gyrus)	Perception, vision, reading, speech
41	Primary auditory cortex	Heschl's gyri and superior temporal gyrus	Hearing
42	Secondary auditory cortex	Heschl's gyri and superior temporal gyrus	Hearing
43	Gustatory cortex	Insular cortex, frontoparietal operculum	Taste
44	Broca's area; lateral premotor cortex	Inferior frontal gyrus (frontal operculum)	Speech, movement planning
45	Prefrontal association cortex	Inferior frontal gyrus (frontal operculum)	Thought, cognition, planning behavior
46	Prefrontal association cortex (dorsolateral prefrontal cortex)	Middle frontal gyrus	Thought, cognition, planning behavior, aspects of eye movement control
47[a]	Prefrontal association cortex	Inferior frontal gyrus (frontal operculum)	Thought, cognition, planning behavior

Source: Martin, JH. 1996. *Neuroanatomy Text and Atlas.* McGraw-Hill, New York.

[a]Areas 13–16 and 48–52 are either not visible on the surface of the human cortex or are present only in other mammalian species, so are not shown in Figure 2.15.

ing both sensory and motor functions. However, they also carry out more specialized functions relating to the organs of the head (Table 2.5). Examination of the cranial nerves provides crucial information about the functioning of the nervous system, as we will discuss in the next chapter.

In addition to the cranial nerve nuclei and pathways, the brainstem is tightly packed with numerous other important nuclei and white matter tracts. All information passing between the cerebral hemispheres and the spinal cord must pass through the brainstem. Therefore, lesions in the brainstem can have a devastating effect on sensory and motor function. In addition, the brainstem contains nuclei that play important roles in the motor system; nuclei that produce nausea and vomiting in response to certain chemicals; modulatory nuclei containing the neurotransmitters norepinephrine, serotonin, dopamine, and acetylcholine (see Table 2.2), which project widely throughout the CNS; nuclear areas involved in pain modulation; and nuclei controlling heart rate, blood pressure, and respiration, among other functions.

An important region of the brainstem that contains many of these nuclei is the **reticular formation**. Named for the network-like appearance of its fibers in histological sections, the reticular formation extends throughout the central portions of the brainstem from the medulla to the midbrain. The more caudal portions of the reticular formation in the medulla and lower pons tend to be involved mainly in motor and autonomic functions. The rostral reticular formation in the upper pons and midbrain plays an important role in regulating the **level of consciousness**, influencing higher areas through modulation of thalamic and cortical activity (Figure 2.23). Thus, lesions that affect the pontomesencephalic reticular formation can cause lethargy and coma.

Cortical, thalamic, and other forebrain networks are also important in maintaining consciousness. Therefore, the level of consciousness can also be impaired in bilateral lesions of the thalami or in bilateral (or large unilateral)

TABLE 2.5 Overview of the Cranial Nerves

NERVE	NAME	FUNCTIONS
CN I	Olfactory nerve	Olfaction
CN II	Optic nerve	Vision
CN III	Oculomotor nerve	Extraocular muscles, except those innervated by CN IV and VI; parasympathetics to pupil constrictor and to ciliary muscles of lens for near vision
CN IV	Trochlear nerve	Superior oblique muscle; causes the eye to move downward and to rotate inward (depression and intorsion)
CN V	Trigeminal nerve	Sensations of touch, pain, temperature, vibration, and joint position for the face, mouth, nasal sinuses, and meninges; muscles of mastication; tensor tympani muscle
CN VI	Abducens nerve	Lateral rectus muscle; causes abduction (outward movement) of the eye
CN VII	Facial nerve	Muscles of facial expression; also stapedius muscle and part of digastric; taste from anterior two-thirds of tongue; sensation from a region near the ear; parasympathetics causing lacrimation and supplying the submandibular and sublingual salivary glands
CN VIII	Vestibulocochlear nerve	Hearing; vestibular sensation
CN IX	Glossopharyngeal nerve	Stylopharyngeus muscle; taste from posterior one-third of tongue; sensation from posterior pharynx, and from a region near the ear; chemo- and baroreceptors of the carotid body; parasympathetics to the parotid gland
CN X	Vagus nerve	Pharyngeal muscles (swallowing); laryngeal muscles (voicebox); parasympathetics to the heart, lungs, and digestive tract up to the splenic flexure; taste from epiglottis and pharynx; sensation from the pharynx, posterior meninges, and a region near the ear; aortic arch chemo- and baroreceptors
CN XI	Spinal accessory nerve	Sternomastoid muscle; upper part of the trapezius muscle
CN XII	Hypoglossal nerve	Intrinsic muscles of the tongue

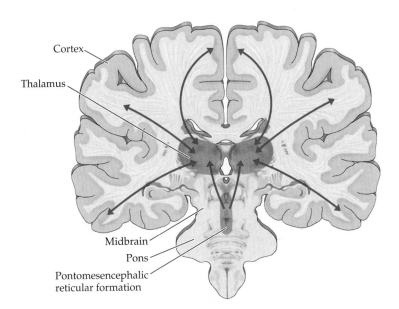

lesions of the cerebral hemispheres. Note that mass lesions above the brainstem often cause impaired consciousness indirectly when they exert pressure on the brainstem through mass effect, thus distorting or compressing the reticular formation and thalamus.

Limbic System

Several structures in the brain are referred to collectively as the **limbic system** because they are located near the medial edge or fringe (*limbus* in Latin) of the cerebral cortex (**Figure 2.24**). These structures have evolved, from a system devoted mainly to olfaction in simpler animals, to perform diverse functions, including the regulation of emotions, memory, appetitive drives, and autonomic and neuroendocrine control. The limbic system includes certain cortical areas located in the medial and anterior temporal lobes (see Figure 2.24A), anterior insula (see Figure 2.24B), inferior medial frontal lobes, and cingulate gyri. It also includes deeper structures, such as the **hippocampal formation** and the **amygdala**, located within the medial temporal lobes (see Figure 2.24A), several nuclei in the medial thalamus, hypothalamus, basal ganglia, septal area, and brainstem. These areas are interconnected by a variety of pathways, including the **fornix**—a paired, arch-shaped white matter structure that connects the hippocampal formation to the hypothalamus and septal nuclei (see Figure 2.24A).

Lesions in the limbic system can cause deficits in the consolidation of immediate recall into longer-term memories. Thus, patients with lesions in these areas may have no trouble recalling remote events but have difficulty forming new memories. In addition, limbic dysfunction can cause behavioral changes and may underlie a number of psychiatric disorders. Finally, epileptic seizures most commonly arise from the limbic structures of the medial temporal lobe, resulting in seizures that may begin with emotions such as fear, memory distortions such as déjà vu, or olfactory hallucinations.

Association Cortex

In addition to the primary motor and sensory areas described earlier, the cerebral cortex contains a large amount of **association cortex** (**Figure 2.25**), which

FIGURE 2.24 **Limbic System Structures** Components of the limbic system in the diencephalon and brainstem are not shown. (A) Medial view, with brainstem removed. (B) Lateral view, with Sylvian fissure opened using retractors.

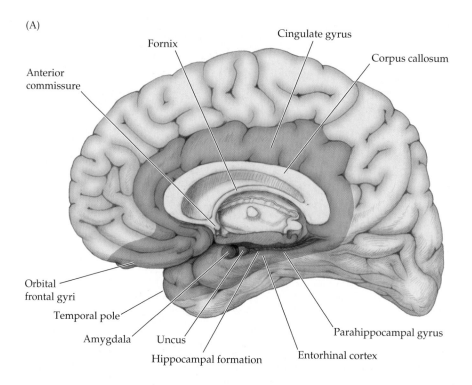

(A)

Fornix

Cingulate gyrus

Corpus callosum

Anterior commissure

Orbital frontal gyri

Temporal pole

Amygdala

Uncus

Hippocampal formation

Entorhinal cortex

Parahippocampal gyrus

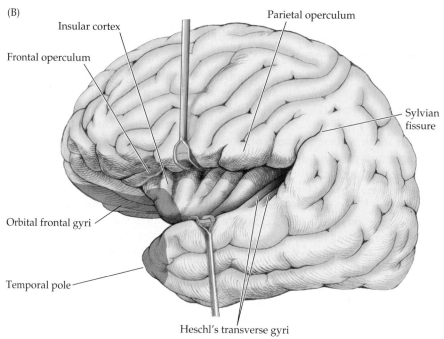

(B)

Insular cortex

Frontal operculum

Parietal operculum

Sylvian fissure

Orbital frontal gyri

Temporal pole

Heschl's transverse gyri

carries out higher-order information processing. In **unimodal** association cortex, higher-order processing takes place mostly for a single sensory or motor modality. Unimodal association cortex is usually located adjacent to a primary motor or sensory area; for example, unimodal visual association cortex is located adjacent to the primary visual cortex, and unimodal motor association cortex (premotor cortex and supplementary motor area) is located adjacent to the primary motor cortex (see Figure 2.25). **Heteromodal** association cortex is involved in integrating functions from multiple sensory and/or motor modalities. We will now briefly review the functions of several clinically important

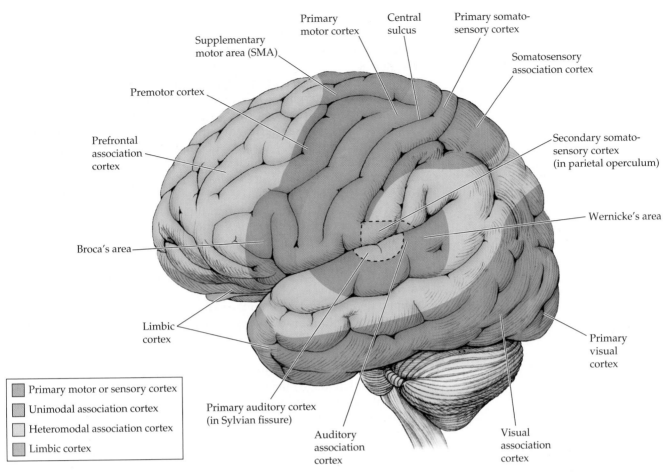

Primary motor cortex

Supplementary motor area (SMA)

Central sulcus

Primary somato-sensory cortex

Premotor cortex

Somatosensory association cortex

Prefrontal association cortex

Secondary somato-sensory cortex (in parietal operculum)

Wernicke's area

Broca's area

Limbic cortex

Primary visual cortex

■ Primary motor or sensory cortex
□ Unimodal association cortex
□ Heteromodal association cortex
■ Limbic cortex

Primary auditory cortex (in Sylvian fissure)

Auditory association cortex

Visual association cortex

FIGURE 2.25 Association Cortex
Lateral view of left hemisphere showing main areas of primary sensory and motor cortex, unimodal association cortex, and heteromodal association cortex.

areas of the association cortex, particularly since we will describe how to examine these structures with the Mental Status exam in the next chapter.

Language is usually perceived first by the primary auditory cortex in the superior temporal lobe when we are listening to speech or by the primary visual cortex in the occipital lobes when we are reading. From here, cortical–cortical association fibers convey information to **Wernicke's area** in the **dominant** (usually left) **hemisphere** (see Figure 2.25). Lesions in Wernicke's area cause deficits in language comprehension, also sometimes called **receptive** or **sensory aphasia**, or **Wernicke's aphasia**. **Broca's area** is located in the frontal lobe, also in the left hemisphere, adjacent to the areas of primary motor cortex involved in moving the lips, tongue, face, and larynx. Lesions in Broca's area cause deficits in the production of language, with relative sparing of language comprehension. This is called **expressive** or **motor aphasia**, or **Broca's aphasia**, which can be remembered by the mnemonic "Broca's broken *boca*" (*boca* means "mouth" in Spanish).

MNEMONIC

The parietal lobe is divided by the intraparietal sulcus into a superior and inferior parietal lobule (see Figures 2.11A,D). Lesions in the inferior parietal lobule in the left hemisphere can produce an interesting constellation of abnormalities, including difficulty with calculations, right–left confusion, inability to identify fingers by name (**finger agnosia**), and difficulties with written language. This group of abnormalities is called **Gerstmann's syndrome**.

Performance of complex motor tasks, such as brushing one's teeth or throwing a baseball, requires higher-order planning before the primary motor cortex can be activated. Motor planning appears to be distributed in many different areas of cortex. Thus, diffuse lesions of the cortex, or sometimes more focal lesions affecting the frontal or left parietal lobe, can produce abnormalities in motor conceptualization, planning, and execution called **apraxia**.

The parietal lobes also play an important role in spatial awareness. Thus, lesions in the parietal lobe, especially in the **nondominant** (usually right) **hemisphere**, often cause a distortion of perceived space and neglect of the contralateral side. For example, right parietal lesions can cause left **hemineglect**. With this syndrome, patients will often ignore objects in their left visual field, but they may see them if their attention is strongly drawn to that side. They may draw a clock face without filling in any numbers on the left side of the clock. They may also be completely unaware of the left side of their body, thinking, for example, that their left arm belongs to someone else, and they may be unaware of any weakness or other deficits on that side. Unawareness of a deficit is called **anosognosia** (from the Greek *a*, "lack"; *nosos*, "disease"; *gnosis*, "knowledge").

Patients may also display a phenomenon called **extinction**, in which a tactile or visual stimulus is perceived normally when it is presented to one side only, but when it is presented on the side opposite the lesion *simultaneously* with an identical stimulus on the normal side, the patient neglects the stimulus on the side opposite the lesion. These severe abnormalities in spatial orientation and awareness are less common in lesions of the dominant (usually left) parietal lobe, possibly because the dominant hemisphere is more specialized for language than it is for visuospatial functions.

The frontal lobes are the largest hemispheres and contain vast areas of association cortex (see Figure 2.25). Lesions in the frontal lobes cause a variety of disorders in personality and cognitive functioning. **Frontal release signs** are "primitive" reflexes that are normal in infants, such as **grasp**, **root**, **suck**, and **snout reflexes**, but that can also be seen in adults with frontal lobe lesions. In addition, patients with frontal lobe lesions may have particular difficulty when asked to perform a sequence of actions repeatedly or to change from one activity to another. In doing these tasks they tend to **perseverate**, meaning that they repeat a single action over and over without moving on to the next one. Personality changes with frontal lobe lesions may include impaired judgment, a cheerful lack of concern about one's illness, inappropriate joking, and other **disinhibited** behaviors. Other patients with frontal lobe lesions may be **abulic** (opposite of ebullient), with a tendency to stare passively and to respond to commands only after a long delay. Frontal lesions can also cause a characteristic unsteady **magnetic gait**, in which the feet shuffle close to the floor, and **urinary incontinence**.

Lesions in the visual association cortex in the parieto-occipital and inferior temporal lobes can produce a variety of interesting phenomena, including **prosopagnosia** (inability to recognize faces), **achromatopsia** (inability to recognize colors), **palinopsia** (persistence or reappearance of an object viewed earlier), and other phenomena. Seizures in the visual association cortex can cause elaborate visual hallucinations.

Blood Supply to the Brain and Spinal Cord

There are two pairs of arteries that carry all the blood supply to the brain and one pair of draining veins (**Figure 2.26A,B**). The **internal carotid arteries** form the anterior blood supply, and the **vertebral arteries**, which join together in a single **basilar artery**, form the posterior blood supply (see Figure 2.26A). The anterior and posterior blood supplies from the carotid and vertebrobasilar systems, respectively, join together in an anastomotic ring at the base of the brain

(A)

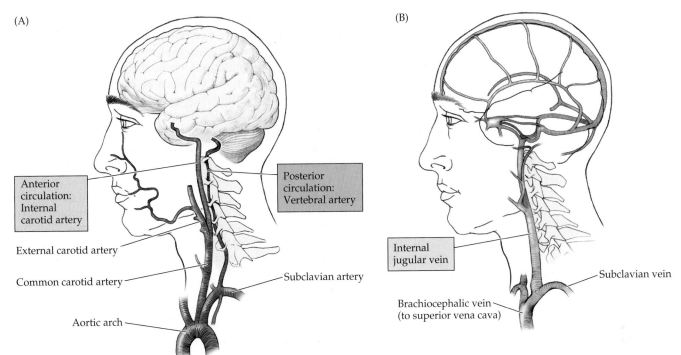

Anterior circulation: Internal carotid artery

Posterior circulation: Vertebral artery

External carotid artery

Common carotid artery

Subclavian artery

Aortic arch

(B)

Internal jugular vein

Subclavian vein

Brachiocephalic vein (to superior vena cava)

FIGURE 2.26 Blood Supply to the Brain (A) Arterial supply is provided by the internal carotid arteries (anterior circulation) and vertebral arteries (posterior circulation). (B) Venous drainage is provided by the internal jugular veins. (C) The circle of Willis. View of the base of the brain from below, showing main arteries arising from the anastomotic ring formed by connections between the anterior (carotid) and posterior (vertebrobasilar) circulations.

(C)

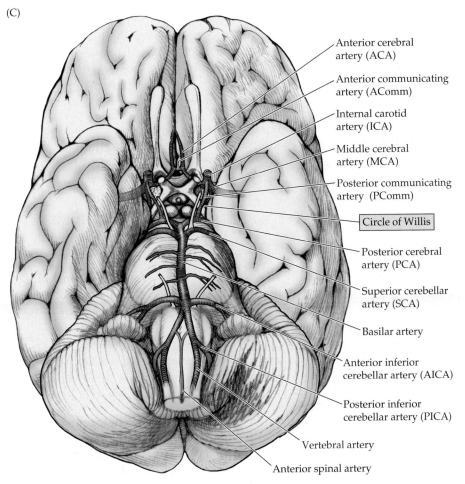

Anterior cerebral artery (ACA)

Anterior communicating artery (AComm)

Internal carotid artery (ICA)

Middle cerebral artery (MCA)

Posterior communicating artery (PComm)

Circle of Willis

Posterior cerebral artery (PCA)

Superior cerebellar artery (SCA)

Basilar artery

Anterior inferior cerebellar artery (AICA)

Posterior inferior cerebellar artery (PICA)

Vertebral artery

Anterior spinal artery

called the **circle of Willis** (see Figure 2.26C). The main arteries supplying the cerebral hemispheres arise from the circle of Willis. Ordinarily, however, the **anterior** and **middle cerebral arteries** derive their main blood supply from the internal carotid arteries (anterior circulation), while the **posterior cerebral arteries** derive their main supply from the vertebrobasilar system (posterior circulation). The main arteries supplying the brainstem and cerebellum also arise from the vertebral and basilar arteries. These include the **superior, anterior inferior**, and **posterior inferior cerebellar arteries**. Venous drainage for the brain is provided almost entirely by the **internal jugular veins** (see Figure 2.26B).

The spinal cord receives its blood supply from the **anterior spinal artery**, which runs along the ventral surface of the cord in the midline and from the paired **posterior spinal arteries**, which run along the right and left dorsal surfaces of the cord (see Figure 6.5). The anterior and posterior spinal arteries are supplied in the cervical region mainly by branches arising from the vertebral arteries (see Figure 2.26C). In the thoracic and lumbar regions, the spinal arteries are supplied by radicular arteries arising from the aorta.

Conclusions

In this chapter we have reviewed the overall structure and organization of the nervous system. In addition, we have discussed in general terms the functions of each of the main areas in the brain, spinal cord, and peripheral nervous system, and we have briefly reviewed the blood supply to the brain. This overview should provide a framework upon which important details will be filled in and reinforced through clinical cases in the chapters that follow. It also provides the necessary background for the next chapter, in which we will introduce the neurologic exam and explore what each portion of the exam can teach us about neuroanatomy. Thus, you, the reader, have completed the rough sketch and now have before you an undertaking that will require innumerable finer and finer brush strokes, each subtly enhancing the final composition.

References

Billings-Gagliardi S, Mazor KM. 2009. Effects of review on medical students' recall of different types of neuroanatomical content. *Acad Med.* 84 (10 Suppl): S34–37.

Carpenter MB. 1991. *Core Text of Neuroanatomy.* 4th Ed. Williams & Wilkins, Baltimore, MD.

Cooper JR, Bloom FE, Roth RH. 2002. *The Biochemical Basis of Neuropharmacology.* 8th Ed. Oxford, New York.

Gorman DG, Unutzer J. 1993. Brodmann's "missing" numbers. *Neurology* 43: 226–227.

Jones EG. 2007. *The Thalamus.* Cambridge University Press, Cambridge, UK.

Kandel ER, Schwartz JH, Jessell TM (eds.). 2000. *Principles of Neural Science.* 4th Ed. McGraw-Hill, New York.

Martin JH. 2003. *Neuroanatomy Text and Atlas.* 3rd Ed. McGraw-Hill, New York.

Mesulam MM (ed.). 2000. *Principles of Behavioral Neurology.* 2nd Ed. Oxford University Press, New York.

Purves D, Augustine GJ, Fitzpatrick D, Hall WC, LaMantia A-S, McNamara JO, White LE (eds.). 2007. *Neuroscience.* 4th Ed. Sinauer, Sunderland, MA.

Steriade M, Jones EG, McCormick DA. 1997. *Thalamus.* Elsevier, Oxford, England.

Zilles K, Amunts K. 2010. Centenary of Brodmann's map—conception and fate. *Nature Rev Neurosci.* 11 (2): 139–145.

CONTENTS

Chapter 3

The Neurologic Exam as a Lesson in Neuroanatomy

Throughout this book, we will encounter case presentations that include findings from the neurologic exam such as the following: "A 37-year-old woman suddenly developed pain and numbness in her right shoulder and fingers. Her general physical exam was unremarkable. A neurologic exam showed that her mental status and cranial nerves were normal; motor exam was notable for diminished right triceps strength; reflexes were absent in the right triceps muscle; coordination and gait were normal; and sensation was normal except for diminished pain and temperature sensation in her right index and middle fingers." Performing the neurologic exam carefully and presenting findings clearly are crucial to accurately diagnosing and effectively treating patients.

In this chapter we will learn about each part of the neurologic exam and its basis in functional neuroanatomy.

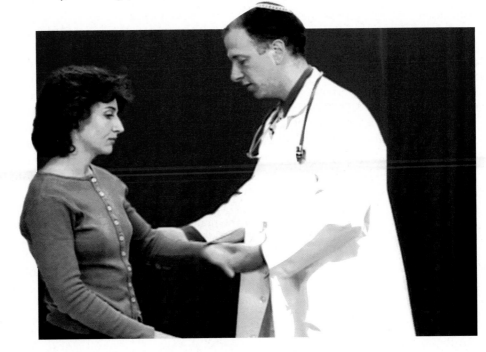

Overview of the Neurologic Exam

THE NEUROLOGIC EXAM as a diagnostic tool gained mythical proportions in the pre-CT/MRI era, when great clinicians could pinpoint a lesion in the nervous system with often astounding accuracy. Decisions for surgery and other interventions were frequently made entirely on the basis of the neurologic history and physical findings. Today, with the availability of modern imaging techniques, the neurologic exam takes on a new and equally important role in diagnosis and management. Rather than serving as an end in and of itself, the neurologic exam today is a critical way station in the clinical decision-making process. Does the patient who just collapsed on the street have cardiac disease or an intracranial bleed? Is the patient with leg weakness and numbness suffering from degenerative joint disease or from impending spinal cord compression? Does the patient with nausea and vomiting need a gastroenterology consult, a head CT, or emergency interventions to lower dangerously elevated intracranial pressure? These—and many similar questions that frequently arise for health care providers in all subspecialties—can quickly be answered by a carefully performed neurologic exam.

We will study the neurologic exam in this chapter with two goals in mind. First, since most of the remainder of the book is based on clinical case presentations, it is important for you to become familiar with the neurologic exam and how to interpret normal or abnormal findings. Second, we will use the neurologic exam in conjunction with the material presented in Chapter 2 to attain an overview of neuroanatomical function and clinical localization.

Although there are some variations in assessment style among clinicians, we describe the neurologic exam here in a fairly conventional format consisting of the following six subdivisions:

1. Mental status
2. Cranial nerves
3. Motor exam
4. Reflexes
5. Coordination and gait
6. Sensory exam

Table 3.1 provides a more detailed outline of the neurologic exam as it will be presented here.

As we discussed in Chapter 1, the neurologic exam is just one part of patient evaluation. Although we discuss it here separately, the neurologic exam should always be performed and interpreted in the context of a more general assessment. This includes the patient history, general physical exam (which includes the neurologic exam as one part) and finally, other diagnostic tests such as radiological studies and blood tests. Review the neurologic information that can be learned from the general physical exam, as discussed in Chapter 1. Some parts of the general exam that have special neurologic significance are summarized in Table 3.2.

A unique feature of the neurologic exam is that it tests *function*. Each part of the neurologic exam should be used to *titrate* the patient's level of function. To do this, the experienced clinician employs several tests for each part of the exam, ranging from the easiest to the most difficult. It is crucial to then record the tests a patient can or cannot perform. This allows comparison with subsequent examinations so that changes in the patient's status can be determined accurately.

In the first section of this chapter we will describe how each part of the neurologic exam is performed. We will include discussions of *what is being*

REVIEW EXERCISE

For each part of the neurologic exam listed in Table 3.1, use the information from Chapter 2 to imagine what part of the nervous system is being tested and how the test could be performed.

TABLE 3.1 Outline of the Neurologic Exam

I. MENTAL STATUS

1. Level of alertness, attention, and cooperation
2. Orientation
3. Memory
   Recent memory
   Remote memory
4. Language
   Spontaneous speech
   Comprehension
   Naming
   Repetition
   Reading
   Writing
5. Calculations, right–left confusion, finger agnosia, agraphia
6. Apraxia
7. Neglect and constructions
8. Sequencing tasks and frontal release signs
9. Logic and abstraction
10. Delusions and hallucinations
11. Mood

II. CRANIAL NERVES

1. Olfaction (CN I)
2. Ophthalmoscopic exam (CN II)
3. Vision (CN II)
4. Pupillary responses (CN II, III)
5. Extraocular movements (CN III, IV, VI)
6. Facial sensation and muscles of mastication (CN V)
7. Muscles of facial expression and taste (CN VII)
8. Hearing and vestibular sense (CN VIII)
9. Palate elevation and gag reflex (CN IX, X)
10. Muscles of articulation (CN V, VII, IX, X, XII)
11. Sternocleidomastoid and trapezius muscles (CN XI)
12. Tongue muscles (CN XII)

III. MOTOR EXAM

1. Observation
   Involuntary movements, tremor, hypokinesia

2. Inspection
   Muscle wasting, fasciculations
3. Palpation
   Tenderness, fasciculations
4. Muscle tone
5. Functional testing
   Drift
   Fine finger movements
   Rapid toe tapping
6. Strength of individual muscle groups

IV. REFLEXES

1. Deep tendon reflexes
2. Plantar response
3. Reflexes tested in special situations
   Suspected spinal cord damage
   Frontal release signs
   Posturing

V. COORDINATION AND GAIT

1. Appendicular coordination
   Rapid alternating movements
   Finger–nose–finger test
   Heel–shin test
   Overshoot
2. Romberg test
3. Gait
   Ordinary gait
   Tandem gait
   Forced gait

VI. SENSORY EXAM

1. Primary sensation—asymmetry, sensory level
   Pain (sharp vs. dull)
   Temperature (cold vs. warm)
   Vibration and joint position sense
   Light touch and two-point discrimination
2. Cortical sensation
   Graphesthesia
   Stereognosis
3. Extinction

tested, correlating clinical findings with neuroanatomy. Much of the material in this first section is also covered in **neuroexam.com** (discussed in the following section), where it is supplemented by video demonstrations.

The remaining sections of this chapter describe the usefulness of the neurologic exam in a variety of special circumstances. Since impairment on any portion of the exam may interfere with the ability to test many other functions, examination strategies and limitations in these situations are discussed. Coma is a special situation in which the neurologic exam is pivotal in evaluation; therefore, a separate section is devoted to this topic. We also briefly discuss the neurologic exam in brain death and in patients with conversion disorder, malingering, and related disorders, before presenting a minimal screening form of the neurologic exam at the conclusion of the chapter.

TABLE 3.2 Parts of the General Physical Exam with Special Neurologic Significance

Vital signs, including orthostatics

Ophthalmoscopic exam

Signs of cranial trauma

Bruits

Cardiac exam, including murmurs, irregular heartbeat

Meningismus

Straight-leg raising

Rectal tone

Dermatologic exam

Note: See also Chapter 1.

The material presented here may be difficult to absorb fully on the first pass, either in print or on the Web. However, by referring back to it while attempting to localize lesions in the clinical cases, you will gradually become proficient in understanding both the neurologic exam and neuroanatomy.

neuroexam.com

A picture—better yet, a moving picture—is worth a thousand words. It is difficult to adequately describe in print the technique used in performing many aspects of the neurologic exam. Therefore, we have provided an interactive website, **neuroexam.com**, which demonstrates through brief, streaming video clips how to perform each part of the neurologic exam. The website also includes much of the text contained in the next section, "The Neurologic Exam: Examination Technique and What Is Being Tested." Therefore, if you have access to the Internet, we suggest that you now browse over to **neuroexam.com** while reading the next section (text from this section also appears on the website). Note, however, that the second half of this chapter, covering the neurologic exam in a variety of special situations, including coma, does not appear on the website.

Some may prefer to read from the book and view video segments selectively. Therefore, in this chapter and elsewhere in the book when discussing the neurologic exam we will provide a cross-reference to the relevant video segment on **neuroexam.com**. The segments can be accessed directly through the "Video Menu" on the website. Alternatively, rather than viewing individual segments, the complete exam can be viewed from start to finish by watching *The NeuroExam Video* (Sinauer, Sunderland, MA).

The Neurologic Exam: Examination Technique and What Is Being Tested

In this section we will describe the neurologic examination (see Table 3.1) and what is being tested by each part of the exam. Like the general physical examination, different parts of the neurologic exam may be done in greater or less detail, depending on the clinical suspicion that a particular lesion is present. For example, an emergency neurologic exam in a comatose patient can be performed in less than 2 minutes (see the "Coma Exam" section later in this chapter). In a patient where the suspicion for focal findings on neurological exam is low, a "screening exam" (as described at the end of this chapter) can be completed in the office setting in about 10 minutes. In contrast, in a patient with unusual findings, where the diagnosis is uncertain, detailed testing may require up to an hour. Sometimes, certain parts of the exam can be combined, or performed in a slightly different order, to minimize the number of times the patient has to change positions. Understanding how to best tailor the exam to the clinical situation comes with experience and practice. The accomplished clinician uses the neurologic exam as a flexible tool. In experienced hands, the neurologic exam remains an unparalleled method both to screen for unsuspected lesions and to test hypotheses for localization.

1. Mental Status

The mental status exam has many different versions. The structure provided here follows a fairly standard format and is organized around the anatomy of the brain, as discussed in the section "Association Cortex" in Chapter 2 (see also Chapters 14, 18, and 19). Thus, we begin with tests that involve global brain function and that determine how well we will be able to perform the rest of the exam (level of alertness, attention, and cooperation). We next ask a few standard questions that make for easy comparisons between different patients or different exams of the same patient (orientation). This part of the exam is followed by testing of limbic and global functions (memory); testing of dominant (usually left) hemispheric language functions (language); and additional testing for left parietal dysfunction (Gerstmann's syndrome), right parietal dysfunction (neglect and constructions), and frontal dysfunction (sequencing tasks and frontal release signs). Finally, we conclude with a few more tests that are less localizing but provide important clues about brain dysfunction (apraxia, logic and abstraction, delusions, hallucinations, and mood).

LEVEL OF ALERTNESS, ATTENTION, AND COOPERATION Be as specific as possible in documenting the level of alertness, making note of what the patient can or cannot do in response to which stimuli (see the section "Coma Exam" later in this chapter). One can test attention by seeing whether the patient can remain focused on a simple task, such as spelling a short word forward and backward (W-O-R-L-D/D-L-R-O-W is a standard), repeating a string of integers forward and backward (**digit span**), or naming the months forward and then backward (see **neuroexam.com Video 4**). Normal digit span is six or more forward and four or more backward, depending slightly on the patient's age and education. It normally takes up to twice as long to recite months backward as forward. Note that these tests of attention depend on language, memory, and some logic functions as well. Degree of cooperation should be noted, especially if it is abnormal, since this will influence many aspects of the exam.

■ *What Is Being Tested?* Level of consciousness is severely impaired in damage to the brainstem reticular formation and in bilateral lesions of the thalami or cerebral hemispheres (see Figure 2.23). It may also be mildly impaired in unilateral cortical or thalamic lesions. Toxic or metabolic factors are also common causes of impaired consciousness because of their effects on the structures mentioned here (see KCC 14.2).

Generalized impaired attention and cooperation are relatively nonspecific abnormalities that can occur in many different focal brain lesions, in diffuse abnormalities such as dementia or encephalitis, and in behavioral or mood disorders (see KCC 19.14–19.16).

ORIENTATION: A CAVEAT TO THOSE WHO WRITE "A&O×3" Ask for the patient's full name, the location, and the date, and note the exact response (see **neuroexam.com Video 5**). It is common practice to use brief phrases in clinical notes such as "alert and oriented" or "alert and oriented to person, place, and time"—abbreviated as "A&O×3"—as a substitute for documenting the full mental status exam. Given realistic time constraints, it is probably reasonable in non-neurologic patients with *normal* mental status to write "A&O×3," as long as the meaning is clear. For patients with compromised mental status, however, it is very important to *document specifically the questions they were asked and how they answered*. This is really the only way to detect changes in a patient's mental status when different doctors are following a patient. For example, for the orientation section on a patient, Harry Smith, you *should* write the following:

Name: "Harry Smith"

Location: "Hospital," but does not know which one

Date: "1942," and does not know month, date, or season

You should *never* write instead: "The patient was A&O×2," since this is ambiguous and makes it hard to know what the patient's true mental status was at the time of the exam.

■ *What Is Being Tested?* The main usefulness of this set of questions is that it is so standard. It mostly tests recent and longer-term memory (see below), but as in all other parts of the exam, the response is also influenced by level of alertness, attentiveness, and language capabilities.

MEMORY It is striking that many patients are able to discuss details of their history reasonably well, and appear to have intact memory in casual discussion, yet have significant memory deficits when explicitly tested. There is, therefore, no substitute for specifically testing memory on mental status exam.

Recent memory. Ask the patient to recall three items or a brief story after a delay of 3 to 5 minutes (see **neuroexam.com Video 6**). Be sure the information has been registered by asking the patient to repeat it immediately before initiating the delay. Provide distractions during the delay to prevent the patient from rehearsing the items repeatedly. A timer, such as a digital watch alarm, should be used to provide a consistent interval from patient to patient—and to prevent the examiner from forgetting to ask for the test items!

Remote memory. Ask the patient about historical or verifiable personal events (see **neuroexam.com Video 7**).

■ *What Is Being Tested?* Memory can be impaired on many different timescales. Impaired ability to register and recall something within a few seconds after it was said is an abnormality that blends into the category of impaired attention discussed earlier. If immediate recall is intact, then difficulty with recall after about 1 to 5 minutes usually signifies damage to the limbic memory structures located in the medial temporal lobes and medial diencephalon (see Figure 2.24; see also KCC 18.1). Dysfunction of these structures causes two characteristic forms of amnesia, which usually coexist. **Anterograde amnesia**, is difficulty remembering new facts and events occurring after lesion onset, and **retrograde amnesia**, is impaired memory of events for a period of time immediately before lesion onset, with relative sparing of earlier memories. Loss of memory which does not fit the typical anterograde/retrograde pattern may signify damage to areas other than the medial temporal and medial diencephalic structures and can also occur in psychogenic amnesia.

LANGUAGE Language, like memory, may seem intact during casual conversation, even when substantial deficits are present. Explicit testing of language is, therefore, a mandatory part of the mental status exam.

1. **Spontaneous speech**. Note the patient's fluency, including phrase length, rate, and abundance of spontaneous speech (see **neuroexam.com Video 8**). Also note tonal modulation and **paraphasic errors** (inappropriately substituted words or syllables), **neologisms** (nonexistent, invented words), or errors in grammar.
2. **Comprehension**. Can the patient understand simple questions and commands? Comprehension of grammatical structure should be tested as well; for example, "Mike was shot by John. Is John dead?" (See **neuroexam.com Video 9**.)
3. **Naming**. Ask the patient to name some easy objects (e.g., pen, watch, tie, etc.) and some more difficult-to-name objects (e.g., fingernail, belt buckle, stethoscope, etc.) objects (see **neuroexam.com Video 10**). Naming parts of objects is often more difficult and should also be tested. Write down what was said to enable follow-up comparisons.
4. **Repetition**. Can the patient repeat single words and sentences (a standard is "no ifs, ands, or buts")? Again, titrate function using tasks that

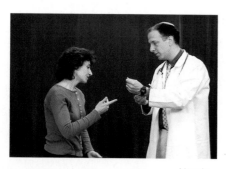

Naming

range from easy to difficult, and write down what the patient says. (See **neuroexam.com Video 11**.)

5. **Reading**. Ask the patient to read aloud single words, a brief passage, and the front page of a newspaper and test for comprehension.
6. **Writing**. Ask the patient to write their name and write a sentence. (See **neuroexam.com Video 12**.)

Reading and writing

■ *What Is Being Tested?* Different kinds of language abnormalities are caused by lesions in the dominant (usually left) frontal lobe, including Broca's area; the left temporal and parietal lobes, including Wernicke's area (see Figure 2.25); subcortical white matter and gray matter structures, including the thalamus and caudate nucleus; as well as the nondominant hemisphere. For further details regarding the neuroanatomy of specific language disorders, see Chapters 2 and 19.

CALCULATIONS, RIGHT–LEFT CONFUSION, FINGER AGNOSIA, AGRAPHIA Impairment of all four of these functions in an otherwise intact patient is referred to as **Gerstmann's syndrome**. Since Gerstmann's syndrome is caused by lesions in the dominant parietal lobe, aphasia is often (but not always) present as well, which can make the diagnosis difficult or impossible. Each of the individual components of Gerstmann's syndrome is poorly localizing on its own, but they are worth documenting as part of the assessment of overall cognitive function:

1. **Calculations**. Can the patient do simple addition, subtraction, and so on? (See **neuroexam.com Video 13**.)
2. **Right–left confusion**. Can the patient identify right and left body parts?
3. **Finger agnosia**. Can the patient name and identify each digit?
4. **Agraphia**. Can the patient write their name and a sentence?

These functions are often tested as part of language assessment (discussed earlier). Right–left confusion and finger agnosia can both be quickly screened for with the classic command, "Touch your right ear with your left thumb." (See **neuroexam.com Video 14**.)

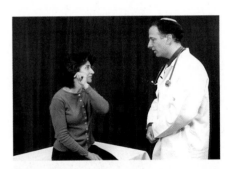

Right ear left thumb

■ *What Is Being Tested?* As we have noted, abnormality of all four of these functions that is out of proportion to other cognitive deficits is strongly localizing to the dominant (usually left) parietal lobe. Otherwise, each of the individual abnormalities can be seen in many different lesions and may be present in individuals with impaired attention, language, praxis (see the next section), constructions, logic and abstraction, and so on.

APRAXIA The term **apraxia** will be used here to mean inability to follow a motor command, when this inability is not due to a primary motor deficit or a language impairment. It is caused by a deficit in higher-order planning or conceptualization of the motor task. You can test for apraxia by asking the patient to do complex tasks, using commands such as "Pretend to comb you hair" or "Pretend to strike a match and blow it out" and so on (see **neuroexam.com Video 15**). Patients with apraxia perform awkward movements that only minimally resemble those requested, despite having intact comprehension and an otherwise normal motor exam. This kind of apraxia is sometimes called **ideomotor apraxia**. In some patients, rather than affecting the distal extremities, apraxia can involve primarily the mouth and face, or movements of the whole body, such as walking or turning around.

Praxis

Unfortunately, the term "apraxia" has also been attached to a variety of other abnormalities—for example, "constructional apraxia" in patients who have visuospatial difficulty drawing complex figures, "ocular apraxia" in patients who have difficulty directing their gaze, "dressing apraxia" in patients who have difficulty getting dressed, and so on. It is unclear at present whether these various types of "apraxia" are related in some way or are caused by completely different mechanisms.

■ *What Is Being Tested?* Although apraxia indicates brain dysfunction, it can be caused by lesions in any of many different regions, so exact localization is often difficult. Apraxia is commonly present in lesions affecting the language areas and adjacent structures of the dominant hemisphere. This can make it challenging to prove that the deficit is apraxia rather than impaired language comprehension. The distinction can often be made by asking patients to perform a task; and then if the patient fails, demonstrate several tasks and ask them to choose the correct one.

NEGLECT AND CONSTRUCTIONS Hemineglect is an abnormality in attention to one side of the universe that is not due to a primary sensory or motor disturbance. In sensory neglect, patients ignore visual, somatosensory, or auditory stimuli on the affected side, despite intact primary sensation (see KCC 19.9). This can often be demonstrated by testing for **extinction on double simultaneous stimulation**. Thus, patients may detect a stimulus when the affected side is tested alone, but when stimuli are presented simultaneously on both sides, only the stimulus on the unaffected side may be detected. In motor neglect, normal strength may be present; however, the patient does not move the affected limb unless attention is strongly directed toward it. Sensory and motor neglect are usually tested as part of the visual, auditory, somatosensory, and motor exams (which will be described in this chapter). During the reading and writing portions of the language exam, patients may be noted to neglect one side of the page.

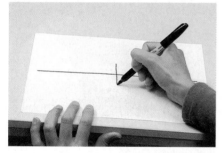

Neglect drawing tests

Copy drawing

During the mental status exam, there are certain other aspects of neglect which should be screened for. Patients should be asked, "Is anything wrong with you right now?" because patients with **anosognosia** may be strikingly unaware of severe deficits on the affected side of their body. For example, some patients with acute stroke who are completely paralyzed on the left side believe there is nothing wrong and may even be perplexed about why they are in the hospital. Some patients do not even comprehend that affected limbs belong to them. In addition, certain drawing tasks, such as asking the patient to bisect a line or draw a clock face, can demonstrate neglect (see **neuroexam.com Video 16**). **Construction tasks** that involve drawing complex figures or manipulating blocks or other objects in space may be abnormal as a result of neglect or other visuospatial impairments (see **neuroexam.com Video 17**). However, constructional abilities can also be abnormal because of other cognitive difficulties, such as impaired sequencing (see the next section) or apraxia.

■ *What Is Being Tested?* Hemineglect is most common in lesions of the right (nondominant) parietal lobe, causing patients to neglect the left side. Left-sided neglect can also occasionally be seen in right frontal lesions, right thalamic or basal ganglia lesions, and, rarely, in lesions of the right midbrain. In left parietal lesions a much milder neglect is usually seen, affecting the patient's right side.

Abnormal constructions demonstrating neglect can occur with right parietal lesions. Other abnormalities in constructions can occur as well, as a result of lesions in many other parts of the brain. Generally, however, impaired visuospatial function is more severe with damage to the nondominant (right) hemisphere (see also KCC 19.9 and KCC 19.10).

SEQUENCING TASKS AND FRONTAL RELEASE SIGNS Patients with frontal lobe dysfunction may have particular difficulty in changing from one action to the next when asked to perform a repeated sequence of actions. For example, when asked to continue drawing a silhouette pattern of alternating triangles and squares, they may get stuck on one shape and keep drawing it (Figure 3.1; see also **neuroexam.com Video 20**). This phenomenon is called **perseveration**. The **Luria manual sequencing task**, in which the patient is asked to tap the table with a fist, open palm, and side of an open hand and then to repeat the sequence as quickly as possible, is also a useful test for perseveration (see **neuroexam.com Video 19**). Another common finding is **motor impersistence**, a form of distractibility in which patients only briefly sustain a motor action in response to a command such as "Raise your arms" or "Look to the right." Ability to suppress inappropriate behaviors can be tested by the **Auditory Go-No-Go Test**, in which the patient moves a finger in response to one tap on the table but must keep it still in response to two taps (see **neuroexam.com Video 21**). Additional support for frontal lobe pathology comes from the presence of **frontal release signs** such as the grasp reflex (see **neuroexam.com Video 18**), described in the section on reflexes later in this chapter. Patients with frontal lobe lesions may also exhibit very slow responses termed **abulia** or changes in personality and judgment based on consecutive exams or reported by family members.

■ *What Is Being Tested?* The constellation of abnormalities just described helps localize lesions to the frontal lobes (see "Association Cortex" in Chapter 2; see also KCC 19.11).

LOGIC AND ABSTRACTION Can the patient solve simple problems such as the following (see **neuroexam.com Video 22**): "If Mary is taller than Jane and Jane is taller than Ann, who's the tallest?" How does the patient interpret proverbs such as "Don't cry over spilled milk"? How well can they comprehend similarities such as "How are a car and an airplane alike?" How well can they generalize and complete a series—for example, "Continue the following: AZ BY CX D_"? A more detailed evaluation can be done, when indicated, using formal, neuropsychological testing batteries. Educational background must always be taken into account in interpretation of these tests.

■ *What Is Being Tested?* These functions can be abnormal in damage to a variety of brain areas involving higher-order association cortex and are not well localized.

DELUSIONS AND HALLUCINATIONS Does the patient have any delusional thought processes? Does the patient have auditory or visual hallucinations? Ask questions such as "Do you ever hear things that other people don't hear or see things that other people don't see?" "Do you feel that someone is watching you or trying to hurt you?" or "Do you have any special abilities or powers?" (See **neuroexam.com Video 23**.)

■ *What Is Being Tested?* These abnormalities can be seen in toxic or metabolic abnormalities and other causes of diffuse brain dysfunction as well as in primary psychiatric disorders (see KCC 18.3). In addition, abnormal sensory phenomena can be caused by focal lesions or seizures in

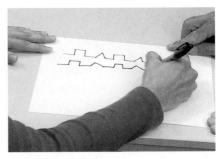

Written alternating sequencing test

Manual alternating sequencing test

Auditory go-no-go

visual, somatosensory, or auditory cortex, and thought disorders can be caused by lesions in the association cortex and limbic system.

MOOD Does the patient have signs of depression, anxiety, or mania? Signs of **major depression** include depressed mood, changes in eating and sleeping patterns, loss of energy and initiative, low self-esteem, poor concentration, lack of enjoyment of previously pleasurable activities, and self-destructive or suicidal thoughts and behavior. Anxiety disorders are characterized by preoccupation with worrisome thoughts. Mania causes patients to be abnormally active and cognitively disorganized.

■ *What Is Being Tested?* These disorders are often considered psychiatric in origin and may be due to imbalances in neurotransmitter systems in several different areas of the brain (see KCC 18.3). However, features of these disorders are also seen in focal brain lesions and in toxic or metabolic abnormalities such as thyroid dysfunction.

Some of the most difficult and interesting diagnostic dilemmas arise because of overlap and confusion between psychiatric and neurologic disorders. Thus, depressed patients with somatization or conversion disorders (which are discussed later in the chapter) often have complaints such as pain, numbness, weakness, or even seizure-like activity and are therefore referred to neurologists for evaluation. Likewise, neurologic disorders such as brain tumors, strokes, metabolic derangements, encephalitis, vasculitis, and so on can produce confusional states or bizarre behavior that may be misinterpreted as psychiatric in origin.

2. Cranial Nerves

Perhaps more than any other part of the neurologic exam, cranial nerve testing can raise "red flags" that suggest specific neurologic dysfunction rather than a systemic disorder. For example, many medical conditions cause lethargy, unsteadiness, headaches, or dizziness. However, any of these symptoms together with cranial nerve abnormalities strongly suggests brainstem dysfunction as the cause (see Chapters 12–14). Careful testing of the cranial nerves can therefore reveal crucial information to help pinpoint disorders in the nervous system. While learning to test the cranial nerves, refer to Figure 2.22 and Table 2.5.

OLFACTION (CN I) Can the patient smell coffee or soap with each nostril? (See **neuroexam.com Video 24**.) Do not use noxious odors, since they may stimulate pain fibers from CN V. CN I is often not tested unless specific pathology, such as a subfrontal brain tumor, is suspected.

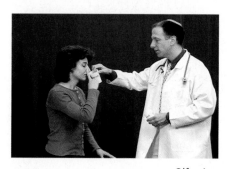

Olfaction

■ *What Is Being Tested?* Impairment can be due to nasal obstruction, damage to the olfactory nerves in the nasal mucosa, damage to the nerves as they cross the cribriform plate, or intracranial lesions affecting the olfactory bulbs (see Figure 2.11C).

OPHTHALMOSCOPIC EXAM (CN II) Examine both retinas carefully with an ophthalmoscope (see **neuroexam.com Video 25**).

■ *What Is Being Tested?* This exam allows direct visualization of damage to the retina or retinal vessels, optic nerve atrophic changes, papilledema (see KCC 5.3), and other important abnormalities.

Ophthalmoscopic exam

VISION (CN II)

1. **Visual acuity**. Test visual acuity for each eye separately (by covering one eye at a time), using an eye chart.

2. **Color vision**. Test each eye separately for ability to distinguish colors. Test for **red desaturation**, a sign of subtle asymmetry in optic nerve function (as seen, for example, in optic neuritis described in KCC 11.4), by asking the patient to cover each eye alternately while looking at a red object and to report any relative dullness of the color as viewed through either eye (see **neuroexam.com Video 26**).

3. **Visual fields**. Test visual fields for each eye by asking the patient to fixate straight ahead and to report when a finger can be seen moving in each quadrant. Alternatively, ask the patient to report how many fingers are being shown in each quadrant (see **neuroexam.com Video 27**). More precise mapping of visual fields can be done in the laboratory for patients who will be followed over time (see KCC 11.2). In comatose or uncooperative patients (discussed later in this chapter), visual fields can be tested roughly using blink-to-threat.

4. **Visual extinction**. Test for visual extinction on **double simultaneous stimulation** by asking patients how many fingers they see when fingers are presented to both sides at the same time (see **neuroexam.com Video 27**). In visual extinction, a form of hemineglect, patients do not report seeing the fingers on the affected (usually left) side of the visual field, although they can see fingers when they are presented to that side alone.

Red desaturation

Visual fields

■ *What Is Being Tested?* Damage anywhere in the visual pathway from the eye to the visual cortex can cause specific deficits in the visual fields of one or both eyes (see Figure 11.15). It is important to note that some visual information from each eye crosses to the opposite side at the **optic chiasm**. Therefore, lesions in front of the optic chiasm (eye, optic nerve) cause visual deficits in one eye, while lesions behind the optic chiasm (optic tract, thalamus, white matter, visual cortex) cause visual field deficits that are similar for both eyes.

Visual hemineglect or extinction is usually caused by contralateral parietal lesions and less often by frontal or thalamic lesions. Neglect is usually more robust in lesions of the right hemisphere (see KCC 19.9).

PUPILLARY RESPONSES (CN II, III) First, record the pupil size and shape at rest. Next, note the **direct response**, meaning constriction of the illuminated pupil, as well as the **consensual response**, meaning constriction of the opposite pupil (see **neuroexam.com Video 29**).

In an **afferent pupillary defect** there is a decreased direct response caused by decreased visual function (CN II) in one eye. However, there is spared pupillary constriction (CN III) when elicited through the consensual response. This can be demonstrated with the **swinging flashlight test**, in which the light is moved back and forth between the eyes every 2 to 3 seconds (see **neuroexam.com Video 30**). The afferent pupillary defect becomes obvious when the flashlight is moved from the normal to the affected eye and the affected pupil *dilates* instead of constricting in response to light. Brief oscillations of pupillary size called **hippus** occur normally in response to light and should not be confused with an afferent pupillary defect.

Finally, test the pupillary response to **accommodation** (also called the **near response**). Normally, the pupils constrict while fixating on an object being moved toward the eyes (see **neuroexam.com Video 31**).

Pupil light reflex

■ *What Is Being Tested?*

1. **Direct response (pupil illuminated)**. The direct response is impaired in lesions of the ipsilateral optic nerve, the pretectal area, the ipsi-

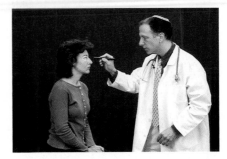

Swinging flashlight test

lateral parasympathetics traveling in CN III, or the pupillary constrictor muscle of the iris (see Figure 13.8).

2. **Consensual response (contralateral pupil illuminated)**. The consensual response is impaired in lesions of the contralateral optic nerve, the pretectal area, the ipsilateral parasympathetics traveling in CN III, or the pupillary constrictor muscle (see Figure 13.8).

3. **Accommodation (response to looking at something moving toward the eye)**. Accommodation is impaired in lesions of the ipsilateral optic nerve, the ipsilateral parasympathetics traveling in CN III, the pupillary constrictor muscle, or in bilateral lesions of the pathways from the optic tracts to the visual cortex. Accommodation to near stimuli is spared in lesions of the pretectal area that may impair the pupillary light response, a condition called "light-near dissociation."

See Chapter 13 for further details on the neuroanatomy of pupillary reflexes and pupillary abnormalities (see KCC 13.5).

EXTRAOCULAR MOVEMENTS (CN III, IV, VI) Check **extraocular movements** (eye movements) by having the patient look in all directions without moving their head. While doing this, ask them if they experience any double vision. Test **smooth pursuit** by having the patient follow an object moved across their full range of horizontal and vertical eye movements (see **neuroexam.com Video 32**).

Test **convergence movements** by having the patient fixate on an object as it is moved slowly toward a point right between the patient's eyes.

Also, observe the eyes at rest to see if there are any abnormalities such as spontaneous nystagmus (which we discuss shortly) or **dysconjugate gaze** (eyes not both fixated on the same point), resulting in **diplopia** (double vision).

Saccades are eye movements used to rapidly refixate from one object to another. The examiner can test saccades by holding up two widely spaced targets (such as the examiner's thumb on one hand and index finger on the other) and asking the patient to look back and forth between the targets (e.g., by saying, "Now look at my finger . . . thumb . . . finger . . . thumb" (see **neuroexam.com Video 33**).

Test **optokinetic nystagmus** (**OKN**) by moving a strip with parallel stripes on it in front of the patient's eyes and asking them to watch the stripes go by (see **neuroexam.com Video 34**). Normally, rhythmic eye movements called **nystagmus** occur, consisting of an alternating **slow phase**, with slow pursuit movements in the direction of strip movement, and a **rapid phase**, with quick, saccadic refixations back to midline. OKN testing can be very useful in detecting subtle abnormalities or asymmetries in saccadic or smooth pursuit eye movements.

In comatose or severely lethargic patients, eye movements can also be evaluated with **oculocephalic** or **caloric testing**. (See "Coma Exam," later in this chapter; see also **neuroexam.com Video 35**.)

■ *What Is Being Tested?* Careful testing can often identify abnormalities in individual muscles or in particular cranial nerves (oculomotor, trochlear, or abducens)—in their course from the brainstem to the orbit, in the brainstem nuclei, or finally, in the higher-order centers and pathways in the cortex and brainstem that control eye movements (review Table 2.5; see also Chapter 13 for more details). Spontaneous nystagmus can indicate toxic or metabolic conditions such as drug overdose, alcohol intoxication, or peripheral or central vestibular dysfunction (see "Hearing and Vestibular Sense (CN VIII)," below).

Smooth pursuit

Saccades

OKN

Oculocephalic testing

FACIAL SENSATION AND MUSCLES OF MASTICATION (CN V) Test facial sensation using a cotton wisp and a sharp object. Also test for **tactile extinction**, using double simultaneous stimulation (see above). The **corneal reflex**, which involves both CN V and CN VII, is tested by touching each cornea gently with a cotton wisp and observing any asymmetries in the blink response (see **neuroexam.com Video 37**).

Feel the masseter muscles during jaw clench (see **neuroexam.com Video 38**). Test for a **jaw jerk reflex** by gently tapping on the jaw with the mouth slightly open (see **neuroexam.com Video 39**).

■ *What Is Being Tested?* Facial sensation can be impaired by lesions of the trigeminal nerve (CN V), the trigeminal sensory nuclei in the brainstem, or ascending sensory pathways to the thalamus and somatosensory cortex in the postcentral gyrus (see Figures 7.9A,B and 12.8). The corneal reflex is mediated by polysynaptic connections in the brainstem between the trigeminal (CN V) and facial (CN VII) nerves and can be impaired by lesions anywhere in this circuit (see KCC 12.4).

Extinction in the presence of intact primary sensation is usually caused by right parietal lesions.

Weakness of the muscles of mastication can be due to lesions in the upper motor neuron pathways synapsing onto the trigeminal (CN V) motor nucleus, in the lower motor neurons of the trigeminal motor nucleus in the pons, in the peripheral nerve as it exits the brainstem to reach the muscles of mastication, in the neuromuscular junction or in the muscles themselves.

Presence of a jaw jerk reflex is abnormal, especially if it is prominent. It is a sign of hyperreflexia associated with lesions of upper motor neuron pathways projecting to the trigeminal motor nucleus. Both the afferent and the efferent limbs of the jaw jerk reflex are mediated by CN V (see KCC 12.4).

MUSCLES OF FACIAL EXPRESSION AND TASTE (CN VII) Look for asymmetry in facial shape or in depth of furrows such as the nasolabial fold. Also look for asymmetries in spontaneous facial expressions and blinking. Facial weakness may be difficult to detect in cases where it occurs bilaterally, also known as **facial diplegia**, because the facial weakness is symmetrical. Ask patients to smile, puff out their cheeks, clench their eyes tight, wrinkle their brow, and so on (see **neuroexam.com Video 40**). Old photographs of the patient can often aid your recognition of subtle changes.

Check taste with sugar, salt, or lemon juice on cotton swabs applied to the lateral aspect of each side of the tongue (see **neuroexam.com Video 41**). Like olfaction, taste is often tested only when specific pathology is suspected, such as in lesions of the facial nerve or in lesions of the gustatory nucleus (nucleus solitarius).

■ *What Is Being Tested?* Facial weakness can be caused by lesions of upper motor neurons in the contralateral motor cortex or descending CNS pathways, lower motor neurons in the ipsilateral facial nerve nucleus (CN VII) or exiting nerve fibers, the neuromuscular junction, or the face muscles. Note that the upper motor neurons for the *upper face* (the upper portions of the orbicularis oculi and the frontalis muscles of the forehead) project to the facial nuclei *bilaterally* (see Figure 12.13). Therefore, **upper motor neuron lesions** such as in stroke cause contralateral face weakness, sparing the forehead, while **lower motor neuron lesions** such as in facial nerve injury typically cause weakness involving the whole ipsilateral face.

Unilateral deficits in taste can occur in lesions of the lateral medulla involving the nucleus solitarius or in lesions of the facial nerve.

Corneal reflex

Masseter

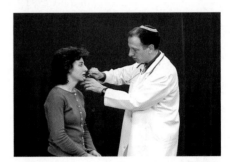

Jaw jerk reflex

Facial muscles

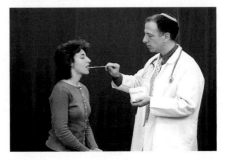

Taste

Hearing

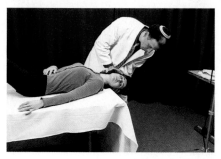

Positional vertigo

HEARING AND VESTIBULAR SENSE (CN VIII) Can the patient hear fingers rubbed together or words whispered just outside of the auditory canal and identify which ear hears the sound? (See **neuroexam.com Video 42**.) A tuning fork can be used to distinguish neural from mechanical conductive hearing problems (see KCC 12.5). Vestibular sense is generally not specifically tested except in the following important situations:

1. **Patients with vertigo**. Certain maneuvers can help distinguish central from peripheral lesions (see KCC 12.6; see also **neuroexam.com Video 43**).
2. **Patients with limitations of horizontal or vertical gaze**. Testing the **vestibulo-ocular reflex** can help localize the lesion (see Chapter 13). The vestibulo-ocular reflex can be tested either using the **oculocephalic maneuver**, in which the eyes are held open and the head is turned rapidly either from side to side or up and down, or by using **caloric testing**, in which cold or warm water is instilled into one ear, producing asymmetrical stimulation of the semicircular canals. See the section "Coma Exam" later in this chapter for further details of these tests and their significance; see also KCC 12.6.
3. **Patients in coma**. The vestibulo-ocular reflex is often the only way to test eye movements in these patients (see the discussion of comatose patients later in this chapter).

■ *What Is Being Tested?* Hearing loss can result from lesions in the acoustic and mechanical elements of the ear, the neural elements of the cochlea, or the acoustic nerve (CN VIII) (see Figure 12.15). After the hearing pathways enter the brainstem, they cross over at multiple levels and ascend *bilaterally* to the thalamus and auditory cortex (see Figure 12.16). Therefore, clinically significant unilateral hearing loss is invariably caused by peripheral neural or mechanical lesions.

Abnormalities in vestibular testing can be associated with lesions in the vestibular apparatus of the inner ear (see Figure 12.15), the vestibular portion of CN VIII, the vestibular nuclei in the brainstem, the cerebellum, or pathways in the brainstem (such as the medial longitudinal fasciculus) that connect the vestibular and oculomotor systems (see Figure 12.19). Further details are provided in Chapter 12 and in the section "Coma exam," later in this chapter.

PALATE ELEVATION AND GAG REFLEX (CN IX, X) Does the palate elevate symmetrically when the patient says, "Aah" (see **neuroexam.com Video 44**)? Does the patient gag when the posterior pharynx is brushed? The **gag reflex** needs to be tested only in patients with suspected brainstem pathology, impaired consciousness, or impaired swallowing.

■ *What Is Being Tested?* Palate elevation and the gag reflex are impaired in lesions involving CN IX, CN X, the neuromuscular junction, or the pharyngeal muscles.

MUSCLES OF ARTICULATION (CN V, VII, IX, X, XII) Is the patient's speech hoarse, slurred, quiet, breathy, nasal, low or high pitched, or otherwise unusual (see **neuroexam.com Video 45**)? It is often important to ask whether the patient's speech has changed from baseline. Note that **dysarthria** (see KCC 12.8) means abnormal pronunciation of speech ("slurred speech"), and should be distinguished from **aphasia**, which is an abnormality in language production or comprehension.

■ *What Is Being Tested?* Abnormal articulation of speech can occur in lesions involving the muscles of articulation, the neuromuscular junction, or the peripheral or central portions of CN V, VII, IX, X, or XII. Furthermore, speech production can be abnormal as a result of lesions in the motor cortex, cerebellum, basal ganglia, or descending pathways to the brainstem.

STERNOCLEIDOMASTOID AND TRAPEZIUS MUSCLES (CN XI) Ask the patient to shrug their shoulders, turn their head in both directions, and raise their head from the bed, flexing forward against the force of your hands (see **neuroexam.com Video 46**).

■ *What Is Being Tested?* Weakness in the sternocleidomastoid or trapezius muscles can be caused by lesions in the muscles, neuromuscular junction, or lower motor neurons of the accessory spinal nerve (CN XI) (see KCC 12.7). Unilateral upper motor neuron lesions in the cortex or descending pathways cause contralateral weakness of the trapezius, with relative sparing of sternocleidomastoid strength. This may be remembered by analogy to upper motor neuron facial lesions sparing the upper portion of the face. When sternocleidomastoid weakness is present with upper motor neuron lesions, there is weakness of head turning away from the side of the lesion (see also KCC 13.10).

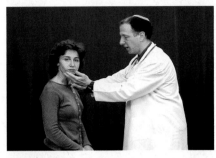

CN XI testing

MNEMONIC

TONGUE MUSCLES (CN XII) Note any atrophy or **fasciculations** (spontaneous, quivering movements) of the tongue while it is resting on the floor of the mouth. Ask the patient to stick their tongue straight out and note whether it curves to one side or the other (see **neuroexam.com Video 47**). Ask the patient to move their tongue from side-to-side and to push it forcefully against the inside of each cheek.

■ *What Is Being Tested?* Fasciculations and atrophy are signs of lower motor neuron lesions (Table 3.3). **Unilateral tongue weakness causes the tongue to deviate toward the weak side**. Tongue weakness can result from lesions of the tongue muscles, the neuromuscular junction, the lower motor neurons of the hypoglossal nerve (CN XII), or the upper motor neurons originating in the motor cortex. Lesions of the motor cortex cause contralateral tongue weakness.

Tongue protrusion

3. Motor Exam

The motor exam has several steps, including (1) observation, (2) inspection, (3) palpation, (4) muscle tone testing, (5) functional testing, and (6) strength testing of individual muscle groups. We now discuss each of these steps in turn.

OBSERVATION Carefully observe the patient to detect any twitches, tremors, or other involuntary movements, as well as any unusual paucity of movement (see KCC 15.2 and KCC 16.1). Note also the patient's posture.

Atrophy? Fasciculations?

TABLE 3.3 Signs of Upper Motor Neuron (UMN) and Lower Motor Neuron (LMN) Lesions

SIGN	UMN LESIONS	LMN LESIONS
Weakness	Yes	Yes
Atrophy	No[a]	Yes
Fasciculations	No	Yes
Reflexes	Increased[b]	Decreased
Tone	Increased[b]	Decreased

[a] Mild atrophy may develop as a result of disuse.
[b] With acute upper motor neuron lesions, reflexes and tone may be decreased.

Upper extremity tone

Lower extremity tone

Drift

Rapid hand movements

Rapid foot tapping

■ *What Is Being Tested?* Involuntary movements and tremors are commonly associated with lesions of the basal ganglia or cerebellum (see KCC 15.2 and KCC 16.1). Tremors can also occasionally be seen with peripheral nerve lesions.

INSPECTION Inspect several individual muscles to see if muscle wasting, hypertrophy, or fasciculations are present (see **neuroexam.com Video 48**). The best muscles to look at for fasciculations in generalized lower motor neuron disorders are the intrinsic hand muscles, shoulder girdle, and thigh.

PALPATION In cases of suspected myositis, palpate the muscles to see if there is tenderness.

MUSCLE TONE TESTING Next, test muscle tone. Ask the patient to relax, and then passively move each limb at several joints to get a feeling for any resistance or rigidity that may be present (see **neuroexam.com Videos 49** and **50**).

■ *What Is Being Tested?* Many parts of the motor exam can help distinguish between **upper motor neuron** and **lower motor neuron** lesions (see Chapters 2 and 6). Recall that upper motor neurons project via the corticospinal tract to lower motor neurons located in the anterior horn of the spinal cord. Signs of lower motor neuron lesions (see Table 3.3) include weakness, atrophy, fasciculations, and **hyporeflexia** (reduced reflexes). (See the section "Reflexes," later in this chapter.) Signs of upper motor neuron lesions include weakness, **hyperreflexia** (increased reflexes), and increased tone. The hyperreflexia and increased tone seen with corticospinal lesions is apparently caused by damage to pathways that travel in close association with the corticospinal tract rather than directly by damage to the corticospinal tract itself. Note that with *acute* upper motor neuron lesions there is often *flaccid paralysis*, with decreased tone and decreased reflexes. With time (hours to weeks), increased tone and hyperreflexia usually develop.

Increased tone can occur in upper motor neuron lesions but can also occur in basal ganglia dysfunction (see KCC 16.1). In addition, slow or awkward fine finger movements or toe tapping in the absence of weakness can signify a subtle abnormality of the corticospinal pathways, but these findings can also occur in lesions of the cerebellum or basal ganglia.

FUNCTIONAL TESTING Before formally testing strength in each muscle, it is useful to do a few general functional tests that help detect subtle abnormalities. Check for **drift** by having the patient hold up both arms simultaneously or both legs simultaneously and close their eyes (see **neuroexam.com Video 51**). Check **fine movements** by testing rapid finger tapping, rapid hand pronation–supination (as in screwing in a light bulb), rapid hand tapping, and rapid foot tapping against the floor or other object (see **neuroexam.com Videos 52** and **53**). Tests for subtle weakness are discussed further in KCC 6.4.

STRENGTH OF INDIVIDUAL MUSCLE GROUPS Patterns of weakness can help localize a lesion to a particular cortical or white matter region, spinal cord level, nerve root, peripheral nerve, or muscle.

Test the strength of each muscle group and record it in a systematic fashion. It is wise to pair the testing of each muscle group immediately with testing of its contralateral counterpart to enhance detection of any asymmetries.

Muscle strength is often rated on a scale of 0/5 to 5/5 as follows:

0/5: No contraction

1/5: Muscle flicker, but no movement

2/5: Movement possible, but not against gravity (test the joint in its horizontal plane)

3/5: Movement possible against gravity, but not against resistance by the examiner

4/5: Movement possible against some resistance by the examiner (sometimes this category is subdivided further into 4–/5, 4/5, and 4+/5)

5/5: Normal strength

While testing muscle strength, it is important to keep in mind anatomical information such as which nerves, nerve roots, and brain areas control each muscle and to allow this information to guide the exam (see **neuroexam.com Videos 54–57**). Also, compare proximal versus distal weakness because these features can sometimes suggest muscle versus nerve disease, respectively.

■ *What Is Being Tested?* A detailed discussion of patterns of muscle weakness and localization is provided in KCC 6.3 and in Chapters 8 and 9. The actions tested are also demonstrated through video clips on the website **neuroexam.com**. In Tables 3.4 and 3.5, we briefly summarize some of the main actions, muscle groups, peripheral nerves, and nerve roots tested during the motor exam.

Upper extremity strength

Lower extremity strength

TABLE 3.4 Upper Extremity Strength Testing

ACTION	MUSCLES	NERVES	NERVE ROOTS
Finger extension at metacarpophalangeal joints	Extensor digitorum, extensor indicis, extensor digiti minimi	Radial nerve (posterior interosseous nerve)	C7, C8
Thumb abduction in plane of palm	Abductor pollicis longus	Radial nerve (posterior interosseous nerve)	**C7**, C8
Finger abduction	Dorsal interossei, abductor digiti minimi	Ulnar nerve	C8, **T1**
Finger and thumb adduction in plane of palm	Adductor pollicis, palmar interossei	Ulnar nerve	C8, **T1**
Thumb opposition	Opponens pollicis	Median nerve	C8, **T1**
Thumb abduction perpendicular to plane of palm	Abductor pollicis brevis	Median nerve	C8, **T1**
Flexion at distal interphalangeal joints, digits 2, 3	Flexor digitorum profundus to digits 2, 3	Median nerve	C7, **C8**
Flexion at distal interphalangeal joints, digits 4, 5	Flexor digitorum profundus to digits 4, 5	Ulnar nerve	C7, **C8**
Wrist flexion and hand abduction	Flexor carpi radialis	Median nerve	C6, C7
Wrist flexion and hand adduction	Flexor carpi ulnaris	Ulnar nerve	C7, **C8**, T1
Wrist extension and hand abduction	Extensor carpi radialis	Radial nerve	C5, **C6**
Elbow flexion (with forearm supinated)	Biceps, brachialis	Musculocutaneous nerve	C5, C6
Elbow extension	Triceps	Radial nerve	C6, **C7**, C8
Arm abduction at shoulder	Deltoid	Axillary nerve	**C5**, C6

Note: When one nerve root is more important than the others, it is shown in boldface.

TABLE 3.5 Lower Extremity Strength Testing

ACTION	MUSCLES	NERVES	NERVE ROOTS
Hip flexion	Iliopsoas	Femoral nerve and L1–L3 nerve roots	L1, L2, L3, L4
Knee extension	Quadriceps	Femoral nerve	L2, **L3**, **L4**
Knee flexion	Hamstrings (semitendinosus, semimembranosus, biceps femoris)	Sciatic nerve	L5, **S1**, S2
Leg abduction	Gluteus medius, gluteus minimus, tensor fasciae latae	Superior gluteal nerve	**L4**, **L5**, S1
Leg adduction	Obturator externus, adductor longus, magnus, and brevis; gracilis	Obturator nerve	**L2**, **L3**, L4
Toe dorsiflexion	Extensor hallucis longus, extensor digitorum longus	Deep peroneal nerve	**L5**, S1
Foot dorsiflexion	Tibialis anterior	Deep peroneal nerve	L4, L5
Foot plantar flexion	Triceps surae (gastrocnemius, soleus)	Tibial nerve	S1, S2
Foot eversion	Peroneus longus, peroneus brevis	Superficial peroneal nerve	L5, S1
Foot inversion	Tibialis posterior	Tibial nerve	L4, L5

Note: When one nerve root is more important than the others, it is shown in boldface.

4. Reflexes

The **deep tendon reflexes** and **plantar response** should be checked in all patients. Certain other reflexes should also be tested in special situations, as described in the sections that follow.

Deep tendon reflexes

DEEP TENDON REFLEXES Check the deep tendon reflexes using impulses from a reflex hammer to stretch the muscle and tendon (see **neuroexam.com Video 58**). The limbs should be in a relaxed and **symmetrical** position, since these factors can influence reflex amplitude. As in muscle strength testing, it is important to compare each reflex immediately with its contralateral counterpart so that any asymmetries can be detected. If you cannot elicit a reflex, you can sometimes bring it out by certain **reinforcement** procedures. For example, have the patient gently contract the muscle being tested by instructing the patient to raise the limb very slightly, or to concentrate on forcefully contracting muscles in a different part of the body just at the moment when the reflex is tested. When reflexes are very brisk, clonus is sometimes seen. **Clonus** is a repetitive vibratory contraction of the muscle that occurs in response to muscle and tendon stretch. Deep tendon reflexes are often rated according to the following scale:

0:	Absent reflex
1+:	Trace, or seen only with reinforcement
2+:	Normal
3+:	Brisk
4+:	Nonsustained clonus (i.e., repetitive vibratory movements)
5+:	Sustained clonus

Deep tendon reflexes are normal if they are 1^+, 2^+, or 3^+ unless they are *asymmetrical* or there is a dramatic difference between the arms and the legs. Reflexes rated as 0, 4^+, or 5^+ are usually considered abnormal. In addition to clonus, other signs of hyperreflexia include **spreading** of reflexes to other muscles not directly being tested and **crossed adduction** of the opposite leg when the medial aspect of the knee is tapped. **Hoffmann's sign** indicates heightened reflexes involving the finger flexor muscles. You can elicit this sign by holding the patient's middle finger loosely and flicking the fingernail downward, causing the finger to

Finger flexors

(A) Normal plantar response

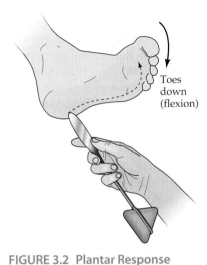

Toes down (flexion)

(B) Extensor plantar response (Babinski's sign)

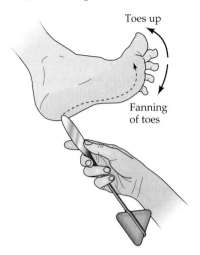

Toes up

Fanning of toes

FIGURE 3.2 Plantar Response

NERVE REFLEX	MAIN SPINAL ROOTS INVOLVED
TABLE 3.6 Deep Tendon Reflexes	
Biceps	C5, C6
Brachioradialis	C6
Triceps	C7
Patellar	L4
Achilles tendon	S1

rebound slightly into extension (see **neuroexam.com Video 60**). If the thumb flexes and adducts in response, Hoffmann's sign is present.

■ *What Is Being Tested?* Deep tendon reflexes (see Figure 2.21) may be diminished by abnormalities in muscles, sensory neurons, lower motor neurons, and the neuromuscular junction; acute upper motor neuron lesions; and mechanical factors such as joint disease. Abnormally increased reflexes are associated with upper motor neuron lesions (see Table 3.3). Note that deep tendon reflexes can be influenced by age, metabolic factors such as thyroid dysfunction or electrolyte abnormalities, and the anxiety level of the patient. The main spinal nerve roots involved in testing the deep tendon reflexes are summarized in Table 3.6.

PLANTAR RESPONSE Test the plantar response by scraping an object across the sole of the foot beginning from the heel, moving forward toward the small toe, and then arcing medially toward the big toe (Figure 3.2; see also **neuroexam.com Video 59**). The normal response is downward contraction of the toes. The abnormal response, called **Babinski's sign**, is characterized by an upgoing big toe and fanning outward of the other toes. In some patients the toes are "silent," moving neither up nor down. If the toes are downgoing on one side and silent on the other, the silent side is considered abnormal. The presence of Babinski's sign in an adult is always abnormal, but it is often present in infants, up to the age of about 1 year.

■ *What Is Being Tested?* Babinski's sign is associated with upper motor neuron lesions anywhere along the corticospinal tract. Note that it may not be possible to elicit Babinski's sign if there is severe weakness of the toe extensors.

A concise way to document reflexes at the biceps, triceps, brachioradialis, patellar, and Achilles tendons, as well as the plantar responses, is to illustrate them with a stick figure. An example for a normal patient is shown in Figure 3.3.

REFLEXES TESTED IN SPECIAL SITUATIONS Additional reflexes are tested in special situations such as coma, spinal cord injury, frontal lobe dysfunction, and neurodegenerative disorders.

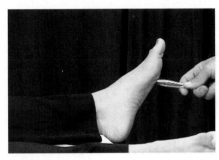

Plantar response

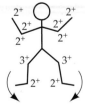

FIGURE 3.3 **Reflex Stick Figure** Commonly tested deep tendon reflexes and plantar responses are summarized pictorially using stick figures, a widespread clinical shorthand.

TABLE 3.7   Additional Reflexes for Localizing Spinal Cord Lesions

REFLEX	SPINAL NERVE ROOTS INVOLVED
Abdominal cutaneous reflexes	
Above umbilicus	T8–T10
Below umbilicus	T10–T12
Cremasteric reflex	L1–L2
Bulbocavernosus reflex	S2–S4
Anal wink	S2–S4

With suspected **spinal cord damage**, absence of certain normal reflexes can help localize the level and ascertain the severity of the lesion (Table 3.7). Elicit **abdominal cutaneous reflexes** by scraping the abdominal skin on each side, above and below the umbilicus, and observing the abdominal muscle contractions. Elicit the **cremasteric reflex** in males by scraping the upper inner thigh and observing ascent of the testicle. The **bulbocavernosus reflex** is contraction of the rectal sphincter in response to pressure on the bulbocavernosus muscle. In males this response can be elicited by compressing the glans penis; in females, since a Foley catheter is usually present in the urethra in this clinical setting, traction on the Foley catheter can elicit the response. **Anal wink** is contraction of the rectal sphincter in response to a sharp stimulus in the perianal area.

Frontal lobe lesions in adults can cause the re-emergence of certain primitive reflexes that are normally present in infants but are pathological in adults. These so-called **frontal release signs** include the **grasp**, **snout**, **root**, and **suck reflexes** (see **neuroexam.com Video 18**, for a demonstration of the grasp reflex). Frontal release signs are sometimes tested for during the mental status exam when there is suspicion of frontal lobe pathology. Two other reflexes deserve mention, although they are less specific and can be seen in a wide variety of neurodegenerative conditions. Elicit the **glabellar response** by tapping with a finger repeatedly in the midline between the eyes and asking the patient to keep their eyes open. The normal patient may blink a few times, but the response then extinguishes. The abnormal response—continued blinking with each tap (Myerson's sign)—is most commonly seen in neurodegenerative movement disorders such as Parkinson's disease. In the **palmomental reflex**, scraping the hypothenar eminence causes ipsilateral contraction of the mentalis muscles of the chin. This response is very nonspecific, and is present in some normal individuals.

In patients with damage to descending motor pathways, **posturing** can sometimes be seen. Posturing consists of complex reflexes involving brainstem and spinal cord circuitry, as described in the section "Coma Exam," later in this chapter.

■ *What Is Being Tested?* Special reflexes tested in spinal cord lesions help localize the level of the damage (see Table 3.7; see also Table 8.1 and Figure 8.4). Frontal release signs support the localization of lesions to the frontal lobes (see KCC 19.11). The localizing value of posturing will be discussed later in this chapter.

5. Coordination and Gait

Coordination and gait are usually described under a separate section because cerebellar disorders can disrupt coordination or gait while leaving other motor functions relatively intact. There is much overlap, however, between the systems being examined in this section and those examined in the earlier, general motor exam section as well as in other parts of the exam. Keep in mind that disturbances of coordination and gait can be caused by lesions in many systems other than the cerebellum.

The term **ataxia** is often used to describe the abnormal movements seen in coordination disorders (see KCC 15.2). In ataxia there are medium- to large-amplitude involuntary movements with an irregular oscillatory quality superimposed on and interfering with the normal smooth trajectory of movement (see Figure 3.4B). **Overshoot** is also commonly seen as part of ataxic movements and is sometimes referred to as **past pointing** when target-oriented movements are being discussed. Another feature of coordination disorders is **dysdiadochokinesia**—that is, abnormal alternating movements.

Cerebellar lesions can cause different kinds of coordination problems depending on their location. One important distinction is between truncal ataxia and appendicular ataxia (see KCC 15.2). **Appendicular ataxia** affects movements of the extremities and is usually caused by lesions of the cerebellar hemispheres and associated pathways (see Figure 15.3). **Truncal ataxia** affects the proximal musculature, especially that involved in gait stability, and is caused by midline damage to the cerebellar vermis and associated pathways.

Finger–nose–finger

APPENDICULAR COORDINATION Fine movements of the hands and feet, as discussed earlier as part of the general motor exam, should be tested. **Rapid alternating movements**, such as wiping one palm alternately with the palm and dorsum of the other hand, should be tested as well (see **neuroexam.com Video 62**). Perhaps the most popular test of coordination, however, is the **finger–nose–finger test**, in which the patient is asked to alternately touch their nose and the examiner's finger as quickly as possible (**Figure 3.4**; see also **neuroexam.com Video 64**). Ataxia is best revealed if the examiner's finger is held at the extreme of the patient's reach and if the examiner's finger is occasionally moved suddenly to a different location. Test for overshoot by having the patient raise both arms suddenly from their lap to the level of your

Overshoot

(A)

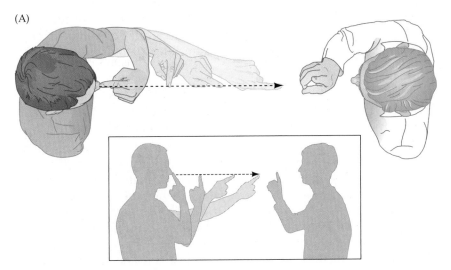

(B)

FIGURE 3.4 Finger–Nose–Finger Test (A) Normal patient. (B) Patient with ataxia.

Precision finger tap

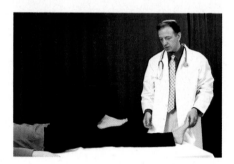

Heel–shin test

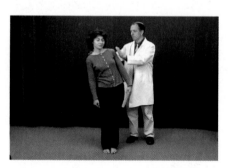

Romberg test

hand (see **neuroexam.com Video 66**). In addition, you can apply pressure to the patient's outstretched arms and then suddenly release it. To test the accuracy of movements in a way that requires very little strength, you can draw a line on the crease of the patient's thumb and then ask the patient to touch the line repeatedly with the tip of their forefinger (see **neuroexam.com Video 63**). This test can help distinguish between irregular, wavering movements caused by limb weakness and abnormal movements caused by ataxia.

Similar tests can be done with the legs. In the **heel–shin test** the patient is asked to touch the heel of one foot to the opposite knee and then to drag their heel in a straight line all the way down the front of their shin and back up again. In order to eliminate the effect of gravity in moving the heel down the shin, this test should always be done in the supine position (see **neuroexam.com Video 65**). Testing for ataxia is discussed further in KCC 15.2.

■ *What Is Being Tested?* Normal performance of these motor tasks depends on the integrated functioning of multiple sensory and motor subsystems. These include position sense pathways, visual pathways, lower motor neurons, upper motor neurons, the basal ganglia, and the cerebellum. Thus, in order to convincingly demonstrate that abnormalities are due to a cerebellar lesion, one must first test for normal joint position sense, vision, strength, and reflexes and confirm the absence of involuntary movements caused by basal ganglia lesions. As already mentioned, *appendicular ataxia* is usually caused by lesions of the cerebellar hemispheres and associated pathways, while *truncal ataxia* (see the following two sections, on the Romberg test and on gait) is often caused by damage to the midline cerebellar vermis and associated pathways (see Figures 15.3 and 15.9).

ROMBERG TEST Ask the patient to stand with their feet together (touching each other). Then ask the patient to close their eyes. Remain close at hand in case the patient begins to sway or fall (see **neuroexam.com Video 67**).

■ *What Is Being Tested?* With the eyes open, three sensory systems provide input to the cerebellum to maintain truncal stability. These are vision, proprioception, and vestibular sense. If there is a mild lesion in the vestibular or proprioception systems, when the eyes are open the patient is usually able to compensate and remain stable. When the patient closes their eyes, however, visual input is removed and instability can be brought out. If there is a more severe proprioceptive or vestibular lesion or if there is a midline cerebellar lesion causing truncal instability, the patient will be unable to maintain this position even with their eyes open.

Note that instability can also be seen with lesions in other parts of the nervous system, such as upper or lower motor neurons or the basal ganglia, so these other systems should be tested separately in the remainder of the exam.

GAIT A patient's gait can be difficult to describe in a reproducible fashion. Observe the patient walking toward you and away from you in an open area with plenty of room. Note **stance** (how far apart the feet are), posture, stability, how high the feet are raised off the floor, trajectory of leg swing and whether there is **circumduction** (an abnormal arced trajectory in the medial to lateral direction), leg stiffness and degree of knee bending, arm swing, tendency to fall or swerve in any particular direction, rate and speed, difficulty initiating or stopping gait, and any involuntary movements that are brought out by walking. Turns should also be observed closely. When following a patient over several visits, it may be useful to time the patient walking a fixed distance and to count the number of steps taken and the number

Gait

of steps required to turn around. The patient's ability to rise from a chair with or without assistance should also be recorded.

To bring out abnormalities in gait and balance, ask the patient to do more difficult maneuvers. Test **tandem gait** by asking the patient to walk a straight line while touching the heel of one foot to the toe of the other with each step (see **neuroexam.com Video 68**). Patients with *truncal ataxia* caused by damage to the cerebellar vermis (see Figure 15.3) or associated pathways will have particular difficulty with this task, since they tend to have a wide-based, unsteady gait and become more unsteady when attempting to keep their feet close together. To bring out subtle gait abnormalities or asymmetries, it may be appropriate in some cases to ask patients to perform so-called **forced gait** testing by asking them to walk on their heels, their toes, or the insides or outsides of their feet; to stand or hop on one leg; or to walk up stairs (see **neuroexam.com Video 69**).

Gait apraxia is a perplexing (and somewhat controversial) abnormality in which the patient is able to carry out all of the movements required for gait normally when lying down but is unable to walk in the standing position. It is thought to be associated with frontal disorders or normal pressure hydrocephalus (see KCC 5.7).

■ *What Is Being Tested?* As with tests of appendicular coordination, gait involves multiple sensory and motor systems. These include vision, proprioception, vestibular sense, lower motor neurons, upper motor neurons, basal ganglia, the cerebellum, and higher-order motor planning systems in the association cortex. Once again, it is important to test each of these systems for normal function before concluding that a gait disturbance is caused by a cerebellar lesion. Localization and diagnosis of gait disorders is described further in KCC 6.5 and Table 6.6.

6. Sensory Exam

The sensory exam relies to a large extent on the ability or willingness of patients to report what they are feeling. It can therefore often be the most difficult part of the exam to interpret with certainty. Tests should be performed in all extremities, as well as on the face and trunk, with the patient's eyes closed or covered to improve objectivity.

PRIMARY SENSATION, ASYMMETRY, SENSORY LEVEL Light touch is best tested with a cotton-tipped swab, but a light finger touch will often suffice, as long as care is taken to make the stimulus fairly reproducible. You can test the relative sharpness of **pain sensation** by randomly alternating stimuli with the sharp or dull end of a safety pin or broken wooden swab (always use a new pin or stick for each patient; see **neuroexam.com Video 70**). **Temperature sensation** can be tested with a cool piece of metal such as a tuning fork (see **neuroexam.com Video 71**). Test **vibration sense** by placing a vibrating tuning fork on the ball of the patient's right or left large toe or fingers and asking them to report when the vibration stops (see **neuroexam.com Video 72**). Classically, a low-frequency tuning fork (128 or even 64 Hz) is used to test vibration, although some claim that higher-frequency tuning forks (256 or even 512 Hz) may be used. Take care not to place the tuning fork on a bone, since bones conduct the vibration to much more proximal sites, where they can be detected by nerves far from the location being tested. Test **joint position sense** by moving one of the patient's fingers or toes upward and downward and asking the patient to report which way it moves (see **neuroexam.com Video 73**). Hold the digit lightly by the sides while doing this so that tactile inputs don't provide significant clues to the direction of movement. The digit should be moved very slightly because normal individuals can detect move-

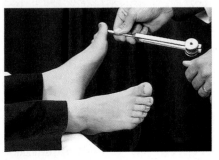

Vibration sense

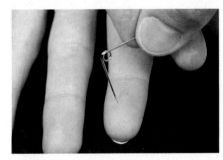

Temperature sensation

Pinprick

Joint position sense

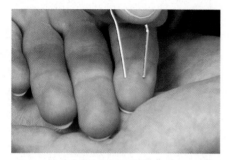

Two-point discrimination

Graphesthesia

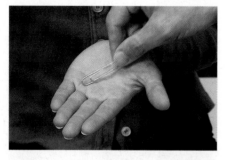

Stereognosis

Tactile extinction

ments that are barely perceptible by eye. **Two-point discrimination** can be tested with a special pair of calipers or with a bent paper clip, alternating randomly between touching the patient with one or both points (see **neuroexam.com Video 74**). The minimal separation (in millimeters) at which the patient can distinguish these stimuli should be recorded in each extremity and compared from side to side.

As in other parts of the exam, the patient's deficits, as well as the anatomy of the nerves, nerve roots, and central pathways, should be used to guide the exam (see Chapters 7–9). Comparisons should be made from one side of the body to the other and from proximal to distal on each extremity. Note especially if there is a **sensory level** corresponding to a particular spinal segment (see Figure 8.4) below which sensation abruptly changes, since such a change may indicate a spinal cord lesion requiring emergency intervention. Whenever there are uncertainties in the sensory exam or other parts of the neurologic exam, a good strategy is to repeat the relevant portions of the exam several times.

CORTICAL SENSATION, INCLUDING EXTINCTION Higher-order aspects of sensation, or **cortical sensation**, should be tested as well. To test **graphesthesia**, ask patients to close their eyes and identify letters or numbers that are being traced onto their palm or the tip of their finger (see **neuroexam.com Video 75**). To test **stereognosis**, ask patients to close their eyes and identify various objects by touch, using one hand at a time (see **neuroexam.com Video 76**). Test also for **tactile extinction** on double simultaneous stimulation (as described earlier; see **neuroexam.com Video 77**). Note that graphesthesia, stereognosis, and extinction cannot reliably be tested for unless primary sensation is intact bilaterally.

■ *What Is Being Tested?* Somatosensory deficits can be caused by lesions in peripheral nerves, nerve roots, the posterior columns or anterolateral sensory systems in the spinal cord or brainstem, the thalamus, or sensory cortex. Recall that position and vibration sense ascend in the posterior column pathway and cross over in the medulla, while pain and temperature sense cross over shortly after entering the spinal cord and then ascend in the anterolateral pathway (see Figures 2.13, 2.18, and 2.19). Intact primary sensation with deficits in cortical sensation such as agraphesthesia or astereognosis suggests a lesion in the contralateral sensory cortex. Note, however, that severe cortical lesions can cause deficits in primary sensation as well. Extinction with intact primary sensation is a form of hemineglect that is most commonly associated with lesions of the right parietal lobe. Like other forms of neglect, extinction can also occasionally be seen in right frontal or subcortical lesions, or in left hemisphere lesions causing mild right hemineglect.

The pattern of sensory loss can provide important information that helps localize lesions to particular nerves, nerve roots, and regions of the spinal cord, brainstem, thalamus, or cortex (see KCC 7.3 and Chapters 8 and 9).

The Neurologic Exam as a Flexible Tool

In actual clinical practice, the neurologic exam just described is hardly ever performed in its entirety from start to finish. As we mentioned earlier, with experience, one learns how to perform a screening exam covering the most important parts of the assessment and then focuses on relevant portions of the exam in greater detail, depending on the clinical situation. In the sections that follow, we will discuss some alternative exam strategies and limitations in interpretation in certain situations as well as examination techniques used for the following special circumstances: coma, brain death, and somatoform psychiatric disorders.

Exam Limitations and Strategies

One of the more challenging aspects of the neurologic exam is that impairments in one area can affect the patient's ability to perform other parts of the exam. For example, if a patient is not fully alert, attentive, and cooperative, and does not have intact language functions during the mental status exam, then detailed testing of each individual muscle group will not be possible during the motor exam. Therefore, the examiner must be prepared to modify the exam appropriately on the basis of the patient's limitations.

In addition to strategies for working around limitations, it is important to know when parts of the exam cannot be done reliably in different situations. Documentation of the patient's limitations and how they affected the exam are essential. For example, in a patient with impaired attention, it is preferable to write "joint position sense testing was unreliable due to impaired attention" rather than "joint position sense was poor."

We will now describe a few specific strategies for examining patients with deficits. In mild to moderate **impairment of alertness or attention**, most aspects of the exam can be done with repeated stimulation of the patient. Patience is required. In severe impairment, the various strategies outlined later for the comatose patient are employed. With the **uncooperative patient**, different strategies are used depending on the situation. These include careful observation of spontaneous speech and movements and, when appropriate, some techniques from the coma exam (see the next section).

With mild to moderate **impairment in language comprehension**, patients can often understand some simple questions or commands and sometimes have an easier time understanding gestures or demonstrations of the action desired. For impaired expression, the patient can be asked yes–no or multiple-choice questions. For testing memory, several objects can be hidden around the room and the patient can be asked to find them after a delay. Again, patience and creativity are essential.

With **sensory and motor neglect**, some patients will show an improvement when the head is turned toward the affected side. With motor neglect, some patients show improved performance when their attention is repeatedly directed to the affected side, in response to pain, or during activities requiring bimanual coordination. When performing the motor exam, patients with **apraxia** may initially appear weak, but show improved strength if the movement is demonstrated or if their limbs are moved through the desired motion before allowing them to continue on their own. Sometimes, midline and appendicular motor functions are differentially affected. For example, some patients can close their eyes and stick out their tongue on command but cannot squeeze the examiner's hand. In **motor impersistence**, estimates of limb weakness or of gaze deficits must be made based on any deficits in the brief movements achieved by the patient.

In patients with bilateral **deafness**, communication can usually be achieved by writing or by using a sign language interpreter, and nonauditory language comprehension can still be tested. In patients who are **unable to speak** due to motor deficits or who are intubated, expressive communication can often be achieved by providing the patient with a pen and paper, computer keyboard, or other device or code, including eye movements or blinking for patients with severe motor impairment (see KCC 14.1).

Numerous additional strategies can be employed for examining patients in special situations, some of which we will describe in the sections that follow.

REVIEW EXERCISE

Now that we have discussed each part of the neurologic exam, review Table 3.1 once again. For each heading in the table, review in your mind (1) how that part of the exam is performed, and (2) what neuroanatomical systems are being tested.

Coma Exam

The coma exam is much simpler than the exam in the awake patient. The same general format is used; however, the exam is shorter because many things cannot be tested without the patient's cooperation. An advantage of the short length of the coma exam is that it can be done quickly in emergency situations, thus enabling a rapid assessment of neurologic status that often yields critical information for patient care.

For patients who are neither fully awake nor fully unconscious, various elements of the coma exam and the exam in the awake patient should be combined. Similarly, for the uncooperative but awake patient, some elements of the coma exam may be useful. Review the overall format of the coma exam shown in Table 3.8. Since many aspects of the coma exam are the same as in the general neurologic exam, these elements will be described more briefly here; refer to the previous discussion and earlier sections titled "What Is Being Tested?" for details.

General Physical Exam

As always, a good general physical exam should be performed in the comatose patient. Particular attention should be paid to parts of the exam that could reveal the cause of coma (see Table 3.2), such as vital signs, airway status, signs of cranial trauma (Table 3.9), nuchal rigidity (see KCC 5.9), and so on. If trauma is suspected, the neck should be immobilized with a rigid cervical collar.

TABLE 3.8 Outline of the Neurologic Exam in the Comatose Patient

I. MENTAL STATUS

Document level of consciousness with a *specific statement* of what the patient did in response to particular stimuli

II. CRANIAL NERVES

1. Ophthalmoscopic exam (CN II)
2. Vision (CN II)
   Blink-to-threat
3. Pupillary responses (CN II, III)
4. Extraocular movements and vestibulo-ocular reflex (CN III, IV, VI, VIII)
   Spontaneous extraocular movements
   Nystagmus
   Dysconjugate gaze
   Oculocephalic maneuver (doll's eyes test)
   Caloric testing
5. Corneal reflex, facial asymmetry, grimace response (CN V, VII)
6. Gag reflex (CN IX, X)

III. SENSORY AND IV. MOTOR EXAM

1. Spontaneous movements
2. Withdrawal from a painful stimulus

V. REFLEXES

1. Deep tendon reflexes
2. Plantar response
3. Posturing reflexes
4. Special reflexes in cases of suspected spinal cord lesions (see Table 3.7)

VI. COORDINATION AND GAIT

Usually not testable

TABLE 3.9 Important External Signs of Cranial Trauma

NAME	DESCRIPTION
Bony step-off	Palpable discontinuity in shape of skull due to displaced fracture
CSF rhinorrhea	Exudation of cerebrospinal fluid from the nose due to base-of-skull fracture usually involving the ethmoid bone
CSF otorrhea	Exudation of cerebrospinal fluid from the ear due to base-of-skull fracture usually involving the temporal bone
Hemotympanum	Dark purple blood visible behind the tympanic membrane caused by base-of-skull fracture usually involving the temporal bone
Battle's sign	Dark purple ecchymoses visible in the skin overlying the mastoid processes due to base-of-skull fracture and blood exudation into the subcutaneous tissues
Raccoon eyes	Dark purple ecchymoses visible in the skin around the eyes due to base-of-skull fracture and blood exudation into the subcutaneous tissues

1. Mental Status

Coma was defined in the classic work of Plum and Posner as unarousable unresponsiveness in which the patient lies with eyes closed. There is a wide continuum of levels of consciousness between coma and the fully awake state. A variety of more poorly defined terms are sometimes used to describe different states along this continuum, such as lethargy, stupor, obtundation, semicoma, and so on. Use of these terms without further details can be confusing to other physicians when they read the chart and try to assess the patient's progress. Therefore, it is essential to document the patient's level of alertness with a *specific statement* of what the patient did in response to particular stimuli. For example:

"The patient opens eyes and turns toward voice but obeys no verbal commands," or

"The patient responds only to painful sternal rub by moving the right arm and grimacing," or

"The patient is unresponsive to voice and sternal rub."

Level of consciousness is often the only part of the mental status exam that can be performed in these patients. Recall that consciousness can be impaired by damage to the brainstem reticular formation and in bilateral lesions of the thalami or cerebral hemispheres (see Figure 2.23). It may also be mildly impaired in unilateral cortical or thalamic lesions. Toxic or metabolic factors are also common causes of impaired consciousness because of their effects on these structures (see KCC 14.2 and KCC 19.14–19.16).

A number of conditions can be mistaken for coma, but are actually quite different from coma in anatomy and pathophysiology. Large lesions involving the frontal lobes or their connections can cause a condition of extreme abulia resembling coma called **akinetic mutism** (see Table 14.3). In this state the patient has profoundly decreased initiative and minimal responsiveness, but the eyes are usually open and there may be occasional normal-appearing movements. **Catatonia** is another condition in which there is profoundly decreased responsiveness due to psychiatric illness. In **locked-in syndrome**, consciousness and sensation may be normal, but the patient is unable to move

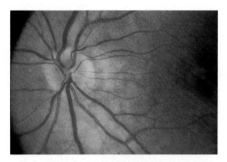

Ophthalmoscopic exam

Blink-to-threat

Pupil light reflex

Oculocephalic testing

MNEMONIC

because of a lesion in the brainstem motor pathways or because of peripheral neuromuscular blockade (see KCC 14.1).

2. Cranial Nerves

Examination of the cranial nerves provides crucial information about brainstem dysfunction as the possible cause of coma.

OPHTHALMOSCOPIC EXAM (CN II) Examine both retinas carefully with an ophthalmoscope, looking especially for papilledema, which suggests elevated intracranial pressure (see KCC 5.3; see also **neuroexam.com Video 25**).

VISION (CN II) If the patient cannot cooperate with visual field testing, the **blink-to-threat** test can be used to roughly map visual fields. Observe whether the patient blinks in response to moving your hand rapidly toward their eyes from different directions (see **neuroexam.com Video 28**).

PUPILLARY RESPONSES (CN II, III) This is one of the most important parts of the exam in patients with impaired consciousness. Pupil size and responsiveness can be a helpful guide to the cause of coma (see Table 14.5; see also **neuroexam.com Video 29**). Although many exceptions exist, toxic or metabolic causes of coma often produce normal-sized, reactive pupils. Asymmetrical or bilateral dilated, unresponsive ("blown") pupils can indicate transtentorial herniation (see KCC 5.4) or other disorders affecting the midbrain. Bilateral small but responsive pupils are often seen in pontine lesions. Bilateral pinpoint pupils are seen in opiate overdosage.

EXTRAOCULAR MOVEMENTS AND VESTIBULO-OCULAR REFLEX (CN III, IV, VI, VIII) Check for spontaneous extraocular movements, nystagmus, dysconjugate gaze, or fixed deviation of the eyes in a particular direction (see **neuroexam.com Videos 32–35**). Optokinetic nystagmus (OKN) can sometimes be useful in eliciting eye movements and testing vision in uncooperative patients, but it is often suppressed when consciousness is impaired. If the patient cannot follow commands to move their eyes, the vestibulo-ocular reflex can be used to test whether brainstem eye movement pathways are intact (see Figures 12.19 and 13.12). Elicit the **oculocephalic reflex** by holding the eyes open and rotating the head from side to side or up and down. These maneuvers obviously should not be performed in cases of head or neck injury until possible cervical spinal trauma has been ruled out by appropriate radiological studies. Presence of the oculocephalic reflex is sometimes called **doll's eyes**, since the eyes move in the direction opposite to head movements. Note that in awake patients, doll's eyes are usually not present. This is because visual fixation and voluntary eye movements mask the reflex. Thus, the absence of doll's eyes suggests brainstem dysfunction in the comatose patient but can be normal in the awake patient.

Another, more potent stimulus of the vestibulo-ocular reflex is **caloric stimulation**. Always examine the external auditory canal first with an otoscope. Then, with the patient lying in the supine position and the head elevated at 30°, infuse ice water into one ear. If the brainstem vestibulo-ocular reflex pathways are intact, this will produce nystagmus, with the fast phase directed opposite to the side of the cold water infusion. A useful mnemonic for interpreting the results of this test is **COWS** (**C**old **O**pposite, **W**arm **S**ame). It is important to note, however, that in comatose patients the fast phase is often absent, and all that is then observed is slow, tonic deviation of the eyes *toward* the cold water. After waiting 5 to 10 minutes for equilibration, try the second ear to drive the eyes in the opposite direction.

CORNEAL REFLEX, FACIAL ASYMMETRY, AND GRIMACE RESPONSE (CN V, VII) Look for facial asymmetry when the patient is at rest and for asymmetrical spontaneous blinking or grimacing. Test the **corneal reflex** by touching each cornea gently with a cotton wisp (see KCC 12.4; see also **neuroexam.com Video 37**). Facial grimacing should also be observed in response to pain during the sensory and motor exams (to be discussed shortly) or in response to firm pressure applied to each orbital ridge.

Corneal reflex

GAG REFLEX (CN IX, X) Test the **gag reflex** by touching the posterior pharynx on each side with a cotton swab. In intubated patients the endotracheal tube can be shaken slightly to elicit a gag reflex. It is also helpful to ask personnel who were present at the time of intubation whether a gag reflex was observed or whether a gag or cough reflex occurs during suctioning of the tracheal passages.

3. Sensory Exam and 4. Motor Exam

Look for spontaneous movements of all extremities. Test resting muscle tone. You can assess asymmetries in tone by raising each limb and letting it fall to the bed (see **neuroexam.com Videos 49** and **50**).

Test each limb for **withdrawal from a painful stimulus,** such as nail bed pressure or skin pinch. Several different responses are possible, depending on the severity of damage to the nervous system. Beginning with the least severe, a lethargic but otherwise intact patient may wake up and shout at the examiner. Thus, out of consideration to the patient, painful stimuli should be used only when absolutely necessary. A more lethargic patient may not wake up fully but may **localize** the stimulus by using another limb to attempt to stop the stimulus in addition to withdrawing the limb from the pain. Grimacing provides additional evidence that the pain sensory pathways are functioning. More severely impaired patients may simply withdraw the limb from the pain. The examiner must be careful to distinguish **purposeful withdrawal** from posturing (discussed in the next section). Also, even vegetative patients who are incapable of purposeful movements may turn and orient their head towards a painful stimulus. Finally, if the pain sensory pathways or motor pathways for the limb are not functional, there may be no response.

5. Reflexes

Test deep tendon reflexes and the plantar response as in the awake patient (see Table 3.6; see also **neuroexam.com Videos 58** and **59**).

Posturing reflexes can be seen in patients with damage to the descending upper motor neuron pathways. These reflexes depend on brainstem and spinal circuitry and are often seen in severe lesions associated with coma. Sherrington studied the effects of lesions at various levels in the brainstem on postural reflexes in the cat. In the **decorticate** preparation, the brainstem was transected above the level of the red nuclei; in the **decerebrate** preparation, the transection was performed below the red nuclei. In response to a painful stimulus the decorticate cat flexes its upper limbs and extends its lower limbs, while the decerebrate cat extends both upper and lower limbs. In humans, the anatomical interpretation of flexor or extensor posturing has not been localized to particular brainstem structures. Thus, although the terms "decorticate posturing" and "decerebrate posturing" are sometimes used with patients, it is probably more correct instead to use the terms **flexor posturing** and **extensor posturing** and to mention which limb is involved. Although there are many exceptions, in humans as well as animals, flexor (decorticate) posturing tends to occur with lesions higher in the neuraxis, at the midbrain or above, whereas

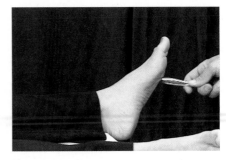

Plantar response

FIGURE 3.5 **Posturing and Triple Flexion** (A) Flexor (decorticate) posturing. (B) Extensor (decerebrate) posturing. (C) Triple flexion.

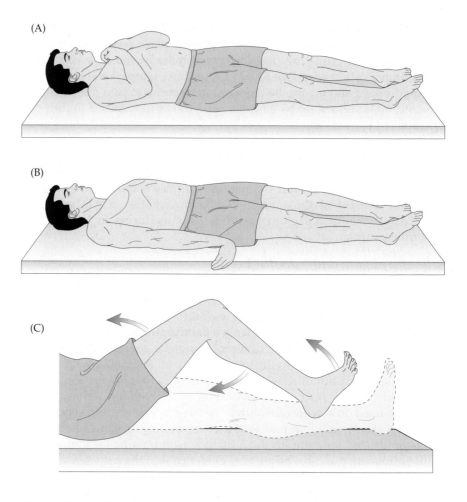

MNEMONIC

extensor (decerebrate) posturing tends to occur with more severe lesions extending lower down in the brainstem. (Mnemonic: in decorticate posturing the lesion is higher, and flexed arms point up toward the cortex; in decerebrate posturing the lesion is lower, and extended arms point down.) Extensor posturing may carry a slightly worse prognosis.

Flexor posturing of the upper extremities is shown in Figure 3.5A. In extensor posturing of the upper extremities, the arms are extended and rotated inward, as shown in Figure 3.5B. Extensor posturing of the lower extremities often accompanies either flexor or extensor upper extremity posturing, as is shown in Figure 3.5A,B. These reflexes depend on brainstem function. Their presence thus suggests damage to descending motor pathways, with some brainstem function left intact. They can occur unilaterally or bilaterally and can be different on the two sides. It is important to distinguish these reflexes from purposeful withdrawal. You can make this distinction by pinching the skin on the extensor and flexor sides of the limb and noting the direction of movement. For example, in flexor posturing the arm flexes even when the flexor side of the arm is pinched, thus moving *toward* the painful stimulus. In purposeful withdrawal, in contrast, the movement is always *away from* the painful stimulus. In addition, purposeful withdrawal often involves limb *abduction* at the shoulder or hip, which is not seen in posturing reflexes.

A flexion reflex in the lower extremity can sometimes be seen as well. This is called **triple flexion** because it involves flexion at the thigh and knee and dorsiflexion at the ankle (see Figure 3.5C). In contrast to the other postural reflexes already mentioned, triple flexion does not require brainstem function and depends only on spinal cord circuitry. Once again, you can distinguish this reflex from purposeful withdrawal by pinching the dorsal and ventral aspects of the leg or foot.

In patients with suspected spinal cord injuries, the special reflexes listed in Table 3.7 may further aid in localization.

6. Coordination and Gait

These are usually not testable in this setting.

Brain Death

The definition of **brain death** is irreversible lack of brain function. The exact criteria used for brain death depend on the hospital; however, the mainstay of the evaluation is the neurologic exam. Generally speaking, there must be no evidence of brain function, including the brainstem. In addition to conducting the usual neurologic exam, in order to ensure that no brainstem function is present, the examiner does caloric testing as well as an **apnea test**, in which lack of spontaneous respirations without the ventilator must be demonstrated despite certain standard changes in blood pH or pCO_2. In the United States, a patient with posturing reflexes involving the brainstem (see Figure 3.5) does not meet brain death criteria, although a patient with only triple flexion and deep tendon reflexes may. Reversible causes such as hypoxia, hypoglycemia, hypothermia, drug overdose, and so on must be tested for. At least two separate brain death examinations should be done to confirm the diagnosis. If part of the clinical evaluation is inconclusive, then confirmatory tests are done— for example, an angiogram that demonstrates no flow to the brain or an EEG that demonstrates electrocerebral inactivity. These tests, however, play a confirmatory role only, and the diagnosis of brain death remains a clinical one. Specific practice parameters for determining brain death have been published by the American Academy of Neurology, and by similar organizations in other countries (see this chapter's References section).

Conversion Disorder, Malingering, and Related Disorders

A number of disorders can mimic neurologic illness but in fact are psychiatric in origin. We have already discussed, under the mood portion of the mental status exam, how difficult this distinction can be. One such disorder is **conversion disorder**, in which psychiatric illness causes the patient to have sensory or motor deficits without a corresponding focal lesion in the nervous system. In **somatization disorder**, patients have multiple somatic complaints that change over time. In these two disorders patients are not consciously "faking" their symptoms, and they usually believe that they have a nonpsychiatric illness. It is essential to avoid being judgmental regarding such patients, since they usually suffer distress and functional impairment from their condition that is equal to or even worse than that of patients with identifiable lesions. Other terms that are sometimes used for these types of disorders include *hypochondriasis* and *hysteria*. Some other related examples include *psychogenic amnesia* and *psychogenic coma*.

In a second, much less common class of disorders, patients do have conscious control over their symptoms, and they are intentionally using them for an ulterior motive. In **factitious disorder** (formerly known in severe cases as **Munchausen syndrome**) the ulterior motive is internal to the patient. It is believed that these patients feign illness, including neurologic illness, because they gain some form of emotional pleasure from assuming the role of patient. In **malingering**, the ulterior motive involves some *external gain* for the patient, such as avoiding work, obtaining disability benefits, or the like.

While it is difficult to distinguish these disorders from neurologic illness, it can often be even more difficult to distinguish these disorders from each other, and sometimes there is overlap. There is an unfortunate tendency to dismiss all such patients as "fakers." However, these patients may be severely impaired by their illnesses, and they deserve psychiatric care to help them recover and to help avoid future confusion with neurologic disease. In addition, overhasty labeling of a patient's symptoms as psychiatric in origin without appropriate investigation can lead to misdiagnosis, particularly when the neurologic findings are subtle or when neurologic and psychogenic diseases overlap, as is often the case.

The most important tool for identifying these patients and for ruling out focal lesions in the nervous system is a thorough knowledge of the neurologic exam and of neuroanatomy. Of the many techniques that can be used, only a few of the most clear-cut methods are described here (see the references at the end of this chapter for more complete discussion):

Hand-dropping test in pseudocoma When a patient is truly in coma and their hand is released directly above their face, their hand should strike their face on its way down.

Saccadic eye movements in pseudocoma Saccades should not be present in coma. Note, however, that they may be present if the patient is *locked in* (see KCC 14.1) or if they are experiencing sleep paralysis, as is seen in narcolepsy (see Chapter 14).

Variable resistance A patient with psychogenic weakness of a limb may vary their resistance over a wide range up to normal strength when their strength is tested by variable resistance from the examiner. This must be distinguished from a similar condition called paratonia, often seen in frontal lobe lesions (see KCC 19.11)

Hoover test In unilateral leg weakness, palpate the contralateral gastrocnemius while the patient tries to raise the affected leg off the bed. In normal individuals the contralateral gastrocnemius is used to exert force against the bed, and it should contract. Lack of gastrocnemius contraction demonstrates lack of effort.

Unconscious movements Patients with psychogenic paralysis may be observed to move the affected limbs during sleep, while being transferred onto a stretcher, or in other situations when distracted.

Midline change in vibration sense Loss of vibration sense on only one side of the sternum or one side of the skull is nonphysiological, since vibration is readily conducted through the bone to the contralateral side.

Other neuroanatomical inconsistencies can sometimes be detected if the examiner uses common sense, a little experience, and a thorough neurologic exam in which uncertain portions of the exam are repeated. It is important, once again, to be cautious because lesions in the nervous system often present with atypical or unusual symptoms and signs for a given location, leading to apparent inconsistencies in the exam. Finally, it should be noted that a substantial number of patients have disorders in which neither a clear-cut neurologic nor a psychiatric diagnosis can be made. Patients of this kind are in danger of being treated by *neither* neurologists nor psychiatrists, who both may consider them out of their area of expertise. A better approach would be for these patients to be followed by *both* neurologists and psychiatrists until either a clear-cut diagnosis emerges or they respond to empiric therapy.

TABLE 3.10 Minimal Screening Neurologic Exam[a]

PART OF EXAM	TESTS
Mental status	Level of alertness and orientation. Assess attention using months forward/backward. Immediate registration and delayed recall of 3 objects for 4 minutes (timed). Naming of watch parts. Note behavior, language, affect, etc., while taking history.
Cranial nerves	Pupil light reflexes. Ophthalmoscopic exam. Visual fields, including extinction testing. Horizontal and vertical smooth pursuit eye movements. Facial sensation to light touch including extinction testing. Facial symmetry during emotional smile. Hearing of finger rub bilaterally. Palate elevation. Note quality of voice during remainder of exam. Head turning and shoulder shrug against resistance. Tongue protrusion.
Motor exam	Drift. Rapid hand and foot tapping. Upper and lower extremity tone. Strength in several proximal and distal muscles in the upper and lower extremities bilaterally (e.g., finger extensors, finger abductors, wrist extensors, biceps, triceps, deltoids, iliopsoas, quadriceps, foot and toe dorsiflexors, and knee flexors).
Reflexes	Bilateral biceps, brachioradialis, patellar, Achilles tendon, and plantar reflexes.
Coordination and gait	Finger–nose–finger and heel–shin tests bilaterally. Gait and tandem gait.
Sensory exam	Light touch in hands and feet, including extinction testing. Pin prick or temperature testing in feet bilaterally. Vibration and joint position sense in feet bilaterally.

[a]Duration = 5 to 10 minutes.

The Screening Neurologic Exam

When evaluating patients, it is useful to be proficient in a brief form of the neurologic exam that can be performed in less than 10 minutes. While performing this abbreviated exam, it is essential to be highly vigilant for any suggestion of a subtle abnormality. Any suspicious part of the examination should then be repeated and evaluated more carefully with more detailed tests, including those described earlier in this chapter. In addition, any suspicions that arise from the patient history should be investigated more closely. For example, a patient with visual complaints should have a detailed visual exam. A patient with complaints of weakness should have a detailed motor exam.

There is no single standard for a screening neurologic exam. However, the items listed in Table 3.10 may be useful as a minimal starting point from which more detailed testing should be done when appropriate.

Conclusions

In this chapter we have reviewed techniques for performing the neurologic exam in many different situations. We have seen how the exam can be adapted to test patients who are awake and cooperative; comatose; or suffering from psychiatric disorders, malingering, or any combination of other impairments. In addition, we have begun to explore the neuroanatomical systems being tested and the effects of disease on function.

The lessons learned here will serve as the basis for understanding the clinical cases in Chapters 5 through 19 and will help localize lesions to particular areas of the nervous system. Once the clinical suspicion of a lesion has been raised as a result of the history and physical exam, several important decisions need to be made. Depending on the type and location of the lesion that is suspected, the options include emergency surgical or medical therapy; less urgent therapy; or further investigations, including blood tests, cerebrospinal fluid tests, electrophysiological studies, or neuroradiological imaging.

The clinician is guided in these difficult decisions by the information obtained from the history and the exam. For example, the decisions to do neuroimaging, which method to use, and what areas of the nervous system to study are based on conclusions made about the probable location and nature of the lesion derived from the history and physical exam. In the next chapter we will discuss the use of neuroimaging and its applications to understanding clinical cases in the context of the complete patient assessment.

References

Aids to the Examination of the Peripheral Nervous System. 1986. Baillière Tindall on behalf of the Guarantors of Brain, London.

Bickley LS (ed.). 2008. *Bates' Guide to Physical Examination and History Taking.* 10th Ed. Lippincott-Raven, Philadelphia.

Blumenfeld H. 2001. *The NeuroExam Video.* Sinauer, Sunderland, MA.

Blumenfeld H. 2009. The neurological examination of consciousness. In *The Neurology of Consciousness*, S Laureys and G Tononi (eds.), Chapters 15–30. Academic Press, New York.

Brazis PW, Masdeu JC, Biller J. 2001. *Localization in Clinical Neurology.* 4th Ed. Lippincott Williams & Wilkins, Boston.

Devinsky O, Feldmann E. 1988. *Examination of the Cranial and Peripheral Nerves.* Churchill Livingstone, New York.

Gilman S. 2000. *Clinical Examination of the Nervous System.* McGraw-Hill, New York.

Goldberg S. 2004. *The Four-Minute Neurological Exam.* MedMaster, Miami, FL.

Haerer AF. 2005. *DeJong's The Neurologic Examination.* 6th Ed. Lippincott, Philadelphia.

Lanska, DJ. 2006. Functional weakness and sensory loss. *Semin Neurol* 26 (3): 297–309.

Patten J. 1995. *Neurological Differential Diagnosis: An Illustrated Approach.* 2nd Ed. Springer Verlag, London.

Plum F, Saper CB, Schiff N, Posner JB. 2007. *The Diagnosis of Stupor and Coma.* 4th Ed. Oxford University Press, New York.

Quality Standards Subcommittee, American Academy of Neurology. 1995. Practice parameters for determining brain death in adults (summary statement). *Neurology* 45 (5): 1012–1014 (reaffirmed January, 2007).

Ross RT. 2006. *How to Examine the Nervous System.* 4th Ed. Humana Press, Totawa, NJ.

Strub RL, Black FW. 2000. *The Mental Status Examination in Neurology.* 4th Ed. FA Davis, Philadelphia.

Wijdicks EFM. 2001. The Diagnosis of Brain Death. *N Engl J Med* 344: 1215.

CONTENTS

Chapter 4

Introduction to Clinical Neuroradiology

Advances in clinical imaging, particularly in neuroradiology, are some of the most exciting recent developments in medicine. One patient, a 52-year-old woman, suddenly developed left-sided weakness and increased reflexes. Although her initial head CT as well as T1-weighted, T2-weighted and FLAIR MRI scans were normal, diffusion-weighted MRI sequences revealed an infarct involving the right motor cortex. An MR angiogram and carotid Doppler studies suggested severe narrowing of the right internal carotid artery. Based on these findings, a surgical procedure was performed to open the narrowing in her carotid artery, and she subsequently did well.

In this chapter, we will learn about current neuroimaging techniques and their clinical applications.

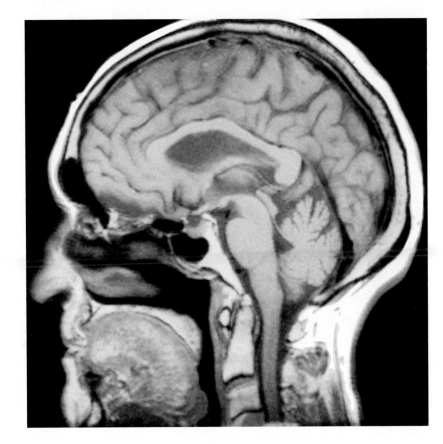

Introduction

MODERN TECHNIQUES OF NEUROIMAGING have revolutionized both clinical practice and neuroscience research. This chapter will focus on the three imaging modalities most commonly used in clinical practice: computerized tomography (CT), magnetic resonance imaging (MRI), and neuroangiography (including ultrasound, magnetic resonance angiography [MRA], and CT angiography). We will also briefly touch on functional imaging modalities such as positron emission tomography (PET), single photon emission computerized tomography (SPECT), and functional MRI (fMRI).

Imaging Planes

Most CT and MRI scan images are two-dimensional "slices" through the brain. The **imaging planes** used are similar to the horizontal (axial), coronal, and sagittal planes described in Chapter 2 (see Figure 2.5). However, the angle of the axial slices in CT scans is sometimes adjusted by a few degrees off the true axial plane (**Figure 4.1**). This adjustment enables the whole brain to be covered using fewer slices and decreases radiation exposure to the eyes. In MRI scans, the axial slices are usually true horizontal slices, although this may vary slightly depending on the institution. **Scout images**, also called **localizer images**, such as that shown in Figure 4.1 should be included on all CT and MRI films so that the exact angle of scanning can be documented. Although in practice a slight angulation off the horizontal plane does not greatly affect the appearance of images, it should be kept in mind, especially when carefully comparing for differences between scans.

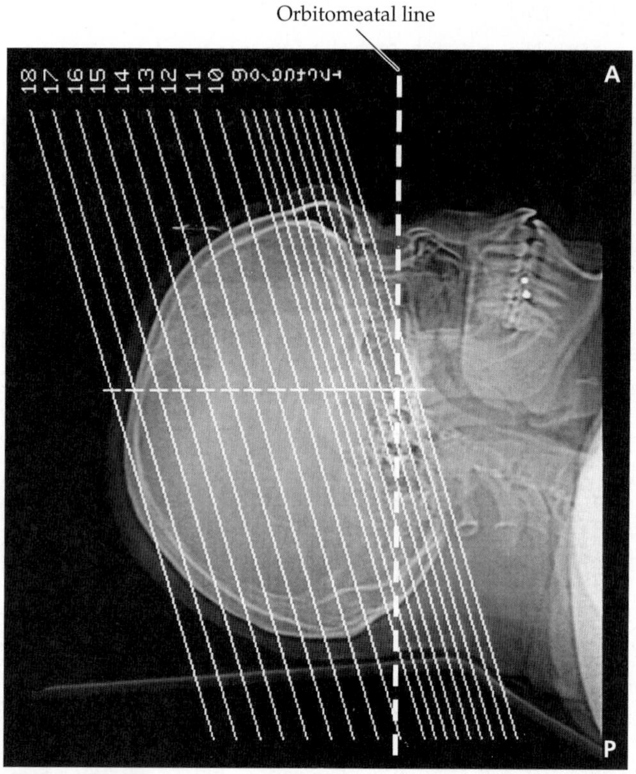

FIGURE 4.1 **CT Scout (Localizer) Image** Lateral view of skull is shown with imaging planes indicated by lines. The true horizontal plane is approximated by the orbitomeatal line, while the typical CT imaging plane is angled slightly upward anteriorly.

Computerized Tomography

Computerized tomography (**CT**) was developed directly from conventional X-ray technology and therefore shares many of the same principles. Like conventional X-ray radiographs, CT scans measure **density** of the tissues being studied. There are really only two differences from conventional X-rays:

1. Rather than taking one view, the X-ray beam is rotated around the patient to take many different views of a single **slice** of the patient; hence the term "tomography" (from the Greek *tomos*, meaning "section").

2. The X-ray data acquired in this way are then reconstructed by a computer to obtain a detailed image of all the structures in the slice (including soft tissues, liquid, and air, as well as bone); hence the term "computerized."

Advances in CT technology have now made it possible to acquire multiple CT slices simultaneously. For simplicity, in this discussion we will first describe single-slice CT. The patient lies on a special table, which moves the patient in small steps through the scanner to obtain many horizontal slices. The scanner is shaped like a large ring (**Figure 4.2**). At each stop of the table, a thin beam of X-rays is scanned through the patient from many different points around the ring and picked up by detectors on the opposite side of the ring. As a single beam of X-rays passes through the patient in a CT scanner, it is partially absorbed by the tissues it encounters. The amount of energy absorbed depends on the *density* of the tissues traversed.

Since the X-ray beam is passed through the patient from many different directions, crossing and recrossing the same structures from different angles, enough information is obtained for the computer to calculate the density for every point within the horizontal slice. These densities are then displayed as an image that looks like a cross section through the head (Figure 4.3). Recently, **spiral (helical) CT** scanners have been developed that can acquire data continuously as the patient moves through the scanner ring, without requiring stops. In addition, instead of acquiring single slices, up to 256 rows of detectors are now being used so that multiple overlapping slices can be obtained. These technical advances reduce patient radiation exposure and also greatly improve the resolution and speed of CT scanning.

As in conventional X-rays, dense structures like bone or other calcifications appear white in CT scans, and less dense materials such as air appear black (see Figure 4.3). The terms **hyperdense** and **hypodense** are frequently used to refer to brighter and darker areas, respectively, on CT scans. Structures of intermediate density similar to that of brain tissue appear gray and are called **isodense**. Cerebrospinal fluid (CSF) is dark gray, and fat tissue (seen subcutaneously just outside the skull) appears nearly black. Since fat is less dense than water, white matter (which has a high myelin content) appears slightly darker than the cellular gray matter (which has a high water content).

Density in CT scans is often expressed in Hounsfield units (HU). The HU scale is based on the following values: HU = 0 for water and HU = –1000 for air. Table 4.1 lists the HU numbers for several materials commonly imaged with CT scans. Take a few moments to review the series of normal CT scan

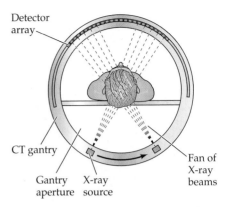

FIGURE 4.2 Schematic Diagram of CT Scanning Gantry

REVIEW EXERCISE

Cover the labels on the CT images in the Neuroradiological Atlas (see Figure 4.12) and name as many structures as possible.

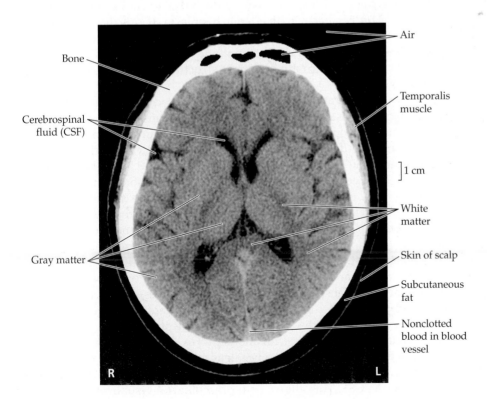

FIGURE 4.3 Typical CT Scan Image Image was acquired in an axial plane, without intravenous contrast. See also Table 4.1.

MNEMONIC

TABLE 4.1 CT Hounsfield Units (HU) for Commonly Scanned Tissues

TISSUE	HU
Air	−1000 to −600
Fat	−100 to −60
Water	0
CSF	8–18
White matter	30–34
Gray matter	37–41
Freshly congealed blood	50–100
Bone	600–2000

images in Figure 4.12, in the Neuroradiological Atlas at the end of this chapter, and become acquainted with normal anatomy as seen in CT scans. Note that in horizontal (axial) sections through the brain (see Figures 4.11C, 4.12J,K, and 4.13J,K), several important gyri form a sideways "T" shape, facilitating localization of the central sulcus. The superior frontal gyrus and the superior parietal lobule form the top of the "T," and the stem of the "T" is formed by the pre- and postcentral gyri together.

CT scans can be used to visualize a variety of different intracranial abnormalities. The appearance of **hemorrhage** on CT depends on how recently it occurred (see Figure 5.19). Fresh intracranial hemorrhage coagulates nearly immediately and therefore shows up on CT scans as hyperdense areas relative to brain. With typical image display settings, fresh hemorrhage may appear about as white as bone, although the actual HU is significantly lower (see Table 4.1). As the clot is broken down, after about a week it becomes isodense with brain tissue, and after 2 to 3 weeks it becomes hypodense (see Figure 5.19).

Acute **cerebral infarctions** often cannot be seen by CT scanning in the first 6 to 12 hours after the event. Subsequently, cell death and edema lead to an area of hypodensity seen in the distribution of the artery that has been occluded, along with some distortion of the local anatomy due to the edema (see Images 10.6A,B and 10.8A–C). Over weeks to months, the brain tissue surrounding the infarct may shrink, producing a local area of prominent sulci or enlarged ventricles. Persistent areas of hypodensity in the brain tissue may be present as a result of gliosis and of brain necrosis with replacement by CSF.

Neoplasms may appear hypodense, hyperdense, or isodense depending on the type and stage (see Images 15.2, 19.5A,B, and 19.7A). They may contain areas of calcification, hemorrhage, or fluid-filled cysts. Neoplasms may produce surrounding edema that is hypodense. Intravenous contrast dye (discussed shortly) is often helpful in visualizing neoplasms.

Mass effect is anything that distorts the brain's usual anatomy by displacement. This can occur with edema, neoplasm, hemorrhage, and other conditions. It can be detected on CT by localized compression of the ventricles, effacement of sulci, or distortion of other brain structures seen, for example, in subfalcine herniation of brain structures across the midline (see Image 5.6A,B).

Intravenous contrast material is sometimes used in CT scanning, especially to facilitate visualization of suspected neoplasms or brain abscess. The contrast material contains iodine, which is denser than brain and will therefore appear hyperdense (white) in areas of increased vascularity or breakdown of the blood–brain barrier (see Images 15.2 and 19.5A,B). Images without intravenous contrast are also obtained for comparison with the contrast images. Review the contrast CT image in Figure 4.4 and identify the structures that normally take up contrast by comparison with the noncontrast images. These structures include arteries, venous sinuses, the choroid plexus, and dura. In suspected intracranial hemorrhage, it is very important to obtain a **noncontrast** CT scan. This is because small hemorrhages often appear on CT as whitish areas at the base of the brain, which could be masked by the normal hyperdense contrast material in blood vessels and meninges at the brain base. Another important application of intravenous contrast is in CT angiography (CTA), discussed later in this chapter.

CT scanning is combined with another form of contrast enhancement in **myelography**. In this technique, a needle is introduced into the CSF, usually by lumbar puncture (see KCC 5.10), and an iodinated contrast dye is introduced into the CSF. This allows better radiological visualization of nerve roots and of abnormal impingements on the spinal CSF space—caused, for example, by a herniated intervertebral disc. In conventional myelography,

on ADC, while T2-shine-through from old stroke appears *bright* on DWI but is *normal or bright* on ADC imaging.

A number of other factors affect the intensity of MRI signals. **Paramagnetic** substances, such as iron in blood degradation products, can cause either relatively bright or dark signals, depending on the exact circumstances (see Tables 4.3 and 4.4). Similarly, although calcium in cortical bone appears dark, calcium bound to protein in brain deposits can appear bright on MRI. When special pulse sequences for **magnetic susceptibility** are used, even minute amounts of hemosiderin from an old hemorrhage can be detected as black areas on the scan. The paramagnetic substance **gadolinium** is used in MRI for intravenous contrast. In analogy with CT, gadolinium contrast injection produces a bright signal in regions of increased vascularity or breakdown of the blood–brain barrier. Gadolinium contrast agents are much less nephrotoxic and less prone to cause allergic reactions than are the iodinated contrast agents used in CT, although gadolinium has been implicated in a condition called nephrogenic systemic fibrosis seen in patients with renal failure.

Another factor that can affect MRI signals is **flow artifact** in blood vessels and, to a lesser extent, in CSF. Flow artifact occurs as a result of protons moving rapidly into or out of the area being imaged. Some protons appear or disappear between the time that the excitation and recording radio frequency pulses occur. This change can cause increases or decreases in MRI signal intensity, depending on the rate and direction of flow and on the pulse sequence used. Magnetic resonance angiography (MRA) takes advantage of these effects to create images of arterial blood flow (see the next section, "Neuroangiography"). MRI scans are also distorted by artifact in patients with metallic implants in the skull. Metallic fragments in the eye; pacemakers, cochlear implants, metallic heart valves, and older aneurysm clips can be moved or damaged by the powerful MRI magnet, and therefore will preclude MRI scanning in certain patients.

Intracranial hemorrhage undergoes a characteristic series of changes on MRI images over time (Table 4.4). In simple terms, on both T1- and T2-weighted images, acute hemorrhage may be difficult to see because it is gray and resembles CSF. Subacute hemorrhage contains methemoglobin, causing it to appear white. Chronic hemorrhage contains dark areas resulting from hemosiderin deposition. Usually, the center of the hemorrhage has a different composition from the periphery so that, particularly for older hemorrhages, there is a characteristic bright center with a dark rim. Eventually, the center of the hemorrhage may resorb, forming a fluid-filled cavity that is dark on T1-weighted images and bright on T2-weighted images (not shown in Table 4.4).

We will now briefly summarize the above discussion (see Tables 4.3 and 4.4) and provide some examples. On MRI imaging, abnormal areas of increased fluid such as cysts, infarcts, edema, gliosis, or demyelination appear dark on T1-weighted images (see Images 7.5A and 19.6A) and bright on T2-weighted images (see Images 6.2A–C, 7.7A,B, and 10.5A,B). Regions of inflammation or neoplasms often enhance with intravenous gadolinium (see Images 12.1A–C, 15.3A,B, and 18.2A–C). Hemorrhage is difficult to see acutely but becomes bright on subsequent imaging (see Image 14.9A,B) and later often develops both bright and dark regions (see Table

TABLE 4.4 MRI Appearance of Intracranial Hemorrhage

TIME SINCE HEMORRHAGE	T1-WEIGHTED	T2-WEIGHTED
Acute: first 6–24 hours (intracellular oxyhemoglobin)	Gray	Light gray
Early subacute: 1–5 days (intracellular deoxyhemoglobin)	Gray	Dark gray
Middle subacute: 3–7 days (intracellular methemoglobin)	White	Dark gray
Late subacute: 3–30 days (extracellular methemoglobin)	White	White
Chronic: >14 days (hemosiderin, mainly on outer rim)	Dark gray	Black

Note: The actual sequence of changes in the appearance of hemorrhage on MRI scans can be fairly complicated and variable, depending on individual scanners.

FIGURE 4.7 **Coronal T1-weighted MRI Image with Intravenous Gadolinium Contrast** This MRI image demonstrates enhancement of arteries, venous sinuses, choroid plexus, and dura mater.

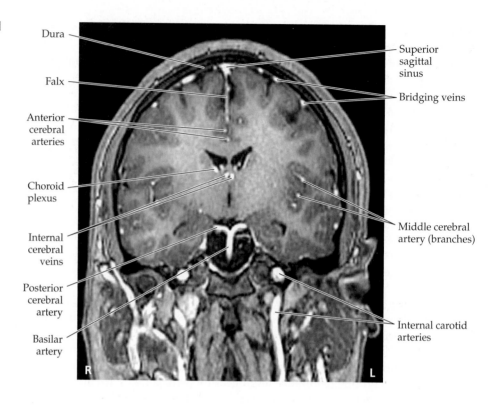

4.4). New pulse sequences continually improve the clinical usefulness of MRI. Diffusion-weighted MRI allows visualization of acute cerebral infarcts far earlier than with conventional MRI (see Images 14.2A,B and 14.7A,B).

Take a few moments now to review the series of normal T1-weighted MRI scan images in the Neuroradiological Atlas at the end of this chapter (see Figures 4.13–4.15) to become acquainted with normal anatomy as seen by MRI. The images in Figure 4.13 were obtained in approximately the same imaging planes as the CT images in Figure 4.12 (although the angle of the slices is somewhat more horizontal for the MRI images). Note the markedly superior tissue contrast and anatomical detail of the MRI images. In addition, note that the administration of intravenous gadolinium (Figure 4.7) causes enhancement of the arteries, venous sinuses, choroid plexus, and dura.

MRI scans are sometimes reformatted into three-dimensional surface representations. An example is shown in Figure 4.8A. This method enables the detection of subtle abnormalities in sulcal morphology that may not be appreciated in two-dimensional sections. In addition, it can be helpful to represent functional neuroimaging data on a three-dimensional representation of the brain surface (Figure 4.8B).

A variety of additional specialized and constantly improving MRI pulse sequences and techniques exist that are beyond the scope of this discussion, and are reviewed in the references listed at the end of the chapter. A few highlights include **magnetic resonance spectroscopy** (**MRS**), measuring brain neurotransmitters and other biochemicals, which has some clinical applications in evaluating brain tumors and regions of epileptic seizure onset; **diffusion tensor imaging** (**DTI**), allowing sensitive assessment of white matter pathways based on water diffusion constrained by axon fibers (Figure 4.9); **functional MRI** (**fMRI**) discussed later in this chapter; and a variety of techniques to measure blood volume or flow in cerebral tissue (e.g., arterial spin labeled MRI) or flow in cerebral vessels (discussed in next section).

(A)

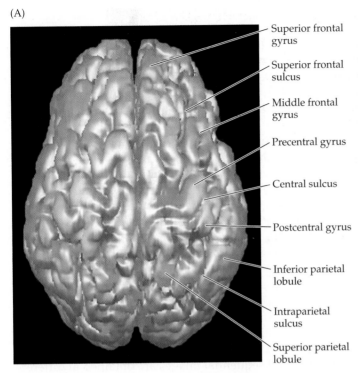

- Superior frontal gyrus
- Superior frontal sulcus
- Middle frontal gyrus
- Precentral gyrus
- Central sulcus
- Postcentral gyrus
- Inferior parietal lobule
- Intraparietal sulcus
- Superior parietal lobule

(B)

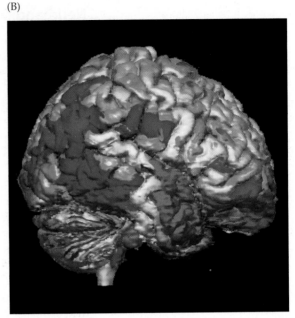

FIGURE 4.8 Three-Dimensional Surface Reconstructions (A) This reconstruction of a normal MRI scan demonstrates anatomy of sulci and gyri. (B) Ictal-interictal SPECT difference imaging. The patient was injected intravenously with Tc99 HM-PAO during a seizure, and a SPECT scan was then performed (ictal SPECT). A second SPECT scan was performed at another time, when the patient was not having a seizure (interictal SPECT). The interictal SPECT was then subtracted from the ictal SPECT, and the results are displayed as a three-dimensional surface reconstruction of the patient's MRI scan. Increased cerebral perfusion is shown as a red–yellow color, and decreased perfusion is shown as blue–green. This scan helped localize the region of seizure onset in this patient (red–yellow) to the right temporal lobe, allowing successful surgical treatment to be performed. (A Courtesy of Rik Stokking; B Courtesy of George Zubal, Susan Spencer, Dennis Spencer, Rik Stokking, Colin Studholme, and Hal Blumenfeld, Yale University School of Medicine.)

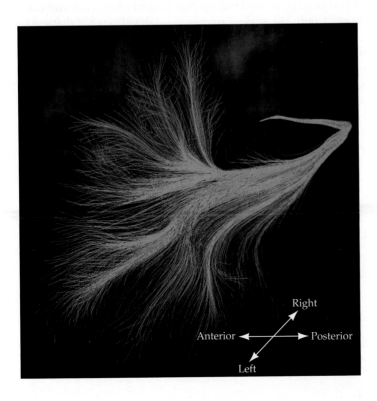

Right

Anterior ◄──────► Posterior

Left

FIGURE 4.9 DTI: Diffusion Tensor Tractography Fiber tracts through anterior corpus callosum are demonstrated originating from seed point in right frontal lobe, and projecting to widespread areas of the left frontal cortex. This technique traces the direction of maximal water diffusion, constrained by white matter microarchitecture, to follow major fiber tracts through the brain. (With permission from Jackowscki et al. 2005. *Medical Image Analysis* 9: 427–440.)

Neuroangiography

Cerebral angiography is one of the oldest neuroradiological techniques, and its role has therefore evolved in many ways. Before the availability of CT and MRI, neuroangiography was often used to detect slight distortions in the patterns of blood vessels suggestive of intracranial mass lesions. These subtle angiographic changes were combined with findings from other techniques no longer used for this purpose (plain skull films, pneumoencephalography, and EEG) to provide circumstantial evidence for intracranial lesions that today can easily be visualized with CT or MRI.

Now that CT and MRI are widely available, neuroangiography is used mainly to visualize lesions of the blood vessels themselves, rather than to provide indirect information about surrounding structures. Lesions optimally seen by angiography include atherosclerotic plaques and other vessel narrowings, aneurysms, and arteriovenous malformations. Angiography is also sometimes used during planning of neurosurgery to assess the vascular anatomy of tumors. In addition to diagnostic uses, neuroangiography can also be used therapeutically, as we will discuss at the end of this section.

Unlike CT and MRI, angiography is an invasive procedure. Under local anesthesia, a catheter is inserted, usually in the femoral artery (Figure 4.10), and threaded up the aorta under continuous X-ray guidance (fluoroscopy). Radio-opaque iodinated contrast material is then injected into the carotid and vertebral arteries on both sides, and sequential images are obtained at different times during the injection and runoff. The arteries are thus well visualized early in the series of images, while the veins are better visualized late. Images are generally taken from several different views to optimally visualize the vessels of both the anterior and posterior circulations. To become familiar with normal vascular anatomy as seen by conventional angiography, review the series of normal angiograms seen in Figures 4.16 and 4.17.

A number of less invasive means for visualizing blood vessels and assessing flow have been developed. These include Doppler ultrasound, magnetic resonance angiography, and spiral CT angiography. Because these noninvasive methods have shown continual improvements over time, conventional angiography has gradually become an uncommon procedure for diagnostic purposes. Nevertheless, conventional angiography remains the "gold standard" and is still used in cases where the diagnosis based on noninvasive tests remains unclear.

Doppler ultrasound can be used to measure flow and lumen diameter of large blood vessels in the head and neck. It is most useful for assessing the

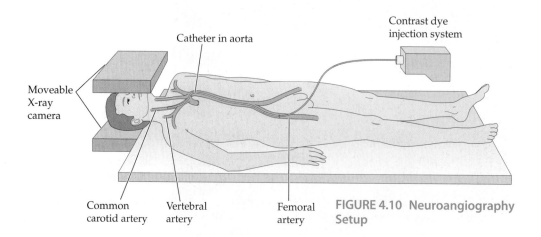

Contrast dye injection system

Catheter in aorta

Moveable X-ray camera

Common carotid artery

Vertebral artery

Femoral artery

FIGURE 4.10 Neuroangiography Setup

proximal portions of the internal carotid, where it can aid in decisions for possible surgery in carotid stenosis (see KCC 10.5). Transcranial Doppler can be used to evaluate flow in the proximal middle cerebral, anterior cerebral, posterior cerebral, vertebral, and basilar arteries. Smaller, more distal branches cannot be assessed. Transcranial Doppler is most often used in the intensive care unit to detect vasospasm following subarachnoid hemorrhage (see KCC 5.6) or, occasionally, as an ancillary test in brain death evaluation (see Chapter 3). Ultrasound usually cannot detect aneurysms or other vascular abnormalities.

Magnetic resonance angiography (**MRA**) takes advantage of the change in magnetic resonance signal that occurs in areas of flow as a result of the movement of protons into and out of the region being assessed between the time that the radio frequency excitation pulse is emitted and the time that the signal is collected. The direction and speed of flow to be detected can be selected via the computer. In some MRA studies, bolus injections of gadolinium are used to improve contrast. A series of normal MRA images of the intracranial vessels is shown in Figure 4.18. These images should be compared to the conventional angiographic images in Figures 4.16 and 4.17. Note that with MRA, although the major vessels can be seen, the smaller, more distal branches cannot. Like conventional angiography, MRA can also be used to image potential sites of narrowing or other pathology of the carotid and vertebral arteries in the neck (Figure 4.19), or even as they arise from the aortic arch (Figure 4.20A). MRA is used mainly to detect regions of decreased or absent arterial blood flow caused by atherosclerotic narrowing, thrombosis, or dissection. In addition, MRA can be useful for detecting some aneurysms and other vascular abnormalities. Venous flow can be visualized using **magnetic resonance venography** (**MRV**). An example of MRV used to detect venous sinus thrombosis is shown in Image 10.13C,D.

Spiral CT angiography (**CTA**) is a method in which rapid injection of intravenous contrast is used together with helical CT scan techniques to quickly obtain images of blood vessels. Pictures of the blood vessels are reconstructed in three dimensions by means of a computer. Information obtained from CTA can sometimes supplement that obtained from MRA, and CTA can be performed in patients in which MRA is contraindicated (e.g., patients with pacemakers). Like MRA, CTA can also often be used to detect carotid stenosis, aneurysms, or vascular malformations.

Although noninvasive techniques have further narrowed the role of angiography in diagnosis, a new role has meanwhile emerged for neuroangiography in invasive functional testing and therapy. This field is known as **interventional neuroradiology**. For example, in the angiogram **Wada test**, a sedative medication (typically amobarbital) is selectively infused into each carotid artery while the patient is awake. This test can help localize the side of language and memory function and is useful in planning neurosurgery (see KCC 18.2). Brain aneurysms and arteriovenous malformations, which can cause massive intracranial hemorrhage (see Chapter 5), can often be clotted off and rendered harmless by filling them with gluelike material or tiny metal coils via the angiography catheter. Angioplasty, in which a balloon is used to dilate a narrowed blood vessel, is being investigated as an alternative to surgery in patients with carotid artery stenosis. Finally, therapeutic trials are under way in acute stroke in which thrombolytic agents are infused via the catheter directly at the site of the clot to try to reestablish perfusion.

Functional Neuroimaging

In clinical situations, the most important radiological information usually comes from a structural anatomical image of the abnormality produced by one of the techniques already described. Sometimes, however, it is also useful to assess the physiological function of the structures in question. Several techniques can be used for measuring various aspects of brain function and, as we will discuss in this section, exciting clinical applications are beginning to emerge. Research applications of functional neuroimaging have seen a remarkable growth in recent years, giving rise to speculation that with these techniques we may soon be able to virtually "peer into someone's head and see what they are thinking."

The original method for measuring brain activity was the **electroencephalogram** (**EEG**). In this technique, an array of electrodes is applied to the surface of the scalp and connected to amplifiers to detect the weak electrical signals transmitted from the brain through the skull. The normal EEG pattern consists of waveforms of various frequencies that vary with the level of alertness of the patient. Abnormalities in large regions of brain produce abnormal or asymmetrical waveforms that can be detected with EEG. The sensitivity and spatial resolution of EEG in detecting focal brain lesions is poor, however, compared to modern neuroimaging methods. EEG remains very useful today in evaluating patients for epileptic (seizure-producing) brain activity (see KCC 18.2) or for detecting widespread abnormalities in brain function (see KCC 19.15 and KCC 19.16). With **evoked potentials**, a method similar to EEG, brain electrical signals are recorded in response to specific stimuli. Additional refinements in electrical measurements of brain function have been developed. These include quantitative EEG analysis and magnetoencephalography (MEG), which uses a superconducting quantum interference device (SQUID) to detect the very weak magnetic signals from the brain. The possible clinical role of these techniques has been a topic of ongoing investigation.

Other methods of functional assessment depend on brain metabolic activity and blood flow. Brain metabolic activity is used as an indirect measure of brain electrical activity, or neuronal firing. Increased local neuronal firing causes increased brain metabolism, which in turn leads to increased local blood flow and increased turnover of local blood volume. Techniques that can produce images based on blood flow or dynamic blood volume include: Xenon regional cerebral blood flow mapping (Xe rCBF), positron emission tomography (PET), **Tc99m single photon emission computerized tomography** (**SPECT**), dynamic contrast functional MRI (perfusion MRI), and arterial spin labeling MRI (ASL MRI). Brain metabolism can be measured by MRI-based techniques or, more commonly, by **fluoro-deoxyglucose positron emission tomography** (**FDG-PET**). Blood flow, blood volume, and the rate of oxygen metabolism all contribute to the signal intensity measured with **blood oxygen level–dependent functional MRI** (**BOLD fMRI**). FDG-PET, SPECT, and BOLD fMRI are the most commonly used techniques to evaluate cerebral blood flow and metabolism for clinical purposes.

Brain PET and SPECT imaging are typically performed by the **nuclear medicine** department. **FDG-PET** scans are used to map local brain consumption of glucose. FDG-PET is often used in patients with dementia or epilepsy to localize regions of abnormal glucose metabolism (see Image 18.5G–J). They may also be useful for distinguishing metabolically active recurrent brain tumors from radiation-induced necrosis. **SPECT** scans are occasionally used to generate a map of regional cerebral blood flow at rest but are less sensitive for most forms of pathology than FDG-PET. **Ictal SPECT** is used clinically to help local-

(A) Broca's area   (B) Wernicke's area   (C) Hand sensorimotor

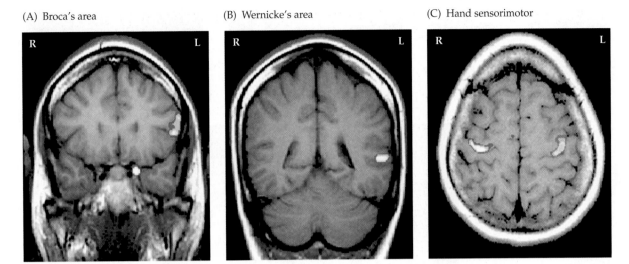

FIGURE 4.11 BOLD fMRI Mapping of Sensorimotor Cortex and Language Cortex (A,B) Language mapping. The subject was asked to read words and determine whether they were real words or nonsense syllables. This task caused activation of both Broca's area (A) and Wernicke's area (B). (C) Sensorimotor mapping. The subject squeezed a rubber ball repeatedly with his left hand, causing activation of the right sensorimotor hand cortex (reddish orange), and with his right hand, causing activation of the left sensorimotor hand cortex (blue). (Compare to Figure 4.13J.) (Courtesy of R. Todd Constable, Yale University School of Medicine.)

ize the region of seizure onset (see KCC 18.2) by measuring cerebral blood flow to indirectly map local electrical brain activity during seizures. An example is shown in Figure 4.8B. Use of new, specific neuroreceptor ligands in PET and SPECT have shown some promise for localizing and monitoring brain abnormalities in neurodegenerative disorders and in epilepsy.

The development of **functional MRI** (**fMRI**) has caused a virtual explosion in neuroscience research and is continually advancing our knowledge of localized brain function. In addition, some potential clinical applications are beginning to emerge. BOLD fMRI is being investigated as a method to help plan neurosurgery by allowing the neurosurgeon to know in advance the locations of vital regions of brain function. As shown in Figure 4.11, BOLD fMRI can be used to localize regions of sensory-motor function and language function. Eventually, with further investigation, this method may replace the angiogram Wada test (see KCC 18.2). As time goes on, additional clinical applications of these powerful functional neuroimaging methods are likely to be found.

Conclusions

Neuroradiology plays an essential role in the diagnosis, and sometimes treatment, of patients with disease of the nervous system. However, it is the role of the clinician to decide, on the basis of the history and physical exam, what the most likely diagnoses are so that neuroradiological methods are used appropriately. The clinician must first determine if a neuroradiological study is needed at all. Then the history and physical exam are used to formulate hypotheses about both **localization** and the **pathophysiology** of lesions. Using this information, the clinician can decide whether CT, MRI, angiography, or other methods of evaluation are most appropriate, as well as which regions of the nervous system should be the focus so that an optimal study is obtained. By combining the clinical history, examination, and other methods of assessment together with the powerful neuroradiological methods available today, clinicians are able to offer an ever-increasing number of patients accurate neurologic diagnosis and appropriate treatment.

NEURORADIOLOGICAL ATLAS

FIGURE 4.12 CT Images   Unenhanced axial CT images with major structures labeled.

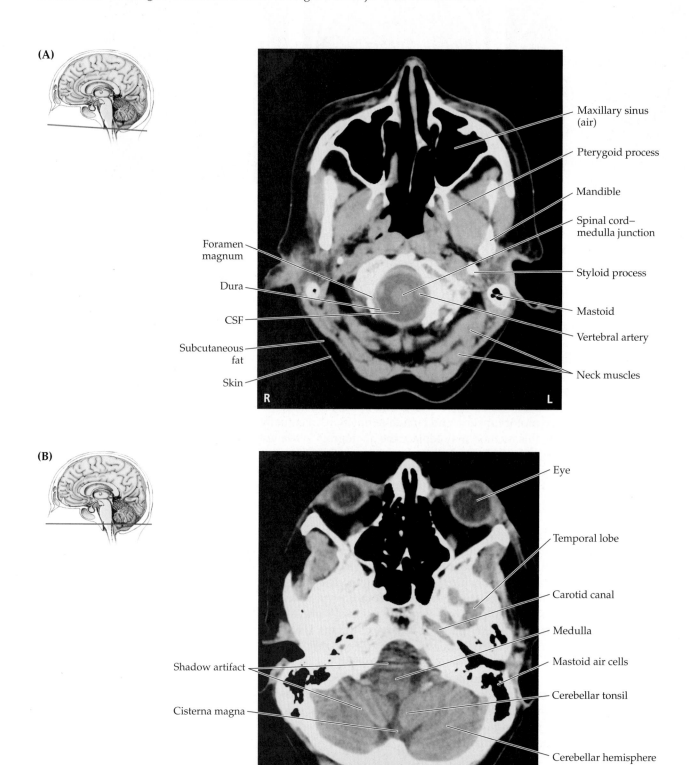

CT, AXIAL

(C)

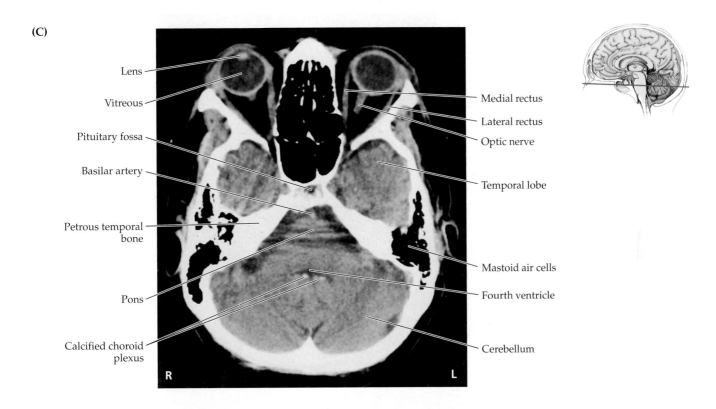

Lens
Vitreous
Pituitary fossa
Basilar artery
Petrous temporal bone
Pons
Calcified choroid plexus

Medial rectus
Lateral rectus
Optic nerve
Temporal lobe
Mastoid air cells
Fourth ventricle
Cerebellum

(D)

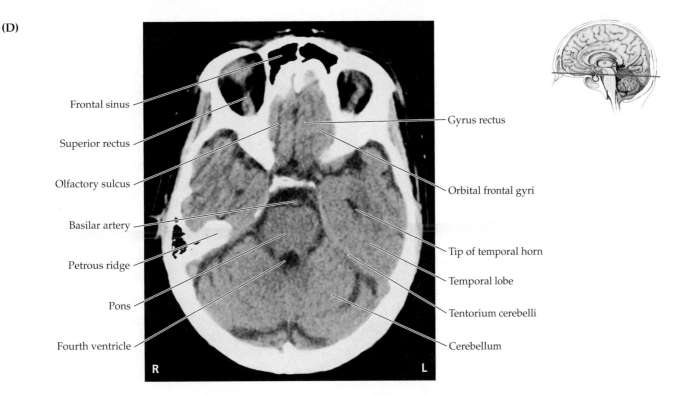

Frontal sinus
Superior rectus
Olfactory sulcus
Basilar artery
Petrous ridge
Pons
Fourth ventricle

Gyrus rectus
Orbital frontal gyri
Tip of temporal horn
Temporal lobe
Tentorium cerebelli
Cerebellum

NEURORADIOLOGICAL ATLAS

FIGURE 4.12 (*continued*)

(E)

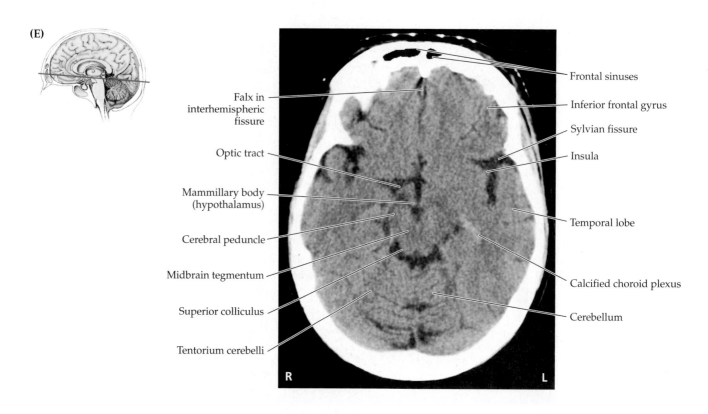

Falx in interhemispheric fissure

Optic tract

Mammillary body (hypothalamus)

Cerebral peduncle

Midbrain tegmentum

Superior colliculus

Tentorium cerebelli

Frontal sinuses

Inferior frontal gyrus

Sylvian fissure

Insula

Temporal lobe

Calcified choroid plexus

Cerebellum

R   L

(F)

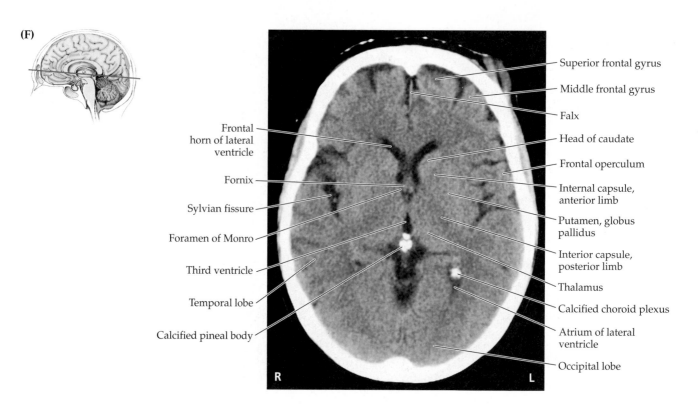

Frontal horn of lateral ventricle

Fornix

Sylvian fissure

Foramen of Monro

Third ventricle

Temporal lobe

Calcified pineal body

Superior frontal gyrus

Middle frontal gyrus

Falx

Head of caudate

Frontal operculum

Internal capsule, anterior limb

Putamen, globus pallidus

Interior capsule, posterior limb

Thalamus

Calcified choroid plexus

Atrium of lateral ventricle

Occipital lobe

R   L

CT, AXIAL

(G)

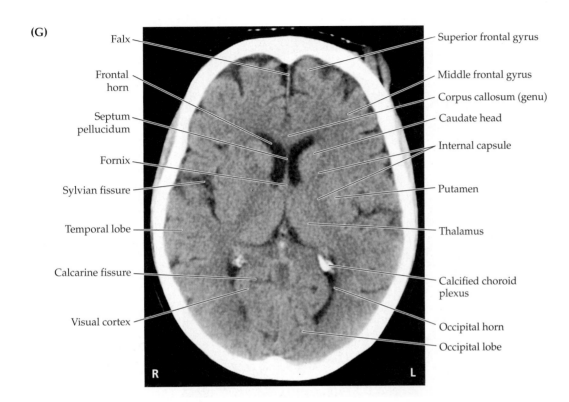

Falx
Frontal horn
Septum pellucidum
Fornix
Sylvian fissure
Temporal lobe
Calcarine fissure
Visual cortex

Superior frontal gyrus
Middle frontal gyrus
Corpus callosum (genu)
Caudate head
Internal capsule
Putamen
Thalamus
Calcified choroid plexus
Occipital horn
Occipital lobe

R   L

(H)

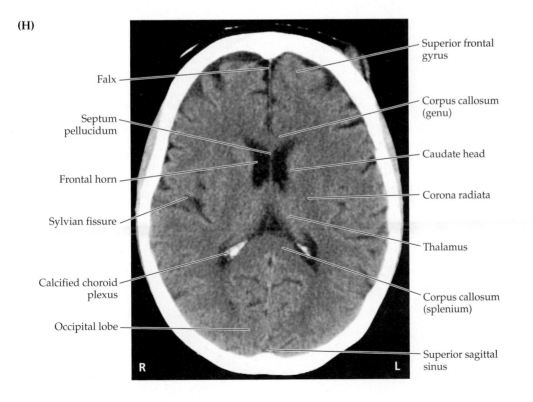

Falx
Septum pellucidum
Frontal horn
Sylvian fissure
Calcified choroid plexus
Occipital lobe

Superior frontal gyrus
Corpus callosum (genu)
Caudate head
Corona radiata
Thalamus
Corpus callosum (splenium)
Superior sagittal sinus

R   L

NEURORADIOLOGICAL ATLAS

FIGURE 4.12 (*continued*)

(I)

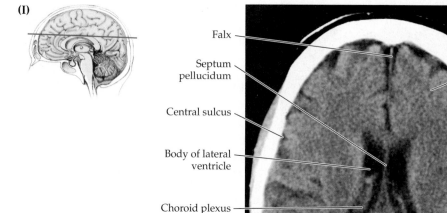

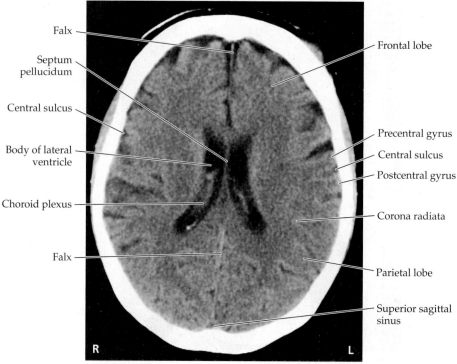

Falx

Septum
pellucidum

Central sulcus

Body of lateral
ventricle

Choroid plexus

Falx

Frontal lobe

Precentral gyrus

Central sulcus

Postcentral gyrus

Corona radiata

Parietal lobe

Superior sagittal
sinus

R   L

(J)

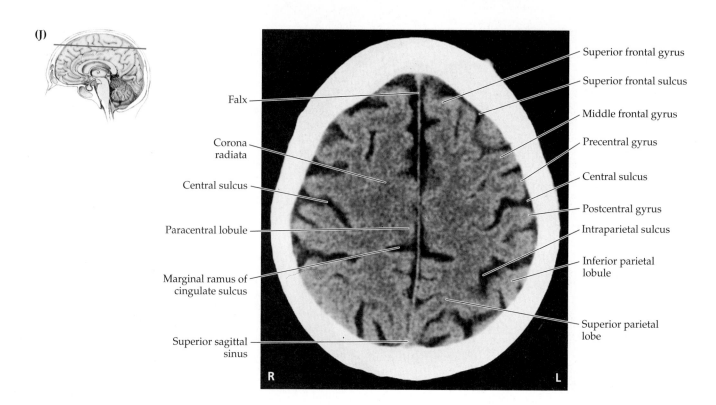

Falx

Corona
radiata

Central sulcus

Paracentral lobule

Marginal ramus of
cingulate sulcus

Superior sagittal
sinus

Superior frontal gyrus

Superior frontal sulcus

Middle frontal gyrus

Precentral gyrus

Central sulcus

Postcentral gyrus

Intraparietal sulcus

Inferior parietal
lobule

Superior parietal
lobe

R   L

CT, AXIAL

(K)

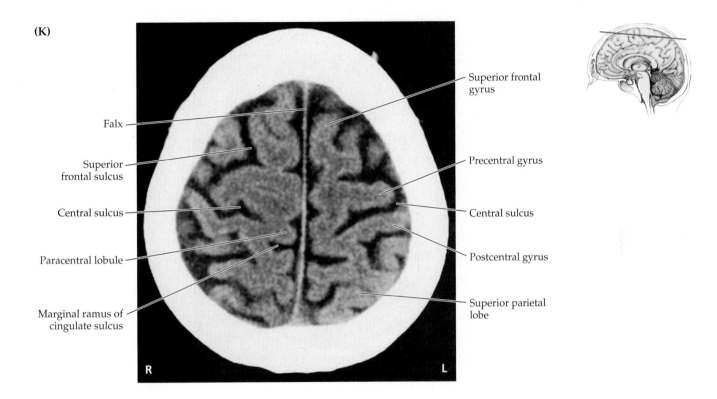

Falx

Superior frontal sulcus

Central sulcus

Paracentral lobule

Marginal ramus of cingulate sulcus

Superior frontal gyrus

Precentral gyrus

Central sulcus

Postcentral gyrus

Superior parietal lobe

R

L

FIGURE 4.13 MRI: Axial T1-Weighted Images Unenhanced axial MRI images
with major structures labeled. TR = 500, TE = 11.

(A)

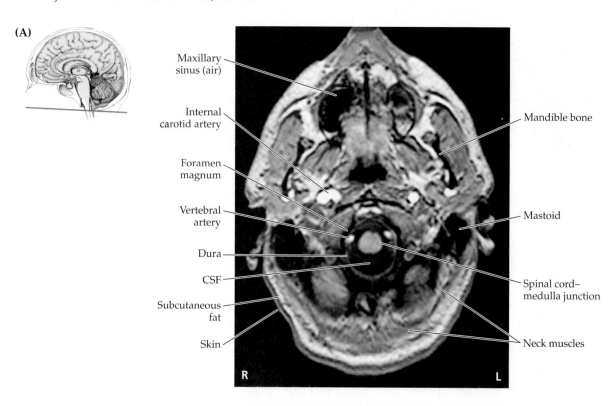

Maxillary sinus (air)

Internal carotid artery

Foramen magnum

Vertebral artery

Dura

CSF

Subcutaneous fat

Skin

Mandible bone

Mastoid

Spinal cord–medulla junction

Neck muscles

R   L

(B)

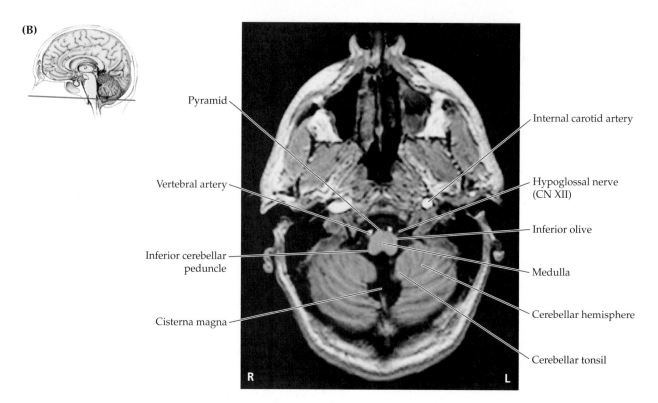

Pyramid

Vertebral artery

Inferior cerebellar peduncle

Cisterna magna

Internal carotid artery

Hypoglossal nerve (CN XII)

Inferior olive

Medulla

Cerebellar hemisphere

Cerebellar tonsil

R   L

MRI, AXIAL T1–WEIGHTED

(C)

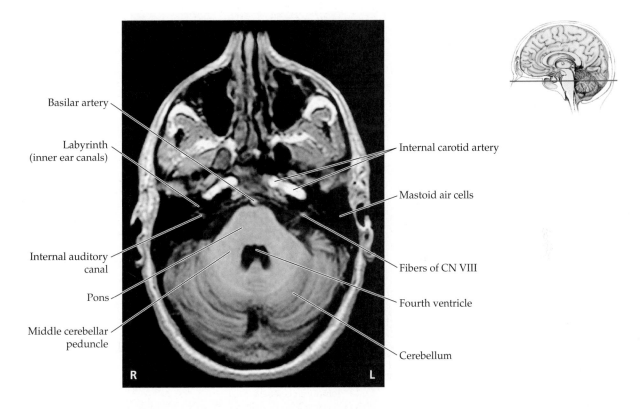

Basilar artery

Labyrinth
(inner ear canals)

Internal carotid artery

Mastoid air cells

Internal auditory
canal

Pons

Fibers of CN VIII

Fourth ventricle

Middle cerebellar
peduncle

Cerebellum

R   L

(D)

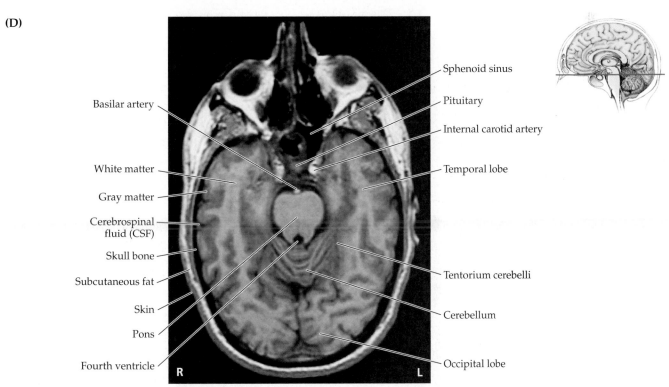

Basilar artery

Sphenoid sinus

Pituitary

Internal carotid artery

White matter

Gray matter

Temporal lobe

Cerebrospinal
fluid (CSF)

Skull bone

Subcutaneous fat

Skin

Tentorium cerebelli

Pons

Fourth ventricle

Cerebellum

Occipital lobe

R   L

NEURORADIOLOGICAL ATLAS

FIGURE 4.13 (*continued*)

(E)

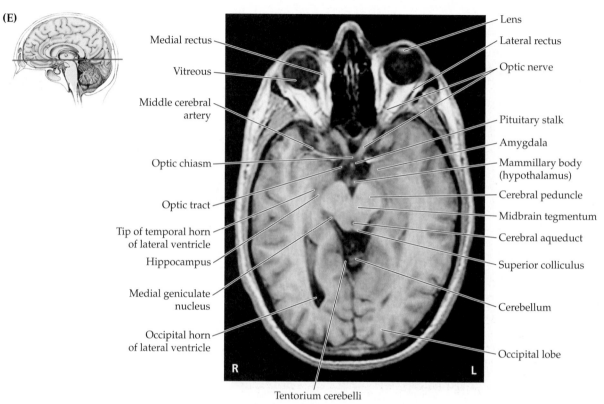

Medial rectus

Vitreous

Middle cerebral artery

Optic chiasm

Optic tract

Tip of temporal horn of lateral ventricle

Hippocampus

Medial geniculate nucleus

Occipital horn of lateral ventricle

Lens

Lateral rectus

Optic nerve

Pituitary stalk

Amygdala

Mammillary body (hypothalamus)

Cerebral peduncle

Midbrain tegmentum

Cerebral aqueduct

Superior colliculus

Cerebellum

Occipital lobe

Tentorium cerebelli

R   L

(F)

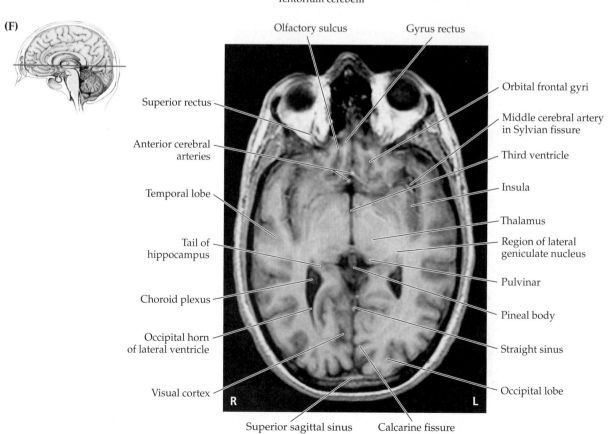

Olfactory sulcus

Gyrus rectus

Superior rectus

Anterior cerebral arteries

Temporal lobe

Tail of hippocampus

Choroid plexus

Occipital horn of lateral ventricle

Visual cortex

Orbital frontal gyri

Middle cerebral artery in Sylvian fissure

Third ventricle

Insula

Thalamus

Region of lateral geniculate nucleus

Pulvinar

Pineal body

Straight sinus

Occipital lobe

Superior sagittal sinus

Calcarine fissure

R   L

MRI, AXIAL T1–WEIGHTED

(G)

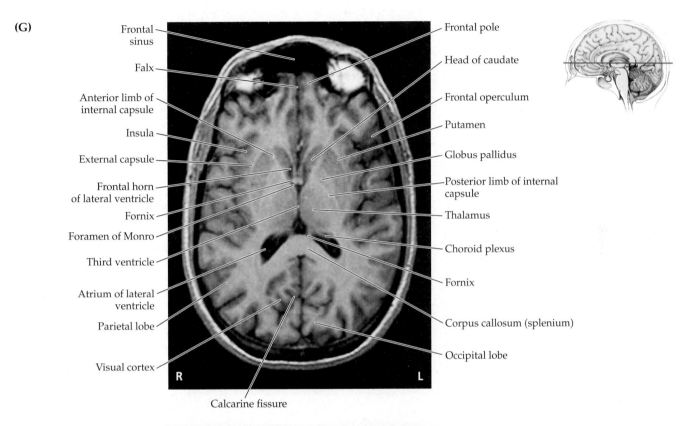

Frontal sinus — Frontal pole
Falx — Head of caudate
Anterior limb of internal capsule — Frontal operculum
Insula — Putamen
External capsule — Globus pallidus
Frontal horn of lateral ventricle — Posterior limb of internal capsule
Fornix — Thalamus
Foramen of Monro — Choroid plexus
Third ventricle — Fornix
Atrium of lateral ventricle — Corpus callosum (splenium)
Parietal lobe — Occipital lobe
Visual cortex

R L

Calcarine fissure

(H)

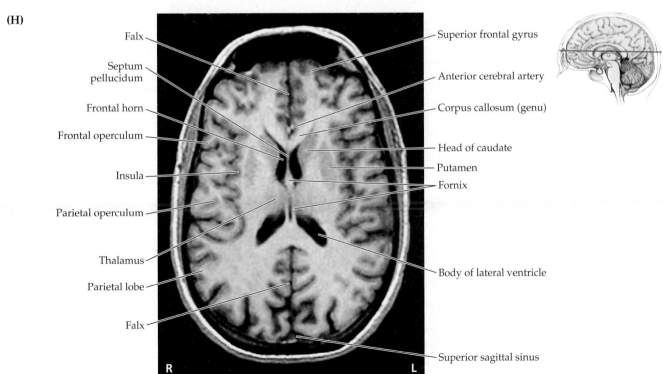

Falx — Superior frontal gyrus
Septum pellucidum — Anterior cerebral artery
Frontal horn — Corpus callosum (genu)
Frontal operculum — Head of caudate
Insula — Putamen
Parietal operculum — Fornix
Thalamus — Body of lateral ventricle
Parietal lobe
Falx

R L

Superior sagittal sinus

FIGURE 4.13 (*continued*)

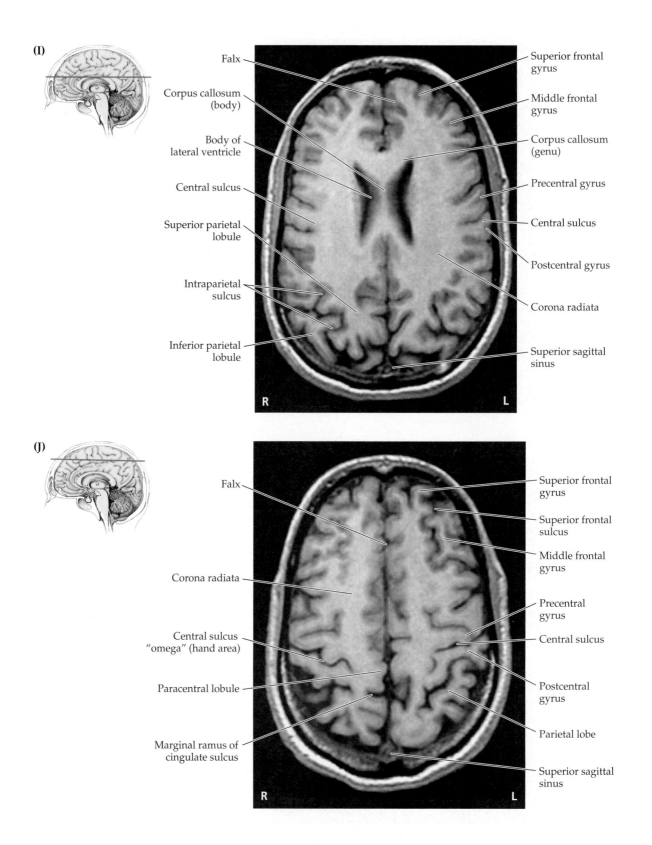

(I)

Falx

Corpus callosum (body)

Body of lateral ventricle

Central sulcus

Superior parietal lobule

Intraparietal sulcus

Inferior parietal lobule

Superior frontal gyrus

Middle frontal gyrus

Corpus callosum (genu)

Precentral gyrus

Central sulcus

Postcentral gyrus

Corona radiata

Superior sagittal sinus

R   L

(J)

Falx

Corona radiata

Central sulcus "omega" (hand area)

Paracentral lobule

Marginal ramus of cingulate sulcus

Superior frontal gyrus

Superior frontal sulcus

Middle frontal gyrus

Precentral gyrus

Central sulcus

Postcentral gyrus

Parietal lobe

Superior sagittal sinus

R   L

MRI, AXIAL T1–WEIGHTED

(K)

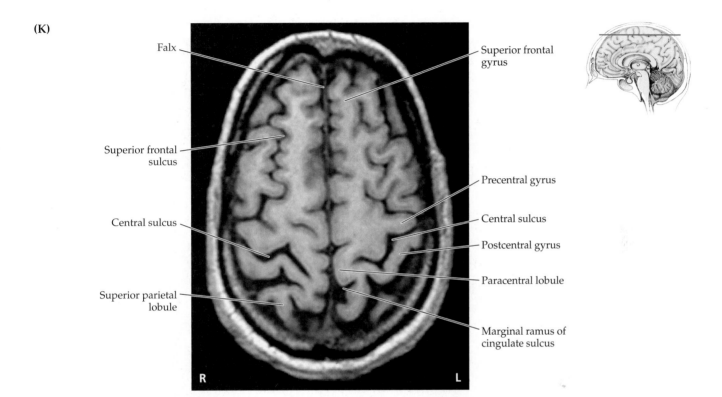

Falx

Superior frontal gyrus

Superior frontal sulcus

Central sulcus

Superior parietal lobule

Precentral gyrus

Central sulcus

Postcentral gyrus

Paracentral lobule

Marginal ramus of cingulate sulcus

R

L

NEURORADIOLOGICAL ATLAS

FIGURE 4.14 MRI: Coronal T1-Weighted Images Unenhanced coronal MRI images with major structures labeled. Images acquired using 3-D SPGR sequence with TR = 23, TE = 4.

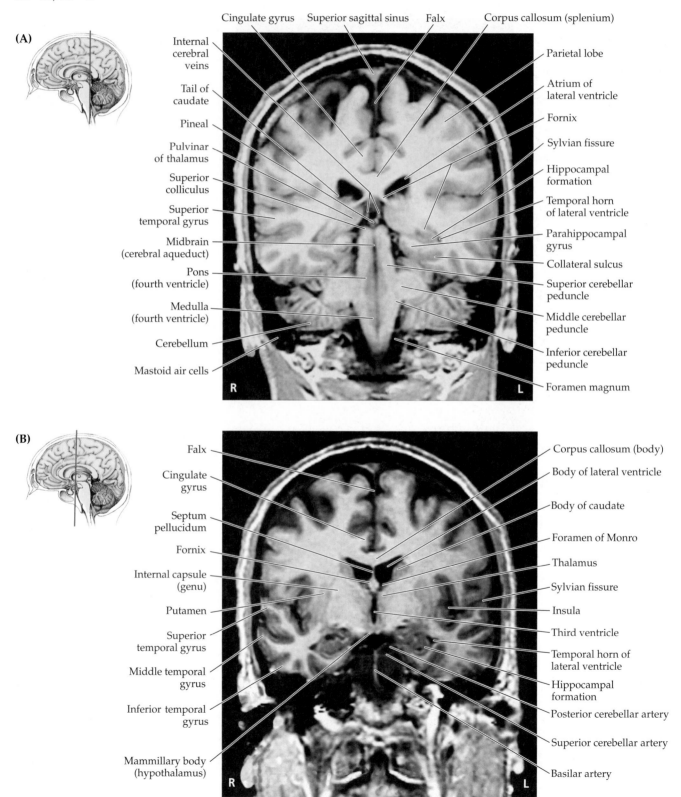

MRI, CORONAL T1–WEIGHTED

(C)

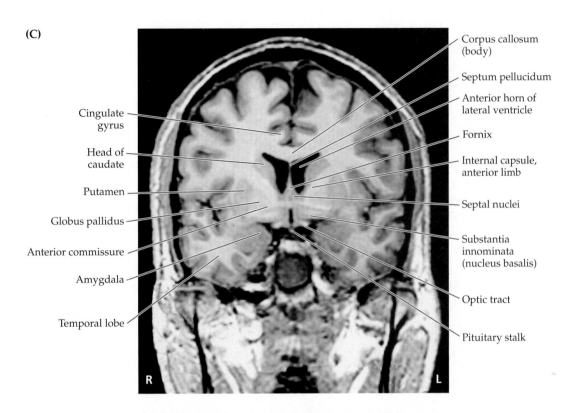

Cingulate gyrus

Head of caudate

Putamen

Globus pallidus

Anterior commissure

Amygdala

Temporal lobe

Corpus callosum (body)

Septum pellucidum

Anterior horn of lateral ventricle

Fornix

Internal capsule, anterior limb

Septal nuclei

Substantia innominata (nucleus basalis)

Optic tract

Pituitary stalk

R   L

(D)

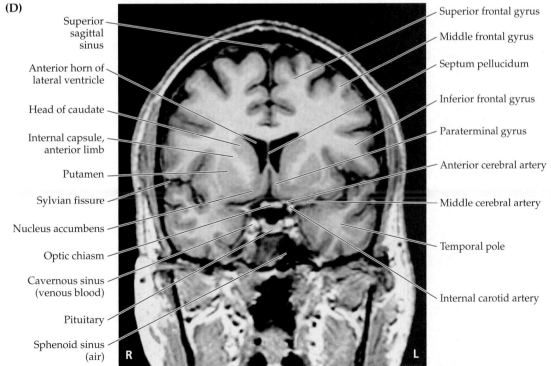

Superior sagittal sinus

Anterior horn of lateral ventricle

Head of caudate

Internal capsule, anterior limb

Putamen

Sylvian fissure

Nucleus accumbens

Optic chiasm

Cavernous sinus (venous blood)

Pituitary

Sphenoid sinus (air)

Superior frontal gyrus

Middle frontal gyrus

Septum pellucidum

Inferior frontal gyrus

Paraterminal gyrus

Anterior cerebral artery

Middle cerebral artery

Temporal pole

Internal carotid artery

R   L

NEURORADIOLOGICAL ATLAS

FIGURE 4.15 MRI: Sagittal T1-Weighted Images Unenhanced sagittal MRI images with major structures labeled. TR = 600, TE = 12.

(A)

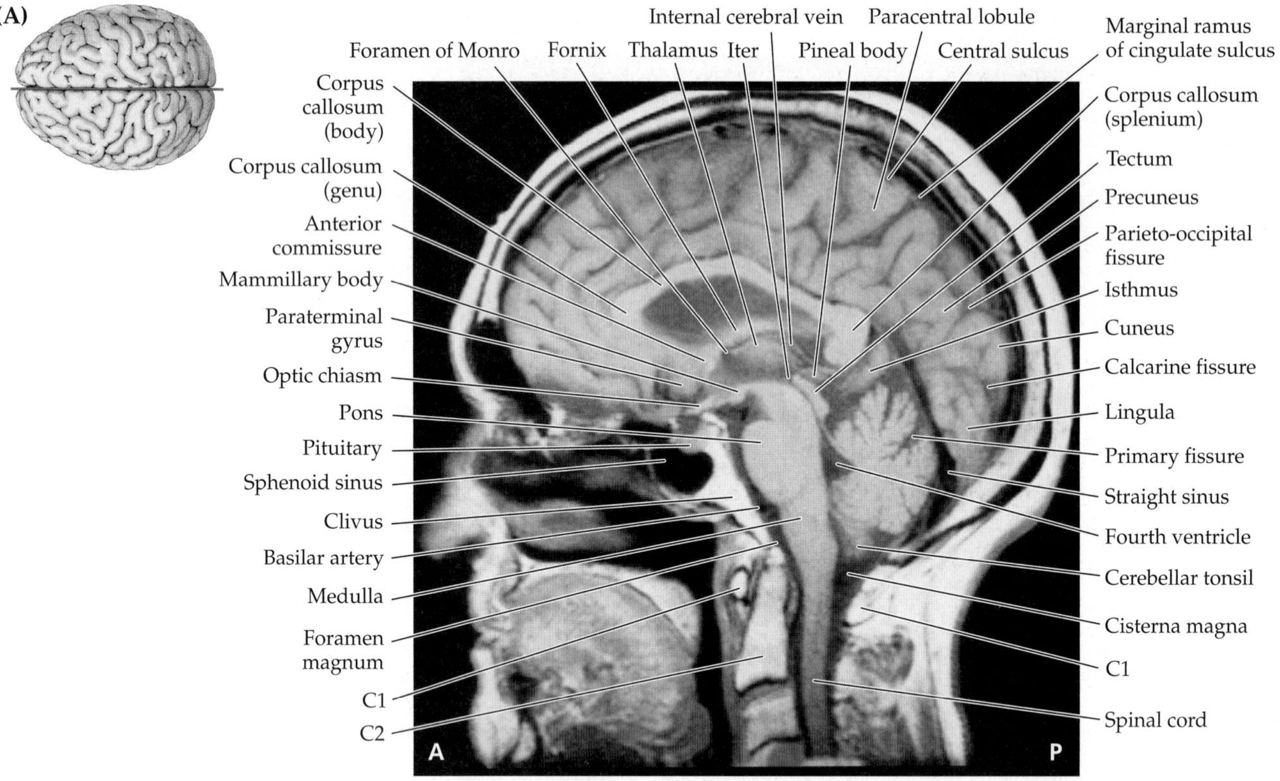
Internal cerebral vein
Paracentral lobule
Foramen of Monro
Fornix
Thalamus
Iter
Pineal body
Central sulcus
Marginal ramus of cingulate sulcus
Corpus callosum (body)
Corpus callosum (genu)
Anterior commissure
Mammillary body
Paraterminal gyrus
Optic chiasm
Pons
Pituitary
Sphenoid sinus
Clivus
Basilar artery
Medulla
Foramen magnum
C1
C2
Corpus callosum (splenium)
Tectum
Precuneus
Parieto-occipital fissure
Isthmus
Cuneus
Calcarine fissure
Lingula
Primary fissure
Straight sinus
Fourth ventricle
Cerebellar tonsil
Cisterna magna
C1
Spinal cord
A
P

(B)

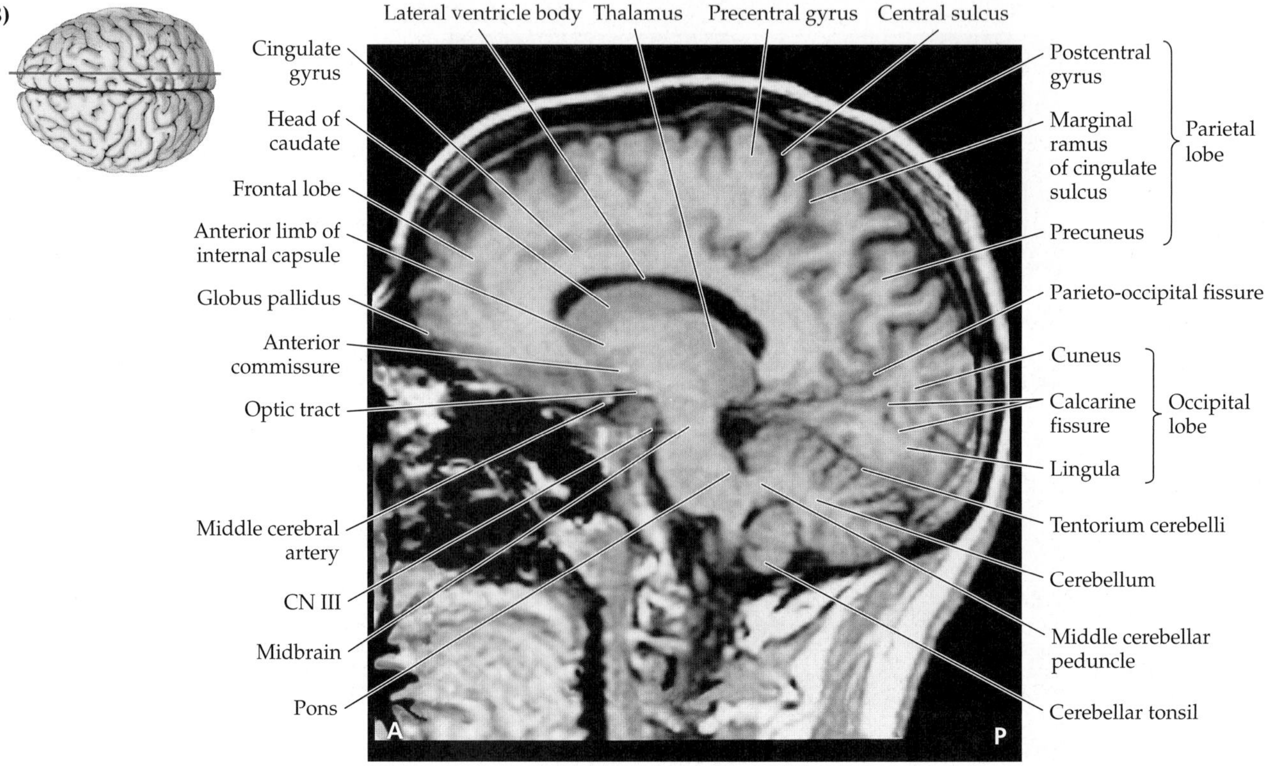
Lateral ventricle body
Thalamus
Precentral gyrus
Central sulcus
Cingulate gyrus
Head of caudate
Frontal lobe
Anterior limb of internal capsule
Globus pallidus
Anterior commissure
Optic tract
Middle cerebral artery
CN III
Midbrain
Pons
Postcentral gyrus
Marginal ramus of cingulate sulcus
Precuneus
Parietal lobe
Parieto-occipital fissure
Cuneus
Calcarine fissure
Lingula
Occipital lobe
Tentorium cerebelli
Cerebellum
Middle cerebellar peduncle
Cerebellar tonsil
A
P

MRI, SAGGITAL T1–WEIGHTED

(C)

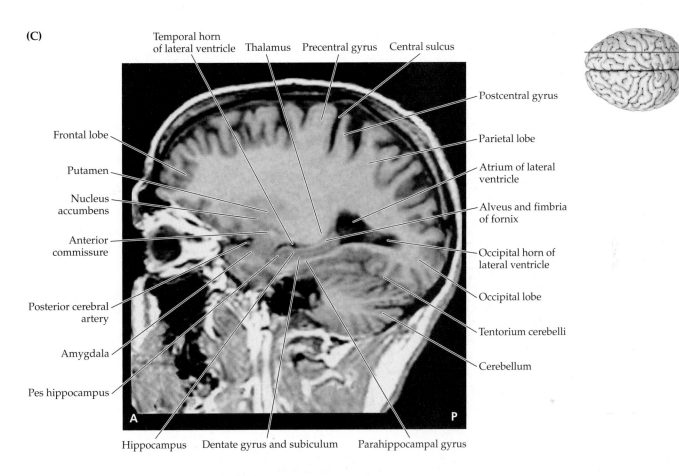

Temporal horn of lateral ventricle
Thalamus
Precentral gyrus
Central sulcus
Postcentral gyrus
Frontal lobe
Parietal lobe
Putamen
Atrium of lateral ventricle
Nucleus accumbens
Alveus and fimbria of fornix
Anterior commissure
Occipital horn of lateral ventricle
Posterior cerebral artery
Occipital lobe
Amygdala
Tentorium cerebelli
Pes hippocampus
Cerebellum
Hippocampus
Dentate gyrus and subiculum
Parahippocampal gyrus
A
P

(D)

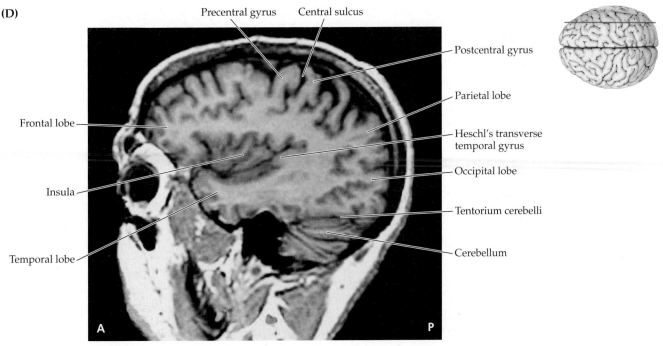

Precentral gyrus
Central sulcus
Postcentral gyrus
Frontal lobe
Parietal lobe
Heschl's transverse temporal gyrus
Insula
Occipital lobe
Tentorium cerebelli
Temporal lobe
Cerebellum
A
P

NEURORADIOLOGICAL ATLAS

FIGURE 4.16 Angiographic Images: Anterior Circulation
(A) Anterior–posterior view following injection of left internal carotid artery, demonstrating filling of left anterior and middle cerebral arteries (ACA, MCA). (B) Close-up view of recurrent artery of Heubner arising from anterior cerebral artery and lenticulostriate arteries arising from middle cerebral artery. (C)

Lateral view following injection of right internal carotid artery, demonstrating filling of right anterior and middle cerebral arteries. The Sylvian triangle is formed by hairpin loops that are made by branches of the middle cerebral artery as they pass from the insula over the operculum and onto the cortical surface.

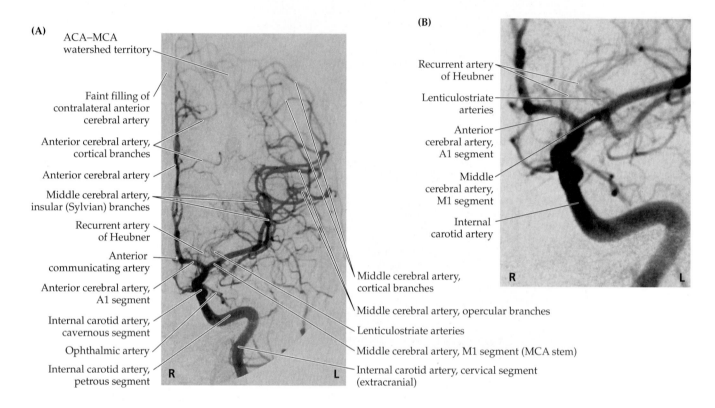

(A)

ACA–MCA watershed territory

Faint filling of contralateral anterior cerebral artery

Anterior cerebral artery, cortical branches

Anterior cerebral artery

Middle cerebral artery, insular (Sylvian) branches

Recurrent artery of Heubner

Anterior communicating artery

Anterior cerebral artery, A1 segment

Internal carotid artery, cavernous segment

Ophthalmic artery

Internal carotid artery, petrous segment

R   L

Middle cerebral artery, cortical branches

Middle cerebral artery, opercular branches

Lenticulostriate arteries

Middle cerebral artery, M1 segment (MCA stem)

Internal carotid artery, cervical segment (extracranial)

(B)

Recurrent artery of Heubner

Lenticulostriate arteries

Anterior cerebral artery, A1 segment

Middle cerebral artery, M1 segment

Internal carotid artery

R   L

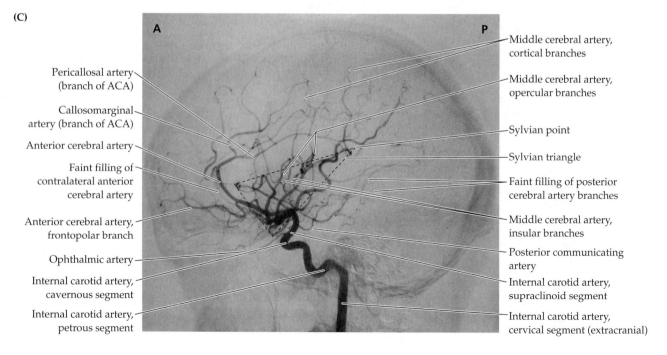

(C)

A   P

Pericallosal artery (branch of ACA)

Callosomarginal artery (branch of ACA)

Anterior cerebral artery

Faint filling of contralateral anterior cerebral artery

Anterior cerebral artery, frontopolar branch

Ophthalmic artery

Internal carotid artery, cavernous segment

Internal carotid artery, petrous segment

Middle cerebral artery, cortical branches

Middle cerebral artery, opercular branches

Sylvian point

Sylvian triangle

Faint filling of posterior cerebral artery branches

Middle cerebral artery, insular branches

Posterior communicating artery

Internal carotid artery, supraclinoid segment

Internal carotid artery, cervical segment (extracranial)

ANGIOGRAPHY

FIGURE 4.17 Angiographic Images: Posterior Circulation (A) Anterior–posterior view following injection of left vertebral artery. Reflux into the right vertebral artery can be seen. (B) Lateral view following injection of left vertebral artery.

(A)

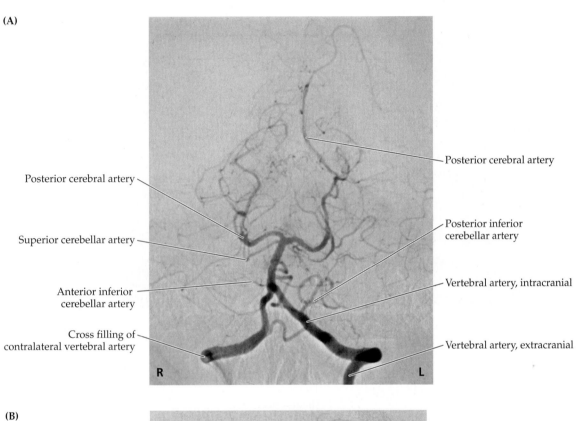

Posterior cerebral artery

Superior cerebellar artery

Anterior inferior cerebellar artery

Cross filling of contralateral vertebral artery

Posterior cerebral artery

Posterior inferior cerebellar artery

Vertebral artery, intracranial

Vertebral artery, extracranial

R   L

(B)

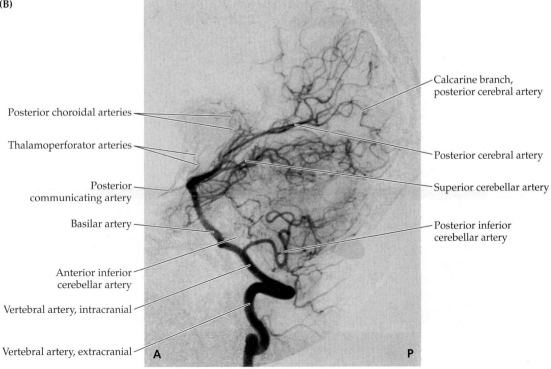

Posterior choroidal arteries

Thalamoperforator arteries

Posterior communicating artery

Basilar artery

Anterior inferior cerebellar artery

Vertebral artery, intracranial

Vertebral artery, extracranial

Calcarine branch, posterior cerebral artery

Posterior cerebral artery

Superior cerebellar artery

Posterior inferior cerebellar artery

A   P

NEURORADIOLOGICAL ATLAS

FIGURE 4.18 MRA Images: Intracranial Circulation (A) Superior view of circle of Willis. (B) Lateral view. Compare to Figures 4.16 and 4.17.

(A)

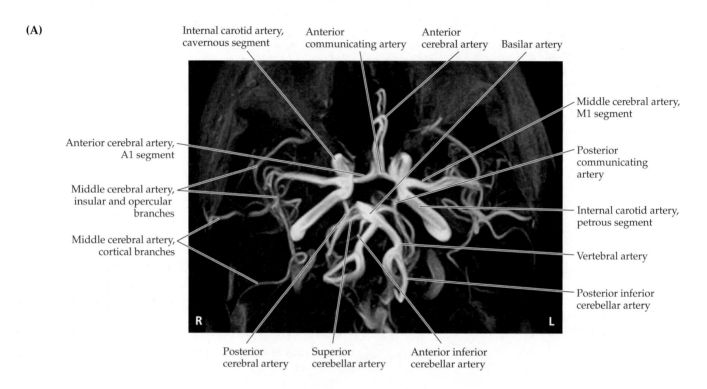

Internal carotid artery, cavernous segment
Anterior communicating artery
Anterior cerebral artery
Basilar artery
Middle cerebral artery, M1 segment
Posterior communicating artery
Internal carotid artery, petrous segment
Vertebral artery
Posterior inferior cerebellar artery
Anterior cerebral artery, A1 segment
Middle cerebral artery, insular and opercular branches
Middle cerebral artery, cortical branches
Posterior cerebral artery
Superior cerebellar artery
Anterior inferior cerebellar artery
R
L

(B)

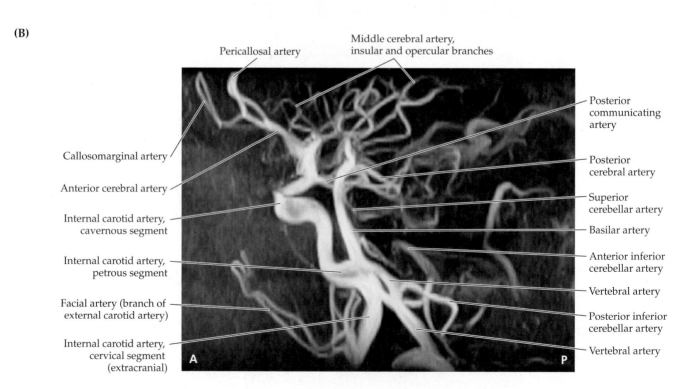

Pericallosal artery
Middle cerebral artery, insular and opercular branches
Posterior communicating artery
Callosomarginal artery
Anterior cerebral artery
Internal carotid artery, cavernous segment
Internal carotid artery, petrous segment
Facial artery (branch of external carotid artery)
Internal carotid artery, cervical segment (extracranial)
Posterior cerebral artery
Superior cerebellar artery
Basilar artery
Anterior inferior cerebellar artery
Vertebral artery
Posterior inferior cerebellar artery
Vertebral artery
A
P

MAGNETIC RESONANCE ANGIOGRAPHY

FIGURE 4.19 MRA Images: Neck Vessels (A) Lateral view. (B) Close-up view of left carotid bifurcation. (C) Anterior–posterior view.

(A)

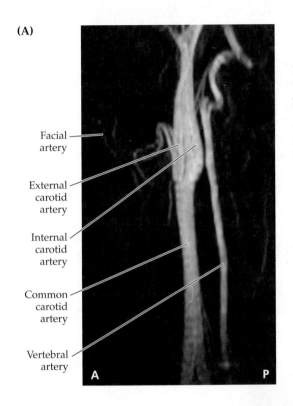

Facial artery

External carotid artery

Internal carotid artery

Common carotid artery

Vertebral artery

A   P

(B)

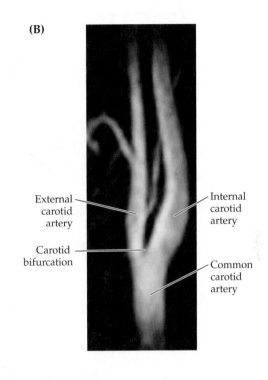

External carotid artery

Carotid bifurcation

Internal carotid artery

Common carotid artery

(C)

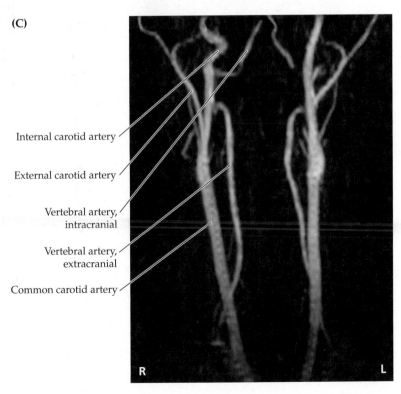

Internal carotid artery

External carotid artery

Vertebral artery, intracranial

Vertebral artery, extracranial

Common carotid artery

R   L

NEURORADIOLOGICAL ATLAS

FIGURE 4.20 MRA Images: Origins of Carotid and Vertebral Arteries
Anterior–posterior view with successive slices progressing from anterior (A) to posterior (B). (A) Origin of common carotid arteries from aortic arch and brachiocephalic artery. (B) Origin of vertebral arteries from subclavian arteries.

(A)

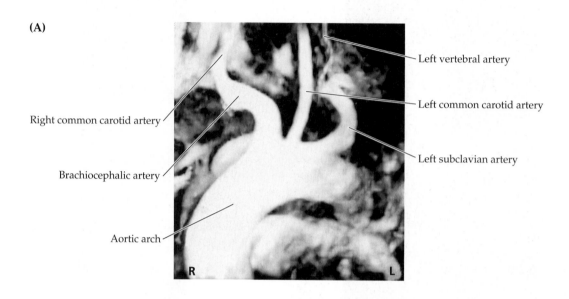

Left vertebral artery

Left common carotid artery

Right common carotid artery

Left subclavian artery

Brachiocephalic artery

Aortic arch

(B)

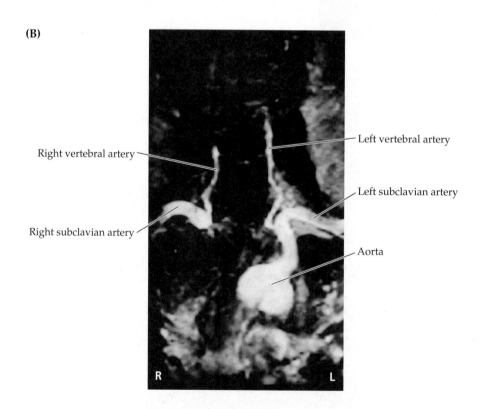

Right vertebral artery

Left vertebral artery

Left subclavian artery

Right subclavian artery

Aorta

References

General

Grossman RI, Yousem DM. 2003. *Neuroradiology: The Requisites.* Mosby, Elsevier, Philadelphia.

Hathout G. 2008. *Clinical Neuroradiology: A Case-Based Approach.* Cambridge University Press, New York.

Osborn AG, Blaser SI, Salzman KL. 2004. *Diagnostic Imaging: Brain.* Saunders, Philadelphia.

Wolbarst AB. 2005. *Physics of Radiology.* 2nd Ed. Chapters 1, 42. Appleton & Lange, Norwalk, CT.

CT

Hu H. 1999. Multi-slice helical CT: Scan and reconstruction. *Med Phys* 26: 5–18.

Seeram E. 2008. *Computed Tomography: Physical Principles, Clinical Applications and Quality Control (Contemporary Imaging Techniques).* 3rd Ed. Saunders, Philadelphia.

MRI

Bushong SC. 2003. *Magnetic Resonance Imaging: Physical and Biological Principles.* 3rd Ed. Elsevier Health Sciences.

Edelman RR, Zlatkin MB, Hesselink JR (eds.). 2005. *Clinical Magnetic Resonance Imaging.* 3rd Ed. Saunders, Philadelphia.

Moseley ME, Liu C, Rodriguez S, Brosnan T. 2009. Advances in magnetic resonance neuroimaging. *Neurol Clin* 27 (1): 1–19.

Vlaardingerbroek MT, den Boer JA. 2004. *Magnetic Resonance Imaging: Theory and Practice.* 3rd Ed. Springer, Berlin.

Weishaupt D, Koechli VD, Marincek B. 2008. *How does MRI work? An Introduction to the Physics and Function of Magnetic Resonance Imaging.* 2nd Ed. Springer, Berlin.

Neuroangiography

Babikian VL, Wechsler LR, Higashida RT (eds.). 2003. *Imaging Cerebrovascular Disease.* Butterworth-Heinemann, Oxford.

Borden NM. 2006. *3D Angiographic Atlas of Neurovascular Anatomy and Pathology.* Cambridge University Press, New York.

Hurst RW, Rosenwasser RH. 2007. *Interventional Neuroradiology.* Informa Healthcare, New York.

Morris PP. 2006. *Practical Neuroangiography.* 2nd Ed. Lippincott, Williams & Wilkins, Baltimore.

Rubin GD, Rofsky NM. 2008. *CT and MR Angiography: Comprehensive Vascular Assessment.* Lippincott, Williams & Wilkins, Baltimore.

Schneider G, Prince MR, Meaney JFM, Ho VB. 2005. *Magnetic Resonance Angiography: Techniques, Indications and Practical Applications.* Springer, Berlin.

Wakhloo AK, Deleo MJ 3rd, Brown MM. 2009. Advances in interventional neuroradiology. *Stroke.* 40 (5):e305–312.

Functional Neuroimaging

Barrington SF, Maisey MN, Wahl RL. 2005. *Atlas of Clinical Positron Emission Tomography.* Oxford University Press, Oxford.

Holodny AI. 2008. *Functional Neuroimaging: A Clinical Approach.* Informa Healthcare, New York.

Huettel SA, Song AW, McCarthy G. 2009. *Functional Magnetic Resonance Imaging.* 2nd Ed. Sinauer, Sunderland, MA.

Toga AW, Mazziotta JC, Frackowiak RSJ (eds.). 2000. *Brain Mapping: The Trilogy.* 3 vols. Academic Press, San Diego.

Valk PE, Delbeke D, Bailey DL, Townsend DW. 2006. *Positron Emission Tomography: Clinical Practice.* Springer, Berlin.

Van Heertum RL, Ichise M. 2009. *Functional Cerebral SPECT and PET Imaging.* 4th Ed. Lippincott Williams & Wilkins, Philadelphia.

CONTENTS

Chapter 5

Brain and Environs: Cranium, Ventricles, and Meninges

After a domestic altercation in which he fell down a flight of cement stairs and injured his head, a 51-year-old man was arrested and taken to prison. He had been conscious and smelled of alcohol when the police had arrested him, but the next morning in prison he was found unresponsive and thrashing aimlessly in his cell. His left pupil was dilated. His right side was paralyzed and had brisk reflexes.

This case illustrates how head injury can cause abnormal shifts among various compartments in the head, including the cranial vault, ventricles, and meninges. In this chapter, we will learn about the normal anatomy and function of each of these compartments as well as clinical consequences of injury or illness.

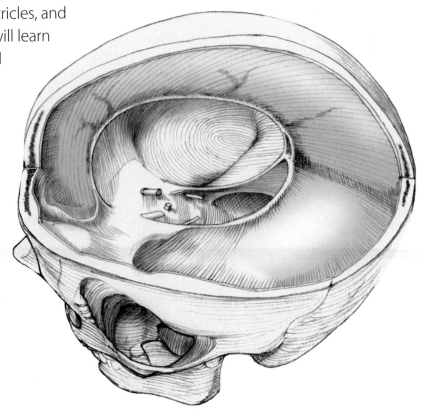

ANATOMICAL AND CLINICAL REVIEW

IN THE SECTIONS THAT FOLLOW, we will briefly discuss the brain in relation to its local environment, including the skull, meninges, blood vessels, and cerebrospinal fluid. In addition, we will summarize several important clinical abnormalities that involve these structures, including headache, intracranial mass lesions, elevated intracranial pressure, brain herniation, intracranial hemorrhage, hydrocephalus, brain tumors, and infections of the nervous system. Because this is the first chapter in the book that contains clinical cases, we introduce many Key Clinical Concepts that will be used not just here, but throughout the remainder of the book. For now, you can skim these Key Clinical Concepts (KCC 5.1–5.11) briefly. Later, while you're thinking through cases and attempting to make diagnoses, it will be useful to refer back to these sections in more detail.

Cranial Vault and Meninges

The brain is encased in several protective layers that cushion it from trauma (Figure 5.1). Beneath the skin and subcutaneous tissues lie the hard bones that form the skull. The skull has many foramina, or holes, which allow the cranial nerves, spinal cord, and blood vessels to enter and leave the intracranial cavity. We will review these foramina in greater detail in Chapter 12, but for now it is important to recognize the largest foramen at the base of the skull: the **foramen magnum** (Figure 5.2). The point where the spinal cord meets the medulla, the **cervicomedullary junction**, occurs at the level of the foramen magnum (see Figures 2.2C and 5.10). You should be able to easily identify the foramen magnum and the other major foramina at the base of the skull on a CT scan (Figure 5.3).

On the inner surface of the skull, several ridges of bone divide the base of the cranial cavity into different compartments, or fossae (see Figure 5.2B and

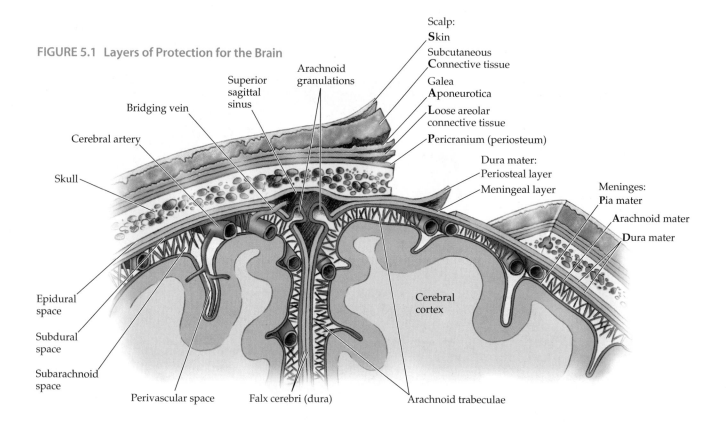

FIGURE 5.1 Layers of Protection for the Brain

(A)

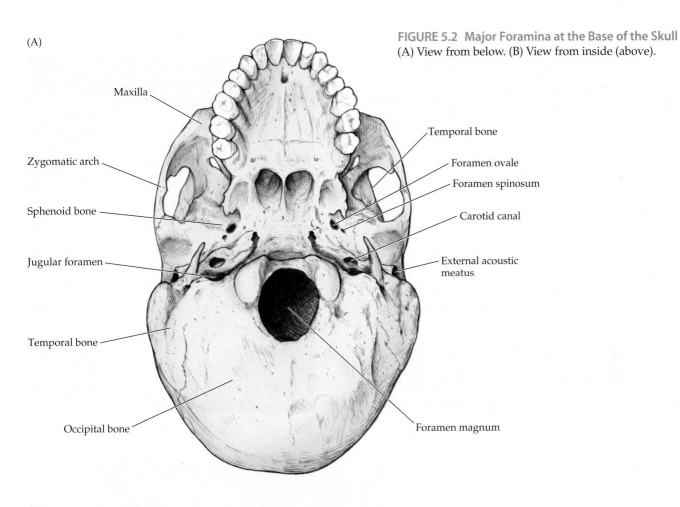

Maxilla

Zygomatic arch

Sphenoid bone

Jugular foramen

Temporal bone

Occipital bone

Temporal bone

Foramen ovale

Foramen spinosum

Carotid canal

External acoustic meatus

Foramen magnum

FIGURE 5.2 Major Foramina at the Base of the Skull
(A) View from below. (B) View from inside (above).

(B)

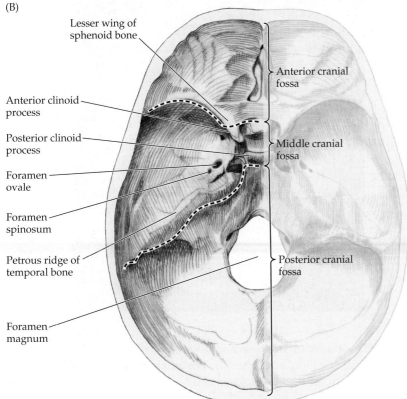

Lesser wing of sphenoid bone

Anterior clinoid process

Posterior clinoid process

Foramen ovale

Foramen spinosum

Petrous ridge of temporal bone

Foramen magnum

Anterior cranial fossa

Middle cranial fossa

Posterior cranial fossa

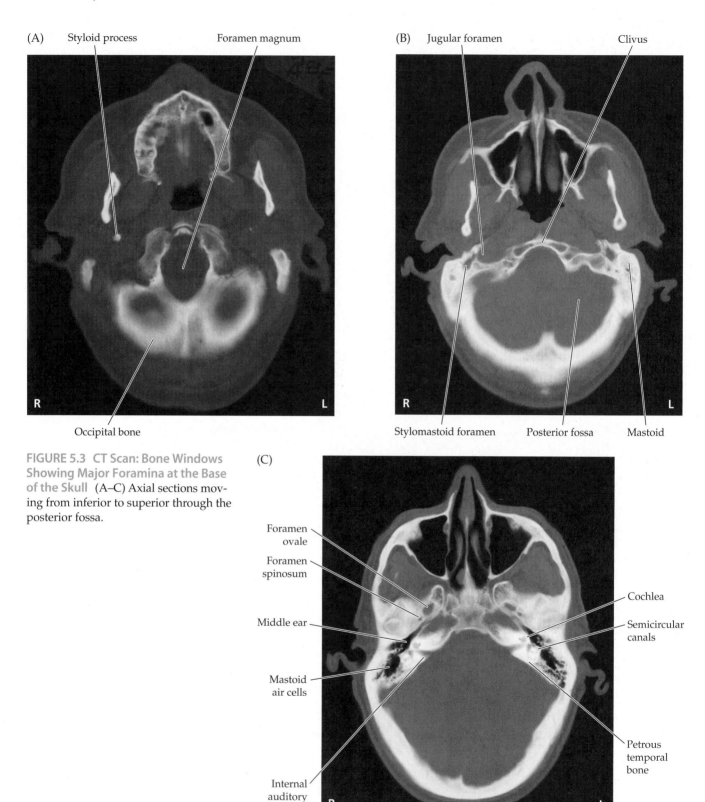

(A)
Styloid process   Foramen magnum

R   L

Occipital bone

(B)
Jugular foramen   Clivus

R   L

Stylomastoid foramen   Posterior fossa   Mastoid

FIGURE 5.3 CT Scan: Bone Windows Showing Major Foramina at the Base of the Skull (A–C) Axial sections moving from inferior to superior through the posterior fossa.

(C)

Foramen ovale

Foramen spinosum

Middle ear

Mastoid air cells

Internal auditory meatus

Cochlea

Semicircular canals

Petrous temporal bone

R   L

Figure 5.4). The **anterior fossa** on each side contains the frontal lobe. The **middle fossa** contains the temporal lobe. The **posterior fossa** contains the cerebellum and brainstem. The anterior fossa is divided from the middle fossa by the lesser wing of the sphenoid bone. The middle fossa is divided from the posterior fossa by the petrous ridge of the temporal bone as well as by a

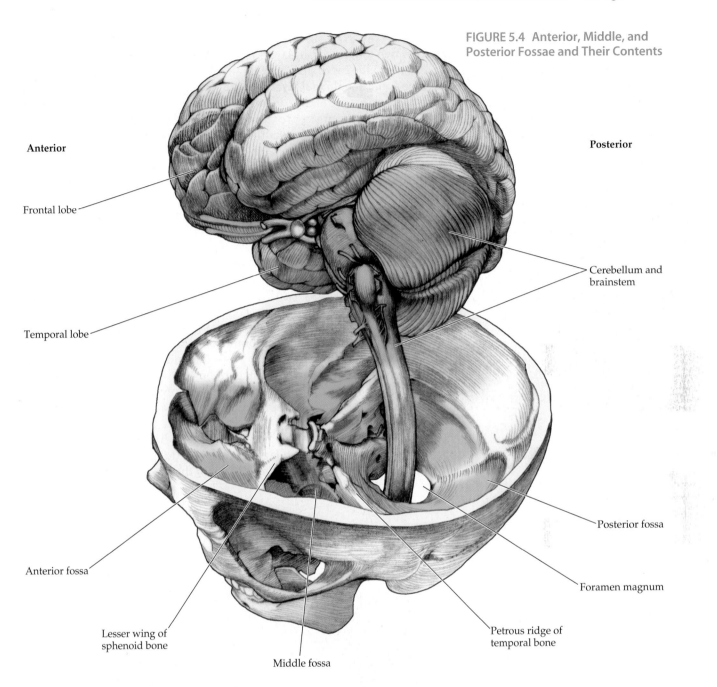

FIGURE 5.4 Anterior, Middle, and Posterior Fossae and Their Contents

Anterior

Posterior

Frontal lobe

Temporal lobe

Cerebellum and brainstem

Posterior fossa

Anterior fossa

Foramen magnum

Lesser wing of sphenoid bone

Petrous ridge of temporal bone

Middle fossa

sheet of meninges, which will be described next. These fossae can also be identified on CT and MRI scans (see Figures 4.12A–D and 4.13A–D).

The final layers of protection within the skull and surrounding the brain are the **meninges** and cerebral spinal fluid (see Figure 5.1). The three layers of meninges from inside to outside are

1. **Pia**
2. **Arachnoid**
3. **Dura**

Thus, a mnemonic for the meningeal layers is **PAD**. The term "mater" (meaning "mother") is sometimes added after these names—for example, pia mater, arachnoid mater, and dura mater. A mnemonic for the layers of the scalp (**SCALP**) is shown in Figure 5.1.

MNEMONIC

Moving now from outside to inside, the **dura**, meaning "hard," is composed of two tough, fibrous layers (see Figure 5.1). The outer **periosteal layer** is adherent to the inner surface of the skull. This outer layer of dura is fused with the inner **meningeal layer** of dura, except in a few places where the inner layer forms folds that descend far into the cranial cavity (see Figure 5.1). This mainly occurs in two places. The first is the **falx cerebri**, a flat sheet of dura that is suspended from the roof of the cranium and separates the right and left cerebral hemispheres, running in the interhemispheric fissure (Figure 5.5; see also Figure 5.1). The second is the **tentorium cerebelli**, a tent-like sheet of dura that covers the upper surface of the cerebellum (see Figure 5.5 and Figure 5.6).

The tentorium cerebelli, together with the petrous portions of the temporal bones, divide the posterior fossa from the rest of the cranial vault. The portion of the intracranial cavity above the tentorium is referred to as **supratentorial**; that below is called **infratentorial**. To understand the relationship between the tentorium cerebelli and the other intracranial structures and how the tentorium is truly "tent" shaped, review the CT scan images in Figure 4.12D,E and the MRI images in Figures 4.13D,E and 4.15A–D. Note that the occipital lobes and part of the temporal lobes rest on the upper surface of the tentorium. Recall that the midbrain connects the cerebral hemispheres with the brainstem and cerebellum. Thus, the midbrain can be seen to pass through an important narrow opening in the tentorium cerebelli, the **tentorial incisura**, also called the **tentorial notch** (see Figure 5.6).

The **arachnoid** is a wispy, "spidery" meningeal layer that adheres to the inner surface of the dura. Within the arachnoid, the cerebrospinal fluid percolates over the surface of the brain (see Figures 5.1, 5.10). The innermost meningeal layer is a very thin layer of cells called the **pia**. Unlike the arachnoid, the pia adheres closely to the surface of the brain and follows it along

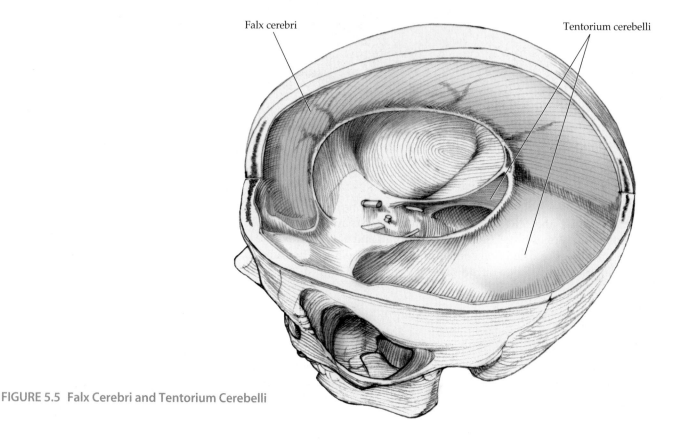

Falx cerebri

Tentorium cerebelli

FIGURE 5.5 Falx Cerebri and Tentorium Cerebelli

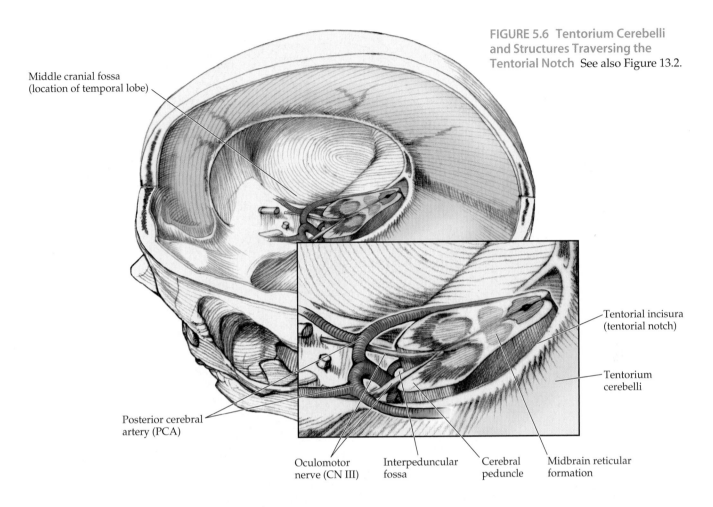

FIGURE 5.6 Tentorium Cerebelli and Structures Traversing the Tentorial Notch See also Figure 13.2.

Middle cranial fossa (location of temporal lobe)

Tentorial incisura (tentorial notch)

Tentorium cerebelli

Posterior cerebral artery (PCA)

Oculomotor nerve (CN III)

Interpeduncular fossa

Cerebral peduncle

Midbrain reticular formation

all the gyri and into the depths of the sulci. The pia also surrounds the initial portion of each blood vessel as it penetrates the brain surface, forming a perivascular space (Virchow–Robin space), and then fuses with the blood vessel wall (see Figure 5.1).

The meninges form three spaces or potential spaces with clinical significance, listed here from outside to inside (see Figure 5.1):

1. **Epidural space**
2. **Subdural space**
3. **Subarachnoid space**

Each of these spaces contains some important blood vessels that can give rise to hemorrhage (see KCC 5.6). The **epidural space** is a potential space located between the inner surface of the skull and the tightly adherent dura (see Figure 5.1). The **middle meningeal artery** enters the skull through the **foramen spinosum** (see Figures 5.2 and 5.3C) and runs in the epidural space between the dura and the skull (Figure 5.7). Grooves, which are formed by this artery and its many branches, can often be seen on the inner surface of the skull. Note that the middle *meningeal* artery is a branch of the external carotid artery (see Figure 2.26A) and supplies the dura, while the middle *cerebral* artery is a branch of the internal carotid artery and supplies the brain (see Figure 2.26C).

The **subdural space** is a potential space between the inner layer of dura and the loosely adherent arachnoid (see Figure 5.1). The **bridging veins** traverse the subdural space. These veins drain the cerebral hemispheres and pass through the subdural space en route to several large **dural venous sinuses** (Fig-

FIGURE 5.7 Middle Meningeal Artery
Shown emerging from the foramen spinosum and running in epidural space between dura and skull.

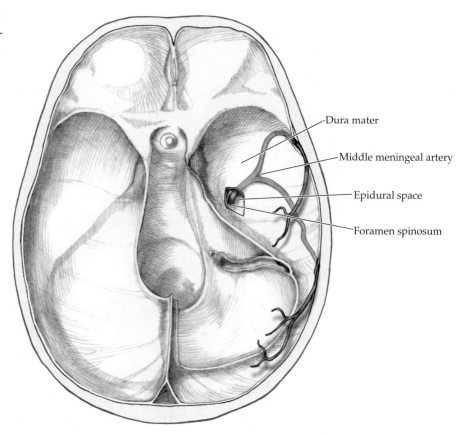

Dura mater

Middle meningeal artery

Epidural space

Foramen spinosum

ure 5.8 and Figure 5.9; see also Figure 5.1). Dural sinuses are large venous channels that lie enclosed within the two layers of dura. The dural sinuses drain blood mainly via the **sigmoid sinuses** to reach the **internal jugular veins**.

The cerebrospinal fluid-filled space between the arachnoid and the pia is called the **subarachnoid space** (see Figure 5.1 and Figure 5.10). In addition to cerebrospinal fluid, the major arteries of the brain also travel within the subarachnoid space and then send smaller penetrating branches inward through the pia.

As the spinal cord exits through the foramen magnum and continues downward through the bony spinal canal, it is enveloped by the same three meningeal layers (see Figure 5.10). The only significant difference is the layer of **epidural fat** in the spinal canal between the dura and periosteum (see Figure 8.2C,D); in the cranium, both layers of dura adhere tightly to bone (see Figure 5.1).

Ventricles and Cerebrospinal Fluid

During early development, the neural tube forms several cavities within the brain called **ventricles** (see Figure 2.2). The ventricles contain **cerebrospinal fluid** (**CSF**), which is produced mainly by a specialized vascular structure called the **choroid plexus** that lies inside the ventricles (see Figure 5.10). The inner walls of the ventricles are lined with a layer of **ependymal cells**, and the blood vessels of the choroid plexus are lined with similar-appearing cuboidal cells called **choroid epithelial cells** (see Figure 5.13C). There are two lateral ventricles (one inside each cerebral hemisphere), a third ventricle located within the diencephalon, and the fourth ventricle, which is surrounded by the pons, medulla, and cerebellum (**Figure 5.11** and **Table 5.1**).

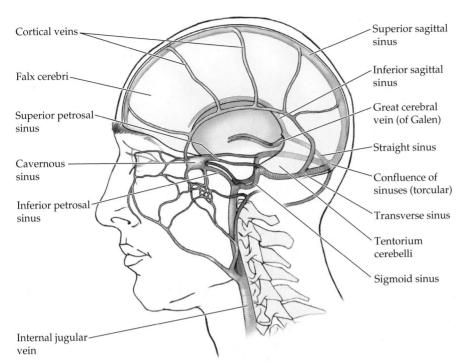

Cortical veins

Falx cerebri

Superior petrosal sinus

Cavernous sinus

Inferior petrosal sinus

Internal jugular vein

Superior sagittal sinus

Inferior sagittal sinus

Great cerebral vein (of Galen)

Straight sinus

Confluence of sinuses (torcular)

Transverse sinus

Tentorium cerebelli

Sigmoid sinus

FIGURE 5.8 Dural Venous Sinuses and Internal Jugular Veins Lateral view.

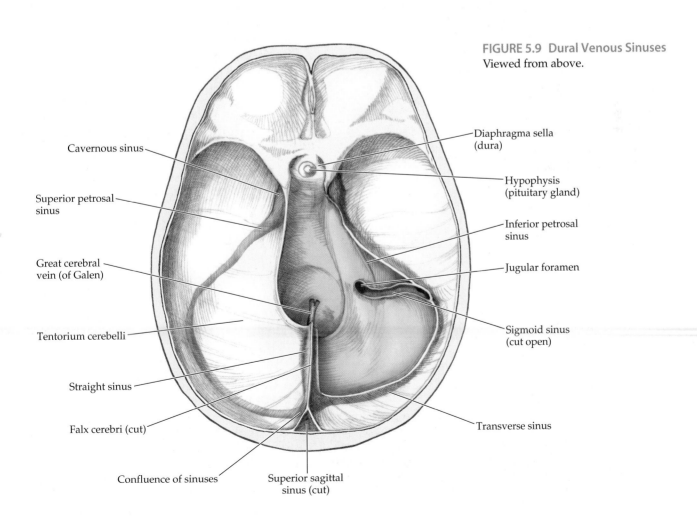

Cavernous sinus

Superior petrosal sinus

Great cerebral vein (of Galen)

Tentorium cerebelli

Straight sinus

Falx cerebri (cut)

Confluence of sinuses

Superior sagittal sinus (cut)

Diaphragma sella (dura)

Hypophysis (pituitary gland)

Inferior petrosal sinus

Jugular foramen

Sigmoid sinus (cut open)

Transverse sinus

FIGURE 5.9 Dural Venous Sinuses Viewed from above.

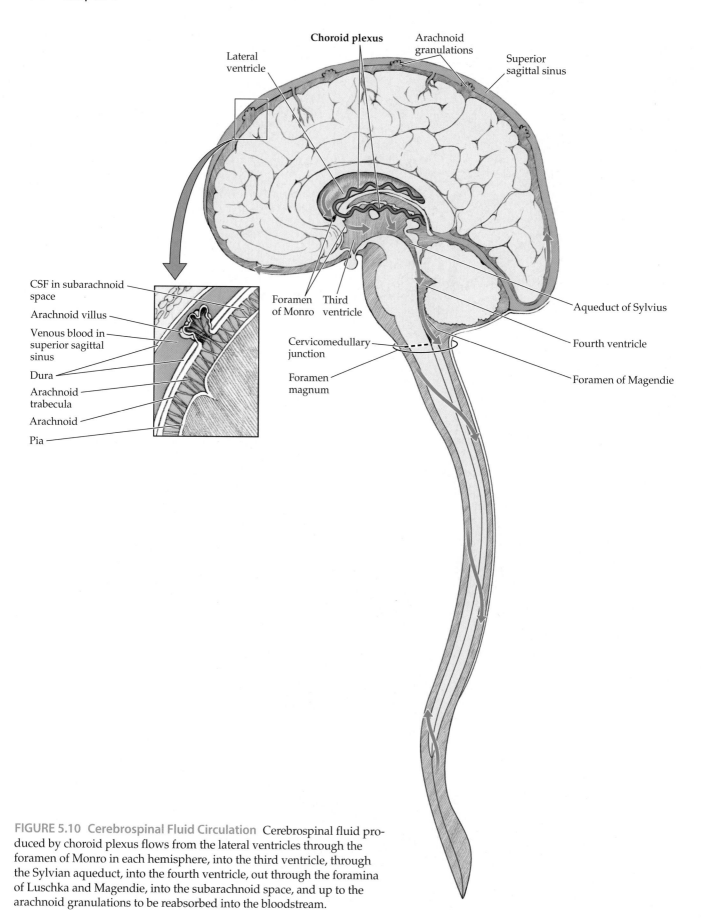

FIGURE 5.10 Cerebrospinal Fluid Circulation Cerebrospinal fluid produced by choroid plexus flows from the lateral ventricles through the foramen of Monro in each hemisphere, into the third ventricle, through the Sylvian aqueduct, into the fourth ventricle, out through the foramina of Luschka and Magendie, into the subarachnoid space, and up to the arachnoid granulations to be reabsorbed into the bloodstream.

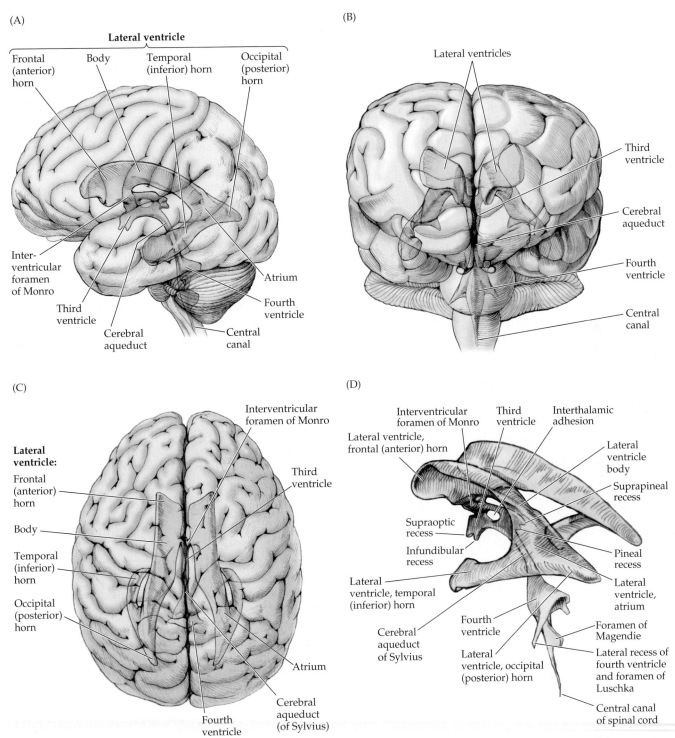

(A)

Lateral ventricle

Frontal (anterior) horn

Body

Temporal (inferior) horn

Occipital (posterior) horn

Interventricular foramen of Monro

Third ventricle

Cerebral aqueduct

Central canal

Atrium

Fourth ventricle

(B)

Lateral ventricles

Third ventricle

Cerebral aqueduct

Fourth ventricle

Central canal

(C)

Interventricular foramen of Monro

Third ventricle

Lateral ventricle:

Frontal (anterior) horn

Body

Temporal (inferior) horn

Occipital (posterior) horn

Atrium

Cerebral aqueduct (of Sylvius)

Fourth ventricle

(D)

Interventricular foramen of Monro

Third ventricle

Interthalamic adhesion

Lateral ventricle, frontal (anterior) horn

Lateral ventricle body

Suprapineal recess

Supraoptic recess

Infundibular recess

Pineal recess

Lateral ventricle, temporal (inferior) horn

Cerebral aqueduct of Sylvius

Fourth ventricle

Lateral ventricle, occipital (posterior) horn

Lateral ventricle, atrium

Foramen of Magendie

Lateral recess of fourth ventricle and foramen of Luschka

Central canal of spinal cord

FIGURE 5.11 Brain Ventricles (A) Ventricles viewed from lateral surface of brain. (B) Ventricles viewed from anterior surface of brain. (C) Ventricles viewed from superior surface of brain. (D) Details of ventricular structure.

The largest of the ventricles are the two **lateral ventricles** (see Figure 5.11), formerly called the first and second ventricles. The lateral ventricles have extensions called horns that are named after the lobes or after the direction in which they extend (see Table 5.1 and Figure 5.11). The **frontal horn**, also known as the **anterior horn**, of the lateral ventricle extends anteriorly from the **body** of the lateral ventricle into the frontal lobe. By definition, the frontal horn begins anterior to the foramen of Monro, described below. The

TABLE 5.1 Brain Ventricles	
VENTRICLE	**LOCATION**
Lateral ventricles	Within the cerebral hemisphere
Frontal (anterior) horn	Begins anterior to the interventricular foramen of Monro and extends into the frontal lobe
Body	Posterior to the interventricular foramen of Monro, within the frontal and parietal lobes
Atrium (trigone)	Area of convergence of the occipital horn, the temporal horn, and the body of the lateral ventricle
Occipital (posterior) horn	Extends from the atrium posteriorly into the occipital lobe
Temporal (inferior) horn	Extends from the atrium inferiorly into the temporal lobe
Third ventricle	Within the thalamus and hypothalamus
Fourth ventricle	Within the pons, medulla, and cerebellum

REVIEW EXERCISE

Cover the labels in Figure 5.10 and name each space or foramen that CSF travels through en route from the choroid plexus of the lateral ventricles to the arachnoid granulations.

Next, use the CT and MRI images from the "Neuroradiological Atlas" (see Figures 4.12–4.15) to identify each of the structures listed in the left column of Table 5.1 and to identify the foramen of Monro and the cerebral aqueduct.

MNEMONIC

body of the lateral ventricle merges posteriorly with the **atrium** or **trigone**. The atrium connects three parts of the lateral ventricle—the body; the **occipital horn**, also known as the **posterior horn**, which extends back into the occipital lobe; and the **temporal horn**, also known as the **inferior horn**, which extends inferiorly and anteriorly into the temporal lobe.

There are several **C-shaped structures** in the brain, including the caudate nucleus, corpus callosum, fornix, and stria terminalis. These structures follow the C-shaped curve of the lateral ventricles. The spatial relationships between these structures are discussed at the end of this chapter in the "Brief Anatomical Study Guide" and "A Scuba Expedition through the Brain."

The lateral ventricles communicate with the **third ventricle** (see Table 5.1) via the **interventricular foramen of Monro** (see Figures 5.10 and 5.11). The walls of the third ventricle are formed by the thalamus and hypothalamus. The third ventricle communicates with the **fourth ventricle** via the **cerebral aqueduct**, also called the **aqueduct of Sylvius**, which travels through the midbrain (see Figures 5.10 and 5.11). The roof of the fourth ventricle is formed by the cerebellum, and the floor is formed by the pons and medulla.

Cerebrospinal fluid leaves the ventricular system via several foramina in the fourth ventricle—the *lateral foramina of Luschka* and the *midline foramen of Magendie* (see Figure 5.11D). Cerebrospinal fluid then percolates around the brain and spinal cord in the subarachnoid space and is ultimately reabsorbed by the **arachnoid granulations** (see Figures 5.1 and 5.10) into the dural venous sinuses, and thus back into the bloodstream. The normal total volume of cerebrospinal fluid in an adult is about **150 cc**. It is produced by the choroid plexus at a rate of **20 cc/hour**, or about 500 cc/day.

The subarachnoid space widens in a few areas to form larger CSF collections called **cisterns**. The following cisterns come up fairly often in clinical practice (**Figure 5.12**):

- Perimesencephalic cisterns
    - Ambient cistern (cisterna ambiens)
    - Quadrigeminal cistern (cisterna quadrigemina)
    - Interpeduncular cistern
- Prepontine cistern (pontine cistern)
- Cisterna magna
- Lumbar cistern

The **ambient cistern** is located lateral to the midbrain; the **quadrigeminal cistern** is posterior to the midbrain, beneath the posterior portion of the corpus callosum (see Figure 5.12). The name "quadrigeminal" comes from the four bumps of the superior and inferior colliculi (see Figures 2.22B and 5.12). The **interpeduncular cistern**, also sometimes called the **interpeduncular fossa**, is located on the ventral surface of the midbrain, between the cerebral peduncles (see Figure 5.6). Note that the third nerve exits the midbrain through the interpeduncular fossa. The **prepontine cistern** is located just ventral to the pons. It contains the basilar artery and the sixth nerves (see Figures 2.22A and 2.26C) as they ascend from the pontomedullary junction up along the clivus (see Figure 5.3B). The **cisterna magna**, also known as the cerebellomedullary cistern, is the largest cistern and is located beneath the cerebellum near the foramen magnum (see Figure 5.12; see also Figure 4.15A). Finally, the **lumbar cistern**, located in the lumbar portions of the spinal column, contains the cauda equina (see Figure 2.8) and is the region from which cerebrospinal fluid is obtained during a **lumbar puncture**, or spinal tap (see KCC 5.10).

Blood–Brain Barrier

Anatomists discovered in the 1800s that when a colored dye is injected into the bloodstream of an animal, all of its organs become stained except the brain. The reason is that the capillary wall endothelial cells in most of the body are separated from each other by clefts, or fenestrations, allowing relatively free passage of fluids and solute molecules (Figure 5.13A). In the brain, however, capillary endothelial cells are linked by **tight junctions** (Figure 5.13B), and substances entering or leaving the brain must travel through the endothelial cells,

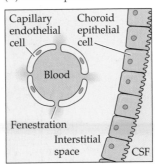

Perimesencephalic cisterns:
Interpeduncular cistern
Quadrigeminal
and ambient cisterns

Prepontine cistern

Cisterna magna

Lumbar cistern

FIGURE 5.12 Principal CSF Cisterns in the Subarachnoid Space

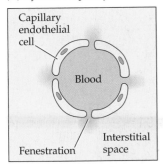 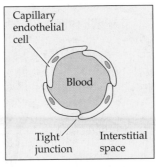 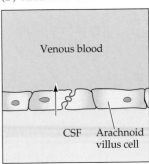

(A) Systemic capillary

Capillary endothelial cell

Blood

Fenestration

Interstitial space

(B) Brain capillary

Capillary endothelial cell

Blood

Tight junction

Interstitial space

(C) Choroid plexus

Capillary endothelial cell

Choroid epithelial cell

Blood

Fenestration

Interstitial space

CSF

(D) Arachnoid villi

Venous blood

CSF

Arachnoid villus cell

FIGURE 5.13 Blood–Brain and Blood–CSF Barriers (A) Typical fenestrated capillary lying outside the nervous system, allowing the passage of water and solutes. (B) Brain capillary with tight junctions between endothelial cells, forming the blood–brain barrier. Cellular transport across the endothelial layer is required for the passage of water-soluble substances between blood and brain. (C) The choroid plexus capillary allows the passage of water and solutes, but choroid plexus epithelial cells form the blood–CSF barrier, requiring cellular transport for passage. (D) The arachnoid villus cell carries out one-way bulk flow of CSF from the subarachnoid space to venous sinuses via giant vacuoles.

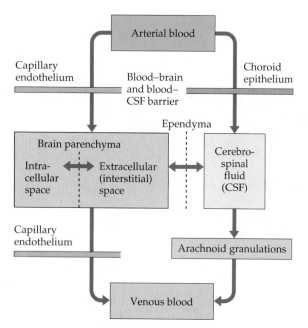

FIGURE 5.14 Fluid Compartments of the Nervous System Blood–brain and blood–CSF barriers separate arterial blood from brain parenchyma and CSF. Substances pass relatively freely between brain parenchyma interstitial space and CSF.

mostly by active transport processes. These endothelial cells and the tight junctions between them form the **blood–brain barrier**.

A similar selective barrier exists between the choroid plexus and the CSF (Figure 5.13C), sometimes referred to as the blood–CSF barrier. The capillaries of the choroid plexus are freely permeable, but the **choroid epithelial cells** form a barrier between the capillaries and the CSF. Lipid-soluble substances, including O_2 and CO_2, permeate readily across the cell membranes of the blood–brain and blood–CSF barriers. However, most other substances must be conveyed in both directions through specialized transport systems, including active transport, facilitated diffusion, ion exchange, and ion channels. In contrast to these vascular barriers, substances can pass relatively freely across the ependymal layer between the CSF and brain parenchyma (Figure 5.14). CSF is reabsorbed at the arachnoid granulations, where **arachnoid villus cells** mediate one-way bulk transport of CSF through giant vacuoles large enough to engulf entire red blood cells (Figure 5.13D).

Because synaptic transmission depends largely on chemical communication between neurons, the blood–brain and blood–CSF barriers protect brain function from most of the fluctuations in blood chemistry that occur on a continual basis. However, in certain specialized brain regions known as the **circumventricular organs**, the blood–brain barrier is interrupted, allowing the brain to respond to changes in the chemical milieu of the remainder of the body and to secrete modulatory neuropeptides into the bloodstream (Figure 5.15). Best known among these are the **median eminence** and the **neurohypophysis**, which are involved in the regulation and release of pituitary hormones (see Chapter 17).

The **area postrema** is the only paired circumventricular organ, and it is located along the caudal wall of the fourth ventricle in the medulla. Also known as the **chemotactic trigger zone**, it is involved in detecting circulating toxins that cause vomiting. The other circumventricular organs have less

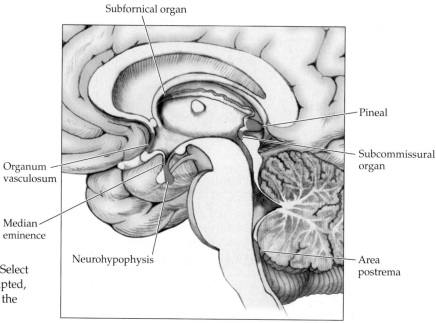

FIGURE 5.15 Circumventricular Organs Select regions where blood–brain barrier is interrupted, allowing chemical communication between the brain and the systemic circulation.

known clinical relevance, and will be mentioned briefly only for the sake of completeness. The organum vasculosum of the lamina terminalis may have neuroendocrine functions, the subfornical organ may regulate fluid balance, the **pineal** may be involved in melatonin-related circadian rhythms, and the function of the subcommissural organ is not known.

Brain tumors, infections, trauma and other disorders can disrupt the blood–brain barrier, resulting in extravasation of fluids into the interstitial space (see Figure 5.14). This excessive extracellular fluid is called **vasogenic edema**. Cellular damage—for example, in cerebral infarction—can cause excessive intracellular fluid accumulation within brain cells, a condition known as **cytotoxic edema**. Both kinds of edema often occur simultaneously.

This concludes our anatomical review of the cranial vault, ventricles, and meninges. The sections that follow (KCC 5.1–5.11) introduce several clinical concepts that will be referred to frequently throughout this book. Therefore, you may choose to skim these sections only briefly now and move on to Clinical Cases 5.1–5.10 to solidify your knowledge of the anatomical material in this chapter.

5.1 KEY CLINICAL CONCEPT

HEADACHE

Headache is one of the most common neurologic symptoms. Although usually benign, it occasionally signals life-threatening conditions. Interestingly, there are no pain receptors in the brain parenchyma itself. Therefore, headache is caused by mechanical traction, inflammation, or irritation of other structures in the head that are innervated, including the blood vessels, meninges, scalp, and skull. The supratentorial dura (most of the intracranial cavity) is innervated by the trigeminal nerve (CN V), while the dura of the posterior fossa is innervated mainly by CN X, but also by CN IX and the first three cervical nerves. The side of the headache often, but not always, corresponds to the side of pathology. Most headaches can be classified as either a vascular headache or a tension headache (Table 5.2). A diverse list of other causes of headache is also found in Table 5.2, organized roughly according to the "Arrowhead of Neurologic Differential Diagnosis" represented in Figure 1.1.

The term **vascular headache** is used to mean migraine as well as the less common but closely related disorder called cluster headache. The pathophysiology of vascular headache is not fully understood, but it is thought to involve inflammatory, autonomic, serotonergic, neuroendocrine, and other influences on blood vessel caliber in the head, leading to headache and other associated symptoms. In **migraine**, about 75% of patients have a positive family history, suggesting a genetic basis. Symptoms may be provoked by certain foods, stress, eye strain, the menstrual cycle, changes in sleep pattern, and a variety of other triggers. Migraine often is preceded by an **aura** or warning symptoms, classically involving visual blurring, shimmering, scintillating distortions, or **fortification scotoma**—a characteristic region of visual loss bordered by zigzagging lines resembling the walls of a fort. The headache is often unilateral, but if it is always on the same side, an MRI scan is warranted to exclude a vascular malformation or other lesion as a trigger for the headaches. The pain is often throbbing and may be exacerbated by light (photophobia), sound (phonophobia), or sudden head movement. Nausea and vomiting may occur, and the scalp may be tender to the touch. Duration is typically 30 minutes to up to 24 hours, and relief often occurs after sleeping. The severity of migraine headaches ranges from mild to very

TABLE 5.2 Differential Diagnosis of Headache

Vascular headache
  Migraine
  Cluster headache
Tension headache
Other causes[a]
  Acute trauma
  Intracranial hemorrhage
  Cerebral infarct
  Carotid or vertebral artery
    dissection
  Venous sinus thrombosis
  Post-ictal headache
  Hydrocephalus
  Pseudotumor cerebri
  Low CSF pressure
  Toxic or metabolic derangements
  Meningitis
  Epidural abscess
  Vasculitis
  Trigeminal or occipital neuralgia
  Neoplasm
  Disorders of the eyes, ears,
    sinuses, teeth, joints, or scalp

[a]Following format of Figure 1.1

severe in different individuals, and migraines may recur from once every few years to up to several times per week.

Complicated migraine may be accompanied by a variety of transient focal neurologic deficits (see KCC 10.3), including sensory phenomena, motor deficits (e.g., hemiplegia), visual loss, brainstem findings in **basilar migraine**, and impaired eye movements in **ophthalmoplegic migraine**. Migraine as a cause of these deficits should be a diagnosis of exclusion and should be accepted only in the setting of recurrent episodes and only after appropriate tests have been done to exclude cerebrovascular disease, epilepsy, or other disorders.

Treatment of migraine is often quite effective. Acute attacks usually respond to nonsteroidal anti-inflammatory drugs, anti-emetics, triptans (serotonin agonists), ergot derivatives, or other medications and to resting in a dark, quiet room. Preventive measures include avoiding triggers when possible and, for patients who have frequent attacks, treatment with prophylactic agents such as beta-blockers, topiramate, valproate, calcium channel blockers such as flunarizine, tricyclic antidepressants, or nonsteroidal anti-inflammatory drugs.

Cluster headache is less than one-tenth as common as migraine. It occurs about five times more often in males than in females. Typically, clusters of headaches occur from once to several times per day every day over a few weeks and then vanish for several months. Headache pain is extremely severe, often described as a steady, boring sensation behind one eye, lasting from about 30 to 90 minutes. It is usually accompanied by unilateral autonomic symptoms such as tearing, eye redness, Horner's syndrome (see KCC 13.5), unilateral flushing, sweating, and nasal congestion. Treatment is similar to that for migraine. In addition, inhaled oxygen is often effective in aborting attacks.

Tension headache, recently renamed **tension-type headache**, is a steady dull ache, sometimes described as a bandlike sensation. Although possibly related to excessive contraction of scalp and neck muscles, the pathophysiological distinction between tension headache and migraine has been questioned. Tension-type headache includes the common type of mild to moderate headache that most individuals experience from time to time, lasting up to a few hours. However, some patients have tension-type headaches that occur continuously every day for years. This chronic form of headache is commonly associated with psychological stress, but it is often unclear which is cause and which is effect. Chronic daily headache of this kind is also commonly seen in **posttraumatic headache**. Treatment for tension-type headache includes muscle relaxation techniques, nonsteroidal anti-inflammatory drugs, other analgesics, and tricyclic antidepressants.

It is important for clinicians to be familiar with the causes of headache listed in Table 5.2 since, with many of these disorders, diagnosis and intervention can be potentially lifesaving. We will mention only a few salient points here; the specific disorders will be discussed in greater detail in the sections and chapters that follow. Sudden "explosive" onset of severe headache should always be taken seriously. A CT scan should be done urgently to see if a subarachnoid hemorrhage has occurred (see KCC 5.6; see also Figure 5.19F). It is less well recognized that headache is common in cerebral ischemia and infarction (see KCC 10.4) and in the post-ictal period following seizures (see KCC 18.2). Low CSF pressure can occur spontaneously or following lumbar puncture (see KCC 5.10), resulting in headache that is worse while standing up and better while lying down. In contrast, in disorders such as neoplasms that can increase intracranial pressure, the headache may be worse when lying down during the night (see KCC 5.3).

Headache accompanied by fever or signs of meningeal irritation, such as stiff neck and sensitivity to light (see Table 5.6), should be evaluated and treated immediately for possible infectious meningitis, since patients with this condition can deteriorate rapidly if untreated.

Idiopathic intracranial hypertension or **pseudotumor cerebri** is a condition of unknown cause characterized by headache and elevated intracranial pressure (see KCC 5.3) with no mass lesion. It is most common in adolescent females, and it is treated with acetazolamide or, when severe, with shunting procedures (see KCC 5.7). **Temporal arteritis**, also called giant cell arteritis, is an important, treatable cause of headache. In this disorder, seen most commonly in elderly individuals, vasculitis affects the temporal arteries and other vessels, including those supplying the eye. The temporal artery is characteristically enlarged and firm. Diagnosis is made by measurement of the blood **erythrocyte sedimentation rate** (**ESR**) and by temporal artery biopsy. Prompt diagnosis and treatment with steroids is essential to prevent possible vision loss. ■

5.2 KEY CLINICAL CONCEPT

INTRACRANIAL MASS LESIONS

Anything abnormal that occupies volume within the cranial vault functions as a mass. Examples include tumor, hemorrhage, abscess, edema, hydrocephalus, and other disorders. Intracranial mass lesions can cause neurologic symptoms and signs by the following mechanisms:

1. Compression and destruction of adjacent regions of the brain can cause neurologic abnormalities.
2. A mass located within the cranial vault can raise the **intracranial pressure**, which causes certain characteristic symptoms and signs.
3. Mass lesions can displace nervous system structures so severely that they are shifted from one compartment into another—a situation called **herniation**.

In this section we will discuss local effects of the mass itself in the brain. In KCC 5.3 and 5.4 we will discuss elevated intracranial pressure and herniation.

Mass lesions can cause both local tissue damage and remote effects through mechanical distortion of adjacent structures. **Mass effect** is a descriptive term used for any distortion of normal brain geometry due to a mass lesion. Mass effect can be as subtle as a mild flattening, or **effacement**, of sulci next to a lesion, seen on MRI scan but producing no symptoms. Depending on location and size, a mass can produce neurologic abnormalities due to local damage. For example, a lesion located in the primary motor cortex will cause contralateral weakness. If the mass distorts or irritates blood vessels or meninges, it may cause headache (see KCC 5.1). Compression of blood vessels can also cause ischemic infarction, and erosion through blood vessel walls can cause hemorrhage.

Disruption of the blood–brain barrier results in extravasation of fluid into the extracellular space, producing vasogenic edema (see Figures 5.13 and 5.14). Compression of the ventricular system can obstruct CSF flow, producing hydrocephalus (see KCC 5.7). Lesions can provoke abnormal electrical discharges in the cerebral cortex, resulting in seizures (see KCC 18.2). In addition, remote effects may result from functional changes in regions receiving important synaptic connections from the damaged areas. Large masses can produce dramatic **midline shift** of brain structures away from the side of the lesion. Displacement and stretching of the upper brainstem impairs function of the reticular activating systems (see Figure 2.23), causing impaired consciousness and, ultimately, coma. The pineal calcification (see Figure 4.12F) is a useful landmark for measuring extent of midline shift at the level of the upper brainstem. The amount of pineal shift has been shown to correlate with impairment of consciousness. In the extreme, mass effect causes brain structures to shift from one compartment into another, leading to herniation (see KCC 5.4). ■

5.3 KEY CLINICAL CONCEPT

ELEVATED INTRACRANIAL PRESSURE

The contents of the intracranial space are confined by the hard walls of the bony skull. Of the three residents of this cavity—cerebrospinal fluid, blood, and brain tissue—not one is compressible (although they can be deformed). Therefore, whenever there is a space-occupying or mass lesion within the skull, something must leave the skull to accommodate the extra volume (Figure 5.16). Smaller lesions can be accommodated by a decrease in intracranial CSF and blood without causing much rise in intracranial pressure (flat part of the curve in Figure 5.16). Larger lesions overcome this compensatory mechanism, and the intracranial pressure eventually begins to rise steeply. This can ultimately lead to herniation (see KCC 5.4) and death (the rightmost part of the curve in Figure 5.16).

Severely elevated intracranial pressure can cause decreased cerebral blood flow and brain ischemia. Cerebral blood flow depends on **cerebral perfusion pressure**, which is defined as the **mean arterial pressure minus the intracranial pressure** (**CPP = MAP – ICP**). Therefore, as the intracranial pressure increases, cerebral perfusion pressure decreases. **Autoregulation** of cerebral vessel caliber can compensate for modest reductions in cerebral perfusion

FIGURE 5.16 Intracranial Pressure versus Intracranial Mass Volume Small intracranial masses can be compensated for by reductions in intracranial CSF and blood volume. Larger masses lead to a steep rise in intracranial pressure, causing reduced cerebral perfusion and, ultimately, herniation. Note that the volume of CSF has been exaggerated in the equilibrium state (as seen in cerebral atrophy) for illustrative purposes.

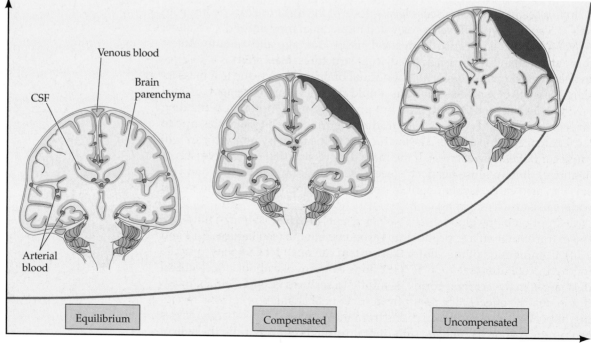

Intracranial pressure

Venous blood

Brain parenchyma

CSF

Arterial blood

Equilibrium

Compensated

Uncompensated

Intracranial mass volume

pressure, leading to relatively stable cerebral blood flow. However, large increases in intracranial pressure can exceed the capacity of autoregulation, leading to reduced cerebral blood flow and brain ischemia. Depending on the type of lesion, intracranial pressure can change suddenly or more slowly, over a period of days to weeks. If left untreated, severely elevated intracranial pressure causes irreversible brain damage and death, sometimes within a few hours.

It is essential for clinicians to recognize the **symptoms and signs of elevated intracranial pressure** (Table 5.3) so that appropriate treatment can be instituted without delay. Let's discuss each of the symptoms and signs in Table 5.3 in turn. **Headache** associated with intracranial mass lesions is often worse in the morning, since brain edema increases overnight from the effects of gravity on a patient in the reclining position. **Altered mental status**, especially *irritability and depressed level of alertness and attention*, is often the most important indicator of elevated intracranial pressure. The mechanisms giving rise to **nausea and vomiting** in elevated intracranial pressure are not known. Vomiting occasionally occurs suddenly and without much nausea. This is called **projectile vomiting**.

Elevated intracranial pressure is transmitted through the subarachnoid space to the optic nerve sheath, obstructing axonal transport and venous return in the optic nerve. Ophthalmoscopic exam (see **neuroexam.com Video 25**) may, therefore, reveal **papilledema**, in which there is engorgement and elevation of the optic disc, sometimes accompanied by retinal hemorrhages (Figure 5.17). This classic sign of elevated intracranial pressure takes several hours to days to develop and is *often not present in the acute setting*. Transient or permanent optic nerve injury can occur in association with papilledema, leading to visual blurring or **visual loss**. Areas of decreased vision most commonly include an **increased blind spot** or a **concentric visual field deficit** affect-

TABLE 5.3 Common Symptoms and Signs of Elevated Intracranial Pressure
SYMPTOM OR SIGN
Headache
Altered mental status, especially *irritability and depressed level of alertness and attention*[a]
Nausea and vomiting
Papilledema
Visual loss
Diplopia (double vision)
Cushing's triad: hypertension, bradycardia, and irregular respirations

[a]This is often the most important indicator of elevated intracranial pressure.

Ophthalmoscopic exam

(A)

(B)

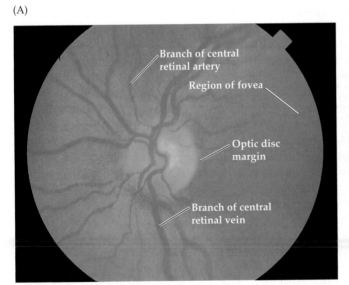

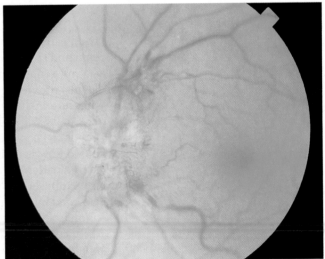

FIGURE 5.17 Papilledema (A) Funduscopic view of retina from a normal subject (left eye). Note sharp margins of the optic disc. (B) Papilledema in a patient with elevated intracranial pressure (left eye). This was a 43-year-old man who developed headaches, visual blurring, and horizontal diplopia. Lumbar puncture revealed an opening pressure of 40 cm H2O. Magnetic resonance venogram revealed a bilateral sigmoid sinus venous thrombosis (see KCC 10.7). Blood studies showed an elevated anticardiolipin antibody (see KCC 10.4). The patient was treated successfully with chronic oral anticoagulation.

TABLE 5.4   Treatment Measures for Elevated Intracranial Pressure

INTERVENTION	TIME TO ONSET OF EFFECT	PROPOSED MECHANISM/COMMENTS
Elevate head of bed 30°, and maintain head straight to avoid obstructing jugular venous return.	Immediate	Promotes venous drainage
Intubate and hyperventilate to pCO_2 of 25–30 mm Hg.	30 seconds	Causes cerebral vasoconstriction
IV mannitol 1 g/kg bolus, then 0.25 g/kg every 6 hours, (or, alternatively, hypertonic saline) aiming for serum Na^+ >138 mEq/L and osmolarity 300–310 mOsm/L while maintaining normal volume status and normal blood pressure. Furosemide may also be added.	5 minutes	Promotes removal of edema and other fluids from CNS while maintaining cerebral perfusion
Ventricular drainage	Minutes	Removal of CSF decreases intracranial pressure
If other measures fail, try barbiturate-induced coma.	1 hour	Causes cerebral vasoconstriction and reduced metabolic demands
Hemicraniectomy (removal of skull overlying mass lesion)	Immediate	Decompresses intracranial cavity Experimental
Steroids	Hours	Reduces cerebral edema, possibly by strengthening blood–brain barrier May also work by other mechanisms Often used with brain tumors Not shown to improve outcome in acute head trauma, stroke, or hemorrhage

ing mainly the peripheral margins of the visual field (see Figure 11.16A). **Diplopia** (double vision) can occur as a result of downward traction on CN VI (see Figure 12.3A), causing unilateral or bilateral abducens nerve palsies (see KCC 13.4). Finally, **Cushing's triad**—hypertension, bradycardia, and irregular respirations—is another classic sign of elevated intracranial pressure. Hypertension may be a reflex mechanism to maintain cerebral perfusion pressure, bradycardia may be a reflex response to the hypertension, and irregular respirations are caused by impaired brainstem function (see Figure 14.17). In reality, a variety of changes in vital signs other than Cushing's triad can be seen as a result of brainstem dysfunction, including hypotension and tachycardia.

The goal of treating elevated intracranial pressure is to reduce it to safe levels, providing time to treat the underlying disorder. **Normal intracranial pressure** in adults is **less than 20 cm H_2O**, or **less than 15 mm Hg** (**torr**) (1 cm H_2O = 0.735 mm Hg; 1 mm Hg = 1.36 cm H_2O). Another critical goal of therapy is to keep cerebral perfusion pressure above 50 mm Hg so that cerebral blood flow is maintained. Intracranial pressure can be measured in clinically stable patients during lumbar puncture (see KCC 5.10); however, lumbar puncture should not be performed in patients suspected of having severely elevated intracranial pressure, since this carries a risk of precipitating herniation (see KCC 5.4).

In critically ill patients, intracranial pressure can be monitored continuously with a ventricular drain, intraparenchymal monitor, subarachnoid bolt, or a variety of other devices placed neurosurgically within the cranium and connected to a pressure transducer. Measures to lower intracranial pressure can then be instituted as described in Table 5.4 while the results are being monitored. Note that treatment of elevated intracranial pressure is somewhat controversial, and authorities differ as to which measures are actually beneficial (the list in Table 5.4 does not reflect this controversy). It should be reemphasized that all of these measures are temporary, and they are best used to buy time while treatment of the underlying cause of the elevated intracranial pressure is instituted. ■

5.4 KEY CLINICAL CONCEPT

BRAIN HERNIATION SYNDROMES

As discussed in KCC 5.2, intracranial tumors, hemorrhage, edema, and other masses cause a displacement of intracranial structures called mass effect. **Herniation** occurs when mass effect is severe enough to push intracranial structures from one compartment into another. Herniation between different compartments is associated with certain distinct clinical features. Some authors argue that herniation does not actually cause these clinical features and that it is merely an epiphenomenon. Nevertheless, herniation remains a clinically useful concept and we will therefore discuss it in more traditional terms in this section.

The three most clinically important herniation syndromes are caused by herniation through the tentorial notch (transtentorial herniation), herniation centrally and downward (central herniation), and herniation under the falx cerebri (subfalcine herniation) (Figure 5.18).

Transtentorial Herniation

Transtentorial herniation (or, simply, **tentorial herniation**) is herniation of the medial temporal lobe, especially the **uncus** (**uncal herniation**), inferiorly through the tentorial notch (see Figures 5.6 and 5.18). Uncal herniation is heralded by the **clinical triad of a "blown" pupil, hemiplegia, and coma**. Compression of the oculomotor nerve (CN III) (see Figure 5.6), usually ipsilateral to the lesion, produces first a dilated, unresponsive pupil (a **blown pupil**), and, later, impairment of eye movements. *In uncal herniation, the dilated pupil is ipsilateral to the lesion in 85% of cases.*

Compression of the cerebral peduncles (see Figure 5.6) can cause **hemiplegia** (paralysis of half of the body). The relationship of the hemiplegia to the side of the lesion is more complicated than the pupil findings. Recall that the corticospinal tract crosses to the opposite side as it descends through the medulla into the spinal cord at the pyramidal decussation (see Figure 2.16). Thus, often the hemiplegia is contralateral to the lesion either because of uncal herniation compressing the ipsilateral corticospinal tract in the mid-

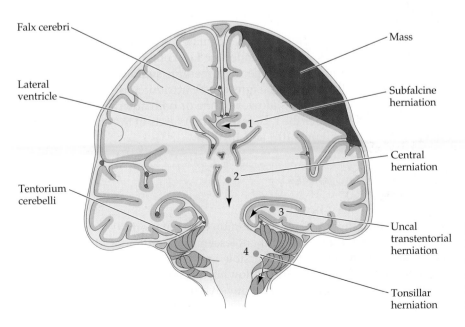

Falx cerebri

Lateral ventricle

Tentorium cerebelli

Mass

Subfalcine herniation

Central herniation

Uncal transtentorial herniation

Tonsillar herniation

FIGURE 5.18 Herniation Syndromes Coronal section through brain and skull demonstrating subfalcine herniation (1), central herniation (2), transtentorial herniation (3), and tonsillar herniation (4), resulting from a mass lesion.

brain, or because of a direct effect of the lesion on the ipsilateral motor cortex, or because of both. However, sometimes in uncal herniation, the midbrain is pushed all the way over until it is compressed by the opposite side of the tentorial notch (see Figure 5.6). In these cases the contralateral corticospinal tract is compressed, producing hemiplegia that is ipsilateral to the lesion. This is called **Kernohan's phenomenon**.

Distortion of the midbrain reticular formation leads to decreased level of consciousness and, ultimately, to **coma** (see KCC 14.2). In addition, the posterior cerebral arteries may be compressed as they pass upward through the tentorial notch (see Figure 5.6). The result can be infarction in the posterior cerebral artery territory (see Figure 10.5). Uncal transtentorial herniation can be unilateral or bilateral and is caused by supratentorial mass lesions. Occasionally, large mass lesions in the posterior fossa can cause *upward* transtentorial herniation.

Central Herniation

Central herniation is central downward displacement of the brainstem (see Figure 5.18). Central herniation can be caused by any lesion associated with elevated intracranial pressure, including hydrocephalus or diffuse cerebral edema. Mild central herniation causes traction on the abducens nerve (CN VI) during its long course over the clivus (see Figures 2.22A, 5.3B, 12.3A, and 13.4), producing lateral rectus palsy (see Table 2.5), which may be unilateral or bilateral. Larger supratentorial masses or elevated intracranial pressure can produce significant central herniation through the tentorial opening, resulting in bilateral uncal herniation (as described already). With severe elevations in intracranial pressure, large supratentorial mass lesions, or mass lesions in the posterior fossa, central herniation can progress downward through the foramen magnum (see Figure 5.18).

Herniation of the cerebellar tonsils downward through the foramen magnum is called **tonsillar herniation** (see Figure 5.18). This condition is associated with compression of the medulla and usually leads to respiratory arrest, blood pressure instability, and death. Some studies have called into question the pathophysiological importance of central herniation, and there is controversy over whether central herniation is seen only as a postmortem phenomenon.

Subfalcine Herniation

Unilateral mass lesions can cause the cingulate gyrus (see Figure 2.11B) and other brain structures to herniate under the falx cerebri (see Figure 5.5) from one side of the cranium to the other. The result is **subfalcine herniation** (see Figure 5.18). Usually, no clinical signs can be attributed directly to the subfalcine herniation. Sometimes, however, one or both anterior cerebral arteries can be occluded under the falx, leading to infarcts in the anterior cerebral artery territory (see Figures 10.4 and 10.5). ■

5.5 KEY CLINICAL CONCEPT

HEAD TRAUMA

Head trauma is, unfortunately, a common cause of morbidity and mortality, especially in the young adult and adolescent population. Mild head trauma causes **concussion**, defined as reversible impairment of neurologic function for a period of minutes to hours following a head injury. The mechanism of concussion is unknown, but it may involve transient diffuse neuronal dysfunction. CT and MRI scans are normal. Clinical features of concussion include loss of consciousness; "seeing stars," followed by headache, dizziness, and, occasionally, nausea; and vomiting. Some of these features may result from migraine-like phenomena (see KCC 5.1) triggered by head injury. Occasionally, head trauma is accompanied by anterograde and retrograde amnesia (see KCC 18.1) for a period of several hours surrounding the injury. Recovery is usually complete, although occasional patients develop **postconcussive syndrome** even after relatively minor trauma, with headaches, lethargy, mental dullness, and other symptoms lasting up to several months after the accident. Even relatively mild head injuries can occasionally cause dissection of the carotid or vertebral arteries (see KCC 10.6), resulting in transient ischemic attacks or cerebral infarcts.

More severe head trauma can cause permanent injury to the brain through various mechanisms, including **diffuse axonal shear injury**, which causes widespread or patchy damage to the white matter and cranial nerves; **petechial hemorrhages**, or small spots of blood in the white matter; larger **intracranial hemorrhages** (see Figure 5.19; see also KCC 5.6); **cerebral contusion** (see Figure 5.21); and direct tissue injury by **penetrating trauma** such as gunshot wounds or open skull fracture. **Cerebral edema** may occur as well, with or without other injuries, contributing to elevated intracranial pressure in head injury.

In addition to the neurologic exam, certain signs on general physical examination may provide clues of significant head trauma (see Table 3.9). In all head injuries, the **spine** should be evaluated carefully because the same mechanism of injury may cause an unnoticed unstable spinal fracture. Detailed X-rays or CT of the spine are usually necessary in head injuries, especially when the patient is not fully responsive. Symptoms of intracranial hemorrhage resulting from head injury sometimes occur after a delay of up to several hours following the event (see KCC 5.6). Therefore, all patients with neurologic deficits, even when transient, should undergo a **CT scan** and should be observed closely for signs of deterioration during the first 24 hours following head injury. Patients with more serious head injuries should be treated as discussed in KCC 5.3 and KCC 5.6. ∎

5.6 KEY CLINICAL CONCEPT

INTRACRANIAL HEMORRHAGE

Intracranial hemorrhage can be **traumatic** or **atraumatic**. It can occur in several different compartments within the cranial vault (Figure 5.19). Intracranial hemorrhages are classified according to location and are often abbreviated as follows:

1. Epidural hematoma (EDH)
2. Subdural hematoma (SDH)
3. Subarachnoid hemorrhage (SAH)
4. Intracerebral or intraparenchymal hemorrhage (ICH)

Specific traumatic or atraumatic causes are most common in each of these locations, as indicated in the discussion that follows.

Epidural Hematoma

Location: In the tight potential space between the dura and the skull.

Usual cause: **Rupture of the middle meningeal artery** (see Figure 5.7) due to **fracture of the temporal bone** by head trauma.

Clinical features and radiological appearance: Rapidly expanding hemorrhage under arterial pressure peels the dura away from the inner surface of the

FIGURE 5.19 Types of Intracranial Hemorrhage Demonstrated by CT Scans

(A) Epidural hematoma

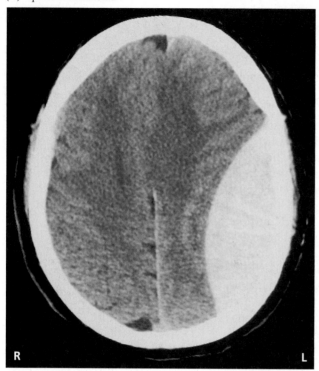

(B) Acute subdural hematoma

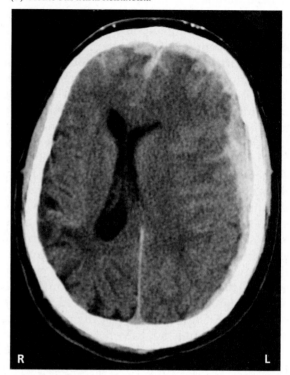

(C) Isodense subdural hematoma

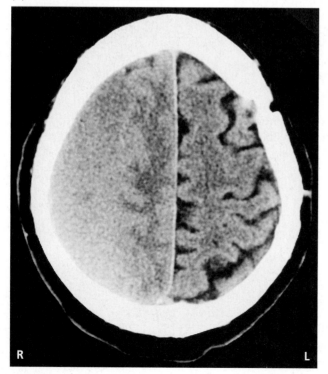

(D) Chronic subdural hematoma

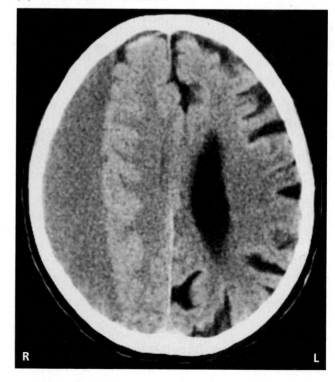

(E) Hematocrit effect resulting from mixed acute
    and chronic subdural blood

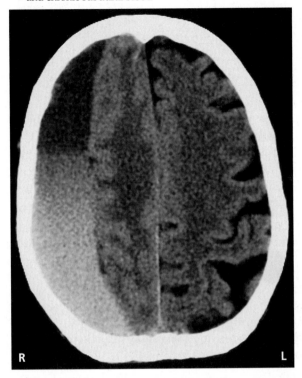

(F) Subarachnoid hemorrhage

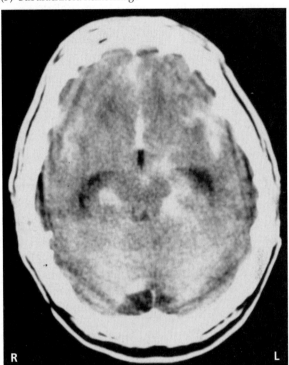

(G) Contusion

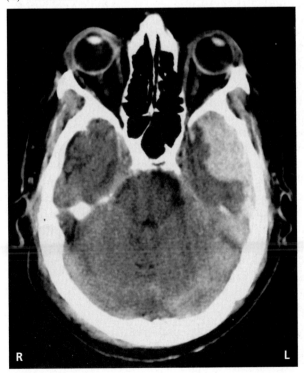

(H) Intraparenchymal (basal ganglia) hemorrhage

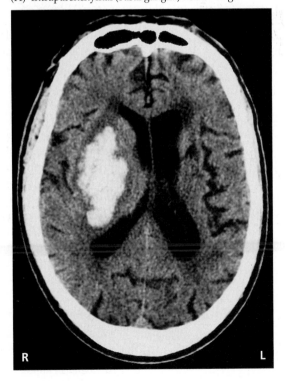

skull, forming a **lens-shaped biconvex** hematoma that often does not spread past the cranial sutures where the dura is tightly apposed to the skull (see Figure 5.19A). Initially the patient may have no symptoms (**lucid interval**). However, within a few hours the hematoma begins to compress brain tissue, often causing elevated intracranial pressure (see KCC 5.3) and, ultimately, herniation (see KCC 5.4) and death unless treated surgically.

Subdural Hematoma

Location: In the potential space between the dura and the loosely adherent arachnoid.

Usual cause: **Rupture of the bridging veins**, which are particularly vulnerable to shear injury as they cross from the arachnoid into the dura (see Figure 5.1).

Clinical features: Venous blood dissects relatively easily between the dura and the arachnoid, spreading out over a large area and forming a **crescent-shaped** hematoma. Two types of subdural hematoma—chronic and acute—are distinguished by different clinical features.

CHRONIC SUBDURAL HEMATOMA Often seen in elderly patients, in whom atrophy allows the brain to move more freely within the cranial vault, thus making the bridging veins more susceptible to shear injury. This type of hematoma (see Figure 5.19D) may be seen with minimal or no known history of trauma. Oozing slowly, venous blood collects over a period of weeks to months, allowing the brain to accommodate and therefore causing vague symptoms such as headache, cognitive impairment, and unsteady gait. In addition, **focal dysfunction of the underlying cortex may result in focal neurologic deficits** and, occasionally, focal seizures.

ACUTE SUBDURAL HEMATOMA For a significant subdural hematoma to occur immediately after an injury, the impact velocity must be quite high. Therefore, acute subdural hematoma is usually associated with other serious injuries, such as traumatic subarachnoid hemorrhage and brain contusion (to be discussed shortly). The prognosis is thus usually worse than with chronic subdural hematoma or even epidural hematoma.

Radiological appearance: Subdural hematomas are typically crescent shaped and spread over a large area (see Figures 5.19B–E). Density depends on the age of the blood. Recall that acute blood is **hyperdense** (see Figure 5.19B) and therefore bright on CT scan (see Chapter 4). After 1 to 2 weeks, the clot begins to liquefy and may appear **isodense** (see Figure 5.19C). If there is no further bleeding, after 3 to 4 weeks the hematoma will be completely liquefied and will appear uniformly **hypodense** (see Figure 5.19D). However, if there is continued occasional bleeding, there will be a **mixed density** appearance resulting from liquefied chronic blood mixed with clotted hyperdense blood. Sometimes, with mixed-density hematomas, the denser acute blood settles to the bottom, giving a characteristic **hematocrit effect** (see Figure 5.19E). Subdural hematoma is treated by surgical evacuation, except for small to moderate-sized chronic subdural hematomas, which, depending on the severity of symptoms, can be followed clinically because some will resolve spontaneously.

Subarachnoid Hemorrhage

Location: In the CSF-filled space between the arachnoid and the pia, which contains the major blood vessels of the brain (see Figure 5.1).

Radiological appearance: Unlike subdural hematoma, blood can be seen on CT to track down into the sulci following the contours of the pia (see Figure 5.19F).

Usual cause: Subarachnoid hemorrhage is seen in two clinical settings: nontraumatic (spontaneous) and traumatic.

Nontraumatic (Spontaneous) Subarachnoid Hemorrhage

Spontaneous subarachnoid hemorrhage usually presents with a sudden catastrophic headache. Patients may describe this as the **"worst headache of my life"** or as feeling like the head is suddenly about to explode. In the vast majority of cases (75%–80%), spontaneous subarachnoid hemorrhage occurs as a result of rupture of an arterial **aneurysm** in the subarachnoid space. Less often (4%–5% of cases), it results from bleeding of an **arteriovenous malformation** and from other rarer or unknown causes. Risk factors for intracranial aneurysm include atherosclerotic disease, congenital anomalies in cerebral blood vessels, polycystic kidney disease, and connective tissue disorders such as Marfan's syndrome.

Saccular aneurisms, or **berry aneurysms**, usually arise from arterial branch points near the circle of Willis (**Figure 5.20** and see Figure 2.26C). These aneurisms are balloon-like outpouchings of the vessel wall that typically have a **neck** connected to the parent vessel and a fragile **dome** that can rupture. Over 85% occur in the anterior circulation (carotid artery and its branches). The most common locations, listed in descending order, are the anterior communicating artery (**AComm**, about **30%**), posterior communicating artery (**PComm**, about **25%**), and middle cerebral artery (**MCA**, about **20%**). Saccular aneurisms can also occur in the branches of the posterior cir-

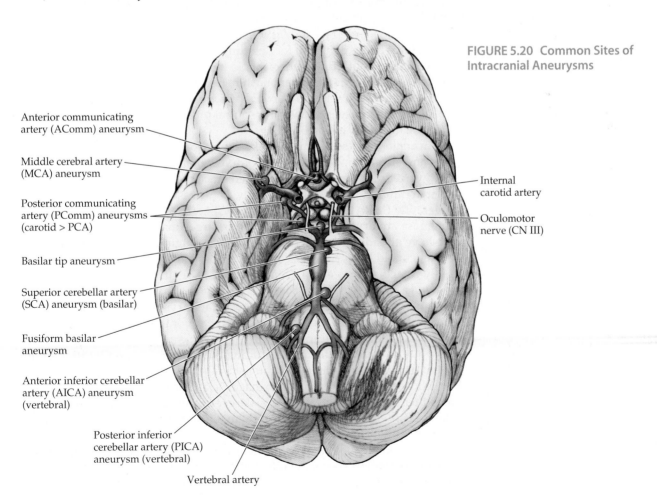

FIGURE 5.20 Common Sites of Intracranial Aneurysms

Anterior communicating artery (AComm) aneurysm

Middle cerebral artery (MCA) aneurysm

Posterior communicating artery (PComm) aneurysms (carotid > PCA)

Basilar tip aneurysm

Superior cerebellar artery (SCA) aneurysm (basilar)

Fusiform basilar aneurysm

Anterior inferior cerebellar artery (AICA) aneurysm (vertebral)

Posterior inferior cerebellar artery (PICA) aneurysm (vertebral)

Vertebral artery

Internal carotid artery

Oculomotor nerve (CN III)

culation (**vertebrobasilar system**, about **15%**). Occasionally the main vessel itself becomes dilated, forming a **fusiform aneurysm** (see Figure 5.20), which is less prone to rupture than saccular aneurysms. Aside from symptoms caused by rupture, large unruptured aneurysms can occasionally present with symptoms due to mass effect or the compression of adjacent structures. An important example is **PComm aneurysm** arising from the internal carotid artery (see Figure 5.20), which can cause a **painful third-nerve palsy** (see Figures 5.6 and 13.2; see also KCC 13.2). The PComm junction with the posterior cerebral artery can also give rise to aneurysms but does so much less commonly than the PComm junction with the carotid.

Risk factors for aneurysmal rupture include hypertension, cigarette smoking, alcohol consumption, and situations causing sudden elevation in blood pressure. The clinical effects of subarachnoid hemorrhage can range from headache and meningeal irritation (see Table 5.6), causing nuchal rigidity and photophobia, to cranial nerve and other focal neurologic deficits, to impaired consciousness, coma, and death. Perhaps 25% of all patients with subarachnoid hemorrhage die in the immediate aftermath of the event and thus never reach the hospital for treatment. The overall mortality of subarachnoid hemorrhage is about 50%; however, the prognosis is better in mild cases. In subarachnoid hemorrhage due to ruptured aneurysm, the risk of rebleeding is 4% on the first day and 20% in the first 2 weeks. Therefore, prompt diagnosis and early treatment of aneurysmal subarachnoid hemorrhage is essential.

CT scan performed within the first 3 days after rupture can detect the hemorrhage in over 95% of cases. It is important to perform a **CT scan without contrast** because both subarachnoid blood and contrast material appear white on the scan (see Figures 4.4 and 5.19F), making it difficult to see a small hemorrhage. CT is better than MRI for detecting acute subarachnoid hemorrhage, although after about 2 days subarachnoid hemorrhage may no longer be visible on CT (see Chapter 4). **Lumbar puncture** (see KCC 5.10) should be performed in suspected subarachnoid hemorrhage with a negative CT but not with a positive CT because increased transmural pressure across the aneurysm can occasionally precipitate rebleeding.

An **angiogram** (see Chapter 4) is next performed to define the exact location and size of the aneurysm. A four-vessel angiogram (both carotids and both vertebrals) should be performed since multiple aneurysms are often seen in different vessels. Increasingly, noninvasive magnetic resonance angiography (MRA) or CT angiography (CTA; see Chapter 4) are also used to diagnose aneurysms, particularly those larger than 2mm–3mm in size. As we have already mentioned, it is crucial to diagnose and treat aneurysmal subarachnoid hemorrhage as quickly as possible to prevent a second, potentially cataclysmic event. Treatment consists of either neurosurgical placement of a clip across the neck of the aneurysm or interventional neuroradiology to place detachable coils within the aneurysm. Both approaches can successfully prevent rebleeding and have favorable long-term outcomes. The choice between clipping or coiling aneurysms is an evolving area and depends on multiple factors, including the shape, size, and location of the aneurysm and the patient's overall medical condition.

Following subarachnoid hemorrhage, delayed **cerebral vasospasm** occurs in about half of all patients, with a peak severity about 1 week after the hemorrhage. This can lead to cerebral ischemia or infarction. Vasospasm is often treated with **"triple H" therapy**, consisting of induced hypertension, hypervolemia, and hemodilution in the intensive care unit. Moderate hypervolemia and hemodilution is usually initiated at the time of hemorrhage diagnosis in an effort to prevent delayed vasospasm. Induced hypertension can be done more safely after the aneurysm has been clipped or coiled, which is another reason that early treatment is preferred. Administration of the calcium channel blocker nimodipine early after the hemorrhage can also improve the outcome, although the mechanism is uncertain because calcium channel blockers do not improve vasospasm angiographically. Cases of refractory vasospasm can be treated with interventional neuroradiology procedures such as balloon angioplasty and local injection of the vasodilator papaverine.

Traumatic Subarachnoid Hemorrhage

Traumatic subarachnoid hemorrhage, which is caused by bleeding into the CSF from damaged blood vessels associated with cerebral contusions and other traumatic injuries, is actually more common than spontaneous subarachnoid hemorrhage. Like spontaneous subarachnoid hemorrhage, it is usually associated with severe headache due to meningeal irritation from blood in the CSF. Deficits are usually related to the presence of other cerebral injuries. Unlike in aneurysmal subarachnoid hemorrhage, vasospasm is not usually seen. An example of traumatic subarachnoid hemorrhage can be seen in Figure 5.19B, which mainly shows acute subdural blood, but also demonstrates some subarachnoid blood located within the sulci.

Intracerebral or Intraparenchymal Hemorrhage

Location: Within the brain parenchyma in the cerebral hemispheres, brainstem, cerebellum, or spinal cord.

Usual cause: Once again, this type of hemorrhage may be traumatic or nontraumatic.

Traumatic Intracerebral or Intraparenchymal Hemorrhage

Contusions of the cerebral hemispheres occur in regions where cortical gyri abut the ridges of the bony skull (**Figure 5.21**, and see Figures 5.2 and 5.4). Thus, contusions are most common at the temporal and frontal poles (see Figure 5.19G). Interestingly, they are less common at the occipital poles.

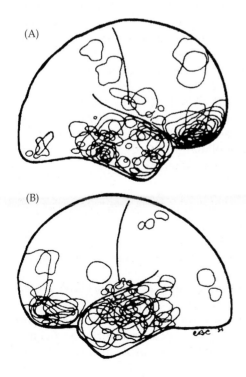

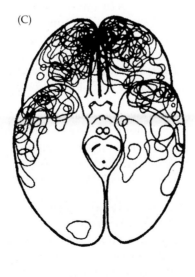

FIGURE 5.21 Common Sites of Cerebral Contusion Composite drawings of contusion size and shape from a series of 40 cases. (A) Right hemisphere; (B) Left hemisphere; (C) Inferior view. (From Courville CB. 1937. *Pathology of the Nervous System*, Part 4. Pacific, Mountain View, CA.)

Contusions occur on the side of the impact (**coup injury**) as well as on the side opposite the impact (**contrecoup injury**) because of rebound of the brain against the skull. Shearing forces can produce areas of bleeding in the white matter as well, including small petechial or larger confluent intraparenchymal hemorrhage. Severe injuries often are accompanied by a combination of contusion, subarachnoid hemorrhage, and subdural hemorrhage. When a combination of intraparenchymal, subarachnoid, and acute subdural hemorrhage is seen on CT scan, head trauma can be assumed.

Nontraumatic Intracerebral or Intraparenchymal Hemorrhage

There are many different **causes of intraparenchymal hemorrhage**, including hypertension, brain tumors, secondary hemorrhage after ischemic infarction, vascular malformations, blood coagulation abnormalities, infections, vessel fragility caused by deposition of amyloid protein in the blood vessel wall (amyloid angiopathy), vasculitis, mycotic (infectious) aneurysms in the setting of endocarditis, and so on.

Hypertensive hemorrhage is the most common cause, and it tends to involve small, penetrating blood vessels (see Figure 5.19H). The pathogenesis is uncertain, but it may be related to chronic pathologic effects of hypertension on the small vessels, such as the lenticulostriate arteries (see Figures 4.16B and 10.7), including **lipohyalinosis** and **microaneurysms of Charcot–Bouchard**. The **most common locations** for hypertensive hemorrhage, in decreasing order of frequency, are the **basal ganglia** (usually the putamen), **thalamus**, **cerebellum**, and **pons**. Some hemorrhages may involve the ventricles, either by extending from adjacent parenchyma or by arising from blood vessels in the ventricles themselves. These are described as **intraventricular extension** of an intraparenchymal hemorrhage, or as **intraventricular hemorrhage**, respectively. Unlike aneurysmal hemorrhage, the rebleeding rate for hypertensive hemorrhage is low, although the hematoma often continues to enlarge, causing worsening clinical status for several hours after onset. **Edema** also gradually develops in the tissue surrounding the hemorrhage, causing a slow clinical worsening that reaches its peak about 3 days after onset.

In **lobar hemorrhage**, bleeding involves the occipital, parietal, temporal, or frontal lobe. The most common cause of lobar hemorrhage is probably **amyloid** (congophilic) **angiopathy**. In this condition, deposits of amyloid in the vessel wall of older patients (usually >50 years old) cause vascular fragility. Unlike hypertensive hemorrhage, in amyloid angiopathy the hemorrhages tend to be recurrent or multiple, and they are often more superficial in location. Transient symptoms resembling transient ischemic attack (see KCC 10.3) or seizures can occur in amyloid angiopathy for weeks or months preceding hemorrhage. Lobar hemorrhage can also be seen in hypertension.

Certain **vascular malformations** are another important cause of intracranial hemorrhage. Vascular malformations are classified as:

1. Arteriovenous malformations
2. Cavernomas (also called cavernous hemangiomas, cavernous angiomas, or cavernous malformations)
3. Capillary telangiectasias (capillary angiomas)
4. Developmental venous anomalies (venous angiomas, venous malformations)

Of these, only arteriovenous malformations and cavernomas have a high likelihood of causing intracranial hemorrhage.

Arteriovenous malformations (**AVMs**) are congenital abnormalities in which there are abnormal direct connections between arteries and veins, often forming a tangle of abnormal blood vessels visible as flow voids on

MRI scan, but best seen on conventional angiography (see Image 11.5A,B and 11.5C). Size can range from a few centimeters to half the brain. Aside from sudden severe symptoms from intracranial hemorrhage, patients also commonly present with seizures or with migraine-like headaches in the absence of hemorrhage. Hemorrhage is usually intraparenchymal, but it can extend to the intraventricular or subarachnoid space as well. The risk of rebleeding is about 1% to 4% per year, much lower than in aneurysmal hemorrhage (discussed earlier). Treatments for AVM, which depend on clinical status, size, and location of the lesions, include neurosurgical removal, intravascular embolization, and stereotactic radiosurgery (see KCC 16.4).

Cavernomas are abnormally dilated vascular cavities lined by only one layer of vascular endothelium. They are not visible on conventional angiography, but with the advent of MRI the diagnosis of cavernous malformations has increased dramatically. They have a characteristic MRI appearance, with a central 1 cm to 2 cm core of increased signal on T1 or T2, surrounded by a dark rim on T2-weighted sequences because of the presence of hemosiderin (see Table 4.4). Some patients have multiple cavernomas, and a familial autosomal dominant form of this disorder exists. Patients often present with seizures. Risk of hemorrhage is between 0.1% and 2.7% per lesion per year. Some studies have reported that the risk of hemorrhage increases after an initial bleed and may be higher (5% per year with 30% per year rebleed rate) for cavernomas located in the brainstem. Although still evolving, clinical criteria for operating on cavernomas often include clinically significant hemorrhage or seizures, infratentorial location, and favorable operative location near the surface of the brain.

Capillary telangiectasias are small regions of abnormally dilated capillaries that rarely give rise to intracranial hemorrhage. **Developmental venous anomalies** are dilated veins usually visible on MRI scans as a single flow void extending to the brain surface. They are most often an incidental finding on MRI and are not known to cause any clinical symptoms themselves, but they can sometimes be seen in association with cavernomas.

Extracranial Hemorrhage

Head trauma can also cause hemorrhage in the inner ear, called hemotympanum; hemorrhage in subcutaneous tissues, resulting in Battle's sign; or "raccoon eyes" (see Table 3.9). Scalp hemorrhage can cause profuse bleeding. Hemorrhage in the loose space between the external periosteum and galea aponeurotica (see Figure 5.1) can produce a "goose egg," or **subgaleal hemorrhage**. In newborns, bleeding during delivery can occur between the skull and external periosteum (pericranium), called a **cephalohematoma**, which can occasionally be quite large. ▪

5.7 KEY CLINICAL CONCEPT

HYDROCEPHALUS

Hydrocephalus (meaning "water in the head") is caused by excess CSF in the intracranial cavity. This condition can result from (1) excess CSF production, (2) obstruction of flow at any point in the ventricles or subarachnoid space, or (3) decrease in reabsorption via the arachnoid granulations.

Excess CSF production is quite rare as a cause of hydrocephalus; it is seen only in certain tumors, such as choroid plexus papilloma. **Obstruction of CSF flow** is a common cause of hydrocephalus and can be produced by obstruction of the ventricular system by tumors, intraparenchymal hemorrhage, other masses, and congenital malformations. This can occur anywhere along the path of CSF flow (see Figures 5.10 and 5.11), but especially at narrow points such as the foramen of Monro, the cerebral aqueduct, or the fourth ventricle. Obstruction can also occur outside the ventricles in the subarachnoid space as a result of debris or adhesions from prior hemorrhage, infection, or inflammation. **Decreased CSF reabsorption** can cause hydrocephalus when the arachnoid granulations are damaged or clogged. Decreased reabsorption at the arachnoid granulations is difficult to distinguish clinically from obstruction of CSF flow in the subarachnoid space and often has similar causes (i.e., prior hemorrhage, infection, inflammation, etc.). For this reason, in clinical practice, hydrocephalus is often divided into two categories:

1. **Communicating hydrocephalus** is caused by impaired CSF reabsorption in the arachnoid granulations, obstruction of flow in the subarachnoid space, or (rarely) by excess CSF production.
2. **Noncommunicating hydrocephalus** is caused by obstruction of flow within the ventricular system.

The main symptoms and signs of hydrocephalus are similar to those of any other cause of elevated intracranial pressure (see KCC 5.2 and KCC 5.3) and can be acute or chronic, depending on how quickly the hydrocephalus develops. These symptoms and signs include headache, nausea, vomiting, cognitive impairment, decreased level of consciousness, papilledema, decreased vision, and sixth-nerve palsies. In addition, ventricular dilation in hydrocephalus may compress descending white matter pathways from the frontal lobes, leading to frontal lobe–like abnormalities including an unsteady **magnetic gait** (feet barely leave the floor) and incontinence. In neonatal hydrocephalus, when the cranial sutures have not yet fused, the skull expands to reduce elevated intracranial pressure, resulting in increased head circumference. A bulging anterior fontanelle is also an important sign of elevated intracranial pressure in infants.

It is important to recognize the eye movement abnormalities associated with hydrocephalus. In mild or slowly developing cases, only a sixth-nerve palsy may be seen, which causes incomplete or slow abduction of the eye in the horizontal direction. Interestingly, hydrocephalus may affect the sixth nerve of one or both eyes. When hydrocephalus is more severe, inward deviation of one or both eyes may be present at rest. Also, in more severe or rapidly developing cases, dilation of the **suprapineal recess** (see Figure 5.11D) of the posterior third ventricle can push downward onto the collicular plate of the midbrain, producing **Parinaud's syndrome**. This syndrome is described in greater detail in Chapter 13 (see KCC 13.9). The important abnormality to be aware of for now is limited vertical gaze, especially in the upward direction. Particularly in children with acute hydrocephalus, the ominous "setting sun" sign, consisting of bilateral deviation of the eyes downward and inward, may be seen. These abnormalities often reverse after treatment.

Treatment of hydrocephalus usually involves a procedure that allows cerebrospinal fluid to bypass the obstruction and drain from the ventricles. An **external ventricular drain** (also called ventriculostomy) works by draining the fluid from the lateral ventricles into a bag outside of the head. A more permanent treatment is a **ventriculoperitoneal shunt**, in which the shunt tubing passes from the lateral ventricle out of the skull and is then tunneled under the skin to drain into the peritoneal cavity of the abdomen. A valve prevents flow of fluid in the reverse direction, from abdomen to ventricle.

Endoscopic neurosurgery has recently been gaining in popularity for treatment of hydrocephalus and other disorders. In this minimally invasive approach, a narrow tube (cannula) is introduced into the cranium or spine through a small incision. Surgical procedures are then performed by passing instruments and a viewing endoscope through the cannula. Endoscopic neurosurgery can be used for treating obstructive hydrocephalus and intraventricular mass lesions. The third ventricle can be accessed by passing the endoscope through the right frontal lobe into the right lateral ventricle and through the foramen of Monro. Endoscopic **third ventriculostomy** is used as an alternative to shunting. In this procedure an endoscope is passed through the foramen of Monro into the third ventricle. A blunt-tipped instrument is then used to make a perforation in the floor of the third ventricle just in front of the mamillary bodies and behind the pituitary infundibulum (see Figure 17.2), allowing CSF to drain into the interpeduncular cistern (see Figure 5.12). A fold of the arachnoid known as Liliequist's membrane located in the suprasellar region can sometimes prevent adequate CSF drainage if it is not perforated during this procedure. Endoscopic neurosurgery can also be used to access pituitary tumors and other lesions in the sellar/suprasellar region via a transsphenoidal approach (see KCC 17.1) and has even been used to perform minimally invasive surgery of the spine (see KCC 8.5).

In closing, two other forms of hydrocephalus should be mentioned. **Normal pressure hydrocephalus**, a condition sometimes seen in elderly individuals, is characterized by chronically dilated ventricles. Patients with normal pressure hydrocephalus typically present with the **clinical triad of gait difficulties**, **urinary incontinence**, and **mental decline**. Measurements of CSF pressure in normal pressure hydrocephalus are usually not elevated; however, some studies have shown pressure elevations to occur only intermittently. Although the exact mechanism is not known, normal pressure hydrocephalus is thought to be a form of communicating hydrocephalus with impaired CSF reabsorption at the arachnoid villi. Some patients improve dramatically, particularly with regard to gait, after large-volume CSF removal by lumbar puncture or in a more permanent manner following ventriculoperitoneal shunting.

Hydrocephalus ex vacuo is simply a descriptive term and is not itself responsible for any pathology. It refers to excess cerebrospinal fluid in a region where brain tissue was lost as a result of stroke, surgery, atrophy, trauma, or other insult. ■

5.8 KEY CLINICAL CONCEPT

BRAIN TUMORS

There are two broad categories of brain tumors. **Primary CNS tumors** arise from abnormal proliferation of cells originating in the nervous system. **Metastatic tumors** arise from neoplasms originating elsewhere in the body that spread to the brain. The relative incidence reported for primary and metastatic brain tumors varies depending on the methodology used to select patients. In most series, however, metastases are 5 to 10 times more common than all primary CNS tumors combined. Common primary brain tumors are listed in Table 5.5. Gliomas and meningiomas are the most common, followed by pituitary adenoma, schwannoma, and lymphoma.

In adults, about 70% of tumors are supratentorial and 30% are infratentorial, while in pediatric patients the reverse is true, with about 70% of tumors located in the posterior fossa and 30% supratentorially. The most common brain tumors in children are astrocytoma and medulloblastoma, followed by ependymoma. Since pediatric brain tumors are often in the posterior fossa, they tend to cause hydrocephalus (see KCC 5.7) through compression and obstruction of the fourth ventricle or aqueduct of Sylvius.

TABLE 5.5 Primary Brain Tumors

TYPE OF TUMOR	PERCENT OF TOTAL
Glioma	33
Glioblastoma multiforme	20
Astrocytoma grades I and II	5
Astrocytoma grade III	3
Oligodendroglioma	2
Ependymoma	2
Other mixed or unclassified	1
Meningioma	33
Pituitary adenoma	12
Schwannoma	9
Lymphoma	3
Embryonal/primitive/ medulloblastoma	1
Other*	9

Note: Metastases are 5 to 10 times more common than primary brain tumors.

*Includes choroid plexus, neuroepithelial, pineal parenchymal, hemangioblastoma, hemangioma, craniopharyngioma, germ cell, and other tumors.

Source: CBTRUS (2008): *Statistical Report: Primary Brain Tumors in the United States,* 2000–2004 (http://www.cbtrus.org).

Symptoms of brain tumors depend on the location, size, and rate of growth of the tumor. Headache and other signs of elevated intracranial pressure (see Table 5.3) are common at presentation. Some tumors may present with seizures or with focal symptoms and signs, depending on the location of the tumor. The tumors most commonly associated with seizures are low-grade gliomas and meningiomas.

Brain tumors are considered benign if they do not infiltrate or disseminate widely through the nervous system and malignant if they have the potential to spread. Unlike systemic malignancies, however, malignant brain tumors only rarely undergo metastatic spread outside of the central nervous system. In addition, so-called benign brain tumors may be incurable—and ultimately fatal if they grow in vital areas of the brain where surgical excision is not possible.

Treatments for brain tumors depend on the histological type, location, and size of the tumor. Surgical removal of as much tumor as possible without causing serious deficits is usually pursued. Recent data suggest that the extent of surgical resection should be greater than 90% to have a positive effect on outcome. Next, depending on the lesion, **radiation therapy** and/or **chemotherapy** may be beneficial. **Steroids** are often used to reduce edema and swelling. Small, benign-looking tumors may simply be followed with serial MRI scans, depending on the clinical situation, especially in elderly patients.

Gliomas are subdivided into several different types (see Table 5.5). Glial tumors arising from astrocytes are called **astrocytomas**. Astrocytomas are usually classified by the World Health Organization (WHO) grading system, in which the most benign is grade I and the most malignant is grade IV, or **glioblastoma multiforme**. Unfortunately, glioblastoma is relatively common and usually leads to death within 1 to 2 years despite maximal resection, radiation, and chemotherapy.

Meningiomas arise from the arachnoid villus cells and occur, in order of decreasing frequency, over the lateral convexities, in the falx, and along the basal regions of the cranium. They grow quite slowly and usually appear on CT and MRI scans as homogeneous enhancing areas that arise from the meningeal layers. In female patients, there may be an association between meningiomas and breast cancer, although the pathophysiological link is uncertain. Meningiomas are treated by local excision. Five to 10% of meningiomas behave in an atypical or malignant fashion.

Pituitary adenomas can cause endocrine disturbances or compress the optic chiasm, usually resulting in a bitemporal visual field defect (see Figure 11.15C). Other lesions can arise in this area as well (including meningioma, craniopharyngioma, hypothalamic glioma, and others). Pituitary adenomas are discussed further in Chapter 17 (see KCC 17.1). Treatment with dopaminergic agonists will often inhibit function and shrink the size of prolactinomas, the most common type of pituitary adenoma. If this is ineffective, transsphenoidal resection is performed (see KCC 17.1).

Schwannomas, which are most common on CN VIII, are discussed in Chapter 12 (see KCC 12.5).

Lymphoma of the central nervous system has been on the rise in recent years, only in part attributable to the increase in human immunodeficiency virus (HIV). This neoplasm arises from B lymphocytes and commonly involves regions adjacent to ventricles, so it is occasionally diagnosed by CSF cytology. It can often be controlled for several years with chemotherapy and radiation therapy and currently has a median survival rate of close to four years.

Pineal region tumors are relatively uncommon (less than 1% of CNS neoplasms) and include pinealomas (pineocytoma and pineoblastoma), germinoma, and rarely, teratoma or glioma. Tumors in this region may obstruct the cerebral aqueduct, causing hydrocephalus (see KCC 5.7) or may compress the dorsal midbrain, causing Parinaud's syndrome (see KCC 13.9).

Brain metastases can occur with numerous tumor types. The three most common are lung or breast carcinoma and melanoma. Some brain metastases have a tendency to hemorrhage, including melanoma, renal cell carcinoma, thyroid carcinoma, and choriocarcinoma. Although lung cancer metastases do not often hemorrhage, the most common tumor causing brain hemorrhage is lung cancer, simply because the incidence of lung cancer and metastases to the brain is so high. When solitary brain metastases occur, outcome is improved by complete surgical excision, when possible. Surgical treatment of more than one brain metastasis is controversial, and multiple or unresectable brain metastases are usually treated with radiation therapy.

In pediatric patients, the most common brain tumors are posterior fossa astrocytoma, medulloblastoma, and ependymoma. **Cerebellar astrocytoma** is a type of grade I astrocytoma that can often be cured by surgical resection. **Medulloblastoma** and **ependymoma** of the posterior fossa have worse prognoses, although long-term survival is possible following a combination of surgery, radiation, and chemotherapy. Medulloblastoma occurs before age 10 about 90% of the time, while cerebellar astrocytomas occur mainly between ages 2 and 20.

Paraneoplastic syndromes are relatively rare neurologic disorders caused by remote effects of cancer in the body, leading to an abnormal autoimmune response. Examples of paraneoplastic syndromes include limbic or brainstem encephalitis, cerebellar Purkinje cell loss, spinal cord anterior horn cell loss, neuropathy, impaired neuromuscular transmission (Lambert–Eaton syndrome), and opsoclonus myoclonus, which is characterized by irregular jerking movements of the eyes and limbs. Tumors that most often cause paraneoplastic syndromes include small cell lung carcinoma, breast cancer, and ovarian cancer. Testing for specific antibodies that cross-react with tumor cells may be helpful in the diagnosis. ∎

5.9 KEY CLINICAL CONCEPT

INFECTIOUS DISORDERS OF THE NERVOUS SYSTEM

Like all other parts of the body, the nervous system can be affected by a variety of infectious pathogens, including bacteria, viruses, parasites, fungi, and prions. In this section we will briefly review the diagnosis and treatment of common infections of the nervous system.

Bacterial Infections

INFECTIONS CAUSED BY COCCI AND BACILLI Important bacterial infections of the nervous system caused by cocci and bacilli include bacterial meningitis, brain abscess, and epidural abscess. Bacteria most often gain access to the nervous system through the bloodstream, and they frequently originate from an infection elsewhere in the body, such as the respiratory tract or heart valves (endocarditis). In addition, infections can spread by direct extension from the oronasal passages. Finally, trauma or surgery can introduce bacteria into the nervous system from the skin.

Infectious meningitis is an infection of the cerebrospinal fluid in the subarachnoid space. It can be caused by bacteria, viruses, fungi, or parasites. Ex-

TABLE 5.6 Signs and Symptoms of Meningeal Irritation

Headache

Lethargy

Sensitivity to light (photophobia) and noise (phonophobia)

Fever

Nuchal rigidity (stiff neck): unable to touch chin to chest

    Kernig's sign: pain in hamstrings when knees are straightened with hips flexed

    Brudzinski's sign: flexion at neck causes hips to flex

cept for in elderly, very young, or immunocompromised patients, infectious meningitis is usually heralded by marked signs and symptoms of **meningeal irritation**, or **meningismus**. These **meningeal signs** can also be seen in subarachnoid hemorrhage, carcinomatous meningitis, and chemical meningitis. Common features of meningeal irritation include headache, lethargy, sensitivity to light (photophobia) and noise (phonophobia), fever, and nuchal rigidity (Table 5.6). In **nuchal rigidity**, the neck muscles contract involuntarily, resulting in resistance to active or passive neck flexion, accompanied by neck pain.

Depending on the cause of meningeal irritation, onset of symptoms may be gradual, over a period of weeks to months in the case of some fungal or parasitic infections, or symptoms may progress rapidly, within hours in the case of many bacterial infections. Diagnosis is made by clinical evaluation and by **sampling of cerebrospinal fluid by lumbar puncture** (see KCC 5.10; Table 5.7). A head CT should always be performed before lumbar puncture is considered because removal of CSF in the presence of a mass lesion can occasionally precipitate herniation. However, *antibacterial therapy should not be delayed* while these procedures are performed, since bacterial meningitis can be rapidly fatal if left untreated. A single dose of intravenous antibiotics prior to CT and lumbar puncture may not reduce diagnostic yield, particularly now that bacterial antigen and PCR (polymerase chain reaction) analysis of CSF is available. We will discuss bacterial meningitis in this section, and we will discuss meningitis caused by other pathogens (see Tables 5.7 and 5.9) in the sections that follow.

In acute **bacterial meningitis**, cerebrospinal fluid typically has a high white blood cell count with a polymorphonuclear predominance, high protein, and low glucose (see Table 5.7). Bacteria can sometimes be identified microscopically by Gram stain, bacterial antigen tests, PCR analysis, or cultures of CSF. The most common pathogens in bacterial meningitis depend on the patient's age (Table 5.8). Treatment, therefore, also depends on age, as shown in

TABLE 5.7 Cerebrospinal Fluid Profiles in Normal Adults and in Those with Infectious Meningitis

CONDITION	WHITE BLOOD CELLS (PER MM^3)	PROTEIN (MG/DL)	GLUCOSE (MG/DL)	COMMENTS
Normal (adults)	<5–10, lymphocytes only	15–45	50–100	If traumatic lumbar puncture,[a] expect 1 additional white blood cell for every ~700 red blood cells.
Acute bacterial meningitis	100–5000, usually polymorphonuclear leukocytes	100–1000	Reduced, <40	In patients with hyperglycemia, CSF glucose is abnormal if <50% of serum glucose.
Viral meningitis or "aseptic" meningitis (see Table 5.9)	10–300, usually lymphocytes	50–100	Normal	Glucose is occasionally reduced in herpes, mumps, and lymphocytic choriomeningitis virus.
Herpes meningoencephalitis	0–500, usually lymphocytes	50–100	Normal or reduced	Red blood cells or xanthochromia may be present.
Tuberculous meningitis or cryptococcal meningitis	10–200, usually lymphocytes	100–200	Reduced, <50	

[a]See also KCC 5.10 for discussion of lumbar puncture technique and interpretation of red blood cell count in CSF.

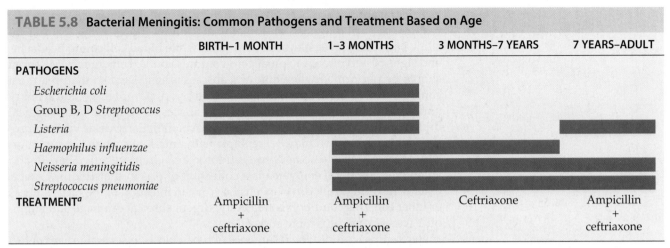

TABLE 5.8 Bacterial Meningitis: Common Pathogens and Treatment Based on Age

	BIRTH–1 MONTH	1–3 MONTHS	3 MONTHS–7 YEARS	7 YEARS–ADULT
PATHOGENS				
Escherichia coli	▬▬▬	▬		
Group B, D *Streptococcus*	▬▬▬	▬		
Listeria	▬▬▬	▬		▬▬▬
Haemophilus influenzae		▬▬	▬▬▬	
Neisseria meningitidis		▬▬	▬▬▬	▬▬▬
Streptococcus pneumoniae		▬▬	▬▬▬	▬▬▬
TREATMENT[a]	Ampicillin + ceftriaxone	Ampicillin + ceftriaxone	Ceftriaxone	Ampicillin + ceftriaxone

[a](1) There is some evidence that treatment of adult patients with *S. pneumonaie* meningitis and of pediatric patients at risk for *H. influenzae* meningitis with dexamethasone prior to antibiotics may improve outcome. Vaccination has markedly decreased the incidence of *H. influenzae* in recent years.

(2) Close household contacts of patients with *N. meningitidis* or *H. influenzae* meningitis should be treated prophylactically with oral rifampicin.

(3) Patients who are elderly or immunocompromised or who have had head trauma or neurosurgery, are also susceptible to *E. coli*, *Klebsiella*, *Pseudomonas*, *Staph. aureus*, *Staph. epidermidis*, and others. Therefore, additional antibiotics are often used in these populations.

(4) If herpes simplex meningoencephalitis is suspected, acyclovir should be added.

Table 5.8. Note that antibiotic therapy is constantly evolving, and current treatment recommendations for bacterial meningitis (and other infections discussed below) should always be checked at the local institution. It is essential to begin treatment as rapidly as possible in cases of suspected bacterial meningitis because these patients can deteriorate very quickly. Therapy should not be postponed during patient transport or while awaiting the results or performance of diagnostic tests. Complications of bacterial meningitis include seizures, cranial neuropathies, cerebral edema, hydrocephalus, herniation, cerebral infarcts, and death. Following recovery from bacterial meningitis in young children, it is important to screen for hearing loss, which, if treated early by cochlear implant, can lead to improved long-term hearing and language function.

Brain abscess is another important bacterial infection of the nervous system. Brain abscess presents as an expanding intracranial mass lesion, much like a brain tumor, but often with a more rapid course. Common presenting features include headache, lethargy, fever, nuchal rigidity, nausea, vomiting, seizures, and focal signs determined by the location of the abscess. Of note, fever is absent in about 40% of cases, and peripheral white blood cell count is not elevated in about 20% of cases, making the diagnosis of infection more difficult. The ESR (erythrocyte sedimentation rate) is usually elevated. Common infecting organisms include streptococci, *Bacteroides*, enterobacteriaceae, *Staphylococcus aureus*, and rarely, *Nocardia*. More than one organism is often present. Patients who are clinically stable and have abscesses less than about 2.5 cm in diameter, and in whom an infectious organism has been identified from another source in the body, can be treated with antibiotics and careful observation. Patients with larger abscesses (at risk for catastrophic rupture) or with clinical signs of mass effect or progressive deterioration should be treated with stereotactic needle aspiration (see KCC 16.4) or surgical removal, as well as antibiotics. Another important cause of brain abscess other than bacteria is the parasite *Toxoplasma gondii*, discussed along with HIV later in this section.

Epidural abscess can occasionally occur, especially in the spinal canal, and requires prompt diagnosis and treatment. Common presenting features include back pain, fever, elevated peripheral white blood cell count, headache, and signs of nerve root or spinal cord compression. An emergency MRI scan should be performed when epidural abscess is suspected so that treatment can be initiated before spinal cord compression, paraparesis, and urinary and fecal incontinence occur. Epidural abscess is treated with surgical drainage and antibiotics (nafcillin and ceftriaxone). Early cases without progressive deterioration may be treatable with antibiotics alone. Common organisms are *Staphylococcus aureus*, streptococci, Gram-negative bacilli, and anaerobes. **Subdural empyema** is a collection of pus in the subdural space, usually resulting from direct extension from an infection of the nasal sinuses or inner ear. This condition is treated by urgent surgical drainage and antibiotics (ceftriaxone plus metronidazole).

In recent years, with the resurgence of tuberculosis in several urban centers in the United States, **tuberculous meningitis** has become more common as well. Headache, lethargy, and meningeal signs (see Table 5.6) usually appear over the course of several weeks. There is often an inflammatory response in the basal cisterns of the brain, which can affect the circle of Willis vessels, resulting in infarcts. If untreated, coma, hydrocephalus, and death ensue. Tuberculous involvement of the epidural space and vertebral bones, called Pott's disease, can also occur. Populations at risk are intravenous drug abusers, patients with HIV, and people from areas where tuberculosis is endemic.

The meningeal involvement results from reactivation of previous tuberculosis infection, and signs of pulmonary tuberculosis are often not present at the time of presentation. Cerebrospinal fluid shows an elevated white blood cell count with lymphocyte predominance (see Table 5.7), elevated protein, and low glucose. Early on, there may be a polymorphonuclear predominance. The *Mycobacterium tuberculosis* organisms that cause this infection are often not visible microscopically in the CSF. The diagnosis can be confirmed by culture, which takes several weeks, or thanks to recent scientific advances, by polymerase chain reaction (PCR). Tuberculous meningitis is treated with a combination of isoniazid, rifampicin, ethambutol, and pyrazinamide.

As we will see in the sections that follow, **lymphocyte-predominant meningitis**, or "aseptic" meningitis, can have numerous other causes in addition to tuberculosis (Table 5.9), but it is most commonly viral in origin. Rarely, **cat scratch disease**, caused by the bacillus *Bartonella henselae*, can produce headaches, mental status changes, and seizures. MRI may show T2 bright areas, and CSF is either normal or shows lymphocyte-predominant meningitis.

INFECTIONS CAUSED BY SPIROCHETES The two most important spirochetal infections of the nervous system are neurosyphilis and Lyme disease. **Neurosyphilis** was fairly common in the pre-penicillin era and has had a resurgence in recent years, possibly related to HIV. Syphilis, formerly called lues, is caused by the spirochete *Treponema pallidum*. It is transmitted sexually, and various stages occur at different times after primary infection. In primary syphilis, painless skin lesions called chancres appear at the site of infection about 1 month after exposure. In secondary syphilis, more diffuse skin lesions appear within approximately 6 months, characteristically including the palms and soles. In tertiary syphilis, neurologic manifestations are often present.

Meningeal involvement can cause **aseptic meningitis** (see Table 5.9), sometimes with associated cranial nerve palsies, especially involving the optic, facial, and vestibulocochlear nerves. Later stages of neurologic involvement can occur following a latency of about 4 to 15 years. These are classified as meningovascular syphilis, general paresis, and tabes dorsalis. In **meningovascular syphilis**, chronic meningeal involvement causes an arteritis, typically involving medium-sized vessels, that results in diffuse white matter infarcts. If untreated, this condition eventually leads to **general paresis**, in which the accumulation of lesions causes dementia, behavioral changes, delusions of grandeur, psychosis, and diffuse upper motor neuron–type weakness. In another variant that often coexists with general paresis, patients with **tabes dorsalis** have involvement of the spinal cord dorsal roots, especially in the lumbosacral region, resulting in degeneration of the dorsal columns. Therefore, these patients have sensory loss in the lower extremities, sensory ataxia (see KCC 15.2) with a characteristic high-stepping tabetic gait (see KCC 6.5), and incontinence (see KCC 7.5). Other associated features include Argyll Robertson pupils (see KCC 13.5) and optic atrophy.

Diagnosis of neurosyphilis is based on blood tests for treponemes (FTA-ABS or MHA-TP), together with cerebrospinal fluid showing **lymphocyte-predominant meningitis**. So-called nontreponemal blood tests (RPR or VDRL) may be positive or negative in neurosyphilis, but cerebrospinal fluid VDRL is usually positive. Neurosyphilis is treated with intravenous penicillin G, and serial lumbar punctures should be performed to monitor the response to therapy.

Lyme disease is caused by the spirochete *Borrelia burgdorferi*, carried by *Ixodes* species of deer tick, which are endemic to certain areas of the United States, Europe, and Australia. The disease is named for the town of Lyme, Connecticut, where the disorder was first described. Primary infection is often heralded by a characteristic raised rash, called erythema chronicum migrans, which gradually shifts its location and enlarges over days to weeks. In some cases, neurologic manifestations occur. These usually appear after a delay of several weeks, and include a lymphocyte-predominant meningitis (see Table 5.9) or mild meningoencephalitis, characterized by meningeal signs and emotional changes, with impaired memory and concentration.

Other features that may be present include cranial neuropathies (especially of the facial nerve), peripheral neuropathies, and, rarely, spinal cord involvement. Non-neurologic manifestations include arthritis and cardiac conduction abnormalities. Lyme disease is diagnosed by typical clinical features, lumbar puncture, and serological testing. Untreated cases can eventually show white matter abnormalities on MRI scan. Lyme disease with neurologic involvement is treated with intravenous ceftriaxone.

Viral Infections

Viral meningitis tends to be less fulminant than bacterial meningitis, and recovery usually occurs spontaneously within 1 to 2 weeks. Patients present with headache, fever, lethargy, nuchal rigidity, and other signs of meningeal irritation (see Table 5.6). Common causes include enteroviruses such as echovirus, coxsackievirus, and mumps virus. Often the causative agent is not identified. There is no specific treatment for most viral infections of the nervous system, except for herpes and HIV.

Cerebrospinal fluid in viral meningitis shows an elevated white blood cell count with a lymphocytic predominance, normal or mildly elevated protein, and normal glucose. In the early stages, a polymorphonuclear predomi-

TABLE 5.9 Differential Diagnosis of Lymphocyte-Predominant "Aseptic" Meningitis

Viral infections (numerous, including HIV)

Partially treated bacterial meningitis

Tuberculous meningitis

Cryptococcal meningitis and other fungal infections

Parameningeal infection (e.g., epidural abscess)

Postinfectious encephalomyelitis

Postvaccination encephalomyelitis

Myelitis

Lyme disease

Neurosyphilis

Parasitic infections (eosinophils may also be present)

Carcinomatous (or other neoplastic) meningitis

Central nervous system vasculitis

Sarcoidosis

Venous sinus thrombosis

Subarachnoid hemorrhage several days previously

Drug reaction

Chemical irritation (e.g., from contrast material injected in CSF)

Note: See also Table 5.7.

nance may be present. The differential diagnosis of **lymphocytic**, or **lymphocyte-predominant meningitis**, of the kind seen with viral meningitis is broad, occasionally making the diagnosis difficult (see Table 5.9; see also Table 5.7). Viral meningitis and other types of lymphocyte-predominant meningitis are sometimes called **aseptic meningitis** to distinguish them from bacterial meningitis. We have already discussed several other conditions characterized by lymphocyte-predominant meningitis, including tuberculous meningitis, neurosyphilis, and CNS Lyme disease.

When viral infections involve the brain parenchyma, they are called **viral encephalitis**. Unlike typical cases of viral meningitis, the clinical manifestations of viral encephalitis are often quite severe. The meninges are often also involved, resulting in **meningoencephalitis**. The most common cause of viral encephalitis is **herpes simplex virus** type 1 (type 2 also occasionally causes encephalitis). As discussed in Chapter 18, the herpes simplex virus has a tropism for limbic cortex. Patients often present with bizarre psychotic behavior, confusion, lethargy, headache, fever, meningeal signs, and seizures. Focal signs such as anosmia, hemiparesis, memory loss, and aphasia may be present as well. Herpes simplex encephalitis causes necrosis of unilateral or bilateral temporal and frontal structures often visible on MRI scan. Untreated, it usually progresses within days to coma and death. Therefore, it is essential to initiate therapy promptly with acyclovir.

Other characteristic features include periodic sharp waves seen over one or both temporal lobes on EEG. Cerebrospinal fluid shows a lymphocytic or mixed lymphocytic–polymorphonuclear predominance, with elevated protein and normal glucose (see Table 5.7). When necrosis is prominent, the CSF red blood cell count may be elevated and the glucose level decreased. The virus is difficult to culture from cerebrospinal fluid, but it can be identified in most cases by polymerase chain reaction (PCR).

There are a variety of other causes of viral encephalitis, but unfortunately none of these have a specific treatment. Prognosis depends on the causative agent. In addition, **postinfectious encephalitis** can occur, usually several days after a viral infection, with diffuse autoimmune demyelination of the central nervous system. Prognosis is variable. Measles is occasionally associated with a delayed, slowly progressive fatal encephalitis called **subacute sclerosing panencephalitis**. Fortunately, the incidence of this disorder has dropped markedly since the introduction of the measles vaccine.

Herpes zoster, or shingles, is an infection caused by the same virus as chickenpox (varicella-zoster virus). The primary symptom is a painful rash conforming to nerve root distributions and will be discussed further in KCC 8.3.

Viral infections of the nervous system are also a common cause of **transverse myelitis** (see KCC 7.2). Important viral causes of myelitis include enteroviruses (such as coxsackie and poliomyelitis), varicella-zoster virus, HIV, or, less commonly, Epstein–Barr virus, cytomegalovirus, herpes simplex, rabies, or Japanese B virus. The virus HTLV-1 causes a more chronic type of spinal cord disease called HTLV-I associated myelopathy, or tropical spastic paraparesis.

HIV-ASSOCIATED DISORDERS OF THE NERVOUS SYSTEM **Human immunodeficiency virus** (**HIV**) can increase susceptibility to numerous infectious disorders of the nervous system, including viral, bacterial, fungal, and parasitic infections. HIV itself can cause an aseptic meningitis at the time of seroconversion. This condition is sometimes associated with cranial neuropathies, especially involving the facial nerve (see KCC 12.3). **HIV-associated neurocognitive disorder** (**HAND**) is a common neurologic manifestation of HIV, with increased frequency late in the course of the illness. Treatment with anti-retroviral agents (highly active antiretroviral therapy, or HAART), can cause some improvement in AIDS-related dementia.

In addition to having an effect on the brain, the HIV virus has been associated with involvement of the spinal cord, peripheral nerves, and muscles (myelopathy, neuropathy, and myopathy, respectively). Other viral infections in patients with HIV include encephalitis caused by **herpes simplex virus**, **varicella-zoster virus**, or **cytomegalovirus**. Cytomegalovirus can also cause retinitis, which responds to ganciclovir, and a polyradiculitis that often involves the cauda equina (see KCC 8.4). **Progressive multifocal leukoencephalopathy** (**PML**) can occur in patients with AIDS or other immunodeficiency states. This disorder is caused by a papovavirus called the JC virus and results in gradual demyelination of the brain, usually leading to death within 3 to 6 months. Survival time has improved to about 11 months for patients on HAART. MRI shows T2 bright white matter abnormalities, especially in the posterior regions of the brain. When PML occurs in HAART-treated patients with immune reconstitution inflammatory syndrome (IRIS), MRI may show contrast enhancing lesions. JC virus has also been associated with cerebellar atrophy caused by granule cell neuronopathy.

Important bacterial infections of the nervous system in patients with HIV include tuberculous meningitis and neurosyphilis (discussed earlier). There is some evidence that neurosyphilis may have a more rapid or atypical course in patients with HIV. In the category of fungal infections, **cryptococcal meningitis** is common and should be suspected in all HIV-positive patients with chronic headache. Diagnosis is by lumbar puncture (see KCC 5.10). Cerebrospinal fluid may show an elevated white blood cell count with lymphocytic predominance (see Table 5.7), but it is occasionally normal. The presence of cryptococcal antigen in the CSF should therefore be checked. The organisms can also sometimes be identified by India ink stain. Cryptococcal meningitis is treated with intravenous amphotericin B, followed by oral fluconazole. Milder cases may be treated with fluconazole alone. Severe cases can cause progressive obtundation, neuropathies, seizures, hydrocephalus, and death.

A common parasitic infection of the nervous system in patients with HIV is **toxoplasmosis**. Central nervous system toxoplasmosis is caused by reactivation of infection with the parasite *Toxoplasma gondii*. Initial exposure is from cysts in cat feces or undercooked meat, and is usually asymptomatic. In patients with AIDS or other causes of immunosuppression, the *Toxoplasma* infection becomes reactivated and spreads to the central nervous system, forming brain abscesses visible on MRI scans as ring-enhancing lesions: a nonenhancing center (dark on T1) surrounded by a ring of enhancement after administration of gadolinium. Edema often causes mass effect leading to compression of adjacent structures. Common presenting features are seizures, headache, fever, lymphocytic predominant meningitis, and focal signs, depending on the location of the lesions. Serological tests for toxoplasmosis are unreliable because much of the general population has been exposed (30% in the United States, 80% in France). The diagnosis can be established by PCR of CSF with approximately 50% sensitivity. Toxoplasmosis is the most common cause of intracranial mass lesions in patients with HIV. Therefore, when typical-appearing lesions are found on MRI scan in these patients, they are usually treated empirically with pyrimethamine and sulfadiazine for approximately 2 weeks, and a follow-up MRI scan is then performed. If there is improvement, therapy is continued. If not, a brain biopsy is recommended to establish the diagnosis.

Patients with AIDS are at increased risk for **primary central nervous system lymphoma** (see KCC 5.8). This B cell lymphoma can appear radiologically similar to toxoplasmosis, and it is the second most common cause of intracranial mass lesions in patients with HIV. CNS lymphoma is diagnosed by brain biopsy. Treatment with steroids and radiation therapy may be of some benefit, although the prognosis is much worse than for primary central nervous system lymphoma without HIV. Kaposi's sarcoma has only very rarely been reported to metastasize to the central nervous system.

Parasitic Infections

Parasitic infections that can involve the nervous system include cysticercosis, toxoplasmosis, malaria, African sleeping sickness (caused by the parasite *Trypanosoma brucei*), amebiasis, rickettsial illnesses, hydatid cysts (echinococcosis) and schistosomiasis. In this section we will discuss cysticercosis. We discussed toxoplasmosis earlier, in the section on HIV. Further information on the other disorders, which are relatively uncommon in North America, is available from the references listed at the end of this chapter.

Cysticercosis is caused by ingestion of the eggs of the pork tapeworm *Taenia solium*, found predominantly in Latin America and in certain regions of Africa, Asia, and Europe. The organism migrates through the bloodstream to the whole body, forming multiple small cysts in the muscles, eyes, and central nervous system. Seizures are a common result. Other common features are headache, nausea, vomiting, lymphocytic meningitis, and focal deficits, depending on cyst location. The spinal cord can also occasionally be involved. Obstruction of the ventricular system by cysts can cause hydrocephalus. CT scans in active infection typically show multiple small, 1 cm to 2 cm cysts in the brain parenchyma, with surrounding edema. The organisms eventually die, leaving numerous 1 mm to 3 mm calcifications scattered throughout the brain ("brain sand").

Cysticercosis is diagnosed by history in appropriate populations, by typical radiologic appearance, and by antibody tests of the serum and cerebrospinal fluid. Sometimes eosinophilia, parasites in the stool, and soft tissue calcifications on X-rays may be present as well. In questionable cases, biopsy may be necessary. The condition is treated with albendazole.

Fungal Infections

Fungal infections of the central nervous system are uncommon in immunocompetent hosts, but they can occur occasionally. Cryptococcal meningitis was discussed earlier, in the section on HIV. Aspergillosis (caused by the fungus *Aspergillus*) and candidiasis (caused by *Candida*) can involve the brain parenchyma and are usually accompanied by an intense inflammatory response. Other fungi that can infect the brain parenchyma or meninges include *Histoplasma*, *Coccidioides*, and *Blastomyces*. *Aspergillus* can also occasionally spread from the nasal passages to the orbital apex, causing an orbital apex syndrome (see KCC 13.7). An important and potentially fatal fungal infection to be aware of is mucormycosis, which occurs mainly in diabetics in the rhinocerebral form and also involves the orbital apex. Rhinocerebral mucormycosis causes ophthalmoplegia, facial numbness, visual loss, and facial weakness, with a typical violet coloration of the tips of the eyelids. Most fungal infections can be diagnosed only by biopsy, which should be pursued aggressively because early treatment is essential. Mucormycosis is treated with amphotericin B. Steroids can exacerbate fungal infections and should be avoided when a fungal infection is suspected.

Prion-Related Illnesses

In recent years, a novel, protein-based infectious agent called the **prion** has been identified in certain neurologic disorders. Prions are unique in their ability to transmit illnesses from one animal to another, despite the fact that they apparently do not contain DNA or RNA. Pathologically, diffuse degeneration of the brain and spinal cord occurs, with multiple vacuoles resulting in a spongiform appearance. Human prion-related illnesses include Creutzfeldt–Jakob disease, Gerstmann–Sträussler–Scheinker disease, kuru, and fatal familial insomnia. These disorders are all relatively rare. The most common, Creutzfeldt–Jakob disease, has an incidence of approximately one new case per million individuals per year.

Typical presenting features of **Creutzfeldt–Jakob disease** are rapidly progressive dementia, an exaggerated startle response, myoclonus, visual distortions or hallucinations, and ataxia. EEG often shows periodic sharp wave complexes, especially late in the course of the illness. CSF usually shows increased 14-3-3 protein. MRI may show a characteristic increased signal in the basal ganglia and cortical ribbon on diffusion weighted imaging (DWI). Unfortunately, there is currently no treatment. Progressive neurologic deterioration and death usually occur within 6 to 12 months. Prion-related illnesses can occur in an inherited pattern, or they can be transmitted by exposure to infected tissues with an incubation period of 2 to 25 years (prions were formerly misnamed "slow viruses"). A recent cluster of atypical cases of Creutzfeldt–Jakob disease may have been caused by ingestion of cattle infected with bovine spongiform encephalopathy (mad cow disease) in Britain. ■

5.10 KEY CLINICAL CONCEPT

LUMBAR PUNCTURE

Lumbar puncture is an important procedure that provides direct access to the subarachnoid space of the lumbar cistern (Figure 5.22). It can be used to obtain samples of CSF, measure CSF pressure, to remove CSF in cases of suspected normal pressure hydrocephalus, and occasionally to introduce drugs (such as antibiotics or cancer chemotherapy) or radiological contrast material (in myelography; see Chapter 4) into the CSF. Before a lumbar puncture is performed, the patient should be evaluated for evidence of elevated intracranial pressure (see KCC 5.3), and the safest practice is to perform a CT scan first to avoid risk of herniation. In addition, caution should be used in cases of impaired coagulation because of the risk of iatrogenic spinal epidural hematoma, which can compress the cauda equina.

The lumbar puncture procedure is performed with sterile technique under local anesthesia (see Figure 5.22). A hollow spinal needle is introduced through the skin with a stylet occluding the lumen to prevent the introduction of skin cells into CSF during needle insertion. The needle passes through subcutaneous tissues, ligaments of the spinal column, dura, and arachnoid, to finally encounter CSF in the subarachnoid space of the lumbar cistern. Note that the lumbar cistern is normally in direct communication with CSF in the ventricles and CSF flowing over the surface of the brain (see Figures 5.10, 5.12). The procedure may be done in the lying or seated position. A manometer tube is used to measure CSF pressure. Pressure measurements are more reliable in the lying position (see Figure 5.22A) because in the seated position the entire column of CSF in the spinal canal adds to the pressure measured in the lumbar cistern. Normal CSF pressure in adults is less than 20 cm H2O (see KCC 5.3).

(A)

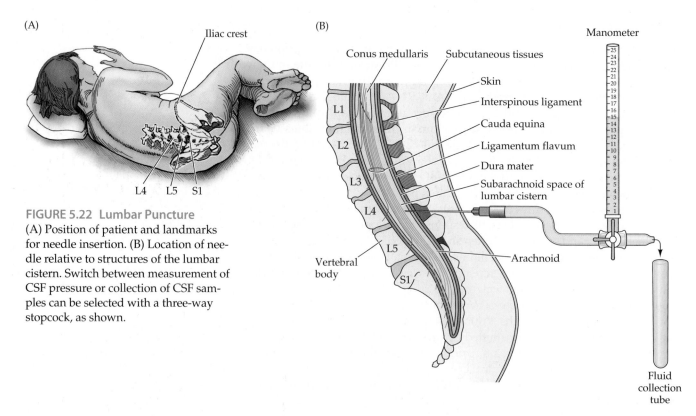

Iliac crest

L4 L5 S1

(B)

Conus medullaris   Subcutaneous tissues

Manometer

Skin

Interspinous ligament

Cauda equina

Ligamentum flavum

Dura mater

Subarachnoid space of
lumbar cistern

Arachnoid

Vertebral
body

Fluid
collection
tube

FIGURE 5.22 Lumbar Puncture
(A) Position of patient and landmarks
for needle insertion. (B) Location of needle relative to structures of the lumbar
cistern. Switch between measurement of
CSF pressure or collection of CSF samples can be selected with a three-way
stopcock, as shown.

Note that the bottom portion of the spinal cord, or **conus medullaris**, ends
at about the L1 or L2 level of the vertebral bones, and the nerve roots continue downward into the lumbar cistern, forming the **cauda equina**, meaning
"horse's tail" (see Figure 5.22B). To avoid hitting the spinal cord, the spinal
needle is generally inserted at the space between the L4 or L5 vertebral
bones. As the tip of the needle enters the subarachnoid space, the nerve
roots are usually harmlessly displaced. The posterior iliac crest serves as a
landmark to determine the approximate level of the L4–L5 interspace.

The opening CSF pressure should be measured and recorded (see Figure
5.22B). CSF samples are then collected and sent for numerous studies, including cell count, protein, glucose, and microbiological testing. The CSF
samples are collected in different tubes, which should be numbered sequentially, since the order in which CSF is taken can affect cell count, as we will
discuss shortly. CSF profiles under normal conditions and in several infectious disorders are listed in Table 5.7, and causes of lymphocytic-predominant meningitis are listed in Table 5.9.

Normally, red blood cells are not present in the CSF. Red blood cells in the
CSF can indicate subarachnoid hemorrhage (see KCC 5.6), hemorrhagic herpes encephalitis (see KCC 5.9), or they may simply have been introduced by
damage to blood vessels caused by the spinal needle at the time of lumbar
puncture, referred to as a **traumatic tap**. A traumatic tap can often be distinguished from pathological subarachnoid blood by the following guidelines:
(1) The number of red blood cells usually decreases from the first to last
tubes of CSF collected in a traumatic tap, but not in subarachnoid hemorrhage. (2) If the CSF is centrifuged, the supernatant may have a yellowish, or
xanthochromic, appearance as a result of red blood cell lysis if hemorrhage
occurred several hours previously, but no xanthochromia should be present
in a traumatic tap if the CSF is centrifuged immediately after collection. A
traumatic tap may also confound the analysis of CSF white blood cell count
because white blood cells are introduced as well. In this case the ratio of red

blood cells to white blood cells, and the white blood cell differential count, should be the same in the CSF as in the patient's peripheral blood smear. As a general guideline, in a traumatic tap *one white blood cell is introduced into the CSF for every 700 red blood cells*. Large amounts of hemorrhage in the CSF from any cause can also sometimes result in reduced CSF glucose or elevated protein.

In addition to diagnosing infection or hemorrhage, lumbar puncture can be useful for obtaining cytological specimens for the diagnosis of **neoplastic** or **carcinomatous, meningitis**, and it can be useful for immunologic testing such as detection of **oligoclonal bands** in suspected multiple sclerosis (see KCC 6.6). ■

5.11 KEY CLINICAL CONCEPT

CRANIOTOMY

Archaeological evidence suggests that some forms of craniotomy have been practiced since prehistoric times. In the twentieth century, neurosurgery entered the modern era, with a dramatic decline in operative morbidity and mortality associated with more refined techniques and sterile operating methods. Craniotomy now provides access to the intracranial contents for a wide variety of therapeutic and diagnostic procedures that can be performed relatively safely by skilled neurosurgeons.

In craniotomy, the initial step is to position the head to provide optimal access to the desired structures. The hair is shaved, the scalp is cleaned with iodine solution, and an incision is then made to expose the skull. **Burr holes** are drilled through the skull at several locations in the operative field, and care is taken not to enter the dura. The burr holes are then joined together with a small saw so that a piece of bone called a **bone flap** can be removed, exposing the dura. The bone flap is usually saved so that it can be replaced after the procedure. The dura is then carefully opened and folded back, providing access to the intracranial contents. In addition to the usual monitoring performed during surgery, in neurosurgery the anesthesiologist must be constantly aware of interventions that will affect intracranial pressure and cerebral blood flow (see KCC 5.3). A lumbar drain is sometimes used to remove controlled amounts of CSF and to enhance mechanical brain relaxation. When the procedure is complete, the dura is closed, the bone flap replaced, and the scalp sutured.

Different neurosurgical approaches are used to gain access to different regions of the intracranial cavity. The most common approaches are pterional, temporal, frontal, and suboccipital craniotomy. The pterion is the region over the temple where the frontal, parietal, temporal, and sphenoid bones meet. In **pterional craniotomy** this region of skull is removed, providing access to the inferior frontotemporal lobes. This approach is used for surgery on anterior circulation and basilar tip aneurysms, the cavernous sinus, and suprasellar tumors. In **temporal craniotomy** a more lateral approach is used for operating on the temporal lobe, including surgery to resect temporal lobe seizure foci (see KCC 18.2), and for decompression of most intracranial hematomas in head trauma (see KCC 5.6). **Frontal craniotomy** is used to operate on frontal lobe lesions such as tumors. **Suboccipital craniotomy** provides access to posterior fossa structures such as the cerebellopontine angle (see KCC 12.5), vertebral artery, brainstem, and lower cranial nerves. Other specialized approaches, including the **transsphenoidal approach**, in which the pituitary region is reached via the nasal passages (see KCC 17.1), are used in special situations. In image-guided **stereotactic procedures** (see KCC

16.4), an instrument is introduced through a small burr hole and directed to a specific target deep within the brain. Recently, minimally invasive **endoscopic neurosurgery**, also known as **neuroendoscopy**, has grown in popularity as an alternative to open craniotomy. We have already discussed endoscopic third ventriculostomy (see KCC 5.7). The usual approach for neuroendoscopic procedures to reach the intraventricular structures is through the right lateral ventricle and foramen of Monro, although a transsphenoidal approach (see KCC 17.1) can also be used to reach the sellar and suprasellar regions. Other applications of this emerging technique include endoscopic biopsy of masses within or along the walls of the ventricles as well as removal and drainage of cysts, hemorrhage, or abscesses. ■

CLINICAL CASES

CASE 5.1   AN ELDERLY MAN WITH HEADACHES AND UNSTEADY GAIT

CHIEF COMPLAINT

An 82-year-old man came to his physician with right-sided headaches and difficulty walking.

HISTORY

The patient was in a motor vehicle accident 3 months ago. He did not hit his head or lose consciousness, but the car was "demolished." He was examined in the emergency room and no abnormalities were found, so he was sent home. Ever since, he has complained of **generalized fatigue** and **right-sided headaches**, especially over the past 2 weeks. The headaches are worse at night, often keeping him awake. He has also recently stumbled several times while walking because of left leg weakness.

PHYSICAL EXAMINATION

General appearance: A thin, elderly man in no acute distress.

Vital signs: T = 97°F, P = 86, BP = 146/80, R = 18.

Head: No evidence of trauma.

Neck: Supple with no bruits.

Lungs: Clear.

Heart: Regular rate with no murmurs, gallops, or rubs.

Abdomen: Normal bowel sounds; soft, nontender.

Extremities: Normal.

Neurologic exam:

 MENTAL STATUS: Alert and oriented × 3. Speech fluent, with good naming and repetition. Able to count down from 100 by sevens (a test of attention and calculation skills).

 CRANIAL NERVES (SEE TABLE 2.5): Pupils equal round and reactive to light (CN II, III). Extraocular movements intact (CN III, IV, VI). **Visual fields** (CN II) full, but with **extinction on the left to double simultaneous stimulation**. Facial sensation intact (CN V). Face symmetrical (CN VII). Normal gag (CN IX, X). Normal sternomastoid strength (CN XI). Tongue midline (CN XII).

 MOTOR: **Pronator drift of left arm** (see KCC 6.4). **Mild left hemiparesis (4⁺/5 strength throughout left arm and leg).** Normal strength on the right side.

 SENSORY: Intact except for **left extinction on double simultaneous stimulation**.

 REFLEXES:

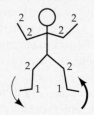

 GAIT: Takes short steps, but with good speed. Able to perform tandem gait.

 COORDINATION: Not tested.

LOCALIZATION AND DIFFERENTIAL DIAGNOSIS

1. On the basis of the symptoms and signs shown in **bold** above, what is the general location of the patient's lesion? Is it intracranial or extracranial? Is it on the left side or the right side?

2. Given the recent history of a motor vehicle accident in this elderly man and the fact that the headaches are worse at night, what is the most likely diagnosis? What are some other possibilities?

Discussion

1. The key symptoms and signs in this case are:

   - **Left hemiparesis and left Babinski's sign**
   - **Visual and tactile extinction on the left**
   - **Right-sided headaches**
   - **Generalized fatigue**

     The patient has mild left arm and leg weakness with a left Babinski's sign, suggesting an upper motor neuron lesion (see Table 3.3) in the corticospinal system anywhere from the right motor cortex to the left cervical spine (see Figure 2.16). The visual and tactile extinction in this patient are forms of left hemineglect (see KCC 19.9), which is most commonly seen in right parietal lesions but can also be seen in right frontal or subcortical lesions. These findings are compatible with a right intracranial, rather than a left spinal cord localization. Generalized fatigue is a nonspecific complaint, but in association with right-sided headaches it suggests a possible right intracranial mass lesion (see KCC 5.1–5.3). *The most likely clinical localization is a right hemisphere cortical and/or subcortical lesion affecting corticospinal and attentional pathways.*

2. This is an elderly patient with 3 months of headaches and generalized fatigue following a motor vehicle accident, who more recently developed a left hemiparesis. Chronic vague symptoms of this type with stepwise worsening in an elderly patient following minor head injury, are very suggestive of chronic subdural hematoma (see KCC 5.6). As a result of his multiple falls, there may be some more recent blood in the hematoma as well. Another possible but less likely diagnosis would be several small right-sided intracerebral hemorrhages over the past few months, caused by hypertension, tumor, vascular malformation, amyloid angiopathy, coagulation disorder, or trauma (contusions). In addition, it is possible that the car accident was a coincidence, or it may even have been an effect of the lesion rather than its cause. This would be the case if the patient had a preexisting right hemisphere lesion such as a brain tumor, infarct, demyelination, or infection that impaired driving skills.

Neuroimaging

The patient's physician decided to order a **head CT** (Image 5.1, page 172).

1. How old is the blood in the hematoma seen on the head CT (see Figure 5.19 and Chapter 4)?

2. Cover the labels on Image 5.1 and identify as many structures as possible on the CT scan.

3. Between what meningeal layers is the hemorrhage located, and what is the name of this kind of hematoma?

Discussion

1. The fluid collection is hypodense relative to the brain, but more dense than CSF. It generally takes about 2 to 3 weeks for a hematoma to become hypodense in appearance.

2. See labels on Image 5.1.

3. A large, right-sided fluid collection is seen located between the brain and the skull. The crescent shape is characteristic of a subdural hematoma (see KCC 5.6), which follows the contours of the brain as it expands in the potential

space between the dura and the loosely adherent arachnoid. Since it is hypodense in appearance, this is a chronic subdural hematoma.

Note that the underlying brain parenchyma on the right side is significantly compressed by the hematoma, including the area of the central sulcus and the frontal and parietal lobes. These structures can be better distinguished on the left side of the brain. Also note that there is a small amount of midline shift, amounting to mild subfalcine herniation (see KCC 5.4). It is remarkable that the patient was not more severely impaired, given the appearance of the CT scan. This case demonstrates the ability of the brain to adapt to large lesions that expand slowly over a long period of time.

Clinical Course

The patient was admitted to the hospital and taken to the operating room for drainage of the hematoma. Two burr holes were drilled through the skull, one in the frontal area and a second in the parietal area. The dura was opened through the burr holes, releasing "crankcase oil–colored" (dark purple) fluid under a large amount of pressure. There was no evidence of any acute blood. The subdural space was irrigated with saline until there was no further oil-colored fluid.

The patient's hemiparesis showed immediate improvement after surgery. Follow-up exam in the physician's office 3 weeks later showed 5/5 strength, intact vision, intact somatic sensation, no hemineglect, and toes downgoing bilaterally. The patient had no further complaints of headaches or generalized fatigue.

CASE 5.1 AN ELDERLY MAN WITH HEADACHES AND UNSTEADY GAIT

IMAGE 5.1 Head CT Showing Chronic Subdural Hematoma A hypodense right chronic subdural hematoma is present. Compare to Figure 4.12J.

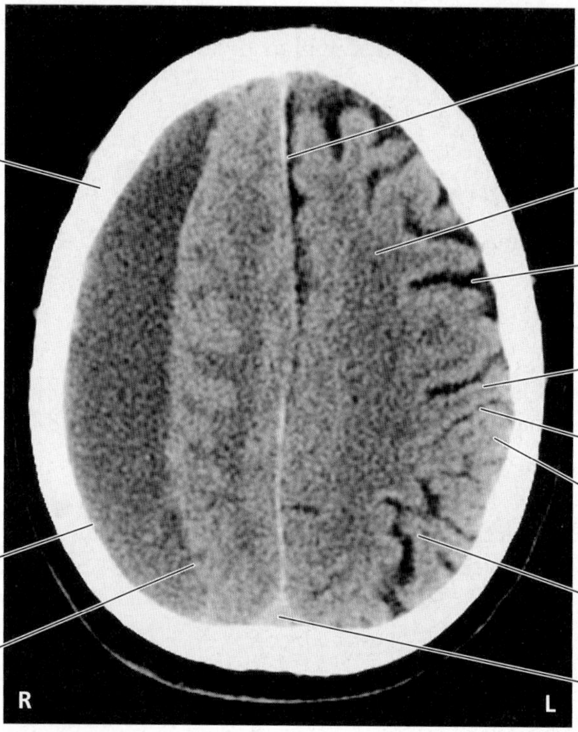

CASE 5.2   ALTERED MENTAL STATUS FOLLOWING HEAD INJURY

CHIEF COMPLAINT

A 67-year-old man was found at the bottom of a flight of stairs, lethargic and smelling of alcohol.

HISTORY

Little information could be obtained from the patient. He was discovered at the base of a staircase intoxicated and with a posterior scalp laceration and was therefore brought to the emergency room.

INITIAL PHYSICAL EXAMINATION

General appearance: An unkempt man lying on a stretcher.

Vital signs: T = 98°F, P = 90, BP = 176/89, R = 20.

Head: **Scalp laceration** over right occipital area; normal tympanic membranes (see Table 3.9).

Neck: In cervical stabilization collar put on by emergency personnel.

Lungs: Clear.

Heart: Regular rate with no murmurs, gallops, or rubs.

Abdomen: Normal bowel sounds; soft, nontender.

Extremities: No edema, normal pulses.

Rectal: Normal tone, heme negative.

Neurologic exam:

   MENTAL STATUS: **Lethargic** but arousable, with **garbled speech**. Stated his full name but **did not know his location or the date. Did not recall what happened to him,** saying "I'm all right." Followed simple commands.

   MOTOR: Able to move all four extremities.

LOCALIZATION AND DIFFERENTIAL DIAGNOSIS

1. Dysfunction of which regions of the nervous system could produce the mild to moderately depressed level of consciousness seen in this patient?

2. In broad terms, what are the two most likely causes of this patient's altered mental status?

CLINICAL COURSE IN THE EMERGENCY ROOM

The initial impression of the emergency room personnel was that the patient was intoxicated with alcohol and perhaps had a mild concussion, both of which were likely to improve with observation over the next few hours. The patient was sent for X-rays of the cervical spine and chest. His blood alcohol level came back at 325 mg/dl (>100 mg/dl can cause intoxication, although chronic users can develop tolerance). While at the radiology department, he became **uncooperative and combative**, moving too much for the X-rays to be done. He then became **increasingly somnolent** and developed irregular respirations, requiring emergency intubation and mechanical ventilation. A second, rapid but more detailed neurologic exam was done, and the patient was taken for an emergency head CT scan.

FOLLOW-UP PHYSICAL EXAMINATION

Vital signs: P = 95, BP = 184/90.

The remainder of the general exam was unchanged.

Neurologic exam:

   MENTAL STATUS: **Unresponsive except for movement to painful stimuli**.

   CRANIAL NERVES: Pupils 3 mm, constricting to 2 mm bilaterally. Oculocephalic maneuvers (see Chapter 3) were not done because of the cervical collar. **Corneal reflex** present on the left, **but absent on the right**.

   SENSORY AND MOTOR: Moved left arm and leg in response to painful stimulation. **Right arm and leg did not move in response to pain**.

   PLANTAR REFLEXES: **No response on the right, upgoing on the left**.

   COORDINATION AND GAIT: Not tested.

Discussion

1. The patient was lethargic and disoriented. Little other information is provided; however, he presumably had no other major abnormalities. Depressed level of alertness and attention can be seen with diffuse bilateral cerebral dysfunction or with dysfunction of the brainstem and diencephalic reticular activating systems (see Figure 2.23).

2. Although there are many possible causes of altered mental status that should be considered, there are two obvious causes in this patient. One is alcohol intoxication. The second is head trauma (see KCC 5.5). There are many additional possible causes of altered mental status in this patient (see KCC 19.15; see also Table 19.14) including other toxic or metabolic derangements, seizures, infection, infarct, or brain tumor.

CASE 5.2 ALTERED MENTAL STATUS FOLLOWING HEAD INJURY (continued)

LOCALIZATION AND DIFFERENTIAL DIAGNOSIS

1. The neurologic exam was incomplete, but it suggests **depressed consciousness** together with **right hemiplegia** (unilateral paralysis) and/or right sensory loss. Decreased function in which structures could cause this picture? What is the most likely cause of this patient's sudden deterioration?

2. In what location(s) might we expect to see hemorrhage on CT scan in a patient with acute head trauma (see KCC 5.5, 5.6)? To answer this question, complete the following table. For each location, identify the inner and outer layers that confine the space (see Figure 5.1), and state whether it is an actual space or a potential space.

Locations of Hemorrhage Associated with Head Trauma			
NAME	LOCATION	POTENTIAL OR ACTUAL SPACE	BOUNDED BY
Intraventricular hemorrhage			
Contusion with intracerebral or intraparenchymal hemorrhage			
Subarachnoid hemorrhage			
Subdural hematoma			
Epidural hematoma			
Subgaleal hematoma			

Discussion

1. The key symptoms and signs in this case are:

- **Unresponsiveness except to painful stimuli**
- **Absent right corneal reflex, and no right arm or leg movement in response to pain, with plantar response absent on the right and upgoing on the left**

Depressed consciousness can be caused by dysfunction of brainstem–diencephalic activating systems or of bilateral cerebral cortices (see Figure 2.23). Absent right face, arm, and leg movements in response to pain could be explained by impaired function in the motor pathways that begin in the left motor cortex (see Figures 2.13, 2.16), and/or by right body sensory loss (see Figure 2.19). The absent right plantar response is also compatible with corticospinal dysfunction because reflexes are sometimes depressed rather than increased in acute upper motor neuron lesions (see Table 3.3).

Although unilateral arm and leg paralysis can be caused by lesions above or below the foramen magnum, the presence of hemiplegia, together with impaired consciousness, strongly suggests pathology within the cranial vault. The absent right corneal reflex also supports intracranial localization. Intracranial lesions of the left brainstem or left cerebral hemisphere can cause right hemiplegia. There is some evidence of right brain dysfunction as well, with an upgoing toe in the left foot; this suggests either that the lesion is so large that it compromises structures across the midline or that a second lesion is present.

In summary, one possible localization is the upper brainstem, with involvement of the pontomesencephalic reticular formation and of the left (more so than the right) corticospinal and corticobulbar tracts. Because of the cervical collar, eye movements could not be tested easily; such testing might have helped further with localization (see **neuroexam.com Video**

35). Another possibility is a large intracranial lesion affecting the left motor cortex or descending white matter pathways and also compressing the upper brainstem–diencephalic junction through mass effect and transtentorial herniation (see KCC 5.4).

The external evidence of head trauma (scalp laceration) is most suggestive of a rapidly expanding lesion in the left cranial cavity, which could impinge on the left hemisphere corticospinal system in both the cortex and the white matter. It could also impair consciousness through midline shift, distorting the reticular formation, and through elevated intracranial pressure. The most likely causes of an expanding lesion following trauma would be epidural hematoma, acute subdural hematoma, cerebral contusion, or cerebral edema (see KCC 5.5, 5.6). Although less likely, additional possibilities for the deterioration seen in this patient would include intracranial hemorrhage from a preexisting lesion such as a tumor, vascular malformation, or aneurysm; ischemic infarction of the left cerebral hemisphere or brainstem; hydrocephalus; or delayed absorption of an ingested toxin (although the last two possibilities should not by themselves cause hemiplegia).

Oculocephalic testing

2. See Table 5.10 and Figure 5.1.

Neuroimaging

In patients of this kind, there are several serious pitfalls in making the diagnosis of intracranial pathology. First, alcohol intoxication obscures the clinical picture, making it difficult to recognize mental status changes from causes other than alcohol. Second, the restless agitation and combativeness present in this patient are frequently seen as a sign of acute worsening of intracranial hypertension or hydrocephalus (see KCC 5.3, 5.7) but can easily be mistaken for an assaultive, intoxicated personality. Therefore, intoxicated patients require extra vigilance, and if a clear trend toward neurologic improvement is not seen, an urgent CT scan of the head is warranted. Our patient had an emergency **head CT** (Image 5.2A–F, pages 177–179) and was then rushed immediately to the operating room. The images in Image 5.2A–C were taken just before surgery; those in Image 5.2D–F were taken 1 year later.

1. How old is the hemorrhage in Image 5.2A–C?

2. What kind of hemorrhages are present in this patient? (Hint: All the hemorrhages listed in Table 5.10 can be seen except for epidural hematoma.)

TABLE 5.10 Locations of Hemorrhage Associated with Head Trauma

NAME	LOCATION	POTENTIAL OR ACTUAL SPACE	BOUNDED BY
Intraventricular hemorrhage	Ventricles (CSF)	Actual	Walls of the ventricles
Contusion with intracerebral or intraparenchymal hemorrhage	Cerebral hemispheres or elsewhere in brain parenchyma	Potential	Brain tissue
Subarachnoid hemorrhage	Subarachnoid space (CSF)	Actual	Pia and arachnoid
Subdural hematoma	Subdural space	Potential	Arachnoid and dura
Epidural hematoma	Epidural space	Potential	Dura and skull
Subgaleal hematoma	Subaponeurotic space (loose areolar connective tissue)	Potential	Periosteum of the skull and galea aponeurotica

Discussion

1. All of the hemorrhage appears quite hyperdense, and on radiological grounds it is less than about 1 week old (see Chapter 4). On the basis of the clinical story, the blood is probably only a few hours old.

2. A thin, crescent-shaped hematoma on the left side extends over a large region, consistent with an acute subdural hematoma (see KCC 5.6). In addition, some blood can be seen to extend down into the sulci (see Image 5.2C). Recall that the pia follows the brain surface down into the sulci, while the arachnoid does not (see Figure 5.1). The blood in the sulci must therefore be in the subarachnoid space, representing subarachnoid hemorrhage. Large, confluent areas of hemorrhage are present in the left temporal and inferior frontal poles, consistent with cerebral contusion (see Image 5.2A,B). Note that on the right posterior scalp is an area of soft tissue swelling and subgaleal hemorrhage. One could draw an imaginary "line of force" from the right posterior scalp injury straight through to the left frontotemporal contusion. This is a classic **coup contrecoup** injury, in which a blow to one side of the head is accompanied by deceleration injury on the opposite side of the brain as it bangs against the inner surface of the skull. The frontal and temporal poles are especially susceptible to contusion where they abut the bony ridges of the anterior and middle cranial fossae (see Figure 5.21). A small amount of intraventricular blood can be seen layering the right occipital horn (see Image 5.2B), which can be better appreciated by comparison to the follow-up study (see Image 5.2E). Finally, there is an abnormally bright band of hyperdensity between the cerebellum and medial temporo-occipital lobes (see Image 5.2A). Recall that the dura forms a fold extending into the cranial vault between these structures called the tentorium cerebelli (see Figure 5.6). This is therefore an acute subdural hematoma of the tentorium cerebelli, which can be seen on other images (not shown) to be a direct extension of the subdural hematoma on the left cerebral convexity. Also not shown are the CT image bone windows that revealed a nondisplaced right occipital skull fracture.

   Severe mass effect produced by hemorrhage and edema can also be seen on this scan. The **midline shift at the level of the pineal calcification** is a good indicator of how much the reticular formation is distorted at the midbrain–diencephalic junction. Our patient has approximately 11 mm of rightward pineal shift (see Image 5.2B). More than 10 mm of shift is usually associated with profound coma. The midbrain appears elongated in the anterior to posterior dimension and somewhat flattened from side to side (see Image 5.2A). The left uncus and medial temporal lobe can be seen to extend across the region normally delineated by the tentorial edge and to press up against the midbrain. This is consistent with early left uncal transtentorial herniation (see KCC 5.4). Marked mass effect is also demonstrated by the near-complete effacement of the basal cisterns (see Image 5.2A) as compared with after recovery (see Image 5.2D). The left lateral ventricle and sulci are also nearly completely obliterated (see Image 5.2A–C). The right ventricle, in contrast, is somewhat dilated because of partial obstruction of CSF flow and consequent mild hydrocephalus. The left hemisphere appears swollen and somewhat hypodense (see Image 5.2B,C), consistent with diffuse cerebral edema.

Clinical Course

Emergency measures were instituted to lower intracranial pressure (see KCC 5.3), including intravenous administration of the hyperosmolar agent mannitol and adjustment of the ventilator settings for hyperventilation, as

CASE 5.2   ALTERED MENTAL STATUS FOLLOWING HEAD INJURY

IMAGE 5.2A–F Head CT Images Showing Acute Sub-dural Hematoma and Recovery (A–C) Acute left fron-totemporal and tentorial subdural hematoma. A left fron-totemporal contusion and subarachnoid blood are present as well. (D–F) Follow-up CT scans 1 year later.

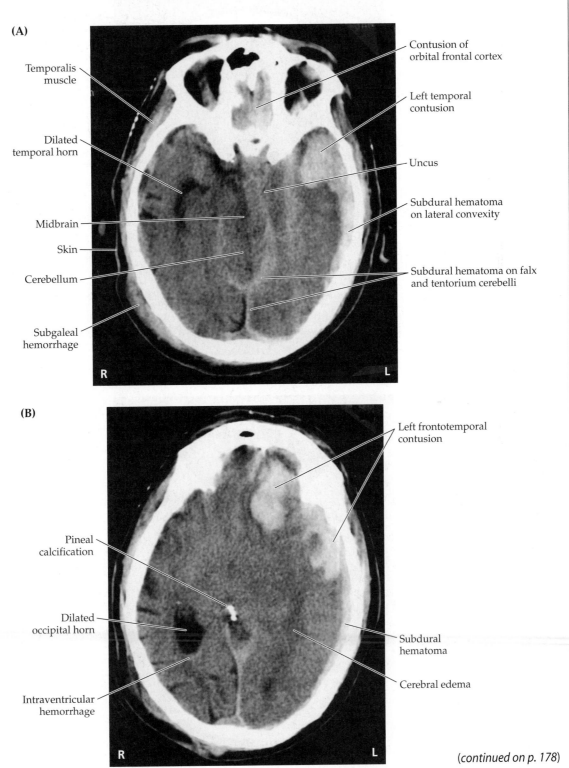

(A)

Temporalis muscle

Dilated temporal horn

Midbrain

Skin

Cerebellum

Subgaleal hemorrhage

Contusion of orbital frontal cortex

Left temporal contusion

Uncus

Subdural hematoma on lateral convexity

Subdural hematoma on falx and tentorium cerebelli

R   L

(B)

Pineal calcification

Dilated occipital horn

Intraventricular hemorrhage

Left frontotemporal contusion

Subdural hematoma

Cerebral edema

R   L

(continued on p. 178)

CASE 5.2 (*continued*)

(C)

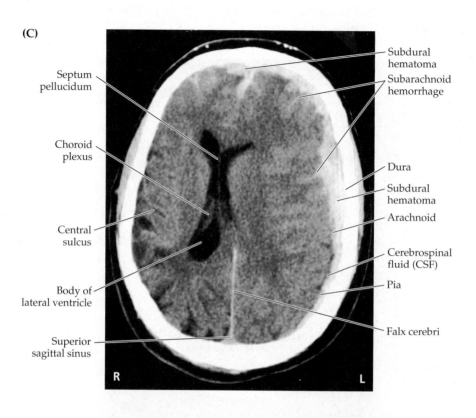

Septum pellucidum

Choroid plexus

Central sulcus

Body of lateral ventricle

Superior sagittal sinus

Subdural hematoma

Subarachnoid hemorrhage

Dura

Subdural hematoma

Arachnoid

Cerebrospinal fluid (CSF)

Pia

Falx cerebri

R   L

(D)

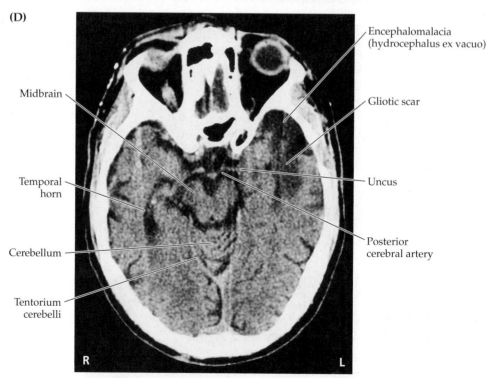

Midbrain

Temporal horn

Cerebellum

Tentorium cerebelli

Encephalomalacia (hydrocephalus ex vacuo)

Gliotic scar

Uncus

Posterior cerebral artery

R   L

CASE 5.2 (continued)

(E)

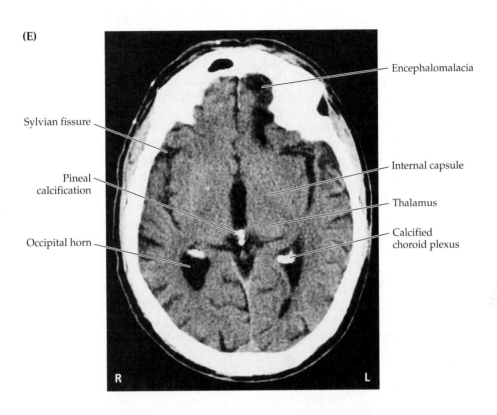

Sylvian fissure

Pineal
calcification

Occipital horn

Encephalomalacia

Internal capsule

Thalamus

Calcified
choroid plexus

(F)

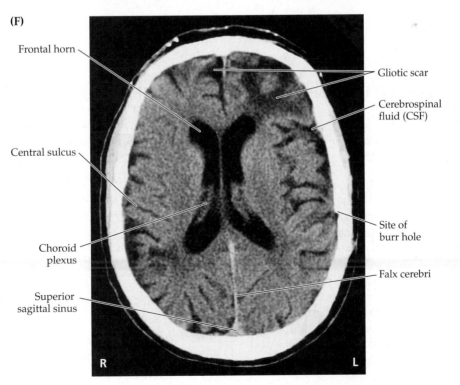

Frontal horn

Central sulcus

Choroid
plexus

Superior
sagittal sinus

Gliotic scar

Cerebrospinal
fluid (CSF)

Site of
burr hole

Falx cerebri

the patient was taken directly to the operating room. During surgery an incision was made in the left scalp, and a large **bone flap** (see KCC 5.11) was removed, exposing the underlying dura, which appeared tense and blue. The dura was opened, revealing a large, freshly clotted subdural hematoma, which was then evacuated. Findings also included significant subarachnoid blood, swelling, and contusions of the left temporal and frontal poles. To further decompress the intracranial cavity, the severely contused portions of the anterior left temporal lobe as well as the anterior inferior left frontal lobe, were removed, and a large left frontal intraparenchymal hematoma was evacuated. The dura was closed with sutures, the bone flap was replaced, and the scalp closed.

In the intensive care unit, 3 hours after surgery, the patient remained intubated but was awake, with full extraocular movements and equal and reactive pupils. He was able to squeeze the examiner's hand or wiggle his toes bilaterally on verbal command, with greater strength on the left side. His left plantar reflex was downgoing and right plantar silent. After a prolonged recovery period in the hospital and at a rehabilitation facility, the patient returned home, was able to function fully independently, and soon found his way back to the neighborhood bar. Follow-up head CT scan done 1 year later during another episode of intoxication (see Image 5.2D–F) showed complete resolution of all hemorrhage and mass effect, with residual hypodensities present in the region of the left temporal and frontal poles representing glial scar and tissue loss (hydrocephalus ex vacuo). This case is somewhat unusual in showing such dramatic recovery despite the severe appearance of the patient's preoperative CT scan with multiple contusions, acute subdural blood, and marked distortion of the brainstem. He was fortunate to have had this acute deterioration occur in a place where emergency neurosurgery could be undertaken without delay but unfortunate to have a continued addiction to alcohol.

CASE 5.3 DELAYED UNRESPONSIVENESS AFTER HEAD INJURY

CHIEF COMPLAINT

A 51-year-old man was found to be progressively unresponsive the morning after a head injury.

HISTORY

The night before admission, the patient fell down a flight of cement stairs at 12:00 A.M., following a domestic altercation. He struck his left temporal area and lost consciousness for about 15 minutes. By the time the police and ambulance arrived, however, he was fully awake, smelled of alcohol, and refused medical treatment. He was arrested on domestic violence charges and spent the night in prison. In the morning the guards came to summon him for a court appearance and found him **difficult to arouse, thrashing about incoherently.** He had vomited and defecated in the cell overnight. An ambulance was called, and he was brought to the emergency room for evaluation.

PHYSICAL EXAMINATION

General appearance: Disheveled-appearing man lying on a stretcher.

Vital signs: T = 98°F, P = 96, BP = 150/100, R = 28.

Head: Left forehead abrasion. No Battle's sign, raccoon eyes, CSF otorrhea, or rhinorrhea. Tympanic membranes normal (see Table 3.9).

Neck: In cervical stabilization collar put on by emergency personnel.

Lungs: Clear.

Heart: Regular rate with no murmurs, gallops, or rubs.

Abdomen: Normal.

Extremities: Normal.

Rectal: Normal tone, heme negative.

CASE 5.3 DELAYED UNRESPONSIVENESS AFTER HEAD INJURY (continued)

Neurologic exam:

MENTAL STATUS: **Unresponsive** to commands. Not speaking. Occasionally thrashed on stretcher in **agitated**, semipurposeful fashion.

CRANIAL NERVES: **Left pupil 5 mm, fixed** (no response to light). Right pupil 2 mm, constricting to 1 mm in response to light. Oculocephalic maneuvers not done because of cervical collar. Gag reflex present.

SENSORY AND MOTOR: Left arm and left leg moved spontaneously. Withdrew left arm and leg purposefully from painful stimulation. **Right arm and leg did not move**, even in response to pain.

REFLEXES:

COORDINATION AND GAIT: Not tested.

Shortly after arriving in the emergency room the patient developed respiratory distress and was therefore intubated. He subsequently stopped moving his left side as well and became completely unresponsive.

LOCALIZATION AND DIFFERENTIAL DIAGNOSIS

1. For each of the symptoms and signs appearing in **bold** above, review the neuroanatomical pathways involved (see Figures 2.16, 2.23; Tables 2.5, 3.3). In what single brain region do these pathways intersect? Which herniation syndrome (see KCC 5.4) will cause compression of this brain region? On which side would you expect to find the cause of the compression?

2. Given the history of acute head injury with delayed and progressive unresponsiveness over the course of several hours, what is the most likely diagnosis? What are some other possibilities?

Discussion

1. The key symptoms and signs in this case are:

   - **Agitation and decreased level of consciousness**
   - **Left pupil fixed and dilated**
   - **Right hemiplegia, right hyperreflexia, and right Babinski's sign**

   Impairment of consciousness could be caused by lesions affecting the brainstem–diencephalic reticular activating systems (either directly, or through compression and distortion from nearby lesions; see Figure 2.23) or by lesions affecting the cerebral hemispheres bilaterally. Pupillary constriction is mediated by parasympathetic fibers that travel with the oculomotor nerve (CN III) arising from the midbrain (see Figures 2.9, 2.22; Table 2.5). Therefore, a fixed, dilated left pupil could be caused by a lesion involving the left midbrain, CN III as it exits the brainstem and travels toward the eye, or the pupillary constrictor muscle itself. The combination of right-sided weakness and increased reflexes suggests an upper motor neuron lesion (see Table 3.3) somewhere in the corticospinal pathway beginning in the left motor cortex and ending in the right spinal cord (see Figure 2.16). These three pathways or systems intersect in the midbrain, producing the triad of coma, a "blown" pupil, and hemiplegia seen in uncal herniation (see KCC 5.4; Figure 5.18). The dilated pupil on the left suggests that the lesion is on the left side of the intracranial cavity, compressing the left midbrain and left CN III.

   The most likely *clinical localization* is left uncal herniation compressing the left midbrain.

2. The patient suffered a significant blow to the left temporal area. Recall that fracture of the temporal bone can lacerate the middle meningeal artery, resulting in epidural hematoma (see Figure 5.7; KCC 5.6). In addition, as can

sometimes be seen with epidural hematomas, the patient initially regained consciousness during a **lucid interval** lasting several hours and only later lapsed back into unconsciousness, probably because of progressive expansion of the hematoma. Recall that epidural hematomas are arterial in origin and therefore can cause rapid deterioration due to expansion. We have already seen in Case 5.2 that acute subdural hematoma, brain contusion, and edema can also sometimes cause progressive deterioration, so these diagnoses should be considered here as well. Finally, it is possible but less likely that during the night the patient had a cerebral infarct, perhaps due to posttraumatic vertebral artery dissection (see KCC 10.6), or that he had an intracerebral hemorrhage caused by something not directly related to the head injury, such as hypertension, brain tumor, or vascular malformation (see KCC 5.6).

Neuroimaging

Because of the change in this patient's clinical status, an urgent **head CT** was done (Image 5.3A–D, pages 184–185).

1. How recent is the hemorrhage seen on the CT scan (see Chapter 4)?

2. On the basis of its appearance, what kind of hemorrhage is this, and where is it located (see KCC 5.6)?

3. Identify the fracture on the bone windows (see Image 5.3D). What bone does it involve?

4. Describe the mass effect (see KCC 5.2) caused by the hematoma and any herniation (see KCC 5.4) that can be identified on the CT.

Discussion

1. The hemorrhage is hyperdense and therefore occurred within the past week (see Chapter 4).

2. A large, lens-shaped fluid collection is seen along the inner surface of the left temporal and parietal bones. Note that this hematoma has a biconvex shape and is limited anteriorly by the coronal suture line (see Image 5.3B,C) where the periosteal layer of dura forms a tight insertion. These features are characteristic of an epidural hematoma formed by high-pressure arterial blood dissecting between the dura and bone (see KCC 5.6).

3. A fracture of the left temporal bone can be seen (see Image 5.3D). On higher cuts (not shown), the fracture was seen to extend into the left parietal bone as well.

4. There is extensive midline shift caused by the hematoma, with displacement of the entire brain to the right, compressing the left lateral ventricle (see Image 5.3A,B). The left anteromedial temporal lobe, including the uncus, extends over the tentorial edge and compresses the left side of the midbrain (see Image 5.3A), constituting left transtentorial herniation (see KCC 5.4). In addition, the cingulate gyrus is forced under the falx (see Image 5.3B), causing subfalcine herniation (see Figure 5.18).

Clinical Course

Because of his rapid decline, the patient was taken immediately from the CT scanner to the operating room. Emergency measures to lower intracranial pressure (see KCC 5.3; Table 5.3), including hyperventilation and intravenous mannitol, were initiated en route. In the operating room, an incision was made from just above the left ear up to the crest of the head, revealing a large linear fracture in the temporal and parietal bones. A craniotomy was performed (see KCC 5.11), and a large fresh clot of blood was evacuated from the epidural space. Several tears in the middle meningeal artery were identified (see Figure 5.7), so the artery was coagulated down to the skull base. The patient had a prolonged hospital course postoperatively, but he gradually regained consciousness and the ability to ambulate and use his right hand. A follow-up CT scan performed a week and a half after the injury showed remarkable recovery of the normal intracranial anatomy following removal of the hematoma. This case demonstrates the importance of being able to recognize the neuroanatomical features of uncal herniation so that therapeutic measures can be initiated promptly, while recovery is still possible.

CASE 5.4 HEADACHE AND PROGRESSIVE LEFT-SIDED WEAKNESS

MINICASE
A 52-year-old lumber executive developed mild difficulty running, initially stubbing his left big toe, which progressed to constant left leg weakness over the course of 6 months. He also complained of **headaches**. Examination was normal except for **decreased left nasolabial fold** and **4⁺/5 weakness in the left triceps and left leg**.

LOCALIZATION AND DIFFERENTIAL DIAGNOSIS
1. On the basis of the limited information provided, is the lesion intracranial or extracranial (see Figures 2.13, 2.16)? Which side is the lesion on?
2. What is the most likely diagnosis? What are some other possibilities?

Discussion

1. The key symptoms and signs in this case are:

   - **Headaches**
   - **Left hemiparesis**

   Weakness affecting the left face, arm, and leg together is caused by lesions of the corticobulbar (face) and corticospinal (arm and leg) pathways originating in the right motor cortex (see Figures 2.13, 2.16). The lesion must be at or above the level of the pons, since the facial nerve (CN VII) nucleus is in the pons and exits the brainstem at the pontomedullary junction (see Figure 2.22; Table 2.5); lesions below this point would not cause face weakness. The presence of headaches further supports an intracranial localization.

   The most likely *clinical localization* is a right intracranial lesion affecting the corticobulbar and corticospinal tracts at or above the level of the pons.

2. Headaches can have many causes (see KCC 5.1). New-onset headaches for a man in his 50s, together with a progressive neurologic deficit over a 6-month period, suggests brain tumor (see KCC 5.8), most commonly meningioma, glioblastoma multiforme, or metastasis. Other, less likely possibilities include infection, demyelination, multiple small hemorrhages or infarcts, or vasculitis.

Clinical Course

An MRI scan was consistent with the primary brain tumor called glioblastoma multiforme (see KCC 5.8) involving mainly the right hemisphere. The diagnosis was confirmed by biopsy. The patient was treated with surgical resection of as much tumor as possible, followed by radiation therapy*. Nev-

---

*Today, chemotherapy with agents such as temodar is typically used in combination with surgery and radiation therapy for glioblastoma. This has successfully prolonged life expectancy by a few months, but the overall prognosis unfortunately remains poor.

CASE 5.3 DELAYED UNRESPONSIVENESS AFTER HEAD INJURY

IMAGE 5.3A–D Head CT Showing Epidural Hematoma (A–C) Left epidural hematoma causing left uncal transtentorial herniation. Axial images progressing from inferior to superior. (D) Bone windows demonstrating left temporal bone fracture.

(A)

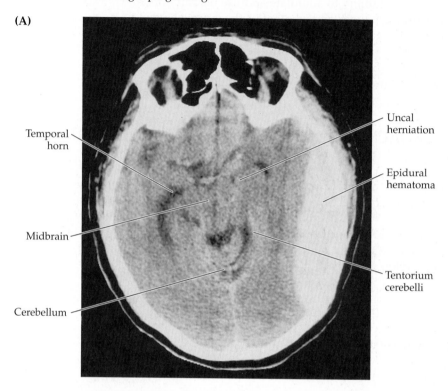

Temporal horn

Midbrain

Cerebellum

Uncal herniation

Epidural hematoma

Tentorium cerebelli

(B)

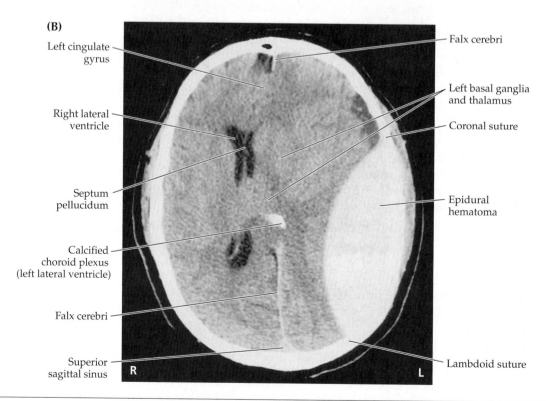

Left cingulate gyrus

Right lateral ventricle

Septum pellucidum

Calcified choroid plexus (left lateral ventricle)

Falx cerebri

Superior sagittal sinus

Falx cerebri

Left basal ganglia and thalamus

Coronal suture

Epidural hematoma

Lambdoid suture

R   L

CASE 5.3 (*continued*)

(C)

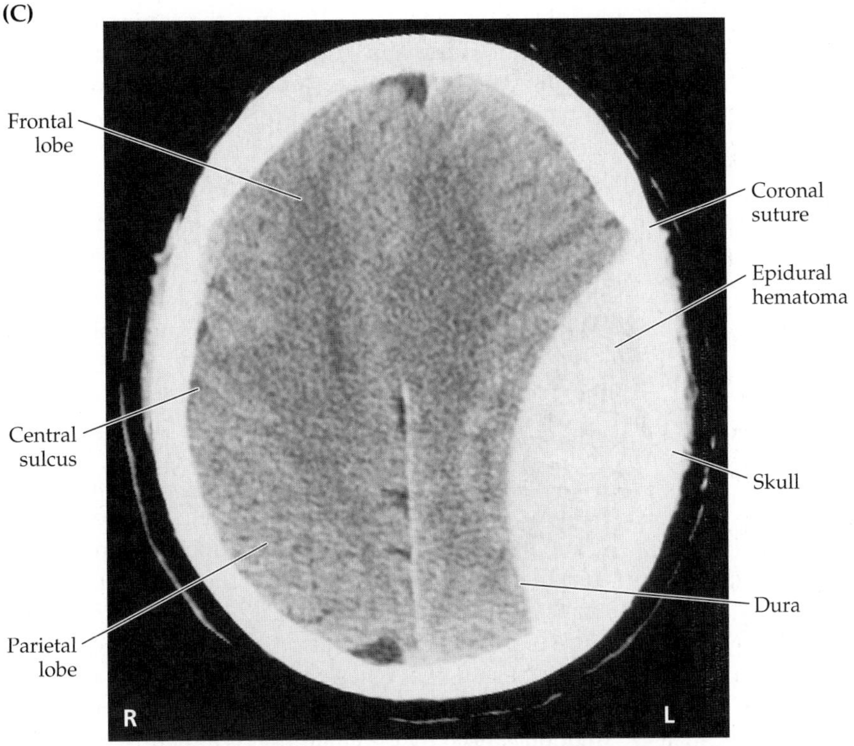

Frontal lobe

Coronal suture

Epidural hematoma

Central sulcus

Skull

Parietal lobe

Dura

R   L

(D)

Left temporal bone fracture

R   L

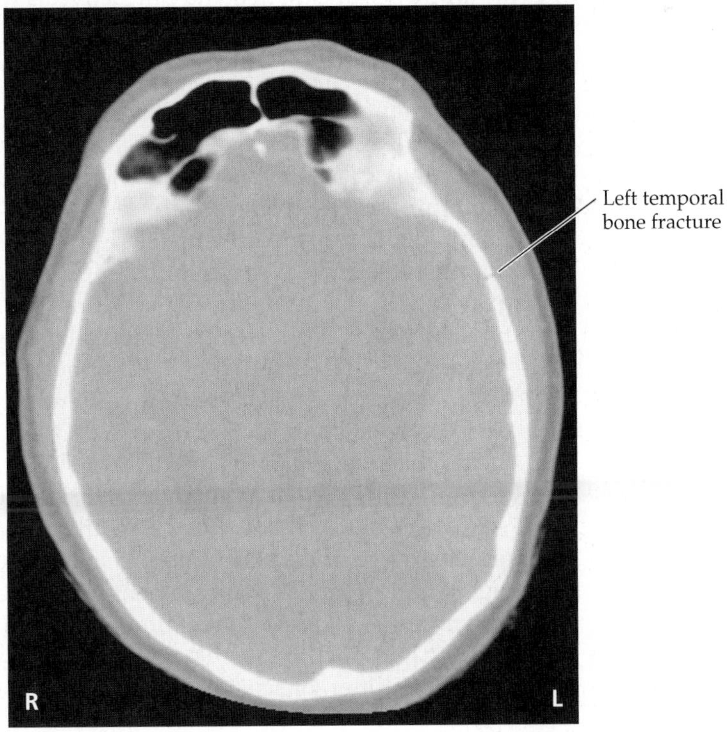

ertheless, about 1 year after his first symptoms he developed gradually worsening headaches, lethargy, and left-sided weakness. MRI revealed recurrent tumor. He was treated with high-dose steroids (dexamethasone) and intravenous mannitol (see KCC 5.3) but continued to worsen. On the day before he died, at 3:00 P.M. he was described as sleepy but arousable and "oriented × 3." Pupils were 4 mm and briskly responsive bilaterally. He was hemiplegic on the left. At 10:00 P.M. the same day, he was unresponsive even to deep pain. His right pupil was 7 mm and unreactive, left pupil 4 mm. A little later the pupils were 7 mm bilaterally and fixed. The dexamethasone and mannitol were increased further, with no improvement. Given his untreatable condition, the family requested no resuscitation measures, and the patient stopped breathing the next day.

1. Impaired function in which brainstem region could explain the progressive hemiplegia, lapse in consciousness, and pupillary abnormalities seen just before death in this patient?

2. What herniation syndrome causes compression of this brain region?

Discussion

1. The midbrain (see Figures 2.16, 2.22A, 2.23; Tables 2.5, 3.3).

2. This patient had a long-standing left hemiparesis caused by a right hemispheric mass lesion, so it is difficult to invoke a herniation syndrome to explain the hemiplegia. However, impaired consciousness with a dilated, unreactive pupil is characteristic of the midbrain compression seen with uncal transtentorial herniation (see Figure 5.6; KCC 5.4). Elevated intracranial pressure (see KCC 5.3) may also have contributed to the impaired level of consciousness. Thus, when the patient became unresponsive, with a dilated, unreactive right pupil, he was probably undergoing right transtentorial herniation caused by the enlarging right hemisphere mass. Eventually, the midbrain was compromised bilaterally, as evidenced by the fact that the left pupil became dilated and unreactive as well.

Pathology

The patient's family requested an autopsy to confirm the cause of death (Image 5.4A–D, pages 188–189). The brain weighed 1420 g (normal weight is 1250–1400 g) and appeared swollen and edematous. On surface examination, prominent grooves were seen on the inferior medial surfaces of the temporal lobes, especially on the right side, located 1.0 cm from the uncal tip on the right and 0.4 cm from the uncal tip on the left (see Image 5.4A), consistent with bilateral uncal herniation, more severe on the right side. The midbrain appeared crowded at this level and deformed from right to left (see Image 5.4B). The right CN III appeared flattened for about 1.0 cm of its extent in the area adjacent to the right uncus (see Image 5.4A).

Coronal sections (see Image 5.4C,D) revealed a necrotic mass centered near the leg region of the motor strip in the right hemisphere, with mild pallor and dramatic enlargement of the right hemisphere white matter consistent with edema. The gyri appeared flattened from being pressed against the inner surface of the skull, and the sulci were effaced. The right uncus could again be seen to have a prominent groove caused by the tentorial edge, and the midbrain–diencephalic junction appeared deformed and compressed (see Image 5.4C). The right cingulate gyrus was shifted 1.0 cm from right to left of midline, consistent with subfalcine herniation. A brownish area was present in the right medial occipital lobe, involving gray and white matter surrounding the calcarine fissure (see Image 5.4D). This is consistent with an ischemic infarct in the territory of the right posterior cerebral artery, with subsequent petechial hemorrhage. Recall that the posterior cerebral artery passes through the tentorial notch and can therefore sometimes be compressed during transtentorial herniation, causing infarcts (see Figures 5.6, 10.5; KCC 5.4).

Transverse section through the midbrain (see Image 5.4B) showed it to be markedly elongated in the anterior to posterior dimension and distorted by compression from right to left. There was an irregular area of dark brown pigmentation in the center of the midbrain. This finding is called a **Duret–Bernard hemorrhage** and can be seen with severe compression of the midbrain and other brainstem areas during transtentorial herniation.

CASE 5.5 SUDDEN COMA AND BILATERAL POSTURING DURING INTRAVENOUS ANTICOAGULATION

MINICASE

A 61-year-old woman presented to the emergency room with 2 hours of left face and arm weakness. On examination at 8:00 A.M. she was fully alert, with mild left face and arm weakness completely sparing the left leg. She had a history of atrial fibrillation, which can cause blood clots to form in the cardiac atria as a result of stasis (see KCC 10.4). For this reason, she was taking the oral anticoagulant warfarin (Coumadin). However, blood tests in the emergency room showed that she was not adequately anticoagulated on her current dosage. It was felt that she most likely had an embolus from the heart to the right side of the brain. A head CT within 3 hours of onset was unremarkable. This finding was consistent with the diagnosis, since it can take from 6 to 24 hours for an acute stroke to become visible on CT (see Chapter 4). She was therefore admitted to the hospital and started on the intravenous anticoagulant heparin in order to rapidly achieve a therapeutic range of anticoagulation (note that today this patient would also have been a candidate for acute tPA administration; see KCC 10.4). The patient remained stable throughout the day, but at 10:00 P.M. she was suddenly found to be **unarousable**. Exam was notable for **fixed, midsized pupils, no extraocular movements, and bilateral extensor (decerebrate) posturing of both arms and legs**. She was emergently intubated because of shallow respirations.

LOCALIZATION AND DIFFERENTIAL DIAGNOSIS

1. On the basis of the findings shown in **bold** above, where is the lesion located (see Figures 2.16, 2.22, 2.23, 3.5; Table 2.5)?
2. What is the most likely diagnosis?

Discussion

1. The key symptoms and signs in this case are:
   - **Unarousable**
   - **Bilateral fixed pupils and absent eye movements**
   - **Bilateral extensor posturing**

   Coma, together with absent eye movements and absent pupillary reflexes, suggests severe brainstem dysfunction involving the ascending activating systems (see Figure 2.23) as well as CNs III, IV, and VI (see Figure 2.22; Table 2.5). Bilateral extensor posturing signifies that the corticospinal tracts are involved bilaterally (see Figure 2.16) but that some brainstem function remains, to enable this abnormal reflex (see Figure 3.5B).

2. Given the recent increase in the patient's anticoagulation therapy, the most likely diagnosis is hemorrhage. The hemorrhage could have occurred in the brainstem itself, or a large hemorrhage above the brainstem could be compressing the brainstem bilaterally through downward herniation (see KCC

CASE 5.4 HEADACHE AND PROGRESSIVE LEFT-SIDED WEAKNESS

IMAGE 5.4A–D Pathologic Specimens Showing Herniations and PCA Hemorrhagic Infarcts
(A) Inferior surface view showing effects of uncal transtentorial herniation on CN III. Herniation is more severe on the right side. (B) Transverse section revealing distortion of the midbrain and Duret–Bernard hemor-rhages. (C) Coronal section showing evidence of uncal and subfalcine herniation. Necrotic tumor mass can be seen in the right hemisphere. (D) Coronal sections through occipital lobes with posterior cerebral artery (PCA) territory hemorrhagic infarct caused by compression of the PCA in the tentorial notch.

(A)

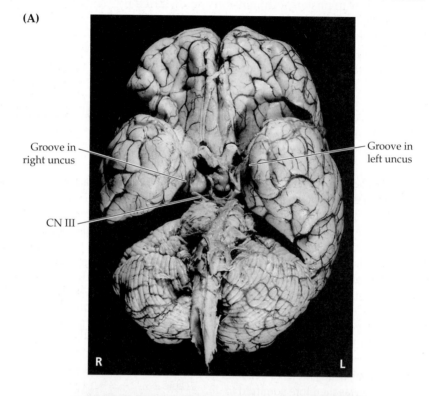

Groove in right uncus

Groove in left uncus

CN III

R   L

(B)

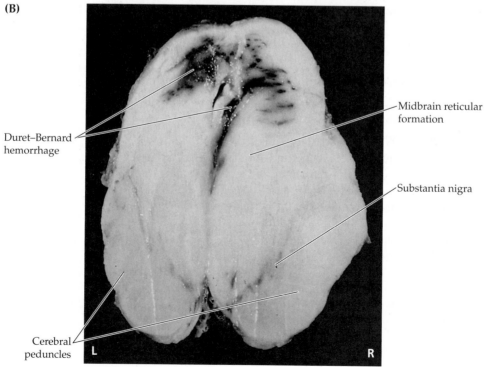

Duret–Bernard hemorrhage

Midbrain reticular formation

Substantia nigra

Cerebral peduncles

L   R

CASE 5.4 (continued)

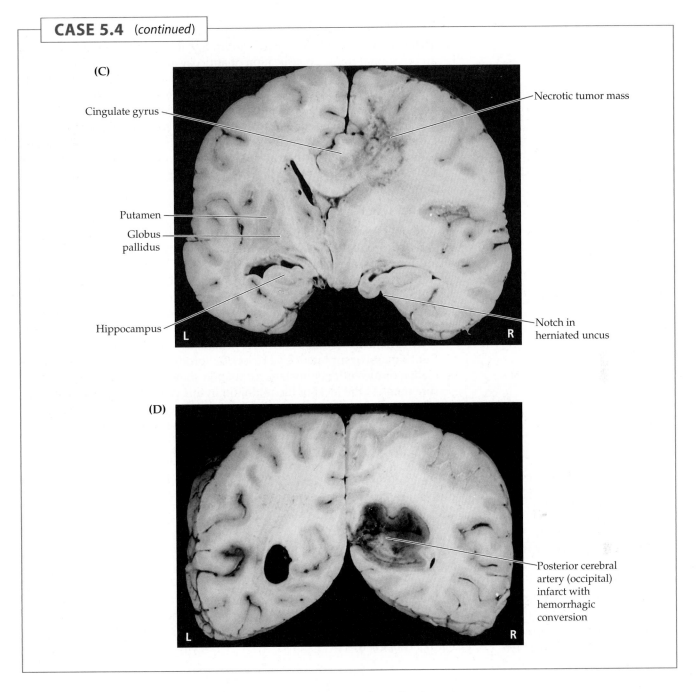

(C)

Cingulate gyrus

Necrotic tumor mass

Putamen

Globus pallidus

Hippocampus

L

R

Notch in herniated uncus

(D)

L

R

Posterior cerebral artery (occipital) infarct with hemorrhagic conversion

5.4; Figure 5.18). Another possibility, given her history of atrial fibrillation and recent possible embolic events, is a large brainstem infarct.

Neuroimaging

After intubation, the patient was taken for an emergency **head CT** (Image 5.5A–H, pages 191–194). The initial CT on admission is shown in Image 5.5A,C,G,E. Repeat CT after her change in status is shown in Image 5.5B,D,F,H.

1. What kind of hemorrhage is seen on these CT scans (see KCC 5.6)?

2. Identify the foramen magnum, medulla, and cerebellar tonsils (see Image 5.5A–D), as well as the midbrain and uncus bilaterally (see Image 5.5E,F).

3. What kinds of herniation occurred in this patient (see KCC 5.4; Figure 5.18)?

Discussion

1. This patient has had a massive hemorrhage in the region of initial infarct, an occasional tragic complication of anticoagulation therapy. The hemorrhage is centered deep in the right hemisphere but is so large that essentially the whole brain is compressed. The blood has extended into the ventricular system, and several pockets of blood with CSF–blood levels can be seen (see Image 5.5H). The cortical sulci and gyri are completely effaced by mass effect, the left ventricle appears dilated from obstructive hydrocephalus, and there is substantial midline shift. On the basis of the large volume of the bleed and the patient's poor level of neurologic function, her prognosis is quite poor.

2. See labels on Image 5.5A–H.

3. Note that before the hemorrhage, there was a large amount of CSF-filled space surrounding the cervicomedullary junction and caudal medulla at the level of the foramen magnum (see Image 5.5A,C). The cerebellar tonsils can be seen to extend into the cisterna magna at a distance somewhat removed from the medulla. After the hemorrhage, the cerebellar tonsils have been forced downward and inward, compressing the caudal medulla and extending downward through the foramen magnum (see Image 5.5B,D). Thus, **tonsillar herniation** has occurred. In this patient, tonsillar herniation has resulted from a massive supratentorial lesion that produced downward **central herniation** of the entire brainstem, a situation sometimes referred to as a **pressure cone**. At the level of the midbrain in this patient, there was also complete effacement of the basal cisterns and bilateral midbrain compression due to **bilateral uncal transtentorial herniation** (see Image 5.5E,F).

Clinical Course

In view of the patient's grim prognosis, the family decided not to pursue heroic measures, and the patient died the next day.

CASE 5.6 SEVERE HEAD INJURY

MINICASE

An 80-year-old man was found lying on the rocks under a 6-foot-high wall near the shoreline. He was conscious and speaking for a brief time before lapsing into a coma. On initial examination he had a right scalp abrasion; 6 mm right pupil and 5 mm left pupil, both nonreactive to light; no corneal reflexes; flexor (decorticate) posturing of the upper extremities to pain (see Figure 3.5A); and upgoing plantar responses bilaterally.

Neuroimaging

The patient was brought to the emergency room, where an emergency **head CT** was done, shown in Image 5.6A,B, page 196.

1. What kind of hemorrhage is present (see KCC 5.6)?

2. Severe herniation of what kind can be seen (see KCC 5.4; Figure 5.18)?

Discussion

1. A large, crescent-shaped, hyperdense fluid collection is seen between the right hemisphere and the skull, consistent with acute subdural hematoma. Note that some areas of decreased density are present toward the top of the

hematoma, likely representing CSF (darker areas) and nonclotted blood or blood mixed with CSF (gray areas). Small amounts of blood can also be seen extending into the sulci, consistent with subarachnoid hemorrhage. There is also a small amount of blood in the right lateral ventricle (see Image 5.6A).

2. Marked subfalcine herniation is present, with extrusion of a substantial amount of the right hemisphere under the falx. In addition to the findings already described, lower images on the CT demonstrated right transtentorial herniation and a right frontotemporal fracture.

CASE 5.5 SUDDEN COMA AND BILATERAL POSTURING DURING INTRAVENOUS ANTICOAGULATION

IMAGE 5.5A–H CT Scan Images Showing Intracranial Hemorrhage Axial CT images progressing from inferior to superior. Baseline scans (A, C, E, G) compared with scans showing catastrophic intracranial hemorrhage with tonsillar and bilateral uncal herniation (B, D, F, H).

(A) Baseline scan

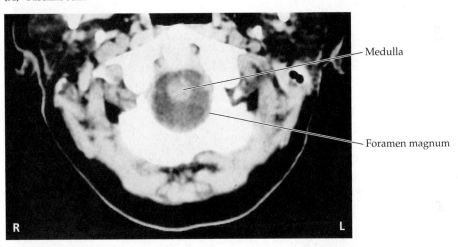

Medulla

Foramen magnum

(B) After hemorrhage

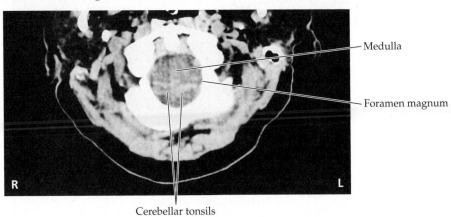

Medulla

Foramen magnum

Cerebellar tonsils

(*continued on p. 192*)

CASE 5.5 (*continued*)

(C) Baseline scan

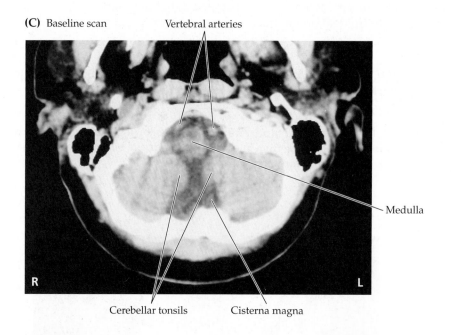

Vertebral arteries

Medulla

Cerebellar tonsils   Cisterna magna

(D) After hemorrhage

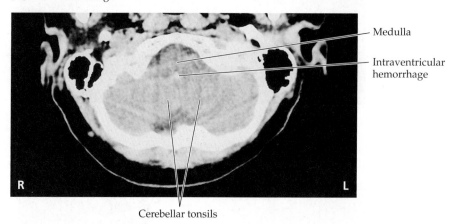

Medulla

Intraventricular hemorrhage

Cerebellar tonsils

CASE 5.5 (*continued*)

(E) Baseline scan

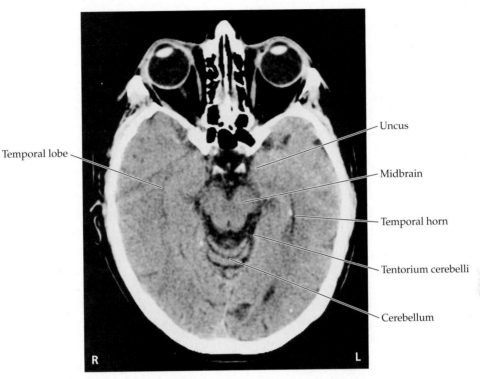

Temporal lobe

Uncus

Midbrain

Temporal horn

Tentorium cerebelli

Cerebellum

R L

(F) After hemorrhage

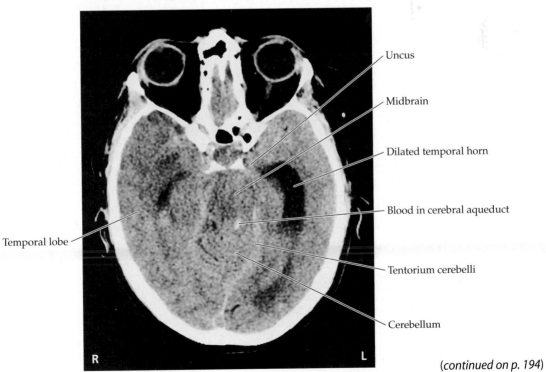

Uncus

Midbrain

Dilated temporal horn

Blood in cerebral aqueduct

Temporal lobe

Tentorium cerebelli

Cerebellum

R L

(*continued on p. 194*)

CASE 5.5 (*continued*)

(G) Baseline scan

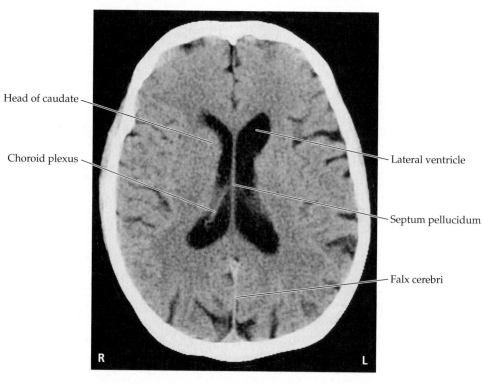

Head of caudate

Choroid plexus

Lateral ventricle

Septum pellucidum

Falx cerebri

R   L

(H) After hemorrhage

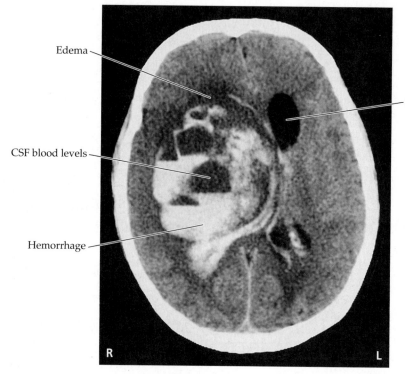

Edema

Dilated lateral ventricle

CSF blood levels

Hemorrhage

R   L

Clinical Course

The patient was taken immediately to the operating room, and the subdural hematoma was evacuated. Unfortunately, he did not improve postoperatively. A follow-up CT scan was done the next day (see Image 5.6B).

The subdural hematoma was no longer present, and there was no longer subfalcine herniation or midline shift. However, there were now two strips of hypodensity located on either side of the falx. These represent ischemic infarctions in the territory of the anterior cerebral arteries bilaterally (see Figure 10.5). Some spots of hemorrhage were located within these infarcts, suggesting hypoperfusion causing infarction, followed by reperfusion with hemorrhage into necrotic areas. The anterior cerebral arteries (see Figure 2.26C) probably were trapped under the falx by the severe degree of subfalcine herniation, leading to infarction and, later, reperfusion injury once the herniation was relieved.

The patient continued to decline medically and neurologically over the following days and eventually died 8 days after admission.

CASE 5.7 A CHILD WITH HEADACHES, NAUSEA, AND DIPLOPIA

CHIEF COMPLAINT

An 11-year-old girl was brought to the pediatrician's office because of worsening headaches, nausea, and diplopia during the past week.

HISTORY

The patient was healthy until 1 week ago, when she developed persistent bifrontal **headaches and nausea**. Both symptoms **gradually worsened**, and for the past 2 days she has had multiple episodes of **vomiting**. Four or 5 days ago she also noticed **horizontal diplopia when looking to the left**. She denied any other symptoms. Her birth and developmental history were unremarkable, according to the parents. At the time of her appointment she was a sixth-grade honor student.

PHYSICAL EXAMINATION

Vital signs: T = 98.8°F, P = 76, BP = 120/68, R = 16.

Head circumference: 54 cm (75th percentile for age).

Neck: Supple with no bruits.

Lungs: Clear.

Heart: Regular rate with no murmurs, gallops, or rubs.

Abdomen: Normal bowel sounds; soft, nontender.

Extremities: Normal.

Neurologic exam:

  MENTAL STATUS: Alert and oriented × 3. Normal language and praxis.

  CRANIAL NERVES: Pupils 5 mm, constricting to 3 mm with light bilaterally. Ophthalmoscopic exam: **Bilateral papilledema** (see Figure 5.17). Extraocular movements: **Left eye did not fully abduct on left lateral gaze.** Otherwise, intact horizontal and vertical eye movements and intact convergence. Visual fields full. Facial sensation intact. Intact corneal reflex bilaterally. Face symmetric. Hearing normal to finger rub. Normal gag and normal palate elevation. Normal sternomastoid strength. Tongue midline.

  MOTOR: No pronator drift. Normal tone. Power 5/5 throughout.

  SENSORY: Intact light touch, pinprick, joint position sense, and vibration sense.

  REFLEXES:

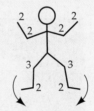

  COORDINATION: Normal.

  GAIT: Normal.

LOCALIZATION AND DIFFERENTIAL DIAGNOSIS

1. Horizontal diplopia with incomplete abduction of the left eye on left lateral gaze can be caused by dysfunction of which cranial nerve or extraocular muscle (see Table 2.5; Figure 2.22)?

2. Headaches, nausea, vomiting, papilledema, and horizontal diplopia are signs of what disorder of the intracranial space (see KCC 5.3)? Given this patient's age and the gradually worsening symptoms over the course of a week, what are some possibilities for the diagnosis (see KCC 5.1, 5.7–5.9)?

CASE 5.6 SEVERE HEAD INJURY

IMAGE 5.6A,B CT Scan Images Showing Subfalcine Herniation and Anterior Cerebral Artery Infarcts (A) Subfalcine herniation caused by acute right subdural hematoma. (B) Follow-up scan 1 day later (following surgical removal of the hematoma), showing bilateral anterior cerebral artery infarcts.

(A)

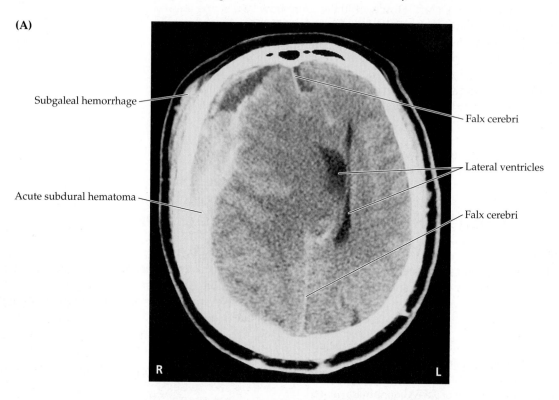

Subgaleal hemorrhage

Falx cerebri

Lateral ventricles

Acute subdural hematoma

Falx cerebri

R   L

(B)

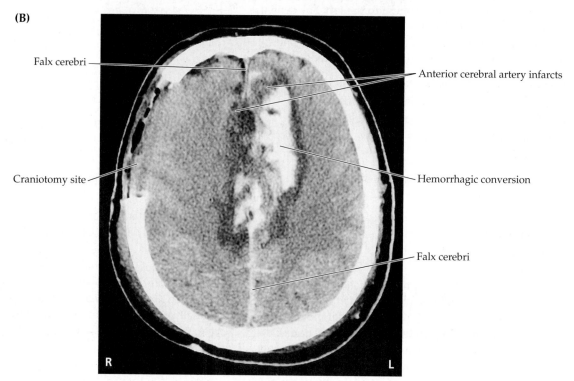

Falx cerebri

Anterior cerebral artery infarcts

Craniotomy site

Hemorrhagic conversion

Falx cerebri

R   L

Discussion

1. The key symptoms and signs in this case are:
   - **Headaches, nausea, and papilledema**
   - **Horizontal diplopia and incomplete abduction of the left eye when looking to the left**

   Dysfunction of the left abducens nerve (CN VI) or the left lateral rectus muscle could cause incomplete abduction of the left eye on left lateral gaze, and horizontal diplopia (see KCC 13.4).

2. This patient has worrisome signs of elevated intracranial pressure (see KCC 5.3). Interestingly, the abducens palsy produced by mildly elevated intracranial pressure can be unilateral (on either side) and may become bilateral with more severe pressure elevations.

   The progressively worsening signs of elevated intracranial pressure in this patient could be caused by a mass lesion such as a brain tumor (see KCC 5.8), hydrocephalus (see KCC 5.7), or idiopathic intracranial hypertension (pseudotumor cerebri; see KCC 5.1). Other, less likely possibilities include a slowly developing intracranial infection (see KCC 5.9) or perhaps a coagulation disorder causing sagittal sinus thrombosis (see Chapter 10).

Neuroimaging

On the basis of the patient's symptoms and signs, the pediatrician decided to admit her to the hospital and ordered an **MRI scan** to be performed the same day (Image 5.7A–C, pages 199–200).

1. Are these T1-, T2-, or proton density–weighted images (see Figure 4.6)? Are they horizontal, coronal, or sagittal (see Figure 2.5)?

2. Cover the labels in Images 5.7A and B and identify as many structures as possible. In particular, review the pathway of CSF flow (see Figures 5.10, 5.11) from synthesis to reabsorption by identifying the following: choroid plexus, occipital and frontal horns of the lateral ventricles, region of foramen of Monro, third ventricle, cerebral aqueduct (of Sylvius), fourth ventricle, cisterna magna, foramen magnum, and finally, superior sagittal sinus.

3. Gadolinium was given intravenously for the image shown in Image 5.7B but not for that shown in Image 5.7A. Describe the location of the mass lesion. Does the mass enhance with gadolinium?

4. What is the cause of the elevated intracranial pressure (see KCC 5.3) in this patient? Which ventricles appear dilated, and why?

Discussion

1. The TR (repetition time) and TE (echo time) are both relatively short, so these are T1-weighted images (see Chapter 4). Recall that in T1-weighted images CSF appears dark, and white matter appears relatively bright compared to gray matter, thus resembling a true brain section. Image 5.7A is a sagittal section near the midline; Images 5.7B and C are horizontal images.

2. See labels on Images 5.7A and B. Compare to Figures 4.13 and 4.15.

3. A roundish mass lesion approximately 2 cm in diameter can be seen in the posterior third ventricle, located between the thalami (see Image 5.7B) and extending slightly into the rostral midbrain (see Image 5.7A). The mass appears bright in Image 5.7B but dark in Image 5.7A, demonstrating enhancement with gadolinium. Recall that enhancement (see Chapter 4) sug-

gests increased vascularity, or breakdown of the blood–brain barrier that can be seen with inflammation, tissue damage, or tumors.

4. This patient has elevated intracranial pressure caused by hydrocephalus (see KCC 5.7). The mass lesion bulges into the posterior portion of the third ventricle and blocks the **iter**, or opening, of the aqueduct of Sylvius, preventing CSF outflow into the fourth ventricle (see Figures 5.10, 5.11). The lateral ventricles and the third ventricle are therefore dilated, but the fourth ventricle is not. In the sagittal view (see Image 5.7A) the corpus callosum appears somewhat thinned out and ballooned upward from the dilated ventricles. Note also that there are some regions of decreased intensity in the white matter adjacent to the frontal and occipital horns (see Image 5.7B). This represents **transependymal absorption of CSF** from the ventricles into the white matter and is a sign that hydrocephalus has developed relatively recently and is severe. In addition, if the patient had had hydrocephalus in infancy before closure of the cranial sutures, she may have had an enlarged head circumference; however, this patient's head circumference was normal, again suggesting the hydrocephalus had developed recently.

Clinical Course

The patient was taken to the operating room the day after admission for placement of a right **ventriculoperitoneal shunt** (see KCC 5.7). A small incision was made in the scalp over the right frontal area. A second small incision was made in the abdominal skin and extended down to the peritoneum. The shunt tubing was passed under the skin through the subcutaneous tissues extending from the scalp incision down to the abdomen. A hole was drilled through the skull, the dura was opened, and a catheter was passed through the right frontal lobe into the right lateral ventricle at a depth of approximately 6 cm from the surface. There was good flow of clear CSF under markedly elevated pressure. The catheter was connected to the shunt tubing through the scalp incision. Good flow was observed from the distal end of the tubing at the abdominal incision before it was inserted into the peritoneal cavity, and both incisions were closed with sutures. The shunt system contains a one-way flow valve to prevent fluid from traveling in the wrong direction. Postoperatively, the patient's headaches and nausea resolved immediately, and her sixth-nerve paresis resolved more slowly but recovered completely by 2 months after surgery.

CSF marker studies and cytology can often be helpful in investigation of pineal region tumors but did not provide a diagnosis for this patient. Therefore, one week after admission, she was taken back to the operating room for a biopsy of the mass lesion. Because of its location deep within the brain adjacent to the midbrain, open surgical resection or biopsy would have been risky. Therefore, a **stereotactic needle biopsy** (see KCC 16.4) was performed. Note that alternatively, this patient could have been treated by **endoscopic neurosurgery** (see KCC 5.11), which allows both treatment of the hydrocephalus by third ventriculostomy and biopsy of the mass in the same procedure.

The results of the biopsy in our patient showed that she had a primitive neuroectodermal tumor (PNET) of the pineal region, also called a pineoblastoma (see KCC 5.8). This is an uncommon brain tumor that often responds well to treatment, but can eventually be fatal. The patient was treated with radiation and chemotherapy and returned to school several months later. A follow-up MRI scan 8 months after initial presentation showed nearly complete disappearance of the tumor with therapy (see Image 5.7C). The hydro-

CASE 5.7 A CHILD WITH HEADACHES, NAUSEA, AND DIPLOPIA

IMAGE 5.7A–C MRI Scan Images Showing Pineal Tumor and Recovery T1-weighted images with TR = 450, TE = 11. (A) Sagittal image. (B) Axial image with intravenous gadolinium, showing pineal tumor obstructing iter (opening of cerebral aqueduct) and causing noncommunicating hydrocephalus. (C) Follow-up axial images with gadolinium 8 months after treatment.

(A)

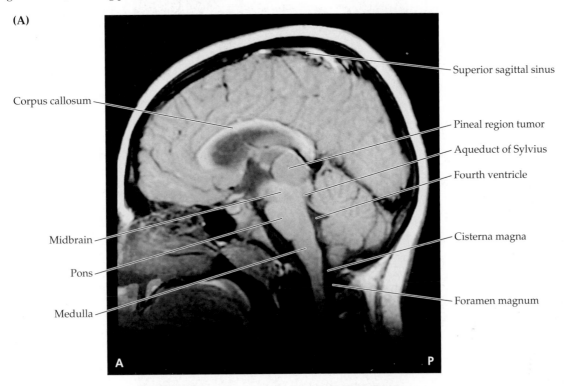

(B)

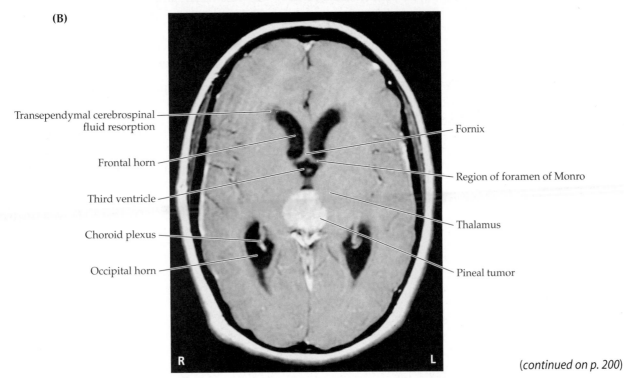

(*continued on p. 200*)

CASE 5.7 (*continued*)

(C)

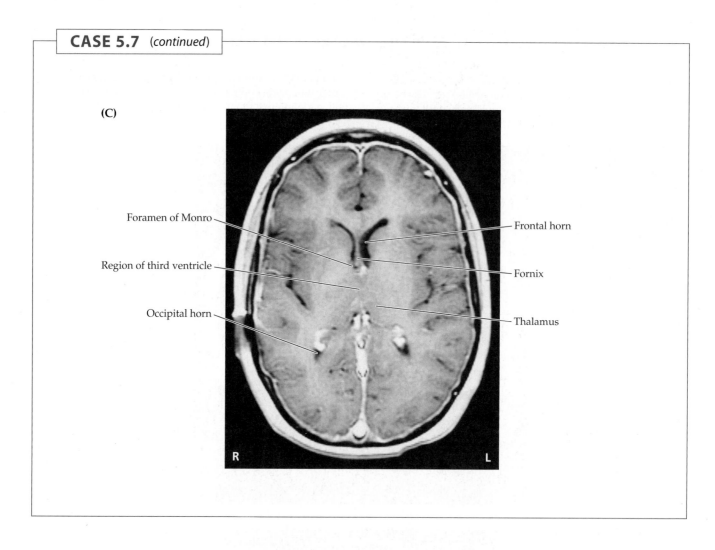

Foramen of Monro

Region of third ventricle

Occipital horn

Frontal horn

Fornix

Thalamus

R   L

cephalus had also completely resolved. At 3-year follow-up, she continued to do well, with no further growth of the tumor.

RELATED CASE. An MRI scan from another patient with hydrocephalus from obstruction at the iter is shown in Image 5.7D–H, pages 201–203. This patient was a 42-year-old Portuguese-speaking man brought to the emergency room by his girlfriend after he became increasingly difficult to arouse, as well as agitated and confused over the course of 1 day. He had a past history of seizures. A CT scan showed multiple small calcifications consistent with CNS cysticercosis (see KCC 5.9). Note the presence of a cyst obstructing the iter visible on the MRI scan in the horizontal (see Images 5.7D–F), sagittal (see Image 5.7G, and coronal (see Image 5.7H) views. He was treated by ventriculoperitoneal shunting (see KCC 5.7) and appropriate antiparasitic medication and made a full recovery.

CASE 5.7 *RELATED CASE*

IMAGE 5.7D–H MRI Scan Images Showing Cysticercosis Obstructing Iter and Causing Noncommunicating Hydrocephalus T1-weighted images. (D–F) Axial images. (G) Sagittal image. (H) Coronal image.

(D)

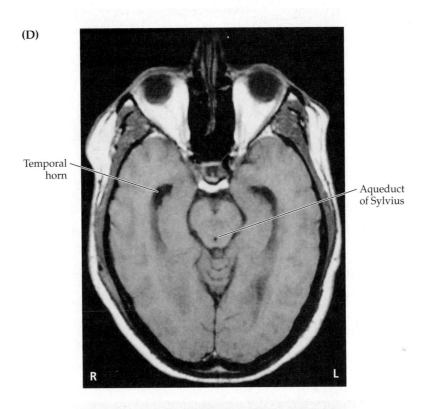

(E)

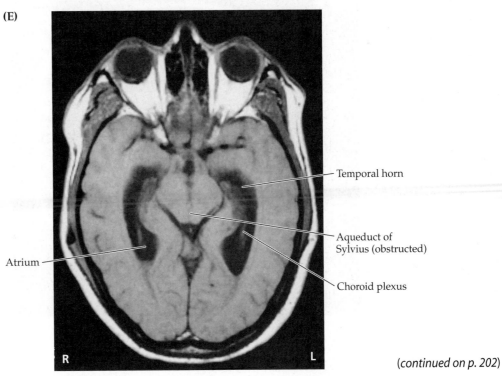

(*continued on p. 202*)

CASE 5.7 *RELATED CASE* *(continued)*

(F)

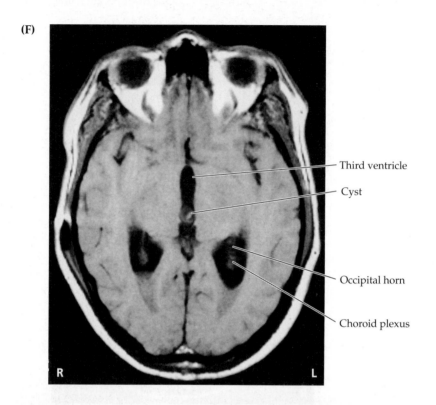

Third ventricle

Cyst

Occipital horn

Choroid plexus

(G)

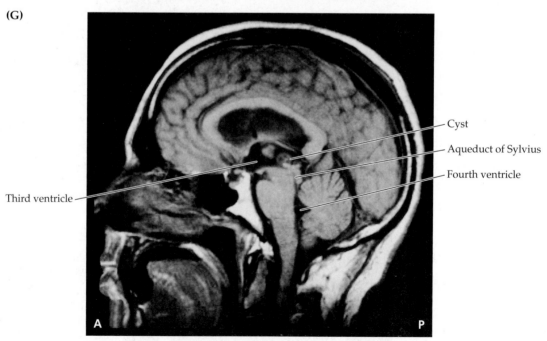

Cyst

Aqueduct of Sylvius

Fourth ventricle

Third ventricle

CASE 5.7 *RELATED CASE* (*continued*)

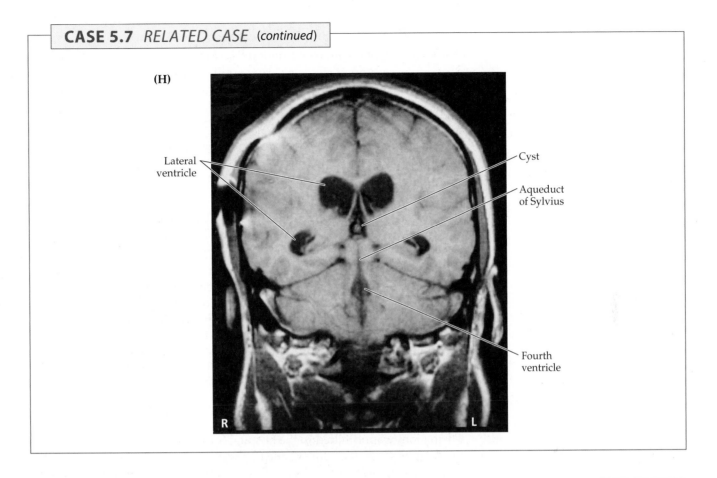

(H)

Lateral ventricle

Cyst

Aqueduct of Sylvius

Fourth ventricle

R   L

CASE 5.8 HEADACHES AND PROGRESSIVE VISUAL LOSS

MINICASE

A 51-year-old man came to an ophthalmology appointment complaining of 8 months of progressive **visual loss** and **headaches**. He was found to have bilateral mild **papilledema**, with some pallor of the right optic disc, visual fields with **enlarged blind spots**, and **concentric loss of the peripheral visual fields in both eyes** (he could see only the center of the visual field with either eye; see Figure 11.16A). The remainder of his neurologic exam was normal.

LOCALIZATION AND DIFFERENTIAL DIAGNOSIS

1. Headaches, papilledema, and visual loss of this kind are seen in what syndrome?

2. What is the appropriate test to perform next?

Discussion

1. Headaches with papilledema are worrisome signs of elevated intracranial pressure (see KCC 5.3). Concentric visual loss can be caused by certain ophthalmologic conditions but also by chronic or intermittent elevated intracranial pressure (see Table 5.3). Elevated intracranial pressure compresses the optic nerve, causing damage especially to the more superficial fibers nearest the nerve sheath, thus leading to concentric visual loss. Elevated intracranial pressure in this patient present over several months would most likely be caused by hydrocephalus (see KCC 5.7) or a mass lesion (see KCC 5.2), but it could also be caused by pseudotumor cerebri, sagittal sinus thrombosis, or other disorders.

2. The best test to perform next would be an MRI scan.

Neuroimaging

The attending ophthalmologist ordered an **MRI scan** (Image 5.8A–C, page 206).

1. Cover the labels on Image 5.8A–C and identify the plane of section, image type (T1-weighted, T2-weighted, or FLAIR), and the labeled structures for each image (see Figures 2.5, 4.6). Again, review the pathway of CSF flow through the ventricles (see Figures 5.10, 5.11).

2. Which ventricles are dilated, and what does this suggest about the site of the obstruction?

Discussion

1. All images have relatively short TR and TE and are therefore T1-weighted (see Chapter 4). Image 5.8A is coronal; 5.8B, axial; and 5.8C, sagittal.

2. The lateral ventricles are markedly dilated, while the third and fourth ventricles are not, suggesting obstruction bilaterally at the foramina of Monro. In fact, a small mass lesion can be identified in the anterior third ventricle, just underneath the fornix, obstructing the foramina of Monro in Image 5.8A. This is the typical location and appearance of a colloid cyst of the third ventricle. Although histologically benign, this tumor causes symptoms mainly through intermittent hydrocephalus, which can sometimes become rapidly fatal (see KCC 5.7). Note the prominent dilation of the lateral ventricles bilaterally (see Image 5.8A–C).

Clinical Course

The patient was admitted to the hospital for surgery. An incision was made in the right scalp, and a large bone flap was carefully removed, exposing the dura over the right hemisphere and a portion of the superior sagittal sinus in the midline. The dura was opened in a longitudinal fashion to the right of the midline. A large bridging vein (see Figure 5.1) was encountered entering the superior sagittal sinus from the right side, and care was taken not to disrupt it. The dura was folded back, revealing the right cerebral cortex and the falx cerebri in the midline. The cortex was gently retracted laterally off the falx along the medial surface of the hemisphere, allowing the surgeon to peer down onto the top surface of the corpus callosum. The pericallosal vessels (see Figure 4.16) were carefully moved to either side, and a 2 cm longitudinal incision was made in the corpus callosum, providing access to both lateral ventricles, with the septum pellucidum visible in the midline. An operating microscope was used for the remainder of the procedure. The colloid cyst—easily visible through the right foramen of Monro—was gently aspirated, with care taken to avoid damaging the fornix. At the end of the procedure, all visible colloid cyst was removed, and excellent flow of irrigant was attained through both foramina of Monro. The dura was then closed and the bone flap replaced before closing the scalp wound.

Postoperatively, the patient did quite well, with no further headaches and no further worsening of—or perhaps even some mild improvement in—his vision over the following months. An MRI done 1 week after surgery showed marked improvement in the hydrocephalus. The hole made in the corpus callosum caused no functional deficits because of its relatively small size (see KCC 19.8).

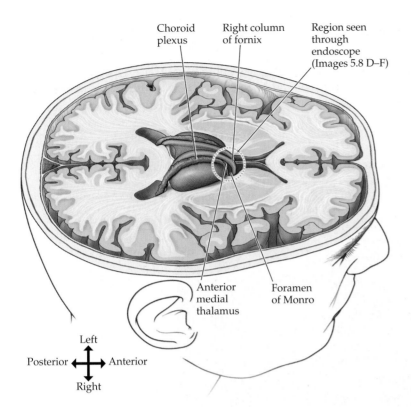

Choroid plexus

Right column of fornix

Region seen through endoscope (Images 5.8 D–F)

Anterior medial thalamus

Foramen of Monro

Left

Posterior ◄──► Anterior

Right

FIGURE 5.23 View of Right Foramen of Monro Region as seen Through Endoscope Top half of brain has been removed to provide perspective.

RELATED CASE. A 36-year-old man saw his primary care doctor for progressively worsening headaches. Neurological examination was normal, but MRI revealed enlargement of the bilateral lateral ventricles (similar to, but not as severe as Image 5.8D–F, pages 207–208), and a roundish mass lesion obstructing the bilateral interventricular foramina of Monro. The patient was referred to a neurosurgeon and underwent **endoscopic neurosurgery** (see KCC 5.7). Using a small incision and burr hole the endoscope was inserted through the right frontal lobe into the right lateral ventricle providing a view of the right foramen of Monro as shown in Figure 5.23 and Image 5.8D. A round mass was visible obstructing the right foramen of Monro (see Image 5.8D). The capsule of the cyst was opened revealing gelatinous colloid material (see Image 5.8E). This patient's colloid cyst was removed entirely through the endoscope, which restored patency of the foramen of Monro restoring CSF flow to the third ventricle (see Image 5.8F). The patient made a rapid recovery and was discharged home with no further problems.

CASE 5.8   HEADACHES AND PROGRESSIVE VISUAL LOSS

IMAGE 5.8A–C MRI Scan Images Showing Colloid Cyst Causing Obstructive Hydrocephalus Colloid cyst of the third ventricle obstructing foramina of Monro and causing noncommunicating obstructive hydrocephalus.

T1-weighted images with TR = 450, TE = 11. (A) Coronal (B) axial, and (C) sagittal views. The coronal scan (A) was performed with intravenous gadolinium.

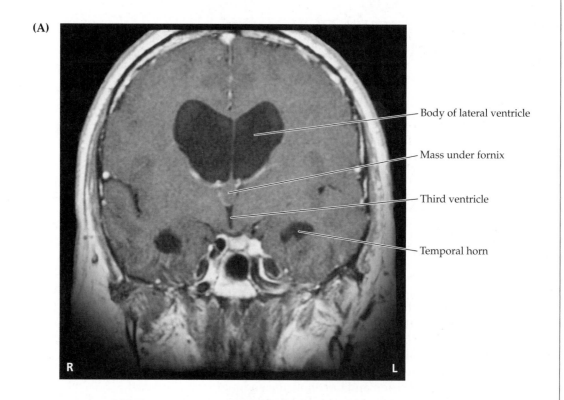

(A)

Body of lateral ventricle

Mass under fornix

Third ventricle

Temporal horn

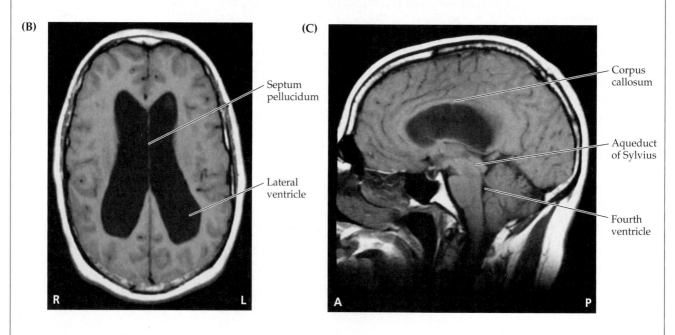

(B)

Septum pellucidum

Lateral ventricle

(C)

Corpus callosum

Aqueduct of Sylvius

Fourth ventricle

CASE 5.8 *RELATED CASE*

IMAGE 5.8D–F **Endoscopic Neurosurgical Treatment of Colloid Cyst of the Third Ventricle** See Figure 5.23 for perspective on structures visible through endoscope. (D) View of right foramen of Monro obstructed by colloid cyst. (E) Colloid material extruded once capsule was opened.

(F) Following removal of colloid cyst, patency of the foramen of Monro was restored, enabling the hypothalamus to be seen forming the floor of the third ventricle. Case and images courtesy of Drs. Howard Weiner, Jonathan Roth and David Harter, NYU School of Medicine, New York.

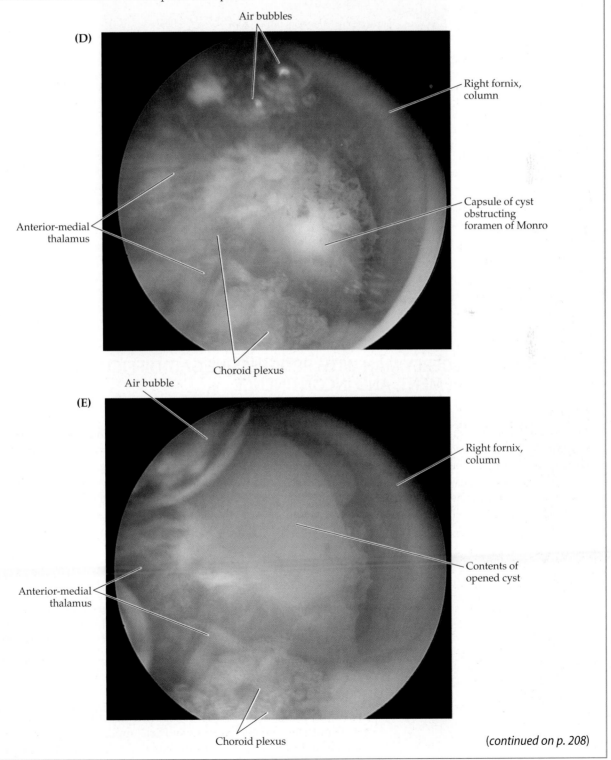

(D)

Air bubbles

Right fornix, column

Capsule of cyst obstructing foramen of Monro

Anterior-medial thalamus

Choroid plexus

(E)

Air bubble

Right fornix, column

Contents of opened cyst

Anterior-medial thalamus

Choroid plexus

(continued on p. 208)

CASE 5.8 *RELATED CASE* (continued)

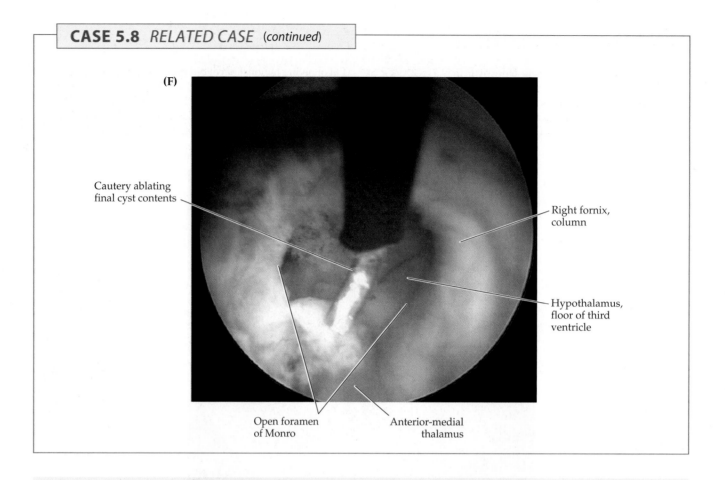

(F)

Cautery ablating
final cyst contents

Right fornix,
column

Hypothalamus,
floor of third
ventricle

Open foramen
of Monro

Anterior-medial
thalamus

CASE 5.9 AN ELDERLY MAN WITH PROGRESSIVE GAIT DIFFICULTY, COGNITIVE IMPAIRMENT, AND INCONTINENCE

MINICASE

A 76-year-old man was admitted to the hospital because of **progressive gait unsteadiness**, **memory difficulty**, and **incontinence**. His gait unsteadiness developed over the course of about 1 year, beginning with a shuffling stride and difficulty rising from a chair. This unsteadiness progressed until he required a cane and, eventually assistance from another person in order to ambulate. Urinary incontinence began 4 months prior to admission, and his family noted the onset of memory problems at around the same time. Examination on admission was notable for recall of only one of three objects at 5 minutes and an unsteady, shuffling gait, the patient barely lifting his feet off the floor.

Neuroimaging

A **head CT** was performed (Image 5.9A–D, pages 210–211).

1. Cover the labels on Image 5.9A–D and identify as many structures as possible on each image. Describe the appearance of the ventricles. Is this appearance due to generalized brain atrophy or to hydrocephalus (see KCC 5.7)?

2. What syndrome fits the clinical history, exam, and CT scan of this patient?

Discussion

1. See the labels on Image 5.9A–D for specific structures. The lateral ventricles, third ventricle, and even fourth ventricle appear enlarged. Note that in patients with brain atrophy, both the sulci and the ventricles are proportion-

ately increased in size. In hydrocephalus, however, the ventricles are increased out of proportion to the amount of sulcal prominence. In our patient the sulci are slightly prominent, while the ventricles are markedly enlarged, making this as a case of hydrocephalus (see KCC 5.7). This diagnosis can be best appreciated if you look near the top of the brain (see Image 5.9D), where the sulci are not especially prominent in this patient. Patients with generalized brain atrophy usually have enlarged sulci in this area.

2. This patient has the clinical triad of shuffling "magnetic gait," incontinence, and mental decline, together with a head CT strongly supporting the diagnosis of normal pressure hydrocephalus (see KCC 5.7).

Clinical Course

The patient was evaluated before and after CSF removal by lumbar puncture (see KCC 5.10). Two large-volume lumbar punctures were performed—one on hospital day 1, and one on hospital day 12—with 45 cc and 33 cc of CSF removed, respectively. Locomotion was evaluated before and after each lumbar puncture (see table) as well as on the intervening days (not shown). After each lumbar puncture, the patient showed dramatic improvement in the speed and stability of his gait as well as in his ability to stand up from a lying position. The improvement lasted for several days after each lumbar puncture, but then his condition gradually worsened again. It is not known why a single removal of 30–40 cc of CSF by lumbar puncture should have an effect persisting longer than the time it takes to replace this volume of CSF (a few hours). It has been suggested that this extended effect may result from a small tear in the dura created by the lumbar puncture that continues to leak for a period of time.

Effects of CSF Removal via Lumbar Puncture (LP) on Locomotion

	DAY 1, PRE-LP	DAY 2, 24 HOURS POST-LP	DAY 12, PRE-LP	DAY 12, 4 HOURS POST-LP
Time to walk 15 feet (s; average of 3 trials)	10.5	6.5	14	8
Number of steps to walk 15 feet (average of 3 trials)	10	6	—	—
Number of steps to turn 180°	9	10	—	—
Time from lying to standing (s; 1 trial)	28	14	42	15
Subjective evaluation	Unsteady, shuffling	Improved stability	Very unsteady	Improved stability

Because of the patient's excellent but temporary response to lumbar puncture and the clinical setting consistent with normal pressure hydrocephalus, a right ventriculoperitoneal shunt was placed (see KCC 5.7). The result was lasting improvement. When the patient was seen in the office 7 months later, his gait was more stable, he lifted his feet higher off the floor, and he walked 15 feet in 9 seconds. He no longer had urinary incontinence. His memory and attention remained moderately impaired, however. For example, he recalled only three of six objects after 5 minutes.

CASE 5.9   AN ELDERLY MAN WITH PROGRESSIVE GAIT DIFFICULTY, COGNITIVE IMPAIRMENT, AND INCONTINENCE

IMAGE 5.9A–D Head CT Showing Dilated Ventricles Typical of Normal Pressure Hydrocephalus

(A–D) Axial images progressing from the inferior to superior direction.

(A)

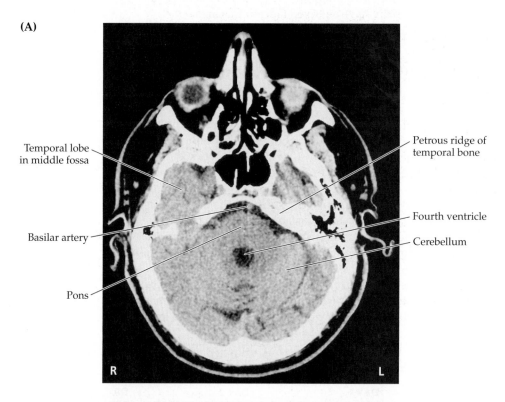

Temporal lobe in middle fossa

Petrous ridge of temporal bone

Basilar artery

Fourth ventricle

Cerebellum

Pons

R   L

(B)

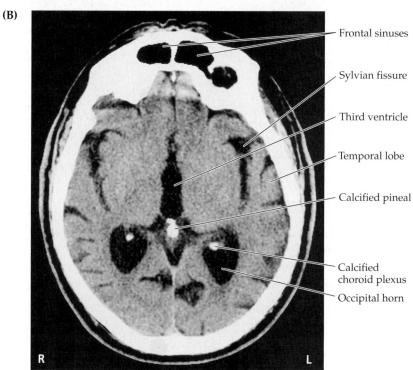

Frontal sinuses

Sylvian fissure

Third ventricle

Temporal lobe

Calcified pineal

Calcified choroid plexus

Occipital horn

R   L

(C)

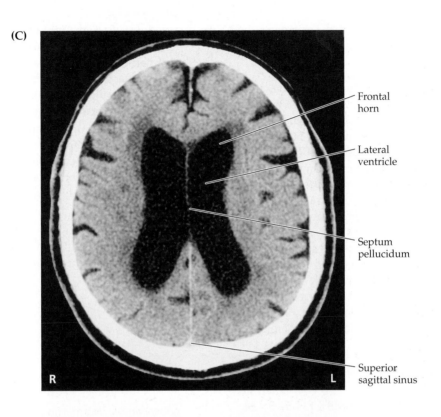

Frontal
horn

Lateral
ventricle

Septum
pellucidum

Superior
sagittal sinus

R   L

(D)

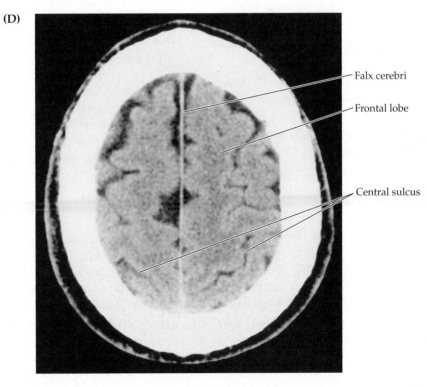

Falx cerebri

Frontal lobe

Central sulcus

CASE 5.10   A YOUNG MAN WITH HEADACHE, FEVER, CONFUSION, AND STIFF NECK

CHIEF COMPLAINT

A 28-year-old man was brought to the emergency room with 1 day of worsening headache, fever, confusion, and stiff neck.

HISTORY

Previously healthy, the patient awoke on the day of admission at about 4:00 A.M. with chills and body aches. By noon he was breathing quickly and had nausea and vomiting, but then he felt better and took a nap. He awoke at 3:00 P.M. with a mid-frontal **headache, photophobia,** and **stiff neck with a temperature of 102°F**. He took some acetaminophen and went to a party that evening. However, by 9:00 P.M. he appeared **confused**, talking about "ragtime" and "skirt party pick me up." His girlfriend then brought him to the emergency room by taxicab.

The patient had no known recent exposure to anyone ill and no HIV risk factors. He had visited a friend in Mississippi the previous week but had not recently traveled abroad, and he had not had any recent insect bites or rashes. He had no exotic pets. He worked as an air force pilot. He was not taking any medications and had no known drug allergies.

PHYSICAL EXAM

General appearance: Acutely ill–appearing young man lying on a stretcher.

Vital signs: **T = 101.7°F**, P = 110, BP = 136/84, R = 24.

Head: Normal tympanic membranes bilaterally. Dry oral mucosa. No oral thrush.

Neck: **Marked nuchal rigidity**. No bruits. No adenopathy.

Lungs: Clear.

Heart: Regular rate with no murmurs, gallops, or rubs.

Abdomen: Normal bowel sounds; soft, nontender.

Extremities: Normal.

Skin: **Nonblanching, purplish 2 to 3 mm petechiae scattered on arms, legs, and chest.**

Neurologic exam:

MENTAL STATUS: Lethargic **but arousable. Normal speech. Oriented to correct month and year, but not to date or name of hospital. The patient said he is a pilot for Northwest, but his *father* is actually a pilot for that airline. He did not recall where he lives. He spelled "w-o-r-l-d" forward but not backward.**

CRANIAL NERVES: Pupils 6 mm, constricting to 3 mm with light bilaterally. Ophthalmoscopic exam: Normal fundi bilaterally. Extraocular movements intact. Visual fields full. Facial sensation intact. Face symmetric. Hearing normal to finger rub bilaterally. Palate elevation normal. Tongue midline.

MOTOR: Normal tone. Power 5/5 throughout, with intermittent cooperation.

SENSORY: Intact light touch, pinch.

REFLEXES:

COORDINATION: Normal finger–nose–finger testing.

GAIT: Not tested.

LOCALIZATION AND DIFFERENTIAL DIAGNOSIS

1. Irritation of which structure(s) is suggested by the signs and symptoms seen in this patient? What is the most likely diagnosis?

2. What are the most likely pathogens, given this patient's age and history? What are some other possibilities?

3. What diagnostic test should be done, and what treatment should be started without delay?

Discussion

1. Headache, fever, photophobia, and nuchal rigidity are strongly suggestive of meningeal irritation (see Table 5.6). The lethargy and confusion suggest a process causing diffuse cerebral dysfunction as well (encephalopathy; see KCC 19.15). The acute progressive time course of this patient's presentation strongly suggests acute bacterial meningitis or possibly viral meningoencephalitis.

2. Acute bacterial meningitis in immunocompetent adults is usually caused by *Streptococcus pneumoniae* or *Neisseria meningitidis,* but it can also be caused by *Listeria monocytogenes* (see Table 5.8). The purplish skin lesions are suggestive of *N. meningitidis,* although they can also be seen in rickettsial infections. Viral meningoencephalitis, including herpes encephalitis, should also

be considered. Postinfectious meningitis or other less common forms of meningitis are also possible.

3. Patients with acute infectious meningitis can deteriorate within a matter of hours or even minutes, so rapid evaluation and treatment are essential. A head CT and lumbar puncture (see KCC 5.10) should be performed to narrow the diagnosis and help determine the most appropriate therapy (see Tables 5.7, 5.8). However, antimicrobial treatment should not be delayed if for any reason a lumbar puncture cannot be performed immediately.

Initial Clinical Course

The patient was started on intravenous ceftriaxone and ampicillin. A head CT was normal, and lumbar puncture (see table) was performed successfully about 20 minutes after starting antibiotics. Blood toxicology screen was negative, and other routine blood tests were unremarkable. Chest X-ray was normal.

Results of Lumbar Puncture

TUBE NUMBER	RED BLOOD CELLS (PER MM³)	WHITE BLOOD CELLS (PER MM³)	PERCENT PMNS[a]	PERCENT LYMPHOCYTES
1	230	3280	100	0
4	220	2030	99	1

[a]PMNs = polymorphonuclear leukocytes.
Protein: 714 mg/dl; **Glucose:** <20 mg/dl.

1. From what space is CSF removed to perform lumbar puncture, and by which two meningeal layers is the space bound (see Figure 5.22B)? What landmark is used to enter the lumbar cistern at the appropriate level (see Figure 5.22A)?

2. What do the results of the CSF analysis in this patient suggest (compare Table 5.7)?

Discussion

1. In lumbar puncture, CSF is removed from the subarachnoid space of the lumbar cistern, bounded by the pia and the arachnoid (see Figures 5.1, 5.22B). The iliac crest is used as a landmark to enter the lumbar cistern at the L4–L5 interspace, well below the conus medullaris of the spinal cord (see KCC 5.10; Figure 5.22).

2. The CSF in this patient shows a very high protein level, low glucose level, and very high white blood cell count, consisting mostly of polymorphonuclear cells, consistent with acute bacterial meningitis (see Table 5.7).

    The CSF cultures grew no organisms. However, blood cultures performed prior to the administration of antibiotics grew *Neisseria meningitidis*. By 1 day after admission the patient's fevers had stopped and his mental status cleared. He was treated with a course of intravenous antibiotics, made an excellent recovery, and was discharged 7 days later with no sequelae.

Additional Cases

Related cases for the following topics can be found in other chapters: **herniation** (Case 10.10); **intracranial hemorrhage** (Cases 10.1, 10.13, 14.9, 19.3, 19.4); **aneurysm** (Cases 10.1 and 13.1); **arteriovenous malformation (AVM)** (Case

11.5); **hydrocephalus** (Case 15.3); **brain tumors** (Cases 7.4, 11.3, 11.4, 12.2, 12.3, 12.5–12.7, 13.9, 15.2, 17.1, 18.2, 18.4, 18.5, 19.7); and **infectious disorders of the nervous system** (Cases 8.4, 16.1, 19.10). Other relevant cases can be found using the **Case Index** located at the end of this book.

Brief Anatomical Study Guide

1. In this chapter, we have discussed the anatomy of the meninges (see Figure 5.1), including the **pia**, **arachnoid**, and **dura**, as well as the major dural infoldings (see Figures 5.5 and 5.6).

2. The cranial cavity is composed of **anterior**, **middle**, and **posterior fossae**, which contain specific brain structures (see Figures 5.2–5.4).

3. The **blood–brain barrier** is formed by brain capillary endothelial cells sealed together by **tight junctions** (see Figures 5.13B and 5.14).

4. To gain a three-dimensional understanding of the ventricular system and its spatial relationships to adjacent structures, review Figures 5.10 and 5.11 as well as the Neurological Atlas MRI images in Figures 4.13–4.15. Several **C-shaped structures** follow the curve of the lateral ventricles. These include the caudate nucleus, corpus callosum, and fornix (see Figure 4.15). We will now discuss the spatial relationships of these structures—both as a review of the anatomy of the ventricular system and to prepare you for the next section, "A Scuba Expedition through the Brain."

5. The **caudate nucleus** and **thalamus** bulge inward from the lateral walls of the **lateral ventricles** (see Figures 4.13 and 4.14). The caudate nucleus forms a C-shaped structure that lies along the wall of the C-shaped lateral ventricle in all planes of section (see Figures 4.14 and 16.4).

6. The **septum pellucidum** is a thin membrane that separates the two lateral ventricles in the midline. The septum pellucidum hangs from the **corpus callosum**, another C-shaped structure, which forms the roof of most parts of the lateral ventricles (see Figures 4.14 and 4.15).

7. Dangling from the bottom of the septum pellucidum, the **fornix** (see Figure 18.13) again forms a C-shaped structure that parallels the curve of the lateral ventricles. The fornix is composed of a pair of archlike bundles of myelinated axons that connect structures in the temporal lobes to the hypothalamus and basal forebrain (see Figure 18.9).

8. The **hippocampal formation**, a structure involved in memory and other limbic functions (see Chapter 18), lies on the floor and the medial wall of the temporal horn of the lateral ventricles (see Figure 4.14).

9. The **interventricular foramina of Monro** are bounded medially and superiorly by the fornix, laterally by the thalami, and inferiorly by the **anterior commissure** (a white matter tract connecting structures in the two temporal lobes) (see Figures 16.2, 16.4, and 18.9A).

10. The **third ventricle** is bounded laterally by the thalami and hypothalamus; superiorly by the fornix; inferiorly by the hypothalamus; anteriorly by the anterior commissure, fornix, lamina terminalis, and hypothalamus; and posteriorly by the posterior commissure, pineal region, and hypothalamus (see Figures 2.11B, 4.14B, and 5.11).

Brief Anatomical Study Guide (continued)

11. The **cerebral aqueduct** (**of Sylvius**), a thin canal of cerebrospinal fluid, is located entirely within the midbrain gray matter (see Figure 5.10).

12. The **fourth ventricle** is a pyramid-shaped cavity with its base resting on the dorsal aspect of the pons and medulla and its apex covered by the cerebellum (see Figures 5.10 and 5.11).

13. Note also the locations of the major **CSF cisterns** in the subarachnoid space (see Figure 5.12).

14. Finally, recall that the meninges make a "PAD" composed, from inside to outside, of the **pia**, **arachnoid**, and **dura** (see Figure 5.1).

15. CSF is produced in the **choroid plexus** and reabsorbed in the **arachnoid granulations** (see Figures 5.1 and 5.10).

A Scuba Expedition through the Brain

Imagine your colleague has lost his memory just before the neuroanatomy final exam. Fortunately, a mutual friend owns a miniaturization ray and special electronic apparatus that, when plugged directly into the hippocampus, can immediately retrieve any lost memories. To help your friend, you bravely don your scuba gear and allow yourself to be miniaturized and injected via lumbar puncture into your friend's lumbar cistern (see Figure 5.22), taking along a set of your friend's MRI scans as a map (see Figures 4.13–4.15). *Your mission: Find the hippocampus.*

Looking around as you swim in the cerebrospinal fluid, you see many wispy, spiderweb-like strands extending inward from the (1._____). You are in the (2._____) space (see Figure 5.1), bounded externally by the (3._____) and internally by the (4._____). As you swim upward, you notice long, rope-like strings descending all around you in the lumbar cistern; for some reason, they remind you of a horse's tail. This is the (5._____), consisting of spinal (6._____) roots. Soon you see a thick, gleaming, pinkish-white structure: the (7._____). It has sensory roots entering its (8._____) surface and motor roots exiting its (9._____) surface.

As you swim up the spinal canal in the subarachnoid space, eventually you come to a large, ring-shaped hole leading into the cranial cavity. This entryway is the (10._____), bringing you into the cisterna (11._____). Above you is a pinkish-gray structure with lots of bumps on it that seems to be furiously calculating coordination and other operations. It is the (12._____). You peek around to the sides at the lateral foramina of (13._____), but then you decide to swim straight up along the dorsal surface of the medulla and under the cerebellum to enter the midline foramen of (14._____). Suddenly, you feel a rush of clear cerebrospinal fluid against your face as it flows out into the subarachnoid space, and you need to kick a little harder with your flippers. You have entered a large cavity called the (15._____) ventricle. You let yourself sink to the ventral floor of the fourth ventricle and take a few steps forward. Beneath your flippers, the floor of the fourth ventricle is formed initially by the (16._____) and then, as you progress farther rostrally, by the (17._____).

Looking upward, you direct your headlight toward a large structure forming the roof of the cavern. This is the (18._____). You decide to swim farther rostrally to enter a narrow tunnel called the (19._____). To pass through this tunnel you have to wriggle your shoulders against the walls, which are part of the (20._____), and you continue to swim forward against the CSF current. Finally, you pop out of the top end of the tunnel and find yourself sinking down toward the floor of another cavity, called the (21._____) ventricle. As you sink downward, you look to the left and to the right, and you see walls made up first by the (22._____) and later by the (23._____). You decide not to let yourself sink to the bottom, which is also formed by the (23._____) but instead, you look up at the roof and see two parallel, gleaming white arches running from back to front, making up the roof of the third ventricle. This is the (24._____). You kick with your flippers and swim up toward the front of the third ventricle, where you see two holes, called the foramina of (25._____). You choose the foramen on the right, climb up onto the threshold of the hole, and find yourself standing on the (26._____) commissure, with your left (medial) hand leaning up against the (27._____) and your right (lateral) hand leaning up against the (28._____). You swim forward and upward and find that you have entered another very large cavity, the right (29._____) ventricle. As you continue to swim forward in this cavity, you are entering the (30._____) horn of the lateral ventricle. Looking upward, you see a whitish "hard body" forming the roof, the (31._____). You soon reach a dead end, also formed by the (31._____) as it curves downward around the most anterior portions of the lateral ventricle.

You turn around 180° and start swimming back caudally toward the other portions of the lateral ventricle. You soon pass the foramen of (32._____) again and have to kick a little harder to avoid being sucked back down into the (33._____) ventricle. You have now entered the (34._____) of the right lateral ventricle. On your right side is a translucent membrane forming the medial wall of the ventricle, called the (35._____). It extends downward from the (36._____), which still forms the roof of the lateral ventricle, to the (37._____) on the medial floor. As you shine your powerful searchlight through this membrane, you can barely make out another, nearly identical cavity on the other side of the brain, the left (38._____) ventricle. Looking laterally (to your left), you see a large gray matter structure bulging inward into the ventricle and forming its lateral wall. Checking your map (see Figures 4.13–4.15), you realize this is the (39._____) nucleus. As you swim farther back, another gray matter structure called the (40._____) bulges inward from the lateral wall. Suddenly you notice that your foot has become ensnared in a pulsatile tangle of blood vessels that seems to be secreting a clear fluid. This is the (41._____), which you had noticed all along the way inside the ventricles but had so far managed to avoid. Carefully you disentangle your flipper and soon enter another part of the ventricle, called the (42._____).

From here you seem to have three choices of where to swim next. You could turn around and swim forward into the (43._____) of the lateral ventricle or you could swim downward into the (44._____) horn, but instead you choose to continue swimming toward the back of the brain and enter the (45._____) horn. However, you soon meet another dead end when you find yourself well within the (46._____) lobe. So you turn around and start walking forward, but then—WHOA! Suddenly you are sliding down a steep slope into the (47._____) horn. You stand up, brush yourself off, and lean your left hand against the medial wall. You look around and decide you must be deep within the (48._____) lobe. Then you look down at your feet and at your left hand, and, what do you know, you're finally standing right on top of the (49._____). Congratulations! You have succeeded in reaching your goal and in preserving many valuable memories.

Answers are located on page 220.

References

General

Greenberg MS, Arredondo N, Duckworth M, Nichols T. 2005. *Handbook of Neurosurgery.* 6th Ed. Thieme, New York.

Laterra J, Goldstein GW. 2000. Ventricular organization of cerebrospinal fluid: blood-brain barrier, brain edema, and hydrocephalus. In Kandel ER, Schwartz JH, Jessell TM (eds.). *Principles of Neural Science.* 4th Ed., Appendix Part B. McGraw-Hill, New York.

Moore AJ, Newell DW (eds.). 2005. Neurosurgery Principles and Practice. Springer, London.

Rengachary S, Ellenbogen R (eds.). 2004. *Principles of Neurosurgery.* Mosby, 2nd Ed.

Headache

Detsky ME, McDonald DR, Baerlocher MO, Tomlinson GA, McCrory DC, Booth CM. 2006. Does this patient with headache have a migraine or need neuroimaging? *JAMA* 296 (10): 1274–1283.

Goadsby PJ, Silberstein SD, Lipton RB (eds.). 2002. *Headache in Clinical Practice.* Informa HealthCare.

Quality Standards Subcommittee of the American Academy of Neurology. 2000. Practice parameter: Evidence-based guidelines for migraine headache (an evidence-based review). *Neurology* 55: 754–762.

Rose G, Edmeads J, Dodick D. 2007. *Critical Decisions in Headache Management.* 2nd Ed. Hamilton: B. C. Decker, Malden, MA.

Silberstein SD. 2008. Treatment recommendations for migraine. *Nat Clin Pract Neurol* 4 (9): 482–489.

Elevated Intracranial Pressure

Bershad EM, Humphreis WE, Suarez JI. 2008. Intracranial Hypertension. *Semin Neurol* 28 (5): 690–702.

Kuroiwa T, Baethmann A, Czernicki Z, et al. 2004. *Brain Edema XII: Proceedings of the 12th International Symposium.* Springer, London.

Raslan A, Bhardwaj A. 2007. Medical management of cerebral edema. *Neurosurg Focus* 15; 22 (5): E12.

Rincon F, Mayer SA. 2008. Clinical review: Critical care management of spontaneous intracerebral hemorrhage. *Crit Care* 12 (6): 237.

Schrader H, Lofgren J, Zwetnow N. 1985. Regional cerebral blood flow and CSF pressures during the Cushing response induced by an infratentorial expanding mass. *Acta Neurol Scand* 72 (3): 273–282.

Brain Herniation Syndromes

Fisher, CM. 1984. Acute brain herniation: A revised concept. *Sem Neurol* 4: 417–421.

Kernohan JW, Woltman HW. 1929. Incisura of the crus due to contralateral brain tumor. *Arch Neurol Psychiatry* 21: 274.

Plum F, Posner JB. 2007. Structural lesions causing stupor and coma. In *Plum and Posner's Diagnosis of Stupor and Coma,* Chapter 3. Oxford University Press, Oxford.

Rhoton AL, Ono M. 1996. Microsurgical anatomy of the region of the tentorial incisura. In *Neurosurgery,* RH Wilkins and SS Rengachary (eds.), 2nd Ed., Vol. 1, Chapter 91. McGraw-Hill, New York.

Ropper AH. 1989. A preliminary MRI study of the geometry of brain displacement and level of consciousness with acute intracranial masses. *Neurology* 39 (5): 622–627.

Ross DA, Olsen, WL, Ross AM, Andrews BT, Pitts LH. 1989. Brain shift, level of consciousness, and restoration of consciousness in patients with acute intracranial hematoma. *J Neurosurg* 71 (4): 498–502.

Head Trauma

Cameron MM. 1978. Chronic subdural haematoma: A review of 114 cases. *J Neurol Neurosurg Psychiatry* 41 (9): 834–839.

Gallagher JP, Browder EJ. 1968. Extradural hematoma: Experience with 167 patients. *J Neurosurg* 29 (1): 1–12.

Haydel MJ. 2005 Clinical decision instruments for CT scanning in minor head injury. *JAMA* 294 (12): 1551–1553.

McKissock W, Taylor JC, Bloom WH, et al. 1960. Extradural hematoma: Observation on 125 cases. *Lancet* 2: 167–172.

Ropper AH, Gorson KC. 2007. Concussion. *N Engl J Med* 356 (2): 166–172.

Servadei F, Compagnone C, Sahuquillo J. 2007. The role of surgery in traumatic brain injury. *Curr Opin Crit Care* 13 (2):163–168.

Unterberg AW, Stover J, Kress B, Kiening KL. 2004. Edema and brain trauma. *Neuroscience* 129 (4): 1021–1029.

Intracranial Hemorrhage

Broderick JP, Brott TG, Duldner JE, Tonsick T, Huster G. 1993. Volume of intracerebral hemorrhage: A powerful and easy-to-use predictor of 30-day mortality. *Stroke* 24 (7): 987–993.

Edlow JA, Caplan LR. 2000. Avoiding pitfalls in the diagnosis of subarachoid hemorrhage. *N Engl J Med* 342 (1): 29–36.

Levine JM. Critical care management of subarachnoid hemorrhage. 2008. *Curr Neurol Neurosci Rep* 8 (6): 518–525.

Molyneux A, Kerr R, Stratton I, Sandercock P, Clarke M, Shrimpton J, Holman R; International Subarachnoid Aneurysm Trial (ISAT) Collaborative Group. 2002. International Subarachnoid Aneurysm Trial (ISAT) of neurosurgical clipping versus endovascular coiling in 2143 patients with ruptured intracranial aneurysms: a randomised trial. *Lancet* 360 (9342): 1267–1274.

Qureshi AI, Tuhrim S, Broderick JP, Batjer HH, Hondo H, Hanley DF. 2001. Spontaneous intracerebral hemorrhage. *N Engl J Med* 344: 1450.

Rost NS, Smith EE, Chang Y, et al. 2008. Prediction of functional outcome in patients with primary intracerebral hemorrhage: the FUNC score. *Stroke 39* (8): 2304–2309.

Segal R, Furmanov A, Umansky F. 2006. Spontaneous intracerebral hemorrhage: to operate or not to operate, that's the question. *Isr Med Assoc J* 8 (11): 815–818.

Suarez JI, Tarr RW, Selman WR. Aneurysmal Subarachnoid Hemorrhage. 2006. *N Engl J Med* 354: 387.

Teunissen LL, Rinkel GJE, Algra A, van Gijn J. 1996. Risk factors for subarachnoid hemorrhage: A systematic review. *Stroke* 27 (3): 544–549.

The Arteriovenous Malformation Study Group. 1999. Arteriovenous malformations of the brain in adults. *N Engl J Med* 340 (23): 1812–1818.

The International Study of Unruptured Intracranial Aneurysms Investigators. 1998. Unruptured intracranial aneurysms—risk of rupture and risks of surgical intervention. *N Engl J Med* 339 (24): 1725–1733.

Hydrocephalus

Adams RD, Fisher CM, Hakim S, Ojemann RG, Sweet WH. 1965. Symptomatic occult hydrocephalus with "normal" cerebrospinal fluid pressure—A treatable syndrome. *N Engl J Med* 273 (3): 117–126.

Hamilton MG. 2009. Treatment of hydrocephalus in adults. *Semin Pediatr Neurol* 16 (1): 34-41.

Milhorat TH. 1996. Hydrocephalus: Pathophysiology and clinical features. In *Neurosurgery*, RH Wilkins and SS Rengachary(eds.), 2nd Ed., Vol. 3. McGraw-Hill, New York.

Rosenberg GA. 2007. Brain edema and disorders of cerebrospinal fluid circulation. In *Neurology in Clinical Practice*, WG Bradley, RB Daroff, J Jankovic, and G Fenichel (eds.), 5th Ed., Vol. 2, Chapter 63. Butterworth-Heinemann, Boston.

Tsakanikas D, Relkin N. 2007. Normal pressure hydrocephalus. *Semin Neurol* 27 (1): 58–65.

Brain Tumors

Baehring JM, Piepmeier JM (eds.). 2007. *Brain Tumors: Practical Guide to Diagnosis and Treatment*. Taylor & Francis, Inc., Oxford.

DeAngenlis LM. 2001. Brain tumors. *N Engl J Med* 344 (2): 114–124.

DeMonte F, Gilbert MR, Mahajan A, McCutcheon IE. 2007. *Tumors of the Brain and Spine*. Springer.

Stupp R et al. 2009. Effects of radiotherapy with concomitant and adjuvant temozolomide versus radiotherapy alone on survival in glioblastoma in a randomised phase III study: 5-year analysis of the EORTC-NCIC trial. *Lancet Oncol* 10 (5): 459–466.

Wen PY, Kesari S. 2008. Malignant gliomas in adults. *N Engl J Med* 359 (5): 492–507.

Infectious Disorders of the Nervous System

Greenberg BM. 2008. Central nervous system infections in the intensive care unit. *Semin Neurol* 28 (5): 682–689.

Minagar A, Commins D, Alexander JS, Hoque R, Chiappelli F, Singer EJ, Nikbin B, Shapshak P. 2008. NeuroAIDS: characteristics and diagnosis of the neurological complications of AIDS. *Mol Diagn Ther* 12 (1): 25–43.

Peltola H, Roine I. 2009. Improving the outcomes in children with bacterial meningitis. *Curr Opin Infect Dis* 22 (3): 250–255.

Ropper AH, Samuels MA. 2009. Infections of the nervous system (bacterial, fungal, spirochetal, parasitic) and sarcoidosis. In *Adams & Victor's Principles of Neurology*, 9th Ed., Chapter 32. McGraw-Hill, New York.

Ropper AH, Samuels MA. 2009. Viral infections of the nervous system, chronic meningitis, and prion diseases. In *Adams & Victor's Principles of Neurology*, 9th Ed., Chapter 33. McGraw-Hill, New York.

Rosenstein NE, Perkins BA, Stevens DS, Popovic T, Hughes JM. 2001. Meningococcal disease. *N Engl J Med* 344 (18): 1378–1388.

van de Beek D, de Gans J, Tunkel AR, Wijdicks EFM. 2006. Community-acquired bacterial meningitis in adults. *N Engl J Med* 354: 44–53.

Ziai WC, Lewin JJ, 3rd. 2008. Update in the diagnosis and management of central nervous system infections. *Neurol Clin* 26 (2): 427–468, viii.

Lumbar Puncture

Ellenby MS, Tegtmeyer K, Lai S, Braner DAV. 2006. Lumbar puncture. *N Engl J Med* 355 (13): e12.

Joffe AR. 2007. Lumbar puncture and brain herniation in acute bacterial meningitis: a review. *J Intensive Care Med* 22 (4): 194–207.

Quality Standards Subcommittee of the American Academy of Neurology. 1993. Practice parameter: lumbar puncture (summary statement). *Neurology* 43 (3 Pt 1): 625–627.

Scuba Expedition through the Brain—Answers

1. arachnoid
2. subarachnoid
3. arachnoid
4. pia
5. cauda equina
6. nerve
7. spinal cord
8. dorsal (or posterior)
9. ventral (or anterior)
10. foramen magnum
11. magna
12. cerebellum
13. Luschka
14. Magendie
15. fourth
16. medulla
17. pons
18. cerebellum
19. cerebral aqueduct (of Sylvius)
20. midbrain (or mesencephalon)
21. third
22. thalamus
23. hypothalamus
24. fornix
25. Monro
26. anterior
27. fornix
28. thalamus
29. lateral
30. anterior (or frontal)
31. corpus callosum
32. Monro
33. third
34. body
35. septum pellucidum
36. corpus callosum
37. fornix
38. lateral
39. caudate
40. thalamus
41. choroid plexus
42. atrium (or trigonum)
43. body
44. temporal (or inferior)
45. occipital (or posterior)
46. occipital
47. temporal (or inferior)
48. temporal
49. hippocampus

CONTENTS

Chapter 6

Corticospinal Tract and Other Motor Pathways

Movement is crucial to our normal functioning, and damage to the motor systems can be profoundly disabling. In this chapter, we will learn of a 74-year-old woman who awoke one morning and suddenly developed slurred speech and paralysis of the entire right side of her body, including her right face, arm, and leg. Her reflexes were brisk on the right side, and she also had a Babinski's sign on the right, but her sensory exam was normal. She was unable to walk or stand without assistance.

To diagnose and treat patients with such symptoms, we must learn about the pathways in the brain and spinal cord that control movement of the body.

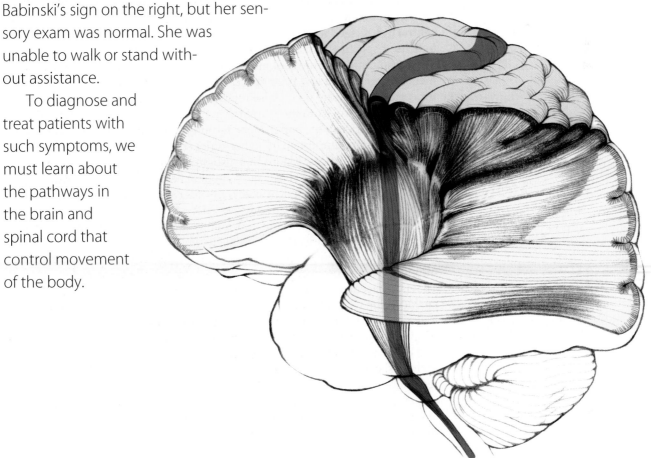

ANATOMICAL AND CLINICAL REVIEW

I N THIS CHAPTER AND THE NEXT, we will focus on the three most important motor and sensory "long tracts" in the nervous system. Familiarity with these three pathways is essential and suffices for full neuroanatomical localization in many clinical cases. These three fundamental pathways and their functions are listed in Table 6.1.

Each of these pathways crosses over, or decussates, to the contralateral side at a specific point in its course. Knowledge of the crossover points is very helpful for localizing lesions. A second clue to localizing lesions often comes from an understanding of the topographical representation of different body parts in these pathways. In these two chapters (6 and 7), we will trace the routes of the main motor and sensory pathways through all levels of the nervous system and review their overall organization and functions, with special emphasis on the spinal cord. We will also briefly discuss other systems that involve the spinal cord, such as the autonomic nervous system, sphincter control mechanisms, and motor pathways other than the corticospinal tract.

Motor Cortex, Sensory Cortex, and Somatotopic Organization

The primary motor cortex and primary somatosensory cortex are shown in Figure 6.1. Recall from Chapter 2 that these areas are located on either side of the **central**, or **Rolandic**, **sulcus**, which divides the frontal lobe from the parietal lobe. The **primary motor cortex** (Brodmann's area 4) is in the **precentral gyrus**, while the primary somatosensory cortex (Brodmann's areas 3, 1, and 2) is in the **postcentral gyrus** (see Figure 6.1). Lesions in these areas cause motor or sensory deficits, respectively, in the contralateral body.

Several important areas of motor association cortex, including the supplementary motor area and premotor cortex, lie just anterior to the primary motor cortex (see Figure 6.1). These regions are involved in higher-order motor planning and project to the primary motor cortex. Similarly, somatosensory association cortex in the parietal lobe receives inputs from primary somatosensory cortex and is important in higher-order sensory processing. Unlike lesions in the primary cortices, lesions in sensory or motor association cortex do not produce severe deficits in basic movement or sensation. Instead, lesions of the association cortex cause deficits in higher-order sensory analysis or motor planning, discussed further in Chapter 19. Interestingly, there are reciprocal connections between primary and association cortex, and sensory and motor areas, as shown in Figure 6.1A.

Functional mapping and lesion studies have demonstrated that the primary motor and somatosensory cortices are **somatotopically organized** (Figure 6.2). That is, adjacent regions on the cortex correspond to adjacent areas on the body surface. The cortical maps are classically depicted by a **motor homunculus** and a **sensory homunculus** (*homunculus* means "little man" in Latin). Since the original description of these homunculi, additional work in both humans and other animals has shown that the somatotopic maps are not as clear-cut and consistent as originally depicted, especially when studied at high-spatial resolution. Multiple fractionated representations exist, more so for motor than for sensory maps. Nevertheless, the homunculi remain a useful concept for understanding the broad strokes of cortical representation, and they are widely used for clinical localization.

As we will see in the sections that follow, somatotopic organization is not confined to the cortex. Rather, motor and sensory pathways maintain a rough

TABLE 6.1 Main Long Tracts of the Nervous System

PATHWAY	FUNCTION
Lateral cortico-spinal tract	Motor
Posterior columns	Sensory (vibration, joint position, fine touch)
Anterolateral pathways	Sensory (pain, temperature, crude touch)

(A)

Premotor cortex

Supplementary motor area

Primary motor cortex

Central sulcus

Primary somatosensory cortex

6 4 3, 1, 2

5, 7

Parietal association cortex

Secondary somatosensory area (in parietal operculum)

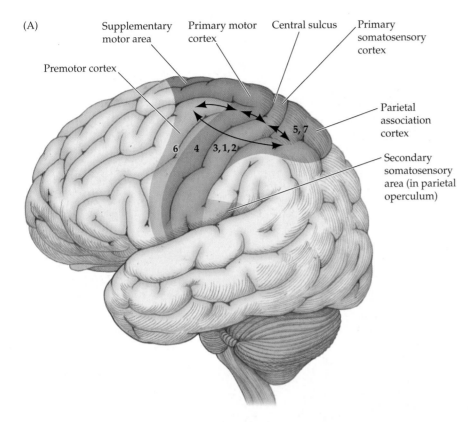

FIGURE 6.1 Motor and Somatosensory Cortical Areas
(A) Lateral view showing primary and association cortical areas of the sensory and motor cortex and reciprocal connections between these regions. Numbers indicate corresponding Brodmann areas (see Figure 2.15). (B) Medial view.

(B)

Supplementary motor area

Primary motor cortex

Primary somatosensory cortex

6 4 3, 1, 2

5, 7

Parietal association cortex

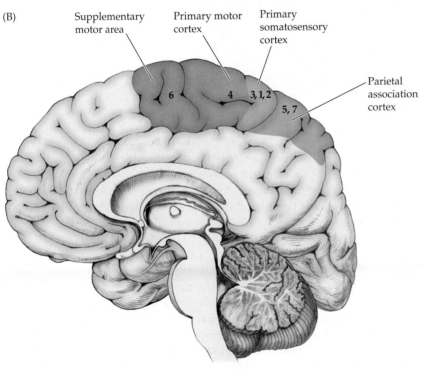

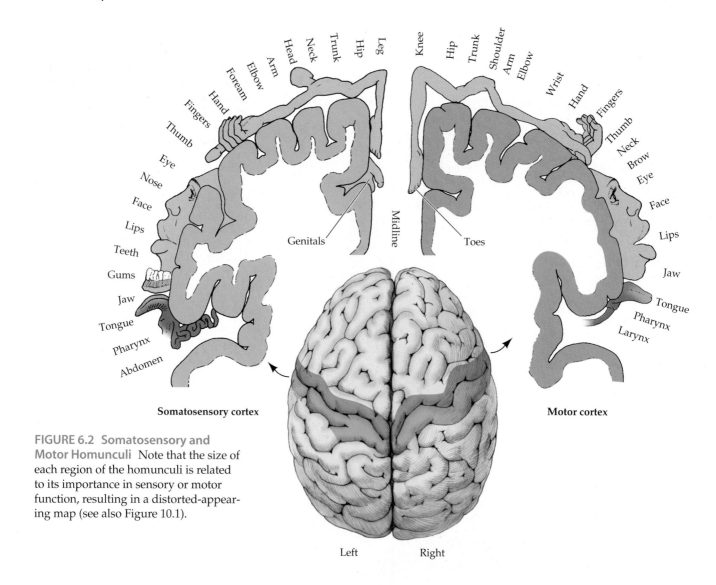

Genitals · Midline · Toes

Somatosensory cortex

Motor cortex

Left   Right

FIGURE 6.2 Somatosensory and Motor Homunculi Note that the size of each region of the homunculi is related to its importance in sensory or motor function, resulting in a distorted-appearing map (see also Figure 10.1).

MNEMONIC

somatotopic organization along their entire length, which can be traced from one level to the next in the nervous system.

A useful generalization for remembering the somatotopic representations is: *The arms are medial to the legs with two exceptions: the primary sensorimotor cortices and the posterior columns* (see, for example, Figures 6.2, 6.10, and 7.3).

Basic Anatomy of the Spinal Cord

The spinal cord contains a butterfly-shaped **central gray matter** surrounded by ascending and descending white matter columns, or funiculi (**Figure 6.3A**). **Sensory neurons** in the **dorsal root ganglia** have axons that bifurcate. One branch conveys sensory information from the periphery, and the other carries this information through the **dorsal nerve root filaments** into the dorsal aspect of the spinal cord. The central gray matter has a **dorsal** (**posterior**) **horn** that is involved mainly in sensory processing, an **intermediate zone** that contains interneurons and certain specialized nuclei (**Table 6.2**), and a **ventral** (**anterior**) **horn** that contains motor neurons. Motor neurons send their axons out of the spinal cord via the **ventral nerve root filaments**. The spinal gray matter can also be divided into nuclei or, using a different nomenclature, into laminae named by Bror Rexed (**Figure 6.3B**; see also Table 6.2), with different functions that

(A)

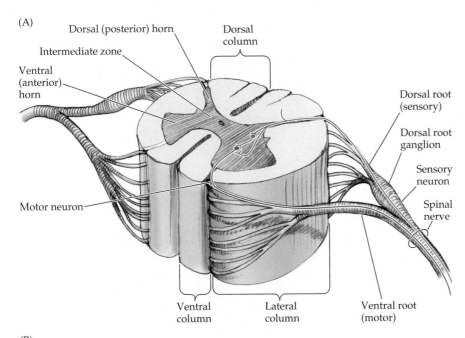

FIGURE 6.3 Basic Spinal Cord Anatomy (A) Gray matter, white matter, and dorsal and ventral roots. (B) Spinal cord nuclei (left) and Rexed's laminae (right). (See also Table 6.2.) (B from DeArmond SJ, Fusco MM, Maynard MD. 1989. *Structure of the Human Brain: A Photographic Atlas*. 3rd Ed. Oxford, New York.)

(B)

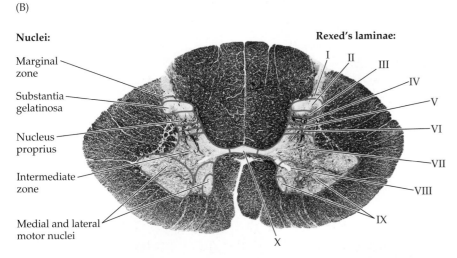

we will discuss in this chapter and in Chapter 7. The spinal cord white matter consists of **dorsal (posterior) columns**, **lateral columns**, and **ventral (anterior) columns** (see Figure 6.3A).

The spinal cord does not appear the same at all levels (**Figure 6.4**). The white matter is thickest in the cervical levels (see Figure 6.4C), where most ascending fibers have already entered the cord and most descending fibers have not yet terminated on their targets, while the sacral cord is mostly gray matter (see Figure 6.4F). In addition, the spinal cord has two enlargements (see Figure 6.4A). The **cervical enlargement** and the **lumbosacral enlargement** give rise to the nerve plexuses for the arms and legs. The spinal cord has more gray matter at the cervical and lumbosacral levels (see Figure 6.4C,E,F) than

TABLE 6.2 Nuclei and Laminae of the Spinal Cord

REGION	NUCLEI	REXED'S LAMINAE
Dorsal horn	Marginal zone	I
Dorsal horn	Substantia gelatinosa	II
Dorsal horn	Nucleus proprius	III, IV
Dorsal horn	Neck of dorsal horn	V
Dorsal horn	Base of dorsal horn	VI
Intermediate zone	Clarke's nucleus, intermediolateral nucleus	VII
Ventral horn	Commissural nucleus	VIII
Ventral horn	Motor nuclei	IX
Gray matter surrounding central canal	Grisea centralis	X

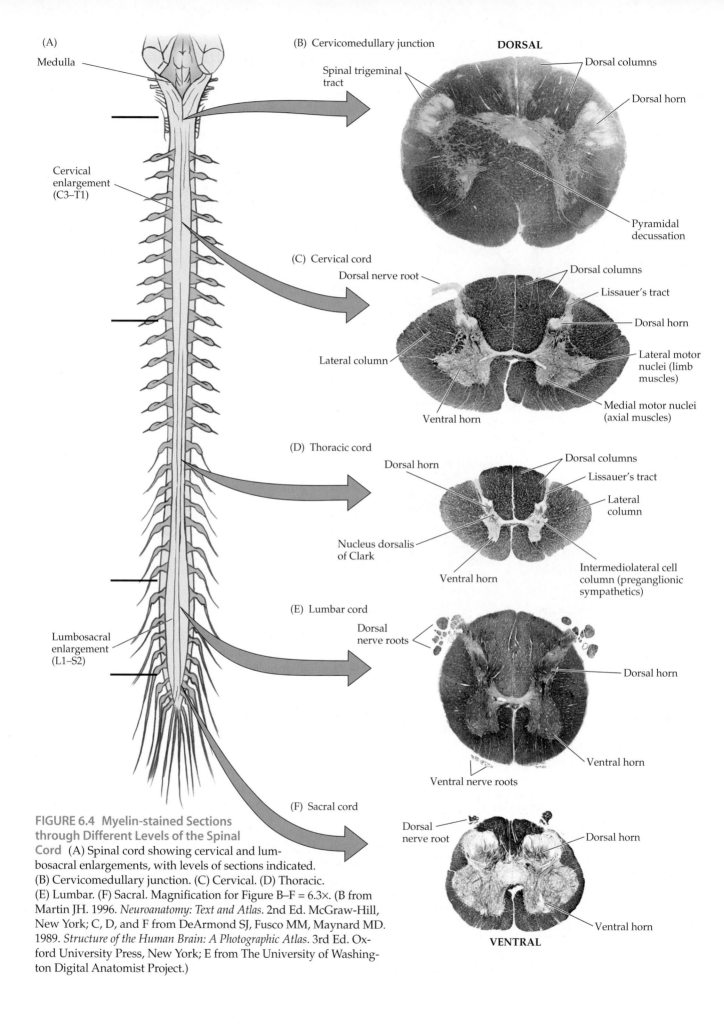

(A)

Medulla

Cervical
enlargement
(C3–T1)

Lumbosacral
enlargement
(L1–S2)

(B) Cervicomedullary junction

DORSAL

Spinal trigeminal
tract

Dorsal columns

Dorsal horn

Pyramidal
decussation

(C) Cervical cord

Dorsal nerve root

Dorsal columns

Lissauer's tract

Dorsal horn

Lateral column

Lateral motor
nuclei (limb
muscles)

Ventral horn

Medial motor nuclei
(axial muscles)

(D) Thoracic cord

Dorsal horn

Dorsal columns

Lissauer's tract

Lateral
column

Nucleus dorsalis
of Clark

Ventral horn

Intermediolateral cell
column (preganglionic
sympathetics)

(E) Lumbar cord

Dorsal
nerve roots

Dorsal horn

Ventral horn

Ventral nerve roots

(F) Sacral cord

Dorsal
nerve root

Dorsal horn

Ventral horn

VENTRAL

FIGURE 6.4 Myelin-stained Sections
through Different Levels of the Spinal
Cord (A) Spinal cord showing cervical and lumbosacral enlargements, with levels of sections indicated.
(B) Cervicomedullary junction. (C) Cervical. (D) Thoracic.
(E) Lumbar. (F) Sacral. Magnification for Figure B–F = 6.3×. (B from
Martin JH. 1996. *Neuroanatomy: Text and Atlas.* 2nd Ed. McGraw-Hill,
New York; C, D, and F from DeArmond SJ, Fusco MM, Maynard MD.
1989. *Structure of the Human Brain: A Photographic Atlas.* 3rd Ed. Oxford University Press, New York; E from The University of Washington Digital Anatomist Project.)

at the thoracic levels (see Figure 6.4D), particularly in the ventral horns, where lower motor neurons for the arms and legs reside. In the thoracic cord, a **lateral horn** is present (see Figure 6.4D) that contains the intermediolateral cell column.

Spinal Cord Blood Supply

The **blood supply to the spinal cord** arises from branches of the vertebral arteries and spinal radicular arteries (Figure 6.5). The vertebral arteries give rise to the **anterior spinal artery** that runs along the ventral surface of the spinal cord (see Figure 2.26C).

In addition, two **posterior spinal arteries** arise from the vertebral or posterior inferior cerebellar arteries and supply the dorsal surface of the cord. The anterior and posterior spinal arteries are variable in prominence at different spinal levels and form a **spinal arterial plexus** that surrounds the spinal cord (see Figure 6.5). Thirty-one segmental arterial branches enter the spinal canal along its length; most of the branches arise from the aorta and supply the meninges. Only six to ten of these reach the spinal cord as **radicular arteries**, arising at variable levels

REVIEW EXERCISE

Cover the labels in Figure 6.3A and Figure 6.4 and name as many structures as possible.

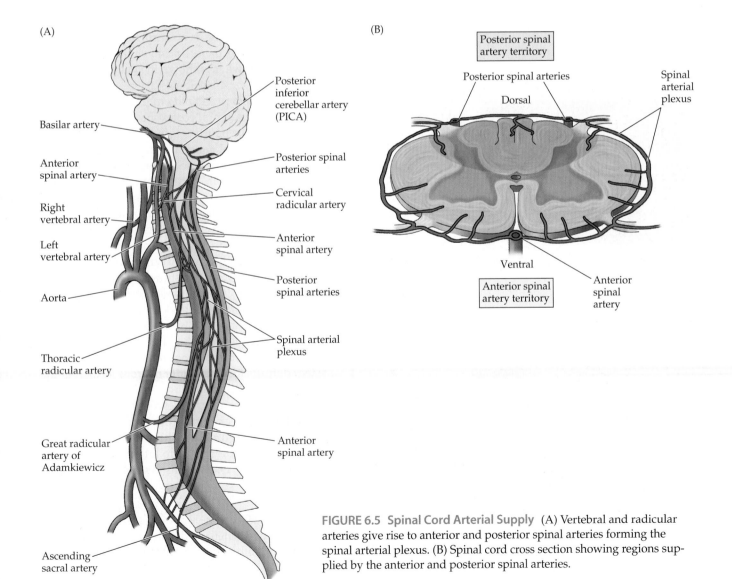

FIGURE 6.5 Spinal Cord Arterial Supply (A) Vertebral and radicular arteries give rise to anterior and posterior spinal arteries forming the spinal arterial plexus. (B) Spinal cord cross section showing regions supplied by the anterior and posterior spinal arteries.

(see Figure 6.5A). There is usually a prominent radicular artery arising from the left side, anywhere from T5 to L3, but usually between T9 and T12. This is called the **great radicular artery of Adamkiewicz**, and it provides the major blood supply to the lumbar and sacral cord.

The mid-thoracic region, at approximately **T4** to **T8**, lies between the lumbar and vertebral arterial supplies and is a **vulnerable zone** of relatively decreased perfusion. This region is most susceptible to infarction during thoracic surgery or other conditions causing decreased aortic pressure. The anterior spinal artery supplies approximately the anterior two-thirds of the cord, including the anterior horns and anterior and lateral white matter columns (see Figure 6.5B). The posterior spinal arteries supply the posterior columns and part of the posterior horns. Venous return from the spinal cord occurs initially via drainage into a plexus of veins located in the epidural space before reaching systemic circulation. These epidural veins, called **Batson's plexus** (see Figure 8.2), *do not contain valves*, so elevated intra-abdominal pressure can cause reflux of blood carrying metastatic cells (such as prostate cancer) or pelvic infections into the epidural space.

General Organization of the Motor Systems

Given the extraordinarily refined movements that can be carried out by a musician, gymnast, or surgeon, it should come as no surprise that motor systems form an elaborate network of multiple, hierarchically organized feedback loops. A summary of motor system pathways is shown in Figure 6.6. Only the most important loops are depicted, and sensory inputs have been omitted. Recall that the **cerebellum** and **basal ganglia** participate in important feedback loops, in which they project back to the cerebral cortex via the **thalamus**, and do not themselves project to lower motor neurons (see Figure 2.17). We will discuss the roles of the cerebellum and basal ganglia in motor control further in Chapters 15 and 16. Within the **cerebral cortex** itself, there are numerous important circuits for motor control. For example, circuits involving association cortex regions such as the supplementary motor area, premotor cortex, and parietal association cortex are crucial to the planning and formulation of motor activities (see Figure 6.1). As we will discuss in Chapter 19, lesions of these regions of association cortex can lead to **apraxia**, a condition characterized by a deficit in higher-order motor planning and execution despite normal strength (**neuroexam.com Video 15**). Although not shown in Figure 6.6, **sensory inputs** clearly also play an essential role in motor control and participate in motor circuits and feedback loops that range from the level of the spinal cord (see Figure 2.21) to the cerebral cortex (see Figure 6.1).

Recall also that **upper motor neurons** carry motor system outputs to **lower motor neurons** located in the spinal cord and brainstem, which, in turn, project to muscles in the periphery. Descending upper motor neuron pathways arise from the cerebral cortex and brainstem (see Figure 6.6). These descending motor pathways can be divided into **lateral motor systems** and **medial motor systems** based on their location in

Praxis

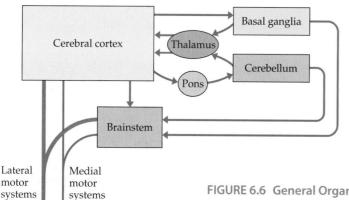

FIGURE 6.6 **General Organization of the Motor Systems** Multiple feedback loops involve the cortex, brainstem, cerebellum, basal ganglia, and thalamus. Motor output pathways (medial and lateral motor systems) arise from the cortex and from the brainstem. See also Figure 2.17.

the spinal cord. Lateral motor systems travel in the lateral columns of the spinal cord and synapse on the more lateral groups of ventral horn motor neurons and interneurons (Figure 6.7). Medial motor systems travel in the anteromedial spinal cord columns to synapse on medial ventral horn motor neurons and interneurons.

The two lateral motor systems are the **lateral corticospinal tract** and the **rubrospinal tract** (Table 6.3). These

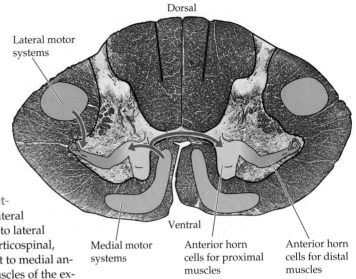

FIGURE 6.7 Somatotopic Organization of Medial and Lateral Motor System Projections to Anterior Horn Cells Lateral motor systems (corticospinal and rubrospinal tracts) project to lateral anterior horn cells, while medial motor systems (anterior corticospinal, vestibulospinal, reticulospinal, and tectospinal tracts) project to medial anterior horn cells. Lateral anterior horn cells control distal muscles of the extremities, while medial anterior horn cells control proximal trunk muscles. (Spinal cord section from DeArmond SJ, Fusco MM, Maynard MD. 1989. *Structure of the Human Brain: A Photographic Atlas.* 3rd Ed. Oxford University Press, New York.)

TABLE 6.3 Lateral and Medial Descending Motor Systems

TRACT	SITE OF ORIGIN	SITE OF DECUSSATION (WHERE RELEVANT)	LEVELS OF TERMINATION	FUNCTION
LATERAL MOTOR SYSTEMS				
Lateral corticospinal tract	Primary motor cortex and other frontal and parietal areas	Pyramidal decussation, at the cervicomedullary junction	Entire cord (predominantly at cervical and lumbosacral enlargements)	Movement of contralateral limbs
Rubrospinal tract	Red nucleus, magnocellular division	Ventral tegmental decussation, in the midbrain	Cervical cord	Movement of contralateral limbs (function is uncertain in humans)
MEDIAL MOTOR SYSTEMS				
Anterior corticospinal tract	Primary motor cortex and supplementary motor area	—	Cervical and upper thoracic cord	Control of bilateral axial and girdle muscles
Vestibulospinal tracts (VSTs)[a]	Medial VST: medial and inferior vestibular nuclei; lateral VST: lateral vestibular nucleus	—	Medial VST: Cervical and upper thoracic cord; Lateral VST: entire cord	Medial VST: positioning of head and neck; lateral VST: balance
Reticulospinal tracts	Pontine and medullary reticular formation	—	Entire cord	Automatic posture and gait-related movements
Tectospinal tract	Superior colliculus	Dorsal tegmental decussation, in the midbrain	Cervical cord	Coordination of head and eye movement (uncertain in humans)

[a]Despite their names, both medial and lateral VSTs are medial motor systems.

pathways control the movement of the extremities (see Figure 6.7). The lateral corticospinal tract in particular is essential for rapid, dextrous movements at individual digits or joints. Both of these pathways cross over from their site of origin and descend in the contralateral lateral spinal cord to control the contralateral extremities (see Table 6.3; Figure 6.11A,B).

The four medial motor systems are the **anterior corticospinal tract** (see Figure 6.11C), the **vestibulospinal tracts** (see Figure 6.11D), the **reticulospinal tracts**, and the **tectospinal tract** (see Figure 6.11E; Table 6.3). These pathways control the proximal axial and girdle muscles involved in postural tone, balance, orienting movements of the head and neck, and automatic gait-related movements (see Figure 6.7). The medial motor systems descend ipsilaterally or bilaterally. Some extend only to the upper few cervical segments (see Table 6.3; Figure 6.11C–E).

The medial motor systems tend to terminate on interneurons that project to both sides of the spinal cord, controlling movements that involve multiple bilateral spinal segments. Thus, unilateral lesions of the medial motor systems produce no obvious deficits. In contrast, lesions of the lateral corticospinal tract produce dramatic deficits (see the next section). The rubrospinal tract in humans is small, and its clinical importance is uncertain, but it may participate in taking over functions after corticospinal injury. It may also play a role in flexor (decorticate) posturing of the upper extremities (see Figure 3.5A), which is typically seen with lesions above the level of the red nuclei, in which the rubrospinal tract is spared.

Lateral Corticospinal Tract

The corticospinal tract—more specifically, the **lateral corticospinal tract**—is the most clinically important descending motor pathway in the nervous system. This pathway controls movement of the extremities, and lesions along its course produce characteristic deficits that often enable precise clinical localization. Because of its clinical importance, we will discuss the corticospinal tract in greater detail than the other descending motor systems. Let's follow the course of the corticospinal tract from cerebral cortex to spinal cord (Figure 6.8). Over half of the corticospinal tract fibers originate in the primary motor cortex (Brodmann's area 4) of the precentral gyrus. The remainder arise from the premotor and supplementary motor areas (area 6) or from the parietal lobe (areas 3, 1, 2, 5, and 7) (Figure 6.9A). The primary motor cortex neurons contributing to the corticospinal tract are located mostly in cortical layer 5 (see Figure 2.14B). Layer 5 pyramidal cell projections synapse directly onto motor neurons in the ventral horn of the spinal cord as well as onto spinal interneurons. About 3% of corticospinal neurons are giant pyramidal cells called **Betz cells**, which are the largest neurons in the human nervous system.

Axons from the cerebral cortex enter the upper portions of the cerebral white matter, or **corona radiata** (see Figure 4.13J), and descend toward the internal capsule (see Figure 4.13G). In addition to the corticospinal tract, the cerebral white matter conveys bidirectional information between different cortical areas, and between cortex and deep structures such as the basal ganglia, thalamus, and brainstem (Figure 6.9B). These white matter pathways form a fanlike structure as they enter the internal capsule, which condenses down to fewer

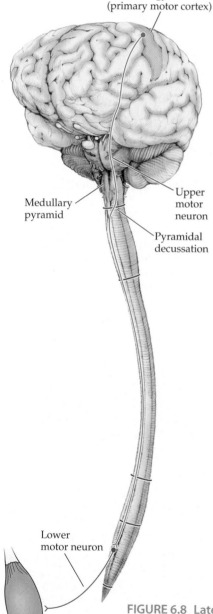

Precentral gyrus
(primary motor cortex)

Medullary
pyramid

Upper
motor
neuron

Pyramidal
decussation

Lower
motor neuron

Skeletal
muscle

FIGURE 6.8 Lateral Corticospinal Tract Upper motor neuron in the primary motor cortex (precentral gyrus) sends an axon downward to cross over at the pyramidal decussation. The axon then continues downward in the contralateral spinal cord before synapsing on the lower motor neuron in the anterior horn.

(A)

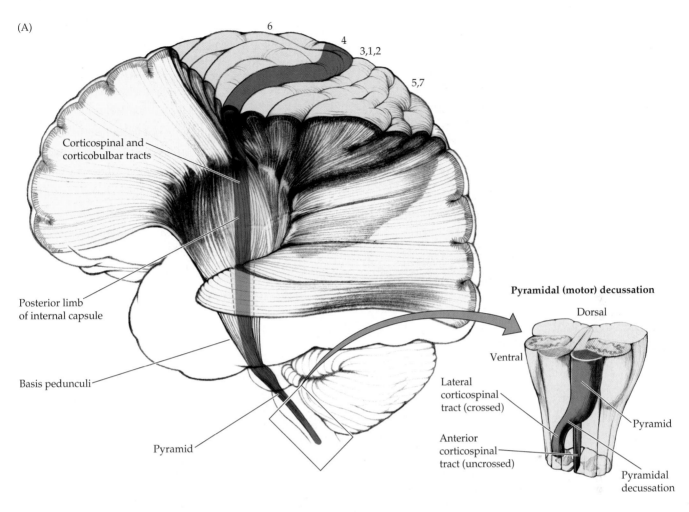

6

4

3,1,2

5,7

Corticospinal and
corticobulbar tracts

Posterior limb
of internal capsule

Basis pedunculi

Pyramid

Pyramidal (motor) decussation

Dorsal

Ventral

Lateral
corticospinal
tract (crossed)

Anterior
corticospinal
tract (uncrossed)

Pyramid

Pyramidal
decussation

(B)

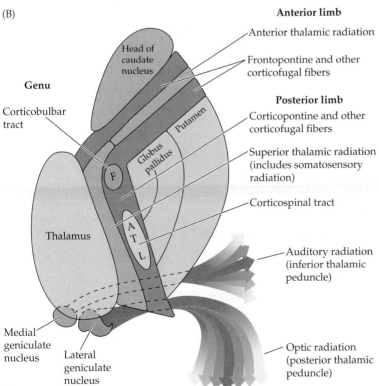

Anterior limb

Anterior thalamic radiation

Frontopontine and other
corticofugal fibers

Head of
caudate
nucleus

Genu

Corticobulbar
tract

Putamen

Globus
pallidus

F

Posterior limb

Corticopontine and other
corticofugal fibers

Superior thalamic radiation
(includes somatosensory
radiation)

Corticospinal tract

Thalamus

A
T
L

Auditory radiation
(inferior thalamic
peduncle)

Medial
geniculate
nucleus

Lateral
geniculate
nucleus

Optic radiation
(posterior thalamic
peduncle)

FIGURE 6.9 Internal Capsule (A) Three-dimensional representation of the internal capsule. Corticospinal and corticobulbar fibers arising from the primary motor cortex and adjacent regions form part of the internal capsule. Corticobulbar fibers project to lower motor neurons in the brainstem. About 85% of corticospinal fibers cross over at the pyramidal decussation to form the lateral corticospinal tract, while the remaining fibers form the anterior corticospinal tract. (B) Horizontal section through internal capsule showing the anterior limb, genu, and posterior limb in relation to the thalamus, head of the caudate, and lentiform nucleus (putamen and globus pallidus). Major fiber pathways of the internal capsule are indicated.

(A) Forebrain

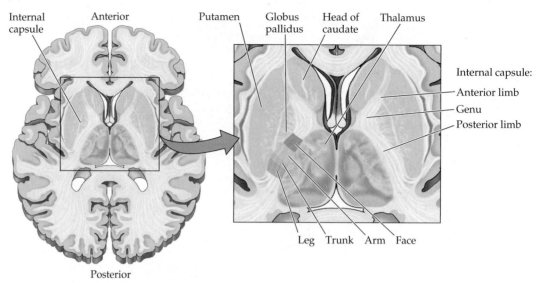

(B) Midbrain

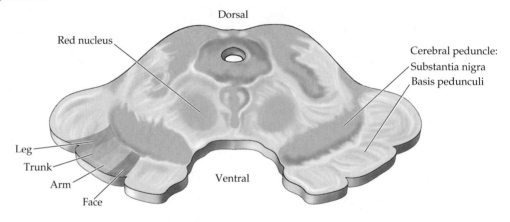

(C) Spinal cord

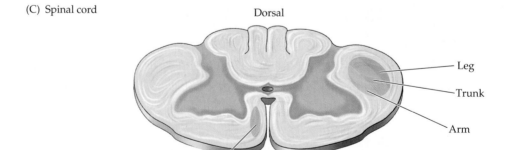

FIGURE 6.10 Somatotopic Organization of Corticobulbar and Corticospinal Tracts (A) Horizontal section through the internal capsule. (B) Midbrain. (C) Spinal cord.

and fewer fibers as connections to different subcortical structures are made (see Figure 6.9A).

The **internal capsule** is best appreciated in horizontal brain sections (Figure 6.10A), in which the right and left internal capsules look like arrowheads or two letter *V*s, with their points facing inward. Note that the thalamus and caudate nucleus are always medial to the internal capsule, while the globus pallidus and putamen are always lateral to the internal capsule. There are three parts to the internal capsule: anterior limb, posterior limb, and genu. Note that the **anterior limb** of the internal capsule separates the head of the caudate from the globus pallidus and putamen, while the **posterior limb** separates the thalamus from the globus pallidus and putamen (see also Figures 16.2 and 16.3). The **genu** ("knee" in Latin) is at the transition between the anterior and posterior limbs, at the level of the foramen of Monro. The **corticospinal tract lies in the posterior limb** of the internal capsule. The **somatotopic map** is preserved in the internal capsule, so motor fibers for the face are most anterior, and those for the arm and leg are progressively more posterior (see Figure 6.10A). Fibers projecting from the cortex to the brainstem, including motor fibers for the face, are called **corticobulbar** instead of corticospinal because they project from the cortex to the brainstem, or "bulb." Despite the somatotopic arrangement, the fibers of the internal capsule are compact enough that lesions at this level often produce weakness of the entire contralateral body (face, arm, and leg) (see KCC 6.3; see also Figure 6.14A). However, occasionally capsular lesions can also produce more selective motor deficits. Additional details of the fibers carried by the internal capsule other than the corticobulbar and corticospinal tracts are shown in Figure 6.9B.

The internal capsule continues into the midbrain **cerebral peduncles**, meaning, literally, "feet of the brain" (Figure 6.10B). The white matter is located in the ventral portion of the cerebral peduncles and is called the **basis pedunculi**. The **middle one-third of the basis pedunculi** contains corticobulbar and corticospinal fibers with the face, arm, and leg axons arranged from medial to lateral, respectively (see Figure 6.10B). The other portions of the basis pedunculi contain primarily corticopontine fibers (see Chapter 15).

The corticospinal tract fibers next descend through the ventral pons, where they form somewhat scattered fascicles (Figure 6.11A). These collect on the ventral surface of the medulla to form the **medullary pyramids** (see Figures 6.8 and 6.11A). For this reason the corticospinal tract is sometimes referred to as the **pyramidal tract** (this terminology, though widely used, is somewhat imprecise since the pyramids include reticulospinal and other brainstem pathways in addition to the corticospinal tract). The transition from medulla to spinal cord is called the **cervicomedullary junction**, which occurs at the level of the foramen magnum (see Figure 5.10). At this point about 85% of the pyramidal tract fibers cross over in the **pyramidal decussation** to enter the lateral white matter columns of the spinal cord, forming the lateral corticospinal tract (see Figures 6.8 and 6.11A). A somatotopic representation is present in the lateral corticospinal tract, with fibers that control the upper extremity located medial to those that control the lower extremity (Figure 6.10C). Finally, the axons of the lateral corticospinal tract enter the spinal cord central gray matter to synapse onto anterior horn cells (see Figures 6.7, 6.8, and 6.11A). The remaining ~15% of corticospinal fibers continue into the spinal cord ipsilaterally, without crossing, and enter the anterior white matter columns to form the anterior corticospinal tract (see Figure 6.9A and Figure 6.11C).

In addition to the lateral corticospinal tract, the other lateral and medial descending motor systems (see Table 6.3) are shown in Figure 6.11 as well. They include the rubrospinal, anterior corticospinal, tectospinal, reticulospinal, and vestibulospinal tracts.

MNEMONIC

REVIEW EXERCISE

Lesions of the corticospinal tract occurring above the pyramidal decussation produce contralateral weakness, whereas lesions below the pyramidal decussation produce ipsilateral weakness (see Figures 6.8 and 6.11A). For a lesion in each of the following locations, state whether the weakness would be ipsilateral or contralateral to the lesion: cortex, internal capsule, midbrain, pons, medulla, and spinal cord.

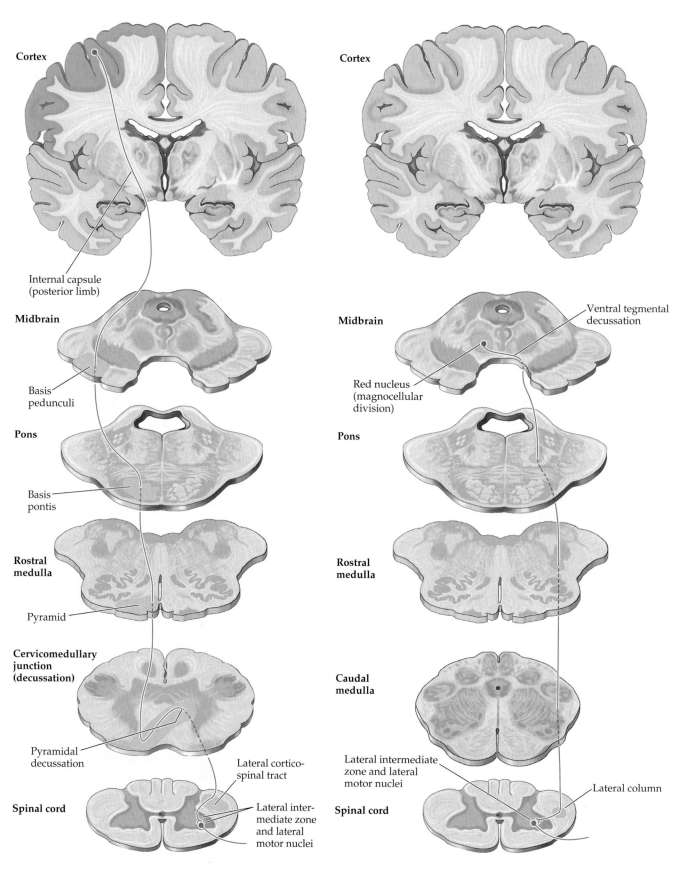

Cortex

Internal capsule
(posterior limb)

Midbrain

Basis
pedunculi

Pons

Basis
pontis

Rostral
medulla

Pyramid

Cervicomedullary
junction
(decussation)

Pyramidal
decussation

Spinal cord

Lateral cortico-
spinal tract

Lateral inter-
mediate zone
and lateral
motor nuclei

Cortex

Ventral tegmental
decussation

Midbrain

Red nucleus
(magnocellular
division)

Pons

Rostral
medulla

Caudal
medulla

Lateral intermediate
zone and lateral
motor nuclei

Lateral column

Spinal cord

(A) **Lateral corticospinal tract**

(B) **Rubrospinal tract**

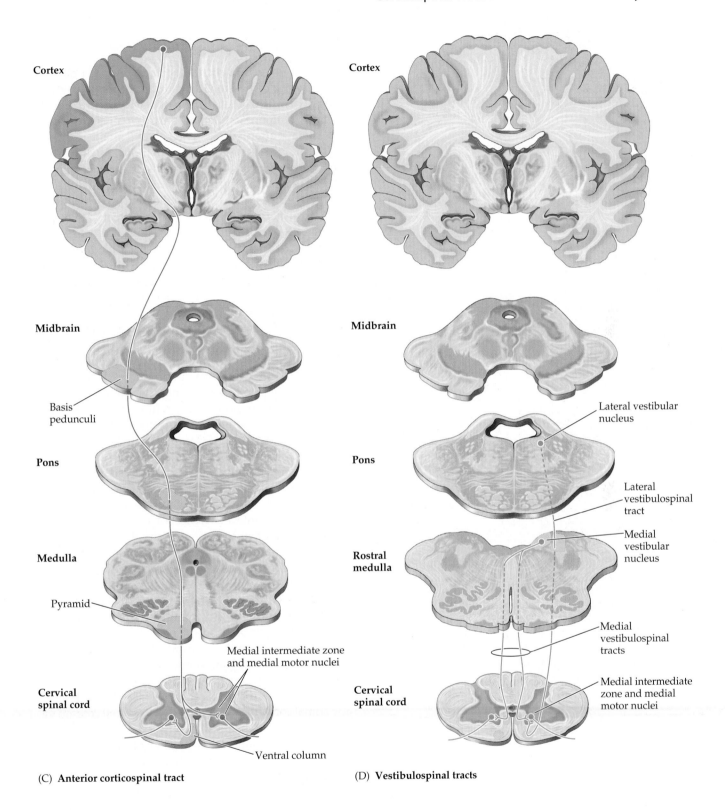

Cortex

Cortex

Midbrain

Midbrain

Basis pedunculi

Lateral vestibular nucleus

Pons

Pons

Lateral vestibulospinal tract

Medulla

Rostral medulla

Medial vestibular nucleus

Pyramid

Medial intermediate zone and medial motor nuclei

Medial vestibulospinal tracts

Cervical spinal cord

Cervical spinal cord

Medial intermediate zone and medial motor nuclei

Ventral column

(C) **Anterior corticospinal tract**

(D) **Vestibulospinal tracts**

FIGURE 6.11 **Descending Motor Pathways** Illustrated on this page, overleaf, and page 238 (see also Table 6.3).

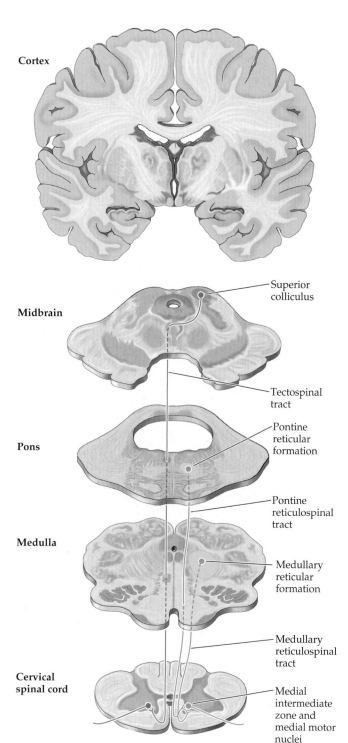

Cortex

Midbrain

Superior
colliculus

Tectospinal
tract

Pons

Pontine
reticular
formation

Pontine
reticulospinal
tract

Medulla

Medullary
reticular
formation

Medullary
reticulospinal
tract

Cervical
spinal cord

Medial
intermediate
zone and
medial motor
nuclei

(E) **Tectospinal tract and reticulospinal tract**

FIGURE 6.11   (*continued*)

Autonomic Nervous System

In contrast to the somatic motor pathways described in the preceding discussion, the **autonomic nervous system** generally controls more automatic and visceral bodily functions. Autonomic efferents are different anatomically from **somatic efferents** (Figure 6.12). In somatic efferents, anterior horn cells or cranial nerve motor nuclei project directly from the central nervous system to skeletal muscle (see Figure 6.12A). In **autonomic efferents**, there is a peripheral synapse located in a ganglion interposed between the central nervous system and the effector gland or smooth muscle (see Figure 6.12B,C). There are sensory inputs to the autonomic nervous system located both centrally and in the periphery. However, the autonomic nervous system itself consists of only efferent pathways and is therefore discussed here with other motor systems.

The autonomic nervous system has two main divisions (Figure 6.13). The **sympathetic**, or **thoracolumbar**, **division** arises from T1 to L2, or T1 to L3, spinal levels and is involved mainly in such "fight-or-flight" functions as increasing heart rate and blood pressure, bronchodilation, and increasing pupil size. The **parasympathetic**, or **craniosacral**, **division**, in contrast, arises from cranial nerve nuclei and from S2 through S4 and is involved in "rest and digest" functions, such as increasing gastric secretions and peristalsis, slowing the heart rate, and decreasing pupil size. The **enteric nervous system**, considered a third autonomic division, consists of a neural plexus, lying within the walls of the gut, which is involved in controlling peristalsis and gastrointestinal secretions.

Preganglionic neurons of the sympathetic division are located in the **intermediolateral cell column** in lamina VII of spinal cord levels T1 to L2 or T1 to L3 (see Figures 6.4D, 6.12B, and 6.13). There are two sets of sympathetic ganglia. The paired **paravertebral ganglia** form a chain called the **sympathetic trunk** (or sympathetic chain) running all the way from cervical to sacral levels on each side of the spinal cord. The sympathetic trunk allows sympathetic efferents, which exit only at thoracolumbar levels, to reach other parts of the body as well. For example, sympathetics are provided to the head and neck by the upper thoracic spinal cord (T1–T3) intermediolateral cell column via three sympathetic chain ganglia named the **superior**, **middle** (often absent), and **inferior** (stellate) **cervical ganglia** (see Figure 13.10). The other sympathetic ganglia are unpaired **prevertebral ganglia**, located in the celiac plexus surrounding the aorta and include the celiac ganglion, superior mesenteric ganglion, and inferior mesenteric ganglion. Axons of preganglionic sympathetic neurons thus have a fairly short distance to travel, while axons of **postganglionic neurons** travel a long distance to reach effector organs (see Figures 6.12B and 6.13). In contrast, parasympathetic preganglionic fibers must travel a long distance to reach the **terminal ganglia** located within or near the effector organs (see Figures 6.12C and 6.13). Parasympathetic preganglionic fibers arise from **cranial nerve parasympathetic**

(A) Somatic efferent

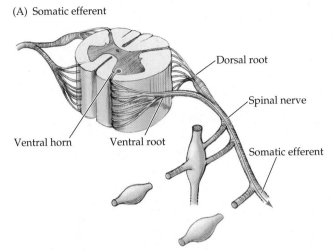

(B) Sympathetic efferent

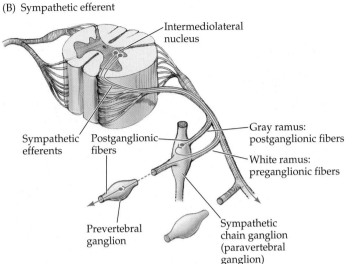

FIGURE 6.12 Somatic and Autonomic Efferents
(A) Somatic efferent arising from an anterior horn cell.
(B) Sympathetic efferents arising from the intermediolateral
nucleus. (C) Parasympathetic efferent arising from sacral
parasympathetic nuclei.

(C) Parasympathetic efferent

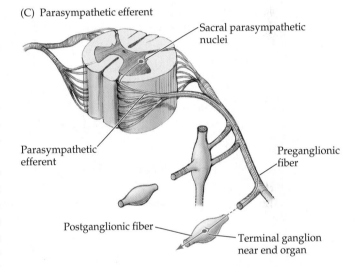

nuclei (see Figures 12.5 and 12.6) and from the **sacral parasympathetic nuclei** located in the lateral gray matter of S2, S3, and S4, in a location similar to that of the intermediolateral cell column (see Figure 6.12C).

The sympathetic and parasympathetic systems also differ in terms of their **neurotransmitters** (see Chapter 2; see also Figures 6.12 and 6.13). **Sympathetic postganglionic** neurons release predominantly **norepinephrine** onto end organs. **Parasympathetic postganglionic** neurons release predominantly **acetylcholine**, activating **muscarinic cholinergic receptors** on end organs. **Preganglionic neurons** in both sympathetic and parasympathetic ganglia release **acetylcholine**, activating **nicotinic cholinergic receptors** (see Figures 6.12 and 6.13). Noradrenergic (α_1, α_2, β_1, β_2, β_3) and muscarinic cholinergic (M_1, M_2, M_3) receptor subtypes mediate different actions of these transmitters on end organs (see the References section at the end of this chapter). In addition, various peptides and other substances (such as ATP and adenosine) are released at autonomic synapses. One notable exception to the norepinephrine/sympathetic, acetylcholine/parasympathetic rule for postganglionic neurotransmitters is the sweat glands, which are innervated by sympathetic postganglionic neurons that release acetylcholine (see Figure 6.13).*

Sympathetic and parasympathetic outflow are controlled both directly and indirectly by higher centers, including the hypothalamus (see Chapter 17), brainstem nuclei such as the nucleus solitarius (see Chapter 12), the amygdala, and several regions of limbic cortex (see Chapter 18). Autonomic responses are also regulated by afferent sensory information, including signals from internal receptors such as chemoreceptors, osmoreceptors, thermoreceptors, and baroreceptors.

——————— Acetylcholine
——————— Norepinephrine

*For the interested reader, this point of relatively minor clinical significance can be remembered by noting that botulinum toxin, a blocker of cholinergic transmission, has recently been shown to be an effective treatment for hyperhidrosis (excessive sweating) when injected locally into the skin of the axilla.

MNEMONIC

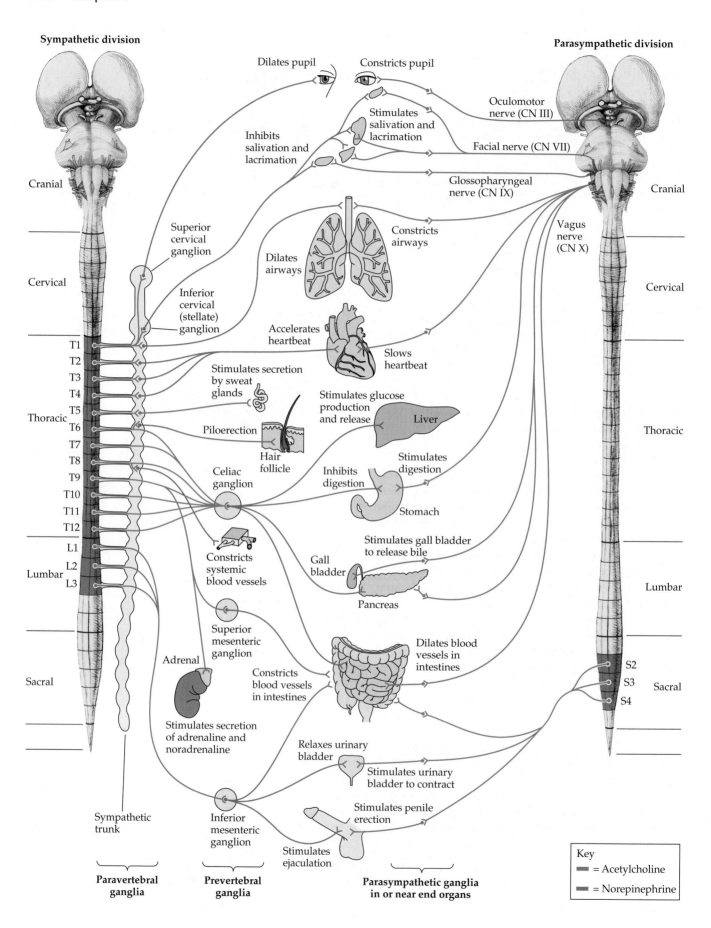

Sympathetic division

Dilates pupil

Constricts pupil

Oculomotor nerve (CN III)

Stimulates salivation and lacrimation

Inhibits salivation and lacrimation

Facial nerve (CN VII)

Glossopharyngeal nerve (CN IX)

Parasympathetic division

Vagus nerve (CN X)

Cranial

Superior cervical ganglion

Inferior cervical (stellate) ganglion

Constricts airways

Dilates airways

Cervical

Cervical

Accelerates heartbeat

Slows heartbeat

T1
T2
T3
T4
T5
T6
T7
T8
T9
T10
T11
T12

Stimulates secretion by sweat glands

Stimulates glucose production and release

Liver

Piloerection

Hair follicle

Celiac ganglion

Inhibits digestion

Stimulates digestion

Stomach

Thoracic

Thoracic

L1
L2
L3

Constricts systemic blood vessels

Stimulates gall bladder to release bile

Gall bladder

Pancreas

Lumbar

Lumbar

Superior mesenteric ganglion

Adrenal

Constricts blood vessels in intestines

Dilates blood vessels in intestines

Stimulates secretion of adrenaline and noradrenaline

S2
S3
S4

Sacral

Sacral

Relaxes urinary bladder

Stimulates urinary bladder to contract

Sympathetic trunk

Inferior mesenteric ganglion

Stimulates penile erection

Stimulates ejaculation

Paravertebral ganglia

Prevertebral ganglia

Parasympathetic ganglia in or near end organs

Key
= Acetylcholine
= Norepinephrine

◀ **FIGURE 6.13 Sympathetic and Parasympathetic Divisions of the Autonomic Nervous System** Sympathetic outputs (left) arise from thoracolumbar spinal cord segments and synapse in paravertebral and prevertebral ganglia. Parasympathetic outputs (right) arise from craniosacral regions and synapse in ganglia in or near effector organs.

KEY CLINICAL CONCEPT

6.1 UPPER MOTOR NEURON VERSUS LOWER MOTOR NEURON LESIONS

The concept of upper motor neuron versus lower motor neuron lesions is very useful in clinical localization (see Figure 6.8). **Upper motor neurons** of the corticospinal tract project from the cerebral cortex to lower motor neurons located in the anterior horn of the spinal cord. **Lower motor neurons**, in turn, project via peripheral nerves to skeletal muscle. An identical concept of upper and lower motor neurons applies to the corticobulbar tract and cranial nerve motor nuclei.

Signs of **lower motor neuron lesions** include muscle weakness, atrophy, fasciculations, decreased tone, and hyporeflexia (Table 6.4; see also **neuroexam.com Video 48**). **Fasciculations** are abnormal muscle twitches caused by spontaneous activity in groups of muscle cells. An example of a benign fasciculation not associated with motor neuron damage is the eyelid twitching often experienced during periods of fatigue, caffeine excess, and eye strain (such as while reading too much neuroanatomy in one night). Signs of **upper motor neuron lesions** include muscle weakness and a combination of increased tone (see **neuroexam.com Videos 49** and **50**) and hyperreflexia sometimes referred to as **spasticity**. Additional abnormal reflexes seen with upper motor neuron lesions, such as **Babinski's sign** (see Figure 3.2), Hoffmann's sign, posturing, and so on, are discussed in Chapter 3 (see also **neuroexam.com Videos 58–60**). Note that with acute upper motor neuron lesions, flaccid paralysis with decreased tone and decreased reflexes may initially occur and gradually, over hours or even months, develop into spastic paresis. Spinal shock is an example of this process (see KCC 7.2).

Increased tone and hyperreflexia do not occur in experimental animals when a selective lesion is made in the corticospinal tract alone. It has therefore been hypothesized that spasticity is caused by damage to descending inhibitory pathways that travel closely with the corticospinal tract, rather than by damage to the corticospinal tract itself. Loss of these descending inhibitory influences may lead to increased excitability of motor neurons in the anterior horn, resulting in brisk reflexes and increased tone. Although this hypothesis fits available data, it has yet to be definitely proven in humans. ■

Atrophy? Fasciculations?

Lower extremity tone

TABLE 6.4 Signs of Upper Motor Neuron and Lower Motor Neuron Lesions		
SIGN	UPPER MOTOR NEURON LESIONS	LOWER MOTOR NEURON LESIONS
Weakness	Yes	Yes
Atrophy	No[a]	Yes
Fasciculations	No	Yes
Reflexes	Increased[b]	Decreased
Tone	Increased[b]	Decreased

[a]Mild atrophy may develop due to disuse.

[b]With acute upper motor lesions, reflexes and tone may be decreased.

TABLE 6.5 Terms Commonly Used to Describe Weakness

TERM	DEFINITION	EXAMPLE	CLINICAL SYMPTOMS
DENOTING SEVERITY			
Paresis	Weakness (partial paralysis)	Hemiparesis	Weakness of one side of body (face, arm, and leg)
-plegia	No movement[a]	Hemiplegia	No movement of one side of body (face, arm, and leg)
Paralysis	No movement[a]	Leg paralysis	No movement of the leg
Palsy	Imprecise term for weakness or no movement	Facial palsy	Weakness or paralysis of face muscles
DENOTING LOCATION			
Hemi-	One side of body (face, arm, and leg)	Hemiplegia	No movement of one side of body (face, arm, and leg)
Para-	Both legs	Paraparesis	Weakness of both legs
Mono-	One limb	Monoparesis	Weakness of one limb (arm or leg)
Di-	Both sides of body equally affected	Facial diplegia	Symmetrical facial weakness
Quadri- or tetra-	All four limbs	Quadriplegia (tetraplegia)	Paralysis of all four limbs

[a]In upper motor neuron lesions causing paralysis or plegia there is no voluntary movement of the limb, but reflexes may be present.

KEY CLINICAL CONCEPT

6.2 TERMS USED TO DESCRIBE WEAKNESS

Weakness is one of the most important functional consequences of both upper and lower motor neuron lesions. Various terms are used in clinical practice to describe both the severity and distribution of weakness (Table 6.5). We will discuss localization of these different patterns of weakness in the next section. ■

KEY CLINICAL CONCEPT

6.3 WEAKNESS PATTERNS AND LOCALIZATION

Weakness can be caused by lesions or dysfunction at any level in the motor system, including the association and limbic cortices involved in volitional or motivational control of movement, the upper motor neurons of the corticospinal tract anywhere from cortex to spinal cord, the lower motor neurons anywhere from anterior horn to peripheral nerve, the neuromuscular junction, the muscles, and the mechanical functions of joints and tendons. The process of localizing lesions, as outlined in the sections that follow, requires that you identify the correct motor system level, side, and specific neuroanatomical structures affected. In the illustrations in this section, lesions are shown in red and deficits are shown in purple.

Unilateral Face, Arm, and Leg Weakness or Paralysis

OTHER NAMES: Hemiparesis or hemiplegia.

1. With No Associated Sensory Deficits Figure 6.14A

OTHER NAMES: Pure motor hemiparesis.

LOCATIONS RULED OUT: Unlikely to be cortical because the lesion would have to involve the entire motor strip, in which case sensory involvement is hard to avoid. Unlikely to be muscle or peripheral nerve because in that case coincidental involvement of the face, arm and leg, all on one side of the body, would be required. Not the spinal cord or medulla because in that case the face would be spared.

REVIEW EXERCISE

Cover the right two columns in Table 6.4. For each sign, state whether it is present, increased, or decreased with upper motor neuron lesions and with lower motor neuron lesions.

FIGURE 6.14A Pure Hemiparesis ▶

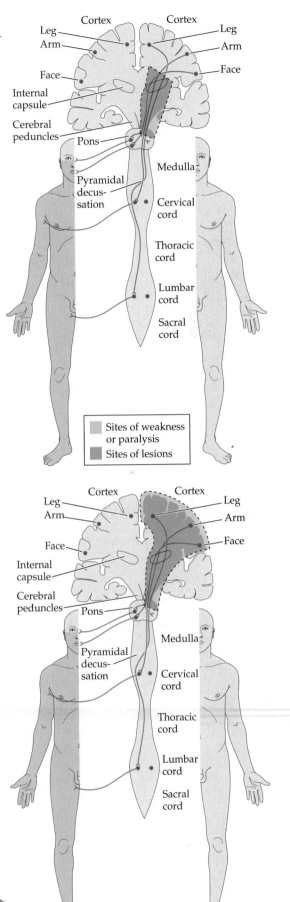

LOCATIONS RULED IN: Corticospinal and corticobulbar tract fibers below the cortex and above the medulla: corona radiate, posterior limb of the internal capsule, basis pontis, or middle third of the cerebral peduncle.

SIDE OF LESION: Contralateral to weakness (above the pyramidal decussation).

COMMON CAUSES: Lacunar infarct of the internal capsule (lenticulostriate branches of the middle cerebral artery or anterior choroidal artery; see Figures 10.7 and 10.9; see also Table 10.3) or of the pons (median perforating branches of the basilar artery; see Figure 14.21B,C; see also Table 14.8). Infarct of the cerebral peduncle (see Figure 14.21A) is less common. Demyelination, tumor, or abscess in these locations or in the corona radiata can also cause pure motor hemiparesis.

ASSOCIATED FEATURES: Upper motor neuron signs (see Table 6.4) are usually present. Dysarthria (see KCC 12.8) is common, giving rise to the name **dysarthria–pure motor hemiparesis**. Ataxia of the affected side may also occasionally be seen because of involvement of the cerebellar pathways (corticopontine fibers), giving rise to the name **ataxia-hemiparesis** (see Table 10.3; see also KCC 15.2).

2. With Associated Somatosensory, Oculomotor, Visual, or Higher Cortical Deficits Figure 6.14B

LOCATIONS RULED OUT: Unlikely to be below the medulla, for the reasons listed in the previous section.

LOCATIONS RULED IN: Entire primary motor cortex, including face, arm, and leg representations of the precentral gyrus, or corticospinal and corticobulbar tract fibers above the medulla (e.g, thalamocapsular lacune; see Table 10.3). The lesion can usually be further localized on the basis of other associated deficits.

SIDE OF LESION: Contralateral to weakness (above the pyramidal decussation).

ASSOCIATED FEATURES ALLOWING FURTHER LOCALIZATION: In addition to somatosensory, oculomotor, visual, or higher cortical deficits such as aphasia or neglect, there may be dysarthria or ataxia. Upper motor neuron signs are usually present as well.

COMMON CAUSES: Numerous, including infarct, hemorrhage, tumor, trauma, herniation, post-ictal state, and so on.

Unilateral Arm and Leg Weakness or Paralysis
Figure 6.14C

OTHER NAMES: Hemiplegia or hemiparesis sparing the face; brachiocrural plegia or paresis.

LOCATIONS RULED OUT: Unlikely to be the corticospinal tract below the motor cortex above the medulla, because the corticobulbar tract fibers are located very close nearby and the face would thus usually be involved. Unlikely to be muscle or peripheral nerve because in that case coincidental involvement of both the arm and leg on one side of the body would be required. Not below C5 in the cervical cord because in that case some arm muscles would be spared.

FIGURE 6.14B **Hemiparesis with Additional Deficits** ▶

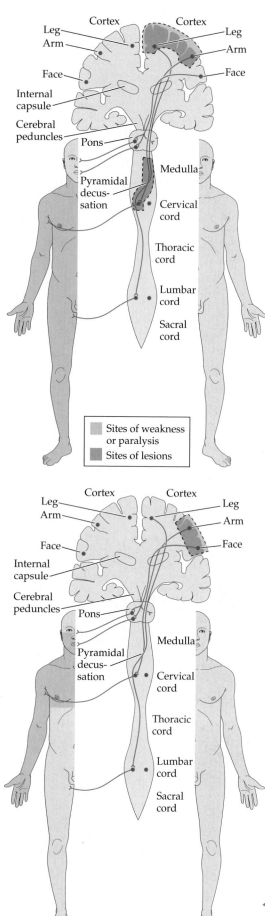

Cortex
Leg
Arm
Face
Internal capsule
Cerebral peduncles
Pons
Pyramidal decussation
Medulla
Cervical cord
Thoracic cord
Lumbar cord
Sacral cord

Sites of weakness or paralysis
Sites of lesions

◀ FIGURE 6.14C Hemiparesis Sparing the Face

LOCATIONS RULED IN: Arm and leg area of the motor cortex; corticospinal tract from the lower medulla to the C5 level of cervical spinal cord.

SIDE OF LESION: Motor cortex or medulla (above the pyramidal decussation): contralateral to weakness. Cervical spinal cord (below the pyramidal decussation): ipsilateral to weakness.

ASSOCIATED FEATURES ALLOWING FURTHER LOCALIZATION: Upper motor neuron signs are usually present. Cortical lesions sparing the face are often in a watershed distribution, and they affect proximal more than distal muscles ("man in the barrel" syndrome; see KCC 10.2). Cortical lesions may be associated with aphasia (see KCC 19.6) or hemineglect (see KCC 19.9). In medial medullary lesions there may be loss of vibration and joint position sense on the same side as the weakness and tongue weakness on the opposite side (see Figure 14.21D; Table 14.7). In lesions extending to the lateral medulla, the lateral medullary syndrome may be present (see KCC 14.3; Table 14.7). In lesions of the spinal cord, the Brown–Séquard syndrome may be present (see KCC 7.4). High cervical lesions may involve the spinal trigeminal nucleus and tract (see Figure 12.8), causing decreased facial sensation.

COMMON CAUSES: Watershed infarct (anterior cerebral–middle cerebral watershed), medial or combined medial and lateral medullary infarcts, multiple sclerosis, lateral trauma, or compression of the cervical spinal cord. Occasionally, infarcts of the posterior limb of the internal capsule that are removed from the genu (see Figure 6.9B) may cause contralateral hemiparesis sparing the face.

Unilateral Face and Arm Weakness or Paralysis
Figure 6.14D

OTHER NAMES: Faciobrachial paresis or plegia.

LOCATIONS RULED OUT: Unlikely to be muscle or peripheral nerve because in that case coincidental involvement of the face and arm would be required. Uncommon (but not impossible) in lesions at the internal capsule or below because the corticobulbar and corticospinal tracts are fairly compact, resulting in leg involvement with most lesions.

LOCATIONS RULED IN: Face and arm areas of the primary motor cortex, over the lateral frontal convexity.

SIDE OF LESION: Contralateral to weakness (above the pyramidal decussation).

ASSOCIATED FEATURES ALLOWING FURTHER LOCALIZATION: Upper motor neuron signs and dysarthria are usually present. In dominant-hemisphere lesions, Broca's aphasia is common (see KCC 19.4). In nondominant-hemisphere lesions, hemineglect may occasionally be present (see KCC 19.9). Sensory loss can occur if the lesion extends into the parietal lobe (see KCC 7.3).

COMMON CAUSES: Middle cerebral artery superior division infarct is the classic cause (see Figures 10.1 and 10.5). Tumor, abscess, or other lesions may also occur in this location.

◀ FIGURE 6.14D Unilateral Face and Arm Weakness

FIGURE 6.14E **Brachial Monoparesis** ▶

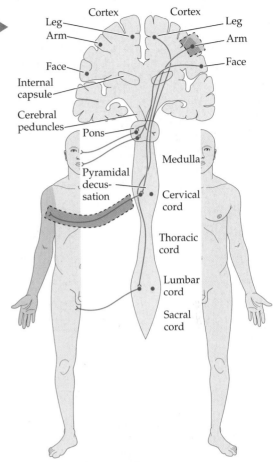

Unilateral Arm Weakness or Paralysis   Figure 6.14E

OTHER NAMES: Brachial monoparesis or monoplegia; there are specific names for different weakness patterns associated with peripheral nerve injuries (see Table 8.1; KCC 9.1).

LOCATIONS RULED OUT: Unlikely anywhere along the corticospinal tract (internal capsule, brainstem, spinal cord), because in that case the face and/or lower extremity would also likely be involved. Rare cases of foramen magnum tumors may initially affect one arm.

LOCATIONS RULED IN: Arm area of the primary motor cortex or peripheral nerves supplying the arm.

SIDE OF LESION: Motor cortex: contralateral to weakness. Peripheral nerves: ipsilateral to weakness.

ASSOCIATED FEATURES ALLOWING FURTHER LOCALIZATION:

Motor cortex lesions: There may be associated upper motor neuron signs, cortical sensory loss, aphasia (see KCC 19.6), or subtle involvement of the face or leg. Occasionally, none of these are present. The weakness pattern may be incompatible with a lesion of peripheral nerves (see Table 8.1; see also KCC 9.1). For example, marked weakness of all finger, hand, and wrist muscles with no sensory loss and normal proximal strength does not occur with peripheral nerve lesions.

Peripheral nerve lesions: There may be associated lower motor neuron signs. Weakness and sensory loss may be compatible with a known pattern for a peripheral nerve lesion (see Table 8.1; see also KCC 9.1).

COMMON CAUSES:

Motor cortex lesion: Infarct of a small cortical branch of the middle cerebral artery, or a small tumor, abscess, or the like.

Peripheral nerve lesion: Compression injury, diabetic neuropathy, and so on (see KCC 8.3 and KCC 9.1).

Unilateral Leg Weakness or Paralysis   Figure 6.14F

OTHER NAMES: Crural monoparesis or monoplegia; there are specific names for weakness patterns associated with various peripheral nerve or spinal cord lesions (see KCC 7.4 and KCC 9.1; see also Table 8.1).

LOCATIONS RULED OUT: Unlikely to be in the corticospinal tract above the upper thoracic cord (internal capsule, brainstem, cervical spinal cord) because in that case the face and/or upper extremity would usually also be involved. Rarely, cervical cord tumors can initially cause leg weakness only.

LOCATIONS RULED IN: Leg area of the primary motor cortex along the medial surface of the frontal lobe, lateral corticospinal tract below T1 in the spinal cord, or peripheral nerves supplying the leg.

SIDE OF LESION: Motor cortex: contralateral to weakness. Spinal cord or peripheral nerves: ipsilateral to weakness.

ASSOCIATED FEATURES ALLOWING FURTHER LOCALIZATION:

Motor cortex lesions: There may be associated upper motor neuron signs, cortical sensory loss, frontal lobe signs such as a

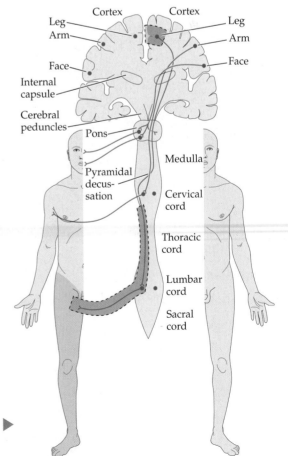

FIGURE 6.14F **Crural Monoparesis** ▶

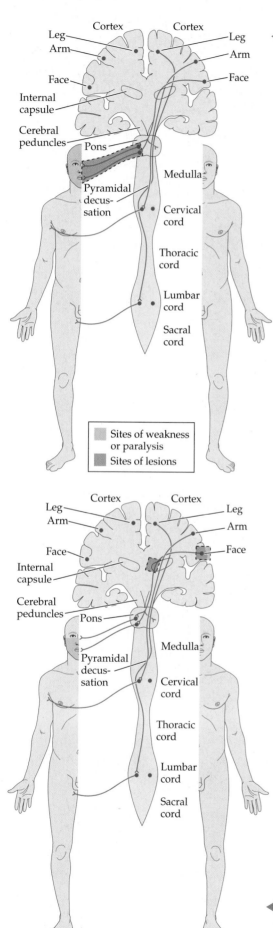

Sites of weakness or paralysis

Sites of lesions

◀ FIGURE 6.14G Facial Weakness (Lower Motor Neuron–Type)

grasp reflex, or subtle involvement of the arm or face. Occasionally, none of these are present. The weakness pattern may be incompatible with a lesion of peripheral nerves—for example, diffuse weakness of all muscles in one leg.

Spinal cord lesions: There may be associated upper motor neuron signs, a Brown–Séquard syndrome (see KCC 7.4), a sensory level, or some subtle spasticity of the contralateral leg. Sphincter function may be involved (see KCC 7.5). The weakness pattern may be incompatible with a lesion of peripheral nerves (see Table 8.1; KCC 9.1).

Peripheral nerve lesions: There may be associated lower motor neuron signs. Weakness and sensory loss may be compatible with a known pattern for a peripheral nerve lesion (see Chapters 8 and 9).

COMMON CAUSES:

Motor cortex lesion: Infarct in the anterior cerebral artery territory, or a small tumor, abscess, or the like.

Spinal cord lesion: Unilateral cord trauma, compression by tumor, or multiple sclerosis.

Peripheral nerve lesion: Compression injury, diabetic neuropathy, and so on.

Unilateral Facial Weakness or Paralysis   Figure 6.14G,H

OTHER NAMES: Bell's palsy (peripheral nerve); isolated facial weakness.

LOCATIONS RULED OUT: Unlikely with lesions below the rostral medulla.

LOCATIONS RULED IN: Common—peripheral facial nerve (CN VII). Uncommon—lesions in the face area of the primary motor cortex or in the genu of the internal capsule (usually, lesions in these locations cause arm and leg involvement as well); facial nucleus and exiting nerve fascicles in the pons or rostral lateral medulla.

SIDE OF LESION: Facial nerve or nucleus—ipsilateral to weakness. Motor cortex or internal capsule—contralateral to weakness.

ASSOCIATED FEATURES ALLOWING FURTHER LOCALIZATION:

Facial nerve or nucleus lesions (lower motor neuron; see Figure 6.14G): The forehead and orbicularis oculi are not spared (see Figure 12.13, lesion B). With facial nerve lesions (e.g., Bell's palsy), there may be hyperacusis, decreased taste, decreased lacrimation, and pain behind the ear on the affected side (see KCC 12.3). In facial nucleus lesions in the pons there are usually deficits associated with damage to nearby nuclei and pathways, such as CN VI, CN V, or the corticospinal tract (see Figures 12.11 and 14.21C; see also Table 14.8). In rostral lateral medullary lesions, a lateral medullary syndrome will be present. Interestingly, some authors report the forehead to be spared (upper motor neuron pattern) in facial weakness caused by medullary lesions.

Motor cortex or capsular genu lesions (upper motor neuron; see Figure 6.14H): The forehead is relatively spared (see Figure

◀ FIGURE 6.14H Facial Weakness (Upper Motor Neuron–Type)

FIGURE 6.14I Brachial Diparesis ▶

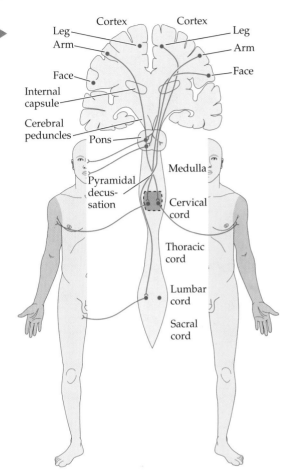

12.13, lesion A). Dysarthria and unilateral tongue weakness are common. There may be subtle arm involvement. In cortical lesions, sensory loss or aphasia may be present.

COMMON CAUSES: Facial nerve: Bell's palsy, trauma, surgery. Motor cortex, capsular genu, pons, or medulla: infarct.

To summarize, only cases of isolated facial weakness of a lower motor neuron pattern, possibly with some hyperacusis, loss of taste, dry eye, or retroauricular pain, can be localized with certainty to the peripheral facial nerve. The presence of sensory loss or any other cranial nerve or motor abnormalities requires evaluation for a central nervous system lesion.

Note that facial weakness may be difficult to detect in cases in which it occurs bilaterally, called **facial diplegia**, since the weakness is symmetrical. Causes include motor neuron disease (see KCC 6.7), bilateral peripheral nerve lesions (such as in Guillain–Barré syndrome, sarcoidosis, Lyme disease, or bilateral Bell's palsy), or bilateral white matter abnormalities caused by ischemia or demyelination (such as in pseudobulbar palsy).

Bilateral Arm Weakness or Paralysis Figure 6.14I

OTHER NAMES: Brachial diplegia.

LOCATIONS RULED OUT: Unlikely to be in the corticospinal tracts because in that case the face and/or legs would also be involved.

LOCATIONS RULED IN: Medial fibers of both lateral corticospinal tracts (see Figure 6.10C); bilateral cervical spine ventral horn cells; peripheral nerve or muscle disorders affecting both arms.

ASSOCIATED FEATURES ALLOWING FURTHER LOCALIZATION: A central cord syndrome or anterior cord syndrome may be present (see KCC 7.4).

COMMON CAUSES: Central cord syndrome: syringomyelia, intrinsic spinal cord tumor, myelitis. Anterior cord syndrome: anterior spinal artery infarct, trauma, myelitis. Peripheral nerve: bilateral carpal tunnel syndrome, or disc herniations.

Bilateral Leg Weakness or Paralysis Figure 6.14J

OTHER NAMES: Paraparesis or paraplegia.

LOCATIONS RULED OUT: Unlikely to be in the corticospinal tracts above the upper thoracic cord (internal capsule, brainstem, cervical spinal cord) because in that case the face and/or upper extremities would also be involved. Rarely, cervical cord tumors can initially cause bilateral leg weakness without arm involvement.

LOCATIONS RULED IN: Bilateral leg areas of the primary motor cortex along the medial surface of the frontal lobes; lateral corticospinal tracts below T1 in the spinal cord; cauda equina syndrome or other peripheral nerve or muscle disorders affecting both legs.

ASSOCIATED FEATURES ALLOWING FURTHER LOCALIZATION:

Bilateral medial frontal lesions: Upper motor neuron signs may be present. There may also be frontal lobe dysfunction (see KCC 19.11), including confusion, apathy, grasp reflexes, and incontinence.

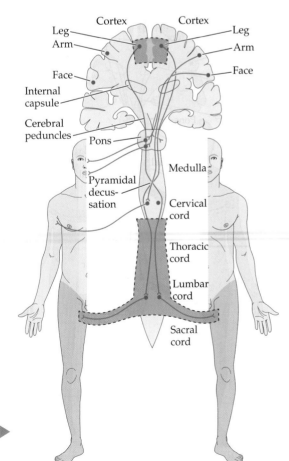

FIGURE 6.14J Paraparesis ▶

TABLE 6.6 *(continued)*

NAME	LOCALIZATION	DESCRIPTION OF GAIT ABNORMALITIES	COMMON CAUSES
DYSKINETIC GAIT (see KCC 16.1, 16.3)	Subthalamic nucleus, or other regions of basal ganglia	Unilateral or bilateral dancelike (choreic), flinging (ballistic), or writhing (athetoid) movements occur during walking and may be accompanied by some unsteadiness.	Huntington's disease; infarct of subthalamic nucleus or striatum; side effect of levodopa; other familial or drug-induced dyskinesias
TABETIC GAIT (see KCC 7.4)	Posterior columns or sensory nerve fibers	High-stepping, foot-flapping gait, with particular difficulty walking in the dark or on uneven surfaces. Patients sway and fall in attempts to stand with feet together and eyes closed (Romberg sign).	Posterior cord syndrome; severe sensory neuropathy
PARETIC GAIT (see KCC 8.3, 9.1)	Nerve roots, peripheral nerves, neuromuscular junction, or muscles	Exact appearance depends on location of lesion. With proximal hip weakness there may be a waddling, Trendelenburg gait. Severe thigh weakness may cause sudden knee buckling. Foot drop can cause a high-stepping, slapping gait, with frequent tripping.	Numerous peripheral nerve and muscle disorders
PAINFUL (ANTALGIC) GAIT	Peripheral nerve or orthopedic injury	Pain may be obvious based on patient's report or facial expression. Patients tend to avoid putting pressure on affected limb.	Herniated disc; peripheral neuropathy; muscle strain; contusions; fractures
ORTHOPEDIC GAIT DISORDER	Bones, joints, tendons, ligaments, and muscles	Depends on nature and location of the disorder. Peripheral nerve injury or spinal cord–related deficits may be present as well.	Arthritis; fractures; dislocations; contractures; soft tissue injuries
FUNCTIONAL GAIT DISORDER	Psychologically based	Can be hard to diagnose. Sometimes patients say they have poor balance, yet spontaneously perform highly destabilizing swaying movements while walking, without ever falling.	Conversion disorder; factitious disorder

KEY CLINICAL CONCEPT

6.6 MULTIPLE SCLEROSIS

Multiple sclerosis is an autoimmune inflammatory disorder affecting central nervous system myelin. The cause is unknown, although there is mounting evidence that T lymphocytes may be triggered by a combination of genetic and environmental factors to react against oligodendroglial myelin. Myelin in the peripheral nervous system is not affected. Discrete plaques of demyelination and inflammatory response can appear and disappear in multiple locations in the central nervous system over time, eventually forming sclerotic glial scars. Demyelination causes slowed conduction velocity, dispersion or loss of coherence of action potential volleys, and ultimately conduction block. Because dispersion increases with temperature, some patients have worse symptoms when they are warm. In addition to demyelination, recent studies have shown that some axons may be destroyed as well in multiple sclerosis plaques.

Prevalence is about 0.1% in the United States, with a higher worldwide prevalence in whites from northern climates, and about a 2:1 female-to-male ratio. Lifetime risk of developing multiple sclerosis goes up to 3–5% if a first-degree relative is affected. Peak age of onset is 20 to 40 years. Onset before age 10 or after 60 years of age is rare but not unheard of.

The classic clinical definition of multiple sclerosis is *two or more deficits separated in neuroanatomical space and time*. In practice, the **diagnosis** is based on the presence of typical clinical features, together with MRI evidence of white matter lesions, slowed conduction velocities on evoked potentials, and the presence of oligoclonal bands in cerebrospinal fluid obtained by lumbar puncture (see KCC 5.10). **Oligoclonal bands** are abnormal discrete bands seen on CSF gel electrophoresis. They result from the synthesis of large amounts of relatively homogeneous immunoglobulin by individual plasma cell clones in the cerebrospinal fluid (CSF). Oligoclonal bands are present in over 85% of patients with clinically definite multiple sclerosis, but they can be seen in about 8% of patients with other disorders as well. CSF with more than 50 white blood cells or with non-lymphocytes is unusual for multiple sclerosis. MRI findings suggestive of multiple sclerosis include multiple T2-bright areas, representing demyelinative plaques located in the white matter. The plaques tend to extend into the white matter from periventricular locations (resulting in "Dawson's fingers"), and they occur in both supratentorial and infratentorial structures. Acute plaques may enhance with gadolinium. The clinical features described here need not all be present to make the diagnosis of multiple sclerosis. In unusual or atypical cases of suspected multiple sclerosis it is essential to test for other inflammatory, infectious, neoplastic, hereditary, and degenerative conditions that can have similar clinical features.

Multiple sclerosis can affect numerous systems (Table 6.7). When patients first present with obvious symptoms, more subtle previous episodes can often be elicited in retrospect. About 50% of patients with a single episode of optic neuritis (see KCC 11.4) or transverse myelitis (see Table 7.4) subsequently develop multiple sclerosis. The course of multiple sclerosis is usually **relapsing–remitting** at the onset but may eventually evolve into a more refractory **chronic progressive** phase. Although severity of the disease varies from patient to patient, with good medical and neurologic care many patients with multiple sclerosis can now be expected to live a normal or near-normal life span despite their disability. Current therapies include a variety of immunomodulatory agents. Acute exacerbations are often treated with high-dose steroids, which can speed recovery in the short term but do not affect the overall course of the disease. For relapsing–remitting multiple sclerosis, the "first line" treatments are beta-interferon and copolymer (glatiramer acetate), which have a moderate effect on preventing exacerbations and delaying progression. "Second line" treatments for breakthrough relapsing disease include monoclonal antibodies such as natalizumab, rituximab, or campath and chemotherapeutic agents such as cyclophosphamide and mitoxantrone. There are currently no proven neuroprotective or neuoregenerative therapies for the more progressive phase of the disease, but this is an area of active investigation. It is also important to emphasize that multidisciplinary treatment of symptoms such as spasticity; pain; impaired bowel, bladder and sexual function; diplopia; dysphagia; and psychiatric manifestations can greatly improve a patient's quality of life. ■

TABLE 6.7 Common Symptoms and Signs in Chronic Multiple Sclerosis

FUNCTIONAL SYSTEM	PERCENT FREQUENCY
MOTOR	
Muscle weakness	65–100
Spasticity	73–100
Reflexes (hyperreflexia, Babinski's sign, absent abdominal cutaneous reflexes)	62–98
SENSORY	
Impairment of vibratory/position sense	48–82
Impairment of pain, temperature, or touch sense	16–72
Pain (moderate to severe)	11–37
Lhermitte's sign	1–42
CEREBELLAR	
Ataxia (limb/gait/truncal)	37–78
Tremor	36–81
Nystagmus (brainstem or cerebellar)	54–73
Dysarthria (brainstem or cerebellar)	29–62
CRANIAL NERVE/BRAINSTEM	
Vision affected	27–55
Ocular disturbances (excluding nystagmus)	18–39
Cranial nerves V, VII, VIII	5–52
Bulbar signs	9–49
Vertigo	7–27
AUTONOMIC	
Bladder dysfunction	49–93
Bowel dysfunction	39–64
Sexual dysfunction	33–59
Others (sweating and vascular abnormalities)	38–43
PSYCHIATRIC	
Depression	8–55
Euphoria	4–18
Cognitive abnormalities	11–59
MISCELLANEOUS	
Fatigue	59–85

Source: Rowland LP (ed.). 2000. *Merritt's Textbook of Neurology.* 10th Ed., Table 13.1. Lippincott Williams and Wilkins, Baltimore, MD.

KEY CLINICAL CONCEPT

6.7 MOTOR NEURON DISEASE

Several relatively uncommon disorders can selectively affect upper motor neurons, lower motor neurons, or both, producing motor deficits without sensory abnormalities or other findings. Most of these disorders are degenerative conditions referred to collectively as **motor neuron disease**. The classic example of motor neuron disease is **amyotrophic lateral sclerosis**, also known as **ALS** or **Lou Gehrig's disease**. ALS is characterized by gradually progressive degeneration of both upper motor neurons and lower motor neurons, leading eventually to respiratory failure and death. ALS has an incidence of 1 to 3 per 100,000 and is slightly more common in men, by a factor of about 1.5. Usual age of onset is in the 50s and 60s, although early-onset cases can be seen. Most cases occur sporadically, but there are also inherited forms that can have autosomal dominant, recessive, or X-linked transmission.

Initial symptoms are usually weakness or clumsiness, which *often begins focally* and then spreads to involve adjacent muscle groups. Painful muscle cramping and fasciculations are also common. Some patients present initially with predominantly bulbar complaints, such as dysarthria and dysphagia, or with respiratory symptoms. On neurologic examination, patients with ALS have weakness, with upper motor neuron findings such as increased tone and brisk reflexes, as well as lower motor neuron findings such as atrophy and fasciculations (see Table 6.4), sometimes best appreciated in the tongue muscles. A head droop is often present because of weakness of the neck muscles. Some patients have uncontrollable bouts of laughter or crying without the usual accompanying emotions, a finding known as pseudobulbar affect (see KCC 12.8). Sensory exam and mental status are typically normal. The extraocular muscles tend to be relatively spared. As the disease progresses, some patients can communicate only through eye movements. Electromyography (see KCC 9.2) shows evidence of muscle denervation and reinnervation in two or more extremities or body segments (for example head and arm, trunk and leg, arm and leg, etc.).

Unfortunately, there is no cure for this tragic disorder at present, and median survival from onset is 23 to 52 months. Riluzole, a blocker of glutamate release, has been shown to prolong survival by several months, and experimental trials with other agents are under way. Education of the patient and family about this disorder and implementation of a comprehensive program of medical and psychosocial services are imperative.

In evaluating patients with suspected ALS, it is important to test for other disorders that can, rarely, cause similar clinical findings. These include lead toxicity, dysproteinemia, thyroid dysfunction, vitamin B_{12} deficiency, vasculitis, paraneoplastic syndromes (see KCC 5.8), hexosaminidase A deficiency, multifocal motor neuropathy with conduction block, and other disorders. Cervical spine compression can occasionally produce a mixture of upper motor neuron signs (cord compression) and lower motor neuron signs in the upper extremities (nerve root compression). A cervical MRI is helpful to rule this out.

Some motor neuron disorders affect primarily upper motor neurons or primarily lower motor neurons. **Primary lateral sclerosis** is an example of an upper motor neuron disease, while **spinal muscular atrophy** affects lower motor neurons. Spinal muscular atrophy occurring in infancy is known as Werdnig–Hoffmann disease and usually leads to death by the second year of life. Much progress has been made recently in understanding the molecular basis of motor neuron disorders, offering some hope that we will have effective treatments for these devastating disorders in the near future. ■

CLINICAL CASES

CASE 6.1 SUDDEN ONSET OF RIGHT HAND WEAKNESS

CHIEF COMPLAINT

A 64-year-old man developed right hand weakness following cardiac arrest.

HISTORY

The patient had a history of hypertension and cigarette use but was otherwise healthy until the day of admission, when he suddenly collapsed in church. Family members at the scene administered immediate CPR, and when the ambulance arrived the patient received electrical defibrillation and promptly regained a normal cardiac rhythm. He was admitted to the cardiac intensive care unit and was found to have episodes of rapid atrial fibrillation. Several days after admission he was noted to have **weakness of the right hand**, and a neurology consult was requested.

PHYSICAL EXAMINATION

Vital signs: T = 98°F, P = 100, BP = 130/60.

Neck: Supple with no bruits.

Lungs: Clear.

Heart: Irregular rhythm, with a soft systolic murmur.

Abdomen: Normal bowel sounds; soft, nontender.

Extremities: Normal.

Neurologic exam:

  MENTAL STATUS: Alert and oriented × 3. Language fluent, with intact naming, repetition, and reading. Able to recall 3/3 objects after 5 minutes.

CRANIAL NERVES: Normal, including no facial weakness.

MOTOR: Power 5/5 throughout, except for right hand and wrist: **Right wrist flexion, extension, and hand grip 3/5. Right finger extension, abduction, adduction, and thumb opposition 0/5.**

REFLEXES:

COORDINATION AND GAIT: Not tested.

SENSORY: Intact light touch, pinprick, joint position, and vibration sense. No extinction on double simultaneous stimulation.

LOCALIZATION AND DIFFERENTIAL DIAGNOSIS

1. On the basis of the symptoms and signs shown in **bold** above, where is the lesion?

2. Given the relatively acute onset of the deficits and the presence of atrial fibrillation, what is the most likely diagnosis? What are some other possibilities?

Discussion

1. The key symptom and sign in this case is:

  • **Isolated right-sided weakness of wrist flexion and extension, finger flexion, extension, abduction, and adduction, and thumb opposition**

    Isolated hand weakness can be caused by a lesion in the primary motor cortex or in peripheral nerves (see KCC 6.3; Figure 6.14E). There are no upper or lower motor neuron signs to help localization in this patient. However, weakness of all finger, hand, and wrist muscles with no sensory loss and no proximal weakness cannot be explained by a peripheral nerve lesion (see KCC 8.3, 9.1; Table 8.1).

    The most likely *clinical localization* is left precentral gyrus, primary motor cortex, hand area.

2. Given the presence of cardiac disease including atrial fibrillation and the relatively acute onset of deficits, the most likely diagnosis is an embolic infarct (see KCC 10.4). An infarct of the left precentral gyrus hand area would be caused by occlusion of a small cortical branch of the left middle cerebral artery, superior division (see Figures 10.5, 10.6). Some other, much less likely causes of a lesion in the cortex in this setting include a small cortical hemorrhage, brain abscess, or tumor.

Clinical Course and Neuroimaging

A **head CT** (Image 6.1A–C, pages 258–259) showed an area of hypodensity representing a cortical infarct in the left precentral gyrus in approximately the hand area (compare to Figure 6.2). Additional workup revealed that the patient had a dilated heart and continued to be at risk for atrial fibrillation and recurrent emboli. He was therefore treated with anticoagulation therapy and antiarrhythmic drugs. By 10 days after the initial consult, his right hand strength had improved substantially, with wrist flexors and extensors 5/5 strength, finger flexors $4^+/5$, and finger extensors $4^-/5$.

CASE 6.2 SUDDEN ONSET OF LEFT FOOT WEAKNESS

CHIEF COMPLAINT

An 81-year-old woman presented to the emergency room because of left foot weakness.

HISTORY

The patient was previously healthy except for a history of hypertension and diabetes. On the morning of admission, as she got out of bed, she noticed difficulty when she first put her left foot on the floor. As she tried to walk, she felt that she was **dragging her left foot**. Nevertheless, she continued her usual morning activities, using a chair for support. Later in the morning, when the gait difficulty persisted, she called her children, who brought her to the emergency room. She had no other complaints except for a mild **right frontal headache**.

PHYSICAL EXAMINATION

Vital signs: Not recorded on admission.

Neck: Supple with no bruits.

Lungs: Clear.

Heart: Regular rate, with soft systolic murmur.

Abdomen: Benign, with normal bowel sounds.

Extremities: Normal.

Neurologic exam:

MENTAL STATUS: Alert and oriented × 3. Speech fluent, with intact naming and comprehension.

CRANIAL NERVES: Normal, including no facial weakness.

MOTOR: No pronator drift. Normal tone. Power 5/5 throughout, except for left foot and leg: **Left iliopsoas and hamstrings $4^+/5$, left ankle dorsiflexion and extensor hallucis longus 4/5.**

REFLEXES:

COORDINATION: Normal except for **slowing of heel-to-shin testing with left leg**.

GAIT: Not tested.

SENSORY: Intact light touch, temperature, joint position, and vibration sense.

LOCALIZATION AND DIFFERENTIAL DIAGNOSIS

1. On the basis of the symptoms and signs shown in **bold** above, where is the lesion?

2. Given the sudden onset of symptoms in an elderly patient with diabetes and hypertension, what is the most likely diagnosis? What are some other possibilities?

Discussion

1. The key symptoms and signs in this case are:

   • **Isolated left-sided weakness and slowness of the iliopsoas, hamstrings, ankle dorsiflexors, and extensor hallucis longus**

   • **Right frontal headache**

   Isolated leg weakness can be caused by a lesion in the primary motor cortex, spinal cord, or peripheral nerves (see KCC 6.3; Figure 6.14F). There are no upper or lower motor neuron signs to help localization. However, weakness in both a femoral and sciatic nerve distribution with no sensory loss (see Table 8.1; KCC 8.3, 9.1) makes a peripheral nerve or spinal cord lesion less likely. In addition, the right frontal headache suggests a cranial localization (see KCC 5.1).

   The most likely clinical localization is: right precentral gyrus, primary motor cortex, leg area.

2. Given the presence of diabetes and hypertension and the relatively acute onset of deficits, the most likely diagnosis is an embolic infarct (see KCC 10.4). An infarct of the right precentral gyrus leg area would be caused by occlusion of a cortical branch of the right anterior cerebral artery (see Figure 10.5). Some other, less likely causes of a cortical lesion in this setting include a small hemorrhage, brain abscess, or tumor. A spinal cord lesion or motor neuron disease is unlikely but possible.

Clinical Course and Neuroimaging

A **head MRI** (Image 6.2A–C, pages 260–261) showed increased T2 signal representing an infarct in the right primary motor cortex leg area. The strength in the patient's right foot gradually improved to the $4^+/5$ to $5/5$ range by the time she was discharged home. A variety of tests (see KCC 10.4) showed no obvious cause for the stroke. She was therefore entered into an experimental trial for strokes of unknown cause comparing treatment with aspirin to warfarin (Coumadin). (The trial ultimately showed no difference in clinical effect between these treatments.)

CASE 6.3 SUDDEN ONSET OF RIGHT FACE WEAKNESS

CHIEF COMPLAINT

A 62-year-old man came to the emergency room with right facial weakness.

HISTORY

The patient awoke in the morning with a "funny feeling" in his right eye and thought he might have conjunctivitis. He looked in the mirror and noticed his right eyebrow drooping. He also thought his speech sounded slightly slurred, so he called his wife to confirm this. She told him to go to the emergency room, and he complied. Past medical history was notable for diabetes.

PHYSICAL EXAMINATION

Vital signs: T = 97°F, P = 80, BP = 160/80, R = 18.

Neck: Supple with no bruits.

Lungs: Clear.

Heart: Regular rate and rhythm, with no murmurs, gallops, or rubs.

Abdomen: Normal bowel sounds; soft, nontender.

Extremities: Normal.

Neurologic exam:

MENTAL STATUS: Alert and oriented × 3. Speech fluent, with intact naming and repetition. Recalled 3/3 words after 5 minutes.

CRANIAL NERVES: Pupils 4 mm, constricting to 3 mm with light bilaterally. Visual fields full. Extraocular movements intact. Normal optokinetic nystagmus bilaterally. Intact pinprick sense, light touch sense, and graphesthesia in V_1, V_2, and V_3. Intact corneal reflex bilaterally. **Right eyebrow slightly depressed. Right lower face showed delay of movements with smile.** Taste on both sides of tongue intact in response to mustard or sweet preserves on a cotton swab. Hearing normal to finger rub. Normal gag and normal palate elevation. **Speech** sounded normal. (The patient felt it was still **mildly slurred**, but better than earlier in the day). Normal sternomastoid strength. Tongue midline.

MOTOR: No pronator drift. However, with pronation testing there was **trace curling of the right fingertips** (see KCC 6.4), with palms upward and eyes closed, that was not seen on the left side. Normal tone. Normal finger and toe taps. Power 5/5 throughout.

REFLEXES:

COORDINATION AND GAIT: Normal on finger-to-nose and heel-to-shin testing bilaterally. Normal reverse tandem gait. No Romberg sign.

SENSORY: Intact light touch, pinprick, and vibration sense and graphesthesia.

LOCALIZATION AND DIFFERENTIAL DIAGNOSIS

1. On the basis of the symptoms and signs shown in **bold** above, where is the lesion?

2. What is the most likely diagnosis? What are some other possibilities?

CASE 6.1 SUDDEN ONSET OF RIGHT HAND WEAKNESS

IMAGE 6.1A–C Infarct in Left Precentral Gyrus Hand Area (A–C) Head CT with sequentially higher horizontal slices. The infarct in the left precentral gyrus hand area is visible in (B) and (C). (Compare to the normal CT in the Neuroradiological Atlas, Figure 4.12.)

(A)

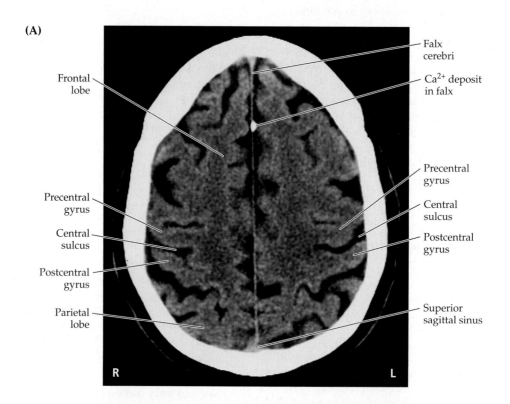

- Frontal lobe
- Precentral gyrus
- Central sulcus
- Postcentral gyrus
- Parietal lobe
- Falx cerebri
- Ca²⁺ deposit in falx
- Precentral gyrus
- Central sulcus
- Postcentral gyrus
- Superior sagittal sinus

R L

(B)

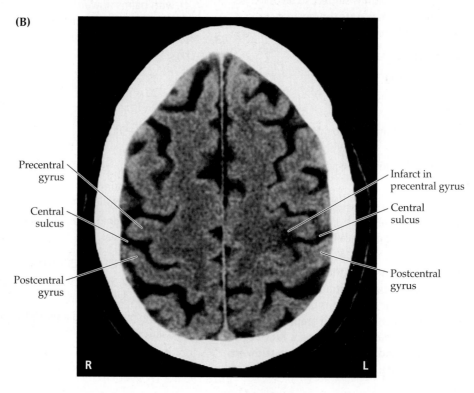

- Precentral gyrus
- Central sulcus
- Postcentral gyrus
- Infarct in precentral gyrus
- Central sulcus
- Postcentral gyrus

R L

CASE 6.1 *(continued)*

(C)

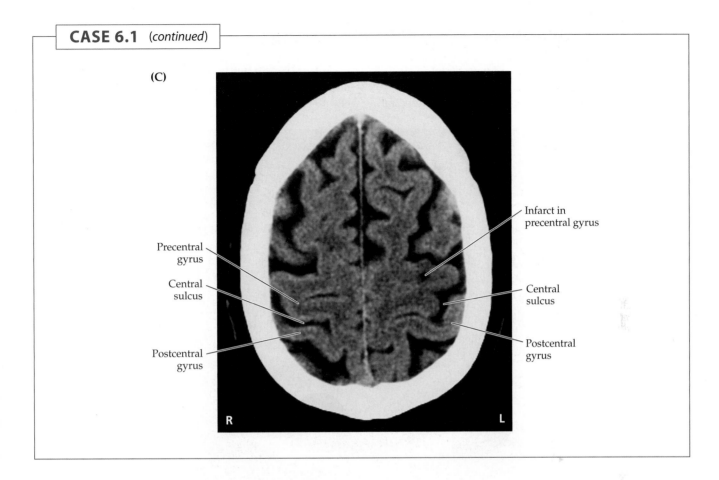

Precentral gyrus

Central sulcus

Postcentral gyrus

Infarct in precentral gyrus

Central sulcus

Postcentral gyrus

R   L

Discussion

1. The key symptoms and signs in this case are:

   - **Right eyebrow slightly depressed; right lower face showed delay of movements with smile; speech mildly slurred**

   - **Trace curling of the right fingertips**

   Unilateral facial weakness without other significant deficits is most commonly caused by peripheral lesions of the facial nerve (see KCC 6.3; Figure 6.14G). However, it can occasionally be seen in cortical lesions or in lesions of the internal capsule genu (see Figures 6.10A, 6.14H). Our patient's main problem was right facial weakness, but he also had other subtle neurological findings. The presence of mild dysarthria and finger curling suggests minor involvement of the corticobulbar and corticospinal tracts, respectively. Thus, the lesion is most likely located in the left motor cortex face area (see Figure 6.14H), with slight impingement on adjacent structures or in the left capsular genu. One unusual feature of this case is the slight involvement of the eyebrow, which is normally spared in upper motor neuron–type facial weakness.

2. Given the presence of diabetes and the relatively acute onset of deficits, the most likely diagnosis is an embolic infarct (see KCC 10.4). An infarct of the left precentral gyrus face area would be caused by occlusion of a cortical branch of the left middle cerebral artery (see Figure 10.6). Some other, less likely causes of a cortical lesion in this setting include a small hemorrhage, brain abscess, or tumor.

CASE 6.2 SUDDEN ONSET OF LEFT FOOT WEAKNESS

IMAGE 6.2A–C Infarct in Right Precentral Gyrus Leg Area (A–C) Sequentially higher axial (horizontal) T2-weighted MRI images. The infarct in the right precentral gyrus leg area is visible in (B) and (C). (Compare to the normal MRIs in the Neuroradiological Atlas, Figures 4.6B and 4.14.)

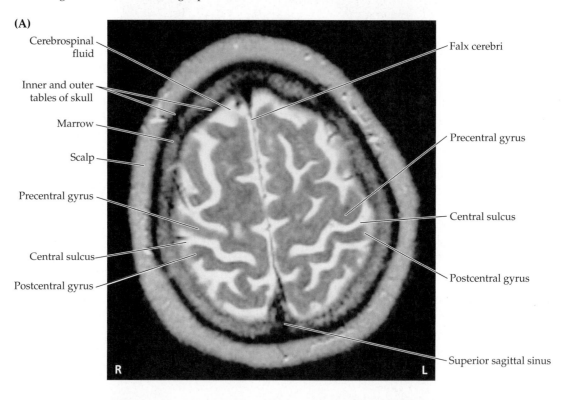

(A)

Cerebrospinal fluid

Inner and outer tables of skull

Marrow

Scalp

Precentral gyrus

Central sulcus

Postcentral gyrus

Falx cerebri

Precentral gyrus

Central sulcus

Postcentral gyrus

Superior sagittal sinus

R   L

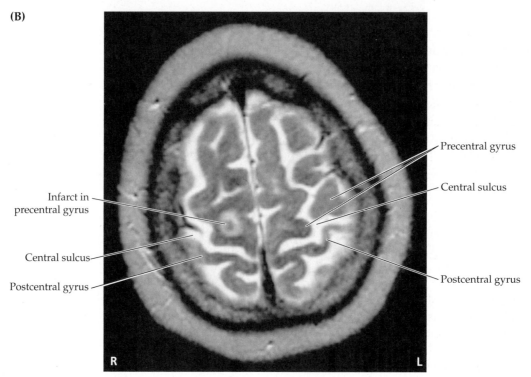

(B)

Infarct in precentral gyrus

Central sulcus

Postcentral gyrus

Precentral gyrus

Central sulcus

Postcentral gyrus

R   L

CASE 6.2 *(continued)*

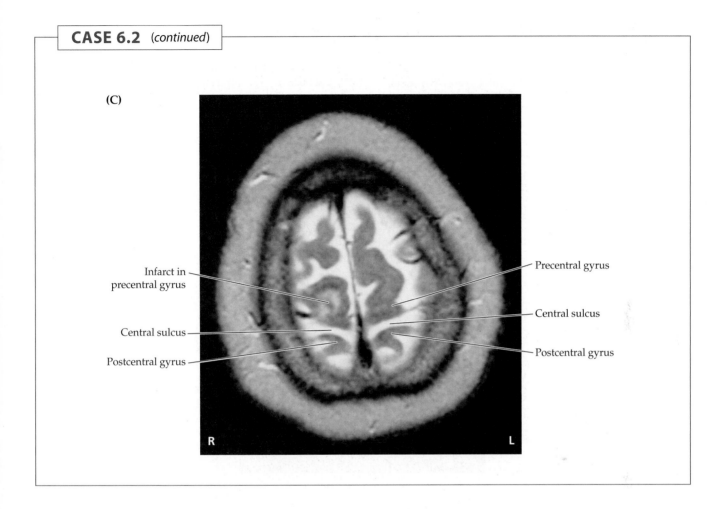

(C)

Infarct in precentral gyrus

Precentral gyrus

Central sulcus

Central sulcus

Postcentral gyrus

Postcentral gyrus

R   L

Clinical Course and Neuroimaging

T1- and T2-weighted MRI sequences done in the emergency room were normal. However, a **diffusion-weighted MRI** (see Chapter 4) done at the same time revealed several small areas of decreased diffusion (increased signal) in the left precentral gyrus (Image 6.3, page 262). The **"omega"** (or inverted omega) in the central sulcus, a landmark usually corresponding to the hand area on axial MRI sections (see Figures 4.11C, 4.13J), can be identified in Image 6.3B. Note that the diffusion changes are lateral to this, suggesting that they are in the face motor cortex.

Within a few hours of his arrival in the emergency room, the trace curling of this patient's right fingertips on pronation testing disappeared. He was admitted to the hospital, and by the next day the patient and his wife no longer thought that his speech was slurred. However, his right face remained mildly weak. Investigations for an embolic source (see KCC 10.4) were negative. He was discharged home on aspirin to reduce his risk of future strokes.

CASE 6.3 SUDDEN ONSET OF RIGHT FACE WEAKNESS

IMAGE 6.3A,B Infarct in Left Precentral Gyrus Face Area Diffusion-weighted MRI with (A) and (B) sequentially higher horizontal slices. The infarct is visible in the left precentral gyrus face area, just lateral to the omega-shaped bend in the central sulcus where the hand area is usually located.

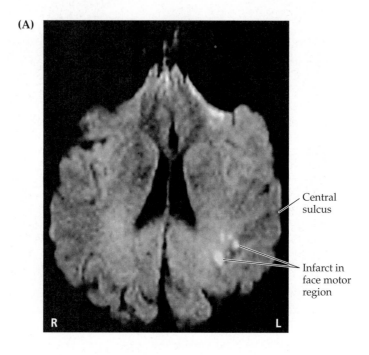

(A)

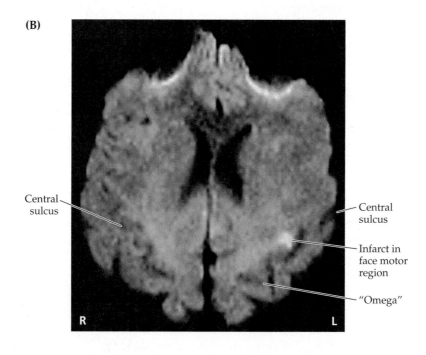

(B)

CASE 6.4 PURE MOTOR HEMIPARESIS I

CHIEF COMPLAINT

A 31-year-old woman developed left face, arm, and leg weakness.

HISTORY

Three days prior to admission, while on a business trip, the patient noticed some difficulty walking, **veering slightly to the left and bumping into corners and walls on her left side**. The next day she had some **stuttering of her speech**, which subsequently resolved. She returned home, and on the morning of admission she noticed that her **left arm and hand were somewhat weak and clumsy**. She did not have any sensory symptoms, visual problems, headaches, or changes in bowel or bladder function. Her symptoms worsened in a warm meeting room and improved with a cold shower.

PHYSICAL EXAMINATION

Vital signs: T = 98.4°F, P = 85, BP = 132/81, R = 20.

Neck: Supple.

Lungs: Clear.

Heart: Regular rate with no murmurs.

Abdomen: Normal bowel sounds; soft, nontender.

Extremities: Normal.

Neurologic exam:

  MENTAL STATUS: Alert and oriented × 3. Recalled 2/3 words after 5 minutes. Speech fluent, with intact comprehension and repetition. No neglect on drawing a clock face or on line cancellation tasks. Normal abstraction and judgment.

CRANIAL NERVES: Normal except for a **decreased left nasolabial fold**. No dysarthria on exam. Fundi normal.

MOTOR: No pronator drift. **Tone slightly increased in left arm. Rapid finger tapping slower on the left. Power 4⁺/5 in left deltoid, triceps, iliopsoas, quadriceps, and hamstrings.** Power otherwise 5/5 throughout.

REFLEXES:

COORDINATION: Normal on finger-to-nose testing bilaterally.

GAIT: **Tends to veer to the left, especially with eyes closed. Decreased arm swing on the left. Unsteady tandem gait, falling to the left.**

SENSORY: Intact light touch, pinprick, temperature, vibration, and joint position sense.

LOCALIZATION AND DIFFERENTIAL DIAGNOSIS

1. On the basis of the symptoms and signs shown in **bold** above, where is the lesion?

2. What is the most likely diagnosis? What are some other possibilities?

Discussion

1. The key symptoms and signs in this case are:

   • Left face, arm, and leg weakness, clumsiness, slowness, increased tone, hyperreflexia, and equivocal Babinski's sign

   • Dysarthria

   • Unsteady gait, falling to the left, with decreased left arm swing

   This patient has pure motor hemiparesis, with left face, arm, and leg upper motor neuron–type weakness and no sensory or cortical deficits such as neglect, aphasia, or other cognitive or visual disorders. Pure motor hemiparesis can be caused by lesions of the contralateral corticobulbar and corticospinal tracts, most commonly in the internal capsule or pons (see KCC 6.3; Figure 6.14A). Dysarthria (no longer present at time of exam but reported by history) can be caused by lesions in numerous locations (see KCC 12.8) but often accompanies pure motor hemiparesis, giving rise to the term "dysarthria-hemiparesis." Similarly, the unsteady gait could be caused by numerous lesions (see KCC 6.5) but is most easily explained by the patient's spastic left hemiparesis.

   The most likely *clinical localization* is right corticobulbar and corticospinal tracts in the posterior limb of the internal capsule or ventral pons.

2. Pure motor hemiparesis is usually caused by lacunar infarction (see Table 10.3) of the contralateral internal capsule or pons. However, given that the patient is a woman in her 30s with no vascular risk factors whose symptoms

worsen with warm temperature, the possibility that this is the first episode of multiple sclerosis should be seriously considered (see KCC 6.6). Other, less likely possibilities include a small tumor, abscess, or hemorrhage in the right internal capsule, cerebral peduncle, or ventral pons.

Clinical Course and Neuroimaging

A **head MRI** (Image 6.4A,B, page 266) showed increased T2 signal in the posterior limb of the right internal capsule. Note that there was enhancement with gadolinium, signifying breakdown of the blood–brain barrier. This is often seen with inflammatory lesions such as demyelinative plaques (see KCC 6.6), but it can also be seen a few days after infarcts. There were a few additional areas of increased T2 signal adjacent to the left frontal horn, suggesting possible prior episodes of demyelination.

An extensive workup for thromboembolic, demyelinative, inflammatory, infectious, and neoplastic disorders was done. All results were negative, except for two oligoclonal bands in the patient's cerebrospinal fluid (see KCC 6.6). Her left-sided weakness was improving by the time of discharge, 1 week after admission. Fortunately, she continued to improve and had no further problems over the following year. At a follow-up appointment 15 months after admission, her neurologic exam was entirely normal except for slight slowness of the left hand on testing of rapid alternating movements and 3+ reflexes on the left side compared to 2+ on the right. Her diagnosis remained uncertain, and she was subsequently followed periodically for the possible development of recurrent neurological signs.

CASE 6.5 PURE MOTOR HEMIPARESIS II

CHIEF COMPLAINT

A 74-year-old woman developed right face, arm, and leg weakness.

HISTORY

The patient was residing in a rehabilitation facility while recovering from an infection. She was doing well until one morning, when she suddenly developed **slurred speech and right-sided weakness**. An emergency neurology consultation was called. Medical history was notable for hypertension, coronary artery disease, and recent onset of atrial fibrillation.

PHYSICAL EXAMINATION

Vitals: T = 99.3°F, P = 84, BP = 110/70, R = 18.

Neck: No bruits.

Lungs: Clear.

Heart: Regular rate with no murmurs.

Abdomen: Normal bowel sounds; soft.

Neurologic exam:

MENTAL STATUS Alert and oriented × 3. Intact comprehension, repetition, and reading. Intact calculations.

CRANIAL NERVES: Pupils equal round and reactive to light. Visual fields full. Extraocular movements intact. Corneal reflex present bilaterally. **Decreased right nasolabial fold. Weak movements of the right face, but with only mild forehead involvement.** Gag reflex present, but with **decreased palate movement on the right. Speech slurred and dysarthric.** Normal sternomastoid strength. **Rightward tongue deviation.**

MOTOR: **Tone flaccid on right side,** normal on left. **Marked right hemiparesis, with power 2/5 in right deltoid, 0/5 in right triceps, biceps, and hand muscles. Power 2/5 in right iliopsoas and quadriceps and 0/5 in right foot.**

REFLEXES:

COORDINATION: Normal on the left; too weak to test on the right.

GAIT: Unable to stand.

SENSORY: Intact light touch, pinprick, and joint position sense. No extinction.

LOCALIZATION AND DIFFERENTIAL DIAGNOSIS

1. On the basis of the symptoms and signs shown in **bold** above, where is the lesion?

2. What is the most likely diagnosis? What are some other possibilities?

Discussion

1. The key symptoms and signs in this case are:

   - **Right face, arm, and leg weakness, hyperreflexia, and Babinski's sign**
   - **Dysarthria, decreased right palate movement, rightward tongue deviation**

     Like the patient in the previous case, this patient has pure motor hemiparesis with dysarthria (dysarthria-hemiparesis). The reflexes are decreased on the left side, probably from a chronic neuropathy (see KCC 8.1), and the reflexes on the right are relatively increased, confirming that this is an upper motor neuron lesion. Again, there are no sensory or cortical deficits such as neglect, aphasia, or other cognitive or visual disorders. Pure motor hemiparesis can be caused by lesions of the contralateral corticobulbar and corticospinal tracts, most commonly in the internal capsule or pons (see KCC 6.3; Figure 6.14A). Control of right CN IX and X (right palate movement) and right CN XII (rightward tongue deviation) have been affected as well, also probably because of a contralateral corticobulbar tract lesion.

     The most likely *clinical localization* is left corticobulbar and corticospinal tracts in the posterior limb of the internal capsule or ventral pons.

2. Pure motor hemiparesis is usually caused by lacunar infarction (see Table 10.3) of the contralateral internal capsule or pons. This patient has multiple vascular risk factors, and lacunar infarction from small-vessel disease (see KCC 10.4) is the most likely diagnosis. Given the history of atrial fibrillation, an embolus to a small perforating vessel of the left internal capsule or pons should be considered as well.

Clinical Course and Neuroimaging

The patient was transferred to a nearby hospital and started on anticoagulation therapy because of her history of atrial fibrillation. An **MRI scan** (Image 6.5, page 269) showed a T2 bright area in the left ventral pons consistent with lacunar infarction (see Figure 14.21B). A magnetic resonance angiogram (see Chapter 4) showed no significant narrowing of the circle of Willis vessels. The patient's hemiparesis had improved only slightly by the time of discharge, with power in the right arm and leg 3/5 proximally and 0/5 distally. Over the next year she had continued problems with dysarthria and swallowing difficulties, ultimately requiring a feeding tube for nutrition.

CASE 6.4   PURE MOTOR HEMIPARESIS I

IMAGE 6.4A,B   Lesion in Posterior Limb of Right Internal Capsule   MRI of the brain. (A) Axial (horizontal) proton density–weighted section. (B) Coronal T1-weighted section after intravenous administration of gadolinium contrast.

(A)

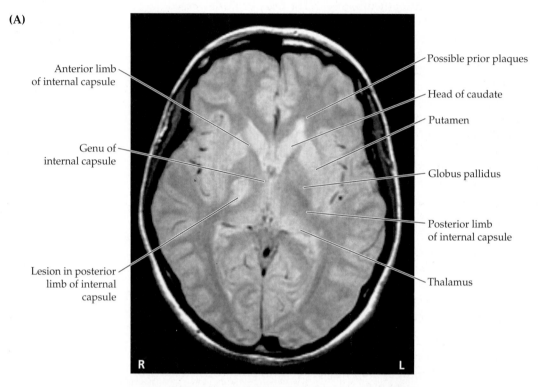

Anterior limb of internal capsule

Genu of internal capsule

Lesion in posterior limb of internal capsule

Possible prior plaques

Head of caudate

Putamen

Globus pallidus

Posterior limb of internal capsule

Thalamus

R   L

(B)

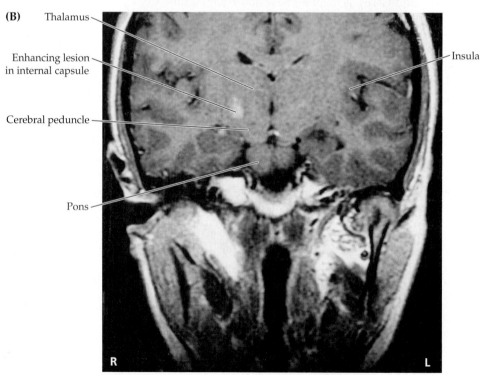

Thalamus

Enhancing lesion in internal capsule

Cerebral peduncle

Pons

Insula

R   L

CASE 6.6 PROGRESSIVE WEAKNESS, MUSCLE TWITCHING, AND CRAMPS

CHIEF COMPLAINT

A 52-year-old right-handed man was referred to a neurologist for evaluation of weakness and difficulty walking.

HISTORY

The patient first noticed gait difficulty 6 months prior to the appointment. He felt "off balance" and over the next 2 months developed difficulty raising his feet off the floor while seated in a chair. A few months later his **leg weakness** had become worse, making it difficult to walk downstairs. In addition, his **arms and hands had become weak**, making it difficult for him to carry out his work as a carpenter. He also noticed constant **twitching of his arm and leg muscles** and painful **cramps in his legs**. He did not complain of diplopia, dysarthria, or dysphagia. There was no history of trauma or neck pain, and no history of toxin exposure. Family history was negative. Initial evaluation by his primary care physician included an MRI of the cervical spine, which was normal.

PHYSICAL EXAMINATION

Vitals: T = 98°F, P = 96, BP = 120/70, R = 18.

Neck: No bruits.

Lungs: Clear.

Heart: Regular rate with no murmurs.

Abdomen: Normal.

Extremities: Normal.

Neurologic exam:

  MENTAL STATUS: Alert and oriented × 3. Intact naming, comprehension, and repetition.

  CRANIAL NERVES: Pupils equal, round, and reactive to light. Extraocular movements intact. No nystagmus. No ptosis. Facial sensation intact. No facial weakness or asymmetry. Hearing good bilaterally. Tongue and palate midline, with no tongue fasciculations.

MOTOR: **Increased tone in bilateral lower extremities. Nearly continuous fasciculations in all four extremities. Atrophy present in the left hand interosseous muscles and bilateral foot intrinsic muscles. Weakness present bilaterally (see the table below), affecting the legs more than the arms and the right side slightly more than the left.**

REFLEXES: **Hoffmann's sign present bilaterally. Jaw jerk reflex present.**

COORDINATION: Finger-to-nose, heel-to-shin, and rapid alternating movements normal.

GAIT: **Required assistance to walk. Paraparetic and spastic gait.**

SENSORY: Intact light touch, pinprick, vibration, and joint position sense. No extinction.

LOCALIZATION AND DIFFERENTIAL DIAGNOSIS

1. Which of the symptoms and signs shown in **bold** above and in table below, are upper motor neuron signs? Which are lower motor neuron signs (see Table 6.4)? Could these findings be caused by cervical cord compression?

2. What is the most likely diagnosis? What are some other possibilities?

Results of Strength Testing						
ARM			WRIST		FINGER	
DELTOID	BICEPS	TRICEPS	EXTENSORS	FLEXORS	EXTENSORS	GRIP
R	4/5	5/5	5/5	4⁺/5	5/5	4⁺/5	4⁺/5
L	4/5	5/5	5/5	5⁻/5	5/5	4⁺/5	5⁻/5

	NECK		LEG				
	FLEXORS	EXTENSORS	ILIOPSOAS	HAMSTRINGS	DORSIFLEXION	PLANTARFLEXION	QUADRICEPS
R	5/5	5/5	4/5	4⁺/5	3⁻/5	5⁻/5	5/5
L	5/5	5/5	4⁺/5	4⁺/5	3⁻/5	5⁻/5	5/5

Discussion

1. As Table 6.4 indicates, weakness is both an upper motor neuron and a lower motor neuron sign. Fasciculations and atrophy are lower motor neuron signs. Increased tone, spastic gait, and hyperreflexia including Babinski's sign, Hoffmann's sign, and jaw jerk reflex are upper motor neuron signs.

   Cervical cord compression can cause upper motor neuron findings in the arms and legs, as well as lower motor neurons findings in the arms due to local nerve root compression. In this patient, however, lower motor neuron findings of fasciculations and atrophy were present in the lower extremities as well. In addition, a jaw jerk reflex was present, which is a sign of hyperreflexia produced by upper motor neuron dysfunction in the corticobulbar pathways well above the level of the cervical cord. Thus, this patient's findings result from diffuse upper and lower motor neuron dysfunction extending from the brain to the lumbosacral spinal cord.

2. Focal weakness progressing to become more diffuse with upper and lower motor neuron signs, muscle cramping, and fasciculations with no sensory deficits form the classic presentation of amyotrophic lateral sclerosis (see KCC 6.7). Also possible, but far less likely, are paraneoplastic motor neuron disease, hexosaminidase deficiency, lead toxicity, or a variety of other disorders listed in KCC 6.7.

Clinical Course

The diagnosis of probable ALS and its implications were discussed at length with the patient and his family. He decided to try treatment with riluzole and was enrolled in a comprehensive rehabilitation program. Additional testing included an EMG (see KCC 9.2), which showed evidence of denervation and reinnervation in all four extremities, compatible with the diagnosis of ALS. Other tests included serum protein electrophoresis, serum B_{12}, folate, and complete blood count, all of which were normal.

The patient was followed over the following months as his symptoms worsened. Within 1 year after initial symptoms he was wheelchair bound because of increasing weakness. By approximately 2 years into the illness, at his last follow-up appointment, he had developed profound dysarthria, dysphagia, and tongue atrophy with fasciculations, and 0/5 to 4^-/5 strength in his extremities. He continued to live at home with family supports and hospice care, and he died a few weeks later of respiratory failure.

Additional Cases

In other chapters, related cases can be found for the following topics: **weakness caused by corticospinal** and **corticobulbar dysfunction** (see Cases 5.1–5.4, 7.3, 7.4, 10.2, 10.4, 10.5, 10.7–10.9, 10.11, 10.12, 12.8, 13.7, 14.1–14.3, 14.5, 14.6, 18.3). Other relevant cases can be found using the **Case Index** located at the end of the book.

CASE 6.5 PURE MOTOR HEMIPARESIS II

IMAGE 6.5 **Left Pontine Basis Infarct** T-2 weighted axial (horizontal) head MRI showing infarct in the left ventral pons.

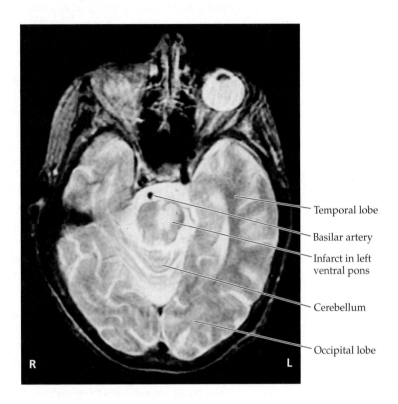

Temporal lobe

Basilar artery

Infarct in left ventral pons

Cerebellum

Occipital lobe

R   L

Brief Anatomical Study Guide

1. Motor and sensory pathways are **somatotopically organized**, with the cortical representations for the face located lateral to the hand and with the leg represented most medially (see Figure 6.2).

2. The **spinal cord** has **dorsal sensory roots**, **ventral motor roots**, **central gray matter**, and surrounding **white matter columns** (see Figure 6.3). The appearance of the spinal cord varies at different levels and is thickest at the **cervical enlargement** and **lumbosacral enlargement**, where nerves for the arms and legs, respectively, arise (see Figure 6.4). **Blood supply** for the spinal cord derives from the **anterior** and **posterior spinal arteries** (see Figure 6.5).

3. The **lateral corticospinal tract** is the most clinically important pathway in the nervous system, and knowledge of its anatomy is sufficient to localize many neurologic disorders (see Figures 6.8 and 6.11A). The lateral corticospinal tract originates mainly in the primary motor cortex of the **precentral gyrus**, descends through the **posterior limb of the internal capsule** (see Figure 6.10), down through the **cerebral peduncle** in the midbrain, and passes through the ventral pons to form fiber bundles along the

Brief Anatomical Study Guide (continued)

ventral medulla called the **pyramids** (see Figures 6.11A and 2.22A). The lateral corticospinal tract crosses to the opposite side at the **pyramidal decussation** located at the junction between the medulla and spinal cord—an essential piece of information for localizing lesions (see Figures 6.8, 6.11A, and 6.14). It then continues in the lateral spinal cord white matter to synapse onto motor neurons in the spinal cord **anterior** (**ventral**) **horn**.

4. Motor neurons projecting from the motor cortex to the spinal cord are called **upper motor neurons**; those projecting from the spinal cord to the muscles are called **lower motor neurons** (see Figure 6.8). **Upper motor neuron versus lower motor neuron signs** (see Table 6.4) often have important implications for determining whether patients are suffering from lesions of the central nervous system versus peripheral nerves.

5. **Patterns of weakness** can also be very useful for localizing lesions (see Figure 6.14).

6. Although the lateral corticospinal tract is clinically the most important, there are several additional descending motor pathways. Descending motor pathways are organized into **lateral motor systems**, such as the lateral corticospinal tract involved in **limb control**, and **medial motor systems**, which are involved in controlling **proximal trunk muscles** (see Table 6.3; see also Figures 6.6 and 6.11).

7. The **autonomic nervous system** generally controls homeostatic body functions that are not under voluntary control and has two main divisions (see Figures 6.12 and 6.13). The **sympathetic division** is involved in "fight-or-flight" functions such as increased heart rate and blood pressure and uses **norepinephrine** as its neurotransmitter on end organs. The **parasympathetic division** subserves "rest and digest" functions such as increased salivation and peristalsis, using **acetylcholine** as its peripheral neurotransmitter. Sympathetic (**thoracolumbar**) efferents arise from the **intermediolateral cell column** of the thoracic and upper lumbar spinal cord and synapse in paravertebral and prevertebral ganglia en route to their targets. Parasympathetic (**craniosacral**) efferents arise from brainstem and sacral spinal nuclei, synapsing in ganglia located in or near their end organs.

References

General References

Carpenter MB. 1991. *Core Text of Neuroanatomy*. 4th Ed., Chapters 3, 4, 9. Williams & Wilkins, Baltimore, MD.

Kandel ER, Schwartz JH, Jessell TM (eds.). 2000. *Principles of Neural Science*. 4th Ed., Chapters 33–38, 47. McGraw-Hill, New York.

Martin JH. 2003. *Neuroanatomy: Text and Atlas*. 3rd Ed., Chapters 5, 10, 11. McGraw-Hill, NY.

Purves D, Augustine GJ, Fitzpatrick D, Katz LC, LaMantia A-S, McNamara JO, Williams SM (eds.). 2008. *Neuroscience*. 4th Ed., Chapters 16, 17, 21. Sinauer, Sunderland, MA.

Motor Cortex, Sensory Cortex, and Somatotopic Organization

Iwata M. 1984. Unilateral palatal paralysis caused by lesion in the corticobulbar tract. *Arch Neurol*. 41 (7): 782–784.

Kurata K. 1992. Somatotopy in the human supplementary motor area. *Trends Neurosci* 15 (5): 159–160.

Nii Y, Uematsu S, Lesser R, Gordon B. 1996. Does the central sulcus divide motor and sensory function? Cortical mapping of the human hand areas as revealed by electrical stimulation through subdural grid electrodes. *Neurology* 46 (2): 360–367.

Penfield W, Boldrey E. 1937. Somatic motor and sensory representation in the cerebral cortex of man as studied by electrical stimulation. *Brain* 60: 389–443.

Penfield W, Rasmussen T. 1950. The Cerebral Cortex of Man: A Clinical Study of Localization of Function. Macmillan, New York.

Sanes JN, Donoghue JP, Thangaraj V, Edelmann RR, Warach S. 1995. Shared neural substrates controlling hand movements in human motor cortex. *Science* 268 (5218): 1775–1778.

Tharin S, Golby A. 2007. Functional brain mapping and its applications to neurosurgery. *Neurosurgery.* 60 (4 Suppl 2): 185–201.

Lateral Corticospinal Tract

Davidoff RA. 1990. The pyramidal tract. *Neurology* 40 (2): 332–339.

Dum RP, Strick PL. 1991. The origin of corticospinal projections from the premotor areas in the frontal lobe. *J Neurosci* 11 (3): 667–689.

Lemon RN. 2008. Descending pathways in motor control. *Annu Rev Neurosci.* 31: 195–218.

Nathan PW, Smith MC, Deacon P. 1990. The corticospinal tract in man: Course and location of fibers at different segmental levels. *Brain* 113 (Pt. 2): 303–324.

Other Medial and Lateral Descending Motor Systems

Muto N, Kakei S, Shinoda Y. 1996. Morphology of single axons of tectospinal neurons in the upper cervical spinal cord. *J Comp Neurol* 372 (1): 9–26.

Nathan PW, Smith MC. 1982. The rubrospinal and central tegmental tracts in man. *Brain* 105 (Pt. 2): 223–269.

Nathan PW, Smith M, Deacon P. 1996. Vestibulospinal, reticulospinal and descending propriospinal nerve fibres in man. *Brain* 119 (Pt. 6): 1809–1883.

Nudo RJ, Sutherland DP, Masterton RB. 1993. Inter- and intra-laminar distribution of tectospinal neurons in 23 mammals. *Brain Behav Evol* 42 (1): 1–23.

Teroa S, Takahashi M, Li M, Hashizume Y, Ikeda H, Mitsuma T, Sobue G. 1996. Selective loss of small myelinated fibers in the lateral corticospinal tract due to midbrain infarction. *Neurology* 47 (2): 558–591.

Autonomic Nervous System

Cooper JR, Roth RF, Bloom FE. 2002. *Biochemical Basis of Neuropharmacology*. 8th Ed. Oxford, New York.

Goldstein DS, Robertson D, Esler M, Straus SE, Eisenhofer G. 2002. Dysautonomias: clinical disorders of the autonomic nervous system. *Ann Intern Med.* 137 (9): 753–763.

Robertson D, Low PA, Polinsky RJ (eds.). 1996. *Primer on the Autonomic Nervous System*. Academic Press, San Diego.

Wilson-Pauwels L, Stewart PA, Akesson EJ. 1997. *Autonomic Nerves*. B C Decker, Malden, MA.

Upper Motor Neuron versus Lower Motor Neuron Lesions

Phillips CG, Landau WM. 1990. Clinical neuropathology VIII. Upper and lower motor neurons: The little old synecdoche that works. *Neurology* 40 (6): 884–886.

Young RR. 1994. Spasticity: A review. *Neurology* 44 (Suppl 9): S12–S20.

Pure Motor Hemiparesis

Fisher CM, Curry HD. 1965. Pure motor hemiplegia of vascular origin. *Arch Neurol* 13: 30.

Gait Disorders

Boonstra TA, van der Kooij H, Munneke M, Bloem BR. 2008. Gait disorders and balance disturbances in Parkinson's disease: clinical update and pathophysiology. *Curr Opin Neurol.* 21 (4): 461–471.

Jankovic J, Nutt JG, Sudarsky L. 2001. Classification, diagnosis, and etiology of gait disorders. *Adv Neurol.* 87: 119–133.

Snijders AH, van de Warrenburg BP, Giladi N, Bloem BR. 2007. Neurological gait disorders in elderly people: clinical approach and classification. *Lancet Neurol.* 6 (1): 63–74.

Multiple Sclerosis

Agrawal SM, Yong VW. 2007. Immunopathogenesis of multiple sclerosis. *Int Rev Neurobiol.* 79: 99–126.

Bartt RE. 2006. Multiple sclerosis, natalizumab therapy, and progressive multifocal leukoencephalopathy. *Curr Opin Neurol.* Aug; 19 (4): 341–349.

Carmosino MJ, Brousseau KM, Arciniegas DB, Corboy JR. 2005. Initial evaluations for multiple sclerosis in a university multiple sclerosis center: outcomes and role of magnetic resonance imaging in referral. *Arch Neurol.* 62 (4): 585–590.

DeAngelis T, Lublin F. 2008. Multiple sclerosis: new treatment trials and emerging therapeutic targets. *Curr Opin Neurol.* 21 (3): 261–271.

Miller DH, Weinshenker BG, Filippi M, Banwell BL, Cohen JA, Freedman MS, Galetta SL, Hutchinson M, et al. 2008. Differential diagnosis of suspected multiple sclerosis: a consensus approach. *Mult Scler.* 14 (9): 1157–1174.

Olek MJ, Dawson DM. 2000. Multiple sclerosis and other inflammatory demyelinating diseases of the central nervous system. In *Neurology in Clinical Practice*, WG Bradley, RB Daroff, GB Fenichel, and J Jankovic (eds.), 4th Ed., Chapter 60. Butterworth-Heinemann, Boston.

Rudick RA, Fisher E, Lee JC, Simon J, Jacobs L. 1999. Use of the brain parenchymal fraction to measure whole brain atrophy in relapsing-remitting MS. *Neurology* 53 (8): 1698–1704.

Motor Neuron Disease

Bromberg MB. 2002. Diagnostic criteria and outcome measurement of amyotrophic lateral sclerosis. *Adv Neurol.* 88: 53–62.

Brooks B. 1994. El Escorial World Federation of Neurology criteria for the diagnosis of amyotrophic lateral sclerosis. Subcommittee on Motor Neuron Diseases/Amyotrophic Lateral Sclerosis of the World Federation of Neurology Research Group on Neuromuscular Diseases and the El Escorial "Clinical limits of amyotrophic lateral sclerosis." *Journal of the Neurological Sciences.* 124: 96–107.

Brooks BR. 2009. Managing amyotrophic lateral sclerosis: slowing disease progression and improving patient quality of life. *Ann Neurol.* 65 Suppl 1: S17–23.

Gordon PH, Cheng B, Katz IB, Mitsumoto B, Rowland LP. 2009. Clinical features that distinguish PLS, upper motor neuron-dominant ALS, and typical ALS. *Neurology* 72: 1948–1952.

Miller R, Rosenberg J, Gelinas D, et al. 1999. Practice parameter: the care of the patient with ALS (an evidence-based review): report of the quality standards subcommittee of the American Academy of Neurology: ALS Practice Parameters Task Force. *Neurology* 52: 1311–1323.

Rothstein JD. 2009. Current hypotheses for the underlying biology of amyotrophic lateral sclerosis. *Ann Neurol.* 65 Suppl 1: S3–S9.

Rowland LP, Shneider NA. 2001. Amyotrophic lateral sclerosis. *N Engl J Med.* 344 (22):1688–1700.

Strober JB, Tennekoon GI. 1999. Progressive spinal muscular atrophies. *J Child Neurol* 14 (11): 691–695.

CONTENTS

Chapter 7

Somatosensory Pathways

A 71-year-old woman developed gradually worsening numbness and tingling in her right leg, along with weakness in her left leg. She also experienced occasional urinary incontinence. A neurologic examination revealed decreased pinprick sensation on her right side below the level of the umbilicus and decreased vibration and joint position sense in the left foot. Her left leg had brisk reflexes and was slightly weak.

In this patient, these complex sensory and motor deficits arose from a single lesion. In this chapter, we will learn about the pathways for sensations such as touch, pain, and position sense of the limbs, and we will use this knowledge to accurately localize lesions of these pathways in clinical cases.

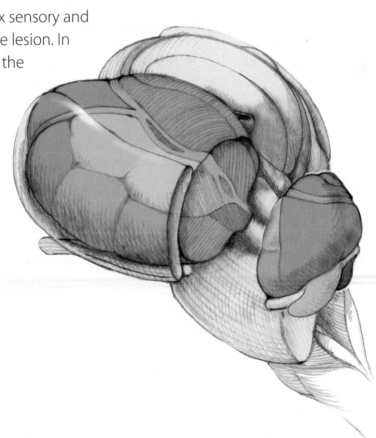

ANATOMICAL AND CLINICAL REVIEW

IN CHAPTER 6, WE DISCUSSED the anatomy of the corticospinal tract and other descending motor pathways. In this chapter we will discuss the other two major "long tracts" of the nervous system (Table 7.1). These are the somatosensory pathways: the **posterior column–medial lemniscal system** and the **anterolateral systems**. Like the corticospinal tract, these pathways are somatotopically organized (see Figure 6.2). Understanding the functions and points of decussation of the three major long tracts (see Table 7.1) is fundamental to clinical neuroanatomical localization.

In the sections that follow, we will learn to use the anatomy of the three major long tracts to localize lesions in the nervous system. We will discuss common disorders of the spinal cord and other locations that affect these pathways. In addition, brainstem and spinal cord mechanisms of pain modulation will be addressed. The organization of the thalamus, serving as the major relay for sensory and other information traveling to the cortex, will be reviewed as well. Finally, we will discuss the roles of sensory and motor pathways in bowel, bladder, and sexual function.

Main Somatosensory Pathways

The term **somatosensory** generally refers to bodily sensations of touch, pain, temperature, vibration, and **proprioception** (limb or joint position sense). There are two main pathways for somatic sensation (see Table 7.1 and Figures 7.1 and 7.2):

- The **posterior column–medial lemniscal pathway** conveys proprioception, vibration sense, and fine, discriminative touch (see Figure 7.1).
- The **anterolateral pathways** include the **spinothalamic tract** and other associated tracts that convey pain, temperature sense, and crude touch (see Figure 7.2).

Since some aspects of touch sensation are carried by both pathways, touch sensation is not eliminated in isolated lesions to either pathway.

Four types of sensory neuron fibers are classified according to axon diameter (Table 7.2). These different fiber types have specialized peripheral receptors that subserve different sensory modalities. Larger-diameter, myelinated axons conduct faster than smaller-diameter or unmyelinated axons.

Sensory neuron cell bodies are located in the **dorsal root ganglia** (see Figures 7.1 and 7.2). Each dorsal root ganglion cell has a stem axon that bifurcates, resulting in one long process that conveys sensory information from the periphery and a second process that carries information into the spinal cord through the dorsal nerve roots. A peripheral region innervated by sensory fibers from a single nerve root level is called a **dermatome**. The dermatomes for the different spinal levels form a map over the surface of the body (see Figure 8.4) that can be useful in localizing lesions of the nerve roots or spinal cord. In Chapters 8 and 9 we will discuss localization based on dermatome and peripheral nerve patterns of sensory and motor loss. In this chapter we will focus on the central course of the somatosensory pathways in the spinal cord and brain. Just as our knowledge that the corticospinal tract crosses over at the pyramidal decussation helps us localize CNS lesions (see Figure 6.11A), it is equally

TABLE 7.1   Main Long Tracts of the Nervous System

PATHWAY(S)	FUNCTION	NAME (AND LEVEL) OF DECUSSATION
Lateral corticospinal tract	Motor	Pyramidal decussation (cervico-medullary junction)
Posterior column–medial lemniscal pathway	Sensory (vibration, joint position, fine touch)	Internal arcuate fibers (lower medulla)
Anterolateral pathways	Sensory (pain, temperature, crude touch)	Anterior commissure (spinal cord)

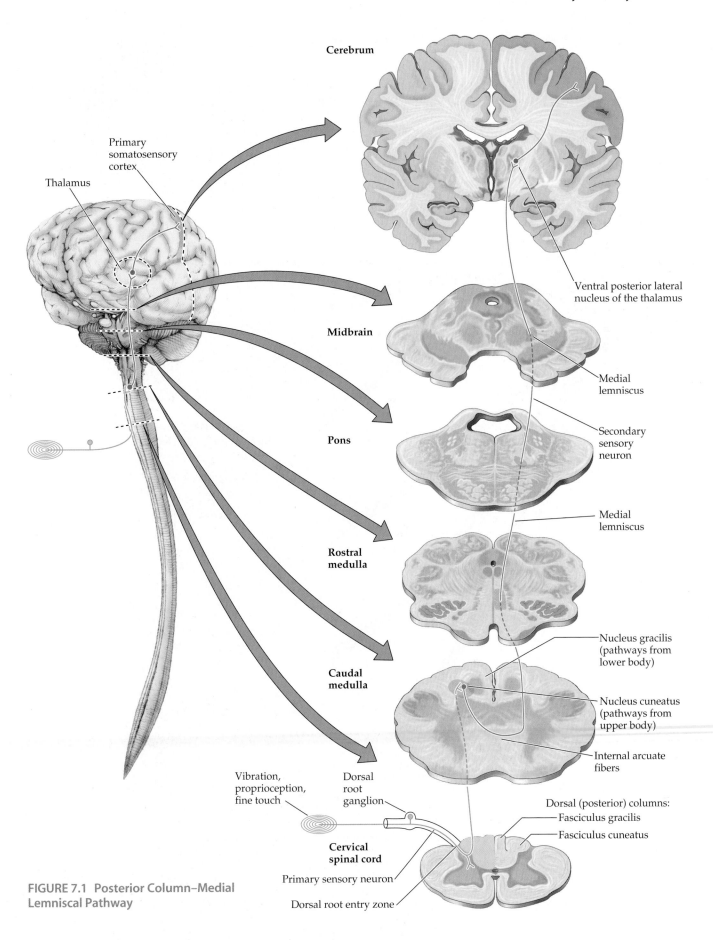

Cerebrum

Primary
somatosensory
cortex

Thalamus

Midbrain

Pons

Rostral
medulla

Caudal
medulla

Vibration,
proprioception,
fine touch

Dorsal
root
ganglion

**Cervical
spinal cord**

Primary sensory neuron

Dorsal root entry zone

Ventral posterior lateral
nucleus of the thalamus

Medial
lemniscus

Secondary
sensory
neuron

Medial
lemniscus

Nucleus gracilis
(pathways from
lower body)

Nucleus cuneatus
(pathways from
upper body)

Internal arcuate
fibers

Dorsal (posterior) columns:
Fasciculus gracilis
Fasciculus cuneatus

**FIGURE 7.1 Posterior Column–Medial
Lemniscal Pathway**

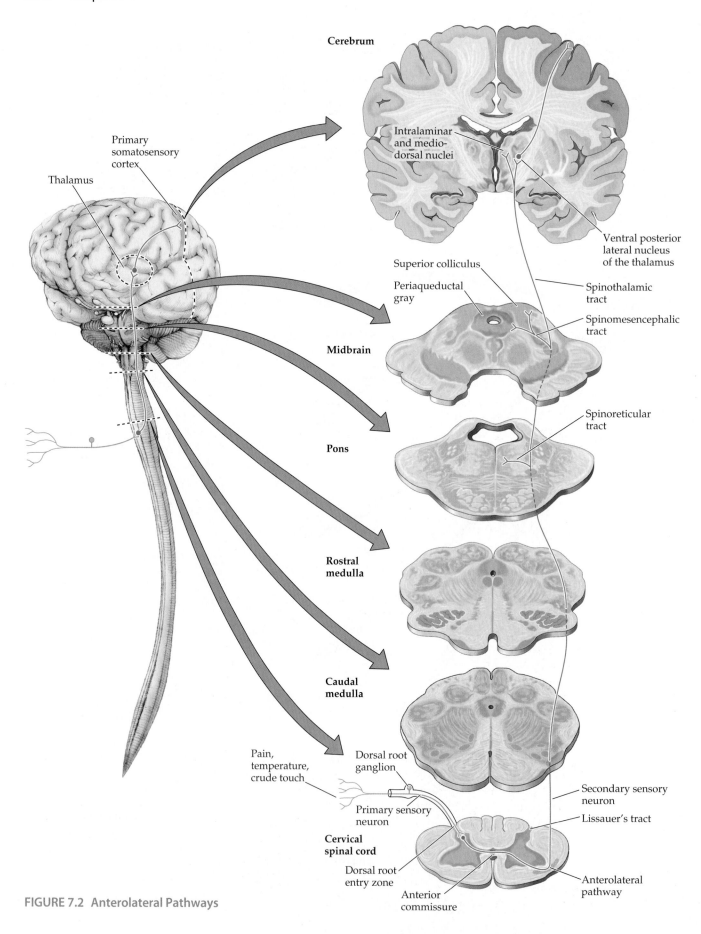

FIGURE 7.2 Anterolateral Pathways

NAME	ALTERNATE NAME	FIBER DIAMETER (μm)	MYELINATED	RECEPTORS	SENSORY MODALITIES
A-α	I	13–20	Yes	Muscle spindle	Proprioception
				Golgi tendon organ	Proprioception
A-β	II	6–12	Yes	Muscle spindle	Proprioception
				Meissner's corpuscle	Superficial touch
				Merkel's receptor	Superficial touch
				Pacinian corpuscle	Deep touch, vibration
				Ruffini ending	Deep touch, vibration
				Hair receptor	Touch, vibration
A-δ	III	1–5	Yes	Bare nerve ending	Pain
				Bare nerve ending	Temperature (cool)
				Bare nerve ending	Itch
C	IV	0.2–1.5	No	Bare nerve ending	Pain
				Bare nerve ending	Temperature (warm)
				Bare nerve ending	Itch

TABLE 7.2 Sensory Neuron Fiber Types

important to know the points of decussation of the two main somatosensory pathways (see Table 7.1; Figures 7.1 and 7.2). We will therefore now trace the course of these pathways from spinal cord to primary somatosensory cortex.

Posterior Column–Medial Lemniscal Pathway

Large-diameter, myelinated axons carrying information about proprioception, vibration sense, and fine touch enter the spinal cord via the medial portion of the dorsal root entry zone (see Figure 7.1). Many of these axons then enter the ipsilateral **posterior columns** to ascend all the way to the **posterior column nuclei** in the medulla. In addition, some axon collaterals enter the spinal cord central gray matter to synapse onto interneurons and motor neurons (see Figure 2.21). It is easier to remember the **somatotopic organization** of the posterior columns (Figure 7.3) if you picture fibers adding on laterally from higher levels as the posterior columns ascend. Thus, the medial portion, called the **gracile fasciculus** (gracile means "thin") carries information from the legs and lower trunk. The more lateral **cuneate fasciculus** (cuneate means "wedge shaped") carries information from the upper trunk above about T6, and from the arms and neck. The first-order sensory neurons that have axons in the gracile and cuneate fasciculi (also called fasciculus gracilis and fasciculus cuneatus) synapse onto second-order neurons in the **nucleus gracilis** and **nucleus cuneatus**, respectively (see Figure 7.1).

MNEMONIC

Axons of these second-order neurons decussate as **internal arcuate fibers** and then form the **medial lemniscus** on the other side of the medulla (see Figure 14.5). The medial lemniscus initially has a vertical orientation and then comes to occupy a progressively more lateral and inclined position as it ascends in the brainstem (see Figures 7.1, 14.3, and 14.4). The somatotopic organization of the medial lemniscus assumes a vertical position in the medulla such that the feet are represented more ventrally ("the little person stands up") and then inclines again in the pons and midbrain such that the arms are represented more medially and legs more laterally ("the little person lies down"). Note the reversal in somatotopic orientation: for the medial lemniscus in the pons and

MNEMONIC

FIGURE 7.3 Somatotopic Organization of Posterior Column and Anterolateral Pathways in the Spinal Cord

FIGURE 7.3 **Somatotopic Organization of Posterior Column and Anterolateral Pathways in the Spinal Cord** Compare to somatotopic organization of the lateral corticospinal tract shown in Figure 6.10C. (Spinal section from DeArmond SJ, Fusco MM, Maynard MD. 1989. *Structure of the Human Brain: A Photographic Atlas.* 3rd Ed. Oxford University Press, New York.)

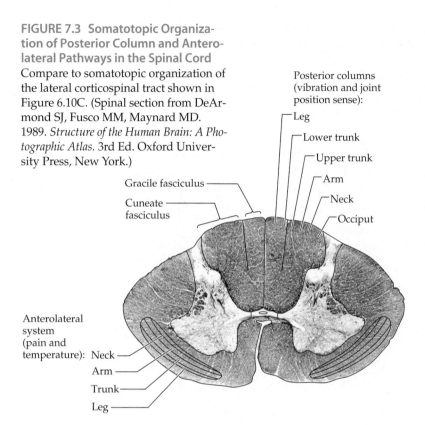

Posterior columns (vibration and joint position sense):
- Leg
- Lower trunk
- Upper trunk
- Arm
- Neck
- Occiput

Gracile fasciculus
Cuneate fasciculus

Anterolateral system (pain and temperature):
- Neck
- Arm
- Trunk
- Leg

MNEMONIC

midbrain the feet are lateral, while in the posterior columns the feet are medial. Recall that most somatotopic maps represent the upper body medially and lower body laterally, with the notable exceptions of the posterior columns (see Figure 7.3) and primary sensory-motor cortices (see Figure 6.2).

The next major synapse occurs when the medial lemniscus axons terminate in the **ventral posterior lateral nucleus (VPL)** of the thalamus. The neurons of the VPL then project through the **posterior limb of the internal capsule** in the **thalamic somatosensory radiations** (see Figure 6.9B) to reach the **primary somatosensory cortex** (see Figure 6.1, areas 3, 1, and 2) in the postcentral gyrus (see Figure 7.1). As we will discuss in Chapter 12 (see Figure 12.8), an analogous pathway called the trigeminal lemniscus conveys touch sensation for the face via the ventral posterior medial nucleus of the thalamus (VPM) to the somatosensory cortex. Synaptic inputs to the primary somatosensory cortex from both the face and body occur mainly in cortical layer IV and the deep portions of layer III, with some inputs also reaching layer VI (see Figure 2.14).

Spinothalamic Tract and Other Anterolateral Pathways

Smaller-diameter and unmyelinated axons carrying information about pain and temperature sense also enter the spinal cord via the dorsal root entry zone (see Figure 7.2). However, these axons make their first synapses immediately in the gray matter of the spinal cord, mainly in the dorsal horn **marginal zone (lamina I)** and deeper in the dorsal horn, in **lamina V** (see Figure 6.3B; Table 6.2). Some axon collaterals ascend or descend for a few segments in **Lissauer's tract** before entering the central gray (see Figures 6.4 and 7.2). Axons from the second-order sensory neurons in the central gray cross over in the **spinal cord anterior (ventral) commissure** to ascend in the anterolateral white matter. It should be noted that it takes **two to three spinal segments** for the decussating fibers to reach the opposite side, so a lateral cord lesion will affect contralateral pain and temperature sensation beginning a few segments below the level of the lesion. The anterolateral pathways in the spinal cord have a **somatotopic organization** (see Figure 7.3) in which the feet are most laterally represented. To help you remember this organization, picture fibers from the anterior commissure adding on medially as the anterolateral pathways ascend in the spinal cord. This somatotopic organization, with arms more medial and legs more lateral, is preserved as the anterolateral pathways pass through the brainstem. When they reach the medulla, the anterolateral pathways are located laterally, running in the groove between the inferior olives and the inferior cerebellar peduncles (see Figures 7.2 and 14.5). They then enter the pontine tegmentum to lie just lateral to the medial lemniscus in the pons and midbrain (see Figures 14.3 and 14.4).

The anterolateral pathways consist of three tracts (see Figure 7.2): the spinothalamic, spinoreticular, and spinomesencephalic tracts. The **spinothalamic tract** is the best known and mediates discriminative aspects of pain and temperature sensation, such as location and intensity of the stimulus. Like the posterior column–medial lemniscal pathway, a major relay for the spinothalamic tract is in the **ventral posterior lateral nucleus** (**VPL**) of the thalamus (see Figure 7.2). However, the terminations of the spinothalamic tract and the posterior column–medial lemniscal pathway reach separate neurons within the VPL. From the VPL, information travelling in the spinothalamic tract is again conveyed via the thalamic somatosensory radiations (see Figure 6.9B) to the primary somatosensory cortex (see Figure 6.1, areas 3, 1, and 2) in the postcentral gyrus (see Figure 7.2). Pain and temperature sensation for the face is carried by an analogous pathway called the trigeminothalamic tract, to be discussed further in Chapter 12 (see Figure 12.8).

There are also spinothalamic projections to other thalamic nuclei, including **intralaminar thalamic nuclei** (central lateral nucleus) and medial thalamic nuclei such as the **mediodorsal nuclei** (see Figure 7.2). These projections probably participate together with the **spinoreticular tract** in a phylogenetically older pain pathway responsible for conveying the emotional and arousal aspects of pain. The spinoreticular tract terminates on the medullary–pontine reticular formation, which in turn projects to the intralaminar thalamic nuclei (centromedian nucleus). Unlike the VPL, which projects specifically in a somatotopic fashion to the primary sensory cortex, the intralaminar nuclei project diffusely to the entire cerebral cortex and are thought to be involved in behavioral arousal (see the section on the thalamus later in this chapter).

The **spinomesencephalic tract** projects to the midbrain periaqueductal gray matter and the superior colliculi (see Figure 7.2). The **periaqueductal gray** participates in central modulation of pain, as we will discuss shortly.

The spinothalamic and spinomesencephalic tracts arise mainly from spinal cord laminae I and V, while the spinoreticular tract arises diffusely from intermediate zone and ventral horn laminae 6 through 8 (see Figure 6.3). In addition to pain and temperature, the anterolateral pathways can convey some crude touch sensation, therefore, touch sensation is not lost when the posterior columns are damaged.

To summarize, if you step on a thumbtack with your left foot, your spinothalamic tract enables you to realize "something sharp is puncturing the sole of my left foot"; your spinothalamic intralaminar projections and spinoreticular tract cause you to feel "ouch, that hurts"; and your spinomesencephalic tract leads to pain modulation, allowing you eventually to think "aah, that feels better."

A summary of spinal cord sensory and motor pathways is shown in Figure 7.4. The sensory pathways shown on the left are discussed in this chapter; the spinocerebellar tracts are covered in Chapter 15; and the motor pathways shown on the right are discussed in Chapter 6 (see Figure 6.11). As we will see in KCC 7.4, clinical syndromes of the spinal cord provide a practical review of regional spinal cord anatomy. We will now discuss the continuation of the somatosensory pathways, from the thalamus to the cerebral cortex.

REVIEW EXERCISE

At what level does the decussation occur for the posterior column–medial lemniscal pathway (see Figure 7.1) and for the anterolateral pathways (see Figure 7.2)? Suppose that a patient has a lesion of the left half of the spinal cord. Which side of the patient's body will have decreased vibration and joint position sense below the level of the lesion? Which side of the patient's body will have decreased pain and temperature sensation below the level of the lesion? Which side will have weakness below the level of the lesion (see Figure 6.8)? Which side will these deficits be on if the lesion is in the left cerebral cortex?

MNEMONIC

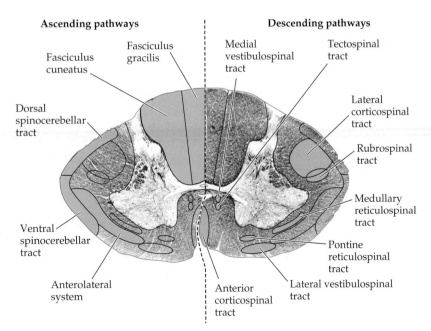

FIGURE 7.4 Spinal Cord Sensory and Motor Spinal Cord Pathways This diagram summarizes the sensory and motor spinal cord pathways discussed in Chapters 6 and 7. (Spinal section from DeArmond SJ, Fusco MM, Maynard MD. 1989. *Structure of the Human Brain: A Photographic Atlas.* 3rd Ed. Oxford University Press, New York.)

Ascending pathways

Fasciculus gracilis
Fasciculus cuneatus
Dorsal spinocerebellar tract
Ventral spinocerebellar tract
Anterolateral system

Descending pathways

Medial vestibulospinal tract
Tectospinal tract
Lateral corticospinal tract
Rubrospinal tract
Medullary reticulospinal tract
Pontine reticulospinal tract
Lateral vestibulospinal tract
Anterior corticospinal tract

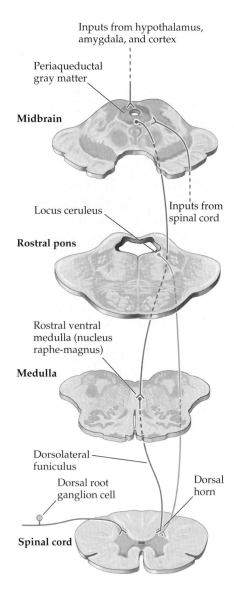

Inputs from hypothalamus, amygdala, and cortex

Periaqueductal gray matter

Midbrain

Locus ceruleus

Rostral pons

Inputs from spinal cord

Rostral ventral medulla (nucleus raphe-magnus)

Medulla

Dorsolateral funiculus

Dorsal root ganglion cell

Dorsal horn

Spinal cord

FIGURE 7.5 Central Pathways Involved in Pain Modulation

Somatosensory Cortex

From the thalamic VPL and VPM nuclei, somatosensory information is conveyed to the **primary somatosensory cortex** in the postcentral gyrus, which includes Brodmann's areas 3, 1, and 2 (see Figures 7.1, 7.2, and 6.1). Like the primary motor cortex, the primary somatosensory cortex is somatotopically organized, with the face represented most laterally and the leg most medially (see Figure 6.2). Information from the primary somatosensory cortex is conveyed to the **secondary somatosensory association cortex** located within the Sylvian fissure, along its superior margin in a region called the parietal operculum (see Figure 6.1). The secondary somatosensory cortex is also organized somatotopically. Further processing of somatosensory information occurs in association cortex of the **superior parietal lobule**, including Brodmann's areas 5 and 7 (see Figure 6.1). Primary somatosensory cortex and somatosensory association cortex also have extensive connections with the motor cortex. Lesions of the somatosensory cortex and adjacent regions produce characteristic deficits referred to as **cortical sensory loss** (see KCC 7.3).

Central Modulation of Pain

Pain modulation involves interactions between local circuits at the level of the spinal cord dorsal horn and long-range modulatory inputs (Figure 7.5). In a mechanism called the **gate control theory**, sensory inputs from large-diameter, nonpain A-β fibers (see Table 7.2) reduce pain transmission through the dorsal horn. Thus, for example, transcutaneous electrical nerve stimulation (TENS) devices work to reduce chronic pain by activating A-β fibers. This is also why shaking your hand after striking your thumb with a hammer temporarily helps the pain. The **periaqueductal gray** receives inputs from the hypothalamus, amygdala, and cortex, and it inhibits pain transmission in the dorsal horn via a relay in a region at the pontomedullary junction called the **rostral ventral medulla (RVM)** (see Figure 7.5). This region includes serotonergic (5-HT) neurons of the raphe nuclei (see Figure 14.12) that project to the spinal cord, modulating pain in the dorsal horn. The RVM also sends inputs mediated by the neuropeptide substance P to the locus ceruleus (see Figure 14.11), which in turn sends noradrenergic (NE) projections to modulate pain in the spinal cord dorsal horn (see Figure 7.5). Histamine also contributes to modulation of pain through H3 receptors.

   Opiate medications such as morphine are likely to exert their analgesic effects through receptors found in widespread locations throughout the nervous system, including receptors located on peripheral nerves and neurons in the spinal dorsal horn. However, opiate receptors and **endogenous opiate peptides**, such as **enkephalin**, **β-endorphin**, and **dynorphin**, are found in particularly high concentrations at key points in the pain modulatory pathways. Thus, enkephalin- and dynorphin-containing neurons are concentrated in the periaqueductal gray, RVM, and spinal cord dorsal horn, while β-endorphin–containing neurons are concentrated in regions of the hypothalamus that project to the periaqueductal gray.

The Thalamus

The **thalamus** (meaning "inner chamber" or "bedroom" in Greek) is an important processing station in the center of the brain. Nearly all pathways that project to the cerebral cortex do so via synaptic relays in the thalamus. The thalamus is often thought of as the major sensory relay station, and it is therefore appropriate to introduce thalamic networks in this chapter. However, in addition to sensory information, the thalamus also conveys nearly all other inputs to the cortex, including motor inputs from the cerebellum and basal ganglia (see Chapters 15 and 16), limbic inputs (see Chapter 18), widespread modulatory inputs involved in behavioral arousal and sleep–wake cycles (see Chapter 14), and other inputs. We will introduce the thalamus in some detail here, both because of its relevance to sensory processing discussed in this chapter and because thalamic nuclei are important in several subsequent chapters.

Some thalamic nuclei have specific topographical projections to restricted cortical areas, while others project more diffusely. Thalamic nuclei typically receive dense reciprocal feedback connections from the cortical areas to which they project. In fact, corticothalamic projections outnumber thalamocortical projections.

As mentioned in Chapter 2, the thalamus is part of the **diencephalon**, together with the hypothalamus and epithalamus. The diencephalon is located just rostral to the midbrain (see Figure 2.2). The **hypothalamus**, located immediately ventral to the thalamus, is discussed in Chapter 17; the **epithalamus** consists of several small nuclei including the habenula, parts of the pretectum, and the pineal body. In horizontal sections, the thalami are visible as deep, gray matter structures shaped somewhat like eggs, with their posterior ends angled outward, forming an inverted *V* (see Figures 4.13, 6.9B, 6.10A, and 16.2). The thalamus is divided into a **medial nuclear group**, **lateral nuclear group**, and **anterior nuclear group** by a Y-shaped white matter structure called the **internal medullary lamina** (Figure 7.6). Nuclei located within the internal medullary lamina itself are called the **intralaminar nuclei**. The **midline thalamic nuclei** are an additional thin collection of nuclei lying adjacent to the third ventricle, several of which are continuous with and functionally very similar to the intralaminar nuclei. Finally, the **thalamic reticular nucleus** (to be distinguished from the reticular nuclei of the brainstem) forms an extensive but thin sheet enveloping

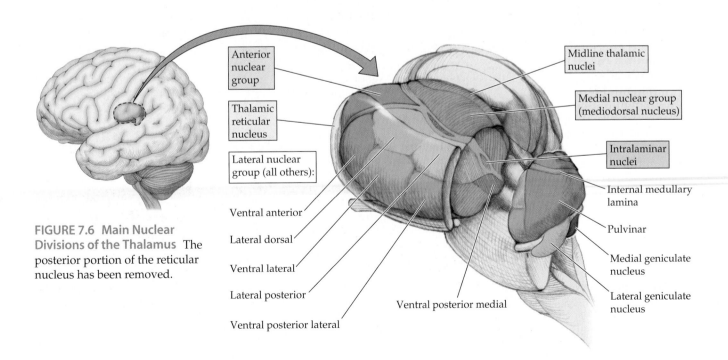

FIGURE 7.6 Main Nuclear Divisions of the Thalamus The posterior portion of the reticular nucleus has been removed.

the lateral aspect of the thalamus. We will now discuss three main categories of thalamic nuclei (Table 7.3):

1. Relay nuclei
2. Intralaminar nuclei
3. Reticular nucleus

Relay Nuclei

Most of the thalamus is made up of **relay nuclei**, which receive inputs from numerous pathways and then project to the cortex. In addition, relay nuclei receive massive reciprocal connections back from the cortex. Projections of relay nuclei to the cortex may be fairly localized to specific cortical regions or more diffuse (see Table 7.3).

SPECIFIC THALAMIC RELAY NUCLEI Among the thalamic relay nuclei, projections to the primary sensory and motor areas tend to be the most localized. These specific relay nuclei lie mainly in the lateral thalamus. All sensory modalities, with the exception of olfaction, have specific relays in the lateral

---

TABLE 7.3 Major Thalamic Nuclei

NUCLEI[a]	MAIN INPUTS[b]	MAIN OUTPUTS	DIFFUSENESS OF PROJECTIONS TO CORTEX[c]	PROPOSED FUNCTIONS
RELAY NUCLEI				
Lateral nuclear group				
Ventral posterior lateral nucleus (VPL)	Medial lemniscus, spinothalamic tract	Somatosensory cortex	+	Relays somatosensory spinal inputs to cortex
Ventral posteromedial nucleus (VPM)	Trigeminal lemniscus, trigeminothalamic tract, taste inputs	Somatosensory and taste cortex	+	Relays somatosensory cranial nerve inputs and taste to cortex
Lateral geniculate nucleus (LGN)	Retina	Primary visual cortex	+	Relays visual inputs to cortex
Medial geniculate nucleus (MGN)	Inferior colliculus	Primary auditory cortex	+	Relays auditory inputs to cortex
Ventral lateral nucleus (VL)	Internal globus pallidus, deep cerebellar nuclei, substantia nigra pars reticulata	Motor, premotor, and supplementary motor cortex	+	Relays basal ganglia and cerebellar inputs to cortex
Ventral anterior nucleus (VA)	Substantia nigra pars reticulata, internal globus pallidus, deep cerebellar nuclei	Widespread to frontal lobe, including prefrontal, premotor, motor, and supplementary motor cortex	+++	Relays basal ganglia and cerebellar inputs to cortex
Pulvinar	Tectum (extrageniculate visual pathway), other sensory inputs	Parietotemporooccipital association	++	Behavioral orientation toward relevant visual and other stimuli
Lateral dorsal nucleus	See anterior nucleus	—	++	Functions with anterior nuclei
Lateral posterior nucleus	See pulvinar	—	++	Functions with pulvinar
Ventral medial nucleus	Midbrain reticular formation	Widespread to cortex	+++	May help maintain alert, conscious state

thalamus en route to their primary cortical areas (see Table 7.3; Figures 7.7 and 7.8). For example, as we have discussed, somatosensory pathways from the spinal cord and cranial nerves relay in the **ventral posterior lateral** (**VPL**) and **ventral posterior medial** (**VPM**) nuclei, respectively. The VPL and VPM in turn project to the primary somatosensory cortex. Visual information is relayed in the **lateral geniculate nucleus** (**LGN**), as we will discuss further in Chapter 11, and auditory information is relayed in the **medial geniculate nucleus** (**MGN**), as we will discuss in Chapter 12 (see also Figure 6.9B). A useful mnemonic for these two nuclei is lateral light (vision), medial music (audition). Motor pathways leaving the cerebellum and basal ganglia (see Figures 2.17, 15.9, and 16.6) also have specific thalamic relays in the **ventral lateral nucleus** (**VL**) en route to the motor, premotor, and supplementary motor cortex (see Table 7.3; Figures 7.7 and 7.8). Even some limbic pathways (see Chapter 18) have fairly specific cortical projections, such as those carried by the **anterior nuclear group** to the anterior cingulate cortex. The anterior thalamic nuclear group forms a prominent bulge in the anterior superior thalamus (see Figure 7.6).

MNEMONIC

TABLE 7.3 (*continued*)

NUCLEI[a]	MAIN INPUTS[b]	MAIN OUTPUTS	DIFFUSENESS OF PROJECTIONS TO CORTEX[c]	PROPOSED FUNCTIONS
Medial nuclear group				
Mediodorsal nucleus (MD)	Amygdala, olfactory cortex, limbic basal ganglia	Frontal cortex	++	Limbic pathways, major relay to frontal cortex
Anterior nuclear group				
Anterior nucleus	Mammillary body, hippocampal formation	Cingulate gyrus	+	Limbic pathways
Midline thalamic nuclei Paraventricular, parataenial, interanteromedial, intermediodorsal, rhomboid, reuniens (medial ventral)	Hypothalamus, basal forebrain, amygdala, hippocampus	Amygdala, hippocampus, limbic cortex	++	Limbic pathways
INTRALAMINAR NUCLEI				
Rostral intralaminar nuclei Central medial nucleus Paracentral nucleus Central lateral nucleus	Deep cerebellar nuclei, globus pallidus, brainstem ascending reticular activating systems (ARAS), sensory pathways	Cerebral cortex, striatum	+++	Maintain alert consciousness; motor relay for basal ganglia and cerebellum
Caudal intralaminar nuclei **Centromedian nucleus** Parafascicular nucleus	Globus pallidus, ARAS, sensory pathways	Striatum, cerebral cortex	+++	Motor relay for basal ganglia
RETICULAR NUCLEUS	Cerebral cortex, thalamic relay and intralaminar nuclei, ARAS	Thalamic relay and intralaminar nuclei, ARAS	None	Regulates state of other thalamic nuclei

[a]The most well known and clinically relevant nuclei are shown in **bold**. Some additional, smaller nuclei have not been included.

[b]In addition to the inputs listed, all thalamic nuclei receive reciprocal inputs from the cortex and from the thalamic reticular nucleus. Modulatory cholinergic, noradrenergic, serotonergic, and histaminergic inputs also reach most thalamic nuclei (see Chapter 14).

[c]+ represents least diffuse (specific thalamic relay nuclei); ++ represents moderately diffuse; +++ represents most diffuse.

FIGURE 7.7 Noncortical Inputs to the Thalamus Main noncortical inputs to the different thalamic nuclei are shown. Cortical connections are shown in Figure 7.8. See Table 7.3 for additional details. Ant., anterior nuclear group; In, intralaminar nuclei; LD, lateral dorsal nucleus; LGN, lateral geniculate nucleus; LP, lateral posterior nucleus; MD, mediodorsal nucleus; MGN, medial geniculate nucleus; VA, ventral anterior nucleus; VLc, ventral lateral nucleus, pars caudalis; VLo, ventral lateral nucleus, pars oralis; VPL ventral posterior lateral nucleus; VPM, ventral posterior medial nucleus.

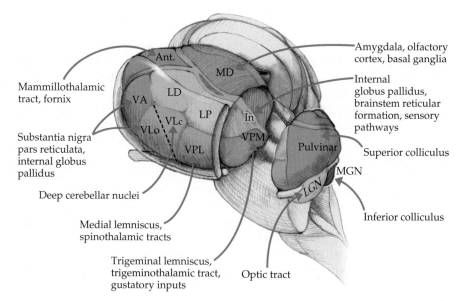

WIDELY PROJECTING (NONSPECIFIC) THALAMIC RELAY NUCLEI Many thalamic nuclei have more widespread cortical projections (see Table 7.3; Figures 7.7 and 7.8). For example, visual and other sensory inputs to the **pulvinar** are relayed to large regions of the parietal, temporal, and occipital association cortex involved in behavioral orientation toward relevant stimuli. The pulvinar ("couch" or "cushion" in Latin) is a large, pillow-shaped nucleus that occupies most of the posterior thalamus (see Figure 7.6). Diffuse relays of limbic inputs and other information involved in cognitive functions occur in the **mediodorsal nucleus (MD)**, as well as in the midline and intralaminar thalamic nuclei. The mediodorsal nucleus, sometimes called the dorsomedial nucleus, forms a large bulge lying medial to the internal medullary lamina, best seen in coronal sections (see Figure 16.4). The MD serves as the major thalamic relay for information traveling to the frontal association cortex (see Figure 7.8; see also Chapters 16, 18, and 19). Other examples of widely projecting thalamic nuclei are listed in Table 7.3.

Intralaminar Nuclei

The **intralaminar nuclei** lie within the internal medullary lamina (see Figure 7.6). Like the relay nuclei, they receive inputs from numerous pathways and have reciprocal connections with the cortex. They are sometimes classified along with other "nonspecific" relay nuclei. However, we have placed them in a separate category here because, unlike relay nuclei, their main inputs and outputs are from the basal ganglia. Intralaminar nuclei can be divided into two functional regions (see Table 7.3): The **caudal intralaminar nuclei** include the large **centromedian nucleus** and are involved mainly in basal ganglia circuitry (see Chapter 16); the **rostral intralaminar nuclei** also have input and output connections with the basal ganglia. In addition, the rostral intralaminar nuclei appear to have an important role in relaying inputs from the **ascending reticular activating systems (ARAS)** to the cortex, maintaining the alert, conscious state (see Figure 2.23 and Chapter 14).

Reticular Nucleus

The **reticular nucleus** forms a thin sheet located just lateral to the rest of the thalamus and just medial to the internal capsule (see Figures 7.6 and 16.4D). It should not be confused with the similarly named *reticular formation*, located in the brainstem (see Chapter 14). The reticular nucleus is the only nucleus of the

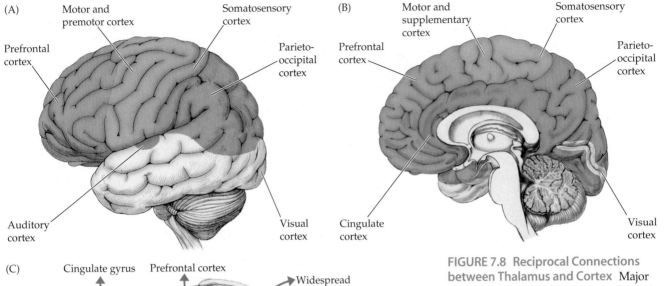

FIGURE 7.8 Reciprocal Connections between Thalamus and Cortex Major connections between thalamic nuclei and cortical areas are shown using corresponding colors. (A) Cortex, lateral view. (B) Cortex, medial view. (C) Thalamus. See Table 7.3 for additional details. Abbreviations are the same as in Figure 7.7.

thalamus that does not project to the cortex. Instead, it receives inputs mainly from other thalamic nuclei and the cortex and then projects back to the thalamus. The reticular nucleus consists of an almost pure population of inhibitory GABAergic neurons. This composition, together with its connections with the entire thalamus, make it well suited to regulate thalamic activity. In addition to cortical and thalamic inputs, other inputs to the reticular nucleus arising from the brainstem reticular activating systems and the basal forebrain may participate in modulating the state of alertness and attention (see Chapters 14 and 19).

In summary, the thalamus has major reciprocal connections with all regions of the cerebral cortex. It contains many different nuclei with different functions. These nuclei convey information from other parts of the nervous system, as well as from the periphery to the cortex.

> ## REVIEW EXERCISE
>
> Name the thalamic nuclei that are most important in relaying the following information to the cortex: somatosensory input, somatosensory face input, visual input, auditory input, basal ganglia and cerebellar input, limbic input (see the boldfaced entries in Table 7.3).

KEY CLINICAL CONCEPT

7.1 PARESTHESIAS

In addition to negative symptoms of sensory loss, lesions of the somatosensory pathways can cause abnormal positive sensory phenomena called **paresthesias**.

Both the character and the location of these abnormal sensations reported by the patient can have localizing value. In lesions of the posterior column–medial lemniscal pathways, patients commonly describe a tingling, numb sensation; a feeling of a tight, bandlike sensation around the trunk or limbs; or a sensation similar to gauze on the fingers when trying to palpate objects. In lesions of the anterolateral pathways, there is often sharp, burning, or searing pain. Lesions of the parietal lobe or primary sensory cortex may cause contralateral numb tingling, although pain can also be prominent. Lesions of the thalamus can cause severe contralateral pain, called **Dejerine–Roussy syndrome**. Lesions of the cervical spine may be accompanied by **Lhermitte's sign**, an electricity-like sensation running down the back and into the extremities upon neck flexion. Lesions of nerve roots often produce **radicular pain** (see KCC 8.3) that radiates down the limb in a dermatomal distribution, is accompanied by numbness and tingling, and is provoked by movements that stretch the nerve root. Peripheral nerve lesions, similarly, often cause pain, numbness, and tingling in the sensory distribution of the nerve.

In addition to "paresthesia," other common terms for sensory abnormalities include **dysesthesia** (unpleasant, abnormal sensation), **allodynia** (painful sensations provoked by normally nonpainful stimuli such as light touch), and **hyperpathia** or **hyperalgesia** (enhanced pain to normally painful stimuli). The term **hypesthesia** means decreased sensation, but it may be misinterpreted (does the "hyp-" refer to "hyper" or "hypo"?) and is thus best avoided. ■

KEY CLINICAL CONCEPT

7.2 SPINAL CORD LESIONS

Spinal cord lesions are a major source of disability because they affect motor, sensory, and autonomic pathways. Suspected dysfunction of the spinal cord must therefore be treated as a medical emergency in an attempt to prevent irreversible damage. The most common causes of spinal cord dysfunction are extrinsic compression due to degenerative disease of the spine, trauma, and metastatic cancer. Additional causes of spinal cord dysfunction are listed in Table 7.4.

Symptoms and signs of spinal cord dysfunction may be obvious when a **sensory level** and **motor dysfunction** corresponding to the level of the lesion are present (see KCC 7.4). Reflex abnormalities, including **abnormal sphincteric function**, can also help confirm the diagnosis (see Tables 3.6 and 3.7; KCC 7.5). In more subtle cases, however, minor sensory or motor changes, back or neck pain, or fever (in epidural abscess) may be the only clues warning of incipient disastrous loss of function. Clinicians should therefore have a low threshold for ordering an urgent MRI scan in cases of suspected spinal cord lesions. In this regard, note that although a sensory level (see KCC 7.4) can suggest the spinal level of a lesion, sometimes the lesion may actually be much higher in the spinal cord. Therefore, higher spinal levels, such as the thoracic and cervical cord, often must be imaged as well, even in cases of suspected lumbosacral pathology.

In acute, severe lesions such as **trauma**, there is often initially a phase of **spinal shock** characterized by flaccid paralysis below the lesion, loss of tendon reflexes, decreased sympathetic outflow to vascular smooth muscle causing moderately decreased blood pressure, and absent sphincteric reflexes and tone. Over the course of weeks to months, spasticity and upper motor neuron signs usually develop. Some sphincteric and erectile reflexes may return, although often without voluntary control. Acute traumatic spinal cord lesions may have improved outcome if treated within the first 8 hours with high doses of steroids.

TABLE 7.4 Differential Diagnosis[a] of Spinal Cord Dysfunction

Trauma or mechanical
    Contusion
    Compression
    Disc herniation
    Degenerative disorders of
      vertebral bones
    Disc embolus
Vascular (see Figure 6.5)
    Anterior spinal artery infarct
    Watershed infarct
    Spinal dural AVM (arteriovenous
      malformation)
    Epidural hematoma
Nutritional deficiency
    Vitamin B_{12}
    Vitamin E
Epidural abscess
Infectious myelitis
    Viral, including HIV
    Lyme disease

    Tertiary syphilis
    Tropical spastic paraparesis
    Schistosomiasis
Inflammatory myelitis
    Multiple sclerosis
    Lupus
    Postinfectious myelitis
Neoplasms
    Epidural metastasis
    Meningioma
    Schwannoma
    Carcinomatous meningitis
    Astrocytoma
    Ependymoma
    Hemangioblastoma
Degenerative/developmental
    Spina bifida
    Chiari malformation
    Syringomyelia

[a]Following format of Figure 1.1.

Chronic **myelopathy** (spinal cord dysfunction) is often seen with **degenerative disorders of the spine**, most commonly in the cervical or lumbar regions. Because both the spine and nerve roots are often compressed, there can be a combination of upper and lower motor neuron signs mimicking motor neuron disease in some cases.

In cord compression caused by **tumors**, it is essential to institute radiation and/or surgical intervention promptly to prevent **irreversible loss of ambulation**. An approximate rule of thumb is that 80% of patients treated for metastatic spinal cord compression after they lose ambulation will remain permanently nonambulatory, while 80% of patients treated before losing ambulation remain ambulatory for the rest of their lives. Metastatic spread to the epidural space is by far the most common cause of neoplastic spinal cord compression, but primary spinal cord tumors can also be seen (see Table 7.4).

Infarction of the spinal cord is usually due to anterior spinal artery occlusion, leading to an anterior cord syndrome (see KCC 7.4). Common causes are trauma, aortic dissection, thromboemboli, and disc emboli (intervertebral disc material that enters local circulation due to trauma). Watershed infarction of the spinal cord is typically at the mid-thoracic vulnerable zone, as we have already discussed (see Figure 6.5). **Spinal dural AVM** can lead to permanent or transient episodes of spinal cord dysfunction and may be challenging to diagnose without appropriate imaging studies, often including angiography.

Myelitis, which can be infectious or inflammatory in etiology, is another important and common cause of spinal cord dysfunction (see Table 7.4; see also KCC 5.9 and 6.6). Patients with myelitis usually present with spinal cord dysfunction that develops relatively quickly, over the course of hours to days. MRI often shows T2 bright areas, and CSF has elevated white blood cell count—usually lymphocytic-predominant, depending on the cause. **Epidural abscess** (see KCC 5.9) can cause irreversible damage to the spinal cord if not diagnosed and treated promptly. ■

Graphesthesia

KEY CLINICAL CONCEPT

7.3 SENSORY LOSS: PATTERNS AND LOCALIZATION

Sensory loss can be caused by lesions anywhere in the somatosensory pathways (see Figures 7.1 and 7.2), including peripheral nerves, nerve roots, posterior column–medial lemniscal and anterolateral pathways, the thalamus, thalamocortical white matter pathways, and the primary somatosensory cortex. In this section and in KCC 7.4, we will review localizing patterns of sensory loss and associated deficits. Sensory loss in the face is discussed further in KCC 12.2. In the illustrations for this section, lesions are indicated in pink, while regions of sensory loss are indicated in purple. Neurons in the posterior column/medial lemniscal system (vibration and position sense) are shown in red, while those in the anterolateral or trigeminothalamic systems (pain and temperature sensation) are shown in blue.

Primary Somatosensory Cortex Figure 7.9A

Deficit is contralateral to the lesion. Despite the depiction in Figure 7.9A, sensory loss does not begin neatly at the midline, and various subregions may be differentially affected, depending on lesion size and location. Discriminative touch and joint position sense are often most severely affected (see **neuroexam.com Videos 73** and **74**), but all modalities may be involved. Sometimes all primary modalities are relatively spared, but a pattern called **cortical sensory loss** is present, with extinction, or decreased stereognosis, and graphesthesia (see Chapter 3; **neuroexam.com Videos 75–77**). Associated deficits from involvement of adjacent cortical areas may include upper motor neuron–type weakness (see Table 6.4), visual field deficits, or aphasia (see KCC 19.6).

Stereognosis

FIGURE 7.9 Patterns of Sensory Loss in Lesions of the Brain or Peripheral Nerves Lesions are shown in red; regions of sensory loss are shown in purple.

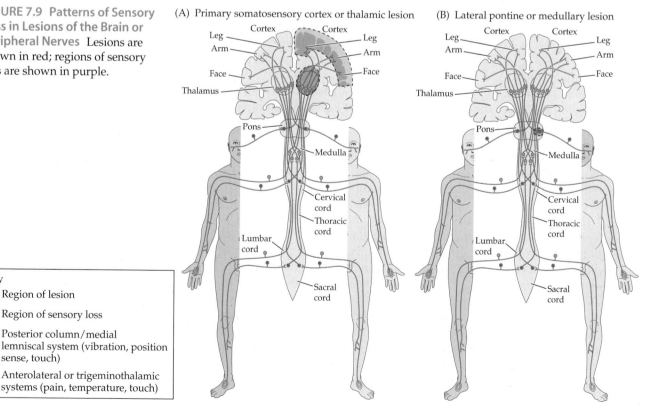

(A) Primary somatosensory cortex or thalamic lesion

(B) Lateral pontine or medullary lesion

Key
- Region of lesion
- Region of sensory loss
- — Posterior column/medial lemniscal system (vibration, position sense, touch)
- — Anterolateral or trigeminothalamic systems (pain, temperature, touch)

Thalamic Ventral Posterior Lateral (VPL) and Ventral Posterior Medial (VPM) Nuclei or Thalamic Somatosensory Radiations

Figure 7.9A

Deficit is contralateral to the lesion. As with lesions of the primary somatosensory cortex, sensory loss does not begin neatly at the midline, and various subregions may be differentially affected. The deficit may be more noticeable in the face, hand (particularly in the lips and fingertips), and foot than in the trunk or proximal extremities. All sensory modalities may be involved, sometimes with no motor deficit. Larger lesions may be accompanied by hemiparesis or hemianopia caused by involvement of the internal capsule; lateral geniculate; or optic radiations. Lesions of the thalamic somatosensory radiations can also cause contralateral hemisensory loss, which is associated with hemiparesis because of the involvement of adjacent corticobulbar and corticospinal fibers (see Figure 6.9B). Less commonly, lesions in other locations, such as the midbrain or upper pons, can cause contralateral somatosensory deficits involving the face, arm, and leg.

Lateral Pons or Lateral Medulla Figure 7.9B

The lesion involves anterolateral pathways and the spinal trigeminal nucleus on the same side. It causes loss of pain and temperature sensation in the body opposite the lesion, and loss of pain and temperature sensation in the face on the same side as the lesion. Associated deficits of the lateral pontine and the lateral medullary syndromes are discussed in KCC 14.3 (see Table 14.7).

Medial Medulla Figure 7.9C

The lesion involves the medial lemniscus, causing contralateral loss of vibration and joint position sense. Associated deficits of the medial medullary syndrome are discussed in Chapter 14 (see Table 14.7).

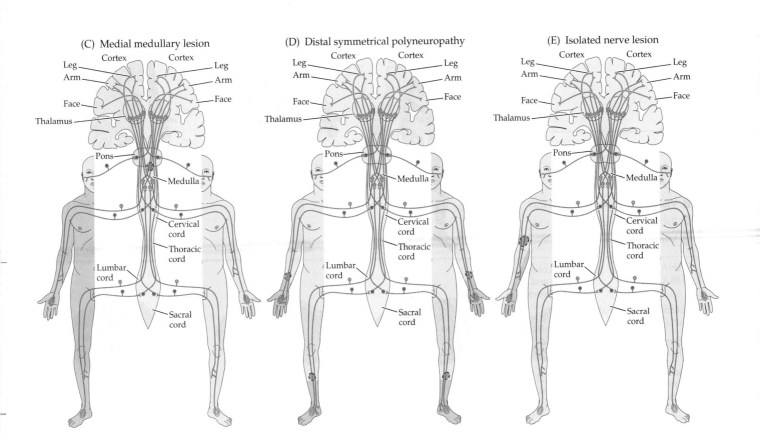

CLINICAL CASES

CASE 7.1 SUDDEN ONSET OF RIGHT ARM NUMBNESS

CHIEF COMPLAINT

An 81-year-old right-handed man came to the emergency room because of right arm numbness and mild language difficulties.

HISTORY

The patient had not seen a doctor in many years. Past medical history was notable for hypertension, diabetes, and angina. At 6:30 P.M. on the day of admission, he suddenly became **confused** and **had difficulty combining words correctly into sentences**. He also complained that he **could not feel things with his right arm** and that **it felt numb**. Furthermore, he had some difficulties using his right hand, saying he could not "**grape with it**." Finally, he also had a vague **blurring of vision** that he could not explain in greater detail.

PHYSICAL EXAMINATION

Vital signs: T = 98.7°F, P = 60, BP = 169/79, R = 12.

Neck: No bruits.

Lungs: Clear.

Heart: Regular rate with no murmurs.

Abdomen: Normal bowel sounds; soft, nontender.

Neurologic exam:

  MENTAL STATUS: Alert and oriented × 3. Speech **fluent, but with occasional paraphasic errors, substituting some letters in words incorrectly** (see "History" above when the patient described use of his right hand). Good naming and repetition. No right–left confusion or finger agnosia. Was able to do simple calculations. Had difficulty with reading and writing.

  CRANIAL NERVES: Pupils 2 mm, constricting to 1.5 mm in response to light. Poor cooperation on visual field testing, but seemed to have **difficulty at times seeing fingers on his right side**. Extraocular movements full. Facial sensation intact to light touch and pinprick. Symmetrical smile. Gag reflex present. Normal sternomastoid strength. Tongue midline.

  MOTOR: **Mild right pronator drift.** Normal tone. 5/5 power throughout.

  REFLEXES:

  COORDINATION: Normal on finger-to-nose and heel-to-shin testing.

  GAIT: Normal on limited testing.

  SENSORY: Intact light touch and pinprick sense. Decreased vibration sense in both toes. **Graphesthesia and stereognosis** normal in the left hand but **absent in the right hand. Occasional extinction in the right hand** on double simultaneous stimulation.

LOCALIZATION AND DIFFERENTIAL DIAGNOSIS

1. On the basis of the symptoms and signs shown in **bold** above, where is the lesion?

2. What is the most likely diagnosis? What are some other possibilities?

Discussion

1. The key symptoms and signs in this case are:

   • **Right arm numbness, extinction, agraphesthesia, and astereognosis with preserved primary sensory modalities**

   • **Mild, fluent aphasia**

   • **Difficulty at times seeing fingers on his right side**

   • **Right pronator drift**

   The patient has a pattern of cortical sensory loss in the right arm consistent with a lesion in the left postcentral gyrus, primary somatosensory cortex, arm area (see KCC 7.3). The loss of vibration sense in the toes is probably not relevant to his current problem and is likely due to longstanding distal symmetrical polyneuropathy (see KCC 8.1), commonly seen in diabetics (this conclusion is also supported by the patient's generally diminished reflexes). The mild fluent aphasia can be explained by a lesion in the dominant (left) parietal lobe (see Chapter 19), as can his right-sided visual difficulties (see Chapter 11). The right pronator drift suggests some mild involvement of corticospinal fibers to the arm, arising from the nearby motor cortex.

The most likely *clinical localization* is left postcentral gyrus, primary somatosensory cortex, arm area, and some adjacent left parietal cortex.

2. The patient is an elderly man with hypertension, diabetes, and cardiac disease, with sudden onset of deficits suggesting embolic stroke (see KCC 10.4). An infarct of the left postcentral gyrus arm area and adjacent parietal cortex would be caused by occlusion of a cortical branch of the left middle cerebral artery (see Figure 10.6). Some other, less likely causes of a cortical lesion in this setting include a small hemorrhage, brain abscess, or tumor. A left subdural hematoma, should also be considered, since it could cause several mild left hemisphere deficits, as seen in this patient.

Clinical Course and Neuroimaging

A head MRI was done (Image 7.1A,B, page 300), showing an area of increased T2 signal consistent with an infarct in the left postcentral gyrus and adjacent parietal lobe. The patient was admitted to the hospital, and an embolic workup (see KCC 10.4) was significant only for regional wall movement abnormalities on his echocardiogram, suggesting prior cardiac ischemia. His language improved back to normal within 5 days, and his right-hand sensation also recovered. He was discharged home on aspirin.

RELATED CASE. A 58-year-old man was sent to the emergency room for suspected myocardial infarction. He had a past history of hypertension and elevated cholesterol. On the day of admission he rose from a chair and suddenly felt transient numbness in his right arm followed by persistent numbness in his right axilla and over his right shoulder. He was admitted to the medicine service where a cardiac evaluation was negative, and neurology was consulted. Neurological exam was normal except for decreased sensation to all modalities over his right shoulder. MRI revealed an acute infarct in the left postcentral gyrus (Image 7.1C,D, page 301). Note that the central sulcus on CT and axial MRI scans has an "inverted omega" shaped portion (see Figure 4.13J), which usually represents the hand sensory and motor cortex. Therefore, this patient's infarct was located just medial to the hand somatosensory cortex in a region consistent with his shoulder numbness (see Figure 6.2; Image 7.1C,D). The patient underwent an embolic evaluation (see KCC 10.4), which was negative, and was discharged home on aspirin and a cholesterol lowering medication to decrease his risk of recurrent stroke.

CASE 7.1 SUDDEN ONSET OF RIGHT ARM NUMBNESS

IMAGE 7.1A,B Infarct in Left Postcentral Gyrus Somatosensory Cortex and Adjacent Parietal Lobe   T2-weighted axial (horizontal) MRI sections. (A) More inferior. (B) More superior.

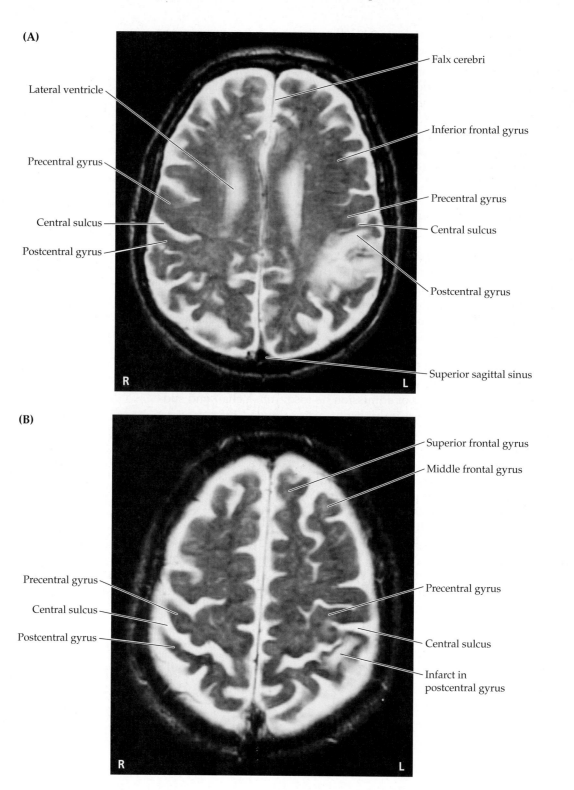

CASE 7.1 *RELATED CASE*

IMAGE 7.1C,D **Left Postcentral Gyrus Infarct** MRI scan. (C) Diffusion-weighted axial MRI image showing acute infarct in the left postcentral gyrus just medial to the "omega-shaped" hand region of the central sulcus. (D) Axial FLAIR image.

(C)

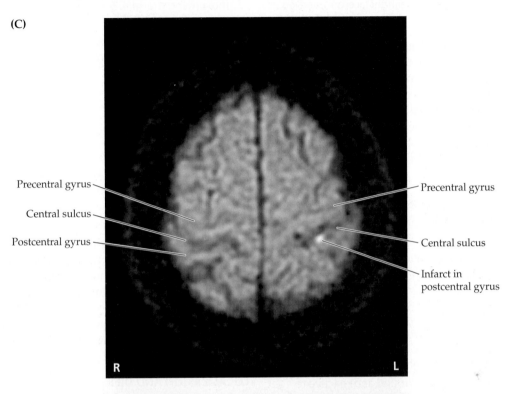

(D)

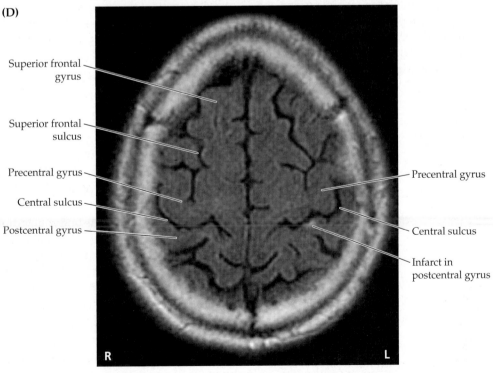

CASE 7.2   SUDDEN ONSET OF RIGHT FACE, ARM, AND LEG NUMBNESS

CHIEF COMPLAINT

A 62-year-old, right-handed woman came to a medical clinic because of 2 days of numbness in her right face, arm, and leg.

HISTORY

On the morning prior to presentation the patient awoke with **decreased sensation in her right face and arm** "as though they were asleep." She had no difficulties with language, motor skills, or vision. Her symptoms continued until the next morning, when she also noticed **loss of sensation in her right foot**. She became concerned and came to her medical clinic for evaluation. Past medical history was notable for hypertension, cigarette smoking, and depression. Her father had a stroke at age 64.

PHYSICAL EXAMINATION

Vital signs: T = 97.4°F, P = 80, BP = 198/114, R = 16.

Neck: Supple with no bruits.

Lungs: Clear.

Heart: Regular rate with no murmurs, gallops, or rubs.

Abdomen: Normal bowel sounds; soft.

Extremities: No edema.

Neurologic exam:

MENTAL STATUS: Alert and oriented × 3. Recalled 3/3 words after 5 minutes. Language normal. Normal finger naming and calculations. No right–left confusion.

CRANIAL NERVES: Pupils equal round and reactive to light. Fundi with A-V nicking (arteriovenous nicking is an ophthalmoscopic sign of chronic hypertension), but normal otherwise. Visual fields full. Extraocular movements intact. **Decreased pinprick, temperature, and light touch sensation in the right face**, especially near the mouth. Facial strength normal. Articulation and palate movements normal. Sternomastoid strength normal. Tongue midline.

MOTOR: No drift. Normal tone. Normal rapid alternating movements. 5/5 power throughout.

REFLEXES:

COORDINATION: Normal on finger-to-nose and heel-to-shin testing.

GAIT: Normal. Tandem gait normal.

SENSORY: **Decreased pinprick, temperature, light touch, and vibration sense on the right half of the body**, especially in the right hand and foot, and less so in the trunk. **Two-point discrimination 15 mm in the right index finger**, compared with 4 mm in the left index finger (using a ruler and the ends of a paper clip). Normal graphesthesia. No extinction on double simultaneous stimulation.

LOCALIZATION AND DIFFERENTIAL DIAGNOSIS

1. On the basis of the symptoms and signs shown in **bold** above, where is the lesion?
2. What is the most likely diagnosis? What are some other possibilities?

Discussion

1. The key symptoms and signs in this case are:

   • **Decreased pinprick, temperature, vibration, and light touch sensation in the right face and body, with decreased two-point discrimination in the right hand**

   Decreased primary sensation in half of the body, including the face, with no other deficits is sometimes thought to be psychologically based because there may be no objective findings on exam. However, this condition can be caused by lesions in the ventral posterior medial and ventral posterior lateral nuclei of the contralateral thalamus (see KCC 7.3).

   The most likely *clinical localization* is left ventral posterior thalamus.

2. Given the relatively sudden onset of deficits, and the history of hypertension, cigarette smoking, and family history of stroke, the most likely diagnosis is ischemic infarct of the left thalamus. Thalamic infarcts are usually caused by occlusion of small penetrating arteries such as the lenticulostriate (middle cerebral), anterior choroidal (internal carotid), or thalamoper-

forator (posterior cerebral) branches, resulting in lacunar infarction (see KCC 10.4; Table 10.3). Hemorrhage is also not uncommon in this location. Other possibilities include tumor or abscess.

Clinical Course and Neuroimaging

The patient was sent from the medical clinic to the emergency room, where a head CT showed a small hypodensity in the left thalamus. This was confirmed with a **head MRI** (Image 7.2, page 304), showing increased T2 and proton density signal in the lateral thalamus consistent with a lacunar infarct. She was admitted to the neurology stroke service, and embolic workup did not reveal a cause for her infarct. She was started on medication to better control her hypertension and was entered into a clinical trial for patients with stroke of unknown cause, comparing aspirin to warfarin (Coumadin) in preventing stroke recurrence (the trial later revealed no clinical difference between these treatments, so aspirin is currently favored). Her numbness gradually improved and was completely gone by 5 days after admission.

CASE 7.3 A FALL CAUSING PARAPLEGIA AND A SENSORY LEVEL

MINICASE

A 24-year-old man was drinking heavily with some friends on the Fourth of July weekend and fell from a second-story balcony. He struck his back on a hard object on the way down and landed in a seated position. He immediately noticed **complete loss of movement and sensation in his legs**. In the emergency room, exam was notable for **flaccid tone and 0/5 strength throughout the lower extremities, decreased rectal tone, absent bulbocavernosus reflex, and a bilateral T10 sensory level to pinprick, touch, vibration, and joint position sense** (Figure 7.13, see Figure 8.4).

LOCALIZATION AND DIFFERENTIAL DIAGNOSIS

1. On the basis of the symptoms and signs shown in **bold** above, where is the lesion?

2. What is the most likely diagnosis?

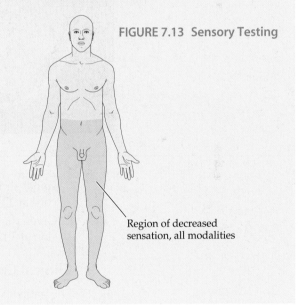

FIGURE 7.13 Sensory Testing

Region of decreased sensation, all modalities

Discussion

1. The key symptoms and signs in this case are:

- **Bilateral leg flaccid paralysis**
- **Decreased rectal tone and absent bulbocavernosus reflex**
- **T10 sensory level to all modalities**

Bilateral leg weakness can be caused by lesions of the bilateral medial frontal lobes, the spinal cord, nerve roots, peripheral nerves, or muscles (see KCC 6.3). However, complete paralysis of both legs with no weakness elsewhere, along with decreased rectal tone and reflexes (see KCC 7.4, 7.5; see also Tables 3.5, 3.7), strongly suggests a spinal cord or cauda equina lesion. The T10 sensory level localizes the lesion more specifically to the T10 level of the spinal cord (cauda equina cannot cause deficits above L1 or L2). The absent bulbocavernosus reflex supports an acute spinal cord injury causing

CASE 7.2 SUDDEN ONSET OF RIGHT FACE, ARM, AND LEG NUMBNESS

IMAGE 7.2 Lacunar Infarct in Left Thalamus Region of VPL and VPM
Proton density-weighted axial (horizontal) MRI section through thalamus and basal ganglia.

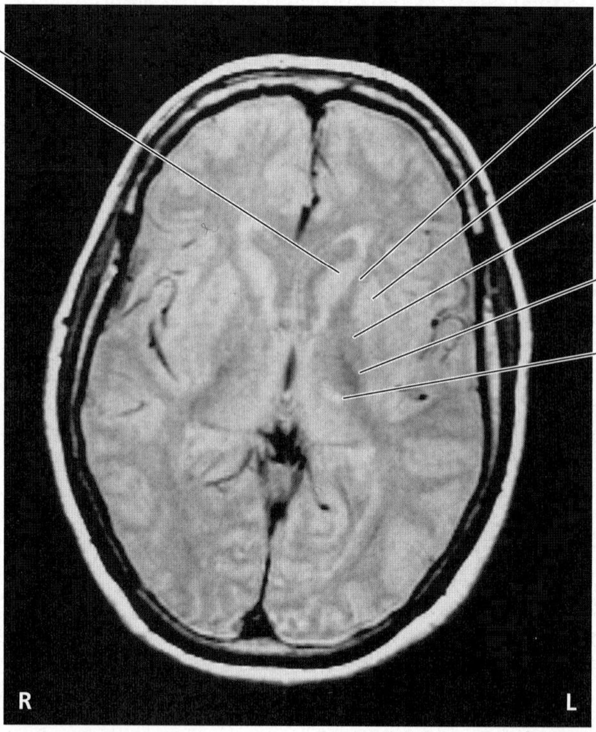

spinal shock (see KCC 7.2, 7.5). The most likely *clinical localization* is spinal cord at T10 level (approximately T9 vertebral bone—see Figure 8.1).

2. Given the history of a significant fall, a fractured spine with cord compression at T10 is likely.

Clinical Course and Neuroimaging

The patient was given high-dose intravenous methylprednisolone (a steroid) in a standard protocol for acute spinal cord injuries. Plain radiographs were done, followed by a **CT of the spine** (Image 7.3A,B, page 306), showing burst fractures of the T9 and T10 vertebral bones leading to near complete obliteration of the spinal canal. Lateral radiographs showed anterior displacement of T9 on T10 by about 4 cm, further compressing the spinal canal. The patient was admitted to the intensive care unit and placed on a special bed for spinal stabilization. Because the spinal compression seemed complete, there was little hope for recovery of function. However, he was taken to the operating room to mechanically stabilize his thoracic spine, which is functionally important for maintaining a seated posture. A T6–T11 vertebral fusion was done with bone grafts and metal rods. The patient remained paraplegic at the time of transfer to a spinal cord rehabilitation center 2 weeks after admission.

CASE 7.4 LEFT LEG WEAKNESS AND RIGHT LEG NUMBNESS

CHIEF COMPLAINT

A 71-year-old woman was referred to a neurologist with 10 months of progressive gait difficulty, right leg numbness, and urinary problems.

HISTORY

The patient was in good health, walking 3 to 4 miles per day until about 10 months prior to admission, when she first noticed mild gait unsteadiness and bilateral leg stiffness. She felt her feet were not fully under her control. Her left leg gradually became weaker than her right, with occasional left leg buckling when she walked. Meanwhile, her **right leg developed progressive numbness and tingling**, and she had intermittent **left-sided thoracic back pain**. Finally, she had **increasing urinary frequency**, **occasional urinary incontinence**, **and difficulty completing a bowel movement** despite several laxatives.

PHYSICAL EXAMINATION

Vital signs: T = 96.3°F, P = 78, BP = 136/76.

Neck: Supple.

Lungs: Clear.

Breasts: Normal.

Heart: Regular rate with no murmurs.

Abdomen: Soft; no masses.

Extremities: No edema.

Rectal: Normal tone; however, patient **could not voluntarily contract anal sphincter**. Stool was negative for occult blood.

Neurologic exam:

MENTAL STATUS: Alert and oriented × 3. Mildly anxious but otherwise normal.

CRANIAL NERVES: Intact.

MOTOR: Upper extremities: Normal bulk and tone; 5/5 strength throughout. Lower extremities: Normal bulk; **tone increased in left leg**; 5/5 strength throughout, except **4/5 strength in the left iliopsoas**.

REFLEXES:

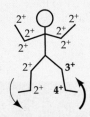

COORDINATION: Normal on finger-to-nose and heel-to-shin testing.

GAIT: **Stiff-legged and unsteady.**

SENSORY: **Pinprick sensation decreased on the right side below the umbilicus** (Figure 7.14). **Vibration and joint position sense decreased in the left foot and leg.** Otherwise intact.

LOCALIZATION AND DIFFERENTIAL DIAGNOSIS

1. On the basis of the symptoms and signs shown in **bold** above, where is the lesion?
2. What is the most likely diagnosis? What are some other possibilities?

FIGURE 7.14 Pinprick Testing

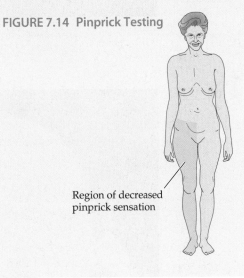

Region of decreased pinprick sensation

Discussion

1. The key symptoms and signs in this case are:

   • Left leg weakness, increased tone, hyperreflexia, and Babinski's sign

   • Decreased vibration and joint position sense in the left foot and leg

   • Decreased pinprick sensation on the right side below the umbilicus, with right leg numb, tingling paresthesias

   • Left-sided thoracic back pain

   • Stiff-legged, unsteady gait

   • Impaired bowel and bladder function

CASE 7.3 A FALL CAUSING PARAPLEGIA AND A SENSORY LEVEL

IMAGE 7.3A,B T9, T10 Vertebral Fractures with Obliteration of Spinal Canal Axial CT of the spine. (A) Image at level of T10 vertebral body shows smashed T10 vertebral bone with obliteration of spinal canal and spinal cord at that level. (B) Image slightly lower, at level of T11, showing normal appearance of spinal canal.

(A)

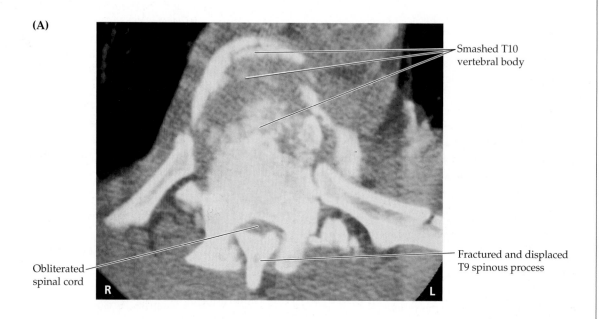

Smashed T10 vertebral body

Obliterated spinal cord

Fractured and displaced T9 spinous process

R   L

(B)

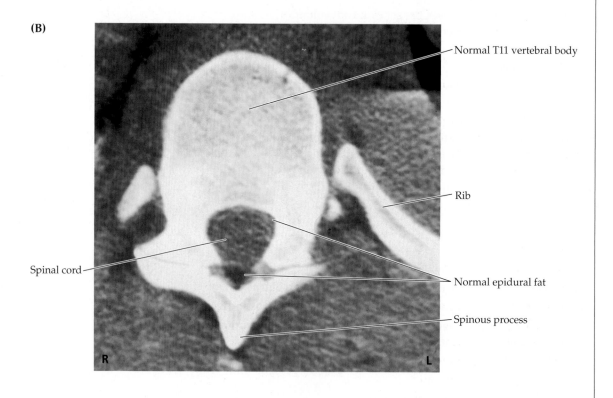

Normal T11 vertebral body

Rib

Spinal cord

Normal epidural fat

Spinous process

R   L

Unilateral weakness and upper motor neuron signs (see Table 6.4) in the leg could be caused by a lesion in the ipsilateral spinal cord or contralateral motor cortex (see KCC 6.3). The associated sensory abnormalities and left thoracic pain support a left hemicord lesion at approximately T9 or T10 (see Figure 8.4), producing a Brown–Séquard syndrome (see KCC 7.4). The presence of bowel and bladder dysfunction suggests some bilateral involvement of the cord (see KCC 7.5), as does the spastic gait (see KCC 6.5), which sounds as if it may have involved both legs.

The most likely *clinical localization* is left T9 or T10 hemicord lesion, with possible mild involvement of the right cord as well.

2. Given the gradual, progressive onset of deficits over several months in an elderly woman and the presence of local pain, the most likely diagnosis is a tumor compressing the thoracic spinal cord from the left side (see KCC 7.2; Table 7.4). Other, less likely possibilities include arthritic bony changes compressing the spinal cord or multiple sclerosis.

Clinical Course and Neuroimaging

The patient underwent a **spine MRI** (Image 7.4A,B, page 308), which showed a lesion arising from the dura consistent with a meningioma compressing the spinal cord from the left side at the T9 level. She was started on steroids and admitted to the hospital. Meningiomas are histologically benign (see KCC 5.8) and can often be treated effectively by local resection. She was therefore taken to the operating room, and a T8–T9 posterior laminectomy (see KCC 8.5) was performed. Once the dural sac was entered, a tannish mass was found on the left side and carefully dissected away from the spinal cord and nerve roots. Pathologic analysis was consistent with meningioma. The patient did well postoperatively, regaining pain and temperature sensation on the right and position and vibration sense on the left and experiencing improved ambulation and sphincter control.

CASE 7.5 SENSORY LOSS OVER BOTH SHOULDERS

MINICASE

A 46-year-old man was in a motor vehicle accident at age 18 years, sustaining C2 and C3 fractures. These fractures caused quadriparesis that gradually improved. In recent years, however, he has had **increasing difficulty walking**. He has also developed **pain and numbness in the shoulders and arms, more severe on the left side**. Exam was notable for increased **tone in all extremities and 3/5 to 4/5 power throughout**. These exam findings were slightly worse than in previous years. He had **brisk reflexes and bilateral upgoing toes**. Gait was **shuffling and slow**. He had **decreased pinprick sensation in the left arm from the shoulder down, and over the right shoulder** (Figure 7.15). Vibration sense was intact.

LOCALIZATION AND DIFFERENTIAL DIAGNOSIS

1. On the basis of the symptoms and signs shown in **bold** above, where is the lesion?

2. What is the most likely diagnosis? What are some other possibilities?

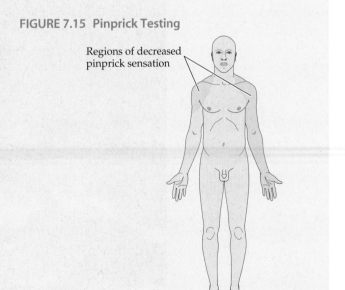

FIGURE 7.15 Pinprick Testing

Regions of decreased pinprick sensation

CASE 7.4 LEFT LEG WEAKNESS AND RIGHT LEG NUMBNESS

IMAGE 7.4A,B Intradural Mass Compatible with Meningioma Compressing Left Spinal Cord at T9 MRI scan. (A) Sagittal T1-weighted image with intravenous gadolinium, showing a homogeneous enhancing mass lesion within spinal canal at T9 level. (B) Axial T2-weighted image at T9 level, showing mass located within the dural sac but outside the spinal cord, compressing cord from the left side.

(A)

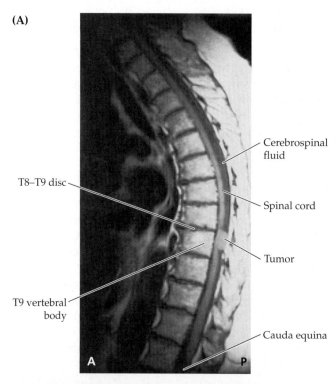

Cerebrospinal fluid

T8–T9 disc

Spinal cord

Tumor

T9 vertebral body

Cauda equina

(B)

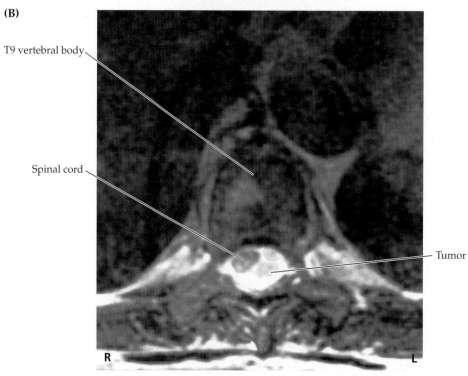

T9 vertebral body

Spinal cord

Tumor

Discussion

1. The key symptoms and signs in this case are:

   - Decreased pinprick sensation and painful numb paresthesias in bilateral shoulders and in the left arm
   - Weakness, increased tone, and hyperreflexia in all extremities
   - Slow, shuffling gait

   This patient has a region of suspended sensory loss to pinprick in a "cape" distribution over both shoulders consistent with a cervical central cord syndrome affecting approximately the C5 dermatomes bilaterally, and C6–T1 on the left (see KCC 7.4; see also Figure 8.4). A cervical central cord lesion could also explain the diffuse weakness with upper motor neuron signs and abnormal gait, since large central cord lesions can cause bilateral corticospinal dysfunction.

   The most likely *clinical localization* is central cord lesion above approximately C5.

2. Given the past history of spinal cord trauma and delayed onset of a central cord syndrome, the most likely diagnosis is posttraumatic syringomyelia. **Syringomyelia** is a fluid-filled cavity in the spinal cord, which can occur with spinal cord tumors, with congenital abnormalities of the craniocervical junction, or following trauma. **Posttraumatic syringomyelia** is a delayed sequela of about 1% of spinal cord injuries, with symptoms beginning from a few months to up to 30 years after the injury (9 years on average). The pathophysiological mechanism is unknown, but patients present with a central cord syndrome that can progressively worsen or spontaneously stabilize. Surgical decompression has been tried, but the long-term benefits are uncertain. Aside from posttraumatic syringomyelia, other, less likely diagnoses in this patient include central cord involvement by multiple sclerosis or an intrinsic spinal cord tumor such as astrocytoma or ependymoma (see Table 7.4).

Clinical Course and Neuroimaging

A **cervical spine MRI** (Image 7.5, page 310) showed a T1 dark cavity consistent with fluid in the center of the spinal cord representing a syrinx extending from C3 to C4. The patient underwent decompressive surgery and did well in the immediate postoperative period but was subsequently lost to follow-up.

CASE 7.6 BODY TINGLING AND UNSTEADY GAIT

MINICASE

A 37-year-old woman came to the emergency room complaining of **tingling and numbness of her arms, legs, and trunk** of 1 week's duration. She also complained of **clumsiness when walking** quickly or up stairs and of **decreased dexterity in her hands**. She had no other complaints except for an occasional, brief, **electricity-like sensation** that she felt beginning in her spine and running down her arms and legs. This unusual sensation was brought on by neck movement. Exam was notable for **dysmetria** on finger-to-nose testing, which was markedly **worse with the eyes closed**. Tandem gait was somewhat **unsteady**. A dramatic **Romberg sign** was present. There was **profound loss of vibration sense**, absent in the toes, an-

kles, and knees and reduced in the knuckles. **Joint position sense** was moderately **decreased** in the toes and slightly decreased in the fingers. Pinprick and temperature sensation were intact, but testing elicited a **girdle-like band of increased sensation** extending from the lower chest to the abdomen. The remainder of the exam was normal.

LOCALIZATION AND DIFFERENTIAL DIAGNOSIS

1. On the basis of the symptoms and signs shown in **bold** above, where is the lesion?

2. What is the most likely diagnosis? What are some other possibilities?

CASE 7.5 SENSORY LOSS OVER BOTH SHOULDERS

IMAGE 7.5 **Fluid-Filled Syrinx in Cervical Spinal Cord at C3, C4 Levels** T1-weighted sagittal MRI.

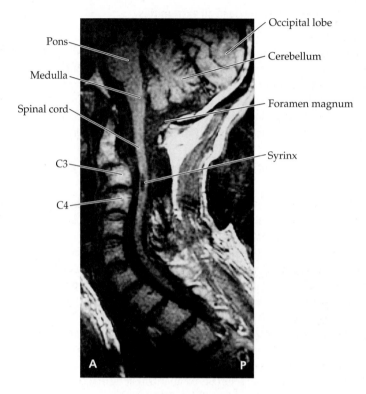

Pons —
Medulla —
Spinal cord —
C3 —
C4 —

— Occipital lobe
— Cerebellum
— Foramen magnum
— Syrinx

A   P

Discussion

1. The key symptoms and signs in this case are:

   - **Loss of vibration and joint position sense in the arms and legs bilaterally**
   - **Clumsy, unsteady gait, with Romberg sign**
   - **Decreased dexterity and dysmetria of the hands, especially with eyes closed**
   - **Paresthesias: Diffuse tingling and numbness, Lhermitte's sign (KCC 7.1), and girdle-like band of hyperpathia around the abdomen**

   Loss of vibration and joint position sense in all extremities with no other associated deficits can be caused by a lesion of the posterior columns in the cervical spinal cord (see KCC 7.4) or by a large fiber neuropathy (see Table 7.2; KCC 8.1). The gait unsteadiness and Romberg sign (see KCC 6.5) can also be explained by the proprioceptive deficit, as can impaired coordination of the hands that is worse with the eyes closed. Lhermitte's sign (see KCC 7.1) suggests that the lesion is in the cervical spine.

   The most likely *clinical localization* is posterior columns of the cervical spinal cord, extending above approximately C5 (arms are involved).

2. Given the relatively rapid onset of deficits localizable to central nervous system white matter tracts in a young woman, these symptoms could represent the first episode of multiple sclerosis (see KCC 6.6), presenting as myelitis (see KCC 7.2). Alternatively, it could represent myelitis (see Table 7.4) from another cause, such as infectious or postinfectious myelitis. Other conditions causing prominent posterior column involvement include tabes

dorsalis of tertiary syphilis and vitamin B_{12} deficiency (tabes dorsalis actually affects mainly the dorsal roots, with secondary degeneration of the posterior columns).

Clinical Course and Neuroimaging

A cervical **spine MRI** (Image 7.6, page 312) showed increased T2 signal in the cervical posterior cord. Additional history was elicited of a paternal uncle with multiple sclerosis (see KCC 6.6). In addition, the patient had had a respiratory infection during the preceding month. Additional tests were negative, except for two oligoclonal bands (see KCC 6.6) present in the cerebrospinal fluid. The presumed diagnosis was either postinfectious myelitis or myelitis as a precursor to multiple sclerosis. The patient's gait unsteadiness and loss of vibration sense were improved but still present at the time of discharge.

CASE 7.7 HAND WEAKNESS, PINPRICK SENSORY LEVEL, AND URINARY RETENTION

MINICASE

A 26-year-old woman was pushing a shopping cart and suddenly developed neck pain, arm pain, and **bilateral hand weakness**. Shortly afterward she found that she had **urinary retention**, being unable to void voluntarily, as well as **fecal incontinence**. She went to the emergency room, where her exam was notable for **upper extremities with bilateral decreased tone, 3/5 triceps strength (C7), and 0/5 grip and finger extensors bilaterally (C8–T1), with absent triceps reflexes.** In addition, she had a **sensory level (see KCC 7.4) to pinprick and temperature sensation** as shown in Figure 7.16, with preserved vibration and joint position sense. **Rectal tone was absent, and she required urinary catheterization to void.**

LOCALIZATION AND DIFFERENTIAL DIAGNOSIS

1. Is the bilateral hand weakness caused by an upper motor neuron or a lower motor neuron lesion (see Table 6.4)? A lesion in what location can cause such weakness (see KCC 6.3)?

2. The reduced sensory level to pinprick occurs at what spinal level (see Figure 8.4)?

3. Lesions in what location can cause acute urinary retention and fecal incontinence with absent rectal tone (see KCC 7.5)?

4. On the basis of the symptoms and signs described above, where is the patient's lesion? What are some possible causes?

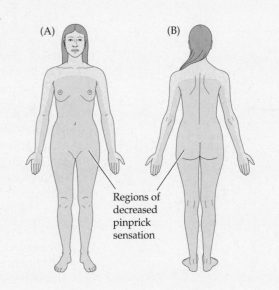

FIGURE 7.16 Pinprick Testing

Discussion

1. The key symptoms and signs in this case are:

- **Bilateral hand and triceps weakness, hypotonia, and absent triceps reflexes**
- **Sensory level to pinprick and temperature**
- **Urinary retention, fecal incontinence, and absent rectal tone**

The patient has weakness, absent reflexes, and hypotonia at C7–T1, suggesting either a lower motor neuron lesion or an acute upper motor neuron

CASE 7.6   BODY TINGLING AND UNSTEADY GAIT

IMAGE 7.6 Region of Hyperintensity in the Posterior (Dorsal) Cervical Spinal Cord at the C3, C4, and C5 Levels Compatible with Demyelination T2-weighted sagittal MRI of the spine.

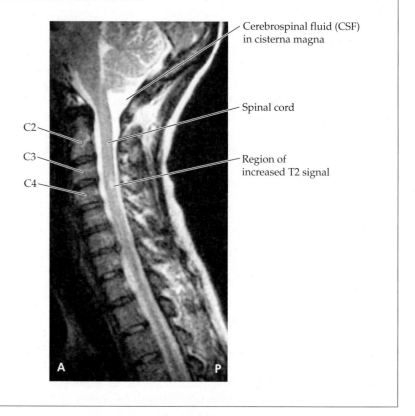

Cerebrospinal fluid (CSF) in cisterna magna

Spinal cord

Region of increased T2 signal

C2

C3

C4

A   P

lesion (see Table 6.4). Bilateral hand (C8–T1) and triceps (C7) weakness can be caused by bilateral anterior horn cell lesions at this level (see KCC 6.3; Figure 6.14I).

2. The reduced sensory level to pinprick occurs at spinal level C7.

3. Urinary retention caused by flaccid bladder with preserved hyperactive urinary sphincter reflexes can be caused by acute lesions below the pontine micturition center and above the conus medullaris (see KCC 7.5). Fecal incontinence with absent rectal tone can be seen in cauda equina or acute spinal cord lesions. Therefore, the most likely localization is an acute spinal cord lesion.

4. C7–T1 lower motor neuron weakness, a C7 sensory level to pinprick with preserved vibration and position sense, and urinary retention with fecal incontinence suggest an anterior cord syndrome (see KCC 7.4) at approximately the C7 to T1 levels. Some possible causes include anterior spinal artery infarct, infectious or inflammatory myelitis, or possibly undisclosed trauma (see Table 7.4).

Clinical Course and Neuroimaging

A **cervical spine MRI** (Image 7.7A,B, page 314) showed increased T2 signal in the anterior cervical spinal cord extending from C5 to T1. A very extensive workup for inflammatory, infectious, and embolic disorders was carried out but was negative. The patient's strength and urinary retention gradually improved, and on a follow-up examination 1 year later she had recovered 4/5 strength in both hands. She continued, however, to have bothersome neck and shoulder pain requiring treatment in a comprehensive pain management program.

Additional Cases

Other chapters include related cases of **somatosensory deficits** (see Cases 8.1, 8.5–8.7, 8.11, 9.1, 9.3, 9.4, 9.6–9.10, 10.2, 10.8, 10.10, 10.11, 10.13, 12.2, 12.3, 13.5, 14.1, 14.4, 15.4, and 19.6). Additional relevant cases can be found using the **Case Index** located at the end of this book.

Brief Anatomical Study Guide

1. This chapter completes our discussion of the three most clinically important long tracts in the nervous system: the **lateral corticospinal tract** (see Chapter 6), the **posterior column–medial lemniscal pathway**, and the **anterolateral pathways** (see Table 7.1). The posterior column–medial lemniscal pathway subserves the sensory modalities of **proprioception**, **vibration sense**, and **fine touch** and has its **major decussation in the lower medulla**. The anterolateral pathways convey the sensory modalities of **pain**, **temperature**, and **crude touch** and have their **major decussation in the spinal cord**.

2. In the posterior column–medial lemniscal pathway (proprioception, vibration sense, and fine touch), myelinated fibers enter the spinal cord dorsal roots and ascend in the posterior column **gracile fasciculus (legs)** or **cuneate fasciculus (arms)**, to synapse in the **nucleus gracilis** or the **nucleus cuneatus** of the caudal medulla (see Figure 7.1).

3. Second-order sensory neurons then **cross to the opposite side of the medulla** in the **internal arcuate fibers** to form the **medial lemniscus**, which ascends to the **ventral posterior lateral (VPL) nucleus** of the thalamus. Tertiary sensory neurons from the VPL project to the **primary somatosensory cortex** (see Figures 7.1; see also Figure 6.1) via the posterior limb of the internal capsule.

4. The anterolateral system **spinothalamic pathway** mediates discriminative aspects of pain and temperature sensation, while other anterolateral pathways are more involved in the arousal, affective, and modulatory aspects of pain perception. In the spinothalamic pathway, small myelinated and unmyelinated fibers enter the spinal cord dorsal roots and synapse in the **dorsal horn marginal zone** and other layers (see Figure 7.2; see also Figure 6.3B).

5. Secondary sensory neurons in the dorsal horn have axons that **cross to the opposite side of the spinal cord** and then ascend in the ventral lateral part of the spinal cord and brainstem to reach the **VPL** of the thalamus. Once again, tertiary sensory neurons from the VPL then project to the **primary somatosensory cortex** (see Figure 7.2; see also Figure 6.1) via the posterior limb of the internal capsule.

6. Like the lateral corticospinal tract, the somatosensory pathways are **somatotopically organized** (see Figure 7.3; see also Figure 6.2). Spinal cord pathways discussed in Chapters 6 and 7 are summarized in Figure 7.4.

7. The **thalamus** is the major subcortical relay station for signals traveling to all regions of the cortex. In addition to the VPL and VPM, which relay somatosensory information, the thalamus contains many other nuclei (see Table 7.3; Figures 7.6–7.8) that convey sensory, motor, limbic, associative, modulatory, or other signals to different cortical areas. The most clinically important nuclei of the thalamus are shown in bold text in Table 7.3.

Mariani C, Cislaghi MG, Barbieri S, Filizzdo F, DiPalma F, Farina E, D'Aliberti G, Scarlato G. 1991. The natural history and results of surgery in 50 cases of syringomyelia. *J Neurol* 238 (8): 433–438.

Miko I, Gould R, Wolf S, Afifi S. 2009. Acute spinal cord injury. *Int Anesthesiol Clin* 47 (1): 37–54.

Schneider RC. 1955. The syndrome of acute anterior spinal cord injury. *J Neurosurg* 12: 95–122.

Schneider RC, Cherry G, Pantek H. 1954. The syndrome of acute central cervical spinal cord injury. *J Neurosurg* 11: 546–573.

Vernon JD, Chir B, Silver JR, Ohry A. 1982. Post-traumatic syringomyelia. *Paraplegia* 20 (6): 339–364.

Anatomy of Bowel, Bladder, and Sexual Function

Beckre HD, Stenzi A, Wallwiener D, Zittel TT. 2005. *Urinary and Fecal Incontinence: An Interdisciplinary Approach.* Springer.

Craggs MD, Balasubramaniam AV, Chung EA, Emmanuel AV. 2006. Aberrant reflexes and function of the pelvic organs following spinal cord injury in man. *Auton Neurosci* 126–127: 355–370.

Fowler C, DasGupta R. 2007. Neurological causes of bladder, bowel, and sexual dysfunction. In *Neurology in Clinical Practice: Principles of Diagnosis and Management,* WG Bradley, RB Daroff, GM Fenichel and CD Marsden (eds.), 5th Ed., Chapter 31. Butterworth-Heinemann, Boston.

Sung VW. 2009. Urinary incontinence. *Med Health R I* 92 (1): 16–19.

Wald A. 2007. Fecal incontinence in adults. *N Engl J Med* 356: 1648.

CONTENTS

Chapter 8

Spinal Nerve Roots

A 38-year-old man was thrown violently to the ground by an explosion while he was working on a road. While in the hospital recovering from burn injuries, he noticed mild back pain with a numb "pins and needles" sensation running down his left leg into the sole and pinky toe of his left foot. He was unable to stand on his toes with his left foot, and his left Achilles tendon reflex was absent. This patient's symptoms illustrate the kinds of sensory and motor deficits associated with spinal nerve root damage.

In this chapter, we will learn about the anatomical course of exiting spinal nerve roots, nerve root functions, and the clinical consequences of injury to these structures.

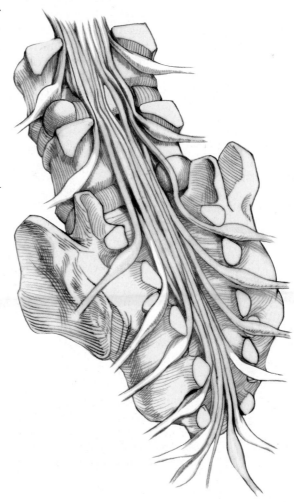

ANATOMICAL AND CLINICAL REVIEW

(A)

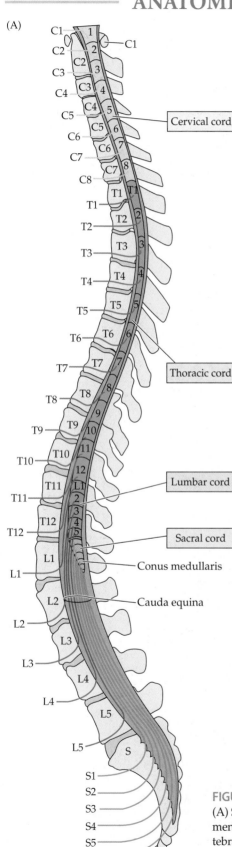

I N THE PRECEDING TWO CHAPTERS, we studied the three main motor and sensory pathways in the central nervous system (see Table 7.1). We will now follow these somatic sensory and somatic motor pathways into the peripheral nervous system (see Figure 2.1) to explore the anatomy and clinical disorders of the peripheral nerves. In this chapter, we will concentrate on the anatomy of spinal nerve roots and their relation to the vertebral structures, regions of innervation, and common clinical disorders. Peripheral autonomic functions were already discussed in Chapter 6. In Chapter 9, we will follow the nerves further into the periphery, and we will discuss the brachial and lumbosacral plexuses and peripheral nerve branches.

Segmental Organization of the Nervous System

Like their invertebrate ancestors, humans retain some degree of segmental organization, especially in the spinal cord. There are **8 cervical (C1–C8), 12 thoracic (T1–T12), 5 lumbar (L1–L5), 5 sacral (S1–S5)**, and **1 coccygeal (Co1) spinal segments** (Figure 8.1). During development, the bones of the spine continue to grow after the spinal cord has reached its full size. Therefore, in adults the spinal cord normally ends with the **conus medullaris** at the level of the **L1 or L2 vertebral bones**. The nerve roots (see Figure 8.1; see also Figure 6.3) travel downward to reach their exit points at the appropriate level. Below the L1 or L2 vertebral bones, the spinal canal contains nerve roots with no spinal cord, forming the **cauda equina**, meaning "horse's tail" (see Figures 8.1A and 8.3C). The conus medullaris tapers into the **filum terminale**, a thin strand of connective tissue running in the center of the cauda equina. The roots of the cauda equina are organized such that the most centrally located roots are from the most caudal segments of the spinal cord (see Figure 8.3C).

(B)

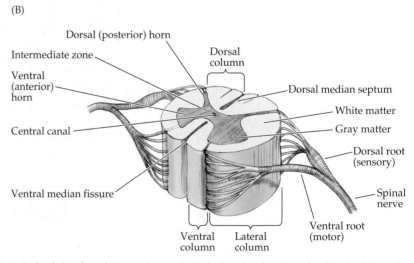

FIGURE 8.1 **Spinal Cord and Nerve Roots in Relation to the Vertebral Spinal Canal** (A) Sagittal view. The cervical (C5–T1) and lumbosacral (L1–S3) spinal cord enlargements supply nerves to the arms and legs, respectively. At the level of the L1 or L2 vertebral bones the spinal cord ends, and nerve roots continue as the cauda equina. (B) Motor (ventral) roots and sensory (dorsal) roots join at each segment to form spinal nerves.

Left and right sensory and motor **nerve roots** arise from each segment of the spinal cord (see Figure 8.1B) except for the C1 segment, which has no sensory roots. As discussed in Chapter 6, the **cervical enlargement** (C5–T1) gives rise to the nerve roots for the arms, and the **lumbosacral enlargement** (L1–S3) gives rise to the nerve roots for the legs (see Figure 8.1A). A short distance from the spinal cord, before leaving the spinal canal, the sensory and motor nerve roots fuse to form a single **mixed spinal nerve** for each segment (see Figure 8.1B). The spinal nerves then further fuse and intermingle peripherally to form plexuses and nerve branches, as will be discussed in Chapter 9. In this chapter we will focus on the innervation provided by single spinal segments and the distinct clinical disorders associated with damage to individual spinal nerve roots.

Nerve Roots in Relation to Vertebral Bones, Discs, and Ligaments

The vertebral bones function both as the central mechanical support for the body and as protection for the spinal cord. Each vertebral bone has a sturdy cylindrical **vertebral body** located anteriorly (Figure 8.2A,B). The vertebral bodies are separated from each other by connective tissue **intervertebral discs**, consisting of a central **nucleus pulposus** surrounded by a capsule called the **annulus fibrosus** (Figure 8.2C). Posteriorly, the neural elements are surrounded by an arch of bone formed by the **pedicles, transverse processes, laminae,** and **spinous processes** (see Figure 8.2B). The **superior and inferior articular processes** or **facet joints** form additional points of mechanical contact between adjacent vertebrae (see Figure 8.2A,B).

The spinal cord runs through the **spinal canal** (**vertebral foramen**) and is surrounded by pia, arachnoid, and dura mater (see Figure 8.2). As the dura exits the skull at the foramen magnum, the inner layer continues and the outer layer becomes indistinguishable from periosteum (see Figure 5.10). Unlike in the cranium, there is a layer of **epidural fat** between the dura and the periosteum in the spinal canal (see Figure 8.2C, and Figure 8.2D), which is a useful landmark on MRI scans. In addition, there is a valveless meshwork of epidural veins called **Batson's plexus** that is thought to play a role in the spread of metastatic cancers and infections in the epidural space. The elastic **ligamentum flavum** is particularly prominent in cervical and lumbar regions and can sometimes become hypertrophied and contribute to spinal cord or nerve root compression.

The nerve roots exit the spinal canal via the **neural** (**intervertebral**) **foramina** (see Figure 8.2A,D and Figure 8.3). **Disc herniations** (see KCC 8.3) are most common at the cervical and lumbosacral levels. An understanding of the anatomy of the nerve roots and discs should make clear the following important rule of thumb: **For both cervical and lumbosacral disc herniations, the nerve root involved usually corresponds to the lower of the adjacent two vertebrae.** For example, a C5–C6 disc herniation usually produces a C6 radiculopathy, an L5–S1 disc usually produces an S1 radiculopathy, and so on. The explanation for this rule is different for cervical versus lumbosacral discs, as we will now discuss.

MNEMONIC

Thoracic, lumbar, and sacral nerve roots exit below the correspondingly numbered vertebral bone (see Figure 8.1). Cervical nerve roots, on the other hand, exit above the correspondingly numbered vertebral bone—except for C8, which has no corresponding vertebral bone and exits between C7 and T1. Cervical nerve roots have a fairly horizontal course as they emerge from the dural or **thecal sac** near the intervertebral disc and exit through the intervertebral foramen (see Figure 8.3B). Cervical discs are usually constrained by the **posterior longitudinal ligament** to herniate laterally toward the nerve root, rather than centrally toward the spinal cord. Thus, in the cervical cord the nerve root involved usually corresponds to the lower vertebral bone of the disc space (see Figure 8.3A).

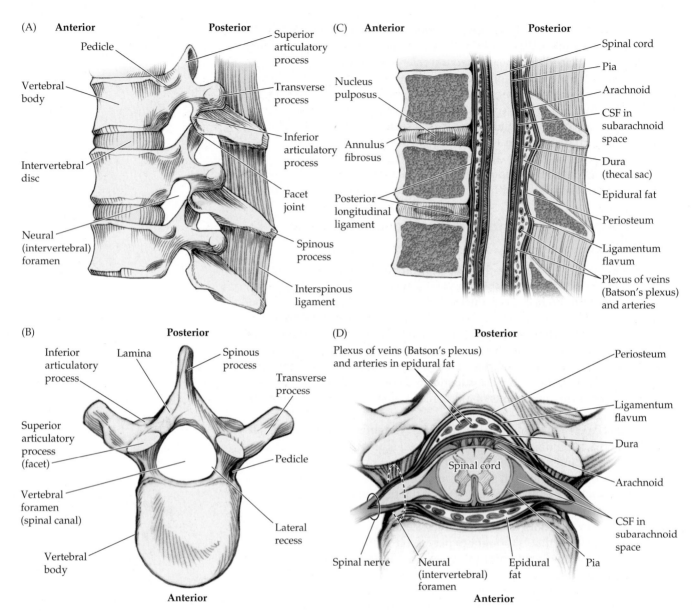

FIGURE 8.2 Vertebral Bones, Meninges, and Other Tissues
(A) Lateral view of vertebral spinal bones. (B) View of a vertebral bone from above. (C) Sagittal section of spinal column including vertebral bones, discs, ligaments, spinal cord, meninges, epidural fat, and blood vessels. (D) Axial section of spinal column and its contents.

Unlike cervical nerve roots, lumbar and sacral nerve roots must travel downward several levels before they exit the spinal canal (see Figure 8.1). In addition, the intervertebral foramina of the lumbosacral spine are such that the nerve roots exit some distance *above* the intervertebral discs (see Figures 8.2 and 8.3B). As they are about to exit, the nerve roots move into the **lateral recess** of the spinal canal (see Figure 8.2B), and it is at this point that they are closest to the disc (see Figure 8.3B). Thus, **posterolateral disc herniations** in the lumbosacral spine usually impinge on nerve roots on their way to exit beneath the next lower vertebral bone, which corresponds to the number of the nerve root involved (see Figure 8.3B,C).

Occasionally, a **far lateral disc herniation** will reach the nerve root exiting at that level, resulting in impingement of the next higher nerve root. For example, a far lateral L5–S1 disc herniation can cause an L5 radiculopathy (see Figure 8.3C). In addition, a **central disc herniation** at the level of the cauda equina can impinge on nerve roots lower than the level of herniation, or it can compress the spinal cord if it occurs above L1.

FIGURE 8.3 Relation of Cervical and Lumbosacral Nerve Roots to Intervertebral Discs (A) Cervical disc herniation usually compresses the nerve root exiting at that level. This corresponds to the number of the lower vertebral bone at that interspace. (B) Lumbosacral disc herniation usually spares the nerve root exiting at that level and compresses the nerve root exiting at the next level down. However, this again corresponds to the number of the lower vertebral bone at the level of the herniation. (C) Far lateral lumbosacral disc herniation affects the nerve root exiting at that level, and central lumbosacral disc herniation can cause cauda equina syndrome (see KCC 8.4).

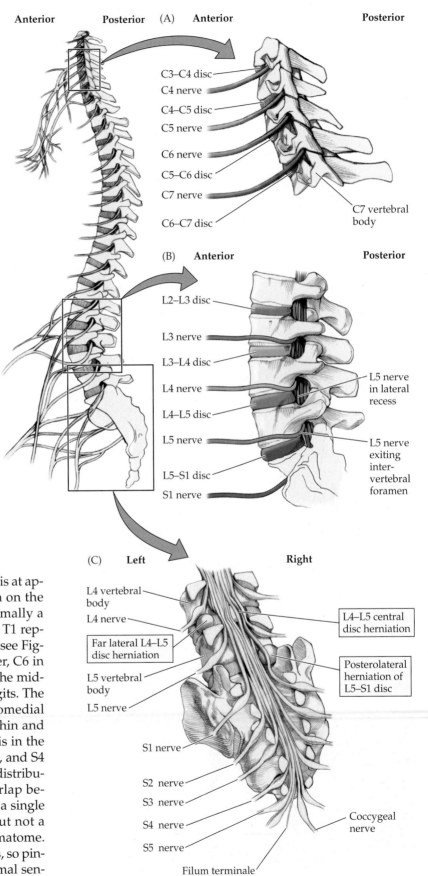

Dermatomes and Myotomes

The sensory region of skin innervated by a nerve root is called a **dermatome** (Figure 8.4). Interestingly, dermatome maps vary somewhat from one text to the next. This variation is likely due to differences in both the methods of testing and individual patients studied. Nevertheless, some familiarity with the usual locations of dermatomes can be very helpful clinically for localizing lesions.

Sensation for the face is provided by the trigeminal nerve (see Figure 12.7), while most of the remainder of the head is supplied by C2 (via the greater and lesser occipital nerves). The nipples are usually at the T4 level, while the umbilicus is at approximately T10. When testing sensation on the chest or back, remember that there is normally a skip between C4 and T2, with C5 through T1 represented mainly on the upper extremities (see Figure 8.4B). C5 is represented in the shoulder, C6 in the lateral arm and first two digits, C7 in the middle digit, and C8 in the fourth and fifth digits. The L4 representation extends over the anteromedial shin, L5 extends down the anterolateral shin and dorsum of the foot to the big toe, and S1 is in the small toe, lateral foot, sole, and calf. S2, S3, and S4 innervate the perineal area in a saddlelike distribution. Note that there is considerable overlap between adjacent dermatomes, so lesions of a single nerve root ordinarily cause a decrease but not a complete loss of sensation in a given dermatome. There may be less overlap for smaller fibers, so pinprick is a more sensitive test for dermatomal sensory loss than touch.

(A)

(B)

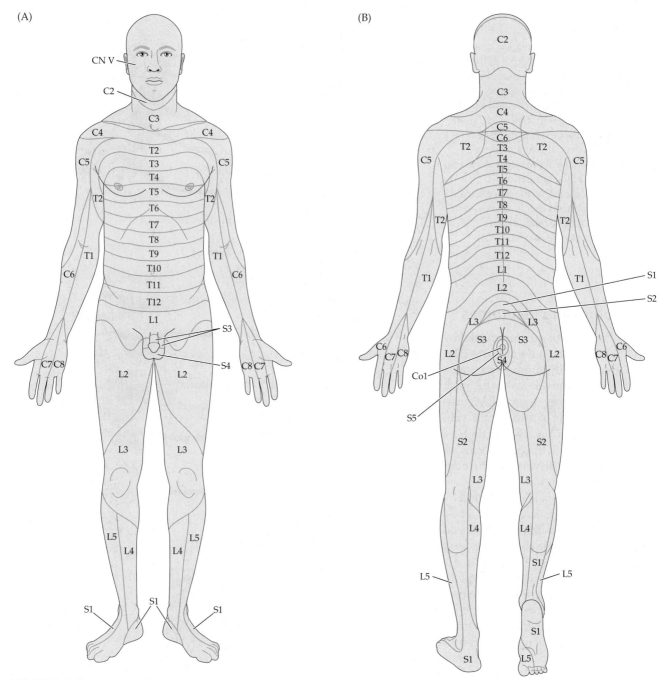

FIGURE 8.4 Dermatomes

Upper extremity strength

The muscles innervated by a single nerve root constitute a **myotome**. The segmental innervation of the muscles is summarized in Table 8.1. This table also includes a summary of nerve roots supplying each peripheral nerve and the main functions of each muscle (see Chapter 9 for further details). Strength testing and reflexes for individual muscles, nerves, and nerve roots is discussed in Chapter 3 (see Tables 3.4–3.7) and is demonstrated on **neuroexam.com Videos 54–60**.

TABLE 8.1 Summary of Peripheral Nerves, Muscles, and Nerve Roots in the Upper and Lower Extremities

NERVE	MUSCLE(S)	FUNCTION OF THE MUSCLE(S)	NERVE ROOTS[a]
Spinal accessory nerve	Trapezius	Elevates shoulder/arm and fixes scapula	CN XI, C3, C4
Phrenic nerve	Diaphragm	Inspiration	C3, C4, C5
Dorsal scapular nerve	Rhomboids	Draw scapula up and in	C4, **C5**
	Levator scapulae	Elevates scapula	C3, C4, C5
Long (Bell's) thoracic nerve	Serratus anterior	Fixes scapula on arm raise	C5, C6, C7
Lateral pectoral nerve	Pectoralis major (clavicular head)	Pulls shoulder forward	**C5**, C6
Medial pectoral nerve	Pectoralis major (sternal head)	Adducts and medially rotates arm	C6, **C7**, C8, T1
	Pectoralis minor	Depresses scapula and pulls shoulder forward	C6, C7, C8
Suprascapular nerve	Supraspinatus	Abducts humerus from 0° to 15°	**C5**, C6
	Infraspinatus	Externally rotates humerus	**C5**, C6
Subscapular nerve	Subscapularis	Internally rotates humerus	C5, C6
	Teres major	Adducts and internally rotates humerus	C5, C6, C7
Thoracodorsal nerve	Latissimus dorsi	Adducts and internally rotates humerus	C6, **C7**, C8
Axillary nerve	Teres minor	Adducts and externally rotates humerus	C5, C6
	Deltoid	Abducts humerus beyond 15°	**C5**, C6
Musculocutaneous nerve	Biceps brachii	Flexes and supinates arm and forearm	C5, C6
	Brachialis	Flexes forearm	C5, C6
	Coracobrachialis	Flexes and adducts arm	C6, **C7**
Radial nerve	Triceps	Extends forearm	C6, **C7**, C8
	Brachioradialis	Flexes forearm	C5, **C6**
	Extensor carpi radialis (longus and brevis)	Extend wrist, abduct hand	C5, **C6**
Posterior interosseus nerve (branch of radial nerve)	Supinator	Supinates forearm	C6, C7
	Extensor carpi ulnaris	Extends wrist, adducts hand	**C7**, C8
	Extensor digitorum communis	Extends fingers (test at metacarpophalangeal joints)	**C7**, C8
	Extensor digiti quinti	Extends little finger	**C7**, C8
	Abductor pollicis longus	Abducts thumb in plane of palm	**C7**, C8
	Extensor pollicis (longus and brevis)	Extends thumb	**C7**, C8
	Extensor indicis proprius	Extends index finger	**C7**, C8
Median nerve	Pronator teres	Pronates and flexes forearm	C6, C7
	Flexor carpi radialis	Flexes wrist, abducts hand	C6, C7
	Palmaris longus	Flexes wrist	C7, **C8**, T1
	Flexor digitorum superficialis	Flexes metacarpophalangeal and proximal interphalangeal joints	C7, **C8**, T1
Muscles affected in carpal tunnel syndrome	Lumbricals (I, II)	For second and third digits, flexes metacarpophalangeal joints, extends other joints	C8, **T1**
	Opponens pollicis	Flexes, opposes thumb	C8, **T1**
	Abductor pollicis brevis	Abducts thumb perpendicular to plane of palm	C8, **T1**
	Flexor pollicis brevis (superficial head)	Flexes first phalanx of thumb	C8, **T1**

(continued on p. 326)

TABLE 8.1 (*continued*)

NERVE	MUSCLE(S)	FUNCTION OF THE MUSCLE(S)	NERVE ROOTS[a]
Anterior interosseous nerve (branch of median nerve)	Flexor digitorum profundus (digits 2, 3)	Flexes second and third fingers (best tested in distal phalanges)	C7, C8
	Flexor pollicis longus	Flexes distal phalanx of thumb	C7, **C8**
	Pronator quadratus	Pronates forearm	C7, **C8**, **T1**
Ulnar nerve	Flexor carpi ulnaris	Flexes wrist, adducts hand	C7, **C8**, T1
	Flexor digitorum profundus (digits 4, 5)	Flexes fourth and fifth fingers (best tested in distal phalanges)	C7, **C8**
	Lumbricals (III, IV)	For fourth and fifth digits, flex metacarpophalangeal joints, extend other joints	C8, **T1**
	Palmar interossei	Adduct fingers, flex metacarpophalangeal joints, extend other joints	C8, **T1**
	Dorsal interossei	Abduct fingers, flex metacarpophalangeal joints, extend other joints	C8, **T1**
	Flexor pollicis brevis (deep head)	Flexes and adducts thumb	C8, **T1**
	Adductor pollicis	Adducts thumb	C8, **T1**
	Muscles of hypothenar eminence Opponens digiti minimi	Internally rotates fifth finger	C8, **T1**
	Abductor digiti minimi	Abducts fifth finger	C8, **T1**
	Flexor digiti minimi	Flexes fifth finger at metacarpophalangeal joint	C8, **T1**
Obturator nerve	Obturator externus	Adducts and outwardly rotates leg	**L2**, **L3**, L4
	Adductor longus	Adducts thigh	**L2**, **L3**, L4
	Adductor magnus	Adducts thigh	**L2**, **L3**, L4
	Adductor brevis	Adducts thigh	**L2**, **L3**, L4
	Gracilis	Adducts thigh	**L2**, **L3**, L4
Femoral nerve	*Iliopsoas muscle* Iliacus	Flexes leg at hip	L1, L2, **L3**
	Psoas	Flexes leg at hip	L2, **L3**, L4
	Quadriceps femoris Rectus femoris	Extends leg at knee, flexes hip	L2, **L3**, **L4**
	Vastus lateralis	Extends leg at knee	L2, **L3**, **L4**
	Vastus intermedius	Extends leg at knee	L2, **L3**, **L4**
	Vastus medialis	Extends leg at knee	L2, **L3**, **L4**
	Pectineus	Adducts thigh	**L2**, **L3**, L4
	Sartorius	Inwardly rotates leg, flexes hip and knee	**L2**, **L3**, L4
Sciatic nerve	Adductor magnus	Adducts thigh	L4, L5, S1
	Hamstring muscles Semitendinosus	Flexes knee, medially rotates thigh, extends hip	L5, **S1**, S2
	Semimembranosus	Flexes knee, medially rotates thigh, extends hip	L5, **S1**, S2
	Biceps femoris	Flexes knee, extends hip	L5, **S1**, S2
Tibial nerve (branch of sciatic nerve)	*Triceps surae muscles* Gastrocnemius	Plantar flexes foot	S1, S2
	Soleus	Plantar flexes foot	S1, S2
	Popliteus	Plantar flexes foot	L4, L5, S1
	Tibialis posterior	Plantar flexes and inverts foot	L4, L5
	Plantaris	Spreads, brings together, flexes proximal phalanges	L4, L5, S1
	Flexor digitorum longus	Flexes distal phalanges, aids plantar flexion	L5, S1, S2

TABLE 8.1 (*continued*)

NERVE	MUSCLE(S)	FUNCTION OF THE MUSCLE(S)	NERVE ROOTS[a]
	Flexor hallucis longus	Flexes great toes, aids plantar flexion	L5, **S1**, **S2**
	Small foot muscles	Cup sole	S1, S2
Superficial peroneal nerve (branch of sciatic nerve)	Peroneus longus	Plantar flexes and everts foot	L5, S1
	Peroneus brevis	Plantar flexes and everts foot	L5, S1
Deep peroneal nerve (branch of sciatic nerve)	Tibialis anterior	Dorsiflexes and inverts foot	L4, L5
	Extensor digitorum longus	Extends phalanges, dorsiflexes foot	**L5**, S1
	Extensor hallucis longus	Extends great toe, aids dorsiflexion	**L5**, S1
	Peroneus tertius	Plantar flexes foot in pronation	L4, **L5**, S1
	Extensor digitorum brevis	Extends toes	L5, S1
Superior gluteal nerve	Gluteus medius	Abducts and medially rotates thigh	**L4**, **L5**, S1
	Gluteus minimus	Abducts and medially rotates thigh	**L4**, **L5**, S1
	Tensor fasciae latae	Abducts and medially rotates thigh	**L4**, **L5**, S1
Inferior gluteal nerve	Gluteus maximus	Extends, abducts, and laterally rotates thigh, extends lower trunk	**L5**, **S1**, S2

Source: Modified and reproduced with permission from Devinsky O and Feldmann E. 1988. *Examination of the Cranial and Peripheral Nerves.* Churchill Livingstone, New York.

[a]Bold text indicates the most important nerve roots, where applicable.

KEY CLINICAL CONCEPT

8.1 DISORDERS OF NERVE, NEUROMUSCULAR JUNCTION, AND MUSCLE

A variety of disorders can affect the peripheral nervous system at multiple levels. This text focuses on neuroanatomical localization, so in this chapter and in Chapter 9 we concentrate mainly on localized disorders of the spinal nerve roots, nerve plexuses, or individual nerve branches. Here, we will place these disorders in the wider context of peripheral nervous system disease so that a more complete differential diagnosis can be formulated.

Disorders of the peripheral nervous system can often be distinguished from central nervous system dysfunction by the anatomical pattern of sensory or motor deficits (see KCC 6.3 and KCC 7.3; Table 8.1). In addition, presence of lower motor neuron signs (see KCC 6.1) such as atrophy, fasciculations, decreased tone, or hyporeflexia suggests peripheral nervous system dysfunction, as do paresthesias in a peripheral nerve distribution (see KCC 7.1). When the location of lesions in the central versus peripheral nervous system remains uncertain based on the history and physical examination, diagnostic tests such as neuroimaging studies (see Chapter 4), blood tests, CSF analysis, and electrodiagnostic studies (see KCC 9.2) can be helpful. Disorders of the peripheral nervous system can be produced by a large number of mechanical, toxic, metabolic, infectious, autoimmune, inflammatory, degenerative, and congenital causes that are beyond the scope of this text (see the References at the end of this chapter for additional details). We will now briefly discuss several common disorders of nerve, neuromuscular junction, and muscle.

Common Neuropathies

Neuropathy is a general term meaning nerve disorder. The site of pathology can be in the axons, myelin, or both and can affect large-diameter fibers, small-diameter fibers, or both. Usually, neuropathies affect both sensory and motor fibers in the nerve, although one or the other may be preferentially involved. Damage can be reversible or permanent. The location of neuropathy can be focal (**mononeuropathy**), multifocal (**mononeuropathy multiplex**), or generalized (**polyneuropathy**). Neuropathy affecting the spinal nerve roots is called **radiculopathy**, which we will discuss in greater detail in KCC 8.3. Like neuropathies, motor neuron disorders (see KCC 6.7) can also cause lower motor neuron–type weakness but motor neuron disorders do not cause sensory involvement.

Important causes of neuropathy include diabetes; mechanical causes; infectious disorders such as Lyme disease, HIV, CMV, varicella-zoster virus, or hepatitis B (see KCC 5.9); toxins; malnutrition; immune disorders such as Guillain–Barré syndrome; and hereditary neuropathies such as **Charcot-Marie-Tooth disease**, among others. We will only discuss a few of the more common causes of neuropathy here.

Diabetic neuropathy is produced by a number of mechanisms, including compromise of the microvascular blood supply of the peripheral nerves (other possible mechanisms include oxidative stress, autoimmunity, and neurotrophic and biochemical disturbances). The most common pattern of diabetic neuropathy is **distal symmetrical polyneuropathy**, which results in a characteristic **glove and stocking** pattern of sensory loss (see Figure 7.9D). Mononeuropathies are also relatively common in diabetes. Acute diabetic mononeuropathy can affect any cranial or spinal nerve but is most common in CN III and the femoral and sciatic nerves. Onset is usually fairly sudden, and sensorimotor deficits in the nerve distribution may be accompanied by painful paresthesias. There is often partial or complete recovery over the course of weeks to months after onset.

Mechanical causes of nerve injury include **extrinsic compression**, **traction**, **laceration**, or **entrapment** by intrinsic structures such as bone or connective tissue. Mild mechanical disruption of a nerve causes **neurapraxia**, temporary impairment of nerve conduction that usually resolves within hours to weeks. More severe injury can interrupt the axons, leading to **Wallerian degeneration** (degeneration of axons and myelin) distal to the site of injury. As long as the structural elements of the nerve are intact, **axonal regeneration** may occur at a rate of about 1 mm/day (a little more than 1 in./month). Occasional long-term complications include incomplete or aberrant reinnervation and the **complex regional pain syndrome**. Complex regional pain syndrome **Type 1**, also called **reflex sympathetic dystrophy**, is more common and follows an injury without specific nerve damage, while **Type 2**, also called **causalgia**, follows damage to a specific nerve. Both types are characterized by intense local burning pain accompanied by edema, sweating, and changes in skin blood supply. In some situations, when peripheral nerves are severed or otherwise disrupted they can be reanastomosed surgically. In addition, some entrapment syndromes may be amenable to surgical decompression. Painful paresthesias associated with neuropathies of all causes are often treated with medications such as anticonvulsants, serotonin-norepinephrine reuptake inhibitors, or tricyclic antidepressants. Common mechanical neuropathies are discussed further in KCC 8.3 and KCC 9.1.

Guillain–Barré syndrome, also known as **acute inflammatory demyelinating polyneuropathy** (**AIDP**), is an important form of neuropathy caused by immune-mediated demyelination of peripheral nerves. Onset typically occurs 1 to 2 weeks following a viral illness, *Campylobacter jejuni* enteritis, HIV infection, or other infections. Presentation is with progressive weakness, **areflexia**, and tingling paresthesias of the hands and feet, with motor involvement typically much more severe than sensory involvement. Symptoms usually reach their worst point 1 to 3 weeks after onset; recovery occurs over many months. Diagnosis is based on typical clinical presentation, cerebrospinal fluid (CSF) demonstrating **elevated protein** without a significantly elevated white blood cell count, and EMG/nerve conduction studies compatible with **demyelination** (see KCC 9.2). Recovery occurs more quickly when patients are treated with **plasmapheresis** or **intravenous immunoglobulin** therapy. In severe cases, patients require intubation and mechanical ventilation. Autonomic dysfunction can be prominent in some cases, requiring careful monitoring. With good supportive care and immune therapy, the majority of patients enjoy complete or near-complete recovery, although about 20% of patients have some residual weakness 1 year after onset.

Common Disorders of the Neuromuscular Junction

Impaired neuromuscular transmission can lead to motor weakness without sensory deficits. Causes include myasthenia gravis, neuromuscular blocking agents and other drugs, Lambert–Eaton myasthenic syndrome (usually paraneoplastic), and botulism.

Myasthenia gravis is an immune-mediated disorder in which there are circulating antibodies against the postsynaptic nicotinic acetylcholine receptors at the neuromuscular junction of skeletal muscle cells. The disorder can sometimes be accompanied by other autoimmune phenomena such as hypothyroidism, lupus, rheumatoid arthritis, and vitiligo. Myasthenia gravis has a bimodal age-related onset, with onset in the second or third decades more common in women and onset in the sixth or seventh decades more common in men. The prevalence is 50 to 125 cases per million. Clinical features include generalized symmetrical weakness, especially of proximal limb muscles, neck muscles, the diaphragm, and eye muscles. Involvement of bulbar muscles can cause facial weakness, a nasal-sounding voice, and dysphagia (see KCC 12.8). Reflexes and sensory exam are normal. Characteristically, **weakness becomes more severe with repeated use of a muscle** or during the course of the day. About 15% of cases have weakness involving only the extraocular muscles and eyelids, a condition called **ocular myasthenia**.

Diagnosis of myasthenia gravis is based on clinical features, and several diagnostic tests, including the ice pack test, repetitive nerve stimulation, measurement of anti-acetylcholine antibodies or muscle specific receptor tyrosine kinase (MuSK) antibodies, single fiber EMG, and chest CT or MRI. The **ice pack** test, which can be performed in patients with ptosis, is administered by placing a bag of ice on the closed eyelids for 2 minutes, and reevaluating for improvement in the ptosis (possibly due to reduced cholinesterase function at lower temperatures). Formerly, the **Tensilon test**, using a short-acting acetylcholinesterase inhibitor (edrophonium) was administered at bedside while observing clinical effects on involved muscles. Commercial manufacture of this agent was discontinued in 2008, however, making future availability of this test uncertain. Clinical response to intermediate-acting acetylcholinesterase inhibitors, such as neostigmine, may also be diagnostically helpful in some cases. Compound motor action potential measurement (see KCC 9.2) with **repetitive nerve stimulation at a rate of 3 per second** often produces a characteristic decrement in amplitude in myasthenia and is considered positive if there is a decrement greater than 10%. Single-fiber EMG is more sensitive (about 90%) but is not specific for myasthenia.

Anti-acetylcholine receptor antibodies (AchR-Ab) are positive in about 85% of cases of generalized myasthenia but in only about 50% of cases of ocular myasthenia. About half of the patients with generalized myasthenia who are AchR-Ab negative have positive serology for **muscle specific receptor tyrosine kinase antibodies (MuSK-Ab)**.

About 12% of patients with myasthenia have a **thymoma**, a tumor of the thymus gland, and many others have thymic hyperplasia, so CT or MRI of the chest should be performed. In addition, testing for associated conditions such as thyroid disease and other immune disorders is appropriate.

Myasthenia gravis is treated by immune therapy. Anticholinesterase medications are also helpful to relieve symptoms. **Pyridostigmine (Mestinon)** is a long-acting cholinesterase inhibitor, with onset of action beginning about 30 minutes after oral administration and duration of about 2 hours. Patients' doses are individually titrated but should not ordinarily exceed about 120 mg every 3 hours, since excess anticholinesterase can actually worsen weakness. Most patients in the age range of adolescence to 60 years are treated surgically with **thymectomy** (whether a thymoma is present or not), as this usually leads to improvement by unclear mechanisms, likely involving a reduced autoimmune response. Use of thymectomy outside this age range or in patients with ocular myasthenia is more controversial but has been used in some cases. Thymectomy should be performed at a time when patients are relatively clinically stable in order to minimize complications in the perioperative period. Short-term immunotherapy with **plasmapheresis** or **intravenous immune globulin** (**IVIg**) can be helpful, particularly when patients are in **myasthenic crisis** requiring intubation, experiencing other severe worsening in symptoms, or in preparation for elective surgery. Longer-term **immunosuppressive agents**, including steroids, azathioprine, mycophenolate, and cyclosporine, are also typically prescribed.

Common Muscle Disorders

Muscle disorders, or **myopathies**, produce weakness that is typically more severe proximally than distally, without loss of sensation or reflexes. Common causes of myopathy include thyroid disease, malnutrition, toxins, viral infections, dermatomyositis, polymyositis, and muscular dystrophy. **Dermatomyositis** and **polymyositis** are immune-mediated inflammatory myopathies. The blood creatinine phosphokinase (CPK) is typically elevated, and electromyography (EMG) studies (see KCC 9.2) are compatible with myopathy. In dermatomyositis there is a characteristic violet-colored skin rash, typically involving the extensor surface of the knuckles and other joints. Although numerous other forms exist, **Duchenne muscular dystrophy** is the most common form of muscular dystrophy. Transmitted by X-linked inheritance, it affects male children and causes progressive proximal weakness. The abnormal protein (dystrophin) has been identified, providing hope for a cure in the near future. ■

KEY CLINICAL CONCEPT

8.2 BACK PAIN

Back pain is one of the most common reasons that people seek medical attention. In this chapter we focus on back pain caused by nerve root disorders; however, it is important to briefly review causes of back pain in general. Table 8.2 is a partial list intended to emphasize the diverse nature of conditions that can be associated with back pain. Many of the diagnoses listed can be elucidated on the basis of a careful history and physical exam. Musculoskeletal causes are most common in all age groups. However, in individuals with onset of back pain over age 50, a neoplasm should be suspected. Back pain in a younger person that worsens with exertion and improves with rest is usually caused by a musculoskeletal problem, including disc herniation in some cases (see KCC 8.3). Symptoms and signs of a radiculopathy (see

TABLE 8.2 Differential Diagnosis of Back Pain

TRAUMA/MECHANICAL	Disc herniation; spondylolysis; vertebral fracture; arthritis; muscle strain/ligament sprain; soft tissue injury
VASCULAR	Spinal arteriovenous malformation; spinal cord infarct; subarachnoid hemorrhage; spinal epidural hematoma
INFECTIOUS/ INFLAMMATORY/ NEOPLASTIC	Osteomyelitis; arachnoiditis; spinal epidural abscess; myositis; cytomegalovirus radiculitis; muscle aches in viral illness; Guillain–Barré syndrome; primary or metastatic neoplasms (extradural, extramedullary, or intramedullary)
DEGENERATIVE/ DEVELOPMENTAL	Scoliosis; degenerative joint disease; amyotrophic lateral sclerosis
REFERRED/OTHER (NON-NEUROLOGIC)	Normal pregnancy; ectopic pregnancy; menses; urinary tract infection; pyelonephritis; renal stone; retroperitoneal abscess; retroperitoneal hematoma; retroperitoneal tumor; pancreatitis; aortic aneurysm; aortic dissection; angina; myocardial infarction; pulmonary embolism

KCC 8.3) should be sought. Back pain in any age group that progressively worsens or does not improve over time should be evaluated with appropriate imaging studies (usually an MRI of the spine). In addition, one should never neglect to evaluate bowel, bladder, and sexual function in patients with back pain, so that irreversible loss of function can be prevented (see KCC 7.2 and KCC 8.4). Several clarifying definitions for degenerative disorder of the spine are given in Table 8.3. ■

KEY CLINICAL CONCEPT

8.3 RADICULOPATHY

Sensory or motor dysfunction caused by pathology of a nerve root is called **radiculopathy**. (Neuropathies in general are discussed in KCC 8.1; radiculopathy is a specific subtype involving nerve roots.) Radiculopathy is often asso-

TABLE 8.3 Clarifying Definitions for Degenerative Disorders of the Spine

DISORDER	DEFINITION
SPONDYLOLYSIS	A general term for degenerative disorders of the spine. From the Greek *spondylos*, meaning "vertebra."
SPONDYLOLYSIS	Fractures that appear in the interarticular portion of the vertebral bone, between the facet joints (see Figure 8.2A). *Lysis* means "loosening" in Greek.
SPONDYLOLISTHESIS	Displacement of a vertebral body relative to the vertebral body beneath it. Includes **anterolisthesis** or **retrolisthesis**, meaning "anterior" or "posterior" displacement of the upper vertebral bone, respectively. Anterolisthesis often coexists with spondylolysis. In Greek, *olisthesis* means "slipping and falling."
OSTEOPHYTES	Bony spurs that form on regions of apposition between adjacent vertebrae because of chronic degeneration. From the Greek *osteo*, meaning "bone," plus *phyton*, meaning "plant" or "outgrowth."
SPINAL STENOSIS	Congenital or acquired narrowing of the spinal canal.

TABLE 8.4 Common Causes of Radiculopathy
Disc herniation
Osteophytes
Spinal stenosis
Trauma
Diabetes
Epidural abscess
Epidural metastases
Meningeal carcinomatosis
Nerve sheath tumors (schwannomas and neurofibromas)
Guillain–Barré syndrome
Herpes zoster (shingles)
Lyme disease
Cytomegalovirus
Idiopathic neuritis

ciated with a burning, tingling pain that **radiates** or shoots down a limb in the dermatome of the affected nerve root (see Figure 8.4). There may be loss of reflexes and motor strength in a radicular distribution (see Tables 3.4–3.7 and 8.1). Chronic radiculopathy can result in atrophy and fasciculations (see KCC 6.1). Sensation may be diminished if a single dermatome is involved, but because of overlap from adjacent dermatomes, sensation is usually not absent. Testing with pinprick is more sensitive than touch for detecting radicular sensory loss. Relatively mild or recent-onset radiculopathy can cause sensory changes without motor deficits. T1 radiculopathy can interrupt the sympathetic pathway to the cervical sympathetic ganglia (see Figure 6.13), resulting in Horner's syndrome (see KCC 13.5). Involvement of multiple nerve roots below L1 can result in a cauda equina syndrome (see KCC 8.4).

Common causes of radiculopathy are listed in Table 8.4. The most common cause by far is intervertebral **disc herniation**, which occurs when part or all of the nucleus pulposus extrudes through a tear in the annulus fibrosus, often causing root compression (see Figures 8.2C and 8.3). It usually occurs without any recent history of traumatic injury, but it can occasionally be caused or exacerbated by trauma. Disc herniation as a cause of radiculopathy is common for the **C6**, **C7**, **L5**, and **S1 nerve roots** and less common at other levels. Lumbosacral radiculopathies are about two to three times as common as cervical radiculopathies. Thoracic disc herniations are less common, since this region of the spinal column is less mobile and fixed by the rib cage. Patients with intervertebral disc herniation typically present with back or neck pain, as well as sensorimotor symptoms in a radicular distribution. As the spine degenerates over time, bony **osteophytes** form (see Table 8.3). Osteophytes, together with disc material, may contribute to narrowing of the intervertebral foramina or may protrude more centrally into the canal, causing spinal stenosis and chronic injury to the spinal cord (myelopathy).

The **straight-leg raising test** can be helpful in the diagnosis of mechanical nerve root compression in the lumbosacral region (Figure 8.5A). In this test the patient lies supine and the examiner slowly elevates the patient's leg at an increasing angle to the table while keeping the leg straight at the knee

FIGURE 8.5 Straight Leg Raising and Spine Percussion Tests (A) Straight leg raising or crossed straight leg raising may reproduce typical radicular symptoms. (B) Pain on percussion of the spine may indicate metastatic, infectious, or other disorders of the vertebral bones.

(B)

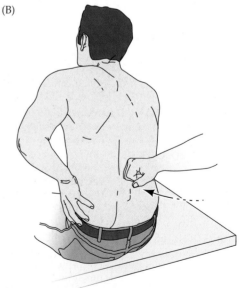

(A)

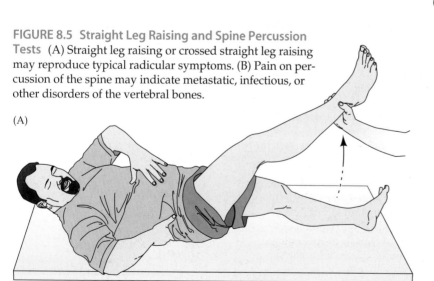

Pain on straight leg raising

Pain on percussion of spine

joint. This provides traction on the nerve roots, and the test is considered positive if it reproduces the patient's typical radicular pain and paresthesias. A response to less than 10° or more than 60°F of straight-leg raising is probably not caused by root compression. In the **crossed straight-leg raising test**, elevating the asymptomatic leg causes typical symptoms in the symptomatic leg. The crossed straight-leg raising test has a specificity of over 90% for lumbosacral nerve root compression. Radicular symptoms may also be increased by the Valsalva maneuver (e.g., coughing, sneezing, straining). In cervical radiculopathy, radicular symptoms may be increased by flexing or turning of the head toward the affected side, likely because of increased narrowing of the intervertebral foramina by these movements. Pain on **percussion of the spine** (Figure 8.5B) may indicate metastatic disease, epidural abscess, osteomyelitis, or other disorders of the vertebral bones, although this sign can be absent in these conditions.

Back pain that is persistent, progressively worsens, or occurs in an older individual, in a patient with prior history of neoplastic disease, or where there is a possibility of epidural abscess, should always be evaluated with a neuroimaging study. An MRI of the spine is usually the test of choice (see Chapter 4). It is important, however, to carefully interpret the MRI in the context of the history and physical examination, since incidental disc bulges and other degenerative changes of the spine are common findings even in individuals without symptoms. In some cases, CT-myelography (see Chapter 4) can help define abnormalities that are not well visualized on MRI. When diagnostic uncertainty remains, EMG and nerve conduction studies (see KCC 9.2) may be helpful.

Other causes of radiculopathy are listed in Table 8.4. **Spinal stenosis**, meaning "narrowing of the spinal canal," can arise congenitally; gradually, as the result of degenerative processes; or by a combination of both factors. **Lumbar stenosis** may result in **neurogenic claudication**, in which bilateral leg pains and weakness occur with ambulation. **Cervical stenosis** can cause a mixture of radicular and long tract signs. Trauma produces radiculopathy through root compression, traction, or **avulsion** of nerve roots off the spinal cord. Diabetic neuropathy can occasionally involve nerve roots, particularly at thoracic levels, producing abdominal pain. **Epidural metastases** most commonly occur in the vertebral bodies, but they can extend laterally to compress nerve roots. Spread of cancer cells such as adenocarcinoma, lymphoma, medulloblastoma, and glioblastoma within the cerebrospinal fluid can involve the nerve roots.

Many causes of radiculopathy are similar to those causing neuropathy in general (see KCC 8.1) but may have an increased tendency to involve the nerve roots. For example, some autoimmune disorders, such as **Guillain–Barré syndrome**, have a predilection for nerve roots. Reactivation of latent **varicella-zoster virus** (chickenpox virus) in dorsal root ganglia produces the painful blistering lesions of **herpes zoster**, or **shingles**. These occur in a dermatomal distribution, associated with sensory and, less commonly, motor loss in the affected nerve roots. Herpes zoster is most common in thoracic dermatomes but can occur anywhere. Treatment with oral antiviral agents such as valacyclovir, famciclovir, or acyclovir can shorten the duration of blistering lesions. Severe pain, referred to as postherpetic neuralgia, can persist after the blistering eruption and is shortened by treatment with antiviral treatment. When herpes zoster occurs in the ophthalmic division of the trigeminal nerve, it can threaten vision, so prompt treatment is critical. **Lyme disease**, a tick-borne illness caused by the spirochete *Borrelia burgdorferi*, can cause radiculopathies. **Cytomegalovirus polyradiculopathy** can be seen in patients with HIV infection, most commonly in the lumbosacral

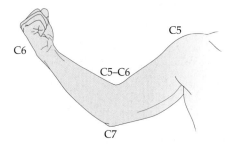

FIGURE 8.6 Three Roots to Remember in the Arm C5 mediates arm abduction at the shoulder; C5 and C6 mediate flexion at the elbow and the biceps reflex; C6 mediates wrist extension; C7 mediates elbow extension and the triceps reflex (see also Table 8.5).

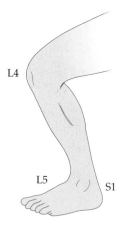

FIGURE 8.7 Three Roots to Remember in the Leg L4 mediates leg extension at the knee and the patellar tendon reflex; L5 mediates dorsiflexion at the ankle; S1 mediates plantar flexion at the ankle and the Achilles tendon reflex (see also Table 8.6).

roots. A milder form of radiculopathy can also be caused by HIV itself. Dumbbell-shaped nerve sheath tumors, such as **schwannomas** and **neurofibromas** (in neurofibromatosis), can occasionally occur in a neural foramen, producing radiculopathy. ∎

Simplification: Three Nerve Roots to Remember in the Arm

For practical purposes, the most clinically important nerve roots in the arm are **C5**, **C6**, and **C7**. It is important to be familiar with the reflexes and the motor and sensory functions associated with these nerve roots, as summarized in Figure 8.6 and Table 8.5. When examining patients, it is helpful to have memorized at least one muscle that gets its major innervation from each of these three nerve roots. In addition to the nerve roots listed in Table 8.5, it is also worth knowing that C8 radiculopathy accounts for about 6% of cervical radiculopathies, is usually caused by C7–T1 disc herniation, and is associated with weakness of the intrinsic hand muscles and decreased sensation in the fourth and fifth digits and the medial forearm. About 20% of all cervical radiculopathies involve two or more cervical levels.

Simplification: Three Nerve Roots to Remember in the Leg

The most clinically important nerve roots in the leg are **L4**, **L5**, and **S1**. Reflexes and the motor and sensory functions associated with L4, L5, and S1 are summarized in Figure 8.7 and Table 8.6. As with the cervical nerve roots, it is helpful, when examining patients, to have memorized at least one muscle that gets its major innervation from each of these nerve roots.

KEY CLINICAL CONCEPT

8.4 CAUDA EQUINA SYNDROME

Impaired function of multiple nerve roots below L1 or L2 is called **cauda equina syndrome**. If the deficits begin at the S2 roots and below, there may be no obvious leg weakness. Sensory loss in an S2 to S5 distribution (see Figure 8.4) is sometimes called **saddle anesthesia**. Involvement of the S2, S3, and S4 nerve roots can produce a distended atonic bladder with urinary retention or overflow incontinence (see KCC 7.5), constipation, decreased rectal tone,

TABLE 8.5 Three Important Nerve Roots in the Arm					
NERVE ROOT	**MAIN WEAKNESS**[a]	**REFLEX DECREASED**[a]	**REGION OF SENSORY ABNORMALITY**[b]	**USUAL DISC INVOLVED**	**APPROXIMATE PERCENTAGE OF CERVICAL RADICULOPATHIES**
C5	Deltoid, infraspinatus, biceps	Biceps, pectoralis	Shoulder, upper lateral arm	C4–C5	7%
C6	Wrist extensors, biceps	Biceps, brachioradialis	First and second fingers, lateral forearm	C5–C6	18%
C7	Triceps	Triceps	Third finger	C6–C7	46%

[a]See Figure 8.6.
[b]See Figure 8.4.

TABLE 8.6 Three Important Nerve Roots in the Leg

NERVE ROOT	MAIN WEAKNESS[a]	REFLEX DECREASED[a]	REGION OF SENSORY ABNORMALITY[b]	USUAL DISC INVOLVED	APPROXIMATE PERCENTAGE OF LUMBOSACRAL RADICULOPATHIES
L4	Iliopsoas, quadriceps	Patellar tendon (knee jerk)	Knee, medial lower leg	L3–L4	3%–10%
L5	Foot dorsiflexion, big toe extension, foot eversion, inversion	None	Dorsum of foot, big toe	L4–L5	40%–45%
S1	Foot plantar flexion	Achilles tendon (ankle jerk)	Lateral foot, small toe, sole	L5–S1	45%–50%

[a]See Figure 8.7.

[b]See Figure 8.4.

fecal incontinence, and loss of erections. It is essential to detect and treat cauda equina syndrome promptly to avoid irreversible deficits. Cauda equina syndrome can sometimes be difficult to differentiate from **conus medullaris syndrome**, in which similar deficits occur as the result of a lesion in the sacral segments of the spinal cord (see Figure 8.1). **Causes of cauda equina syndrome** include compression by a central disc herniation (see Figure 8.3C), epidural metastases, schwannoma, meningioma, neoplastic meningitis, trauma, epidural abscess, arachnoiditis, and cytomegalovirus polyradiculitis. ◼

KEY CLINICAL CONCEPT

8.5 COMMON SURGICAL APPROACHES TO THE SPINE

Most patients with radiculopathy caused by disc herniation recover within a few months without surgery. Indications for urgent surgery include the rare instances in which cord compression or cauda equina syndrome occurs. Semiurgent surgery is indicated in patients with progressive or severe motor deficits or in the occasional patient with intolerable, medically intractable pain. Elective surgery is contemplated when a clear radiculopathy is present and conservative measures such as rest, physical therapy, and traction have been tried for 1 to 3 months but were ineffective.

In the cervical spine, surgical options include a **posterior approach** with **laminectomy**, meaning removal of the lamina over affected levels (see Figure 8.2B), combined with **discectomy** to remove herniated disc material, and **foraminotomy** to widen the lateral recess through which the nerve root passes just before it exits the intervertebral foramen. An **anterior approach** can also be used in the cervical spine. In this procedure, an incision is made in the anterior neck and the dissection is carried down to the vertebral bodies. The anterior approach provides direct access to the discs without traversing the spinal canal and also allows mechanical **fusion** of adjacent vertebral bodies, usually using a bone graft. An anterior approach is also often favored in cases of thoracic disc herniation, which is rare. In the lumbar spine, a posterior approach is generally used. Sometimes a variety of hardware is implanted to increase mechanical stability. ◼

CLINICAL CASES

CASE 8.1 UNILATERAL NECK PAIN AND TINGLING NUMBNESS IN THE THUMB AND INDEX FINGER

CHIEF COMPLAINT

A 30-year-old man came to his physician's office because of 4 weeks of left-sided neck and arm pain and tingling.

HISTORY

Previous history was unremarkable except for some minor sports injuries. These included a skiing accident 2 years ago in which he struck the left side of his neck and had local pain lasting about 3 weeks, and a backward fall during a softball game 1 year ago in which he struck his occiput without loss of consciousness but had some confusion for about 30 minutes. Four weeks ago he awoke one morning with severe **left neck and shoulder pain with tingling radiating down into the first and second fingers of the left hand** (thumb and index finger). The symptoms improved slightly after a few days but then recurred about 4 days ago, making sleep difficult. Over-the-counter pain medications helped but did not eliminate the pain. He did not notice any weakness, numbness, change in bowel or bladder function (see KCC 7.5), or Lhermitte's sign (see KCC 7.1).

PHYSICAL EXAMINATION

Vital signs: T = 98°F, P = 72, BP = 140/80.

Neck: Supple; no tenderness.

Lungs: Clear.

Heart: Regular rate.

Abdomen: Soft.

Extremities: Normal.

Neurologic exam:

MENTAL STATUS: Alert and oriented × 3. Normal language. Gave detailed history.

CRANIAL NERVES: Intact.

MOTOR: Normal tone. 5/5 power throughout, except for **$4^+/5$ power in the left biceps, brachioradialis, and wrist extensors**.

REFLEXES:

COORDINATION: Normal on finger-to-nose and heel-to-shin testing.

GAIT: Normal. Tandem gait normal. No Romberg sign.

SENSORY: Intact light touch, vibration, and joint position sense. **Mildly decreased pinprick sensation in the left first and second fingers. Two-point discrimination 4–5 mm in the left index finger**, compared to 3 mm in the right index finger (using a ruler and the ends of a paper clip).

LOCALIZATION AND DIFFERENTIAL DIAGNOSIS

On the basis of the symptoms and signs shown in **bold** above, where is the lesion? What is the most likely diagnosis?

Discussion

The key symptoms and signs in this case are:

- **Left neck and shoulder pain with tingling radiating down into the first and second fingers, accompanied by decreased pinprick and discriminative sensation**
- **Weakness of the left biceps, brachioradialis, and wrist extensors**
- **Absent left biceps and brachioradialis reflexes**

The patient has a typical left C6 radiculopathy (see KCC 8.3; Figures 8.4, 8.6; Tables 8.1, 8.5). It is very likely that the diagnosis in this setting is a left C5–C6 disc herniation (see Table 8.4 for less common causes).

Clinical Course and Neuroimaging

The patient underwent a cervical spine MRI (Image 8.1A,B, page 338), which confirmed a left C5–C6 disc herniation. He was referred to a neurosurgeon to discuss possible surgery to relieve the nerve root compression. The patient opted to wait for a few more weeks and to undergo surgery if his symptoms did not resolve. He used a hard cervical collar for traction, avoided physical exertion, and continued taking nonsteroidal anti-inflammatory pain medication. One month later, his symptoms had resolved completely. His exam was normal except for a diminished left biceps reflex and barely detectable left biceps weakness, which continued to improve with physical therapy.

REVIEW EXERCISE

Using Figure 8.3A and Image 8.1A,B, explain why a C5–C6 disc herniation usually causes a C6 radiculopathy.

CASE 8.2 UNILATERAL OCCIPITAL AND NECK PAIN

MINICASE

A 74-year-old man with a past history of bladder carcinoma developed **left-sided occipital and neck pain** over the course of 2 weeks. Exam was normal except for **questionable altered sensation over the left occipital area**. Head CT and cervical X-rays were normal. A bone scan and MRI of the cervical spine were therefore done.

LOCALIZATION AND DIFFERENTIAL DIAGNOSIS

On the basis of the symptoms and signs shown in **bold** above, where is the lesion? What is the most likely diagnosis?

Discussion

The key symptoms and signs in this case are:

- **Pain and sensory changes over the left occipital area**

The combination of pain and sensory loss in this territory suggests a peripheral nerve lesion. Sensation in the occipital scalp is provided by C2 (see Figure 8.4), which gives rise to the greater and lesser occipital nerves.

The most likely *clinical localization* is left C2 nerve root or left cipital nerves. **Occipital neuralgia** (similar to trigeminal neuralgia; see KCC 12.2) is a relatively common cause of unilateral occipital pain, which is sometimes accompanied by altered sensation. Given the patient's history of bladder cancer, another possible diagnosis is epidural metastasis compressing the left C2 nerve root or left occipital nerves. Less likely possibilities include degenerative disease of the spine or the other diagnoses listed in Table 8.4.

Clinical Course and Neuroimaging

A **cervical spine MRI** (Image 8.2, page 339) was done, revealing a left cervical mass involving C2. The patient had a CT-guided needle biopsy of the mass, and pathology revealed metastatic transitional cell bladder carcinoma. Radiation therapy was instituted, but the patient gradually deteriorated and was ultimately referred for hospice care.

CASE 8.3 UNILATERAL SHOULDER PAIN AND WEAKNESS

MINICASE

A 50-year-old man with a past history of multiple high school and college football injuries was in a motor vehicle accident and developed **left shoulder pain and numbness** that occasionally radiated down the left arm into the thumb and was increased by neck extension. Exam was normal except for **4/5 deltoid power on the left** and **decreased pinprick sensation in the left shoulder**.

LOCALIZATION AND DIFFERENTIAL DIAGNOSIS

On the basis of the symptoms and signs shown in **bold** above, where is the lesion? What is the most likely diagnosis?

CASE 8.1   UNILATERAL NECK PAIN AND TINGLING NUMBNESS IN THE THUMB AND INDEX FINGER

IMAGE 8.1A,B   Herniated C5–C6 Intervertebral Disc Obliterating Left C6 Neural (Intervertebral) Foramen
MRI of the cervical spine. (A) Sagittal T1-weighted image showing herniated C5–C6 intervertebral disc. (B) Axial T2-weighted image at level of herniated C5–C6 intervertebral disc showing that the disc obliterates the left C6 neural foramen.

(A)

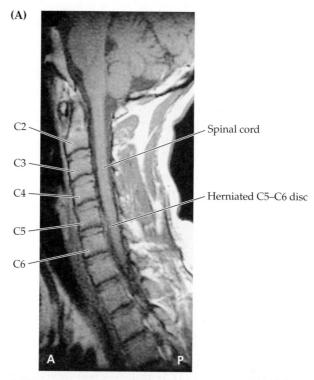

(B)

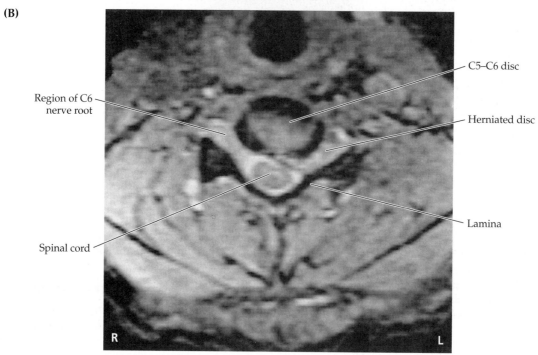

Discussion

The key symptoms and signs in this case are:

- **Left shoulder pain and decreased sensation**
- **Left deltoid weakness**

Sensation to the shoulder and motor innervation of the deltoid muscle is provided by the axillary nerve, which gets its predominant supply from the C5 nerve root (see Table 8.1; Figure 8.6). The most likely diagnosis is therefore a left C5 radiculopathy caused by C4–C5 disc herniation or osteophytes. Other, less likely causes of a C5 radiculopathy are listed in Table 8.4. An axillary neuropathy should also be considered (see Table 9.1). In addition, another diagnosis to be considered is rotator cuff tear, an injury to tendons and ligaments that can cause weakness of abduction and external rotation at the shoulder. However, such an injury would not explain this patient's sensory changes.

Clinical Course

The patient underwent a cervical spine MRI, which demonstrated bony osteophytes (see Table 8.3) causing narrowing of the intervertebral neural foramina at C4–C5 (not shown). He was taken to the operating room for laminectomy and decompression of the neural foramina (see KCC 8.5) and had a good postoperative recovery.

CASE 8.2 UNILATERAL OCCIPITAL AND NECK PAIN

IMAGE 8.2 Metastatic Bladder Carcinoma Encasing Left C2 Nerve Root
Axial T1-weighted MRI of the cervical spine. Compare to Figure 4.12A.

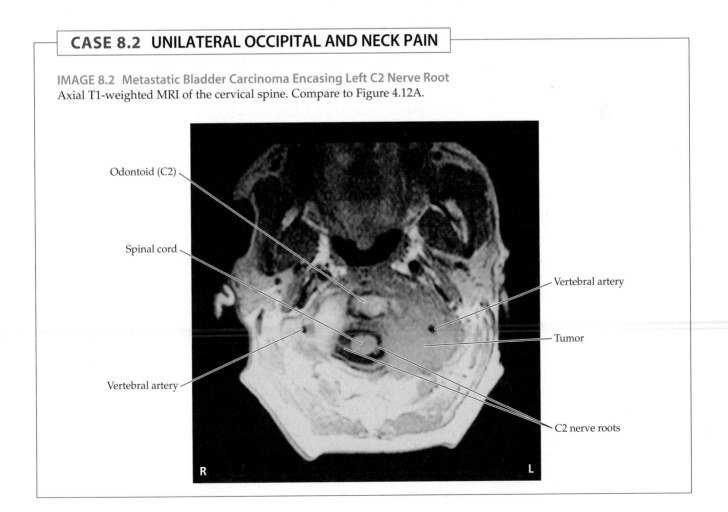

CASE 8.4 BLISTERS, PAIN, AND WEAKNESS IN THE LEFT ARM

MINICASE

A 68-year-old man awoke one morning with a painful, blistering rash on his left shoulder and arm, accompanied by left arm weakness and numbness that progressed over the next week. He was eventually **unable to raise his left arm** but had good hand strength. He visited his family doctor, and on exam he had reddish blisters, some of which were scaled over, in the distribution shown in Figure 8.8. In addition, he had **2/5 left deltoid strength, 3/5 strength in the left arm external rotation, and 4/5 strength in the left biceps and brachioradialis. Left biceps and brachioradialis reflexes were absent**, while other reflexes were 2⁺. There was **decreased pinprick sensation in the same distribution as the rash** (see Figure 8.8). The remainder of the exam was unremarkable.

LOCALIZATION AND DIFFERENTIAL DIAGNOSIS

On the basis of the symptoms and signs shown in **bold** above, where is the lesion? What is the most likely diagnosis?

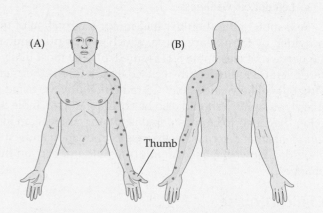

(A)   (B)

Thumb

FIGURE 8.8 Region of Skin Blistering and Decreased Pinprick Sensation Compare to Figure 8.4.

Discussion

The key symptoms and signs in this case are:

- Blistering rash and decreased pinprick sensation in the left shoulder and arm
- Weak left deltoid, arm external rotation, biceps, and brachioradialis, with absent left biceps and brachioradialis reflexes

The herpetic skin lesions, together with sensory loss in a left C5–C6 distribution, make the most likely diagnosis herpes zoster (shingles) of the left C5 and C6 nerve roots (see KCC 8.3). Although more common in thoracic dermatomes, herpes zoster can occur in other dermatomes as well and can occasionally cause weakness. The muscle weakness and reflex loss in this patient are also consistent with C5 and C6 involvement, although C5 seems to be more severely affected, since wrist extension (C6) was not weak (see Tables 8.1, 8.5; Figure 8.6).

Clinical Course

A lumbar puncture was done. The cerebrospinal fluid (CSF) was normal except for the presence of 224 white blood cells per cubic millimeter (normal is 0–5; see KCC 5.9, 5.10; Table 5.7) with 89% lymphocytes. Viral cultures from CSF and from the skin lesions were negative; however, a polymerase chain reaction (PCR) test on CSF was positive for varicella-zoster virus. The patient was treated with intravenous and, later, oral acyclovir. A cervical spine MRI was negative, as was careful examination for cranial nerve involvement. When the patient was seen 3 months later in follow-up, the rash had resolved. The pain and weakness in his left arm were improving but still a significant problem.

CASE 8.5 UNILATERAL SHOULDER PAIN AND NUMBNESS IN THE INDEX AND MIDDLE FINGERS

MINICASE

A 37-year-old physician had sudden onset of sharp **pain in her right shoulder**, followed by tingling and numbness radiating down her arm into her hand. She noticed that her **right second and third fingers felt numb**. She tried to comb her hair, but her right arm kept bending at the elbow, causing her hand to fall onto her head. She was fairly certain of the diagnosis at this point and scheduled an appointment with an orthopedic colleague. Her exam was notable only for **3/5 strength in the right triceps, 4⁺/5 strength in the right finger extensors, an absent right triceps reflex, and decreased sensation to pinprick and light touch in the right second finger and lateral half of the third finger** (Figure 8.9).

LOCALIZATION AND DIFFERENTIAL DIAGNOSIS

On the basis of the symptoms and signs shown in **bold** above, where is the lesion? What is the most likely diagnosis?

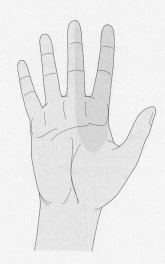

FIGURE 8.9 Region of Decreased Sensation
Compare to Figure 8.4.

Discussion

The key symptoms and signs in this case are:

- **Weakness of the right triceps and finger extensors, with absent right triceps reflex**
- **Right shoulder pain with paresthesias and decreased sensation in the right second finger and lateral half of the third finger**

The weakness, reflex loss, and decreased sensation in this patient are consistent with a right C7 radiculopathy (see Tables 8.1, 8.5; Figures 8.4, 8.6). Note that, as mentioned previously, dermatomal patterns can vary slightly; however, C7 radiculopathy usually affects the middle finger and may involve the index finger as well. The most likely diagnosis is right C7 nerve root compression caused by right C6–C7 disc herniation. Other, less common possibilities are listed in Table 8.4.

Clinical Course

An MRI of the cervical spine revealed a C6–C7 disc herniation impinging on the right C7 nerve root (see Image 8.1A,B for a similar scan at a different level). She initially elected to pursue conservative therapy, but continued to have pain and developed some atrophy of the right triceps muscle. Therefore, she underwent a partial laminectomy with a C6–C7 discectomy (see KCC 8.5). Postoperatively, her triceps strength recovered fully, although she continued to have some pain and numbness in the right arm and hand over the next few months.

CASE 8.6 UNILATERAL NECK PAIN, HAND WEAKNESS, AND NUMBNESS IN THE RING AND LITTLE FINGERS

MINICASE

A 34-year-old cardiothoracic surgeon developed **left neck and shoulder pain**, with numbness and tingling radiating down the ulnar aspect of his arm into the fourth and fifth fingers. On exam he had some **weakness of the intrinsic muscles of the left hand** and **decreased sensation to pinprick and light touch over the left fourth and fifth fingers** (Figure 8.10). The remainder of the exam was normal.

LOCALIZATION AND DIFFERENTIAL DIAGNOSIS

On the basis of the symptoms and signs shown in **bold** above, where is the lesion? What is the most likely diagnosis?

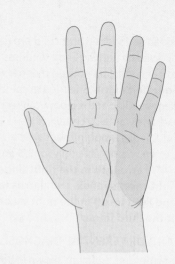

FIGURE 8.10 Region of Decreased Sensation
Compare to Figure 8.4.

Discussion

The key symptoms and signs in this case are:

- **Weakness of the intrinsic muscles of the left hand**
- **Left neck and shoulder pain, with paresthesias and decreased sensation in the left fourth and fifth fingers**

It would have been helpful for localization if additional details of the motor exam had been provided. Weakness of hand intrinsic muscles (lumbricals, interossei) can be caused by lesions of the ulnar nerve, median nerve, lower trunk of the brachial plexus (see Chapter 9), C8, or T1 nerve roots (see Table 8.1). In addition, the distribution of abnormal sensation in this patient is consistent with a lesion of the ulnar nerve, lower trunk of the brachial plexus (C8, T1; see Chapter 9), or C8 nerve root (see Figure 8.4). Neck and shoulder pain suggests a radiculopathy. Therefore, a left C8 radiculopathy caused by leftward C7–T1 disc herniation is the most likely diagnosis (see also Table 8.4); however, an ulnar neuropathy or lower brachial plexus lesion should also be considered.

Clinical Course

An MRI showed a C7–T1 disc herniation (see Image 8.1A,B for a similar scan at a different level). At the time of laminectomy, some free disc fragments were found compressing the left C8 nerve root and were removed. Postoperatively, the patient's pain was resolved, and his hand strength recovered fully.

CASE 8.7 PAIN AND NUMBNESS IN THE MEDIAL ARM

MINICASE

A 66-year-old executive had been suffering for 2 years with **pain and numbness in his left shoulder and medial arm**. Exam was notable for **decreased sensation to light touch in the left medial arm and forearm** (Figure 8.11) and was otherwise unremarkable. Several MRIs and a CT myelogram suggested possible neural compression at multiple levels, including C6–C7, C7–T1, and T1–T2.

LOCALIZATION AND DIFFERENTIAL DIAGNOSIS

On the basis of the symptoms and signs shown in **bold** above, where is the lesion? What is the most likely diagnosis?

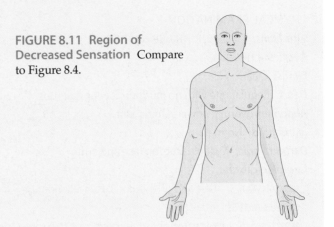

FIGURE 8.11 Region of Decreased Sensation Compare to Figure 8.4.

Discussion

The key symptoms and signs in this case are:

- **Left shoulder pain, with decreased sensation in medial arm and forearm**

This patient had sensory changes in the left T1 dermatome (see Figure 8.4). Thoracic radiculopathy, though uncommon, can occasionally be seen as a result of disc herniation or other disorders listed in Table 8.4. It should be noted that this patient had sensory findings only, with no motor involvement (see Table 8.1), which can occur with incomplete nerve root lesions (see KCC 8.3). It should also be noted that Horner's syndrome, which sometimes can be seen in T1 radiculopathy, was not present in this case.

Clinical Course

Because the imaging studies did not provide a definite level, yet the patient had refractory pain, he was taken to the operating room for laminectomy and exploration of the left C7, C8, and T1 nerve roots. At the time of surgery the left T1 nerve root was found to be compressed by T1–T2 disc fragments, which were removed. This case illustrates that MRI or CT findings suggesting nerve root compression need to be interpreted in the context of clinical symptoms and signs, since asymptomatic radiological abnormalities are common. Postoperatively, the patient's pain improved markedly.

CASE 8.8 LOW BACK PAIN RADIATING TO THE SOLE OF THE FOOT AND THE SMALL TOE

CHIEF COMPLAINT

Following an accident, a 38-year-old man developed difficulty walking and low back pain radiating to the lateral sole of his left foot.

HISTORY

The patient was working on a road when he was injured by an explosion. He suffered severe burns requiring plastic surgery.

In addition, he experienced low back pain with **numbness and "pins and needles"** running down his left leg into the **sole and lateral aspect of the left foot, including the small toe**. He had some trouble walking, mostly because of pain, but he also noticed **difficulty standing on his toes with the left foot**. He denied changes in bowel, bladder, or erectile function.

(continued on p. 344)

CASE 8.8 *(continued)*

PHYSICAL EXAMINATION

Vital signs: T = 98°F, P = 80, BP = 112/80.

Neck: Supple.

Lungs: Clear.

Heart: Regular rate with no murmurs, gallops, or rubs.

Abdomen: Normal bowel sounds; soft.

Extremities: Normal.

Dermatologic: Multiple scars on face and arms.

Neurologic exam:

MENTAL STATUS: Alert and oriented × 3. Fluent language.

CRANIAL NERVES: Intact.

MOTOR: 5/5 power throughout, except for **4/5 power in the left gastrocnemius and hamstrings** (semitendinosus, semimembranosus, and biceps femoris).

REFLEXES:

COORDINATION: Normal on finger-to-nose testing.

GAIT: Slow and painful. **Unable to stand on toes of left foot.**

SENSORY: Intact except for **decreased light touch and pin-prick sensation in the left lateral calf, left lateral foot including the small toe, and sole of the left foot** (Figure 8.12).

LOCALIZATION AND DIFFERENTIAL DIAGNOSIS

On the basis of the symptoms and signs shown in **bold** above, where is the lesion? What is the most likely diagnosis?

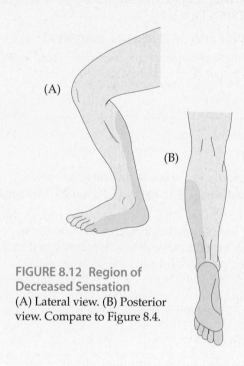

FIGURE 8.12 Region of Decreased Sensation (A) Lateral view. (B) Posterior view. Compare to Figure 8.4.

REVIEW EXERCISE

Trace the course of the L4, L5, and S1 nerve roots in Image 8.8A–G (see also Figure 8.3B,C) to demonstrate why an L4–L5 disc herniation usually causes an L5 radiculopathy and an L5–S1 disc herniation usually causes an S1 radiculopathy.

Discussion

The key symptoms and signs in this case are:

- Weakness of the left gastrocnemius and hamstrings, with absent left Achilles tendon reflex
- Paresthesias and decreased sensation in the left lateral calf, lateral foot including the small toe, and sole

The weakness, reflex loss, and sensory changes in this patient are consistent with a left S1 radiculopathy (see Tables 8.1, 8.6; Figures 8.4, 8.7). The most likely diagnosis is a left posterolateral L5–S1 disc herniation compressing the left S1 nerve root (see also Table 8.4 for other possibilities).

Clinical Course and Neuroimaging

A **spine MRI** was performed and showed a left L5–S1 disc herniation (Image 8.8A–G, pages 346–347). The patient's symptoms did not improve, and he was therefore taken to the operating room for a laminectomy, which revealed a herniated L5–S1 disc with a free disc fragment compressing the left S1 nerve root in the lateral recess. The fragment was removed, and the patient did well until 1 year later, when he had recurrent pain in the same distribution and mild calf weakness. Repeat MRI showed scar tissue surrounding the left S1 nerve root. This could not be treated surgically, and he was therefore treated with pain medications and local steroid injections with only partial relief.

RELATED CASE. Image 8.8H (page 348) shows an example from a different patient of a myelogram (see Chapter 4) demonstrating bilateral L5 nerve root compression by a herniated L4–L5 disc.

CASE 8.9 UNILATERAL THIGH WEAKNESS WITH PAIN RADIATING TO THE ANTERIOR SHIN

MINICASE

A 76-year-old man suffered for 1 year with **relentless pain and numbness radiating from his right buttock down the anterior thigh into the shin. Exam was notable for 4⁻/5 right quadriceps strength, 4⁺/5 right iliopsoas strength, absent right patellar reflex, and decreased pinprick sensation in the right shin and medial calf** (Figure 8.13).

LOCALIZATION AND DIFFERENTIAL DIAGNOSIS

On the basis of the symptoms and signs shown in **bold** above, where is the lesion? What is the most likely diagnosis?

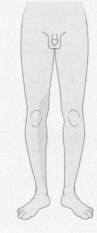

FIGURE 8.13 Region of Decreased Sensation Compare to Figure 8.4.

Discussion

The key symptoms and signs in this case are:

- Weakness of the right quadriceps and iliopsoas, with absent right patellar reflex
- Paresthesias and decreased sensation in the anterior thigh, shin, and medial calf

The pattern of weakness, reflex loss, and sensory changes is compatible with a right femoral neuropathy or L4 radiculopathy (see Tables 8.1, 8.6; Figures 8.4, 8.7; see also Table 9.3). **Lesions of the femoral nerve can sometimes be distinguished from an L4 radiculopathy by testing for weakness of thigh adduction**, which may be present in L4 radiculopathy but not femoral neuropathy (see Table 8.1). Unfortunately, thigh adduction testing was not documented in this patient. An L2 or L3 radiculopathy could also be considered in this patient; however, these radiculopathies do not usually produce sensory changes extending below the knee, and they are also much less common than an L4 radiculopathy. The most likely diagnosis is therefore right femoral neuropathy or right posterolateral L3–L4 disc herniation compressing the right L4 nerve root (see also Table 8.4 for other possibilities).

Clinical Course

Interestingly, rather than an L3–L4 posterolateral disc herniation, this patient's MRI (not shown) revealed a herniated right L4–L5 disc extending far upward and laterally to compress the right L4 nerve root (review Figure 8.3C). Following laminectomy and removal of the herniated disc material, the patient had complete resolution of the pain, and his right leg strength improved.

CASE 8.8 LOW BACK PAIN RADIATING TO THE SOLE OF THE FOOT AND THE SMALL TOE

IMAGE 8.8A,B L5–S1 Posterolateral Disc Herniation Compressing Left S1 Nerve Root in the Lateral Recess T1-weighted MRI of the spine. (A) Parasagittal view slightly to the left of midline, showing herniated L5-S1 intervertebral disc. (B) Parasagittal view, farther to the left of midline, showing neural foramina. Note that the foramen for the L5 root (L5–S1) is not obstructed.

(A)

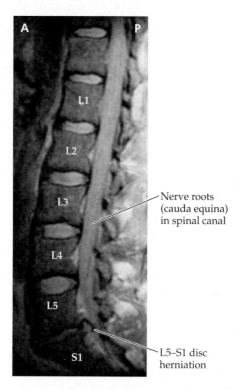

(B)

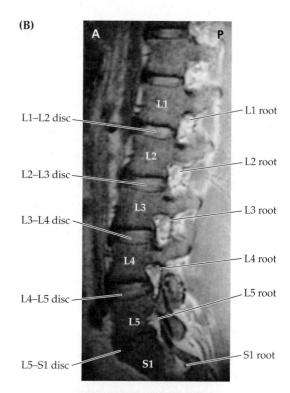

IMAGE 8.8C–G Axial Sections Showing Posterolateral Disc Herniation Compressing Left S1 Nerve Root in the Lateral Recess T1-weighted MRI of the spine. (C) Midsagittal view showing herniated L5–S1 intervertebral disc (as in Image 8.8A), with levels of axial sections in D–G indicated. Sections D–G proceed from rostral to caudal. (D) Axial section at level of L5 vertebral body showing L5 nerve roots in lateral recesses. (E) Axial section at level of L5 neural foramen, above the level of the L5–S1 intervertebral disc (compare to Figure 8.3B,C). (F) Axial section at level of L5–S1 intervertebral disc showing herniation of disc into left lateral recess compressing left S1 nerve root (compare to Figure 8.3B,C). (G) Axial section at level of S1 body showing S1 nerve roots in lateral recess below the level of compression.

(C)

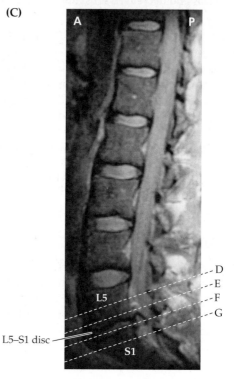

CASE 8.8 (*continued*)

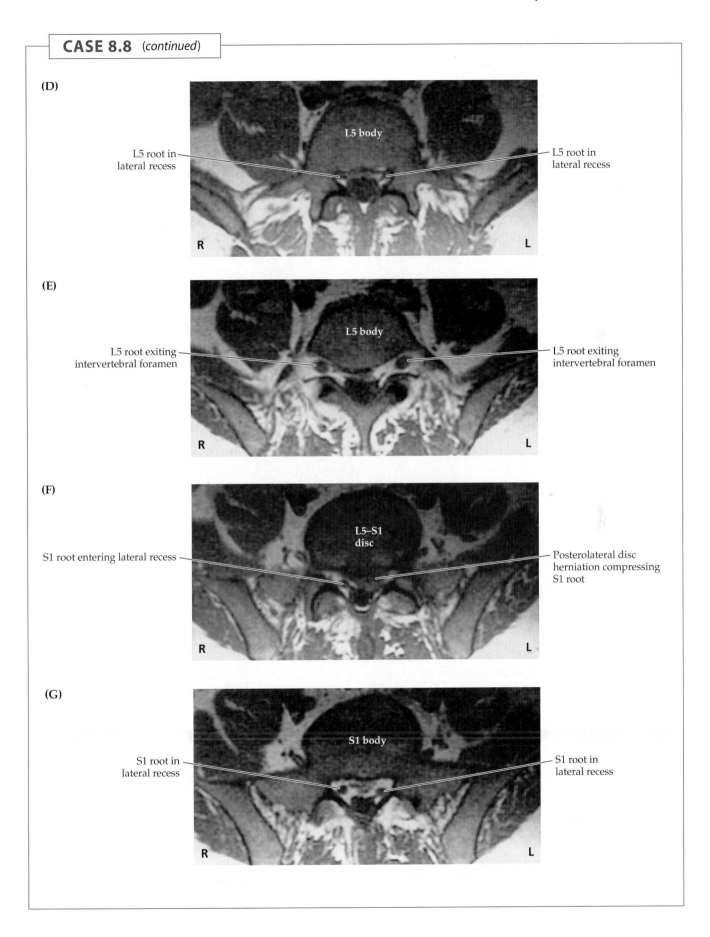

(D)

L5 body

L5 root in lateral recess

L5 root in lateral recess

R   L

(E)

L5 body

L5 root exiting intervertebral foramen

L5 root exiting intervertebral foramen

R   L

(F)

L5–S1 disc

S1 root entering lateral recess

Posterolateral disc herniation compressing S1 root

R   L

(G)

S1 body

S1 root in lateral recess

S1 root in lateral recess

R   L

CASE 8.8 *RELATED CASE*

IMAGE 8.8H **Example of Myelogram Showing Bilateral L5 Nerve Root Compression at the Level of the L4–L5 Intervertebral Disc** An anterior–posterior plain X-ray film is shown after introduction of myelographic contrast material into the subarachnoid space (see Chapter 4; see also KCC 5.10). The normal L3, L4, S1, and S2 nerve root sleeves can be visualized in relation to vertebral bones (compare to Figure 8.3). The L5 nerve root sleeves are truncated bilaterally because of L4–L5 intervertebral disc herniation.

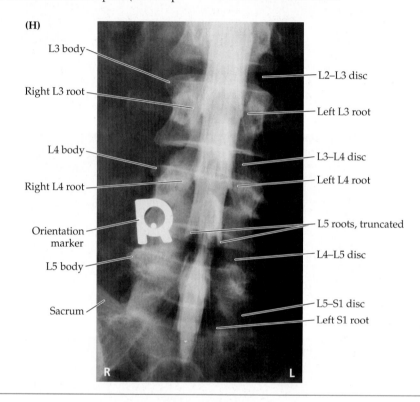

(H)

L3 body
Right L3 root
L4 body
Right L4 root
Orientation marker
L5 body
Sacrum

L2–L3 disc
Left L3 root
L3–L4 disc
Left L4 root
L5 roots, truncated
L4–L5 disc
L5–S1 disc
Left S1 root

R   L

CASE 8.10 LOW BACK PAIN RADIATING TO THE BIG TOE

MINICASE

A 57-year-old man with low back pain for over 20 years tripped over a door ledge and had a sudden increase in **right-sided low back pain radiating down his leg to the right big toe**. He had some difficulty walking because of the pain, causing him to visit the emergency room several times over the next 3 months, where his exam was notable for **3/5 power in the right extensor hallucis longus and tibialis anterior, 4⁺/5 power in the right foot invertors and evertors**, normal reflexes, and **decreased pinprick sensation in the right anterolateral calf and dorsum of the foot** (Figure 8.14). **Straight-leg raising** (see Figure 8.5A) beyond 30° on the left side had no effect, but on the right side **reproduced the patient's usual pain**.

LOCALIZATION AND DIFFERENTIAL DIAGNOSIS

On the basis of the symptoms and signs shown in **bold** above, where is the lesion? What is the most likely diagnosis?

FIGURE 8.14 **Region of Decreased Sensation** Compare to Figure 8.4.

Discussion

The key symptoms and signs in this case are:

- **Weakness of the right extensor hallucis longus, tibialis anterior, and right foot invertors and evertors**
- **Pain radiating to the right big toe reproduced by straight-leg raising, with decreased sensation in the anterolateral calf and dorsum of the foot**

This patient has typical radicular pain, sensory loss, and weakness compatible with a right L5 radiculopathy (see Tables 8.1, 8.6; Figures 8.4, 8.7). A peroneal nerve palsy (see Tables 8.1, 9.3) can also produce similar decreased sensation and foot drop but does not cause painful paresthesias with straight-leg raising. In addition, lesions of the peroneal nerve can sometimes be distinguished from an L5 radiculopathy by testing for weakness of foot inversion, which may be present in L5 radiculopathy but not in peroneal nerve palsy (see Table 8.1).

The most likely diagnosis is therefore right posterolateral L4–L5 disc herniation compressing the right L5 nerve root (see also Table 8.4 for other possibilities).

Clinical Course

An MRI showed a herniated L4–L5 disc compressing the right L5 nerve root (see Image 8.8F for a similar scan at a different level). Surgery was discussed with the patient; however, he was lost to follow-up.

CASE 8.11 SADDLE ANESTHESIA WITH LOSS OF SPHINCTERIC AND ERECTILE FUNCTION

CHIEF COMPLAINT

A 39-year-old man came to the emergency room with 10 days of bilateral gluteal pain, numbness, and sphincteric dysfunction.

HISTORY

Ten days prior to admission the patient was doing heavy labor with concrete when he coughed and felt a sudden "pop" followed by sharp **pain in the gluteal region bilaterally**. The pain was only partly relieved by over-the-counter pain medications. During the following days he noticed that he **had no erections**, even upon awakening. In addition, he noticed a **loss of sensation over his genitals and buttocks**. When he sat down it felt as though he was "on air" because he could not feel the seat. He also became **constipated** and did not have any bowel movements for 10 days, despite frequent attempts. **Urination was also difficult**, and when he felt discomfort from bladder distention, he applied pressure over his lower abdomen to initiate flow. Because of increasing problems with urinary retention, he finally came to the emergency room.

PHYSICAL EXAMINATION

Vital signs: T = 98.6°F, P = 60, BP = 130/80, R = 16.
Neck: Supple with no bruits.

Lungs: Clear.
Heart: Regular rate with no murmurs, gallops, or rubs.
Abdomen: Normal bowel sounds; soft. **Firm, distended bladder palpable in lower abdomen above pubic bone.**
Extremities: No edema.
Rectal: **Rectal tone flaccid.**
Neurologic exam:
  MENTAL STATUS: Alert and oriented × 3. Normal language.
  CRANIAL NERVES: Intact II–XII.
  MOTOR: Normal bulk and tone. 5/5 power throughout.
  REFLEXES: **No anal wink. Only trace bulbocavernosus reflex** (see Table 3.7). Cremasteric reflex was present.
  COORDINATION: Normal on finger-to-nose and heel-to-shin testing.
  GAIT: Normal.

(continued on p. 350)

CASE 8.11 *(continued)*

SENSORY: **Decreased pinprick and light touch sensation in a saddle distribution, including the genitals, perianal area, buttocks, and upper posterior thighs** (Figure 8.15). Pinprick, light touch, vibration, and joint position sense were normal in all other areas.

POSTVOID RESIDUAL VOLUME (see KCC 7.5): The patient was catheterized after attempting voluntary urination, and **1300 cc** of urine were obtained (normal volume is less than 100 cc).

LOCALIZATION AND DIFFERENTIAL DIAGNOSIS

On the basis of the symptoms and signs shown in **bold** above, where is the lesion? What is the most likely diagnosis?

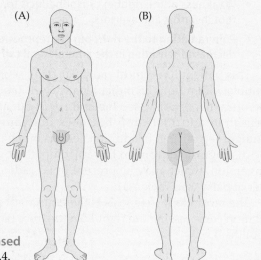

FIGURE 8.15 Region of Decreased Sensation Compare to Figure 8.4.

Discussion

The key symptoms and signs in this case are:

- **Pain in the gluteal region bilaterally with loss of sensation in a saddle distribution over the genitals and buttocks**
- **Constipation, urinary retention, loss of erections, loss of rectal tone, no anal wink, and loss of bulbocavernosus reflex**

Impaired bowel, bladder, and sexual function can be caused by bilateral lesions of the cerebral hemispheres, spinal cord, conus medullaris, cauda equina, or peripheral nerves (see KCC 7.5). The intact cremasteric reflex suggests that function of the L1–L2 nerve roots is preserved (see Table 3.7), and the normal lower extremity strength suggests preserved function down through S1. Meanwhile, the region of pain and sensory loss is in the bilateral S2 through S5 or coccygeal dermatomes (see Figure 8.4), suggesting a lesion of the lower cauda equina or conus medullaris.

The most likely *clinical localization* is: cauda equina S2 through S5 roots, or conus medullaris.

Given the sudden onset of symptoms, a central disc herniation is likely as the diagnosis. Since the lower nerve roots are located more medially in the cauda equina (see Figure 8.3C), a central disc herniation tends to compress the lower nerve roots. For other, less likely causes of cauda equina syndrome in this patient, see KCC 8.4. Possible lesions of the conus medullaris include intrinsic tumors such as ependymoma or astrocytoma; metastatic lesions; demyelinating processes; and sarcoidosis.

Clinical Course and Neuroimaging

The patient immediately underwent a **spinal CT/myelogram** (Image 8.11A–C, pages 352–353), which showed an L5–S1 central disc herniation compressing the cauda equina. An urgent laminectomy was therefore performed. A large mass of disc material was found compressing the thecal sac from the anterior aspect at the L5–S1 intervertebral level. Following decompression, the patient's pain improved, and he regained some gluteal sensation; however, he still had urinary retention and required intermittent catheterization at the end of his 11-day hospital stay.

Additional Cases

Other chapters include related cases for the following topics: **peripheral nerve disorders** (Cases 9.1–9.14); **distal symmetric polyneuropathy** (Cases 6.5 and 10.3); and **cranial neuropathy** (Cases 12.2–12.7, 13.1–13.3, and 13.5). Other relevant cases can be found using the **Case Index** located at the end of this book.

Brief Anatomical Study Guide

1. In this chapter we have discussed the segmental innervation of the body provided by **dorsal sensory** and **ventral motor nerve roots** that exit the spinal cord at **cervical, thoracic, lumbar, and sacral** levels (see Figure 8.1A) and fuse to form mixed spinal nerves (see Figure 8.1B). Because the vertebral bones outgrow the spinal cord during development, the lower roots continue **below the L1 or L2 vertebral bones** as the **cauda equina** (see Figure 8.1). The sensory regions innervated by nerve roots are called **dermatomes** (see Figure 8.4), and motor territories of nerve roots are called **myotomes**.

2. The most common cause of nerve root dysfunction, or **radiculopathy**, is **intervertebral disc herniation** at the cervical or lumbosacral levels (see Figures 8.2 and 8.3). The nerve root involved usually corresponds to the **vertebral body below the level of the herniated disc**. For example, an L5–S1 disc herniation usually causes an S1 radiculopathy.

3. The **three most clinically relevant arm and leg nerve roots are C5, C6, and C7, and L4, L5, and S1, respectively**. A summary of the sensory and motor functions of these nerve roots is provided in Tables 8.5 and 8.6 and Figures 8.6 and 8.7.

CASE 8.11   SADDLE ANESTHESIA WITH LOSS OF SPHINCTERIC AND ERECTILE FUNCTION

IMAGE 8.11A–C   Large Posterior L5–S1 Disc Herniation Compressing Cauda Equina   Spinal CT-myelogram (see Chapter 4). Sections A–C proceed from rostral to caudal. (A) Section at level of L4–L5 intervertebral disc showing normal contrast agent–filled CSF space and cauda equina at this level. (B) Massive L5–S1 interverte- bral disc herniation, obliterating spinal canal and causing complete block of contrast dye, at level of cauda equina. (C) Section at level of S1 vertebral body showing normal-appearing nerve roots, spinal canal, and other structures. Compare to Figure 8.3C.

(A)

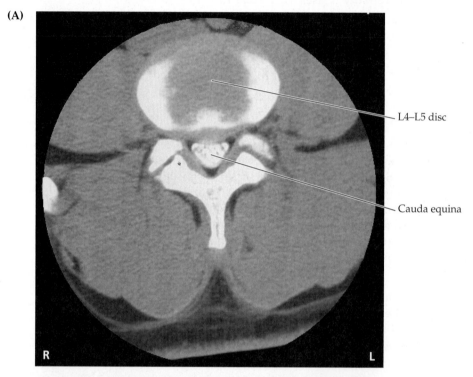

L4–L5 disc

Cauda equina

(B)

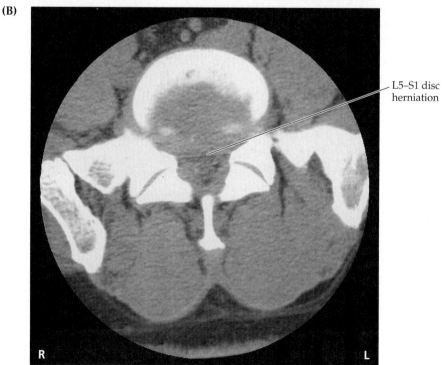

L5–S1 disc herniation

CASE 8.11 (*continued*)

(C)

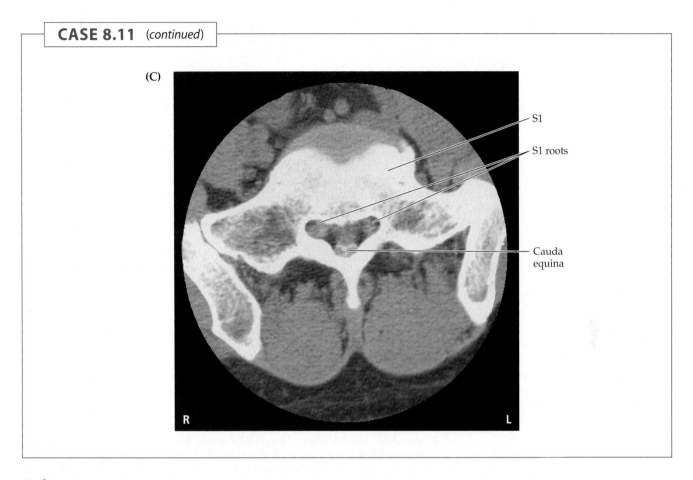

References

References

General References

Aids to the Examination of the Peripheral Nervous System. 4th Ed. 2000. W.B. Saunders on behalf of the Guarantors of Brain, Edinburgh.

Bell R, Balderston RA, Garfin SR, Herkowitz HN, Eismont FJ. 2006. *Rothman-Simeone The Spine.* 5th Ed. Saunders, Philadelphia.

Brazis PW, Masdeu JC, Biller J. 2006. The localization of spinal nerve and root lesions. In *Localization in Clinical Neurology,* 5th Ed., Chapter 3. Lippincott Williams & Wilkins.

Devinsky O, Feldmann E. 1988. *Examination of the Cranial and Peripheral Nerves.* Churchill Livingstone, New York.

Greenberg MS. 2006. *Handbook of Neurosurgery.* 6th Ed. Thieme, New York.

Neuromuscular Disorders

Amato A, Russell J. 2008. *Neuromuscular Disorders.* McGraw-Hill, Columbus, OH.

Deymeer FS (ed.). 2000. Neuromuscular Diseases: From Basic Mechanisms to Clinical Management. *Monogr Clin Neurosci,* Vol. 18.

Gold, R, Schneider-Gold, C. 2008. Current and future standards in treatment of myasthenia gravis. *Neurotherapeutics.* 5 (4): 535–541.

Preston DC, Shapiro BE. 2005. *Electromyography and Neuromuscular Disorders: Clinical-Electrophysiologic Correlations.* 2nd Ed. Butterworth-Heinemann, Boston.

Back Pain

Carragee EJ. 2005. Clinical practice. Persistent low back pain. *N Engl J Med* 352 (18): 1891–1898.

Chou R, Fu R, Carrino JA, Deyo RA. 2009. Imaging strategies for low-back pain: systematic review and meta-analysis. *Lancet* 373 (9662): 463–472.

Peul WC, van Houwelingen HC, van den Hout WB, Brand R, Eekhof JAH, Tans JTJ, Thomeer RTWM, Koes BW. 2007. Surgery versus prolonged conservative treatment for sciatica. *N Engl J Med* 356: 2245.

Cervical Radiculopathy

Brown S, Guthmann R, Hitchcock K, Davis JD. 2009. Clinical inquiries. Which treatments are effective for cervical radiculopathy? *J Fam Pract* 58 (2): 97–99.

Carette S, Fehlings MG. 2005. Cervical radiculopathy. *N Engl J Med* 353: 392.

Nasca RJ. 2009. Cervical radiculopathy: current diagnostic and treatment options. *J Surg Orthop Adv* 18 (1): 13–18.

Poletti CE, Sweet WH. 1990. Entrapment of the C2 root and ganglion by the atlanto-epistrophic ligament: clinical syndrome and surgical anatomy. *Neurosurgery* 27 (2): 288–291.

Samuraki M, Yoshita M, Yamada M. 2005. MRI of segmental zoster paresis. *Neurology* 64 (7): 1138.

Tanaka N, Fujimoto Y, An HS, Ikuta Y, Yasuda M. 2000. The anatomic relation among the nerve roots, intervertebral foramina, and intervertebral discs of the cervical spine. *Spine* 25 (3): 286–291.

Thoracic Radiculopathy

Kanno H, Aizawa T, Tanaka Y, Hoshikawa T, Ozawa H, Itoi E, Kokubun S. 2009. T1 radiculopathy caused by intervertebral disc herniation: symptomatic and neurological features. *J Orthop Sci* 14 (1): 103–106.

Kumar R, Buckley TF. 1986. First thoracic disc protrusion. *Spine* 11 (5): 499–501.

Kumar R, Cowie RA. 1992. Second thoracic disc protrusion. *Spine* 17 (1): 120–121.

Levin KH. 1999. Neurologic manifestations of compressive radiculopathy of the first thoracic root. *Neurology* 53 (5): 1149–1151.

McCormick WE, Will SF, Benzel EC. 2000. Surgery for thoracic disc disease. Complication avoidance: overview and management. *Neurosurg Focus*. 9(4):e13.

Sellman MS, Mayer RF. 1988. Thoracoabdominal radiculopathy. *Southern Med J* 81(2): 199–201.

Lumbosacral Radiculopathy

Chou R, Fu R, Carrino JA, Deyo RA. 2009. Imaging strategies for low-back pain: systematic review and meta-analysis. *Lancet* 373 (9662): 463–472.

Cohen MS, Wall EJ, Olmarker K, Rydevik BL, Garfin SR. 1992. Anatomy of the spinal nerve roots in the lumbar and lower thoracic spine. In *The Spine*, RH Rothman and FA Simeone (eds.), 3rd Ed., Chapter 4. Saunders, Philadelphia.

Goldstein B. 2002. Anatomic issues related to cervical and lumbosacral radiculopathy. *Phys Med Rehabil Clin N Am* 13 (3): 423–437.

Madigan L, Vaccaro AR, Spector LR, Milam RA. 2009. Management of symptomatic lumbar degenerative disk disease. *J Am Acad Orthop Surg* 17 (2): 102–111.

Tarulli AW, Raynor EM. 2007. Lumbosacral radiculopathy. *Neurol Clin* 25 (2): 387–405.

Cauda Equina Syndrome

Ahn UM, Ahn NU, Buchowski JM, Garrett ES, Sieber AN, Kostuik JP. 2000. Cauda equina syndrome secondary to lumbar disc herniation: a meta-analysis of surgical outcomes. *Spine* 25 (12): 1515–1522.

Gindin RA, Volcan IJ. 1978. Rupture of the intervertebral disc producing cauda equina syndrome. *Am Surg* 44 (9): 585–593.

Lavy C, James A, Wilson-MacDonald J, Fairbank J. 2009. Cauda equina syndrome. *BMJ* 338.

McCarthy MJ, Aylott CE, Grevitt MP, Hegarty J. 2007. Cauda equina syndrome: factors affecting long-term functional and sphincteric outcome. *Spine* 32 (2): 207–216.

CONTENTS

Chapter 9

Major Plexuses and Peripheral Nerves

A 3-week-old infant was not moving her left arm normally. She was a large baby and had endured significant traction on her left shoulder during delivery. Her left arm had decreased tone and appeared internally rotated. She was able to extend her left arm at the elbow and could open and close her hand, but she could not abduct her left arm at the shoulder or flex it at the elbow. The left biceps reflex was absent.

In this chapter, we will learn the sensory and motor functions of the major nerves in the arms and legs. As we shall see, this patient's symptoms are characteristic of injury to specific nerves.

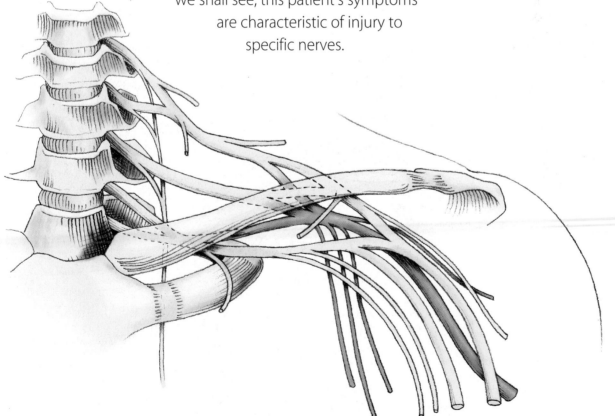

ANATOMICAL AND CLINICAL REVIEW

IN THIS CHAPTER, we will discuss the functions of the brachial plexus, lumbosacral plexus, and nerve branches that arise from them. Knowledge of the motor and sensory territories of the spinal nerve roots (see Chapter 8), major plexuses, and peripheral nerves can be very useful clinically in identifying specific nerve lesions and in distinguishing them from lesions of the central nervous system. Disorders specifically affecting motor neurons were discussed in Chapter 6 (see KCC 6.7). Nerve roots and radiculopathies were discussed in Chapter 8, where we also introduced peripheral nerve and neuromuscular disorders in general (see KCC 8.1). Here we will discuss the most important peripheral nerves in the upper and lower extremities, as well as common localized plexus and nerve syndromes.

Brachial Plexus and Lumbosacral Plexus

The **brachial plexus** is formed by nerve roots arising from the cervical enlargement at **C5**, **C6**, **C7**, **C8**, and **T1** (Figure 9.1). These nerve roots provide the major sensory and motor innervation for the upper extremities. The nerves of the

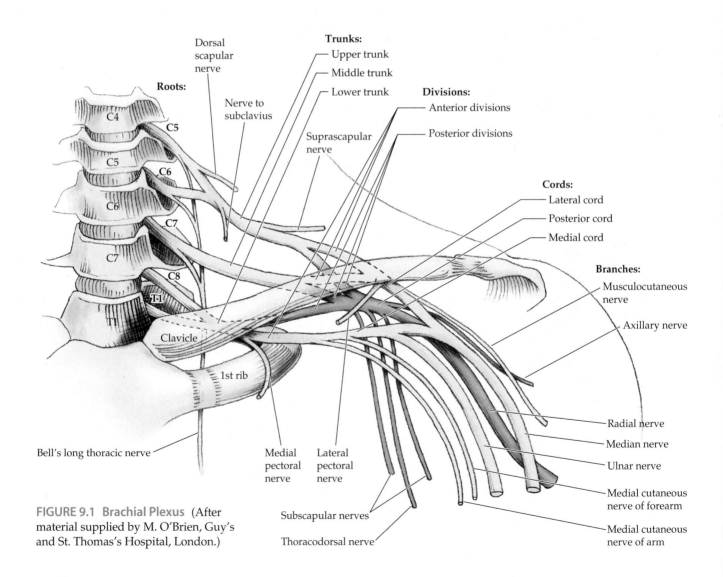

FIGURE 9.1 Brachial Plexus (After material supplied by M. O'Brien, Guy's and St. Thomas's Hospital, London.)

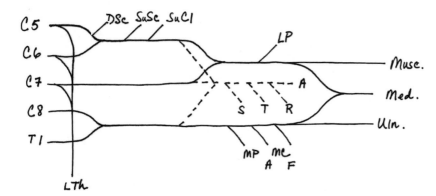

FIGURE 9.2 **Brachial Plexus: Simplified Schematic** Lth = Bell's long thoracic nerve; DSc = dorsal scapular nerve; SuSc = suprascapular nerve; SuCl = nerve to subclavius; LP = lateral pectoral nerve; A = axillary nerve; R = radial nerve; T = thoracodorsal nerve; S = subscapular nerve; MP = medial pectoral nerve; MC,A = medial cutaneous nerve of arm; MC,F = medial cutaneous nerve of forearm; Musc. = musculocutaneous nerve; Med. = median nerve; Uln. = ulnar nerve.

brachial plexus are so clinically important that it is worth committing the structure of the brachial plexus to memory. A simplified schematic can be helpful in this regard (**Figure 9.2**). The parts of the brachial plexus can be remembered by this mnemonic: Robert (**r**oots) Taylor (**t**runks) Drinks (**d**ivisions) Cold (**c**ords) Beer (**b**ranches). It is also important to know the muscles innervated by each of the nerve branches (see Table 8.1). The five most clinically important nerve branches arising from the brachial plexus are the radial, median, ulnar, musculocutaneous, and axillary nerves (see the next section). A few more mnemonics may be helpful: The nerve branches of the posterior cord can be memorized with the mnemonic **STAR** or **ARTS** (**A**xillary, **R**adial, **T**horacodorsal, **S**ubscapularis). The muscles innervated by the musculocutaneous nerve are represented by the mnemonic **BBC** (**B**iceps, **B**rachialis, **C**oracobrachialis).

The **lumbosacral plexus** arises from **L1**, **L2**, **L3**, **L4**, **L5**, **S1**, **S2**, **S3**, and **S4** at the lumbosacral enlargement and provides innervation to the lower extremities and pelvis (**Figure 9.3**). Once again, a simplified schematic may be helpful (**Figure 9.4**). The muscles innervated by each of the lumbosacral nerve branches should be reviewed (see Table 8.1). The most clinically important nerve branches arising from the lumbosacral plexus are the femoral, obturator, sciatic, tibial, and peroneal nerves, as will be described shortly.

There is also a plexus formed by branches of CN XII and C1 through C5 called the **cervical plexus**, which supplies mainly the neck muscles. We will not discuss this plexus further except to mention that the **phrenic nerve**, which supplies the diaphragm, arises from **C3**, **C4**, and **C5**.

Regions of sensory innervation by cutaneous nerve branches are shown in

MNEMONIC

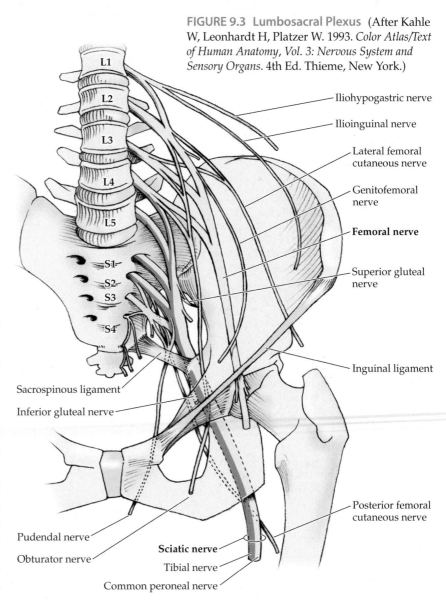

FIGURE 9.3 **Lumbosacral Plexus** (After Kahle W, Leonhardt H, Platzer W. 1993. *Color Atlas/Text of Human Anatomy, Vol. 3: Nervous System and Sensory Organs.* 4th Ed. Thieme, New York.)

Iliohypogastric nerve

Ilioinguinal nerve

Lateral femoral cutaneous nerve

Genitofemoral nerve

Femoral nerve

Superior gluteal nerve

Inguinal ligament

Posterior femoral cutaneous nerve

Sacrospinous ligament

Inferior gluteal nerve

Pudendal nerve

Obturator nerve

Sciatic nerve

Tibial nerve

Common peroneal nerve

FIGURE 9.4 Lumbosacral Plexus: Simplified Schematic IlHyp = iliohypogastric nerve; IlIng = ilioinguinal nerve; GF = genitofemoral nerve; LFC = lateral femoral cutaneous nerve; F = femoral nerve; Obt = obturator nerve; Saph = saphenous nerve; SG = superior gluteal nerve; IG = inferior gluteal nerve; Sc = sciatic nerve; CP = common peroneal nerve; SP = superficial peroneal nerve; DP = deep peroneal nerve; T = tibial nerve; Sur = sural nerve; PFC = posterior femoral cutaneous nerve; Pud = pudendal nerve.

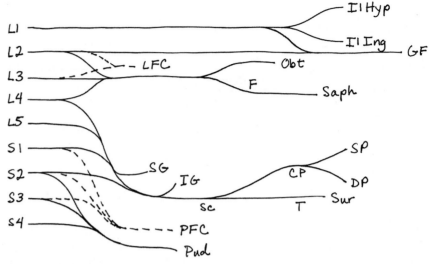

TABLE 9.1 Five Important Nerves in the Arm

NERVE	MOTOR FUNCTIONS	REGION OF SENSORY LOSS WITH NEUROPATHY
Radial	Extension at all arm, wrist, and proximal finger joints below the shoulder; forearm supination; thumb abduction in plane of palm	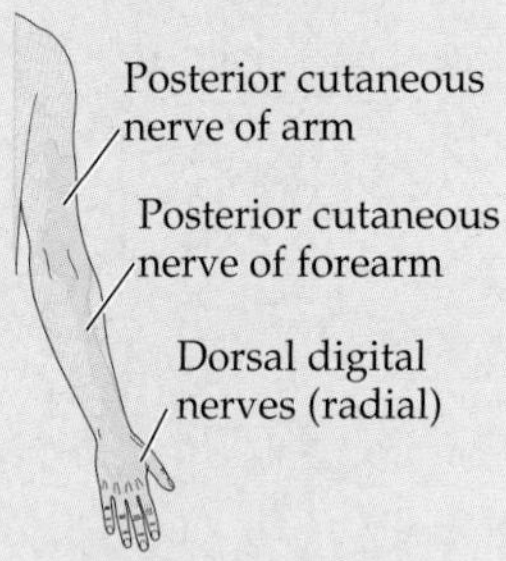 Posterior cutaneous nerve of arm; Posterior cutaneous nerve of forearm; Dorsal digital nerves (radial)
Median	Thumb flexion and opposition, flexion of digits 2 and 3, wrist flexion and abduction, forearm pronation	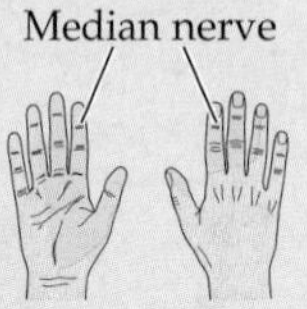 Median nerve
Ulnar	Finger adduction and abduction other than thumb; thumb adduction; flexion of digits 4 and 5; wrist flexion and adduction	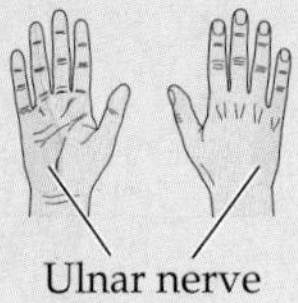 Ulnar nerve
Axillary	Abduction of arm at shoulder beyond first 15°	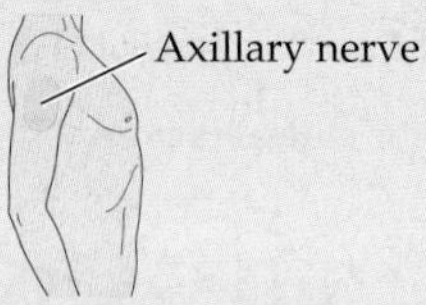 Axillary nerve
Musculo-cutaneous	Flexion of arm at elbow, supination of forearm	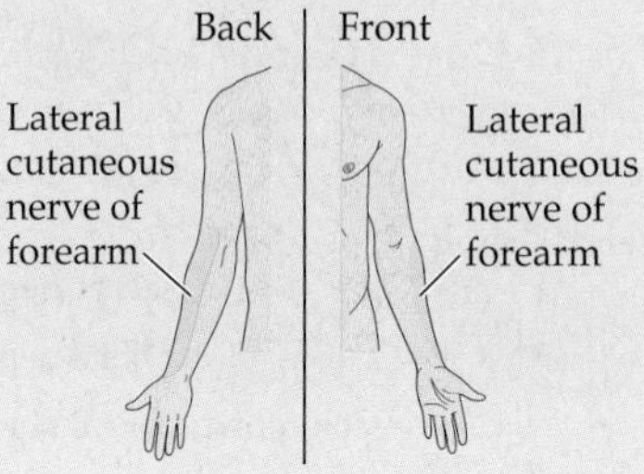 Back / Front; Lateral cutaneous nerve of forearm; Lateral cutaneous nerve of forearm

REVIEW EXERCISE

1. Practice drawing the simplified schematic of the brachial plexus shown in Figure 9.2.

2. Practice drawing the simplified schematic of the lumbosacral plexus shown in Figure 9.4.

Figure 9.5. The actual area of sensory loss following a nerve injury is somewhat smaller than the territories shown because of overlap from adjacent nerves. Compare this figure to the dermatomal sensory distribution of nerve roots shown in Figure 8.4.

Simplification: Five Nerves to Remember in the Arm

It is most clinically important to be familiar with the functions of the **radial**, **median**, **ulnar**, **axillary**, and **musculocutaneous nerves** in the arm. **Table 9.1** summarizes the motor and sensory functions of these nerves, and they are demonstrated on **neuroexam.com Videos 54** and **55**. Additional details are found in Table 8.1. In general, the radial nerve is important for extension of all joints in the arm and proximal fingers, the median nerve is important for the thumb side of the hand and wrist, and the ulnar nerve is important for the pinky side of the hand and wrist. Note that (1) the sensory terri-

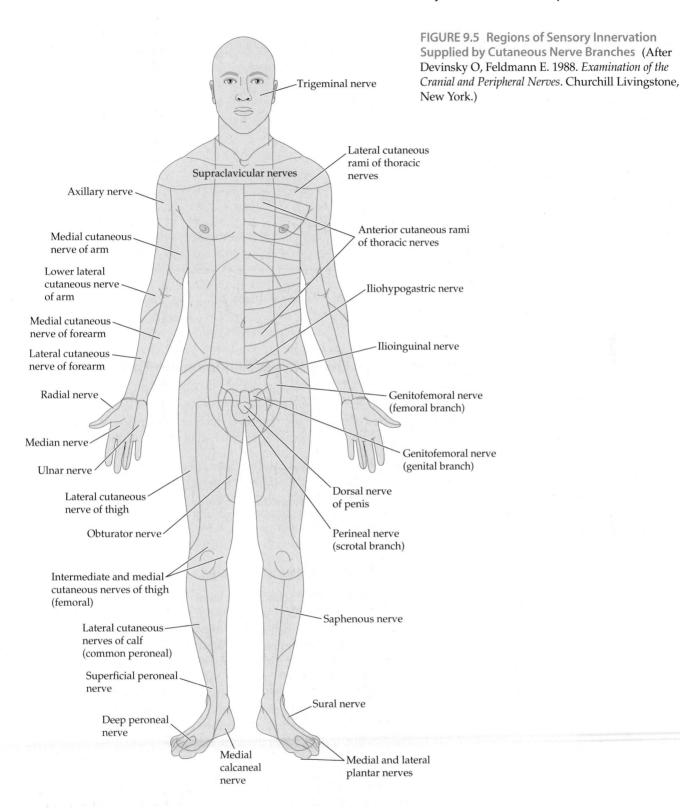

Trigeminal nerve

Supraclavicular nerves

Axillary nerve

Medial cutaneous nerve of arm

Lower lateral cutaneous nerve of arm

Medial cutaneous nerve of forearm

Lateral cutaneous nerve of forearm

Radial nerve

Median nerve

Ulnar nerve

Lateral cutaneous nerve of thigh

Obturator nerve

Intermediate and medial cutaneous nerves of thigh (femoral)

Lateral cutaneous nerves of calf (common peroneal)

Superficial peroneal nerve

Deep peroneal nerve

Medial calcaneal nerve

Lateral cutaneous rami of thoracic nerves

Anterior cutaneous rami of thoracic nerves

Iliohypogastric nerve

Ilioinguinal nerve

Genitofemoral nerve (femoral branch)

Genitofemoral nerve (genital branch)

Dorsal nerve of penis

Perineal nerve (scrotal branch)

Saphenous nerve

Sural nerve

Medial and lateral plantar nerves

FIGURE 9.5 Regions of Sensory Innervation Supplied by Cutaneous Nerve Branches (After Devinsky O, Feldmann E. 1988. *Examination of the Cranial and Peripheral Nerves.* Churchill Livingstone, New York.)

tories shown in Table 9.1 are smaller than in Figure 9.5 because adjacent nerves overlap somewhat, and (2) Table 9.1 shows regions of *sensory loss* with nerve injury rather than the whole region innervated by the nerve. Finger flexion is best tested at the distal interphalangeal joints (see **neuroexam.com Video 55**), where the flexor digitorum profundus (median nerve for digits 2 and 3; ulnar nerve for digits 4 and 5) acts without significant contributions from other muscles (Table 9.2).

TABLE 9.2   Muscles Contributing to Flexion and Extension at Finger Joints Other than the Thumb

MUSCLE	NERVE	FLEXION[a]			EXTENSION[a]		
		MCP	PIP	DIP	MCP	PIP	DIP
Flexor digitorum profundus	Median (2nd, 3rd digits), ulnar (4th, 5th digits)	X	X	**X**			
Flexor digitorum superficialis	Median	X	**X**				
Flexor digiti minimi (fifth digit)	Ulnar	X					
Lumbricals	Median (2nd, 3rd digits), ulnar (4th, 5th digits)	**X**			**X**	**X**	
Palmar and dorsal interossei	Ulnar	X			X	X	
Extensor digitorum	Radial				**X**	X	X
Extensor indicis (second digit)	Radial				X	X	X
Extensor digiti minimi (fifth digit)	Radial				X	X	X

Note: **Bold** text indicates the most important muscles.

[a]MCP, metacarpophalangeal joint; PIP, proximal interphalangeal joint; DIP, distal interphalangeal joint.

Simplification: Three Nerves Acting on the Thumb

MNEMONIC

Different thumb muscles are innervated by the radial, ulnar, and median nerves. It is easiest to remember these by the mnemonic RUM (**R**adial, **U**lnar, **M**edian), as shown in Figure 9.6. Thumb abduction in the plane of the palm (abductor pollicis longus) is mediated by the **R**adial nerve, adduction (adductor pollicis) by the **U**lnar nerve, and opposition (opponens pollicis) and flexion (flexor pollicis longus and superficial head of the flexor pollicis brevis) by the **M**edian nerve. It should also be recalled that thumb abduction perpendicular to the palm (see Table 3.4; **neuroexam.com Video 55**) is mediated by the abductor pollicis brevis, which is innervated by the median nerve after it passes through the carpal tunnel.

Intrinsic and Extrinsic Hand Muscles

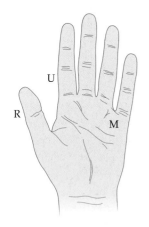

MNEMONIC

The **intrinsic hand muscles** include the muscles of the **thenar eminence** at the base of the thumb (opponens pollicis, abductor pollicis brevis, flexor pollicis brevis, adductor pollicis), the muscles of the **hypothenar eminence** at the base of the pinky finger (opponens digiti minimi, abductor digiti minimi, flexor digiti minimi), the **lumbricals**, and the **interossei**. Intrinsic hand muscles are innervated by the ulnar nerve, except for the LOAF (**L**umbricals I and II, **O**pponens pollicis, **A**bductor pollicis brevis, **F**lexor pollicis brevis—superficial head) muscles, which are innervated by the median nerve after it passes through the carpal tunnel. All intrinsic hand muscles are supplied by **C8** and **T1** (see Table 8.1).

In addition to the intrinsic hand muscles, **extrinsic muscles** in the forearm are important for finger movements (see Table 8.1). Intrinsic and extrinsic muscles contributing to flexion and extension at finger joints other than the thumb are summarized in Table 9.2. As we have already mentioned, it should be clear from this table that the flexor digitorum profundus (median nerve for digits 2 and 3; ulnar nerve for digits 4 and 5) is best tested at the distal interphalangeal joints, since other muscles participate in flexion at the other joints. Similarly, the extensor digitorum (radial nerve and C7) is best tested at the metacarpophalangeal joints (see Table 9.2). This is because other muscles, most notably the lumbricals, are predominantly responsible for finger extension at the proximal and distal interphalangeal joints (median nerve for 2nd and 3rd digits; ulnar nerve for 4th

FIGURE 9.6   Three Nerves Acting on the Thumb   The radial nerve abducts the thumb in the plane of the palm. The ulnar nerve adducts the thumb in the plane of the palm. The median nerve opposes the thumb. Note also that the abductor pollicis brevis (median nerve) abducts the thumb perpendicular to the plane of the palm (not shown).

and 5th digits). See Tables 8.1 and 9.1 for muscles contributing to finger adduction, abduction, and opposition. Note, for example, that the **palmar interossei** adduct the fingers, while the **dorsal interossei** abduct them.

Simplification: Five Nerves to Remember in the Leg

It is most clinically important to be familiar with the functions of the **femoral**, **obturator**, **sciatic**, **tibial**, and **peroneal nerves** in the leg. Table 9.3 summarizes the motor and sensory functions of these nerves, and they are demonstrated on **neuroexam.com Videos 56** and **57**. Table 8.1 provides additional details. Note again that the sensory territories shown in Table 9.3 are smaller than in Figure 9.5, since here we are interested in regions of sensory loss.

The tibial and common peroneal nerves are the two most important branches of the sciatic nerve. The **hamstring muscles** (semitendinosus, semimembranosus, and biceps femoris) are innervated by the sciatic nerve itself before it divides into the tibial and common peroneal nerves. The common peroneal nerve divides further to give rise to the **superficial** and **deep peroneal nerves** (see Figures 9.3 and 9.4; Table 9.3).

REVIEW EXERCISE

1. Turn back to Tables 3.4–3.6 in Chapter 3, where we discussed strength and reflex testing (see also **neuroexam.com Videos 54–58**). In these tables, cover all columns except for the left-most column. For each action or reflex, list the appropriate muscle, nerves, and nerve roots being tested (refer to Table 8.1).

2. In Tables 9.1 and 9.3, cover the columns showing the regions of sensory loss and sketch the region of sensory loss for each of the five nerves in the arm and the leg.

TABLE 9.3 **Important Nerves in the Leg**

NERVE	MOTOR FUNCTIONS	REGION OF SENSORY NERVE LOSS WITH NEUROPATHY
Femoral	Leg flexion at the hip, leg extension at the knee	
Obturator	Adduction of the thigh	
Sciatic	Leg flexion at the knee (see also tibial and peroneal nerves, in column at left)	
Tibial	Foot plantar flexion and inversion, toe flexion	
Superficial peroneal	Foot eversion	
Deep peroneal	Foot dorsiflexion, toe extension	

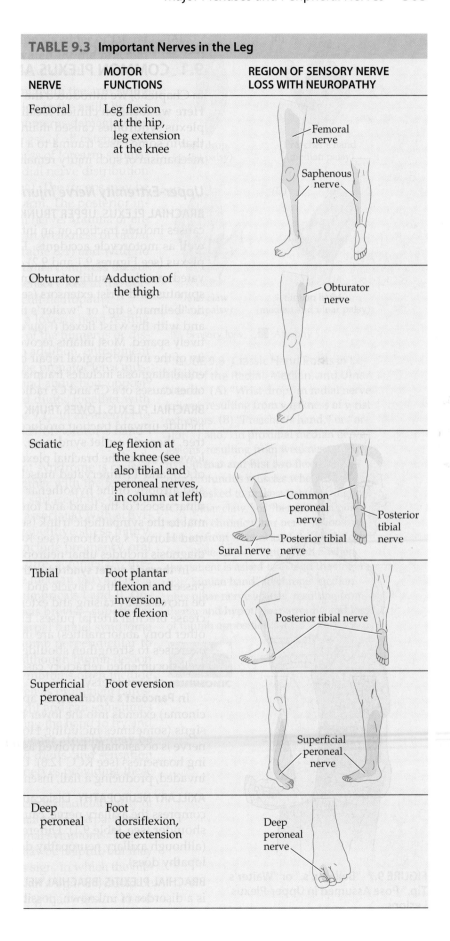

1 minute. Treatments include immobilization by a removable wrist splint, steroid injections, and surgical decompression of the carpal tunnel.

ULNAR NEUROPATHY The medial epicondyle of the elbow carries the name "funny bone" because mild trauma to the ulnar nerve as it passes over the elbow in the **ulnar groove** (between the olecranon and medial epicondyle) produces transient paresthesias (see KCC 7.1) in an ulnar distribution. A common cause of ulnar neuropathy is entrapment (see KCC 8.1) at the elbow in the cubital canal, which lies in the region of the ulnar groove. This condition is sometimes called tardy ulnar palsy, a delayed result of a posttraumatic, degenerative, or congenital increased carrying angle at the elbow. The ulnar nerve can also be damaged acutely by fractures of the medial epicondyle, or it may be compressed by a habit of resting the elbows on a hard table.

Findings include weakness of wrist flexion and adduction, finger adduction and abduction, and flexion of the fourth and fifth digits, together with sensory loss and paresthesias in an ulnar distribution (see Table 9.1). As with most neuropathies, motor findings may be absent in mild cases. Severe cases may include atrophy and fasciculations in the hypothenar eminence. Because of weak lumbricals at the fourth and fifth digits, these fingers may assume a characteristic "ulnar claw" or "benediction posture" (Figure 9.8C). Differential diagnosis includes C8 and T1 radiculopathy, Pancoast's syndrome, thoracic outlet syndrome, or other lesions of the brachial plexus inferior trunk or medial cord (see Figures 9.1 and 9.2). Unlike ulnar neuropathy, these conditions sometimes produce a Horner's syndrome, sensory changes in the T1 dermatome of the upper medial arm (see Figure 8.4) or involvement of hand muscles innervated by the median nerve. Entrapment in the cubital canal at the elbow can be treated surgically by translocation of the ulnar nerve to the flexor side of the elbow.

Compression of the ulnar nerve in the hand as it passes over the hamate bone in Guyon's canal can occur from prolonged leaning forward while cycling. The result is weakness of finger adduction and abduction without sensory loss because the cutaneous branches of the ulnar nerve are given off more proximally.

Combination of chronic median and ulnar nerve lesions leads to thenar and hypothenar atrophy with lack of thumb opposition, resulting in a "simian hand," or "monkey's paw" (Figure 9.8D).

Lower-Extremity Nerve Injuries

FEMORAL NEUROPATHY The femoral nerve can occasionally be injured during pelvic surgery or compressed by a retroperitoneal hematoma or pelvic mass. Abnormalities include weakness of thigh flexion and knee extension, loss of the patellar reflex, and sensory loss in the anterior thigh (see Table 9.3). Differential diagnosis includes L3 or L4 radiculopathy. L3 or L4 radiculopathy, however, may include weakness of thigh adduction (obturator nerve), a feature not associated with femoral neuropathy (see Table 8.1).

SCIATIC NEUROPATHY Causes of sciatic neuropathy include posterior hip dislocation, acetabular fracture, and intramuscular injection placed too medially and inferiorly in the buttocks. There is weakness of all foot and ankle muscles and of knee flexion, loss of the Achilles tendon reflex, and sensory loss in the foot and lateral leg below the knee (see Table 9.3). Differential diagnosis includes lesions in the foot area of the motor cortex (see KCC 6.3; Figure 6.14F).

The term "sciatica" is vague and refers to all disorders causing painful paresthesias in a sciatic distribution. The most common cause is compression of lumbosacral roots by disc material and osteophytes (see KCC 8.3). Rarely, the sciatic nerve may be compressed more distally by muscular or skeletal elements.

PERONEAL NERVE PALSY As the common peroneal nerve passes around the fibular head near the skin surface, it is vulnerable to laceration, stretch injury by forcible foot inversion, or compression by tight stockings, a cast, crossed legs, or trauma. In peroneal nerve palsy, there is **foot drop**, with weakness of foot dorsiflexion and eversion, and sensory loss over the dorsolateral foot and shin. Most patients recover spontaneously when the mechanical cause is removed. A foot brace may improve function if foot drop is significant. Differential diagnosis includes L5 radiculopathy. However, L5 radiculopathy includes weakness of foot dorsiflexion, eversion, and inversion, while in peroneal palsy, foot inversion is normally spared because this function can be carried out by the tibialis posterior (innervated by the tibial nerve) (see Table 9.3; see also Table 8.1).

OBTURATOR NERVE PALSY The obturator nerve (originating from L2–L4; see Figure 9.3 and Table 9.3) can be compressed in women during complicated delivery or occasionally during pelvic trauma or surgery. Deficits include gait instability due to weakness of the leg adductor muscles and pain and numbness in the medial thigh.

MERALGIA PARESTHETICA The lateral femoral cutaneous nerve (which originates in L2 and L3; see Figures 9.3 and 9.4) can be entrapped as it passes under the inguinal ligament or fascia lata, producing paresthesias and loss of sensation in the lateral thigh (see Figure 9.5). This entrapment syndrome includes no motor involvement or reflex changes. Common causes include obesity, pregnancy, weight loss, or heavy equipment belts, and symptoms may be worse after prolonged walking, standing, or sitting. Differential diagnosis includes L2 or L3 radiculopathy, although, unlike meralgia paresthetica, these conditions usually include motor changes or decreased patellar reflex. Symptoms most often resolve spontaneously or by avoidance of mechanical precipitants; however, surgical decompression is occasionally attempted in refractory cases.

MORTON'S METATARSALGIA Tight-fitting shoes can compress the digital nerves, especially of the third and fourth toes, producing patches of numbness and paresthesias.

In conclusion, familiarity with the patterns of sensory and motor loss in the common plexus and nerve syndromes discussed in this section, as well as those seen in radiculopathies (see KCC 8.3) and other disorders of nerves, muscles, and the neuromuscular junction (see KCC 8.1), can greatly aid in localizing neurological deficits and in distinguishing disorders of the peripheral and central divisions of the nervous system. When the diagnosis remains uncertain, electrodiagnostic tests can often be helpful, as discussed in the next section. ◼

KEY CLINICAL CONCEPT

9.2 ELECTROMYOGRAPHY (EMG) AND NERVE CONDUCTION STUDIES

Electromyography (EMG) and nerve conduction studies are valuable diagnostic tools that can help localize and determine the causes of nerve and muscle disorders. In **nerve conduction studies**, stimulating electrodes are placed on the skin overlying a nerve, and recording electrodes are placed either at a different point along the same nerve or overlying a muscle innervated by the nerve (Figure 9.9A–C). When a stimulus is given to the nerve, a **compound motor action potential** (**CMAP**) can be recorded over the belly of a muscle innervated by that nerve, resulting from the summated electrical activity of muscle cells (Figure 9.9D). If a distal nerve branch with purely sensory function is used for

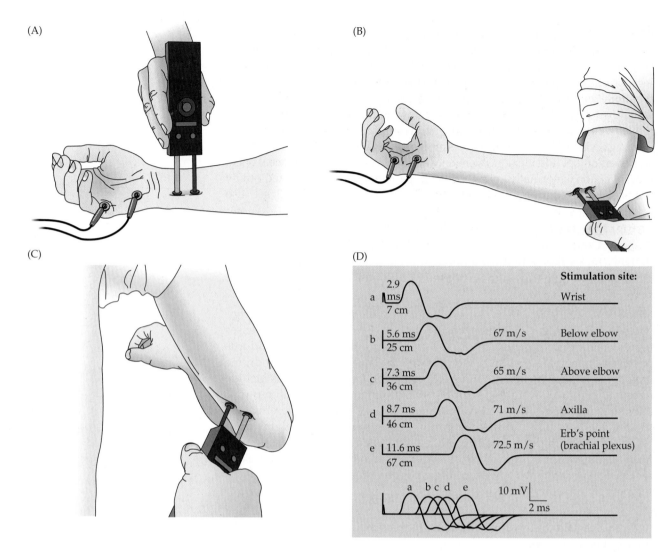

FIGURE 9.9 Nerve Conduction Study
The compound motor action potential (CMAP) of the ulnar nerve was recorded in this example from a normal subject at several stimulus locations. (A) Stimulation of the ulnar nerve at the wrist and recording of CMAP from the abductor digiti quinti in the hypothenar eminence. (B) Ulnar nerve stimulation just distal to the ulnar groove. (C) Ulnar nerve stimulation just proximal to the ulnar groove. (D) CMAP recordings for different stimulation sites. For each trace, the distance from the recording electrode (in centimeters) and the latency to CMAP onset (in milliseconds) are indicated to the left. Conduction velocities between adjacent stimulation sites (in meters per second) were calculated by dividing the distance between the two stimulation sites by the difference in motor latencies for the two sites. (After Rajesh KS, Thompson LL. 1989. *The Electromyographer's Handbook*. 2nd Ed. Little, Brown, Boston.)

recording or stimulation and a second set of stimulating or recording electrodes is placed somewhere along the nerve, a **compound sensory nerve action potential** (**SNAP**) can be recorded over the nerve, resulting from summated electrical activity in the sensory neuron axons of the nerve.

Lesions proximal to the dorsal root ganglia will leave the cell bodies and distal axons of sensory neurons intact (see Figure 8.1B). Therefore, SNAPs will be preserved. In contrast, proximal lesions of motor nerve roots will cause degeneration of the distal motor neuron axons, reducing or abolishing CMAPs. There are standard values for SNAP and CMAP latencies or conduction velocities for each major nerve when stimulated at various points along its course. These values allow nerve conduction studies to be used to determine if there is evidence for slowed nerve conduction—for example, in the case of **demyelination** (see KCC 8.1). In addition, there are standard values for SNAP amplitudes. Decreased SNAP amplitudes suggest that conduction in some axons of the nerve has been interrupted, as is the case in **axonal damage**.

CMAP studies can be used to evaluate the function of the neuromuscular junction by the use of **repetitive stimulation**. Slow, repetitive stimulation (2–3 Hz) depletes presynaptic stores of acetylcholine; faster repetitive stimulation (>5 Hz) increases presynaptic calcium, facilitating neurotransmitter release. Under normal conditions, repetitive stimulation does not significantly affect

CMAP amplitude because there is a "safety factor," meaning that every presynaptic action potential results in a postsynaptic potential well above the threshold needed to produce a muscle cell action potential. Under pathologic conditions, however, failures in neuromuscular transmission occur. Therefore, for example in **myasthenia gravis** (see KCC 8.1), in which there is a decrease in postsynaptic acetylcholine receptors on muscle cells, slow repetitive stimulation results in a gradual **decrement** in CMAP amplitude. Decrement of >10% is considered abnormal. In **Lambert–Eaton myasthenic syndrome** and in **botulism**, in which there is decreased presynaptic neurotransmitter release, fast repetitive stimulation (or active volitional muscle contraction) causes CMAPs to **increment** in amplitude from an abnormally low starting point.

In **electromyography** (**EMG**), an electrode is inserted directly into a muscle, and motor unit action potentials (MUPs) are recorded from muscle cells. The EMG pattern provides information useful in distinguishing weakness caused by **neuropathic disorders** (nerve or motor disease) from that caused by **myopathic disorders** (muscle disease). In neuropathic disorders, **increased spontaneous activity** (fibrillation potentials and positive sharp waves) often is recorded on EMG and is sometimes also visible on physical examination as **fasciculations** (see KCC 6.1). Fasciculations and other forms of spontaneous activity can occur due to chronic deinnervation of muscle cells. Deinnervation also causes adjacent motor axons to sprout and reinnervate a larger region, resulting in abnormally large motor units (a motor unit consists of all the muscle cells innervated by a single motor neuron axon). Therefore, with neuropathic disorders, MUPs are of abnormally large amplitude and duration. Reduced MUP amplitude and duration suggests a **myopathic** disorder is present.

When a muscle is voluntarily contracted, the EMG normally shows a pattern of continuous firing of motor units referred to as a normal **recruitment pattern**. In neuropathic disorders, the recruitment pattern has normal amplitude but shows interrupted firing, since some motor units are not successfully activated. This phenomenon is referred to as decreased, reduced, or incomplete recruitment. In myopathic disorders, the recruitment pattern is continuous or even increased (since more motor units need to be activated for a given force), but the amplitude is often decreased. ■

CLINICAL CASES

CASE 9.1 COMPLETE PARALYSIS AND LOSS OF SENSATION IN ONE ARM

CHIEF COMPLAINT

A 60-year-old man with a history of lung cancer gradually developed severe pain, weakness, and numbness in his right arm.

HISTORY

The patient smoked for 34 years. Two years ago he was diagnosed with lung cancer and underwent a right upper-lobe lung resection followed by radiation and chemotherapy. Six months ago he developed shooting **pain and swelling of the right arm**. He gradually **lost all strength and sensation in the entire right arm up to the shoulder** but continued to have severe burning pain. Past medical history was notable for right eye surgery following an assault with a baseball bat 20 years ago.

PHYSICAL EXAMINATION

Vital signs: T = 99.4°F, P = 110, BP = 130/80.

Neck: Supple; no tenderness.

Lungs: Clear.

Heart: Regular rate with no murmurs, gallops, or rubs.

Abdomen: Normal bowel sounds; soft, no masses.

Extremities: Right arm swollen, with two firm, 5 cm discolored masses—one in the right axilla and one in the upper right chest wall. Also marked clubbing of the fingers bilaterally.

(continued on p. 370)

CASE 9.1 *(continued)*

Neurologic exam:

MENTAL STATUS: Alert and oriented × 3.

CRANIAL NERVES: Intact, except for the right eye, which had an irregular pupil and diminished vision in both the lateral and medial fields.

MOTOR: Normal tone except for the **right arm**, which was **flaccid**. 5/5 power throughout, except for **0/5 power in the right shoulder, arm, and hand**.

REFLEXES:

```
        0        2⁺
          O
        0      2⁺
              2⁺
        0      2⁺

      2⁺      2⁺

       2⁺   2⁺
```

COORDINATION: Normal on finger-to-nose (except unable to test right arm) and heel-to-shin testing.

GAIT: Normal.

SENSORY: **Absent light touch, pinprick, and vibration sense in the entire right arm up to the deltoid** (Figure 9.10). Sensation otherwise normal.

LOCALIZATION AND DIFFERENTIAL DIAGNOSIS

1. On the basis of the symptoms and signs shown in **bold** above, where is the lesion?

2. What is the most likely diagnosis?

3. This patient had an abnormal right eye due to prior trauma. If his right eye had been normal previously, what additional finding might be present on exam that would help with the localization?

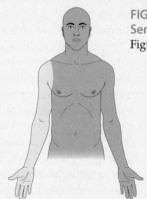

FIGURE 9.10 Region of Sensory Loss Compare to Figures 8.4 and 9.5.

Discussion

1. The key symptoms and signs in this case are:

   - **Paralysis, with decreased tone and absent reflexes in the entire right arm and hand**

   - **Absent light touch, pinprick, and vibration sense in the entire right arm up to the deltoid**

   - **Pain and swelling in the right arm**

   Weakness in one arm can be caused by peripheral nerve lesions or by lesions in the arm region of the motor cortex (see KCC 6.3; Figure 6.14E). However, it is unlikely that a cortical lesion would produce complete paralysis and sensory loss in the entire arm ending sharply at the shoulder, with no face or leg weakness at all. In addition, no single peripheral nerve lesion could produce this pattern. Therefore, the lesion must involve the entire right brachial plexus, or all right-sided nerve roots from C5 through T1.

2. The history of right apical lung tumor supports the possibility of a lesion invading the right brachial plexus from below (see the description of Pancoast's syndrome in KCC 9.1), as does the presence of swelling in the arm, suggesting obstruction of venous return. Because of the past history of radiation therapy, **radiation plexitis** is another possibility, in which numbness and sometimes weakness can develop in a limb months to years after treatment due to radiation-induced nerve injury.

3. A lesion of the proximal portion of the lower brachial plexus involving the T1 nerve root can cause a Horner's syndrome (see Figures 6.13, 13.10; KCC 13.5). This condition could be difficult to appreciate in this patient because of his prior history of right eye trauma.

Clinical Course and Neuroimaging

A **brachial plexus MRI** (Image 9.1, page 372) showed extensive invasion of the apical lung mass into the region of the right brachial plexus. The cancer in this patient, unfortunately, was no longer amenable to treatment. However, his pain was managed by a multidisciplinary team using oral, intravenous, and epidural medications to provide adequate pain relief.

CASE 9.2 A NEWBORN WITH WEAKNESS IN ONE ARM

MINICASE

A 3-week-old infant girl was brought to the pediatrician because of left arm weakness. She was born at 42 weeks (2 weeks past the due date) weighing 10 pounds 11 ounces, and the delivery was complicated by shoulder dystocia (difficulty delivering the shoulder) resulting in significant traction on the left neck and shoulder during delivery. Left arm weakness was noted at birth that improved slightly but was still present at the appointment. Exam was normal except for the **left upper extremity**, which had **decreased tone and lay internally rotated at the infant's side with decreased spontaneous movements. She did not abduct the left arm or flex it at the elbow** but did have spontaneous opening and closing of the hand with normal grip strength, normal elbow extension, and some wrist flexion. **The left biceps reflex was absent**, and other reflexes were 2+ throughout.

LOCALIZATION AND DIFFERENTIAL DIAGNOSIS

On the basis of the symptoms and signs shown in **bold** above, where is the lesion? What is the most likely diagnosis?

Discussion

The key symptoms and signs in this case are:

- **Weakness of left arm external rotation, abduction, and elbow flexion, with decreased tone and absent biceps reflex**

The patient has typical findings consistent with a left brachial plexus upper trunk injury (Erb–Duchenne palsy), affecting C5 and C6 innervated muscles, caused by left shoulder traction at birth (KCC 9.1).

Clinical Course

A physical therapy program was initiated to preserve range of motion during recovery. At age 7 weeks the patient was able to lift her arm off the table and had some external rotation of the arm, as well as slight spontaneous elbow flexion. The left biceps reflex was still absent. By age 4 months she was able to reach for objects well with either hand, although she preferred to use her right hand, and she had 4+/5 left biceps strength when pulled to a seated position. Continued improvement was anticipated.

CASE 9.1 COMPLETE PARALYSIS AND LOSS OF SENSATION IN ONE ARM

IMAGE 9.1 **Right Apical Lung Cancer Extending into the Region of the Brachial Plexus** T1-weighted coronal MRI scan of the chest.

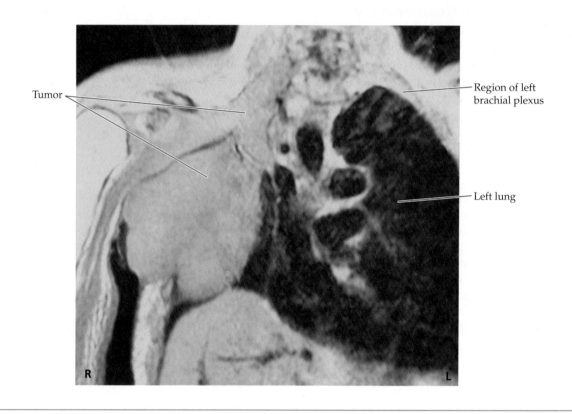

Tumor

Region of left brachial plexus

Left lung

R L

CASE 9.3 A BLOW TO THE MEDIAL ARM CAUSING HAND WEAKNESS AND NUMBNESS

MINICASE

A 38-year-old alcoholic man was seen to fall, catching his right arm on a garbage can. He was brought to the emergency room. He had an **abrasion and tenderness of the right upper medial arm**. Neurologic exam was normal except for **4/5 strength of right thumb opposition, second and third finger flexors, and wrist flexion and abduction. There was also decreased pinprick and light touch sense along the lateral surface of the right hand and first, second, and third fingers** (Figure 9.11).

LOCALIZATION AND DIFFERENTIAL DIAGNOSIS

On the basis of the symptoms and signs shown in **bold** above, where is the lesion? What is the most likely diagnosis?

FIGURE 9.11 Region of Sensory Loss

Discussion

The key symptoms and signs in this case are:

- **Weakness of right thumb opposition, second and third finger flexors, and wrist flexion and abduction**
- **Decreased pinprick and light touch sense along the lateral surface of the right distal forearm, hand, and first, second, and third fingers**
- **Abrasion and tenderness of the right upper medial arm**

The pattern of weakness and sensory loss in this patient is consistent with a median nerve injury (see Table 9.1; Figure 9.5; see also Table 8.1). The most likely cause is compression of the nerve in the upper medial arm, as evidenced by the tenderness in this area and the mechanism of injury.

Clinical Course

X-rays of the right arm revealed no fractures. The patient was discharged from the emergency room and did not return for follow-up.

CASE 9.4 NOCTURNAL PAIN AND TINGLING IN THE THUMB, POINTER, AND MIDDLE FINGER

MINICASE

A 38-year-old man who works in a cola factory developed **pain and tingling in his right thumb, index, and middle fingers** over the past 2 months that occasionally radiates into the right arm and forearm. His symptoms are worse at night or when the arm is relaxed. He has also noticed some **decreased sensation of the fingertips of the same fingers** while buttoning his shirt. Exam was notable only for obesity and **4⁺/5 weakness of the right opponens pollicis and decreased pinprick sensation in the palmar aspect of the right first, second, and third fingers, sparing the thenar area** (Figure 9.12). Tinel's and Phalen's signs were not present.

LOCALIZATION AND DIFFERENTIAL DIAGNOSIS

On the basis of the symptoms and signs shown in **bold** above, where is the lesion? What is the most likely diagnosis?

FIGURE 9.12 Region of Sensory Loss

Discussion

The key symptoms and signs in this case are:

- **Mild weakness of the right opponens pollicis**
- **Pain, tingling, and decreased pinprick sensation in the palmar aspect of the right first, second, and third fingers, sparing the thenar area**

The opponens pollicis is supplied by the median nerve (Table 9.1; see also Table 8.1). The fact that wrist flexion and abduction are spared, as is sensation over the thenar area, suggests a lesion of the median nerve after the branch to the flexor carpi radialis and the palmar cutaneous branch are given off, such as in the carpal tunnel (see KCC 9.1). Carpal tunnel syndrome of the right wrist is thus the most likely diagnosis. Note that the abductor pollicus brevis is usually affected in carpal tunnel syndrome (see KCC 9.1) but was not specifically tested in this patient. Other, less likely possibilities include a mild right C6 and C7 radiculopathy or a more proximal right median nerve lesion.

Clinical Course

Thyroid function tests and routine blood chemistries were normal. The patient was given a removable splint to hold his wrist in slight extension at night, and his symptoms gradually improved.

CASE 9.5 HAND AND WRIST WEAKNESS AFTER A FALL

MINICASE

A 20-year-old male waiter tripped while working at a restaurant and broke his fall by extending his left hand onto a table. That night he had pain in his left arm that resolved by the next day, but he then noticed **weakness of the left hand and wrist** and came to the emergency room. Exam was normal except for **3/5 strength in the left wrist extensors, finger extensors, and thumb abduction in the plane of the palm**, and 4/5 **strength in forearm supination**. Strength in all other muscles was intact, as was sensation.

LOCALIZATION AND DIFFERENTIAL DIAGNOSIS

On the basis of the symptoms and signs shown in **bold** above, where is the lesion? What is the most likely diagnosis?

Discussion

The key symptoms and signs in this case are:

- **Weakness of left forearm supination, wrist extensors, finger extensors, and abduction of the thumb in the plane of the palm**

These muscles are all supplied by the radial nerve (Table 9.1; see also Table 8.1). In particular, the facts that the triceps was spared and that there was no sensory loss suggest a lesion of the posterior interosseous nerve, a purely motor branch of the radial nerve. The posterior interosseous branch of the radial nerve was apparently injured during the fall, with the exact mechanism being unclear (see KCC 9.1).

Clinical Course

X-rays of the left arm did not reveal a fracture. The patient was given a splint to avoid developing contracture deformities and was followed by an occupational therapist as his strength gradually recovered. An EMG done 2 months after the injury was consistent with a lesion of the left radial nerve distal to the fibers innervating the triceps. By 4 months after the injury, strength had returned to $4^+/5$ in the affected muscles and was continuing to improve gradually. (Note: In posterior interosseous nerve injuries, the extensor carpi radialis is usually spared, so extension of the wrist in the radial direction is preserved. This was not tested for in this case.)

CASE 9.6 NUMBNESS AND TINGLING IN THE PINKY AND RING FINGER

MINICASE

A 32-year-old computer programmer developed 2 months of worsening **tingling and numbness in his left fifth digit, in the medial aspect of his left fourth digit, and along the medial surface of his left hand and forearm**. The symptoms were worse upon awakening in the morning and were exacerbated after resting his elbows on a hard surface. Exam was normal except for **4/5 strength in left fifth finger abduction, and decreased pinprick sensation in the left fifth digit and the medial half of the left fourth digit** (Figure 9.13). Symptoms were not worsened by arm abduction plus elevation.

LOCALIZATION AND DIFFERENTIAL DIAGNOSIS

On the basis of the symptoms and signs shown in **bold** above, where is the lesion? What is the most likely diagnosis?

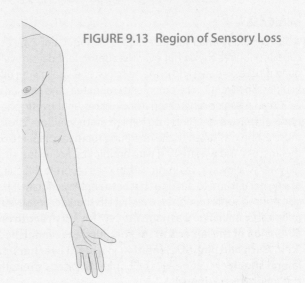

FIGURE 9.13 Region of Sensory Loss

Discussion

The key symptoms and signs in this case are:

- **Weakness of left fifth finger abduction**
- **Paresthesias and decreased pinprick sensation in the left fifth digit and the medial half of the left fourth digit**

The sensory and motor deficits in this patient could be caused by mild dysfunction in the ulnar nerve, the brachial plexus lower trunk (e.g., thoracic outlet syndrome), the brachial plexus medial cord, or the C8 or T1 nerve roots (see Table 9.1; KCC 9.1; see also Table 8.1). The fact that the symptoms are worse after pressure on the elbow, are not accompanied by neck pain (common in cervical radiculopathy; see KCC 8.3), and are not exacerbated by arm abduction plus elevation (characteristic of thoracic outlet syndrome) suggests, but does not prove, that the ulnar nerve is the culprit.

Clinical Course

Nerve conduction studies (see KCC 9.2; Figure 9.9) were normal except in the ulnar nerves. To measure conduction velocity in the ulnar nerve, a recording electrode was placed on the skin over the belly of the abductor digiti minimi muscle and a stimulating electrode was placed on the skin at various points on the arm over the course of the ulnar nerve. When a stimulus was given to the nerve, a compound motor action potential (CMAP) could be recorded over the muscle. Distal conduction velocity in this patient was normal when the ulnar nerve was stimulated below the medial epicondyle of the elbow (see Figure 9.9A,B), but was decreased when the ulnar nerve was stimulated just above the elbow (see Figure 9.9C), suggesting a conduction problem at the elbow.

The nerve conduction studies showed that both ulnar nerves were affected, although only the left one had produced symptoms. Nerve conduction studies of the median nerves were normal bilaterally. The patient was given elbow pads to wear while sleeping or while working at the computer, and he was instructed to avoid resting on his elbows. Two months later his paresthesias had improved markedly, strength in his finger abductors was normal, and he had only mildly decreased sensation in a left ulnar distribution.

CASE 9.7 SHOULDER WEAKNESS AND NUMBNESS AFTER STRANGULATION

MINICASE

A 39-year-old woman was assaulted by strangulation. She escaped serious asphyxia but over the subsequent days developed progressive swelling in the neck area and swallowing difficulties. She went to the emergency room ten days later and a CT scan revealed a hematoma and abscess in the retropharyngeal space extending into her left sternocleidomastoid muscle. She was taken to the operating room for drainage of the abscess and was treated with antibiotics. Following surgery her swallowing was improved, but she reported difficulty raising her left arm to put on a shirt or to apply deodorant. She also noticed some numbness over her left shoulder area. Neurological examination was normal except for **3/5 strength on abduction of the left arm at the shoulder**, and **diminished light touch and pinprick sensation in a patch over her left lateral shoulder** (Figure 9.14). Of note, left biceps strength and reflexes were normal.

LOCALIZATION AND DIFFERENTIAL DIAGNOSIS

On the basis of the symptoms and signs shown in **bold** above, where is the lesion? What is the most likely diagnosis?

FIGURE 9.14 Region of Sensory Loss

Discussion

The key symptoms and signs in this case are:

- **Weakness of left arm abduction**
- **Decreased sensation to light touch and pinprick on left lateral shoulder**

Weakness of arm abduction and shoulder numbness could be caused by a left C5 radiculopathy (see Figure 8.4; Table 8.5) or by a left axillary nerve injury (see Table 9.1; Figure 9.5). Sparing of the biceps, which receives important innervations from the C5 nerve root, makes a C5 radiculopathy less likely. Therefore, the most likely diagnosis is a left axillary nerve injury (see KCC 9.1). Axillary nerve injury is usually caused by dislocation or traction of the shoulder, and can also been seen as a complication of surgery in the neck and shoulder region.

Clinical Course

The nerve injury was discussed with the patient, and follow-up was advised since in some cases when prompt recovery does not occur, axillary nerve injury can benefit from treatment by nerve grafting or decompressive surgery. However, after a course of intravenous antibiotics in the hospital, the patient was discharged home and did not return for follow-up appointments.

CASE 9.8 UNILATERAL THIGH PAIN, WEAKNESS, AND NUMBNESS IN A DIABETIC

MINICASE

A 45-year-old man spent several weeks in the intensive care unit for diabetic ketoacidosis and severe bilateral pneumonia. When he finally stabilized and was transferred to a regular hospital floor, he noticed **weakness and numbness in the left leg, with numbness and tingling over the anterior thigh down to the medial calf above the foot**. A neurology consult was called, and on exam he had **4/5 strength in the left iliopsoas and quadriceps**, with preserved strength in all other muscle groups, including the thigh adductors. There was **decreased pinprick sensation in the left anterior thigh and medial calf** (Figure 9.15). Reflexes were normal and symmetrical except for an **absent left patellar reflex**.

LOCALIZATION AND DIFFERENTIAL DIAGNOSIS

On the basis of the symptoms and signs shown in **bold** above, where is the lesion? What is the most likely diagnosis?

FIGURE 9.15 Region of Sensory Loss

Discussion

The key symptoms and signs in this case are:

- **Weakness of the left iliopsoas and quadriceps with absent patellar reflex**
- **Paresthesias and decreased pinprick sensation in the left anterior thigh and medial calf**

The pattern of weakness, reflex loss, and sensory changes in this patient could be caused by an L4 radiculopathy or a femoral neuropathy (see Table 9.3; see also Tables 8.1, 8.6; Figure 8.4). The fact that thigh adduction is spared suggests a femoral neuropathy, since L4 contributes significantly to both the obturator and femoral nerves. The most likely diagnosis is left femoral neuropathy caused by diabetes (see KCC 8.1, 9.1). Less likely, insertion of a femoral vein catheter during the patient's stay in the intensive care unit may have caused an undetected hematoma compressing the femoral nerve. Compare this case to Case 8.9.

Clinical Course

The patient gradually recovered strength in the left leg. When seen 1 year later, he had 5/5 power in all muscle groups, but he had persistent sensory loss in a left femoral nerve distribution and an absent left patellar reflex.

CASE 9.9 TINGLING AND PARALYSIS OF THE FOOT AFTER A FALL

CHIEF COMPLAINT

A 30-year-old woman presented to the emergency room after a fall with **tingling and paralysis of the right foot**.

HISTORY

Two days ago the patient slipped on a wet floor in the supermarket and fell backward, landing on her back. She initially noticed no symptoms, but she awoke at 3:00 A.M. to feed her 2-month-old baby and was **unable to move her right foot**. She also had a **tingling sensation in her right lateral lower leg and foot**. These symptoms did not resolve over the next 2 days, so she came to the emergency room. There was no back pain, and there were no bowel or bladder symptoms.

(continued on p. 378)

CASE 9.9 *(continued)*

PHYSICAL EXAMINATION

Vital signs: T = 98°F, P = 84, BP = 136/68.

Neck: Supple with no bruits.

Lungs: Clear.

Heart: Regular rate.

Abdomen: Soft.

Extremities: Normal.

Back and spine: No tenderness.

Rectal: Normal tone with no masses.

Neurologic exam:

MENTAL STATUS: Alert and oriented × 3. Fluent speech.

CRANIAL NERVES: Intact.

MOTOR: No drift. Normal tone, except for **diminished tone in the right foot**. 5/5 power throughout, except for **0/5 power in the right tibialis anterior, extensor hallucis longus, foot invertors, foot evertors, and gastrocnemius, and 3/5 power in the right hamstrings.**

REFLEXES:

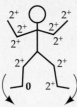

FIGURE 9.16 Region of Sensory Loss

COORDINATION: Normal on finger-to-nose testing.

GAIT: **Flailing movements of the right foot while raising it off the floor with each step.**

SENSORY: **Decreased light touch, pinprick, vibration, and joint position sense in the right lateral calf and in the entire right foot** (Figure 9.16)**.** Sensation otherwise normal.

LOCALIZATION AND DIFFERENTIAL DIAGNOSIS

On the basis of the symptoms and signs shown in **bold** above, where is the lesion? What is the most likely diagnosis?

Discussion

The key symptoms and signs in this case are:

- **Weakness of the right tibialis anterior, extensor hallucis longus, foot invertors, foot evertors, gastrocnemius, and hamstrings, with decreased right foot tone and absent right Achilles tendon reflex**
- **Paresthesias and decreased light touch, pinprick, vibration, and joint position sense in the right foot and lateral calf**

Right leg weakness, decreased tone, and hyporeflexia could be caused by a peripheral nerve lesion in the right leg or by an acute upper motor neuron lesion in the motor cortex or spinal cord (see KCC 6.1, 6.3; Figure 6.14F). The pattern of sensory loss is not consistent with a spinal cord lesion (see KCC 7.4) or cortical lesion (see KCC 7.3), but it does match that of a sciatic nerve lesion (see KCC 9.1; Table 9.3), as do the details of the motor exam. The most likely diagnosis is right sciatic neuropathy, probably caused by the fall, although the exact mechanism of injury is unclear.

Clinical Course and Neuroimaging

X-rays of the lumbosacral spine and pelvis revealed no fractures. A **lumbar plexus MRI** (Image 9.9A,B, page 380) revealed T2 bright signal in the right sciatic nerve, consistent with edema. The patient was treated with physical therapy and a right foot brace to aid ambulation. An EMG done 1 week after the onset of symptoms revealed inability to activate the muscles of the right sciatic nerve, including the gastrocnemius, tibialis anterior, flexor hallucis brevis, and medial hamstrings. A follow-up study was not done. The patient's strength gradually improved, and by 5 months after onset she felt almost back to normal. On exam 1 year after onset, she had 4/5 to 4+/5 strength in the right tibialis anterior, foot evertors, and invertors. Gastrocnemius strength was 5/5 bilaterally, and she had recovered sensation and the Achilles tendon reflex in the right foot.

CASE 9.10 A LEG INJURY RESULTING IN FOOT DROP

MINICASE

A 27-year-old man slipped on a wet tile floor and twisted his right foot toward the left, resulting in acute foot pain, followed by weakness. He was seen in the emergency room and had **0/5 power in his right tibialis anterior and extensor hallucis longus and 3/5 power in his right foot evertors.** Power was otherwise 5/5, including the right foot invertors and gastrocnemius. He had **decreased sensation to pinprick on the dorsum of the right foot, which was especially pronounced in the web space between the first and second toes** (Figure 9.17).

LOCALIZATION AND DIFFERENTIAL DIAGNOSIS

On the basis of the symptoms and signs shown in **bold** above, where is the lesion? What is the most likely diagnosis?

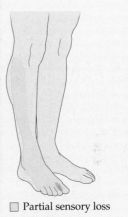

FIGURE 9.17 Region of Sensory Loss

☐ Partial sensory loss
■ Complete sensory loss

Discussion

The key symptoms and signs in this case are:

- **Weakness in the right tibialis anterior and extensor hallucis longus; moderate weakness of the right foot evertors**
- **Decreased sensation to pinprick on the dorsum of the right foot, especially between the first and second toes**

The pattern of weakness and sensory loss is consistent with injury to the common peroneal nerve (see Table 9.3; KCC 9.1; see also Table 8.1), which most likely occurred during the fall. Note that the deep peroneal nerve (tibialis anterior, extensor hallucis longus, sensation between first and second toes) appears to be more severely involved in this patient than the superficial peroneal nerve (foot eversion, sensation on dorsal foot and lateral shin). In addition, this should be distinguished from an L5 radiculopathy, in which there may also be weakness of foot inversion, not seen in the present case (compare to Case 8.10).

Clinical Course

An EMG (see KCC 9.2) 2 days after the injury showed abnormally low recruitment of motor unit action potentials in the right tibialis anterior, extensor hallucis longus, extensor digitorum brevis, and peroneus muscles, suggesting a neuropathic process. Motor nerve conduction studies showed decreased amplitudes when the right peroneal nerve was stimulated just above the fibular neck and normal amplitudes when stimulated just below the fibular neck, suggesting nerve injury at the fibular neck. The patient gradually improved over the following months.

CASE 9.9 TINGLING AND PARALYSIS OF THE FOOT AFTER A FALL

IMAGE 9.9A,B Abnormal Bright Signal in the Right Sciatic Nerve, Compatible with Sciatic Neuropathy MRI of the lumbar plexus. (A) Coronal T2-weighted MRI showing the plane of section for B. (B) Axial T2-weighted section showing abnormal bright signal in the right sciatic nerve as it passes dorsal to the femur.

(A)

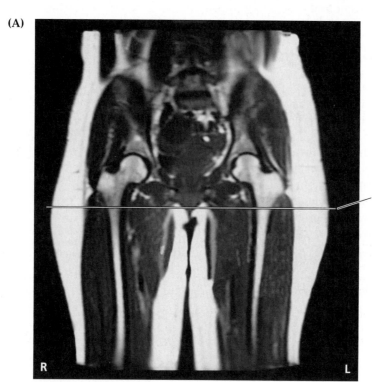

Plane of section for B

(B) Femur

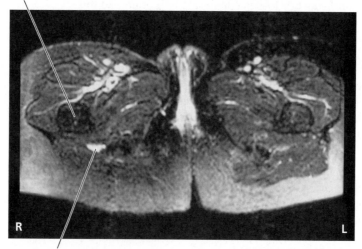

Bright signal in sciatic nerve

CASE 9.11 LATERAL THIGH PAIN AND NUMBNESS AFTER PREGNANCY

MINICASE

Two days after giving birth, a 24-year-old woman developed burning **pain and numbness in the right lateral thigh**, which was worsened by walking. Exam was normal except for a patch of **decreased sensation to light touch, pinprick, and cold on the right lateral thigh** (Figure 9.18). Importantly, her reflexes and motor strength were normal.

LOCALIZATION AND DIFFERENTIAL DIAGNOSIS

On the basis of the symptoms and signs shown in **bold** above, where is the lesion? What is the most likely diagnosis?

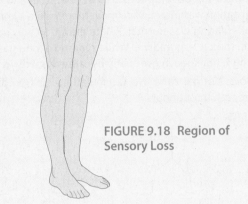

FIGURE 9.18 Region of Sensory Loss

Discussion

The key symptoms and signs in this case are:

- **Pain, paresthesias, and decreased sensation to light touch, pinprick, and cold on the right lateral thigh**

A purely sensory disorder affecting the lateral thigh is consistent with dysfunction of the lateral femoral cutaneous nerve, or meralgia paresthetica (see Figure 9.5; KCC 9.1). An L2 or L3 radiculopathy could be considered; however, there were no motor deficits, reflex abnormalities, or back pain to support this possibility.

Clinical Course

The patient was reassured that her symptoms were caused by injury to a sensory nerve that would likely improve with time. Her symptoms gradually resolved over the following 5 months and required no specific treatment.

CASE 9.12 DYSARTHRIA, PTOSIS, AND DECREASED EXERCISE TOLERANCE

CHIEF COMPLAINT

A 35-year-old woman saw a neurologist because of worsening dysarthria and muscle fatigue.

HISTORY

The patient worked as a nurse, and over the course of four months she noticed that at the end of her dictations, she had profound **difficulty enunciating her words**. This was **most apparent at the end of her work day**. Also, toward the end of her work day she had **difficulty producing a full smile**. Her symptoms disappeared with rest. She also noticed some mild neck discomfort and felt that it was **difficult at times to hold her head up**. In addition, she had **reduced exercise tolerance**, becoming short of breath sooner than previously when using the treadmill at her gym.

PHYSICAL EXAMINATION

Vital signs: T = 98°F, P = 80, BP = 90/70.

Neck: Supple, no bruits.

Lungs: Clear.

Heart: Regular rate.

Abdomen: Soft.

Extremities: Normal.

Neurologic exam:

MENTAL STATUS: Alert and oriented × 3. Fluent speech. Recalled 3/3 words after 5 minutes. Normal calculations.

CRANIAL NERVES: Intact visual fields and acuity. Pupils equal and reactive to light and accommodation. Extraocular movements intact with no nystagmus. **On prolonged**

(continued on p. 382)

CASE 9.12 *(continued)*

upgaze she developed ptosis of the left eyelid. Facial sensation was intact. Face movements were symmetrical. Hearing was normal. Palate elevation was normal and tongue was midline. **While reading a long passage aloud, her speech gradually became dysarthric.**

MOTOR: Normal tone. No fasciculations or tremor. 5/5 strength throughout.

REFLEXES:

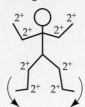

COORDINATION: Normal finger-to-nose and heel-to-shin testing.

SENSORY: Normal pinprick, temperature, vibration, and joint position sense. No extinction.

LOCALIZATION AND DIFFERENTIAL DIAGNOSIS

1. Based on the symptoms and signs shown in **bold** above, where is the lesion?
2. What is the most likely diagnosis?

Discussion

The key symptoms and signs in this case are:

- Dysarthria, worse at the end of prolonged speech
- Difficulty producing a full smile at the end of the day
- Ptosis of the left eye on prolonged upgaze
- Difficulty holding head up at the end of the day
- Reduced strength and shortness of breath on treadmill

1. Dysarthria can be caused by cranial nerve or upper motor neuron disorders (see KCC 12.8), as can facial weakness during smiling (see Figure 12.13). Ptosis can result from disorders of CN III supplying the levator palpebrae superior or from Horner's syndrome (see KCC 13.6). Weakness of the neck muscles, legs, and respiratory muscles, likewise, can have central or peripheral causes. However, the weakness in these multiple locations, in the absence of any sensory findings, would be unusual for a neuropathy. In addition, there are no upper motor neuron signs to suggest a multifocal disorder affecting the central nervous system. Therefore, a peripheral disorder affecting the neuromuscular junction or muscles involved in speech, eyelid elevation, respiratory muscles, and proximal respiratory, neck, and leg muscles is most likely.

2. This pattern of weakness with no sensory loss and intact reflexes, together with the fact that the weakness was worse toward the end of the day or with repeated use of the muscles, is most suggestive of myasthenia gravis (see KCC 8.1). Other possible causes of diffuse, slowly progressive weakness without sensory loss, reflex loss, or upper motor neuron signs include Lambert–Eaton syndrome and myopathic disorders (see KCC 8.1).

Diagnostic Studies and Clinical Course

The neurologist performed a "Tensilon" (edrophonium) test (see KCC 8.1) by evaluating the patient's ability to read a long passage aloud. Her dysarthria was markedly reduced after administration of edrophonium, so the test was considered positive. Acetylcholine receptor antibodies were also positive at 1.73 nmol/L (normal is less than 0.3 nmol/L). **Repetitive stimulation** (see KCC 9.2) of the ulnar nerve at 3 stimuli per second produced a 23%

decrement of the CMAP amplitude recorded over the abductor digiti minimi muscle (Image 9.12, page 385). Decrement of greater than 10% is considered abnormal and supports the diagnosis of myasthenia gravis (see KCC 9.2). Pulmonary function tests were normal. The patient underwent a chest CT, which revealed a 7×5 cm lobulated mass in the right anterior mediastinum extending over to the right side of the pericardium, consistent with a thymoma (see KCC 8.1). She was treated with the anticholinesterase medication pyridostigmine (Mestinon) and underwent surgical resection of the thymic mass, which was confirmed histologically to be a thymoma. Following surgery, her dysarthria and fatigue resolved completely, and she had a normal neurologic exam, including no dysarthria and no ptosis, even after prolonged upgaze.

CASE 9.13 GENERALIZED WEAKNESS AND AREFLEXIA

CHIEF COMPLAINT

A 70-year-old woman came to the emergency room because of progressive weakness, gait difficulty and shortness of breath.

HISTORY

The patient was well until about 2 weeks previously when she developed intermittent diarrhea. About 8 or 9 days prior to admission she noticed **weakness in her arms and legs**, and had a few falls. She was sent for physical therapy, but her weakness progressed so that for the past 4 days she was **unable to walk**. She also noticed **tingling in her feet and finger tips**. Finally, she developed **breathing problems** and came to the hospital.

PHYSICAL EXAMINATION

Vital signs: T = 98.2°F, P = 74, BP = 132/74, RR = 20.

Bedside pulmonary function tests: **Vital capacity 1.6 L** (normal is greater than ~3.5 L for adult females, ~4.5 L for males); **Negative inspiratory force –35 cm H$_2$O** (normal is larger than –80 cm H$_2$O).

Neck: Supple, no bruits.

Lungs: Clear.

Heart: Regular rate, no murmurs.

Abdomen: Normal bowel sounds, soft, nontender.

Extremities: Normal, no edema

Neurologic exam:

  MENTAL STATUS: Alert and oriented × 3. Normal language, attention and memory.

CRANIAL NERVES: Visual fields intact. Pupils equal and reactive to light. Extraocular movements intact. Facial sensation intact V$_1$ –V$_3$. **Facial movements appeared weak bilaterally. Palate and pharyngeal movements appeared weak. Weak shoulder shrug.** Tongue normal.

MOTOR: Normal bulk and tone. **Power in bilateral deltoids was 4/5, biceps 3/5, triceps 4$^-$/5, wrist extensors 3/5, finger abductors 4$^-$/5, hip flexors 3/5, knee flexors 4$^+$/5, foot dorsiflexors 4$^-$/5.**

REFLEXES:

GAIT: **Unable to stand unsupported.**

SENSORY: **Decreased vibration sense in both feet up to the ankles.** Pinprick intact.

LOCALIZATION AND DIFFERENTIAL DIAGNOSIS

1. Based on the symptoms and signs shown in **bold** above, where is the lesion?

2. What is the most likely diagnosis?

Discussion

The key symptoms and signs in this case are:

- **Progressive weakness of the face, palate, arms and legs leading to inability to walk**
- **Breathing difficulty, with reduced vital capacity and inspiratory force**
- **Absent reflexes**
- **Tingling in her fingertips and feet, and decreased vibration sense in both feet**

1. Weakness affecting bilateral cranial nerves, arms, legs and breathing muscles could be seen in a central disorder such as a brainstem lesion involving corticobulbar and corticospinal tracts, in widespread lower motor neuron disease, or in a diffuse disorder of the peripheral nerves (polyneuropathy), neuromuscular junctions, or muscles themselves (see KCC 6.3, Generalized Weakness or Paralysis). Cranial nerve involvement suggests the lesion is not in the spinal cord. An upper motor neuron lesion above the spinal cord is also unlikely since there is no hyperrflexia despite presence of deficits for over a week. In addition, the absent reflexes along with bilateral distal sensory involvement is not compatible with lower motor neuron disease, neuromuscular or muscle disorders since these conditions are not associated with sensory deficits, and do not usually cause reflex loss unless weakness is profound (dropped reflexes are typically due to sensory nerve involvement). Therefore, the most likely localization is a diffuse symmetric polyneuropathy.

2. Neuropathy can have many causes, but when it occurs days to weeks following an acute illness, with progressive motor greater than sensory involvement and absent reflexes, the most likely diagnosis is Guillain–Barré syndrome (see KCC 8.1), also known as acute inflammatory demyelinating polyneuropathy (AIDP). There are a few other causes of generalized rapidly progressive predominantly motor weakness including myasthenia gravis, heavy metal or organophosphate toxicity, diphtheria, botulism, Lyme polyradiculitis, porphyria, poliomyelitis, and tick paralysis. However, these other disorders can usually be distinguished from Guillain–Barré syndrome based on clinical features (see KCC 8.1) and diagnostic tests including nerve conduction studies and lumbar puncture (see KCC 5.10, 9.2).

Diagnostic Studies and Clinical Course

Although there appeared to be cranial nerve involvement, since the patient was elderly and had some degenerative disease of the spine, an urgent MRI of the cervical spine was performed to rule out spinal cord compression as the cause of her progressive weakness and sensory loss. The MRI did not reveal cord compression, enabling a **lumbar puncture** (see KCC 5.10) to be performed safely. Cerebral spinal fluid showed 2 white blood cells/mm^3 (67% lymphocytes and 31% monocytes), 2 red blood cells/mm^3, normal glucose of 69 mg/dl, and elevated protein of 115 mg/dl (see Table 5.7). This elevated CSF protein with normal cell count is referred to as **albuminocytologic dissociation** and is evidence of the autoimmune response against nerves and nerve roots characteristically seen in Guillain–Barré syndrome. **Nerve conductions studies** (see KCC 9.2) revealed markedly reduced conduction velocities, and somewhat reduced compound motor action potential (CMAP) amplitudes, compatible with greater demyelination than axonal loss, and confirmed the diagnosis of Guillain–Barré syndrome.

The patient was treated with a course of intravenous immunoglobulin daily for five days, but by the second hospital day, her vital capacity had dropped to 800cc and negative inspiratory force to –20 mm H$_2$O, so she was electively intubated. Her strength gradually improved, and by hospital day 5 she was successfully extubated and was able to breathe on her own. By the time of discharge from the hospital, 9 days after admission, her strength was 4/5 to 4$^+$/5 in the upper and lower extremities, and she no longer had diminished vibration sense in the feet. She was last seen as an outpatient seven months later and had normal nerve conduction velocities, and a normal neurological exam including 5/5 strength throughout, 2$^+$ reflexes and a normal gait.

CASE 9.12 DYSARTHRIA, PTOSIS, AND DECREASED EXERCISE TOLERANCE

IMAGE 9.12 Decrement on Repetitive Stimulation
Repetitive stimulation testing was performed by stimulating the ulnar nerve of the right arm and recording the CMAP over the right abductor digiti minimi muscle (see Figure 9.9; KCC 9.2). Stimulation was repeated at 3 per second for a total of 9 stimuli. Successive stimuli are shown displaced sequentially to the right (3 ms displacement per stimulus) to allow comparison of amplitudes of successive CMAPs: The first CMAP in the series is displayed farthest to the left, and the ninth CMAP is displayed farthest to the right. There was a decrement in CMAP amplitude of 23%.

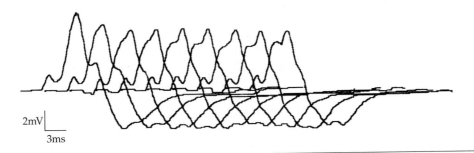

2mV | 3ms

CASE 9.14 MYSTERIOUS WEAKNESS AFTER DINNER*

MINICASE

One evening after dinner, a 58 year old woman began to have **swallowing difficulties** and **double vision**. Over the next few hours she developed **ptosis, facial weakness**, and **difficulty breathing** so her family brought her to the emergency room. Examination was notable for normal mental status, **severely limited horizontal and vertical eye movements with normally reactive pupils, facial diplegia, dysarthria, weakness of the bilateral arms and legs worse proximally than distally, 1$^+$ patellar reflexes but otherwise undetectable deep tendon reflexes**, and a normal sensory exam. **Vital capacity was 600 cc** (normal is greater than ~3.5 L for an adult woman) so she was intubated.

LOCALIZATION AND DIFFERENTIAL DIAGNOSIS

On the basis of the symptoms and signs shown in **bold** above, where is the lesion? What are some possibilities for the diagnosis?

---

*A description of this patient was published previously in Shapiro BE, Soto O, Shafqat S and Blumenfeld H. 1997. *Muscle and Nerve*, 20: 100–102.

Discussion

The key symptoms and signs in this case are:

- Diplopia, limited horizontal and vertical eye movements, ptosis, facial diplegia, dysarthria and dysphagia
- Proximal arm and leg weakness
- Breathing difficulty, with reduced vital capacity
- Diminished reflexes

Generalized weakness without sensory loss can be caused by a brainstem lesion (see KCC 14.1) or by a widespread disorder of the lower motor neurons, peripheral nerves, neuromuscular junctions, or muscles (see KCC 6.3, Generalized Weakness or Paralysis). Proximal greater than distal weakness suggests a muscle or neuromuscular junction disorder, but the diminished reflexes suggest a possible acute polyneuropathy such as Guillain–Barré syndrome (see KCC 8.1). Another possibility is an acute upper motor neuron lesion, which can sometimes be associated with decrease rather than increased reflexes (see Table 6.4). In summary, the differential diagnosis is large and includes an acute brainstem lesion such as infarct or hemorrhage, and rapidly progressive peripheral disorders such as Guillain–Barré syndrome, myasthenia gravis, heavy metal or organophosphate toxicity, botulism, porphyria, poliomyelitis, and tick paralysis (see KCC 8.1).

Clinical Course

The patient was intubated and admitted to the intensive care unit. MRI with MRA did not reveal any brainstem abnormalities, routine blood tests and cerebrospinal fluid were normal. The family and patient repeatedly denied any possible toxin exposure. **Nerve conduction studies** demonstrated normal conduction velocities but markedly diminished compound motor action potential (CMAP) amplitudes (see KCC 9.2). In addition, fast repetitive stimulation or strong voluntary muscle contraction caused CMAP amplitudes to increase 2- to 3-fold. Increment of this kind is usually seen in presynaptic disorders of the neuromuscular junction, such as Lambert-Eaton myasthenic syndrome or botulism (see KCC 8.1; 9.2).

On further questioning, the family admitted to operating a home canning operation. They revealed that the patient had prepared a spaghetti dinner using their canned tomato sauce on the day of admission. Stool specimens and residual sauce brought in by the family were positive for botulinum toxin type B. She was treated with botulinum antitoxin, and had a prolonged course in the intensive care unit, but eventually recovered fully. When she was extubated and able to talk again, she explained that she had opened a dented can, and the sauce did not smell right, so she tried to feed it to the dog. The dog refused to eat it (wise choice), but the patient tasted some of the raw sauce which seemed alright, so she then cooked it and fed the whole family a spaghetti dinner. Fortunately, she cooked the sauce long enough to inactivate the toxin.

Additional Cases

Other chapters describe related cases for the following topics: **radiculopathy** (Cases 8.1–8.11); **distal symmetric polyneuropathy** (Cases 6.5 and 10.3); and **cranial neuropathy** (Cases 12.2–12.7, 13.1–13.3, and 13.5). Other relevant cases can also be found using the **Case Index** located at the end of this book.

Brief Anatomical Study Guide

1. The **brachial plexus** arises from **C5 through T1** (see Figure 9.2), while the **lumbosacral plexus** arises from **L1 through S4** (see Figure 9.4).

2. The most clinically important nerves in the upper extremity are the **radial, median, ulnar, axillary, and musculocutaneous nerves**; Table 9.1 summarizes the sensory and motor functions of these nerves.

3. The most important nerves in the lower extremity are the **femoral, obturator, sciatic, tibial, and peroneal nerves**; Table 9.3 summarizes the sensory and motor functions of these nerves.

References

General References

Aids to the Examination of the Peripheral Nervous System. 1986. Bailliere Tindall on behalf of the Guarantors of Brain, London.

Dawson DM, Hallett M, Wilbourn AJ, Campbell WW, Terrono AL, Millender LH. 1999. *Entrapment Neuropathies*. 3rd Ed. Lippincott Williams & Wilkins, New York.

Devinsky O, Feldmann E. 1988. *Examination of the Cranial and Peripheral Nerves*. Churchill Livingstone, New York.

Deymeer FS (ed.). 2000. *Neuromuscular Disease: From Basic Mechanisms to Clinical Management*. (*Monogr Clin Neurosci*, vol. 18). S Karger AG, New York.

Massey EW. 1998. Sensory Mononeuropathies. *Seminars in Neurology* 8 (2): 177–183.

Moore KL, Dalley AF. 2005. *Clinically Oriented Anatomy*. 5th Ed. Lippincott Williams & Wilkins, Philadelphia.

Preston DC, Shapiro BE. 2005. *Electromyography and Neuromuscular Disorders: Clinical–Electrophysiologic Correlations*. 2nd Ed. Butterworth-Heinemann, Boston.

Salter RB. 1999. *Textbook of Disorders and Injuries of the Musculoskeletal System*. 3rd Ed. Williams & Wilkins, Baltimore, MD.

Upper Extremity

Anto C, Aradhya P. 1996. Clinical diagnosis of peripheral nerve compression in the upper extremity. *Orthop Clin North Am* 27 (2): 227–236.

Colbert SH, Mackinnon SE. 2008. Nerve compressions in the upper extremity. *Mo Med* 105 (6): 527–535.

Brachial Plexus

Arcasoy SM, Jett JR. 1997. Superior pulmonary sulcus tumors and Pancoast's syndrome. *N Engl J Med* 337 (19): 1370–1376.

Blaauw G, Muhlig RS, Vredeveld JW. 2008. Management of brachial plexus injuries. *Adv Tech Stand Neurosurg* 33: 201–231.

Kawai H, Kawabata. 2000. *Brachial Plexus Palsy*. World Science Publishing Company.

Sandmire HF, DeMott RK. 2008. Newborn brachial plexus palsy. *J Obstet Gynaecol* 28 (6): 567–572.

Zafeiriou DI, Psychogiou K. 2008. Obstetrical brachial plexus palsy. *Pediatr Neurol* 38 (4): 235–242.

Median Nerve

Katz JN, Simmons BP. 2002. Carpal Tunnel Syndrome. *N Engl J Med* 346: 1807.

Phalen GS, Kendrick JI. 1957. Compression neuropathy of the median nerve in the carpal tunnel. *JAMA* 164: 524.

Wertsch JJ, Melvin J. 1982. Median nerve anatomy and entrapment syndromes: A review. *Arch Phys Med Rehabil* 63 (12): 623–627.

Radial Nerve

Kleinert JM, Mehta S. 1996. Radial nerve entrapment. *Orthop Clin North Am* 27 (2): 305–315.

Massey EW, Pleet AB. 1978. Handcuffs and Cheiralgia Paresthetica. *Neurology* 28 (12): 1312–1313.

Ulnar Nerve

Khoo D, Carmichael SW, Spinner RJ. 1996. Ulnar nerve anatomy and compression. *Orthop Clin North Am* 27 (2): 317–338.

Shea JD, McClain EJ. 1969. Ulnar-nerve compression syndromes at and below the wrist. *J Bone Joint Surg* 51 (6): 1095–1103.

Vanderpool DW, Chalmers J, Lamb DW, Whiston TB. 1968. Peripheral compression lesions of the ulnar nerve. *J Bone Joint Surg* 50 (4): 792–803.

Sciatic Nerve

Fassler PR, Swiontkowski MF, Kilroy AW, Routt ML, Jr. 1993. Injury of the sciatic nerve associated with acetubular fracture. *J Bone Joint Surg (Am)* 75 (8): 1157–1166.

Johnson ME, Foster L, DeLee JC. 2008. Neurologic and vascular injuries associated with knee ligament injuries. *Am J Sports Med* 36 (12): 2448–2462.

Peroneal Nerve

Berry H, Richardson PM. 1976. Common peroneal nerve palsy: A clinical and electrophysiological review. *J Neurol Neurosurg Psychiatry* 39 (12): 1162–1171.

Masakado Y, Kawakami M, Suzuki K, Abe L, Ota T, Kimura A. 2008. Clinical neurophysiology in the diagnosis of peroneal nerve palsy. *Keio J Med* 57 (2): 84–89.

Meralgia Paresthetica

Harney D, Patijn J. 2007. Meralgia paresthetica: diagnosis and management strategies. *Pain Med* 8 (8): 669–677.

Kitchen C, Simpson J. 1972. Meralgia paresthetica: A review of 67 patients. *Acta Neurol Scand* 48 (5): 547–555.

Nouraei SA, Anand B, Spink G, O'Neill KS. 2007. A novel approach to the diagnosis and management of meralgia paresthetica. *Neurosurgery* 60 (4): 696–-700.

Sarala PK, Nishihara T, Oh SJ. 1979. Meralgia paresthetica: Electrophysiologic study. *Arch Phys Med Rehabil* 60 (1): 30–31.

CONTENTS

Chapter 10

Cerebral Hemispheres and Vascular Supply

A 45-year-old man with a history of cigarette smoking and hypertension staggered into a diner grunting incoherently, then tripped and fell to the floor. The manager called an ambulance, and in the emergency room the patient was found to have severe right face and arm weakness. Six days later, he was still able to utter only a few barely articulate words, but he could follow many simple commands and answered yes/no questions appropriately.

Each area of the cerebral cortex belongs to a specific vascular "territory," and this patient's symptoms are typical of cortical damage to one such territory. To diagnose and treat patients with these disorders, we must learn about the local functions of different parts of the cerebrum and their associated blood supply—the focus of this chapter.

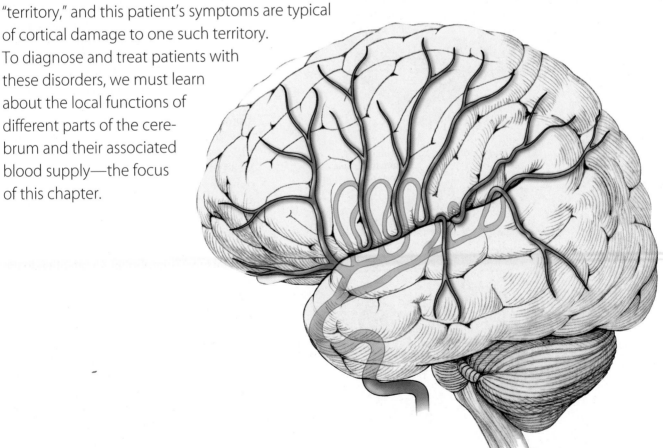

ANATOMICAL AND CLINICAL REVIEW

W̲HEN WE THINK OF THE HUMAN BRAIN, we usually envision the cerebral hemispheres. In this chapter, we will review the functional anatomy of the cerebral hemispheres by studying their blood supply and the clinical–anatomical correlations that can be made when the blood supply is transiently or permanently disrupted. We will discuss the blood supply and functional anatomy of the brainstem and cerebellum in Chapters 14 and 15; the spinal cord blood supply was discussed in Chapter 6. Understanding the blood supply of the brain provides a useful review of regional brain anatomy, since blood vessel territories typically overlap several spatially adjacent functional systems. In addition, knowledge of blood vessel territories is clinically useful, since it enables the localization of common stroke syndromes on clinical grounds, allowing the prompt initiation of proper diagnostic and therapeutic interventions.

Review of Main Functional Areas of Cerebral Cortex

We will now briefly review the main functional areas of the cerebral cortex that are commonly affected by cerebral infarcts (**Figure 10.1**). Additional details can be found in other chapters that discuss motor, somatosensory, visual, and association cortex at greater length (see Chapters 2, 6, 7, 11, and 19.)

Recall that the face and hand areas of the sensorimotor homunculi are on the lateral convexities, while the leg areas are in the interhemispheric fissure (see Figure 6.2). In the dominant, usually left hemisphere, Broca's area lies in the inferior frontal gyrus, just anterior to the articulatory areas of the primary motor cortex, a location well suited for planning the articulatory program (see Figure 10.1A; see also Figure 19.2 and KCC 19.4). Meanwhile, Wernicke's area lies in the superior temporal gyrus, adjacent to the primary auditory cortex (see also Figure 12.16) and is involved in language processing.

Association cortex in the nondominant, usually right, hemisphere (especially the right parietal lobe) is important for attention to the contralateral body and space. Primary visual cortex for the contralateral visual hemifield lies along the calcarine fissure of the occipital lobe (see Figure 10.1B; see also Figure 11.15). The optic radiations, white matter pathways carrying visual information from the thalamus to the visual cortex, pass under the parietal and temporal cortex (see Figure 10.1A); they can be damaged in infarcts of these lobes, causing contralateral visual field deficits.

FIGURE 10.1 Some Important Functional Areas of the Cerebral Cortex
(A) Lateral view of left hemisphere.
(B) Medial view of right hemisphere.

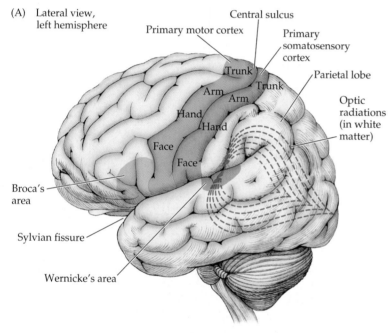

(A)   Lateral view, left hemisphere

Central sulcus
Primary motor cortex
Primary somatosensory cortex
Parietal lobe
Optic radiations (in white matter)
Trunk
Trunk
Arm
Arm
Hand
Hand
Face
Face
Broca's area
Sylvian fissure
Wernicke's area

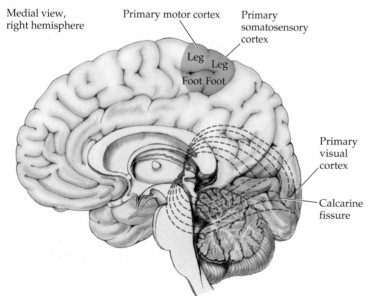

(B)   Medial view, right hemisphere

Primary motor cortex
Primary somatosensory cortex
Leg
Leg
Foot Foot
Primary visual cortex
Calcarine fissure

Circle of Willis: Anterior and Posterior Circulations

The arterial supply to the cerebral hemispheres is derived from the **anterior circulation** provided by the bilaterally paired **internal carotid arteries**, as well as by the **posterior circulation** provided by the bilateral **vertebral arteries** (Figure 10.2). The anterior circulation arises from the common carotid arteries originating at the aorta or brachiocephalic arteries (see Figure 4.20A). At the carotid bifurcation, the common carotid splits, forming the internal carotid and external carotid arteries (see Figure 4.19). The vertebral arteries, which supply the posterior circulation, arise from the subclavian arteries (see Figure 4.20B) and then ascend (see Figure 4.19) through foramina in the transverse processes of the cervical vertebrae (foramina transversaria; see Figure 10.2B) before entering the foramen magnum and joining to form the basilar artery. These anterior and posterior circulations meet in an anastomotic ring called the **circle of Willis**, from which all major cerebral vessels arise (Figure 10.3). The circle of Willis provides abundant opportunities for collateral flow; however, anatomical variants are common, and a complete full-caliber ring is present in only approximately 34% of individuals. The main arteries supplying the cerebral hemispheres are the anterior, middle, and posterior cerebral arteries. The **anterior cerebral arteries** (**ACAs**) and **middle cerebral arteries** (**MCAs**) are the terminal branches of the internal carotid arteries. The anterior cerebral arteries anastomose anteriorly at the **anterior communicating artery** (**AComm**). The anterior and posterior circulations are linked to each other via the **posterior communicating arteries** (**PComms**), which connect the internal carotids to the posterior cerebral arteries, thereby joining the ante-

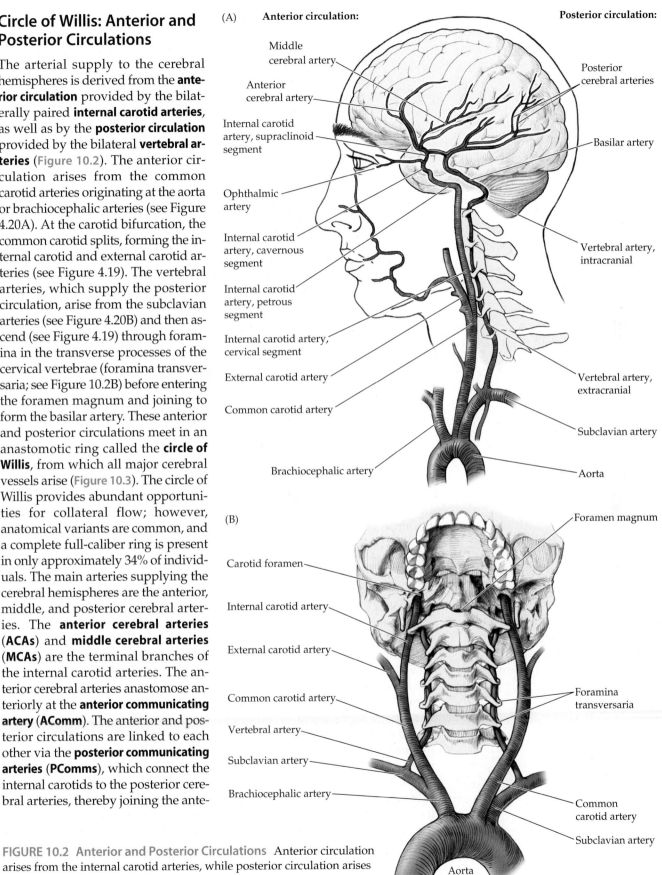

(A)   **Anterior circulation:**   **Posterior circulation:**

Middle cerebral artery
Anterior cerebral artery
Internal carotid artery, supraclinoid segment
Ophthalmic artery
Internal carotid artery, cavernous segment
Internal carotid artery, petrous segment
Internal carotid artery, cervical segment
External carotid artery
Common carotid artery
Brachiocephalic artery

Posterior cerebral arteries
Basilar artery
Vertebral artery, intracranial
Vertebral artery, extracranial
Subclavian artery
Aorta

(B)

Carotid foramen
Internal carotid artery
External carotid artery
Common carotid artery
Vertebral artery
Subclavian artery
Brachiocephalic artery

Foramen magnum
Foramina transversaria
Common carotid artery
Subclavian artery
Aorta

FIGURE 10.2 Anterior and Posterior Circulations Anterior circulation arises from the internal carotid arteries, while posterior circulation arises from the vertebral arteries.

FIGURE 10.3 Circle of Willis and Its Main Branches

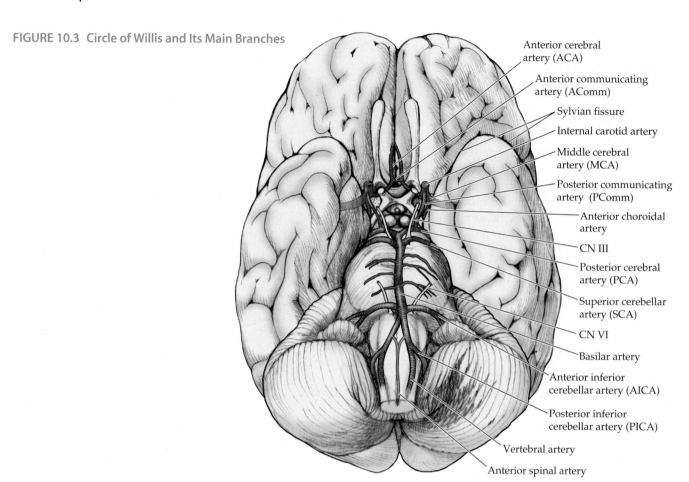

Anterior cerebral
artery (ACA)

Anterior communicating
artery (AComm)

Sylvian fissure

Internal carotid artery

Middle cerebral
artery (MCA)

Posterior communicating
artery (PComm)

Anterior choroidal
artery

CN III

Posterior cerebral
artery (PCA)

Superior cerebellar
artery (SCA)

CN VI

Basilar artery

Anterior inferior
cerebellar artery (AICA)

Posterior inferior
cerebellar artery (PICA)

Vertebral artery

Anterior spinal artery

REVIEW EXERCISE

In the conventional angiograms and magnetic resonance angiograms of Figures 4.16–4.18, cover the labels and identify as many structures as possible. Identify especially the internal carotid arteries, vertebral arteries, basilar artery, ACA, MCA, PCA, AComm, and PComm.

MNEMONIC

rior and posterior circulations. The **posterior cerebral arteries (PCAs)** arise from the top of the basilar artery, which in turn is formed by the convergence of the two vertebral arteries. In addition to the posterior cerebral arteries, several branches to the brainstem and cerebellum arise from the vertebrobasilar system, as we will discuss in Chapters 14 and 15.

The internal carotid artery has several named segments during its course (see Figure 10.2). These can be well visualized in the angiogram images in Figures 4.16A,C and 4.18B. First comes the relatively vertical **cervical segment** in the neck, followed by a sharp horizontal bend as the internal carotid enters the temporal bone as the **petrous segment**. Next comes the **cavernous segment** as the internal carotid begins an *S*-shaped turn, also known as the **carotid siphon**, within the cavernous sinus (see Figure 13.11). It then passes the anterior clinoid process (see Figure 5.2B) to pierce the dura and bends posteriorly to enter the subarachnoid space as the **supraclinoid**, or **intracranial segment** (see Figure 4.16C). Although there are several smaller branches, the main branches of the supraclinoid internal carotid artery can be remembered by the mnemonic **OPAAM** (if you can remember "OPAAM"), which stands for the **O**phthalmic, **P**osterior communicating, **A**nterior choroidal, **A**nterior cerebral, and **M**iddle cerebral arteries. The **ophthalmic artery** usually arises from the bend in the internal carotid just after it enters the dura (see Figure 4.16A,C). The ophthalmic artery enters the optic foramen with the optic nerve and provides the main blood supply to the retina.

Sometimes, in an alternative nomenclature, the terms **A1**, **M1**, and **P1** are used for the initial segments of the ACA, MCA, and PCA, respectively, and second- and third-order branches are referred to as A2, A3, and so on.

Anatomy and Vascular Territories of the Three Main Cerebral Arteries

The three main cerebral arteries (ACA, MCA, and PCA) give rise to numerous branches that travel in the subarachnoid space over the surface of the brain and into the sulci. Small, penetrating branches arise from these vessels to supply the superficial portions of the brain, including the cortex and underlying white matter. The deep structures of the brain, such as basal ganglia, thalamus, and internal capsule, are supplied by small, penetrating branches that arise from the initial segments of the main cerebral arteries near the circle of Willis at the base of the brain. We will now review the vascular territories of the superficial and deep structures of the cerebral hemispheres.

Vascular Territories of the Superficial Cerebral Structures

The **anterior cerebral artery** passes forward to travel in the interhemispheric fissure as it sweeps back and over the corpus callosum (Figure 10.4). Two major branches commonly seen are the pericallosal and callosomarginal arteries (see Figures 4.16C and 4.18B). The anterior cerebral artery thus supplies most of the cortex on the anterior medial surface of the brain, from the frontal to the anterior parietal lobes (Figure 10.5), usually including the medial sensorimotor cortex (see Figure 10.1B).

The **middle cerebral artery** turns laterally to enter the depths of the Sylvian fissure (see Figure 10.3). Within the Sylvian fissure it usually bifurcates into the **superior division** and the **inferior division** (Figure 10.6). This is somewhat variable, and sometimes there are three or even four main branches of the middle cerebral artery. The branches of the middle cerebral artery form loops as they pass over the insula and then around and over the operculum to exit the Sylvian fissure onto

REVIEW EXERCISE

What are the three main arteries of the cerebral hemispheres (see Figure 10.2)? Which arise from the anterior circulation and which arise from the posterior circulation in most individuals?

FIGURE 10.4 Anterior Cerebral Artery (ACA) and Posterior Cerebral Artery (PCA) Simplified course of the main ACA and PCA branches is shown on the medial surface of the brain.

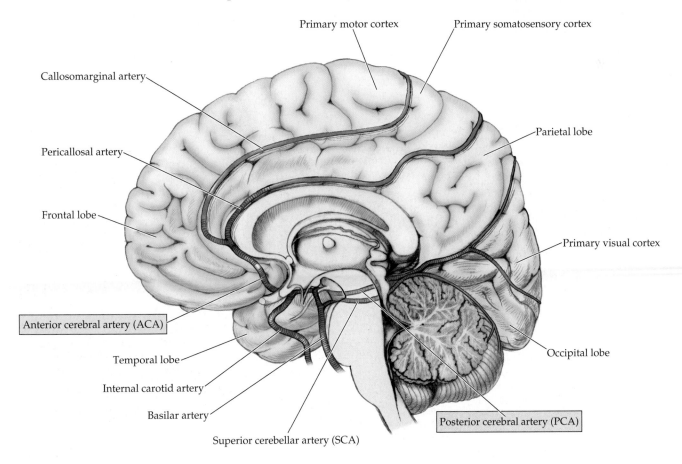

Primary motor cortex

Primary somatosensory cortex

Callosomarginal artery

Parietal lobe

Pericallosal artery

Frontal lobe

Primary visual cortex

Anterior cerebral artery (ACA)

Temporal lobe

Occipital lobe

Internal carotid artery

Posterior cerebral artery (PCA)

Basilar artery

Superior cerebellar artery (SCA)

(A)

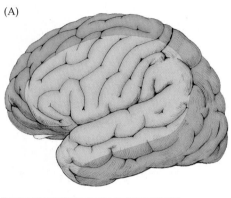

(B)

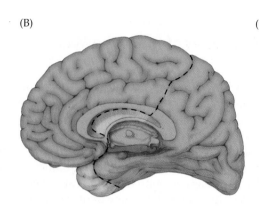

(C)

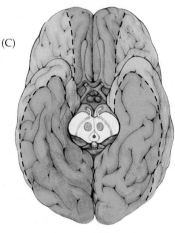

Key
- Anterior cerebral artery
- Middle cerebral artery
- Posterior cerebral artery

FIGURE 10.5 Regions of Cortex Supplied by the Anterior Cerebral Artery (ACA), Middle Cerebral Artery (MCA), and Posterior Cerebral Artery (PCA) (A) Lateral view. (B) Medial view. (C) Inferior view.

REVIEW EXERCISE

Draw the vascular territories for the three main cerebral arteries (ACA, MCA, and PCA) on the brain surfaces below. Refer to Figure 10.1 to deduce the deficits seen in infarcts of each of these three territories. This information has many important clinical applications (see KCC 10.1).

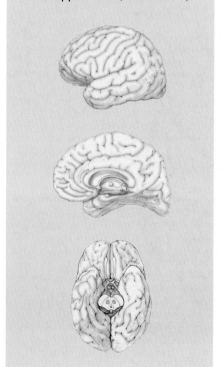

the lateral convexity (see Figures 4.16A,C and 4.18B). The superior division supplies the cortex above the Sylvian fissure, including the lateral frontal lobe and usually including the peri-Rolandic cortex (see Figure 10.6). The inferior division supplies the cortex below the Sylvian fissure, including the lateral temporal lobe and a variable portion of the parietal lobe. The middle cerebral artery thus supplies most of the cortex on the dorsolateral convexity of the brain (see Figure 10.5).

The **posterior cerebral artery** curves back after arising from the top of the basilar and sends branches over the inferior and medial temporal lobes and over the medial occipital cortex (see Figures 4.17, 4.18, and 10.4). The posterior cerebral artery territory therefore includes the inferior and medial temporal and occipital cortex (see Figure 10.5).

Vascular Territories of the Deep Cerebral Structures

The most important penetrating vessels at the base of the brain are the **lenticulostriate arteries**. These small vessels arise from the initial portions of the mid-

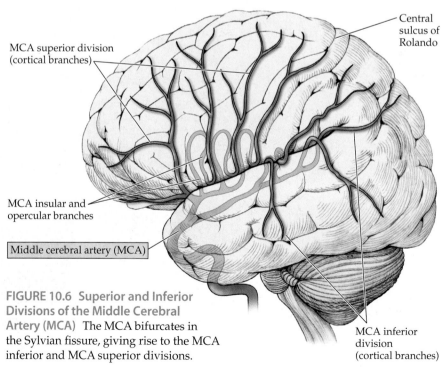

FIGURE 10.6 Superior and Inferior Divisions of the Middle Cerebral Artery (MCA) The MCA bifurcates in the Sylvian fissure, giving rise to the MCA inferior and MCA superior divisions.

MCA superior division (cortical branches)

Central sulcus of Rolando

MCA insular and opercular branches

Middle cerebral artery (MCA)

MCA inferior division (cortical branches)

dle cerebral artery before it enters the Sylvian fissure (Figure 10.7), and they penetrate the anterior perforated substance (see Figure 2.11C) to supply large regions of the basal ganglia and internal capsule (Figure 10.8). In hypertension, the lenticulostriate arteries and other similar small vessels are particularly prone to narrowing, which can lead to lacunar infarction (see KCC 10.4), as well as to rupture, causing intracerebral hemorrhage (see KCC 5.6).

Other small vessels supply deep structures as well, with some variability (see Figures 10.7 and 10.8). The **anterior choroidal artery** arises from the internal carotid artery (see Figure 10.3). Its territory includes portions of the globus pallidus, putamen, thalamus (sometimes involving part of the lateral geniculate nucleus), and the posterior limb of the internal capsule (see Figure 6.9B) extending up to the lateral ventricle (see Figures 10.8 and 10.9B). Recall that the posterior limb of the internal capsule contains important motor pathways through the corticobulbar and corticospinal tracts (see Figures 6.9 and 6.10). Thus, lacunar infarction in either the lenticulostriate or anterior choroidal territories often causes contralateral hemiparesis. The **recurrent artery of Heubner** comes off the initial portion of the anterior cerebral artery to supply portions of the head of the caudate, anterior putamen, globus pallidus, and internal capsule (see Figures 10.7–10.9). Other variable branches may also come off the initial portions of the anterior cerebral arteries to supply deep structures. Small, penetrating arteries that arise from the proximal posterior cerebral arteries near the top of the basilar artery include the **thalamoperforator arteries** (see Figure 10.8) (as well as the thalamogeniculate and posterior choroidal arteries), which supply the thalamus and sometimes extend to a portion of the posterior limb of the internal capsule. As we will see in Chapter 14, small, penetrating vessels arising from the top of the basilar artery also supply the midbrain (see Figure 14.21A).

The superficial and deep territories of the main cerebral arteries are summarized in coronal and axial sections in Figure 10.9.

REVIEW EXERCISE

In the angiographic images shown in Figures 4.16B and 4.17B, identify (1) the lenticulostriate arteries arising from the middle cerebral artery, (2) the recurrent artery of Heubner arising from the anterior cerebral artery, and (3) the thalamoperforator and posterior choroidal arteries arising from the posterior cerebral arteries.

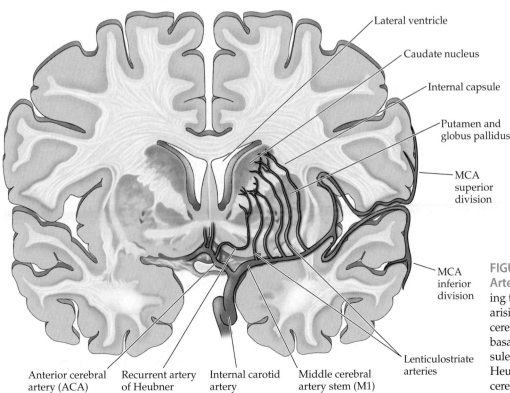

Lateral ventricle

Caudate nucleus

Internal capsule

Putamen and globus pallidus

MCA superior division

MCA inferior division

Lenticulostriate arteries

Anterior cerebral artery (ACA)

Recurrent artery of Heubner

Internal carotid artery

Middle cerebral artery stem (M1)

FIGURE 10.7 Lenticulostriate Arteries Coronal section showing the lenticulostriate arteries arising from the proximal middle cerebral artery and supplying the basal ganglia and internal capsule. The recurrent artery of Heubner arises from the anterior cerebral artery.

FIGURE 10.8 Blood Supply to Deep Cerebral Structures (A) Blood vessels supplying the basal ganglia and thalamus. (B) Blood supply to the internal capsule and globus pallidus.

(A) Blood vessels supplying the basal ganglia and thalamus

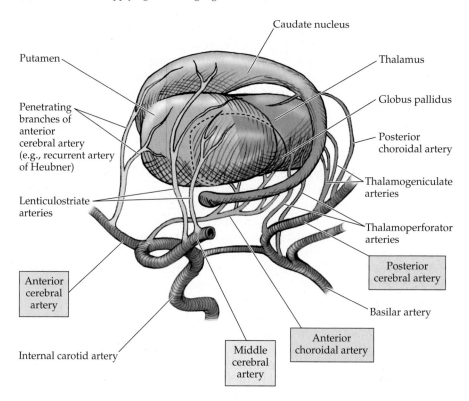

(B) Blood supply to the internal capsule and globus pallidus

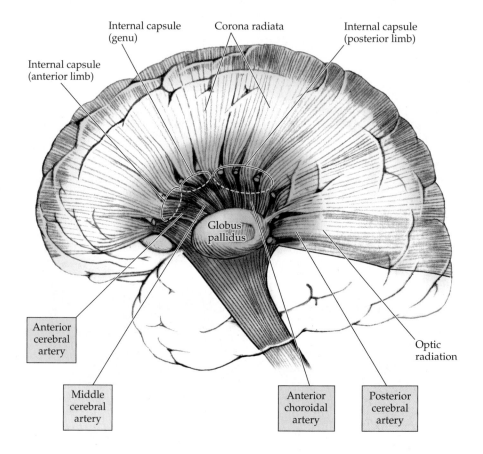

(A)

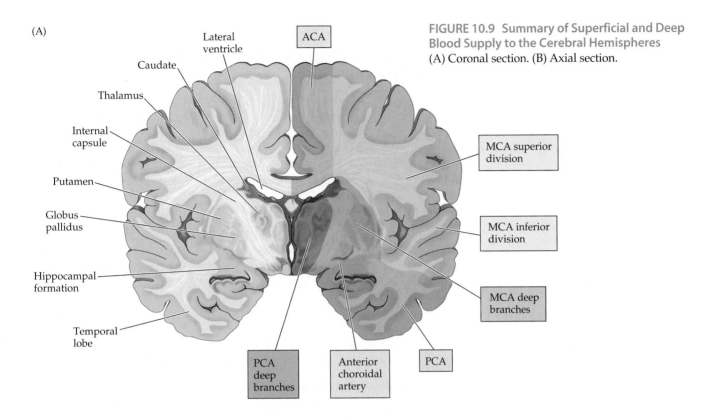

FIGURE 10.9 Summary of Superficial and Deep Blood Supply to the Cerebral Hemispheres (A) Coronal section. (B) Axial section.

(B)

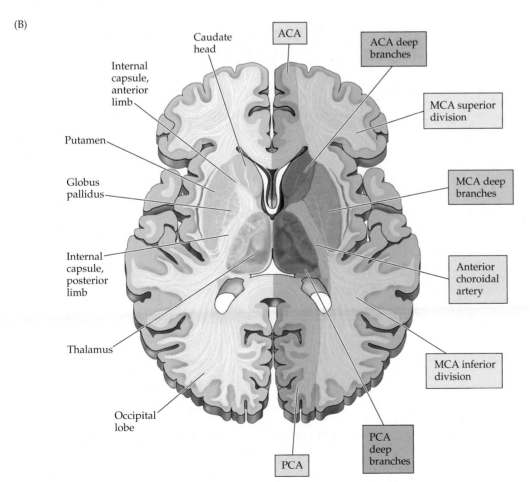

KEY CLINICAL CONCEPT

10.1 CLINICAL SYNDROMES OF THE THREE CEREBRAL ARTERIES

Recognition of the classic syndromes produced by infarcts of the middle cerebral artery (MCA), anterior cerebral artery (ACA), and posterior cerebral artery (PCA) territories remains a cornerstone of neurologic assessment and continues to play an important role in evaluating patients with acute stroke. We discuss localization of these syndromes here; stroke pathophysiology and clinical management are discussed in KCC 10.3 and KCC 10.4.

Middle Cerebral Artery (MCA)

Infarcts and ischemic events are more common in the middle cerebral artery than in the anterior or posterior cerebral arteries, at least in part because of the relatively large territory supplied by the middle cerebral artery. MCA infarcts occur in the following three general regions (see Figures 10.6, 10.7, and 10.9):

1. Superior division
2. Inferior division
3. Deep territory

Proximal MCA occlusions affecting all three of these regions are called MCA stem infarcts. The most common deficits seen with infarcts of left or right MCA territories are summarized in Table 10.1. Knowledge of the deficits associated with each of these territories is clinically useful since

TABLE 10.1 Major Clinical Syndromes of the MCA, ACA, and PCA Territories

LOCATION OF INFARCT	AFFECTED TERRITORY	DEFICITS[a]
Left MCA superior division		Right face and arm weakness of the upper motor neuron type and a nonfluent, or Broca's, aphasia. In some cases there may also be some right face and arm cortical-type sensory loss.
Left MCA inferior division		Fluent, or Wernicke's, aphasia and a right visual field deficit. There may also be some right face and arm cortical-type sensory loss. Motor findings are usually absent, and patients may initially seem confused or crazy but otherwise intact, unless carefully examined. Some mild right-sided weakness may be present, especially at the onset of symptoms.
Left MCA deep territory		Right pure motor hemiparesis of the upper motor neuron type. Larger infarcts may produce "cortical" deficits, such as aphasia as well.
Left MCA stem		Combination of the above, with right hemiplegia, right hemianesthesia, right homonymous hemianopia, and global aphasia. There is often a left gaze preference, especially at the onset, caused by damage to left hemisphere cortical areas important for driving the eyes to the right.

TABLE 10.1 (*continued*)

LOCATION OF INFARCT	AFFECTED TERRITORY	DEFICITS[a]
Right MCA superior division		Left face and arm weakness of the upper motor neuron type. Left hemineglect is present to a variable extent. In some cases, there may also be some left face and arm cortical-type sensory loss.
Right MCA inferior division		Profound left hemineglect. Left visual field and somatosensory deficits are often present; however, these may be difficult to test convincingly because of the neglect. Motor neglect with decreased voluntary or spontaneous initiation of movements on the left side can also occur. However, even patients with left motor neglect usually have normal strength on the left side, as evidenced by occasional spontaneous movements or purposeful withdrawal from pain. Some mild, left-sided weakness may be present. There is often a right gaze preference, especially at onset.
Right MCA deep territory	L ... R	Left pure motor hemiparesis of the upper motor neuron type. Larger infarcts may produce "cortical" deficits, such as left hemineglect as well.
Right MCA stem	L ... R	Combination of the above, with left hemiplegia, left hemianesthesia, left homonymous hemianopia, and profound left hemineglect. There is usually a right gaze preference, especially at the onset, caused by damage to right hemisphere cortical areas important for driving the eyes to the left.
Left ACA	L ... R	Right leg weakness of the upper motor neuron type and right leg cortical-type sensory loss. Grasp reflex, frontal lobe behavioral abnormalities, and transcortical aphasia can also be seen. Larger infarcts may cause right hemiplegia.
Right ACA	L ... R	Left leg weakness of the upper motor neuron type and left leg cortical-type sensory loss. Grasp reflex, frontal lobe behavioral abnormalities, and left hemineglect can also be seen. Larger infarcts may cause left hemiplegia.
Left PCA	L ... R	Right homonymous hemianopia. Extension to the splenium of the corpus callosum can cause alexia without agraphia. Larger infarcts, including the thalamus and internal capsule, may cause aphasia, right hemisensory loss, and right hemiparesis.
Right PCA	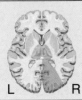 L ... R	Left homonymous hemianopia. Larger infarcts including the thalamus and internal capsule may cause left hemisensory loss and left hemiparesis.

[a]Compare regions of infarcts to Figure 10.1.

MCA infarcts are relatively common. Deficits such as aphasia, hemineglect, hemianopia, and face–arm or face–arm–leg sensorimotor loss are described further in KCC 6.3, 7.3, 11.2, 19.4, 19.5, and 19.9. Large MCA territory infarcts often have a **gaze preference toward the side of the lesion** (see Figures 13.14 and 13.15), especially in the acute period, shortly after onset.

Other combinations not listed in Table 10.1, such as superior plus inferior division infarcts sparing deep territories, or superior division plus deep territories, can occasionally occur. In addition, there are sometimes partial or overlapping syndromes. Smaller cortical infarcts can also occur within one territory, producing more focal deficits, such as monoparesis (see KCC 6.3; Figure 6.14E,F).

Small, deep infarcts involving penetrating branches of the MCA or other vessels are called lacunes, as we will discuss in KCC 10.4. Certain characteristic lacunar syndromes (see Table 10.3) can often be distinguished on clinical grounds from infarcts involving large blood vessel territories (see Table 10.1).

Anterior Cerebral Artery (ACA)

ACA infarcts typically produce upper motor neuron-type weakness and cortical-type sensory loss (see KCC 7.3) affecting the **contralateral leg more than the arm or face** (see Table 10.1; Figure 10.1B). Larger ACA strokes may cause contralateral hemiplegia, at least initially. Dominant ACA strokes sometimes are associated with transcortical motor aphasia (see KCC 19.6), and nondominant ACA strokes can produce contralateral neglect (see KCC 19.9). There may also be a variable degree of frontal lobe dysfunction depending, in part, on the size of the infarct. Such dysfunction may include a grasp reflex, impaired judgment, flat affect, apraxia, abulia, and incontinence (see KCC 19.11). Sometimes damage to the supplementary motor area and other regions in the frontal lobe leads to an unusual "alien hand syndrome" characterized by semiautomatic movements of the contralateral arm that are not under voluntary control.

Posterior Cerebral Artery (PCA)

PCA infarcts typically cause a **contralateral homonymous hemianopia** (see Table 10.1; Figure 11.15). Smaller infarcts that do not involve the whole PCA territory may cause smaller homonymous visual field defects. Sometimes the small, penetrating vessels that come off the proximal PCA are involved, leading to infarcts in the thalamus or posterior limb of the internal capsule. The result can be a contralateral sensory loss; contralateral hemiparesis; or even thalamic aphasia (see KCC 19.6) if the infarct is in the dominant (usually left) hemisphere, thereby mimicking features of MCA infarcts. PCA infarcts that involve the left occipital cortex and the splenium of the corpus callosum can produce alexia without agraphia (see KCC 19.7).

Small, perforating vessels arising from the proximal PCAs at the top of the basilar artery (see Figure 4.17B) supply the midbrain. Vascular syndromes of this region of the brainstem will be discussed in Chapter 14 (see KCC 14.3). ■

KEY CLINICAL CONCEPT

10.2 WATERSHED INFARCTS

When a cerebral artery is occluded, ischemia or infarction occurs in the territory supplied by that vessel, with regions near other vessels relatively spared. In contrast, when the blood supply to two adjacent cerebral arteries

---

REVIEW EXERCISE

Cover all but the left-most column in Table 10.1. For each vessel territory, describe the regions of the brain involved (see Figures 10.5 and 10.9) and the expected deficits (see Figure 10.1).

FIGURE 10.10 Watershed Zones for the Major Cerebral Arteries
(A) Coronal section. (B) Axial section. Compare to Figure 10.9.

(A)

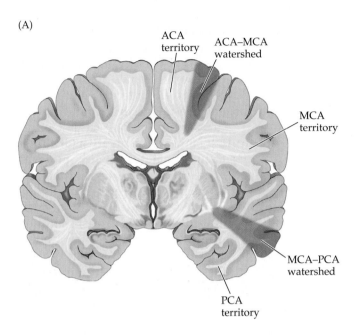

(B)

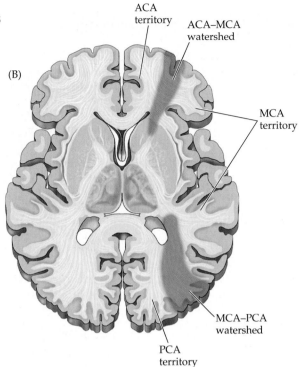

is compromised, the regions between the two vessels are most susceptible to ischemia and infarction. These regions between cerebral arteries are called **watershed zones** (Figure 10.10). Bilateral watershed infarcts in both the ACA–MCA and MCA–PCA watershed zones can occur with severe drops in systemic blood pressure. A sudden occlusion of an internal carotid artery or a drop in blood pressure in a patient with carotid stenosis (KCC 10.5) can cause an ACA–MCA watershed infarct, since the MCA and ACA are both fed by the carotid (see Figure 10.2).

Watershed infarcts can produce proximal arm and leg weakness ("man in the barrel" syndrome) because the regions of homunculus involved often include the trunk and proximal limbs (see Figure 10.1A). In the dominant hemisphere, watershed infarcts can cause transcortical aphasia syndromes (see KCC 19.6). MCA–PCA watershed infarcts can cause disturbances of higher-order visual processing (see KCC 19.12). In addition to watershed infarcts between the superficial territories of different cerebral vessels, watershed infarcts can also occasionally occur between the superficial and deep territories of the MCA (see Figure 10.9). ■

KEY CLINICAL CONCEPT

10.3 TRANSIENT ISCHEMIC ATTACK AND OTHER TRANSIENT NEUROLOGIC EPISODES

Transient neurologic episodes are a common diagnostic problem. Symptoms and signs may be positive or negative, and they can be motor, somatosensory, visual, auditory, olfactory, kinesthetic, emotional, or cognitive in nature. Some causes of transient neurologic episodes are listed in Table 10.2. Of these, the **most common causes** are transient ischemic attack, migraine, seizures, and other non-neurologic conditions such as cardiac arrhythmia or hypoglycemia.

TABLE 10.2 Differential Diagnosis of Transient Neurologic Episodes[a]

CATEGORY	EXAMPLES
Structural/mechanical	Intermittent compression of spinal cord or peripheral nerves; Chiari malformation; platybasia
Vascular	Transient ischemic attack (TIA); migraine; arteriovenous malformation; amyloid angiopathy
Epileptic seizures	
CSF flow–related	Colloid cyst of the third ventricle
Genetic	Hypokalemic or hyperkalemic periodic paralysis; episodic ataxias
Toxic/metabolic	Medication-related; toxin exposure; hypoglycemia; carcinoid; pheochromocytoma
Infectious/inflammatory	Encephalitis; multiple sclerosis
Movement disorders	Chorea; dystonia; tic disorders
Psychogenic	Panic attacks; dissociative disorders; somatization
Other	Benign paroxysmal positional vertigo; trigeminal neuralgia; narcolepsy
Other non-neurologic	Retinal detachment; shoulder dislocation; angina; cardiac arrhythmias; hypotension; hypoglycemia; peripheral vascular disease; breath-holding spells

[a]Following format of Figure 1.1.

In this chapter we will be concerned primarily with transient neurologic episodes caused by cerebrovascular disease. A **transient ischemic attack**, or **TIA**, was classically defined as a neurologic deficit lasting less than 24 hours, caused by temporary brain ischemia. This concept has been revised in recent years for several reasons. First, although some transient ischemic deficits last longer, the more **typical duration for a TIA is about 10 minutes**. Second, improved imaging technology and animal studies suggest that ischemic deficits lasting more than about 10 minutes probably produce at least some permanent cell death in the involved region of the brain. TIAs lasting more than an hour, in fact, are usually small infarcts. On the other hand, despite the appearance of a small infarct on an MRI scan, complete functional recovery can sometimes occur within 1 day. The concept of a TIA remains useful, at the very least, as an important warning sign for a potentially larger ischemic injury to the brain.

TIAs are a **neurological emergency** akin to acute coronary disease or unstable angina. Approximately 15% of patients with TIAs will have a stroke causing persistent deficits within 3 months, and about half of these strokes occur within the first 48 hours. Therefore, all patients who experience a TIA should be urgently admitted and evaluated for treatable causes of ischemic cerebrovascular disease (see next section).

Several mechanisms for TIAs have been proposed, each of which may occur in different situations. One possibility is that an embolus temporarily occludes the blood vessel but then dissolves, allowing return of blood flow before permanent damage occurs. Other possibilities include in situ thrombus formation on the blood vessel wall and/or vasospasm leading to temporary narrowing of the blood vessel lumen.

TIAs must be distinguished clinically from the other kinds of transient episodes listed in Table 10.2. When transient deficits occur in a typical vascular pattern (see KCC 10.1; see also Chapter 14), especially in a patient with stroke risk factors (see KCC 10.4), TIA must be high on the list of possible diagnoses, and an appropriate workup should be pursued. Other diagnoses that can commonly mimic TIAs include focal seizures (see KCC 18.2) and migraine (see KCC 5.1). Interestingly, episodes of hypoglycemia can sometimes produce transient focal neurologic deficits, especially in the elderly.

Transient loss of consciousness *without other focal features* is a special case of transient neurologic dysfunction. The most common cause by far is cardiogenic **syncope** including vasovagal transient episodes of hypotension ("fainting"), arrhythmias, and other non-neurologic causes. Neurologic causes are responsible for less than 5 to 10% of cases of syncope and include seizures (see KCC 18.2), other causes of coma listed in Table 14.4, and rarely, TIA of the posterior circulation affecting the brainstem reticular activating systems (see KCC 14.3). ■

10.4 ISCHEMIC STROKE: MECHANISMS AND TREATMENT

Stroke is the third leading cause of death in the United States and a major cause of permanent disability. Recent improvements in the acute diagnostic and therapeutic management of stroke have made it clear that stroke should be handled as an emergency along lines similar to cardiac emergencies. **Stroke** refers to both hemorrhagic events, such as intracerebral or subarachnoid hemorrhage, and to ischemic infarction of the brain. Sometimes ischemic strokes can cause blood vessels to become fragile and rupture, leading to secondary **hemorrhagic conversion**. Ischemic stroke is discussed in this section, and nontraumatic intracranial hemorrhage is discussed in KCC 5.6.

Mechanisms of Ischemic Stroke

Ischemic stroke occurs when inadequate blood supply to a region of the brain lasts for enough time to cause infarction (death) of brain tissue. There are numerous mechanisms for ischemic stroke. In clinical practice, a distinction is often made between embolic and thrombotic infarcts. In an **embolic infarct**, a piece of material (usually a blood clot) is formed in one place and then travels through the bloodstream to suddenly lodge in and occlude a blood vessel supplying the brain. In a **thrombotic infarct**, a blood clot is formed locally on the blood vessel wall, usually at the site of an underlying atherosclerotic plaque, causing the vessel to occlude. Embolic infarcts are considered to occur suddenly, with maximal deficits at onset, while thrombotic infarcts may have a more stuttering course. In reality, this distinction is not often easy to make on clinical grounds alone.

Another important distinction is between large-vessel and small-vessel infarcts. **Large-vessel infarcts** involve the major blood vessels on the surface of the brain, such as the middle cerebral artery and its main branches (see KCC 10.1). Large-vessel infarcts are most often caused by emboli, although thrombosis can also occasionally occur, especially in large proximal vessels such as the vertebral, basilar (see KCC 14.3), and carotid arteries. **Small-vessel infarcts** involve the small, penetrating vessels that supply deep structures. In the cerebral hemispheres these include the basal ganglia, thalamus, and internal capsule (see Figures 10.7 and 10.8), while in the brainstem these include the medial portions of the midbrain, pons, and medulla (see Figures 14.18 and 14.20). Small-vessel infarcts are sometimes also called **lacunar infarcts** because they resemble small lakes or cavities when the brain is examined on pathologic section.

In embolic infarcts, the goal is to determine the **source of the embolus** so that future strokes may be prevented. Emboli are most commonly composed of thrombotic (blood clot) material. In **cardioembolic infarcts**, the embolus originates in the heart. Cardioembolic infarcts occur in conditions such as **atrial fibrillation**, in which thrombi form in the fibrillating left atrial appendage; **myocardial infarction**, in which thrombi form on hypokinetic or akinetic regions of infarcted myocardium; and **valvular disease** or mechanical prostheses, in which thrombi form on the valve leaflets or prosthetic parts. **Artery-to-artery emboli** can also occur. These include emboli arising from a stenosed segment of the internal carotid artery (see KCC 10.5), vertebral stenosis, or an ectatic dilated basilar artery. **Dissection** of the carotid or vertebral arteries (see KCC 10.6) often results in thrombus formation, which can embolize to the brain. In addition, atherosclerotic disease of the aortic arch is increasingly being recognized as an important potential source of artery-to-artery thromboembolic material. Occasionally, a **patent foramen ovale** can allow a thromboembolus formed in the venous system to bypass the lungs and pass directly from the right to the left side of the heart, reaching the brain.

Aside from thrombus, emboli composed of other materials can, less commonly, lead to stroke. Examples include **air emboli** in deep-sea divers or iatrogenic introduction of air into the circulation; **septic emboli** in bacterial endocarditis, which can lead to mycotic aneurysms and hemorrhage; **fat or cholesterol emboli** in trauma to long bones or to arterial walls; proteinaceous emboli in **marantic endocarditis**; **disc emboli** in cervical trauma; **amniotic fluid emboli** during childbirth; platelet aggregates; and foreign materials introduced into the circulation (such as talc or other contaminants of illicit intravenous drugs).

Lacunar infarcts are usually associated with **small-vessel disease** caused by chronic hypertension. In hypertension, small penetrating vessels can become occluded by a several pathologic processes. Atherosclerotic disease, in situ thrombosis, small emboli, or hypertension-related changes in the vessel wall, known as **lipohyalinosis**, can lead to occlusion of small intracranial blood vessels. In addition, abnormalities of the parent vessel wall such as thrombosis, atheroma formation, or dissection can occlude the openings to one or more small vessels. Numerous characteristic **lacunar syndromes** have been described, and some of the more common ones are listed in Table 10.3. The clinical features of these lacunar syndromes can help localize infarcts and can help distinguish them from large-vessel infarcts (see Table 10.1). **Pure motor hemiparesis** (see Table 10.3) was already discussed in KCC 6.3 (see also Figure 6.14A). In **ataxic hemiparesis**, the ataxia (see KCC 15.2) is caused by damage to proprioceptive or cerebellar circuitry rather than by damage to the cerebellum itself. **Thalamic lacunes** can cause contralateral somatosensory deficits (see KCC 7.3; Figure 7.9A), sometimes followed by a thalamic pain syndrome. **Basal ganglia lacunes** can occasionally cause move-

TABLE 10.3 Common Lacunar Syndromes

SYNDROME	CLINICAL FEATURES	POSSIBLE LOCATIONS FOR INFARCT	POSSIBLE VESSELS INVOLVED
Pure motor hemiparesis or dysarthria hemiparesis	Unilateral face, arm, and leg upper motor neuron–type weakness, with dysarthria	Posterior limb of internal capsule (common)	Lenticulostriate arteries (common), anterior choroidal artery, or perforating branches of posterior cerebral artery
		Ventral pons (common)	Ventral penetrating branches of basilar artery
		Corona radiata	Small middle cerebral artery branches
		Cerebral peduncle	Small proximal posterior cerebral artery branches
Ataxic hemiparesis	Same as pure motor hemiparesis, but with ataxia on same side as weakness	Same as pure motor hemiparesis	Same as pure motor hemiparesis
Pure sensory stroke (thalamic lacune)	Sensory loss to all primary modalities in the contralateral face and body	Ventral posterior lateral nucleus of the thalamus (VPL)	Thalamoperforator branches of the posterior cerebral artery
Sensorimotor stroke (thalamocapsular lacune)	Combination of thalamic lacune and pure motor hemiparesis	Posterior limb of the internal capsule, and either thalamic VPL or thalamic somatosensory radiation	Thalamoperforator branches of the posterior cerebral artery, or lenticulostriate arteries
Basal ganglia lacune	Usually asymptomatic, but may cause hemiballismus (see KCC 16.1)	Caudate, putamen, globus pallidus, or subthalamic nucleus	Lenticulostriate, anterior choroidal, or Heubner's arteries

ment disorders such as hemiballismus (see KCC 16.1). For additional details on less common lacunar syndromes, see the references at the end of this chapter.

Cortical vs. subcortical lesions can sometimes be differentiated clinically based on the absence or presence of so called **cortical signs**, including aphasia (see KCC 19.6), neglect (see KCC 19.9), homonymous visual field defects (see KCC 11.2), and cortical sensory loss (see KCC 7.3). However, each of these deficits can be seen in some cases of subcortical lesions as well. Presence of a typical lacunar syndrome (see Table 10.3), such as pure motor hemiparesis, suggests that a subcortical lesion is present. Clinical differentiation between **hemispheric vs. brainstem lesions** will be discussed in KCC 14.3.

In addition to **focal neurological deficits** (see KCC 10.1, 10.2, 11.3, and 14.3), ischemic stroke can be associated with **headache** or, less commonly, **seizures**. Headache (see KCC 5.1) occurs in 25% to 30% of ischemic strokes. When headache is unilateral, it is more commonly on the side of the infarct, although exceptions do occur. Headache may be more common for posterior than for anterior circulation infarcts. In addition, headaches or neck pain are often seen in dissection of the carotid or vertebral arteries (see KCC 10.6). Seizures (see KCC 18.2) occur in 3% to 10% of stroke patients, usually sometime after the stroke but occasionally as the presenting abnormality.

To summarize:

- Emboli usually cause large-vessel infarcts involving cerebral or (less commonly) cerebellar cortex, with sudden onset of maximal deficits.

- Lacunes are small-vessel infarcts usually seen in chronic hypertension, commonly affecting the deep white matter and nuclei of the cerebral hemispheres and brainstem.

- Thrombosis occasionally occurs in large proximal vessels, such as vertebral, basilar, and carotid arteries, and may also contribute to large or small vessel infarction.

Stroke Risk Factors

Certain patients are at increased risk for vascular disease, including ischemic stroke. When the history is being taken, patients should be asked if they have the following common vascular risk factors: hypertension, diabetes, hypercholesterolemia, cigarette smoking, family history, or prior history of stroke or other vascular disease (Table 10.4). In addition, certain cardiac disorders are important risk factors for stroke, especially atrial fibrillation, mechanical valves or other valvular abnormalities, patent foramen ovale (mentioned in the preceding section), and a severely decreased ejection fraction.

Less commonly, several other systemic medical conditions may affect the coagulation pathways or work through other mechanisms to increase both thrombotic and embolic infarcts (Table 10.5). These hypercoagulable states also increase the risk for venous thrombosis (see KCC 10.7).

Ischemic stroke is relatively uncommon in young individuals because the cumulative effects of the major stroke risk factors (see Table 10.4) tend to worsen with age. When stroke does occur in a **young patient**, conditions such as arterial dissection (see KCC 10.6), patent foramen ovale, or the disorders listed in Table 10.5 should be considered in addition to the usual causes.

Treatment and Diagnostic Workup of Ischemic Stroke and TIA

ACUTE MANAGEMENT There is a growing movement to treat stroke or TIA as an acute medical emergency or **brain attack**, similar to heart attack. Prompt

TABLE 10.4 Common Stroke Risk Factors

Hypertension

Diabetes

Hypercholesterolemia

Cigarette smoking

Positive family history

Cardiac disease (valvular disease, atrial fibrillation, patent foramen ovale, low ejection fraction)

Prior history of stroke or other vascular disease

TABLE 10.5 Medical Conditions Leading to Hypercoagulability

Protein S deficiency

Protein C deficiency

Antithrombin III deficiency

Other hereditary coagulation factor disorders

Dehydration

Adenocarcinoma

Surgery, trauma, childbirth

Disseminated intravascular coagulation

Antiphospholipid antibody syndromes

Vasculitis (effects on vessel wall in addition to hypercoagulability):

    Temporal arteritis

    Primary central nervous system vasculitis (granulomatous angiitis)

    Systemic lupus erythematosus

    Polyarteritis nodosa

    Wegener's granulomatosis

    Other rheumatological disorders

    Infections

    Neoplasms

Sickle cell disease

Polycythemia

Leukemia

Cryoglobulinemia

Homocystinuria (increases risk of atherosclerosis)

medical attention allows early therapeutic interventions that improve outcome. When the history and exam suggest a possible ischemic event, an imaging study of the brain should be done immediately to rule out hemorrhage. In most emergency rooms, CT scans can be performed more quickly than MRI scans, and a CT will suffice for this purpose. Remember that an infarct will often not be visible on the initial CT scan, especially if it is done within a few hours of symptom onset; however, a hemorrhage will almost always be visible (see Chapter 4). Meanwhile, routine blood chemistries, cell counts, and coagulation studies should be sent. Specialized coagulation studies can be sent when clinically appropriate (see Table 10.5), such as in a young individual with no known vascular risk factors.

Once a hemorrhage has been ruled out by CT, patients may be eligible for treatment with the thrombolytic agent **tissue plasminogen activator** (**tPA**). Intravenous tPA was originally demonstrated to improve the chances of a good functional outcome if given within 3 hours of stroke onset; however, recent studies show that patients may benefit even if the drug is given *within 4.5 hours of onset*. Nevertheless, it is likely that the earlier patients are treated, the better. tPA does carry some increased risk of intracranial and systemic hemorrhage, which can occasionally be life threatening. There are standard recommendations and contraindications to giving tPA, including evidence or history of intracranial hemorrhage, AVM or aneurysm, active internal bleeding, abnormal platelet or coagulation studies on blood tests, and uncontrolled hypertension, among others (see the References at the end of the chapter for details). Following administration of tPA, patients are typically watched closely in an intensive care unit setting for at least 24 hours before transfer to the regular patient floor.

In patients with stroke who are not eligible for tPA or in patients who have had a TIA, acute administration of the antiplatelet agent **aspirin** can reduce the risk of early recurrent stroke. Heparin anticoagulation, though commonly used in the past, has not proven effective in the routine treatment of acute stroke and is no longer recommended. Several studies have shown that the increased risk of hemorrhagic conversion outweighs any benefits of intravenous heparin. There may still be a role for heparin in a few situations, such as to prevent further embolization in patients who have an arterial dissection or atrial fibrillation, but this remains controversial. Other acute interventions being investigated include intra-arterial thrombolysis and mechanical clot extraction, which may allow extension of the therapeutic window out to as much as 8 hours after stroke onset. These procedures are performed by catheterization of the occluded vessel (see Figure 4.9), which allows direct administration of the thrombolytic agent to the thrombus or use of a specialized device to pull the clot out of the vessel. In addition, several "neuroprotectant" compounds are being tested that may preserve brain tissue before irreversible cell damage occurs. They include antioxidants, calcium-channel blockers, glutamate receptor antagonists, and antagonists towards cellular receptors that modulate inflammation. Angioplasty and stenting of stenosed vertebral, carotid, and intracranial vessels are being tried along the lines of similar procedures in coronary arteries.

Some of the most effective measures for treating acute stroke involve good general supportive care. Patients should be given intravenous fluids to correct hydration status, maintain cerebral perfusion, and prevent hypotension. **Hypoglycemia** or **hyperglycemia** should be quickly corrected in acute stroke because these conditions are known to worsen infarctions, possibly by increasing local tissue acidosis and blood–brain barrier permeability. Good nursing care, prevention of deep vein thrombosis, and early mobilization can all help improve recovery. For these reasons, patients are best managed in an inpatient setting, and some studies suggest improved outcomes when patients are treated on specialized stroke units.

DIAGNOSTIC EVALUATION The diagnostic evaluation begins with the patient history and exam, including questions about stroke risk factors (see Table 10.4), and continues with a number of diagnostic tests, which we will now discuss. **Blood flow in the major cranial and neck vessels** should be assessed with Doppler ultrasound and/or magnetic resonance angiography (MRA; see Figures 4.18 and 4.19) or CT angiography (CTA). *This is particularly important in suspected internal carotid artery stenosis*, since carotid endarterectomy may be required (see KCC 10.5). Conventional angiography is occasionally needed in cases where the degree of stenosis is uncertain based on these noninvasive tests.

The possibility of a cardioembolic source should be investigated with an **electrocardiogram**, to look for evidence of cardiac ischemia or arrhythmias, and an **echocardiogram**, to look for structural abnormalities or thrombi. Some practitioners perform a 24-hour Holter monitor test to look for arrhythmias. Studies have shown that *patients with atrial fibrillation are at increased risk of embolic stroke* and that this risk is significantly reduced when they are treated with warfarin (Coumadin) oral anticoagulation. Blood tests for cardiac enzymes to detect myocardial infarction are also performed on most patients admitted for acute stroke. As already mentioned, young patients with stroke or other patients with suggestive history should also be evaluated for the less common conditions listed in Table 10.5.

MEDIUM- TO LONG-TERM MANAGEMENT Common medium-term complications of ischemic stroke include hemorrhagic conversion, seizures, and delayed swelling. In patients with large MCA infarcts, substantial edema and mass effect may develop over the first 3 to 4 days. Measures for lowering intracranial pressure (see KCC 5.3) may be necessary in an effort to prevent herniation (see KCC 5.4) and death. One investigational therapeutic measure for such patients is **hemicraniectomy**, in which a portion of skull is temporarily removed over the region of swelling and is later replaced after the danger of herniation has passed. Similarly, large cerebellar infarcts can cause mass effect in the posterior fossa, which may require surgical decompression in some cases.

Because stroke often occurs in conjunction with other serious medical disorders, a multidisciplinary treatment approach is appropriate. Careful attention to medical complications, high quality nursing care, and a comprehensive rehabilitation program can substantially reduce morbidity and mortality. **Recovery and rehabilitation from stroke** is a remarkable process with variable outcome. Functional neuroimaging studies have demonstrated that over time other brain areas can "take over" the functions previously carried out by the infarcted regions of brain tissue. This process can occur very quickly, within days in some individuals, while function in other patients continues to gradually improve for up to approximately 1 year.

Preventive measures are most important in reducing the incidence of recurrent stroke. Modifiable **risk factors** such as hypertension, smoking, and hypercholesterolemia should be addressed. Hydroxymethylglutaryl-coenzyme A (HMG-CoA) reductase inhibitors, or **statins**, may also have benefits in reducing stroke risk beyond their effects on cholesterol levels, for example, by reducing inflammation and increasing nitric oxide synthase activity. In addition, antiplatelet drugs, such as **aspirin**, have been shown to reduce the risk of ischemic stroke recurrence. ■

Carotid auscultation

10.5 CAROTID STENOSIS

Atherosclerotic disease commonly leads to stenosis of the internal carotid artery just beyond the carotid bifurcation (see Figures 4.19B and 10.2). Thrombi formed on a stenotic internal carotid artery can embolize distally, giving rise to TIAs or infarcts of various carotid branches, especially the MCA, ACA, and ophthalmic artery. Carotid stenosis is thus associated with MCA territory symptoms such as contralateral face–arm or face–arm–leg weakness, contralateral sensory changes, contralateral visual field defects, aphasia, or neglect. In addition, there may be ophthalmic artery symptoms such as ipsilateral monocular visual loss classically known as **amaurosis fugax** (see KCC 11.3) and ACA territory symptoms such as contralateral leg weakness.

Carotid stenosis can sometimes be detected on physical examination as a whooshing sound, or **bruit**, that continues into diastole and is best heard with the bell of the stethoscope applied lightly just below the angle of the jaw (see **neuroexam.com Video 2**). The severity of carotid stenosis can usually be estimated noninvasively with Doppler ultrasound and MRA (or CTA), so that conventional angiography is only occasionally needed as the "gold standard."

While carotid stenosis can be clinically silent, patients who have carotid stenosis along with transient monocular blindness on the same side, or TIAs or strokes causing symptoms on the contralateral side, are considered to have **symptomatic carotid stenosis**. The mainstay of treatment for symptomatic carotid stenosis is **carotid endarterectomy**. In this procedure the carotid artery is exposed surgically and temporarily clamped. A longitudinal incision is made in the artery, and atheromatous material is shelled out from the internal carotid lumen, eliminating the stenosis. Carotid endarterectomy has been compared prospectively with medical therapy in patients who had a stroke or TIA on the side of a >70% stenosis of the internal carotid artery. Over 2 years of follow-up in one major trial (the North American Symptomatic Carotid Endarterectomy Trial), the rate of stroke on the side of the stenosis was 26% in the medically treated group, compared with 9% in the group that underwent endarterectomy. Studies have also suggested that in less severe (50% to 70%) carotid stenosis with symptoms or severe carotid stenosis even without symptoms, outcome may be improved by surgery, although the magnitude of the benefit is less than in symptomatic carotid stenosis. The possible role of **angioplasty and stenting** for carotid stenosis is still under investigation and presently is reserved mainly for those patients with high surgical risk for traditional carotid endarterectomy.

Sometimes an internal carotid artery can gradually or suddenly become 100% occluded, causing infarcts in the MCA, ACA, or ACA–MCA watershed territories (see KCC 10.1 and KCC 10.2). **Carotid occlusion** may be completely asymptomatic if there is adequate collateral flow via the anterior or posterior communicating arteries (see Figures 10.2 and 10.3). The occlusion usually occurs just beyond the carotid bifurcation, and the vessel then becomes filled with thrombus up to the level of the ophthalmic artery, which is perfused by collateral flow. Emboli may become dislodged from the top of the thrombus and cause TIAs or strokes. In contrast to carotid stenosis, endarterectomy is not usually performed in cases of 100% carotid occlusion because of the risk of dislodging more emboli and because the procedure has no proven benefit. Because of the different therapeutic implications, it is thus critical to distinguish severe carotid stenosis from carotid occlusion.

Another mechanism for infarction with carotid stenosis is a sudden drop in systemic blood pressure, leading to infarction in the ACA–MCA watershed territory (see KCC 10.2). ◼

KEY CLINICAL CONCEPT

10.6 DISSECTION OF THE CAROTID OR VERTEBRAL ARTERIES

Head or neck trauma and sometimes even minor events such as a cough or sneeze can cause a small tear to form on the intimal surface of the carotid or vertebral arteries. This may allow blood to burrow into the vessel wall, producing a **dissection**. A flap then protrudes into the vessel lumen, under which thrombus forms that can embolize distally. Patients with a dissection may describe feeling or hearing a pop at the onset. In **carotid dissection** they may hear a turbulent sound with each heartbeat and have an ipsilateral Horner's syndrome (see KCC 13.5) and pain over the eye. In **vertebral dissection**, there is often posterior neck and occipital pain. TIAs or infarcts occur in the anterior circulation with carotid dissection and in the posterior circulation with vertebral dissection. There may be a delay of hours to up to several weeks between onset of dissection and ischemic events. Diagnosis is usually by MRI/MRA or CTA of the neck, which show vessel irregularity, narrowing, and sometimes visualization of a **false lumen** in the vessel wall adjacent to the true lumen. Dissection is most often treated with intravenous heparin anticoagulation followed by oral warfarin (Coumadin) anticoagulation to prevent thromboembolic events. The required duration of therapy has not been adequately studied, but most practitioners continue anticoagulation for several months and perform follow-up MRAs to ensure adequate vessel patency before discontinuing anticoagulation. Sometimes dissection, particularly of the vertebral artery, leads to formation of a pseudoaneurysm that may, rarely, rupture, causing subarachnoid hemorrhage. ■

Venous Drainage of the Cerebral Hemispheres

Like the arterial system, the venous drainage of the brain has both superficial and deep territories. The **superficial veins** drain mainly into the **superior sagittal sinus** and the **cavernous sinus**, while the **deep veins** drain into the **great vein of Galen** (Figure 10.11). Ultimately, nearly all venous drainage for the brain reaches the **internal jugular veins**. As we discussed in Chapter 5, the major cerebral venous sinuses lie enclosed within folds of the two layers of dura (see Figure 5.1).

The superior sagittal sinus sweeps back and drains into the two **transverse sinuses** (see Figure 10.11A,B). Each of these sinuses turns downward to become a **sigmoid sinus** that exits the skull through the **jugular foramen**, forming the internal jugular vein. The **cavernous sinus** is a plexus of veins located on either side of the sella turcica. The internal carotid artery and cranial nerves III, IV, V_1, V_2, and VI all pass through the cavernous sinus (see also Figure 13.11). The cavernous sinus drains via the **superior petrosal sinus** into the transverse sinus and via the **inferior petrosal sinus** into the internal jugular vein (see Figure 10.11A,B). The deep structures drain into the **internal cerebral veins** (see Figure 4.7), the **basal veins of Rosenthal**, and other veins to reach the **great cerebral vein of Galen** (see Figure 10.11B,D). The great vein of Galen enters the dura of the tentorium and is joined by the **inferior sagittal sinus** to form the **straight sinus**, or **sinus rectus** (see Figure 10.11B; see also Figure 4.15A).

The **confluence of the sinuses**, also known as the **torcular Herophili** (or more simply as the **torcular**), occurs where the superior sagittal, straight, and occipital sinuses join together and are drained by the transverse sinuses (see Figure 10.11A,B). The torcular is often shaped in such a manner that most blood from the superior sagittal sinus enters the right transverse sinus, while most blood from the straight sinus enters the left transverse sinus.

> **REVIEW EXERCISE**
>
> Which sinuses receive most of the venous blood from the cortical surface? What vein receives most of the venous blood from deep structures (see Figure 10.11C,D)?

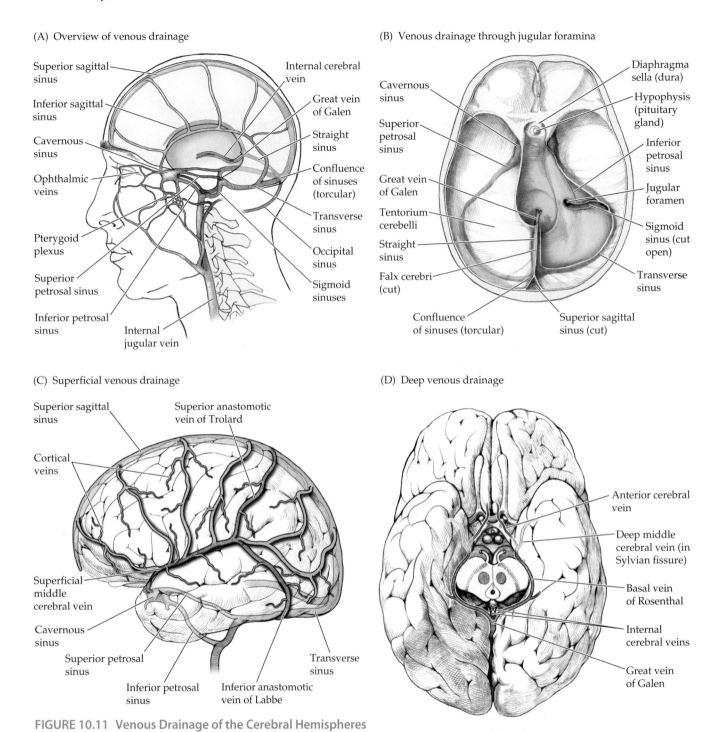

(A) Overview of venous drainage

Superior sagittal sinus

Inferior sagittal sinus

Cavernous sinus

Ophthalmic veins

Pterygoid plexus

Superior petrosal sinus

Inferior petrosal sinus

Internal jugular vein

Internal cerebral vein

Great vein of Galen

Straight sinus

Confluence of sinuses (torcular)

Transverse sinus

Occipital sinus

Sigmoid sinuses

(B) Venous drainage through jugular foramina

Cavernous sinus

Superior petrosal sinus

Great vein of Galen

Tentorium cerebelli

Straight sinus

Falx cerebri (cut)

Confluence of sinuses (torcular)

Superior sagittal sinus (cut)

Diaphragma sella (dura)

Hypophysis (pituitary gland)

Inferior petrosal sinus

Jugular foramen

Sigmoid sinus (cut open)

Transverse sinus

(C) Superficial venous drainage

Superior sagittal sinus

Superior anastomotic vein of Trolard

Cortical veins

Superficial middle cerebral vein

Cavernous sinus

Superior petrosal sinus

Inferior petrosal sinus

Inferior anastomotic vein of Labbe

Transverse sinus

(D) Deep venous drainage

Anterior cerebral vein

Deep middle cerebral vein (in Sylvian fissure)

Basal vein of Rosenthal

Internal cerebral veins

Great vein of Galen

FIGURE 10.11 Venous Drainage of the Cerebral Hemispheres

Although **cortical veins** are highly variable, a few major, fairly constant veins include the **inferior anastomotic vein of Labbe**, which drains into the transverse sinus, the **superior anastomotic vein of Trolard**, which drains into the superior sagittal sinus, and the **superficial middle cerebral vein**, which drains into the cavernous sinus (see Figure 10.11C). The **anterior cerebral veins** and **deep middle cerebral veins** drain into the **basal veins of Rosenthal**, which then join the internal cerebral veins to form the great vein of Galen (see Figure 10.11D).

KEY CLINICAL CONCEPT

10.7 SAGITTAL SINUS THROMBOSIS

Sagittal sinus thrombosis is often associated with one of the hypercoagulable states listed in Table 10.5. It occurs with increased frequency in pregnant women and within the first few weeks post partum. Obstruction of venous drainage usually causes elevated intracranial pressure (see KCC 5.3). Back pressure in cortical veins can cause parasagittal hemorrhages. In addition, the increased venous pressure can decrease cerebral perfusion, leading to infarcts. Seizures are common. Patients often have headaches and papilledema, and they may have depressed level of consciousness. The superior sagittal sinus can normally be seen as a triangular region on axial CT and MRI scan images (see Figures 4.12 and 4.13). The sinus normally fills with intravenous contrast (see Figure 4.4), but in sagittal sinus thrombosus there may be a central, darker-filling defect, called the **empty delta sign**. More subtle radiological signs of sagittal sinus thrombosus include increased density of the sagittal sinus on CT due to coagulated blood (see Table 4.1), or increased T1 signal on MRI (see Table 4.4). In suspected sagittal sinus thrombosus, regardless of whether these subtle radiological findings are present, a more definitive study should be performed, such as magnetic resonance venography (MRV) or a conventional angiogram. Treatment usually involves anticoagulation therapy, although this is controversial when hemorrhage has occurred. Seizures (see KCC 18.2) and elevated intracranial pressure (see KCC 5.3) should be treated as well, when present.

Venous thrombosis can also occur less commonly in other intracranial venous sinuses, in the deep cerebral veins, or in a major cortical vein, leading to infarcts or hemorrhage in the territories of these vessels. ■

CLINICAL CASES

CASE 10.1 SUDDEN-ONSET WORST HEADACHE OF LIFE

MINICASE

A 68-year-old man suddenly developed "**the worst headache of my life**." He had a history of severe diffuse atherosclerosis, including coronary artery disease and peripheral vascular disease requiring multiple bypass operations. He was also a heavy smoker for over 40 years. On the morning of admission he was walking in the hallway at home, and at 10:00 A.M. he suddenly developed an explosive headache worse than anything he had ever experienced. The headache began in the bifrontal area and over the next few minutes spread all over his head and down his neck. He denied nausea, vomiting, loss of consciousness, or vision changes. Examination was unremarkable except for **mild nuchal rigidity**.

LOCALIZATION AND DIFFERENTIAL DIAGNOSIS

1. What diagnosis should be suspected in cases with this kind of clinical presentation (see Chapter 5)?

2. What is the most common cause of this disorder, and what vessels are most commonly affected?

3. What is the significance of the nuchal rigidity?

4. What tests should be performed?

Discussion

1. Sudden onset of severe headache that is worse than any experienced previously should be considered a subarachnoid hemorrhage until proven otherwise (see KCC 5.1, 5.6).

2. In about 80% of cases, spontaneous subarachnoid hemorrhage is caused by rupture of an arterial aneurysm in the subarachnoid space. The most com-

mon locations for aneurysms are the origin of the anterior communicating artery, posterior communicating artery, or bifurcation points of the middle cerebral artery (see Figure 5.20).

3. Nuchal rigidity is often a sign of meningeal irritation (see Table 5.6) caused by inflammation, infection, or hemorrhage in the subarachnoid space.

4. In suspected subarachnoid hemorrhage an emergency head CT should be performed. Rarely, the CT is negative in subarachnoid hemorrhage, so if the clinical situation suggests subarachnoid hemorrhage, a lumbar puncture should then be performed (see KCC 5.10). Lumbar puncture should not be performed if the CT is positive, especially since lumbar puncture can occasionally cause aneurysmal rupture by increasing the pressure across the wall of the aneurysm. Once the diagnosis of subarachnoid hemorrhage has been confirmed, an angiogram should be performed to localize the aneurysm so that it can promptly be treated by surgical clipping or endovascular occlusion (see KCC 5.6).

Neuroimaging

The patient underwent an emergency **head CT** (Image 10.1A, page 419), which demonstrated regions of hyperdensity in the subarachnoid space consistent with diffuse subarachnoid hemorrhage layering in the interhemispheric fissure, in the Sylvian fissures, and around the brainstem. In addition, the lateral ventricles appeared mildly dilated, compatible with hydrocephalus. This can occur from impaired CSF flow in the subarachnoid space because of the hemorrhage (see KCC 5.7). Next the patient was taken for an **angiogram** (see Image 10.1B–D).

1. What blood vessel was injected with contrast dye to produce Image 10.1B–D? Compare this to the normal angiogram in Figure 4.16. Cover the labels on Image 10.1 and identify the internal carotid, anterior cerebral, and middle cerebral arteries (including the lenticulostriate arteries). Note the presence of a saccular aneurysm measuring approximately 1 cm in diameter.

2. What is the site of origin of the aneurysm?

Discussion and Clinical Course

1. Injection of the left carotid artery caused opacity of the left internal and external carotid arteries, of the left middle cerebral artery (including the lenticulostriate arteries), and, through the anterior communicating artery, of the anterior cerebral arteries bilaterally.

2. In the oblique view (see Image 10.1D), the aneurysm can be clearly seen to arise in the region of the anterior communicating artery.

    The patient underwent craniotomy (see KCC 5.11) and surgical clipping of the aneurysm and subsequently recovered fully.

RELATED CASE. Intracranial aneurysms are increasingly treated using a less invasive approach, by filling the aneurysm with detachable coils introduced via an angiogram catheter (see KCC 5.6). An example is shown in Image 10.1E,F, page 421. This was a 71 year old man who had a mass adjacent to his left internal carotid artery, discovered as an "incidental finding" on head CT performed for other reasons. A CT angiogram and later a conventional angiogram (see Image 10.1E) demonstrated an aneurysm arising from the left internal carotid artery, in the region of the left posterior communicating artery (PComm). The aneurysm caused no symptoms in this patient, but because of the risk of hemorrhage, he underwent an interventional neuroradiological procedure to fill the aneurysm with metal coils (see Image 10.1F). This caused a clot to form within the aneurysm so that it no longer posed a significant danger to the patient.

CASE 10.2 LEFT LEG WEAKNESS AND LEFT ALIEN HAND SYNDROME

CHIEF COMPLAINT

A 67-year-old woman suddenly developed left leg weakness and difficulty using her left hand.

HISTORY

Past history was notable for hypertension, peripheral vascular disease, and smoking one pack per day for 40 years. On the morning of admission, after finishing breakfast the patient tried to stand up and suddenly found she could not support her weight. She fell against a door, scraping her left side, but managed to reach a telephone and call an ambulance.

PHYSICAL EXAMINATION

Vital signs: T = 98°F, P = 76, BP = 140/90, R = 14.

Neck: Supple with no bruits.

Lungs: Clear.

Heart: Regular rate.

Abdomen: Soft.

Extremities: A few abrasions present on the left arm and leg.

Neurologic exam:

MENTAL STATUS: Alert and oriented × 3. The patient seemed **unaware at times of any weakness on her left side and did not complain about her abrasions.** Language was fluent.

CRANIAL NERVES: Normal, except for a **minimally decreased left nasolabial fold and mild dysarthria.**

MOTOR: Power 5/5 throughout, except for **1/5 to 2/5 strength in the left leg, both proximally and distally, and 4/5 strength in the proximal left arm.**

REFLEXES:

COORDINATION AND GAIT: Not tested.

SENSORY: **There was inconsistent decreased response to pinprick on the left side.**

CLINICAL COURSE

The patient's weakness temporarily worsened, so that by 2 days after admission she had 0/5 strength in both the left leg and arm. She also had **extinction on the left side to double simultaneous tactile stimulation.** One month later the patient had recovered 3/5 strength in the left arm but continued to have 0/5 strength in the left leg. Interestingly, she felt that her **left arm was "out of control."** Her left arm would occasionally grab onto things without her being aware of it, and she then had to use her right arm to release its **grasp.** She could not localize her left arm in space and had difficulty using the left hand to perform voluntary activities, with marked **motor impersistence.** When distracted, however, she could use both hands to perform certain automatic overlearned behaviors, such as folding a piece of paper in half.

LOCALIZATION AND DIFFERENTIAL DIAGNOSIS

1. On the basis of the symptoms and signs shown in **bold** above, where is the lesion?
2. What is the most likely diagnosis, and what are some other possibilities?

Discussion

1. The key symptoms and signs in this case are:

   • **Profound weakness of the left leg, with mild weakness of the left arm and face, mild dysarthria, left leg hyperreflexia, and Babinski's sign**

   • **Left grasp reflex and motor impersistence**

   • **Left arm "out of control"**

   • **Unawareness of left-sided weakness and abrasions, decreased response to pinprick on the left, tactile extinction on the left**

   Upper motor neuron–type weakness of the left leg could be caused by a lesion in the right primary motor cortex leg area, or the left thoracolumbar spinal cord (see KCC 6.3; Figure 6.14F). However, since there was also mild dysarthria and some left face and arm weakness, there must also have been some milder involvement of corticobulbar and corticospinal fibers for the face and arm, ruling out a spinal cord lesion. A grasp reflex and motor impersistence suggest a frontal lobe lesion (see KCC 19.11). The unusual behavior of her left arm is compatible with an alien hand syndrome (see KCC 10.1), sometimes seen in lesions of the supplementary motor area (see Figure 6.1). Anosognosia and contralateral neglect (see KCC 19.9) can be seen in nondominant (usually right) hemisphere lesions, especially in the parietal lobe, but sometimes also in the frontal lobe. In addition, some of the patient's apparent left-sided weakness may have been due to left motor neglect rather than to actual weakness, as suggested by her preserved ability to use her left hand well at times.

The most likely *clinical localization* is right primary motor cortex foot area, supplementary motor area, and other adjacent regions of the right frontal or parietal lobes.

2. Given the sudden onset of the deficits, together with the patient's age, history of hypertension, smoking, and peripheral vascular disease, the most likely diagnosis is an embolic stroke (see KCC 10.4). An infarct of the right medial frontal lobe including foot motor cortex and supplementary motor area would be caused by occlusion of the right anterior cerebral artery (see KCC 10.1). Another, less likely cause of a lesion in the cortex with this time course would be a hemorrhage. Because the deficits improved over time, a tumor or infection is unlikely as the cause.

Neuroimaging

A head CT scan done shortly after admission suggested probable right anterior cerebral artery infarct. Follow-up **head CT scan** 1 month after admission (Image 10.2A,B, page 422) confirmed the presence of a hypodense area on the anterior medial aspect of the right hemisphere consistent with a right anterior cerebral artery infarct (compare to Figures 10.4, 10.5, and 10.9). Note that today this patient may have been a candidate for treatment with tPA if she had reached medical attention quickly enough.

CASE 10.3 DECREASED VISION ON ONE SIDE

CHIEF COMPLAINT

A 63-year-old woman went to an ophthalmologist because of episodes of decreased vision in her "right eye" and headaches.

HISTORY

Past medical history was notable for diabetes, elevated cholesterol, and coronary artery disease. About 5 or 6 weeks ago the patient began having **episodes of sudden "blurry wavy" vision. She believed this was mostly in the right eye**, but she did not try looking with one eye at a time. The episodes would last 15 to 20 minutes, occurring three to four times per week, and were accompanied by a **severe left retro-**

orbital headache. She was able to recognize faces during the episodes but had difficulty reading. She denied any other symptoms. Two days ago an episode began that resulted in persistent decreased vision on the right.

PHYSICAL EXAMINATION

Vital signs: T = 98.6°F, P = 84, BP = 180/78, R = 20.

Neck: Supple with no bruits.

Lungs: Clear.

Heart: Regular rate with no murmurs.

Abdomen: Soft, nontender.

CASE 10.3 *(continued)*

Neurologic exam:

MENTAL STATUS: Alert and oriented × 3. Speech fluent.

CRANIAL NERVES: Pupils 3 mm, constricting to 2 mm bilaterally. Normal fundi. Visual acuity 20/30 right eye, 20/25 left eye. Visual field testing (see KCC 11.2) revealed a **right homonymous hemianopia** (Figure 10.12). Extraocular movements intact. Facial sensation intact to light touch and pinprick. Face symmetrical. Normal palate elevation. Normal shoulder shrug. Tongue midline.

MOTOR: No drift. Normal tone. 5/5 power throughout.

REFLEXES:

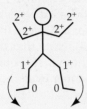

COORDINATION: Normal on finger-to-nose and heel-to-shin testing.

GAIT: Normal.

SENSORY: Intact pinprick and joint position sense. Pinprick and vibration sense diminished in both feet (likely related to diabetic neuropathy).

LOCALIZATION AND DIFFERENTIAL DIAGNOSIS

1. On the basis of the symptoms and signs shown in **bold** above, where is the lesion? What is the most likely diagnosis, and what are some other possibilities?

2. Is this patient a candidate for treatment with intravenous tPA (see KCC 10.4)?

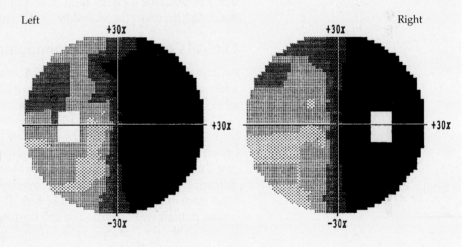

FIGURE 10.12 Automated Visual Field Mapping Showing Right Homonymous Hemianopia See KCC 11.2 for discussion of automated visual field mapping.

Discussion

The key symptoms and signs in this case are:

- **Right homonymous hemianopia**
- **Left retro-orbital headache**

As we will discuss in Chapter 11, a right homonymous hemianopia can be caused by a lesion in the left hemisphere visual pathways anywhere from the left optic tract to the left primary visual cortex (see Figure 11.15). The transient episodes of 15 to 20 minutes of decreased right-sided vision occurring over several weeks, followed by a persistent deficit, are suggestive of TIAs (see KCC 10.3) preceding a cerebral infarct (see KCC 10.4). In addition, the patient's age and history of diabetes, hypercholesterolemia, and coronary artery disease raise the suspicion for cerebrovascular disease as the cause. Following the visual pathways from front to back, the optic tract and lateral geniculate nucleus of the thalamus are supplied by multiple small vessels. Infarcts of the optic tract are rare, and infarcts of the lateral geniculate nucleus are usually accompanied by damage to the adjacent internal capsule, causing contralateral hemiparesis. Infarcts of the entire optic radiation can occur, with large MCA infarcts causing a hemianopia; however, this would also result in a contralateral hemiplegia and other deficits. Thus, the most common cause of hemianopia without other deficits is infarction of the primary visual cortex caused by occlusion of the posterior cerebral artery. The patient's left retro-orbital headache is also consistent with left PCA disease (see KCC 10.4). Occasionally, proximal PCA occlusion can involve small penetrating vessels (see Figures 10.8, 10.9), resulting in infarction of the thalamus or internal capsule as well. However, this patient did not have somatosensory or motor deficits, so the most proximal segment of the PCA must have been spared. Other, less likely diagnoses in this setting include hemorrhage, tumor, abscess, or demyelination in the left occipital cortex.

The most likely *clinical localization* and *diagnosis* is: Left primary visual cortex lesion caused by left posterior cerebral artery infarct.

Because the patient developed persistent deficits 2 days before coming to medical attention, she would not be candidate for tPA, since the current time window is 4.5 hours after onset (see KCC 10.4). Note that the patient described her vision loss as occurring in the right eye. It is common for patients to describe a visual field defect in this manner even though in reality the defect involves the visual fields of both eyes.

Clinical Course and Neuroimaging

An initial CT scan suggested left PCA infarct, and a follow-up **head MRI scan** a few days later (Image 10.3A–D, page 423–424) confirmed the presence of a left PCA infarct involving the left primary visual cortex. Note the presence of T2 bright signal in Image 10.3A, consistent with increased water content from edema and necrosis. In addition, on the T1-weighted images (see Image 10.3C) there were some bright areas consistent with methemoglobin (see Table 4.4) resulting from petechial hemorrhagic conversion. A more significant hemorrhage would have resulted in more dramatic evidence of a frank hematoma seen on both T1- and T2-weighted images.

The patient was admitted to the hospital and testing was done, including a cardiac Holter monitor, an echocardiogram, and Doppler studies of the neck vessels to look for a source for the embolus (see KCC 10.4). These tests were negative; however, a magnetic resonance angiogram revealed multiple stenoses of the cerebral vessels compatible with diffuse intracranial atherosclerotic disease. It was therefore felt that the patient most likely had an artery-to-artery embolus, or a PCA thrombosis caused by severe atherosclerotic disease. Formerly, this patient would have been treated with oral anticoagulation, however, more recent studies have shown no clinical benefit of warfarin over antiplatelet agents like aspirin, which is currently the treatment of choice. Her right hemianopia did not improve, but over time she learned to adapt to her deficit, with improved reading, and she used extra caution to avoid bumping into objects on her right side.

CASE 10.1 SUDDEN-ONSET WORST HEADACHE OF LIFE

IMAGE 10.1A–D Subarachnoid Hemorrhage Caused by Aneurysmal Bleeding (A) Head CT axial image demonstrating subarachnoid hemorrhage (SAH) and hydrocephalus. (B) Angiogram, anterior–posterior view. (C) Angiogram, lateral view. (D) Angiogram, oblique view.

(A)

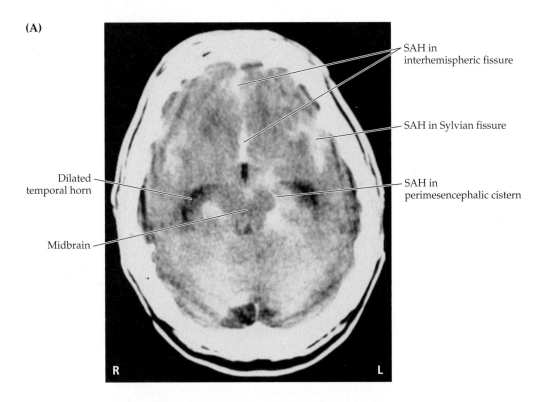

SAH in interhemispheric fissure

SAH in Sylvian fissure

Dilated temporal horn

SAH in perimesencephalic cistern

Midbrain

(B)

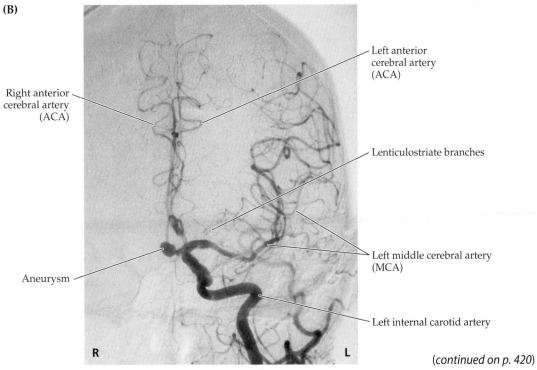

Left anterior cerebral artery (ACA)

Right anterior cerebral artery (ACA)

Lenticulostriate branches

Left middle cerebral artery (MCA)

Aneurysm

Left internal carotid artery

(*continued on p. 420*)

CASE 10.1 *(continued)*

(C)

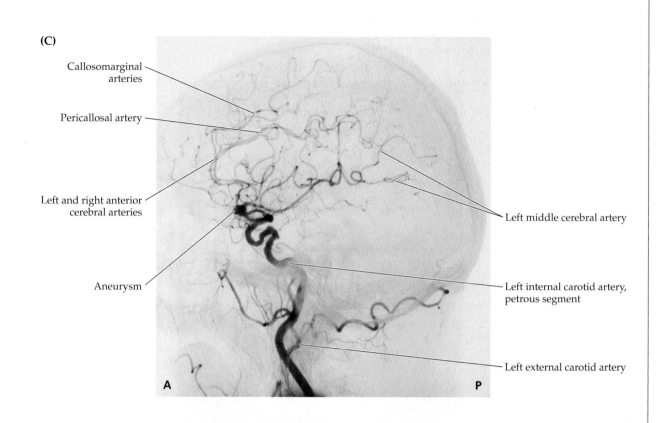

Callosomarginal arteries

Pericallosal artery

Left and right anterior cerebral arteries

Aneurysm

Left middle cerebral artery

Left internal carotid artery, petrous segment

Left external carotid artery

A   P

(D)

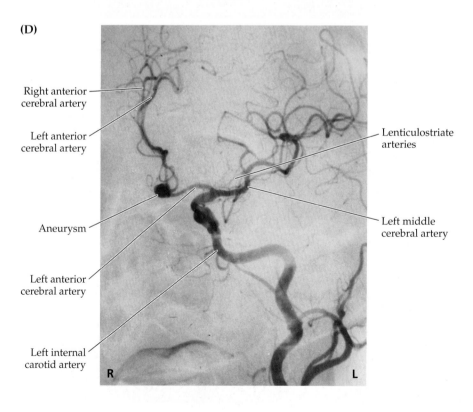

Right anterior cerebral artery

Left anterior cerebral artery

Lenticulostriate arteries

Aneurysm

Left middle cerebral artery

Left anterior cerebral artery

Left internal carotid artery

R   L

CASE 10.1 *RELATED CASE*

IMAGE 10.1E,F PComm Aneurysm Treated by Coiling
(E) Angiogram, anterior–posterior view after injection of
left internal carotid artery. (F) Repeat angiogram after
treatment by coiling no longer shows flow of contrast
within the aneurysm.

(E)

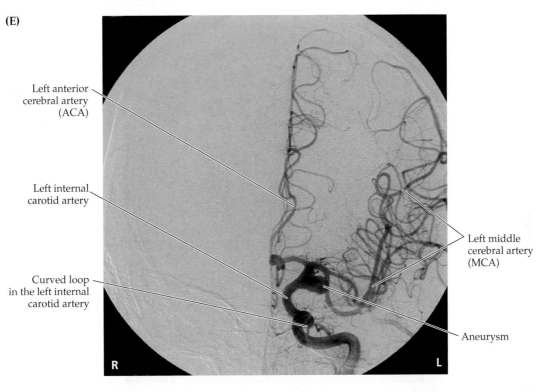

Left anterior
cerebral artery
(ACA)

Left internal
carotid artery

Curved loop
in the left internal
carotid artery

Left middle
cerebral artery
(MCA)

Aneurysm

R   L

(F)

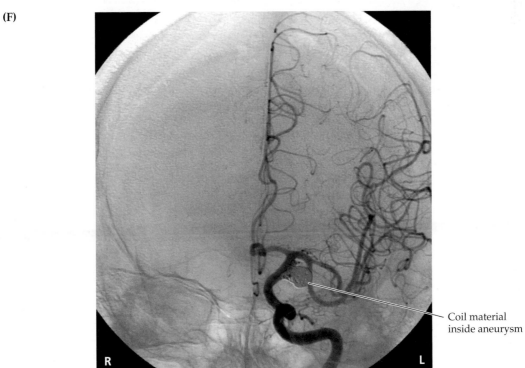

Coil material
inside aneurysm

R   L

CASE 10.2 LEFT LEG WEAKNESS AND LEFT ALIEN HAND SYNDROME

IMAGE 10.2A,B Right Anterior Cerebral Artery (ACA) Infarct (A,B) Axial CT images progressing from inferior to superior. The central sulcus was located by following it down from higher sections (not shown; compare to Figure 4.12).

(A)

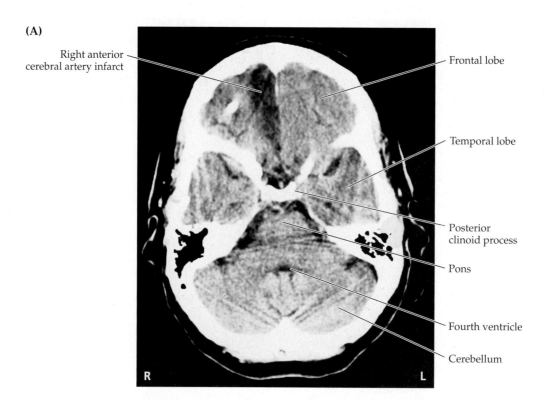

Right anterior cerebral artery infarct

Frontal lobe

Temporal lobe

Posterior clinoid process

Pons

Fourth ventricle

Cerebellum

R　　　L

(B)

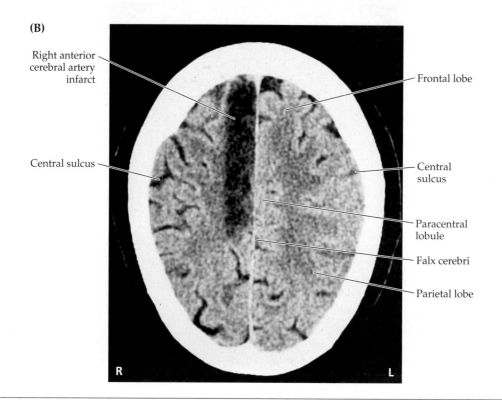

Right anterior cerebral artery infarct

Central sulcus

Frontal lobe

Central sulcus

Paracentral lobule

Falx cerebri

Parietal lobe

R　　　L

CASE 10.3 DECREASED VISION ON ONE SIDE

IMAGE 10.3A–D **Left Posterior Cerebral Artery (PCA) Infarct** MRI of the brain. (A,B) Axial T2-weighted images proceeding from inferior to superior. (C) Parasagittal T1-weighted image of left hemisphere showing bright areas consistent with petechial hemorrhage in the region of the PCA infarct. (D) Parasagittal T1-weighted image of normal right hemisphere from the same patient for comparison, showing locations of calcarine and parieto-occipital fissures.

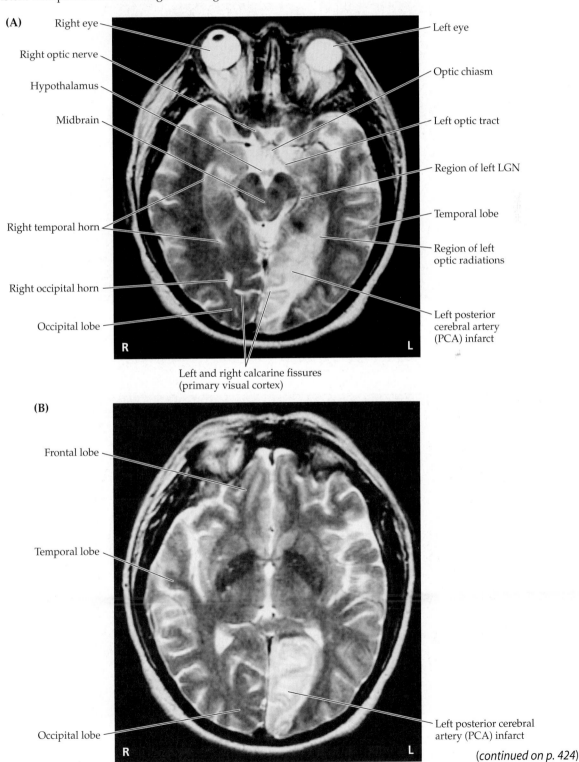

(A)

Right eye
Right optic nerve
Hypothalamus
Midbrain
Right temporal horn
Right occipital horn
Occipital lobe
R   L

Left eye
Optic chiasm
Left optic tract
Region of left LGN
Temporal lobe
Region of left optic radiations
Left posterior cerebral artery (PCA) infarct

Left and right calcarine fissures (primary visual cortex)

(B)

Frontal lobe
Temporal lobe
Occipital lobe
R   L

Left posterior cerebral artery (PCA) infarct

(continued on p. 424)

CASE 10.3 (*continued*)

(C)

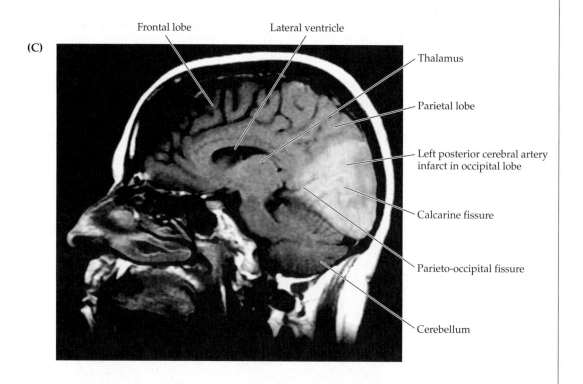

Frontal lobe · Lateral ventricle · Thalamus · Parietal lobe · Left posterior cerebral artery infarct in occipital lobe · Calcarine fissure · Parieto-occipital fissure · Cerebellum

(D)

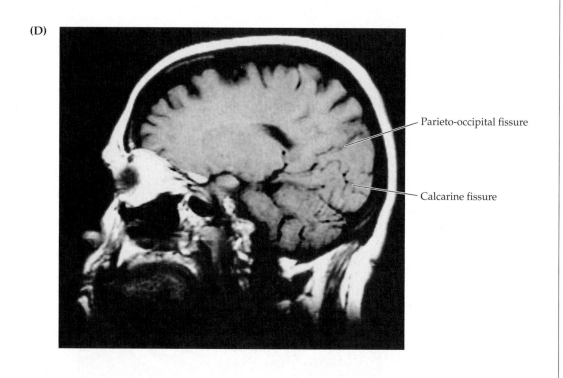

Parieto-occipital fissure · Calcarine fissure

CASE 10.4 TRANSIENT EPISODES OF LEFT EYE BLURRINESS OR RIGHT HAND WEAKNESS

MINICASE

A 71-year-old right-handed man with a long history of cigarette smoking and hypertension had an episode 5 months before admission of **right hand weakness and speech difficulty, "mixing up words."** Since then, he has had several episodes, lasting a few minutes each, of **dim, blurry vision in the left eye**. Finally, he fell on three separate occasions when his **right leg suddenly gave out**, most recently on the day of admis-

sion. Examination was normal except for a **high-pitched bruit audible over the left carotid artery**.

LOCALIZATION AND DIFFERENTIAL DIAGNOSIS

1. What is the most likely cause of this patient's transient neurologic episodes? What are some other possibilities?
2. For each type of episode shown in **bold** above, identify a branch of the internal carotid artery that could be responsible for the patient's symptoms.

Discussion

1. Given the patient's age and history of smoking and hypertension, atherosclerotic cerebrovascular disease is likely. The episodes lasted for a few minutes each and fit with a vascular anatomical pattern (see the localization discussion that follows), suggesting TIAs (see KCC 10.3). In addition, the **left carotid bruit** is a "smoking gun" that heightens the suspicion further for TIAs caused by left internal carotid stenosis. Table 10.2 lists other, less likely possibilities.

2. The three types of episodes in this case are:

   - **Right hand weakness and speech difficulty, "mixing up words":** Symptomatic branch of internal carotid = left MCA superior division (see KCC 10.1)

   - **Right leg weakness:** Symptomatic branch of internal carotid = left ACA (see KCC 10.1)

   - **Decreased vision in the left eye:** Symptomatic branch of internal carotid = left ophthalmic artery (see KCC 10.5, 11.3)

   Interestingly, unlike the case in this patient, internal carotid stenosis often causes TIAs mainly in one carotid branch, resulting in recurrent, nearly stereotyped episodes. For example, a patient may have several episodes of contralateral hand weakness, numbness, and tingling or several episodes of transient monocular visual loss.

Clinical Course

Carotid Doppler studies showed a very tight stenosis of the left internal carotid artery. This was confirmed by MRA. The patient underwent an endarterectomy of the left internal carotid artery. In this procedure the carotid is temporarily cross-clamped, an incision is made in the artery, atherosclerotic plaque is carefully shelled out, and the artery wall is then stitched back together (see KCC 10.5). A large atheromatous plaque was removed in this patient, and when it was examined pathologically, it was found to have a residual lumen of only 0.1 cm in diameter. The patient did very well postoperatively, with no further episodes of weakness or vision changes.

RELATED CASES. Typical MRA findings in a different patient with critical stenosis of the right internal carotid artery are shown in Image 10.4A,B, page 426. This patient presented with two episodes, lasting 5 minutes each, of left hand numbness, tingling, and a feeling like it was not part of her body. A pathologic specimen from yet another patient removed at the time of carotid endarterectomy is shown in Image 10.4C, page 426.

CASE 10.4 *RELATED CASE*

IMAGE 10.4A,B Internal Carotid Artery Stenosis
Magnetic resonance angiogram (MRA) of the carotid arteries from a patient with TIAs consisting of left-hand numbness. (A) Right carotid MRA showing a "skip lesion" consistent with severe stenosis of the right internal carotid artery just beyond the carotid bifurcation. (B) Left carotid MRA showing normal flow.

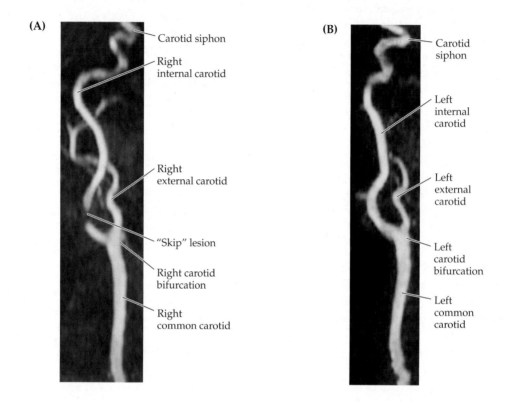

(A)
- Carotid siphon
- Right internal carotid
- Right external carotid
- "Skip" lesion
- Right carotid bifurcation
- Right common carotid

(B)
- Carotid siphon
- Left internal carotid
- Left external carotid
- Left carotid bifurcation
- Left common carotid

IMAGE 10.4C Serial Pathologic Sections of Severe Internal Carotid Artery Stenosis Caused by Atherosclerosis

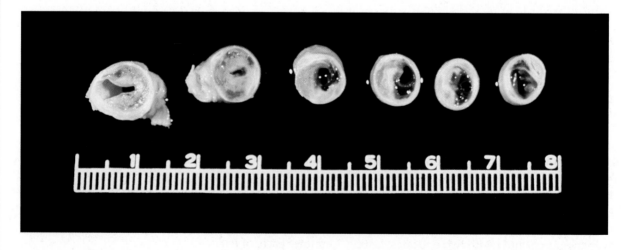

CASE 10.5 NONFLUENT APHASIA WITH RIGHT FACE AND ARM WEAKNESS

CHIEF COMPLAINT

A 45-year-old man was brought to the emergency room because of right face and arm weakness and inability to speak.

HISTORY

The patient had a past history of alcohol use, cigarette smoking, and hypertension. He ate breakfast at the same diner every day but had not shown up there for 2 days. On the morning of admission, he staggered back into the diner grunting incoherently, tripped, and fell on the floor. The manager noticed that the patient was having trouble moving his right arm, so he called an ambulance.

PHYSICAL EXAMINATION

Vital signs: T = 97°F, P = 88, BP = 218/116, R = 16.

Neck: Supple with no bruits.

Lungs: Clear.

Cardiac: Regular rate with no murmurs. S4 gallop.

Abdomen: Soft, nontender.

Extremities: No edema.

Neurologic exam:

MENTAL STATUS: Alert. **Grunted only, producing no words. Followed no commands except to close eyes or open mouth. Obeyed gestures to raise arms or legs.**

CRANIAL NERVES: Pupils 3 mm, constricting to 2 mm bilaterally. Preserved blink to threat bilaterally. Extraocular movements intact. **Decreased right nasolabial fold at rest and decreased movements of right face, sparing forehead.**

MOTOR: **No right arm movement except for slight flexion toward a painful stimulus. Able to raise right leg off bed, but not against resistance.** Good, purposeful movements of left arm and leg against resistance.

REFLEXES:

COORDINATION AND GAIT: Not tested.

SENSORY: Grimaced in response to pinch in all extremities.

INITIAL CLINICAL COURSE

The patient's global aphasia quickly evolved into a Broca's aphasia (see KCC 19.4). By 6 days after admission he was still **able to utter only a few barely articulate words** and he **could not repeat, but he could follow many simple commands** and answer yes/no questions appropriately.

LOCALIZATION AND DIFFERENTIAL DIAGNOSIS

1. On the basis of the symptoms and signs shown in **bold** above, where is the lesion?

2. What is the most likely diagnosis, and what are some other possibilities?

3. Is this patient a candidate for treatment with intravenous tPA for acute ischemic stroke?

Discussion

1. The key symptoms and signs in this case are:

   - **Decreased movements of right face (sparing forehead), profound right arm weakness, and mild right leg weakness**
   - **Nonfluent (Broca's) aphasia**

   Unilateral face and arm weakness are usually caused by a lesion in the contralateral face and arm areas of the motor cortex (see KCC 6.3; Figure 16.14). In support of this, the pattern of facial weakness fits with an upper motor neuron lesion. Note that there was no hyperreflexia on the right side, but this is not unusual in an acute upper motor neuron lesion. The slight right arm flexion toward the painful stimulus likely represents a fragment of flexor posturing (see Figure 3.5A). Another possible localization would be a deep lesion involving the internal capsule; however, the leg is only mildly involved, making this less likely. In addition, the Broca's aphasia in this patient is compatible with a lesion involving the left frontal cortex. Note that the patient initially had a global aphasia, which evolved to Broca's aphasia. This is a pattern commonly seen with large acute lesions involving Broca's area and adjacent regions of the left frontal lobe (see KCC 19.4). Note also that the intact visual fields tested by blink to threat are very helpful for ruling out a more posterior lesion.

The most likely *clinical localization* is left primary motor cortex face and arm areas, Broca's area, and adjacent left frontal cortex.

2. The patient was not elderly; however, given the history of hypertension and cigarette smoking and the anatomical distribution of his deficits, the most likely diagnosis is a left MCA superior division infarct (see KCC 10.1). Other, less likely diagnoses include a hemorrhage, tumor, or abscess affecting the left frontal lobe.

3. Because the patient had not been seen for 2 days before staggering into the diner, the true time of onset is not known. Therefore, he does not meet criteria for onset with 4.5 hours and is not a tPA candidate.

Clinical Course and Neuroimaging

A head CT done in the emergency room showed a left MCA superior division infarct that appeared more than 24 hours old. The patient was admitted and an embolic workup was done (see KCC 10.4). A **brain MRI scan** done 4 days after admission confirmed the left MCA superior division infarct (Image 10.5A,B, page 431). Note the presence of a T2 bright area consistent with increased water from edema and necrosis in the territory of the left MCA superior division (see Figures 10.5, 10.6, 10.9; KCC 10.1). The sulci appear effaced when compared to the same region in the opposite hemisphere, representing mass effect. The infarct can be seen to involve the left frontal operculum (Broca's area) and the face and arm motor areas on the lateral convexity of the frontal lobe. However, the leg motor area, lying superiorly in the interhemispheric fissure, was spared, as was the posterior limb of the internal capsule (see Figures 10.1, 10.9B).

Carotid Doppler studies and magnetic resonance angiography showed no flow in a segment of the left internal carotid artery. Because of the critical importance of distinguishing carotid occlusion from a tight stenosis (see KCC 10.5), a conventional angiogram was performed that confirmed carotid occlusion (see Image 10.10 for angiographic findings in carotid occlusion). Thus, the patient most likely had an embolus that formed in the occluded left internal carotid and migrated upward to the left middle cerebral artery superior division. He was discharged on oral anticoagulation therapy to try to prevent further emboli. Note that embolic infarcts can also occur with carotid stenosis (rather than occlusion), in which case endarterectomy may be indicated (see KCC 10.5).

As already noted, the patient's global aphasia quickly evolved into a Broca's aphasia, but he continued to have some difficulty comprehending more complex commands. His right leg strength recovered fully, and his right arm strength also improved. By the sixth hospital day he was able to extend his fingers and move most muscles in the right arm against gravity but not against resistance.

CASE 10.6 "TALKING RAGTIME"

CHIEF COMPLAINT

A 64-year-old, right-handed man with a history of schizophrenia suddenly began talking nonsense and repeating himself over and over again.

HISTORY

The patient had a history of chronic schizophrenia, with occasional auditory hallucinations and paranoid delusions treated with antipsychotic medications. He also had hypertension and a recent episode of chest pain. On the day of admission he was last seen in his usual state of health at 12:00 noon. His wife went out shopping, and when she returned at 4:00 P.M., he was seated at the kitchen table **saying meaningless syllables and irrelevant phrases over and over and would not respond to any of her questions**. She also noticed that his right arm was hanging down at his side, and she called an ambulance.

PHYSICAL EXAMINATION

Vital signs: T = 98.4°F, P = 67, BP = 153/82, R = 20.

Neck: No bruits.

Lungs: Clear.

Heart: Regular rate with no murmurs, gallops, or rubs.

Abdomen: Soft, nontender.

Extremities: Mild ankle edema.

Rectal: Normal.

Neurologic exam:

MENTAL STATUS: Alert. Mildly agitated. **Responded to all questions by saying only "Yup, yup" or "I don't know" repeatedly. Followed no commands, named no objects, and could not repeat.** Occasionally asked, "What time is it?" or repeated other **irrelevant phonemes**, such as "Jillian" or "lexi" multiple times. Did not follow written commands. Mimicked some gestures, but only when shown in the left visual field.

CRANIAL NERVES: Pupils 4 mm, constricting to 2 mm bilaterally. Normal fundi. **Blink to threat was present only on the left.** Extraocular movements intact. Facial grimace symmetrical. No dysarthria.

MOTOR: **Slightly increased tone in the right arm.** Moved all four extremities and was **able to hold his right arm over his head**, **but not as readily** as the left arm.

REFLEXES:

COORDINATION: Not tested.

GAIT: Stood cautiously and took short, tentative steps without support.

SENSORY: Grimaced and withdrew in response to pinprick in all extremities. **Grimaced more in response to pinprick on the left side than the right.**

LOCALIZATION AND DIFFERENTIAL DIAGNOSIS

1. On the basis of the symptoms and signs shown in **bold** above, where is the lesion?

2. What is the most likely diagnosis, and what are some other possibilities?

3. Is this patient a candidate for treatment with tPA?

Discussion

1. The key symptoms and signs in this case are:

   - Fluent aphasia with impaired comprehension and repetition
   - Blink to threat present only on the left
   - Greater grimace in response to pinprick on the left side than the right
   - Slightly increased tone in the right arm, with right Babinski's sign

   This patient had a Wernicke's aphasia (see KCC 19.5), with fluent, meaningless speech together with severely impaired comprehension and repetition. This syndrome suggests a lesion in the left temporoparietal cortex, including Wernicke's. The right visual field defect could also be explained by a lesion in this area if the optic radiations were involved (see Figure 10.1A). Movements on the right side are relatively spared, suggesting that the primary motor cortex was not directly involved. However, the right sensory loss, mild right-sided upper motor neuron findings, and possible mild right neglect (as evidenced by the patient's reluctance to use his right arm) suggest involvement of the left parietal lobe and left primary sensory cortex, with mild impingement on corticospinal tract fibers.

The most likely *clinical localization* is left temporal and parietal lobes, including Wernicke's area, optic radiations, and somatosensory cortex.

2. The anatomical distribution of the lesion fits with a left MCA inferior division infarct (see KCC 10.1). This diagnosis is further supported by the patient's age, history of hypertension, and possible cardiac disease (chest pain). Other, less likely possibilities include hemorrhage, abscess, or tumor in the left temporoparietal region.

    Interestingly, this patient had a history of schizophrenia, which can also cause nonsensical speech and disregard for verbal questions or commands. The other neurologic findings enabled the diagnosis of a neurologic rather than a psychiatric disorder in this case. However, in left MCA inferior division infarcts, sensory and motor deficits are often absent, making the diagnosis more difficult. Once again, visual field testing is essential to detect posterior lesions, although the results may be hard to interpret even with blink to threat if the patient is sufficiently agitated.

3. The patient was last witnessed to be at his normal baseline 4 hours before onset, so theoretically the patient could have been treated with tPA within the 4.5 hour window if he had immediately arrived at the emergency room and everything had occurred in an expedited manner. In reality, by the time this patient arrived at the hospital, the 4.5 hour time window had just passed and he was no longer eligible.

Clinical Course and Neuroimaging

Head CT in the emergency room showed a region of hypodensity in the left temporoparietal lobes, and follow-up **head CT** 2 days after admission confirmed a left MCA inferior division infarct (Image 10.6A,B, page 433). Note the sulcal effacement in the region of hypodensity in the left temporal and parietal lobes. The stroke can be seen to include Wernicke's area (see Figure 10.1A), the optic radiations, and left parietal cortex but to spare the precentral gyrus and internal capsule. Recall that the margin between the MCA superior and inferior divisions is somewhat variable in the parietal lobe. For example, in inferior division infarcts, less of the parietal lobe is often involved than in this case.

The patient was admitted for a workup, which revealed no embolic source (see KCC 10.4). By the second hospital day he had nearly normal movements on the right side and no longer had an upgoing toe on the right. His aphasia improved somewhat, with more varied and intermittently coherent spontaneous speech. He was fluent but continued to have severely impaired language comprehension and decreased vision on the right side. The long-term treatment plan in a patient of this kind should include primary risk factor management including antihypertensive and cholesterol-lowering therapy, an anti-platelet agent such as aspirin, and occupational and speech therapy.

CASE 10.5   NONFLUENT APHASIA WITH RIGHT FACE AND ARM WEAKNESS

IMAGE 10.5A,B Left Middle Cerebral Artery (MCA) Superior Division Infarct
T-2 weighted MRI axial sections, with A and B progressing from inferior to superior.

(A)

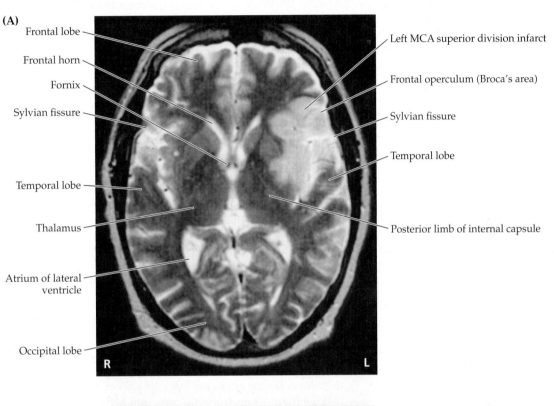

Frontal lobe

Frontal horn

Fornix

Sylvian fissure

Temporal lobe

Thalamus

Atrium of lateral ventricle

Occipital lobe

Left MCA superior division infarct

Frontal operculum (Broca's area)

Sylvian fissure

Temporal lobe

Posterior limb of internal capsule

R   L

(B)

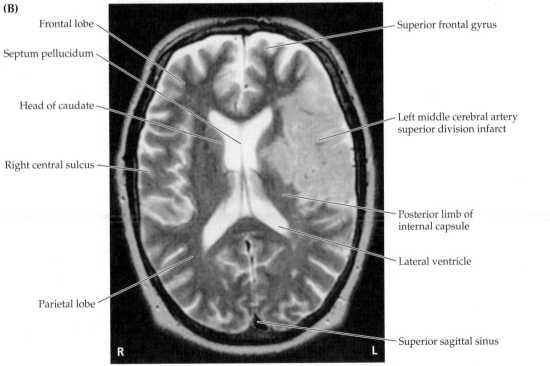

Frontal lobe

Septum pellucidum

Head of caudate

Right central sulcus

Parietal lobe

Superior frontal gyrus

Left middle cerebral artery superior division infarct

Posterior limb of internal capsule

Lateral ventricle

Superior sagittal sinus

R   L

CASE 10.7 DYSARTHRIA AND HEMIPARESIS

MINICASE

An 84-year-old woman with a history of hypertension and diabetes had two episodes of slurred speech and right-sided weakness on two consecutive days, and on the third day she developed persistent **dysarthria and right hemiplegia**. Exam was normal except for **right facial weakness sparing the forehead, dysarthria, decreased right-sided tone, 0/5 power in the right arm and leg, and right upgoing plantar response***.

LOCALIZATION AND DIFFERENTIAL DIAGNOSIS

1. On the basis of the symptoms and signs shown in **bold** above, where is the lesion?
2. What is the most likely diagnosis, and what are some other possibilities?

---

*This patient was seen before tPA was in routine clinical use but otherwise may have been a candidate for thrombolysis, depending on timing and other eligibility criteria.

Discussion

1. The key symptoms and signs in this case are:

   - **Dysarthria and right face, arm, and leg paralysis with a right Babinski's sign**

   This patient has pure motor hemiplegia, without sensory abnormalities or cortical signs such as aphasia or neglect. This can be localized to the contralateral corticobulbar and corticospinal tracts, most commonly in the internal capsule or ventral pons (see KCC 6.3; Figure 6.14A). Dysarthria is commonly present as well, giving rise to the term "dysarthria hemiparesis" (see Table 10.3). In addition, the facial weakness is of the upper motor neuron type, since the forehead is spared. The right Babinski's sign also supports an upper motor neuron lesion.

   The most likely *clinical localization* is left corticobulbar and corticospinal tracts in the internal capsule or ventral pons.

2. Given the patient's age and the history of hypertension and diabetes, an ischemic infarct is the most likely diagnosis. Internal capsule infarcts are most commonly caused by occlusion of the lenticulostriate arteries, which take their origin from the proximal MCA and supply the deep MCA territories (see Figures 4.16B, 10.7–10.9). Infarcts of this kind are often called lacunes (see KCC 10.4). Other vessels that can cause internal capsule lacunes include the anterior choroidal artery and, less commonly, small penetrating vessels from the proximal PCA. Infarcts in the ventral pons can be caused by occlusion of the small paramedian perforators arising from the basilar artery (see Figures 14.18–14.20). Aside from a lacunar infarct, other possibilities include hemorrhage, tumor, abscess, or demyelination in the left posterior limb of the internal capsule, ventral pons, corona radiata, or cerebral peduncle (compare to Cases 6.4 and 6.5).

Clinical Course and Neuroimaging

Head CT in the emergency room showed a hypodensity in the left internal capsule, which appeared more clearly on a follow-up **head CT** 10 days later (Image 10.7, page 434). Note that lacunar infarcts are often smaller than the one seen in this patient, and an infarct of this size is sometimes called a giant lacune. The patient was admitted to the hospital, and an MRA revealed bilateral severe stenosis of the middle cerebral arteries. It was felt that thrombus or atheroma forming along the wall of the vessel might have occluded several lenticulostriate vessels coming off the proximal left middle cerebral artery (see Figures 4.16B, 10.7). Patients of this kind are currently managed with anti-platelet agents and primary risk factor management, although at some centers an investigational procedure such as endovascular stenting might be considered. During the hospitalization, she had no further progression of her deficits, but she continued to have severe right-sided weakness.

CASE 10.6 "TALKING RAGTIME"

IMAGE 10.6A,B Left Middle Cerebral Artery (MCA) Inferior Division Infarct
CT scan axial sections, with A and B progressing from inferior to superior.

(A)

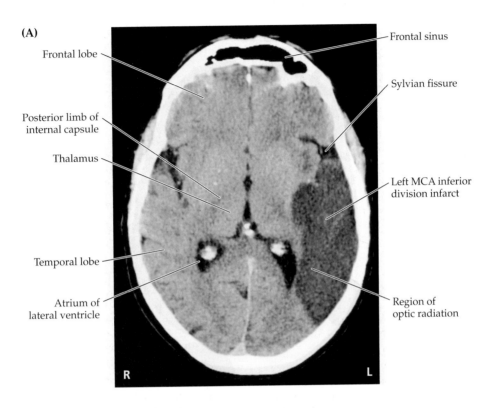

Frontal lobe

Posterior limb of
internal capsule

Thalamus

Temporal lobe

Atrium of
lateral ventricle

Frontal sinus

Sylvian fissure

Left MCA inferior
division infarct

Region of
optic radiation

R   L

(B)

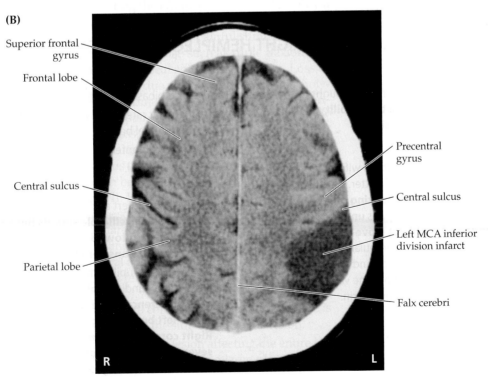

Superior frontal
gyrus

Frontal lobe

Central sulcus

Parietal lobe

Precentral
gyrus

Central sulcus

Left MCA inferior
division infarct

Falx cerebri

R   L

2. The patient's age and history of hypertension and cardiac disease suggest possible cerebrovascular disease. A left MCA stem infarct could produce all of the deficits described above (see Table 10.1). Other possibilities include massive left hemisphere hemorrhage or, less likely given the time course, an abscess or tumor.

Clinical Course and Neuroimaging

Initial head CT within a few hours of onset was negative, except for a hyperdensity in the proximal left MCA, consistent with a blood clot. An EKG revealed atrial fibrillation, suggesting that an embolus had formed in the left atrium and traveled to the left MCA stem. Thrombolysis and other acute interventions were not available at the time this patient was admitted. A repeat **head CT** 1 day after admission again showed a hyperdensity in the left MCA stem but also showed a massive area of hypodensity consistent with infarction of the entire left MCA territory (Image 10.8A–C, pages 441–442). Note that the infarct involved both the superficial and the deep MCA territories (compare to Figures 10.8 and 10.9) while sparing the thalamus, inferior temporal lobe, and medial occipitoparietal cortex (PCA territory), in addition to sparing the medial frontoparietal cortex (ACA territory). Note also that the sulci over the left hemisphere appear effaced, and there is considerable left-to-right shift under the falx in the region of the midline and deformation of the midbrain in the region of the tentorium, which suggests incipient uncal herniation (see KCC 5.4).

Three days after admission the patient became increasingly somnolent, and a repeat CT scan showed increased swelling of the left hemisphere, with about 1 cm of left-to-right midline shift and some effacement of the basal cisterns. The patient was intubated and treated with intravenous mannitol in an attempt to decrease brain swelling by osmotic diuresis (see KCC 5.3). However, by the fourth hospital day the patient was unresponsive, exhibiting bilateral extensor posturing of the arms and legs (see Figure 3.5B). The family had a living will written previously by the patient stating that he did not want extreme measures to be taken to sustain his life if he had an illness with poor prognosis for good functional recovery. He was therefore extubated and given pain medications, and he died the next day with his family at the bedside.

RELATED CASE. An 86-year-old woman with a history of hypertension, paroxysmal atrial fibrillation, aortic valve replacement and hyperlipidemia was observed at her nursing home to suddenly develop right-sided weakness and inability to speak at 11:45 A.M. She was not taking oral anticoagulation. The patient arrived at the emergency room at 12:30 P.M. and on neurological exam she had global aphasia, no blink to threat on the right side, a left gaze preference, right facial droop, 0/5 strength in the right arm and leg, no response to painful stimuli on the right side, and a Babinski sign on the right. Head CT showed no hemorrhage or other acute changes aside from increased density in the left middle cerebral artery possibly representing a blood clot. In summary, this patient had a very similar clinical presentation to the patient in Case 10.8, consistent with a left MCA stem infarct, affecting the left MCA superior, inferior, and deep territories (see Table 10.1).

Unlike the patient in Case 10.8, this patient arrived at an emergency room equipped to administer acute thrombolysis. She was treated with intravenous tPA shortly after completing her initial evaluation in the emergency room, and was then transferred to the intensive care unit. By hospital day 2, she had 2/5 strength in the right upper extremity and 3/5 strength in the right lower extremity. **MRI scan** with diffusion weighted imaging (DWI) performed over 24 hours after onset showed only scattered areas of increased signal in the left MCA territory (Image 10.8D,E, page 443). These findings were compatible with some infarction in the left MCA territory, but with remarkable sparing of many regions likely due to reperfusion. The patient continued to improve and by the time of discharge from the hospital ten days after admission she was fully oriented, had normal naming, mildly dysarthric speech, mild right facial droop, 3/5 strength in her right arm and 4/5 strength in her right leg.

CASE 10.9 LEFT FACE AND ARM WEAKNESS

MINICASE

A 91-year-old, right-handed woman with a history of paroxysmal atrial fibrillation called her daughter one morning because she was **unable to get her arm through the left sleeve of her dress.** The patient's **speech sounded slightly slurred** over the telephone, so her daughter called an ambulance. Examination was notable for **left facial weakness sparing the forehead, mild dysarthria, left arm pronator drift, 4/5 strength in the left arm, and brisk 3⁺ reflexes in the left arm** compared to 2⁺ reflexes in the right arm. In addition, there was **occasional extinction on the left side to double simultaneous visual or tactile stimulation.** The remainder of the exam was essentially normal, including intact visual fields and normal leg strength.

LOCALIZATION AND DIFFERENTIAL DIAGNOSIS

1. On the basis of the symptoms and signs shown in **bold** above, where is the lesion?

2. What is the most likely diagnosis, and what are some other possibilities?

Discussion

1. The key symptoms and signs in this case are:

   - **Left facial weakness sparing the forehead, left arm weakness, and hyperreflexia**

   - **Mild dysarthria**

   - **Occasional extinction on the left side to double simultaneous visual or tactile stimulation**

   Unilateral face and arm weakness are usually caused by contralateral lesions of the face and arm motor cortex (see KCC 6.3; Figure 6.14D). Dysarthria can be caused by numerous lesions (see KCC 12.8), including the face and mouth motor cortex. Extinction suggests mild hemineglect (see KCC 19.9), which is most often caused by right parietal lesions but can also be seen in right frontal lesions.

   The most likely *clinical localization* is right primary motor cortex face and arm areas and adjacent right frontal lobe.

2. Given the patient's history of atrial fibrillation, her age, the acute onset, and the typical clinical pattern fitting the anatomical distribution of the right middle cerebral artery superior division (see Table 10.1), the most likely diagnosis is right MCA superior division infarct. Other, less likely possibilities include hemorrhage, infection, or tumor located in the right frontal lobe.

Clinical Course

The exact time of deficit onset in this case was not known, so the patient was not a candidate for tPA thrombolysis. Initial head CT showed a subtle hypodensity in the right frontal lobe. Follow-up head CT 4 days later confirmed a right MCA superior division infarct that appeared very similar to the infarct in Case 10.5 (see Image 10.5A,B), except on the other side of the brain. As already noted, the patient had a known history of paroxysmal atrial fibrillation. Despite the increased bleeding risk at her age, the patient was treated with intravenous heparin and later switched over to oral anticoagulation, which she tolerated well. During a stay at an inpatient rehabilitation facility, she gained partial recovery of strength and function in the left arm.

CASE 10.10 LEFT HEMINEGLECT

MINICASE

A 61-year-old, left-handed security guard had an **episode of left hand tingling lasting less than an hour** that was reported to medical staff by a friend. The next day he was at the grocery store buying a lottery ticket and reportedly slumped briefly to the floor. He **denied that anything was wrong** but said, "They called an ambulance because they said I had a stroke." On examination he was **unaware of having any deficits** and wanted to go home. He had **profound left visual neglect**, describing only the curtains to the far right in a picture of a complex visual scene and reading only the right two words on each line of a magazine article. When trying to write, he **moved the pen in the air off to the right of the page**. He had **no blink to threat on the left**, a **marked right gaze preference**, **and mildly decreased left nasolabial fold**. Spontaneous **movements were decreased on the left side**, but with provocation he was able to achieve 4/5 strength in the left arm and leg. He was able to feel touch on the left side but had extinction on the left to double simultaneous tactile stimulation. Reflexes were slightly brisker on the left.*

LOCALIZATION AND DIFFERENTIAL DIAGNOSIS

1. What is the significance of the transient episode of hand tingling?
2. On the basis of the symptoms and signs shown in **bold** above, where is the lesion?
3. What is the most likely diagnosis, and what are some other possibilities?

---

*This patient was seen before tPA was in routine clinical use but otherwise might have been a candidate.

Discussion

1. The key symptoms and signs in this case are:

   - Anosognosia, left visual neglect, extinction on the left to double simultaneous tactile stimulation, moving the hand to the right of the page, and decreased spontaneous movements on the left side
   - Right gaze preference
   - No blink to threat on the left
   - Decreased spontaneous movements on the left side, with mildly decreased left nasolabial fold, and slightly brisker reflexes on the left

   The transient episode of hand tingling occurring the day before the onset of a fixed deficit is suggestive of a TIA forewarning an ischemic stroke. Left hand tingling could be caused by compromised flow in the right middle cerebral artery, most commonly from a cardiogenic embolus, or from a right carotid stenosis. Other possible causes of transient neurologic symptoms are listed in Table 10.2.

2. This patient exhibits several forms of neglect (see Chapter 19). In addition to anosognosia, he has left sensory neglect to visual and tactile stimuli, as well as left motor neglect. These features are most commonly seen in patients with nondominant (usually right) parietal lobe lesions but can also occasionally be seen with lesions in the right frontal lobe and in other locations. The right gaze preference further supports a right hemisphere frontal or parietal localization. However, decreased blink to threat is ordinarily caused by damage to primary visual pathways and not by neglect. Therefore, the decreased blink to threat on the left suggests that the lesion is more posteriorly located, possibly involving the optic radiations as they travel be-

neath the right temporal and parietal lobes (see Figure 10.1A). Mild corticobulbar and corticospinal findings can also be seen in parietal lesions, especially acutely (see KCC 10.1).

The most likely *clinical localization* is right temporoparietal lobe, including the optic radiations.

3. Given the sudden onset of the deficits and the patient's age, the most likely diagnosis is TIA followed by ischemic infarct. The right temporoparietal lobe is supplied by the right MCA inferior division (see Table 10.1; Figures 10.7, 10.9). Another possibility is that the initial episode was a focal seizure and that the patient had a tumor, hemorrhage, or infection with deficits previously neglected, that became more severe on the day of admission.

Clinical Course and Neuroimaging

Head CT on the day of admission showed a mild hypodensity in the right temporoparietal area. Follow-up head CT 10 days later confirmed right MCA inferior division infarct that appeared very similar to Image 10.6A,B (see Case 10.6), except on the other side of the brain. Carotid Doppler studies and an MRA suggested occlusion versus critical stenosis of the right carotid artery (see KCC 10.5), so a conventional **cerebral angiogram** was done (see Chapter 4). Injection of the right carotid showed occlusion of the right common carotid artery (Image 10.10A,B, page 444). Injection of the left carotid resulted in cross-filling to the right ACA and right MCA via the anterior communicating artery (see Image 10.10B). Therefore, the patient most likely had a right carotid occlusion, possibly on the day prior to admission, followed by thrombus formation in the carotid, with embolization to the right MCA inferior division. This is similar to the cause of the infarct in Case 10.5 and should be contrasted with the case of carotid stenosis seen in Case 10.4.

The patient was treated with intravenous and, later, oral anticoagulation to try to reduce the risk of further emboli from his carotid. By 3 days after admission he was able to look voluntarily to the left, strength was normal on the left side when he was motivated, and reflexes were symmetrical. He still had decreased blink to threat on the left, and occasional (1/3 trials) left extinction on double simultaneous tactile stimulation. Coumadin (warfarin) was eventually stopped, and in follow-up 1 year later the patient had a normal exam except for a left visual field cut (not precisely mapped out by the examiner).

CASE 10.11 LEFT HEMINEGLECT, HEMIPLEGIA, AND HEMIANOPIA

MINICASE

A 62-year-old, right-handed woman with a history of hypertension, hyperthyroidism, and atrial fibrillation awoke early one morning with pain behind her right eye. She tried to walk to the bathroom but fell at the doorway. Her family later found her on the floor, **unable to move her left side**, and as they called the ambulance she kept repeating "Do not call anyone" because **she believed nothing was wrong**. On exam, when **shown her left hand and asked what it was**, **she replied**, **"Someone's hand."** When asked whose hand it was, she replied, "The doctor's." She had **no blink to threat on the left and no voluntary gaze to the left past the midline, and** there was marked weakness of the lower portion of the left face. Strength was 0/5 in the left arm and leg, the left plantar response was upgoing, and there was no response to pinprick on the left side.

LOCALIZATION AND DIFFERENTIAL DIAGNOSIS

1. On the basis of the symptoms and signs shown in **bold** above, where is the lesion?

2. What is the most likely diagnosis, and what are some other possibilities?

Discussion

1. The key symptoms and signs in this case are:
   - Anosognosia, hemiasomatognosia
   - Left face, arm, and leg plegia with left Babinski's sign
   - No blink to threat on the left
   - No voluntary gaze to the left past the midline
   - No response to pinprick on the left side

   This patient clearly has a large lesion of the right hemisphere, including the entire corticobulbar and corticospinal systems, the retrochiasmal visual pathways, the somatosensory systems, and the pathways responsible for personal awareness and for directing gaze into the contralateral hemifield. This is similar to Case 10.8, but since the right hemisphere is involved, instead of aphasia the patient has loss of awareness of illness and of her entire left body. She has all the deficits of Cases 10.9 and 10.10, plus a left hemiplegia and hemisensory loss. In summary, she must have a large lesion affecting the entire right cerebral cortex and/or all right hemisphere subcortical pathways.

2. Given her age, the sudden onset of deficits, and the history of hypertension and atrial fibrillation, the most likely diagnosis is a right MCA stem infarct (see Table 10.1). A large right hemisphere hemorrhage is also possible. Less likely, given the time course, is a large abscess or tumor.

Initial Clinical Course

The patient was admitted for further evaluation and treatment. Note that because exact time of onset was not known, she was not a candidate for tPA. Head CT on admission was compatible with early right MCA stem infarct, and this was confirmed by MRI and MRA. Two days after admission the patient became **increasingly difficult to arouse**, and she eventually developed a **dilated right pupil** with **flexor posturing in the right arm and an upgoing toe on the right side**.

1. What clinical syndrome do the new findings in **bold** constitute, and what is its localization?

2. What are some possible causes in this patient?

Discussion

1. The combination of impaired consciousness, dilated right pupil, and new right corticospinal findings is compatible with right uncal transtentorial herniation (see KCC 5.4). These three findings would be caused by compression of the midbrain reticular activating system, right CN III, and left cerebral peduncle (Kernohan's phenomenon; see KCC 5.4), respectively.

2. Right uncal herniation would be caused by an expanding mass lesion in the right cranial cavity. Possible causes in this setting would include increased swelling and edema from the infarct, or hemorrhagic conversion.

Clinical Course and Neuroimaging

An urgent head CT demonstrated increased swelling of the right hemisphere, with right-to-left midline shift and effacement of the interpeduncular cistern. The patient was intubated, hyperventilated, and given intravenous mannitol, which led to temporary improvement. Over the next 2 days an intracranial pressure (ICP) monitoring bolt was used, together with the neurologic exam, to gauge the response to escalating ICP-lowering measures (see KCC 5.3). On hospital day 4, however, the patient's right pupil became dilated, her ICP rose, and she developed bradycardia and hypertension (Cushing response; see Table 5.3), which did not respond to mannitol. Therefore, after discussion with the family, she was taken to the operating room for an investigational procedure called a hemicraniectomy (see KCC 10.4), in which a large piece of skull is temporarily removed to decompress the underlying brain (Image 10.11A,B, page 445). After a long, complicated hospital course and an inpatient rehabilitation stay, she was eventually discharged home with her family. On follow-up exam 2 months after presentation, she was soft-spoken and somewhat lethargic, and she had a persistent left hemiplegia and left hemianopia. However, she knew the correct month and year, and she was able to write her name and identify family members. Follow-up head CT after replacement of the bone flap showed resolution of the brain swelling (see Image 10.11B).

CASE 10.8 GLOBAL APHASIA, RIGHT HEMIPLEGIA, AND HEMIANOPIA

IMAGE 10.8A–C Left Middle Cerebral Artery (MCA) Stem Infarct, Causing Significant Mass Effect Head CT scan axial sections, with A–C progressing from inferior to superior.

(A)

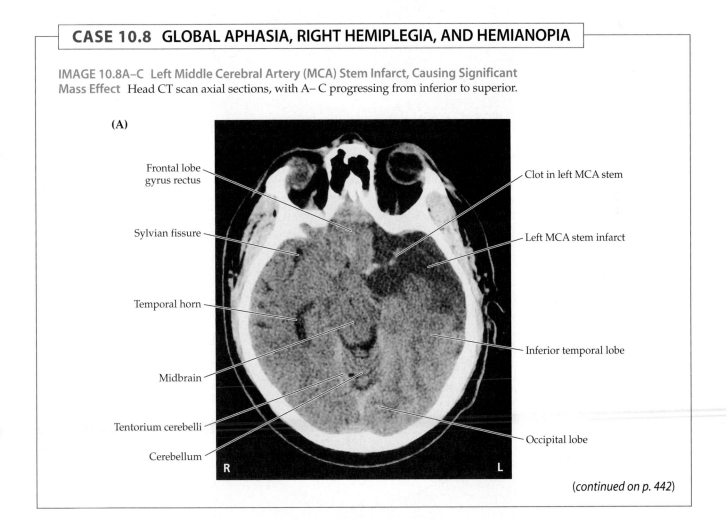

(*continued on p. 442*)

CASE 10.8 (continued)

(B)

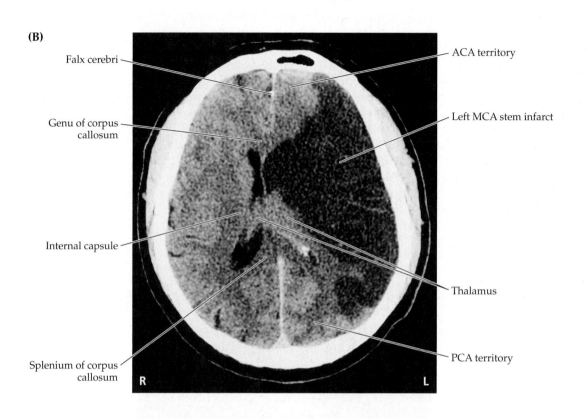

Falx cerebri

Genu of corpus callosum

Internal capsule

Splenium of corpus callosum

ACA territory

Left MCA stem infarct

Thalamus

PCA territory

R L

(C)

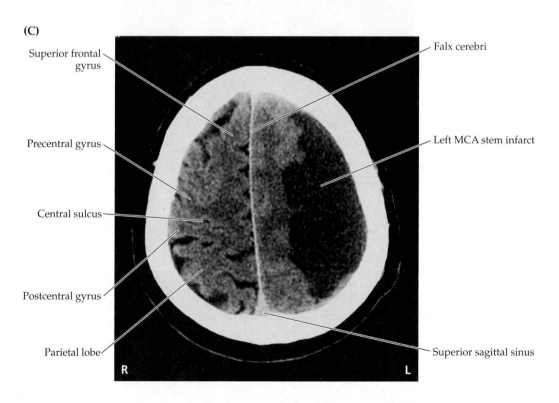

Superior frontal gyrus

Precentral gyrus

Central sulcus

Postcentral gyrus

Parietal lobe

Falx cerebri

Left MCA stem infarct

Superior sagittal sinus

R L

CASE 10.8 *(continued)*

IMAGE 10.8D,E **Left Middle Cerebral Artery (MCA) Stem Occlusion, Treated with tPA** MRI scan axial diffusion weighted images (DWI); D,E progressing from inferior to superior, showing patch areas of damaged and spared cortex.

(D)

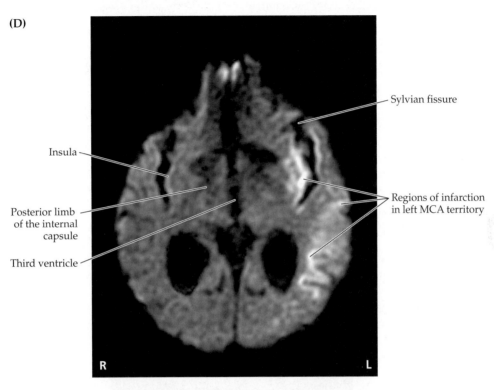

(E)

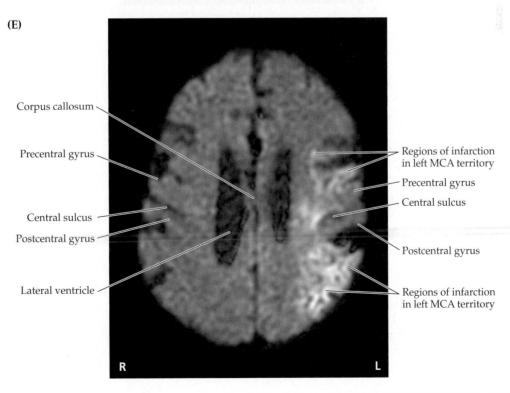

CASE 10.10 LEFT HEMINEGLECT

IMAGE 10.10A,B Right Carotid Occlusion (A) Injection of right carotid showing occlusion of right common carotid artery. (B) Injection of left carotid showing cross-filling via the anterior communicating artery (AComm) to the right anterior cerebral artery (ACA) and the right middle cerebral artery (MCA). A1 = initial segment of ACA, proximal to the first major branch at the AComm.

(A)

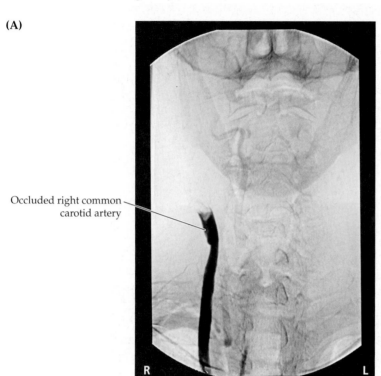

Occluded right common carotid artery

R   L

(B)

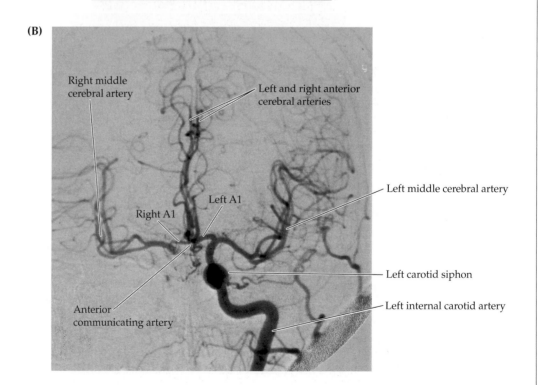

Right middle cerebral artery

Left and right anterior cerebral arteries

Left middle cerebral artery

Right A1

Left A1

Left carotid siphon

Left internal carotid artery

Anterior communicating artery

CASE 10.11 LEFT HEMINEGLECT, HEMIPLEGIA, AND HEMIANOPIA

IMAGE 10.11A,B Right Middle Cerebral Artery (MCA) Stem Infarct, Treated with Hemicraniectomy Head CT axial images. (A) Scan performed four days after admission, shortly after hemicraniectomy. A large swollen infarct is present in the right MCA territory with areas of increased density consistent with petichial hemorrhage.

Removal of the overlying skull prevented fatal uncal herniation. A ventriculostomy was also temporarily placed in the left lateral ventricle to prevent hydrocephalus (see KCC 5.7). (B) Follow-up scan six weeks later. The infarct is no longer swollen, and the skull bone flap has been replaced.

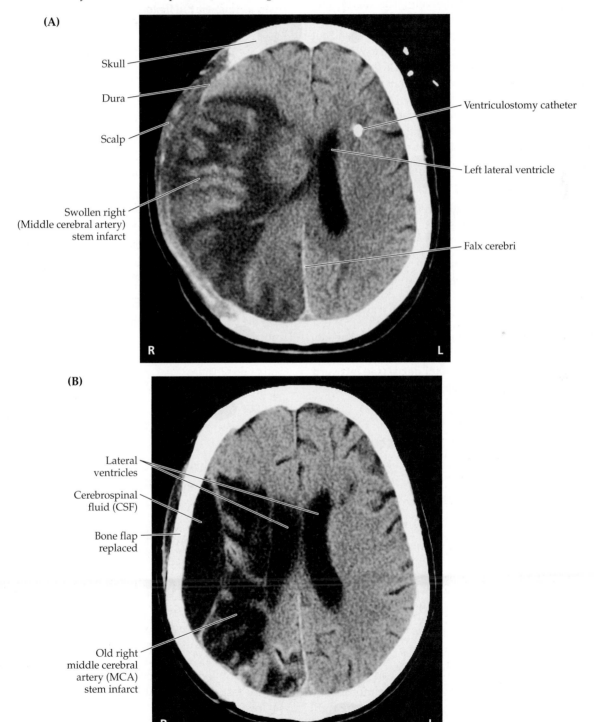

(A)

Skull

Dura

Scalp

Swollen right (Middle cerebral artery) stem infarct

Ventriculostomy catheter

Left lateral ventricle

Falx cerebri

R L

(B)

Lateral ventricles

Cerebrospinal fluid (CSF)

Bone flap replaced

Old right middle cerebral artery (MCA) stem infarct

R L

CASE 10.12 UNILATERAL PROXIMAL ARM AND LEG WEAKNESS

CHIEF COMPLAINT

A 52-year-old right-handed woman went to her physician the morning after developing **difficulty raising her left arm**.

HISTORY

Past history was notable for hypertension and heavy cigarette smoking. After supper on the evening prior to admission the patient tried to reach for a cup of coffee with her left hand but was **unable to raise her left arm**. As she turned to walk away, this movement caused her left arm to flop up in the air slightly and knock the coffee on the floor. She did not make much of this and went to sleep. The next morning, while shopping in the supermarket with her husband, she noticed that she could not raise her left arm to take items off the shelves. On the way home they stopped at her physician's office.

PHYSICAL EXAMINATION

Vital signs: T = 99.3°F, P = 84, BP = 140/70, R = 18.

Neck: Supple, with a **right carotid bruit continuing into diastole**.

Lungs: Clear.

Cardiac: Regular rate, with an S4 gallop.

Abdomen: Normal bowel sounds; soft.

Extremities: Normal.

Neurologic exam:

MENTAL STATUS: Alert and oriented × 3. Normal language. Normal simple calculations. Drew a clockface normally.

CRANIAL NERVES: Normal, except for **decreased leftward fast phases of optokinetic nystagmus** (see Chapter 13).

MOTOR: Normal fine finger movements. Power was 5/5 throughout on the right side. **Left arm power was as follows: shrug 4$^+$/5, deltoid 4$^-$/5, triceps 4/5, biceps 4$^+$/5, wrist extensors 5/5, fingers 5/5. Left leg power was as follows: hip flexors 4/5**, hip extensors 5/5, thigh adductors 5/5, thigh abductors 5/5, distal muscles 5/5.

REFLEXES:

COORDINATION: Slowed finger-to-nose testing in the left arm because of weakness.

GAIT: **Tended to veer to the left. Fell to the left on attempted tandem (heel-to-toe) walking.**

SENSORY: Intact light touch, pinprick, joint position sense, and vibration sense. Normal graphesthesia, no extinction on double simultaneous stimulation.

LOCALIZATION AND DIFFERENTIAL DIAGNOSIS

1. On the basis of the symptoms and signs shown in **bold** above, where is the lesion?

2. What is the most likely diagnosis, and what are some other possibilities?

Discussion

1. The key symptoms and signs in this case are:

   - **Weakness of the proximal left arm and leg, with left hyperreflexia and Babinski's sign**
   - **Unsteady gait, veering to the left**
   - **Decreased leftward fast phases of optokinetic nystagmus**
   - **Right carotid bruit**

   This patient has unilateral proximal arm and leg weakness of the upper motor neuron type (see KCC 6.1), sparing the face. This pattern of weakness, sometimes called "man in the barrel" syndrome, is consistent with a lesion in the contralateral motor cortex proximal arm–trunk–proximal leg areas (see Figure 10.1). Damage to this region can be caused by ACA–MCA watershed infarcts (see KCC 10.2). Gait unsteadiness with veering to the left can result from lesions in many locations (see KCC 6.5), including the right leg motor cortex. Leftward fast phases of optokinetic nystagmus can be impaired by right frontal lobe lesions.

2. The patient's vascular risk factors include hypertension and cigarette smoking. Meanwhile, the presence of a right carotid bruit suggests right carotid

stenosis. In this setting, decreased right carotid perfusion could occur if the systemic blood pressure suddenly decreased or if the stenosis suddenly worsened—for example, from thrombus formation. The most likely diagnosis is therefore decreased right carotid perfusion, resulting in watershed infarct in the right ACA–MCA territory, including the right motor cortex proximal arm and leg areas and the right frontal lobe. Other possibilities include another type of cortical lesion in the same location, such as hemorrhage, tumor, or abscess.

Clinical Course and Neuroimaging

The patient's physician sent her to the emergency room. Because her deficits had begun more than 4.5 hours previously, she was not a candidate for thrombolysis, but she was admitted to the hospital for further evaluation and treatment. A diffusion-weighted **MRI scan** (see Chapter 4) revealed an acute infarct in the right ACA–MCA watershed territory (Image 10.12A, page 450), which was confirmed 2 days later by conventional MRI (Image 10.12B, page 450; compare to Figure 10.10). MRA and carotid Doppler studies revealed a high-grade stenosis of the right internal carotid artery just above the carotid bifurcation. The patient therefore underwent a right carotid endarterectomy (see KCC 10.5). Pathologic examination of the endarterectomy specimen revealed severe atherosclerosis with superimposed mural thrombus, resulting in a 90% stenosis of the lumen. This finding suggests that thrombus could have temporarily occluded the artery at one time, causing the watershed infarct in the distal carotid territory. After surgery, her strength gradually improved, and she was treated with aspirin to reduce recurrent stroke risk. When seen in follow-up 5 weeks after surgery, she had normal strength throughout, except for trace 4$^+$/5 weakness of the left deltoid and iliopsoas.

CASE 10.13 RIGHT FRONTAL HEADACHE AND LEFT ARM NUMBNESS IN A WOMAN WITH GASTRIC CARCINOMA

CHIEF COMPLAINT

A 75-year-old right-handed woman was admitted for gastric carcinoma and then developed a right frontal headache with left arm numbness and weakness.

HISTORY

Two weeks prior to admission, the patient noticed difficulty eating. She was admitted to the hospital on the general surgery service when a large mass was found in her abdomen, and an endoscopic biopsy revealed gastric carcinoma. On the evening after admission, the nurse found her **lying on her left arm in an awkward position.** The patient complained of a **right frontal headache and left arm numbness.** The surgical intern found that she had left-sided weakness, and a neurology consult was called.

PHYSICAL EXAMINATION

Vital signs: T = 97.4°F, P = 80, BP = 130/80.

Neck: Supple with no bruits.

Lungs: Clear.

Heart: Regular rate with no murmurs.

Abdomen: Normal bowel sounds. An approximately 15 cm mass was palpable in the midabdomen, with mild tenderness.

Extremities: No edema.

Neurologic exam:

MENTAL STATUS: Alert and oriented × 3. Language normal. Able to recall 1/3 objects after 5 minutes, and 2/3 objects with prompting.

CRANIAL NERVES: Pupils 3 mm, constricting to 1 mm bilaterally. Normal fundi. Visual field full, but with **extinction on the left side to double simultaneous stimulation.** Extraocular movements full, but with a **right gaze preference. Facial sensation mildly decreased on the left side to light touch and pinprick. Mild left facial weakness, sparing the forehead.** Normal hearing. Normal gag, palate elevation, and articulation. Tongue midline.

MOTOR: **Left pronator drift.** Power 5/5 throughout on the right side. **Left arm strength 3/5 to 4/5.** Left iliopsoas and quadriceps 5/5, and **left extensor hallucis longus 4$^+$/5.**

(continued on p. 448)

CASE 10.13 *(continued)*

REFLEXES:

2^+ 2^+
2^+ 2^+
2^+ 2^+
2^+ 2^+

COORDINATION: Normal rapid alternating movements on the right. Left side not tested.

GAIT: Not tested.

SENSORY: Mildly decreased light touch, pinprick, temperature, vibration, and joint position sense on the left side. Dramatic extinction on the left to double simultaneous stimulation. Decreased stereognosis and graphesthesia in the left hand.

LOCALIZATION AND DIFFERENTIAL DIAGNOSIS

1. On the basis of the symptoms and signs shown in **bold** above, where is the lesion?

2. What is the most likely diagnosis, and what are some other possibilities?

Discussion

1. The key symptoms and signs in this case are:

   - **Right frontal headache**
   - **Weakness of the left face and arm more than the leg, with left Babinski's sign**
   - **Mildly decreased light touch, pinprick, temperature, vibration, and joint position sense on the left side, with decreased left stereognosis and graphesthesia**
   - **Left visual and tactile extinction**

   Weakness in the left face and arm that is greater than in the leg is usually caused by a lesion of the right face and arm motor cortex (see KCC 6.3; Figure 6.14D). The Babinski's sign supports the presence of an upper motor neuron lesion. Impaired primary and cortical sensation on the left side suggests a lesion in the right somatosensory cortex. The left neglect implies that the lesion may extend into parietal or, less likely, frontal association cortex. Right frontal headache has numerous possible causes (see KCC 5.1), including a lesion in the right hemisphere.

   The most likely *clinical localization* is right primary motor cortex face and arm areas, right somatosensory cortex, and right parietal association cortex.

2. Given the patient's age, the sudden onset of deficits, and the hypercoagulability associated with carcinoma (see Table 10.5), the most likely diagnosis is ischemic stroke. Although the findings do not neatly fit a right MCA superior or inferior division infarct (see KCC 10.1), they might be explained by an infarct overlapping these territories. A hemorrhage in this area could also explain her deficits. Other possibilities include an abscess or a tumor such as brain metastasis, especially given this patient's history. It should be noted that in about 10% of brain tumors, symptoms develop rapidly, in a "stroke-like" manner.

Initial Clinical Course

A **head CT** (Image 10.13A,B, page 451) showed hemorrhage with surrounding edema in the right parietal lobe extending to the face and arm regions of the precentral gyrus. The initial impression was hemorrhage into a brain metastasis, or cerebral infarct with hemorrhagic conversion. An MRI scan with gadolinium and an embolic workup were planned to investigate these possibilities. However, shortly after her head CT the patient suddenly became unresponsive and had left facial twitching, consistent with a **seizure** (see KCC 18.2). She was treated with intravenous anticonvulsants (diazepam and phenytoin) and she improved, although she had two more brief seizures over the next day and remained **difficult to arouse**. Repeat head CT showed no change in the bleed, and **head CT with intravenous contrast** (see Image 10.13B) did not show any enhancing lesions consistent with metastases. On careful review of the contrast CT, an empty delta sign was noted in the superior sagittal sinus (see Image 10.13B).

1. What is the significance of the empty delta sign?

2. What possible diagnosis should now be considered, and what tests should be done to investigate this possibility?

Discussion

1. The sagittal sinus normally fills uniformly with contrast, and a relatively dark region in the middle suggests a filling defect, possibly due to a blood clot (see Image 10.13B). Note that in retrospect, there was a suggestion of dense material (bright signal) in the sagittal sinus on the noncontrast scan as well (see Image 10.13A).

2. Given this finding, superior sagittal sinus thrombosis should be strongly suspected (see KCC 10.7). Other features consistent with (but not specific to) superior sagittal thrombosis include the headache, parasagittal hemorrhage, depressed level of consciousness, and seizures. The empty delta sign is suggestive but is not conclusive evidence for this diagnosis, so additional tests, such as a magnetic resonance venogram or conventional angiography, should be done.

Clinical Course and Neuroimaging

A **magnetic resonance venogram** (Image 10.13C, page 452) showed no appreciable flow in the superior sagittal sinus. This image should be compared to the normal MR venogram from another patient (Image 10.13D, page 452). Despite her hemorrhage, the patient was treated with low-level anticoagulation using subcutaneous heparin to prevent further thrombosis. She spent 3 weeks in inpatient rehabilitation, with improvement in her left arm strength and ambulation, and eventually underwent abdominal surgery for her gastric carcinoma.

CASE 10.12 UNILATERAL PROXIMAL ARM AND LEG WEAKNESS

IMAGE 10.12A,B Right ACA–MCA Watershed Infarct
(A) Coronal view from diffusion weighted MRI on the day of admission showing acute right ACA–MCA water-shed infarct. (B) Axial view from conventional T2 weighted MRI 2 days later, confirming infarct in this distribution.

(A)

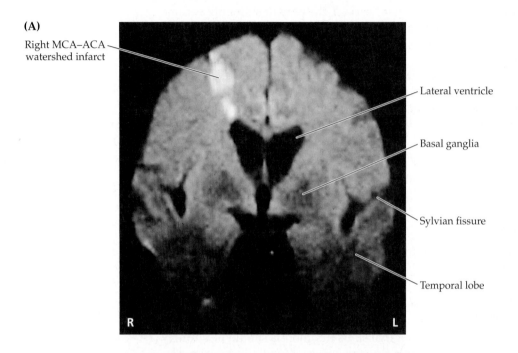

Right MCA–ACA watershed infarct

Lateral ventricle

Basal ganglia

Sylvian fissure

Temporal lobe

R   L

(B)

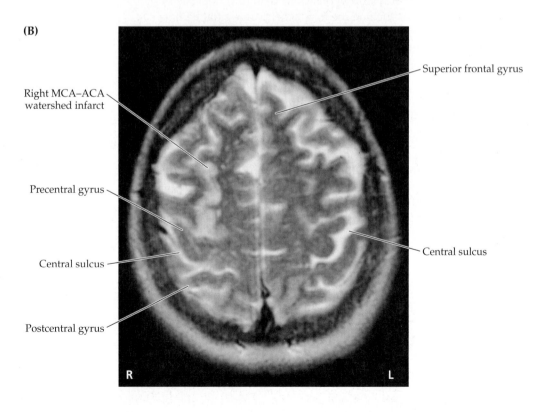

Superior frontal gyrus

Right MCA–ACA watershed infarct

Precentral gyrus

Central sulcus

Central sulcus

Postcentral gyrus

R   L

CASE 10.13 RIGHT FRONTAL HEADACHE AND LEFT ARM NUMBNESS IN A WOMAN WITH GASTRIC CARCINOMA

IMAGE 10.13A,B Right Parietal Hemorrhage and Empty Delta Sign (A) Noncontrast scan showing right parietal hemorrhage with surrounding edema. (B) Repeat scan with intravenous contrast, showing empty delta sign.

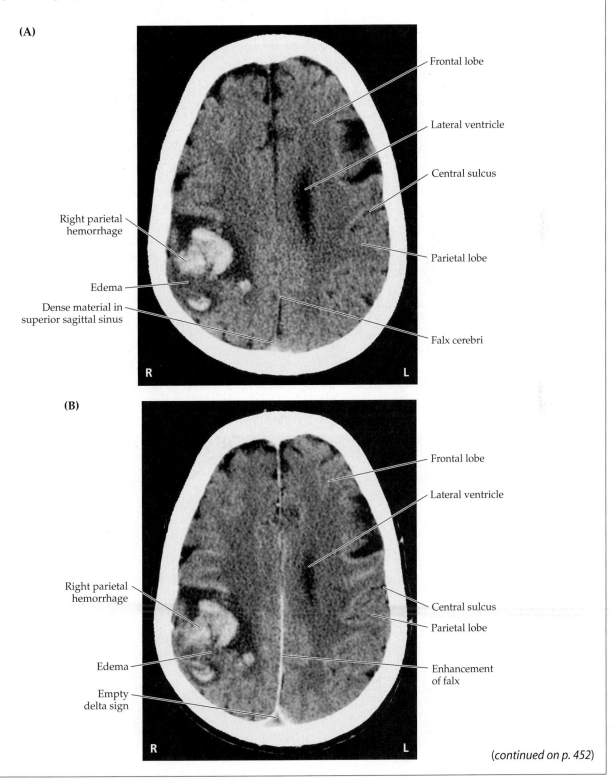

(A)

- Frontal lobe
- Lateral ventricle
- Central sulcus
- Parietal lobe
- Falx cerebri

Right parietal hemorrhage
Edema
Dense material in superior sagittal sinus

R　　L

(B)

- Frontal lobe
- Lateral ventricle
- Central sulcus
- Parietal lobe
- Enhancement of falx

Right parietal hemorrhage
Edema
Empty delta sign

R　　L

(continued on p. 452)

CASE 10.13 (continued)

IMAGE 10.13C,D Superior Sagittal Sinus Thrombosis
Magnetic resonance venogram (MRV). (C) MRV from patient in Case 10.13, showing absence of flow in the superior sagittal sinus, likely due to thrombosis. (D) MRV in a normal patient, showing normal flow in the superior sagittal sinus and cortical veins. Note that the transverse sinuses are not well seen in both C and D because their initial portions near the torcular are being viewed end-on and their more lateral portions are truncated by the imaging method. Compare to Figure 10.11.

(C)

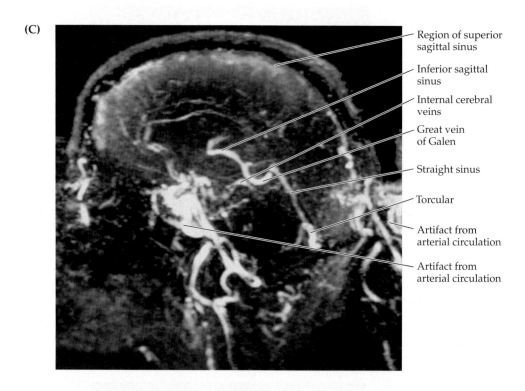

Region of superior sagittal sinus

Inferior sagittal sinus

Internal cerebral veins

Great vein of Galen

Straight sinus

Torcular

Artifact from arterial circulation

Artifact from arterial circulation

(D)

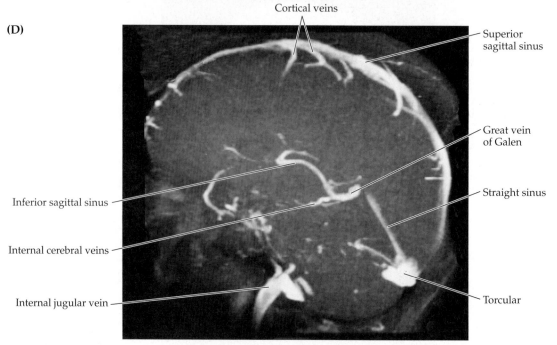

Cortical veins

Superior sagittal sinus

Great vein of Galen

Straight sinus

Torcular

Inferior sagittal sinus

Internal cerebral veins

Internal jugular vein

Additional Cases

Related cases can be found for the following: **cerebral infarct**, or **TIA** (Cases 5.5, 6.1–6.3, 6.5, 7.1, 7.2, 11.1, 11.6, 13.7, 14.1–14.8, 15.1, 18.3, 19.1–19.4, 19.6, 19.8, 19.9); **aneurysm** (Case 13.1); **arteriovenous malformation** (**AVM**) (Case 11.5); **dissection** (Case 13.6); and **intracranial hemorrhage** (Cases 5.1–5.3, 5.5, 5.6, 14.9, 19.3, 19.4). Other relevant cases can be found using the **Case Index** located at the end of the book.

Brief Anatomical Study Guide

1. The **main functions of the cerebral hemispheres** are summarized in Figure 10.1. The three main cerebral arteries are the **anterior cerebral artery** (**ACA**), **middle cerebral artery** (**MCA**), and **posterior cerebral artery** (**PCA**). The ACA and MCA arise from the **anterior circulation**, or **carotid territory** of the **circle of Willis**, while the PCA arises from the **posterior circulation**, or **vertebrobasilar system** (see Figures 10.2 and 10.3).

2. The **ACA** supplies the medial frontal and medial parietal lobes, including the sensorimotor cortex for the lower extremity (see Figure 10.4). The **PCA** supplies the occipital lobes, including the primary visual cortex, and the medial-inferior temporal lobes. The **MCA** supplies the entire lateral surface of the cerebral hemispheres, including the face and arm sensorimotor cortex and many regions of association cortex (see Figure 10.5).

3. The MCA territory has the following three major subdivisions (see Table 10.1): the **MCA superior division** supplies most of the cortex above the Sylvian fissure, including the lateral frontal cortex and the face and arm peri-Rolandic cortex. The **MCA inferior division** supplies the cortex of the lateral temporal and parieto-occipital lobes below the Sylvian fissure. The **MCA deep territory** includes the internal capsule and much of the basal ganglia (see Figures 10.7–10.9).

4. The **ACA** also has a **deep territory**, including portions of the anterior basal ganglia and internal capsule, while the **deep territory of the PCA** includes the thalamus, midbrain, and variable parts of the posterior internal capsule (see Figures 10.8 and 14.21A). The superficial and deep territories of all three major cerebral vessels are summarized in Figures 10.5 and 10.9. Deficits caused by occlusion of the three main cerebral arteries or their branches are summarized in Table 10.1.

5. Blockage of a cerebral artery or its branches often causes an infarct in a specific vascular territory. Infarcts can also occur through another mechanism when the systemic blood pressure drops or when a parent vessel (such as the carotid artery) supplying more than one major cerebral vessel (e.g., the ACA and MCA) becomes blocked, resulting in infarction of the most distal territories of overlap of these vessels. These territories are called **watershed zones** (see Figure 10.10).

6. Venous drainage of the cerebral hemispheres occurs through a system of superficial and deep cerebral veins. The superficial veins drain mainly into the **superior sagittal sinus** and **cavernous sinus**, while the deep veins drain into the **great vein of Galen** (see Figure 10.11). Ultimately, all venous drainage for the brain reaches the **internal jugular veins**, mostly via the **transverse and sigmoid sinuses**.

References

General References

Bogousslavsky J, Regli F. 1990. Anterior cerebral artery territory infarction in the Lausanne Stroke Registry: Clinical and etiologic patterns. *Arch Neurol* 47 (2): 144–150.

Brust JCM. 1998. Anterior cerebral artery. In *Stroke: Pathophysiology, Diagnosis, and Management*. 3rd Ed., HJM Barnett, JP Mohr, BM Stein (eds.), Chapter 18. Churchill Livingstone, New York.

Caplan L. 2009. *Caplan's Stroke: A Clinical Approach*. Saunders, Philadelphia.

Damasio H. 1987. Vascular territories defined by computed tomography. In *Cerebral Blood Flow: Physiologic and Clinical Aspects*, JH Wood (ed.), Chapter 20. McGraw-Hill, New York.

Klatka LA, Depper MH, Marini AM. 1998. Infarction in the territory of the anterior cerebral artery. *Neurology* 51 (2): 620–622.

Osborn AG, Tong KA. 1995. *Handbook of Neuroradiology: Brain and Skull*. 2nd Ed. Mosby, St. Louis, MO.

Renfro MB, Day AL, Rhoton AL. 1997. The extracranial and intracranial vessels: Normal anatomy and variations. In *Cerebrovascular Disease*, HH Batjer (ed.), Chapter 1. Lippincott-Raven, Philadelphia.

Tatu L, Moulin T, Bogousslavsky J, Duvernay H. 1998. Arterial territories of the human brain: Cerebral hemispheres. *Neurology* 50 (60): 1699–1708.

Warlow CP, van Gijn J, Dennis MS, Wardlaw JM. 2008. *Stroke: Practical Management*. 3rd Ed. Wiley-Blackwell, Malden, MA.

Middle Cerebral Artery

Caplan L, Babilkian V, Helgason C, Hier DB, DeWitt D, Patel D, Stein R. 1985. Occlusive disease of the middle cerebral artery. *Neurology* 35 (7): 975–982.

Heinsius T, Bogousslavsky J, Van Melle G. 1998. Large infarcts in the middle cerebral artery territory: Etiology and outcome patterns. *Neurology* 50 (2): 341–350.

Lhermitte F, Gautier JC, Derouesne C. 1970. Nature of occlusions of the middle cerebral artery. *Neurology* 20 (1): 82–88.

Mohr JP, Lazar RM, Marshall RS, Gautier JC. 1998. Middle cerebral artery disease. In *Stroke: Pathophysiology, Diagnosis, and Management*. 3rd Ed., HJM Barnett, JP Mohr, BM Stein (eds.), Chapter 19. Churchill Livingstone, New York.

Moulin T et al. 1996. Early CT signs in acute middle cerebral artery infarction: Predictive value for subsequent infarct locations and outcome. *Neurology* 47 (2): 366–374.

Waddington MM, Ring BA. 1968. Syndromes of occlusions of middle cerebral artery branches. *Brain* 91 (4): 685–696.

Posterior Cerebral Artery

Chambers BR, Brooder RJ, Donnan GA. 1991. Proximal posterior cerebral artery occlusion simulating middle cerebral artery occlusion. *Neurology* 41 (3): 385–390.

DeRenzi E, Zambolin A, Crisi G. 1987. The pattern of neuropsychological impairment associated with left posterior cerebral artery infarcts. *Brain* 110 (Pt. 5): 1099–1116.

Finelli PF. 2008. Neuroimaging in acute posterior cerebral artery infarction. *Neurologist* 14 (3): 170–180.

Hayman LA, Berman SA, Hinck VC. 1981. Correlation of CT cerebral vascular territories with function: II. Posterior cerebral artery. *Am J Neuroradiol* 2: 219–225.

Mohr JP, Pessin MS. 1998. Posterior cerebral artery disease. In *Stroke: Pathophysiology, Diagnosis, and Management*. 3rd Ed., HJM Barnett, JP Mohr, BM Stein (eds.), Chapter 20. Churchill Livingstone, New York.

Pessin MS, Kwan ES, DeWitt LD, Hedges TR, Gale D, Caplan LR. 1987. Posterior cerebral artery stenosis. *Ann Neurol* 21 (1): 85–89.

Lacunar Infarcts and Other Subcortical Infarcts

Adams HP Jr, Damasio HC, Putman SF, Damasio AR. 1983. Middle cerebral artery occlusion as a cause of isolated subcortical infarction. *Stroke* 14 (6): 948–952.

Boiten J, Lodder J. 1991. Discrete lesions in the sensorimotor control system. A clinical topographical study of lacunar infarcts. *J Neurol Sci* 105 (2): 150–154.

Fisher CM. 1965. Lacunes: Small, deep cerebral infarcts. *Neurology* 15: 774–784.

Fisher CM. 1991. Lacunar infarcts—A review. *Cerebrovasc Dis* 1: 311–320.

Fisher CM, Curry HD. 1965. Pure motor hemiplegia of vascular origin. *Arch Neurol* 13: 30–44.

Ghika J, Bogousslavsky J, Regli F. 1991. Infarcts in the territory of lenticulostriate branches from the middle cerebral artery. Etiological factors and clinical features in 65 cases. *Schweiz Arch Neurol Psychiatr* 142 (1): 5–18.

Lodder J, Barnford J, Kappelle J, Boiten J. 1994. What causes false clinical prediction of small deep infarcts? *Stroke* 25 (1): 86–96.

Melo TP, Bogousslavsky J, van Melle G, Regli F. 1992. Pure motor stroke: A reappraisal. *Neurology* 42 (4): 789–795.

Stroke Mechanisms and Treatment

Adams H, Adams R, Del Zoppo G, Goldstein LB. 2005. Guidelines for the Early Management of Patients with Ischemic Stroke. 2005 Guidelines Update. A Scientific Statement from the Stroke Council of the American Heart Association/American Stroke Association. *Stroke* 36: 916.

Amarenco P, Bogousslavsky J, Callahan A 3rd, et al. 2006. High-dose atorvastatin after stroke or transient ischemic attack. *N Engl J Med* 355: 549–559.

Barnett HJM, Taylor DW, Eliasziw M, et al. 1998. Benefit of carotid endarterectomy in patients with symptomatic moderate or severe stenosis. *N Engl J Med* 339 (20): 1415–1425.

Bruno A, Biller J, Adams HP, Jr, Clarke WR, Woolson RF, Williams LS, Hansen MD. 1999. Acute blood glucose level and outcome from ischemic stroke. *Neurology* 52 (2): 280–284.

Delashaw JB, Broaddus WC, Kassell NF, Halcy EC, Pendleton GA, Vollmer DG, Maggio WW, Grady MS. 1990. Treatment of right cerebral infarction by hemicraniectomy. *Stroke* 21 (6): 874–881.

Easton JD, Saver JL, Albers GW, et al. 2009. Definition and evaluation of transient ischemic attack: a scientific statement for healthcare professionals from the American Heart Association/American Stroke Association Stroke Council. *Stroke* 40: 2276–2293.

Executive Committee for the Asymptomatic Carotid Atherosclerosis Study. 1995. Endarterectomy for asymptomatic carotid artery stenosis. *JAMA* 273: 1421–1428.

Furlan A, Higashida R, Wechsler L, et al. 1999. Intra-arterial prourokinase for acute ischemic stroke. The PROACT II Study: A randomized controlled trial. *JAMA* 282: 2003–2011.

Hacke W, Kaste M, Bluhmki E, et al. 2008. Thrombolysis with alteplase 3 to 4.5 hours after acute ischemic stroke. *N Engl J Med* 359: 1317–1329.

Huttner HB, Jüttler E, Schwab S. 2008. Hemicraniectomy for middle cerebral artery infarction. *Curr Neurol Neurosci Rep.* 8 (6): 526–533.

Josephson SA, Sidney S, Pham TN, Bernstein AL, Johnston SC. 2008. Higher ABCD2 score predicts patients most likely to have true transient ischemic attack. *Stroke* 39 (11): 3096–3098.

Mohr JP, Thompson JL, Lazar RM, Levin B, Sacco RL, Furie KL, Kistler JP, Albers GW et al. 2001. Warfarin-Aspirin Recurrent Stroke Study Group. A comparison of warfarin and aspirin for the prevention of recurrent ischemic stroke. *N Engl J Med* 345 (20): 1444–1451.

NINDS Stroke rt-PA Stroke Study Group. 1995. Tissue plasminogen activator for acute ischemic stroke. *N Engl J Med* 333 (24): 1581–1587.

Sacco RL, Adams R, Albers G, et al. 2006. Guidelines for prevention of stroke in patients with ischemic stroke or transient ischemic attack: a statement for healthcare professionals from the American Heart Association/American Stroke Association Council on Stroke. *Circulation* 113: e409–449.

Watershed Infarcts

Bladin CF, Chambers BR. 1993. Clinical features, pathogenesis, and computed tomographic characteristics of internal watershed infarction. *Stroke* 24 (12): 1925–1932.

Bogousslavsky J, Regli F. 1986. Unilateral watershed cerebral infarcts. *Neurology* 36 (3): 373–377.

Wodarz R. 1980. Watershed infarctions and computed tomography. A topographical study in cases with stenosis or occlusion of the carotid artery. *Neuroradiology* 19 (5): 245–248.

Headache and Stroke

Jorgensen HS, Jespersen HF, Nakayamu H, Raaschou HO, Olsen TS. 1994. Headache in stroke: The Copenhagen stroke study. *Neurology* 44 (10): 1793–1797.

Vestergaard K, Andersen G, Nielsen MI, Jensen TS. 1993. Headache in stroke. *Stroke* 24 (11): 1621–1624.

Cerebral Venous Anatomy and Thrombosis

Agostoni E, Aliprandi A, Longoni M. 2009. Cerebral venous thrombosis. *Expert Rev Neurother* 9 (4): 553–564.

Capra NF, Anderson KV. 1984. Anatomy of the cerebral venous system. In *The Cerebral Venous System and Its Disorders*, JP Kapp, HH Schmidek (eds.), Chapter 1. Grune & Stratton, Orlando, FL.

Einhaupl KM, Villringer A, Meister W, Mehraein S, Gamer C, Pellkofer M, Haberl RL, Pfister HW, Schmeidck P. 1991. Heparin treatment of venous sinus thrombosis. *Lancet* 338 (8767): 597–600.

Isensee C, Reul J, Thron A. 1994. Magnetic resonance imaging of thrombosed dural sinuses. *Stroke* 25 (1): 29–34.

Saadatnia M, Fatehi F, Basiri K, Mousavi SA, Mehr GK. 2009. Cerebral venous sinus thrombosis risk factors. *Int J Stroke* 4 (2): 111–123.

Virapongse C, Cazenave C, Quisling R, Sarwar M, Hunter S. 1987. The empty delta sign: Frequency and significance in 76 cases of dural sinus thrombosis. *Radiology* 162 (3): 779–785.

CONTENTS

Chapter *11*

Visual System

Damage to the visual pathways in a single location can affect one or both eyes, causing significant functional impairments. A 57-year-old man repeatedly visited the emergency room because of headaches. He experienced throbbing bilateral or right occipital pain and saw zigzagging lines in his visual field. He had also recently noticed a vision problem that caused him to frequently bump into objects on his left side. On examination, he was unable to see anything in the left lower quadrant of his visual fields in both eyes.

In this chapter, we will learn about the normal anatomy and function of the neural pathways from the retina to the cortex, and we will use this knowledge to accurately diagnose and localize lesions in clinical cases.

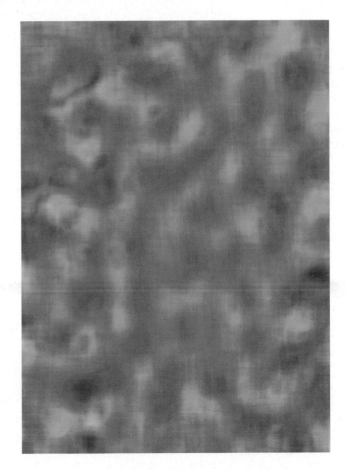

ANATOMICAL AND CLINICAL REVIEW

HUMAN BEINGS ARE highly visual creatures. A greater portion of our brains is devoted to sight than to any other sensory modality. In this chapter we will review the anatomy of the visual pathways from the retina to the lateral geniculate nuclei of the thalamus to the primary visual cortex, and the effects of lesions at various points in this system (see Figure 11.15). Disorders of higher-order visual processing caused by lesions of visual association cortex will be discussed in Chapter 19.

Eyes and Retina

As light enters the eye and passes through the **lens**, it forms an image on the retina that is **inverted and reversed**: Information from the upper visual space is projected onto the lower retina, and the lower visual space projects to the upper retina (Figure 11.1A). Similarly, the right part of visual space projects to the left hemiretina of each eye, and vice versa (Figure 11.1B). The central fixation point for each eye falls onto the **fovea**, which is the region of the retina with the highest visual acuity. The fovea corresponds to the central 1° to 2° of visual space. Despite its relatively small size, information from the fovea is represented by about half of the fibers in the optic nerve and half of the cells in the primary visual cortex. The **macula** is an oval region approximately 3 by 5 millimeters that surrounds the fovea and also has relatively high visual acuity. The macula occupies the central 5° of visual space.

About 15° medial (nasal) to the fovea is the **optic disc**, the region where the axons leaving the retina gather to form the **optic nerve**. There are no photoreceptors over the optic disc. This creates a small blind spot located about 15° lateral (temporal) and slightly inferior to the central fixation point for each eye (Figure 11.1C and Figure 11.2). Note that the blind spots for the two eyes are not superimposed, so there is no functional deficit when both eyes are used. Interestingly, even when one eye is closed our visual analysis pathways appear to "fill in" the blind spot. Thus, we are not usually aware of it unless it is specifically tested for (Figure 11.3).

FIGURE 11.1 **Formation of Images on the Retina** Images on the retina are inverted (A) and reversed (B). In addition, the lack of rods and cones over the optic nerve head (optic disc) creates a blind spot for each eye located about 15° lateral to the central fixation point (C).

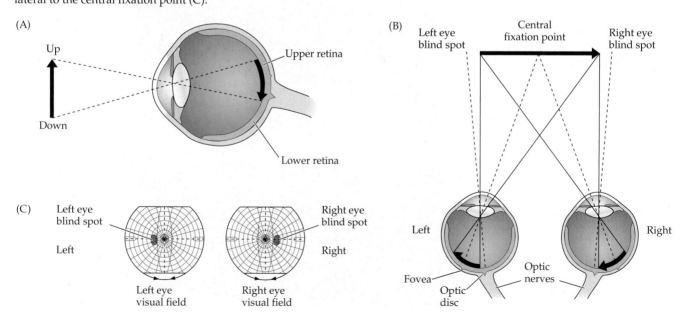

(A) Left eye (O.S.) visual field

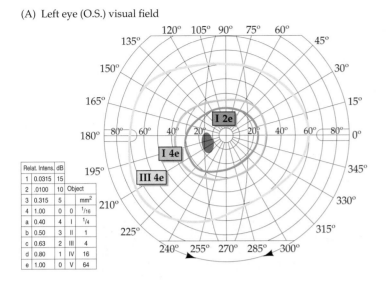

(B) Right eye (O.D.) visual field

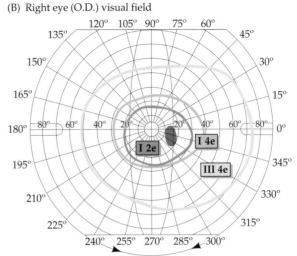

FIGURE 11.2 Normal Visual Fields
Fields were tested by Goldmann perimetry (see KCC 11.2). Test objects of different size and intensity were used. The larger (III = 4 mm^2), and brighter (4e = relative intensity of 1.00) objects have larger visual fields than smaller (I = 1/4 mm^2) and dimmer (2e = relative intensity of 0.01) objects. Blind spots (tested with object I 2e) are also indicated.

Numerous investigators have studied the cellular components of the visual pathways. We will review only the highlights here (see the References section at the end of the chapter for additional details). There are two classes of **photoreceptors** in the retina: rods and cones (Figure 11.4). **Rods** are more numerous than cones by a ratio of about 20:1. However, rods have relatively poor spatial and temporal resolution of visual stimuli, and they do not detect colors. Their main function is for vision in low-level lighting conditions, where they are far more sensitive than cones. In normal daylight the response of most rods is saturated. **Cones** are less numerous overall, but they are much more highly represented in the fovea, where visual acuity is highest. Cones have relatively high spatial and temporal resolution, and they detect colors.

In addition to the photoreceptors, the retina contains several other cellular layers (see Figure 11.4). The photoreceptors form the outermost layer, farthest from the lens. Therefore, light must traverse the entire thickness of the retina to reach them. In the fovea, however, the other layers of the retina are not present, allowing light to reach the photoreceptors without distortion.

The **receptive field** of a neuron in the visual pathway is defined as the portion of the visual field where light causes excitation or inhibition of the cell. Photoreceptors respond to light in their receptive fields and form excitatory or inhibitory synapses onto **bipolar cells**. Bipolar cells, in turn, synapse onto **ganglion cells**, which send their axons into the optic nerve (see Figure 11.4). Unlike

FIGURE 11.3 Demonstration of the Blind Spot for the Left Eye Cover your right eye and fixate on the upper cross with the book held about 15 inches from your eye. Move the book slightly in and out, and the dark circle should disappear when it falls into the blind spot of your left eye. Similarly, cover your right eye and fixate on the lower cross, and the gap in the line should disappear in the left eye's blind spot. (After Hurvich LM. 1981. *Color Vision*. Sinauer, Sunderland, MA.)

FIGURE 11.4 The Retina (A) Spatial relationships between the retina and other structures of the eye. (B) Magnified view of the fovea, where light reaches photoreceptors (rods and cones) without passing through intervening layers. The main cells and cell layers of the retina are indicated.

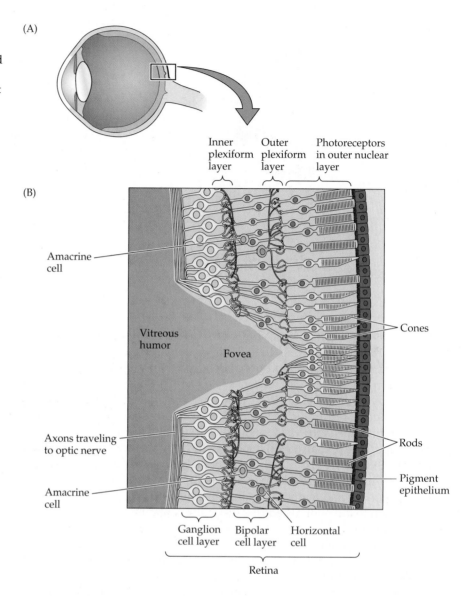

(A)

(B)

Inner plexiform layer

Outer plexiform layer

Photoreceptors in outer nuclear layer

Amacrine cell

Vitreous humor

Fovea

Cones

Axons traveling to optic nerve

Rods

Amacrine cell

Pigment epithelium

Ganglion cell layer

Bipolar cell layer

Horizontal cell

Retina

REVIEW EXERCISE

Name each of the five layers of the retina, proceeding from the outside toward the center of the eye. For each layer, describe the main cell types or synapses that occur (see Figure 11.4).

other neurons, photoreceptors and bipolar cells do not fire action potentials. Instead, information is conveyed along the length of these cells by passive electrical conduction, and they communicate through "nontraditional" synapses that release neurotransmitter in a graded fashion that depends on membrane potential. Ganglion cells, on the other hand, do fire action potentials as they convey information into the optic nerve.

In addition to this direct, or vertical, pathway through the retina, there are also interneurons called **horizontal cells** and **amacrine cells** (see Figure 11.4). These interneurons have lateral inhibitory or excitatory connections with nearby bipolar and ganglion cells. Therefore, a small spot of light on the retina causes excitation (or inhibition) of bipolar and ganglion cells directly in its path and inhibition (or excitation) of the surrounding bipolar and ganglion cells. As a result of these lateral connections, bipolar and ganglion cells have receptive fields with a **center–surround** (concentric) configuration (Figure 11.5).

There are two classes of center–surround cells. **On-center** cells are excited by light in the center of their receptive field and inhibited by light in the surrounding area. Conversely, **off-center** cells are inhibited by light in the center and excited by light in the surrounding area. Beginning with the bipolar cells, many neurons in the visual pathway—including ganglion cells, lateral genic-

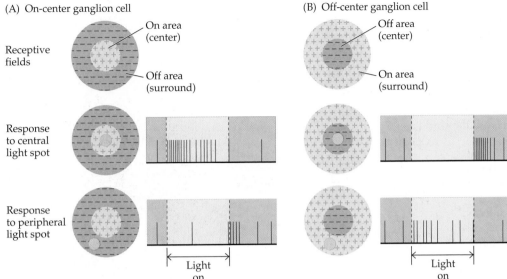

(A) On-center ganglion cell

On area (center)

Receptive fields

Off area (surround)

Response to central light spot

Response to peripheral light spot

Light on

(B) Off-center ganglion cell

Off area (center)

On area (surround)

Response to central light spot

Response to peripheral light spot

Light on

FIGURE 11.5 Retinal Ganglion Cells with Center–Surround (Concentric) Receptive Fields (A) Receptive fields and response patterns of an on-center cell. Duration of the light stimulus and action potential firing pattern, recorded extracellularly, are shown to the right. (B) Receptive fields and response patterns of an off-center cell. (After Kuffler SW. 1953. Discharge patterns and functional organization of mammalian retina. *J Neurophysiol* 16: 37–68.)

ulate neurons, and input neurons of the primary visual cortex—have center–surround receptive fields, which are either on-center or off-center. Beyond the input neurons of the visual cortex, neurons involved in vision have more sophisticated receptive field properties, which will be discussed later.

Retinal ganglion cells can be further classified as **parasol cells** (also known as Pα or A cells), which have large cell bodies, large receptive (and dendritic) fields, and respond best to gross stimulus features and movement, and **midget cells** (also known as Pβ or B cells), which have small cell bodies, small receptive (and dendritic) fields, are more numerous, and are sensitive to fine visual detail and to colors. Parasol cells have larger-diameter fibers and project to the **magnocellular layers** of the lateral geniculate nucleus of the thalamus (see Figure 11.7), while midget cells have smaller-diameter fibers and project to the **parvocellular layers** of the lateral geniculate. There are also other retinal ganglion cells that fit neither class, some of which are sensitive to overall light intensity. Parasol and midget cells can be either on- or off-center cells.

Optic Nerves, Optic Chiasm, and Optic Tracts

The retinal ganglion cells send their axons into the **optic nerve**,* which exits through the orbital apex via the optic canal of the sphenoid bone (see Figure 12.3A,C) to enter the cranial cavity. There is a partial crossing of fibers in the **optic chiasm** (see Figure 11.15). Thus, fibers from the left hemiretinas of both eyes end up in the left **optic tract**, while fibers from the right hemiretinas end up in the right optic tract. Note that in order to accomplish this, the nasal (medial) retinal fibers for each eye, which are responsible for the temporal (lateral) hemifields, cross over in the optic chiasm. Lesions of the optic chiasm therefore often produce **bitemporal (bilateral lateral) visual field defects** (see Figure 11.15C). Lesions of the eye, retina, or optic nerves produce **monocular visual field defects** (see Figure 11.15A,B). Because of the crossover in the optic

---

*The optic nerve is not truly a nerve. Retinal bipolar cells are analogous to primary somatosensory neurons (dorsal root ganglion cells), while retinal ganglion cells are analogous to secondary somatosensory neurons that project to the thalamus. Thus, the pathway formed by the retinal ganglion cell axons actually lies entirely within the central nervous system. However, by convention, the initial portion of this pathway (anterior to the optic chiasm) is called the optic nerve, while the more proximal portion of this pathway (posterior to the optic chiasm) is called the optic tract (see Figure 11.15).

FIGURE 11.6 Geniculate and Extrageniculate Visual Pathways The geniculate (or geniculostriate) pathway relays in the lateral geniculate nucleus (LGN) and continues to the primary visual cortex via the optic radiations (see Figure 11.15). The extrageniculate pathways bypass the LGN via the brachium of the superior colliculus and relay in the pretectal area and superior colliculus. Projections from the pretectal area and superior colliculus then continue to the pulvinar en route to temporoparieto-occipital association cortex. The medial geniculate nucleus (MGN) and inferior colliculus, also shown here, are important relays in the auditory system that will be discussed in Chapter 12.

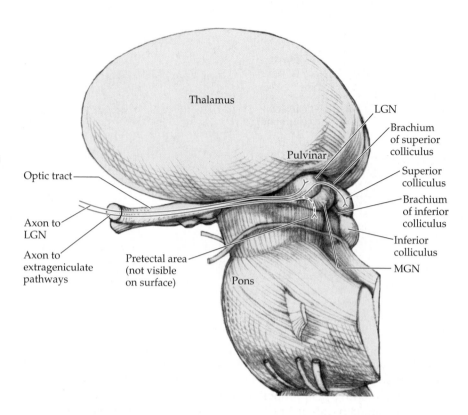

chiasm, lesions proximal to the chiasm (optic tracts, lateral geniculate, optic radiations, or visual cortex) generally produce **homonymous visual field defects**, meaning that the defect occurs in the same portion of the visual field for each eye (see Figure 11.15D,G,H).

The optic chiasm lies on the ventral surface of the brain, beneath the frontal lobes and just in front of the pituitary gland (see Figure 17.2B). It is therefore susceptible to compression by pituitary tumors and other lesions in this vicinity. The optic tracts wrap around the midbrain laterally to reach the lateral geniculate nucleus (LGN) of the thalamus (Figure 11.6).

Lateral Geniculate Nucleus and Extrageniculate Pathways

The axons of retinal ganglion cells in the optic tracts form synapses on neurons in the **lateral geniculate nucleus** (**LGN**) of the thalamus, which in turn project to the primary visual cortex. A minority of fibers in the optic tract bypass the LGN to enter the **brachium of the superior colliculus** (see Figure 11.6). These retinal fibers form the **extrageniculate visual pathways**, which project mainly to the **pretectal area** and **superior colliculus**. As we will see in Chapter 13, the pretectal area is important in the pupillary light reflex and projects to the parasympathetic nuclei controlling the pupils (see Figure 13.8). The superior colliculus and pretectal area are important in directing visual attention and eye movements toward visual stimuli. The superior colliculus and pretectal area therefore project to numerous brainstem areas involved in these functions, as well as to **association cortex** (lateral parietal cortex and frontal eye fields of prefrontal cortex) via relays in the **pulvinar** and lateral posterior nucleus of the thalamus (see Figure 11.6; see also Figures 7.7 and 7.8). Thus, the retino-tecto-pulvinar-extrastriate cortex pathway functions in visual attention and orientation, while the retino-geniculo-striate pathway functions in visual discrimination and perception.

The LGN has six layers, numbered 1 to 6 from ventral to dorsal (Figure 11.7). The first two **magnocellular layers** relay information from parasol cells of the

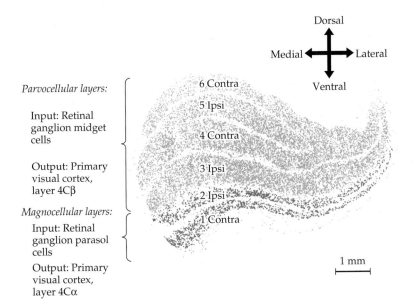

Dorsal

Medial ◄──┼──► Lateral

Ventral

Parvocellular layers:

Input: Retinal ganglion midget cells

Output: Primary visual cortex, layer 4Cβ

Magnocellular layers:

Input: Retinal ganglion parasol cells

Output: Primary visual cortex, layer 4Cα

6 Contra
5 Ipsi
4 Contra
3 Ipsi
2 Ipsi
1 Contra

1 mm

FIGURE 11.7 Layers of the Lateral Geniculate Nucleus The lateral geniculate nucleus (LGN) is composed of six layers, numbered from ventral to dorsal. As shown, the more dorsal parvocellular layers receive inputs from retinal ganglion midget cells and send outputs to visual cortex layer 4Cβ. The more ventral magnocellular layers receive inputs from retinal ganglion parasol cells and send outputs to visual cortex layer 4Cα. Contra = input is from retinal ganglion cells of the contralateral eye; Ipsi = input is from retinal ganglion cells of the ipsilateral eye. (Courtesy of Tim Andrews and Dale Purves, Duke University School of Medicine.)

retina (motion and spatial analysis), while layers 3 through 6, the **parvocellular layers**, relay information from midget cells (detailed form and color). These two pathways through the LGN are sometimes referred to as the **M pathway** (for magnocellular) and **P pathway** (for parvocellular). **Interlaminar neurons** (also called koniocellular neurons) located between the layers of the LGN participate, along with parvocellular neurons, in color vision and possibly serve other functions as well.

The information from the left and right eyes remains segregated even after passing through the LGN. The segregation for each eye is preserved because axons from the ipsilateral and contralateral retinas synapse onto different layers of the LGN (see Figure 11.7). Most neurons of the LGN have on- or off-center–surround receptive fields, similar to retinal neurons (see Figure 11.5). However, some LGN neurons, particularly in the magnocellular layers, are **on/off cells**. These cells detect changes and fire transiently to both on and off stimuli.

Optic Radiations to Primary Visual Cortex

The axons leaving the LGN enter the white matter to sweep over and lateral to the atrium and temporal horn of the lateral ventricle (through the *C* shape of the lateral ventricle) and then back toward the primary visual cortex in the occipital lobe (Figure 11.8). As they do so, these axons fan out over a wide area, forming the **optic radiations**. Axons from the contralateral and ipsilateral retinal layers of the LGN (see Figure 11.7) are intermingled in the optic radiations, so lesions of the optic radiations usually cause homonymous defects affecting the contralateral visual field (see Figure 11.15E–G).

FIGURE 11.8 Optic Radiations Inferior fibers (Meyer's loop) pass through the temporal lobe. Superior fibers of the optic radiations pass through the parietal lobe.

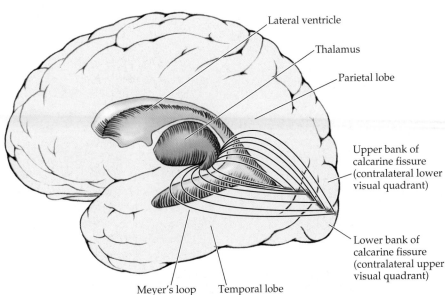

Lateral ventricle

Thalamus

Parietal lobe

Upper bank of calcarine fissure (contralateral lower visual quadrant)

Lower bank of calcarine fissure (contralateral upper visual quadrant)

Meyer's loop   Temporal lobe

The fibers of the inferior optic radiations arc forward into the temporal lobe, forming **Meyer's loop** (see Figures 11.8 and 11.15). The inferior optic radiations carry information from the inferior retina or the superior visual field (see Figure 11.1A). Temporal lobe lesions can therefore cause a **contralateral homonymous superior quadrantanopia** ("pie in the sky" visual field defect) (see Figure 11.15E). Conversely, the upper optic radiations pass under the parietal lobe (see Figure 11.8). Therefore, parietal lobe lesions can sometimes cause a **contralateral homonymous inferior quadrantanopia** ("pie on the floor" visual field defect) (see Figure 11.15F).

The **primary visual cortex** lies on the banks of the **calcarine fissure** in the occipital lobe (Figure 11.9; see also Figures 4.13G and 4.15A). The upper portions of the optic radiations project to the superior bank of the calcarine fissure; the inferior optic radiations terminate on the lower bank (see Figure 11.8). Upper-bank lesions thus cause contralateral inferior quadrant defects, while lower-bank lesions cause contralateral superior quadrant defects (see Figure 11.15I,J). The primary visual cortex, like many other parts of the visual pathways, is **retinotopically organized**. The region of the fovea is represented near the occipital pole, while more peripheral regions of the ipsilateral retinas and contralateral visual fields are represented more anteriorly along the calcarine fissure (see Figure 11.9). Note that despite its small retinal area, because it is the region of highest density of photoreceptors and correspondingly high visual acuity, the fovea has a disproportionate cortical representation, occupying about 50% of the primary visual cortex.

The portion of the medial occipital lobe above the calcarine fissure is named the cuneus (meaning "wedge") and the portion below the calcarine fissure is called the lingula (meaning "little tongue") (see Figure 11.9).

REVIEW EXERCISE

Do the optic radiations pass laterally to the lateral ventricles, or do they remain medial to the lateral ventricles for their entire course? (See Figure 11.8.)

MNEMONIC

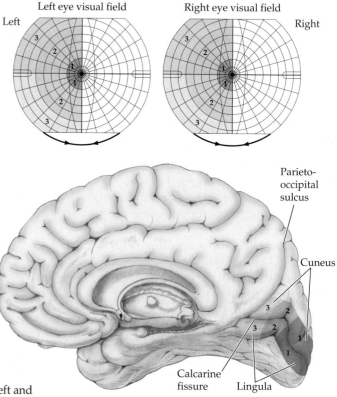

FIGURE 11.9 Retinotopic Map Left visual hemifields of the left and right eyes mapped to the primary visual cortex of the right hemisphere.

Visual Processing in the Neocortex

Review the different cortical layers and their functions (see Figure 2.14). Most input to primary visual cortex arrives at cortical **layer 4**. Because of its functional importance in this region of the brain, layer 4 is relatively thick and is subdivided into sublaminae 4A, 4B, 4Cα, and 4Cβ (Figure 11.10). Layer 4B contains numerous myelinated axon collaterals resulting in the pale-appearing **stria of Gennari**, which is visible in sections of the gray matter even with the naked eye. Because of this distinctive stria (see Figure 11.10), the **primary visual cortex (area 17)** is sometimes referred to as **striate cortex**.

Parallel Channels for Analyzing Motion, Form, and Color

Numerous channels of information undergo parallel processing in the visual system. The three best-characterized channels are for analyzing **motion**, **form**, and **color**. As already discussed, some of the information for these channels is

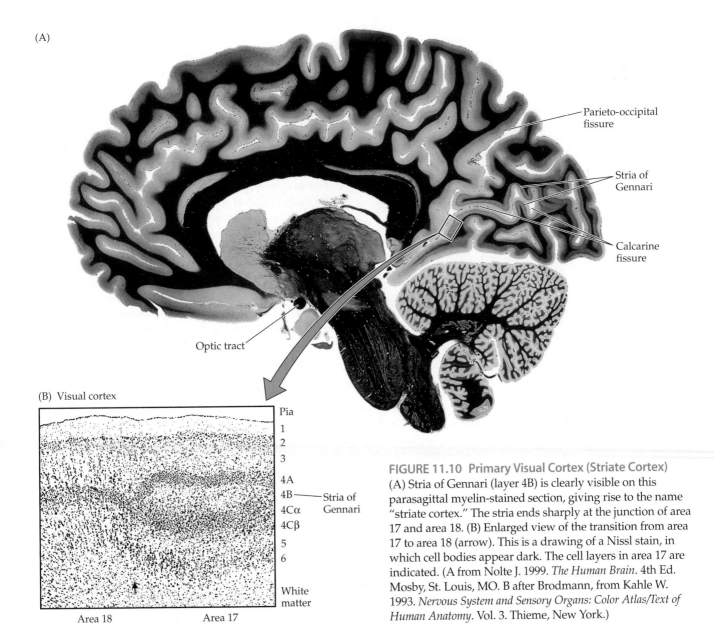

(A)

Parieto-occipital fissure

Stria of Gennari

Calcarine fissure

Optic tract

(B) Visual cortex

Pia
1
2
3
4A
4B —— Stria of Gennari
4Cα
4Cβ
5
6
White matter

Area 18   Area 17

FIGURE 11.10 Primary Visual Cortex (Striate Cortex) (A) Stria of Gennari (layer 4B) is clearly visible on this parasagittal myelin-stained section, giving rise to the name "striate cortex." The stria ends sharply at the junction of area 17 and area 18. (B) Enlarged view of the transition from area 17 to area 18 (arrow). This is a drawing of a Nissl stain, in which cell bodies appear dark. The cell layers in area 17 are indicated. (A from Nolte J. 1999. *The Human Brain*. 4th Ed. Mosby, St. Louis, MO. B after Brodmann, from Kahle W. 1993. *Nervous System and Sensory Organs: Color Atlas/Text of Human Anatomy*. Vol. 3. Thieme, New York.)

(A)

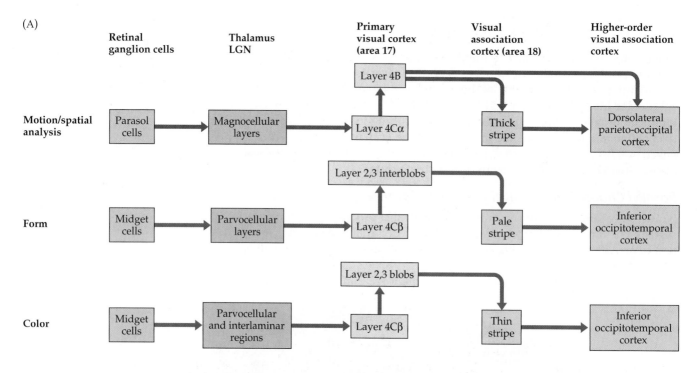

	Retinal ganglion cells	Thalamus LGN	Primary visual cortex (area 17)	Visual association cortex (area 18)	Higher-order visual association cortex
Motion/spatial analysis	Parasol cells	Magnocellular layers	Layer 4B / Layer 4Cα	Thick stripe	Dorsolateral parieto-occipital cortex
Form	Midget cells	Parvocellular layers	Layer 2,3 interblobs / Layer 4Cβ	Pale stripe	Inferior occipitotemporal cortex
Color	Midget cells	Parvocellular and interlaminar regions	Layer 2,3 blobs / Layer 4Cβ	Thin stripe	Inferior occipitotemporal cortex

(B)

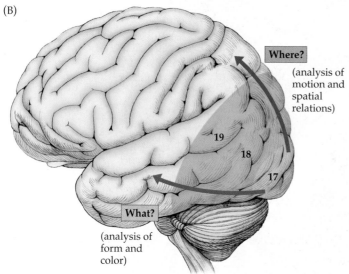

Where?
(analysis of motion and spatial relations)

19
18
17

What?
(analysis of form and color)

FIGURE 11.11 **Visual Processing Pathways** (A) Three parallel channels of visual information processing. (B) Dorsal and ventral streams of higher-order visual processing.

segregated as early as the retinal ganglion cells and LGN. These three channels also project to different layers of the primary visual cortex (**Figure 11.11A**). The magnocellular layers of the LGN, conveying information about movement and gross spatial features, project mainly to **layer 4Cα**. The parvocellular layers of the LGN, carrying fine spatial information, terminate mainly in **layer 4Cβ**. Information about color is also relayed by the parvocellular layers, as well as by the interlaminar zones, to specialized regions of cortical **layers 2 and 3** called **blobs** (see Figure 11.14) because of their appearance on staining with the histochemical marker cytochrome oxidase.

From the primary visual cortex, or area 17, neurons project to **extrastriate** regions of **visual association cortex**, including areas 18, 19, and other regions of the parieto-occipital and occipitotemporal cortex (see Figure 11.11A). In the monkey, the three channels of information processing described here have been shown to project to distinct regions of area 18 named thin stripes, thick stripes, and pale stripes (or interstripes) on the basis of their staining patterns with cytochrome oxidase. From primary and secondary visual cortex (areas 17 and 18), two main streams of higher-order visual processing have been demonstrated in both animals and humans (**Figure 11.11B**). The **dorsal pathways** project to parieto-occipital association cortex. These pathways answer the question "**Where?**" by analyzing motion and spatial relationships between objects as well as between the body and visual stimuli. The **ventral pathways** project to occipitotemporal association cortex. These pathways answer the question "**What?**" by analyzing form, with specific regions identifying colors, faces, letters, and other visual stimuli. The effects of lesions in these two streams of higher-order visual processing will be discussed in Chapter 19.

Ocular Dominance Columns and Orientation Columns

The classic work of David Hubel and Torsten Wiesel in the 1960s demonstrated that the visual cortex has a **columnar organization**. As in the LGN, inputs to the primary visual cortex are segregated on the basis of whether they originated from the contralateral or ipsilateral eye. However, instead of terminating on different layers, inputs from each eye terminate in different alternating bands of cortex, each about 1 millimeter wide, called **ocular dominance columns** (Figure 11.12A; see also Figure 11.14). The original studies by Hubel, Weisel, and collaborators demonstrated the ocular dominance columns using autoradiography and histological techniques on postmortem tissue from animals. Today, it is possible to image the ocular dominance columns and other patterns of cortical activity in living animals (including humans) using intrinsic optical signals related to neural activity, as shown in Figure 11.12A. Using a camera looking down at the pial surface of the brain during presentation of visual stimuli, investigators recorded optical signals from primary visual cortex. A visual stimulus presented to the right eye activated regions shown in white, while a stimulus presented to the left eye activated regions shown in black. This produced a typical pattern of ocular dominance columns consisting of alternating stripes of increased and decreased activation corresponding to inputs from the right eye and left eye, respectively.

The receptive fields of neurons in the primary visual cortex input layers, such as layer 4, are mainly on-center– and off-center–surround cells (Figure

REVIEW EXERCISE

Does analysis of motion and spatial relations take place in the dorsal or ventral stream of visual association cortex? What about analysis of form and color? (See Figure 11.11)

(A)

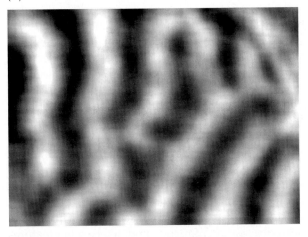

(B)

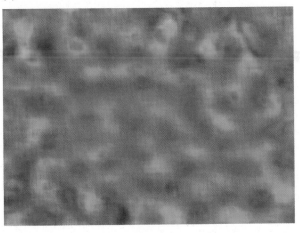

FIGURE 11.12 Intrinsic Optical Signals Demonstrating Ocular Dominance Columns and Orientation Columns View is looking down onto the pial surface of the brain. (A) Ocular dominance columns. Visual stimuli were presented monocularly to the right eye and to the left eye of a macaque monkey. Intrinsic optical signals were then recorded from primary visual cortex. The intrinsic optical signal is correlated with the ratio of oxyhemoglobin to deoxyhemoglobin in the tissue, which in turn is related to the amount of neuronal activity. Regions that receive input mainly from the right eye appear light, while regions that receive input mainly from the left eye appear dark. Dimensions of the region of cortex shown here are 8 × 5 mm. Note that while ocular dominance columns are 1 mm wide in humans, they are only 400 μm wide in the macaque. (B) Orientation columns. Arrays of parallel lines of different orientations were presented binocularly. Regions with a particular orientation selectivity respond most strongly to lines oriented in the preferred direction. Intrinsic optical signals recorded with stimuli of different orientations are represented here by different colors. Line orientation relative to horizontal and color codes, respectively, are: 0° = red; 45° = yellow; 90° = green; 135° = blue. The region of cortex imaged is identical to that shown in (A). (Courtesy of Anna Wang Roe, Vanderbilt University.)

(A) Concentric (center–surround) cell receptive fields

On-center   Off-center

(B) Simple cell receptive fields

(C)

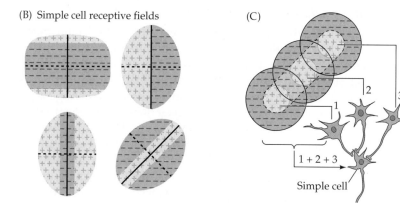

1
2
3

Concentric (center–surround) cells

1 + 2 + 3

Simple cell

FIGURE 11.13 Receptive Fields of Cortical Simple Cells Derived from Integration of Center–Surround Cell Inputs (A) Receptive fields of center–surround cells of retina, LGN, and cortical input layer. (B) Receptive fields of simple cells of primary visual cortex. Exact fields vary, but they always contain orientation-selective regions of excitation and inhibition. (C) Model proposed by Hubel and Wiesel for summation of center–surround inputs to generate the receptive field of a simple cell with orientation selectivity. (After Hubel DH, Wiesel TN. 1962. Receptive fields, binocular interaction and functional architecture in the cat's visual cortex. *J Physiol* (Lond) 160: 106–154.)

11.13A). However, these cells project to other neurons, above and below layer 4, which have more sophisticated receptive fields. **Simple cells** respond to lines or edges that occur at a *specific location* and with a specific angular orientation within their receptive field (Figure 11.13B). Figure 11.13C shows a model for how the summation of activity from several concentric cells with adjacent receptive fields can generate the receptive field of a single simple cell having a specific orientation selectivity. Activity from simple cells, in turn, summates to generate the receptive field properties of **complex cells**, the next level of neuronal processing. Complex cells respond to lines or edges that occur at *any location* in their receptive field, with a specific angular orientation. Several subclasses of these cell types exist, and there are also neurons with even more sophisticated receptive fields.

The orientation selectivity of simple cells, complex cells, and other neurons remains the same within a vertical column of cortex, from pia to white matter. In contrast, as one moves horizontally in the cortex, the orientation selectivity varies continuously, so that a complete 180° sequence of orientation selectivity occurs over a cortical distance of about 1 millimeter (Figure 11.14). The vertical columns of uniform orientation selectivity are called **orientation columns**, which can also be imaged in living animals using optical imaging (Figure 11.12B).

Ocular dominance and orientation columns intersect, so a given region of about 1 square millimeter of cortex will contain a complete sequence of both ocular dominance and orientation columns (see Figure 11.14). Such functional units were referred to by Hubel and Wiesel as **hypercolumns**. More recently, it has been shown that additional columns for functions other than ocular dominance and orientation selectivity also exist in the visual cortex. These functions include direction selectivity, spatial frequency, and, likely, other features of visual perception.

KEY CLINICAL CONCEPT

11.1 ASSESSMENT OF VISUAL DISTURBANCES

The localization and diagnosis of visual disturbances involves two major steps. The first step is a detailed description of the **nature of the visual disturbance**, including its time course and whether positive phenomena such as brightly colored lights or negative phenomena such as regions of decreased vision are present. The second step is a description of the regions of the **visual field** for each eye that are involved. In this section we will discuss the localizing information that can be derived from the nature of the visual disturbance; in the next section we will discuss the visual field defects seen with lesions in specific locations.

Ophthalmoscopic exam

As with other disorders, the evaluation of visual disturbances involves taking a detailed history, followed by a complete exam, which includes examination of the eyes with an ophthalmoscope (see **neuroexam.com Video 25**), as well as testing of visual acuity and of visual fields for each eye (see **neuroexam.com Video 27**). **Visual acuity** is often reported using the **Snellen notation** of 20/X. In this notation, the denominator ("X") is the distance at which a normal individual can see the smallest line of the eye chart seen by the subject at 20 feet. Visual acuity can be impaired by a variety of ophthalmological disorders that are beyond the scope of this text (see the References at the end of the chapter for details). Note that visual field defects do not typically affect visual acuity.

The distinction between a **monocular** or **binocular** visual disturbance is essential for localization. However, patients often describe visual changes as being in one eye, when in reality the left or right visual field is affected for *both* eyes. While describing a transient visual disturbance, patients are sometimes able to recall improvement on covering one eye, suggesting a true monocular disorder. Often it is only by examination of the patient while the problem is still present that this distinction can be confirmed. Similarly, **"blurred" vision** is hard to interpret without further description; it can mean anything from corneal disease to a lesion in the visual cortex. "Blurred" vision can sometimes even be a sign of subtle diplopia, suggesting an ocular motility disorder (see Chapter 13).

Some important terms used to describe visual disturbances are listed in Table 11.1. Visual changes are often divided into positive and negative phenomena. Negative phenomena such as a **scotoma** or a **homonymous visual field defect** (see Table 11.1) can be caused by lesions at various locations in the visual pathways (see Figure 11.15; see KCC 11.2). Patients may be aware of a dark brown, purplish, or white region of their vision where they cannot see. At other times they are unaware of the defect, and the region that cannot be seen is experienced in a manner similar to the physiological blind spot (see Figure 11.2) or to locations behind the head that are normally out of view. Regions of absent vision of this kind are nearly always a result of a lesion of the central visual pathways, while black, dark brown, or purplish scotomas are most often produced by retinal lesions.

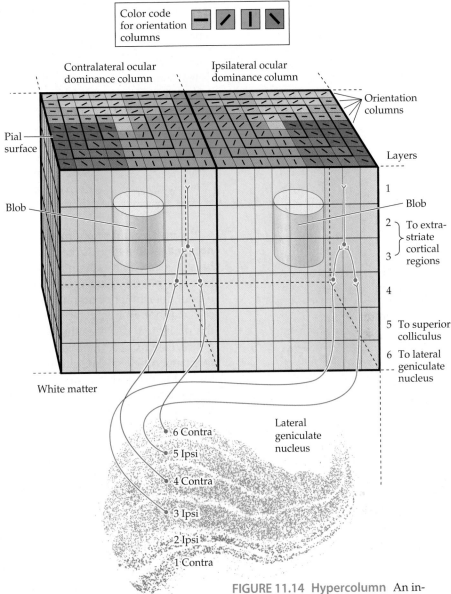

FIGURE 11.14 Hypercolumn An information-processing module of visual cortex containing a complete set of ocular dominance columns, orientation columns, blobs, and interblobs (regions between the blobs). Inputs from the lateral geniculate nucleus are shown as well. Contra = input from contralateral eye; Ipsi= input from ipsilateral eye. (Top illustration after Rosenzweig MR, Breedlove SM, Leiman AL. 2002. *Biological Psychology*. 3rd Ed. Sinauer, Sunderland, MA; bottom courtesy of Tim Andrews and Dale Purves, Duke University School of Medicine.)

TABLE 11.1 Some Terms to Describe Visual Disturbances

TERM	DEFINITION
Scotoma	A circumscribed region of visual loss
Homonymous defect	A visual field defect in the same region for both eyes
Refractive error	Indistinct vision improved by corrective lenses
Photopsias	Bright, unformed flashes, streaks, or balls of light
Phosphenes	Photopsias produced by retinal shear or optic nerve disease
Entopic phenomena	Seeing structures in one's own eye
Illusions	Distortion or misinterpretation of visual perception
Hallucination	Perception of something that is not present

Positive visual phenomena may be simple or formed. **Simple visual phenomena** such as lights, colors, or geometric shapes are caused by disturbances located anywhere from the eye to the primary visual cortex. Important ophthalmological causes of positive phenomena include light flashes in retinal detachment and rainbow-colored halos around objects in acute glaucoma, although positive phenomena can be caused by numerous other ophthalmological diseases that are beyond the scope of this discussion. In migraine (see KCC 5.1), patients may experience visual blurring and scotomas that sometimes have a scintillating appearance or consist of jagged, alternating light and dark zigzag lines called a **fortification scotoma** (because of its resemblance to the fortifications of medieval towns in Europe). These typical migrainous phenomena are thought to be related to transient dysfunction of the primary visual cortex (e.g., zigzagging lines may result from activation of alternating orientation columns). When patients instead experience pulsating colored lights or moving geometric shapes, occipital seizures should be suspected, although occipital seizures may also produce migraine-like visual phenomena at times.

Formed visual hallucinations (see Table 11.1), such as people, animals, or complex scenes, arise from the inferior temporo-occipital visual association cortex. Common causes of formed visual hallucinations include toxic or metabolic disturbances (especially hallucinogens, anticholinergics, and cyclosporin), withdrawal from alcohol or sedatives, focal seizures, complex migraine, neurodegenerative conditions such as Creutzfeldt–Jakob disease or Lewy body disease, narcolepsy, midbrain ischemia (peduncular hallucinosis; see KCC 14.3), or psychiatric disorders. Of note, in psychiatric disorders visual hallucinations are less common than auditory hallucinations, and they usually occur with accompanying sound. Formed visual hallucinations can also appear as a **release phenomenon**. Thus, patients with visual deprivation in part or all of their visual fields caused by either ocular or central nervous system lesions may occasionally see objects, people, or animals in the region of vision loss, especially during the early stages of the deficit. Visual hallucinations that occur in elderly patients as a result of impaired vision have been called Bonnet syndrome. ■

KEY CLINICAL CONCEPT

11.2 LOCALIZATION OF VISUAL FIELD DEFECTS

Once the nature of a visual disturbance has been established (see KCC 11.1), including the time course and other clinical features, such as whether positive, negative, simple, or elaborate visual phenomena are present, the next step is to evaluate which portions of the visual fields are involved. We will first describe the methods used for visual field testing and then discuss the interpretation and localizing value of specific visual field defects.

Visual Field Testing

Basic **visual field testing** can be done at the bedside, using **confrontation testing** (see **neuroexam.com Video 27**). The examiner should test each eye separately by covering one eye at a time. The patient is instructed to look at the examiner's eye while the examiner holds visual stimuli such as fingers or a small cotton-tipped stick halfway between the patient and examiner. In this way, the examiner's visual field is being tested at the same time. The examiner should test each quadrant while watching the patient carefully for central fixation. Moving or wiggling fingers are easier to see but are less sensitive for detecting regions of mildly decreased vision. The examiner should hold up fingers simultaneously on the right and left sides at some point during the exam to test for extinction, a sign of visual neglect (see KCC 19.9). By convention, visual fields are recorded with the right eye on the right side of the page, as if viewing your own visual field (see Figure 11.1). **Blink to threat** can be a useful way of testing crude visual fields in the uncooperative or lethargic patient (see **neuroexam.com Video 28**).

More **formal visual field testing** can be done, when appropriate, using **Goldmann perimetry**, with small lights of different sizes and intensities displayed on a screen in front of the patient (see Figure 11.2). The normal visual field extends about 60° nasally and superiorly and slightly farther inferiorly and temporally. In addition to manually performed Goldmann perimetry, automated computerized perimetry is increasingly being used in some settings. However, automated perimetry usually only tests the central 30° of the visual field.

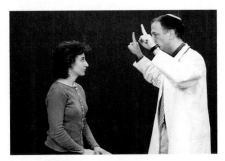

Visual fields

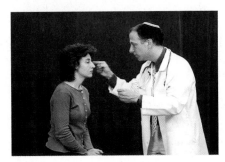

Blink to threat

Visual Field Defects

The position and shape of the scotoma, and whether it affects one eye or both, are the most important pieces of information allowing localization of abnormalities in the visual pathways. **Figure 11.15** summarizes the effects of lesions at various points in the pathways from retina to primary visual cortex that we have been discussing in this chapter and review again here. A lesion of the retina causes a **monocular scotoma** (see Figure 11.15A), with the location, size, and shape depending on the location and extent of the lesion. Common causes include retinal infarcts (see KCC 11.3), hemorrhage, degeneration, or infection. If the lesion is severe enough, the entire retina may be involved, causing **monocular visual loss** (see Figure 11.15B). In addition to retinal disorders, monocular disturbances of vision can result from numerous other diseases of the eye (refer to ophthalmology texts for additional details).

Lesions of the optic nerve cause monocular visual loss or monocular scotomas (see Figure 11.15A,B), which may be partial or incomplete, depending on the severity of the lesion. Common causes include glaucoma, optic neuritis, elevated intracranial pressure, anterior ischemic optic neuropathy, optic glioma, schwannoma, meningioma, and trauma.

The optic chiasm is located near the pituitary gland (see Figure 17.2B) and can be compressed by lesions arising in this area. Damage to the optic chiasm typically causes a **bitemporal hemianopia** (see Figure 11.15C), which is often more asymmetrical than shown in the figure. Common lesions in this area include pituitary adenoma, meningioma, craniopharyngioma, and hypothalamic glioma, although numerous other lesions can also occur in this location.

Retrochiasmal lesions, including lesions of the optic tracts, LGN, optic radiations, or visual cortex, generally cause **homonymous** visual field defects,

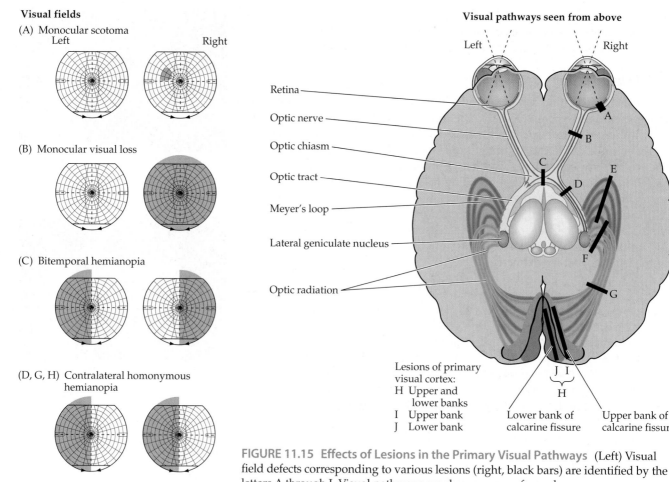

Visual fields

(A) Monocular scotoma

(B) Monocular visual loss

(C) Bitemporal hemianopia

(D, G, H) Contralateral homonymous hemianopia

(E, J) Contralateral superior quadrantanopia

(F, I) Contralateral inferior quadrantanopia

Visual pathways seen from above

Left   Right

Retina

Optic nerve

Optic chiasm

Optic tract

Meyer's loop

Lateral geniculate nucleus

Optic radiation

Lesions of primary visual cortex:
H   Upper and lower banks
I   Upper bank
J   Lower bank

Lower bank of calcarine fissure   Upper bank of calcarine fissure

FIGURE 11.15 Effects of Lesions in the Primary Visual Pathways (Left) Visual field defects corresponding to various lesions (right, black bars) are identified by the letters A through J. Visual pathways are shown as seen from above.

meaning that the same regions of the fields for both eyes are involved. However, since the fibers from each eye are less fully mingled in the optic tract and LGN, visual field defects may not be perfectly congruous for more anterior retrochiasmal lesions, while they are usually perfectly congruous for lesions of the visual cortex.

Lesions of the optic tracts are relatively uncommon and usually cause a **contralateral homonymous hemianopia** (see Figure 11.15D). Possible lesions include tumors, infarct, or demyelination.

Lesions of the lateral geniculate nucleus are also usually associated with a contralateral homonymous hemianopia (see Figure 11.15D), although sometimes more unusual visual field defects can occur, such as a keyhole-shaped sectoranopia. Possible lesions include tumors, infarct, hemorrhage, toxoplasmosis, or other infections.

Lesions of the optic radiations include infarcts, tumors, demyelination, trauma, and hemorrhage. As discussed in the next section, lesions involving the temporal lobe, such as middle cerebral artery (MCA) inferior division infarcts, can interrupt the lower optic radiations as they loop through the temporal lobe (Meyer's loop; see Figure 11.8). Lesions of the temporal lobe therefore cause a **contralateral superior quadrantanopia** (see Figure 11.15E), or **"pie in the sky"** visual defect. Meanwhile, lesions involving the parietal lobe, such as MCA superior division infarcts, can interrupt the upper portions of the optic radiations as they pass through the parietal lobe (see Figure 11.8). Therefore, parietal lesions typically cause a **contralateral inferior quadran-**

tanopia (see Figure 11.15F), or **"pie on the floor"** visual defect. Lesions of the entire optic radiation cause a **contralateral homonymous hemianopia** (see Figure 11.15G).

The primary visual cortex may be damaged by posterior cerebral artery (PCA) infarcts, tumors, hemorrhage, infections, or trauma to the occipital poles. Lesions to the upper bank of the calcarine fissure cause a **contralateral inferior quadrantanopia** (see Figure 11.15I), while lesions to the lower bank cause a **contralateral superior quadrantanopia** (see Figure 11.15J). Damage to the entire primary visual cortex causes a **contralateral homonymous hemianopia** (see Figure 11.15H). Smaller lesions cause homonymous scotomas in the appropriate portion of the contralateral visual field (see Figure 11.9).

Partial lesions of the visual pathways occasionally result in a phenomenon called **macular sparing** (Figure 11.16). This occurs because the fovea has a relatively large representation for its size, beginning in the optic nerve and continuing to the primary visual cortex (see, for example, Figure 11.9). Macular sparing can also occur in visual cortex because either the MCA or the PCA may provide collateral flow to the representation of the macula in the occipital pole (see Figure 10.5). Although the term "macular sparing" is usually used in the context of cortical lesions, other lesions may cause a relative sparing of central vision as well. For example, external compression of the optic nerve, as is seen in elevated intracranial pressure, may cause **concentric visual loss** (constricted visual field; see Figure 11.16A).

Disorders of higher-order visual processing caused by lesions of the visual association cortex are discussed in Chapter 19. ■

MNEMONIC

REVIEW EXERCISE

First, cover all the visual fields in Figure 11.15, and for each lesion marked on the brain (A–J), draw the visual fields expected in a patient with that lesion. Next, cover the brain illustration, and for each visual field defect shown in the figure, state all possible locations where a lesion causing such a deficit could be located.

Examples of macular sparing

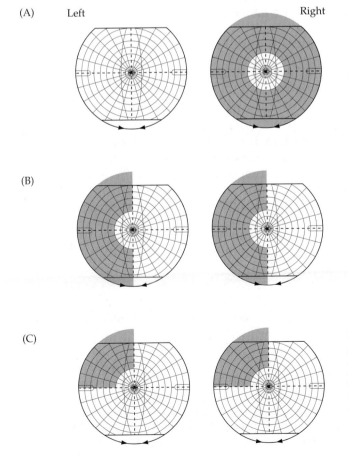
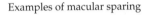

FIGURE 11.16 Examples of Macular Sparing (A) Monocular concentric visual loss caused, for example, by chronically elevated intracranial pressure or retinitis pigmentosa. (B) Left homonymous hemianopia with macular sparing caused, for example, by a right posterior cerebral artery infarct preserving the occipital pole. (C) Left superior quadrantanopia with macular sparing caused, for example, by a lesion of the inferior bank of the right calcarine fissure preserving the occipital pole.

11.3 BLOOD SUPPLY AND ISCHEMIA IN THE VISUAL PATHWAYS

The retina receives its blood supply primarily from branches of the **ophthalmic artery**, which originates just above the genu of the internal carotid artery (see Figure 10.2A). The retinal arteries and veins can be well visualized as they emerge from the optic disc by use of an **ophthalmoscope** (see **neuroexam.com Video 25**; see also Figure 5.17). The three main causes of impaired blood flow in the ophthalmic artery and its branches are (1) **emboli**, often atheromatous material arising from ipsilateral internal carotid stenosis; (2) **stenosis**, usually associated with diabetes, hypertension, or elevated intracranial pressure; and (3) **vasculitis**, as seen, for example, in temporal arteritis.

Central retinal artery occlusion and **branch retinal artery occlusion** can cause infarction of the entire retina or of the affected retinal sector, respectively. The retinal artery usually has two major branches—one covering the upper half, the other covering the lower half of the retina. An **altitudinal** scotoma in one eye can therefore result from occlusion of one of these branches (**Figure 11.17**). Smaller monocular scotomas can also occur from occlusion of smaller branches (see Figure 11.15A). Transient occlusion of the retinal artery caused by emboli results in a transient ischemic attack (TIA) of the retina called **amaurosis fugax**, with "browning out" or loss of vision in one eye for about 10 minutes, sometimes described "like a window shade" moving down or up over the eye. This condition deserves to be worked up just like any other TIA (see KCC 10.3 and KCC 10.4), since it may be a warning sign for an impending retinal or cerebral infarct. A common cause of amaurosis fugax is ipsilateral internal carotid artery stenosis (see KCC 10.5), which causes artery-to-artery emboli.

Impaired blood supply to the anterior optic nerve is called **anterior ischemic optic neuropathy** (**AION**), and is a relatively common cause of sudden vision loss in patients over age 50. The anterior optic nerve is supplied by the **short posterior ciliary arteries**, branches of the ophthalmic artery. One form of AION, referred to as arteritic AOIN, is seen in association with **temporal arteritis** and should be treated promptly with steroids to prevent vision loss (see KCC 5.1). The more common form is **non-arteritic AION**. Risk factors for non-arteritic AION include diabetes, hypertension, elevated cholesterol, and a small cup-to-disc ratio on ophthalmoscopic exam. Onset is classically upon awakening, possibly due to nocturnal hypotension, with painless decreased vision in one eye. The pathophysiology of non-arteritic AION is likely related to atherosclerosis, and the most important current treatment is preventative risk factor reduction.

The optic tracts, optic chiasm, and intracranial segment of the optic nerves receive their blood supply from numerous small branches arising from the proximal portions of the anterior cerebral artery (ACA), middle cerebral artery (MCA), and anterior and posterior communicating arteries. Clinically significant infarcts of these structures are therefore rarely seen.

The lateral geniculate nucleus has a variable blood supply arising from several vessels, including the anterior choroidal artery (branch of the internal carotid), thalamogeniculate artery, and posterior choroidal artery (branches of the posterior cerebral artery). Infarcts of the lateral geniculate generally produce a contralateral homonymous hemianopia (see Figure 11.15D), although more un-

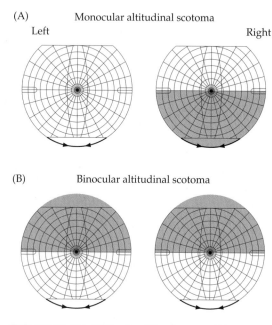

(A) Monocular altitudinal scotoma

Left   Right

(B) Binocular altitudinal scotoma

FIGURE 11.17 Altitudinal Scotomas (A) Monocular altitudinal scotoma caused, for example, by occlusion of the superior branch of the right ophthalmic artery by an embolus. (B) Binocular altitudinal scotoma caused, for example, by bilateral occlusion of the posterior cerebral artery (PCA) branches supplying the lingular gyri (inferior banks of calcarine fissures).

usual field defects can also sometimes be seen, as mentioned earlier. In addition, there may be an associated contralateral hemiparesis or hemisensory loss due to involvement of the nearby posterior limb of the internal capsule and thalamic somatosensory radiations (see Figures 6.9B and 10.8B).

The optic radiations pass through the parietal and temporal lobes, where they may be damaged by infarcts of the superior and inferior divisions of the middle cerebral artery, respectively (see Figure 11.8; see also Figure 10.1 and Table 10.1). Damage to the upper portions of the optic radiations in the parietal lobe cause a contralateral inferior quadrantanopia (see Figure 11.15F), while damage to Meyer's loop in the temporal lobe causes a contralateral superior quadrantanopia (see Figure 11.15E).

The primary visual cortex is supplied by the posterior cerebral artery (PCA; see Figure 10.5). Infarcts of the entire primary visual cortex cause a contralateral homonymous hemianopia (see Figure 11.15H). Smaller infarcts cause smaller contralateral homonymous defects (see Figures 11.15I,J and 11.16C). Sometimes disease of the basilar artery, which supplies both PCAs (see Figure 10.3), can cause bilateral PCA ischemia or infarcts. A bilateral altitudinal scotoma (see Figure 11.17B) is strongly suggestive of vertebrobasilar insufficiency causing bilateral infarcts or TIAs.

The inferior occipitotemporal visual association cortex ("what?" stream; see Figure 11.11B) is supplied by the PCA (see Figure 10.5). The lateral parieto-occipital visual association cortex ("where?" stream) lies in the MCA–PCA watershed territory (see Figures 10.5 and 10.10). Infarcts of the inferior occipitotemporal or dorsolateral parieto-occipital visual association cortex cause characteristic disorders of higher-order visual processing that will be discussed in Chapter 19 (see KCC 19.12). ■

KEY CLINICAL CONCEPT

11.4 OPTIC NEURITIS

Optic neuritis is an inflammatory demyelinating disorder of the optic nerve that is epidemiologically and pathophysiologically related to multiple sclerosis (see KCC 6.6). Like multiple sclerosis, mean age of onset is in the 30s, onset after age 45 is rare, and there is about a 2:1 female-to-male ratio. With careful follow-up, about 50% or more of patients with an isolated episode of optic neuritis will eventually develop multiple sclerosis.

The usual clinical features at onset include eye pain, especially with eye movement, and monocular visual problems. The visual impairment typically includes a monocular **central scotoma** (visual loss in the center of the visual field), decreased visual acuity, and impaired color vision. In severe cases, complete loss of vision in one eye may occur. Ophthalmoscopic exam may reveal a swollen optic disc if the inflammation extends to the fundus, causing **papillitis**, or the fundus may appear normal if the neuritis is entirely **retrobulbar** (behind the eye). Sometimes the disc appears pale, a condition termed **optic disc pallor**; this finding suggests prior episodes of optic neuritis.

A fairly sensitive way to detect impaired cone function in central vision is to test for **red desaturation** by asking the patient to compare the appearance of a bright red object seen with each eye in turn (see **neuroexam.com Video 26**). In patients with past or present optic neuritis, the object will often appear duller in the affected eye. Another very useful way to detect optic nerve dysfunction is to test for an **afferent pupillary defect** using the **swinging flashlight test** (see **neuroexam.com Video 30**; see also KCC 13.5). In addition, **visual evoked potentials** can provide evidence of impaired conduction in the visual pathways. In this test the patient is shown a shifting checkerboard pattern, which elicits a

Swinging flashlight test

voltage waveform that can be detected over the occipital cortex by electrodes placed on the scalp. The normal latency for visual evoked potentials is less than about 115 milliseconds. Prolonged latency with preserved amplitude suggests slowed conduction consistent with demyelination.

Onset of optic neuritis can be acute or slowly progressive over several days to weeks. Recovery usually begins within 2 weeks, and near complete recovery is commonly seen within 6 to 8 weeks and sometimes over a few months. There is often some residual visual loss, which in some cases may be severe, especially after repeated bouts. A second episode occurs in about one-third of cases. Differential diagnosis includes retinal artery occlusion, anterior ischemic optic neuropathy, acute glaucoma, and compressive or infiltrative lesions. Treatment of isolated optic neuritis with high-dose intravenous steroids has been shown to shorten the duration of visual impairment but has no effect on the long-term outcome. MRI of the brain can be used to predict subsequent risk of developing multiple sclerosis, based on the presence or absence of demyelinating lesions.

Most patients with typical clinical features do not require extensive work-up. However, if any atypical features are present, such as age over 45 years, lack of eye pain, bilateral symptoms, or lack of prompt recovery, further investigations are warranted. These include MRI with gadolinium to rule out infiltrative or compressive lesions; blood tests including erythrocyte sedimentation rate, Lyme titer, syphilis serologies, Epstein–Barr virus, human immunodeficiency virus, B_{12}, and folate; serological tests for rheumatological disorders; and, possibly, lumbar puncture. ■

CLINICAL CASES

CASE 11.1   A DARK SPOT SEEN WITH ONE EYE

MINICASE

A 67-year-old man woke up one morning, and when he turned on the bathroom light he noticed a **dark purplish brown spot in the upper part of his vision that disappeared when he covered his right eye**. This did not improve over the following week, so he went to see an ophthalmologist. Initial exam was normal except for a **soft right carotid bruit** and a **scotoma in the upper nasal quadrant of the right eye** on visual field testing (Figure 11.18).

LOCALIZATION AND DIFFERENTIAL DIAGNOSIS

On the basis of the symptoms and signs shown in **bold** above, where is the lesion? What is the most likely diagnosis, and what are some other possibilities? What is the next step in evaluating this patient?

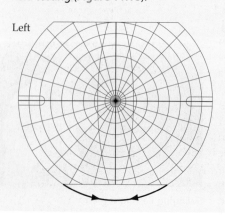
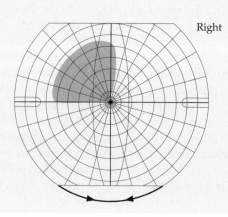

Left   Right

FIGURE 11.18 Visual Fields Fields were evaluated by confrontation testing.

Discussion

The key symptoms and signs in this case are:

- **Scotoma in the upper nasal quadrant of the right eye**
- **Right carotid bruit**

The monocular scotoma seen in this patient could be produced by a lesion of the retina in the right eye (see KCC 11.2; Figure 11.15A). A visual field defect in the upper nasal quadrant would be caused by a lesion in the lower temporal portion of the retina (see Figure 11.1). Given the patient's age and the presence of a right carotid bruit, the most likely diagnosis is an embolus arising from the right internal carotid artery (see KCC 10.5), resulting in a branch retinal artery occlusion in the right eye (see KCC 11.3). Additional possibilities include other disorders of the retina, such as hemorrhage, or a disorder of the optic nerve, such as anterior ischemic optic neuropathy (see KCC 11.3). The next step in the evaluation should be an examination of the retina with an ophthalmoscope.

Clinical Course

A dilated exam of the retina was done, revealing a pale, wedge-shaped area in the inferior, temporal portion of the right retina, consistent with an infarct caused by a branch retinal artery occlusion (see KCC 11.3). The patient was admitted to the hospital for further evaluation of embolic risk factors. Carotid Doppler studies and an MRA showed a tight stenosis of the right internal carotid artery. The patient underwent a right carotid endarterectomy and had no further ischemic events, although the right eye scotoma remained.

CASE 11.2 VISION LOSS IN ONE EYE

MINICASE

A 39-year-old man awoke one morning with **blurred vision in the left eye, as if it were covered with a shaded piece of glass**. This condition progressed over the next few days to **nearly complete loss of vision in the left eye**. He scheduled an appointment with an ophthalmologist, but by the time he was seen, about 2 weeks after the onset of symptoms, the patient felt his vision was 90% back to normal. There was no eye pain, but he did have a generalized headache. On exam, there was slight **pallor of the left optic disc** when viewed with an ophthalmoscope. **When a light was shined into the left pupil, the constriction of both pupils was less** than when a light was shined into the right pupil. This result was best demonstrated with the swinging flashlight test (see KCC 11.4, 13.5). This patient's visual fields are shown in Figure 11.19. The remainder of the examination was normal.

LOCALIZATION AND DIFFERENTIAL DIAGNOSIS

1. On the basis of the symptoms and signs shown in **bold** above, where is the lesion?

2. What is the most likely diagnosis, and what are some other possibilities?

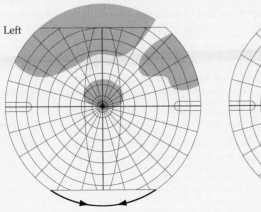

Left

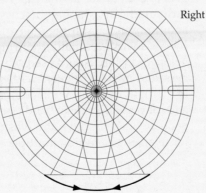

Right

FIGURE 11.19 Visual Fields Fields were evaluated by confrontation testing.

Discussion

1. The key symptoms and signs in this case are:

   - **Monocular visual loss in the left eye, improving to a monocular central scotoma**
   - **Left afferent pupillary defect**
   - **Left optic disc pallor**

   Monocular visual loss or a monocular scotoma can be caused by a lesion in any location anterior to the optic chiasm, including the eye, retina, or optic nerve (see KCC 11.2; Figure 11.15A,B). The afferent pupillary defect (see KCC 11.4, 13.5) in the left eye also supports a lesion in these locations. On ophthalmoscopic exam, the eye and retina were unremarkable, but the left optic disc, which is the entrance point of the optic nerve to the retina (see Figure 11.1), appeared pale, suggesting that there had been a previous lesion in the left optic nerve.

   The most likely *clinical localization* is left optic nerve.

2. Given the patient's relatively young age, the improvement within a few weeks, and the characteristic central scotoma, the most likely diagnosis is optic neuritis of the left eye (see KCC 11.4). Note that red desaturation was not tested but might be expected to be abnormal in this patient (see **neuroexam.com Video 26**). Other possibilities include an ischemic lesion of the optic nerve caused by small-vessel disease (anterior ischemic optic neuropathy), but this is more common in older individuals with diabetes or hypertension (see KCC 11.3). A neoplasm (meningioma, optic glioma, lymphoma, metastasis) or inflammatory process (sarcoid, Lyme disease) compressing or infiltrating the optic nerve is possible, but these conditions do not usually improve spontaneously.

Clinical Course

Visual evoked potentials (see KCC 11.4) showed a latency of 104 ms in the right eye and 148 ms in the left eye (normal is less than 115 ms). An MRI scan showed no infiltrative or compressive lesions of the optic nerve; however, there were several areas of increased T2 signal in the periventricular regions of the brain, suggesting possible demyelination (see KCC 6.6). On further discussion with the patient, some important additional history came to light. Five years previously, he had had a similar episode of visual loss in the left eye that resolved spontaneously in a few weeks and likely represented a prior episode of optic neuritis. This previous episode may account for the optic pallor seen during the current episode, since this finding often takes time to develop. He also admitted to occasionally having tingling in the fingers of both hands and to experiencing times when he felt his gait was uncoordinated. A prolonged discussion was held with the patient and family about possible multiple sclerosis (see KCC 6.6). The diagnosis of multiple sclerosis was eventually confirmed 2 years later, when he unfortunately developed diplopia, decreased sensation in the left arm and leg, gait unsteadiness, and urinary frequency. A repeat MRI scan showed a new T2 bright area in the right medulla, and a lumbar puncture was normal except for the presence of two oligoclonal bands (see KCC 5.10). The patient improved spontaneously within 5 days and was then treated with long-term beta interferon for relapsing remitting multiple sclerosis (see KCC 6.6).

CASE 11.3 MENSTRUAL IRREGULARITY AND BITEMPORAL HEMIANOPIA

MINICASE

A 50-year-old woman went to an ophthalmologist because of several months of worsening vision that had begun to interfere with her driving. Past history was notable for **long-standing menstrual irregularity and infertility**. Exam was normal except for **decreased vision** primarily in the **temporal portions of the visual fields bilaterally** (Figure 11.20).

LOCALIZATION AND DIFFERENTIAL DIAGNOSIS

On the basis of the symptoms and signs shown in **bold** above, where is the lesion? What is the most likely diagnosis, and what are some other possibilities?

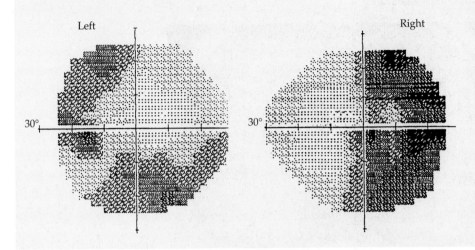

FIGURE 11.20 Visual Fields Fields were evaluated by automated computerized perimetry.

Discussion

The key symptoms and signs in this case are:

- **Bitemporal hemianopia**
- **Long-standing menstrual irregularity and infertility**

Bitemporal hemianopia is usually caused by lesions of the optic chiasm (see KCC 11.2; Figure 11.15C). Note that in this patient, like many patients with optic chiasm lesions, the bitemporal hemiopia is not perfectly symmetrical. The history of a menstrual irregularity and infertility suggests an endocrine disorder, which further supports the diagnosis of a lesion in the vicinity of both the pituitary gland and the optic chiasm. The most likely diagnosis is a pituitary adenoma (see KCC 17.1). Other tumors or masses in the sellar and suprasellar region, including a meningioma, craniopharyngioma, optic or hypothalamic glioma, lymphoma, or sarcoid, could also produce these abnormalities.

Clinical Course and Neuroimaging

A **brain MRI** (Image 11.3A,B, page 482) revealed a homogeneously enhancing mass in the suprasellar region, compressing the optic chiasm. Note that the mass had a "dural tail," suggesting that it arose from the meninges. The patient was referred to a neurosurgeon and was admitted for resection of the mass. The operative approach was through the right frontal and temporal bones. Cerebrospinal fluid was drained from the lumbar cistern, causing the lateral ventricles to collapse partly, thus making the frontal lobes easier to gently lift away from the tumor. The mass was seen to arise from the dura of the diaphragma sellae (see Figure 5.9) and the anterior sphenoid bone. All visible portions of the tumor were resected with careful preservation of the adjacent vasculature, optic nerves, and infundibular stalk. Pathologic sections revealed a meningioma. Following surgery the patient made an excellent recovery, with near complete return of vision in her lateral fields over the course of 6 to 7 months.

CASE 11.3 MENSTRUAL IRREGULARITY AND BITEMPORAL HEMIANOPIA

IMAGE 11.3A,B **Brain MRI Showing a Meningioma of the Suprasellar Region Compressing the Optic Chiasm** (A) Coronal T1-weighted image showing meningioma located just dorsal to the pituitary and compressing the optic chiasm from below. (B) Sagittal T1-weighted image with intravenous gadolinium contrast enhance- ment, showing characteristic features of a meningioma, including relatively uniform contrast enhancement, location adjacent to the meninges, and a tapering extension along the dural surface ("dural tail"). As in (A), the tumor can be seen to compress the optic chiasm from below.

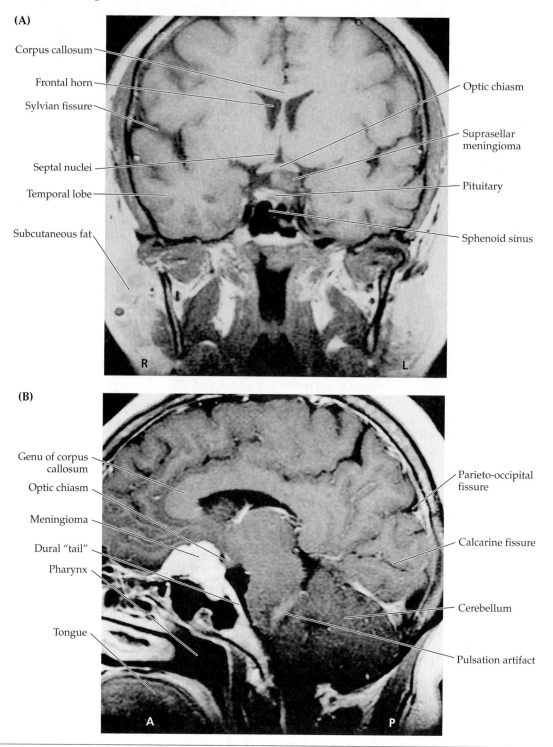

(A)

Corpus callosum

Frontal horn

Sylvian fissure

Septal nuclei

Temporal lobe

Subcutaneous fat

Optic chiasm

Suprasellar meningioma

Pituitary

Sphenoid sinus

R L

(B)

Genu of corpus callosum

Optic chiasm

Meningioma

Dural "tail"

Pharynx

Tongue

Parieto-occipital fissure

Calcarine fissure

Cerebellum

Pulsation artifact

A P

CASE 11.4 HEMIANOPIA AFTER TREATMENT FOR A TEMPORAL LOBE TUMOR

MINICASE

A 29-year-old man was referred to a neuro-ophthalmologist because of **worsening vision in his left visual fields**. Past history was notable for complex partial seizures for 5 or 6 years, beginning with an olfactory aura. One year ago an MRI revealed a left temporal lobe tumor. On resection, the tumor was found to be an oligoastrocytoma, and he was treated with chemotherapy and radiation therapy with an initially good response. Current exam was remarkable only for a **left homonymous hemianopia** (Figure 11.21).

LOCALIZATION AND DIFFERENTIAL DIAGNOSIS

On the basis of the symptoms and signs shown in **bold** above, where is the lesion? What is the most likely diagnosis, and what are some other possibilities?

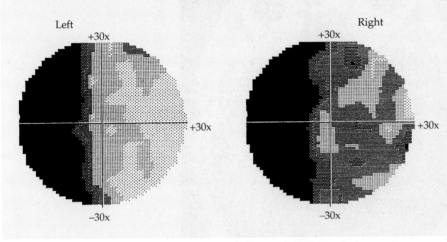

Left   +30x   Right   +30x

+30x   +30x

−30x   −30x

FIGURE 11.21 **Visual Fields** Fields were evaluated by automated computerized perimetry.

Discussion

The key symptoms and signs in this case are:

- **Left homonymous hemianopia**

A left homonymous hemianopia can be caused by a lesion anywhere in the right retrochiasmal visual pathways, including the optic tract, lateral geniculate nucleus, optic radiations, or primary visual cortex (see KCC 11.2; Figure 11.15D,G,H). Given the previous history of a tumor, the most likely diagnosis is a recurrence of the tumor involving one or more of these structures on the right side of the brain. Note that the original tumor was in the left temporal lobe, so we would have to propose recurrence in a region somewhat removed from the original resection. Other, less likely possibilities include delayed radiation-induced necrosis, hemorrhage, infarct, demyelination, or brain abscess.

Clinical Course and Neuroimaging

A **brain MRI** (Image 11.4A,B, page 484) revealed abnormal enhancement of the right optic tract, extending to the right lateral geniculate nucleus. The patient was admitted and treated with high-dose steroids but showed no significant improvement. His vision continued to deteriorate over the next 2 weeks, and he was readmitted for stereotactic biopsy of an enhancing area in the right temporal lobe, which showed malignant astrocytoma. Additional chemotherapy was tried, but he continued to worsen, developing progressive left hemiparesis, and he was ultimately transferred to a hospice for comfort care.

CASE 11.4 HEMIANOPIA AFTER TREATMENT FOR A TEMPORAL LOBE TUMOR

IMAGE 11.4A,B Brain MRI Showing a Tumor Involving the Right Optic Tract T1-weighted axial images with intravenous gadolinium contrast enhancement.

(A) and (B) are adjacent sections progressing from inferior to superior.

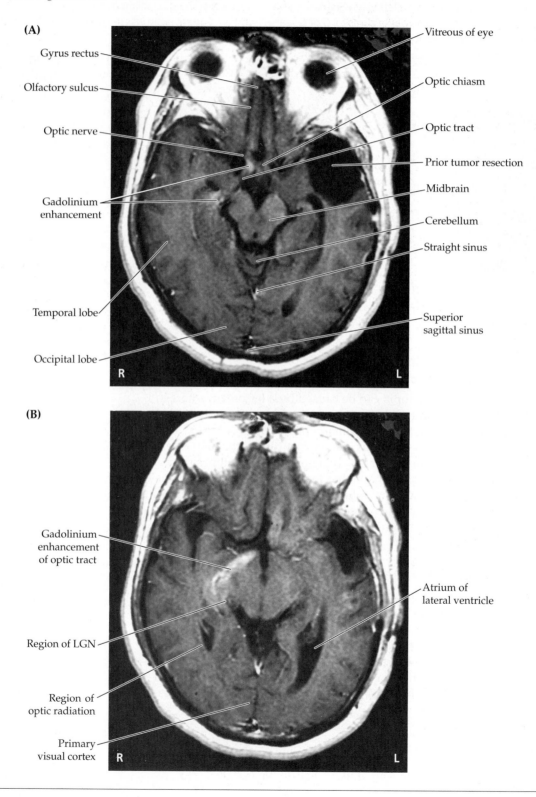

(A)

Gyrus rectus

Olfactory sulcus

Optic nerve

Gadolinium enhancement

Temporal lobe

Occipital lobe

Vitreous of eye

Optic chiasm

Optic tract

Prior tumor resection

Midbrain

Cerebellum

Straight sinus

Superior sagittal sinus

R　L

(B)

Gadolinium enhancement of optic tract

Region of LGN

Region of optic radiation

Primary visual cortex

Atrium of lateral ventricle

R　L

CASE 11.5 VISUAL CHANGES CAUSED BY MIGRAINE HEADACHES?

MINICASE

A 57-year-old, right-handed man visited the emergency room several times because of headaches that had begun 4 months previously. He had **throbbing bilateral or right occipital pain and zigzagging lines** in his field of vision. The headaches were often more severe in the afternoon, and they were relieved by nonsteroidal pain medication (naproxen). He had not previously had headaches of this kind, and he had no family history of migraines. Recently he noticed a vision problem causing him to frequently bump into objects on his left side. He was referred to a neurologist, and examination was normal except for a **left inferior quadrantanopia** (Figure 11.22). The examiner did not listen for a cranial bruit.

LOCALIZATION AND DIFFERENTIAL DIAGNOSIS

On the basis of the symptoms and signs shown in **bold** above, where is the lesion? What is the most likely diagnosis, and what are some other possibilities?

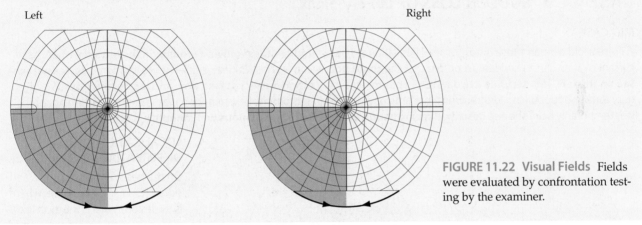

Left   Right

FIGURE 11.22 Visual Fields Fields were evaluated by confrontation testing by the examiner.

Discussion

The key symptoms and signs in this case are:

- **Left inferior homonymous quadrantanopia**
- **Right occipital headaches**

A left inferior homonymous quadrantanopia can be caused by a lesion of the right superior optic radiations or the right superior bank of the calcarine fissure (see Figure 11.15F,I). The right occipital headache also suggests a lesion in the right occipital area. The throbbing quality of the headache and the fortification scotoma with zigzagging lines are typical features of migraine headaches (see KCC 5.1). However, it would be unusual for a man in his late 50s, with no family history, suddenly to develop new-onset migraine. In addition, headache always occurring on the same side of the head is worrisome, and the fixed visual field defect suggests a more permanent deficit. Given his age, a brain tumor such as glioblastoma or brain metastasis should be considered first as the cause of his migraine-like headaches and visual field cut. Other possibilities include hemorrhage, infarct, or abscess. Of note, arteriovenous malformations (AVMs) can cause migraine headaches that recur in the same location (see KCC 5.1, 5.6). They can also cause hemorrhage, ischemia, or infarcts, resulting in focal deficits.

Clinical Course and Neuroimaging

A **brain MRI** (Image 11.5A, page 488 and Image 11.5B, page 489) showed multiple dark flow voids, consistent with a large AVM involving the superior portions of the right occipital lobe. Note that the AVM was located predominantly above the calcarine fissure and therefore affected the contralateral inferior quadrant of the visual field (see Figure 11.15I). An **angiogram** was done to help guide therapy. On injection of the right carotid, the AVM was found to be fed by a large, tortuous, abnormal branch of the right MCA (Image 11.5C, page 489). Injection of the posterior circulation showed that the AVM was also supplied by a similar abnormal branch of the right PCA (not shown). Numerous smaller branches supplied the AVM as well. It was decided that because of its large size and complicated vascular supply, the AVM was not amenable to treatment with surgery or embolization. The patient's headaches resolved spontaneously, and he remained stable when seen in follow-up over the next 4 years.

CASE 11.6 SUDDEN LOSS OF LEFT VISION

MINICASE

A 76-year-old woman with a history of atrial fibrillation and recurrent ovarian cancer suddenly reported that she **could not see on the left**. This occurred five days after abdominal surgery for tumor debulking, during which her anticoagulation was temporarily held. She also **described seeing hands, arms,** **and "waffles" moving around on the left side**, where her vision was lost. Initially, she felt there was some **decreased sensation on her left face, arm**, and **leg** but this quickly resolved. Neurological examination was entirely normal, except for a **left homonymous hemianopia** (Figure 11.23).

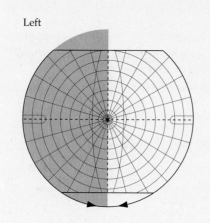
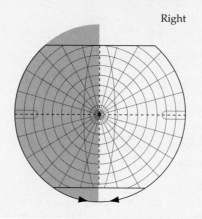

Left

Right

LOCALIZATION AND DIFFERENTIAL DIAGNOSIS

On the basis of the symptoms and signs shown in **bold** above, where is the lesion? What is the most likely diagnosis, and what are some other possibilities?

FIGURE 11.23 **Visual Fields** Fields were evaluated by confrontation testing by the examiner.

Discussion

The key symptoms and signs in this case are:

- Left homonymous hemianopia
- Left-sided visual hallucinations
- Transient sensory loss in the left face, arm, and leg

As we discussed in Case 11.4, left homonymous hemianopia can be caused by a lesion anywhere in the right retrochiasmal visual pathways, including the optic tract, lateral geniculate nucleus, optic radiations, or primary visual cortex (see KCC 11.2; Figure 11.15D,G,H). In this patient, with sudden onset of deficits and a history of atrial fibrillation not being anticoagulated, the most likely cause is embolic infarct (see KCC 10.4). Infarcts of the optic tract and lateral geniculate nucleus are uncommon. An infarct large enough to damage both the temporal and parietal portions of the right optic radiation (see Figures 11.8, 11.15G) would likely involve the entire right middle cerebral artery inferior division, causing severe hemineglect and sensory loss not seen in this patient (see Table 10.1). Therefore, the most likely diagnosis is infarct of the right primary visual cortex in the occipital lobe, caused by right posterior cerebral artery (PCA) embolus (see Table 10.1).

Formed visual hallucinations can be seen in the region of visual loss in patients with acute sensory deprivation, including occipital stroke (see KCC 11.1). Transient left face, arm and leg somatosensory loss could be caused by dysfunction in the right primary somatosensory cortex or thalamus (see Figure 7.9). Given the probable involvement of the right PCA and lack of other deficits, the most likely explanation is transient occlusion of the proximal right PCA by embolus, including branches to the right thalamus (see Figures 10.8, 10.9) with later more permanent occlusion of the more distal right PCA branches to the visual cortex. Other, less likely possibilities include right occipital hemorrhage, right occipital seizures, brain metastases, or brain abscess.

Clinical Course and Neuroimaging

An emergency head CT was performed, which did not reveal hemorrhage or other lesions. Because of her recent surgery she was not eligible for tPA, however, given her history of atrial fibrillation she was treated with intravenous and then oral anticoagulation. An MRI could not be performed since she had a pacemaker, but a **head CT** (Image 11.6A,B, page 490) six days after onset revealed hypodensity in the right medial occipital cortex compatible with infarction in the right PCA territory. Note that some mass effect was present, compressing the right occipital horn of the lateral ventricle (see Image 11.6A). The visual hallucinations stopped over the next few days, but she was left with a left homonymous hemianopia.

Additional Cases

Related cases in other chapters: **localization of lesions in the visual pathways** (Cases 5.1, 5.2, 5.8, 10.3, 10.4, 10.6, 10.8, 10.10, 10.11, 14.8, 17.2, and 19.2); **disorders of higher-order visual processing** (Cases 19.5, 19.6, 19.8, and 19.9). Other relevant cases can be found using the **Case Index** located at the end of the book.

Brief Anatomical Study Guide

1. This chapter covers the **visual pathways** from retina to optic nerve, to optic tract, to lateral geniculate nucleus of the thalamus, to optic radiations, to visual cortex (see Figure 11.15).

2. Throughout these pathways **there are three parallel channels** for process information related to **motion/spatial analysis**, **form**, and **color** (see Figure 11.11).

3. The **retina** (see Figure 11.4) has several different cell types, including photoreceptor **rods**, which are more sensitive to low levels of illumination; **cones,** which have higher spatial and temporal resolution and detect colors; and **ganglion cells**, which form the output layer of the retina.

4. Most retinal ganglion cells can be classified as parasol cells or midget cells. **Parasol cells** have large receptive fields, respond best to gross stimulus features and movement, and project to the **magnocellular layers** of the lateral geniculate nucleus. **Midget cells** are more numerous, have small receptive fields, are sensitive to fine visual detail and colors, and project to the **parvocellular layers** of the lateral geniculate nucleus.

5. The **lateral geniculate nucleus (LGN)** has six layers (see Figure 11.7), with inputs from each eye segregated in different layers. These segregated inputs

Brief Anatomical Study Guide (continued)

are preserved in projections to **layer 4** of the **primary visual cortex (area 17)**, forming **ocular dominance columns** (see Figure 11.12A).

6. Information processing by **visual association cortex (areas 18 and 19)** and **higher-order association cortex** takes place via a **dorsal "where?" stream** dedicated to analysis of motion and spatial relations and a **ventral "what?" stream** involved in analysis of form and color (see Figure 11.11B).

7. Lesions in the primary visual pathways produce characteristic **visual field defects** (see Figure 11.15). Lesions anterior to the optic chiasm produce monocular defects (see Figures 11.15A, 11.16A, and 11.17A). In the optic chiasm the nasal fibers for each eye, carrying visual information from the lateral (temporal) visual hemifields, across to the opposite side. Therefore, visual pathways posterior to the optic chiasm carry information from both eyes corresponding to the contralateral visual field. Lesions of the optic tract (see Figure 11.15D), lateral geniculate nucleus, optic radiations (see Figure 11.8), or visual cortex therefore cause contralateral homonymous ("same named") visual field defects affecting both eyes (see Figures 11.15D–J and 11.16B,C). Lesions of the optic chiasm itself produce bitemporal (bilateral lateral) visual field defects (see Figure 11.15C). Understanding the basic anatomy of visual pathways is essential in clinical diagnosis and differentiation of eye disorders and lesions of the central nervous system.

CASE 11.5 VISUAL CHANGES CAUSED BY MIGRAINE HEADACHES?

IMAGE 11.5A,B Brain MRI Showing an Arteriovenous Malformation (AVM) in the Superior Bank of the Right Calcarine Cortex T1-weighted images. (A) Coronal section. (B) Sagittal section. Dark regions represent flow voids in the AVM.

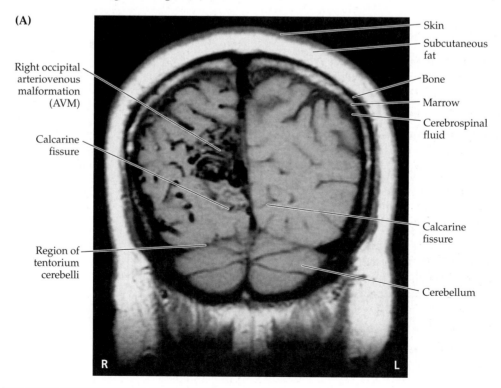

(A)

Right occipital arteriovenous malformation (AVM)

Calcarine fissure

Region of tentorium cerebelli

Skin

Subcutaneous fat

Bone

Marrow

Cerebrospinal fluid

Calcarine fissure

Cerebellum

R   L

CASE 11.5 (continued)

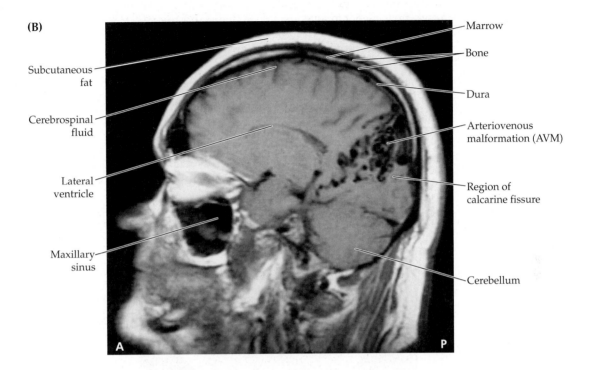

(B)

Subcutaneous fat

Cerebrospinal fluid

Lateral ventricle

Maxillary sinus

Marrow

Bone

Dura

Arteriovenous malformation (AVM)

Region of calcarine fissure

Cerebellum

IMAGE 11.5C Arteriovenous Malformation (AVM) in the Right Superior Occipital Cortex (C) Cerebral angiogram, lateral view, following injection of the right internal carotid artery. The AVM can be seen to fill via abnormal branches of the right middle cerebral artery (MCA).

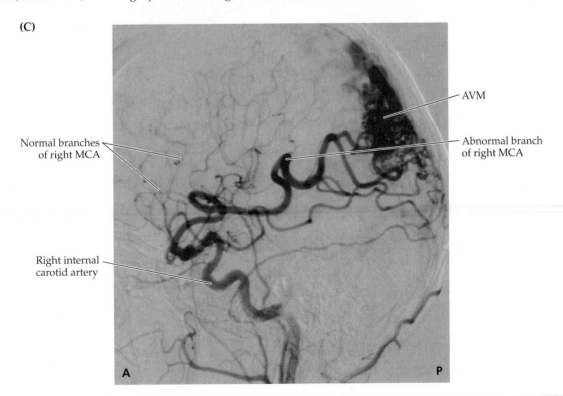

(C)

Normal branches of right MCA

Right internal carotid artery

AVM

Abnormal branch of right MCA

CASE 11.6 SUDDEN LOSS OF LEFT VISION

IMAGE 11.6A,B Right Posterior Cerebral Artery (PCA) Infarct
Head CT images A,B progressing from inferior to superior.

(A)

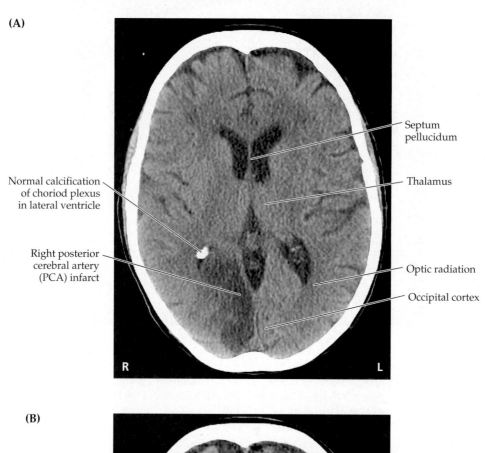

Septum
pellucidum

Thalamus

Normal calcification
of choriod plexus
in lateral ventricle

Right posterior
cerebral artery
(PCA) infarct

Optic radiation

Occipital cortex

R   L

(B)

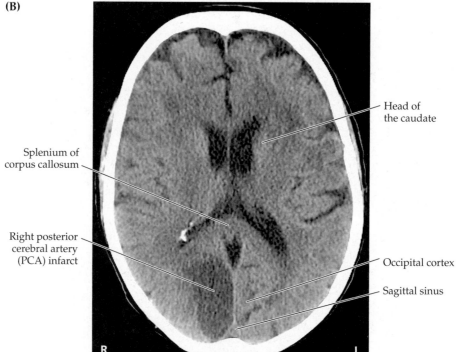

Head of
the caudate

Splenium of
corpus callosum

Right posterior
cerebral artery
(PCA) infarct

Occipital cortex

Sagittal sinus

R   L

References

General References

Chalupa LM, Werner JS (eds.). 2003. *The Visual Neurosciences*. The MIT Press, Cambridge, MA.

Chan, JW. 2007. *Optic Nerve Disorders: Diagnosis and Management*. Springer-Verlag, New York, LLC.

Corbett JJ. 2003. The bedside and office neuro-ophthalmology examination. *Semin Neurol* 23 (1): 63–76.

Kandel ER, Schwartz JH, Jessell TM. 2000. *Principles of Neural Science*. 4th Ed. Chapters 25–29. McGraw Hill, New York.

Sun P, Ueno K, Waggoner RA, Gardner JL, Tanaka K, Cheng K. 2007. A temporal frequency-dependent functional architecture in human V1 revealed by high-resolution fMRI. *Nat Neurosci* 10 (11): 1404–1406.

Volpe NJ, Galetta SL, Liu GT. 2007. *Neuro-Ophthalmology: Diagnosis and Management*. 2nd Ed. Saunders, Philadelphia.

Retinal Artery Occlusion

Biousse V. 1997. Carotid disease and the eye. *Curr Opin Ophthalmol* 8 (6): 16–26.

Burde RM. 1989. Amaurosis fugax. *J Clin Neuroophthalmol* 9 (3): 185–189.

Chen CS, Lee AW. 2008. Management of acute central retinal artery occlusion. *Nat Clin Pract Neurol* 4 (7): 376–383.

Karjalainen K. 1971. Occlusion of the central retinal artery and branch arterioles: A clinical, tonographic and fluorescein angiographic study of 175 patients. *Acta Ophthalmol Suppl* 109: 1–95.

Oshinskie L. 1987. Branch retinal artery occlusion and carotid artery stenosis. *Am J Optom Physiol Optics* 64 (2): 144–149.

Optic Neuritis

Balcer LJ. 2006. Clinical practice. Optic neuritis. *N Engl J Med* 354 (12): 1273–1280.

Hickman SJ. 2007. Optic nerve imaging in multiple sclerosis. *J Neuroimaging* Suppl 1: 42S–45S.

Plant GT. 2008. Optic neuritis and multiple sclerosis. *Curr Opin Neurol* (1): 16–21.

Volpe NJ. 2008. The optic neuritis treatment trial: a definitive answer and profound impact with unexpected results. *Arch Ophthalmol* 126 (7): 996–999.

Xu J, Sun SW, Naismith RT, Snyder AZ, Cross AH, Song SK. 2008. Assessing optic nerve pathology with diffusion MRI: from mouse to human. *NMR Biomed* 21 (9): 928–940.

Optic Nerve Injury Due to Orbital Trauma

Wu N, Yin ZQ, Wang Y. 2008. Traumatic optic neuropathy therapy: an update of clinical and experimental studies. *J Int Med Res* 36 (5): 883–889.

Suprasellar Meningioma

Chicani CF, Miller NR. 2003. Visual outcome in surgically treated suprasellar meningiomas. *J Neuroophthalmol* 23 (1): 3–10.

Ehlers N, Malmros R. 1973. The suprasellar meningioma. A review of the literature and presentation of a series of 31 cases. *Acta Ophthalmol Suppl* 1–74.

Optic Tract Lesions

Savino PJ, Paris M, Schatz NJ, Orr LS, Corbett JJ. 1978. Optic tract syndrome. *Arch Ophthalmol* 96 (4): 656–663.

Lateral Geniculate Nucleus Lesions

Acheson JF, Sanders MD. 1997. *Common Problems in Neuro-Ophthalmology*. Saunders, London.

Miller NR (ed.), Newman NJ, Biousse V. 2004. *Walsh and Hoyt's Clinical Neuro-ophthalmology*. 6th Ed., Vol. 2. Lippincott Williams & Wilkins, Baltimore, MD.

Optic Radiation and Visual Cortex Lesions

Ropper AH, Samuels MA. 2009. *Adams and Victor's Principles of Neurology*. 9th Ed., Chapter 13. McGraw-Hill, New York.

CONTENTS

Brainstem I: Surface Anatomy and Cranial Nerves

A 41-year-old woman noticed that she could not hear anything from the phone receiver when it was on her left ear. Also, over the previous year, she sometimes felt as though the room was spinning slightly when she moved her head. She developed some pain on the left side of her face and decreased taste on the left side of her tongue, as well as a decreased corneal reflex on the left side.

This patient illustrates the complex variety of abnormalities that patients with cranial nerve disorders may experience. In this chapter, we will learn about the origins of cranial nerves in the brainstem, their courses to the periphery, and their various functions.

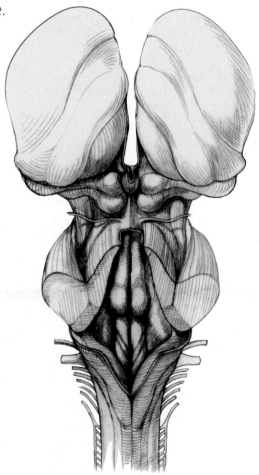

ANATOMICAL AND CLINICAL REVIEW

LOCATED AT THE BASE of the cerebral hemispheres, the **brainstem** is a compact, stalklike structure that carries nearly all information between the brain and the remainder of the body (**Figure 12.1**). This tight space is the corridor to all major sensory, motor, cerebellar, and cranial nerve pathways. However, the brainstem is not simply a conduit for information. It also contains numerous important nuclei of its own, which control the cranial nerves; level of consciousness; cerebellar circuits; muscle tone; posture; and cardiac, respiratory, and numerous other essential functions. If the brain were a city, then the brainstem would be both the Grand Central Station and Central Power Supply packed into one location. Thus, small lesions in the brainstem can result in substantial deficits, often involving multiple motor, sensory, and neuroregulatory modalities.

An intimate knowledge of brainstem anatomy is a powerful clinical tool. Armed with an understanding of brainstem nuclei and pathways, the clinician can intelligently decide on appropriate diagnostic and therapeutic measures for patients with brainstem disorders. Brainstem anatomy is so complex yet so clinically relevant that we devote three full chapters (Chapters 12–14) to understanding it in detail. Thus, in this chapter we will first review the surface features of the brainstem and then discuss the course and functions of each cranial nerve. Next, in Chapter 13 we will focus in greater detail on the cranial nerves and central pathways mediating eye movements and pupillary control. Finally, in Chapter 14 we will study the vascular supply and internal structures of the brainstem, including the major ascending and descending tracts, reticular formation, and other important brainstem nuclei.

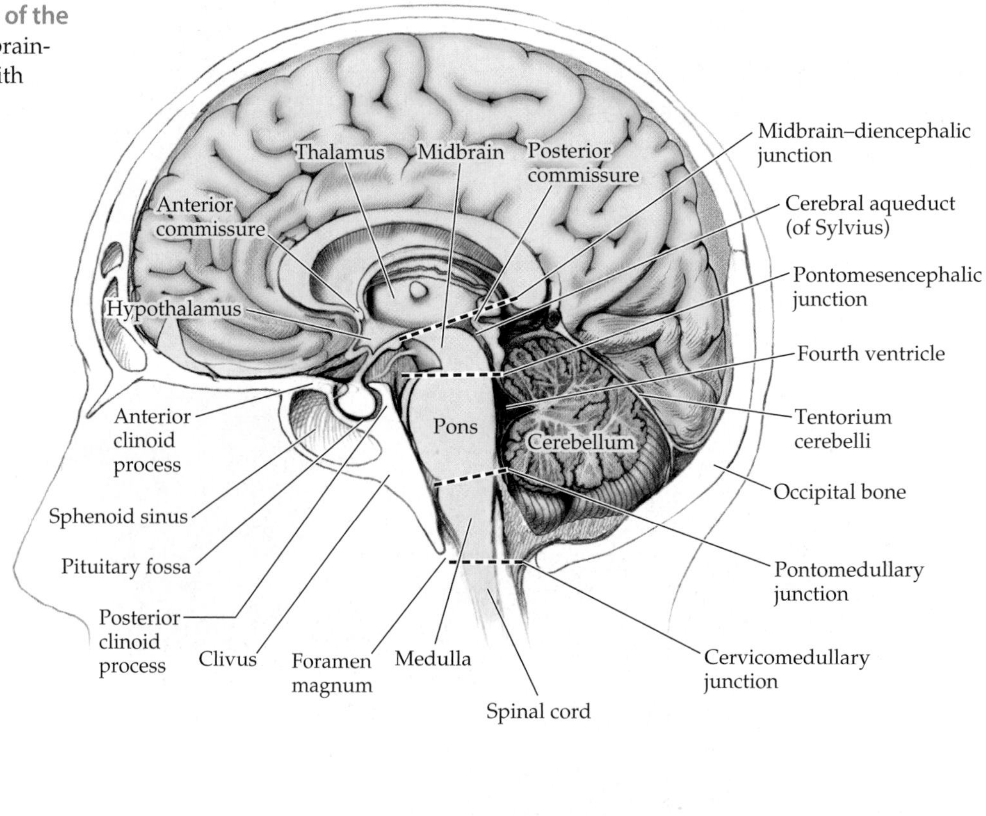

FIGURE 12.1 Midsagittal View of the Brainstem In Situ Parts of the brainstem and spatial relationships with surrounding structures.

TABLE 12.1 Cranial Nerve Names and Main Functions

CN	NAME	MAIN FUNCTION(S)
CN I	Olfactory nerve	Olfaction
CN II	Optic nerve	Vision
CN III	Oculomotor nerve	Eye movements; pupil constriction
CN IV	Trochlear nerve	Eye movements
CN V	Trigeminal nerve	Facial sensation; muscles of mastication
CN VI	Abducens nerve	Eye movements
CN VII	Facial nerve	Muscles of facial expression; taste; lacrimation; salivation
CN VIII	Vestibulocochlear nerve	Hearing; equilibrium sense
CN IX	Glossopharyngeal nerve	Pharyngeal muscles; carotid body reflexes; salivation
CN X	Vagus nerve	Parasympathetics to most organs; laryngeal muscles (voice); pharyngeal muscles (swallowing); aortic arch reflexes
CN XI	Spinal accessory nerve	Head turning (trapezius and sternomastoid muscles)
CN XII	Hypoglossal nerve	Tongue movement

Learning the cranial nerves initially requires some memorization. Over time, however, they become very familiar because of their important clinical relevance. The numbers, names, and main functions of the cranial nerves are listed in Table 12.1. Note that the cranial nerves have both sensory and motor functions. To learn the cranial nerves and their functions, two different review strategies are useful. In one, the **cranial nerves** are listed in numerical sequence and the sensory and motor functions of each nerve are discussed (see Table 12.4). In the second, the different sensory and motor **cranial nerve nuclei** are listed, and the functions and cranial nerves subserved by each nucleus are discussed (see Table 12.3). Both approaches are clinically relevant, and we will use both strategies at various points in these brainstem chapters to integrate knowledge of the peripheral and central course of the cranial nerves.

REVIEW EXERCISE

Cover the two right columns in Table 12.1. For each numbered cranial nerve, provide its name and main functions.

Surface Features of the Brainstem

The **brainstem** consists of the **midbrain**, **pons**, and **medulla** (see Figure 12.1). It lies within the posterior fossa of the cranial cavity. The rostral limit of the brainstem is the **midbrain–diencephalic junction** (see Figure 12.1). Here the brainstem meets thalamus and hypothalamus at the level of the tentorium cerebelli. Midbrain joins pons at the **pontomesencephalic junction**, and pons meets medulla at the **pontomedullary junction**. The caudal limit of the brainstem is the **cervicomedullary junction**, at the level of the foramen magnum and pyramidal decussation (see Figure 12.1, **Figure 12.2A**; see also Figure 6.8). The cerebellum is attached to the dorsal surface of the pons and upper medulla (see Figure 12.1). Although some authors have included the cerebellum or thalamus in the term "brainstem," we adopt common clinical usage here and take brainstem to imply only midbrain, pons, and medulla. We discuss the thalamus and cerebellum at greater length elsewhere (see Chapters 7 and 15).

On the dorsal surface of the **midbrain** are two pairs of bumps called the **superior colliculi** and **inferior colliculi** (**Figure 12.2B**). Together, these form the **tec-**

REVIEW EXERCISE

For the midbrain, pons and medulla in Figure 12.1 or 12.2C, point in the rostral, caudal, dorsal, ventral, superior, inferior, posterior and anterior directions (see definitions in Figure 2.4). How do these differ for points above the midbrain–diencephalic junction?

(A)

ROSTRAL

Superior

Right ⟷ Left

Inferior

Thalamus

Optic nerve (CN II)

Oculomotor nerve (CN III)

Trochlear nerve (CN IV)

Trigeminal nerve (CN V)

Abducens nerve (CN VI)

Facial nerve (CN VII)

Vestibulocochlear nerve (CN VIII)

Glossopharyngeal nerve (CN IX)

Vagus nerve (CN X)

Hypoglossal nerve (CN XII)

Spinal accessory nerve (CN XI)

Optic chiasm

Optic tract

Interpeduncular fossa

Cerebral peduncle

Middle cerebellar peduncle

Cerebellopontine angle

Inferior olive

Pyramid

Pyramidal decussation

Midbrain

Pons

Medulla

Spinal cord

CAUDAL

(B)

ROSTRAL

Superior

Left ⟷ Right

Inferior

Thalamus

Pineal body

Superior colliculus

Inferior colliculus

Anterior medullary velum

Superior cerebellar peduncle

Middle cerebellar peduncle

Inferior cerebellar peduncle

Obex

Dorsal (posterior)
column tubercles:
Nucleus cuneatus
Nucleus gracilis

Dorsal (posterior) columns:
Fasciculus gracilis
Fasciculus cuneatus

Brachium of superior
colliculus

Lateral geniculate nucleus

Medial geniculate nucleus

Brachium of inferior
colliculus

Trochlear nerve (CN IV)

Facial colliculus

Hypoglossal trigone

Glossopharyngeal
nerve (CN IX)

Vagus (CN X)

Vagal trigone

Spinal accessory nerve (CN XI)

Midbrain

Pons

Medulla

Spinal cord

CAUDAL

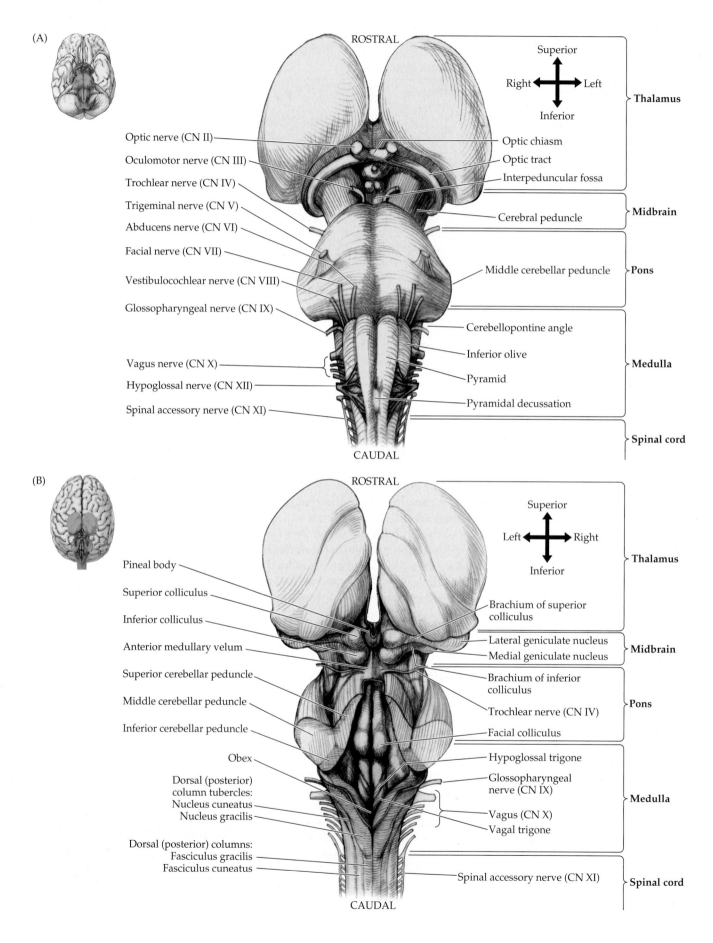

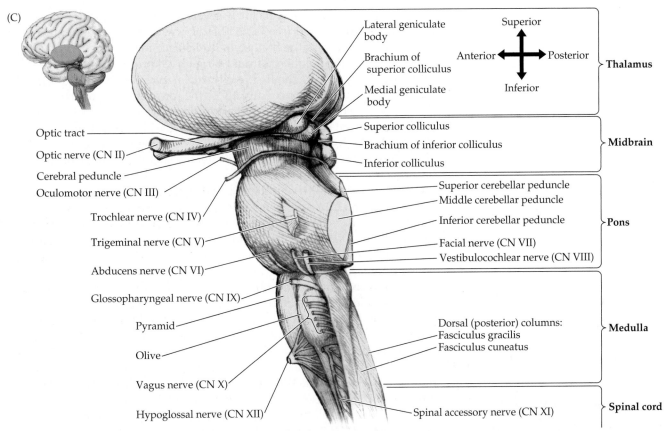

(C)

Lateral geniculate body

Brachium of superior colliculus

Medial geniculate body

Thalamus

Optic tract

Optic nerve (CN II)

Cerebral peduncle

Oculomotor nerve (CN III)

Trochlear nerve (CN IV)

Trigeminal nerve (CN V)

Abducens nerve (CN VI)

Glossopharyngeal nerve (CN IX)

Pyramid

Olive

Vagus nerve (CN X)

Hypoglossal nerve (CN XII)

Superior colliculus

Brachium of inferior colliculus

Inferior colliculus

Midbrain

Superior cerebellar peduncle

Middle cerebellar peduncle

Inferior cerebellar peduncle

Facial nerve (CN VII)

Vestibulocochlear nerve (CN VIII)

Pons

Dorsal (posterior) columns: Fasciculus gracilis Fasciculus cuneatus

Medulla

Spinal accessory nerve (CN XI)

Spinal cord

FIGURE 12.2 Surface Anatomy of the Brainstem and Cranial Nerves (A) Ventral view with cerebral hemispheres removed. (B) Dorsal view with cerebellum removed, exposing floor of the fourth ventricle. (C) Lateral view.

tum (meaning "roof") of the midbrain. The ventral surface of the midbrain is formed by the **cerebral peduncles**, between which lies the **interpeduncular fossa** (see Figure 12.2A; see also Figure 5.6). The **pons** is limited dorsally by the **fourth ventricle** (see Figure 12.1). More dorsolaterally, the pons is attached to the cerebellum by large white matter tracts called the **superior, middle,** and **inferior cerebellar peduncles** (see Figure 12.2B). On the ventral surface of the **medulla**, the **pyramids** can be seen descending from the pontomedullary junction to the **pyramidal decussation** (see Figure 12.2A). It is often useful to divide the medulla into a rostral portion and a caudal portion. In the **rostral medulla** the prominent bulges of the **inferior olivary nuclei** can be seen just lateral to the pyramids (see Figure 12.2A). In the **caudal medulla** the inferior olivary nuclei are no longer seen, but the **posterior columns** and **posterior column nuclei** are visible on the dorsal surface (see Figure 12.2B).

The floor of the fourth ventricle extends from the pons to the rostral half of the medulla. Along the floor of the fourth ventricle, several bumps are visible. These include the **facial colliculi**, formed by the abducens nuclei and fibers of the facial nerve (see Figure 12.2B; see also Figure 14.1C). The **hypoglossal trigone** and **vagal trigone** (see Figure 12.2B) are formed by the hypoglossal nucleus (CN XII) and the dorsal motor nucleus of CN X, respectively. Recall that rostrally, the fourth ventricle joins the cerebral aqueduct, which runs through the midbrain (see Figure 12.1). Caudally, the fourth ventricle drains into the subarachnoid space via the foramina of Luschka (located laterally) and foramen of Magendie (located in the midline). The fourth ventricle ends caudally at the **obex** (see Figure 12.2B), marking the entry to the spinal cord central canal, which in adults is normally closed.

For those who are verbally inclined, several mnemonics exist for the cranial nerve names and numbers (see Table 12.1). However, the best visual mnemonic for the cranial nerves is the brainstem itself, since the cranial nerves emerge roughly in numerical sequence from I through XII proceeding from rostral to caudal (see Figure 12.2). The first two cranial nerves do not emerge from the brainstem, but rather connect directly to the forebrain. The **olfactory nerves** (**CN I**) enter the **olfactory bulbs** and **olfactory tracts**, which run along the ventral surface of the frontal lobes in the **olfactory sulci** (see Figures 18.5 and 18.6). The **optic nerves** (**CN II**) meet at the optic chiasm, forming the optic tracts, which wrap laterally around the midbrain to enter the lateral geniculate nuclei of the thalamus (see Figures 11.6 and 11.15).

Cranial nerves III–XII exit the brainstem either ventrally or ventrolaterally (see Figure 12.2A and Figure 12.2C). The one exception is CN IV, which exits from the dorsal midbrain (see Figure 12.2B). We will see shortly that CN III, VI, and XII, which exit ventrally near the midline, together with CN IV, which exits dorsally, form a distinct functional group innervating somatic motor structures.

The **oculomotor nerves** (**CN III**) emerge ventrally from the interpeduncular fossa of the midbrain (see Figure 12.2A). Note that the oculomotor nerve usually passes between the posterior cerebral artery and the superior cerebellar artery (see Figure 14.18A). As we just mentioned, the **trochlear nerve** (**CN IV**) is exceptional in exiting dorsally from the midbrain (see Figure 12.2B). The fibers of CN IV cross over as they emerge, an arrangement that is also unique to this cranial nerve. The **trigeminal nerve** (**CN V**) exits from the ventrolateral pons (see Figure 12.2A,C). The **abducens nerve** (**CN VI**) exits ventrally, at the pontomedullary junction (see Figure 12.2A,C). Then, proceeding in sequence, the **facial nerve** (**CN VII**), **vestibulocochlear nerve** (**CN VIII**), **glossopharyngeal nerve** (**CN IX**), and **vagus nerve** (**CN X**), exit ventrolaterally from the pontomedullary junction and rostral medulla. The region where CN VII, VIII, and IX exit the brainstem is called the **cerebellopontine angle**. The **spinal accessory nerve** (also known as the **accessory spinal nerve; CN XI**) arises laterally from multiple rootlets along the upper cervical spinal cord. The **hypoglossal nerve** (**CN XII**) exits the medulla ventrally, between the pyramids and inferior olivary nuclei (see Figure 12.2A).

Skull Foramina and Cranial Nerve Exit Points

When we discuss each cranial nerve in the sections that follow, we will describe its course in detail. For now, we will simply introduce the foramina through which the cranial nerves exit the skull (Figure 12.3; Table 12.2).

The olfactory nerves exit via the **cribriform plate**, and the optic nerve via the **optic canal** (see Figure 12.3; Table 12.2). The **superior orbital fissure** transmits several nerves (CN III, IV, VI, and V_1) into the orbit (see Figure 12.3A, C). CN III, IV, and VI mediate eye movements. The ophthalmic division of CN V (CN V_1) conveys sensation for the eye and upper face. The maxillary (CN V_2) and mandibular (CN V_3) divisions of the trigeminal nerve exit via the **foramen rotundum** and **foramen ovale**, respectively, providing sensation to the remainder of the face (see Figure 12.7). CN VII and VIII both exit the cranial cavity via the **internal auditory meatus** to enter the **auditory canal**. CN VIII innervates the inner ear deep within the temporal bone. CN VII exits the skull to reach the muscles of facial expression via the **stylomastoid foramen** (see Figure 12.3B). The **jugular foramen** transmits CN IX, X, and XI (see Figure 12.3A,B). Finally, the hypoglossal nerve (CN XII), controlling tongue movements, exits the skull via its own foramen, the **hypoglossal canal**, which lies just in front of the foramen magnum.

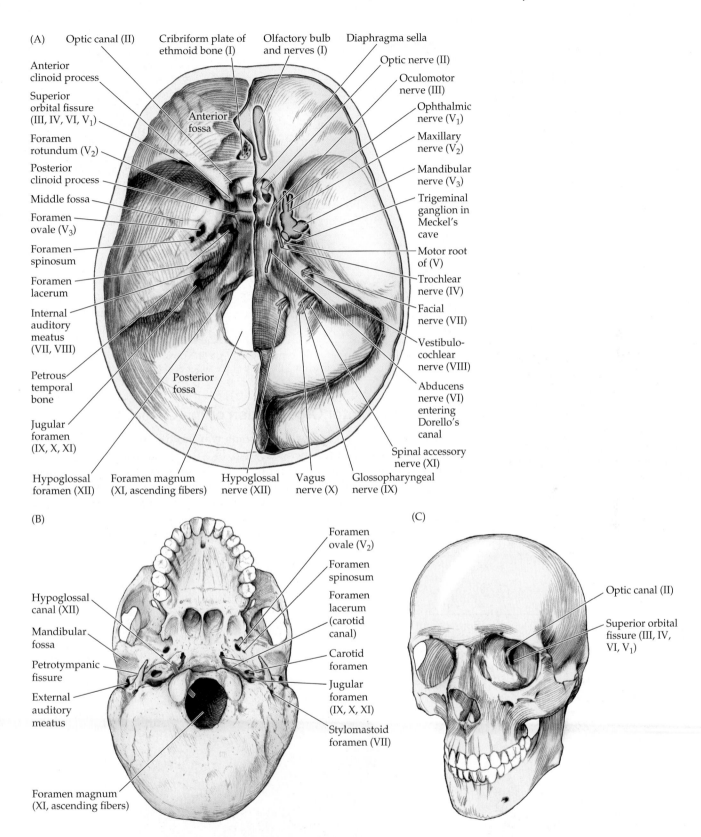

(A) Optic canal (II)   Cribriform plate of ethmoid bone (I)   Olfactory bulb and nerves (I)   Diaphragma sella

Anterior clinoid process

Superior orbital fissure (III, IV, VI, V₁)

Foramen rotundum (V₂)

Posterior clinoid process

Middle fossa

Foramen ovale (V₃)

Foramen spinosum

Foramen lacerum

Internal auditory meatus (VII, VIII)

Petrous temporal bone

Jugular foramen (IX, X, XI)

Anterior fossa

Posterior fossa

Optic nerve (II)

Oculomotor nerve (III)

Ophthalmic nerve (V₁)

Maxillary nerve (V₂)

Mandibular nerve (V₃)

Trigeminal ganglion in Meckel's cave

Motor root of (V)

Trochlear nerve (IV)

Facial nerve (VII)

Vestibulo-cochlear nerve (VIII)

Abducens nerve (VI) entering Dorello's canal

Spinal accessory nerve (XI)

Hypoglossal foramen (XII)   Foramen magnum (XI, ascending fibers)   Hypoglossal nerve (XII)   Vagus nerve (X)   Glossopharyngeal nerve (IX)

(B)

Hypoglossal canal (XII)

Mandibular fossa

Petrotympanic fissure

External auditory meatus

Foramen magnum (XI, ascending fibers)

Foramen ovale (V₂)

Foramen spinosum

Foramen lacerum (carotid canal)

Carotid foramen

Jugular foramen (IX, X, XI)

Stylomastoid foramen (VII)

(C)

Optic canal (II)

Superior orbital fissure (III, IV, VI, V₁)

FIGURE 12.3 Skull Foramina Serving as Cranial Nerve Exit Points (A) Inside view of the base of the skull, seen from above, with cranial nerves shown on the right and exit foramina shown on the left. (B) View of the base of the skull, seen from below. (C) Anterior view of the skull and foramina.

TABLE 12.2 Cranial Nerve Exit Foramina

CN	NAME	EXIT FORAMEN
CN I	Olfactory nerves	Cribriform plate
CN II	Optic nerve	Optic canal
CN III	Oculomotor nerve	Superior orbital fissure
CN IV	Trochlear nerve	Superior orbital fissure
CN V	Trigeminal nerve	V_1: Superior orbital fissure
		V_2: Foramen rotundum
		V_3: Foramen ovale
CN VI	Abducens nerve	Superior orbital fissure[a]
CN VII	Facial nerve	Auditory canal (stylomastoid foramen)
CN VIII	Vestibulocochlear nerve	Auditory canal
CN IX	Glossopharyngeal nerve	Jugular foramen
CN X	Vagus nerve	Jugular foramen
CN XI	Spinal accessory nerve	Jugular foramen (*enters* skull via foramen magnum)
CN XII	Hypoglossal nerve	Hypoglossal foramen (canal)

[a]The abducens nerve first exits the dura through Dorello's canal (see Figure 12.3) and then travels a long distance before exiting the skull at the superior orbital fissure.

FIGURE 12.4 Development of Cranial Nerve Nuclei Sensory and Motor Longitudinal Columns (A) Cross section of human myelencephalon at 45 days showing locations of sensory and motor cranial nerve nuclei functional columns. (B) Adult medulla, with locations of functional columns indicated. Examples of the nuclei in these columns for a section at this level are indicated in parentheses. (A after Tuchman-Duplessis H, Auroux M, and Haegel P. 1974. *Illustrated Human Embryology.* Volume 3. *Nervous System and Endocrine Glands.* Masson & Company, Paris. B after Martin JH. 1996. *Neuroanatomy: Text and Atlas.* 2nd Ed. McGraw-Hill, New York.)

Sensory and Motor Organization of the Cranial Nerves

The cranial nerves are analogous in some ways to the spinal nerves, having both sensory and motor functions. Also, like the spinal cord, motor cranial nerve nuclei are located more ventrally, while sensory cranial nerve nuclei are located more dorsally (**Figure 12.4**). However, cranial nerve sensory and motor functions are more specialized because of the unique anatomy of the head and neck. During embryological development, the cranial nerve nuclei lie adjacent to the ventricular system (see Figure 12.4A). As the nervous sys-

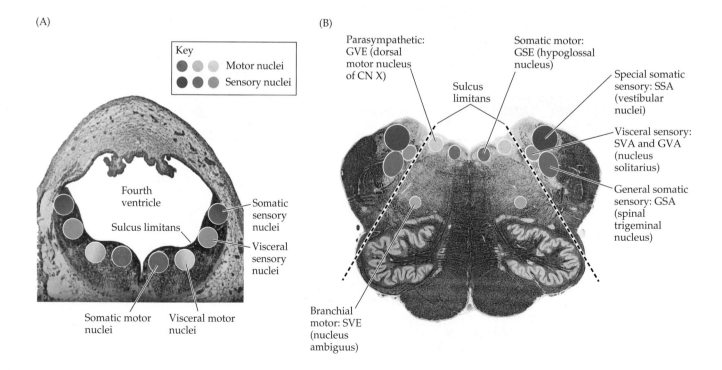

(A)

Key
● ● ● Motor nuclei
● ● ● Sensory nuclei

Fourth ventricle

Sulcus limitans

Somatic sensory nuclei

Visceral sensory nuclei

Somatic motor nuclei

Visceral motor nuclei

(B)

Parasympathetic: GVE (dorsal motor nucleus of CN X)

Somatic motor: GSE (hypoglossal nucleus)

Sulcus limitans

Special somatic sensory: SSA (vestibular nuclei)

Visceral sensory: SVA and GVA (nucleus solitarius)

General somatic sensory: GSA (spinal trigeminal nucleus)

Branchial motor: SVE (nucleus ambiguus)

tem matures, **three motor columns** and **three sensory columns** of cranial nerve nuclei develop that run in an interrupted fashion through the length of the brainstem (see Figure 12.4 and Figure 12.5). Each column subserves a different motor or sensory cranial nerve function, which can be classified as shown in Table 12.3. The color codes for each column used in Figures 12.4 and 12.5 and in Table 12.3 will remain constant throughout this chapter. In another set of terminology described toward the end of this section (and listed in the figures and tables) each column can also be described as general vs. special, somatic vs. visceral, and afferent vs. efferent. Let's review each of these columns in more detail, moving from medial to lateral.

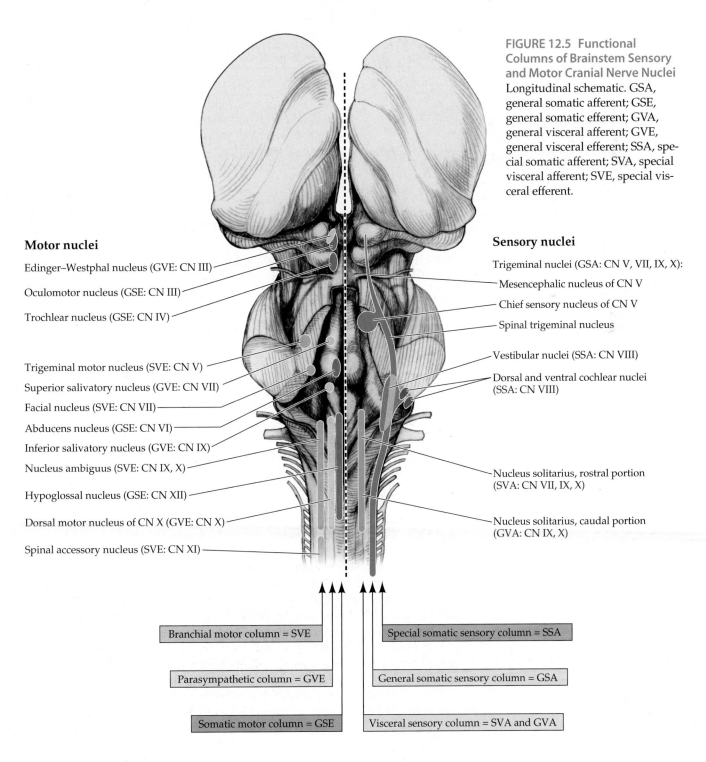

FIGURE 12.5 Functional Columns of Brainstem Sensory and Motor Cranial Nerve Nuclei Longitudinal schematic. GSA, general somatic afferent; GSE, general somatic efferent; GVA, general visceral afferent; GVE, general visceral efferent; SSA, special somatic afferent; SVA, special visceral afferent; SVE, special visceral efferent.

Motor nuclei

Edinger–Westphal nucleus (GVE: CN III)

Oculomotor nucleus (GSE: CN III)

Trochlear nucleus (GSE: CN IV)

Trigeminal motor nucleus (SVE: CN V)

Superior salivatory nucleus (GVE: CN VII)

Facial nucleus (SVE: CN VII)

Abducens nucleus (GSE: CN VI)

Inferior salivatory nucleus (GVE: CN IX)

Nucleus ambiguus (SVE: CN IX, X)

Hypoglossal nucleus (GSE: CN XII)

Dorsal motor nucleus of CN X (GVE: CN X)

Spinal accessory nucleus (SVE: CN XI)

Sensory nuclei

Trigeminal nuclei (GSA: CN V, VII, IX, X):

Mesencephalic nucleus of CN V

Chief sensory nucleus of CN V

Spinal trigeminal nucleus

Vestibular nuclei (SSA: CN VIII)

Dorsal and ventral cochlear nuclei (SSA: CN VIII)

Nucleus solitarius, rostral portion (SVA: CN VII, IX, X)

Nucleus solitarius, caudal portion (GVA: CN IX, X)

Branchial motor column = SVE

Parasympathetic column = GVE

Somatic motor column = GSE

Special somatic sensory column = SSA

General somatic sensory column = GSA

Visceral sensory column = SVA and GVA

TABLE 12.3 Classification of Cranial Nerve Nuclei into Motor and Sensory Columns

CLASSIFICATION	FUNCTION(S)	BRAINSTEM NUCLEI	CRANIAL NERVE(S)
MOTOR			
Somatic motor (general somatic efferent)	Extraocular muscles, intrinsic tongue muscles	Oculomotor, trochlear, abducens, and hypoglossal	CN III, IV, VI, XII
Branchial motor (special visceral efferent)	Muscles of mastication, facial expression, middle ear, pharynx, larynx, sternomastoid, upper portion of trapezius	Motor nucleus of CN V Facial nucleus Nucleus ambiguus Accessory spinal nucleus	CN V CN VII CN IX, X CN XI
Parasympathetic (general visceral efferent)	Parasympathetic innervation of head and thoracoabdominal viscera above splenic flexure	Edinger–Westphal nucleus Superior salivatory nucleus Inferior salivatory nucleus Dorsal motor nucleus of CN X	CN III CN VII CN IX CN X
SENSORY			
Visceral sensory (special visceral afferent)	Taste	Nucleus solitarius (rostral portion, gustatory nucleus)	CN VII, IX, X
(general visceral afferent)	Inputs for control of cardiorespiratory and digestive function	Nucleus solitarius (caudal portion, cardiorespiratory nucleus)	CN IX, X
General somatic sensory (general somatic afferent)	Touch, pain, temperature, position, and vibration sense for face, sinuses, and meninges	Trigeminal nuclei	CN V, VII, IX, X
Special somatic sensory (special somatic afferent)	Olfaction, vision, hearing, vestibular sensation (olfaction and vision do not have sensory nuclei in the brainstem)	Cochlear nuclei, vestibular nuclei	CN VIII

The **somatic motor nuclei** are the oculomotor (CN III), trochlear (CN IV), abducens (CN VI), and hypoglossal (CN XII) nuclei, all of which remain adjacent to the midline (see Figures 12.4 and 12.5; Table 12.3). These nuclei send their fibers to exit the brainstem close to the midline as well (see Figure 12.2). The somatic motor nuclei innervate the extraocular and intrinsic tongue muscles, which are derived embryologically from the **occipital somites**.

The **visceral motor nuclei** (see Figure 12.4A) separate into two different columns of nuclei: branchial motor nuclei and parasympathetic nuclei (see Table 12.3). The **branchial motor nuclei** are the trigeminal motor nucleus (CN V), facial nucleus (CN VII), nucleus ambiguus (CN IX, X), and spinal accessory nucleus (CN XI) (see Table 12.3). During development, the branchial motor nuclei initially lie just lateral to the somatic motor nuclei (see Figure 12.4A), but they gradually migrate ventrolaterally into the tegmentum (Figures 12.4B and 12.5). Both the somatic and branchial motor nuclei innervate *striated muscles*. However, unlike the somatic motor nuclei, the **branchial motor nuclei** innervate muscles derived from the **branchial arches**, including the muscles of mastication, facial expression, middle ear, pharynx, and larynx. Based on the fact that the sternomastoid and upper portion of the trapezius (innervated by CN XI) may derive from somites rather than the branchial arches, some classify the spinal accessory nucleus as somatic or mixed somatic and branchial motor. However, for simplicity, and since the spinal accessory nucleus is located laterally in approximate continuity with the nucleus ambiguus, we will consider the spinal accessory nucleus to be part of the branchial motor column.

The next column comprises the **parasympathetic nuclei** (see Figure 12.5). These are the Edinger–Westphal nucleus (CN III), superior (CN VII) and inferior (CN IX) salivatory nuclei, and the dorsal motor nucleus of the vagus (CN X) (see Table 12.3; see also Figure 12.6). The parasympathetic nuclei do not innervate striated muscle. They provide preganglionic parasympathetic fibers innervating *glands, smooth muscle, and cardiac muscle* of the head, heart, lungs, and digestive tract above the splenic flexure (see also Figure 6.13).

Continuing laterally, the sensory nuclei also form three columns (see Figures 12.4 and 12.5). The **visceral sensory column** contains a single nucleus, the nucleus solitarius, which has two parts. The rostral nucleus solitarius, or gustatory nucleus, receives taste inputs primarily from CN VII, but also from CN IX and X. The caudal nucleus solitarius, or cardiorespiratory nucleus, receives inputs for regulation of cardiac, respiratory, and gastrointestinal function from CN IX and X (see Table 12.3). As discussed in Chapter 14, the nucleus solitarius has other functions as well, such as sleep regulation.

The **general somatosensory nuclei**, or **trigeminal nuclei** (see Table 12.3; Figures 12.4 and 12.5), mediate touch, pain, temperature, position, and vibration sense for the face, sinuses, and meninges. As we will see, sensory inputs to the trigeminal sensory nuclei are mostly from the trigeminal nerve (CN V), but there are also smaller sensory inputs from CN VII, IX, and X (see Figure 12.7B).

By definition, the **special senses** are olfaction, vision, hearing, vestibular sense, and taste. Olfaction and vision do not have their primary sensory nuclei in the brainstem. The brainstem **special somatic sensory nuclei** are the cochlear nuclei mediating hearing (CN VIII) and the vestibular nuclei mediating positional equilibrium (CN VIII) (see Figures 12.4 and 12.5; Table 12.3). Note that although taste is one of the special senses, it has been classified here in the visceral sensory column (nucleus solitarius).

As we mentioned earlier, in another commonly used classification scheme, cranial nerve functions are referred to as either general or special, somatic or visceral, and afferent or efferent. Combinations of these terms result in a total of eight possible categories, such as general somatic efferent (GSE), special visceral afferent (SVA), and so on. However, this scheme still results in the same six longitudinal columns of nuclei (three motor, three sensory) described above (see Figure 12.5; Table 12.3). The reasons are that (1) the somatic motor column is not divided into special and general categories, and (2) the nucleus solitarius contains both the special and general visceral afferents.

Note that most cranial nerve nuclei project to or receive inputs from predominantly one cranial nerve (see Figure 12.5). The three exceptions can be remembered with the mnemonic **SAT** for **S**olitarius, **A**mbiguus, and **T**rigeminal, all of which are long nuclei extending into the medulla (see Figure 12.5).

MNEMONIC

> ## REVIEW EXERCISE
>
> Draw a sketch of the brainstem as in Figure 12.5. Fill in the three columns of motor nuclei and three columns of sensory nuclei. For each nucleus, state its name as well as the cranial nerves that it supplies (see Figure 12.5; Table 12.3). You should be able to draw this sketch from memory.

Functions and Course of the Cranial Nerves

In the sections that follow we will review each of the cranial nerves and their functions in detail. Table 12.4 lists the motor and sensory functions of each cranial nerve. Note that some cranial nerves are **purely motor** (CN III, IV, VI, XI, XII), some are **purely sensory** (CN I, II, VIII), and some have **both motor and sensory** functions (CN V, VII, IX, X). The information contained in Table 12.4 is of central importance to understanding the cranial nerves and should be very familiar to you by the time you complete this chapter. The relevant portions of Table 12.4 will be repeated as we introduce each cranial nerve in the sections that follow.

In addition to describing sensory and motor functions, we will review the course of each cranial nerve from brainstem nuclei to peripheral terminations, including the intracranial course of each cranial nerve, skull exit points (see Table 12.2), cranial nerve branches, and peripheral sensory or parasympathetic ganglia (Table 12.5). As we discuss each cranial nerve, we will also review common clinical disorders associated with it.

TABLE 12.4 **Cranial Nerves: Sensory and Motor Functions**

NERVE	NAME	FUNCTIONAL CATEGORIES	FUNCTIONS
CN I	Olfactory nerve	Special somatic sensory	Olfaction
CN II	Optic nerve	Special somatic sensory	Vision
CN III	Oculomotor nerve	Somatic motor	Levator palpebrae superior and all extraocular muscles, except for superior oblique and lateral rectus
		Parasympathetic	Parasympathetics to pupil constrictor and ciliary muscles for near vision
CN IV	Trochlear nerve	Somatic motor	Superior oblique muscle; causes depression and intorsion of the eye
CN V	Trigeminal nerve	General somatic sensory	Sensations of touch, pain, temperature, joint position, and vibration for the face, mouth, anterior two-thirds of tongue, nasal sinuses, and meninges
		Branchial motor	Muscles of mastication and tensor tympani muscle
CN VI	Abducens nerve	Somatic motor	Lateral rectus muscle; causes abduction of the eye
CN VII	Facial nerve	Branchial motor	Muscles of facial expression, stapedius muscle, and part of digastric muscle
		Parasympathetic	Parasympathetics to lacrimal glands, and to sublingual, submandibular, and all other salivary glands except parotid
		Visceral sensory (special)	Taste from anterior two-thirds of tongue
		General somatic sensory	Sensation from a small region near the external auditory meatus
CN VIII	Vestibulocochlear nerve	Special somatic sensory	Hearing and vestibular sensation
CN IX	Glossopharyngeal nerve	Branchial motor	Stylopharyngeus muscle
		Parasympathetic	Parasympathetics to parotid gland
		General somatic sensory	Sensation from middle ear, region near the external auditory meatus, pharynx, and posterior one-third of tongue
		Visceral sensory (special)	Taste from posterior one-third of tongue
		Visceral sensory (general)	Chemo- and baroreceptors of carotid body
CN X	Vagus nerve	Branchial motor	Pharyngeal muscles (swallowing) and laryngeal muscles (voice box)
		Parasympathetic	Parasympathetics to heart, lungs, and digestive tract down to the splenic flexure
		General somatic sensory	Sensation from pharynx, meninges, and a small region near the external auditory meatus
		Visceral sensory (special)	Taste from epiglottis and pharynx
		Visceral sensory (general)	Chemo- and baroreceptors of the aortic arch
CN XI	Spinal accessory nerve	Branchial motor	Sternomastoid and upper part of trapezius muscle
CN XII	Hypoglossal nerve	Somatic motor	Intrinsic muscles of the tongue

Note: See Table 12.3 and Figure 12.5 for nuclei.

TABLE 12.5 Cranial Nerves: Peripheral Sensory and Parasympathetic Ganglia

NERVE	NAME	PERIPHERAL GANGLIA	FUNCTION(S) OF GANGLIA
CN I	Olfactory nerve	None	—
CN II	Optic nerve	None (retina)	—
CN III	Oculomotor nerve	Ciliary ganglion	Parasympathetics to iris and ciliary muscle
CN IV	Trochlear nerve	None	—
CN V	Trigeminal nerve	Trigeminal ganglion (semilunar or gasserian ganglion)	Primary sensory neuron cell bodies for sensation in the face, mouth, sinuses, and supratentorial meninges
CN VI	Abducens nerve	None	—
CN VII	Facial nerve	Sphenopalatine ganglion (pterygopalatine ganglion)	Parasympathetics to lacrimal glands and nasal mucosa
		Submandibular ganglion	Parasympathetics to submandibular and sublingual salivary glands
		Geniculate ganglion	Primary sensory neuron cell bodies for taste sensation in anterior two-thirds of tongue, and for sensation near outer ear
CN VIII	Vestibulocochlear nerve	Spiral ganglion	Primary sensory neuron cell bodies for hearing
		Scarpa's vestibular ganglion	Primary sensory neuron cell bodies for vestibular sensation
CN IX	Glossopharyngeal nerve	Otic ganglion	Parasympathetics to parotid gland
		Superior (jugular) glosso-pharyngeal ganglion	Primary sensory neuron cell bodies for sensation from middle ear, external auditory meatus, pharynx, and posterior one-third of tongue
		Inferior (petrosal) glosso-pharyngeal ganglion	Primary sensory neuron cell bodies for sensation from middle ear, external auditory meatus, pharynx, posterior one-third of tongue, for taste from posterior tongue, and for carotid body inputs
CN X	Vagus nerve	Parasympathetic ganglia in end organs	Parasympathetics to heart, lungs, and digestive tract to level of splenic flexure
		Superior (jugular) vagal ganglion	Primary sensory neuron cell bodies for sensation from pharynx, outer ear, and infratentorial meninges
		Inferior (nodose) vagal ganglion	Primary sensory neuron cell bodies for laryngeal sensation, for taste from epiglottis, and for reflex inputs from aortic arch receptors and other thoracoabdominal viscera
CN XI	Spinal accessory nerve	None	—
CN XII	Hypoglossal nerve	None	—

CN I: Olfactory Nerve

FUNCTIONAL CATEGORY	FUNCTION
Special somatic sensory	Olfaction

Olfactory stimuli are detected by specialized chemoreceptors on bipolar primary sensory neurons in the olfactory neuroepithelium of the upper nasal cavities. Axons of these neurons travel via short **olfactory nerves** that traverse the **cribriform plate** of the ethmoid bone (see Figure 12.3A; Table 12.2) to synapse in the **olfactory bulbs** (see Figures 18.5 and 18.6). From the olfactory bulbs, information travels via the **olfactory tracts**, which run in the olfactory sulcus between the gyrus rectus and orbital frontal gyri to reach olfactory processing areas, as discussed in Chapter 18. Note that although the olfactory bulbs and tracts are sometimes called CN I, these structures are actually not nerves, but are part of the central nervous system.

Olfaction

KEY CLINICAL CONCEPT

12.1 ANOSMIA (CN I)

Patients with unilateral **anosmia**, or olfactory loss, are rarely aware of the deficit because olfaction in the contralateral nostril can compensate. Therefore, when testing olfaction, the examiner must test each nostril separately (see **neuroexam.com Video 24**). Patients are often aware of bilateral anosmia and may complain of decreased taste because of the important contribution of olfaction to the perception of flavor.

Loss of the sense of smell can be caused by head trauma, which damages the olfactory nerves as they penetrate the cribriform plate of the ethmoid. In addition, viral infections can damage the olfactory neuroepithelium. Obstruction of the nasal passages can impair olfaction. Bilateral anosmia is also common in patients with certain neurodegenerative conditions such as Parkinson's disease and Alzheimer's disease.

Intracranial lesions that occur at the base of the frontal lobes near the olfactory sulci can interfere with olfaction. Possible lesions in this location include meningioma, metastases, basal meningitis or less commonly, **sarcoidosis**, a granulomatous inflammatory disorder that occasionally involves the nervous system, often causing cranial neuropathies. As we will discuss in KCC 19.11, frontal lobe deficits are often difficult to detect clinically, especially with small lesions. Therefore, lesions at the base of the frontal lobes can sometimes grow to a very large size, causing little obvious dysfunction other than anosmia. Large lesions of the olfactory sulcus region (typically meningiomas) can also sometimes produce a condition called Foster Kennedy syndrome, in which there is anosmia together with optic atrophy in one eye (caused by ipsilateral tumor compression) and papilledema in the other eye (caused by elevated intracranial pressure). ∎

CN II: Optic Nerve

FUNCTIONAL CATEGORY	FUNCTION
Special somatic sensory	Vision

As we discussed in Chapter 11, the optic nerve carries visual information from the retina to the lateral geniculate nucleus of the thalamus and to the extrageniculate pathways (see Figures 11.6, 11.15, and 12.2A). The retinal ganglion cells are actually part of the central nervous system, so the optic nerves are, strictly speaking, tracts and not nerves. Nevertheless, by widely accepted convention the portion of the visual pathway in front of the optic chiasm is called the **optic nerve**, and beyond this point it is referred to as the **optic tract**. The optic nerves travel from the orbit to the intracranial cavity via the **optic canal** (see Figure 12.3A,C; Table 12.2). The anatomy and disorders of visual pathways are discussed in greater detail in Chapter 11.

CN III, IV, and VI: Oculomotor, Trochlear, and Abducens Nerves

NERVE	FUNCTIONAL CATEGORY	FUNCTION
CN III	Somatic motor	Levator palpebrae superior and all extraocular muscles, except for superior oblique and lateral rectus
	Parasympathetic	Parasympathetics to pupil constrictor and ciliary muscles for near vision
CN IV	Somatic motor	Superior oblique muscle; causes depression motor and intorsion of the eye
CN VI	Somatic motor	Lateral rectus muscle; causes abduction of the eye

These nerves, which are responsible for controlling the extraocular muscles, will be discussed in detail in Chapter 13. Briefly, CN VI abducts the eye laterally in the horizontal direction; CN IV acts through a trochlea, or pulley-like, structure in the orbit, to rotate the top of the eye medially and move it downward; and CN III subserves all other eye movements. The **oculomotor (CN III)** and **trochlear (CN IV)** nuclei are located in the midbrain, and the **abducens (CN VI)** nucleus is in the pons (see Figures 12.5, 14.3, and 14.4C). Recall that CN III exits the brainstem ventrally in the interpeduncular fossa, CN IV exits dorsally from the inferior tectum, and CN VI exits ventrally at the pontomedullary junction (see Figure 12.2). CN III, IV, and VI then traverse the **cavernous sinus** (see Figure 13.11), and exit the skull via the **superior orbital fissure** (see Figure 12.3A,C; Table 12.2) to reach the muscles of the orbit. CN III also carries parasympathetics to the **pupillary constrictor** and to the **ciliary muscle** of the lens. The preganglionic parasympathetic neurons are located in the Edinger–Westphal nucleus in the midbrain (see Figure 12.5). They synapse in the **ciliary ganglion** located in the orbit (Figure 12.6). Postganglionic parasympathetic fibers then continue to the pupillary constrictor and ciliary muscles.

Other cranial nerve parasympathetics are also summarized in Figure 12.6.

FIGURE 12.6 Summary of Cranial Nerve Parasympathetic Pathways

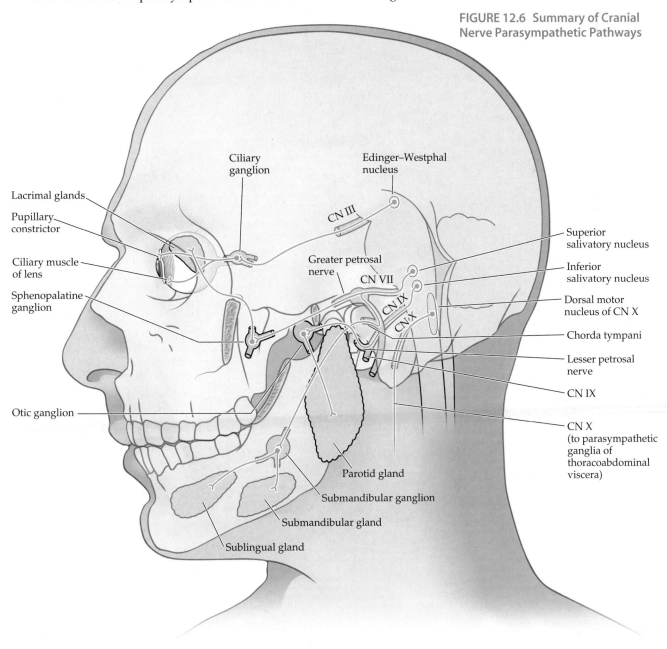

CN V: Trigeminal Nerve

FUNCTIONAL CATEGORY	FUNCTION
General somatic sensory	Sensations of touch, pain, temperature, joint position, and vibration for the face, mouth, anterior two-thirds of tongue, nasal sinuses, and meninges
Branchial motor	Muscles of mastication and tensor tympani muscle

FIGURE 12.7 Trigeminal Nerve (CN V) (A) Summary of trigeminal sensory and motor pathways. (B) General somatic sensory innervation to the face is provided by the trigeminal nerve as well as by CN VII, IX, and X. These inputs all travel to the trigeminal nuclei (see Figure 12.8). The occiput and neck, on the other hand, are supplied by cervical nerve roots C2 and C3. Sensory innervation of the supratentorial dura is provided by CN V (not shown).

The name "trigeminal" was given to this nerve because it has three major branches: the **ophthalmic division (V₁)**, **maxillary division (V₂)**, and **mandibular division (V₃)** (Figure 12.7). The trigeminal nerve provides **sensory** innervation to the face and should be distinguished from the facial nerve, which controls the muscles of facial expression. The trigeminal nerve also has a small

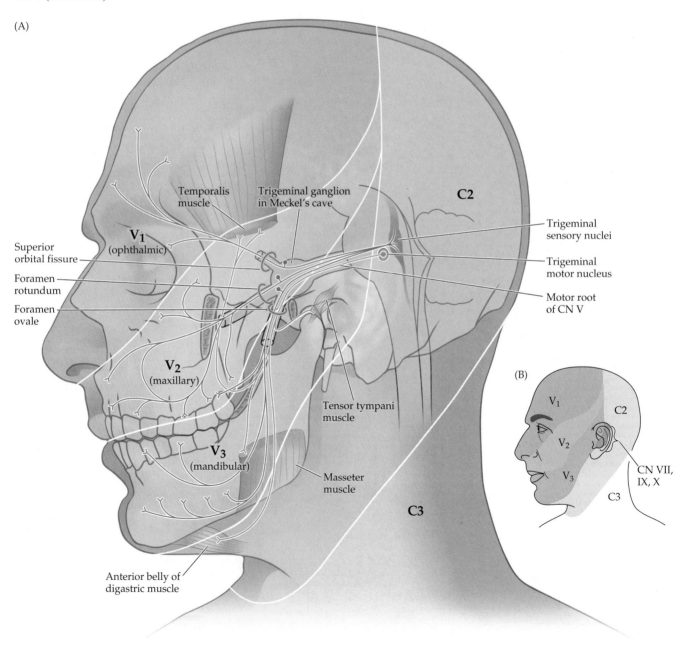

(A)

(B)

branchial **motor root** (see Figure 12.7), which travels with the mandibular division and is responsible for controlling the muscles of mastication and some other smaller muscles.

The trigeminal nerve exits the brainstem from the ventrolateral pons (see Figure 12.2A,C). It then enters a small fossa just posterior and inferolateral to the cavernous sinus called **Meckel's cave**. The **trigeminal ganglion**, also known as the **semilunar** or **gasserian ganglion**, lies in Meckel's cave and is the sensory ganglion of the trigeminal nerve (see Figure 12.7; Table 12.5). The ophthalmic division (V_1) travels through the inferior part of the cavernous sinus to exit the skull via the **superior orbital fissure** (see Figures 12.3A,C, 12.7A; Table 12.2; see also Figure 13.11). The maxillary division (V_2) exits via the **foramen rotundum** and the mandibular division (V_3) via the **foramen ovale**. A mnemonic for the exit points of these three branches is **Single Room Occupancy**, or **SRO** (for **S**uperior, **R**otundum, **O**vale). The sensory territories of V_1, V_2, and V_3 are shown in Figure 12.7B. Recall that sensation to the occiput is conveyed by C2 (see Case 8.2). The trigeminal nerve also provides touch and pain sensation for the nasal sinuses, inside of the nose, mouth, and anterior two-thirds of the tongue. In addition, pain sensation for the **supratentorial dura mater** is supplied by the trigeminal nerve, while the dura of the posterior fossa is innervated by CN X and upper cervical nerve roots.

MNEMONIC

Trigeminal Somatic Sensory Functions

The **trigeminal nuclei** (Figure 12.8 and Figure 12.9) receive **general somatic sensory** inputs from CN V and other cranial nerves (see Table 12.3). The main inputs are carried by CN V and, as we just mentioned, provide sensation for the face, mouth, anterior two-thirds of the tongue, nasal sinuses, and supratentorial dura. Smaller inputs from CN VII, IX, and X provide sensation for part of the external ear (see Figure 12.7B; Table 12.4). In addition, CN IX provides sensation to the middle ear, posterior one-third of the tongue, and pharynx. CN X additionally provides sensation for the infratentorial dura and probably also contributes to pharyngeal sensation (see Table 12.4). As we will now see, the trigeminal sensory systems are analogous to the posterior column–medial lemniscal system and anterolateral systems of the spinal cord (Table 12.6).

The trigeminal nuclear complex runs from the midbrain to the upper cervical spinal cord (see Figure 12.8) and consists of three nuclei: the mesencephalic, chief sensory, and spinal trigeminal nuclei (see Table 12.6). The **chief** (**main** or **principal**) **trigeminal sensory nucleus** and the **spinal trigeminal nucleus** provide sensory systems for the face and head that are analogous to the posterior columns and anterolateral systems, respectively (compare Figures 7.1,

TABLE 12.6 Analagous Trigeminal and Spinal Somatosensory Systems

NUCLEUS	SENSORY MODALITIES	MAIN PATHWAY TO THALAMUS	MAIN THALAMIC NUCLEUS[a]
TRIGEMINAL SENSORY SYSTEMS			
Mesencephalic trigeminal nucleus	Proprioception	—	—
Chief trigeminal sensory nucleus	Fine touch; dental pressure	Trigeminal lemniscus	VPM
Spinal trigeminal nucleus	Crude touch; pain; temperature	Trigeminothalamic tract	VPM
SPINAL SENSORY SYSTEMS			
Posterior column nuclei	Fine touch; proprioception	Medial lemniscus	VPL
Dorsal horn	Crude touch; pain; temperature	Spinothalamic tract	VPL

[a]VPL, ventral posterior lateral nucleus; VPM, ventral posterior medial nucleus.

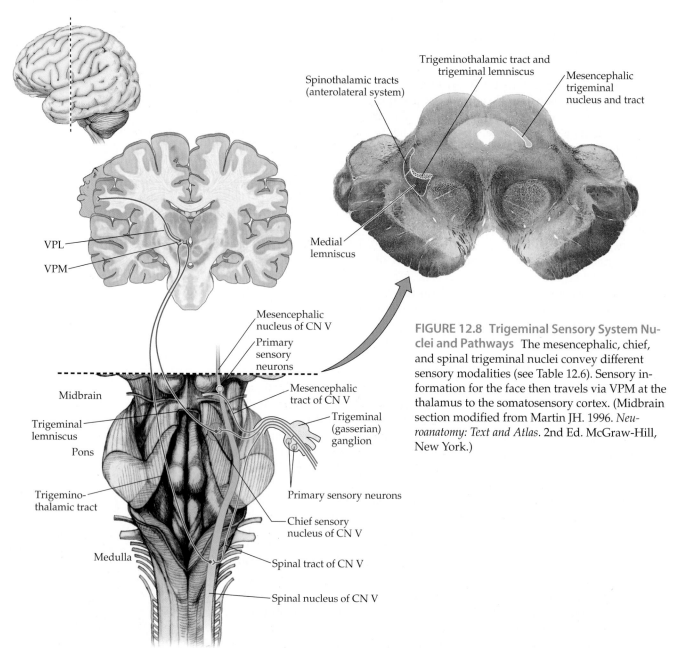

VPL

VPM

Spinothalamic tracts
(anterolateral system)

Trigeminothalamic tract and
trigeminal lemniscus

Mesencephalic
trigeminal
nucleus and tract

Medial
lemniscus

Mesencephalic
nucleus of CN V

Primary
sensory
neurons

Midbrain

Mesencephalic
tract of CN V

Trigeminal
lemniscus

Trigeminal
(gasserian)
ganglion

Pons

Primary sensory neurons

Trigemino-
thalamic tract

Chief sensory
nucleus of CN V

Medulla

Spinal tract of CN V

Spinal nucleus of CN V

FIGURE 12.8 Trigeminal Sensory System Nuclei and Pathways The mesencephalic, chief, and spinal trigeminal nuclei convey different sensory modalities (see Table 12.6). Sensory information for the face then travels via VPM at the thalamus to the somatosensory cortex. (Midbrain section modified from Martin JH. 1996. *Neuroanatomy: Text and Atlas*. 2nd Ed. McGraw-Hill, New York.)

7.2, and 12.8; see Table 12.6). The primary sensory neurons for these trigeminal nuclei lie mainly in the trigeminal ganglion (see Figure 12.8). Some also lie in the peripheral sensory ganglia of CN VII, IX, and X (see Table 12.5).

The **chief trigeminal sensory nucleus**, located in the lateral pons, receives synaptic inputs from large-diameter primary sensory neuron fibers mediating fine touch and dental pressure (see Figure 12.8; see also Figure 14.4B). This nucleus is similar in structure and function to the dorsal column nuclei (see Figures 7.1 and 14.5B). The **trigeminal lemniscus** crosses to the opposite side of the brainstem to ascend with the medial lemniscus toward the thalamus (see Figure 14.3). While the medial lemniscus travels to the ventral posterior lateral nucleus (VPL), the trigeminal lemniscus travels to the **ventral posterior medial nucleus** (**VPM**). From there, tertiary sensory neurons travel to the face area of the primary somatosensory cortex. A second, smaller pathway, called the dorsal trigeminal tract (or dorsal trigeminothalamic tract), travels from the chief trigeminal sensory nucleus to the ipsilateral VPM, without crossing. This pathway appears to convey touch and pressure sensation from the oral cavity, including the teeth.

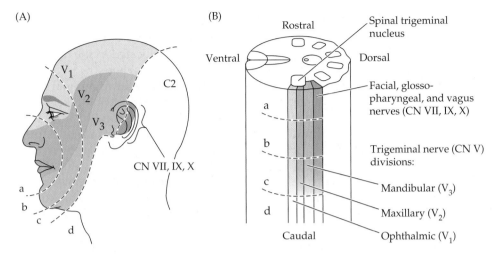

FIGURE 12.9 Somatotopic Maps of the Spinal Trigeminal Nucleus (A) Concentric regions (a–d) emanating from the mouth are indicated as they correspond with rostral to caudal regions of the spinal trigeminal nucleus shown in (B). (B) Regions of spinal trigeminal tract supplying each trigeminal nerve division, as well as CN VII, IX and X. In addition, rostral to caudal regions of the nucleus and tract (a–d) are indicated, corresponding to the concentric regions shown in (A).

The **spinal trigeminal nucleus** is located in the lateral pons and medulla, extending down to the upper cervical spine (see Figure 12.8). Medium- and small-diameter primary sensory fibers conveying crude touch, pain, and temperature sensation enter the lateral pons with the trigeminal nerve and descend in the **spinal trigeminal tract** to synapse in the spinal trigeminal nucleus. Examine the spinal trigeminal nucleus and tract in the serial sections shown in Figures 14.4C and 14.5A–C. It should be clear from these sections that the spinal trigeminal nucleus is the rostral extension of the dorsal horn. Similarly, the spinal trigeminal tract is analogous to Lissauer's tract (see Figures 6.4 and 7.2). Secondary sensory neurons from the spinal trigeminal nucleus cross the brainstem to ascend as the **trigeminothalamic tract** (or ventral trigeminothalamic tract). The trigeminothalamic tract is analogous to the spinothalamic tract (see Table 12.6), and the pathways travel together to the thalamus (see Figures 12.8 and 14.3). Trigeminothalamic tract fibers synapse in the thalamic **ventral posterior medial nucleus** (**VPM**), and tertiary sensory neurons then travel in the internal capsule to the primary somatosensory cortex. Like the anterolateral systems in the spinal cord, there are also pathways from the spinal trigeminal nucleus to intralaminar thalamic nuclei, the reticular formation, and other areas, to mediate the affective and arousal aspects of facial pain.

The spinal trigeminal tract and nucleus are somatotopically organized, with the mandibular division represented dorsally, the ophthalmic division represented ventrally, and the maxillary division in between (see Figure 12.9). In addition, concentric rings form an **"onion skin"**-like representation, with perioral areas represented more rostrally in the nucleus, and areas more removed from the mouth represented more caudally.

The **mesencephalic trigeminal nucleus and tract** run along the lateral edge of the periaqueductal gray matter of the midbrain (see Figure 14.3A,B) and mediate proprioception (see Table 12.6). The neurons of the mesencephalic trigeminal nucleus are the only case in which **primary sensory neurons** lie within the central nervous system instead of in peripheral ganglia (see Figure 12.8). The peripheral processes of these neurons convey proprioceptive input from the muscles of mastication and probably also from the tongue, and from the extraocular muscles. In the monosynaptic **jaw jerk reflex** (see KCC 12.4; **neuroexam.com Video 39**), processes of mesencephalic trigeminal neurons descend to the pons and synapse in the motor trigeminal nucleus (see Figure 12.7A). Ascending and descending fibers form the mesencephalic trigeminal tract (see Figure 12.8; see also Figure 14.3A,B). Other central pathways of the mesencephalic trigeminal nucleus are still under investigation.

REVIEW EXERCISE

1. Cover the second column from the left in Table 12.6. For each nucleus in the left column, provide the sensory modalities served and the location of the nucleus (see Figure 12.8).

2. Which thalamic nucleus is most important for relaying somatosensory information from the face, and which is most important for relaying somatosensory information from the rest of the body to the cortex?

Trigeminal Branchial Motor Functions

The **trigeminal motor nucleus** mediates the **branchial motor** functions of the trigeminal nerve (see Figure 12.7). This nucleus is located in the upper-to-mid pons (see Figures 12.5 and 14.4B), near the level where the trigeminal nerve exits the brainstem. The branchial **motor root** of the trigeminal nerve runs inferomedial to the trigeminal ganglion along the floor of Meckel's cave and then joins V_3 to exit via the foramen ovale (see Figure 12.3A). It then supplies the **muscles of mastication** (see **neuroexam.com Video 38**), including the masseter, temporalis, and medial and lateral pterygoid muscles, as well as several smaller muscles, such as the **tensor tympani**, tensor veli palatini, mylohyoid, and anterior belly of the digastric. The **upper motor neuron control** reaching the trigeminal motor nucleus is predominantly bilateral, so unilateral lesions in the motor cortex or corticobulbar tract usually cause no deficit in jaw movement. Bilateral upper motor neuron lesions, however, can cause hyperreflexia manifested in a brisk jaw jerk reflex (see KCC 12.4).

KEY CLINICAL CONCEPT

12.2 TRIGEMINAL NERVE DISORDERS (CN V)

Disorders of the trigeminal nerve are relatively uncommon, except for **trigeminal neuralgia** (tic douloureux). In this condition, patients experience recurrent episodes of brief severe pain lasting from seconds to a few minutes, most often in the distribution of V_2 or V_3. Attacks usually begin after age 35. Painful episodes are often provoked by chewing, shaving, or touching a specific trigger point on the face. Neurologic exam, including facial sensation, is normal. The cause of trigeminal neuralgia in most cases is unknown. In some cases, compression of the trigeminal nerve by an aberrant vessel has been demonstrated, but the significance of this finding is uncertain. It is important to perform an MRI scan to exclude tumor or other lesions in the region of the trigeminal nerve as the cause. Trigeminal neuralgia can also occur in multiple sclerosis (see KCC 6.6), possibly caused by demyelination in the trigeminal nerve entry zone of the brainstem. Initial treatment of trigeminal neuralgia is with carbamazepine, and alternatives include oxcarbazepine, baclofen, lamotrigine, or pimozide. Refractory cases have been successfully treated with various procedures, including radiofrequency ablation of the Gasserian ganglion, gamma knife, CyberKnife (see KCC 16.4), or surgical microvascular decompression of the trigeminal nerve.

Sensory loss in the distribution of the trigeminal nerve or its branches (see **neuroexam.com Video 36**) can be caused by trauma, metastatic disease—especially in isolated chin or jaw numbness, herpes zoster (see KCC 8.3), aneurysms of the petrous portion of the internal carotid artery (see Figures 4.16C and 12.3A), cavernous sinus or orbital apex disorders (see KCC 13.7), trigeminal or vestibular schwannoma (see KCC 12.5), or sphenoid wing meningioma (see KCC 5.8). Lesions of the trigeminal nuclei in the brainstem cause *ipsilateral* loss of facial sensation to pain and temperature because the primary sensory fibers do not cross before entering the nucleus (see Figure 12.8). Common causes include infarcts (see Chapter 14), demyelination, or other brainstem lesions. Often lesions of the trigeminal nucleus in the pons or medulla also involve the nearby spinothalamic tract (see Figures 7.2, 14.4C, and 14.5A,B). This combination of spinal trigeminal and spinothalamic tract involvement in lateral brainstem lesions leads to a well-recognized pattern with sensory loss to pain and temperature in the face ipsilateral to the lesion, but in the body contralateral to the lesion (see KCC 7.3; Figure 7.9B). ∎

Facial sensation

CN VII: Facial Nerve

FUNCTIONAL CATEGORY	FUNCTION
Branchial motor	Muscles of facial expression, stapedius muscle, and part of digastric muscle
Parasympathetic	Parasympathetics to lacrimal glands, and to sublingual, submandibular, and all other salivary glands except parotid
Visceral sensory (special)	Taste from anterior two-thirds of tongue
General somatic sensory	Sensation from a small region near the external auditory meatus

The main function of the facial nerve is to control the muscles of facial expression (see **neuroexam.com Video 40**); however, it has several other important functions as well. The main nerve trunk carries the **branchial motor** fibers controlling facial expression, while a smaller branch called the **nervus intermedius** carries fibers for the parasympathetic (tears and salivation), visceral sensory (taste), and general somatosensory functions (**Figure 12.10**; see also Figures 12.6 and 12.14).

Facial muscles

The **facial nucleus** is located in the branchial motor column, more caudally in the pons than the trigeminal motor nucleus (see Figure 12.5; see also Figure 14.4B,C). The fascicles of the facial nerve loop dorsally around the abducens nucleus, forming the **facial colliculus** on the floor of the fourth ventricle (see Figure 12.2B and **Figure 12.11**). The nerve then exits the brainstem ventrolaterally at the pontomedullary junction (see Figure 12.2A,C). Upper motor neuron control of the facial nucleus is discussed in KCC 12.3 (see Figure 12.13). Briefly, lesions in the cortex or corticobulbar tracts cause contralateral face weakness that spares the forehead, while lesions of the facial nucleus, nerve fascicles in the brainstem, or peripheral nerve cause ipsilateral weakness of the entire face.

The facial nerve exits the brainstem ventrolaterally at the pontomedullary junction, lateral to CN VI in a region called the **cerebellopontine angle** (see Figure 12.2A,C). It then traverses the subarachnoid space and enters the **internal auditory meatus** (see Figure 12.3A; see also Figure 4.13C) to travel in the **auditory canal** of the **petrous temporal bone** together with the vestibulocochlear nerve (see Figure 12.14). At the **genu** of the facial nerve, the nerve takes a turn posteriorly and inferiorly in the temporal bone to run in the **facial canal**, just medial to the middle ear (see Figures 12.10 and 12.14). The **geniculate ganglion** lies in the genu and contains primary sensory neurons for taste sensation in the anterior two-thirds of the tongue, and for general somatic sensation in a region near the external auditory meatus (see Table 12.5; Figure 12.7B). The main portion of the facial nerve exits the skull at the **stylomastoid foramen** (see Figures 12.3B and 12.10). It then passes through the parotid gland and divides into five major **branchial motor** branches to control the muscles of facial expression: the **temporal, zygomatic, buccal, mandibular,** and **cervical** branches (see Figure 12.10). Other smaller branchial motor branches innervate the **stapedius** (see Figures 12.10 and 12.15), occipitalis, posterior belly of the digastric, and stylohyoid muscles. The cranial nerves controlling the middle ear muscles can be recalled by the mnemonic **T**rigeminal for **T**ensor **T**ympani and **S**eventh for **S**tapedius. Both the tensor tympani and the stapedius dampen movements of the middle ear ossicles (see the section on CN VIII later in this chapter), providing feedback modulation of acoustic signal intensity.

MNEMONIC

The preganglionic **parasympathetic** fibers of the facial nerve originate in the **superior salivatory nucleus** (see Figure 12.10) and are carried by two small branches off the main trunk of the facial nerve. The **greater petrosal nerve** takes off at the genu of the facial nerve (see Figure 12.14) to reach the **sphenopalatine**

FIGURE 12.10 Facial Nerve (CN VII) Summary of facial nerve sensory and motor pathways.

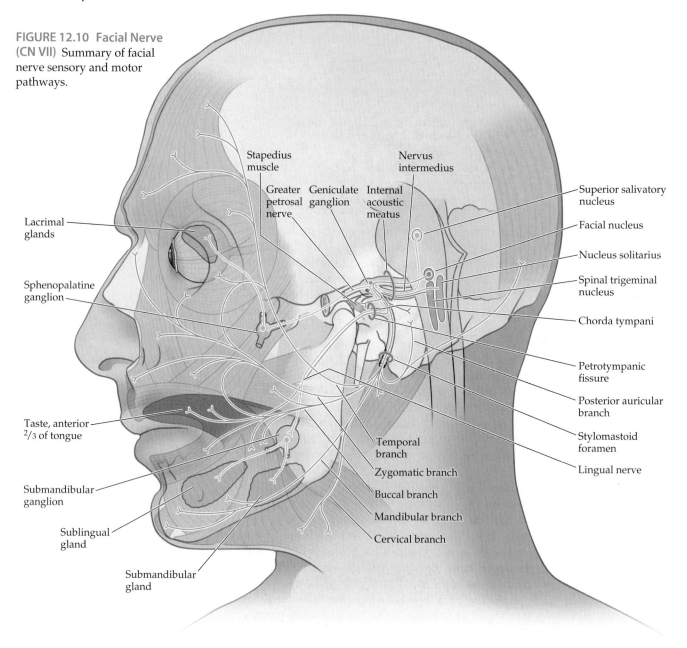

Stapedius muscle

Nervus intermedius

Greater petrosal nerve

Geniculate ganglion

Internal acoustic meatus

Superior salivatory nucleus

Facial nucleus

Nucleus solitarius

Spinal trigeminal nucleus

Chorda tympani

Petrotympanic fissure

Posterior auricular branch

Stylomastoid foramen

Lingual nerve

Lacrimal glands

Sphenopalatine ganglion

Taste, anterior ²⁄₃ of tongue

Submandibular ganglion

Sublingual gland

Submandibular gland

Temporal branch

Zygomatic branch

Buccal branch

Mandibular branch

Cervical branch

Taste

(**pterygopalatine**) **ganglion**, where postganglionic parasympathetic cells project to the lacrimal glands and nasal mucosa (see Figure 12.10). The **chorda tympani** leaves the facial nerve just before the stylomastoid foramen and travels back upward to traverse the middle ear cavity before exiting the skull at the **petrotympanic fissure** (see Figures 12.3B and 12.10), just medial and posterior to the temporomandibular joint. The chorda tympani then joins the **lingual nerve** (a branch of CN V₃) to reach the **submandibular ganglion** (also called the submaxillary ganglion), where postganglionic parasympathetics arise to supply the submandibular (submaxillary) and sublingual salivary glands as well as other minor salivary glands aside from the parotid. Note that the majority (~70%) of saliva production arises from the submandibular salivary glands.

The lingual nerve and chorda tympani also carry **special visceral sensory** fibers mediating **taste sensation** (see **neuroexam.com Video 41**) for the anterior two-thirds of the tongue (see Figure 12.10). The primary sensory taste fibers have their cell bodies in the geniculate ganglion (**Figure 12.12**; see also Figure 12.14 and Table 12.5). These cells synapse onto secondary sensory

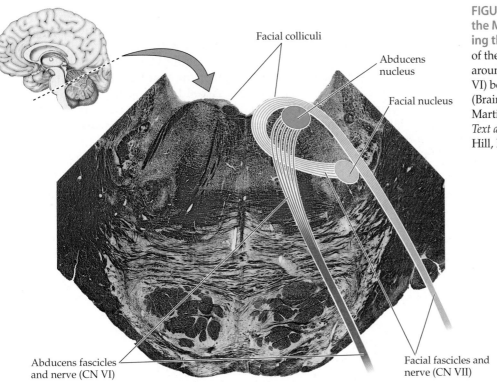

Facial colliculi

Abducens nucleus

Facial nucleus

Abducens fascicles and nerve (CN VI)

Facial fascicles and nerve (CN VII)

FIGURE 12.11 Axial Section of the Mid-to-Lower Pons, Showing the Facial Colliculus Fibers of the facial nerve (CN VII) loop around the abducens nucleus (CN VI) before exiting the brainstem. (Brainstem section modified from Martin JH. 1996. *Neuroanatomy: Text and Atlas*. 2nd Ed. McGraw-Hill, New York.)

FIGURE 12.12 Central Taste Pathways

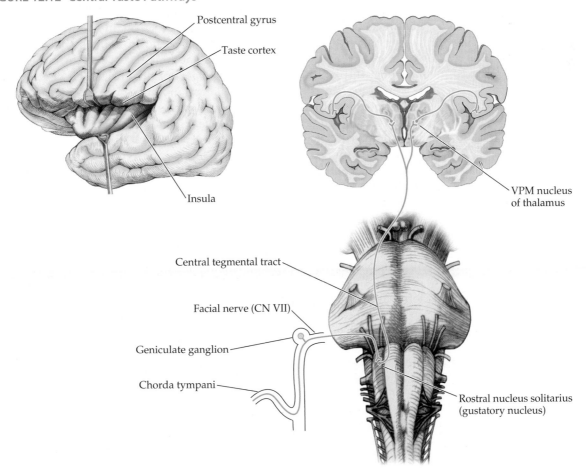

Postcentral gyrus

Taste cortex

Insula

VPM nucleus of thalamus

Central tegmental tract

Facial nerve (CN VII)

Geniculate ganglion

Chorda tympani

Rostral nucleus solitarius (gustatory nucleus)

MNEMONIC

REVIEW EXERCISE

Describe the branchial motor, parasympathetic, visceral sensory, and general somatic sensory functions of the facial nerve. Name the nucleus subserving each function. (See Figure 12.10.)

neurons in the **rostral nucleus solitarius**, also known as the gustatory nucleus. The taste functions of the nucleus solitarius can be remembered since this nucleus looks like a delicious doughnut with cell bodies surrounding a myelinated center in cross sections (see Figure 14.5A). There are also taste inputs for the posterior tongue and pharynx that travel via CN IX and X to enter the rostral nucleus solitarius. Ascending projections continue via the **central tegmental tract** (see Figure 12.12; see also Figures 14.3 and 14.4) to reach tertiary sensory neurons in the **ventral posterior medial nucleus** (**VPM**) of the thalamus. Thalamic neurons from the VPM, in turn, project to the **cortical taste area**, which lies at the inferior margin of the postcentral gyrus adjacent to the tongue somatosensory area and extends into the fronto-parietal operculum and insula (see Figure 12.12). Taste pathways ascend ipsilaterally and likely project to bilateral thalamus and cortex, but the laterality of cortical taste pathways in humans is still under investigation.

Finally, a small branch of the facial nerve provides **general somatic sensation** for a region near the external auditory meatus that lies adjacent to similar regions supplied by CN IX and X (see Figure 12.7B). The somatosensory fibers for CN V, VII, IX, and X all synapse in the trigeminal nuclei (see Figure 12.5; Table 12.3).

KEY CLINICAL CONCEPT

12.3 FACIAL NERVE LESIONS (CN VII)

As we discussed briefly in KCC 6.3, it is clinically important to distinguish between facial weakness caused by upper motor neuron lesions and facial weakness caused by lower motor neuron lesions. Upper motor neurons in the face area of the primary motor cortex control lower motor neurons in the contralateral facial nucleus of the pons (Figure 12.13). In addition, for the superior portions of the face, projections descend from the ipsilateral motor cortex as well as from the contralateral motor cortex. Thus, the lower motor neurons supplying the forehead and part of the orbicularis oculi receive upper motor neuron inputs from bilateral motor cortices. As a result, unilateral **upper motor neuron lesions** tend to spare the forehead and cause only mild contralateral orbicularis oculi weakness resulting in a slightly widened palpebral fissure, or inability to fully bury the eyelash on forced eye closure. In upper motor neuron lesions the weakness affects mainly the inferior portions of the contralateral face (see Figure 12.13, Lesion A). **Lower motor neuron lesions**, in contrast, affect the entire half of the face and do not spare the forehead (see Figure 12.13, Lesion B). Additional clues sometimes present in upper motor neuron–type weakness include neighborhood effects such as hand or arm weakness, sensory loss, aphasia, or dysarthria, none of which are present in lower motor neuron lesions. The connections shown in Figure 12.13 are somewhat oversimplified; in reality, the upper motor neuron corticobulbar fibers controlling the facial nucleus project mainly to pontine interneurons that project, in turn, to lower motor neurons in the facial nucleus.

The most common facial nerve disorder by far is **Bell's palsy**, in which all divisions of the facial nerve are impaired within a few hours or days and then gradually recover. The cause is unknown, although viral or inflammatory mechanisms have been suggested. The most striking feature is unilateral facial weakness of the lower motor neuron type, which can be mild but is often severe (see Figure 12.13, Lesion B). Diagnosis is based on clinical history and exam (see **neuroexam.com Videos 40** and **41**). Patients often initially complain of some retroauricular pain, likely caused by involvement of the general somatosensory component of CN VII (see Figure 12.7B). Hyperacusis can occur because of stapedius muscle weakness (see Figures 12.10 and 12.15). In addi-

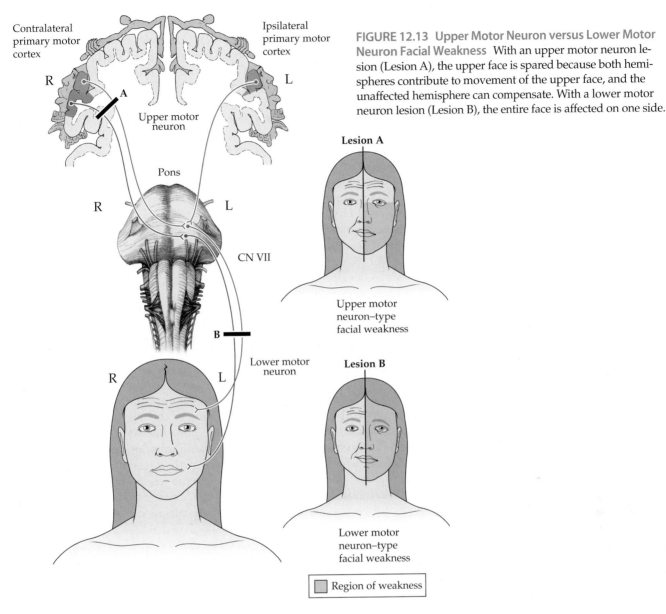

FIGURE 12.13 Upper Motor Neuron versus Lower Motor Neuron Facial Weakness With an upper motor neuron lesion (Lesion A), the upper face is spared because both hemispheres contribute to movement of the upper face, and the unaffected hemisphere can compensate. With a lower motor neuron lesion (Lesion B), the entire face is affected on one side.

tion, patients may suffer from "dry eye," resulting from decreased lacrimation with parasympathetic involvement (see Figure 12.10). Neurologic examination is notable for unilateral lower motor neuron–type facial weakness, sometimes associated with loss of taste on the ipsilateral tongue (test with mustard or sugar applied with a cotton swab; see **neuroexam.com Video 41**). The remainder of the exam should be normal in Bell's palsy. The presence of hand weakness, sensory loss, dysarthria, or aphasia suggests an upper motor neuron lesion. In clinically typical cases, imaging studies are usually normal, however, most practitioners will order an MRI scan to exclude a structural lesion and blood studies including a blood count, glucose, and Lyme titer.

Treatment of Bell's palsy has been controversial, but recent evidence suggests that a 10-day course of oral steroids started soon after onset improves chances for full recovery. The possible role of antiviral agents in treating Bell's palsy remains uncertain. Incomplete eye closure and decreased tearing can cause corneal ulcerations. Therefore, patients should be given lubricating eyedrops and instructions to tape the eye shut at night. About 80% of patients recover fully from Bell's palsy within 3 weeks, although some are left with variable degrees of residual weakness. During recovery, regenerating facial nerve fibers sometimes reach the wrong target. For example, aberrant regeneration of parasympathetic fibers (see Figure 12.6) can result in the phenomenon of "crocodile tears," in which patients experience lacrimation instead of salivation when they see food. Aberrant regeneration of different motor branches of the facial nerve sometimes results in **synkinesis**, meaning abnormal movement together. For example, if the patient is asked to close one eye, the ipsilateral platysma muscle may contract slightly, along with the orbicularis oculi.

In cases of bilateral lower motor neuron–type facial weakness, or if a patient experiences a second episode, a more thorough investigation is warranted. This should include an MRI scan with contrast to look for tumor or other infiltrative disorders, lumbar puncture (see KCC 5.10), and tests for Lyme disease, sarcoidosis, and HIV. In addition to the causes already mentioned, facial nerve injury can occur in head trauma, particularly with fractures of the petrous temporal bone. Facial weakness caused by upper motor neuron lesions is discussed in KCC 6.3. In addition, brainstem lesions can occasionally involve the facial nucleus or exiting nerve fascicles (see Figure 12.11; see also KCC 14.3). ■

Corneal reflex

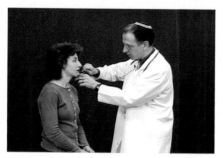

Jaw jerk reflex

KEY CLINICAL CONCEPT

12.4 CORNEAL REFLEX AND JAW JERK REFLEX (CN V, VII)

The **corneal reflex** is elicited by gentle stroking of each cornea with a cotton swab; the response is eye closure (see **neuroexam.com Video 37**). This reflex is mediated by both monosynaptic and polysynaptic pathways. The afferent limb is conveyed by the ophthalmic division of the trigeminal nerve to the chief sensory and spinal trigeminal nuclei. The efferent limb is then carried by the facial nerve to reach the orbicularis oculi muscles causing eye closure. A lesion of the trigeminal sensory pathways, the facial nerve, or their connections causes a decreased corneal reflex in the ipsilateral eye. The corneal reflex is also modulated by inputs from higher centers. Therefore, lesions of sensorimotor cortex and its connections can cause a diminished corneal reflex in the eye contralateral to the lesion.

Since an eye blink response can also be elicited by an object moving toward the eye, when eliciting the corneal response the examiner must take care to ensure that the blink has been elicited by touch rather than a sudden movement toward the eye. In blink to threat (see **neuroexam.com Video 28**) the afferent limb of the reflex is carried by the optic nerve (CN II), while in the corneal reflex, the afferent limb is carried by the trigeminal nerve (CN V).

The **jaw jerk reflex** is elicited by tapping on the chin with the mouth slightly open; the jaw jerks forward in response. The monosynaptic pathway for this reflex consists of primary sensory neurons in the mesencephalic trigeminal nucleus (see Figure 12.8), which send axons to the pons to synapse in the motor trigeminal nucleus. In normal individuals the jaw jerk reflex is minimal or absent (see **neuroexam.com Video 39**). In bilateral upper motor neuron lesions, such as amyotrophic lateral sclerosis (see KCC 6.7) or diffuse white matter disease, the jaw jerk reflex may be brisk. ■

CN VIII: Vestibulocochlear Nerve

FUNCTIONAL CATEGORY	FUNCTION
Special somatic sensory	Hearing and vestibular sensation

This nerve carries the **special somatic sensory** functions of hearing and vestibular sense from the structures of the inner ear. The vestibulocochlear nerve exits the brainstem at the pontomedullary junction just lateral to the facial nerve, in a region called the **cerebellopontine angle** (see Figure 12.2A,C). It then tra-

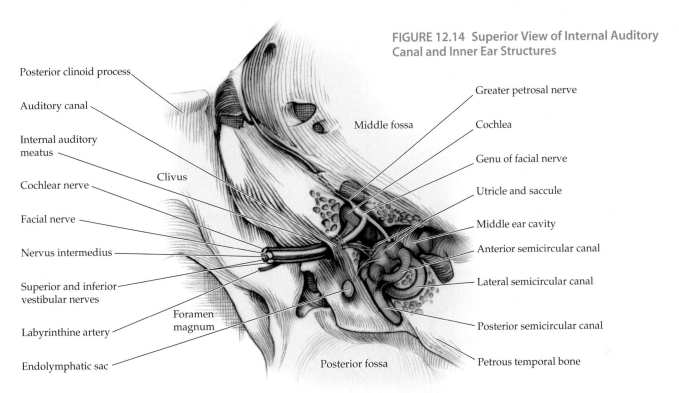

FIGURE 12.14 Superior View of Internal Auditory Canal and Inner Ear Structures

Posterior clinoid process

Auditory canal

Internal auditory meatus

Cochlear nerve

Facial nerve

Nervus intermedius

Superior and inferior vestibular nerves

Labyrinthine artery

Endolymphatic sac

Clivus

Foramen magnum

Middle fossa

Posterior fossa

Greater petrosal nerve

Cochlea

Genu of facial nerve

Utricle and saccule

Middle ear cavity

Anterior semicircular canal

Lateral semicircular canal

Posterior semicircular canal

Petrous temporal bone

verses the subarachnoid space to enter the **internal auditory meatus** (see Figure 12.3A) together with the facial nerve and travels in the **auditory canal** of the petrous temporal bone to reach the cochlea and vestibular organs (**Figure 12.14**; see also Figure 4.13C). In the subsections that follow we will discuss the auditory and vestibular functions of CN VIII, in turn.

Auditory Pathways

Sound waves are transmitted by the **tympanic membrane** and amplified by the middle ear ossicles—the **malleus**, **incus**, and **stapes**—to reach the **oval window** (**Figure 12.15**). The movements of the malleus are dampened by the tensor tympani muscle, and movements of the stapes are dampened by the stapedius in response to loud sounds. From the oval window, vibrations reach the inner ear structures. The **inner ear**, or **labyrinth**, consists of the **cochlea**, **vestibule**, and **semicircular canals** (see Figure 12.15). The labyrinth is composed of a **bony labyrinth**, which is lined with compact bone and contains the **membranous labyrinth**. The bony labyrinth is filled with fluid called **perilymph**, within which the structures of the membranous labyrinth are suspended. Interestingly, perilymph communicates with the subarachnoid space through a small perilymphatic duct (not shown). The membranous labyrinth, in turn, is filled with a fluid of slightly different ionic composition, called **endolymph**. The membranous labyrinth includes the **cochlear duct**, **utricle**, **saccule**, and **semicircular canals** (see Figure 12.15).

Acoustic vibrations from the oval window reach the **scala vestibuli** and are transmitted along the snail-shaped **cochlea** to the end, where it joins the **scala tympani**, and the pressure waves are ultimately relieved at the **round window** back in the wall of the middle ear. The vibrations also reach the **cochlear duct** (scala media) (see Figure 12.15, lower right inset), where mechanoreceptor cilia on the **hair cells** are activated by movement of the **basilar membrane** relative to the stiff **tectorial membrane**. The hair cells form excitatory synapses onto the terminals of primary sensory neurons. These bipolar sensory neurons have their cell bodies in the **spiral ganglion**, located along the central rim of the cochlea, and send their axons into the cochlear

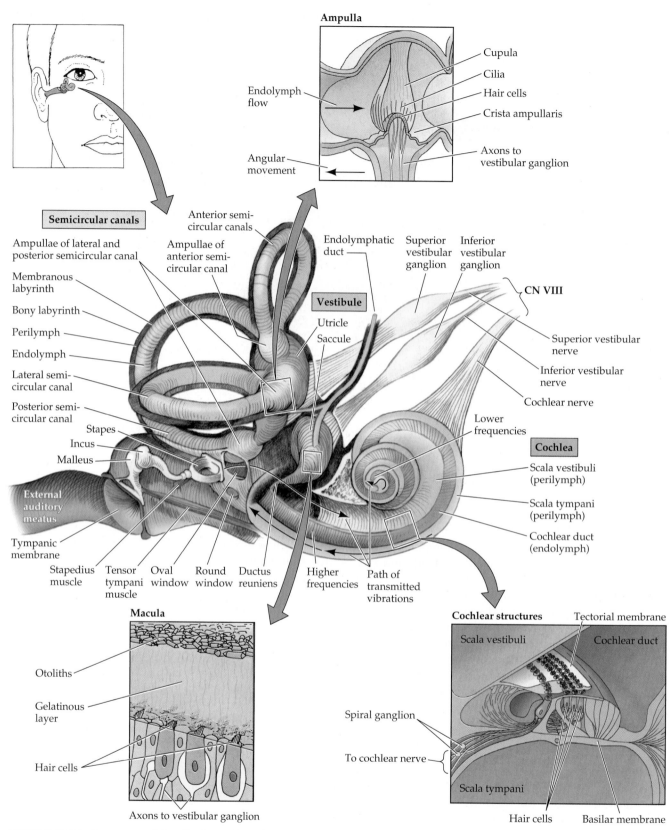

Ampulla

Cupula
Cilia
Hair cells
Crista ampullaris

Endolymph flow

Axons to vestibular ganglion

Angular movement

Semicircular canals

Anterior semi-circular canals

Ampullae of lateral and posterior semicircular canal

Ampullae of anterior semi-circular canal

Endolymphatic duct

Superior vestibular ganglion

Inferior vestibular ganglion

CN VIII

Membranous labyrinth

Vestibule

Bony labyrinth

Utricle

Saccule

Superior vestibular nerve

Perilymph

Inferior vestibular nerve

Endolymph

Cochlear nerve

Lateral semi-circular canal

Lower frequencies

Posterior semi-circular canal

Cochlea

Stapes

Scala vestibuli (perilymph)

Incus

Malleus

Scala tympani (perilymph)

External auditory meatus

Cochlear duct (endolymph)

Tympanic membrane

Stapedius muscle

Tensor tympani muscle

Oval window

Round window

Ductus reuniens

Higher frequencies

Path of transmitted vibrations

Macula

Otoliths

Gelatinous layer

Hair cells

Axons to vestibular ganglion

Cochlear structures

Tectorial membrane

Scala vestibuli

Cochlear duct

Spiral ganglion

To cochlear nerve

Scala tympani

Hair cells

Basilar membrane

FIGURE 12.15 Summary of Vestibular and Cochlear Structures Sound waves enter the external auditory meatus, are transmitted mechanically by the middle ear to the cochlea, and are transduced by hair cells to neural signals carried centrally by the cochlear nerve. The ampullae of the semicircular canals detect angular acceleration, and the maculae of the otolith organs (utricle and saccule) detect linear acceleration and head tilt and transmit this information to the vestibular nerves. (Insets from Rosenzweig MR, Breedlove SM, Leiman AL. 2002. *Biological Psychology*. 3rd Ed. Sinauer, Sunderland, MA.)

nerve (see Figures 12.14 and 12.15). The hair cells of the cochlea, together with their supporting cells, are called the **organ of Corti**. There is a **tonotopic representation** determined by structural width and stiffness along the length of the organ of Corti such that higher-frequency sounds activate hair cells near the oval window, while lower-frequency sounds activate hair cells near the apex of the cochlea (see Figure 12.15).

Let's follow the pathways for hearing centrally, from the cochlear nuclei to the primary auditory cortex (Figure 12.16 and Figure 12.17). Auditory information throughout these pathways is tonotopically organized. Primary sensory neurons in the **spiral ganglion** send their axons in the cochlear division of CN VIII to reach the **dorsal and ventral cochlear nuclei**, which are wrapped around the lateral aspect of the inferior cerebellar peduncle at the pontomedullary junction (see Figures 12.16 and 12.17C). The hearing pathways then ascend through

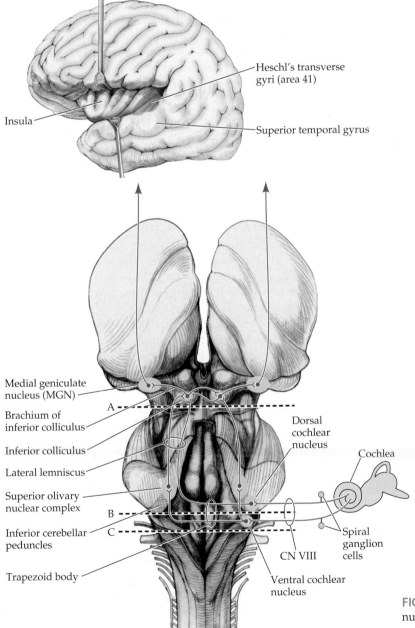

Heschl's transverse gyri (area 41)

Insula

Superior temporal gyrus

Medial geniculate nucleus (MGN)

Brachium of inferior colliculus

Inferior colliculus

Lateral lemniscus

Superior olivary nuclear complex

Inferior cerebellar peduncles

Trapezoid body

Dorsal cochlear nucleus

Cochlea

Spiral ganglion cells

CN VIII

Ventral cochlear nucleus

FIGURE 12.16 Central Auditory Pathways Main nuclei and pathways are shown from the cochlear nerve to the auditory cortex. Levels of sections for Figure 12.17 are indicated.

(A)

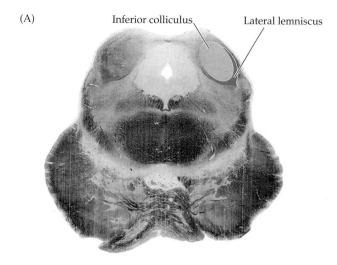

Inferior colliculus
Lateral lemniscus

FIGURE 12.17 Brainstem Sections of Auditory Pathways
Levels of sections are as indicated in Figure 12.16. (A) Midbrain inferior colliculi. (B) Caudal pons at the level of the superior olivary nucleus and trapezoid body. (C) Rostral medulla showing entry of cochlear (auditory) nerve to cochlear nuclei. Myelin-stained cross sections of the human brainstem. (A and B modified from Martin JH. 1996. *Neuroanatomy Text and Atlas*. 2nd Ed. McGraw-Hill, New York. C from The University of Washington Digital Anatomist Project.)

the brainstem *bilaterally* through a series of relays to reach the inferior colliculi, medial geniculate nuclei, and, ultimately, the auditory cortex. Because auditory information from each ear ascends bilaterally in the brainstem, with decussations occurring at multiple levels, *unilateral hearing loss is not seen in lesions in the central nervous system proximal to the cochlear nuclei.*

Fibers from the **dorsal cochlear nucleus** pass dorsal to the inferior cerebellar peduncle, cross the pontine tegmentum, and ascend in the contralateral **lateral lemniscus** (see Figures 12.16 and 12.17A,B). The lateral lemniscus is an important ascending auditory pathway in the pons and lower midbrain that terminates in the inferior colliculus. Many fibers of the **ventral cochlear nucleus** pass ventral to the inferior cerebellar peduncle to synapse bilaterally in the **superior olivary nuclear complex** of the pons (see Figures 12.16 and 12.17B). The superior olivary nuclei appear to function in localizing sounds horizontally in space. Crossing auditory fibers at this level form a white matter structure called the **trapezoid body** (see Figures 12.16 and 12.17B). The trapezoid body is traversed at right angles by the medial lemniscus (see Figure 14.4C).

From the superior olivary nuclear complex, fibers ascend bilaterally in the lateral lemniscus to reach the **inferior colliculi** of the midbrain (see Figures 12.16 and 12.17A). Decussating fibers at the level of the inferior colliculi pass both dorsal and ventral to the cerebral aqueduct. From the inferior colliculi, fibers ascend via the **brachium of the inferior colliculi** to the **medial geniculate nuclei** of the thalamus, which are located just lateral to the superior colliculi of the midbrain (see Figures 11.6, 12.16, and 14.3A). From this thalamic relay, information continues in the **auditory radiations** (see Figure 6.9B) to the primary auditory cortex. The **primary auditory cortex** (Brodmann's area 41) lies on **Heschl's transverse gyri**. We can see these straight, fingerlike gyri in brain specimens by opening the Sylvian fissure and looking at the superior surface of the temporal lobe, just medial to the superior temporal gyrus (see Figure 12.16; see also Figure 4.15D). The nearby areas of cortex in the temporal and parietal lobes are auditory association cortex, including Wernicke's area, which will be discussed in Chapter 19. In addition to the nuclei already mentioned, there are several smaller nuclei in the hearing pathway, including the nuclei of the trapezoid body and the nuclei of the lateral lemniscus.

As noted already, lesions in the central nervous system proximal to the cochlear nuclei do not cause unilateral hearing loss because auditory information crosses bilater-

(B)

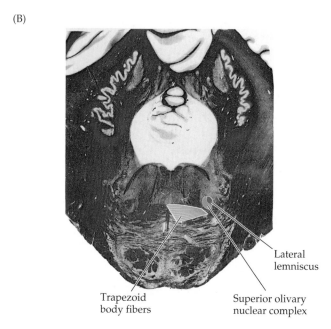

Lateral lemniscus

Trapezoid body fibers

Superior olivary nuclear complex

(C)

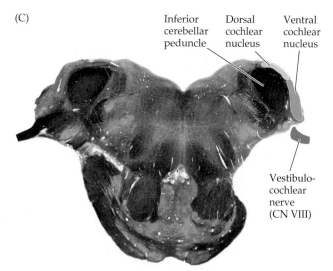

Inferior cerebellar peduncle

Dorsal cochlear nucleus

Ventral cochlear nucleus

Vestibulocochlear nerve (CN VIII)

ally at multiple points in the brainstem. However, auditory information ascending through the brainstem and thalamus to the auditory cortex does contain a relatively greater contribution from the contralateral ear. In auditory seizures, caused by abnormal electrical discharges in the auditory cortex, patients often perceive a tone or roaring sound like an airplane or a train coming from the side opposite the auditory cortex involved. Bilateral damage to the auditory cortex causes cortical deafness (see KCC 19.7).

Efferent feedback pathways from the brainstem to the cochlea in the vestibulocochlear nerve modulate the sensitivity of the hair cells in response to sounds of varying intensities. Similar pathways exist for modulation of vestibular hair cells. Because of these small efferent pathways, some might not consider CN VIII a purely sensory nerve. In addition, reflex pathways from the ventral cochlear nuclei reach the facial and trigeminal motor nuclei to contract the stapedius and tensor tympani muscles. These muscles dampen the response of the middle ear to loud sounds.

Vestibular Pathways

The **vestibular nuclei** are important for adjustment of posture, muscle tone, and eye position in response to movements of the head in space. Not surprisingly, therefore, the vestibular nuclei have intimate connections with the cerebellum, and with the brainstem motor and extraocular systems. In addition, an ascending pathway through the thalamus to the cortex provides an awareness of head position that is integrated with visual and tactile spatial information in the parietal association cortex.

The **semicircular canals** (see Figures 12.14 and 12.15) detect angular acceleration around three orthogonal axes. The spatial orientation of the three semicircular canals can be remembered by imagining the arms of a bodybuilder in three poses (Figure 12.18).

REVIEW EXERCISE

Follow the auditory pathways from the cochlear nerve to the auditory cortex in Figures 12.16 and 12.17.

MNEMONIC

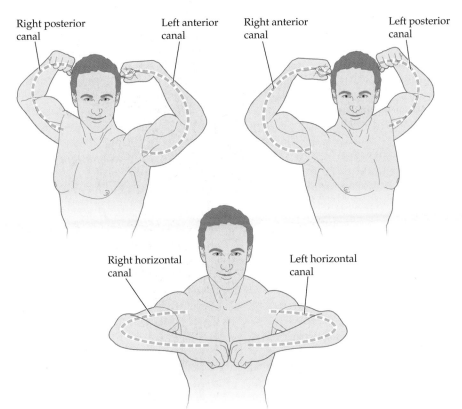

Right posterior canal · Left anterior canal · Right anterior canal · Left posterior canal

Right horizontal canal · Left horizontal canal

FIGURE 12.18 Orientation of Semicircular Canals Images of a bodybuilder in three poses can be helpful to remember the spatial orientations of the semicircular canals.

Rotation of the head around any of these axes causes movement of endolymph through the **ampullae** (see Figure 12.15, upper right inset). This flow deforms the gelatinous **cupula**, within which the mechanoreceptor cilia of hair cells are embedded. The hair cells are located in a ridge within each ampulla, called the **crista ampullaris**. The hair cells activate terminals of bipolar primary sensory neurons that have their cell bodies in the **vestibular ganglia of Scarpa** and send axons into the vestibular nerves (see Figure 12.15). The utricle and saccule contain structures called **maculae** that resemble the cristae ampullaris, but rather than angular acceleration, they detect linear acceleration and head tilt (see Figure 12.15, lower left inset). The macula consists of calcified crystals called **otoliths** sitting on a gelatinous layer within which mechanoreceptor hair cells are embedded. Gravity or other causes of linear acceleration pull on the crystals and activate these hair cells. The **superior vestibular ganglion** receives input from the utricle, anterior saccule, and anterior and lateral semicircular canals. The **inferior vestibular ganglion** receives input from the posterior saccule and posterior semicircular canal.

Primary sensory neurons in the vestibular ganglia (see Figure 12.15) convey information about angular and linear acceleration from the semicircular canals and otolith organs, respectively, through the vestibular division of CN VIII to the vestibular nuclei. There are **four vestibular nuclei** on each side of the brainstem, lying on the lateral floor of the fourth ventricle in the pons and rostral medulla (**Figure 12.19**). These nuclei can also be seen in the

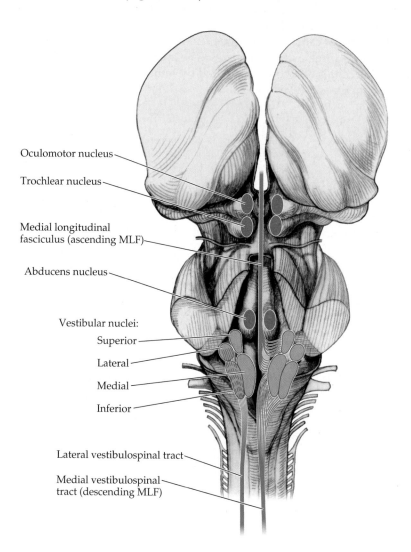

Oculomotor nucleus

Trochlear nucleus

Medial longitudinal fasciculus (ascending MLF)

Abducens nucleus

Vestibular nuclei:

Superior

Lateral

Medial

Inferior

Lateral vestibulospinal tract

Medial vestibulospinal tract (descending MLF)

FIGURE 12.19 Vestibular Nuclei and Vestibular Tracts Connections with the cerebellum and forebrain are not shown.

myelin sections in Figures 14.4C and 14.5A. The **lateral vestibular nucleus** (also called Deiter's nucleus) gives rise to the **lateral vestibulospinal tract**, which, despite its name, is part of the medial descending motor systems (see Table 6.3). The lateral vestibulospinal tract extends throughout the length of the spinal cord and is important in maintaining balance and extensor tone (see Figure 6.11D). The **medial vestibulospinal tract** arises from the **medial vestibular nucleus**, with additional contributions primarily from the **inferior vestibular nucleus**. The medial vestibulospinal tract is also a medial descending motor system, but it extends only to the cervical spine and is important in controlling neck and head position. The medial vestibular nucleus is the largest of the vestibular nuclei. The inferior vestibular nucleus is relatively easy to identify on myelin-stained sections because fibers of the lateral vestibular nucleus traverse the inferior vestibular nucleus as they descend toward the spinal cord (see Figure 12.19), giving the inferior vestibular nucleus a characteristic "checkerboard" appearance (see Figure 14.5A).

MNEMONIC

The **medial longitudinal fasciculus** (**MLF**) is an important pathway that connects the nuclei involved in eye movements to each other and to the vestibular nuclei (see Figure 12.19). The MLF can be identified in the sections in Figures 14.3–14.5 as a heavily myelinated tract running near the midline on each side, just under the oculomotor and trochlear nuclei in the midbrain, and just under the floor of the fourth ventricle in the midline of the pons. Fibers arising from the **medial vestibular nucleus**, with additional contributions mainly from the **superior vestibular nucleus**, ascend in the MLF to the oculomotor, trochlear, and abducens nuclei. This pathway mediates the **vestibulo-ocular reflex**, in which eye movements are adjusted for changes in head position (see **neuroexam.com Video 35**). The function of the MLF in interconnecting the abducens and oculomotor nuclei is discussed in Chapter 13. In another commonly used nomenclature, what we have called the MLF is referred to as the ascending MLF, and the medial vestibulospinal tract is referred to as the descending MLF (see Figure 12.19).

Oculocephalic testing

The vestibular nuclei have numerous important reciprocal connections with the cerebellum. As we will discuss in Chapter 15, vestibular-cerebellar connections occur mainly with the flocculonodular lobes and cerebellar vermis. These regions of the cerebellum are often called the vestibulocerebellum. A small number of primary vestibular sensory neurons bypass the vestibular nuclei and project directly to the vestibulocerebellum.

Ascending pathways from the vestibular nuclei relay in the ventral posterior nucleus of the thalamus to reach the **cerebral cortex**. These pathways are still being investigated in humans; however, one important cortical region for vestibular sensation appears to lie in the parietal association cortex, possibly in Brodmann's area 5, or in the lateral temporoparietal junction and posterior insula.

KEY CLINICAL CONCEPT

12.5 HEARING LOSS (CN VIII)

Unilateral hearing loss can be caused by disorders of the external auditory canal, middle ear, cochlea, eighth nerve, or cochlear nuclei (see Figures 12.14 and 12.15). As we have emphasized already, once the auditory pathways enter the brainstem, information immediately crosses bilaterally at multiple levels (see Figure 12.16). Therefore, unilateral hearing loss is not caused by lesions in the central nervous system proximal to the cochlear nuclei. (Disturbances of higher-order auditory processing and auditory hallucinations are described in KCC 19.7 and 19.13.)

Hearing

Impaired hearing is usually divided into **conductive hearing loss**, caused by abnormalities of the external auditory canal or middle ear, and **sensorineural hearing loss**, usually caused by disorders of the cochlea or eighth nerve. When evaluating a patient for hearing loss, the practitioner should first examine the ears with an otoscope. Hearing can be tested with sounds of different frequencies, such as finger rubbing, whispering, or a ticking watch (see **neuroexam.com Video 42**). Conductive and sensorineural hearing loss can often be distinguished with a simple 256 Hz or 512 Hz tuning fork (some argue that only 512 Hz or higher is useful for testing hearing). In the **Rinne test**, air conduction is compared to bone conduction for each ear. We measure **air conduction** by holding a vibrating tuning fork just outside each ear, and **bone conduction** by placing a tuning fork handle on each mastoid process (see **neuroexam.com Video 42**). Normal individuals hear the tone better by air conduction. In conductive hearing loss, bone conduction is greater than air conduction because bone conduction bypasses problems in the external or middle ear. In sensorineural hearing loss, air conduction is greater than bone conduction in both ears (as in normal hearing); however, hearing is decreased in the affected ear. In the **Weber test** the tuning fork is placed on the vertex of the skull in the midline, and the patient is asked to report the side where the tone sounds louder (see **neuroexam.com Video 42**). Normally, the tone sounds equal on both sides. In sensorineural hearing loss, the tone is quieter on the affected side. In conductive hearing loss, the tone is louder on the affected side, because compensatory neural mechanisms or mechanical factors increase the perceived volume on the side of conduction problem. You can verify this on yourself by producing temporary unilateral conductive hearing loss by closing each ear alternately. If you then hum, the tone should be louder on the occluded side.

Other tests that can help localize the cause of hearing loss include audiometry and brainstem auditory evoked potentials. An MRI scan with fine cuts through the auditory canal should be performed when disorders of the eighth nerve are suspected. Common causes of conductive hearing loss include cerumen in the external auditory canal, otitis, tympanic membrane perforation, and sclerosis of the middle ear ossicles. Causes of sensorineural hearing loss include exposure to loud sounds, meningitis, ototoxic drugs, head trauma, viral infections, aging, Meniere's disease (see KCC 12.6), cerebellopontine angle tumors, and, rarely, internal auditory artery infarct (see KCC 14.3).

Cerebellopontine angle tumors include acoustic neuroma (vestibular schwannoma), meningioma, cerebellar astrocytoma, epidermoid, glomus jugulare, and metastases. The most common tumor by far in this location is **acoustic neuroma**, accounting for about 9% of intracranial neoplasms (see Table 5.6). Mean age of onset is 50 years, and the tumor is nearly always unilateral. The exception is in neurofibromatosis type 2, in which the tumors are bilateral and usually occur by adolescence or early adulthood. This slow-growing tumor develops at the transitional zone between Schwann cells and oligodendrocytes, which occurs at the point where CN VIII enters the internal auditory meatus (see Figures 12.3A and 12.14). The term "acoustic neuroma" is a misnomer because the tumor is actually a **schwannoma**, not a neuroma, and it nearly always arises from the vestibular, not acoustic, division of the eighth nerve. Initially the tumor grows within the bony auditory canal, but then it expands into the cerebellopontine angle (see Figure 12.2A,C). Common early symptoms are unilateral hearing loss, **tinnitus** (ringing in the ear), and unsteadiness. The next cranial nerve to be affected is usually the nearby trigeminal nerve, with facial pain and sensory loss. Often the first sign of trigeminal involvement is a subtle decrease in the corneal reflex (see KCC 12.4). Interestingly, although the vestibular and facial nerves are compressed within the auditory canal, true vertigo is not usually a prominent symptom (although some unsteadiness is common), and the facial nerve does not usually become involved until the tumor is quite large. Eventually there is facial weakness, sometimes with decreased taste sensation on the side of the tumor.

With large tumors, cerebellar and corticospinal pathways are compressed, causing ipsilateral ataxia and contralateral hemiparesis. Impairment of swallowing and the gag reflex (CN IX and X) and unilateral impaired eye movements (CN III and VI) occur only in very large tumors. Ultimately, if left untreated, the tumor will compress the fourth ventricle, causing CSF outflow obstruction, hydrocephalus, herniation, and death. With appropriate clinical evaluation and MRI scanning, acoustic nerve tumors can be detected at an early stage, when they still lie entirely within the auditory canal. Treatment has traditionally been by surgical excision, but more recently there is a shift towards stereotactic radiosurgery (see KCC 16.4) with gamma knife or CyberKnife. Some smaller tumors may be monitored by MRI, especially in older patients, and factors such as age, position, size, hearing and patient preference help determine the choice between radiosurgery vs. surgical exision. Conventional surgery requires a posterior fossa approach, often involving collaboration between a neurosurgeon and otolaryngologist. Surgeons strive to spare facial nerve function during the procedure, and with smaller tumors some hearing may even be spared in the affected ear. Schwannomas can occur on other cranial nerves, as well as on spinal nerve roots, causing radiculopathy or spinal cord compression. **Trigeminal neuroma** is the second most common form of schwannoma affecting the cranial nerves. Schwannomas of the other cranial nerves are very rare. ■

KEY CLINICAL CONCEPT

12.6 DIZZINESS AND VERTIGO (CN VIII)

"Dizziness" is a vague term used by patients to describe many different abnormal sensations. In taking the history, the examiner should clarify whether the patient is referring to **true vertigo**, meaning a spinning sensation of movement, or one of the other meanings of dizziness. Other uses of "dizziness" include light-headedness or faintness, nausea, and unsteadiness on one's feet. True vertigo is more suggestive of vestibular disease than these other symptoms are. However, the situation is complicated by the fact that the other sensations listed here often accompany vertigo, and in some cases they may be the only presenting symptoms of vestibular disease.

Vertigo can be caused by lesions anywhere in the vestibular pathway, from labyrinth, to vestibular nerve, to vestibular nuclei and cerebellum, to parietal cortex. Most cases of vertigo are caused by **peripheral disorders** involving the inner ear, with **central disorders** of the brainstem or cerebellum being less common. It is essential to distinguish these possibilities because some central causes of vertigo, such as incipient brainstem stroke or posterior fossa hemorrhage, require emergency treatment to prevent serious sequelae. In taking the history of a patient with vertigo, it is therefore crucial to ask whether any other symptoms suggestive of posterior fossa disease, such as *diplopia, other visual changes, somatosensory changes, weakness, dysarthria, incoordination, or impaired consciousness,* are present (see Table 14.6). Patients with any of these abnormalities accompanying vertigo should be considered to have posterior fossa disease until proven otherwise and should be treated on an urgent basis. The general physical exam should always include **orthostatic** measurements of blood pressure and pulse in the supine and sitting or standing positions. Normally the systolic blood pressure drops by only about 10 millimeters of mercury, and the pulse increases by about 10 beats per minute when measured a few minutes after going from supine to a seated position with the legs dangling. Substantially greater changes suggest that the patient's symptoms may be caused by hypovolemia, antihypertensive medications, or cardiovascular or autonomic disorders, rather than by a vestibular lesion. In addition, during the general exam the tympanic membranes should be examined with an otoscope. A careful neurologic exam should be done to detect any abnormalities that may suggest a central cause for the vertigo.

Dix–Hallpike (or **Nylén–Bárány**) **positional testing** is a useful part of the exam that can help distinguish peripheral from central causes of vertigo (see **neuroexam.com Video 43**). The patient sits on the bed or examining table. The examiner then supports the head of the patient as the patient lies back with the head turned so that one ear is down and the head extends over the edge of the table. This maneuver should be done rapidly but gently. The patient is asked to keep their eyes open and report any sensations of vertigo, while the examiner looks for nystagmus. This change of position causes maximal stimulation of the posterior semicircular canal of the ear that is down (the anterior semicircular canal of the ear that is up is probably also stimulated) (see Figure 12.18). The maneuver is then repeated with the other ear down.

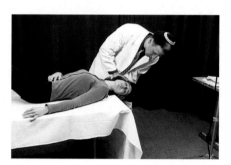

Positional vertigo test

With **peripheral lesions** affecting the inner ear there is usually a delay of 2 to 5 seconds before the onset of nystagmus and vertigo (Table 12.7). The nystagmus is horizontal or rotatory and does not change directions while the patient remains in the same position. Nystagmus and vertigo then fade away within about 30 seconds. If the same maneuver is repeated, there is often adaptation (also called habituation or fatiguing), so that the nystagmus and vertigo are briefer and less intense each time. In contrast, with **central lesions** the nystagmus and vertigo may begin immediately, and there tends to be no adaptation (see Table 12.7). Horizontal or rotatory nystagmus can also be seen with central lesions. However, vertical nystagmus, nystagmus that changes directions while remaining in the same position, or prominent nystagmus in the absence of vertigo is seen only in central, and not in peripheral, lesions.

Let's briefly review a few specific peripheral and central causes of vertigo. **Benign paroxysmal positional vertigo** is possibly the most common cause of true vertigo. Patients experience brief episodes of vertigo lasting for a few seconds and occurring with change of position. When the symptom first occurs, the patient may be dizzy for several hours. However, after the first episode it is usually brief and occurs only with change of position. In some cases, the vertigo may be so intense that patients cannot walk. The proposed

TABLE 12.7 Positional Testing to Distinguish Peripheral from Central Causes of Vertigo and Nystagmus

TYPE OF LESION	ONSET OF NYSTAGMUS	ADAPTATION (HABITUATION)	CHARACTERISTICS OF NYSTAGMUS AND VERTIGO
Peripheral (inner ear)	Delayed	Yes	Horizontal or rotatory, not vertical; does not change directions; prominent nystagmus only if vertigo is present as well.
Central (brainstem or cerebellum)	Immediate or delayed	No	Horizontal, rotatory, or vertical; may change directions; prominent nystagmus may occur in the absence of vertigo.

mechanism for this disorder is the presence of pieces of otolithic debris called otoconia in the semicircular canals (especially the posterior canal) that push against the cupula (see Figure 12.15, inset). Symptoms occur especially when the patient attempts to sleep and lies with the affected ear down, or if the patient turns to the affected side; however, if the patient remains still, the dizziness typically abates. Turning away from the affected ear or sitting up may also provoke symptoms. Treatment by **canalith repositioning maneuvers** to dislodge otolithic debris (Epley maneuver, or Semont liberatory maneuver) is beneficial in most patients. Symptoms may also be improved by adaptation exercises including the Brandt–Daroff procedure (patient sits on edge of bed, lies down sideways with the left ear down until vertigo subsides, and then repeats this ten times on each side) or other forms of vestibular rehabilitation therapy.

Viral infections or idiopathic inflammation of the vestibular ganglia or nerve may cause **vestibular neuritis**, a monophasic illness resulting in several days of intense vertigo and sometimes a feeling of unsteadiness that can last from weeks to months. In **Meniere's disease**, patients have recurrent episodes of vertigo, accompanied by fluctuating and sometimes stepwise, progressive hearing loss and tinnitus. Patients with Meniere's disease also often complain of a full feeling in the ear. The etiology is thought to be excess fluid and pressure in the endolymphatic system (see Figures 12.14 and 12.15). Meniere's disease is most frequently treated with salt restriction and diuretics, although there have been no controlled studies of those therapies. There are multiple surgical procedures that have been effective in some patients; these include vestibular nerve section, labrinthectomy, endolymphatic saculotomy (decompression), and transtympanic gentamycin (to cause permanent loss of vestibular function on the affected side). **Autoimmune inner ear disease**, another important cause of vertigo, can produce symptoms resembling Meniere's disease. **Acoustic neuroma** (vestibular schwannoma) is another cause of hearing loss and tinnitus that can be associated with vertigo (see KCC 12.5). However, unlike Meniere's disease, patients with acoustic neuroma often complain of unsteadiness rather than true vertigo, and they do not usually have discrete episodes.

Common central causes of vertigo include **vertebrobasilar ischemia or infarct**. Involvement of the vestibular nuclei or cerebellum causes vertigo, often with other symptoms and signs of vertebrobasilar disease (see KCC 14.3; Table 14.6). It is essential to recognize this entity so that treatment is not delayed. Similarly, a small **hemorrhage** in the cerebellum or, rarely, in the brainstem may initially present mainly with vertigo and should be treated as soon as possible to prevent catastrophe. Cerebellar hemorrhage that presents initially with some nausea and dizziness, only to rebleed a few hours later, has been called "fatal gastroenteritis." **Encephalitis**, **tumors**, or **demyelination** in the posterior fossa can cause vertigo. In addition, numerous **drugs and toxins**, including alcohol and anticonvulsant medications, cause dysfunction of the vestibular nuclei and cerebellum, producing vertigo along with other symptoms. Ototoxic drugs such as gentamicin cause *bilateral* vestibular dysfunction, which is why they cause unsteadiness of gait and oscillopsia (perception of oscillating vision) rather than true spinning vertigo. Anemia and thyroid disorders can also produce dizziness and should be tested for in all patients complaining of vertigo where etiology is unclear. Other disorders occasionally associated with vertigo include atypical migraines, Lyme disease, and syphilis. Finally, epileptic seizures (see KCC 18.2) are uncommon as a cause of vertigo without other symptoms; however, patients with seizures involving the parietal regions responsible for motion perception may report vertigo as one manifestation of their seizures. ■

CN IX: Glossopharyngeal Nerve

FUNCTIONAL CATEGORY	FUNCTION
Branchial motor	Stylopharyngeus muscle
Parasympathetic	Parasympathetics to parotid gland
General somatic sensory	Sensation from middle ear, region near the external auditory meatus, pharynx, and posterior one-third of tongue
Visceral sensory (special)	Taste from posterior one-third of tongue
Visceral sensory (general)	Chemoreceptors and baroreceptors of carotid body

The glossopharyngeal nerve was named for its role in sensation for the posterior tongue and pharynx; however, it has additional functions as well. It exits the brainstem as several rootlets along the upper ventrolateral medulla, just below the pontomedullary junction and just below CN VIII, between the inferior olive and the inferior cerebellar peduncle (see Figure 12.2A,C). The nerve traverses the subarachnoid space to exit the skull via the **jugular foramen** (see Figure 12.3A,B; Table 12.2).

The **branchial motor** portion of the nerve supplies one muscle, the **stylopharyngeus** (Figure 12.20), which elevates the pharynx during talking and swallowing and contributes (with CN X) to the gag reflex. There is evidence that the glossopharyngeal nerve may provide some innervation to other pharyngeal muscles; however, most pharyngeal muscles are supplied primarily by the vagus (see the next section). The branchial motor component of CN IX arises from the **nucleus ambiguus** in the medulla (see Figure 12.20). "Ambiguus" is Latin for "ambiguous," and this name can be remembered because the nucleus is difficult to discern on conventional stained sections (see Figure 14.5A,B). **Parasympathetic** preganglionic fibers in the glossopharyngeal nerve arise from the **inferior salivatory nucleus** in the pons (see Figure 12.20). These parasympathetic fibers leave the glossopharyngeal nerve via the **tympanic nerve** and then join the **lesser petrosal nerve** to synapse in the **otic ganglion**, providing postganglionic parasympathetics to the **parotid gland**.

The **general visceral sensory** portion of the glossopharyngeal nerve conveys inputs from baroreceptors and chemoreceptors in the **carotid body**. These afferents travel to the **caudal nucleus solitarius** of the medulla, also known as the **cardiorespiratory nucleus** (see Figure 12.20). Glossopharyngeal **special visceral sensation** mediates **taste** for the posterior one-third of the tongue, which reaches the **rostral nucleus solitarius**, or **gustatory nucleus** (see Figures 12.5, 12.12, and 12.20). **General somatic sensory** functions of CN IX are the sensation of touch, pain, and temperature from the posterior one-third of the tongue, pharynx, middle ear, and a region near the external auditory meatus (see Figure 12.7B). The glossopharyngeal nerve has two sensory ganglia located within or just below the jugular foramen (see Table 12.5). General and special **visceral sensation** are conveyed by primary sensory neurons in the **inferior (petrosal) glossopharyngeal ganglion**. **General somatic sensation** is conveyed by primary sensory neurons in both the inferior and **superior (jugular) glossopharyngeal ganglion**.

MNEMONIC

REVIEW EXERCISE

Which cranial nerve contributes to the greater petrosal nerve? Which contributes to the lesser petrosal nerve? (See Figure 12.6.)

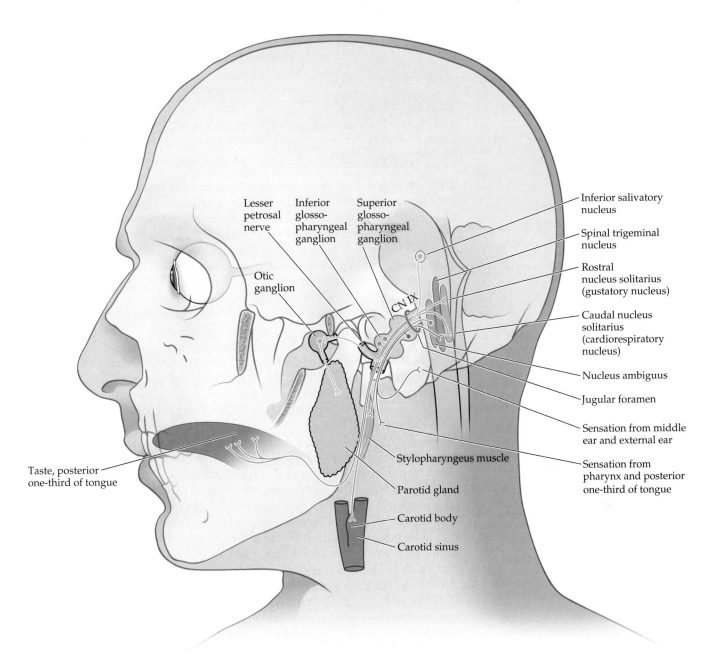

FIGURE 12.20 Glossopharyngeal Nerve (CN IX)
Summary of glossopharyngeal nerve sensory and
motor pathways.

CN X: Vagus Nerve

FUNCTIONAL CATEGORY	FUNCTION
Branchial motor	Pharyngeal muscles (swallowing) and laryngeal muscles (voice box)
Parasympathetic	Parasympathetics to heart, lungs, and digestive tract down to the splenic flexure
General somatic sensory	Sensation from pharynx, meninges, and a small region near the external auditory meatus
Visceral sensory (special)	Taste from epiglottis and pharynx
Visceral sensory (general)	Chemoreceptors and baroreceptors of the aortic arch

The vagus nerve derives its name from the wandering course it takes in providing parasympathetic innervation to organs throughout the body ("vagus" means "wandering" in Latin). Other important functions are also served by the vagus, as we will discuss here. The vagus nerve exits the ventrolateral medulla as several rootlets just below CN IX, between the inferior olive and the inferior cerebellar peduncle (see Figure 12.2A,C). It crosses the subarachnoid space and then leaves the cranial cavity via the **jugular foramen** (see Figures 12.3A,B and 12.21).

The largest part of the vagus nerve provides **parasympathetic** innervation to the heart, lungs, and digestive tract, extending nearly to the splenic flexure (see Figures 6.13 and 12.21). Parasympathetic preganglionic fibers arise from the **dorsal motor nucleus of CN X**, which runs from the rostral to the caudal medulla (see Figure 14.5A,B). The dorsal motor nucleus of CN X forms the **vagal trigone** on the floor of the fourth ventricle, just lateral to the hypoglossal trigone, near the obex (see Figure 12.2B). Postganglionic parasympathetic neurons innervated by the vagus are found in **terminal ganglia** located within or near the effector organs. Recall that parasympathetics to the gastrointestinal tract beyond the splenic flexure—and to the urogenital system—are provided by parasympathetic nuclei in the sacral spinal cord (see Figure 6.13).

The **branchial motor** component of the vagus (Figure 12.21) controls nearly all pharyngeal and upper esophageal muscles (swallowing and gag reflex) and the muscles of the larynx (voice box). The **nucleus ambiguus** supplies branchial motor fibers that travel in the vagus nerve to the muscles of the palate, pharynx, upper esophagus, and larynx, and in the glossopharyngeal nerve (CN IX) to the stylopharyngeus (see Figure 12.20).

A branch of the vagus called the **recurrent laryngeal nerve** (see Figure 12.21) loops back upward from the thoracic cavity to control all intrinsic laryngeal muscles except for the cricothyroid, which is innervated by another branch of the vagus, the **superior laryngeal nerve**. The fibers in the recurrent laryngeal nerve arise from the caudal portion of the nucleus ambiguus. After they exit the brainstem, these fibers travel briefly with CN XI before joining CN X (see the next section). Some texts consider these caudal fibers of the nucleus ambiguus to be part of CN XI and refer to the caudal nucleus ambiguus as the cranial nucleus of CN XI. However, we include these fibers with CN X because they spend the majority of their course traveling with CN X, not CN XI. Upper motor neuron innervation to the nucleus ambiguus controlling the voice and voluntary swallowing is from bilateral motor cortex (see Figure 6.2), except for the palate, which receives unilateral innervation from the contralateral cortex (for example, see Case 6.5).

General somatic sensory fibers of the vagus (see Figure 12.21) supply the pharynx, larynx, meninges of the posterior fossa, and a small region near the external auditory meatus (see Figure 12.7B). Note that below the larynx and pharynx, conscious (general somatic) sensation from the viscera is carried by spinal, not cranial, nerves. However, unconscious, **general visceral sensation** from chemoreceptors and baroreceptors of the aortic arch, cardiorespiratory system, and digestive tract is carried to the brainstem by the vagus nerve.

REVIEW EXERCISE

List the branchial motor, parasympathetic, general somatic sensory, and visceral sensory functions of CN IX and CN X. Name the nucleus subserving each function. (See Figures 12.20 and 12.21.)

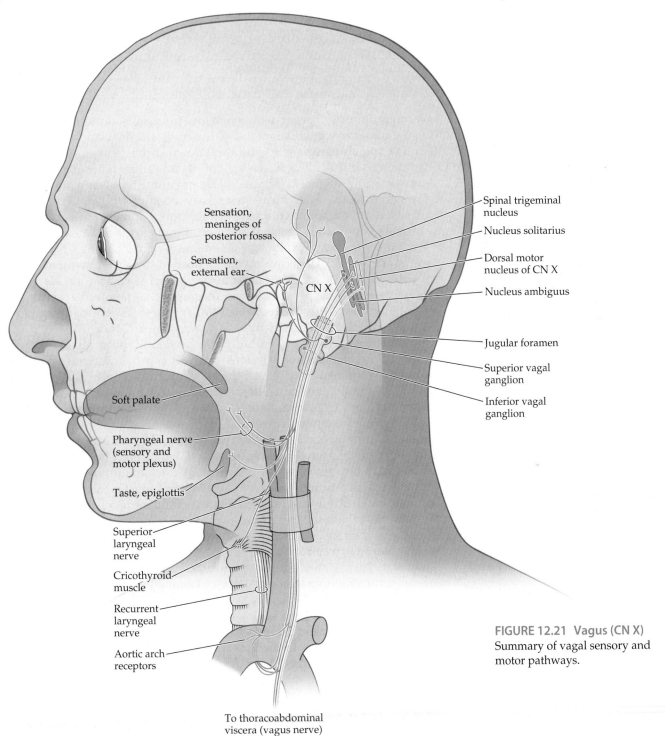

Spinal trigeminal nucleus

Nucleus solitarius

Dorsal motor nucleus of CN X

Nucleus ambiguus

Jugular foramen

Superior vagal ganglion

Inferior vagal ganglion

Sensation, meninges of posterior fossa

Sensation, external ear

CN X

Soft palate

Pharyngeal nerve (sensory and motor plexus)

Taste, epiglottis

Superior laryngeal nerve

Cricothyroid muscle

Recurrent laryngeal nerve

Aortic arch receptors

To thoracoabdominal viscera (vagus nerve)

FIGURE 12.21 Vagus (CN X)
Summary of vagal sensory and motor pathways.

Many of these general visceral afferents reach the **caudal nucleus solitarius** (cardiorespiratory nucleus; see Figures 12.5 and 14.5B). The vagus nerve also contains a small number of **special visceral sensory** fibers that carry **taste** sensation from the epiglottis and posterior pharynx to the **rostral nucleus solitarius** (gustatory nucleus) (see Figures 12.5 and 14.5A).

The primary sensory neuron cell bodies for CN X general and special visceral sensation are located in the **inferior (nodose) vagal ganglion** (Table 12.5), located just below the jugular foramen. Cell bodies for general somatic sensation are located in both the inferior vagal ganglion and the **superior (jugular) vagal ganglion**, which lies within or just below the jugular foramen.

CN XI: Spinal Accessory Nerve

FUNCTIONAL CATEGORY	FUNCTION
Branchial motor	Sternomastoid and upper part of trapezius muscle

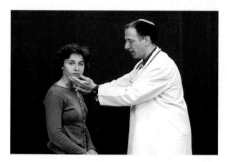

CN XI testing

As its name implies, this nerve does not arise from the brainstem, but rather from the upper five or six segments of the cervical spinal cord (see Figure 12.2). The **spinal accessory nucleus** (also known as the accessory spinal nucleus) protrudes laterally between the dorsal and ventral horns of the spinal cord central gray matter (see Figure 14.5D), providing **branchial motor*** fibers to this nerve. Nerve rootlets leave the spinal accessory nucleus and exit the lateral aspect of the spinal cord between the dorsal and ventral nerve roots just dorsal to the dentate ligament and ascend through the foramen magnum to reach the intracranial cavity (see Figures 12.2A and 12.3A,B). CN XI then exits the cranium again via the **jugular foramen** to supply the **sternomastoid** and upper portions of the **trapezius muscle**. The sternomastoid muscle turns the head toward the opposite side, and the trapezius is involved in elevating the shoulder (see **neuroexam.com Video 46**). The *lower* portions of the trapezius are usually supplied mainly by cervical nerve roots C3 and C4.

Note that the *left* sternomastoid turns the head to the *right*, and vice versa. Therefore, **lower motor neuron lesions** of CN XI may cause some ipsilateral weakness of shoulder shrug or arm elevation, and weakness of head turning *away* from the lesion. In turning the head, other neck muscles can sometimes compensate for the sternomastoid; therefore, in subtle cases, it is best to palpate the sternomastoid with one hand for contractions while the patient attempts turning their head against resistance offered by the examiner's other hand. **Upper motor neuron lesions** can also cause deficits of head turning, toward the side opposite the lesion. Therefore, it is thought that the central pathways for head turning project to the *ipsilateral* spinal accessory nucleus. However, the deficit in head turning to the side opposite the lesion in cortical lesions is often more of a gaze preference than true weakness. With upper motor neuron lesions causing contralateral hemiparesis, the shoulder shrug is also often weak on the side of the hemiparesis.

Before it exits the cranium, the spinal accessory nerve is briefly joined by some fibers arising from the caudal nucleus ambiguus that exit from the lateral medulla adjacent to the vagus nerve. These fibers rejoin the vagus within a few centimeters and ultimately form the recurrent laryngeal nerve. As noted in the previous section, because these fibers travel briefly with CN XI, some textbooks refer to them as the **cranial root of CN XI**. Despite this name, the recurrent laryngeal nerve fibers spend the majority of their course traveling with CN X and can functionally be considered part of the vagus.

CN XII: Hypoglossal Nerve

FUNCTIONAL CATEGORY	FUNCTION
Somatic motor	Intrinsic muscles of the tongue

The hypoglossal nerve exits the ventral medulla as multiple rootlets between the pyramid and inferior olivary nucleus (see Figure 12.2A,C). This

---

*As we have already discussed, some consider the spinal accessory nerve to be somatic or mixed somatic and branchial rather than purely branchial motor, since the sternomastoid and upper trapezius muscles may have somatic embryological origins. For simplicity we have kept CN XI in the branchial motor category, since the spinal accessory nucleus is located laterally, in continuity with the branchial motor column (see Figure 12.5).

nerve exits through its own foramen, the **hypoglossal foramen** (see Figure 12.3A,B), and provides **somatic motor** innervation to all intrinsic and extrinsic tongue muscles except for the palatoglossus, which is supplied by CN X (see **neuroexam.com Video 47**). The **hypoglossal nucleus** is located near the midline on the floor of the fourth ventricle in the medulla (see Figure 14.5A,B), forming the **hypoglossal trigone**, just medial to the dorsal nucleus of CN X (see Figures 12.2B, 12.4B, and 12.5).

Upper motor neurons for tongue movement arise from the tongue region of the primary motor cortex (see Figure 6.2) and travel in corticobulbar pathways that decussate before reaching the hypoglossal nuclei. Therefore, lesions in the primary motor cortex or internal capsule will cause **contralateral tongue weakness**, while lesions of the hypoglossal nucleus, exiting fascicles, or nerve cause **ipsilateral tongue weakness**. Note that unilateral tongue weakness causes the tongue to deviate toward the weak side when it is protruded. Thus, a lesion of the hypoglossal nerve will cause the tongue to deviate toward the side of the lesion.

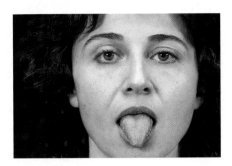

Tongue protrusion

12.7 DISORDERS OF CN IX, X, XI, AND XII

Peripheral lesions of the lower cranial nerves are relatively uncommon. Most disorders of these cranial nerves are associated with central lesions (see Chapter 14). Like all other nerves, however, the lower cranial nerves are occasionally affected by diabetic neuropathy; demyelination; motor neuron disease; and traumatic, inflammatory, neoplastic, toxic, or infectious conditions. Let's briefly discuss a few disorders of the lower cranial nerves.

Glossopharyngeal neuralgia is clinically similar to trigeminal neuralgia but involves the sensory distribution of CN IX, causing episodes of severe throat and ear pain. Injury to the **recurrent laryngeal nerve** (a branch of CN X), which can occur during surgery of the neck (such as carotid endarterectomy, cervical disc surgery, or thyroid surgery) or during cardiac surgery as the left nerve loops around the aorta (see Figure 12.21), produces unilateral vocal cord paralysis and hoarseness (see KCC 12.8). The recurrent laryngeal nerve can also be infiltrated by apical lung tumors during its looping course through the upper thoracic cavity, which produces hoarseness as part of Pancoast's syndrome (see KCC 9.1). **Glomus tumors** are a rare disorder that can affect the lower cranial nerves. Glomus bodies are normal, small epithelioid structures that resemble the carotid bodies histologically, but whose function is unknown. Like the carotid bodies, they are richly innervated by CN IX but are located adjacent to the jugular foramen and along branches of CN IX leading to the middle ear cavity. Tumors arising from the glomus bodies are known by a variety of names, including glomus tumor and **glomus jugulare**. Patients with glomus jugulare often present with impairments of CN IX, X, and XI, resulting from compression of these nerves in the jugular foramen. In addition, the tumor frequently extends to the nearby CN XII and can grow upward to affect CN VIII and VII in the temporal bone. When the tumor grows into the middle ear, it can sometimes be seen on otoscopic exam as a fleshy vascular mass. Treatment is by resection, although radiation therapy is also used in some cases.

Clinical examination of CN IX through XII and the functional effects of lesions of these cranial nerves was discussed in the previous sections and is discussed in further detail in KCC 12.8. ■

12.8 HOARSENESS, DYSARTHRIA, DYSPHAGIA, AND PSEUDOBULBAR AFFECT

Disorders of speech and swallowing can be very disabling, or even fatal in some cases. Causes of these disorders can range from upper motor neuron lesions (corticobulbar pathways) to lower motor neuron lesions to disorders of the neuromuscular junction or muscles themselves, and they can also result from cerebellar or basal ganglia dysfunction. Let's discuss some of the more common causes of each of these disorders.

Voice disorders occur when the larynx and **vocal cords** (more correctly called the vocal folds, or true vocal cords) are not functioning properly. Such malfunction can occur as the result of mechanical factors or neural or muscular disorders. **Hoarseness** of the voice usually results from disorders of the vocal cords causing asynchronous vibratory patterns. Hoarseness is often caused by mechanical factors such as swelling, nodules, polyps, or neoplasms of the vocal cords. **Breathiness** of the voice is caused by paralysis or paresis of the vocal cord(s), which results in incomplete adduction of one or both of the vocal cords and an air leak at the glottis. In common language, breathiness is often called "hoarseness," although this is not strictly correct. Recall that the muscles of the larynx are innervated by the branchial motor portion of CN X. As the recurrent laryngeal nerve loops through the upper thoracic cavity (see Figure 12.21) it can be injured during surgery in the neck or chest, or it can be compressed by apical lung cancer (Pancoast syndrome; see KCC 9.1). Voice disorders can also occur from lesions of CN X as it exits the brainstem, such as glomus jugulare (see KCC 12.7), or of the nucleus ambiguus in the medulla (see Figures 12.20 and 12.21). The most common lesion of the medulla affecting the nucleus ambiguus is a lateral medullary infarct (see Figure 14.21D; Table 14.7). An abnormal, gravelly sounding voice can also occur in Parkinson's disease and other related movement disorders (see KCC 16.2). Spasmodic dysphonia is an uncommon form of dystonia (see KCC 16.1) involving the larynx, presumably resulting from dysfunction of basal ganglia circuitry. Vocal cord lesions or abnormal vocal cord movements can best be evaluated by **fiberoptic laryngoscopy**, in which a flexible scope is used to directly visualize the vocal cords during speech.

Dysarthria is abnormal articulation of speech (see **neuroexam.com Video 45**). Dysarthria should be distinguished from aphasia (see KCC 19.2). Whereas dysarthria is a motor articulatory disorder, aphasia is a disorder of higher cognitive functioning in which language formulation or comprehension is abnormal. Depending on the lesion, aphasia and dysarthria can occur together, or one can occur without the other. Dysarthria can range in severity from mild slurring to unintelligible speech. It can occur in lesions involving the muscles of articulation (jaw, lips, palate, pharynx, and tongue), the neuromuscular junction, or the peripheral or central portions of CN V, VII, IX, X, or XII. In addition, speech articulation can be abnormal because of dysfunction of the motor cortex face area (see Figure 6.2), cerebellum, basal ganglia, or descending corticobulbar pathways to the brainstem. Common causes of dysarthria include infarcts, multiple sclerosis, or other lesions affecting corticobulbar pathways (see KCC 6.3); brainstem lesions; lesions of cerebellar pathways or basal ganglia; toxins (e.g., alcohol); other causes of diffuse encephalopathy; myasthenia gravis; and other disorders of neuromuscular junction, muscle, or peripheral nerves. A few other important but less common specific causes of dysarthria to be aware of include amyotrophic lateral sclerosis (see KCC 6.7), botulism, and Wilson's disease.

Dysphagia is impaired swallowing. Dysphagia can be caused by esophageal strictures, neoplasms, or other local lesions, or it may have a neural or neuromuscular basis. When dysphagia is caused by neural or neuromuscular disorders, it often has the same causes as, and occurs together with, dysarthria (although dysarthria and dysphagia can occur independently as well). **Swallowing** is classically divided into four phases: the **oral preparatory phase** (preparation of the food bolus for swallowing by mastication); **oral phase** (movement of the bolus in an anterior–posterior direction by the oral tongue); **pharyngeal phase** (propulsion of the bolus through the pharynx by base-of-tongue driving force, aided by anterior–superior movement of the larynx); and the **esophageal phase** (opening of the upper esophageal sphincter; esophageal peristalsis; and emptying into the stomach). Thus, dysphagia can be caused by dysfunction of muscles of the tongue, palate, pharynx, epiglottis, larynx, or esophagus; by lesions of CN IX, X, XII, or their nuclei; or by dysfunction at the neuromuscular junction or in the descending corticobulbar pathways.

Impaired oral and pharyngeal swallowing function and impaired reflex closure of the entrance to the trachea by the epiglottis and laryngeal muscles can lead to aspiration of food, and esophageal reflux can lead to aspiration of gastric secretions into the lungs. **Aspiration pneumonia**, caused by impaired swallowing function, is difficult to treat and is a common cause of death in disorders of the nervous system. Pharyngeal reflexes can be tested by the **gag reflex**. This reflex is elicited by stroking of the posterior pharynx with a cotton-tipped swab. The gag reflex is mediated by sensory and motor fibers from both CN IX and X, although CN IX may be more important for the afferent limb, while CN X provides primarily the efferent limb. Although an impaired gag reflex may be indicative of impaired motor or sensory function of the pharynx, its absence or presence is not absolutely predictive of aspiration risk.

A simple way to assess soft palate function is to observe **palate elevation** with a penlight while asking the patient to say, "Aah" (see **neuroexam.com Video 44**). In unilateral lesions of CN X or of the nucleus ambiguus, the uvula and soft palate will deviate toward the normal side, while the soft palate on the abnormal side hangs abnormally low, producing the **stage curtain sign**.

Brainstem nuclei involved in laughing and crying include CN VII, IX, X, and XII. Lesions of corticobulbar pathways in the subcortical white matter or brainstem can occasionally produce a bizarre syndrome called **pseudobulbar affect**. Patients with this syndrome exhibit uncontrollable bouts of laughter or crying without feeling the usual associated emotions of mirth or sadness. Pseudobulbar affect may be likened to an "upper motor neuron" disorder in which there is abnormal reflex activation of laughter and crying circuits in the brainstem, leading to **emotional incontinence**. The term **pseudobulbar palsy** is sometimes used to describe dysarthria and dysphagia caused by lesions not of the brainstem (bulb), but rather of the upper motor neuron fibers in the corticobulbar pathways (hence "pseudo"). Another neurologic cause of episodes of inappropriate laughter is a rare seizure disorder called **gelastic epilepsy**, which is usually associated with hypothalamic lesions (hypothalamic hamartoma) but can occasionally be seen in temporal lobe seizures (see KCC 18.2). ■

Review: Cranial Nerve Combinations

The preceding material is quite detailed, yet, as we will see, it has numerous important clinical applications. Let's review several functional combinations to help consolidate the details of cranial nerve anatomy and to clarify some regional aspects of sensory and motor function. Functional combinations involving the eye muscles will be discussed in Chapter 13.

1. **Sensory and motor innervation of the face:** Sensation is provided by the trigeminal nerve (CN V), while movement of the muscles of facial expression is provided by the facial nerve (CN VII).

2. **Taste and other sensorimotor functions of the tongue and mouth:** The anterior two-thirds and posterior one-third of the tongue are derived from different branchial arches and therefore have different innervation. For the anterior two-thirds of the tongue, taste is provided by the facial nerve (CN VII, chorda tympani), while general somatic sensation is provided by the trigeminal nerve (V_3, mandibular division). For the posterior one-third of the tongue, both taste and general somatic sensation are provided by the glossopharyngeal nerve (CN IX). Taste for the epiglottis and posterior pharynx is provided by the vagus nerve (CN X). General sensation for the teeth, nasal sinuses, and inside of the mouth, above the pharynx and above the posterior one-third of the tongue, is provided by the trigeminal nerve (CN V).

3. **Sensory and motor innervation of the pharynx and larynx:** For the pharyngeal **gag reflex**, general somatic sensation is provided by both the glossopharyngeal and the vagus nerves (CN IX and X), but branchial motor innervation is provided mainly by the vagus (CN X). For the larynx, the vagus provides both sensory and motor innervation. General somatic sensation for organs below the level of the larynx is provided by the spinal nerves.

4. **Sensory and motor innervation of the ear:** General somatic sensation for the middle ear and inner tympanic membrane is provided by the glossopharyngeal nerve (CN IX), while sensation for the external ear and outer surface of the tympanic membrane is provided by the trigeminal (V_3, mandibular branch), facial (CN VII), glossopharyngeal (CN IX), and vagus (CN X) nerves (see Figure 12.7B). Hearing and vestibular sense travel in the vestibulocochlear nerve (CN VIII). Branchial motor innervation for the tensor tympani comes from the trigeminal nerve (CN V), while innervation for the stapedius comes from the facial nerve (CN VII). An aid to remembering this information is the fact that "**t**ensor **t**ympani" and "**t**rigeminal" start with the letter **T**, while "**s**tapedius" and cranial nerve "**s**even" start with **S**. Similarly, the **t**ensor veli palatini is supplied by the **t**rigeminal nerve, while all other muscles of the soft palate are supplied by the vagus.

5. **Sensation from the meninges:** Sensation from the supratentorial dura mater is carried by the trigeminal nerve (CN V), while the dura of the posterior cranial fossa is supplied by the vagus (CN X) and by the upper cervical nerve roots.

6. **General visceral sensation:** Unconscious, general visceral sensation from baroreceptors and chemoreceptors is carried by the glossopharyngeal nerve (CN IX) for the carotid body and sinus and by the vagus nerve (CN X) for the aortic arch and other thoracoabdominal viscera.

7. **Effects of unilateral cortical lesions:** The lower portions of the face (CN VII), soft palate (CN V, X), upper trapezius muscle (CN XI), and tongue (CN XII) receive mainly contralateral input, so they show weakness on the side opposite to cortical or corticobulbar lesions. Effects of unilateral upper motor neuron (UMN) lesions on eye movements are discussed in Chapter 13 (see KCC 13.10). Other cranial nerves do not typically show unilateral deficits with unilateral UMN lesions, although, interestingly, unilateral cortical or corticobulbar tract lesions can cause nonlateralized dysfunction in articulation (dysarthria) and swallowing (dysphagia).

Additional understanding of cranial nerve combinations will be gained through clinical practice.

MNEMONIC

REVIEW EXERCISE

1. For each cranial nerve (I through XII), state its exit point from the brainstem (see Figure 12.2) and its exit point from the cranium (see Figure 12.3, Table 12.2).

2. Now it's time to go for broke: For each cranial nerve, list each sensory and motor category and all associated functions (see Table 12.4).

CLINICAL CASES

CASE 12.1 ANOSMIA AND VISUAL IMPAIRMENT*

MINICASE

A 51-year-old man began having difficulty reading over the course of 5 to 6 weeks. He saw his doctor, and on review of systems it was noted that he had lost his sense of smell about 3 years earlier. On exam, he had a visual acuity of 20/20 in the right eye **and 20/40 in the left eye**. He was **unable to smell coffee or soap with either nostril**.

LOCALIZATION AND DIFFERENTIAL DIAGNOSIS

1. On the basis of the symptoms and signs shown in **bold** above, where is the lesion?
2. What is the most likely diagnosis, and what are some other possibilities?

*This patient was described previously as a case report in *The New England Journal of Medicine*. 1996. 335: 1668–1674.

Discussion

1. The key symptoms and signs in this case are:

   - **Bilateral anosmia**
   - **Difficulty reading and left decreased visual acuity**

   The anosmia could be caused by bilateral lesions of the olfactory mucosa or the olfactory nerves, bulbs, or tracts (see KCC 12.1). Decreased acuity in the left eye is consistent with a disorder in the left eye or the left optic nerve (see KCC 11.2). Deficits of CN I and CN II together suggest a lesion at the base of the frontal lobes, where these two cranial nerves briefly run in proximity (see Figure 18.6) before CN II exits the cranium via the optic canal (see Figure 12.3A). It is also possible that the anosmia is an unrelated incidental finding.

   The most likely *clinical localization* is bilateral orbital frontal areas.

2. Given the prolonged course, a slow-growing tumor at the base of the frontal lobes such as a meningioma should be suspected. Other tumors or chronic inflammatory disorders in this region are also possible, but less likely.

Clinical Course and Neuroimaging

The patient underwent a **brain MRI** (Image 12.1B,C, pages 540–541). For comparison, Image 12.1A (page 540) shows a normal MRI from another patient demonstrating the anatomical structures at the base of the frontal lobes. With the labels covered, identify the olfactory bulbs, olfactory sulcus, gyrus rectus, and cribriform plate. The images of the patient in this case, taken with gadolinium enhancement, are shown in Image 12.1B and C. An enhancing mass at the base of the frontal lobes extends along the dural surface in the region of the olfactory bulbs and erodes through the cribriform plate into the upper nasal passages (see Image 12.1B). The mass also extends back to encase the left optic nerve (see Image 12.1C). On the basis of its appearance and the patient's history, it was felt that this was most likely a meningioma, although the irregular borders of the lesion that appeared to infiltrate adjacent structures were somewhat unusual for a meningioma. A biopsy of the mass was performed through the nose, via a transethmoidal approach. Interestingly, pathology revealed noncaseating granulomas consistent with sarcoidosis (see KCC 12.1). Additional workup supported the diagnosis of sarcoidosis confined to the nervous system. The patient was treated with steroids, and his vision improved in the left eye, but he remained unable to smell.

CASE 12.1 ANOSMIA AND VISUAL IMPAIRMENT

IMAGE 12.1A–C Mass in Orbital Frontal Region T1-weighted MRI images with intravenous gadolinium enhancement. (A) Coronal image from a normal patient showing relation of olfactory bulbs to the frontal lobes and cribriform plate. (B,C) Coronal images from patient in Case 12.1, with B and C progressing from anterior to posterior.

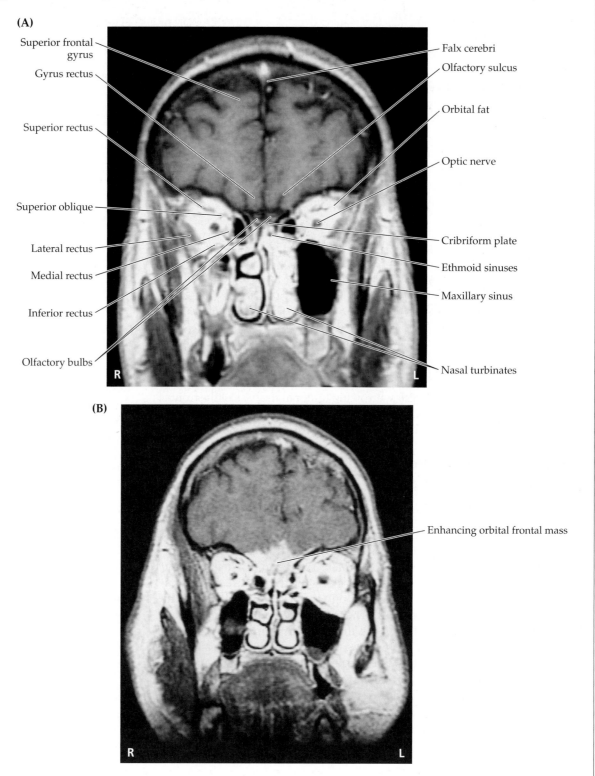

(A)

Superior frontal gyrus —
Gyrus rectus —
Superior rectus —
Superior oblique —
Lateral rectus —
Medial rectus —
Inferior rectus —
Olfactory bulbs —

— Falx cerebri
— Olfactory sulcus
— Orbital fat
— Optic nerve
— Cribriform plate
— Ethmoid sinuses
— Maxillary sinus
— Nasal turbinates

R L

(B)

— Enhancing orbital frontal mass

R L

CASE 12.1 *(continued)*

(C)

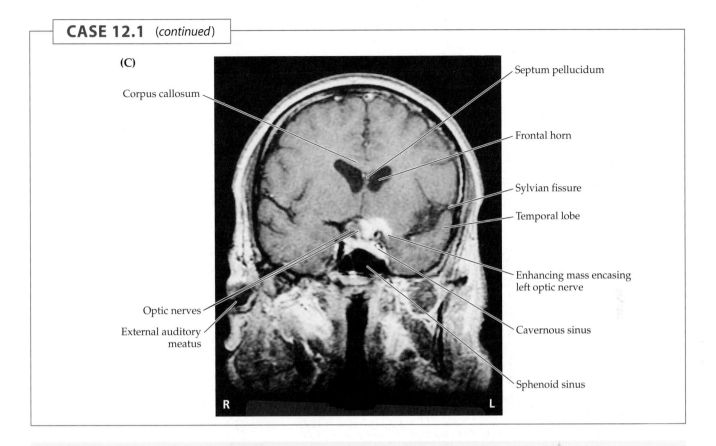

Corpus callosum

Septum pellucidum

Frontal horn

Sylvian fissure

Temporal lobe

Enhancing mass encasing left optic nerve

Cavernous sinus

Optic nerves

External auditory meatus

Sphenoid sinus

R   L

CASE 12.2   CHEEK NUMBNESS AND A BULGING EYE

MINICASE

A 51-year-old woman saw an ophthalmologist because she noticed that her left eye seemed to be bulging out increasingly for the past 3 to 4 years, and she had recently developed **left-sided headaches**. Exam was normal except for an **outward bulging of her left eye (proptosis) and decreased sensation to touch and pinprick over her left cheek.**

LOCALIZATION AND DIFFERENTIAL DIAGNOSIS

1. Which division of which cranial nerve provides sensation to the cheek? Where does this nerve exit the skull?
2. What diagnosis is suggested by the slowly developing left proptosis over the course of several years, together with the cheek sensory deficit and left-sided headache?

Discussion

1. The key symptoms and signs in this case are:

   - **Left-sided headaches**
   - **Left proptosis**
   - **Decreased sensation to touch and pinprick over the left cheek**

   The maxillary division of the trigeminal nerve (CN V_2) provides sensation to the cheek (see Figure 12.7). This branch of the trigeminal nerve exits the skull through the foramen rotundum (see Figures 12.3, 12.7; Table 12.2).

2. The history and exam findings suggest a slow-growing mass lesion such as a meningioma (see KCC 5.8) involving the left foramen rotundum area, causing V_2 sensory loss, and extending into the left orbit, causing proptosis.

Clinical Course and Neuroimaging

The patient underwent an **MRI scan** with gadolinium enhancement (Image 12.2, page 544) that revealed an enhancing mass lying outside the brain in the region of the left foramen rotundum (compare Figure 12.3A) and extending into the left orbit. This appearance was felt to be consistent with a meningioma (see KCC 5.8). Because of concern that the mass would soon lead to impaired vision in the left eye, the patient was referred to a neurosurgeon. A left frontotemporal craniotomy was performed (see KCC 5.11), and a firm, grayish-reddish mass was carefully dissected off the sphenoid wing and removed from the orbit. Pathologic examination confirmed the diagnosis of meningioma. Postoperatively the patient did well, and she had no further problems.

CASE 12.3 JAW NUMBNESS AND EPISODES OF LOSS OF CONSCIOUSNESS

MINICASE

A 24-year-old woman was admitted to the cardiology service after an episode of syncope. Upon further probing, it was found that the patient had had three prior episodes of loss of consciousness over recent years, during which she was **unresponsive for a few minutes**, had some poorly described **shaking movements**, **and was then confused for up to a half an hour**. On review of systems, the patient described a patch of numbness over her left jaw that had been present for approximately 2 years. Exam was normal except for **decreased light touch, pinprick, and temperature sensation in the left jaw and lower face** (Figure 12.22).

LOCALIZATION AND DIFFERENTIAL DIAGNOSIS

1. On the basis of the symptoms and signs shown in **bold** above, where is the lesion?

2. What is the most likely diagnosis, and what are some other possibilities?

FIGURE 12.22 Region of Decreased Sensation

Discussion

1. The key symptoms and signs in this case are:

- **Decreased light touch, pinprick, and temperature sensation in the left jaw and lower face**

- **Episodes of unresponsiveness lasting for a few minutes, with shaking movements, followed by confusion for up to a half an hour**

The patient's sensory loss was in the distribution of the mandibular division of the trigeminal nerve (CN V_3; see Figure 12.7B). A lesion near the foramen ovale or of the mandibular division of CN V could, therefore, explain this deficit (see Figures 12.3A, 12.7A). Brief episodes of unresponsiveness can have numerous causes (see KCC 10.3; Table 10.2). Over 90% of cases are non-neurologic in origin and are caused by transient hypotension (vasovagal syncope), cardiac arrhythmias, or other medical conditions. However, patients with cardiogenic syncope typically recover immediately after the episode ends. Persistent deficits, such as the confusion seen in our patient, suggest a neurologic cause such as seizures (see KCC 18.2), vertebrobasilar transient ischemic attack (see KCC 10.3, 14.3), or vertebrobasilar migraine (see KCC 5.1). The shaking reported in this patient is suggestive of seizures, although a better description would have been helpful. One way to unify this patient's findings into a single diagnosis would be to postulate a mass lesion near the left foramen ovale that extends to the adjacent left medial temporal lobe, causing seizures. We will see in Chapter 18 that the limbic structures of the temporal lobes are especially prone to epileptic seizures.

2. Possible causes of a lesion in the vicinity of the foramen ovale and medial temporal lobe include metastases, meningioma, trigeminal neuroma, aneurysm of the petrous segment of the internal carotid artery, or sarcoidosis (see KCC 12.1, 12.2).

Clinical Course and Neuroimaging

The patient underwent both an **MRI** and a **CT scan** of the head (Image 12.3A–C, pages 545–546). Image 12.3A is an axial proton density–weighted MRI image showing a roundish mass compressing the left medial temporal lobe, lying in the path of CN V in Meckel's cave. Image 12.3B is a coronal T1-weighted MRI image with gadolinium showing enhancement of the mass and extension downward through the foramen ovale. The "dumbbell" shape of this mass, extending through a bony foramen, is typical of a schwannoma (see KCC 12.5). Image 12.3C is an axial CT scan image, using bone windows to demonstrate erosion of the mass through the temporal bone in the region of the left foramen ovale. The mass appeared to lie outside the substance of the brain and was felt to represent a schwannoma (trigeminal neuroma), meningioma, or giant aneurysm. The patient was started on anticonvulsant medications and an angiogram was done, but no aneurysm was visualized. Therefore, she underwent a left frontotemporal craniotomy, and a tannish white mass was identified under the left temporal lobe. The tumor was carefully removed in a 10-hour operation, with care taken not to damage adjacent cranial nerves or blood vessels. Pathologic examination was consistent with a schwannoma. Postoperatively the patient made an excellent recovery, with no further seizures, but she had persistent numbness of the left jaw.

CASE 12.4 ISOLATED FACIAL WEAKNESS

MINICASE

A 26-year-old woman developed **pain behind her left ear one evening**. When she looked in the mirror the next morning, she noticed that her **left face was drooping**. In addition, her **left ear was sensitive to loud sounds**. She saw her physician, who gave her some medication for the pain, but over the next 2 days her **left eye developed a "scratchy" painful sensation**, so she came to the emergency room. Exam was notable for marked **left facial weakness, including the forehead**. Taste was not tested. The remainder of the exam was normal.

LOCALIZATION AND DIFFERENTIAL DIAGNOSIS

1. On the basis of the symptoms and signs shown in **bold** above, where is the lesion?

2. What is the most likely diagnosis, and what are some other possibilities?

Discussion

1. The key symptoms and signs in this case are:

   • **Left retroauricular pain, hyperacusis, and facial weakness including the forehead**

   • **Painful, scratchy sensation in left eye**

   This patient had lower motor neuron–type facial weakness (see Figure 12.13), together with hyperacusis and retroauricular pain on the left side. These findings are compatible with a lesion of the left facial nerve affecting branchial motor and general somatic sensory function (see Table 12.4; Figures 12.7B, 12.10).

   The painful, scratchy eye is a bit of a puzzle. However, patients with a facial nerve lesion may have parasympathetic involvement (see Figure 12.10) causing decreased lacrimation; also, they are often not able to completely close the affected eye, especially while they are sleeping, which can lead to corneal desiccation and corneal ulcers.

CASE 12.3 (continued)

(C)

Erosion of mass
through foramen ovale

Foramen ovale

Foramen spinosum

Foramen magnum

R   L

CASE 12.4 *RELATED CASE*

IMAGE 12.4A–D **Left Temporal Bone Fracture in Region of Facial Canal** Reconstructed sagittal CT scan images through the left temporal bone, with (A) through (D) progressing from medial to lateral.

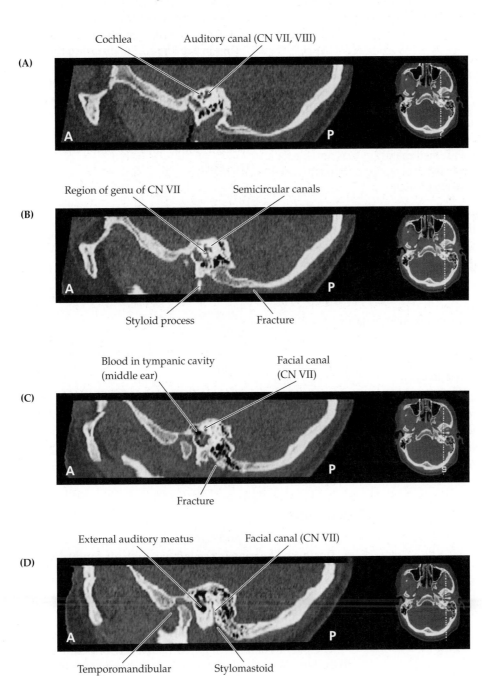

CASE 12.5 HEARING LOSS AND DIZZINESS

CHIEF COMPLAINT

A 41-year-old woman was referred to an otolaryngologist for **dizziness and progressive hearing loss in the left ear**.

HISTORY

One year ago the patient began having episodes of **mild dizziness, which felt like the room was spinning when she moved her head.** Two months ago she noticed **greatly reduced hearing in her left ear**, making it impossible to use the telephone receiver unless it was on her right ear. In addition, she had some **left facial pain** and **decreased taste on the left side of her tongue.** Past medical history was notable for a melanoma resected from the right hip region 6 months previously, with one positive lymph node.

PHYSICAL EXAMINATION

Vital signs: T = 99.1°F, P = 72, BP = 110/80, R = 12.

Ears: Normal otoscopic exam of the external auditory canals and tympanic membranes.

Neck: Supple.

Lungs: Clear.

Heart: Regular rate with no murmurs or gallops.

Abdomen: Benign.

Extremities: No edema.

Dermatologic: No skin lesions.

Neurologic exam:

  MENTAL STATUS: Alert and oriented × 3. Mildly anxious, but otherwise normal.

  CRANIAL NERVES: Pupils equal round and reactive to light. Normal fundi. Visual fields full. Extraocular movements intact. Facial sensation intact to light touch, but **decreased corneal reflex on the left.** Face symmetrical. **Hearing greatly diminished to finger rub or whispering in the left ear. A vibrating tuning fork sounded louder when held just outside the left ear than when the handle was touched to the left mastoid process (air conduction greater than bone conduction).** Taste was not tested. Voice and palate elevation normal. Sternomastoid strength normal. Tongue midline.

MOTOR: Normal tone. 5/5 power throughout.

REFLEXES:

COORDINATION: Normal on finger-to-nose and heel-to-shin testing.

GAIT: Normal.

SENSORY: Intact pinprick, vibration, and joint position sense.

LOCALIZATION AND DIFFERENTIAL DIAGNOSIS

1. On the basis of the symptoms and signs shown in **bold** above, where is the lesion?

2. What is the most likely diagnosis, and what are some other possibilities?

Discussion

1. The key symptoms and signs in this case are:

   • **Decreased hearing in the left ear, with air conduction greater than bone conduction**

   • **Episodes of mild dizziness**

   • **Left facial pain and decreased left corneal reflex**

   • **Decreased taste on the left side of the tongue**

   The patient had hearing loss with a sensorineural pattern that localized to the left cochlea or the left vestibulocochlear nerve (see KCC 12.5). Episodic dizziness can be caused by dysfunction anywhere in the pathways of vestibular sensation, including the labyrinth, vestibular ganglia, CN VIII, vestibular nuclei, cerebellum, or parietal cortex (see KCC 12.6). Given the sensorineural hearing loss, however, the dizziness is probably also caused by a problem in the left inner ear or CN VIII. Similarly, left facial pain (CN V), decreased corneal reflex (CN V_1 or CN VII; see KCC 12.4), and decreased taste (CN VII) could each result from peripheral lesions of the respective cranial nerves or from lesions in the left brainstem. Since unilateral hearing loss must be caused by a lesion outside of the brainstem (see KCC 12.5), the most parsimonious explanation is a lesion in the left cerebellopontine angle, where CN V, VII, and VIII all lie in close proximity (see Figure 12.2A,C).

The most likely *clinical localization* is CN V, VII, and VIII in the left cerebellopontine angle.

2. The most common lesion of the cerebellopontine angle is acoustic neuroma (see KCC 12.5). Our patient in this case recently had a melanoma, so metastases should also be considered, especially since melanoma often metastasizes to the brain. Other, less likely possibilities include meningioma, epidermoid, and glioma. Meniere's disease (see KCC 12.6) could account for hearing loss and dizziness, but not for this patient's abnormalities of CN V and VII.

Clinical Course and Neuroimaging

The otolaryngologist ordered a **brain MRI** with gadolinium and special thin cuts through the region of the internal auditory canal (Image 12.5A,B, page 551). In these T1-weighted images, an enhancing mass can be seen in the left cerebellopontine angle. The mass appears to lie entirely outside of the brainstem and has a lateral knob extending into the left internal auditory meatus in the petrous portion of the temporal bone. These findings are highly suggestive of an acoustic neuroma (vestibular schwannoma; see KCC 12.5).

The patient was referred to a neurosurgeon and admitted for removal of the tumor. As is often the case with this kind of surgery, the procedure was a collaboration between neurosurgery and otolaryngology. The left occipital bone was opened behind the transverse sinus, the dura was opened, and the left cerebellar hemisphere was gently retracted to reveal the tumor. The tumor was carefully dissected away from the adjacent cerebellum; pons; CN V, VII, IX, and X; and branches of the posterior inferior cerebellar artery (see Figure 15.2). The functioning of the facial nerve was monitored continuously during the resection by use of a stimulating electrode placed on CN VII and by EMG (electromyography; see KCC 9.2) leads placed in the orbicularis oculi and labial muscles. Thus, although the facial nerve was severely distorted by the tumor, its function was preserved. CN VIII, however, was sacrificed because it was completely encapsulated by tumor, resulting in unilateral deafness. The pathology report confirmed schwannoma. Post-operatively the patient suffered from vertigo (see KCC 12.6) and had nystagmus at rest for 1 to 2 days, which then resolved. She also had complete left facial paralysis that resolved over the course of several months, and she subsequently did well.

CASE 12.6 HOARSE VOICE FOLLOWING CERVICAL DISC SURGERY

MINICASE

A 38-year-old saleswoman developed left neck and shoulder pain and evaluation revealed a cervical disc herniation, for which she underwent a discectomy and fusion via an anterior approach through the neck (see KCC 8.5). Her symptoms of cervical radiculopathy resolved. However, in the recovery room following surgery she noticed a **marked change in her voice**, which now had a **breathy, "hoarse"** quality. She was reassured that this was a temporary effect of the intubation. Nevertheless, over the next 2 months she continued to have severe breathiness of her voice, making it difficult for her to do her work as a personal shopper. She was referred to an otolaryngologist for evaluation. Conventional neurological exam was normal, aside from the breathiness. Her voice had a soft breathy quality suggesting that an air leak was present in her larynx.

LOCALIZATION AND DIFFERENTIAL DIAGNOSIS

On the basis of the symptoms and signs shown in **bold** above, where is the lesion, and what is the most likely cause?

Discussion

The key symptoms and signs in this case are:

- **Breathy, "hoarse" voice**

Breathiness of the voice (often called hoarseness, although this is not strictly accurate) can be caused by any disorder that prevents complete closure of the vocal folds (true vocal cords) during phonation (see KCC 12.8). Incomplete vocal cord closure can be caused by lesions anywhere in the pathway from the nucleus ambiguus, to the vagus nerve (CN X), to the recurrent laryngeal nerve, to the muscles of the larynx (see Figure 12.21). Given this patient's history of surgery on the left side of her neck, the most likely diagnosis is stretch injury or laceration of the left recurrent laryngeal nerve. Note that injury to the superior laryngeal nerve does not usually cause noticeable deficits, since it only supplies the cricothyroid muscle (a subtle deficit in reaching high notes is occasionally noted by professional singers), and injury of the vagus itself is uncommon during neck surgery because of its deeper location.

Clinical Course and Videostroboscopic Imaging

To confirm the diagnosis, the otolaryngologist performed fiberoptic video imaging of the larynx using a laryngoscope, which can be inserted through the nose or mouth (Image 12.6A–J, page 552). The process of stroboscopy matches the phonatory frequency to a strobe light, and by offsetting the phase slightly, it gives the illusion of slow-motion vibratory cycles of the true vocal cords. This process demonstrated normal movement of the right cord during phonation and during breathing. However, the left cord was paralyzed, and remained in an abducted position. Thus, her left-cord paralysis resulted in incomplete closure of the glottis during phonation and caused this patient's breathiness.

Although recurrent laryngeal nerve injuries sometimes recover over time, this patient was eager to have the problem fixed immediately because of the severity of her deficit, its duration, and the importance of her voice for her work. Therefore, she underwent a procedure in which a precisely carved silastic insert was placed into the left paraglottic space. The insertion was performed while visualizing the cords and testing voice quality until the left cord was restored to a sufficiently medial position to allow normal approximation of the cords during phonation. Follow-up over time showed that her left recurrent laryngeal nerve injury was indeed permanent. However, the procedure enabled an immediate and complete recovery of her normal voice.

CASE 12.5 HEARING LOSS AND DIZZINESS

IMAGE 12.5A,B Left Acoustic Neuroma (Vestibular Schwannoma) Axial T1-weighted MRI images with intravenous gadolinium contrast. (A) and (B) are adjacent sections progressing from inferior to superior.

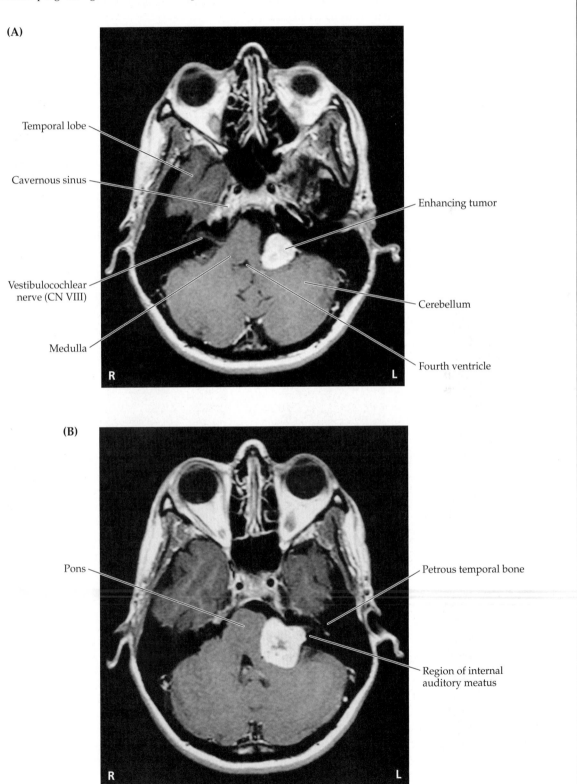

(A)

Temporal lobe

Cavernous sinus

Enhancing tumor

Vestibulocochlear nerve (CN VIII)

Cerebellum

Medulla

Fourth ventricle

R   L

(B)

Pons

Petrous temporal bone

Region of internal auditory meatus

R   L

CASE 12.6 HOARSE VOICE FOLLOWING CERVICAL DISC SURGERY

IMAGE 12.6A–J Left Vocal Cord Paralysis Videostroboscopic images through the laryngoscope during a cycle of phonation (adduction) and inspiration (abduction). (A–G) During phonation, the right cord adducts medially, while the left cord remains immobile, causing an air leak. (H–J) During inspiration, the right cord abducts laterally, while the left cord remains immobile. (Courtesy of Michael Goldrich, Robert Wood Johnson Medical School, University of Medicine and Dentistry of New Jersey.)

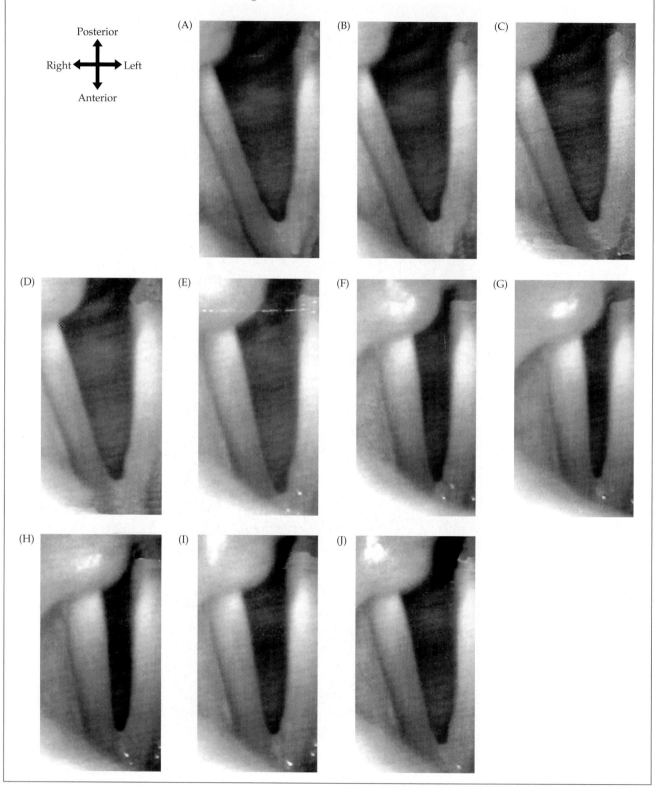

CASE 12.7 HOARSENESS, WITH UNILATERAL WASTING OF THE NECK AND TONGUE MUSCLES

CHIEF COMPLAINT

A 34-year-old man was referred to an otolaryngologist for progressive hoarseness, dysphagia, and weakness of the left sternomastoid and tongue.

HISTORY

Four months prior to presentation, the patient developed a persistent cough and respiratory infection that did not resolve. Soon afterward he noticed **difficulty swallowing** thick foods, and his voice gradually became **hoarse**. Three weeks prior to presentation, he began to have **decreased hearing in the left ear**, some **alteration in taste**, and mild **left-sided headache**. He lost 40 pounds during the 4 months since developing symptoms.

PHYSICAL EXAMINATION

Vital signs: T = 98.1°F, P = 84, BP = 118/86, R = 18.

Neck: Supple; no adenopathy or palpable masses.

Lungs: Clear.

Heart: Regular rate with no gallops or murmurs.

Abdomen: Soft, nontender.

Extremities: Normal.

Neurologic exam:

   MENTAL STATUS: Alert and oriented × 3. Normal language. Gave detailed history.

   CRANIAL NERVES: Pupils 4 mm, constricting to 2 mm bilaterally. Visual fields full. Normal optic discs. Extraocular movements intact. Facial sensation intact to light touch and pinprick. Intact corneal reflexes. **Mildly decreased left nasolabial fold. Decreased hearing to finger rub on the left.** Gag intact. **Uvula deviated to the right with palate elevation. Voice hoarse and breathy in quality. Left trapezius and sternomastoid muscles had fasciculations and strength of 4/5. Tongue had marked asymmetrical atrophy and fasciculations of the left side, with tongue deviating to the left on protrusion. On laryngoscopic examination, the left vocal cord was paralyzed** (see Case 12.6).

MOTOR: No pronator drift. 5/5 power throughout.

REFLEXES:

COORDINATION: Normal on finger-to-nose and heel-to-shin testing.

GAIT: Normal.

SENSORY: Intact light touch, pinprick, and joint position sense.

LOCALIZATION AND DIFFERENTIAL DIAGNOSIS

1. On the basis of the symptoms and signs shown in **bold** above, where is the lesion?

2. What are some possible lesions in this location?

Discussion

1. The key symptoms and signs in this case are:

   - **Difficulty swallowing, decreased left palate movement, hoarseness, and left vocal cord paralysis**

   - **Left trapezius and sternomastoid weakness and fasciculations**

   - **Left tongue deviation, atrophy, and fasciculations**

   - **Decreased hearing in the left ear**

   - **Mildly decreased left nasolabial fold**

   - **Alteration in taste**

   - **Left-sided headache**

   This patient has multiple abnormalities of the cranial nerves on the left side of the head. Although each individual abnormality could be explained by a small brainstem lesion, all of the relevant nuclei could not be involved together without also involving other nearby structures, such as the anterolateral system, inferior cerebellar peduncle, and descending sympathetic pathway (see Figure 14.21D). In addition, as in Case 12.5, the unilateral hearing loss suggests that the lesion lies outside of the brainstem (see KCC 12.5).

Taking each of the above deficits in turn, the swallowing muscles of the pharynx and the left palate are innervated by the left CN X, although CN IX may contribute to the gag reflex as well. A lesion of the left CN X could also explain hoarseness (breathiness) and left vocal cord paralysis, since the larynx is innervated by the vagus as well. Left trapezius and sternomastoid weakness and fasciculations suggest a lower motor neuron lesion (see KCC 6.1) of the left spinal accessory nerve (CN XI). Similarly, deviation of the tongue to the left, together with atrophy and fasciculations, suggests a lower motor neuron lesion of the left hypoglossal nerve (CN XII). Decreased hearing in the left ear can be caused by a lesion in the left external auditory canal, middle ear, cochlea, or vestibulocochlear nerve (CN VIII). Although a decreased left nasolabial fold could be caused by upper motor neuron– or mild lower motor neuron–type weakness, given the other findings, a peripheral lesion of the left facial nerve (CN VII) is more likely. A facial nerve lesion could also explain the alteration in taste. Unilateral headaches can have many causes (see KCC 5.1), but in this setting they support the presence of an intracranial lesion on the left side of the head.

To summarize, this lesion involves CN VII, VIII, IX, X, XI, and XII on the left side. These cranial nerves exit the left lower brainstem and leave the cranium via the internal auditory meatus, jugular foramen, and hypoglossal canal (see Figures 12.2A,C; 12.3A,B; Table 12.2). Note that large lesions of the cerebellopontine angle usually involve CN V (see KCC 12.5 and Case 12.5). Since CN V was spared in this case, it suggests that the lesion lies farther down.

The most likely *clinical localization* is a large lesion lying just outside of the left ventrolateral medulla or in the vicinity of the left internal auditory meatus, jugular foramen, and hypoglossal canal.

2. Possible lesions in this location include meningioma, schwannoma, metastases, granulomatous disease, and glomus tumors (see KCC 12.7).

Clinical Course and Neuroimaging

A **brain MRI** with gadolinium was ordered (Image 12.7A,B, page 557). Image 12.7A shows an axial T1-weighted image demonstrating an enhancing mass filling the left jugular foramen, just behind the left internal carotid artery. Note that the mass extends into the posterior fossa near the region where the left hypoglossal nerve exits the medulla to traverse the subarachnoid space. Image 12.7B is a coronal T2-weighted image revealing that the mass extends up into the left petrous temporal bone to reach the vicinity of the left seventh and eighth cranial nerves. These findings are compatible with a left glomus jugulare tumor (see KCC 12.7).

The patient was initially treated with radiation therapy, resulting in no progression of his deficits. He was actively involved in a speech therapy program to help him work on his voice and swallowing. Four years later, however, he began to have worsening hoarseness, severe left facial weakness, and left mastoid pain. Since his tumor appeared to be growing, surgery was planned. Because glomus tumors tend to be very vascular and can bleed profusely during surgery, he had a preoperative angiogram during which an interventional neuroradiologist (see Chapter 4) embolized as much of the tumor as possible. This procedure was followed by a prolonged operation involving collaboration between two teams of neurosurgeons and otolaryngologists, who accomplished a complete removal of all visible tumor. Unfortunately, on pathologic analysis, rather than the usual benign appearance of glomus tumors, this lesion had malignant features including mitotic figures (indicating active cellular proliferation) and necrosis. He remained stable until 1 year later, when he had worsening left ear pain and an MRI showed recurrence of the tumor. Again, he underwent embolization followed by surgery. However, soon afterward he developed aspiration pneumonia with overwhelming sepsis, and he died five and a half years after the onset of his initial symptoms.

CASE 12.8 UNCONTROLLABLE LAUGHTER, DYSARTHRIA, DYSPHAGIA, AND LEFT-SIDED WEAKNESS*

CHIEF COMPLAINT

A 27-year-old male saxophone player came to the emergency room because of worsening dysarthria, dysphagia, left-sided weakness, and episodes of uncontrollable laughter.

HISTORY

Two and a half years prior to presentation, the patient developed episodes of **left face and mouth pain** precipitated by chewing. One year prior to presentation, he started having **episodes of uncontrollable laughter**, not accompanied by appropriate affect. When he persisted in laughing repeatedly at his girlfriend's father's wake, he was referred to a psychiatrist, who tried behavior modification therapy without benefit. Two to 3 months prior to presentation, he developed increasing difficulty playing the saxophone, and he noticed **slurred speech** and occasional **choking on his food**. He found he also had **an unstable gait, bumping into objects on his left side, difficulty buttoning his shirt with his left hand, and urinary urgency and difficulty initiating urination**.

PHYSICAL EXAMINATION

Vital signs: T = 98°F, P = 72, BP = 130/70, R = 12.

Neck: Supple with no bruits.

Lungs: Clear.

Heart: Regular rate with no gallops or murmurs (but difficult exam because of frequent laughter).

Abdomen: Soft, nontender.

Extremities: Normal.

Neurologic exam:

MENTAL STATUS: Alert and oriented × 3. Recalled 3/3 objects after 5 minutes. Normal language. Normal constructions. Denied depression or anxiety.

CRANIAL NERVES: Pupils 4 mm, constricting to 2 mm bilaterally. Visual fields full. Normal optic discs.

Extraocular movements intact. **Mildly decreased left nasolabial fold.** Normal hearing to finger rub bilaterally. **Absent gag reflex. Speech mildly dysarthric. Had intermittent bouts of uncontrollable laughter about once per minute without accompanying emotion. Mild weakness of head turning to the left.** Tongue midline, with no fasciculations.

MOTOR: **Mild left pronator drift. Slowed finger tapping in the left hand. Tone slightly increased in left lower extremity. Power 4/5 in left deltoid, triceps, wrist extensors, finger extensors, iliopsoas, hamstrings, tibialis anterior, and extensor hallucis longus,** but otherwise 5/5 throughout.

REFLEXES:

COORDINATION: Finger-to-nose and heel-to-shin testing slowed, but without ataxia.

GAIT: **Slightly unsteady, with stiff left lower extremity.**

SENSORY: Intact light touch, pinprick, temperature, vibration, and joint position senses and graphesthesia.

LOCALIZATION AND DIFFERENTIAL DIAGNOSIS

1. On the basis of the symptoms and signs shown in **bold** above, which nerves (see Table 12.4) and which long tracts (see Chapters 6 and 7) are affected by the lesion?

2. In what general region of the nervous system can a single lesion produce all of these findings?

3. What are some possible lesions in this location?

*This patient was described previously as a case report by Shafqat et al. *Neurology.* 1998. 50: 1918–1919.

Discussion

1. The key symptoms and signs in this case are:

- **Episodes of left face and mouth pain**
- **Episodes of uncontrollable laughter not accompanied by appropriate affect**
- **Dysarthria, dysphagia, absent gag reflex**

- Mild weakness of head turning to the left
- Left face arm and leg weakness, with increased tone, hyperreflexia, and gait unsteadiness
- Urinary urgency and difficulty with initiation

The episodes of left face and mouth pain precipitated by chewing were initially suggestive of a trigeminal nerve (CN V) disorder, such as trigeminal neuralgia (see KCC 12.2). However, later findings suggest a lesion affecting central nervous system pathways. These include the development of pseudo-bulbar affect, suggesting a lesion of the corticobulbar pathways (see KCC 12.8), left hemiparesis with upper motor neuron signs compatible with corticobulbar and corticospinal dysfunction (see KCC 6.3), and urinary dysfunction, also compatible with impairment of descending pathways controlling micturition (see KCC 7.5). Dysarthria, dysphagia, and absent gag reflex (CN IX, X) in this patient, along with impaired CN XI function, further support a combination of cranial nerve dysfunction along with involvement of long tracts. These findings suggest a lesion affecting CN V, IX, X, and XI, as well as corticobulbar, corticospinal, and descending sphincteric pathways.

2. A lesion of the brainstem in the region of the pons and medulla could affect these multiple cranial nerves and long tracts. A lesion affecting this many brainstem structures while preserving other brainstem nuclei and pathways would have to be fairly extensive, yet cause patchy involvement.

3. Given the gradual onset of symptoms over several years involving multiple brainstem structures, one possibility would be multiple sclerosis (see KCC 6.6) affecting primarily the brainstem. Other possibilities include a brainstem vascular malformation (see KCC 5.6), a granulomatous disorder such as sarcoidosis (see KCC 12.1), or a slow-growing tumor such as a brainstem glioma or meningioma (see KCC 5.8).

Clinical Course and Neuroimaging

The patient underwent a head CT scan in the emergency room revealing a mass lesion, which was better visualized by **MRI scan** (Image 12.8A,B, page 558). Note the presence of a large mass lying outside of the brain adjacent to the dura and enhancing uniformly with gadolinium, consistent with a meningioma (see KCC 5.8). The mass could be seen to cause severe compression and distortion of the pons and left middle cerebellar peduncle (see Image 12.8A). The patient's relatively mild deficits given this degree of distortion attest to the chronic nature of this lesion. The mass could also be seen to extend into the region of Meckel's cave adjacent to the left cavernous sinus (see Image 12.8B), possibly explaining the patient's early symptoms of left facial pain. The patient underwent a multistage resection, involving preoperative embolization by interventional radiology, and two collaborative operations involving teams of neurosurgeons and otolaryngologists. He made an excellent recovery with minimal deficits. On follow-up examination 1 year later (Image 12.8C, page 559), he still had rare episodes of inappropriate laughter, and he had some mild diplopia that he had developed following surgery, but he was otherwise without deficits. Repeat MRI scan showed near complete removal of the tumor (see Image 12.8C), with only a small portion left where it was adherent to CN IV.

CASE 12.7 HOARSENESS, WITH UNILATERAL WASTING OF THE NECK AND TONGUE MUSCLES

IMAGE 12.7A,B **Left Glomus Jugulare Tumor** (A) Axial T1-weighted MRI image with gadolinium. (B) Coronal T2-weighted MRI image.

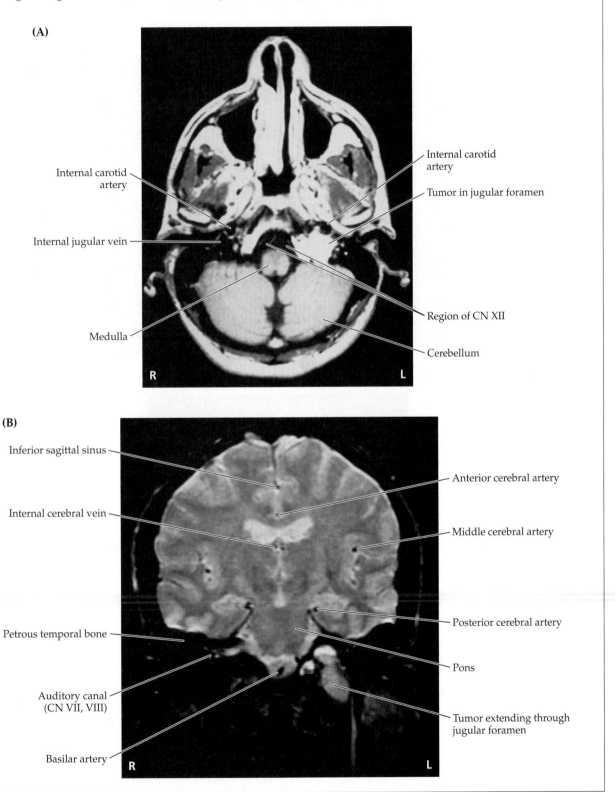

(A)

Internal carotid artery
Internal jugular vein
Medulla

Internal carotid artery
Tumor in jugular foramen
Region of CN XII
Cerebellum

R   L

(B)

Inferior sagittal sinus
Internal cerebral vein
Petrous temporal bone
Auditory canal (CN VII, VIII)
Basilar artery

Anterior cerebral artery
Middle cerebral artery
Posterior cerebral artery
Pons
Tumor extending through jugular foramen

R   L

CASE 12.8 UNCONTROLLABLE LAUGHTER, DYSARTHRIA, DYSPHAGIA, AND LEFT-SIDED WEAKNESS

IMAGE 12.8A–C Meningioma Compressing the Pons T1-weighted MRI images with intravenous gadolinium enhancement. (A) Sagittal view. (B) Axial view. (C) Follow-up MRI axial view 1 year after surgery.

(A)

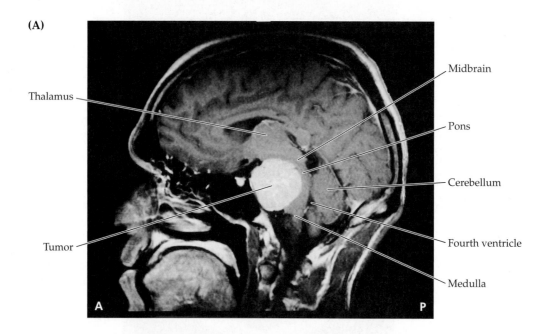

Thalamus

Midbrain

Pons

Cerebellum

Tumor

Fourth ventricle

Medulla

(B)

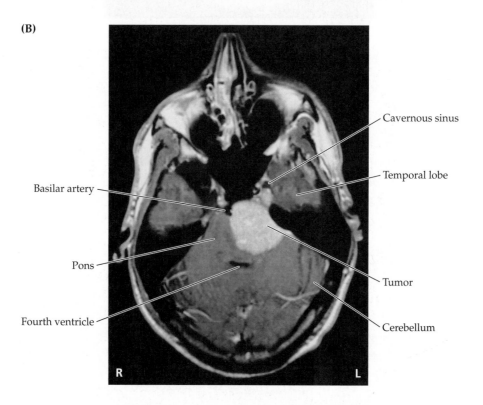

Cavernous sinus

Temporal lobe

Basilar artery

Pons

Tumor

Fourth ventricle

Cerebellum

CASE 12.8 (continued)

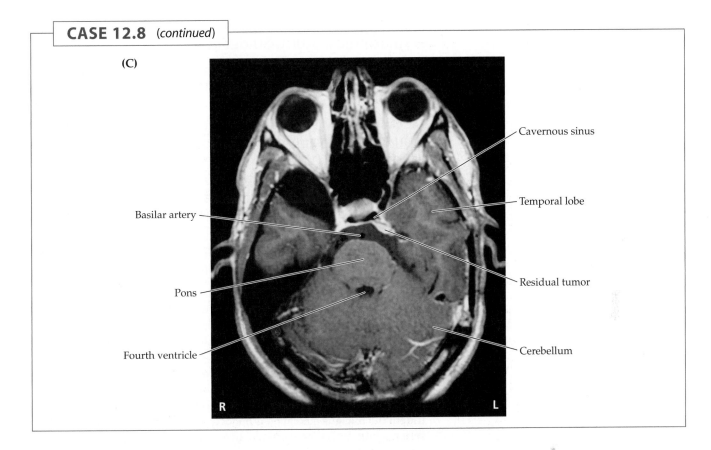

(C)

Cavernous sinus

Temporal lobe

Basilar artery

Residual tumor

Pons

Fourth ventricle

Cerebellum

R   L

Additional Cases

Related cases can be found in other chapters for: **upper or lower motor neuron cranial nerve disorders** (Cases 5.2, 5.3, 5.7, 5.8, 6.3, 6.5, 10.4, 10.5, 10.11, 11.1, 11.2, 13.1–13.3, 13.5, 14.1, 14.4, 14.7, 15.4, 17.2). Other relevant cases can be found using the **Case Index** located at the end of the book.

Brief Anatomical Study Guide

1. The main parts of the **brainstem** are the **midbrain**, **pons**, and **medulla**, as shown in Figure 12.1. The **cranial nerves** (see Table 12.1) exit the brainstem roughly in numerical sequence from rostral to caudal (see Figure 12.2), except for CN I and CN II, which arise from the forebrain. Each cranial nerve exits the skull through a specific foramen, as summarized in Table 12.2 and Figure 12.3.

2. As in the spinal cord gray matter, the cranial nerve nuclei for **motor functions are located more ventrally** in the brainstem, and those for **sensory functions are located more dorsally** (see Figure 12.4).

3. Along the long axis of the brainstem are **three motor columns** and **three sensory columns** of cranial nerve nuclei (see Figures 12.4 and 12.5). The functions, nuclei, and cranial nerves for each of these columns are summarized in Table 12.3. Note that each nucleus can be involved in the motor or sensory functions of one or more cranial nerves. Similarly, each cranial nerve can have both sensory and motor functions, as well as connections with one or more cranial nerve nuclei (see Table 12.4).

Brief Anatomical Study Guide (continued)

4. Some cranial nerves have peripheral **ganglia** that contain either primary sensory neurons or parasympathetic postganglionic neurons (see Table 12.5; Figure 12.6).

5. The **olfactory nerves** (**CN I**) (see Figures 18.5 and 18.6) traverse the cribriform plate of the ethmoid bone to synapse in the **olfactory bulbs** (see Figure 12.3A). Olfactory information then travels to the olfactory cortex (see Chapter 18) via the olfactory tracts.

6. The **optic nerve** (**CN II**) enters the cranial cavity via the optic canal (see Figure 12.3A,C), carrying visual information from the retina to the lateral geniculate nucleus and extrageniculate pathways (see Chapter 11).

7. The **oculomotor** (**CN III**), **trochlear** (**CN IV**), and **abducens** (**CN VI**) nerves are involved in eye movements and pupillary control (CN III) and are discussed at length in Chapter 13.

8. The **trigeminal nerve** (**CN V**) provides sensation for the face, mouth, and meninges of the supratentorial cranial cavity via three major branches: the **ophthalmic division** (V_1), **maxillary division** (V_2), and **mandibular division** (V_3) (see Figure 12.7). These branches enter the skull via the **superior orbital fissure**, **foramen rotundum**, and **foramen ovale**, respectively (see Figures 12.3A, 12.7), with primary sensory neurons located in the **trigeminal ganglion**. The **trigeminal sensory nuclei** include the **mesencephalic trigeminal nucleus** mediating proprioception, the **chief trigeminal sensory nucleus** mediating discriminative touch, and the **spinal trigeminal nucleus** mediating pain and temperature sensation (see Figures 12.5, 12.8; Table 12.6). Trigeminal sensory information travels to the cortex via the **trigeminal lemniscus** and **trigeminothalamic tract**, with a relay in the **VPM** of the thalamus (see Figure 12.8; Table 12.6). The trigeminal nerve also has a small **motor root** that travels with CN V_3, supplying the **muscles of mastication** (see Figure 12.7).

9. The **facial nerve** (**CN VII**) controls the muscles of facial expression (to be distinguished from the trigeminal nerve that mediates facial sensation) via fibers that arise from the **facial nucleus** in the pons (see Figure 12.11). The facial nerve travels along with CN VIII in the auditory canal and then exits the skull via the stylomastoid foramen (see Figures 12.3B and 12.10). **Upper motor neuron control** of the facial nucleus is **bilateral for the upper parts of the face**, so in unilateral upper motor lesions the contralateral side can compensate, resulting in sparing of the upper face muscles (see Figure 12.13). The facial nerve also has sensory fibers that provide taste sensation for the anterior two-thirds of the tongue reaching the nucleus solitarius (see Figure 12.12) and somatic sensation fibers for a region around the outer ear traveling to the trigeminal nuclei (see Figure 12.7B). Sensory cell bodies lie in the **geniculate ganglion**. Parasympathetics arising from the superior salivatory nucleus travel in CN VII via the sphenopalatine ganglion and submandibular ganglion, respectively, to the lacrimal glands and salivary glands (see Figure 12.6).

10. The **vestibulocochlear nerve** (**CN VIII**) carries auditory information from the cochlea to the **dorsal and ventral cochlear nuclei** (see Figures 12.15–12.17). The primary sensory cell bodies lie in the spiral ganglion. Central auditory pathways cross multiple times, so unilateral lesions in the

central nervous system do not cause clinically significant unilateral hearing loss (see Figure 12.16). Information about head position and acceleration is carried by the vestibular portions of CN VIII arising from the **semicircular canals and otolith organs** (see Figure 12.15). Primary cell bodies are in the vestibular ganglia, and this information travels to the **vestibular nuclei** in the brainstem to influence unconscious posture and balance, eye movements, and conscious perception of movement through multiple pathways (see, e.g., Figure 12.19).

11. The **glossopharyngeal nerve** (**CN IX**) exits the skull via the jugular foramen (see Figures 12.3A,B, 12.20; Table 12.2). Motor fibers arising from the **nucleus ambiguus** provide innervation of the stylopharangeus muscle, important for pharynx elevation during speech and swallowing. Sensory fibers from chemoreceptors and baroreceptors in the carotid body reach the caudal nucleus solitarius (cardiorespiratory nucleus). Taste sensory fibers from the posterior one-third of the tongue travel to the rostral nucleus solitarius (gustatory nucleus). Somatic sensation from the posterior tongue, pharynx, middle ear, and external ear travels via CN IX to the trigeminal nuclei. Finally, parasympathetics arising from the inferior salivatory nucleus activate the parotid salivary gland via the otic ganglion.

12. The **vagus nerve** (**CN X**) also has multiple functions, providing parasympathetic innervation for the viscera arising from the **dorsal motor nucleus of CN X** (see Figures 12.5 and 12.21; see also Figure 14.5A,B). In addition, motor fibers of the vagus arising from the **nucleus ambiguus** supply the pharynx (swallowing) and larynx (voice). Sensory fibers from the aortic arch travel to the caudal nucleus solitarius. Sensory fibers for the pharynx, larynx, outer ear, and meninges of the posterior fossa travel to the trigeminal nuclei.

13. The **spinal accessory nerve** (**CN XI**) (see Figure 12.2A,C) arises from the **spinal accessory nucleus** (see Figure 12.5) and innervates the sternomastoid and upper portions of the trapezius muscles. Because of the mechanical attachments of the sternomastoid muscle, lesions of CN XI cause weakness of head turning to the side opposite the lesion.

14. The **hypoglossal nerve** (**CN XII**) (see Figure 12.2A,C) arises from the hypoglossal nucleus (see Figures 12.4 and 12.5) and supplies intrinsic tongue muscles. Hypoglossal nerve lesions cause the tongue to deviate toward the side of the lesion when the tongue is protruded.

References

General References

Bailey BJ (ed.), Johnson JT, Newlands SD, et al. 2006. *Head and Neck Surgery—Otolaryngology*. 4th Ed. Lippincott Williams & Wilkins, Philadelphia.

Cummings CW, Haughey BH, Thomas JR, et al. 2004. *Otolaryngology: Head and Neck Surgery*. 4th Ed. Mosby, New York.

Wilson-Pauwels L, Akesson EJ, Stewart PA. 1988. *Cranial Nerves: Anatomy and Clinical Comments*. B. C. Decker, Toronto, Ontario.

Winn RH (ed.). 2004. *Youmans Neurological Surgery*. 5th Ed., Vols. 1–4. Saunders, Philadelphia.

Cribriform and Suprasellar Meningiomas

Dehdashti AR, Ganna A, Witterick I, Gentili F. 2009. Expanded endoscopic endonasal approach for anterior cranial base and suprasellar lesions: indications and limitations. *Neurosurgery* 64 (4): 677–687.

Paterniti S, Fiore P, Levita A, La Camera A, Cambria S. 1999. Basal meningiomas. A retrospective study of 139 surgical cases. *J Neurosurg Sci* 43 (2): 107–113.

Central Nervous System Sarcoidosis

Joseph FG, Scolding NJ. 2007. Sarcoidosis of the nervous system. *Pract Neurol* 7 (4): 234–244.

Scully RE, Mark EJ, McNeely WF, Ebeling SH (eds.). 1996. Case records of the Massachusetts General Hospital. *N Engl J Med* 335: 1668–1674.

Vinas FC, Rengachary S. 2001. Diagnosis and management of neurosarcoidosis. *J Clin Neurosci* 8 (6): 505–513.

Trigeminal Nerve Lesions

Akhaddar A, El-Mostarchid B, Zrara I, Boucetta M. 2002. Intracranial trigeminal neuroma involving the infratemporal fossa: case report and review of the literature. *Neurosurgery* 50 (3): 633–637.

Avasarala J. 1997. Inflammatory trigeminal sensory neuropathy. *Neurology* 49 (1): 308.

Bartleson JD, Black DF, Swanson JW. 2007. Cranial and facial pain. In WG Bradley, RB Daroff, GM Fenichel, J Jankovic (eds.), *Neurology in Clinical Practice*. 5th Ed., Chapter 21. Butterworth-Heinemann, Boston.

Bordi L, Compton J, Symon L. 1989. Trigeminal neuroma: A report of eleven cases. *Surg Neurol* 31: 272–276.

Gibson RD, Cowan IA. 1989. Giant aneurysm of the petrous carotid artery presenting with facial numbness. *Neuroradiology* 31: 440–441.

Gronseth G, Cruccu G, Alksne J, Argoff C, Brainin M, Burchiel K, Nurmikko T, Zakrzewska JM. 2008. Practice parameter: the diagnostic evaluation and treatment of trigeminal neuralgia (an evidence-based review): report of the Quality Standards Subcommittee of the American Academy of Neurology and the European Federation of Neurological Societies. *Neurology* 71 (15): 1183–1190.

Morantz RA, Kirchner FR, Kishore P. 1976. Aneurysms of the petrous portion of the internal carotid artery. *Surg Neurol* 6: 313–318.

Obermann M, Katsarava Z. 2009. Update on trigeminal neuralgia. *Expert Rev Neurother* 9 (3): 323–329.

Facial Nerve Lesions

Adour KK. 2002. Decompression for Bell's palsy: why I don't do it. *Eur Arch Otorhinolaryngol* 259 (1): 40–47.

Adour KK, Wingerd J, Bell DN, Manning JJ, Hurley JP. 1972. Prednisone treatment for idiopathic facial paralysis (Bell's palsy). *N Engl J Med* 287: 1268–1272.

Gilden DH. 2004. Clinical practice. Bell's Palsy. *N Engl J Med* 351 (13): 1323–1331.

Gilden DH, Tyler KL. 2007. Bell's palsy—is glucocorticoid treatment enough? *N Engl J Med* 357 (16): 1653–1655.

Goudakos JK, Markou KD. 2009. Corticosteroids vs. corticosteroids plus antiviral agents in the treatment of Bell palsy: a systematic review and meta-analysis. *Arch Otolaryngol Head Neck Surg* 135 (6): 558–564.

Guerrissi JO. 1997. Facial nerve paralysis after intratemporal and extratemporal blunt trauma. *J Craniofac Surg* 8 (5): 431–437.

Madhok V, Falk G, Fahey T, Sullivan FM. 2009. Prescribe prednisolone alone for Bell's palsy diagnosed within 72 hours of symptom onset. *BMJ* 338: b255.

Marsh MA, Coker NJ. 1991. Surgical decompression of idiopathic facial palsy. *Otolaryngol Clin North Am* 24: 675–689.

Neimat JS, Hoh BL, McKenna MJ, Rabinov JD, Ogilvy CS. 2005. Aneurysmal expansion presenting as facial weakness: case report and review of the literature. *Neurosurgery* 56 (1): 190.

Sullivan, FM et al. 2007. Early treatment with prednisolone or acyclovir in Bell's palsy. *N Engl J Med* 357 (16): 1598–1607.

Tiemstra JD, Khatkhate N. 2007. Bell's palsy: diagnosis and management. *Am Fam Physician* 76 (7): 997–1002.

Dizziness and Vertigo

Boniver R. 2008. Benign paroxysmal positional vertigo: an overview. *Int Tinnitus J* 14 (2): 159–167.

Boyer FC, Percebois-Macadré L, Regrain E, Lévêque M, Taïar R, Seidermann L, Belassian G, Chays A. 2008. Vestibular rehabilitation therapy. *Neurophysiol Clin* 38 (6): 479–487.

Brandt T, Zwergal A, Strupp M. 2009. Medical treatment of vestibular disorders. *Expert Opin Pharmacother* 10 (10): 1537–1548.

Bronsetin AM, Lempert T (eds). 2007. *Dizziness: A Practical Approach to Diagnosis and Management*. Cambridge University Press, New York.

Chan Y. 2009. Differential diagnosis of dizziness. *Curr Opin Otolaryngol Head Neck Surg* 17 (3): 200–203.

Kerber KA. 2009. Vertigo and dizziness in the emergency department. *Emerg Med Clin North Am* 27 (1): 39–50.

Acoustic Neuroma

Batchelor T. 2007. Cancer in the the nervous system. In WG Bradley, RB Daroff, GM Fenichel, CD Marsden (eds.), *Neurology in Clinical Practice: Principles of Diagnosis and Management*. 5th ed., Vol. 2, Chapter 58. Butterworth-Heinemann, Boston.

Kondziolka D, Lunsford LD. 2008. Future perspectives in acoustic neuroma management. *Prog Neurol Surg* 21: 247–254.

Matthies C, Samii M. 1997. Management of 1000 vestibular schwannomas (acoustic neuromas): Clinical presentation. *Neurosurgery* 40: 1–9.

Pollock BE. 2008. Vestibular schwannoma management: an evidence-based comparison of stereotactic radiosurgery and microsurgical resection. *Prog Neurol Surg* 21: 222–227.

Glomus Jugulare

Ghani GA, Sung Y, Per-Lee JH. 1983. Glomus jugulare tumors—Origin, pathology, and anesthetic considerations. *Anesth Analg* 62: 686–691.

Jackson CG, Kaylie DM, Coppit G, Gardner EK. 2004. Glomus jugulare tumors with intracranial extension. *Neurosurg Focus* 17 (2): E7.

Ramina R, Maniglia JJ, Fernandes YB, Paschoal JR, Pfeilsticker LN, Neto MC, Borges G. 2004. Jugular foramen tumors: diagnosis and treatment. *Neurosurg Focus* 17 (2): E5.

CONTENTS

Chapter 13

Brainstem II: Eye Movements and Pupillary Control

Damage to the eye movement pathways can interfere with normal vision and can also affect pupil and eyelid control. A 48-year-old woman developed gradually worsening left eye pain and double vision over the course of 18 months. Her left pupil appeared dilated and did not constrict in response to light. Her left eye had limited upgaze, downgaze, and medial gaze but normal lateral gaze. In addition, her left upper eyelid drooped about 3 millimeters lower than the right one.

In this chapter we will learn about the brainstem circuits, nerves, and muscles involved in movement of the eyes, eyelids, and pupils, and about the effects of lesions or illness on these functions.

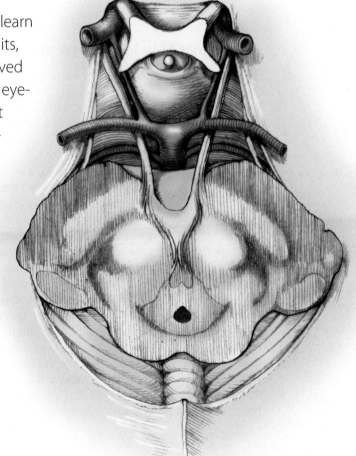

ANATOMICAL AND CLINICAL REVIEW

MOVEMENT OF OUR EYES AND PUPILS occurs continuously, mostly imperceptibly to us, and enables us to maximize the information derived from the relatively small visual area represented in the fovea. Abnormalities of the pupils and eye movements are often warning signs of pathology in the brainstem or cranial nerves and should therefore be evaluated carefully. In this chapter we will discuss the anatomy of both the **extraocular muscles** that cause the eyes to move within the orbits and the **internal ocular muscles** that control pupillary size and lens accommodation. We will also discuss common disorders affecting these systems. Eye movement disorders and pathways are often separated into two levels:

1. **Nuclear and infranuclear pathways** involve the brainstem nuclei of CN III, IV, and VI; the peripheral nerves arising from these nuclei; and the eye movement muscles.
2. **Supranuclear pathways** involve brainstem and forebrain circuits that control eye movements through connections with the nuclei of CN III, IV, and VI.

We will follow this bipartite organization in this chapter. First, we will discuss the peripheral course of CN III, IV, and VI; the muscles innervated by them; and the locations of their brainstem nuclei. Next, we will discuss the central and peripheral pathways involved in pupillary control. We then will discuss CNS supranuclear pathways that control extraocular movements through connections with the nuclei of CN III, IV, and VI. By understanding the anatomy of eye movements and pupillary control, we can often use the neurologic exam to localize a lesion in the central nervous system or the periphery, providing essential guidance to further diagnostic tests and therapeutic interventions.

Extraocular Muscles, Nerves, and Nuclei

The mechanical systems and information processing involved in eye movement control constitute a remarkable design in natural engineering. The muscles, nerves, and nuclei described in this section enable precise, smooth, and rapid eye movements to occur in a synchronized and coordinated fashion.

Extraocular Muscles

There are six **extraocular muscles** for each eye (Figure 13.1). The **lateral rectus**, **medial rectus**, **superior rectus**, and **inferior rectus** muscles move the eye laterally, medially, superiorly, and inferiorly, respectively (see Figure 13.1A). These muscles originate in a common tendinous ring at the orbital apex and insert onto the sclera. In addition to the simple horizontal and vertical eye movements performed by the rectus muscles, there are also **torsional** movements, in which the eye is rotated slightly about its axis. To provide balanced torsional movements, there are two more extraocular muscles: the superior and inferior obliques (see Figure 13.1B). The **superior oblique** muscle originates on the sphenoid bone in the posterior medial orbit and passes anteriorly through the **trochlea**, a pulley-like fibrous loop on the medial superior orbital rim (see Figure 13.1B,D). It then inserts on the superior surface of the eye to produce **intorsion**, meaning movement of the upper pole of the eye inward (see Figure 13.1B). Meanwhile, the **inferior oblique** has no trochlea, but it originates along the anterior medial orbital wall and inserts on the inferior surface of the eye to produce **extorsion**, meaning movement of the upper pole of the eye outward.

(A)

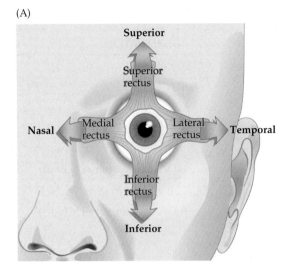

(B)

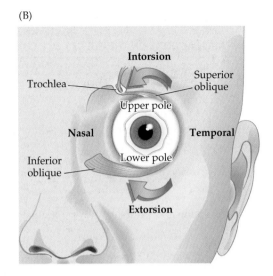

(C)

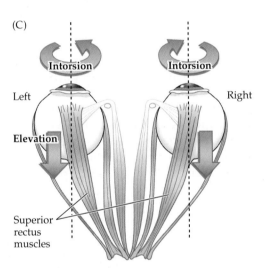

(D)

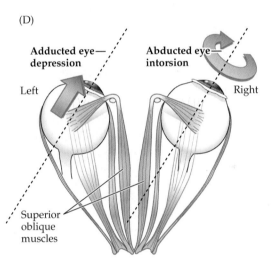

FIGURE 13.1 Extraocular Muscles and Actions (A) Rectus muscles. (B) Superior and inferior oblique muscles. (C) Dual actions of superior rectus muscles. (D) Dual actions of superior oblique muscles.

The movement produced by an extraocular muscle depends on the direction in which the muscle pulls relative to the main axis of the eye (see Figure 13.1C,D). Therefore, as the eyes move by rotating in the orbit, the extraocular muscles can have different actions (Table 13.1). Thus, depending on eye position, the rectus muscles can also produce torsional eye movements, and the oblique muscles can make important contributions to vertical eye movements. For example, when the eyes are directed forward, the superior rectus attaches to the eye at an angle of 23° to the eye's main axis (see Figure 13.1C). Therefore, contraction of the superior rectus causes both elevation and intorsion of the eye. Conversely, contraction of the inferior rectus causes depression and extorsion. If an eye is **abducted** (moving horizontally toward the temple) by 23° so that its axis lines up with the superior rectus muscle, this muscle will now cause a pure elevation movement of the eye. If the eye is **adducted** (moving horizontally toward the nose), the superior rectus has more of an intorsion action. Similarly, as shown in Figure 13.1D, the superior and inferior oblique muscles contribute to vertical movements of the eye. For example, when the eye is adducted (left eye in Figure 13.1D), the superior oblique comes more in line with the axis of the eye and therefore causes depression. Likewise, the inferior oblique causes elevation, especially when the eye is adducted. As the eye is abducted (right eye in Figure 13.1D) the superior oblique becomes more per-

TABLE 13.1 Actions and Innervation of the Extraocular Muscles

MUSCLE	MAIN ACTIONS ON THE EYE	COMMENTS	INNERVATION
Lateral rectus	Abduction (lateral) movement of eye	Abduction = temporal	Abducens nerve (CN VI)
Medial rectus	Adduction (medial) movement of eye	Adduction = nasal	Oculomotor nerve (CN III)
Superior rectus	Elevation and intorsion	Elevation increases with abduction; intorsion increases with adduction	Oculomotor nerve (CN III)
Inferior rectus	Depression and extorsion	Depression increases with abduction; extorsion increases with adduction	Oculomotor nerve (CN III)
Inferior oblique	Elevation and extorsion	Elevation increases with adduction; extorsion increases with abduction	Oculomotor nerve (CN III)
Superior oblique	Depression and intorsion	Depression increases with adduction; intorsion increases with abduction	Trochlear nerve (CN IV)

pendicular to the eye's axis and therefore causes mainly intorsion. Similarly, the inferior oblique causes mainly extorsion when the eye is abducted. The main actions of the extraocular muscles are summarized in Table 13.1.

Other eye muscles that are not extraocular muscles will be discussed in this chapter. These include the **levator palpebrae superior**, which elevates the eyelid; the **pupillary constrictor** and **dilator muscles**, which cause the pupil to become smaller and larger; and the **ciliary muscle**, which adjusts the thickness of the lens in response to viewing distance.

Extraocular Nerves and Nuclei

The oculomotor (CN III), trochlear (CN IV), and abducens (CN VI) nerves pass through the cavernous sinus and then enter the orbit through the superior orbital fissure (see Figure 12.3A,C). The **oculomotor nerve** (**CN III**) supplies all extraocular muscles except the lateral rectus and superior oblique. Shortly after entering the orbit, the oculomotor nerve splits into two major branches. The **superior division** supplies the superior rectus and also innervates the **levator palpebrae**, a muscle important for eyelid elevation. The **inferior division** of the oculomotor nerve supplies the medial rectus, inferior rectus, and inferior oblique muscles. The inferior division of the oculomotor nerve also carries preganglionic parasympathetic fibers to the pupillary constrictor muscles and to the ciliary muscles of the lens (see Figure 12.6). The **trochlear nerve** innervates the superior oblique muscle, and the **abducens nerve** innervates the lateral rectus (see Table 13.1).

Recall that the oculomotor (CN III), trochlear (CN IV), and abducens (CN VI) nuclei, together with the hypoglossal (CN XII) nucleus, constitute the somatic motor column of cranial nerve nuclei (see Figure 12.5; Table 12.3). These nuclei all lie near the midline, adjacent to the ventricular system, and their fibers exit the brainstem ventrally near the midline, with the exception of CN IV, which exits dorsally (see Figure 12.2). Let's discuss each of these nuclei and the intracranial segments of these nerves in more detail.

The **oculomotor nuclei** are located in the upper midbrain at the level of the superior colliculi and red nuclei, just ventral to the periaqueductal gray matter (**Figure 13.2**; see Figure 14.3A). Fascicles of the oculomotor nerve exit the brainstem as CN III in the interpeduncular fossa between the posterior cerebral and superior cerebellar arteries (see Figure 13.2; see also Figures 5.6 and 10.3). The **Edinger–Westphal nuclei**, containing preganglionic parasympathetic fibers,

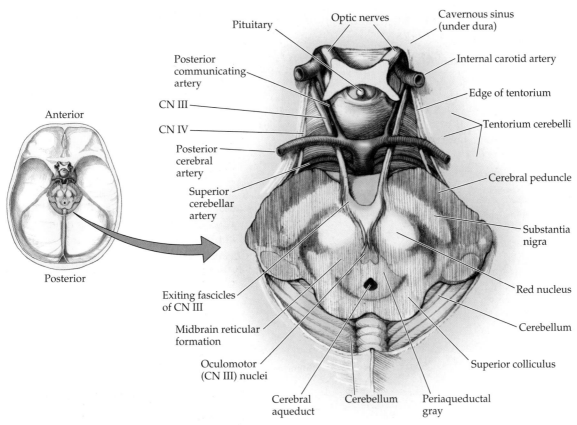

Pituitary

Optic nerves

Cavernous sinus (under dura)

Posterior communicating artery

Internal carotid artery

CN III

Edge of tentorium

CN IV

Tentorium cerebelli

Posterior cerebral artery

Cerebral peduncle

Superior cerebellar artery

Substantia nigra

Exiting fascicles of CN III

Red nucleus

Midbrain reticular formation

Cerebellum

Oculomotor (CN III) nuclei

Superior colliculus

Cerebral aqueduct

Cerebellum

Periaqueductal gray

FIGURE 13.2 Oculomotor Nerves (CN III) in Situ View is from above, with the forebrain removed, showing the oculomotor nuclei in the rostral midbrain and the oculomotor nerves (CN III) in relation to adjacent structures. See also Figure 5.6.

form a *V* shape as they curve over the dorsal aspect of the oculomotor nuclei and fuse anteriorly in the midline (Figure 13.3 and 14.3A). The parasympathetic fibers controlling pupil constriction run in the superficial and medial portion of the oculomotor nerve as it travels in the subarachnoid space, and they are susceptible to compression from aneurysms, particularly arising from the nearby posterior communicating artery (see Figure 13.2). The oculomotor nerve then enters the cavernous sinus and continues to the orbit via the superior orbital fissure, as already described (see Figure 12.3A,C). The oculomotor nucleus actually consists of several subnuclei (see Figure 13.3; Table 13.2). These subnuclei and their connections are of relatively minor clinical significance, except for the following points: (1) Unilateral weakness of the levator palpebrae superior or unilateral pupillary dilation cannot arise from unilateral lesions of the oculomotor nucleus. (2) Lesions of the oculomotor nucleus affect the *contralateral* superior rectus. In addition to the contralateral superior rectus, lesions of the oculomotor nucleus affect the ipsilateral superior rectus since the crossing fibers traverse the oculomotor nucleus before exiting in the third nerve fascicles. *To summarize:* a nuclear lesion of the oculomotor nucleus does not cause unilateral ptosis, unilateral dilated unresponsive pupil, or unilateral superior rectus palsy.

The **trochlear nuclei** are located in the lower midbrain at the level of the inferior colliculi and the decussation of the superior cerebellar peduncle (Figure 13.4; see Figure 14.3B). Like the oculomotor nuclei, they lie just ventral to the periaqueductal gray matter and are bounded ventrally by the fibers of the medial longitudinal fasciculus. The trochlear nerves are the only cranial nerves to exit the brain dorsally (see Figure 12.2B). In addition, unlike any other cranial nerve, the trochlear nerves exit the brainstem in a completely crossed fashion (see Figure 13.4). They travel caudally for a short distance and then cross to the opposite side before exiting at the level of the anterior medullary velum, where they

REVIEW EXERCISE

For each of the six extraocular muscles, list the main action(s) of the muscle and the cranial nerve it is innervated by (refer to Table 13.1 and Figure 13.1). For the cranial nerves CN III, CN IV, and CN VI, list all muscles innervated (intraocular and extraocular) and state the brainstem nuclei from which the fibers arise (refer to Tables 12.3, 12.4, and 13.1 and Figure 12.5).

TABLE 13.2 Subnuclei of the Oculomotor Nucleus (CN III) and Their Functions[a]

SUBNUCLEUS	MUSCLES INNERVATED	SIDE INNERVATED
Dorsal	Inferior rectus	Ipsilateral
Intermediate	Inferior oblique	Ipsilateral
Ventral	Medial rectus	Ipsilateral
Edinger–Westphal (parasympathetic)	Pupillary constrictors and lens ciliary muscles	Bilateral
Central caudal	Levator palpebrae superior	Bilateral
Medial	Superior rectus	Contralateral

[a]Subnuclei are color coded to match Figure 13.3.

are susceptible to compression from cerebellar tumors. The trochlear nerves are very thin and are relatively easily damaged by shear injury from head trauma. They travel through the subarachnoid space along the underside of the tentorium cerebelli (see Figure 13.3) and then enter the cavernous sinus to reach the orbit via the superior orbital fissure, and they innervate the superior oblique muscles (see Figure 13.1).

The **abducens nuclei** lie on the floor of the fourth ventricle under the facial colliculi in the mid-to-lower pons (see Figures 12.2B and 12.11). Abducens fibers travel ventrally to exit at the pontomedullary junction (see Figure 12.2A). The abducens nerve must then follow a long course in the subarachnoid space, ascending between the pons and clivus (see Figure

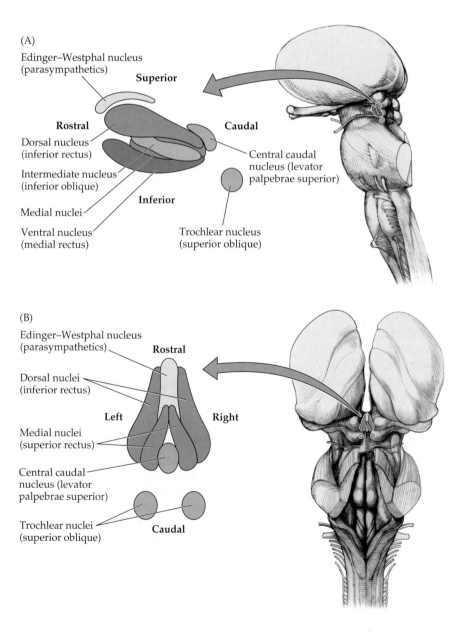

FIGURE 13.3 Oculomotor Nuclear Complex (A) Left lateral view. (B) Dorsal view. The trochlear nuclei are also shown.

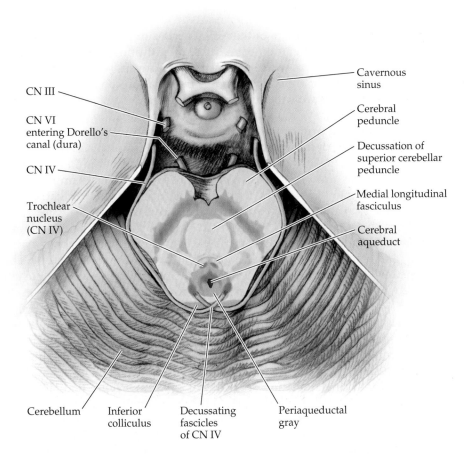

CN III

CN VI entering Dorello's canal (dura)

CN IV

Trochlear nucleus (CN IV)

Cavernous sinus

Cerebral peduncle

Decussation of superior cerebellar peduncle

Medial longitudinal fasciculus

Cerebral aqueduct

Cerebellum   Inferior colliculus   Decussating fascicles of CN IV   Periaqueductal gray

FIGURE 13.4 Trochlear Nerves (CN IV) and Abducens Nerves (CN VI) in Situ View is from above (as in Figure 13.2), with the forebrain and upper midbrain removed, showing the trochlear nuclei in the caudal midbrain and the trochlear nerves (CN IV) in relation to adjacent structures. Also shown are abducens nerves (CN VI) originating from the pontomedullary junction (not visible; see Figure 12.2A), ascending along the clivus, and piercing the dura to enter Dorello's canal.

12.1). The abducens nerve then exits the dura to enter **Dorello's canal**, running between the dura and skull, under the petroclinoid ligament (see Figures 12.3A and 13.4). It next makes a sharp bend as it passes over the petrous tip of the temporal bone to reach the cavernous sinus (see Figure 13.11). This long vertical course may explain why the abducens nerve is highly susceptible to downward traction injury produced by elevated intracranial pressure (see KCC 5.3). After traversing the cavernous sinus, the abducens nerve enters the orbit via the superior orbital fissure to innervate the lateral rectus muscle (see Figure 13.1).

KEY CLINICAL CONCEPT

13.1 DIPLOPIA

Diplopia (double vision) can be caused by abnormalities in several locations. From distal to proximal, these include (1) mechanical problems such as orbital fracture with muscle entrapment; (2) disorders of the extraocular muscles such as thyroid disease, or orbital myositis (orbital pseudotumor); (3) disorders of the neuromuscular junction such as myasthenia gravis; or (4) disorders of CN III, IV, VI and their central pathways. In the following sections we review some disorders of CN III, IV, and VI and their brainstem nuclei that cause diplopia. Diplopia can also occasionally be caused by disorders involving the supranuclear oculomotor pathways such as internuclear ophthalmoplegia (INO) (see KCC 13.8), skew deviation (see KCC 13.3), and ingestion of toxins such as alcohol or anticonvulsant medications.

In taking the history, ask if the diplopia went away when the patient closed or covered one eye. If so, this symptom suggests that the diplopia was caused by an eye movement abnormality. In mild dysconjugate gaze, patients may report only visual blurring, without actual diplopia. Monocular diplopia or polyopia (three or more images) can be caused by ophthalmological disease, disorders of the visual cortex, or psychiatric conditions, but not by eye movement abnormalities. In addition, patients should be asked if the diplopia was worse with near or far objects, when looking up or down or to the left or right, because this information can help localize the cause of diplopia, as we will see later in the chapter.

On examination, eye movements are usually reported in degrees or millimeters from the primary position. When an extraocular muscle is not working properly, **dysconjugate gaze** results, causing diplopia. A helpful rule of thumb is that the *image further from the midline and towards the direction of attempted gaze is always the one seen by the abnormal eye*. For example, when looking at an object to the right, if one eye does not move to the right, then it will form a second image that appears displaced to the right.

The **red glass test** can also be helpful in examining patients with diplopia. A transparent piece of red glass or plastic is held over one eye, usually the right, and a small white light is held directly in front of the patient. The image seen by the right eye is therefore red, and the image seen by the left eye is white. The patient is then asked to follow the light as it is moved to nine different positions of gaze, and to report the locations of the white and red images. Normally the white and red images are fused in all positions of gaze. Examples of the red glass test in different kinds of diplopia are shown in Figures 13.5–13.7. More quantitative methods for measuring diplopia—usually employed by eye movement specialists—are also available.

Abnormal lateral deviation of one eye is called **exotropia**, and abnormal medial deviation is called **esotropia**. Vertical deviation is usually described only with respect to the eye that is higher, and it is called **hypertropia**. A useful test for subtle dysconjugate gaze is to shine a flashlight from directly in front of the patient on both eyes simultaneously and then to examine the position of the reflection of the light on each cornea. Normally the reflection is symmetrical on the two corneas. When an eye is misaligned, the reflected light appears displaced in the opposite direction. Another helpful test for subtle eye muscle weakness is the **cover–uncover test**. Visual input normally helps maintain the eyes yoked in the same direction. Therefore, when an eye is covered while looking in the direction of a weak muscle, it may drift slightly back toward the neutral position. This mild latent weakness present only with an eye covered is called a **phoria** (as in exophoria, esophoria, etc.), in contrast to a tropia.

In young children, because the visual pathways are still developing, congenital eye muscle weakness can produce **strabismus** (dysconjugate gaze) that over time causes suppression of one of the images, resulting in **amblyopia** (decreased vision in one eye). For this reason, early intervention is essential. ∎

KEY CLINICAL CONCEPT

13.2 OCULOMOTOR PALSY (CN III)

Complete disruption of oculomotor nerve function causes paralysis of all extraocular muscles except for the lateral rectus and superior oblique. Therefore, the only remaining movements of the eye are some abduction and some depression and intorsion (see Figure 13.1; Table 13.1). Because of decreased tone in all muscles except the lateral rectus and superior oblique, the eye may come to lie in a **"down and out"** position at rest (Figure 13.5A). In addition, paralysis of the levator palpebrae superior causes the eye to be closed (complete ptosis) unless the upper lid is raised with a finger. The pupil is dilated and unresponsive to light because of involvement of the parasympathetic fibers that run with the oculomotor nerve.

Partial impairment of oculomotor nerve function can cause different combinations of these findings to appear in milder form. For example, the eye

(A) Right oculomotor nerve (CN III) palsy

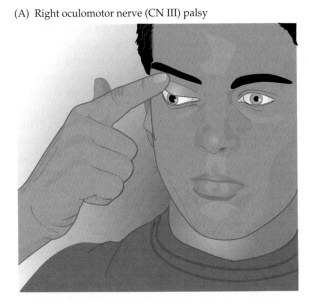

(B)

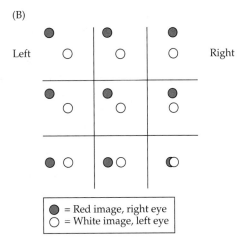

Left   Right

● = Red image, right eye
○ = White image, left eye

FIGURE 13.5 Oculomotor Nerve (CN III) Palsy (A) Appearance of eyes in the presence of a right oculomotor nerve palsy. (B) Results of red glass testing with right oculomotor nerve palsy, drawn as seen from the patient's perspective. Red glass was placed over the right eye.

movements may be impaired, with minimal pupil abnormalities, or the pupil may be involved, with only mild eye movement abnormalities.

When the examiner takes the history of someone with oculomotor palsy, the patient may report that the diplopia is worse when looking at near objects and better when looking at distant objects, since convergence is impaired. Red glass testing in third-nerve palsy generally reveals **diagonal diplopia** that is most severe when looking up and medially with the affected eye (**Figure 13.5B**). Note that the results of red glass testing are drawn as seen by the patient.

Common causes of oculomotor nerve palsy include diabetic neuropathy or other microvascular neuropathy associated with hypertension or hyperlipidemia; and head trauma in which shearing forces damage the nerve. Another important cause of oculomotor palsy is compression of the nerve by intracranial aneurysms, most often arising from the junction of the posterior communicating artery (Pcomm) with the internal carotid artery (see Figures 13.2, 5.6; KCC 5.6). Less commonly, third-nerve palsy can be caused by aneurysms arising from the PComm–posterior cerebral artery (PCA) junction, from the basilar artery–PCA junction, or from the basilar artery–superior cerebellar artery (SCA) junction (see Figure 5.20). The oculomotor nerve can also be damaged by other abnormalities in the subarachnoid space, cavernous sinus, or orbit, such as infection, tumor, or venous thrombosis. Herniation of the medial temporal lobe over the edge of the tentorium cerebelli can compress the oculomotor nerve (see Figure 13.2; see also Figure 5.6). Recall that in addition to an ipsilateral oculomotor palsy, transtentorial uncal herniation also often causes coma and hemiplegia (see KCC 5.4). Ophthalmoplegic migraine is a condition usually seen in children that causes reversible oculomotor nerve palsy. Lesions in the midbrain such as lacunar infarcts (see KCC 10.4), or other infarcts involving the oculomotor nucleus or the exiting nerve fascicles, can also cause an oculomotor palsy (see Figure 14.21A, Table 14.9; see also Table 13.2). In addition, muscle disorders, or disorders of the neuromuscular junction such as myasthenia gravis (see KCC 8.1), can sometimes mimic the eye movement abnormalities and ptosis seen in oculomotor palsy.

Since aneurysms can cause life-threatening intracranial hemorrhage (see KCC 5.6), there should be a high index of suspicion for aneurysms in patients presenting with third-nerve palsy. Aneurysms classically cause a **painful oculomotor palsy that involves the pupil**. The oculomotor palsy may be

REVIEW EXERCISE

How would the eyes appear for a patient with a complete *left* oculomotor nerve palsy? Draw the expected results of red glass testing for this patient. (Compare to Figure 13.5.)

subtle or complete. These patients should be considered to have a **PComm aneurysm** until proven otherwise. A good-quality **emergency CT angiogram (or MRA)** should be done without delay and may be followed a conventional four vessel angiogram in cases where any uncertainty remains. A **painless and complete oculomotor palsy that spares the pupil** is not caused by aneurysms (with rare exceptions); it is usually caused by diabetes or other microvascular neuropathy. The reason is thought to be that the parasympathetic fibers are located near the surface of the nerve, and if the nerve compression is severe enough to cause complete paralysis of the muscles innervated by CN III, then the pupillary fibers should be involved as well. In **partial oculomotor palsy that spares the pupil**, particularly if **pain** is present, then the findings could be caused by partial compression of CN III by an aneurysm, so a CT angiogram should be done.

Lesions of the oculomotor nerve can sometimes affect the superior division or inferior division in isolation. A lesion of the superior division causes weakness of the superior rectus and levator palpebrae superior and is more often caused by a mass in or near the orbit rather than diabetic neuropathy. ■

KEY CLINICAL CONCEPT

13.3 TROCHLEAR PALSY (CN IV)

The trochlear nerve produces **depression and intorsion** of the eye. Therefore, in trochlear nerve palsy there is **vertical diplopia**. If the weakness is severe, the affected eye may show **hypertropia** (Figure 13.6A). There may also be extorsion of the eye, which is not usually visible to the examiner. Patients with trochlear nerve palsy often report that they can improve the diplopia by looking up (chin tuck) and by **tilting the head away from the affected eye** because these maneuvers compensate for the hypertropia and extorsion, respectively (see Figure 13.6A). In addition, recall that the depressing action of the superior oblique is most pronounced when the eye is adducted (see Figure 13.1D; Table 13.1). Therefore, the vertical diplopia is most severe when the affected eye is looking downward and toward the nose. This can be confirmed with red glass testing (Figure 13.6B). To summarize, diagnosing a fourth-nerve

FIGURE 13.6 Trochlear Nerve (CN IV) Palsy (A) Appearance of eyes in the presence of a right trochlear nerve palsy. Hypertropia can be compensated for by tucking the chin and looking up slightly. Extorsion can be compensated for by tilting the head away from the affected eye. (B) Results of red glass testing with right trochlear nerve palsy, drawn as seen from the patient's perspective. Red glass was placed over the right eye. (C) Appearance of a horizontal white line to the patient with a red glass over the right eye.

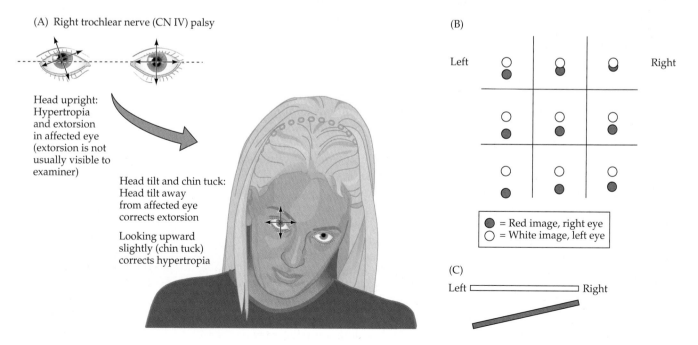

(A) Right trochlear nerve (CN IV) palsy

Head upright: Hypertropia and extorsion in affected eye (extorsion is not usually visible to examiner)

Head tilt and chin tuck: Head tilt away from affected eye corrects extorsion

Looking upward slightly (chin tuck) corrects hypertropia

(B)

Left   Right

● = Red image, right eye
○ = White image, left eye

(C)

Left ▭ Right

palsy usually involves demonstrating typical findings through the following **four steps** (Bielschowsky three step test, plus the "missing step"):

1. The affected eye has hypertropia.
2. Vertical diplopia worsens when the affected eye looks nasally.
3. Vertical diplopia improves with head tilt away from the affected eye.
4. Vertical diplopia worsens with downgaze.

Another test that is sometimes useful is to have the patient look at a horizontal line (or pen). In a trochlear-nerve palsy, the patient will see two lines, with the lower line tilted (see Figure 13.6C). These two lines form an arrowhead with the "point" directed toward the affected side.

MNEMONIC

MNEMONIC

The relationship between compensatory head positions and eye movement abnormalities can easily be remembered because *the head movement is always in the direction of action normally served by the affected muscle.* For example, in a right trochlear palsy the head is held down and tilted to the left; the normal action of the right trochlear nerve is depression and intorsion of the eye.

The trochlear nerve is the most commonly injured cranial nerve in head trauma, probably because of its long course and thin caliber, making it susceptible to shear injury. Other causes of trochlear nerve pathology in the subarachnoid space, cavernous sinus, or orbit include neoplasm, infection, and aneurysms. In many cases the cause remains unknown, and these cases may be caused by microvascular damage to the nerve, especially in patients with diabetes. Vascular or neoplastic disorders within the midbrain or near the tectum (e.g., pineal gland or anterior cerebellum) can also affect the trochlear nuclei or nerve fascicles. Interestingly, **congenital fourth nerve palsy**, a relatively common cause of superior oblique weakness, is often latent for years except for minor head tilt and can later decompensate, leading to diplopia in adulthood.

Other causes of **vertical diplopia** include disorders of extraocular muscles, myasthenia gravis, lesions of the superior division of the oculomotor nerve affecting the superior rectus, and skew deviation. **Skew deviation** is defined as a vertical disparity in the position of the eyes of supranuclear origin. Unlike trochlear palsy, in skew deviation the vertical disparity is typically (but not always) relatively constant in all positions of gaze. Skew deviation can be caused by lesions of the cerebellum, brainstem, or even the inner ear.

Other causes of **head tilt** include cerebellar lesions, meningitis, incipient tonsillar herniation, and torticollis. It is often helpful to look at old photographs to establish whether the head tilt is old or new. ■

> ## REVIEW EXERCISE
>
> How would the eyes appear and what head position would be seen in a patient with a complete *left* trochlear nerve palsy? Draw the expected results of red glass testing for this patient. (Compare to Figure 13.6.)

KEY CLINICAL CONCEPT

13.4 ABDUCENS PALSY (CN VI)

Lesions of the abducens nerve produce **horizontal diplopia**. In some cases esotropia (see KCC 13.1) of the affected eye may be present as well. In contrast to a third-nerve palsy, patients report that the diplopia is better when they are viewing near objects and worse when they are viewing far objects. On examination, the affected eye does not abduct normally (Figure 13.7A). In milder abducens palsy these may simply be **incomplete "burial of the sclera"** on lateral gaze. Diplopia worsens when the patient tries to abduct the affected eye. Again, this can be confirmed by red glass testing (Figure 13.7B). Some patients may tend to turn the head toward the affected eye in an effort to compensate for the diplopia. A subtle sixth-nerve palsy can sometimes be detected by testing bilateral horizontal saccades (see **neuroexam.com Video**

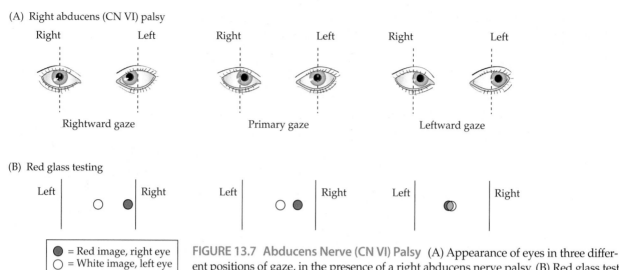

FIGURE 13.7 **Abducens Nerve (CN VI) Palsy** (A) Appearance of eyes in three different positions of gaze, in the presence of a right abducens nerve palsy. (B) Red glass testing in the same three positions of gaze, with right abducens nerve palsy. Red glass was placed over the right eye, and results are drawn as seen from the patient's perspective.

REVIEW EXERCISE

How would the eyes appear in primary gaze for a patient with a complete *left* abducens nerve palsy? Draw the expected results of red glass testing for this patient. (Compare to Figure 13.7.)

33) and observing the slight lag in abduction of the affected eye, or by using the cover–uncover test while the affected eye looks laterally (see KCC 13.1).

Because of its long course along the clivus and over the sharp ridge of the petrous temporal bone (see Figures 13.4, 12.1, 12.2A, and 12.3A), the abducens nerve is particularly susceptible to injury from downward traction caused by elevated intracranial pressure (see KCC 5.3). Abducens palsy is therefore an important early sign of supratentorial or infratentorial tumors, pseudotumor cerebri, hydrocephalus, and other intracranial lesions. The abducens palsy seen with elevated intracranial pressure can be unilateral or bilateral. Other causes of damage to the sixth nerve can occur in the subarachnoid space, cavernous sinus, or orbit. Common disorders include head trauma, infection, neoplasm, inflammation, aneurysms, and cavernous sinus thrombosis. As with oculomotor and trochlear palsies, many cases of acute abducens palsy lack a demonstrated cause, and this condition may result from a microvascular neuropathy like that seen in diabetes.

Pontine infarcts or other disorders affecting the exiting abducens fascicles in the pons (see Figure 12.11) cause weakness of ipsilateral eye abduction resembling a peripheral abducens nerve lesion. However, lesions of the abducens nucleus in the pons produce not a simple abducens palsy but, instead, a horizontal *gaze* palsy in the direction of the lesion. As we will discuss in the section on supranuclear control of eye movements later in this chapter, in a **gaze palsy**, movements of *both* eyes in one direction are decreased (see Figure 13.13B, Lesion 2). In addition, lesions of the abducens nucleus often affect the nearby fibers of CN VII in the facial colliculus, resulting in ipsilateral facial weakness (see Figure 12.11; see also Table 14.8).

Other causes of **horizontal diplopia** include myasthenia gravis and disorders of the extraocular muscles caused by thyroid disease, tumors, inflammation, or orbital trauma. Supranuclear disorders of horizontal gaze and convergence will be discussed later in this chapter. ■

The Pupils and Other Ocular Autonomic Pathways

The pupils are controlled by both parasympathetic and sympathetic pathways. The **parasympathetic** pathways involved in pupillary constriction are shown in Figure 13.8 (see also Figure 12.6). Light entering one eye activates retinal ganglion cells, which project to both optic tracts because of fibers crossing over in the optic chiasm. Fibers in the extrageniculate pathway continue in the brachium of the superior colliculus past the lateral geniculate nucleus to reach the **pretectal area** (see Figure 13.8; see also Figure 11.6) just rostral to the midbrain. After synapsing, axons then continue bilaterally to the **Edinger–Westphal nuclei**, which contain preganglionic parasympathetic neurons. Some of the crossing fibers travel in the **posterior commissure** (see Figure 13.8; see also Figure 12.1). The Edinger–Westphal nuclei lie just dorsal and anterior to the oculomotor (CN III) nuclei near the midline (see Figures 13.3 and 14.3A). Preganglionic parasympathetic fibers travel bilaterally from the Edinger–Westphal nuclei via the oculomotor nerves to reach the **ciliary ganglia** in the orbit (see Figure 13.8). From there, postganglionic parasympathetics continue to the **pupillary constrictor muscles** to cause the pupils to become smaller. Note that a light shone in one eye causes a **direct response** in the same eye and a **consensual response** in the other eye (see **neuroexam.com Video 29**) because information crosses bilaterally at multiple levels.

Pupil light reflex

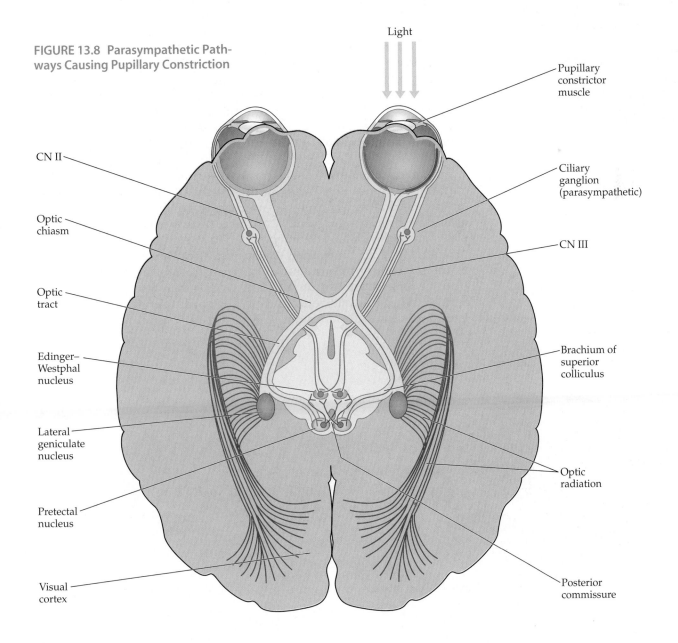

FIGURE 13.8 Parasympathetic Pathways Causing Pupillary Constriction

Light

Pupillary constrictor muscle

CN II

Ciliary ganglion (parasympathetic)

Optic chiasm

CN III

Optic tract

Edinger–Westphal nucleus

Brachium of superior colliculus

Lateral geniculate nucleus

Optic radiation

Pretectal nucleus

Visual cortex

Posterior commissure

Bilateral pupillary constriction also occurs through a slightly different circuit during the **accommodation response** (see **neuroexam.com Video 31**). This response occurs when a visual object moves from far to near, and it has the following three components:

- Pupillary constriction
- Accommodation of the lens ciliary muscle
- Convergence of the eyes

The accommodation response is activated by visual signals relayed to the **visual cortex** (see Figure 13.8). From there, through pathways still under investigation, the pretectal nuclei are again activated, causing bilateral pupillary constriction mediated by the parasympathetic pathways shown in Figure 13.8. Contraction of the **ciliary muscle** of the lens is parasympathetically mediated by the same pathway. Note that the lens is normally under tension from the suspensory ligament (**Figure 13.9A**). The ciliary muscle acts as a sphincter (like the pupillary constrictor), so when it contracts it causes the suspensory ligament to relax, producing a rounder, more convex lens shape (**Figure 13.9B**). Convergence is mediated by mechanisms described later in this chapter.

The **sympathetic** pathway responsible for pupillary dilation is shown in **Figure 13.10**. A descending sympathetic pathway from several hypothalamic nuclei (see Chapter 17) travels in the lateral brainstem and cervical spinal cord to reach thoracic spinal cord levels T1 and T2. This pathway is thought to be in approximately the same location in the brainstem as the spinothalamic tract (see Figure 13.10; see also Figure 7.2) because lesions of the spinothalamic tract tend to be associated with Horner's syndrome (see KCC 13.5 and KCC 13.6). This descending sympathetic pathway activates preganglionic sympathetic neurons in the **intermediolateral cell column** of the upper thoracic cord (see Figure 13.10, inset; see also Figures 6.12B and 6.13). Axons of the preganglionic sympathetic neurons exit the spinal cord via ventral roots T1 and T2 and skirt the apex of the lung before joining the **paravertebral sympathetic chain** via **white rami communicantes**. The axons ascend to synapse in the **superior cervical ganglion**. From there, postganglionic sympathetic fibers ascend through the **carotid plexus** along the walls of the internal carotid artery to the cavernous sinus, ultimately reaching the **pupillary dilator muscle** (see Figures 13.9A and 13.10).

This sympathetic pathway is also important in controlling the smooth muscle of the **superior tarsal muscle** (**Müller's**), which elevates the upper lid, causing a wide-eyed stare in conditions of increased sympathetic outflow. Recall that the levator palpebrae superior is composed of skeletal muscle and also functions

FIGURE 13.9 Actions of the Ciliary Muscle and the Pupillary Muscles (A) When viewing a distant object, the ciliary muscle and the pupillary constrictor muscle are relaxed. (B) When viewing a near object, the ciliary muscle and the pupillary constrictor muscle contract.

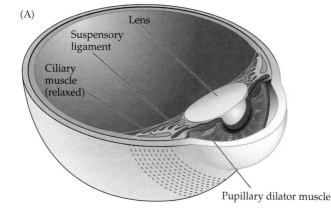

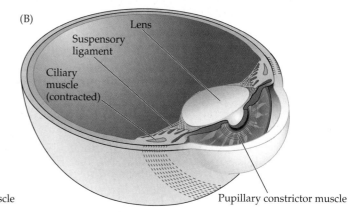

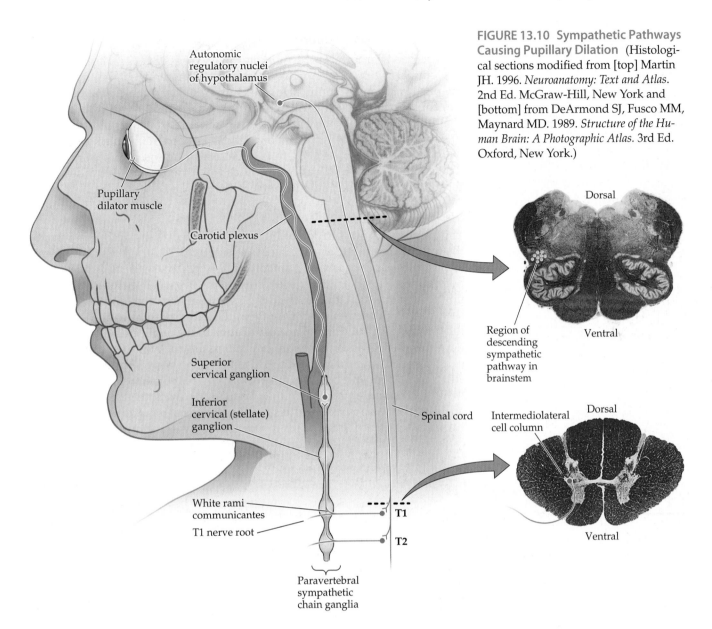

Autonomic
regulatory nuclei
of hypothalamus

Pupillary
dilator muscle

Carotid plexus

Superior
cervical ganglion

Inferior
cervical (stellate)
ganglion

Spinal cord

White rami
communicantes

T1 nerve root

T1

T2

Paravertebral
sympathetic
chain ganglia

Dorsal

Ventral

Region of
descending
sympathetic
pathway in
brainstem

Intermediolateral
cell column

Dorsal

Ventral

FIGURE 13.10 Sympathetic Pathways Causing Pupillary Dilation (Histological sections modified from [top] Martin JH. 1996. *Neuroanatomy: Text and Atlas.* 2nd Ed. McGraw-Hill, New York and [bottom] from DeArmond SJ, Fusco MM, Maynard MD. 1989. *Structure of the Human Brain: A Photographic Atlas.* 3rd Ed. Oxford, New York.)

in eyelid opening under the control of CN III. Sympathetics from the pathway shown in Figure 13.10 also innervate the smooth muscle **orbitalis (Müller's)**, which prevents the eye from sinking back in the orbit, as well as the cutaneous arteries and sweat glands of the face and neck. These various sympathetic functions are impaired in Horner's syndrome (see KCC 13.5 and KCC 13.6).

KEY CLINICAL CONCEPT

13.5 PUPILLARY ABNORMALITIES

Abnormalities of the pupil can be caused by peripheral or central lesions, by sympathetic or parasympathetic lesions, or by disorders of the iris muscle or visual pathways. Pupillary abnormalities can be bilateral or can affect one side only, in which case **anisocoria**, or pupillary asymmetry, is present. In the subsections that follow, we will review the anatomical basis of several important pupillary abnormalities.

REVIEW EXERCISE

Fill in the blanks: Fibers controlling sympathetic outflow descend from the hypothalamus through the (medial or lateral) brainstem and spinal cord to reach preganglionic sympathetic neurons in the _____ cell column. These exit via the _____ and _____ roots and ascend in the sympathetic chain to synapse on neurons in the _____ ganglion. The postganglionic sympathetic fibers then ascend along the carotid plexus to ultimately reach the pupillary _____ muscle of the iris. (See Figure 13.10.)

Oculomotor Nerve Lesion

As discussed earlier under oculomotor palsy (see KCC 13.2), lesions of the efferent parasympathetic pathway from the Edinger–Westphal nucleus to the pupillary constrictor muscle (see Figure 13.8) can cause **impaired pupillary constriction**, resulting in a unilateral **dilated pupil**. When the lesion is complete, the pupil is very large and is sometimes called a "blown pupil." The anisocoria is more obvious in ambient light than in a darkened room (Table 13.3, condition A). There is a decreased or absent direct response when light is shone in the affected eye, as well as a decreased or absent consensual response when light is shone in the opposite eye.

Horner's Syndrome

This important constellation of findings is caused by disruption of sympathetic pathways to the eye and face (see Figure 13.10). The classic syndrome consists of **ptosis**, **miosis**, and **anhidrosis**, as well as several other minor abnormalities. **Ptosis**, or upper eyelid drooping, is caused by loss of innervation to Müller's smooth muscle in the upper lid. **Miosis**, or decreased pupillary size, is caused by loss of sympathetic innervation to the pupillary dilator muscle, resulting in **impaired dilation of the pupil**. Unlike oculomotor nerve lesions, anisocoria is more obvious in the dark than in ambient light (see Table 13.3, condition B). Careful observation can reveal that the pupil still has a direct and consensual constricting response to light. However, there is a **dilation lag** relative to the normal pupil when the light is removed, and pupillary size is reduced. Testing the **ciliospinal reflex** is sometimes useful. In this test, a painful pinch to the neck activates sympathetic outflow, causing pupillary dilation on the normal side but not on the side with

TABLE 13.3 Common Pupil Abnormalities					
LESION OR CONDITION	**DARK ROOM**	**AMBIENT LIGHT**	**DIRECT RESPONSE: LIGHT IN AFFECTED EYE**	**CONSENSUAL RESPONSE: LIGHT IN UNAFFECTED EYE**	**COMMENTS**
A. Left occulomotor nerve lesion					There may be associated ptosis and eye movement abnormalities.
B. Left Horner's syndrome					There is a dilation lag in going from light to dark. Other features of Horner's syndrome (ptosis, anhidrosis) may be present.
C. Left afferent pupillary defect					The swinging flashlight test (**neuroexam.com Video 30**) is useful in subtle cases.
D. Benign essential anisocoria					The same relative anisocoria is present in all lighting conditions. No dilation lag.

Horner's syndrome. **Anhidrosis**, or decreased sweating of the ipsilateral face and neck, is also caused by loss of sympathetic innervation. A simple way to test for this finding is to rub a finger along the skin, which may feel smoother on the affected side because of decreased moisture.

Horner's syndrome can be caused by lesions anywhere along the sympathetic pathway shown in Figure 13.10. Therefore, **possible locations for lesions causing Horner's syndrome** include:

1. Lateral hypothalamus or brainstem (e.g., infarct or hemorrhage; see KCC 14.3)
2. Spinal cord (e.g., trauma)
3. First and second thoracic roots (e.g., apical lung tumor [Pancoast syndrome] or trauma; see KCC 9.1)
4. Sympathetic chain (e.g., tumor or trauma)
5. Carotid plexus (e.g., carotid dissection; see KCC 10.6)
6. Cavernous sinus (e.g., thrombosis, infection, aneurysm, or neoplasm; see KCC 13.7)
7. Orbit (e.g., infection or neoplasm; see KCC 13.7)

Lesions proximal to the superior cervical ganglion (see Figure 13.10), called **preganglionic lesions**, can be distinguished from **postganglionic lesions** by using hydroxyamphetamine eye drops, which stimulate norepinephrine release and dilate the pupil for preganglionic, but not postganglionic, lesions. In addition, postganglionic lesions are not usually associated with anhidrosis because the sympathetics for sudomotor innervation diverge from the oculosympathetic pathway before the superior cervical ganglion. Large bilateral lesions of the pons are sometimes associated with **pontine pupils**, in which both pupils are small but reactive to light. This small pupillary size is probably caused by bilateral disruption of the descending sympathetic pathways.

Afferent Pupillary Defect (Marcus Gunn Pupil)

In this condition the direct response to light in the affected eye is decreased or absent, while the consensual response of the affected eye to light in the opposite eye is normal (see Table 13.3, condition C). The afferent pupillary defect is caused by **decreased sensitivity of the affected eye to light** resulting from lesions of the optic nerve, retina, or eye. Recall that lesions at or behind the optic chiasm would affect inputs from both eyes (see Figure 13.8) and therefore do not generally produce a Marcus Gunn pupil.* A useful way to detect an afferent pupillary defect is with the **swinging flashlight test** (see **neuroexam.com Video 30**). The flashlight is moved back and forth between the eyes every 2 to 3 seconds. The afferent pupillary defect becomes obvious when the flashlight is moved from the normal to the affected eye, and the affected pupil *dilates* in response to light (see Table 13.3, condition C). This abnormal dilation should be distinguished from **hippus**, which is a normal, brief oscillation of the pupil size that sometimes occurs in response to light. It should also be emphasized that optic nerve or retinal disease does not cause anisocoria, since the consensual response still causes both pupils to constrict or dilate together due to crossing fibers at multiple levels (see Figure 13.8), despite any afferent defects on one side (see Table 13.3C).

Swinging flashlight test

---

*Optic tract lesions (an uncommon disorder) can reportedly be associated with a contralateral afferent pupillary defect. In addition, with specialized testing (in which a light is shined onto one-half of the retina at a time), an afferent pupillary defect can be demonstrated in the ipsilateral hemiretina of both eyes for lesions behind the optic chiasm.

Benign (Essential, Physiological) Anisocoria

A slight pupillary asymmetry of less than 0.6 millimeters is seen in 20% of the general population. This asymmetry can vary from one examination to the next, sometimes within a few hours. There are no other associated abnormal findings, such as dilation lag, changes in the asymmetry with lighting conditions, or eye movement abnormalities (see Table 13.3D).

Pharmacological Miosis and Mydriasis

Numerous pharmacological agents can affect pupillary size and can cause confusion in diagnosis, particularly in the comatose patient. Opiates cause bilateral pinpoint pupils, and barbiturate overdose can also cause bilateral small pupils, mimicking pontine lesions. Anticholinergic agents affecting muscarinic receptors, such as scopolamine or atropine, can cause dilated pupils. Pupillary dilation may be unilateral if topical exposure occurs in one eye only, mimicking uncal herniation. When exposure to anticholinergic agents is suspected, 1% pilocarpine eyedrops can be useful because they cause pupillary constriction in parasympathetic lesions but cannot overcome pharmacological muscarinic blockade. In general, diagnosis and treatment of suspected acute brainstem lesions should not be delayed, but the history should include possible drug exposure, and laboratory testing must include a toxicology screen. Pharmacological testing using eyedrops containing cocaine, hydroxyamphetamine, or different strengths of pilocarpine can be useful for establishing the diagnosis in cases of anisocoria with equivocal or subtle findings (see the References at the end of this chapter for additional details).

Light–Near Dissociation

In **light–near dissociation**, the pupils constrict much less in response to light than to accommodation (see **neuroexam.com Videos 29** and **31**). The mechanism for this disparity is not known for certain, and the mechanism may not be the same in different disorders. The classic example of light–near dissociation is the **Argyll Robertson pupil** associated with neurosyphilis, in which, in addition to light–near dissociation, the pupils are also small and irregular. Light–near dissociation can also be seen in diabetes and Adie's myotonic pupil (see the next subsection) and as part of Parinaud's syndrome (see KCC 13.9), which is associated with compression of the dorsal midbrain.

Adie's Myotonic Pupil

This disorder is characterized by degeneration of the ciliary ganglion or postganglionic parasympathetic neurons (see Figure 13.8), resulting in a mid-dilated pupil that reacts poorly to light. Some pupillary constriction can be elicited with the accommodation response, but the pupil then remains constricted and dilates very slowly, a condition described as a tonic or myotonic pupil. The cause is not known.

Midbrain Corectopia

In this relatively rare condition, lesions of the midbrain can sometimes cause an unusual pupillary abnormality in which the pupil assumes an irregular, off-center shape. ■

KEY CLINICAL CONCEPT

13.6 PTOSIS

Eye opening is performed by striated skeletal muscle of the levator palpebrae superior (CN III) together with Müller's smooth muscle in the upper lid (sympathetics). The frontalis muscle of the forehead (CN VII) performs an accessory role. **Eye closure** is performed by the orbicularis oculi muscle (CN VII).

Ptosis, or drooping of the upper eyelid, can be seen in Horner's syndrome, as discussed in KCC 13.5. Other causes of unilateral or bilateral ptosis include oculomotor nerve palsy affecting the levator palpebrae superior, myasthenia gravis, orbital mass, and redundant skin folds associated with aging (pseudoptosis). The ptosis in Horner's syndrome is usually relatively mild, while oculomotor lesions can cause mild or severe ptosis to the point of complete eye closure. The ptosis in myasthenia is classically "fatiguing" and increases following sustained upgaze (see KCC 8.1). Causes of bilateral ptosis or closed eyes without loss of consciousness include nondominant parietal lobe stroke, severe neuromuscular disorders, dorsal lesions of the oculomotor nuclei affecting the central caudal nucleus (see Figure 13.3; see also Table 13.2), and voluntary eye closure associated with photophobia in migraine or meningeal irritation.

Weakness of the orbicularis oculi caused by facial nerve or upper motor neuron lesions (see Figure 12.13) can cause a widened palpebral fissure that may be mistaken for ptosis in the opposite eye. Careful examination of the lids using the irises as a reference point can usually resolve this dilemma: In ptosis, the upper lid comes down farther over the iris in the affected eye (see Figure 13.5A), while in facial weakness, the palpebral fissure is widened mainly because of sagging of the lower lid in the affected eye (see Figure 12.13). ◼

Cavernous Sinus and Orbital Apex

We will now briefly discuss the cavernous sinus and orbital apex because CN III, IV, and VI all pass through this region, and lesions here produce characteristic syndromes that often affect eye movements (see KCC 13.7). The **cavernous sinus** consists of a collection of venous sinusoids located on either side of the pituitary that receives venous blood from the eye and superficial cortex and ultimately drains via several pathways into the internal jugular vein (see Figure 10.11A,B). Like other venous sinuses, the cavernous sinus lies between the periosteal and dural layers of the dura mater. The cavernous sinus surrounds the **carotid siphon** and several important nerves (Figure 13.11; see also Figures 12.3A, 13.2, and 13.4): the **abducens nerve** (**CN VI**), which lies closest to the carotid; and the **oculomotor** (**CN III**), **trochlear** (**CN IV**), and **ophthalmic** (**CN V₁**) nerves, which run in sequence within the lateral wall of the cavernous sinus (see Figure 13.11). These nerves pass forward to enter the orbital apex via the superior orbital fissure (see Figure 12.3A,C). The **maxillary nerve** (**CN V₂**) skirts the lower portion of the cavernous sinus and often runs through it for a short distance before exiting via the foramen rotundum (see Figure 12.3A). **Sympathetic fibers** traveling in the carotid plexus (see Figure 13.10) en route to the pupillary dilator muscle traverse the cavernous sinus as well.

The optic nerve lies just above the cavernous sinus and enters the orbital apex via the optic canal (see Figures 12.3A and 13.2). The **orbital apex** is the region where nearly all nerves, arteries, and veins of the orbit converge before communicating with the intracra-

FIGURE 13.11 Cavernous Sinus Coronal section through the cavernous sinus, showing the locations of important nerves and blood vessels. View is from the front.

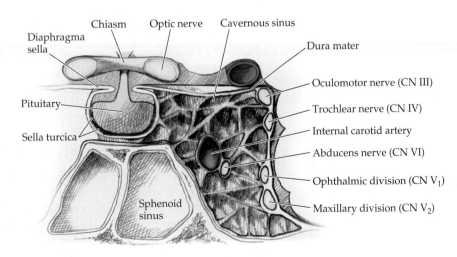

Chiasm
Optic nerve
Cavernous sinus
Diaphragma sella
Dura mater
Pituitary
Oculomotor nerve (CN III)
Trochlear nerve (CN IV)
Internal carotid artery
Sella turcica
Abducens nerve (CN VI)
Ophthalmic division (CN V₁)
Sphenoid sinus
Maxillary division (CN V₂)

nial cavity via the optic canal and superior orbital fissure (see Figure 12.3C). It is important to be familiar with the structures of the cavernous sinus and orbital apex to understand how lesions in these regions can affect multiple cranial nerves (see KCC 13.7).

KEY CLINICAL CONCEPT

13.7 CAVERNOUS SINUS SYNDROME (CN III, IV, VI, V$_1$) AND ORBITAL APEX SYNDROME (CN II, III, IV, VI, V$_1$)

Lesions of the cavernous sinus or orbital apex can affect isolated nerves, or they can affect all the nerves traversing these structures (see Figure 13.11). A complete lesion of the cavernous sinus disrupts CN III, IV, and VI, causing total ophthalmoplegia, usually accompanied by a fixed, dilated pupil. Involvement of CN V$_1$ and variable involvement of V$_2$ causes sensory loss in these divisions of the trigeminal nerve. Horner's syndrome can also occur because of disruption of ocular sympathetics, but this condition may be difficult to appreciate in the setting of a third nerve lesion (sometimes the Horner's syndrome is apparent in the affected eye in the dark; see Table 13.3). Orbital apex lesions produce the same deficits as cavernous sinus syndrome, but they are more likely to involve CN II also, causing visual loss, and often are associated with proptosis, or bulging of the eye, due to mass effect in the orbit. In addition, CN V$_2$ is spared in orbital apex syndrome since it exits the cranium via the foramen rotundum (see Figure 12.3). In both cavernous sinus and orbital apex lesions, partial deficits of the nerves mentioned here often occur when the lesion is less severe. Since the cavernous sinus and orbital apex are contiguous, both structures can be affected by a single lesion. Impaired venous drainage in both disorders can cause vascular engorgement of the orbital structures.

Causes of cavernous sinus syndrome include metastatic tumors, direct extension of nasopharyngeal tumors, meningioma, pituitary tumors or pituitary apoplexy, aneurysms of the intracavernous carotid, cavernous carotid arteriovenous fistula, bacterial infection causing cavernous sinus thrombosis, aseptic thrombosis, idiopathic granulomatous disease (Tolosa–Hunt syndrome), and fungal infections such as aspergillosis or mucormycosis. In cavernous carotid aneurysms or fistulas, the abducens nerve is often involved first because it lies closest to the carotid artery (see Figure 13.11). In pituitary apoplexy (see KCC 17.3), there is hemorrhage within the pituitary gland, often in the setting of a pituitary tumor, which can sometimes extend into the adjacent cavernous sinus. Orbital apex syndrome can be caused by metastatic tumors, orbital cellulitis (bacterial infection), idiopathic granulomatous disease (orbital myositis or pseudotumor), and fungal infections such as aspergillosis. Cavernous sinus and orbital apex syndromes are medical emergencies requiring prompt recognition, diagnosis, and treatment. When MRI scans with contrast enhancement and lumbar puncture do not reveal a specific diagnosis and symptoms are progressive, emergency orbitotomy and biopsy are warranted. ■

Supranuclear Control of Eye Movements

Circuits for the supranuclear control of eye movements extend from the brainstem and cerebellum to the forebrain and exert their influence on the final common output nuclei of CN III, IV, and VI. There appear to be at least three dedicated circuits in the brainstem that feed much of the information from supranuclear control systems to the output nuclei, generating the following:

- Horizontal eye movements
- Vertical eye movements
- Vergence eye movements

We will first discuss the brainstem pathways generating movements in these three directions. Next, for each of these movement directions we will discuss how larger-scale networks, including the cortex, basal ganglia, cerebellum, and vestibular nuclei, generate different types of eye movements for different purposes. These different types of eye movements include the following:

- **Saccades** are rapid eye movements reaching velocities of up to 700° per second (see **neuroexam.com Video 33**). They function to bring targets of interest into the field of view. Vision is transiently suppressed during saccadic eye movements. Saccades are the only type of eye movement that can easily be performed voluntarily, although they can be elicited by reflexes as well.

- **Smooth pursuit** eye movements are not under voluntary control, and they reach velocities of only 100° per second (see **neuroexam.com Video 32**). They allow stable viewing of moving objects.

- **Vergence** eye movements maintain fused fixation by both eyes as targets move toward or away from the viewer (see **neuroexam.com Video 32**). The velocity is about 20° per second.

- **Reflex** eye movements include optokinetic nystagmus (see **neuroexam.com Video 34**) and the vestibulo-ocular reflex (see **neuroexam.com Video 35**). **Nystagmus** is a rhythmic form of reflex eye movements composed of slow eye movements in one direction interrupted repeatedly by fast, saccade-like eye movements in the opposite direction.

Brainstem Circuits for Horizontal Eye Movements

Horizontal eye movements are generated by the lateral rectus and medial rectus muscles, which are controlled by the abducens and oculomotor nuclei, respectively (**Figure 13.12**). As we discussed in Chapter 12, the **medial longitudinal fasciculus** (**MLF**) interconnects the oculomotor, trochlear, abducens, and vestibular nuclei (see Figure 12.19). Through connections in the MLF, eye movements are normally yoked together, resulting in conjugate gaze in all directions. For example, during horizontal eye movements the actions of the abducens and oculomotor nuclei are coordinated through connections in the MLF, as shown in Figure 13.12. Through this circuit, the abducens nucleus does more than just control abduction of the ipsilateral eye. In reality, the abducens

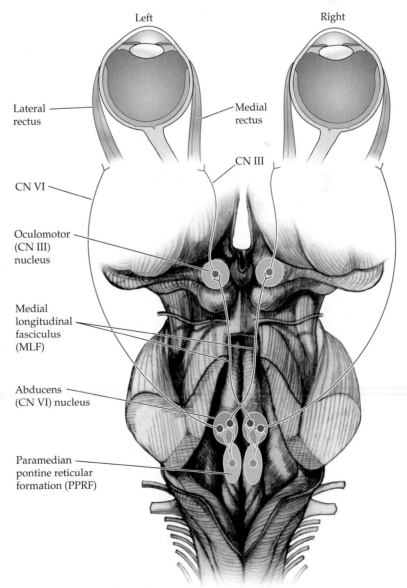

FIGURE 13.12 Brainstem Pathways for Control of Horizontal Eye Movements

Left

Right

Lateral rectus

Medial rectus

CN III

CN VI

Oculomotor (CN III) nucleus

Medial longitudinal fasciculus (MLF)

Abducens (CN VI) nucleus

Paramedian pontine reticular formation (PPRF)

nucleus is a **horizontal gaze center**, controlling horizontal movement of *both* eyes in the direction ipsilateral to the side of the nucleus (see Figure 13.12). Thus, some neurons in the abducens nucleus project to the ipsilateral lateral rectus muscle, while others project via the MLF to the contralateral oculomotor nucleus, which, in turn, activates the contralateral medial rectus muscle.

In the pontine tegmentum near the abducens nucleus, an additional important horizontal gaze center called the **paramedian pontine reticular formation (PPRF)** provides inputs from the cortex and other pathways to the abducens nucleus, resulting in lateral horizontal gaze (see Figure 13.12). As we will discuss later in this chapter, the vestibular nuclei also connect to the extraocular nuclei via the MLF, resulting in vestibulo-ocular reflexes.

KEY CLINICAL CONCEPT

13.8 BRAINSTEM LESIONS AFFECTING HORIZONTAL GAZE

Examination of Figure 13.13 should make clear how different lesions in the brainstem affect horizontal gaze. Lesions of the abducens nerve cause impaired abduction of the ipsilateral eye (see Figure 13.13, Lesion 1; see also KCC 13.4). Lesions of the abducens nerve should be distinguished from lesions of the abducens nucleus, which produce an ipsilateral **lateral gaze palsy** involving both eyes because of the connections through the MLF (see Figure 13.13, Lesion 2). Similarly, lesions of the PPRF cause an ipsilateral lateral gaze palsy (see Figure 13.13, Lesion 3).*

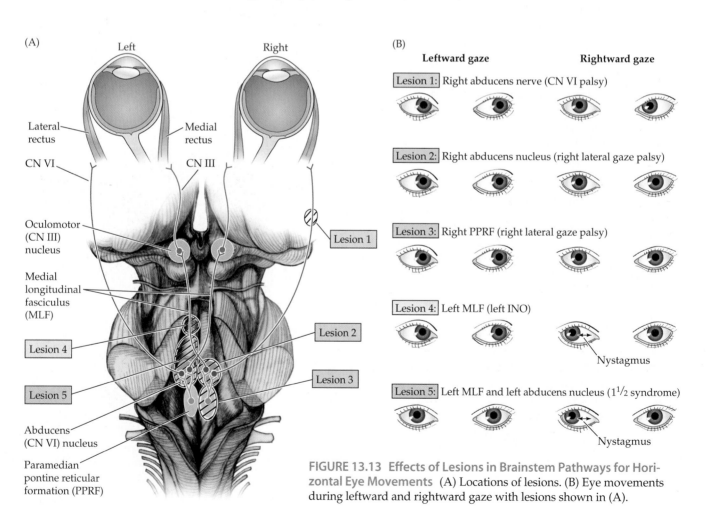

FIGURE 13.13 Effects of Lesions in Brainstem Pathways for Horizontal Eye Movements (A) Locations of lesions. (B) Eye movements during leftward and rightward gaze with lesions shown in (A).

Lesions of the MLF interrupt the input to the medial rectus. Therefore the eye ipsilateral to the lesion does not adduct fully on attempted horizontal gaze (see Figure 13.13, Lesion 4). In addition, for uncertain reasons there is also nystagmus of the opposite eye, possibly because of mechanisms trying to bring the eyes back into alignment. This classic neurologic syndrome produced by an MLF lesion is called an **internuclear ophthalmoplegia** (**INO**). By definition, the side of the INO is the side of the lesion in the MLF. Since the ascending MLF crosses almost immediately after leaving the abducens nucleus (see Figure 13.13) the side of the INO is also the side on which eye adduction is weak. In an INO, eye adduction on the affected side is impaired with horizontal gaze but is often spared during convergence because the inputs to the oculomotor nucleus mediating convergence (see the next section) arise from the pretectal region and hence do not travel in the caudal MLF. Common causes of INO include multiple sclerosis plaques, pontine infarcts, or neoplasms involving the MLF. A subtle INO can sometimes be detected only by testing of horizontal saccades in both directions (see **neuroexam.com video 33**), and observation of a slight lag in the adduction of the eye on the affected side.

Finally, if a lesion involves both the MLF and the adjacent abducens nucleus or PPRF, there is a combination of an ipsilateral INO and an ipsilateral lateral gaze palsy (see Figure 13.13, Lesion 5). Thus, the ipsilateral eye cannot move at all horizontally, and the contralateral eye loses half of its movements, preserving only its ability to abduct, resulting in the quaint name **one-and-a-half syndrome** for this disorder. ■

Brainstem Circuits for Vertical and Vergence Eye Movements

Vertical eye movements are mediated by the superior and inferior rectus and superior and inferior oblique muscles (see Figure 13.1; see also Table 13.1). Brainstem centers controlling vertical eye movements are located in the **rostral midbrain reticular formation and pretectal area**. The **ventral** portion of this region is thought to mediate **downgaze**, while the more **dorsal** region (in the vicinity of the posterior commissure) mediates **upgaze**. One important nucleus that is thought to mediate downgaze is the **rostral interstitial nucleus of the MLF**. Other lesser nuclei in this area that may also play a role include the nucleus of Darkschewitsch and the interstitial nucleus of Cajal. Lesions such as infarcts or tumors (discussed in the next section) of the dorsal parts of the vertical eye movement center cause impaired upgaze, while lesions in the ventral part cause impaired downgaze. Progressive supranuclear palsy (see KCC 16.2) is associated with impaired vertical eye movements and midbrain atrophy. In addition, in locked-in syndrome (see KCC 14.1) large pontine lesions can disrupt the bilateral corticospinal tracts and abducens nuclei, eliminating body movements and horizontal eye movements. However, sometimes the vertical eye movement centers in the midbrain are spared, allowing the patient to communicate entirely through vertical eye movements.

Vertical eye movements are normally closely coordinated with movement of the upper eyelids, except during eye blinks. This coupling is thought to be mediated by the **M-group of neurons** located in the midbrain near the nuclei for vertical eye movements. Eyelid abnormalities seen in Parinaud's syndrome (see next section) likely arise from damage to the M-group of neurons in the rostral midbrain.

*Interestingly, the vestibular nuclei project directly to the abducens nuclei as well as to the PPRF. Therefore, the horizontal gaze palsy with a PPRF lesion (but not an abducens nucleus lesion) may be overcome by vestibular inputs such as the oculocephalic maneuver (see **neuroexam.com Video 35**) or ice water calorics.

REVIEW EXERCISE

1. Cover Figure 13.13B. For each lesion in Figure 13.13A, describe the expected eye positions on leftward and rightward gazes.

2. Cover the labels in Figure 13.13B. State the possible lesion locations and the side involved that would produce the eye movement abnormalities shown.

Convergence of the eyes is produced by the medial recti; **divergence**, by the lateral recti. The exact anatomical locations for centers in the brainstem controlling **vergence** have not been defined, but there appear to be separate pools of neurons in the midbrain reticular formation mediating either convergence or divergence movements. Vergence movements are under the control of descending inputs from the visual pathways in the occipital and parietal cortex and constitute part of the accommodation response discussed previously.

KEY CLINICAL CONCEPT

13.9 PARINAUD'S SYNDROME

Some clinical aspects of the brainstem circuits that control vertical eye movements were discussed in the preceding section and in KCC 13.3. In this section we will discuss an additional syndrome that includes vertical eye movement abnormalities. **Parinaud's syndrome** is a constellation of eye abnormalities usually seen with lesions compressing the dorsal midbrain and pretectal area. The four components of Parinaud's syndrome are:

1. Impairment of **vertical gaze, especially upgaze**. This may be due to compression of the dorsal part of the vertical gaze center (see the preceding section).
2. **Large, irregular pupils** that do not react to light but sometimes may react to near–far accommodation. This **light–near dissociation** (see KCC 13.5) may occur as a result of disruption of optic tract fibers traveling to the Edinger–Westphal nuclei via dorsal pathways including the posterior commissure (see Figure 13.8), while fibers descending from the visual cortex take a different route and are relatively spared.
3. **Eyelid abnormalities** ranging from bilateral lid retraction (Collier's sign) or "tucking" to bilateral ptosis.
4. **Impaired convergence** and sometimes **convergence–retraction nystagmus**, in which the eyes rhythmically converge and retract in the orbits, especially on attempted upgaze.

Common causes of Parinaud's syndrome are pineal region tumors (see KCC 5.8) and hydrocephalus (see KCC 5.7), as well as multiple sclerosis or vascular disease of the midbrain and pretectal area. Hydrocephalus can cause dilation of the suprapineal recess of the third ventricle (see Figure 5.11), which pushes downward onto the collicular plate (tectum) of the midbrain. Thus, hydrocephalus, especially in children, can produce the bilateral **setting-sun sign**, in which the eyes are deviated inward because of bilateral sixth nerve palsies (see KCC 13.4) and downward because of a Parinaud's syndrome. Similar downward and inward deviation of the eyes, sometimes referred to as **"peering at the tip of the nose,"** can be seen in **thalamic hemorrhage**, where the mechanism is not known. ∎

Control of Eye Movements by the Forebrain

Multiple parallel pathways descend from the cerebral cortex to control eye movement circuits in the brainstem; we will mention only a few of the better-known pathways here. Descending cortical pathways travel either directly to the brainstem centers for horizontal, vertical, or convergence eye movements (discussed earlier) or via relays in the midbrain **superior colliculi**.

The best-known cortical area that controls eye movements consists of the **frontal eye fields** (Figure 13.14). Based mainly on animal studies, this was formerly believed to correspond to Brodmann's area 8 (see Figure 2.15). However, recent functional imaging studies have suggested that the frontal eye fields in humans may lie more posteriorly, at the junction between the superior frontal

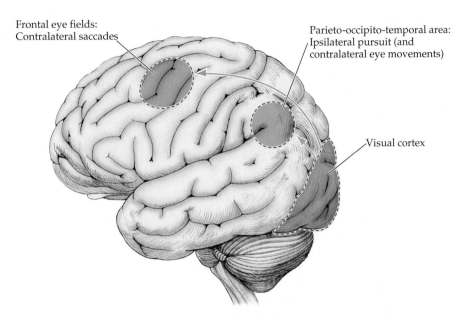

Frontal eye fields:
Contralateral saccades

Parieto-occipito-temporal area:
Ipsilateral pursuit (and
contralateral eye movements)

Visual cortex

FIGURE 13.14 Cortical Regions Important in Eye Movement Control

sulcus and the precentral sulcus, in Brodmann's area 6. Some authors consider the frontal eye fields to overlap the premotor and prefrontal cortices (see Figure 19.11A), reflecting their roles in eye movements and selective attention, respectively. The frontal eye fields generate **saccades** in the **contralateral** direction via connections to the contralateral PPRF (see Figure 13.12). More posterior cortical regions of the **parieto-occipito-temporal cortex** (see Figure 13.14) are primarily responsible for **smooth pursuit** movements in the **ipsilateral** direction, via connections with the vestibular nuclei, cerebellum, and PPRF, as we will discuss in the section on reflex eye movements. The parieto-occipital-temporal cortex may make some contribution to contralateral eye movements as well. Cortical descending control of eye movements is heavily influenced by **visual inputs** arriving at the primary visual cortex and visual association cortex (see Figure 13.14).

The **basal ganglia** also play a role in modulatory control of eye movements, and characteristic disorders of eye movements can be seen in basal ganglia dysfunction (see Chapter 16).

KEY CLINICAL CONCEPT

13.10 RIGHT-WAY EYES AND WRONG-WAY EYES

Lesions of the **cerebral hemispheres** normally impair eye movements in the contralateral direction, often resulting in a **gaze preference toward the side of the lesion**. This gaze preference is typically accompanied by weakness contralateral to the cortical lesion (if the corticospinal pathways are involved), so that the **eyes look away from the side of the weakness** (Figure 13.15A).

Certain clinical situations can cause the eyes to look **toward the side of the weakness**. This condition is called **wrong-way eyes** (Figure 13.15B). Causes of wrong-way eyes include **seizure activity in the cortex**, which can drive the eyes in the contralateral direction because of activation of the frontal eye fields (see Figure 13.14), while also causing abnormal or decreased movements of the contralateral side of the body because of involvement of motor association cortex and other structures. In addition, for unclear reasons, large lesions such as **thalamic hemorrhage** can disrupt the corticospinal pathways of the internal capsule, causing contralateral weakness, yet may also

FIGURE 13.15 Right-Way and Wrong-Way Eyes (A) Right-way eyes caused, for example, by a left cortical lesion affecting corticospinal pathways and frontal eye fields. (B) Wrong-way eyes caused, for example, by a left pontine lesion affecting corticospinal pathways and the PPRF.

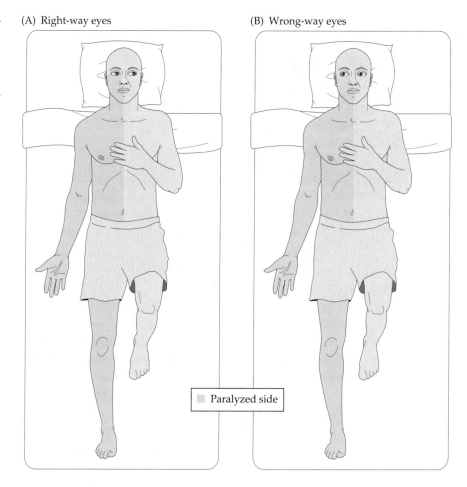

(A) Right-way eyes   (B) Wrong-way eyes

☐ Paralyzed side

cause the eyes to deviate toward the side of the weakness. Lesions in the thalamic region causing wrong-way eyes are usually accompanied by deep coma. Finally, lesions of the **pontine basis and tegmentum** (see Figure 13.13; Lesions 2 and 3; see also Figure 14.21C; Table 14.8) can cause wrong-way eyes because disruption of the corticospinal fibers causes contralateral hemiplegia, while involvement of the abducens nucleus or PPRF causes ipsilateral gaze weakness. ■

Cerebellar, Vestibular, and Spinal Control of Voluntary and Reflex Eye Movements

The cerebellum, vestibular nuclei, and cervical spinal proprioceptors influence ongoing voluntary eye movements and contribute to several forms of reflex eye movements. Two well-described forms of reflex eye movements are optokinetic nystagmus and the vestibulo-ocular reflex (VOR). The examiner can elicit **optokinetic nystagmus (OKN)** in the horizontal direction by moving a thick ribbon with vertical stripes (called an OKN strip) horizontally in front of the eyes (see **neuroexam.com Video 34**). The eyes alternate between smooth pursuit movements in the direction of stripe movement and backup (corrective) saccades opposite the direction of stripe movement in an attempt to stabilize the image. OKN is sometimes called "train nystagmus" because it can be observed in the eyes of fellow passengers as they watch the passing visual scene through an open window.

The **slow phase**, or smooth pursuit phase, of OKN is mediated by the ipsilateral posterior cortex (see Figure 13.14), with connections to the vestibular

OKN

nuclei and flocculonodular lobe of the cerebellum projecting to the PPRF and abducens nuclei (see Figure 13.12). The **fast phase**, or saccadic phase, of OKN is mediated by the frontal eye fields projecting ultimately to the contralateral PPRF (see Figure 13.12). Therefore, lesions of the frontal cortex or anywhere in the saccadic pathways disrupt the fast phases of OKN, while the slow phases are disrupted by lesions in the smooth pursuit pathways. OKN testing is thus useful to check for subtle dysfunction in the eye movement pathways. OKN can also be elicited in the vertical direction.

The **vestibulo-ocular reflex** (**VOR**) stabilizes the eyes on the visual image during head and body movements. Inputs from the vestibular nuclei, especially the medial vestibular nuclei, travel in the MLF to control the extraocular nuclei (see Figure 12.19). You can illustrate the importance of vestibulo-ocular reflexes by first fixating on your finger and turning your head from side to side, noting the relatively stable image of your finger, and comparing this to when you keep your head still and move your finger from side to side at about the same rate. In patients who are comatose and therefore lack visual fixation, the integrity of brainstem circuits mediating the VOR is often tested with the **oculocephalic maneuver** to elicit "doll's eyes" (see **neuroexam.com Video 35**) or with **cold water calorics** (see the section on the coma exam in Chapter 3). In the normal, awake individual, cerebellar circuits involving the flocculus and nodulus (vestibulocerebellum; see Chapter 15) enable visual fixation to overcome the VOR. This is why oculocephalic testing does not produce "doll's eyes" in the normal, awake individual. Similarly, visual fixation can suppress nystagmus evoked by caloric testing. In the **VOR suppression test**, patients fixate on an object moving with the head as it rotates (e.g., place a drinking straw in the patient's mouth and ask them to fixate on the far end of the straw as they turn their head from side to side). The presence of nystagmus indicates cerebellar dysfunction.

Proprioceptive inputs also help stabilize the eyes on the visual image, especially during head and neck movements.

Oculocephalic testing

CLINICAL CASES

CASE 13.1 DOUBLE VISION AND UNILATERAL EYE PAIN

CHIEF COMPLAINT

A 48-year-old woman came to the emergency room with worsening **left eye pain and intermittent double vision**.

HISTORY

Approximately 4 or 5 years previously, the patient had begun to have **left frontal and left retro-orbital headaches** that occurred intermittently at first, and then on an almost daily basis. An MRI scan was reportedly normal, and she was diagnosed with cluster migraine. The headaches continued to occur but were relieved by ibuprofen. One and a half years prior to presentation she began to have intermittent **drooping of the left eyelid and dilation of the left pupil**. She also noticed that her **left eye occasionally drifted to the left**, causing diplopia. This patient was very observant, and she noticed that her **diplopia was worse when looking to the right**. When covering each eye alternately, she reported that the two images did not overlap. In fact, the **image from her left eye appeared to the right and slightly above the image from her right eye**. These symptoms gradually progressed from intermittent to continuous, and her headaches were no longer relieved by up to 12 ibuprofen tablets per day, so she came to the emergency room.

PHYSICAL EXAMINATION

Vital signs: T = 98.7°F, P = 78, BP = 180/90.

(continued on p. 592)

CASE 13.1   *(continued)*

Neck: Supple with no bruits.

Lungs: Clear.

Heart: Regular rate with no murmurs.

Abdomen: Soft, nontender.

Neurologic exam:

MENTAL STATUS: Alert and oriented × 3. Speech fluent, with intact naming and repetition. 3/3 words recalled after 5 minutes.

CRANIAL NERVES: Visual fields full. Fundi normal. Right pupil 4 mm, constricting to 3 mm with direct and consensual light stimulation, and with accommodation. **Left pupil 6 mm, with no direct or consensual response to light and no response to accommodation. Left eye had limited but not absent upgaze, downgaze, and adduction.** Normal left eye abduction. Normal right eye movements. **Left ptosis**, with left palpebral fissure 6 mm and right 9 mm. Corneal reflexes intact. Facial sensation intact. Face symmetrical, other than left ptosis already described. Normal palate and tongue movements.

MOTOR: No drift. Normal tone. 5/5 power throughout.

REFLEXES:

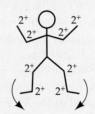

COORDINATION: Normal on finger-to-nose and heel-to-shin testing.

GAIT: Not tested.

SENSORY: Intact light touch, pinprick, vibration, and joint position sense. Normal graphesthesia; no extinction.

LOCALIZATION AND DIFFERENTIAL DIAGNOSIS

1. On the basis of the symptoms and signs shown in **bold** above, where is the lesion?

2. What is the most likely diagnosis, and what are some other possibilities?

Discussion

1. The key symptoms and signs in this case are:

   - **Left frontal and retro-orbital headaches**
   - **History of left eye drifting to the left, and diplopia with image from left eye above and to right of image from right eye, with diplopia worse when looking to the right**
   - **Left eye with limited but not absent upgaze, downgaze, and adduction, left ptosis, and a fixed dilated left pupil**

   This patient has a history of diplopia and findings on exam that are compatible with an oculomotor nerve (CN III) palsy (see KCC 13.2; Figure 13.5). Her left pupil was dilated and unresponsive. Although left eye adduction, upgaze, and downgaze were decreased, they were not absent, suggesting a partial third-nerve palsy. In addition, there was no description of an abnormal eye position at rest. Headaches can have numerous causes (see KCC 5.1); when they are always on the same side, however, an intracranial abnormality on that side should be suspected.

   The most likely *clinical localization* is left oculomotor nerve (CN III).

2. Painful third-nerve palsy should be treated as an aneurysm until proven otherwise. The most common aneurysm causing CN III palsy occurs where the posterior communicating artery (PComm) branches off from the internal carotid (see Figures 5.20, 13.2; KCC 5.6), although aneurysms of the internal carotid, posterior cerebral, and superior cerebellar arteries should also be considered. For other causes of CN III palsy, see KCC 13.2.

Clinical Course and Neuroimaging

A head CT (not shown) revealed an egg-shaped, 1 cm mass near the left edge of the posterior clinoid process (see Figure 12.3A) that enhanced with intravenous contrast. This occurred before CTA was in wide use, so a conventional

cerebral angiogram was performed (Image 13.1, page 594), showing a 1.2 cm aneurysm arising in the region where the PComm branches off the internal carotid (compare to Figure 4.16C). The PComm itself could not be seen on the angiogram. The aneurysm had a well-visualized neck, and its dome pointed posteriorly, along the course of the third nerve. The patient was taken to the operating room and underwent a left frontotemporal craniotomy. The left frontal and temporal lobes were carefully retracted and the left internal carotid artery visualized. Using an operating microscope, the neurosurgeons identified the neck of the aneurysm. The dome of the aneurysm was seen to project posteriorly and inferiorly under the edge of the tentorium cerebelli. A small PComm artery was found to arise from the internal carotid adjacent to the neck of the aneurysm. The neurosurgeons carefully placed a clip across the neck of the aneurysm, being cautious to avoid occluding any other small vessels such as the PComm. The dome of the aneurysm immediately became less tense and could be safely opened with microscissors, and some blood was evacuated, leading to decompression of adjacent structures. Postoperatively the patient did quite well; 1 week after surgery her left pupil was still nonreactive, but she had nearly full movement of the left eye, no diplopia, and no headaches.

CASE 13.2 A DIABETIC WITH HORIZONTAL DIPLOPIA

MINICASE

A 54-year-old man with a history of diabetes awoke one morning with **horizontal diplopia that increased on gaze to the left and decreased on gaze to the right**. He initially had some pain in the left periorbital area, which resolved after a few days. Exam was normal except for **incomplete abduction of the left eye**. He was able to move the left eye slightly past the midline toward the left; however, he was unable to fully "bury the sclera," as he could with the right eye when looking to the right. He had **horizontal diplopia with no vertical component, which was worse on left gaze**.

LOCALIZATION AND DIFFERENTIAL DIAGNOSIS

1. On the basis of the symptoms and signs shown in **bold** above, where is the lesion?
2. What is the most likely diagnosis, and what are some other possibilities?

Discussion

1. The key symptoms and signs in this case are:

   - **Horizontal diplopia, worse on left gaze, with incomplete abduction of the left eye**

     This patient has dysfunction of the left lateral rectus muscle causing dysconjugate horizontal gaze, and diplopia (see Figures 13.1, 13.7, 13.13). Possible causes of incomplete left eye abduction include dysfunction of the left abducens nerve (CN VI) or the lateral rectus muscle, or a mechanical problem in the orbit.

2. Given the patient's history of diabetes and the lack of any other associated findings, the most likely diagnosis is an isolated abducens nerve palsy, caused by microvascular disease. Other possible causes of abducens nerve palsy and horizontal diplopia are discussed in KCC 13.1 and 13.4.

Clinical Course

A head CT was normal. When seen in follow-up 2 weeks later, the patient showed no change, but after 2 months his diplopia was less severe and he was able to bury the sclera almost completely on leftward gaze. Three months after presentation his exam was normal, and he had no symptoms except for occasional transient diplopia when looking in the distance to the left. Review Case 5.7 for another important cause of abducens palsy.

CASE 13.1 DOUBLE VISION AND UNILATERAL EYE PAIN

IMAGE 13.1 Left Posterior Communicating Artery (PComm) Aneurysm
Angiogram from left internal carotid artery injection. Lateral view.

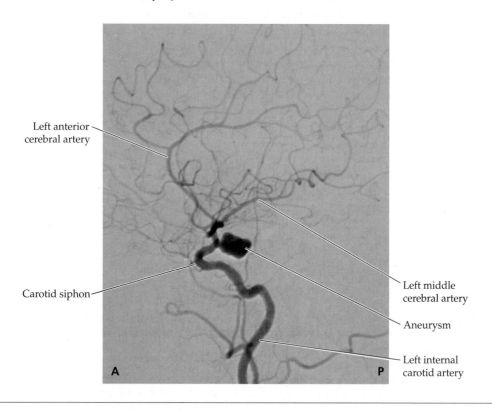

Left anterior
cerebral artery

Left middle
cerebral artery

Carotid siphon

Aneurysm

Left internal
carotid artery

A   P

CASE 13.3 VERTICAL DIPLOPIA

MINICASE

A 74-year-old man awoke one morning with **vertical diplopia**. The diplopia was relieved by covering of either eye and did not vary in severity at different times of the day. He had no history of head trauma. Past medical history was notable only for hypertension. Exam was normal except for a **right hypertropia and incomplete downgaze with the right eye when looking medially**. He was tested with a red glass over the right eye (Figure 13.16) and had **vertical diplopia, with the image from the right eye located below the image from the left eye**. The diplopia **worsened with downward and leftward gaze and improved with leftward head tilt**.

LOCALIZATION AND DIFFERENTIAL DIAGNOSIS

1. On the basis of the symptoms and signs shown in **bold** above, where is the lesion?

2. What is the most likely diagnosis, and what are some other possibilities?

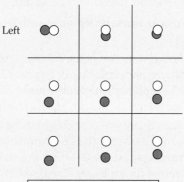

Left

● = Red image, right eye
○ = White image, left eye

FIGURE 13.16 Results of Red Glass Testing Red glass was held over the patient's right eye, and results are drawn as seen from the patient's perspective.

Discussion

1. The key symptoms and signs in this case are:

   • **Right hypertropia, and vertical diplopia worse with downward and leftward gaze and worse with rightward head tilt**

   The findings in this patient are compatible with a right trochlear palsy (see KCC 13.3; Figure 13.6). A disorder of the superior oblique muscle itself is also possible.

2. The most likely cause of the isolated trochlear palsy in this patient is, as in Case 13.2, an idiopathic neuropathy of presumed microvascular origin. For other possible causes of trochlear nerve palsy and vertical diplopia, see KCC 13.1 and KCC 13.3.

Clinical Course

A head CT was normal, Tensilon test (see KCC 8.1) was negative, and hemoglobin A_{1c} (an indicator of average blood glucose) was borderline-elevated at 7.7%, suggesting possible diabetes. The patient initially wore an eye patch over his right eye because this eliminated his diplopia and made it easier for him to function normally. He gradually improved, and 3 months later he no longer needed the eye patch and had diplopia only while reading, which involves looking down.

CASE 13.4 LEFT EYE PAIN AND HORIZONTAL DIPLOPIA

MINICASE

A 27-year-old man with no previous medical problems came to the emergency room because of 1 week of worsening left-sided headaches, left eye pain, and horizontal diplopia. He awoke 1 week prior to presentation with a severe left frontal headache. Two days later the headache had moved to his left eye, and he began noticing horizontal diplopia on rightward gaze. He went to his primary medical doctor and had an MRI scan, which was reportedly normal. Because of continued symptoms, he finally decided to come to the emergency room for further evaluation, where he was seen by a neurology consult resident. Examination was normal except for **mild erythema of the left orbital conjunctiva** and **slightly decreased adduction of the left eye on right lateral gaze**. He had **horizontal diplopia and left eye pain on rightward gaze, and the rightmost image vanished when the left eye was covered**. There was also **minimal horizontal diplopia on leftward gaze, with the leftmost image vanishing when the left eye was covered**.

LOCALIZATION AND DIFFERENTIAL DIAGNOSIS

1. On the basis of the patient's symptoms and his findings on neurologic exam, what is the most likely location for a lesion in this patient causing horizontal diplopia (see KCC 13.1, 13.4)?

2. Given the pain and erythema of the left orbital conjunctiva, what are some possibilities for the diagnosis?

Discussion

1. The key symptoms and signs in this case are:

   • **On right gaze: left eye pain, limited adduction, and horizontal diplopia, with the right image vanishing when the left eye was covered**

   • **On left gaze: mild horizontal diplopia, with the left image vanishing when the left eye was covered**

   • **Pain and erythema of the left orbital conjunctiva**

   The examination reveals bilateral limitations of horizontal eye movement of the left eye, with more difficulty looking to the right. Relatively isolated dysfunction of both the left medial and lateral rectus muscles would be difficult to explain on the basis of nerve or CNS lesions (see Figure 13.1; Table 13.1; KCC

13.1, 13.4). In addition, the eye pain that worsens with movement suggests a possible mechanical cause in the orbit. One possibility would be a lesion that restricts movement of the left lateral rectus muscle, limiting its ability to stretch on right lateral gaze and decreasing its ability to contract on left lateral gaze.

2. The differential diagnosis includes orbital trauma (although there was none based on history); thyroid disease (although pain and subacute onset don't fit with this diagnosis); myasthenia gravis (again, this condition should not cause pain); or, more likely, especially given the pain and erythema, an infectious, inflammatory, or neoplastic disorder such as orbital cellulitis, orbital lymphoma, orbital myositis (orbital pseudotumor), sarcoidosis, Tolosa–Hunt syndrome, fungal infection, or cavernous sinus thrombosis.

Clinical Course and Neuroimaging

In the emergency room the patient underwent a head CT scan. In addition, he had a lumbar puncture (see KCC 5.10), and cerebrospinal fluid studies were normal. On careful review of the CT scan, the left lateral rectus muscle appeared slightly thickened. The patient was therefore admitted, and a **brain MRI** with gadolinium was done (Image 13.4, page 599). The MRI revealed marked enhancement and thickening of the left lateral rectus muscle, compatible with the diagnosis of orbital myositis (orbital pseudotumor), a relatively uncommon inflammatory condition of the extraocular muscles. The patient was treated with oral steroids and discharged home. His symptoms had improved when he was seen 1 week later in the outpatient office, but he was subsequently lost to follow-up because his insurance would not cover visits to a neurologist outside of his program.

CASE 13.5 UNILATERAL HEADACHE, OPHTHALMOPLEGIA, AND FOREHEAD NUMBNESS

CHIEF COMPLAINT

A 24-year-old woman with a history of pituitary adenoma suddenly developed severe headache, left forehead and cheek numbness, and inability to move the left eye.

HISTORY

The patient presented 2 years previously with Cushing's syndrome (see KCC 17.1) and underwent two operations to resect a pituitary adenoma. She did well until 2 weeks before admission, when she started having **left frontal headaches**, especially around the left eye and nose. An MRI scan showed recurrent pituitary adenoma, and radiation therapy was planned. Two days prior to admission, however, she had onset of **horizontal diplopia** and was found by her endocrinologist to have a **left CN VI palsy**. She was admitted to an outside hospital and treated with steroids to try to reduce swelling. However, the next day she had a sudden worsening of her headache, with **pain and numbness involving the left cheek and forehead; a dilated, fixed left pupil; and almost no movement of the left eye.** She was therefore urgently transferred to a tertiary care center for further evaluation and treatment.

PHYSICAL EXAMINATION

General appearance: Crying, anxious young woman in apparent pain.

Vital signs: T = 97°F, P = 92, BP = 172/112.

Neck: Supple with no bruits.

Lungs: Clear.

Heart: Regular rate with no murmurs.

Abdomen: Benign.

Neurologic exam:

MENTAL STATUS: Alert and oriented × 3.

CRANIAL NERVES: Right pupil 3 mm, constricting to 2 mm. **Left pupil 6 mm, with no direct or consensual response to light.** Visual fields full. Fundi normal. Normal movements of the right eye, but the **left eye had no extraocular movements and a marked left ptosis.** She had diplopia in all directions of gaze. **Sensation was slightly decreased to pinprick in the left forehead, eyelid, left bridge of the nose, and upper cheek** (Figure 13.17). Face was symmetrical, aside from the left

CASE 13.5 *(continued)*

ptosis noted above. Shoulder shrug was normal, and tongue was midline.

MOTOR: No pronator drift. 5/5 power throughout.

REFLEXES: Not tested.

COORDINATION: Normal on finger-to-nose testing.

GAIT: Not tested.

SENSORY: Intact light touch and pinprick sensation.

LOCALIZATION AND DIFFERENTIAL DIAGNOSIS

1. On the basis of the symptoms and signs shown in **bold** above, where is the lesion?
2. Given the patient's history of a pituitary adenoma and the sudden onset of her deficits, what is the most likely diagnosis? What are some other possibilities?

FIGURE 13.17 Region of Decreased Sensation

Discussion

1. The key symptoms and signs in this case are:

   - **Initial left abducens palsy, evolving to ophthalmoplegia, ptosis, and a fixed dilated pupil**
   - **Pain, paresthesia, and decreased sensation to pinprick in the left forehead, eyelid, bridge of nose, and upper cheek**

   This patient had dysfunction of the left CN III, IV, VI, and V_1, constituting a left cavernous sinus syndrome (see KCC 13.7; Figure 13.11). Lack of involvement of the optic nerve suggests that the disorder was not in the orbital apex. Interestingly, the initial abnormal finding was an abducens palsy, suggesting a lesion that expanded from medial to lateral within the left cavernous sinus (see Figure 13.11).

   The most likely *clinical localization* is the left cavernous sinus.

2. This patient has a history of recurrent pituitary adenoma. The progress of her symptoms, however, is too rapid to be explained easily by extension of the tumor into the cavernous sinus. The sudden worsening of her symptoms could be explained by hemorrhage into the tumor or pituitary apoplexy (see KCC 17.3). Interestingly, pituitary apoplexy often recurs in patients without a previously known history of adenoma. Other possibilities include cavernous carotid aneurysm or fistula, cavernous sinus thrombosis, or infection (see KCC 13.7).

Clinical Course and Neuroimaging

The tertiary care center to which the patient was transferred had an MRI scanner readily available. Therefore, an **MRI** was performed on an urgent basis because MRI provides better resolution than CT for structures in the pituitary and cavernous sinus region (Image 13.5A,B, page 600). Image 13.5A is a coronal T1-weighted image showing a large hemorrhage extending from the pituitary fossa into the left cavernous sinus. Note that the optic tract was not affected, and more anteriorly (not shown), the optic nerves were also spared. Image 13.5B is an axial T2-weighted image, again showing the hemorrhage. The trigeminal nerve is nicely demonstrated in Image 13.5B and can be seen to enter Meckel's cave on both sides (compare to Figure 12.3A).

Cerebrospinal fluid surrounding the trigeminal ganglion in Meckel's cave appears white on T2-weighted images.

Because of these findings, the patient was taken to the operating room for a transsphenoidal resection (see KCC 17.1) of the hemorrhage and pituitary adenoma. Hemorrhagic tissue was removed from the left cavernous sinus, and hemostatic packing was inserted. By postoperative day 1, the patient had recovered some limited left extraocular movements and slight left pupillary responses. She was transferred back to the hospital closer to her home for the remainder of her recovery.

CASE 13.6 PTOSIS, MIOSIS, AND ANHIDROSIS

MINICASE

A 17-year-old male got into an argument with his sister while intoxicated, and she shot him in the neck with a steel pellet gun. He was brought to the emergency room and treated for a left pneumothorax (collapsed lung). The emergency room physician noticed unequal pupils and called a neurology consult. On exam, the patient had an entry wound at the base of his neck just above the left clavicle (**Figure 13.18**). There was no neck swelling. The right pupil was 4 mm, constricting to 3 mm in response to light, and the **left pupil was 2 mm, constricting to 1.5 mm. The left eyelid had 3 mm of ptosis compared to the right. The left forehead felt smoother than the right, suggesting decreased sweat production.** When the right neck was pinched, the right pupil dilated (**ciliospinal reflex**). **Pinching the left neck caused no change in the left pupil.** The remainder of the exam was normal.

LOCALIZATION AND DIFFERENTIAL DIAGNOSIS

1. On the basis of the symptoms and signs shown in **bold** above, where is the lesion?
2. What is the most likely diagnosis, and what are some other possibilities?

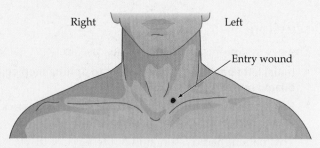

FIGURE 13.18 Location of Entry Wound

Discussion

1. The key symptoms and signs in this case are:

   - **Left ptosis**
   - **Small, reactive left pupil, with decreased ciliospinal reflex**
   - **Decreased left facial sweating**

   This patient has Horner's syndrome (see KCC 13.5, 13.6). Horner's syndrome can be caused by a lesion anywhere in the sympathetic pathway to the eye (see Figure 13.10). However, given the history of a penetrating neck wound, the lesion is probably located in the sympathetic chain or sympathetics of the carotid plexus. The entry wound was low in the neck (see Figure 13.18), making direct injury to the sympathetics in the carotid plexus unlikely (see Figure 13.10). In addition, impaired sweating is more common with preganglionic lesions. Nevertheless, the carotid artery may have been injured low in the neck, causing a carotid dissection (see KCC 10.6) extending superiorly, resulting in a Horner's syndrome. A lesion of the upper thoracic nerve roots or spinal cord is unlikely, given the absence of any other neurologic findings, although the pneumothorax suggests the pellet traversed the region of the lung apex.

2. Possible causes of Horner's syndrome in this pattern include direct traumatic injury to the sympathetic chain in the neck (see Figure 13.10) or carotid dissection (see KCC 10.6).

   In summary, the most likely *clinical localization* is left sympathetic chain in the vicinity of the lower neck or lung apex, or left carotid plexus.

Clinical Course and Neuroimaging

Because of the possibility of a carotid dissection, an angiogram was done (Image 13.6A, page 601; note that the steel pellet would make MRI/MRA dangerous. CT angiography was not available at the time, but might be difficult to interpret due to shadowing artifact). No dissection was seen. The steel pellet took a slightly downward course to lodge behind the carotid artery, at the level of the T1 and T2 nerve root exit points. Injury to the lower cervical or upper thoracic sympathetic trunk during the steel pellet's trajectory through the neck was therefore the most likely diagnosis. Because of the pellet's location adjacent to vital structures, it was not removed. Social services and psychiatry were consulted because of the circumstances of his injury. When seen 1 month later in follow-up the patient was doing well, but his left ptosis and miosis remained essentially unchanged.

RELATED CASE. Image 13.6B (page 601) shows an MRI of the neck from a different patient with ptosis and miosis. This 43-year-old woman struck her head against the steering wheel in a car accident. She was well until 2 weeks later, when she developed pain over her right eye, with right ptosis and miosis. The T1-weighted MRI Image 13.6B done without contrast demonstrates a white crescent of clotted blood (see Table 4.4) in the false lumen of a carotid dissection (see KCC 10.6). In this case, the Horner's syndrome was associated with a carotid dissection. She was treated with anticoagulation for several months to prevent emboli.

CASE 13.4 LEFT EYE PAIN AND HORIZONTAL DIPLOPIA

IMAGE 13.4 Orbital Pseudotumor Involving Left Lateral Rectus Muscle T1-weighted axial MRI image with intravenous gadolinium demonstrating abnormal enhancement and thickening of the left lateral rectus muscle compatible with orbital pseudotumor.

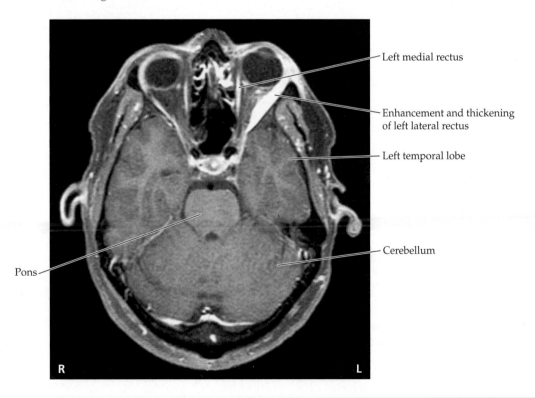

Left medial rectus

Enhancement and thickening of left lateral rectus

Left temporal lobe

Cerebellum

Pons

CASE 13.5 UNILATERAL HEADACHE, OPHTHALMOPLEGIA, AND FOREHEAD NUMBNESS

IMAGE 13.5A,B Pituitary Apoplexy Causing Left Cavernous Sinus Syndrome (A) Coronal T1-weighted image demonstrating hemorrhage extending from the region of the pituitary fossa into the left cavernous sinus. (B) Axial T2-weighted image, again demonstrating hemorrhage in the left cavernous sinus, adjacent to Meckel's cave.

(A)

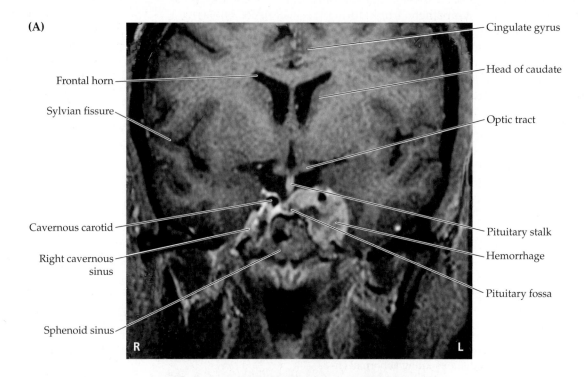

Cingulate gyrus
Head of caudate
Optic tract
Frontal horn
Sylvian fissure
Pituitary stalk
Cavernous carotid
Hemorrhage
Right cavernous sinus
Pituitary fossa
Sphenoid sinus

R   L

(B)

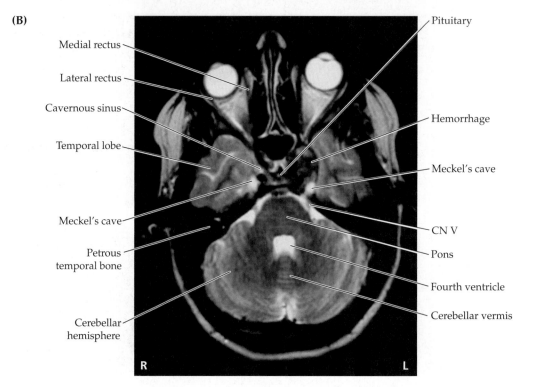

Pituitary
Medial rectus
Lateral rectus
Cavernous sinus
Hemorrhage
Temporal lobe
Meckel's cave
Meckel's cave
CN V
Petrous temporal bone
Pons
Fourth ventricle
Cerebellar hemisphere
Cerebellar vermis

R   L

IMAGE 13.6A Steel Pellet in Region of the Sympathetic Trunk Angiogram from aortic arch injection. Left anterior oblique view. The trajectory made by the steel pellet from the entry site to its final location traverses the region of the left T1 and T2 nerve roots and the junction between the thoracic and cervical portions of the sympathetic trunk (compare to Figure 13.10).

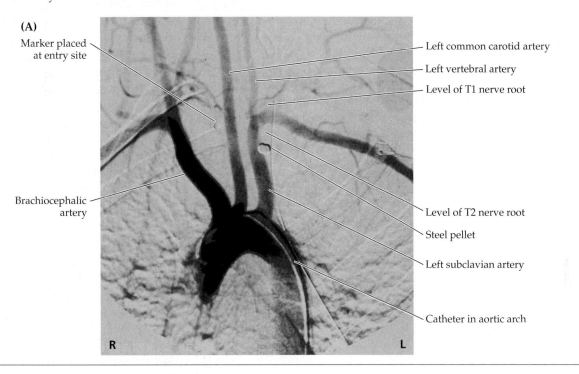

(A)

Marker placed at entry site

Brachiocephalic artery

Left common carotid artery

Left vertebral artery

Level of T1 nerve root

Level of T2 nerve root

Steel pellet

Left subclavian artery

Catheter in aortic arch

R   L

CASE 13.6 *RELATED CASE*

IMAGE 13.6B Right Carotid Dissection Axial, T1-weighted MRI image through the neck and skull base reveals a crescent of blood in the right carotid, consistent with a right carotid dissection.

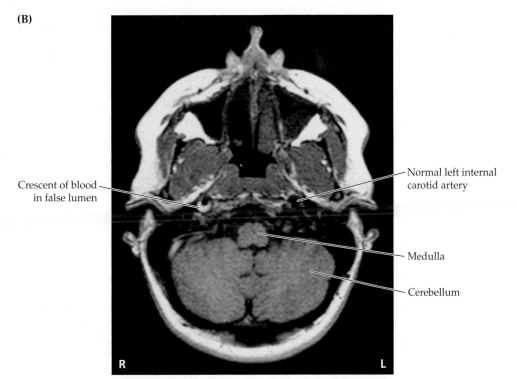

(B)

Crescent of blood in false lumen

Normal left internal carotid artery

Medulla

Cerebellum

R   L

CASE 13.7 WRONG-WAY EYES

MINICASE

A 71-year-old man with a history of diabetes collapsed on the street and was unable to stand up, so he was brought to the emergency room by ambulance. On examination, he was awake but **lethargic**, had a **rightward gaze preference**, and was **unable to move either eye past the midline toward the left**. In addition, he had **weakness of the right lower face** and **2/5 strength in the right arm and leg**, with an **upgoing plantar response on the right**. Examination was otherwise unremarkable.

LOCALIZATION AND DIFFERENTIAL DIAGNOSIS

1. Lesions in which locations can cause the constellation of symptoms and signs shown in **bold** above?
2. Given the patient's age and the time course of his presentation, what is the most likely diagnosis, and what are some other possibilities?

Discussion

1. The key symptoms and signs in this case are:

   - **Lethargy**
   - **Rightward gaze preference, with inability to move either eye past the midline toward the left**
   - **Right face, arm, and leg weakness, with upgoing plantar response on the right**

   This patient has a combination of a left horizontal gaze palsy and right hemiparesis constituting so-called wrong-way eyes (see KCC 13.10; Figure 13.15). Wrong-way eyes can be caused by ongoing seizure activity in one of the cerebral hemispheres, by lesions in the region of the thalami, or by lesions in the pons affecting the corticospinal tract (contralateral hemiparesis) and abducens nucleus (ipsilateral gaze preference). The patient's lethargy suggests possible mild involvement of brainstem activating systems (see Chapter 14). Lesions in the vicinity of the thalami causing wrong-way eyes usually do so in the setting of profound coma, which this patient did not have.

2. Given the patient's age and history of diabetes, the most likely diagnosis is an infarct in the left pons involving the left corticospinal and corticobulbar fibers as well as the left abducens nucleus (or PPRF). A hemorrhage in this location is also a possibility. Another possibility is that the patient could be having ongoing seizure activity (focal status epilepticus) in the left hemisphere, causing the rightward gaze preference and right-sided weakness, but this is less likely, given the history.

Clinical Course and Neuroimaging

The patient's CT scan showed an area of increased density at the top of the basilar artery, so he was taken for an emergency angiogram with a plan for intra-arterial thrombolysis (see KCC 10.4). However, the angiogram revealed a patent basilar artery, and it was concluded that the increased density seen in the basilar artery on CT scan was calcification rather than thrombosis (see Table 4.1). He was therefore admitted to the hospital while an embolic evaluation was pursued (see KCC 10.4). An MRI scan (Image 13.7, page 605) revealed an area of increased T2 signal in the left pons consistent with an infarct from a penetrating vessel arising from the basilar artery (see KCC 14.3; Figure 14.21C; Table 14.8). The patient was found to have paroxysmal atrial fibrillation and was treated with chronic oral anticoagulation and transferred to an inpatient rehabilitation facility. When seen in the office 4 months later, however, he still had severe 3/5 right hemiparesis and a left horizontal gaze palsy.

CASE 13.8 HORIZONTAL DIPLOPIA IN A PATIENT WITH MULTIPLE SCLEROSIS

MINICASE

A 25-year-old woman had a 2-year history of multiple sclerosis with relapsing and remitting episodes of weakness and sensory deficits in her extremities. Two weeks prior to evaluation, she had onset of horizontal diplopia. Compared to previous exams, the new findings consisted of an **abnormality of rightward horizontal gaze**, such that the **left eye did not adduct past the midline** and the **right eye had sustained end gaze nystagmus on abduction**. In contrast, when convergence was tested, the left eye was able to adduct past the midline. Leftward horizontal gaze was normal.

LOCALIZATION AND DIFFERENTIAL DIAGNOSIS

1. Draw the positions of this patient's eyes on leftward horizontal gaze, on rightward horizontal gaze, and during convergence.
2. What is the location of this patient's lesion, and which side is it on?
3. What is the most likely cause?

Discussion

1. The key symptoms and signs in this case are:
   - **Left eye did not adduct past the midline**
   - **Right eye had sustained end gaze nystagmus on abduction**

   The positions of this patient's eyes during leftward and rightward gaze are shown in Figure 13.13B, Lesion 4. Convergence was normal.
2. These findings constitute a left INO localized to the left MLF (see Figure 13.13A, Lesion 4).
3. Given the patient's history, the most likely diagnosis is a multiple sclerosis plaque (see KCC 6.6) in the left MLF.

Initial Clinical Course and Neuroimaging

A **brain MRI** showed a new region of increased T2 signal along the floor of the fourth ventricle in the region of the left MLF (Image 13.8A,B, page 606).

CASE 13.8 (continued)

During the next few days the patient gradually developed problems with leftward horizontal gaze as well, so **neither eye would move past the midline when looking to the left**. On rightward gaze, she continued to have **no adduction of the left eye**. Thus, the only remaining horizontal eye movement was **right eye abduction**, which continued to have **end gaze nystagmus**, as before.

LOCALIZATION AND DIFFERENTIAL DIAGNOSIS

1. Draw the positions of this patient's eyes on leftward horizontal gaze and on rightward horizontal gaze following her clinical worsening.
2. What is the location of this patient's lesion now?

Discussion

1. The key symptoms and signs in this case are:
   - **Inability of either eye to move past the midline when looking to the left**
   - **No adduction of the left eye**
   - **End gaze nystagmus on right eye abduction**

   The positions of this patient's eyes during leftward and rightward gaze are shown in Figure 13.13B, Lesion 5.
2. These findings constitute a left INO plus a left horizontal gaze palsy, also called one-and-a-half syndrome (see KCC 13.8). Most likely her demyelina-

tive plaque in the left MLF enlarged to involve the left abducens nucleus or PPRF (see Figure 13.13A, Lesion 5).

Clinical Course

The patient was treated with steroids, and her eye movements and diplopia gradually improved. On follow-up examination she was able to move her eyes fully in all directions; in tests of saccades to the right, however, her right eye could be seen to move quickly to the right while her left eye moved more slowly to the right, lagging slightly behind. This is a way to test for a mild INO on physical examination (see KCC 13.8) and demonstrated a slight residual deficit in this patient.

CASE 13.9 HEADACHES AND IMPAIRED UPGAZE

MINICASE

A 23-year-old aerospace engineer developed mild **headaches and difficulty looking up** over the course of 3 weeks, so he went to see his family physician. On examination, his **pupils were about 6 mm in diameter bilaterally and had minimal reaction to light**, but they did constrict during accommodation. He was **unable to look upward at all** past the horizontal plane, but he had otherwise full eye movements in other directions. When attempting to look upward, or after closing and opening his eyes, he had **retraction of the upper eyelids and convergence–retraction nystagmus** of both eyes. Examination was otherwise normal.

LOCALIZATION AND DIFFERENTIAL DIAGNOSIS

1. On the basis of the symptoms and signs shown in **bold** above, what syndrome does this patient have affecting his eye movements, and what is the usual localization of this syndrome?

2. What is the most likely diagnosis, and what are some other possibilities?

Discussion

1. The key symptoms and signs in this case are:
   - Headaches
   - Large pupils with minimal reaction to light but preserved reaction to accommodation (light–near dissociation)
   - Inability to look upward
   - Lid retraction and convergence–retraction nystagmus

   The patient has a classic Parinaud's syndrome (see KCC 13.9). Parinaud's syndrome is usually caused by compression of or lesions in the dorsal midbrain and pretectal area. Impaired upgaze is probably caused by dysfunction of the upgaze portion of the vertical gaze center located in the dorsal part of the rostral midbrain reticular formation. Light–near dissociation probably occurs because fibers from the optic tract reaching the Edinger–Westphal nuclei travel via dorsal pathways including the posterior commissure and have been disrupted, while fibers descending from the visual cortex take a different route and are relatively spared (see KCC 13.5; Figure 13.8). Convergence and eyelid abnormalities associated with Parinaud's syndrome also localize to the rostral midbrain-pretectal region. Headaches of the kind seen in this patient can have many causes (see KCC 5.1) but also support the presence of intracranial pathology.

2. Given the presence of a gradually progressive Parinaud's syndrome, the most likely diagnosis in this patient is a pineal region tumor (see KCC 5.8) that has enlarged to compress the dorsal midbrain. In young children before the cranial sutures close, hydrocephalus can present with Parinaud's syndrome as the main abnormality. In an adult, however, if the hydrocephalus were se-

vere enough to cause these findings, then intracranial pressure would likely be severely elevated as well, causing impaired consciousness (see KCC 5.3), which this patient does not have. Therefore, the most likely diagnosis is a primary neoplasm of the pineal region. Another, less likely possibility is that the patient has another lesion in this region, including metastatic tumor, infarct, demyelination, infection, or vascular malformation.

Clinical Course and Neuroimaging

A **brain MRI** with intravenous gadolinium showed an enhancing lesion in the pineal region causing severe compression of the dorsal midbrain and pretectal area (Image 13.9A,B, page 609). He was referred to a neurosurgeon and underwent a biopsy of the lesion performed stereotactically (see KCC 16.4). The pathologic diagnosis was uncertain, revealing either an intermediate-grade pineocytoma or a pineal teratoma. He was treated with multiple cycles of chemotherapy and with radiation therapy, with repeated worsening and temporary improvement in his symptoms, but he remained relatively stable at last follow-up, 2 years after initial presentation.

Additional Cases

Related cases for the following topics can be found in other chapters: **abnormalities of eye movements or pupillary responses** (Cases 5.2–5.7, 10.8, 10.10–10.12, 11.2, 14.1, 14.5–14.8, 15.3, 15.4, 16.2, 16.4, 18.3). Other relevant cases can be found using the **Case Index** located at the end of the book.

CASE 13.7 WRONG-WAY EYES

IMAGE 13.7 Infarct in Left Medial Pontine Basis and Tegmentum Axial, T2-weighted MRI image. An infarct is present in the left pons in the region of the left corticospinal tract, PPRF, and possibly the abducens nucleus.

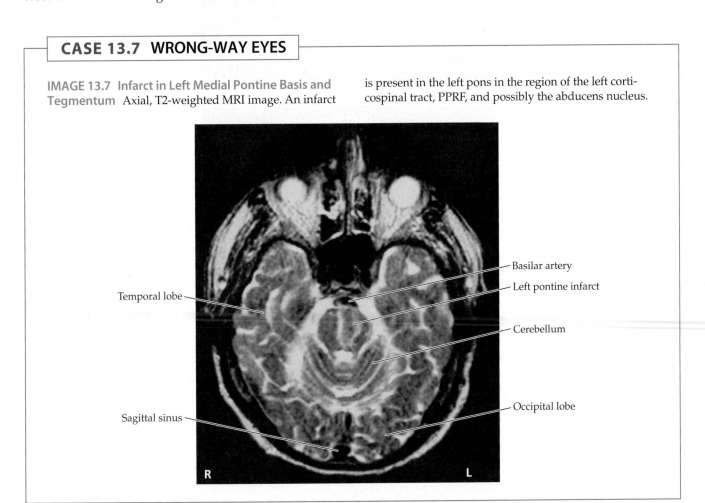

CASE 13.8 HORIZONTAL DIPLOPIA IN A PATIENT WITH MULTIPLE SCLEROSIS

IMAGE 13.8A,B **Plaque in the Left MLF** Axial, proton density–weighted MRI images progressing from inferior (A) to superior (B). A bright plaque is visible in the region of the left MLF along the floor of the fourth ventricle (compare to MLF location shown in Figure 14.4).

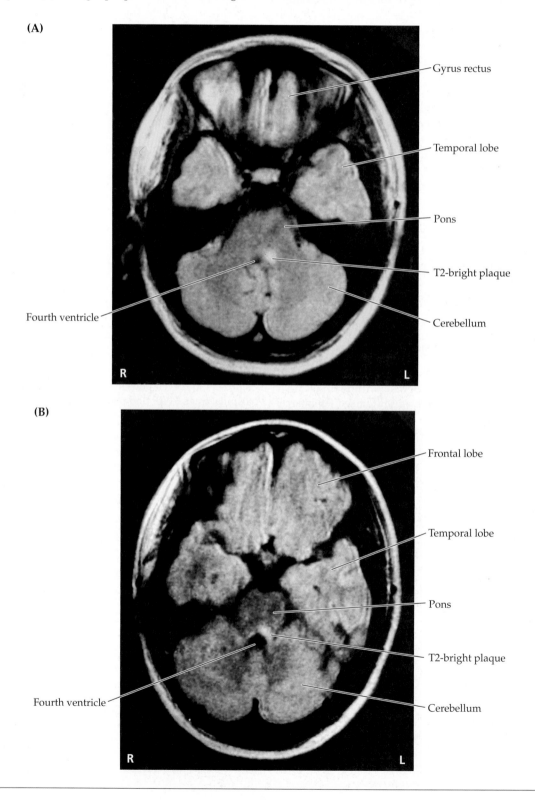

(A)

Gyrus rectus

Temporal lobe

Pons

T2-bright plaque

Fourth ventricle

Cerebellum

R   L

(B)

Frontal lobe

Temporal lobe

Pons

T2-bright plaque

Fourth ventricle

Cerebellum

R   L

Brief Anatomical Study Guide

1. There are **six extraocular muscles** for each eye, each with a different function (see Figure 13.1; see also Table 13.1). The **medial and lateral rectus** muscles move the eye medially and laterally, respectively. The **superior and inferior rectus** and **superior and inferior oblique** muscles are involved in vertical and torsional eye movements.

2. Three cranial nerves control eye movements: The **oculomotor nerve (CN III)** supplies all extraocular muscles except the lateral rectus and superior oblique (see Table 13.1; see also Figure 13.2). In addition, it supplies the **levator palpebrae superior** muscle, which elevates the upper eyelid. The **trochlear nerve (CN IV)** innervates the superior oblique muscle, and the **abducens nerve (CN VI)** innervates the lateral rectus (see Table 13.1; see also Figure 13.4).

3. The **oculomotor nucleus** is located in the rostral midbrain at the level of the superior colliculus and red nucleus (see Figures 13.2 and 14.3A). Fascicles of the oculomotor nerve exit the midbrain ventrally in the interpeduncular fossa between the posterior cerebral and superior cerebellar arteries. The nerve then runs adjacent to the posterior communicating artery, where it is susceptible to compression by aneurysms or by downward herniation of the temporal lobe over the tentorial edge.

4. The **trochlear nucleus** is located in the caudal midbrain at the level of the inferior colliculus and decussation of the superior cerebellar peduncles (see Figure 13.4). Fascicles of the trochlear nerve decussate and exit the brainstem dorsally. The thin trochlear nerve is particularly susceptible to injury from shearing forces in head trauma.

5. The **abducens nucleus**, together with fibers of the facial nerve, forms the facial colliculus on the floor of the fourth ventricle in the midpons (see Figure 12.11). Fascicles of the abducens nerve travel ventrally to exit the brainstem at the pontomedullary junction (see Figure 12.2A). The abducens nerve then ascends a long distance along the clivus and over the petrous ridge (see Figure 13.4; see also Figure 12.1), making it susceptible to injury from downward traction in elevated intracranial pressure.

6. As the oculomotor, trochlear, and abducens nerves (CN III, IV, and VI) exit the cranial cavity, they travel through the **cavernous sinus** (see Figure 13.11) in close association with the ophthalmic division of the trigeminal nerve (CN V$_1$) and then enter the orbit via the **superior orbital fissure** (see Figure 12.3A,C). Lesions in the cavernous sinus or orbital apex can affect multiple cranial nerves involved in eye movement control.

7. The pupils are under both parasympathetic and sympathetic control. **Parasympathetic** pathways mediate **pupillary constriction** via the pathways summarized in Figure 13.8. Preganglionic parasympathetic fibers arise from the **Edinger–Westphal** nucleus. Note that the parasympathetic fibers travel with the oculomotor nerve, so damage to this nerve often produces an abnormally dilated pupil. **Sympathetic** pathways mediate **pupillary dilation** via pathways summarized in Figure 13.10. Sympathetic pathways can be interrupted at multiple levels from the brainstem to the nerve fibers ascending in the neck to the eye, resulting in **Horner's syndrome** (see KCC 13.5 and KCC 13.6).

8. Central pathways for control of eye movements that impinge on the oculo-motor, trochlear, and abducens nuclei are referred to as **supranuclear pathways**. The pathways involve a distributed network including brain-stem, cerebellar, basal ganglia, cortical, and other circuits. For horizontal eye movements, the final common pathway and main **horizontal gaze center** in the brainstem is the **abducens nucleus**, which controls horizon-tal gaze for both the ipsilateral and contralateral eyes through the **medial longitudinal fasciculus (MLF)** (see Figures 13.12 and 13.13). The **parame-dian pontine reticular formation (PPRF)** is another important horizontal gaze center with inputs to the abducens nucleus.

9. **Vertical gaze** is controlled by nuclei in the rostral midbrain and pretectal region, including the **rostral interstitial nucleus of the MLF**. Control of **vergence eye movements** probably arises from nuclei in the midbrain reticular formation. Descending cortical inputs from the **frontal eye fields** (see Figure 13.14) decussate to reach the contralateral PPRF, producing contralateral lateral horizontal gaze. However, descending cortical inputs from more posterior areas of the parieto-occipito-temporal junction tend to cause ipsilateral deviation of gaze.

10. **Nystagmus** refers to rhythmic alternating **slow phases** of eye movements that move in one direction, interrupted by **fast phases** of eye movements in the opposite direction. Nystagmus is abnormal when it occurs at rest (without changing visual or vestibular inputs). Normal nystagmus occurs during attempts to view a visual scene or a series of stripes moving in front of the eyes and is called **optokinetic nystagmus (OKN)**. Vestibular inputs are relayed via the MLF to stabilize the eyes on a visual image during head and body movements. This is called the **vestibulo-ocular reflex (VOR)**.

CASE 13.9   HEADACHES AND IMPAIRED UPGAZE

IMAGE 13.9A,B   Pineal Region Tumor Compressing the Tectum   T1-weighted MRI images with intravenous gadolinium. (A) Axial image demonstrating a large enhancing lesion in the pineal region. (B) Sagittal image showing the enhancing lesion compressing the dorsal midbrain and pretectal area.

(A)

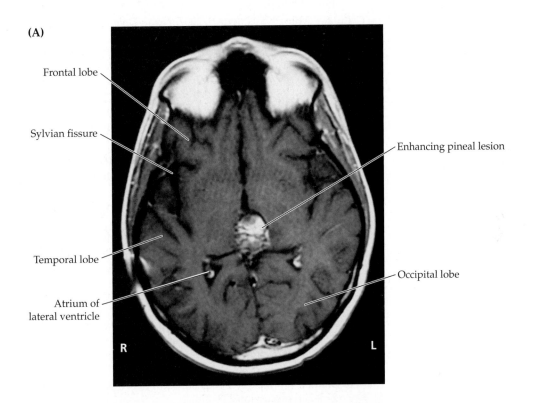

Frontal lobe

Sylvian fissure

Temporal lobe

Atrium of lateral ventricle

Enhancing pineal lesion

Occipital lobe

R   L

(B)

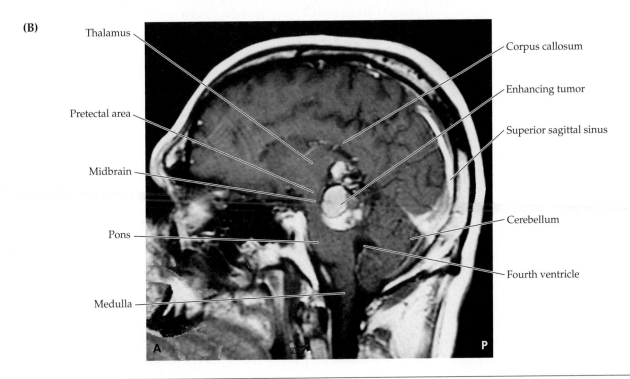

Thalamus

Pretectal area

Midbrain

Pons

Medulla

Corpus callosum

Enhancing tumor

Superior sagittal sinus

Cerebellum

Fourth ventricle

A   P

References

General References

Lee AG, Brazis P, Kline LB. 2009. *Curbside Consultation Neuro-Ophthalmology: Forty-Nine Clinical Questions.* Slack, Inc., Thorofare, NJ.

Leigh RJ, Zee DS. 2006. *The Neurology of Eye Movements.* 4th Ed. Oxford University Press, Oxford, UK.

Liu GT, Volpe NJ, Galetta S. 2007. *Neuro-Ophthalmology: Diagnosis and Management.* Saunders, Philadelphia.

Miller NR, Newman NJ, Hoyt WF (eds.). 2005. *Walsh and Hoyt's Clinical Neuro-Ophthalmology.* 6th Ed. Lippincott Williams & Wilkins, Philadelphia.

Pane A. Burdon B, Miller NR. 2007. The *Neuro-Ophthalmology Survival Guide.* Mosby, London.

Wray SH. 1998. Neuro-ophthalmologic diseases. In *Comprehensive Neurology.* 2nd Ed., RN Rosenberg, DE Pleasure (eds.), Chapter 19. Raven, New York.

Oculomotor, Trochlear, and Abducens Nerve Palsies

Bennett JL, Pelak VS. 2001. Palsies of the third, fourth, and sixth cranial nerves. *Ophthalmol Clin North Am* 14 (1): 169–185, ix.

Brazis PW. 2009. Isolated palsies of cranial nerves III, IV, and VI. *Semin Neurol* 29 (1): 14–28.

Chen PR, Amin-Hanjani S, Albuquerque FC, McDougall C, Zabramski JM, Spetzler RF. 2006. Outcome of oculomotor nerve palsy from posterior communicating artery aneurysms: comparison of clipping and coiling. *Neurosurgery* 58 (6): 1040–1046.

Hamilton SR. 1999. Neuro-ophthalmology of eye-movement disorders. *Curr Opin Ophthalmol* 10 (6): 405–410.

Lee MS, Galetta SL, Volpe NJ, Liu GT. 1999. Sixth nerve palsies in children. *Pediatr Neurol* 20 (1): 49.

Mansour AM, Reinecke RD. 1986. Central trochlear palsy. *Survey Ophthalmol* 30 (5): 279–297.

O'Donnell TJ, Buckley EG. 2006. Sixth nerve palsy. *Compr Ophthalmol Update* 7 (5): 215–221; discussion 223–224.

Richards BW, Jones R, Younge BR. 1992. Causes and prognosis in 4,278 cases of paralysis of the oculomotor, trochlear, and abducens cranial nerves. *Am J Ophthalmol* 113 (5): 489–496.

Sharpe JA, Wong AM, Fouladvand M. 2008. Ocular motor nerve palsies: implications for diagnosis and mechanisms of repair. *Prog Brain Res* 171: 59–66.

Cavernous Sinus and Orbital Apex Syndromes

Bosley TM, Schatz NJ. 1983. Clinical diagnosis of cavernous sinus syndromes. *Neurol Clin* 1 (4): 929–953.

Keane JR. 1996. Cavernous sinus syndrome: Analysis of 151 cases. *Arch Neurol* 53 (10): 967–971.

Linskey ME, Sekhar LN, Hirsch W, Yonas H, Horton JA. 1990. Aneurysms of the intracavernous carotid artery: Clinical presentation, radiographic features, and pathogenesis. *Neurosurgery* 26 (1): 71–79.

Linskey ME, Sekhar LN, Hirsch WL, Yonas H, Horton JA. 1990. Aneurysms of the intracavernous carotid artery: Natural history and indications for treatment. *Neurosurgery* 26 (6): 933–937.

Miller NR. 2007. Diagnosis and management of dural carotid-cavernous sinus fistulas. *Neurosurg Focus* 23 (5): E13.

Nawar RN, AbdelMannan D, Selman WR, Arafah BM. 2008. Pituitary tumor apoplexy: a review. *J Intensive Care Med* 23 (2): 75–90.

Verrees M, Arafah BM, Selman WR. 2004. Pituitary tumor apoplexy: characteristics, treatment, and outcomes. *Neurosurg Focus* 16 (4): E6.

Yeh S, Foroozan R. 2004. Orbital apex syndrome. *Curr Opin Ophthalmol* 15 (6): 490–498.

Horner's Syndrome

Debette S, Leys D. 2009. Cervical-artery dissections: predisposing factors, diagnosis, and outcome. *Lancet* (7): 668–678.

Mokri B, Silbert PL, Schievink WI, Piepgras DG. 1996. Cranial nerve palsy in spontaneous dissection of the extracranial internal carotid artery. *Neurology* 46 (2): 356–359.

Reede DL, Garcon E, Smoker WR, Kardon R. 2008. Horner's syndrome: clinical and radiographic evaluation. *Neuroimaging Clin N Am* 18 (2): 369–385, xi.

Walton KA, Buono LM. 2003. Horner syndrome. *Curr Opin Ophthalmol* 14 (6): 357–363.

Horizontal Gaze Disorders

Tijssen CC. 1994. Contralateral conjugate eye deviation in acute supratentorial lesions. *Stroke* 25 (7): 1516–1519.

Wall M, Wray SH. 1983. The one-and-a-half syndrome—A unilateral disorder of the pontine tegmentum: A study of 20 cases and review of the literature. *Neurology* 33 (8): 971–980.

Vertical Gaze Disorders and Parinaud's Syndrome

Baloh RW, Furman JM, Yee RD. 1985. Dorsal midbrain syndrome. *Neurology* 35 (1): 54–60.

Moffie D, Ongerboer de Visser BW, Stefanko SZ. 1983. Parinaud's syndrome. *J Neurol Sci* 58 (2): 175–183.

Segarra JM. 1970. Cerebral vascular disease and behavior. I. The syndrome of the mesencephalic artery (basilar artery bifurcation). *Arch Neurol* 22 (5): 408–418.

Trojanowski JQ, Wray SH. 1980. Vertical gaze ophthalmoplegia: Selective paralysis of downgaze. *Neurology* 30 (6): 605–610.

CONTENTS

Chapter 14

Brainstem III: Internal Structures and Vascular Supply

A 22-year-old woman felt her neck "snap" during treatment by a chiropractor. As she left the office, she felt dizzy and staggered out to her car, falling to the left. She also developed numbness and tingling on the left side of her face, a hoarse voice, a small left pupil, a drooping left upper eyelid, and decreased sensation on the right side of her body. This patient's symptoms illustrate the complicated deficits that follow interruption of the blood supply to one part of the brainstem. In Chapter 10 we learned about the functions of various cortical regions and their blood supply.

Here we will learn about the functions of brainstem nuclei and pathways and the blood supply that is vital to each brainstem region.

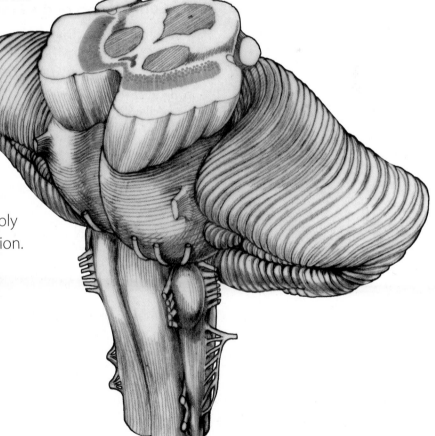

ANATOMICAL AND CLINICAL REVIEW

N O REGION OF THE NERVOUS SYSTEM inspires more awe in the clinician than the brainstem, perhaps because small lesions in this structure can have devastating consequences, or perhaps because of the elegant and complex anatomy of this compact structure juxtaposed between the cerebral hemispheres on the one hand and the spinal cord and cranial nerves on the other. Or finally, perhaps the brainstem inspires respect because it is the most evolutionarily ancient brain region (together with the basal forebrain), harking back to the simpler nervous systems of our reptilian ancestors. Whatever the reason, a thorough knowledge of brainstem anatomy is essential for the clinician's ability to diagnose and treat the often life-threatening disorders of this region of the brain.

In the previous two chapters we discussed the role of the brainstem in cranial nerve function, eye movements, and pupillary control. In this chapter we will examine the inner workings of the brainstem in greater detail, including some important nuclei and white matter pathways. After introducing these structures, we will identify them in the traditional manner by reviewing a series of brainstem sections. Next, several brainstem structures not discussed in other chapters, especially those constituting the reticular formation and related structures, will be examined in greater depth. Finally, we will discuss the vascular territories of the brainstem and the characteristic syndromes that result from damage to particular groups of brainstem structures located in close proximity to one another. Discussion of these vascular syndromes is diagnostically useful and also serves as an excellent review of regional brainstem neuroanatomy.

Main Components of the Brainstem

We can summarize the main components of the brainstem in a simplified manner, as shown in **Figure 14.1**, by using the following four functional groupings:

1. Cranial nerve nuclei and related structures
2. Long tracts
3. Cerebellar circuitry
4. Reticular formation and related structures

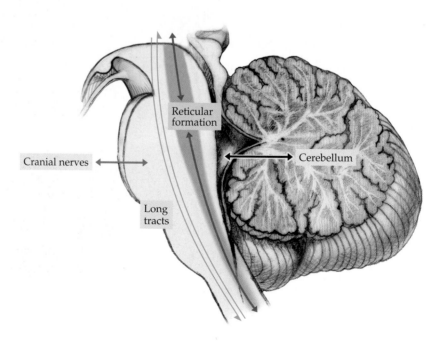

FIGURE 14.1 Main Components of the Brainstem

As would be expected from these functional groupings, brainstem lesions are often associated with cranial nerve abnormalities, long-tract findings, ataxia, and impairments related to reticular formation dysfunction such as impaired level of consciousness and autonomic dysregulation. It can be helpful to think of these four functional groupings when imagining the effects of lesions in different parts of the brainstem.

The cranial nerve nuclei and related structures were discussed in Chapters 12 and 13, the long tracts were discussed in Chapters 6 and 7, and the cerebellar pathways will be discussed in Chapter 15. In this chapter we will review these structures briefly and discuss the brainstem reticular formation and related structures in greater detail. The subcomponents of each grouping are listed in Table 14.1. Before proceeding further, let's identify these structures in the brainstem through the traditional, but so far unparalleled, method of examining stained brainstem sections.

Brainstem Sections

There is no substitute, when learning brainstem anatomy, for examining its detailed structures in stained serial sections of the human brainstem. The sections in Figures 14.3, 14.4, and 14.5 were prepared with a stain for myelin, so that myelin appears dark and gray matter appears light. We will refer to these sections as we review different nuclei, pathways, and functional systems in this chapter. A useful technique for anatomical self-study and review is to follow a given structure or pathway from start to finish as it appears in multiple adjacent sections.

Recall that the brainstem consists of the midbrain, pons, and medulla oblongata (see Figure 12.1). As in the spinal cord, motor nuclei in the brainstem are located more ventrally, while sensory nuclei are located more dorsally. During development, the sensory and motor nuclei are demarcated by the **sulcus limitans** (see Figure 12.4). In the adult, the sulcus limitans is still visible along the lateral wall of the fourth ventricle, dividing **motor nuclei** ventromedially from **sensory nuclei** dorsolaterally.

Other terms sometimes used in discussing the brainstem are "tectum," "tegmentum," and "basis" (**Figure 14.2**). The **tectum**, meaning "roof" in Latin, is obvious only in the midbrain and consists of the superior and inferior colliculi, which lie dorsal to the cerebral aqueduct. The **tegmentum**, meaning "covering," lies ventral to the cerebral aqueduct in the midbrain and ventral to the fourth ventricle in the pons and medulla. The tegmentum makes up the main bulk of the brainstem nuclei and reticular formation, which we will discuss in this chapter. The **basis** is the most ventral portion, where the large collections of fibers making up the corticospinal and corticobulbar tracts lie.

For overall orientation, let's review the brainstem sections briefly, proceeding from rostral to caudal. Functional details of the many structures mentioned here are discussed elsewhere. The **midbrain** is relatively short, and most axial sections cut through either the **superior colliculi**, which are more rostral (**Figure 14.3A**), or the **inferior colliculi**, which are more caudal (**Figure 14.3B**). Sections at these two levels can be distinguished because sections through the superior colliculi also include the **oculomotor nuclei** and **red nuclei** (see Figure 14.3A), while sections through the inferior colliculi also include the **trochlear nuclei** and **brachium conjunctivum** (decussation of the superior cerebellar peduncles)

FIGURE 14.2 **Brainstem Basis, Tegmentum, and Tectum** Axial section through the midbrain at the level of the superior colliculi. (Brainstem section modified from Martin JH. 1996. *Neuroanatomy: Text and Atlas.* 2nd Ed. McGraw-Hill, New York.)

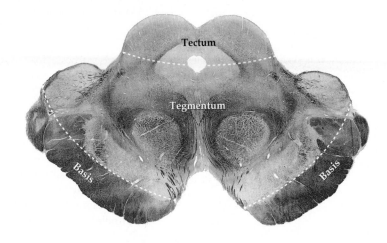

TABLE 14.1 Summary of Brainstem Structures

MAIN FUNCTIONAL GROUPINGS	SUBCOMPONENTS
1. *Cranial nerve nuclei and related structures* (see Chapters 12, 13)	Somatic motor column (GSE)[a] Oculomotor, trochlear, abducens, and hypoglossal nuclei Branchial motor column (SVE) Motor trigeminal, facial, ambiguus, and spinal accessory nuclei Parasympathetic column (GVE) Edinger–Westphal, superior and inferior salivatory nuclei, dorsal motor nucleus of CN X General somatic sensory column (GSA) Trigeminal nuclear complex Special somatic sensory column (SSA) Vestibular nuclei, cochlear nuclei Visceral sensory column Nucleus solitarius: rostral portion (SVA); caudal portion (GVA) Additional nuclei and pathways associated with eye movements Pretectal area; superior colliculus; medial longitudinal fasciculus (MLF); rostral interstitial nucleus of the MLF; convergence center; paramedian pontine reticular formation; accessory hypoglossal nucleus Additional nuclei and pathways associated with hearing Superior olivary nuclear complex; trapezoid body; lateral lemniscus; inferior colliculus Other nuclei and pathways associated with cranial nerve function Reticular formation; central tegmental tract
2. *Long tracts* (see Chapters 6, 7)	Motor pathways Corticospinal and corticobulbar tracts; other descending somatomotor pathways; descending autonomic pathways Somatosensory pathways Posterior column–medial lemniscal system; anterolateral system
3. *Cerebellar circuitry* (see Chapter 15)	Superior, middle, and inferior cerebellar peduncles Pontine nuclei; red nucleus; (parvocellular portion); central tegmental tract; inferior olivary nucleus
4. *Reticular formation and related structures*	Systems with widespread projections Reticular formation; cholinergic nuclei; noradrenergic nuclei; serotonergic nuclei; dopaminergic nuclei; other projecting systems Nuclei involved in sleep regulation Pain modulatory systems Periaqueductal gray; rostral ventral medulla Brainstem motor control systems: somatic, branchial, and autonomic Posture and locomotion (reticular formation; vestibular nuclei; superior colliculi; red nucleus [magnocellular portion]; substantia nigra; pedunculopontine tegmental nucleus); respiration, cough, hiccup, sneeze, shiver, swallow; nausea and vomiting (chemotactic trigger zone); autonomic control, including heart rate and blood pressure; sphincter control, including pontine micturition center

Note: The structures listed here reflect our interest in learning clinically relevant neuroanatomy. There are numerous additional brainstem structures beyond the scope of this text.

[a]GSA, general somatic afferent; GSE, general somatic efferent; GVA, general visceral afferent; GVE, general visceral efferent; SSA, special somatic afferent; SVA, special visceral afferent; SVE, special visceral efferent.

(see Figure 14.3B). Other important landmarks in midbrain sections include the cerebral aqueduct, periaqueductal gray, midbrain reticular formation, medial lemniscus, anterolateral system, and **cerebral peduncles** (= substantia nigra + basis pedunculi [see Figure 14.3]).

In sections through the **pons** (Figure 14.4), the large, thick middle cerebellar peduncles are visible laterally. These large peduncles connecting to the cere-

(A)

FIGURE 14.3 **Myelin-Stained Sections of the Midbrain** Axial planes of section, are shown in the inset. (A) Rostral midbrain (× 3.1). (B) Caudal midbrain (× 3.0). (From Martin JH. 1996. *Neuroanatomy: Text and Atlas*. 2nd Ed. McGraw-Hill, New York.)

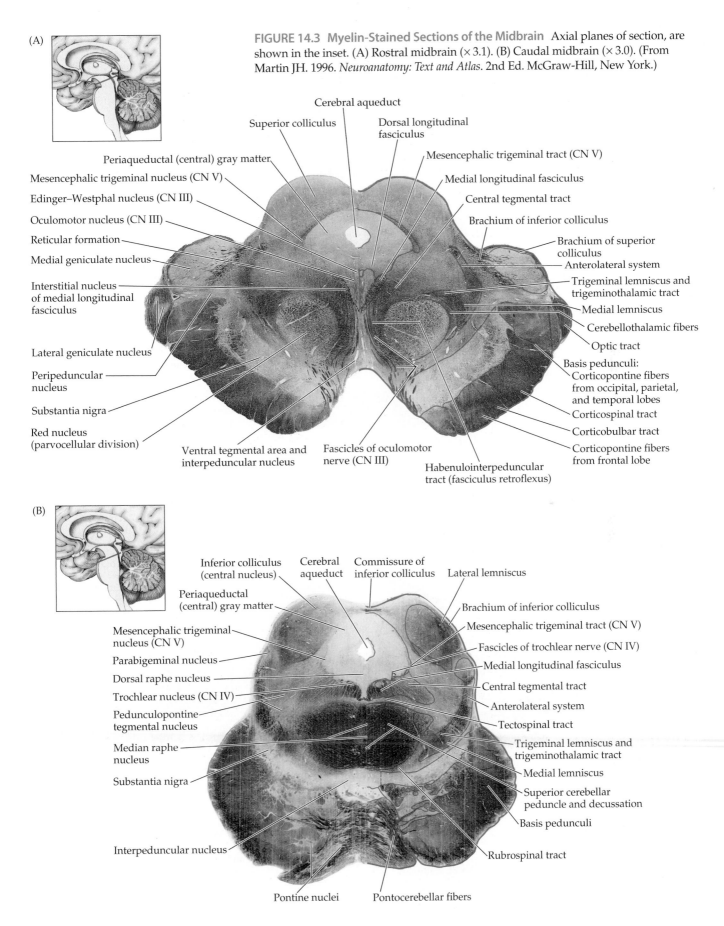

Superior colliculus

Cerebral aqueduct

Dorsal longitudinal fasciculus

Periaqueductal (central) gray matter

Mesencephalic trigeminal tract (CN V)

Mesencephalic trigeminal nucleus (CN V)

Medial longitudinal fasciculus

Edinger–Westphal nucleus (CN III)

Central tegmental tract

Oculomotor nucleus (CN III)

Brachium of inferior colliculus

Reticular formation

Brachium of superior colliculus

Medial geniculate nucleus

Anterolateral system

Interstitial nucleus of medial longitudinal fasciculus

Trigeminal lemniscus and trigeminothalamic tract

Medial lemniscus

Cerebellothalamic fibers

Lateral geniculate nucleus

Optic tract

Peripeduncular nucleus

Basis pedunculi:
Corticopontine fibers from occipital, parietal, and temporal lobes

Substantia nigra

Corticospinal tract

Red nucleus (parvocellular division)

Corticobulbar tract

Ventral tegmental area and interpeduncular nucleus

Fascicles of oculomotor nerve (CN III)

Habenulointerpeduncular tract (fasciculus retroflexus)

Corticopontine fibers from frontal lobe

(B)

Inferior colliculus (central nucleus)

Cerebral aqueduct

Commissure of inferior colliculus

Lateral lemniscus

Periaqueductal (central) gray matter

Brachium of inferior colliculus

Mesencephalic trigeminal nucleus (CN V)

Mesencephalic trigeminal tract (CN V)

Parabigeminal nucleus

Fascicles of trochlear nerve (CN IV)

Dorsal raphe nucleus

Medial longitudinal fasciculus

Trochlear nucleus (CN IV)

Central tegmental tract

Pedunculopontine tegmental nucleus

Anterolateral system

Median raphe nucleus

Tectospinal tract

Trigeminal lemniscus and trigeminothalamic tract

Substantia nigra

Medial lemniscus

Superior cerebellar peduncle and decussation

Basis pedunculi

Interpeduncular nucleus

Rubrospinal tract

Pontine nuclei

Pontocerebellar fibers

(A)

FIGURE 14.4 Myelin-Stained Sections of the Pons Axial planes of section are shown in the insets. (A) Pontomesencephalic junction (× 3.3). (B) Rostral to mid pons (× 3.3). (C) Caudal pons (× 3.0). (A from the University of Washington Digital Anatomist Project; B from DeArmond SJ, Fusco MM, Maynard MD. 1989. *Structure of the Human Brain: A Photographic Atlas*. 3rd Ed. Oxford, New York; C from Martin JH. 1996. *Neuroanatomy: Text and Atlas*. 2nd Ed. McGraw-Hill, New York.)

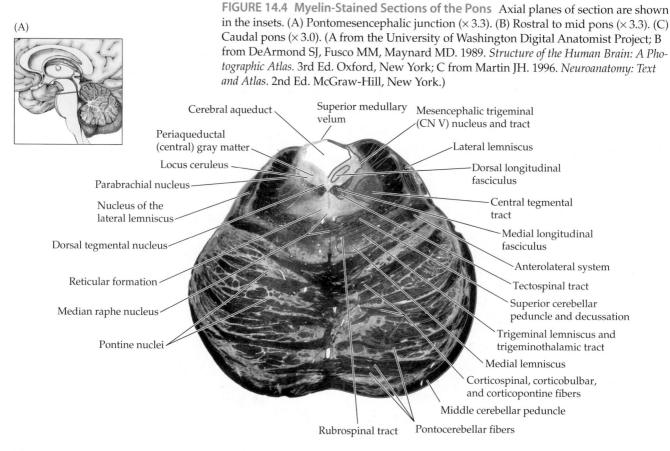

Cerebral aqueduct
Superior medullary velum
Mesencephalic trigeminal (CN V) nucleus and tract
Periaqueductal (central) gray matter
Lateral lemniscus
Locus ceruleus
Dorsal longitudinal fasciculus
Parabrachial nucleus
Central tegmental tract
Nucleus of the lateral lemniscus
Medial longitudinal fasciculus
Dorsal tegmental nucleus
Anterolateral system
Reticular formation
Tectospinal tract
Median raphe nucleus
Superior cerebellar peduncle and decussation
Pontine nuclei
Trigeminal lemniscus and trigeminothalamic tract
Medial lemniscus
Corticospinal, corticobulbar, and corticopontine fibers
Middle cerebellar peduncle
Rubrospinal tract
Pontocerebellar fibers

(B)

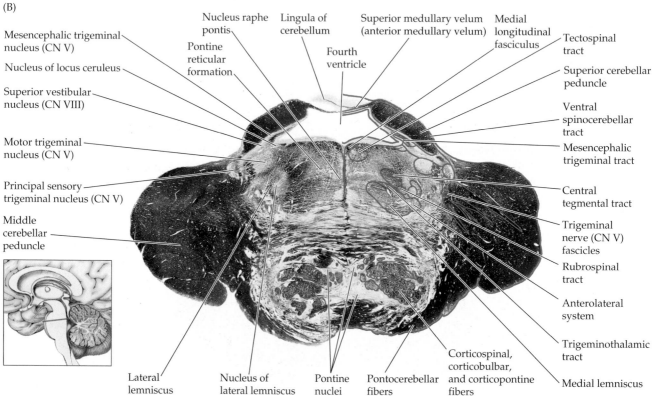

Mesencephalic trigeminal nucleus (CN V)
Nucleus raphe pontis
Lingula of cerebellum
Superior medullary velum (anterior medullary velum)
Medial longitudinal fasciculus
Tectospinal tract
Pontine reticular formation
Nucleus of locus ceruleus
Superior cerebellar peduncle
Superior vestibular nucleus (CN VIII)
Fourth ventricle
Ventral spinocerebellar tract
Motor trigeminal nucleus (CN V)
Mesencephalic trigeminal tract
Principal sensory trigeminal nucleus (CN V)
Central tegmental tract
Middle cerebellar peduncle
Trigeminal nerve (CN V) fascicles
Rubrospinal tract
Anterolateral system
Trigeminothalamic tract
Medial lemniscus
Lateral lemniscus
Nucleus of lateral lemniscus
Pontine nuclei
Pontocerebellar fibers
Corticospinal, corticobulbar, and corticopontine fibers

(C)

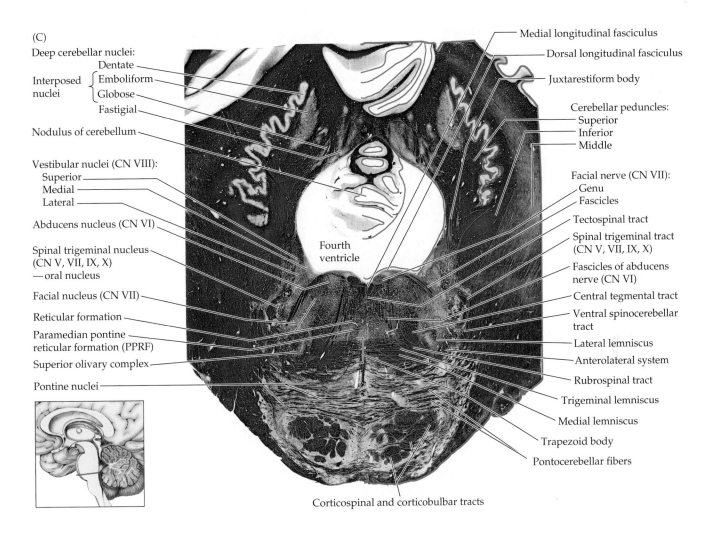

Deep cerebellar nuclei:
- Dentate
- Interposed nuclei {
  - Emboliform
  - Globose
- Fastigial

Nodulus of cerebellum

Vestibular nuclei (CN VIII):
- Superior
- Medial
- Lateral

Abducens nucleus (CN VI)

Spinal trigeminal nucleus (CN V, VII, IX, X) — oral nucleus

Facial nucleus (CN VII)

Reticular formation

Paramedian pontine reticular formation (PPRF)

Superior olivary complex

Pontine nuclei

Fourth ventricle

Medial longitudinal fasciculus

Dorsal longitudinal fasciculus

Juxtarestiform body

Cerebellar peduncles:
- Superior
- Inferior
- Middle

Facial nerve (CN VII):
- Genu
- Fascicles

Tectospinal tract

Spinal trigeminal tract (CN V, VII, IX, X)

Fascicles of abducens nerve (CN VI)

Central tegmental tract

Ventral spinocerebellar tract

Lateral lemniscus

Anterolateral system

Rubrospinal tract

Trigeminal lemniscus

Medial lemniscus

Trapezoid body

Pontocerebellar fibers

Corticospinal and corticobulbar tracts

bellum on either side inspired the name pons, which means "bridge" in Latin. The ventral pons consists of the **basis pontis**, which includes the corticospinal and corticobulbar tracts, as well as the pontine nuclei involved in cerebellar function. The **pontine tegmentum** contains numerous important nuclei and pathways, which we will discuss shortly. The **fourth ventricle** separates the pontine tegmentum from the cerebellum.

As in the midbrain, sections through the medulla can be thought of as rostral or caudal. In sections through the **rostral medulla**, the inferior olivary nucleus is present, as is the fourth ventricle (**Figure 14.5A**). In the **caudal medulla**, the inferior olivary nucleus and fourth ventricle are no longer present, but the posterior columns and posterior column nuclei appear (**Figure 14.5B**). Other important landmarks seen in sections through the medulla include the inferior cerebellar peduncles, pyramidal tracts, anterolateral system, and medial lemniscus. The transition between the medulla and spinal cord is marked by the **pyramidal decussation** (**Figure 14.5C**). The upper cervical spinal cord contains the **accessory spinal nucleus** (**Figure 14.5D**).

Now that we have reviewed the overall architecture of the brainstem, we are ready to discuss each of the four main functional groupings (see Table 14.1) in more detail.

REVIEW EXERCISE

Cover the labels and insets in Figures 14.3–14.5. For each section shown in these figures, identify the level in the brainstem and name as many labeled structures as possible. These structures will be reviewed again in the discussion that follows.

(A)

FIGURE 14.5 Myelin-Stained Sections of the Medulla and Cervical Spine Axial planes of section are shown in the insets. (A) Rostral medulla (× 5.0). (B) Caudal medulla (× 6.9). (C) Cervicomedullary junction (× 9.6). (D) Rostral cervical spinal cord (C2) (× 12.6). (A–C from Martin JH. 1996. *Neuroanatomy: Text and Atlas.* 2nd Ed. Mc-Graw-Hill, New York; D from the University of Washington Digital Anatomist project.)

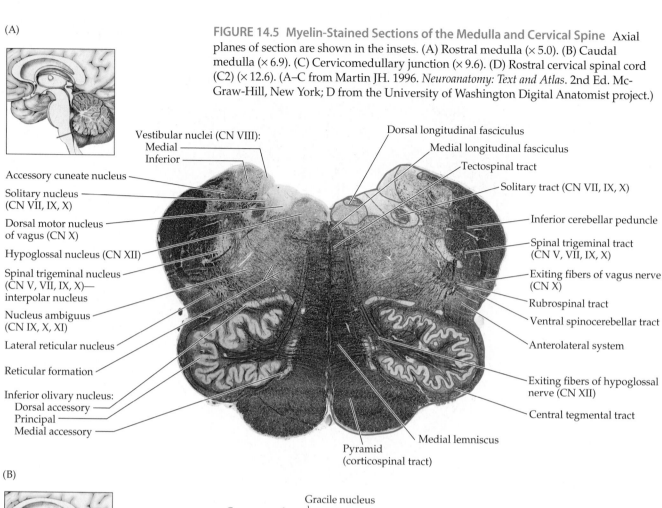

Vestibular nuclei (CN VIII):
Medial
Inferior

Accessory cuneate nucleus

Solitary nucleus
(CN VII, IX, X)

Dorsal motor nucleus
of vagus (CN X)

Hypoglossal nucleus (CN XII)

Spinal trigeminal nucleus
(CN V, VII, IX, X)—
interpolar nucleus

Nucleus ambiguus
(CN IX, X, XI)

Lateral reticular nucleus

Reticular formation

Inferior olivary nucleus:
Dorsal accessory
Principal
Medial accessory

Dorsal longitudinal fasciculus

Medial longitudinal fasciculus

Tectospinal tract

Solitary tract (CN VII, IX, X)

Inferior cerebellar peduncle

Spinal trigeminal tract
(CN V, VII, IX, X)

Exiting fibers of vagus nerve
(CN X)

Rubrospinal tract

Ventral spinocerebellar tract

Anterolateral system

Exiting fibers of hypoglossal
nerve (CN XII)

Central tegmental tract

Medial lemniscus

Pyramid
(corticospinal tract)

(B)

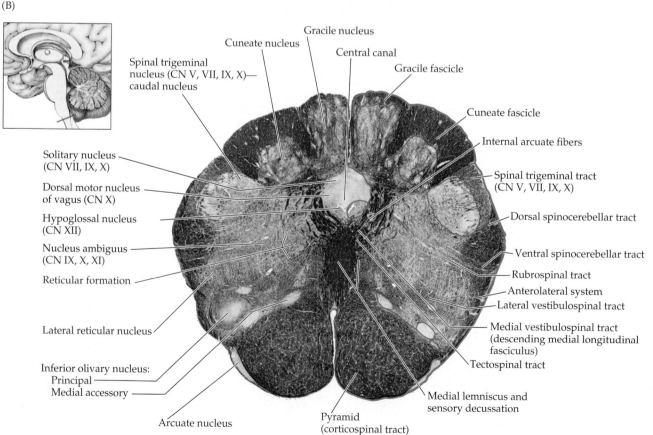

Cuneate nucleus

Gracile nucleus

Central canal

Gracile fascicle

Spinal trigeminal
nucleus (CN V, VII, IX, X)—
caudal nucleus

Cuneate fascicle

Internal arcuate fibers

Solitary nucleus
(CN VII, IX, X)

Dorsal motor nucleus
of vagus (CN X)

Hypoglossal nucleus
(CN XII)

Nucleus ambiguus
(CN IX, X, XI)

Reticular formation

Lateral reticular nucleus

Inferior olivary nucleus:
Principal
Medial accessory

Arcuate nucleus

Pyramid
(corticospinal tract)

Spinal trigeminal tract
(CN V, VII, IX, X)

Dorsal spinocerebellar tract

Ventral spinocerebellar tract

Rubrospinal tract

Anterolateral system

Lateral vestibulospinal tract

Medial vestibulospinal tract
(descending medial longitudinal
fasciculus)

Tectospinal tract

Medial lemniscus and
sensory decussation

(C)

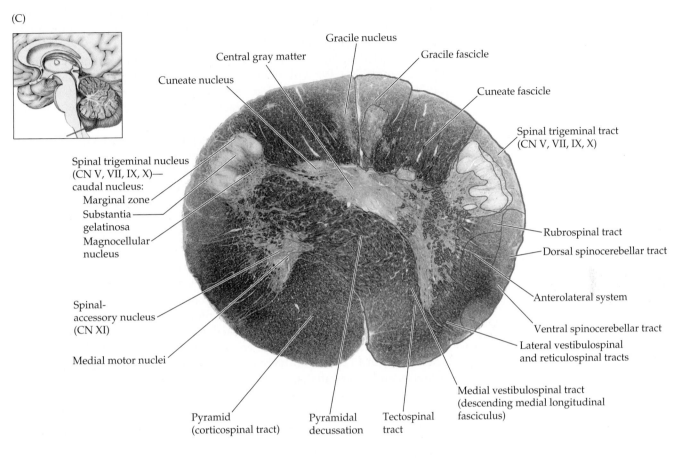

Central gray matter

Gracile nucleus

Gracile fascicle

Cuneate nucleus

Cuneate fascicle

Spinal trigeminal tract
(CN V, VII, IX, X)

Spinal trigeminal nucleus
(CN V, VII, IX, X)—
caudal nucleus:

Marginal zone

Substantia
gelatinosa

Magnocellular
nucleus

Rubrospinal tract

Dorsal spinocerebellar tract

Anterolateral system

Spinal-
accessory nucleus
(CN XI)

Ventral spinocerebellar tract

Lateral vestibulospinal
and reticulospinal tracts

Medial motor nuclei

Medial vestibulospinal tract
(descending medial longitudinal
fasciculus)

Pyramid
(corticospinal tract)

Pyramidal
decussation

Tectospinal
tract

(D)

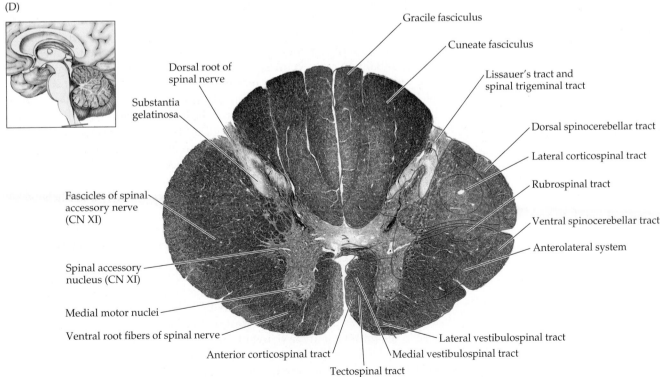

Gracile fasciculus

Cuneate fasciculus

Dorsal root of
spinal nerve

Lissauer's tract and
spinal trigeminal tract

Substantia
gelatinosa

Dorsal spinocerebellar tract

Lateral corticospinal tract

Rubrospinal tract

Fascicles of spinal
accessory nerve
(CN XI)

Ventral spinocerebellar tract

Anterolateral system

Spinal accessory
nucleus (CN XI)

Medial motor nuclei

Lateral vestibulospinal tract

Ventral root fibers of spinal nerve

Medial vestibulospinal tract

Anterior corticospinal tract

Tectospinal tract

Cranial Nerve Nuclei and Related Structures

As discussed in Chapter 12, the cranial nerve motor nuclei are located ventral to the sulcus limitans, and the sensory nuclei dorsal to the sulcus limitans, in an arrangement similar to that of the spinal cord (see Figure 12.4). There are three longitudinal columns of motor nuclei and three columns of sensory nuclei, described in detail in Chapters 12 and 13 (see Figure 12.5; Tables 12.3 and 12.4). We will review these nuclei again briefly here and identify them in the brainstem sections. Knowing the locations of these nuclei in the brainstem is essential for localizing brainstem infarcts, as we will discuss later in this chapter.

The **somatic motor nuclei** (GSE) are the oculomotor, trochlear, abducens, and hypoglossal nuclei, all of which remain adjacent to the midline. The **oculomotor nuclei** (CN III) are in the rostral midbrain, and the **trochlear nuclei** (CN IV) are in the caudal midbrain; both nuclei lie just ventral to the periaqueductal gray (see Figure 14.3A,B). The **medial longitudinal fasciculus (MLF)**, which forms the ventral border of the oculomotor and trochlear nuclei, interconnects these nuclei, the abducens nuclei, and the vestibular nuclei. The **abducens nuclei** (CN VI) help form the facial colliculus on the floor of the fourth ventricle in the mid to lower pons (see Figure 14.4C). As we discussed in Chapter 13, several other brainstem regions are important for the supranuclear control of eye movements, including the paramedian pontine reticular formation (see Figure 13.12), the rostral interstitial nucleus of the MLF, and the convergence center. The **hypoglossal nuclei** (CN XII) form the hypoglossal trigones on the floor of the fourth ventricle in the medulla (see Figure 14.5A; see also Figure 12.2B) and have a longitudinal sausage shape continuing into the caudal medulla (see Figure 14.5B). As can be seen in these sections, three sausage-shaped nuclei bear a constant relation to each other as they run along the floor of the ventricle or central canal of the medulla. These are, from medial to lateral, the hypoglossal, dorsal motor X, and solitary nuclei (see Figure 14.5A,B; see also Figure 12.5).

The **branchial motor nuclei** (SVE) are the trigeminal motor nucleus (CN V), facial nucleus (CN VII), nucleus ambiguus (CN IX, X), and spinal accessory nucleus (CN XI; also known as the accessory spinal nucleus). Recall that the branchial motor nuclei initially lie just lateral to the somatic motor nuclei, but they gradually migrate ventrolaterally into the tegmentum (see Figure 12.4). The **trigeminal motor nucleus** is located in the upper to mid pons (see Figure 14.4B), just ventral to the chief trigeminal sensory nucleus, near the level where the trigeminal nerve exits the brainstem. The **facial nucleus** is located more caudally in the pontine tegmentum (see Figure 14.4C) and gives rise to the looping fibers of the facial colliculus. The **nucleus ambiguus** is difficult to distinguish from the surrounding brainstem (see Figure 14.5A,B). It runs longitudinally through the medulla in a similar position to the facial nucleus. The **spinal accessory nucleus** (also known as the accessory spinal nucleus), as its name implies, is located not in the brainstem, but rather in the upper five segments of the cervical spinal cord (see Figure 14.5D). It protrudes laterally between the dorsal and ventral horns of the spinal cord central gray matter. Recall that some consider the spinal accessory nucleus to be somatic motor (GSE) or mixed (GSE + SVE) rather than branchial motor (SVE).

The **parasympathetic nuclei** (GVE) are the Edinger–Westphal nucleus (CN III), superior (CN VII) and inferior (CN IX) salivatory nuclei, and dorsal motor nucleus of CN X. The **Edinger–Westphal nuclei** form a *V*-shaped cap as they fuse in the midline and curve over the dorsal and rostral aspect of the oculomotor nuclei (see Figure 14.3A; see also Figure 13.3). The **superior and inferior salivatory nuclei** lie in the pontine tegmentum (see Figure 12.5) and do not form discrete, easily visible nuclei on standard sections.* The **dorsal motor nucleus of CN X** runs from the rostral to the caudal medulla, just lateral to the hypoglossal nucleus (see Figure 14.5A,B). It forms the **vagal trigone** on the floor of the fourth ventricle, just lateral to the hypoglossal trigone (see Figure 12.2B).

General somatic sensory (GSA) inputs from the cranial nerves (CN V, VII, IX, and X) all travel to the trigeminal nuclear complex. The **trigeminal nuclear complex** runs from the midbrain to the upper cervical spinal cord (see Figures 14.3–14.5) and consists of three nuclei: the mesencephalic, chief sensory, and spinal trigeminal nuclei (see Table 12.6). The **mesencephalic trigeminal nucleus and tract**, subserving proprioception, run along the lateral edge of the periaqueductal gray matter of the midbrain (see Figures 14.3 and 14.4A). The **chief (main, principal) trigeminal sensory nucleus** is located in the upper to mid pons just dorsolateral to the trigeminal motor nucleus (see Figure 14.4B). The **spinal trigeminal nucleus** and **spinal trigeminal tract** run the length of the lateral pons and medulla (see Figures 14.4C and 14.5A–C). It should be clear from reviewing Figure 14.5C and D that the spinal trigeminal nucleus is the rostral extension of the dorsal horn of the spinal cord. Recall that these systems are analogous, subserving pain and temperature sensation for the face and body (see Table 12.6). Similarly, the chief trigeminal nuclei (see Figure 14.4B) are analogous to the dorsal column nuclei (see Figure 14.5B) in that they both subserve fine discriminative touch.

Special somatic sensory (SSA) inputs for hearing and vestibular sense (CN VIII) reach the cochlear and vestibular nuclei, respectively. The **dorsal and ventral cochlear nuclei** wrap around the lateral aspect of the inferior cerebellar peduncle at the pontomedullary junction (see Figure 12.17C). After entering the brainstem, the hearing pathways decussate at multiple levels, as discussed in Chapter 12. Many fibers from the cochlear nuclei decussate in the trapezoid body in the caudal pons (see Figure 14.4C; see also Figures 12.16, 12.17B). Some fibers synapse in the **superior olivary nuclear complex** of the pons (see Figure 14.4C). Auditory information then ascends in the **lateral lemniscus** (see Figures 14.3B and 14.4) to reach the **inferior colliculus** (see Figure 14.3B). From there, fibers ascend via the **brachium of the inferior colliculi** to the **medial geniculate nuclei** of the thalamus, located just lateral to the superior colliculi of the midbrain (see Figure 14.3A). Information then continues in the auditory radiations to the primary auditory cortex (see Figure 12.16).

There are **four vestibular nuclei—superior, inferior, medial, and lateral**—on each side of the brainstem, lying on the lateral floor of the fourth ventricle, in the pons and rostral medulla (see Figures 14.4B,C, 14.5A; see also Figure 12.19). As we discussed in Chapter 12, the vestibular nuclei convey perception of head position and acceleration to the cerebral cortex via relays in the ventral posterior thalamus. However, the vestibular nuclei are more intimately connected with the cerebellum (see Chapter 15) and local brainstem circuits, and most functions of the vestibular nuclei are not at the level of conscious perception. Thus, the medial and lateral vestibulospinal tracts (see Figure 6.11D) are involved in posture and muscle tone, and they arise primarily from the medial and lateral vestibular nuclei, respectively. The medial vestibular nucleus is the largest of the vestibular nuclei (see Figure 14.5A). The inferior vestibular nucleus is also relatively easy to identify because fibers of the lateral vestibular nucleus traverse the inferior vestibular nucleus as they descend to the spinal

---

*Some authors propose abandoning the names "superior and inferior salivatory nuclei" because, among other reasons, they are too limited: In addition to parasympathetic innervation of the salivary glands, these nuclei innervate the lacrimal glands, the secretory glands of the nasal mucosa, and the cerebral blood vessels. For now, however, these names remain in wide use.

cord, giving the inferior vestibular nucleus a characteristic "checkerboard" appearance on myelin-stained sections (see Figure 14.5A).

As we mentioned earlier in this section, the **medial longitudinal fasciculus (MLF)** is an important pathway, connecting the vestibular nuclei and nuclei involved in extraocular movements (see Figure 12.19). The MLF can be identified as a heavily myelinated tract running near the midline on each side, just under the floor of the fourth ventricle in the midline of the pons (see Figure 14.4) and just under the oculomotor and trochlear nuclei in the midbrain (see Figure 14.3). Fibers arising from the medial vestibular nucleus, with additional contributions mainly from the superior vestibular nucleus, ascend in the MLF to the oculomotor, trochlear, and abducens nuclei, mediating vestibulo-ocular reflexes (see Chapter 13). Finally, as we will discuss in Chapter 15, the vestibular nuclei have numerous important reciprocal connections with the cerebellum, primarily with the inferior cerebellar vermis and flocculonodular lobes.

All **visceral afferents**, whether special or general, travel to the **nucleus solitarius** (see Figure 14.5A,B), located just lateral to the dorsal motor nucleus of CN X. Note that the nucleus solitarius has a unique appearance in myelin sections, with the heavily stained central solitary tract surrounded by the lightly stained tube-shaped (doughnut-shaped in cross section) solitary nucleus. **Special visceral afferents** (SVA) for taste (CN VII, IX, and X) reach the rostral nucleus solitarius, also known as the gustatory nucleus, while **general visceral afferents** (GVA) from the cardiorespiratory (and gastrointestinal) systems (CN IX, and X) reach the caudal nucleus solitarius, also known as the cardiorespiratory nucleus. As discussed in Chapter 12, the taste pathway continues rostrally via the **central tegmental tract** (see Figures 12.12, 14.3, and 14.4) to reach the ventral posterior nucleus (VPM) of the thalamus, which projects to the cortical taste area in the parietal operculum and insula.

Long Tracts

The major long tracts that pass through the brainstem were discussed in Chapters 6 and 7. The major descending motor pathways are summarized in Figure 6.11 and Table 6.3. Recall that the **corticospinal and corticobulbar tracts** travel in the middle third of the cerebral peduncles in the midbrain (see Figure 14.3; see also Figure 6.10B). The other portions of the cerebral peduncles carry predominantly corticopontine fibers involved in cerebellar circuitry (see Chapter 15). The corticospinal fibers continue from the midbrain cerebral peduncles to run through the basis pontis (see Figure 14.4) and then emerge as the pyramids in the ventral medulla (see Figure 14.5A,B). The pyramidal decussation occurs at the cervicomedullary junction (see Figure 14.5C), giving rise to the lateral corticospinal tract (see Figure 14.5D).

The major ascending somatosensory pathways are summarized in Figures 7.1 and 7.2 and Table 7.1. Recall that axons in the **posterior columns**, subserving vibration, joint position sense, and fine touch, synapse onto neurons in the **posterior column nuclei**, consisting of the more medial **nucleus gracilis** for the legs and the more lateral **nucleus cuneatus** for the arms (see Figure 14.5B–D). The posterior column nuclei give rise to the **internal arcuate fibers**, which cross to the opposite side (see Figure 14.5B) and then ascend through the brainstem as the **medial lemniscus** (see Figures 14.3, 14.4, and 14.5A) to reach the VPL (ventral posterior lateral nucleus) of the thalamus (see Figure 7.1). The **anterolateral systems**, including the **spinothalamic tract**, subserve pain, temperature, and crude touch. The anterolateral systems decussate in the spinal cord—not the brainstem—and they assume a fairly fixed, lateral position as they ascend through the brainstem (see Figures 14.3–14.5).

One additional descending pathway that is clinically relevant is the descending sympathetic pathway running through the lateral brainstem in close proximity to the anterolateral systems (see Figure 13.10). Recall that damage to this pathway can cause Horner's syndrome (see KCC 13.5).

KEY CLINICAL CONCEPT

14.1 LOCKED-IN SYNDROME

Patients who have absent motor function but maintain intact sensation and cognition are said to be "locked in." The usual cause is an infarct in the ventral pons (see KCC 14.3) affecting the bilateral corticospinal and corticobulbar tracts. The spinal cord and cranial nerves receive no input from the cortex, and the patient is unable to move. Sensory pathways and the brainstem reticular activating systems are spared. Patients are thus fully aware and able to feel, hear, and understand everything in their environment. This condition can mimic—but should be carefully distinguished from—coma, which is discussed later in this chapter (see KCC 14.2).

As we saw in Chapter 13, vertical eye movements and eyelid elevation are controlled by a region in the tegmentum of the rostral midbrain. Horizontal eye movements, however, depend on pontine circuits (see Figure 13.12). Therefore, locked-in syndrome often spares vertical eye movements and eye opening. Patients with this syndrome can thus communicate using eye movements. Special computer interfaces based on eye movements have been developed for patients with locked-in syndrome. The French editor Jean-Dominique Bauby even wrote an entire book (*Le Scaphandre et le Papillon,* or *The Diving Bell and the Butterfly*) after becoming locked in, spelling out one letter at a time to a transcriptionist by using eye movements. The prognosis is generally poor: About 60% of locked-in patients eventually succumb to respiratory infection or other complications of paralysis. Some patients do regain some motor function over time, however, and, rarely, a near-complete recovery can occur.

In addition to bilateral ventral pontine infarcts, other lesions in the ventral pons, such as hemorrhage, tumor, encephalitis, multiple sclerosis, or central pontine myelinolysis, can also occasionally cause locked-in syndrome. Less commonly, lesions in the bilateral cerebral peduncles of the midbrain, or in the internal capsules, can be the cause. In addition, a locked-in condition can result from severe disorders of motor neurons (see KCC 6.7), peripheral nerves, muscles, or the neuromuscular junction (see KCC 8.1). ■

Cerebellar Circuitry

The cerebellum is discussed in detail in Chapter 15; however, here we will briefly mention a few prominent brainstem structures that participate in cerebellar circuitry. As we will see in Chapter 15, lesions of the cerebellar circuitry produce a characteristic uncoordinated wavering movement abnormality called **ataxia**. Ataxia typically occurs *ipsilateral* to the side of the lesion because cerebellar circuits tend to decussate twice before reaching lower motor neurons.

The cerebellum is connected to the brainstem via three large white matter pathways called the cerebellar peduncles (see Figure 15.3). The **superior cerebellar peduncle** contains mainly cerebellar outputs (see Figure 14.4). The **decussation of the superior cerebellar peduncles** occurs in the midbrain at the level of the inferior colliculi (see Figure 14.3B). Cerebellar output fibers then continue rostrally to reach the **red nucleus** of the midbrain at the level of the superior colliculi (see Figure 14.3A). Other fibers continue rostrally to ultimately influence primary motor cortex and premotor cortex via relays in the ventrolateral nucleus of the thalamus. The **middle cerebellar peduncle** is the largest of the cerebellar peduncles (see Figure 14.4B,C). It contains massive inputs to the cerebellum arising from the **pontine nuclei** scattered through the basis pontis (see Figure 14.4). The pontine nuclei, in turn, receive inputs from the corticopontine fibers of the cerebral peduncles (see Figure 14.3). The **inferior cerebellar peduncle** mainly carries inputs to the cerebellum from the spinal cord (see Figure 14.5A).

In addition to the pontine nuclei, several other brainstem nuclei participate in cerebellar circuitry. We will mention only a few of these here. As we have just discussed, the red nucleus receives inputs from the superior cerebellar peduncle (see Figure 14.3A). The rostral (parvocellular) portion of the red nucleus sends fibers via the central tegmental tract (see Figures 14.3B and 14.4) to reach the inferior olivary nucleus in the rostral medulla (see Figure 14.5A), which in turn sends fibers back to the cerebellum via the inferior cerebellar peduncle. Interruption of this circuit from cerebellum to brainstem and back to cerebellum results in a characteristic, though rare, movement disturbance called palatal myoclonus, which is characterized by continuous, rhythmic clicking movements of the palate. In addition, as mentioned earlier, the vestibular nuclei are intimately interconnected with the cerebellum.

Reticular Formation and Related Structures

The **reticular formation** is a central core of nuclei that run through the entire length of the brainstem (Figure 14.6). It is continuous rostrally with certain diencephalic nuclei and caudally with the intermediate zone of the spinal cord. Simplifying somewhat, we can say that these rostral and caudal extensions highlight the two main functions of the reticular formation. Thus, the **rostral reticular formation** of the mesencephalon and upper pons function together with diencephalic nuclei to maintain an alert conscious state in the forebrain. Meanwhile, the **caudal reticular formation** of the pons and medulla works together with the cranial nerve nuclei and the spinal cord to carry out a variety of important motor, reflex, and autonomic functions. Although there are numerous exceptions, this simplified rostral and caudal conceptualization can be heuristically and clinically useful.

Definitions vary as to which nuclei to include in the reticular formation. The reticular formation is located in the **brainstem tegmentum** (see Figure 14.2). Historically, the term "reticular," meaning net-like or mesh-like in appearance, was first applied to this region in the late 1800s because there were no obvious nuclear divisions when this brain area was viewed with conventional histological staining. In addition, as we will discuss shortly, some neurons in the brainstem tegmentum have very widespread projection patterns, adding to the impression that this structure is diffusely organized. However, over time and with more refined techniques, numerous specific nuclei have been identified, which have quite precisely organized projection patterns. For example, as we will discuss below, the brainstem tegmentum contains nuclei with specific neurotransmitters including acetylcholine, dopamine, norepinephrine, serotonin, and others. In addition, some cranial nerve nuclei, such as the superior and inferior salivatory nuclei, or the nucleus ambiguus, lie buried within the reticular formation, while in other areas, clearly visible subnuclei do appear (for example, the nucleus reticularis gigantocellularis, located in the medial rostral medulla). Also closely associated with but distinct from the reticular formation are regions such as the periaqueductal gray matter in the midbrain, which is involved in pain modulation, and the chemotactic trigger zone (area postrema) in the medulla, involved in causing nausea.

Thus, in the modern era, the term "reticular formation" has come to be used for fewer and fewer structures. Instead, we usually refer to specific nuclei

FIGURE 14.6 Brainstem Reticular Formation Location at multiple levels is shown. The reticular formation merges caudally with the spinal cord intermediate zone and rostrally with the subthalamic region and lateral hypothalamus.

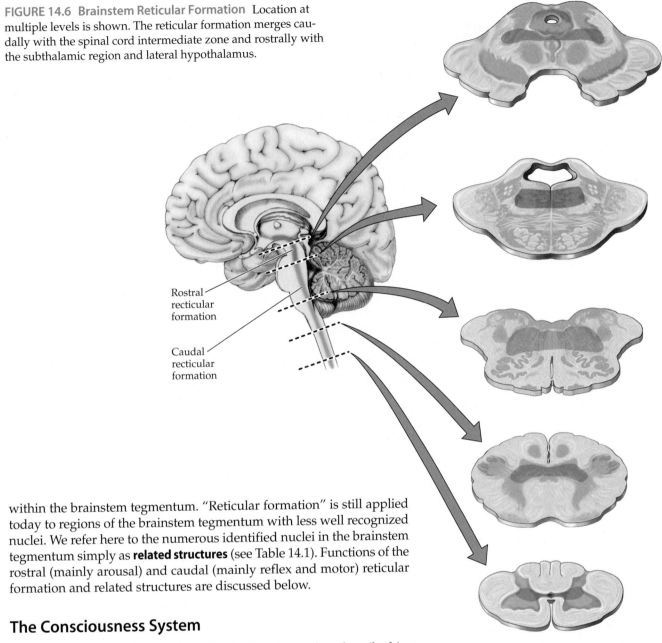

Rostral
recticular
formation

Caudal
recticular
formation

within the brainstem tegmentum. "Reticular formation" is still applied today to regions of the brainstem tegmentum with less well recognized nuclei. We refer here to the numerous identified nuclei in the brainstem tegmentum simply as **related structures** (see Table 14.1). Functions of the rostral (mainly arousal) and caudal (mainly reflex and motor) reticular formation and related structures are discussed below.

The Consciousness System

The main functions of the brain and spinal cord are often described in terms of "systems" such as the motor system, sensory systems, limbic system, etc. We propose that a new term should be introduced to describe the functions related to consciousness. In analogy to other systems, the **"consciousness system"** consists of cortical and subcortical networks in the brain that carry out the major functions of consciousness. Therefore, the consciousness system is formed principally by the medial and lateral frontoparietal association cortex (see Chapter 19) together with arousal circuits in the upper brainstem and diencephalon (see Figure 19.14).

Consciousness can be divided into the **content of consciousness**, comprised of systems mediating sensory, motor, memory, and emotional functions (discussed in other chapters throughout the book); and the **level of consciousness**. The consciousness system regulates the level of consciousness, while other brain systems provide the substrate or content upon which the consciousness system acts (see the figure in the Epilogue at the end of the book). Control of

MNEMONIC

the level of consciousness involves at least three processes, which can be remembered by the mnemonic **AAA** (for **A**lertness, **A**ttention, **A**wareness). The first process, **alertness**, depends on normal functioning of the brainstem and diencephalic arousal circuits and the cortex. The second process is **attention**, which appears to use many of these same circuits, together with additional processing in frontoparietal association cortex and other systems to be discussed in Chapter 19. The third, and most poorly understood, process is that which leads to our subjective and personal experience of **awareness**. Conscious awareness depends on our ability to combine various higher-order forms of sensory, motor, emotional, and mnemonic information from disparate regions of the brain into a unified and efficient summary of mental activity, which can potentially be remembered at a later time.

Here we will focus on widespread projection systems involved in alertness and behavioral arousal, leaving discussion of the higher-order aspects of consciousness (attention, awareness) for Chapter 19. On the basis of animal experiments and human patients studied in the 1930s and 1940s, it was found that lesions in the rostral brainstem reticular formation and medial diencephalon can cause coma, while stimulation of the same regions can lead to behavioral and electrographic arousal from deep anesthesia. Moruzzi and Magoun named these regions of the brain the ascending reticular activating system (ARAS). It has subsequently been found that these systems are not all ascending (descending cortical influences are critical as well) and do not all originate in the reticular formation. Nevertheless, the concept of a fragile region in the upper brainstem–diencephalic junction where focal lesions can cause coma remains clinically useful, although rather than a single "ARAS" there appear to be multiple interconnected arousal systems acting in parallel to maintain normal consciousness.

Where in the brain can a lesion cause coma? Classically, coma is caused either by dysfunction of the **upper brainstem reticular formation and related structures** or by dysfunction of extensive **bilateral regions of the cerebral cortex** (see also KCC 14.2). Lesions in other regions of the brainstem do not typically affect the level of consciousness. For example, lesions located more caudally in the brainstem, such as in the lower pons or medulla (see Figure 14.6), do not affect consciousness. Similarly, in lesions of the ventral midbrain or pons that spare the reticular formation, consciousness is typically spared, as in locked-in syndrome (see KCC 14.1). However, **bilateral lesions of the thalamus**, particularly those involving the medial and intralaminar regions, can also cause coma.

As we will discuss in the next section, **multiple parallel systems** in the upper brainstem, diencephalon, and basal forebrain contribute to maintaining alertness. To summarize briefly, these subcortical arousal systems include the following: (1) **upper brainstem** neurons containing **norepinephrine**, **serotonin**, and **dopamine** which project to both cortical and subcortical forebrain structures; (2) upper brainstem neurons containing **acetylcholine** and **pontomesencephalic reticular formation** neurons possibly containing **glutamate** which project to the thalamus, hypothalamus, and basal forebrain; these subcortical structures in turn have arousal effects through widespread projections to the cerebral cortex; (3) **posterior hypothalamic** neurons containing **histamine** and **orexin** which project to both cortical and subcortical targets; (4) **basal forebrain** neurons containing **acetylcholine**, which project to the cerebral cortex; and (5) neurons in the **rostral thalamic intralaminar nuclei** and **other medial thalamic nuclei**, probably containing glutamate, which project to the cerebral cortex. For illustrative purposes, the projections of one of these systems—the pontomesencephalic reticular formation—are shown in Figure 14.7. The pontomesencephalic reticular formation contributes to arousal by projecting to the thalamus, hypothalamus, and basal forebrain; these structures, in turn, have widespread

REVIEW EXERCISE

Which of the following lesions are usually associated with coma, and which are not?

1. Unilateral frontal cortex

2. Bilateral pontomesencephalic reticular formation

3. Bilateral basis pontis

4. Bilateral medullary reticular formation

5. Bilateral diffuse regions of cortex

6. Bilateral thalami

(Answer: Only 2, 5, and 6 are usually associated with coma.)

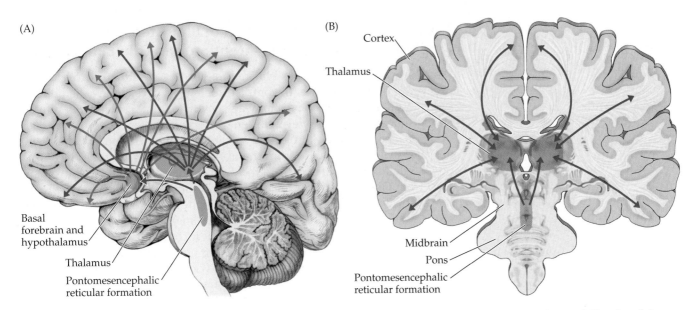

(A)

Basal
forebrain and
hypothalamus

Thalamus

Pontomesencephalic
reticular formation

(B)

Cortex

Thalamus

Midbrain

Pons

Pontomesencephalic
reticular formation

FIGURE 14.7 Arousal Circuits of the Pontomesencephalic Reticular Formation   (A) Midsagittal view; (B) coronal view. Widespread projections to the cortex arise from outputs of the pontomesencephalic reticular formation relayed via the thalamic intralaminar nuclei, basal forebrain, and hypothalamus.

projections to the cerebral cortex (see Figure 14.7). This anatomical arrangement illustrates why lesions of either widespread bilateral regions of cortex, or of the upper brainstem-diencephalic activating systems can cause coma.

What activates this system of behavioral arousal and alertness? The reticular formation and related structures receive inputs from sensory pathways, especially the anterolateral system spinoreticular pathway involved in pain transmission (**Figure 14.8**; see also Figure 7.2). In addition, numerous regions of association cortex and limbic cortex project to the pontomesencephalic reticular formation (as well as to the intralaminar nuclei), so that cognitive processes and emotions, respectively, can lead to an increased level of alertness through this system. The posterior lateral hypothalamus also projects to and activates the arousal systems. Other circuits that may play a role in attentional mechanisms include the superior colliculi, cerebellum, basal ganglia, and thalamic reticular nucleus, as we will discuss in Chapter 19.

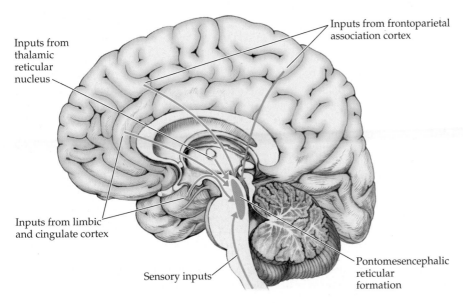

Inputs from
thalamic
reticular
nucleus

Inputs from frontoparietal
association cortex

Inputs from limbic
and cingulate cortex

Sensory inputs

Pontomesencephalic
reticular
formation

FIGURE 14.8 Major Inputs to the Pontomesencephalic Reticular Formation and Related Structures

Widespread Projection Systems of Brainstem and Forebrain: Consciousness, Attention, and Other Functions

Most pathways in the nervous system project from one structure to a limited number of other structures. In contrast, some pathways, often called **diffuse** (or **widespread**) **projection systems**, each emanate from a single region to innervate many structures or even the entire nervous system (Table 14.2). Interestingly, most brainstem projection systems that are directed upward to innervate the forebrain arise from the upper brainstem (midbrain to rostral pons), while those that project to the brainstem, cerebellum, or spinal cord arise from the lower pons or medulla. In addition, some widespread projection systems have their main source outside the brainstem in the hypothalamus (histamine) or basal forebrain (acetylcholine). Together, these projection systems are essential for maintaining an alert conscious state and for regulating attention, the sleep–wake cycle, and emotional balance. In the sections that follow we will discuss the major widespread projection systems—some with identified neurotransmitters, and others for which the neurotransmitters are still under investigation.

TABLE 14.2 Widespread Projection Systems in the Nervous System

PROJECTION SYSTEM	LOCATION(S) OF CELL BODIES	MAIN TARGET(S)	NEUROTRANSMITTER RECEPTOR(S)[a,b]	FUNCTION(S)[c]
Reticular formation	Midbrain and rostral pons	Thalamic intralaminar nuclei, hypothalamus, basal forebrain	Unknown (glutamate?)	Alertness
Intralaminar nuclei	Thalamic intralaminar nuclei	Cortex, striatum	(Glutamate?)	Alertness
Midline thalamic nuclei	Midline thalamic nuclei	Cortex	(Glutamate?)	Alertness
Norepinephrine	Pons: locus ceruleus and lateral tegmental area	Entire CNS	α_{1A-D}, α_{2A-D}, β_{1-3}	Alertness, mood elevation
Dopamine	Midbrain: substantia nigra pars compacta and ventral tegmental area	Striatum, limbic cortex, amygdala, nucleus accumbens, prefrontal cortex	D_{1-5}	Movements, initiative, working memory
Serotonin	Midbrain, pons, and medulla: raphe nuclei	Entire CNS	5-HT_{1A-F}, 5-HT_{2A-C}, 5-HT_{3-7}	Alertness, mood elevation, breathing control
Histamine	Hypothalamus: tuberomammillary nucleus; midbrain: reticular formation	Entire brain	H_{1-3}	Alertness
Orexin (hypocretin)	Posterior lateral hypothalamus	Entire brain	OX_1, OX_2	Alertness, food intake
Acetylcholine	Basal forebrain: nucleus basalis, medial septal nucleus, and nucleus of diagonal band	Cerebral cortex including hippocampus	Muscarinic (M_{1-5}), nicotinic subtypes	Alertness, memory
	Pontomesencephalic region: pedunculopontine nucleus and laterodorsal tegmental nucleus	Thalamus, cerebellum, pons, medulla	Muscarinic (M_{1-5}), nicotinic subtypes	Alertness, memory

[a]Many of the neurons releasing the neuromodulatory transmitters listed here also release a variety of peptides, which are likely to play a neuromodulatory role as well.

[b]Several of the receptor subtypes listed here have been cloned, and additional receptor subtypes are constantly being added.

[c]Functions listed are highly simplified here; see text and References, at the end of this chapter, for additional details.

As we discussed in Chapter 2 (see Table 2.2), neurotransmitters have two general types of functions. One is to mediate communication between neurons through fast excitatory or inhibitory postsynaptic potentials acting in the millisecond time range. The main excitatory and inhibitory neurotransmitters in the central nervous system are glutamate and gamma-aminobutyric acid (GABA), respectively. The second function is **neuromodulation**, generally occurring over slower time scales. Neuromodulation includes a broad range of cellular mechanisms involving signaling cascades that regulate synaptic transmission, neuronal growth, and other functions. Neuromodulation can either facilitate or inhibit the subsequent signaling properties of the neuron. The neurotransmitters of the diffuse projection systems, including acetylcholine, dopamine, norepinephrine, serotonin, and histamine, have a mainly neuromodulatory role in the central nervous system. In addition, a variety of peptides, small molecules, and other as yet unidentified neuromodulatory transmitters participate as well. Depending on the specific receptors present, these transmitters can have a facilitatory or inhibitory effect on neuronal signaling. Some transmitters even have both facilitatory and inhibitory neuromodulatory effects at different synapses or different receptor sites. The functional effects of these transmitters also depend on the regions of the brain being modulated, and they include effects on level of consciousness, the sleep–wake cycle, emotional states, motor behavior, and many other diverse functions. The details of these neuromodulatory actions are beyond the scope of this text (see the References section at the end of this chapter for more information). Here, we will focus instead on the anatomical distribution of these neurotransmitter systems and their functional roles only in the most general terms.

Note that, unlike gross lesions of the pontomesencephalic reticular formation, lesions or pharmacological blockades of the individual projecting neurotransmitter systems discussed in this section do not result in coma. Lesions or blockades of some neurotransmitter systems, especially acetylcholine and histamine, can cause profound confusion and drowsiness, but not coma. Thus, it appears likely that maintenance of the normal awake state does not depend on a single projection system. Rather, it probably depends on intact functioning of multiple anatomical and neurotransmitter systems acting in parallel, including the pontomesencephalic reticular formation and other brainstem projection pathways, bilateral thalamic intralaminar nuclei, and bilateral cerebral cortex. The individual neurotransmitter systems discussed here do play an important role in attentional mechanisms, memory, and emotional states, as will be discussed.

BRAINSTEM RETICULAR FORMATION AND THALAMUS As we have already mentioned, the **pontomesencephalic reticular formation** (or mesopontine reticular formation) projects to the **thalamic intralaminar nuclei** (see Figure 14.7). The neurotransmitters for these projections have not been identified with certainty, although many neurons in this pathway contain glutamate. The thalamic intralaminar nuclei have numerous reciprocal connections with the basal ganglia (see Table 7.3; Figure 7.7). In addition, however, the intralaminar nuclei, particularly the **rostral intralaminar nuclei** (central lateral, paracentral, and central medial nuclei), have widespread projections to the cerebral cortex (see Figure 7.8). These projections, together with other widespread projections from the thalamus, such as those arising from the adjacent **midline thalamic nuclei** (see Figure 7.6; Table 14.2), are thought to be important for maintaining normal alertness. In addition to projections to the intralaminar nuclei, the pontomesencephalic reticular formation projects to the **hypothalamus** and **basal forebrain** (see Figure 14.7). Widespread projections from these regions, in turn, may also participate in the alerting functions of the pontomesencephalic reticular formation.

(A)

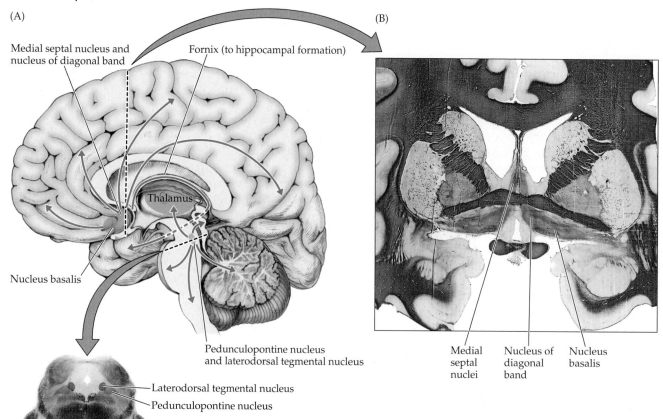

Medial septal nucleus and nucleus of diagonal band

Fornix (to hippocampal formation)

(B)

Thalamus

Nucleus basalis

Pedunculopontine nucleus and laterodorsal tegmental nucleus

Laterodorsal tegmental nucleus
Pedunculopontine nucleus

Medial septal nuclei

Nucleus of diagonal band

Nucleus basalis

FIGURE 14.9 Cholinergic Projection Systems See also Table 14.2. (A and B sections modified from Martin JH. 1996. *Neuroanatomy: Text and Atlas*. 2nd Ed. McGraw-Hill, New York.)

ACETYLCHOLINE Acetylcholine is the major efferent neurotransmitter of the peripheral nervous system, found at the neuromuscular junction, preganglionic autonomic synapses, and postganglionic parasympathetic synapses (see Chapter 6). Cholinergic neurons play a more limited role in the central nervous system, functioning primarily in neuromodulation rather than neurotransmission (recall that the major excitatory neurotransmitter of the central nervous system is glutamate). Neuromodulatory cholinergic neurons with widespread projections are found mainly in two locations (**Figure 14.9**): the pontomesencephalic region of the brainstem and the basal forebrain. Neurons of the brainstem cholinergic projection system are found mainly in the **pedunculopontine tegmental nuclei** and the **laterodorsal tegmental nuclei** (see Figure 14.9). These nuclei are located, respectively, in the lateral portion of the reticular formation and periaqueductal gray, at the junction between the midbrain and pons. Cholinergic projections from this region travel to the thalamus, including the intralaminar nuclei, which in turn project to widespread regions of the cortex (see Figure 14.7). Acetylcholine has different effects on different regions of the thalamus. In addition to its possible role in arousal, the pedunculopontine nucleus has a role in motor systems, and it is sometimes referred to as the mesencephalic locomotor region. Electrical stimulation of this region in animals causes coordinated locomotor movements. In this capacity, the pedunculopontine and laterodorsal tegmental nuclei have extensive connections with the basal ganglia, tectum, deep cerebellar nuclei, pons, medulla, and spinal cord.

Cholinergic inputs to the thalamus generally have an arousal effect mediated indirectly by facilitation of excitatory projections from the thalamus to the cortex. However, direct cholinergic inputs to the cortex do not arise from the brainstem to any significant extent; instead, they come mainly from the basal forebrain (see Figure 14.9). The **nucleus basalis** (of Meynert) contains cholinergic neurons that project to almost the entire cerebral cortex. Cholinergic projections to the hippocampal formation arise from the **medial septal nuclei** and

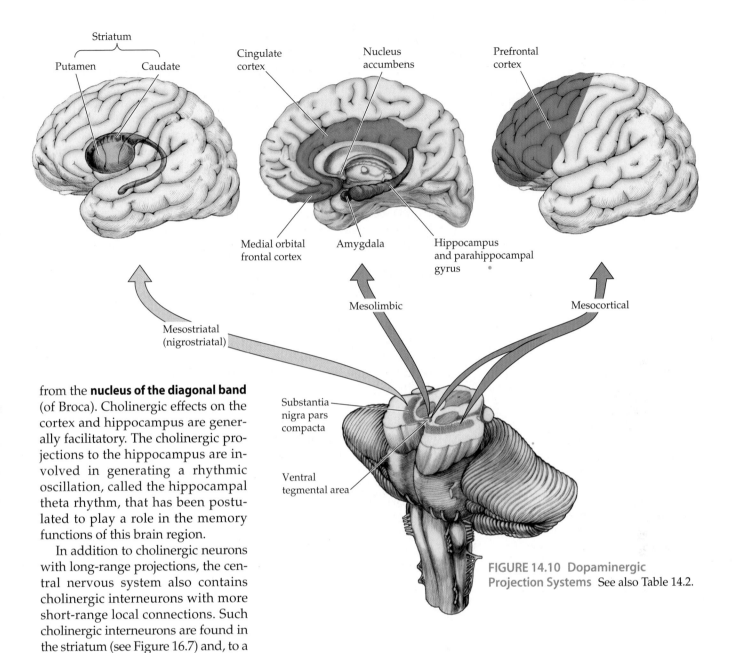

FIGURE 14.10 Dopaminergic Projection Systems See also Table 14.2.

from the **nucleus of the diagonal band** (of Broca). Cholinergic effects on the cortex and hippocampus are generally facilitatory. The cholinergic projections to the hippocampus are involved in generating a rhythmic oscillation, called the hippocampal theta rhythm, that has been postulated to play a role in the memory functions of this brain region.

In addition to cholinergic neurons with long-range projections, the central nervous system also contains cholinergic interneurons with more short-range local connections. Such cholinergic interneurons are found in the striatum (see Figure 16.7) and, to a more limited extent, in the cerebral cortex. There is also a cholinergic projection from the medial habenula to the interpeduncular nucleus.

Classically, the main cholinergic receptor type in the central nervous system is **muscarinic** (see Table 14.2). However, **nicotinic** receptors may play an important role in the central nervous system as well. The main functions of acetylcholine in the central nervous system are the facilitation of attention, memory, and learning. Pharmacological blockade of central cholinergic transmission causes delirium (see KCC 14.2 and KCC 19.15) and memory deficits. Degeneration of cholinergic neurons in the basal forebrain may be one of the mechanisms for memory decline in Alzheimer's disease (see KCC 19.16). The effects of cholinergic blockade on striatal neurons in movement disorders are discussed in Chapter 16.

DOPAMINE Neurons containing **dopamine** are located mainly in the ventral midbrain: in the **substantia nigra pars compacta** and the nearby **ventral tegmental area** (see Figure 14.3 and Figure 14.10). Three projection systems have been described arising from these nuclei in the mesencephalon. The

mesostriatal (nigrostriatal) pathway arises mainly from the substantia nigra pars compacta and projects to the caudate and putamen. Dysfunction of this pathway produces movement disorders such as Parkinson's disease, which is often treated with dopaminergic agonists (see KCC 16.2).

The **mesolimbic** pathway (see Figure 14.10) arises mainly from the ventral tegmental area and projects to limbic structures (see Chapter 18) such as the medial temporal cortex, amygdala, cingulate gyrus, and nucleus accumbens. The mesolimbic pathway plays a major role in reward circuitry and addiction. In addition, overactivity of the mesolimbic pathway is thought to be important in the "positive" symptoms of schizophrenia (see KCC 18.3), such as hallucinations, which often respond to dopaminergic antagonists.

Finally, the **mesocortical** pathway (see Figure 14.10) arises mainly from the ventral tegmental area (as well as from scattered dopaminergic neurons in the vicinity of the substantia nigra) and projects to the prefrontal cortex. Roles for this system have been proposed in frontal lobe functions such as working memory and attentional aspects of motor initiation (see Chapter 19). Damage to the mesocortical dopaminergic pathway may be important for some of the cognitive deficits and hypokinesia seen in Parkinson's disease (see KCC 16.2) and in the "negative" symptoms of schizophrenia (see KCC 18.3). In addition to the dopaminergic projection systems of the substantia nigra pars compacta and ventral tegmental area, dopaminergic neurons with more localized functions are found in the retina, olfactory bulbs, hypothalamus (inhibiting prolactin release; see Chapter 17), and medulla.

NOREPINEPHRINE Neurons containing **norepinephrine** (noradrenaline) were once thought to be located exclusively in the **locus ceruleus**, meaning "blue spot," located near the fourth ventricle in the rostral pons (see Figure 14.4A and Figure 14.11). However, noradrenergic neurons with similar projections to the locus ceruleus are also found scattered in the **lateral tegmental area** of the pons and medulla. Ascending noradrenergic projections from the locus ceruleus and rostral lateral tegmental area reach the entire forebrain (see Figure 14.11).

The effects of norepinephrine on the cortex can be inhibitory or excitatory, but effects on the thalamus are generally excitatory. Some of the better-known receptor types are listed in Table 14.2. Functions of the ascending norepineph-

FIGURE 14.11 Noradrenergic Projection Systems See also Table 14.2. (Brainstem sections modified from [top] the University of Washington Digital Anatomist Project and [bottom] Martin JH. 1996. *Neuroanatomy: Text and Atlas.* 2nd Ed. McGraw-Hill, New York.)

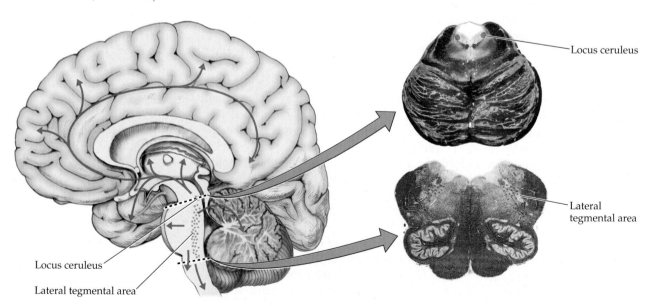

Locus ceruleus

Lateral tegmental area

Locus ceruleus

Lateral tegmental area

rine projection system include modulation of attention, sleep–wake states, and mood. Attention-deficit disorder is often treated with medications that enhance noradrenergic transmission. Firing of locus ceruleus neurons increases in the awake state and decreases dramatically during sleep. However, lesions of the locus ceruleus do not cause somnolence. On the other hand, narcolepsy, a sleep disorder characterized by excessive daytime sleepiness, often responds to treatment with noradrenergic-enhancing medications. Norepinephrine appears to be important, together with serotonin, in central modulation of pain (see Chapter 7) as well as in mood disorders such as depression and manic-depressive disorder and in anxiety disorders including obsessive-compulsive disorder (see KCC 18.3).

The locus ceruleus and lateral tegmental area also supply norepinephrine to the cerebellum, brainstem, and spinal cord. Noradrenergic neurons in the lateral tegmental area of the caudal pons and medulla are involved in sympathetic functions such as blood pressure control. In addition to norepinephrine, the related catecholamine epinephrine (adrenaline) is found in a small number of brainstem neurons. The role of these neurons has not been established, but it may also be related to blood pressure control.

SEROTONIN Neurons containing **serotonin** are found in the **raphe nuclei** of the midbrain, pons, and medulla (**Figure 14.12**). Raphe means "seam" in Greek, and it refers to the midline seam-like appearance of the brainstem in some areas where these nuclei are located (see Figures 14.3B and 14.4). The **rostral raphe nuclei** of the midbrain and rostral pons project to the entire forebrain, including the cortex, thalamus, and basal ganglia (see Figure 14.12). Both excitatory and inhibitory effects of serotonin have been described, even within the same structure. Serotonergic pathways are believed to play a role in several psychiatric syndromes (see also KCC 18.3), including depression, anxiety, obsessive-compulsive disorder, aggressive behavior, and certain eating disorders. In addition, like noradrenergic neurons, serotonergic neurons markedly decrease their firing rate during sleep and are likely involved in arousal. The **caudal raphe nuclei** of the caudal pons and medulla project to the cerebellum, medulla, and spinal cord. Projections to the spinal cord and medulla are involved in pain modulation (see Figure 7.5), breathing, temper-

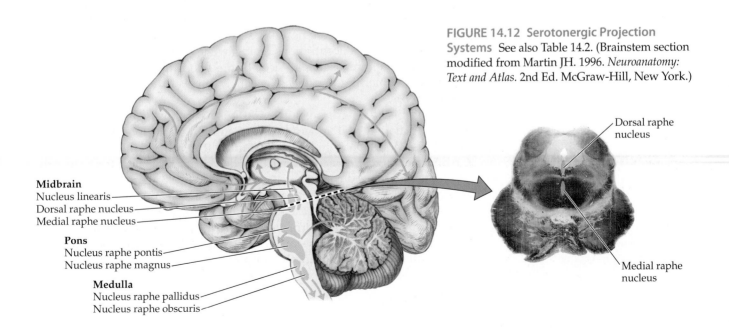

FIGURE 14.12 Serotonergic Projection Systems See also Table 14.2. (Brainstem section modified from Martin JH. 1996. *Neuroanatomy: Text and Atlas.* 2nd Ed. McGraw-Hill, New York.)

Midbrain
Nucleus linearis
Dorsal raphe nucleus
Medial raphe nucleus

Pons
Nucleus raphe pontis
Nucleus raphe magnus

Medulla
Nucleus raphe pallidus
Nucleus raphe obscuris

Dorsal raphe nucleus

Medial raphe nucleus

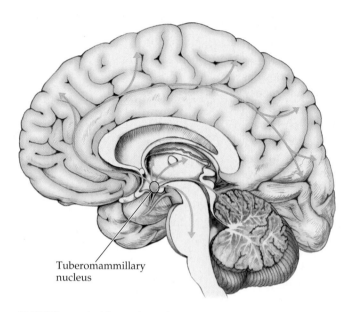

FIGURE 14.13 Histaminergic Projection Systems See also Table 14.2.

Tuberomammillary nucleus

ature regulation, and motor control. Defects in serotonin neurons have been associated with **sudden infant death syndrome** (**SIDS**), possibly caused by impaired arousal in response to hypoventilation. In addition to the raphe nuclei, a small number of serotonergic neurons have been identified in other brainstem regions, including in the area postrema and caudal locus ceruleus, and around the interpeduncular nucleus, as well as in the spinal cord.

HISTAMINE **Histamine** is found mainly in neurons of the posterior hypothalamus in the **tuberomammillary nucleus** (Figure 14.13), although there are some scattered histaminergic neurons in the midbrain reticular formation as well. Most histamine in the body is found outside the nervous system in mast cells, where it plays a role in immune responses and allergic reactions. Histamine-containing neurons were identified only relatively recently in the nervous system. Diffuse histaminergic projections from the tuberomammillary nucleus to the forebrain may be important to maintaining the alert state. Histamine has excitatory effects on thalamic neurons, and both excitatory and inhibitory effects on cortical neurons. Antihistamine medications, used to treat allergies, are thought to cause drowsiness by blocking CNS histamine receptors.

OTHER PROJECTING SYSTEMS In addition to the systems already listed, a variety of other neuromodulatory or projecting pathways that may play a role in alertness, mood regulation, memory, and other functions have been described or are still under investigation. These include peptides and small-molecule neurotransmitters. The peptide **orexin** (also called hypocretin) is produced in neurons of the posterior lateral hypothalamus (see Figure 14.15). Orexinergic neurons project to all major brainstem arousal systems as well as to the cerebral cortex, promoting the awake state. **Adenosine** is another putative neurotransmitter that may be important in mechanisms of alertness. Adenosine receptors are found in both the thalamus and the cortex, and adenosine generally has an inhibitory effect on these structures. The sources of adenosine in the nervous system have not been well characterized. Interestingly, concentrations of adenosine vary in a circadian manner, reaching a maximum just before sleep begins. One important mechanism for the increase in alertness produced by caffeine may be the blockade of adenosine receptors.

The inhibitory neurotransmitter **GABA** is found throughout the nervous system. Although well known for its role in inhibitory local interneurons, GABA also participates in long-range inhibitory projections. For example, GABAergic projection systems have been described in the basal forebrain projecting to widespread cortical areas. In addition, GABAergic neurons in the reticular nucleus of the thalamus project to other thalamic nuclei and to the rostral brainstem reticular formation. These GABAergic projection systems may be crucial for gating information flow in the nervous system and for regulating oscillatory electrophysiological activity underlying sleep and arousal. As we will discuss shortly, another important GABAergic projection system is from hypothalamic neurons, especially in the ventrolateral preoptic area,

REVIEW EXERCISE

Cover Table 14.2 except for the leftmost column. For each widespread projection system, list the locations of the cell bodies and the main targets of their projections.

which inhibit the serotonergic, noradrenergic, histaminergic, and cholinergic arousal systems, thereby promoting deep sleep.

Anatomy of the Sleep–Wake Cycle

The sleep–wake cycle involves a complex interplay of neural circuits, many located in the brainstem. In adult humans, there are five stages of sleep (Figure 14.14). Normally, sleep begins with **stages 1 through 4** of progressively deeper **nonREM** (non-rapid eye movement) sleep. NonREM sleep is followed by **REM** (rapid eye movement) sleep, during which most dreaming typically occurs. The cycle then repeats several times through the night (see Figure 14.14). REM sleep is sometimes called "paradoxical sleep." This name is used because in some ways REM is a deeper stage of sleep than stage 4, while in other ways it more closely resembles the awake state. For example, general muscle tone and brainstem monoaminergic neurotransmission are lower during REM sleep than during any other stage. On the other hand, the electroencephalogram (EEG; see Chapter 4) during REM sleep in some ways resembles that of awake activity (a low-voltage mixture of relatively fast activity), while the EEG of stage 3 or 4 nonREM sleep more closely resembles coma (high-voltage slow activity). It is also easier to awaken an individual from REM sleep than from stage 3 or 4 of nonREM sleep.

Sleep is not, as was once thought, simply a passive process arising from decreased stimulation of the nervous system. Several neural circuits interact to generate sleep, including many of the circuits described earlier in this chapter. Interestingly, whereas classic transection experiments in cats at the midbrain level produced coma, demonstrating the importance of the reticular formation in maintaining the awake state, transection at the level of the lower pons markedly *reduced* sleep in cats. This result suggested the presence of **sleep-promoting regions in the medulla**. These regions have subsequently been postulated to be located in the medullary reticular formation and nucleus solitarius

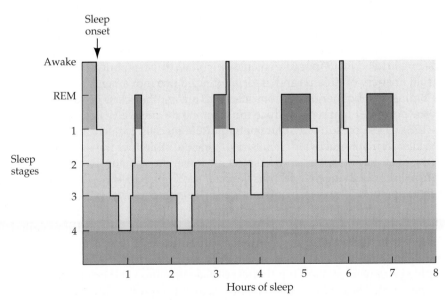

FIGURE 14.14 Sleep Stages Different stages of sleep can be recorded and scored based on polysomnographic monitoring of multiple physiological parameters, including EEG, extraocular movements, body movements, muscle tone, and respirations.

(Figure 14.15A).* In fact, lesions in certain regions of the medulla, as well as in the anterior hypothalamus and basal forebrain, can markedly reduce sleep. These regions are especially important for promoting nonREM sleep. For example, GABAergic neurons in the **ventrolateral preoptic area** (**VLPO**) of the anterior hypothalamus and in nearby regions send inhibitory projections to histaminergic neurons in the posterior hypothalamus (see Figure 14.15A), as well as to the brainstem serotonergic, noradrenergic, dopaminergic, and cholinergic arousal systems. The peptide galanin contributes to this inhibitory pathway, along with GABA. Thus, the VLPO promotes nonREM sleep by inhibiting of the ascending activating systems that project to the forebrain (see Figure 14.15A).

Around the time of World War I, there was a unique epidemic of "encephalitis lethargica" in which patients slept for long periods, sometimes leading to death. Constantin von Economo examined the brains of these patients and found damage in the posterior hypothalamus. More recently, neurons were discovered in the posterior lateral hypothalamus which produce peptides called **orexins** (also known as **hypocretins**). In contrast to the VLPO, the orexin neurons excite the brainstem and hypothalamic arousals systems and are crucial for the awake state. The VLPO inhibits the orexin neurons so that orexin neurons show reduced activity in sleep, further depressing the activity of brainstem and hypothalamic arousal systems (see Figure 14.15A).

A different set of brainstem circuits is thought to control REM sleep (Figure 14.15B). Several classes of **REM-on cells** are located in the pontine reticular formation. During REM sleep, activity in these pontine **GABAergic REM-on cells** together with neurons in the VLPO inhibit monaminergic transmission, particularly norepinephrine (NE) release from the locus ceruleus and lateral tegmental area, and serotonin (5-HT) release from the raphe nuclei (see Figure 14.15B). Activity in orexin neurons is reduced as well. **Noradrenergic and serotonergic** cells show progressively reduced firing during stages 1 through 4 of nonREM sleep, and these neurons are *virtually silent* during REM sleep. This silence leads to removal of inhibition from cholinergic neurons in the pedunculopontine and laterodorsal tegmental nuclei, resulting in increased cholinergic transmission to the thalamus during REM sleep (see Figure 14.15B). These activating changes are thought to underlie the appearance in REM sleep of EEG activity closely resembling the waking state. In addition, brainstem cholinergics produce intermittent waves of activation passing from pons-to-thalamus-to-cortex called PGO (ponto-geniculo-occipital) waves, thought to induce the visual imagery of dreams and associated rapid eye movements.

Brainstem cholinergic neurons also activate another class of REM-on cells in the pons that markedly reduce muscle tone during REM sleep (see Figure 14.15B). These presumably **glutamatergic REM-on cells**, located in the pontine reticular formation, activate circuits involving the inhibitory transmitter glycine in the medulla and spinal cord. As a result, lower motor neurons are inhibited, causing markedly decreased muscle tone (see Figure 14.15B). Lesions or degeneration of this system (sometimes seen as a precursor of parkinsonism or related disorders) abolishes the normal inhibition of motor activity during REM sleep, resulting in complex activities during dreaming, a condition called **REM sleep behavioral disorder**. Although tonic muscle activity is inhibited during REM sleep, there are brief phasic movements during REM sleep, such as rapid eye movements (giving this sleep stage its name) and brief movements of the limbs. These phasic movements occur during the waking state as well and are activated by neurons located in the pontine reticular formation.

Saper and colleagues recently proposed that rapid and complete transitions between sleep and wake, and between REM and nonREM sleep states, occur be-

---

*The precise locations of these sleep-promoting regions in the brainstem have not been confirmed electrophysiologically.

(A) NonREM sleep

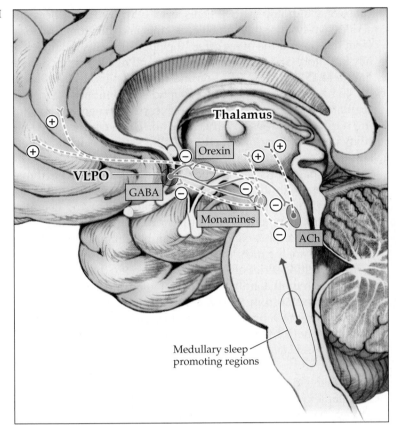

(B) REM sleep

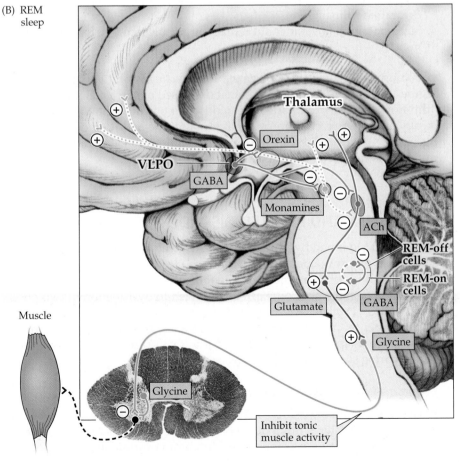

FIGURE 14.15 Brainstem Circuits Important for Sleep Regulation
(A) During nonREM sleep, GABAergic (and galanin) neurons in the ventral lateral preoptic area of the anterior hypothalamus inhibit neurons in the ascending activating systems, including posterior hypothalamic orexin neurons, monamines such as histamine (in the tuberomammillary nucleus, not shown), serotonin, noradrenalin, and dopamine, as well as brainstem cholinergics (Ach). Certain regions of the medulla may also play a role in promoting nonREM sleep. (B) During REM sleep, monamine transmitters, particularly noradrenalin and serotonin, are further reduced. This contributes to increased cholinergic inputs to the thalamus and EEG appearance of arousal. Pontine circuits include mutually inhibitory REM-on and REM-off cells, as well as neurons that inhibit tonic muscle activity during REM dream states. (Spinal cord section modified from DeArmond SJ, Fusco MM, Maynard MD. 1989. *Structure of the Human Brain: A Photographic Atlas.* 3rd Ed. Oxford, New York.)

REVIEW EXERCISE

Are the following brainstem neuro-transmitter systems *activated* or *inactivated* during nonREM sleep? During REM sleep? (See Figure 14.15.)

ACh

NE

5-HT

cause of a "flip-flop switch." This term, borrowed from electronic circuits, describes two systems that inhibit each other and therefore tend to switch strongly from one on and the other off, to the opposite state. Examples of these mutually inhibitory circuits include the VLPO and monaminergic arousal systems (see Figure 14.15A) and the REM-on and REM-off neurons (see Figure 14.15B).

Numerous other transmitters, peptides, and humoral and endocrine factors regulate and are regulated by sleep. Near orexin neurons, melanin concentrating hormone neurons have been found. These neurons have the opposite activity pattern, being most active in REM sleep and least active in the waking state. The suprachiasmatic nucleus of the hypothalamus (see Figure 17.3) receives retinal inputs and is crucial for setting circadian rhythms and synchronizing them with the light–dark cycle. Levels of numerous hormones fluctuate in a circadian pattern. Many naturally occurring chemicals have been investigated as potential sleep-promoting substances (e.g., adenosine, mentioned earlier in this chapter), although there is currently no agreement on which ones are the most important.

In an exciting series of investigations, orexin (hypocretin) cells in the posterior lateral hypothalamus were found to be deficient in both animals and humans that suffer from a sleep disorder called narcolepsy. **Narcolepsy** is characterized by an abnormal tendency to easily enter REM sleep directly from the waking state (compare to the normal sleep stages depicted in Figure 14.14), which is associated with four classic clinical findings:

1. Excessive daytime sleepiness
2. Cataplexy (sudden loss of muscle tone from the awake state, often in response to an emotional stimulus)
3. Hypnagogic (while falling asleep) or hypnopompic (while awaking) dreamlike hallucinations
4. Sleep paralysis (awaking, but remaining unable to move for several minutes)

Because orexin tends to stabilize the awake state by stimulating arousal (see Figure 14.15), it has been proposed that decreased orexin leads to an unstable "flip-flop" switch, provoking repeated transitions in and out of REM. Discovery of the hypocretin/orexin peptides raises the hope that ongoing investigations will explain these phenomena in cellular and molecular terms, which may lead to improved treatments for this disorder.

KEY CLINICAL CONCEPT

14.2 COMA AND RELATED DISORDERS

As classically defined by Plum and Posner, **coma** is unarousable unresponsiveness in which the patient lies with eyes closed. Minimum duration is 1 hour, to distinguish coma from transient disorders of consciousness such as concussion or syncope. Coma can be localized to dysfunction in two possible locations: (1) bilateral widespread regions of the cerebral hemispheres, or (2) the upper brainstem–diencephalic activating systems (or both 1 and 2). Coma can perhaps be best understood if it is contrasted with various similar-appearing states (see Figure 14.16 and Table 14.3).

Coma and other disorders of consciousness

Brain death might be considered an extreme and *irreversible* form of coma. As discussed in Chapter 3, brain death is defined on the basis of clinical examination demonstrating no evidence of forebrain or brainstem function, including no brainstem reflexes (Figure 14.16A; see Table 14.3). Only spinal cord reflexes may persist in brain death. When an EEG is done as a confirmatory test in brain death, it shows "electrocerebral inactivity," or a flat pattern, less than 2 microvolts in amplitude. Cerebral perfusion and metabolism are likewise reduced to zero in brain death.

In **coma**, there is profoundly impaired function of the cerebral cortex and diencephalic–upper brainstem arousal systems (Figure 14.16B). Unlike brain death, many simple or even complex brainstem reflex activities may occur in coma. However, psychologically meaningful or purposeful responses mediated by the cortex are absent (see Table 14.3). For example, in coma a patient may show reflex eye movements (such as the vestibulo-ocular reflex; see Chapter 3, "Coma Exam"), respiratory movements, or posturing (see Figure 3.5). However, there are no purposeful movements such as limb abduction in response to pain, localizing a painful stimulus by touching it with a different limb, or other responses demonstrating volition.

In coma, cerebral metabolism is typically reduced by at least 50%, in agreement with the lack of significant cortical functions. Although coma can be *induced* either by cortical or subcortical pathology, once coma is present, *both* cortical and subcortical arousal systems are depressed (see Figure 14.16B) since these systems are so intimately connected. The EEG (see Chapter 4) is usually abnormal in coma, but it can show many different patterns, including large-amplitude slow waves, burst-suppression, triphasic waves, spindle waves, or even alpha activity (a pattern seen in normal wakefulness). The most consistent abnormality of the EEG in coma is that it is typically monotonous, with little variability over time, unlike the normally varying EEG seen in different sleep stages (see Figure 14.14). **Sleep** differs from coma (see Table 14.3) in that patients in coma are unarousable even with vigorous stimulation and, as already noted, patients in coma do not undergo cyclical variations of state as seen during sleep.

Coma is not generally a permanent condition. Within 2–4 weeks of onset nearly all patients either deteriorate or emerge into other states of less profoundly impaired arousal. Following an initial catastrophic brain insult causing coma (most commonly, trauma or anoxia), some patients may enter a perplexing state in which they regain sleep–wake cycles and other primitive orienting responses and reflexes mediated by the brainstem and diencephalon but remain unconscious. This condition is referred to as a **vegetative state** (Figure 14.16C; see Table 14.3). Vegetative state can also occur in certain end-stage dementias, as well as neurodegenerative or congenital disorders. If duration is longer than 1 month it is called a **persistent vegetative state**. When vegetative state lasts longer than 3 months following non-traumatic causes or longer than 12 months following trauma, prognosis for recovery is very poor.

As in coma, patients in vegetative state have no meaningful responses to stimuli, and they have diffuse cortical dysfunction evidenced by over 50% reduction in cerebral metabolism. However, patients in vegetative state do open their eyes and arouse in response to stimulation, and they may turn their eyes and heads toward auditory or tactile stimuli, presumably through brainstem and diencephalon-mediated pathways. Patients in vegetative state may produce unintelligible sounds and move their limbs, but they do not have meaningful speech or gestures, do not make purposeful movements, do not track visual stimuli, and are incontinent.

Appearance of visual tracking may be one of the earliest signs of emergence into the **minimally conscious state** (Figure 14.16D; see Table 14.3), which can occur as a further stage of recovery from vegetative state or as a primary disorder. In minimally conscious state, patients have some minimal or vari-

(A) Brain death

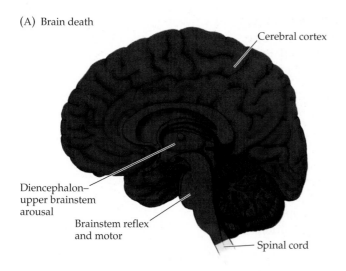

(B) Coma

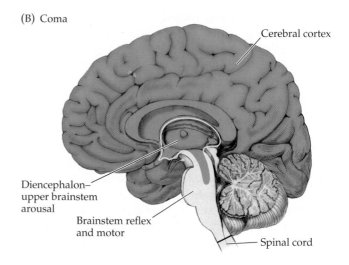

(C) Vegetative state

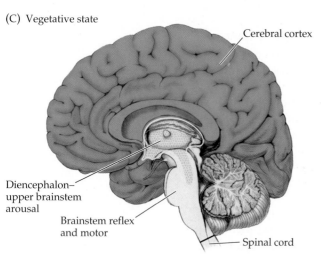

(D) Minimally conscious or better

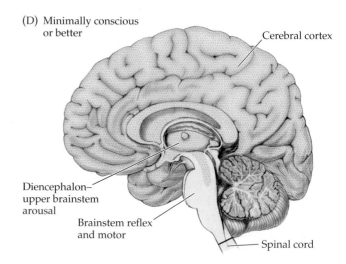

(E) Akinetic mutism, abulia, catatonia

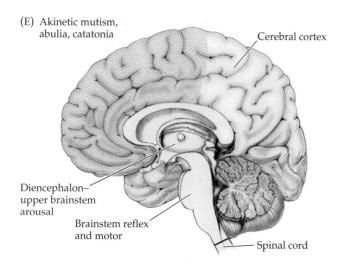

(F) Locked in

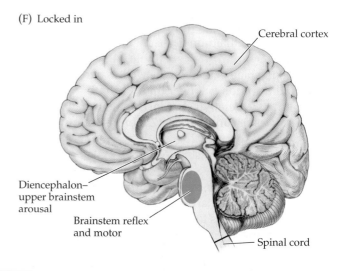

Key
■ Absent function   ■ Severely depressed function   ▦ Variably depressed function

◀ **FIGURE 14.16 Brain Regions Involved In Coma and Related Disorders**
(A) Brain death: All cortical, subcortical, and brainstem function is irreversibly lost.
Spinal cord function may be preserved. No responses can be elicited except for spinal
cord reflexes. (B) Coma: There is severe impairment of cortical function and of the diencephalic–upper brainstem activating systems. Patients are unarousable with eyes
closed and have no purposeful responses, but brainstem reflex activity is present.
(C) Vegetative state: Cortical function is severely impaired, but there is some preserved diencephalic–upper brainstem activating function. Like in coma, patients are
unconscious at all times, with no purposeful responses, but they can open their eyes
spontaneously or with stimulation, and they exhibit primitive orienting responses,
and sleep–wake cycles. (D) Minimally conscious state or better: Impaired function of
the cerebral cortex and diencephalic–upper brainstem activating systems is variable.
Patients exhibit some purposeful responses, along with deficits, depending on the
severity of brain dysfunction. (E) Akinetic mutism, abulia, catatonia: Impaired frontal
and dopaminergic function leads to profound apathy and deficits in response initiation. (F) Locked-in syndrome: Example of lesion in basis pontis eliminating corticospinal and corticobulbar motor output. Sensory function and consciousness are
preserved. (Modified with permission from Blumenfeld H. 2009. The neurological
examination of consciousness. In *The Neurology of Consciousness*, S Laureys and G
Tononi (eds.), Chapter 2, pp. 15–30. Elsevier, Ltd.)

able degree of responsiveness, including the ability to follow simple commands, say single words, or reach for and hold objects. By definition, minimally conscious patients do not have reliable interactive verbal or nonverbal
communication and do not have reliable functional use of objects. Specific
clinical criteria for the vegetative and minimally conscious states have been
established by multisociety task forces (see the References section at the end
of this chapter for more information). Terms such as "coma vigil" or "apallic
syndrome" were used in the past for vegetative and similar states, but they
are imprecise and are not generally used today. Interestingly, recent innova-

TABLE 14.3 Coma and Related States

ANATOMY (SEE FIGURE 14.16)	CEREBRAL CORTEX	DIENCEPHALON– UPPER BRAINSTEM AROUSAL SYSTEMS	BRAINSTEM REFLEX AND MOTOR SYSTEMS	SPINAL CORD CIRCUITS
FUNCTIONS TESTED	PURPOSEFUL RESPONSES TO STIMULI?	BEHAVIORAL AROUSAL, SLEEP- WAKE CYCLES?	BRAINSTEM REFLEXES?	SPINAL CORD REFLEXES?
States of impaired consciousness				
Brain death	No	No	No	Yes
Coma	No	No	Yes	Yes
Vegetative state	No	Yes	Yes	Yes
Minimally conscious state	Yes, at times	Yes	Yes	Yes
Stupor, obtundation, lethargy, delirium	Yes, at times	Variable	Yes	Yes
Status epilepticus	Variable	Variable	Yes	Yes
Akinetic mutism, abulia, catatonia	Yes, at times	Yes	Yes	Yes
Sleep, normal and abnormal	Yes, at times	Yes	Yes	Yes
States resembling impaired consciousness				
Locked-in syndrome	No[a]	Yes	Yes	Yes
Dissociative disorders, somatoform disorders	Yes, at times	Yes	Yes	Yes

[a]Some locked-in patients may have preserved vertical eye movements, eye blinking, or other slight movements under volitional control.

Modified with permission from Blumenfeld H. 2009. The neurological examination of consciousness. In *The Neurology of Consciousness*,
S Laureys and G Tononi (eds.), Chapter 2, pp. 15–30. Elsevier, Ltd.

tive studies have used functional MRI to show that even in vegetative state, some patients retain the ability to follow instructions such as "imagine you are playing tennis" or "imagine you are walking through your house" in a similar manner to normal controls. Further work with this technique may eventually enable improved prediction of prognosis and possibly even some limited form of communication with selected vegetative patients.

There is a wide continuum of levels of consciousness between coma and the fully awake state. A variety of more poorly defined terms—**lethargy**, **obtundation**, **stupor**, **semicoma**, and so on—are sometimes used to describe different states along this continuum. Although definitions for these terms exist, use of these terms alone, without further details, can be confusing to other physicians when they read the chart and try to assess the patient's progress. Therefore, as described in Chapter 3 (section on the coma exam), it is essential to document the patient's level of alertness with a *specific statement* of what the patient did in response to particular stimuli. For example, if pressure applied to a nail bed or to the supraorbital ridge causes a patient to briefly open their eyes, moan, and push away the examiner with one hand before lapsing back into unresponsiveness, the patient is not in a coma. Documenting impaired consciousness and the specific response elicited is the most practical way to follow changes in patients of this kind. Other states along the continuum of impaired alertness, attention, and cognition, such as **delirium** (global confusional states) and **dementia**, are discussed further in Chapter 19 (see KCC 19.14–19.16).

Several states of profound apathy, in the extreme, can resemble coma or vegetative. These include akinetic mutism, abulia, and catatonia (**Figure 14.16E** and see Table 14.3). These disorders have in common the dysfunction of circuits involving the frontal lobes, diencephalon, and ascending dopaminergic projections (see Figure 14.10) important to the initiation of motor and cognitive activity. In **akinetic mutism**, the patient appears fully awake and, unlike patients in vegetative state, such patients visually track the examiner. However, they usually do not respond to any commands. Because the primary deficit is in motor initiation rather than in consciousness, akinetic mutism differs from the minimally conscious state. Akinetic mutism can be viewed as an extreme form of **abulia**, often resulting from frontal lesions, in which patients usually sit passively but may occasionally respond to questions or commands after a long delay. In some patients, abulia or akinetic mutism can be reversed with dopaminergic agonists. Abulia is discussed further in Chapter 19 (see KCC 19.11). **Catatonia** is a similar akinetic state that can occasionally be seen in advanced cases of schizophrenia. Again, frontal-lobe and dopaminergic dysfunction have been implicated. Other, related akinetic–apathetic states include advanced parkinsonism (see KCC 16.2), severe depression, and neuroleptic malignant syndrome.

An important consideration in the differential diagnosis of coma is **status epilepticus**, meaning continuous seizure activity (see KCC 18.2). Often, seizure activity is clinically obvious. However, sometimes only subtle twitching or no motor activity at all is present. Studies in which EEGs were performed indiscriminately in a series of patients in coma revealed that unrecognized status epilepticus was present in up to 20% of cases. Therefore, whenever the cause of coma cannot be found, or when there is a history of seizures, an EEG should be performed promptly so that anticonvulsant therapy can be initiated when needed.

The **locked-in syndrome** discussed in KCC 14.1 can sometimes be mistaken for coma (**Figure 14.16F**; see Table 14.3). Unlike coma, however, these patients are conscious and may be able to communicate through vertical eye movements or eye blinks. Several **psychiatric disorders** can cause patients to appear as if in a coma. In addition to catatonia and severe depression, patients may be unresponsive when in a **dissociative state**, often resulting from severe emotional trauma. **Somatoform disorders** such as conversion disorder, somatization disorder, or factitious disorder can also sometimes produce states resembling coma, sometimes called "pseudocoma." Often these can be distinguished from coma by a carefully performed neurologic exam (see Chapter 3), although in some cases the diagnosis may not be obvious.

Transient loss of consciousness, as we discussed in KCC 10.3, is usually caused by cardiac or other medical conditions and is much less commonly caused by neurological disorders such as seizures or brainstem ischemia.

Clinical Approach to the Patient in Coma

Coma is a neurologic emergency because many causes of coma are reversible if treated promptly but can cause permanent damage that becomes progressively more severe as time goes on. Some important causes of coma are listed in Table 14.4. As we have already mentioned, coma can be localized to either bilateral widespread dysfunction of the cerebral hemispheres, to brainstem–diencephalic dysfunction, or to both. The most common causes of bilateral cerebral dysfunction are global anoxia, other toxic/metabolic disorders, and head trauma. Bilateral ischemic infarcts can also cause coma, usually by involving first one and then the second hemisphere. Brainstem dysfunction causing coma can either be from extrinsic compression from cerebral or cerebellar mass lesions or by intrinsic brainstem lesions, most commonly infarct or hemorrhage.

As in any other emergency situation, when evaluating coma the initial priorities are to ensure that there is an unobstructed **airway**, that the patient is **breathing**, and that there is normal **circulatory function**. When it is clinically appropriate, the patient should be intubated and cardiopulmonary resuscitation should be initiated. Establishing prompt intravenous access is also essential. **Intravenous thiamine, dextrose, and naloxone** should be given immediately, even before laboratory results are obtained*, because thiamine deficiency, hypoglycemia, and opiate overdose are readily treatable causes of coma. Additional doses are needed when these conditions are confirmed. Flumazenil can also be given if benzodiazepine overdose is suspected. Next, a more detailed assessment should be performed, including history, exam, blood tests, and other diagnostic tests, to find specific treatable causes of coma.

Much useful information can be obtained from the **neurologic exam** in the comatose patient, and this exam can usually be completed within a matter of minutes. For example, the size and responsiveness of the pupils (see **neuroexam.com Video 29**) can be a helpful guide to the cause of coma (Table 14.5). Although many exceptions exist, toxic

---

*Infants in coma may have an undiagnosed inborn error of metabolism that could be worsened by giving dextrose, so infants should not be given dextrose unless glucose testing reveals hypoglycemia.

TABLE 14.4 Some Important Causes of Coma[a]

Head trauma
Brainstem ischemia
Diffuse anoxia
Cardiopulmonary arrest
Intracranial hemorrhage
Status epilepticus (or post-ictal state)
Hydrocephalus
Diffuse cerebral edema
Drug or ethanol toxicity
Electrolyte abnormality (e.g., elevated Na^+, Ca^{2+}, Mg^{2+})
Hypoglycemia
Hypothyroidism
Hypoadrenalism
Thiamine deficiency
Hepatic failure
Renal failure
Meningitis
Encephalitis
Brain abscess
Intracranial neoplasm
End-stage neurodegenerative condition
Inborn errors of metabolism
Hereditary endogenous benzodiazepine production
Psychogenic unresponsiveness

[a]Following the format of Figure 1.1.

TABLE 14.5 Typical Pupil Abnormalities in Coma

CAUSE OF COMA	APPEARANCE OF PUPILS
Toxic and metabolic disorders	Normal (usually)
Midbrain lesion or transtentorial herniation	Unilateral or bilateral "blown" pupil[a]
Pontine lesion	Small, responsive to light bilaterally
Opiate overdose	Pinpoint pupils bilaterally

[a]Dilated pupil, unresponsive to light.

Pupil light reflex

and metabolic causes of coma often produce normal-sized, reactive pupils. Asymmetrically or bilaterally dilated, unresponsive ("blown") pupils can indicate midbrain compression and transtentorial herniation (see KCC 5.4). Bilateral small but responsive pupils are often seen in pontine lesions. Bilateral pinpoint pupils are seen in opiate overdose. At this point, please review the section titled "Coma Exam" in Chapter 3, as this information will be useful for the cases in this chapter (as well as in clinical practice).

Blood tests can reveal many toxic and metabolic causes of coma (see Table 14.4). Next, an **emergency head CT** should be obtained so that appropriate neurosurgical or cerebrovascular treatments can be initiated, especially when focal neurologic findings or other evidence of an intracranial lesion is present. Once the danger of herniation has been excluded by head CT, patients with coma of unknown cause should undergo a **lumbar puncture** for CSF analysis (see KCC 5.10). In addition, if the cause of coma remains unknown after this evaluation, an **EEG** should be obtained immediately so that subtle status epilepticus can be detected and treated if present.

Prognosis in coma depends on the specific cause. In diffuse anoxic brain injury, guidelines have been published, including those by Levy and collaborators, that may help predict the outcome on the basis of clinical features at various times after onset (see the References section at the end of this chapter for details). In drug overdose, patients usually recover completely, as long as vital functions are supported adequately during the period of coma. ■

Reticular Formation: Motor, Reflex, and Autonomic Systems

In this chapter we have so far emphasized the role of the reticular formation in modulating alertness, attention, and consciousness. However, many circuits, particularly of the caudal reticular formation, serve crucial functions in motor, reflex, and autonomic function, including basic "life support" systems such as respiration and circulatory control.

Respiration involves a network of control systems acting at multiple levels. Usually, respiratory rhythms occur automatically under the control of circuits in the **medulla**. The importance of the medulla was demonstrated when animals transected at or above the pontomedullary junction continued to breathe. However, other regions of the nervous system have strong modulatory influences on the respiratory pattern. Respiratory rhythms can also be superseded temporarily by voluntary control mediated by the forebrain. Some important brainstem regions involved in respiration are shown in Figure 14.17. There are numerous inputs to respiratory circuits, including peripheral chemoreceptors for blood oxygen level and pH, and stretch receptors in the lungs, many of which project to the cardiorespiratory portion of the **nucleus solitarius**. In addition, there are inputs from central nervous system neurons including chemoreceptors in the medulla, which contain serotonin and stimulate respiration. While the **pre-Bötzinger complex** located in the medulla has been described as a pacemaker for respiration, many other nuclei in the medulla appear to also participate in generating the respiratory rhythms. As shown in Figure 14.17, some nuclei are active during inspiration, while others are active during expiration. Ultimately, these nuclei project to spinal cord lower motor neurons. Cervical spinal segments **C3 to C5** control **phrenic nerve** efferents that contract the diaphragm during inspiration, while thoracic levels control thoracic inspiratory and expiratory muscles.

Lesions of the medulla disrupt respiratory circuits and can cause **respiratory arrest** and death. Other abnormal respiratory patterns are sometimes seen with lesions of the central nervous system. Proceeding from caudal to rostral, lesions of the medulla that do not cause respiratory arrest can lead to **ataxic respi-**

ration, an ominous pattern of very irregular breathing, that may ultimately progress to respiratory arrest. Lesions of the rostral pons (in the medial parabrachial Kölliker–Fuse area located dorsal to the motor nucleus of CN V; see Figure 14.17) can rarely cause a peculiar breathing pattern called **apneustic respiration**, in which the patient has brief 2- to 3-second respiratory pauses at full inspiration. Midbrain lesions, as well as lesions in other regions, may lead to **central neurogenic hyperventilation**. Finally, in **Cheyne–Stokes respiration**, breathing becomes progressively deeper with each breath, then progressively shallower with each breath to the point of apnea. The cycle then repeats, and breathing gradually becomes deeper again, in a continual crescendo–decrescendo pattern. This breathing pattern is not typically harmful in and of itself. Cheyne–Stokes respiration is usually seen in bilateral lesions at or above the level of the upper pons (including lesions of the cerebral cortex), but it can also be seen in mountain climbers sleeping at high altitudes and in medical conditions such as cardiac failure.

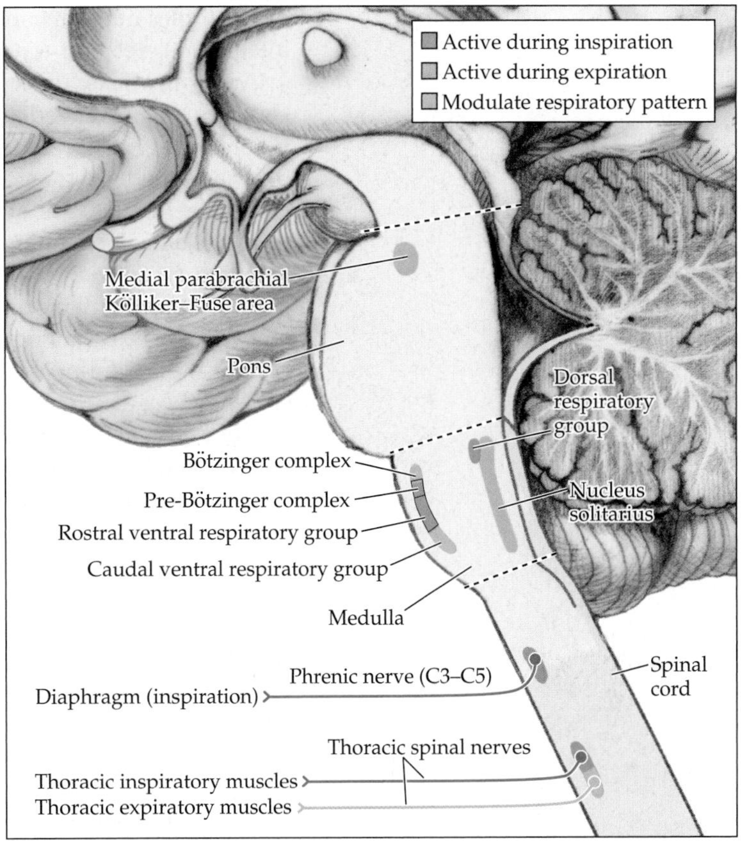

Control of **heart rate and blood pressure** are likewise mediated by circuits at multiple levels in the nervous system. Inputs to the caudal **nucleus solitarius**, also known as the cardiorespiratory nucleus, are again crucial, as are circuits in the nearby medullary reticular formation. The nucleus solitarius receives inputs from baroreceptors in the carotid body and aortic arch via cranial nerves IX and X, respectively (see Figures 12.20 and 12.21). Control of heart rate and blood pressure is then mediated by circuits, many of which project directly from the nucleus solitarius to parasympathetic and sympathetic preganglionic neurons in the brainstem and spinal cord (see Figure 6.13). Presympathetic neurons in the **rostral ventrolateral medulla** project to sympathetic neurons in the spinal cord intermediolateral cell column and are crucial for maintaining normal blood pressure. Interruption of this pathway causes the reduced blood pressure seen in **spinal shock** (see KCC 7.2). Interestingly, the cardiorespiratory portion of the nucleus solitarius also projects rostrally to the forebrain, largely via relays in the parabrachial nucleus of the rostral pons (see Figure 14.4A). Inputs from the nucleus solitarius to the **limbic system** (see Chapter 18) may, for example, be important in mediating emotional responses to altered cardiorespiratory function, and they have been postulated to play a role in triggering panic attacks. Information travels in the other direction as well, so an emotional state manifested as limbic system activity has a strong effect on autonomic function through projections to the brainstem reticular formation. This may explain why an emotional experience can cause your heart to race and your palms to sweat.

The reticular formation is involved in many complex motor tasks. Experimental animals in which higher structures have been disconnected from the brainstem can still perform numerous motor tasks, including orienting toward stimuli, maintaining posture, locomotion, and respiration. **Motor systems arising from the brainstem**, discussed in Chapter 6, include the reticulospinal, vestibulospinal, tectospinal, and rubrospinal tracts (see Figure 6.11). In addition, the substantia nigra and pedunculopontine tegmental nucleus play im-

FIGURE 14.17 **Brainstem Regions Involved in Respiratory Control**

portant roles in basal ganglia circuits, and parts of the reticular formation are integral to cerebellar function.

Abnormal **flexor (decorticate) posturing** or **extensor (decerebrate) posturing** (see Figure 3.5) are mediated largely by brainstem circuits. Regions of the reticular formation adjacent to cranial nerve nuclei are crucial for coordinating activity and mediating reflexes involving the cranial nerves, such as the corneal reflex, eye movements, and many other activities (see Chapters 12 and 13). Behaviors such as **coughing, hiccupping, sneezing, yawning, shivering, gagging, vomiting, swallowing, laughing, and crying** are all heavily dependent on circuits in the pontomedullary reticular formation. Lesions of the brainstem can interfere with these behaviors or cause them to emerge abnormally. For example, patients with pontine infarcts can exhibit abnormal spontaneous shivering, lesions of the medulla can produce hiccups, and lesions of descending white matter pathways can produce abnormal, spontaneous **pseudobulbar laughter and crying** (see KCC 12.8)

The **area postrema**, located along the caudal lateral wall of the fourth ventricle in the medulla, contains a region called the **chemotactic trigger zone** (see Figure 5.15). In this region, the blood–brain barrier is incomplete, allowing endogenous substances or exogenous toxins in the bloodstream to trigger **nausea and vomiting**. Nausea and vomiting can also be triggered by a circuit beginning with the release of serotonin (5-HT) from cells in the stomach and small intestine walls in response to emetic agents. The 5-HT stimulates the endings of afferent fibers traveling with the vagus to reach the nucleus solitarius in the brainstem. Vagal afferents also project to the nearby area postrema. Activation of the area postrema or nucleus solitarius may also play a role in the nausea and vomiting seen in disorders of the vestibular system or cerebellum, and in elevated intracranial pressure, although the mechanisms are still under investigation.

The **pontine micturition center** and other regions of the reticular formation are involved in maintaining sphincter control (see Figure 7.11). As we discussed in Chapter 7, the **periaqueductal gray** functions together with other regions in the brainstem and spinal cord to modulate pain transmission (see Figure 7.5).

Brainstem Vascular Supply

The blood supply to the posterior fossa structures arises from the vertebrobasilar system (Figure 14.18; see also Figure 10.2). The paired **vertebral arteries** arise from the subclavian arteries at the base of the neck and then ascend through the foramina transversaria of cervical vertebrae C6 through C2. They then take a winding course around the lateral aspect of the first cervical vertebra before piercing the dura and entering the cranial cavity via the **foramen magnum**. The vertebral arteries run along the ventral aspect of the medulla and join at the pontomedullary junction to form a single **basilar artery** (see Figure 14.18A). The basilar artery continues rostrally, running along the ventral surface of the pons, before splitting at the pontomesencephalic junction into the two **posterior cerebral arteries**, which connect via the **posterior communicating arteries (Pcomms)** to the internal carotid arteries of the anterior circulation.

The vertebrobasilar system gives rise to multiple branches to provide the blood supply to the brainstem and cerebellum. In addition, most parts of the thalamus, as well as the inferior-medial occipital and temporal lobes, are supplied by the posterior cerebral arteries arising from the top of the basilar artery. The largest branches of the vertebrobasilar system are the **posterior inferior cerebellar artery (PICA)**, **anterior inferior cerebellar artery (AICA)**, **superior cerebellar artery (SCA)**, and **posterior cerebral artery (PCA)** (see Figure 14.18). Note that just as there are three cerebral arteries (ACA, MCA, and PCA) there are also three cerebellar arteries (PICA, AICA, and SCA).

MNEMONIC

(A)

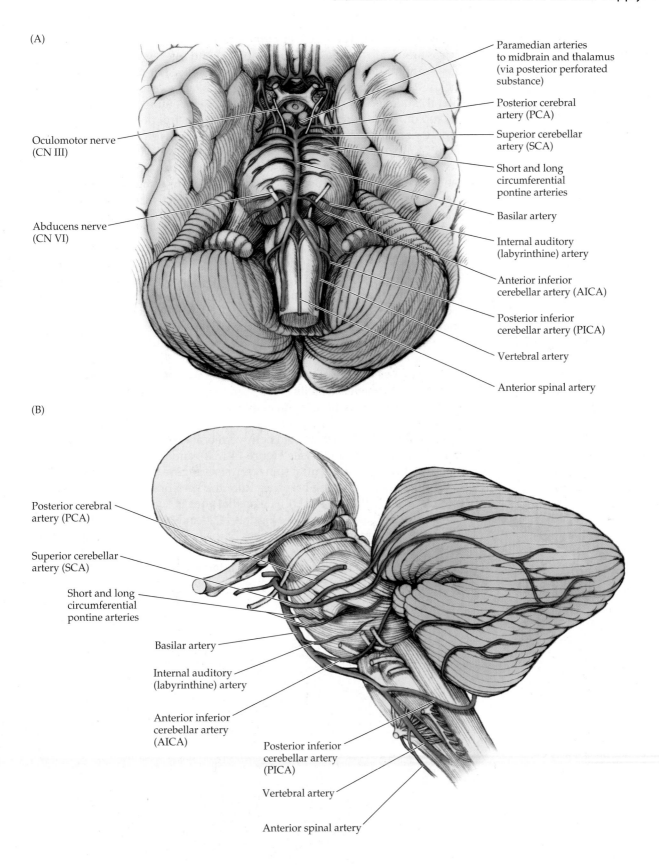

Paramedian arteries
to midbrain and thalamus
(via posterior perforated
substance)

Posterior cerebral
artery (PCA)

Superior cerebellar
artery (SCA)

Short and long
circumferential
pontine arteries

Basilar artery

Internal auditory
(labyrinthine) artery

Anterior inferior
cerebellar artery (AICA)

Posterior inferior
cerebellar artery (PICA)

Vertebral artery

Anterior spinal artery

Oculomotor nerve
(CN III)

Abducens nerve
(CN VI)

(B)

Posterior cerebral
artery (PCA)

Superior cerebellar
artery (SCA)

Short and long
circumferential
pontine arteries

Basilar artery

Internal auditory
(labyrinthine) artery

Anterior inferior
cerebellar artery
(AICA)

Posterior inferior
cerebellar artery
(PICA)

Vertebral artery

Anterior spinal artery

**FIGURE 14.18 Brainstem Blood
Supply** (A) Ventral view. (B) Lateral view.

The **PICA** arises from the vertebral artery at the level of the medulla and wraps around to supply the lateral medulla and inferior cerebellum (see Figure 14.18B). The **AICA** arises from the proximal basilar artery at the level of the caudal pons, usually just after the vertebral arteries fuse, and supplies the lateral caudal pons and a small region of the cerebellum. The **SCA** arises from the top of the basilar artery at the level of the rostral pons and supplies the superior cerebellum, as well as a small region of the rostral laterodorsal pons. The **PCA** arises from the top of the basilar artery as well, just beyond the SCA. Recall that the oculomotor nerves (CN III) usually pass between the SCA and the PCA (see Figure 14.18A). The PCA wraps around the midbrain, supplying it, as well as most of the thalamus, medial occipital lobes, and inferior-medial temporal lobes. The blood supply to the cerebellum is discussed further in Chapter 15.

Several types of smaller branches arise from these main arteries and provide the blood supply to the brainstem (**Figure 14.19**). **Paramedian branches** tend to respect the midline, with individual branches supplying either the right or left paramedian region. They extend a variable distance from the ventral surface of the brainstem, with the longest branches reaching all the way to the ventricle. **Short circumferential arteries** and **long circumferential arteries** (including PICA, AICA, and so on, as well as smaller branches) give rise to penetrating branches that supply the more lateral portions of the brainstem (see Figure 14.19).

The **main vascular territories of the brainstem** are shown in **Figures 14.20** and **14.21**. The **medial medulla** is supplied by paramedian branches of the anterior spinal artery in more caudal regions and by paramedian branches of the vertebral arteries in more rostral regions (see Figure 14.21D). Recall that the anterior spinal artery arises from both vertebral arteries, runs along the ventral surface of the medulla (see Figure 14.18A), and continues out of the cranial vault to supply the ventral spinal cord (see Figure 6.5). The **lateral medulla** is supplied by penetrating branches from the vertebral artery and the PICA (see Figures 14.20 and 14.21D). The **medial pons** is supplied by paramedian branches of the basilar artery (see Figures 14.20 and 14.21B,C). The **lateral pons** is supplied by circumferential branches of the basilar artery. In the more caudal regions, the lateral pons is supplied by the AICA (see Figures 14.20 and 14.21C). The inner ear is supplied by the **internal auditory (labyrinthine) artery** (see Figure 14.18A), which usually arises as a branch of the AICA, but occa-

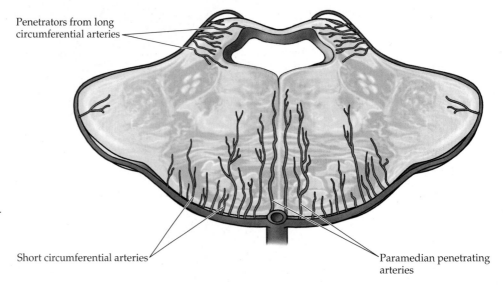

FIGURE 14.19 Penetrating Vessels Supplying the Internal Brainstem Structures Penetrating vessels of the brainstem include the paramedian arteries and penetrators from short and long circumferential arteries.

Penetrators from long circumferential arteries

Short circumferential arteries

Paramedian penetrating arteries

sionally comes directly off the basilar artery. The more rostral lateral pons is supplied mainly by small circumferential branches of the basilar artery called lateral **pontine arteries** (see Figure 14.18). A small variable region of the **superior dorsolateral pons** receives some blood supply from the SCA (see Figures 14.20 and 14.21B), but this artery supplies mainly the superior cerebellum rather than the brainstem. The **midbrain** is supplied by penetrating branches arising from the top of the basilar artery and from the proximal PCAs (see Figures 14.18, 14.20, and 14.21A). Recall that arteries supplying the **thalamus** also arise mainly from the top of the basilar artery and proximal PCAs (see Figure 10.8A). Paramedian branches arising from the top of the basilar enter the interpeduncular fossa to supply the medial midbrain and thalamus (see Figure 14.18A). Sometimes these arteries bifurcate after their origin, giving rise to the so-called arteries of Percheron, which supply the bilateral medial midbrain and thalamus. Occlusion of an artery of Percheron before it bifurcates can lead to bilateral medial midbrain or thalamic infarcts.

Tables 14.7, 14.8, and 14.9 list important structures lying in each of the main brainstem vascular territories. Clinical syndromes associated with these territories will be discussed in the next section.

Because the brainstem is so essential for maintaining consciousness and vital life functions, it is crucial for the clinician to be familiar with the major vascular syndromes of the posterior circulation. We will first discuss general features of vertebrobasilar vascular disease and then review syndromes involving specific vascular territories.

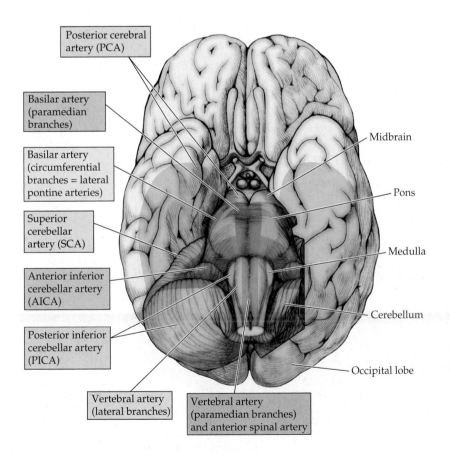

FIGURE 14.20 Brainstem Vascular Territories, Surface View

FIGURE 14.21 **Brainstem Vascular Territories Shown in Axial Sections** Level of sections shown in insets. (A) Midbrain (× 3.1). (B) Rostral pons (× 3.4). (C) Caudal pons (× 3.5). (D) Medulla (× 5.1). (Brainstem sections in A, C, D from Martin JH. 1996. *Neuroanatomy: Text and Atlas.* 2nd Ed. McGraw-Hill, New York. B from DeArmond SJ, Fusco MM, Maynard MD. 1989. *Structure of the Human Brain: A Photographic Atlas.* 3rd Ed. Oxford, New York.)

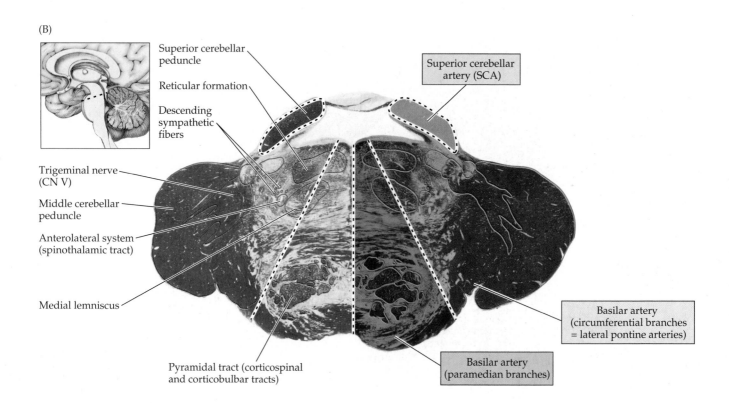

(A)

Reticular formation

Anterolateral system (spinothalamic tract)

SCA and proximal PCA

Descending sympathetic fibers

Medial lemniscus

Proximal PCA

Red nucleus

Substantia nigra

Pyramidal tract (corticospinal and corticobulbar tracts)

Oculomotor nucleus and nerve fascicles (CN III)

Paramedian branches at top basilar artery (interpeduncular fossa)

(B)

Superior cerebellar peduncle

Reticular formation

Descending sympathetic fibers

Superior cerebellar artery (SCA)

Trigeminal nerve (CN V)

Middle cerebellar peduncle

Anterolateral system (spinothalamic tract)

Medial lemniscus

Pyramidal tract (corticospinal and corticobulbar tracts)

Basilar artery (paramedian branches)

Basilar artery (circumferential branches = lateral pontine arteries)

(C)

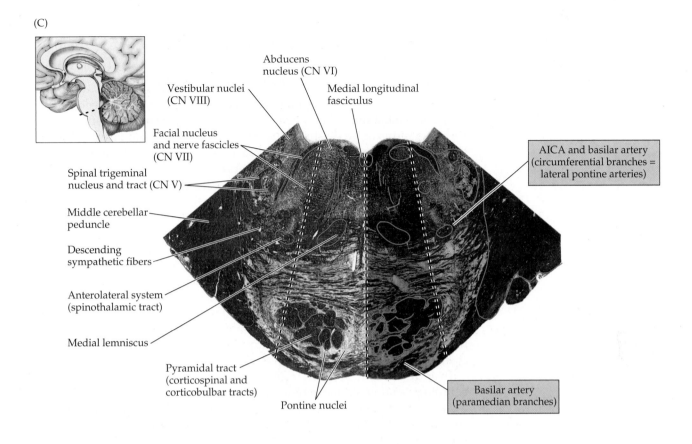

Vestibular nuclei
(CN VIII)

Abducens
nucleus (CN VI)

Medial longitudinal
fasciculus

Facial nucleus
and nerve fascicles
(CN VII)

Spinal trigeminal
nucleus and tract (CN V)

Middle cerebellar
peduncle

Descending
sympathetic fibers

Anterolateral system
(spinothalamic tract)

Medial lemniscus

Pyramidal tract
(corticospinal and
corticobulbar tracts)

Pontine nuclei

AICA and basilar artery
(circumferential branches =
lateral pontine arteries)

Basilar artery
(paramedian branches)

(D)

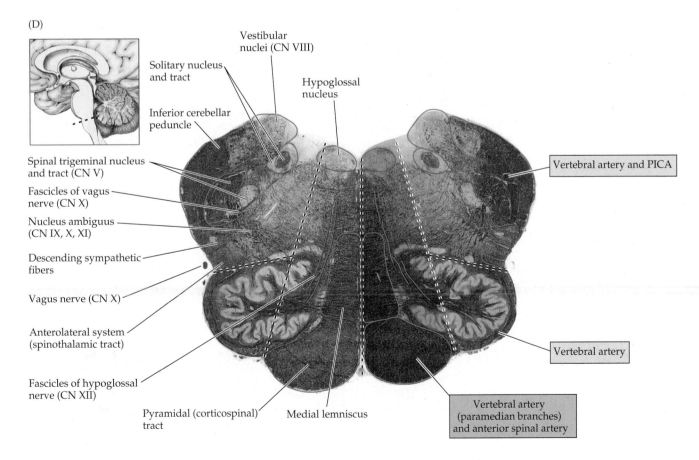

Vestibular
nuclei (CN VIII)

Solitary nucleus
and tract

Hypoglossal
nucleus

Inferior cerebellar
peduncle

Spinal trigeminal nucleus
and tract (CN V)

Fascicles of vagus
nerve (CN X)

Nucleus ambiguus
(CN IX, X, XI)

Descending sympathetic
fibers

Vagus nerve (CN X)

Anterolateral system
(spinothalamic tract)

Fascicles of hypoglossal
nerve (CN XII)

Pyramidal (corticospinal)
tract

Medial lemniscus

Vertebral artery and PICA

Vertebral artery

Vertebral artery
(paramedian branches)
and anterior spinal artery

KEY CLINICAL CONCEPT

14.3 VERTEBROBASILAR VASCULAR DISEASE

General Features of Posterior Circulation Disease

As discussed in Chapter 10, infarcts can occur by a variety of mechanisms, including **embolism**, often of cardiac origin; in situ **thrombosis**, often occurring on a nidus of preexisting atherosclerotic stenosis; and **lacunar disease**, often resulting from small-vessel occlusion in the setting of chronic hypertension. These mechanisms and the others discussed in Chapter 10 (see KCC 10.4) can all occur in the posterior circulation as well, affecting the vertebral and basilar arteries and their branches. Because the vertebrobasilar system supplies posterior fossa structures, including the brainstem, warning signs of vertebrobasilar ischemia can be very ominous. Therefore, when patients report these symptoms, they should always be brought to immediate medical attention in an effort to avoid life-threatening brainstem infarction, coma, and death. Some common **warning signs of vertebrobasilar ischemia** are listed in Table 14.6. These warning signs can be attributed to ischemia of specific anatomical components of the brainstem (see Figure 14.1)—such as the cranial nerve nuclei and their interconnections, long sensory and motor tracts, cerebellar circuits, and the reticular activating systems—or to ischemia of the occipital lobes (PCA territory).

In addition to these clues from the patient's history, certain findings on neurologic examination can often help distinguish brainstem ischemia from ischemia of the cerebral hemispheres. Features strongly suggestive of **brainstem involvement** rather than hemispheric involvement include **crossed signs**, such as decreased sensation on one side of the face and contralateral

TABLE 14.6 Common Warning Signs of Vertebrobasilar Ischemia

SYMPTOM	ISCHEMIC STRUCTURES
Dizziness (vertigo), nausea	Vestibular nuclei, cerebellum, or inner ear
Diplopia, dysconjugate gaze	Supranuclear or infranuclear eye movement pathways (see Chapter 13)
Blurred vision or other visual disturbances	Eye movement pathways, or visual cortex
Incoordination (ataxia)	Cerebellum or cerebellar pathways
Unsteady gait	Cerebellar pathways; long sensory or motor tracts
Dysarthria, dysphagia	Corticobulbar pathways or brainstem cranial nerve nuclei
Numbness and tingling, particularly bilateral or perioral	Long somatosensory pathways or trigeminal system
Hemiparesis, quadriparesis	Corticospinal tract
Somnolence	Pontomesencephalic reticular formation or bilateral thalami
Headache	
Occipital	Posterior fossa meninges and vessels (CN X and cervical roots)
Frontal	Supratentorial meninges and vessels (CN V; PCA is often CN V_1)
Nonlocalized	Supra- and/or infratentorial meninges and vessels

body, or weakness on one side of the face and contralateral body, and **cranial nerve abnormalities**, especially those causing eye movement abnormalities, such as **dysconjugate gaze**, **wrong-way eyes** (see Figure 13.15), **pupillary abnormalities**, or **nystagmus** (see Chapter 13). Features strongly suggestive of **hemispheric involvement** rather than brainstem involvement include aphasia, hemineglect, hemianopia, and seizures. Note that in some situations both brainstem and hemispheric involvement can occur simultaneously, particularly when the PCAs are involved at the top of the basilar artery (see Figures 14.18 and 14.20).

Once brainstem disease is suspected, certain rules of thumb can be helpful in further localizing brainstem vascular disease or other lesions to the midbrain, pons, or medulla. Signs of **midbrain dysfunction** include third-nerve palsy, unilateral or bilateral pupil dilation, ataxia, flexor (decorticate) posturing, and impaired consciousness. Signs of **pontine dysfunction** include bilateral Babinski's signs, generalized weakness, perioral numbness (see Figure 12.9), "salt and pepper" (pins and needles) facial tingling, bilateral upper or lower visual loss or blurring (usually caused by impaired blood flow from the basilar artery to both PCAs), irregular respirations (described earlier in this chapter), ocular bobbing (eyes dip downward quickly and then return gradually to mid position before dipping again), shivering, palatal myoclonus (affecting the central tegmental tract; described earlier in this chapter), abducens palsy or horizontal gaze palsy, bilateral small but reactive pupils (disruption of descending sympathetic fibers), extensor (decerebrate) posturing, and impaired consciousness. Signs of **medullary dysfunction** include vertigo, ataxia, nystagmus, nausea, vomiting, respiratory arrest, autonomic instability, and hiccups. Specific vascular syndromes of different brainstem regions are discussed in greater detail in the next section.

Treatment of vertebrobasilar disease is similar to treatment of ischemic stroke in the anterior circulation and depends on the mechanism of ischemia (see KCC 10.4). As with anterior circulation disease, transient ischemic attacks (TIAs) sometimes provide a warning prior to ischemic infarction (see KCC 10.3). Patients with an initial ischemic event should undergo an evaluation as described in KCC 10.4 to search for a mechanism for the ischemia. Anticoagulation therapy is used to treat thromboembolic disease caused by atrial fibrillation or mechanical cardiac valves. As discussed in KCC 10.6, **vertebral dissection**, often following minor head or neck trauma, is another important source of embolic disease, usually treated with anticoagulation. Occasionally, an ectatic or fusiform basilar artery aneurysm (see Figure 5.20) can form thrombi that embolize intermittently to distal branches. Much more commonly, artherosclerotic disease causes **vertebral stenosis** or **basilar stenosis** (basilar insufficiency) resulting in tenuous waxing and waning brainstem signs that may be sensitive to changes in blood pressure. It is crucial to distinguish vertebral or basilar stenosis from evolving small vessel lacunes, which can also cause waxing and waning symptoms but do not involve stenosis of major blood vessels.

Vertebral thrombosis or, especially, **basilar thrombosis** can be a life-threatening condition due to potential widespread brainstem infarction. As in other forms of acute stroke (see KCC 10.4), systemic administration of the thrombolytic agent tPA can improve outcome if given within 4.5 hours of onset, although there is some increased risk of hemorrhage. Intra-arterial administration of thrombolytic agents locally at the site of the clot using interventional neuroradiological techniques has also been beneficial in the investigational setting. In patients beyond the window for tPA therapy, antiplatelet agents such as aspirin are given acutely. In addition, blood pressure–lowering medications should be used cautiously or avoided to prevent worsening of hy-

REVIEW EXERCISE

Cover the right column in Table 14.6. For each of the common symptoms of vertebrobasilar disease, list the possible structures suffering from ischemia.

poperfusion. In refractory or progressive cases, vertebral and basilar stenosis are often treated with heparin anticoagulation therapy, although such therapy has not been proven effective by randomized trial. Unlike carotid stenosis (see KCC 10.5), vertebral and basilar stenosis have not been successfully treated with endarterectomy, although angioplasty has been tried experimentally with some success. Finally, as with anterior circulation stroke (see KCC 10.4), it is essential to treat the patient with a multidisciplinary approach, with careful attention to other coexisting medical conditions and potential complications, both during the acute stage and during recovery.

SPECIFIC CLINICAL SYNDROMES OF THE VERTEBROBASILAR TERRITORY The discussion in this section focuses on infarcts in several **specific territories of the posterior circulation** (see Tables 14.7–14.9). In addition to being clinically useful, review of these vascular territories and the anatomical structures affected in each territory serves as a useful review of the regional anatomy of the brainstem and can help consolidate knowledge acquired in this and the preceding two chapters. After discussing these focal syndromes, we will review several multifocal or bilateral brainstem syndromes, such as basilar thrombosis, top-of-the-basilar syndrome, and pontine hemorrhage.

Of all the focal brainstem vascular syndromes that we will discuss here (see Tables 14.7–14.9), only two are common: the *lateral medullary syndrome*, usually caused by vertebral thrombosis; and *medial basis pontis infarcts*, usually caused by lacunar disease. Medial medullary syndrome and SCA syndrome are less common. The other syndromes listed are relatively rare when occurring in isolation.

Proceeding from caudal to rostral, **vascular syndromes of the medulla** are listed in Table 14.7 (see Figure 14.21D). The **medial medullary syndrome** is

TABLE 14.7 Focal Vascular Syndromes of the Medulla

REGION	SYNDROME NAME[a]	VASCULAR SUPPLY	STRUCTURE(S)	ANATOMICAL CLINICAL FEATURE(S)
Medial medulla (see Figure 14.21D)	Medial medullary syndrome	Paramedian branches of vertebral and anterior spinal arteries	Pyramidal tract	Contralateral arm or leg weakness
			Medial lemniscus	Contralateral decreased position and vibration sense
			Hypoglossal nucleus and exiting CN XII fascicles	Ipsilateral tongue weakness
Lateral medulla (see Figure 14.21D)	Wallenberg's syndrome (lateral medullary syndrome)	Vertebral artery (more commonly than PICA)	Inferior cerebellar peduncle, vestibular nuclei	Ipsilateral ataxia, vertigo, nystagmus, nausea
			Trigeminal nucleus and tract	Ipsilateral facial decreased pain and temperature sense
			Spinothalamic tract	Contralateral body decreased pain and temperature sense
			Descending sympathetic fibers	Ipsilateral Horner's syndrome
			Nucleus ambiguus	Hoarseness, dysphagia
			Nucleus solitarius	Ipsilateral decreased taste

[a]Eponymous names for vascular syndromes listed in Tables 14.7–14.9 (Weber's syndrome, Claude's syndrome, etc.) need not be memorized, since their exact meanings have varied historically.

caused by occlusion of paramedian branches of the anterior spinal or verte-bral arteries. Infarction of the pyramidal tract results in contralateral arm and leg upper motor neuron weakness sparing the face, which can resemble a cer-vical cord lesion (see Figure 6.14C). Sometimes the contralateral face is in-volved as well, although usually less than the arm and leg. Often, there is ip-silateral tongue weakness from infarction of the exiting CN XII fascicles or, depending on how far the infarct extends from the ventral surface of the medulla, from infarction of the hypoglossal nucleus. Also depending on how far dorsally the infarct extends, there may be contralateral decreased vibra-tion and joint position sense caused by infarction of the medial lemniscus.

Lateral medullary syndrome, or **Wallenberg's syndrome**, is a relatively com-mon brainstem infarct. In addition to being clinically important, this is the one syndrome that students should memorize because understanding the classic clinical features and anatomical structures involved serves as a help-ful reference point for understanding all other brainstem syndromes as well. Because the syndrome affects the lateral tegmentum, motor involvement is usually not prominent, and prognosis is generally good. Lateral medullary syndrome is caused by thrombosis more often than embolus. Vertebral thrombosis is the most common cause. Isolated involvement of the PICA is a less common cause.

The most disabling features of lateral medullary syndrome are ipsilateral ataxia caused by infarction of the inferior cerebellar peduncle and vertigo caused by infarction of the vestibular nuclei (see Figure 14.21D). Unsteady gait, horizontal or rotatory nystagmus, nausea, and vomiting are common associ-ated features. There is often decreased pain and temperature sensation of the ipsilateral face (spinal trigeminal nucleus and tract) and of the contralateral body (spinothalamic tract; see Figure 7.9B). In some cases, facial sensory loss is contrateral, possibly because of the involvement of crossing fibers, or there may be relatively stronger sensation in the ipsilateral face due to abnormally heightened sensitivity or parathesis, particularly shortly after onset. Other vari-ants include sensory loss in just the upper or lower contralateral body, proba-bly due to only partial involvement of the spinothalamic tract. Involvement of the descending sympathetic fibers, which run in the lateral tegmentum of the brainstem near the spinothalamic tracts, causes an ipsilateral Horner's syn-drome (see KCC 13.5; Figure 13.10), with ptosis, miosis, and, less commonly, anhidrosis. Infarction of the nucleus ambiguus and exiting fascicles of CN X causes breathy hoarseness and dysphagia (see KCC 12.8). The gag reflex is often decreased on the side of the lesion, and laryngoscopy shows ipsilateral vocal cord paralysis. Finally, involvement of the nucleus solitarius can occa-sionally be demonstrated by tests for decreased taste sensation on the ipsilat-eral tongue. As already noted, motor involvement is not commonly present. In some cases, however, there may be ipsilateral facial weakness, possibly due to fibers of the facial nerve that loop caudally into the medulla before exiting at the pontomedullary junction. In addition, when infarcts extend somewhat more medially and reach the pyramidal tract, contralateral hemiparesis may be present, and combined lateral and medial medullary infarcts can sometimes occur. An uncommon but interesting manifestation of lateral medullary infarcts is the loss, in some patients, of vertical orientation, making them suddenly feel as if the whole world has turned upside down or sideways.

Many clinical features of lateral medullary syndrome also occur in other le-sions of the lateral brainstem tegmentum, such as AICA syndrome, and some-times in SCA syndrome as well (Table 14.8; see Figure 14.21B,C). The presence of hoarseness or loss of taste sensation helps localize the syndrome to the medulla rather than the pons. In addition, the presence of ipsilateral hearing loss suggests AICA involvement rather than lateral medullary syndrome.

TABLE 14.8   **Focal Vascular Syndromes of the Pons**

REGION	SYNDROME NAME[a]	VASCULAR SUPPLY	STRUCTURE(S)	ANATOMICAL CLINICAL FEATURE(S)
Medial pontine basis (see Figure 14.21B,C)	Dysarthria hemiparesis (pure motor hemiparesis)	Paramedian branches of basilar artery, ventral territory	Corticospinal and corticobulbar tracts	Contralateral face, arm, and leg weakness; dysarthria
	Ataxic hemiparesis	Same as above	Corticospinal and corticobulbar tracts	Contralateral face, arm, and leg weakness; dysarthria
			Pontine nuclei and pontocerebellar fibers	Contralateral ataxia (occasionally, ipsilateral ataxia)
Medial pontine basis and tegmentum (see Figure 14.21C)	Foville's syndrome	Paramedian branches of basilar artery, ventral and dorsal territories	Corticospinal and corticobulbar tracts	Contralateral face, arm, and leg weakness; dysarthria
			Facial colliculus	Ipsilateral face weakness; ipsilateral horizontal gaze palsy
	Pontine wrong-way eyes	Same as above	Corticospinal and corticobulbar tracts	Contralateral face, arm, and leg weakness; dysarthria
			Abducens nucleus or paramedian pontine reticular formation	Ipsilateral horizontal gaze palsy
	Millard–Gubler syndrome	Same as above	Corticospinal and corticobulbar tracts	Contralateral face, arm, and leg weakness; dysarthria
			Fascicles of facial nerve	Ipsilateral face weakness
	Other regions variably involved	Same as above	Medial lemniscus	Contralateral decreased position and vibration sense
			Medial longitudinal fasciculus	Internuclear ophthalmoplegia (INO)
Lateral caudal pons (see Figure 14.21C)	AICA syndrome	AICA	Middle cerebellar peduncle	Ipsilateral ataxia
			Vestibular nuclei	Vertigo, nystagmus
			Trigeminal nucleus and tract	Ipsilateral facial decreased pain and temperature sense
			Spinothalamic tract	Contralateral body decreased pain and temperature sense
			Descending sympathetic fibers	Ipsilateral Horner's syndrome
	Other regions variably involved	Labyrinthine artery	Inner ear	Ipsilateral hearing loss
Dorsolateral rostral pons (see Figure 14.21B)	SCA syndrome	SCA	Superior cerebellar peduncle and cerebellum	Ipsilateral ataxia
			Other lateral tegmental structures (variable)	Variable features of lateral tegmental involvement (see AICA syndrome)

[a]Eponymous names for vascular syndromes listed in Tables 14.7–14.9 (Weber's syndrome, Claude's syndrome, etc.) need not be memorized, since their exact meanings have varied historically.

Vascular syndromes of the pons are listed in Table 14.8 (see Figure 14.21B,C). Like lateral medullary syndrome, **medial pontine syndromes** are also relatively common and clinically important. Because the paramedian pontine perforating vessels tend to respect the midline (see Figure 14.19), unilateral paramedian pontine infarcts often end relatively sharply at the midline. These usually are lacunar infarcts resulting from small-vessel lipohyalinosis in the setting of chronic hypertension, but they can also be caused by microemboli, small-vessel thrombosis, or occlusion of the opening of small penetrating vessels by atherosclerotic disease where they arise from the wall of the basilar artery. Basilar stenosis can also cause paramedian pontine infarcts and may be a precursor to basilar thrombosis. Therefore, an evaluation of the basilar artery with magnetic resonance angiography or CT angiography is mandatory in these patients in the hope of averting this catastrophic outcome. Bilateral infarcts can also occur, often in the setting of basilar thrombosis.

In paramedian pontine infarcts, the extent of involvement of the pons from the ventral surface toward the fourth ventricle varies. The most common distribution involves the **paramedian basis pontis** unilaterally (see Figure 14.21B,C). Infarction of the corticospinal and corticobulbar tracts causes a lacunar syndrome of contralateral face, arm, and leg weakness, together with dysarthria, also known as dysarthria hemiparesis or pure motor hemiparesis (see Figure 6.14A). Recall that dysarthria hemiparesis is also commonly seen with lacunar infarcts of the posterior limb of the internal capsule (see Table 10.3). Involvement of the pontine nuclei and pontocerebellar fibers can cause ataxia, which is usually contralateral (on the same side as the hemiparesis), resulting in the syndrome called ataxic hemiparesis (see Table 14.8). A variant of this condition seen mainly with paramedian basis pontis infarcts is dysarthria–clumsy hand syndrome, in which there is dysarthria along with motor disturbances affecting the contralateral arm more than the leg.

Sometimes pontine infarcts can extend farther into the tegmentum toward the fourth ventricle, resulting in **medial pontine basis and tegmentum** infarcts (see Table 14.8). When the basis pontis is involved together with the facial colliculus (see Figure 14.21C), there is ipsilateral facial weakness, an ipsilateral horizontal gaze palsy (due to involvement of the abducens nucleus or paramedian pontine reticular formation), and contralateral hemiparesis (Foville's syndrome). Recall that ipsilateral horizontal gaze palsy together with contralateral hemiparesis is an example of "wrong-way eyes," in this case caused by a pontine lesion (see KCC 13.10; Figure 13.15). Slightly more laterally placed infarcts that involve the pontine basis and facial nerve fascicles without the abducens nucleus can cause ipsilateral facial weakness and contralateral hemiparesis (Millard–Gubler syndrome). Other regions of the pontine tegmentum that can variably be involved in paramedian infarcts are the medial lemniscus (causing contralateral position and vibration sense loss) and the medial longitudinal fasciculus (causing an internuclear ophthlamoplegia) (see Figure 13.13).

AICA infarcts involve mainly the caudal lateral pons (see Table 14.8; Figure 14.21C). The resulting lateral brainstem tegmentum syndrome may resemble the lateral medullary syndrome in some ways but not others, as noted earlier (presence of hoarseness or loss of taste localizes to the medulla rather than the pons). The labyrinthine artery occasionally comes directly off the basilar artery, but usually it arises as a branch of the AICA. Therefore, in addition to a lateral tegmental syndrome including ipsilateral ataxia, vertigo, nystagmus, pain and temperature sensory loss in the ipsilateral face and contralateral body, and an ipsilateral Horner's syndrome, AICA infarcts can also cause unilateral hearing loss. Sometimes patients with disease of the AICA or with basilar stenosis experience TIAs that include a roaring sound in their ears.

REVIEW EXERCISE

Cover the labels on the left side of Figure 14.21. For each of the following brainstem regions and vascular territories, name the structures affected and the expected deficits:

14.21A: Midbrain basis and tegmentum (PCA and top-of-the-basilar branches) (see Table 14.9)

14.21B,C: Medial pontine basis (paramedian branches of basilar, ventral territory) (see Table 14.8)

14.21C: Lateral inferior pons (AICA) (see Table 14.8)

14.21D: Medial medulla (paramedian branches of vertebral and spinal arteries) (see Table 14.7)

14.21D: Lateral medulla (vertebral artery and PICA) (see Table 14.7)

Isolated infarcts of the rostral lateral pons are uncommon, possibly because of the multiple lateral pontine arteries supplying this region (see Figures 14.18 and 14.20). **SCA infarcts** (see Table 14.8; Figure 14.21B) usually involve mainly the superior cerebellum, causing ipsilateral ataxia (see Chapter 15). A variable region of the rostrolateral pons may also be involved, occasionally causing some features of lateral tegmental syndrome.

Vascular syndromes of the midbrain are listed in Table 14.9 (see Figure 14.21A). **Midbrain infarcts** result from the occlusion of penetrating vessels arising from the top of the basilar artery and proximal PCAs. Infarcts in this territory often occur in the setting of an embolus lodged at the top of the basilar artery (top-of-the-basilar syndrome) causing infarcts in multiple other locations as well, but midbrain infarcts can occasionally be seen in isolation. Midbrain syndromes have been described involving different regions of the basis, tegmentum, or both. Infarction of the cerebral peduncles in the midbrain basis causes contralateral hemiparesis; infarction of the third-nerve nucleus or fascicles causes an ipsilateral third-nerve palsy; and infarction of the red nucleus and fibers of the superior cerebellar peduncle (above the decussation) causes a contralateral tremor and ataxia. Larger infarcts of the midbrain that affect the midbrain reticular formation cause impaired consciousness, although when this occurs, other territories are often involved as well.

In addition to these specific territories (see Tables 14.7–14.9), posterior circulation infarcts can sometimes occur that involve multiple territories. In **basilar thrombosis**, there are often catastrophic bilateral infarctions of multiple regions of the pons and other regions supplied by the basilar artery, including the cerebellum, midbrain, thalamus, and occipital lobes. Basilar thrombosis usually results from thrombosis of a previously narrowed basilar artery in the setting of atherosclerotic disease. Patients often develop multiple cranial nerve abnormalities, long-tract signs, and coma, typically with a poor prognosis.

TABLE 14.9 Focal Vascular Syndromes of the Midbrain[a]

REGION	SYNDROME NAME(S)[a]	VASCULAR SUPPLY	STRUCTURE(S)	ANATOMICAL CLINICAL FEATURE(S)
Midbrain basis (see Figure 14.20A)	Weber's syndrome	Branches of PCA and top of basilar artery	Oculomotor nerve fascicles	Ipsilateral third-nerve palsy
			Cerebral peduncle	Contralateral hemiparesis
Midbrain tegmentum[b] (see Figure 14.20A)	Claude's syndrome	Branches of PCA and top of basilar artery	Oculomotor nerve fascicles	Ipsilateral third-nerve palsy
			Red nucleus, superior cerebellar peduncle fibers	Contralateral ataxia
Midbrain basis and tegmentum[b] (see Figure 14.20A)	Benedikt's syndrome	Branches of PCA and top of basilar artery	Oculomotor nerve fascicles	Ipsilateral third-nerve palsy
			Cerebral peduncle	Contralateral hemiparesis
			Red nucleus, substantia nigra, superior cerebellar peduncle fibers	Contralateral ataxia, tremor, and involuntary movements

[a]Eponymous names for vascular syndromes listed in Tables 14.7–14.9 (Weber's syndrome, Claude's syndrome, etc.) need not be memorized, since their exact meanings have varied historically.

[b]More dorsal infarcts involving the midbrain reticular formation cause impaired consciousness.

Top-of-the-basilar syndrome is usually caused by an embolus that lodges in the distal basilar artery, also causing infarcts of multiple vascular territories. Clinical features include visual disturbances resulting from infarcts of the visual cortex; memory disturbances from infarcts of the bilateral medial thalami or temporal lobes; eye movement abnormalities from infarction of the oculomotor nuclei and third-nerve fascicles in the midbrain; somnolence, delirium, or vivid visual hallucinations ("peduncular hallucinosis") caused by infarction of the midbrain reticular formation; and ataxia resulting from cerebellar infarcts. Of note, corticospinal involvement is often relatively mild in top-of-the-basilar syndrome. Sometimes, as an embolus migrates up the basilar artery toward the top, it occludes various penetrator arteries in the pons, producing a series of transient deficits referred to as the **basilar scrape syndrome**.

Another important vascular syndrome of the brainstem is **pontine hemorrhage**. This is most commonly seen in the setting of chronic hypertension, causing fragility of small, penetrating blood vessels (see KCC 5.6). Pontine hemorrhage usually involves the paramedian branches of the basilar artery, at the junction between the tegmentum and basis pontis. Although small hemorrhages can cause relatively mild deficits, pontine hemorrhages are often large and bilateral, resulting in catastrophic bilateral cranial nerve deficits, long-tract signs, coma, and a poor prognosis. Hemorrhage in other regions of the brainstem is relatively uncommon and is usually caused by vascular malformations rather than hypertension. ■

CLINICAL CASES

CASE 14.1 FACE AND CONTRALATERAL BODY NUMBNESS, HOARSENESS, HORNER'S SYNDROME, AND ATAXIA

CHIEF COMPLAINT

A 22-year-old woman suddenly developed left posterior neck pain, vertigo, ataxia, left facial numbness, and hoarseness after chiropractic neck manipulation.

HISTORY

The patient had been well until 4 months previously, when she injured her neck in a car accident. She saw a chiropractor daily for neck pain. On the day of admission, after her neck was "snapped," she suddenly felt increased **pain in the left posterior neck region**. As she left the chiropractor's office, she felt **dizzy and nauseated** and **staggered** out to her car, falling toward the left. She noticed her **vision bouncing or swaying (oscillopsia)** but had no diplopia. She vomited twice, and when she reached home her husband noticed that her **voice sounded hoarse**. She also felt a **numbness and tingling on the left side of her face**. The symptoms did not improve after a brief nap, so she came to the emergency room.

PHYSICAL EXAMINATION

Vital signs: T = 96°F, P = 60, BP = 126/84.
Neck: No bruits.
Lungs: Clear.

Heart: Regular rate with no murmurs or gallops.
Abdomen: Soft, nontender.
Extremities: Normal.
Neurologic exam:

MENTAL STATUS: Alert and oriented × 3. Normal language. Named months forward and backward with no errors. Recalled 3/3 words after 4 minutes.

CRANIAL NERVES: **Left pupil 2.5 mm, constricting to 2 mm.** Right pupil 3.5 mm, constricting to 2 mm. Visual fields full. **Right-beating horizontal and counterclockwise rotatory nystagmus**, which increased with rightward gaze. Patient reported an associated perception of the **visual field moving back and forth (oscillopsia)**. Extraocular movements full. **Left ptosis. Decreased pinprick and temperature sensation in left ophthalmic, maxillary, and mandibular divisions of CN V** (Figure 14.22). **Decreased left corneal reflex.** Face symmetrical. Taste not tested. Hearing intact. **Voice hoarse. Decreased palate elevation on the left, and decreased left gag reflex.** Normal sternomastoid and trapezius strength. Tongue midline.

(continued on p. 662)

CASE 14.1 *(continued)*

MOTOR: No drift. Normal tone. 5/5 power throughout.

REFLEXES:

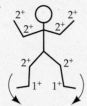

COORDINATION: **Mild ataxia on finger-to-nose testing on the left. Toe tapping on the left was irregular in rhythm (dysrhythmic).**

GAIT: Unable to stand because of severe dizziness.

SENSORY: **Decreased pinprick and temperature sensation in the right limbs and trunk below the neck** (see Figure 14.22). Intact light touch, vibration, and joint position sense.

LOCALIZATION AND DIFFERENTIAL DIAGNOSIS

1. On the basis of the symptoms and signs shown in **bold** above, where is the lesion?
2. Given the sudden onset of deficits and neck pain following neck manipulation, what is the most likely diagnosis? What are some other possibilities?

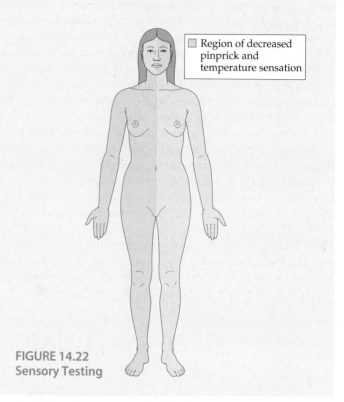

☐ Region of decreased pinprick and temperature sensation

**FIGURE 14.22
Sensory Testing**

Discussion

1. The key symptoms and signs in this case are:
   - **Pain in the left posterior neck region**
   - **Unsteady gait, falling toward the left**
   - **Left ataxia and dysrhythmia**
   - **Dizziness and nausea with right-beating nystagmus**
   - **Decreased pinprick and temperature sensation in the left face**
   - **Decreased left corneal reflex**
   - **Decreased pinprick and temperature sensation in the right limbs and trunk below the neck**
   - **Left ptosis, with small, reactive left pupil**
   - **Hoarseness, with decreased palate elevation on the left and decreased left gag reflex**

   This patient had virtually all the clinical features of lateral medullary syndrome, or Wallenberg's syndrome. Review the structures in the lateral medulla involved, as well as the corresponding deficits that produce this characteristic constellation of findings (see Figure 14.21D; Table 14.7; see also Figure 7.9B).

2. Lateral medullary syndrome is usually caused by thrombosis, most often involving the vertebral artery, and less often the PICA in isolation (see KCC 14.3), resulting in a lateral medullary infarct. Given the patient's recent neck manipulation, neck pain, young age, and lack of other stroke risk factors, vertebral dissection should be strongly considered (see KCC 10.6). Other, much

less likely causes of lateral medullary dysfunction in this patient include hemorrhage into a vascular malformation, abscess, or demyelinating disease.

The most likely *clinical localization* and diagnosis is, therefore, left lateral medullary syndrome, with left lateral medullary infarct caused by left vertebral dissection.

Clinical Course and Neuroimaging

Initial brain CT and conventional **MRI scans** did not show an infarct (Image 14.1A, page 665). However, diffusion-weighted MRI on the day of admission suggested left lateral medullary infarct (not shown), and this was confirmed by follow-up conventional MRI 5 days later (Image 14.1B, page 665). An MRA performed on the day of admission showed loss of flow in the left vertebral artery. Axial T1-weighted MRI sections through the vertebral arteries demonstrated a bright region of thickening in the wall of the left vertebral artery consistent with intramural clot from a left vertebral dissection (Image 14.1C, page 666). The patient was treated with intravenous heparin anticoagulation for a week and then switched to Coumadin (Warfarin). On follow-up 11 days after presentation, she no longer had nausea, vertigo, or nystagmus, and she was able to walk well, with only slight leftward veering on tandem gait. She still had a left Horner's syndrome, mildly decreased pinprick sensation in the left face and right body, and trace appendicular ataxia on the left side.

CASE 14.2   HEMIPARESIS SPARING THE FACE

MINICASE

A 53-year-old man with a history of cigarette smoking and hypercholesterolemia was driving home from the airport at 7:00 A.M. one morning and had a 1-hour episode of **pins and needles in his right perioral area**, **arm**, **and leg**. He reached home, and at 10:00 A.M., while he was walking the dog, these symptoms recurred, together with **difficulty walking and clumsiness and weakness of the right arm and leg**. On exam in the emergency room, he had **decreased tone and 3/5 to 4/5 strength in the right arm and leg**, an upgoing toe on the right, and **decreased vibration and joint position sense** **in the right arm and leg**. There was only a **trace decrease in the right nasolabial fold** at rest, and his smile was symmetrical. Tongue was midline.

LOCALIZATION AND DIFFERENTIAL DIAGNOSIS

1. On the basis of the symptoms and signs shown in **bold** above, what is the most likely location for the lesion?

2. What is the most likely diagnosis?

Discussion

1. The key symptoms and signs in this case are:

   • 3/5 to 4/5 weakness of the right arm and leg, with a right Babinski's sign

   • Trace decreased right nasolabial fold

   • Right body paresthesias, and decreased vibration and joint position sense

   Right arm and leg upper motor neuron weakness sparing the face could be caused by a lesion in the left medulla or in the right cervical spinal cord (see KCC 6.3; Figure 6.14C). Similarly, paresthesias and decreased vibration and joint position sense in the right body could be caused by a lesion in the left brainstem involving the medial lemniscus (see Figure 7.9C) or in the right cervical spinal cord involving the posterior columns (see Figure 7.10E). The fact that some subtle right facial weakness was present makes the spinal cord location less likely. Since the face was nearly spared, however, the lesion is unlikely to lie above the facial nerve exit point (pontomedullary junction) because if it did, more prominent facial involvement would be expected. This leaves the medial medulla as a likely location (see Table 14.7; Figure 14.21D), involving the left corticospinal fibers in the medullary pyramid, and the left medial lemniscus. The lack of tongue motor involvement (hypoglossal nerve) is interesting, but according to the published literature on medial medullary syndrome, the tongue is involved only 50% of the time or less.

The most likely *clinical localization* is left medial medulla, involving the pyramid and medial lemniscus but sparing the hypoglossal nucleus and CN XII fascicles.

2. Given the patient's age and vascular risk factors of cigarette smoking and hypercholesterolemia, the most likely diagnosis is left medial medullary infarction. This is usually caused by occlusion of paramedian branches of the vertebral or anterior spinal arteries (see KCC 14.3; Table 14.7; Figure 14.21D). The right perioral pins and needles experienced by this patient suggests some possible ischemia affecting the spinal trigeminal nucleus or trigeminothalamic fibers (see Figures 12.8, 12.9). Although this symptom is more common in pontine ischemia, it can occasionally be seen in medial medullary syndrome.

Clinical Course and Neuroimaging

Conventional MRI was initially negative, but diffusion-weighted MRI revealed a left medial medullary infarct (Image 14.2A,B, page 667). Several days later this was visible on conventional MRI as well. The patient was admitted for further evaluation, including MRA, echocardiogram, and a Holter monitor, which did not reveal an obvious embolic source (see KCC 10.4). However, an MRA revealed an irregular region of signal loss in the distal left vertebral artery, just prior to the vertebrobasilar junction. Because of the possibility of vertebral dissection (see KCC 10.6) or vertebral stenosis (see KCC 14.3) and the lack of CT angiography at the time this patient was seen, a conventional vertebral angiogram was done. The angiogram confirmed occlusion of the distal left vertebral artery just beyond the left PICA takeoff point but did not reveal a dissection. This suggests that the patient's medial medullary infarct was caused by occlusion of paramedian vessels arising from the distal left vertebral artery (see Figure 14.21D). The vertebral occlusion may have been embolic from an unknown source or could have been caused by thrombosis superimposed on a stenosed atherosclerotic vertebral artery. Currently, patients of this kind are treated with aspirin, although at the time, oral anticoagulation with warfarin was used. His weakness gradually improved, and he was discharged to an inpatient rehabilitation facility to continue his recovery.

CASE 14.1 FACE AND CONTRALATERAL BODY NUMBNESS, HOARSENESS, HORNER'S SYNDROME, AND ATAXIA

IMAGE 14.1A–C Left Laterally Medullary Infarct Caused by Vertebral Dissection Axial MRI images of the brain and cervical spine. (A) T2-weighted MRI done on admission does not show infarct but does show absence of the flow void in the left vertebral artery. (B) Follow-up T2-weighted MRI done 5 days after admission shows increased signal in the left lateral medulla compat-ible with infarction (compare to Figure 14.21D). (C) T1-weighted images of the neck done on the day of admission show the left vertebral artery to have a thickened wall with a bright appearance compatible with dissection and coagulation of blood in the wall of the left vertebral artery (see Table 4.4).

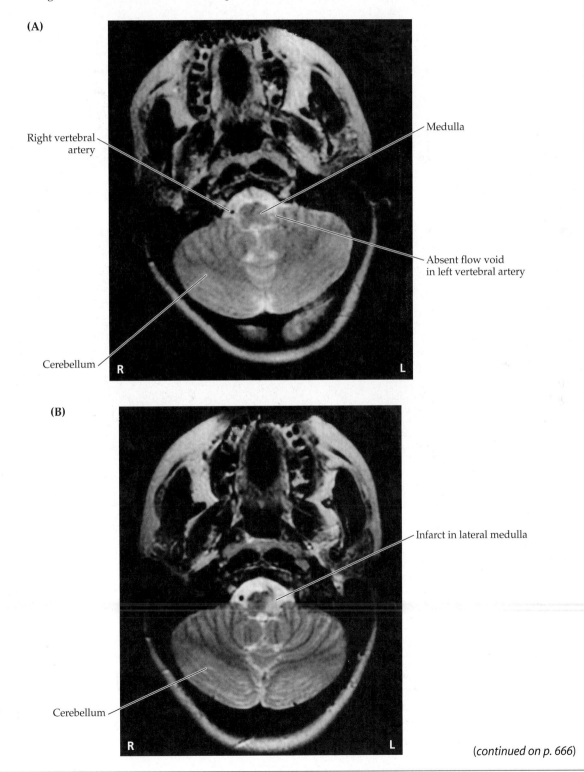

(A)

Right vertebral artery

Medulla

Absent flow void in left vertebral artery

Cerebellum

R   L

(B)

Infarct in lateral medulla

Cerebellum

R   L

(continued on p. 666)

CASE 14.1 *(continued)*

(C)

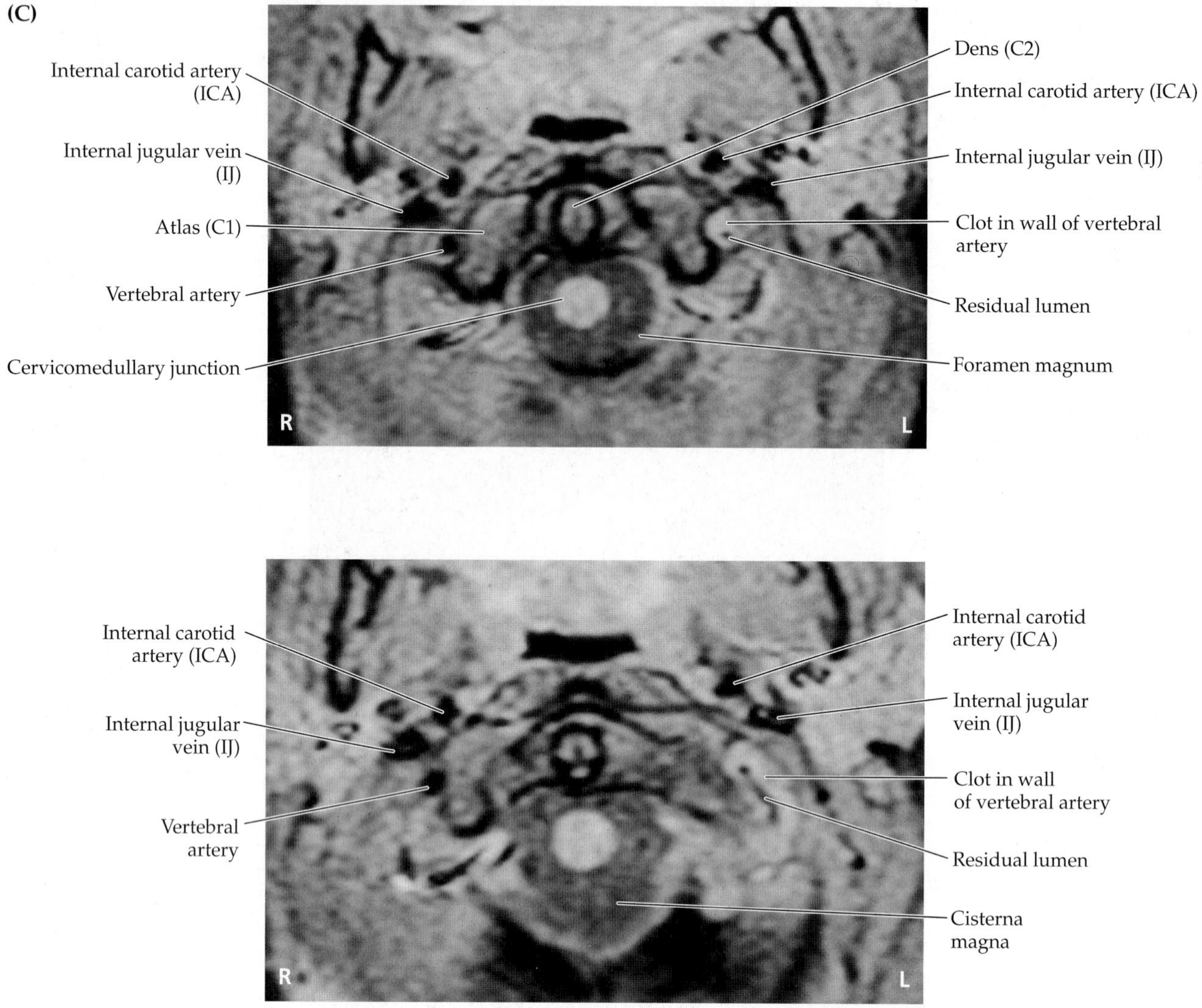

Internal carotid artery (ICA)
Internal jugular vein (IJ)
Atlas (C1)
Vertebral artery
Cervicomedullary junction

Dens (C2)
Internal carotid artery (ICA)
Internal jugular vein (IJ)
Clot in wall of vertebral artery
Residual lumen
Foramen magnum

R   L

Internal carotid artery (ICA)
Internal jugular vein (IJ)
Vertebral artery

Internal carotid artery (ICA)
Internal jugular vein (IJ)
Clot in wall of vertebral artery
Residual lumen
Cisterna magna

R   L

CASE 14.2 HEMIPARESIS SPARING THE FACE

IMAGE 14.2A,B Left Medial Medullary Infarct Diffusion-weighted MRI images of the brain. (A) Axial section through the medulla. (B) Coronal section.

(A)

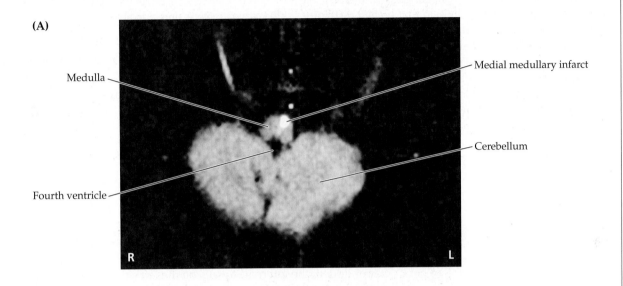

Medulla

Medial medullary infarct

Cerebellum

Fourth ventricle

R

L

(B)

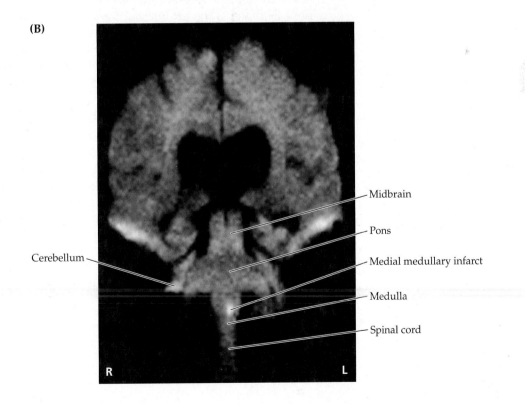

Cerebellum

Midbrain

Pons

Medial medullary infarct

Medulla

Spinal cord

R

L

CASE 14.3 DYSARTHRIA AND HEMIPARESIS

MINICASE

A 48-year-old man with a history of diabetes, hypertension and elevated cholesterol awoke on the day prior to admission with a funny "numb" feeling of his right arm and leg. In explaining what he meant by "numb" he said that he had **difficulty reaching for and holding onto objects with his right hand**. He also had **trouble walking, dragging his right foot**. His wife noted that his **speech sounded slurred and his face looked twisted**. These symptoms gradually became worse, so the next day the patient came to the emergency room. On exam he had **right facial weakness**, and **mildly dysarthric speech** (confirmed by his wife's comparison to baseline). In addition, **his right arm and leg had 2/5 to 4/5 weakness**. He was unable to walk due to weakness. The remainder of his examination was unremarkable.

LOCALIZATION AND DIFFERENTIAL DIAGNOSIS

1. On the basis of the symptoms and signs shown in **bold** above, what is the most likely location for the lesion?
2. What is the most likely diagnosis?

Discussion

1. The key symptoms and signs in this case are:

   - **Dysarthria**
   - **Right face, arm, and leg weakness**

   This patient has dysarthria hemiparesis, or pure motor hemiparesis (see KCC 6.3; Figure 6.14A). The most common localizations for this are the posterior limb of the internal capsule on the left side (see Table 10.3; Case 10.7), or the left basis pontis, involving the corticospinal and corticobulbar tracts (see Table 14.8; Figure 14.21B,C).

2. Given the patient's significant vascular risk factors, the most likely diagnosis is infarct in the left internal capsule or left basis pontis. The gradual progression of symptoms over the course of a day could be consistent with waxing and waning small vessel or lacunar infarction (see KCC 10.4), however, given the patient's young age other diagnoses should also be considered including demyelination, hemorrhage, brain abscess, or tumor.

Clinical Course and Neuroimaging

The patient was admitted to the hospital for further evaluation and treatment. Initial head CT did not show any abnormalities, but **MRI of the brain** revealed an acute infarct in the left basis pontis (Image 14.3, page 671). MRA of the head and neck showed normal flow in the basilar artery and other vessels. Echocardiogram and prolonged electrocardiogram recordings were normal. The patient was treated with aspirin to reduce the risk of recurrent stroke, and discharged to an inpatient rehabilitation hospital. On follow-up examination his strength improved allowing ambulation, and his dysarthria fully resolved, but when seen four years later he remained with 4/5 weakness in the right arm and leg. Compare this case to Case 6.5 and Case 13.7.

CASE 14.4 UNILATERAL FACE NUMBNESS, HEARING LOSS, AND ATAXIA

CHIEF COMPLAINT

A 56-year-old male auto mechanic had 1 month of episodic diplopia and unsteadiness and then suddenly developed persistent right face numbness, hearing loss, and right-sided clumsiness.

HISTORY

Past history was notable for severely elevated cholesterol and cigarette smoking. About 1 month prior to admission, the patient developed **transient episodes** consisting of **light-headedness, nausea, unsteadiness "staggering like I was drunk," diagonal diplopia with the right image higher than the left, perioral numbness, and a generalized headache**.

The episodes were precipitated by his standing up and walking around, lasted 5 or 6 minutes, and occurred up to four or five times per day. The episodes gradually improved over time and nearly stopped.

INITIAL LOCALIZATION, DIFFERENTIAL DIAGNOSIS, AND MANAGEMENT

1. On the basis of the symptoms shown in **bold** above, what general brain region is most likely involved?
2. Given this patient's history, what diagnosis should be seriously considered, and what should be done?

Discussion

1. The symptoms are strongly suggestive of brainstem dysfunction, possibly localized to the pons (see KCC 14.3; Table 14.6). Review each of this patient's symptoms and their localization in Table 14.6.

2. Given the patient's vascular risk factors and the fact that the episodes occur in situations that may lower the patient's systemic blood pressure (standing up), the most likely diagnosis is TIAs in the vertebrobasilar system, possibly caused by basilar stenosis. This is a potentially life-threatening situation, and the patient should be brought immediately to the hospital for evaluation including MRA, and for appropriate treatment.

CASE 14.4 *(continued)*

HISTORY

The patient did not seek medical attention for his symptoms. On a Friday night 3 days prior to admission, the patient abruptly developed **right facial numbness, decreased hearing in the right ear, slurred speech, right hand clumsiness (occasionally dropping things), and unsteady gait**. When he returned to work at the garage on Monday, he had trouble working on the cars, so he finally came to the emergency room.

PHYSICAL EXAMINATION

Vital signs: T = 96.7°F, R = 14.

Orthostatic testing:

  SUPINE: P = 80, BP = 130/80.

  STANDING: P = 88, BP = 122/76.

Neck: Supple with no bruits.

Lungs: Clear.

Heart: Regular rate with no murmurs.

Abdomen: Soft, nontender; normal bowel sounds.

Extremities: Normal.

Neurologic exam:

MENTAL STATUS: Alert and oriented × 3. Naming and repetition intact. Mildly decreased attention; for example, skipped November when naming months backward. Recalled 1/3 words after 5 minutes, but got 3/3 with prompting.

CRANIAL NERVES: Pupils 4 mm, constricting to 2.5 mm bilaterally. Fundi normal. Extraocular movements full, with fine **horizontal nystagmus** (direction of fast phase was not specified). **Light touch and pinprick sensation slightly decreased in right V$_2$ and V$_3$ distribution. Right corneal reflex decreased.** Face symmetrical. **Hearing decreased on the right. On the Weber test** (see KCC 12.5) **sounds were louder on the left. Speech slightly slurred.** Normal palate elevation. Shoulder shrug and sternomastoids normal. Tongue midline.

MOTOR: No drift. Normal tone. 5/5 power throughout.

(continued on p. 670)

CASE 14.4 *(continued)*

REFLEXES:

COORDINATION: **Mild dysmetria on finger-to-nose testing on the right. Right finger tapping and foot tapping were slightly slow and dysrhythmic.**

GAIT: **Slightly wide based. Able to do only two or three steps of tandem gait because of unsteadiness.** When the patient stood or walked during the exam, he did not have symptoms of the transient episodes described above.

SENSORY: Intact light touch, pinprick, vibration, and joint position sense.

LOCALIZATION AND DIFFERENTIAL DIAGNOSIS

1. On the basis of the symptoms and signs shown in **bold** above, what is the most likely location for the lesion? Which blood vessel(s) may be involved?
2. What is the most likely diagnosis?

Discussion

1. The key symptoms and signs in this case are:

   - **Decreased hearing in the right ear, with Weber test also eliciting decreased right hearing**
   - **Light touch and pinprick sensation slightly decreased in right V$_2$ and V$_3$ distribution, with decreased right corneal reflex**
   - **Right dysmetria and dysrhythmia, with slurred speech, horizontal nystagmus, and unsteady, wide-based gait**

   The patient has findings compatible with a right lateral caudal pontine syndrome, most likely caused by a right AICA infarct (see Table 14.8; Figure 14.21C). Decreased hearing of sensorineural origin based on the Weber test (see KCC 12.5; **neuroexam.com Video 42**) may be due to involvement of the labyrinthine artery (see Figure 14.18A); impaired right facial sensation may be due to infarction of the right trigeminal nucleus and tract; and right-sided appendicular ataxia, gait ataxia, slurred speech, and nystagmus may be caused by involvement of the right middle cerebellar peduncle and vestibular nuclei (see Figure 14.21C). The patient also had mildly decreased attention, which is a nonspecific finding that could have many causes, including brainstem ischemia (see KCC 19.14).

   The most likely *clinical localization* is right caudal lateral pons, AICA territory.

2. The patient's vascular risk factors and antecedent episodes of transient symptoms make infarction in the right AICA territory the most likely diagnosis.

Clinical Course and Neuroimaging

The patient was admitted to the hospital for further evaluation. An **MRI** revealed an infarct in the right lateral caudal pons in the territory of the AICA (Image 14.4A, page 672; see also Figure 14.21C). An MRA showed a striking lack of flow in the entire vertebrobasilar system (Image 14.4B, page 672). This finding suggests that the patient most likely had long-standing disease of the posterior circulation and that he developed collateral flow through vessels not visible on the MRA to supply his brainstem. In this tenuous situation, artery-to-artery embolus or thrombosus of the AICA could have caused infarction in the right AICA territory.

Based on current recommendations, this patient would likely be treated with aspirin, however, at the time he was started on warfarin anticoagulation. After 5 days, at the time of discharge, he had no nystagmus, his right hearing was improved, facial sensation was normal except for a small area around the right side of his mouth (see Figure 12.9), he was able to perform tandem gait, and the right arm and leg ataxia was improved but not completely gone. In addition, his inattention (possibly reflecting poor flow in the vertebrobasilar system) had returned to normal.

Five days later he returned to the emergency room with nausea causing decreased eating and drinking, and he had recurrent symptoms of lightheadedness when standing, without other symptoms. Exam was unchanged, except that on orthostatic testing his pulse and blood pressure, respectively, went from 76 and 147/98 supine, to 110 and 124/98 standing. (An increase in heart rate of greater than 10 beats per minute or a decrease in systolic blood pressure of greater than 10 mm Hg is considered abnormal.) Blood tests confirmed that his anticoagulation was adequate, and there was no evidence of bleeding as the cause of orthostasis, so he was admitted for intravenous hydration and observation. Again he did well, gradually tolerated the upright position better, and was discharged. He quit smoking, started taking daily walks, began to follow a low-cholesterol diet, and later began taking a cholesterol-lowering drug. At follow-up 1 month and again 4 months later, he was without symptoms, and exam was normal except for decreased pinprick sensation in a small area around the right lip and trace dysrhythmia of right finger and toe tapping.

CASE 14.3 DYSARTHRIA AND HEMIPARESIS

IMAGE 14.3 **Left Basis Pontis Infarct** Axial diffusion-weighted MRI image (DWI) through the pons showing an acute infarct.

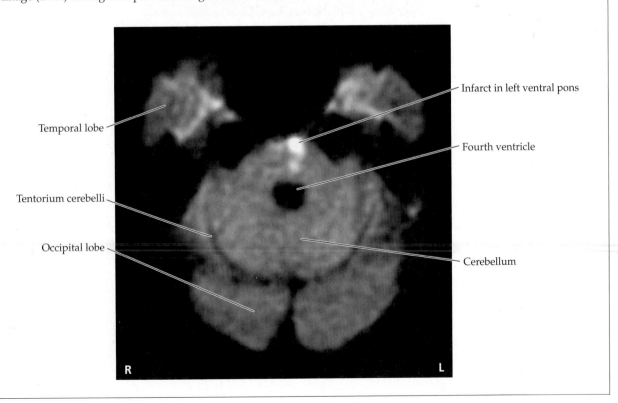

CASE 14.4   UNILATERAL FACE NUMBNESS, HEARING LOSS, AND ATAXIA

IMAGE 14.4A,B   Right AICA Infarct and Basilar Insufficiency (A) Axial T2-weighted MRI image through the pons demonstrating increased signal in the right dorsolateral pons and middle cerebellar peduncle, compatible with right anterior inferior cerebellar artery (AICA) infarct (compare to Figure 14.21C). (B) MRA demonstrating absence of visible flow in the vertebral or basilar arteries. This suggests severe narrowing of the basilar artery.

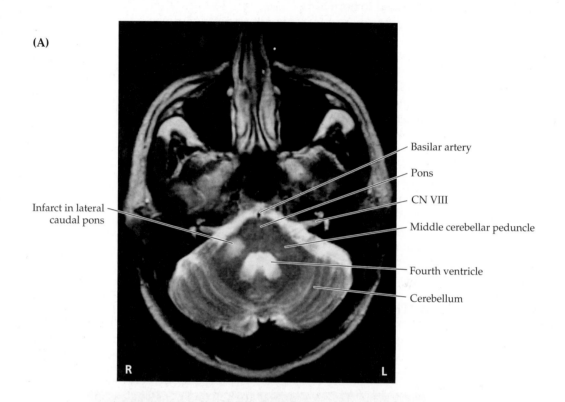

(A)

Infarct in lateral
caudal pons

Basilar artery

Pons

CN VIII

Middle cerebellar peduncle

Fourth ventricle

Cerebellum

R   L

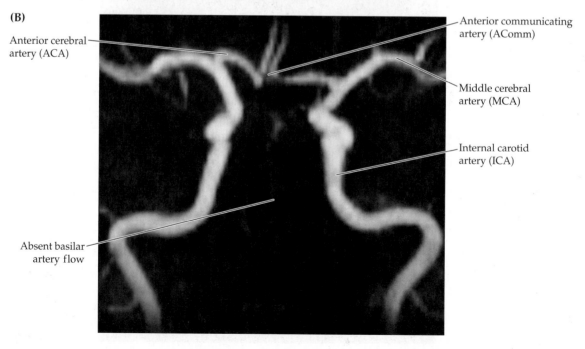

(B)

Anterior cerebral
artery (ACA)

Anterior communicating
artery (AComm)

Middle cerebral
artery (MCA)

Internal carotid
artery (ICA)

Absent basilar
artery flow

CASE 14.5 LOCKED IN

MINICASE

A 52-year-old woman with a history of Crohn's disease was walking in a shopping mall and suddenly had to sit down because she felt sweaty and the left side of her face felt "funny." Her husband noticed a **"lazy eye" on the left**. She was taken to a local hospital, where on initial exam she had **slurred speech** with good comprehension. She also had **weakness and decreased sensation of the left face, arm, and leg**, and she was **ataxic** (side not specified). She was admitted to the hospital for further evaluation.

INITIAL LOCALIZATION AND DIFFERENTIAL DIAGNOSIS

On the basis of the symptoms and signs shown in **bold** above, what is the most likely general location for the lesion? Given this patient's history, what diagnosis should be seriously considered?

Discussion

The symptoms are strongly suggestive of brainstem dysfunction, possibly from vertebrobasilar disease (see KCC 14.3; Table 14.6). Review the localization of each of these symptoms in Table 14.6. Of note, inflammatory bowel disease can sometimes cause hypercoagulability, so incipient thrombosis in the vertebrobasilar system should be strongly considered.

CASE 14.5 *(continued)*

HISTORY

That evening at 9:00 P.M., the patient suddenly had a **respiratory arrest** requiring intubation, and she was found to have decerebrate posturing (see Figure 3.5B). The next morning she was transferred to a tertiary care center. On exam, she was intubated and **unable to move her extremities**, but she was awake and **able to answer yes/no questions appropriately by using eye blinks or vertical eye movements**. She had **no horizontal eye movements, even with oculocephalic maneuvers**, but she did have voluntary vertical eye movements. She also had **ocular bobbing** (fast phase down, slow phase up), **and a skew deviation, with the left eye higher than the right**. There was **no movement of the limbs** in response to commands. In response to pain, **the left arm did not move, the right arm had extensor (decerebrate) posturing, and both legs had triple flexion** (see Figure 3.5C). **Reflexes were absent, and both toes were upgoing**.

LOCALIZATION AND DIFFERENTIAL DIAGNOSIS

1. On the basis of the symptoms and signs shown in **bold** above, what is the most likely localization?

2. What is the most likely cause, and what are some other possibilities? What is the name of this clinical syndrome?

Discussion

1. The key symptoms and signs in this case are:

   - **Respiratory arrest**
   - **No horizontal eye movements**
   - **Ocular bobbing and a skew deviation**
   - **Ability to answer yes/no questions appropriately using eye blinks or vertical eye movements**
   - **No voluntary movements, with only decerebrate posturing, triple flexion, and bilateral Babinski's signs**

   Onset with respiratory arrest suggests possible medullary involvement (see Figure 14.17). The other findings—including lack of horizontal eye movements, ocular bobbing, skew deviation, and bilateral upper motor neuron–type paralysis with extensor posturing—all suggest extensive bilateral involvement of the pons (see KCC 14.3; Figure 14.21B,C). The fact that consciousness and vertical eye movement were preserved suggests that the midbrain was spared.

2. This clinical picture is compatible with locked-in syndrome (see KCC 14.1) caused by extensive bilateral pontine and possibly medullary infarcts, sparing the midbrain.

Clinical Course and Neuroimaging

An **MRI** revealed massive bilateral infarcts of the pons, extending down to the medulla (Image 14.5A,B, page 677). The midbrain (including the midbrain reticular formation) was not involved (Image 14.5C, page 678). An MRA showed absent flow in the vertebrobasilar system similar to that exhibited by the patient in Case 14.4 (see Image 14.4B). The lack of vertebrobasilar flow suggests that this patient had developed a basilar artery thrombosis, but that, unlike the patient in Case 14.4, she did not have sufficient collateral flow to supply much of her brainstem. She was treated with intravenous heparin but made no significant improvements during her hospitalization. Eventually, her left arm also developed decerebrate posturing. The patient and family decided to continue the mechanical ventilator; however, they requested no resuscitation in the event of cardiac arrest. A speech therapist worked with the patient to facilitate communications using picture and letter cards and other devices, and she remained able to communicate but unable to move except for looking up and down or blinking (levator palpebrae superior). She developed several infections that were treated with antibiotics, and when last seen, two and a half months after onset, her exam remained unchanged.

RELATED CASES. Locked-in syndrome (see KCC 14.1) is usually caused by bilateral ventral pontine lesions but, rarely, it can be caused by lesions in other locations. For example, Image 14.5D–F (pages 678–679) shows an MRI scan from a 56-year-old mathematics professor who suddenly became unable to move. Unlike the patient described above, he had preserved horizontal as well as vertical eye movements. MRA showed absent flow in the upper basilar artery, and MRI demonstrated bilateral infarction of the cerebral peduncles (see Image 14.5F). The midbrain tegmentum and the pons were largely spared, explaining his preserved consciousness and ability to make both vertical and horizontal eye movements. He remained in a locked-in state, communicating with eye movements only, for the next year and a half and eventually succumbed to an overwhelming pulmonary infection.

Atherosclerotic basilar stenosis with superimposed basilar thrombosis is a life-threatening neurologic emergency requiring prompt treatment. The two cases of locked-in patients described above, as well as other cases in this chapter, demonstrate the potentially dire consequences of basilar insufficiency. Image 14.5G (page 680) shows a pathology specimen from a patient who died of basilar thrombosis. This patient was a 67-year-old man who, over 2 days, gradually became sedated, weak, and quadriplegic, eventually lapsing into coma with no eye movements. An MRI showed massive infarction of the pons, midbrain (including the reticular formation), thalamus, and cerebellum (all supplied by the basilar artery; see Figures 14.18, 14.20). Postmortem examination revealed severe atherosclerotic narrowing of the basilar artery (see Image 14.5G), with superimposed thrombus.

CASE 14.6 WRONG-WAY EYES, LIMITED UPGAZE, DECREASED RESPONSIVENESS, AND HEMIPARESIS WITH AN AMAZING RECOVERY

CHIEF COMPLAINT

A 53-year-old man was brought to the emergency room with acute onset of **decreased responsiveness and left-sided weakness**.

HISTORY

Past history was notable for cigarette smoking, hypertension, and hypertriglyceridemia. He worked as the chef on a Japanese navy ship. Four days prior to admission, he developed **bilateral frontal and retro-orbital headaches**. At 10:00 A.M. on the day of admission, his headache worsened and he developed **generalized weakness and nausea**. At 10:30 A.M. he had a **transient episode of right-sided weakness** and went to see the ship's doctor. Lunchtime was approaching, however, and he had to return to work. At 2:00 P.M. he had sudden onset of **blurred vision, dysarthria, and gait difficulty**, which rapidly evolved into **decreased responsiveness with left-sided weakness**. Fortunately, the ship was docked at a major city, and the patient was urgently transferred to the emergency room, arriving within a half hour of symptom onset.

PHYSICAL EXAMINATION

Vital signs: T = 98°F, P = 84, BP = 170/90.

Neck: Supple with no bruits.

Lungs: Clear.

Heart: Regular rate; 2/6 systolic murmur heard loudest over the apex.

Abdomen: Soft, nontender.

Extremities: Normal.

Neurologic exam:

MENTAL STATUS: **Lethargic, arousable to voice, but needed repeated stimulation to get a response.** Japanese speaking. Correctly followed simple commands and imitated gestures.

CRANIAL NERVES: Pupils 3 mm, constricting to 2 mm. Intact blink to visual threat bilaterally. Extraocular movements full except for **left gaze preference and inability to move either eye fully to the right, even with oculocephalic maneuvers. Upgaze was also somewhat limited**. Corneal reflexes intact. **Left facial weakness, sparing the forehead. Unable to protrude tongue** on command.

MOTOR: Strong, purposeful movements of the right arm and leg. **Left arm had weak purposeful movements, and left leg had triple flexion only in response to pain.**

PLANTAR RESPONSES: **Upgoing bilaterally.**

OTHER REFLEXES, COORDINATION, GAIT, AND SENSORY EXAM: Not tested.

Toward the end of the exam, the patient became **less responsive** and began having **shivering movements and bilateral right greater than left extensor posturing** (see Figure 3.5B). He was urgently intubated and taken for a head CT and emergency angiogram at 3:00 P.M. (1 hour after symptom onset).

LOCALIZATION AND DIFFERENTIAL DIAGNOSIS

On the basis of the symptoms and signs shown in **bold** above, what is the most likely localization? What is the most likely diagnosis?

Discussion

The key symptoms and signs in this case are:

- **Bilateral frontal and retro-orbital headaches**
- **Nausea**
- **Blurred vision**
- **Generalized weakness; gait difficulty; transient episode of right-sided weakness; weakness of left face sparing the forehead, left arm, and left leg, with left leg triple flexion, and bilateral Babinski's signs; progression to bilateral right greater than left extensor posturing**
- **Decreased responsiveness, progressively worsening**
- **Dysarthria; inability to protrude tongue**
- **Right horizontal gaze palsy; limited upgaze**
- **Shivering movements**

This patient had symptoms and signs suggestive of waxing, waning, and then sudden worsening dysfunction of several bilateral regions supplied by the vertebrobasilar system (see KCC 14.3; Table 14.6). Blurred vision could be the result of occipital lobe involvement or diplopia. The alternating right, then left, and then right hemiparesis with bilateral Babinski's signs is strongly suggestive of basilar artery stenosis affecting the bilateral basis pontis. The presence of wrong-way eyes (see Table 14.8; see also KCC 13.10; Figure 13.15; Case 13.7) at the time of initial examination (left weakness with left gaze preference) suggests extension to the right pontine tegmentum (see Figure 14.21C). Limited upgaze and impaired consciousness suggest impaired function of the midbrain tegmentum. Shivering is characteristic of pontine dysfunction, and the headaches, nausea, and dysarthria are also suggestive of brainstem dysfunction resulting from ischemia (see KCC 14.3; Table 14.6). Given the patient's vascular risk factors, the overall picture is most compatible with evolving basilar thrombosis, possibly in the setting of preexisting basilar stenosis.

Clinical Course and Neuroimaging

The patient's initial **head CT** revealed no infarcts or hemorrhages (Image 14.6A, page 680). Interestingly, there was a region of increased density in the basilar artery with Hounsfield units (HU) of 60 to 70 compatible with clotted blood (see Table 4.1). Intravenous heparin was started and, in an experimental treatment for basilar thrombosis, the patient was then taken for an immediate **angiogram** (Image 14.6B, page 681). When the left vertebral artery was injected with dye, flow stopped after the proximal basilar artery just after the AICAs, and the dye refluxed down into the contralateral right vertebral artery during the injection (see Image 14.6B). Urokinase, a thrombolytic agent, was then infused through a catheter directly into the basilar artery at the region of the occlusion. The result was successful lysis of a clot that had blocked a very narrow region of the basilar artery (see Image 14.6C). After the lysis, the injected dye no longer refluxed significantly into the contralateral (right) vertebral artery, but instead continued distally, demonstrating restored flow in the distal basilar artery, PCA, and SCAs (Image 14.6C, page 681). Follow-up head CT no longer showed a bright clot in the basilar artery, with HU (see Table 4.1) measured at 40 (compare to Case 13.7).

On exam that evening, the patient was alert, following commands, with slightly limited right gaze and upgaze. He had extensor posturing on the left side, purposeful movements on the right, and bilaterally upgoing toes. The next day he had full eye movements except for slightly limited right eye abduction, 4+/5 strength on the left side, and bilateral equivocal plantar responses. He was changed over from heparin to Coumadin, and an MRA just before discharge (2 weeks after onset) showed persistent focal stenosis of the midbasilar artery. MRI showed no evidence of infarction. By the time of discharge, the patient had a completely normal neurologic exam, and he walked out of the hospital with no deficits.

CASE 14.5 LOCKED IN

IMAGE 14.5A–C Bilateral Pontine Basis Infarcts Causing Locked-In Syndrome Axial T2-weighted MRI images. (A) Rostral medulla with small bilateral regions of increased signal, compatible with infarcts. (B) Pons with extensive bilateral infarcts destroying the corticospinal and corticobulbar pathways. (C) Midbrain is spared, including the midbrain reticular formation, allowing preserved consciousness.

(A)

(B)

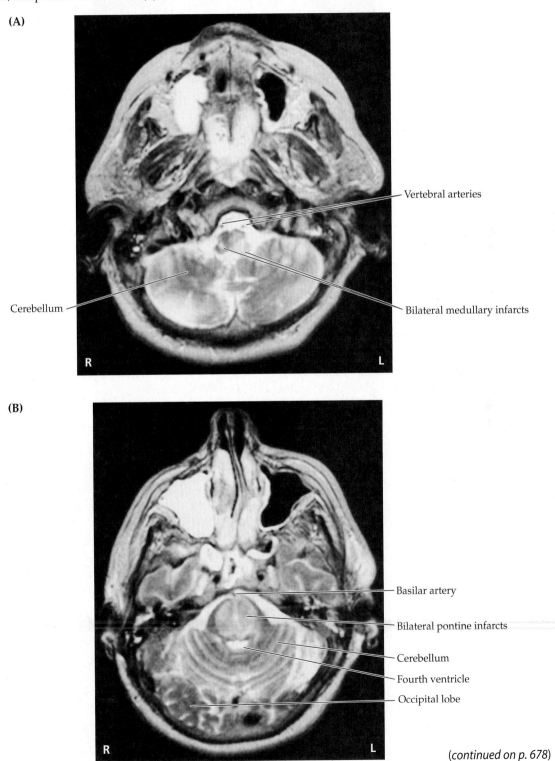

(*continued on p. 678*)

CASE 14.5 *(continued)*

(C)

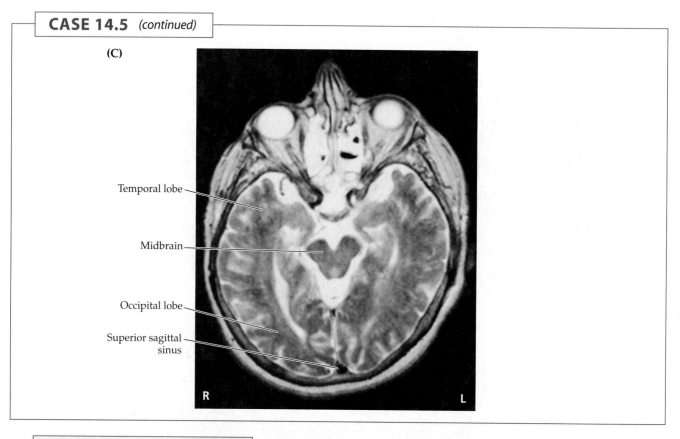

Temporal lobe

Midbrain

Occipital lobe

Superior sagittal sinus

R   L

CASE 14.5 *RELATED CASE*

IMAGE 14.5D–F Bilateral Midbrain Basis Infarcts Causing Locked-In Syndrome Axial T2-weighted MRI images. (D) Medulla. (E) Pons. (F) Midbrain, with bilat- eral T2-bright regions in the cerebral peduncles, compati- ble with infarcts.

(D)

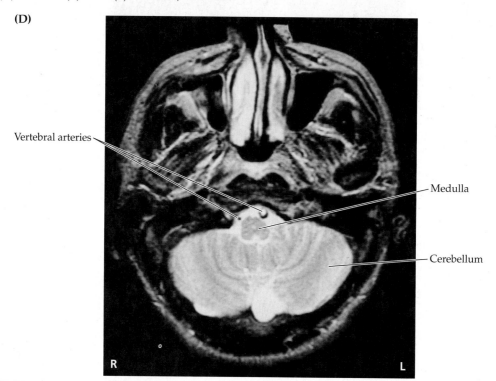

Vertebral arteries

Medulla

Cerebellum

R   L

CASE 14.5 *RELATED CASE* *(continued)*

(E)

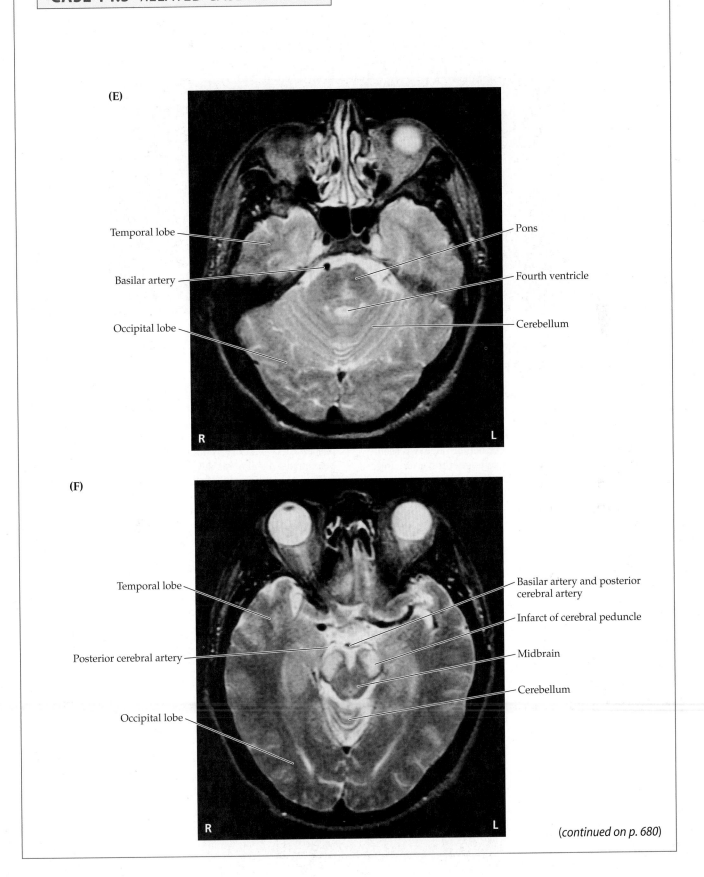

Temporal lobe

Basilar artery

Occipital lobe

Pons

Fourth ventricle

Cerebellum

R   L

(F)

Temporal lobe

Posterior cerebral artery

Occipital lobe

Basilar artery and posterior cerebral artery

Infarct of cerebral peduncle

Midbrain

Cerebellum

R   L

(continued on p. 680)

CASE 14.5 *RELATED CASE* (continued)

IMAGE 14.5G **Basilar Artery Stenosis** Pathology specimen from a patient who died of basilar thrombosis. A severe narrowing of the midbasilar artery can be seen (arrow), caused by atherosclerotic disease. This resulted in the superimposed basilar thrombosis.

(G)

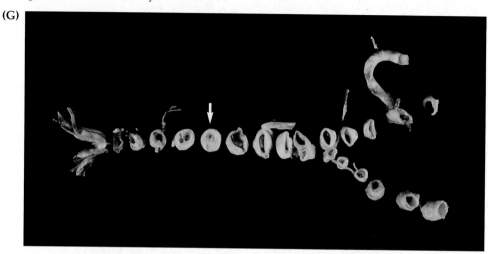

CASE 14.6 **WRONG-WAY EYES, LIMITED UPGAZE, DECREASED RESPONSIVENESS, AND HEMIPARESIS WITH AN AMAZING RECOVERY**

IMAGE 14.6A–C **Basilar Artery Thrombosis Treated with Intra-Arterial Thrombolysis** (A) Head CT on admission showing bright signal in the basilar artery, suggesting thrombosis. No infarct or hemorrhage was seen. (B) Angiogram done following left verterbral artery injection, demonstrating lack of flow in the basilar artery past the anterior inferior cerebellar arteries (AICA). Left anterior oblique view. (C) Repeat angiogram after intra-arterial thrombolysis, demonstrating restored flow in the distal basilar artery territory. A mid-basilar artery narrowing can be seen just distal to the AICAs. Left anterior oblique view following injection of the left vertebral artery, as in B.

(A)

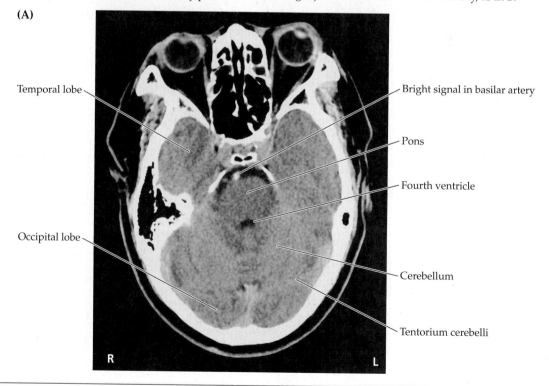

CASE 14.6 *(continued)*

(B)

Anterior inferior cerebellar artery (AICA)

Proximal basilar artery

Vertebral artery

Posterior inferior cerebellar artery (PICA)

Vertebral artery

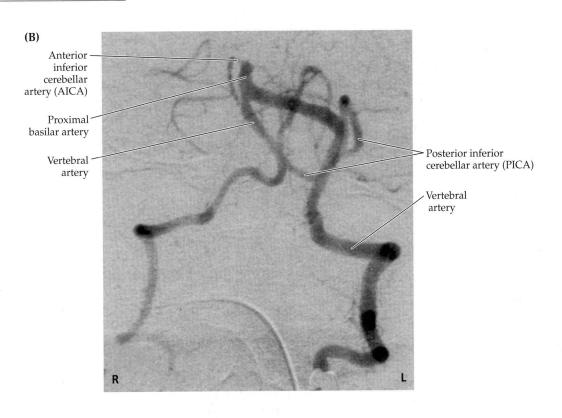

R   L

Distal basilar artery

(C)

Anterior inferior cerebellar artery (AICA)

Proximal basilar artery

Vertebral artery

Posterior cerebral artery (PCA)

Superior cerebellar artery (SCA)

Posterior inferior cerebellar artery (PICA)

Vertebral artery

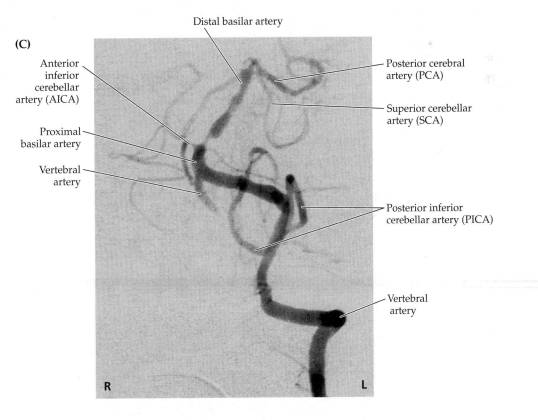

R   L

CASE 14.7 DIPLOPIA AND UNILATERAL ATAXIA

MINICASE

A 72-year-old man with a history of hypertension and hyper-cholesterolemia was watching TV one night and suddenly saw **two faces on the screen diagonally displaced. This diplopia went away when he covered one eye.** In order not to alarm his wife, he quietly made his way to bed, but the next morning the diplopia was unchanged, and he also noticed **gait unsteadiness with staggering to the right.** He tried using a friend's walker, but then also noticed that his **right hand was clumsy.** For example, he had difficulty picking up a credit card from the table with his right hand. On exam, his **left eye would elevate by only 1 mm, adduct by only 2 mm, and depress by only 3 mm** (Figure 14.23B). Left eye abduction was normal. He had diagonal **diplopia,** which was tested with a red glass over the right eye, as shown in Figure 14.23D. There was a **left ptosis,** with the palpebral fissure measuring 4 mm on the left and 9 mm on the right (see Figure 14.23C). The left pupil had a slightly irregular shape (see Figure 14.23A) but reacted normally to light. The patient also had **mild right ataxia on finger-to-nose and heel-to-shin testing, and an unsteady gait, tending to list to the right.** The remainder of the exam was normal, except for the **right plantar response,** which was **equivocal.**

LOCALIZATION AND DIFFERENTIAL DIAGNOSIS

1. What is the cause of the eye movement abnormalities summarized in Image 14.6A–C? (Review KCC 13.1, 13.2.)

2. A lesion in what region of the brain could cause these abnormalities along with right ataxia?

3. What is the most likely diagnosis?

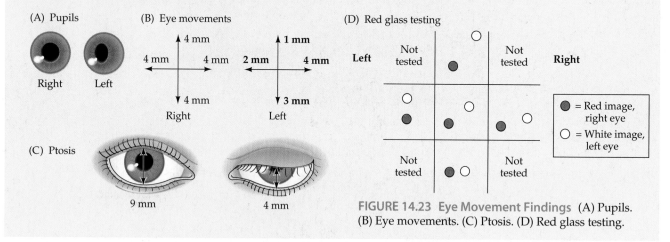

FIGURE 14.23 Eye Movement Findings (A) Pupils. (B) Eye movements. (C) Ptosis. (D) Red glass testing.

Discussion

The key symptoms and signs in this case are:

- Diagonal diplopia, and left eye movements with only 1 mm elevation, 2 mm adduction, 3 mm depression and normal abduction with left ptosis
- Right arm and leg ataxia
- Equivocal right plantar response
- Unsteady gait, tending to list to the right
- Right hand clumsiness

1. The eye movement abnormalities in Figure 14.23 represent a left third-nerve palsy (compare to Figure 13.5). Note that the pupil is not dilated but has an irregular shape. This may represent midbrain corectopia, which can occasionally be seen with midbrain lesions (see KCC 13.5). The fact that elevation of the contralateral eye was normal suggests that the lesion involves the third-nerve fascicles in the midbrain rather than the entire oculomotor nucleus (fibers from the superior rectus subnucleus project contralaterally; see Table 13.2; Figure 13.3).

2. The right-sided ataxia may be caused by involvement of the fibers of the superior cerebellar peduncle in the left midbrain (see Figure 14.21A; Table 14.9). The subtle right plantar abnormality could result from mild involvement of the left cerebral peduncle in the midbrain.

   The most likely *clinical localization* is left midbrain tegmentum, including the oculomotor nerve fascicles and superior cerebellar peduncle fibers (Claude's syndrome; see Table 14.9).

3. Given the sudden onset of symptoms in a man in his seventies with a history of hypertension and elevated cholesterol, the most likely diagnosis is left midbrain infarct, caused by occlusion of the penetrating vessels at the top of the basilar artery and proximal left PCA (see Figures 14.18A, 14.21A). A far less likely possibility is a small hemorrhage in this region.

Clinical Course and Neuroimaging

An MRI revealed subtle increased T2 signal in the midbrain tegmentum (not shown). Diffusion-weighted **MRI** confirmed the presence of an acute infarct in the left medial midbrain tegmentum (Image 14.7A,B, page 686; compare to Figure 14.21A). The patient was admitted to the hospital for further evaluation. MRA was unremarkable, but a Holter monitor revealed intermittent atrial fibrillation, and echocardiogram showed an enlarged left atrium. He was therefore treated with Coumadin (warfarin) anticoagulation and discharged home with a persistent left third-nerve palsy and mild right ataxia.

RELATED CASE. Image 14.7C (page 687) shows an MRI from another patient with a slightly larger midbrain infarct. The infarct involved the right midbrain tegmentum and right cerebral peduncle, resulting in a right CN III palsy, left-sided ataxia, and left hemiparesis (Benedikt's syndrome) (see Table 14.9; Figure 14.21A). Compare Case 14.7 and this related case with Case 13.1.

CASE 14.8 INTERMITTENT MEMORY LOSS, DIPLOPIA, SPARKLING LIGHTS, AND SOMNOLENCE

MINICASE

A 60-year-old retired businesswoman was sent to the emergency room by her physician because of 2 months of worsening episodes of **memory loss**, **sparkling lights**, and **blurry or double vision**. Her past medical history was notable for anticardiolipin (antiphospholipin) antibody syndrome (a condition causing hypercoagulability; see Table 10.5), including elevated anticardiolipin antibodies (IgG 2407, IgM 38, IgA < 10), a first-trimester miscarriage, left subclavian stenosis, and Raynaud's phenomenon. She had been treated in the past with Coumadin, but she elected to take aspirin instead. She also had a long history of brief, 1- to 2-minute, episodes of migraine-like visual scintillations (see KCC 5.1) that she described as "like firecrackers going off in front of my eyes," associated with multiple other vague complaints, including epigastric, pleuritic, and back pain, which were attributed to fibromyalgia or stress. There was a strong family history of migraine. The patient's sister had died 2 months prior to admission, and shortly afterward the patient began having recurrent **episodes of mem-**

ory loss lasting several minutes each, along with worsening of all her other complaints. She was somewhat evasive and vague in describing the episodes, and her physicians initially believed them to be psychiatrically based. Memory lapses included forgetting whether her sister had been buried or cremated and forgetting a real estate deal she had recently completed.

On the day of admission, she saw her physician because of a new complaint she had developed a few days earlier of **blurred and then double vision** to the point where it was difficult for her to stand up and walk. On initial examination she was alert and fluent but had **mildly reduced attention**, able to repeat only 5/7 digits forward, and recalling 2/3 objects after 3 minutes. Her **right pupil was slightly enlarged at 4 mm, constricting sluggishly to 2.5 mm,** and her left pupil was 3 mm, constricting briskly to 2 mm. She had **limited upward gaze with both eyes**. In addition, **medial gaze was reduced**

(continued on p. 684)

CASE 14.8   *(continued)*

with the right eye, and she had a **right ptosis**. Exam was otherwise unremarkable.

Her physician sent her to the emergency room, where she was examined by a neurologist, but by the time she arrived she had normal eye movements, with the only abnormality on exam being a mild residual anisocoria. She was admitted for further evaluation, and the next morning she was found again to have limited upgaze bilaterally, right ptosis, and right limited medial gaze, with a somewhat dilated right pupil. Anticoagulation with intravenous heparin was initiated; however, during the subsequent days she had **waxing and waning somnolence and delirium** to the point of being unarousable at times. At other times, despite being transferred to the intensive care unit, she would wake up, pull out all her intravenous lines, and walk down the hallway to use the bathroom. Her eye movement abnormalities persisted, and in addition

she developed **bilateral ataxia on finger-to-nose testing** (when she was awake enough to cooperate) and **decreased blink to visual threat on the left side**.

LOCALIZATION AND DIFFERENTIAL DIAGNOSIS

1. How can the eye movement abnormalities in this patient be summarized?

2. Dysfunction in what location can cause these eye movement abnormalities along with an impaired level of consciousness and ataxia?

3. Given the history of hypercoagulability and the addition of a left visual field deficit, as well as episodes of memory loss, what vascular syndrome (specify the blood vessels involved) can cause this combination of deficits? (See KCC 14.3.)

Discussion

The key symptoms and signs in this case are:

- **Episodes of memory loss**
- **Mildly reduced attention evolving to waxing and waning somnolence and delirium**
- **Limited upward gaze with both eyes**
- **Double vision with limited right medial and upward gaze, enlarged right pupil, and right ptosis**
- **Bilateral ataxia on finger-to-nose testing**
- **Sparkling lights, blurry vision, and, later, decreased blink to visual threat on the left side**

1. *Summary of eye movement abnormalities*: The patient had limited upgaze bilaterally. In addition, she had findings compatible with right third-nerve dysfunction, including right ptosis, a large right pupil with decreased reactivity to light, and reduced right eye adduction in addition to the limited upgaze.

2. *Localization of eye movement abnormalities, somnolence, and ataxia*: Dysfunction in the midbrain tegmentum (see Figure 14.21A) could cause (1) vertical gaze abnormalities through involvement of the rostral interstitial nucleus of the MLF in the rostral midbrain (see the Chapter 13 section on vertical eye movements); (2) right third-nerve dysfunction through involvement of the right third nerve fascicles or nucleus (a right third-nerve nucleus lesion involving the ipsilateral superior rectus subnucleus, and contralateral crossing fibers, could also cause bilateral impairment of upgaze; see Table 13.2; Figure 13.3); (3) somnolence and delirium through involvement of the reticular formation (see KCC 14.2); and (4) bilateral ataxia through involvement of the superior cerebellar peduncle fibers.

3. *Vascular localization and diagnosis*: Given this patient's history of hyperco-agulability, a thromboembolic disorder is likely. The above deficits can be localized to the midbrain tegmentum, which is supplied by small pene-trating vessels arising from the top of the basilar artery and proximal PCAs (see Figure 14.18A, Figure 14.20, and Figure 14.21A). The left visual field deficit could be explained by an infarct of the right occipital lobe, which is supplied by the right PCA. In addition, the episodes of memory loss could have been caused by TIAs involving the bilateral medial thalami or me-dial temporal lobes (see KCC 18.1), and these structures are also supplied by the PCAs (see Figures 10.5, 10.8). Therefore, the above deficits could all be caused by vascular insufficiency at the top of the basilar artery and prox-imal PCAs, known as top-of-the-basilar sydrome (see KCC 14.3; Figure 14.18). Top-of-the-basilar syndrome is usually caused by an embolus or thrombus lodged at the top of the basilar artery. Another, less likely possi-bility is thrombosis of the more proximal basilar artery, although in that case pontine dysfunction usually occurs, including corticospinal, horizontal gaze, proprioceptive, and other dysfunctions not seen in this patient.

Clinical Course and Neuroimaging

As already noted, despite anticoagulation the patient's condition continued to wax and wane. MRI and **head CT** scans demonstrated multiple bilateral infarcts in the territory of vessels arising from the top of the basilar artery (Image 14.8A,B, page 689), including the midbrain tegmentum, bilateral me-dial thalami, and right occipital lobe. An **angiogram** was performed because of concerns that, given her rheumatological history, she might have CNS vasculitis (see Table 10.5) which would require a different treatment (im-munosuppressive therapy). The angiogram was negative for changes sug-gestive of vasculitis. However, a filling defect was seen at the top of the basi-lar artery, most likely due to a thrombus that had embolized to that location from a remote source such as the heart or, less likely, due to a thrombus that had formed locally (Image 14.8C,D, page 690). Of note, the PCAs did not fill from the basilar artery (compare to Figure 4.17B). However, when the inter-nal carotid arteries were injected (not shown), the PCAs did fill, via the pos-terior communicating arteries, except for the distal right PCA. This result suggests that the midbrain and thalamic infarcts were caused by the occlu-sion of small penetrating vessels arising from the top of the basilar artery and proximal PCAs (see Figure 14.18A), while the right occipital infarct was probably caused by an embolus that broke off from the top of the basilar re-gion and migrated up into the distal right PCA (see Figure 14.20).

During the course of her hospital stay, the patient's condition initially waxed and waned, as already noted, but she later developed right hemiple-gia and lapsed into a coma (see KCC 14.2). A week later, she began to regain some responsiveness to stimulation and to commands, and an MRA re-vealed that some flow had been restored through the distal basilar artery. She was eventually discharged on Coumadin to a rehabilitation facility.

CASE 14.7 DIPLOPIA AND UNILATERAL ATAXIA

IMAGE 14.7A,B Left Midbrain Infarct in Region of Third Nerve Fascicles and Superior Cerebellar Peduncle   Diffusion-weighted MRI images. (A) Axial image through the midbrain. (B) Coronal image.

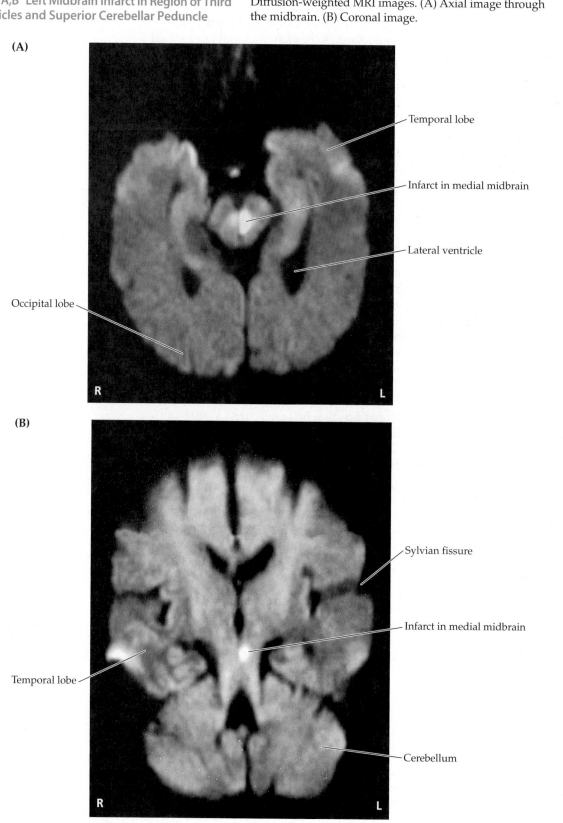

(A)

Temporal lobe

Infarct in medial midbrain

Lateral ventricle

Occipital lobe

R   L

(B)

Sylvian fissure

Infarct in medial midbrain

Temporal lobe

Cerebellum

R   L

CASE 14.7 *RELATED CASE*

IMAGE 14.7C **Right Midbrain Infarct in Region of Cerebral Peduncle, Superior Cerebellar Peduncle,** **Red Nucleus, and Third-Nerve Fascicles** Axial T2-weighted MRI image through the midbrain.

(C)

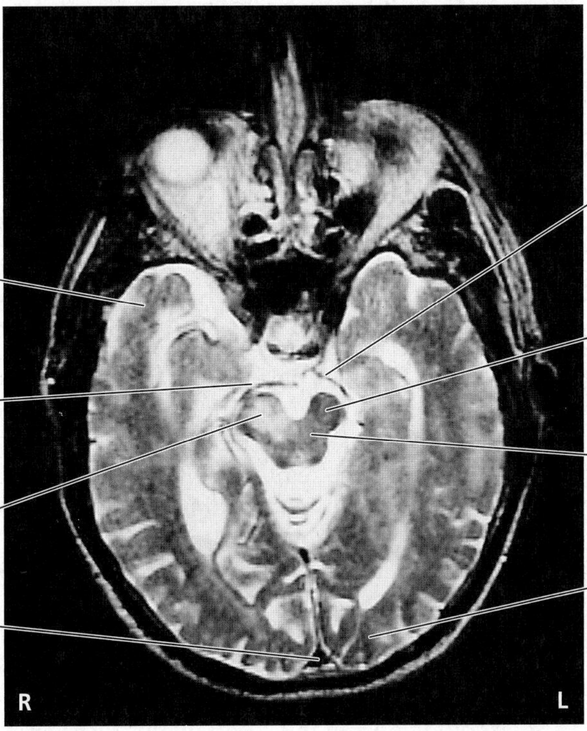

Temporal lobe

Posterior cerebral artery

Infarct

Superior sagittal sinus

Posterior cerebral artery

Cerebral peduncle

Midbrain tegmentum

Occipital lobe

R L

CASE 14.9 INTRACTABLE HICCUPS

MINICASE

A 50-year-old woman developed a **bilateral retro-orbital headache** associated with nasal discharge 2 weeks prior to presentation. She was treated with oral antibiotics for presumed sinusitis, and her symptoms resolved. On review of her past history, she described an episode 14 years previously of vertigo, nystagmus, and dysarthria with a negative workup done in 1977, before MRI was available. Because of these previous symptoms, an MRI scan was scheduled; however, she then suddenly developed **intractable hiccups** lasting 5 days, causing her to come back to her physician's office. General exam and neurologic exam were entirely normal.

LOCALIZATION AND DIFFERENTIAL DIAGNOSIS

Although hiccups (singultus) are usually benign, persistent hiccups can be caused by a variety of systemic or gastrointestinal disorders, as well as by CNS lesions. Lesions in what general region of the CNS are associated with hiccups? Assuming that this patient's previous neurologic symptoms are related to her present complaints, what are some possibilities for the diagnosis?

Discussion

The key symptoms and signs in this case are:

- **Bilateral retro-orbital headache**
- **Intractable hiccups**

Headache can have many causes, but it may be associated with intracranial pathology (see KCC 5.1) Hiccups can be caused by lesions of the posterior fossa, particularly in the medulla (see KCC 14.3). Given the occurrence of symptoms possibly related to brainstem dysfunction many years earlier, a chronic or recurrent lesion of the brainstem, especially the medulla, is the most likely diagnosis. Some possibilities would include demyelination, a low-grade tumor, a small recurrent hemorrhage in an arteriovenous malformation or cavernous angioma, vertebrobasilar migraine, and vasculitis or immune-mediated disorders of the CNS (CNS lupus, Behçet's syndrome, sarcoidosis, etc.).

Clinical Course and Neuroimaging

A **brain MRI** was performed (Image 14.9A,B, page 691). The MRI revealed a small bright region on unenhanced T1-weighted images, consistent with subacute hemorrhage (see Table 4.4), located in the dorsal portion of the rostral pons in the region of the obex. The patient was admitted briefly for observation and then was discharged home with an appointment for an angiogram approximately 1 month later, once the blood had resolved, to look for an arteriovenous malformation. The angiogram was negative, and it was felt that the patient most likely had a cavernous angioma (see KCC 5.6). A follow-up MRI scan 3 to 4 months later revealed resorption of the hemorrhage (see Image 14.9B). Treatment of cavernous angiomas is controversial; however, because of concerns about the potential high risk if another bleed should occur in this location, the decision was made to treat the angioma by surgical resection. The lesion was resected during a long, delicate operation in the posterior fossa. Pathologic examination of the resected tissue confirmed cavernous angioma (see KCC 5.6). She made a complete recovery without any deficits.

Additional Cases

Related cases can be found in other chapters for the following topics: **brainstem internal structures and vascular supply** (Cases 5.2–5.6, 10.3, 10.11, 12.8, 13.7–13.9, 15.4, and 18.3). Other relevant cases can be found using the **Case Index** located at the end of this book.

CASE 14.8 INTERMITTENT MEMORY LOSS, DIPLOPIA, SPARKLING LIGHTS, AND SOMNOLENCE

IMAGE 14.8A,B Infarcts Caused by Top-of-the-Basilar Syndrome Axial CT scan images. (A) Hypodensity in midbrain tegmentem and right occipital lobe compatible with infarcts. (B) Section slightly higher than in A, showing bilateral medial thalamic and right occipital infarcts.

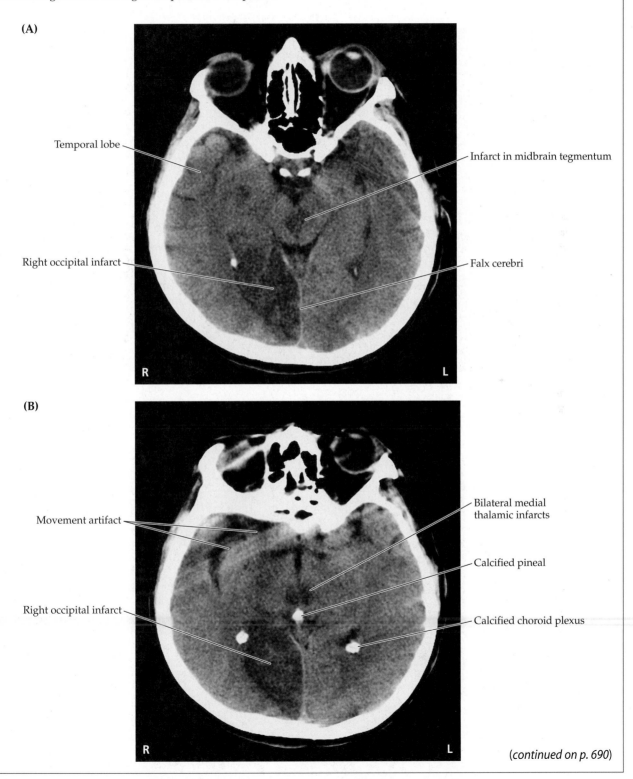

(*continued on p. 690*)

CASE 14.8 (continued)

IMAGE 14.8C,D Top-of-the-Basilar Syndrome
Angiogram following injection of the left vertebral artery.
(C) Left lateral view. Note filling defect in the distal basilar artery and lack of filling of the bilateral posterior cerebral arteries (PCAs). (D) Anteroposterior view (compare to Figure 4.17).

(C)

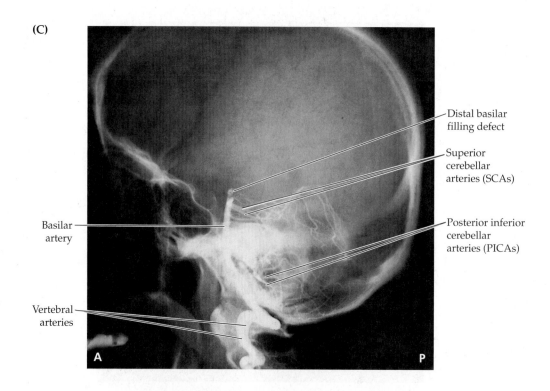

(D)

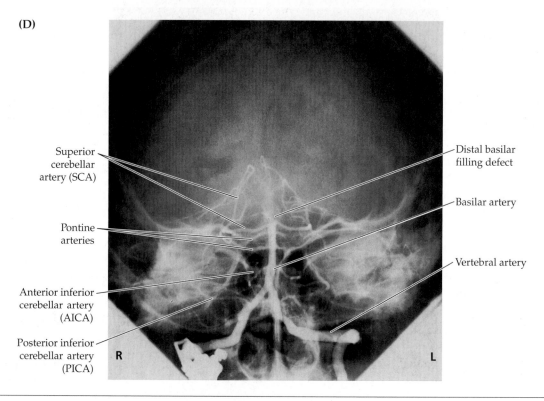

CASE 14.9 INTRACTABLE HICCUPS

IMAGE 14.9A,B Cavernous Angioma in the Rostral Medulla, in Region of the Obex T1-weighted MRI images. (A) Axial image at the time of presentation with bright region in rostral medulla, demonstrating subacute hemorrhage. (B) Sagittal image 3 to 4 months later showing dark cavitated region in rostral medulla compatible with prior hemorrhage.

(A)

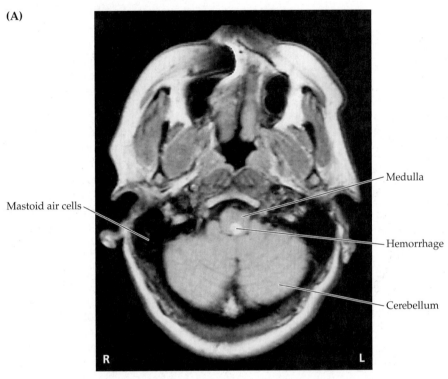

(B)

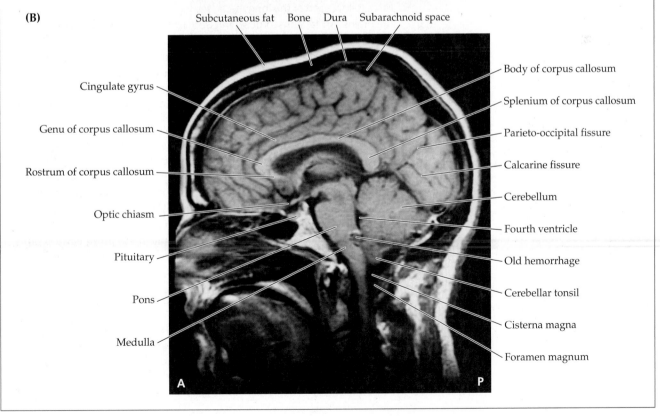

Brief Anatomical Study Guide

In this chapter we focused on the **four main components of internal brainstem structures** shown in Figure 14.1, namely: cranial nerve nuclei and related structures; long tracts; cerebellar circuitry; and reticular formation and related structures. Here we will describe two strategies for reviewing this material. First, we will take a **functional approach** and use the myelin-stained brainstem sections shown in Figures 14.3–14.5 to identify functional pathways and functional groupings of nuclei that are distributed throughout these sections and beyond. Second, we will take a **regional approach** and use the vascular territories shown in Figure 14.21 to identify clinically relevant constellations of deficits that occur with focal brainstem infarcts.

1. To begin the functional review, use Table 14.1 to recall each of the **six functional columns of cranial nerve nuclei**, and follow these columns through the brainstem sections in Figures 14.3–14.5, identifying each of the component nuclei. (All nuclei can be identified on these sections except for the salivatory nuclei and the cochlear nuclei [see also Figures 12.4 and 12.5].) As you trace structures through adjacent brainstem sections, be sure to note their spatial relationships to nearby structures as well.

2. Next, review the **auditory pathway** by following the lateral lemniscus through the sections from the caudal pons to the inferior colliculus (see Figures 14.3 and 14.4), using Figures 12.16 and 12.17 as a guide. Identify the **MLF** in each brainstem section (see Figures 14.3–14.5), and recall its functions (see Figures 12.19 and 13.12). Follow the **taste pathway** from the nucleus solitarius rostrally toward the thalamus by following the central tegmental tract from the medulla up through the midbrain (see also Figure 12.12).

3. Next, follow each of the long tracts (see Figures 6.11A, 7.1, and 7.2) through the brainstem sections, proceeding from rostral to caudal to trace the route of the **corticospinal tract** and from caudal to rostral to trace the **posterior column–medial lemniscal system** and the **anterolateral (spinothalamic) system** (see Figures 14.3–14.5). **Cerebellar circuitry** is discussed in detail in Chapter 15, but as an initial orientation, use the stained brainstem sections in Figures 14.3–14.5 to identify each of the structures listed under cerebellar circuitry in Table 14.1.

4. The **reticular formation** was simplified in this chapter as containing rostral and caudal regions with different primary functions (see Figure 14.6). The **rostral reticular formation**, located in the midbrain and upper pons, includes several widespread projection systems involved mainly in behavioral and cognitive arousal (see Table 14.2; Figures 14.6–14.13). Lesions in the pontomesencephalic reticular formation often cause coma. Other circuits of the reticular formation located mainly in the **more caudal pons and medulla** are important for control of respiration, heart rate, blood pressure, and other autonomic functions, as well as for motor control, posture, muscle tone, locomotion, and a variety of other relatively stereotyped motor activities.

5. By reviewing **brainstem vascular syndromes**, we will complete a regional overview of internal brainstem structures and also enhance our clinical knowledge. The major blood vessels of the posterior circulation are shown

in Figures 14.18–14.21. **Paramedian penetrating branches** supply medial regions of the brainstem, while lateral regions are supplied by penetrators arising from both small-caliber **circumferential branches** (see Figure 14.19) and the larger vessels shown in Figure 14.18.

6. Proceeding from caudal to rostral, identify each vascular territory on the right half of Figure 14.21. Cover the labels and name the vessels that supply each territory (these are also listed in Tables 14.7–14.9). Next, name the structures within each territory that would be affected by occlusion of the blood supply to that territory, and name the associated **clinical symptoms and signs**. By proceeding in this way through the vascular territories of the medulla, pons, and midbrain, you should gain a full appreciation of the elegance of brainstem anatomy and its clinical importance.

References

Anatomical and Clinical Review

Cooper JR, Bloom FE, Roth RH. 2003. *The Biochemical Basis of Neuropharmacology*. 8th Ed. Oxford University Press, New York.

Huguenard JR, McCormick DA. 2007. Thalamic synchrony and dynamic regulation of global forebrain oscillations. *Trends Neurosci* 30 (7): 350–356.

Jones EG (ed.). 2007. *The Thalamus*. 2nd ed. Cambridge University Press, Cambridge, UK.

Stenade H, McCarley RW. 2005. *Brain Control of Wakefulness and Sleep*. 2nd ed. Plenum, New York.

Locked-In Syndrome

Bauby, JD. 1998. *The Diving Bell and the Butterfly*. Knopf Doubleday Publishing Group, New York.

Chia LG. 1991. Locked-in syndrome with bilateral ventral midbrain infarcts. *Neurology* 41 (3): 445–446.

Dollfus P, Milos PL, Chapuis A, Real P, Orenstein M, Soutter JW. 1990. The locked-in syndrome: A review and presentation of two chronic cases. *Paraplegia* 28 (1): 5–16.

Laureys S, Pellas F, Van Eeckhout P, Ghorbel S, Schnakers C, Perrin F, Berré J, Faymonville ME, et al. 2005. The locked-in syndrome : what is it like to be conscious but paralyzed and voiceless? *Prog Brain Res* 150: 495–511.

Patterson JR, Gabois M. 1986. Locked-in syndrome: A review of 139 cases. *Stroke* 17 (4): 758–764.

Reznik M. 1983. Neuropathology in seven cases of locked-in syndrome. *J Neurol Sci* 60 (1): 67–78.

Coma and Related Disorders of Consciousness

Blumenfeld H. 2009. The neurological examination of consciousness. In *The Neurology of Consciousness*. S Laureys, G Tononi (eds.), Chapters 15–30. Elsevier, Academic Press, New York.

Fisher CM. 1969. The neurological examination of the comatose patient. *Acta Neurol Scand Suppl* 45 (Suppl 36): 1–56.

Giacino JT, Ashwal S, Childs N, Cranford R, Jennett B, Katz DI, Kelly JP, Rosenberg JH, et al. 2002. The minimally conscious state: definition and diagnostic criteria. *Neurology* 58: 349–353.

Laureys S, Tononi G (eds). 2008. *The Neurology of Consciousness*. Elsevier, Academic Press, New York.

Lu J, Sherman D, Devor M, Saper CB. 2006. A putative flip-flop switch for control of REM sleep. *Nature* 441 (7093): 589–594.

Posner JB, Plum F, Saper CB, Schiff N. 2008. *The Diagnosis of Stupor and Coma.* 4th Ed. Oxford University Press, Oxford.

The Multi-Society Task Force on PVS. 1994. Medical aspects of the persistent vegetative state (1). The Multi-Society Task Force on PVS. *N Engl J Med* 330: 1499–1508.

Wijdicks EF. 2001. The diagnosis of brain death. *N Engl J Med* 344 (16):1215–1221.

Young GB, Ropper AH, Bolton CF. 1998. *Coma and Impaired Consciousness: A Clinical Perspective.* McGraw-Hill, New York.

General Brainstem Vascular Supply and Vertebrobasilar Disease

Burger KM, Tuhrim S, Naidich TP. 2005. Brainstem vascular stroke anatomy. *Neuroimaging Clin N Am* (2): 297–324.

Mohr JP, Choi D, Grotta J, and Wolf P (eds.). 2004. *Stroke: Pathophysiology, Diagnosis and Management.* 4th Ed. Churchill Livingstone, New York.

Moncayo J, Bogousslavsky J. 2003. Vertebro-basilar syndromes causing oculo-motor disorders. *Curr Opin Neurol* 16 (1): 45–50.

Savitz SI, Caplan LR. 2005. Vertebrobasilar disease. *N Engl J Med* 352 (25): 2618–2626.

Schwarz S, Egelhof T, Schwab S, Hacke W. 1997. Basilar artery embolism: Clinical syndrome and neuroradiologic patterns in patients without permanent occlusion of the basilar artery. *Neurology* 49 (5): 1346–1352.

Silverman IE, Liu GT, Volpe NJ, Galetta SL. 1995. The crossed paralysis: The original brain-stem syndromes of Millard-Gubler, Foville, Weber, and Raymond-Cestan. *Arch Neurol* 52 (6): 635–638.

Tatu L, Moulin T, Bogousslavsky J, Duvernoy H. 1996. Arterial territories of the human brain. *Neurology* 47 (5): 1125–1135.

Tijssen CC. 1994. Contralateral conjugate eye deviation in acute supratentorial lesions. *Stroke* 25 (7): 1516–1519.

Cervical Arterial Dissection

Biller J, Hingtgen WL, Adams HP, Smoker WRK, Godersky JC, Toffol GJ. 1986. Cervicocephalic arterial dissections: A ten-year experience. *Arch Neurol* 43 (12): 1234–1238.

Flis CM, Jäger HR, Sidhu PS. 2007. Carotid and vertebral artery dissections: clinical aspects, imaging features and endovascular treatment. *Eur Radiol* 17 (3): 820–834.

Frumkin LR, Baloh RW. 1990. Wallenberg's syndrome following neck manipulation. *Neurology* 40 (4): 611–615.

Nedeltchev K, Baumgartner RW. 2005. Traumatic cervical artery dissection. *Front Neurol Neurosci* 20: 54–63.

Rizzo L, Crasto SG, Savio D, Veglia S, Davini O, Giraudo M, Cerrato P, De Lucchi R. 2006. Dissection of cervicocephalic arteries: early diagnosis and follow-up with magnetic resonance imaging. *Emerg Radiol* 12 (6): 254–265.

Medullary Infarcts

Bassetti C, Bogousslavsky J, Mattle H, Bernasconi A. 1997. Medial medullary stroke: Report of seven patients and review of the literature. *Neurology* 48 (4): 882–890.

Currier RD, Giles CL, DeJong RN. 1961. Some comments on Wallenberg's lateral medullary syndrome. *Neurology* 11: 778–791.

Katoh M, Kawamoto T. 2000. Bilateral medial medullary infarction. *J Clin Neurosci* 7 (6): 543–545.

Kim JS, Lee JH, Suh DC, Lee MC. 1994. Spectrum of lateral medullary syndrome: Correlation between clinical findings and magnetic resonance imaging in 33 subjects. *Stroke* 25 (7): 1405–1410.

Kim JS, Kim HG, Chung CS. 1995. Medial medullary syndrome: Report of 18 patients and a review of the literature. *Stroke* 26 (9): 1548–1552.

Kitis O, Calli C, Yunten N, Kocaman A, Sirin H. 2004. Wallenberg's lateral medullary syndrome: diffusion-weighted imaging findings. *Acta Radiol* 45 (1): 78–84.

Matsumoto S, Okuda B, Imai T, Kameyama M. 1988. A sensory level on the trunk in lower lateral brainstem lesions. *Neurology* 38 (10): 1515–1519.

Solomon D, Galetta SL, Liu GT. 1995. Possible mechanisms for horizontal gaze deviation and lateropulsion in the lateral medullary syndrome. *J Neuroophthalmol* 15 (1): 26–30.

Toyoda K, Imamura T, Saku Y, Oita J, Ibayashi S, Minematsu K, Yamaguchi T, Fujishima M. 1996. Medial medullary infarction: Analyses of eleven patients. *Neurology* 47 (5): 1141–1147.

Vuilleumier P, Bogousslavsky J, Regli F. 1995. Infarction of the lower brainstem: Clinical, aetiological and MRI–topographical correlations. *Brain* 118 (pt.4): 1013–1025.

Basilar Artery Stenosis and Thrombosis

Archer CR, Horenstein S. 1977. Basilar artery occlusion. *Stroke* 8 (3): 383–390.

Baird TA, Muir KW, Bone I. 2004. Basilar artery occlusion. *Neurocrit Care* (3): 319–329.

Brandt T. Diagnosis and thrombolytic therapy of acute basilar artery occlusion: a review. 2002. *Clin Exp Hypertens* 24 (7-8): 611–622.

Bruckmann H, Ferbert A, del Zoppo GJ, Hacke W, Zeumer H. 1986. Acute vertebral-basilar thrombosis. Angiologic-clinical comparison and therapeutic implications. *Acta Radiol Suppl* 369: 38–42.

Hachinski V. 2007. Intra-arterial thrombolysis for basilar artery thrombosis and stenting for asymptomatic carotid disease: implications and future directions. *Stroke* 38 (2 Suppl): 721–722.

Hankey GJ, Khangure MS, Stewart-Wynne EG. 1988. Detection of basilar artery thrombosis by computed tomography. *Clin Radiol* 39 (2): 140–143.

Idicula TT, Joseph LN. 2007. Neurological complications and aspects of basilar artery occlusive disease. *Neurologist* 13 (6): 363–368.

Kubik CS, Adams RD. 1946. Occlusion of the basilar artery: A clinical and pathological study. *Brain* 69: 73–121.

Schonewille WJ, Wijman CA, Michel P, Rueckert CM, Weimar C, Mattle HP, Engelter ST, Tanne D, et al., on behalf of the BASICS study group. 2009. Treatment and outcomes of acute basilar artery occlusion in the Basilar Artery International Cooperation Study (BASICS): a prospective registry study. *Lancet Neurol* 8 (8): 724–730.

Smith WS. 2007. Intra-arterial thrombolytic therapy for acute basilar occlusion: pro. *Stroke* 38 (2 Suppl): 701–703.

Williams D, Wilson TG. 1975. The diagnosis of the major and minor syndromes of basilar insufficiency. *J Neurol Neurosurg Psychiatry* 39: 741–774.

Top of the Basilar Syndrome

Caplan LR. 1980. "Top of the basilar" syndrome. *Neurology* 30 (1): 72–79.

Mitra S, Ghosh D, Puri R, Parmar VR. 2001. Top-of-the-basilar-artery stroke. *Indian Pediatr* 38 (1): 83–87.

Segarra JM. 1970. Cerebral vascular disease and behavior. I. The syndrome of the mesencephalic artery (basilar artery bifurcation). *Arch Neurol* 22 (5): 408–418.

Pontine Infarcts

Bassetti C, Bogousslavsky J, Barth A, Regli F. 1996. Isolated infarcts of the pons. *Neurology* 46 (1): 165–175.

Ling L, Zhu L, Zeng J, Liao S, Zhang S, Yu J, Yang Z. 2009. Pontine infarction with pure motor hemiparesis or hemiplegia: a prospective study. *BMC Neurol* 9: 25.

Onbas O, Kantarci M, Alper F, Karaca L, Okur A. 2005. Millard-Gubler syndrome: MR findings. *Neuroradiology* 47 (1): 35–37.

Midbrain Infarcts

Kim JS, Kim J. 2005. Pure midbrain infarction: clinical, radiologic, and pathophysiologic findings. *Neurology* 64 (7): 1227–1232.

CONTENTS

Chapter 15

Cerebellum

Cerebellar lesions can cause abnormalities in movement of the body and eyes and can affect balance by disrupting vestibular function. A 13-year-old boy with a lesion in the cerebellum developed gradually worsening headaches, nausea, vomiting, and unsteadiness over the course of 2 months. His headaches were worse at night and were mainly in the left occipital area. A neurologic examination revealed bilateral papilledema, nystagmus, mildly slurred speech, and irregular, ataxic movements that were worse on the left side than the right.

In this chapter we will learn about the anatomy and functions of the cerebellum, including network interactions with other parts of the nervous system, and we will see examples of cases in which these functions have been impaired.

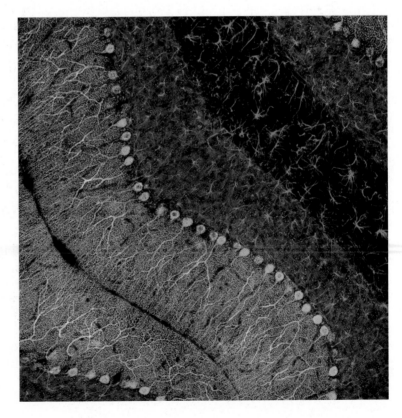

====== ANATOMICAL AND CLINICAL REVIEW ======

T HE CEREBELLUM INTEGRATES massive sensory and other inputs from many regions of the brain and spinal cord. This information is used by the cerebellum to smoothly coordinate ongoing movements and to participate in motor planning. Like the basal ganglia, discussed in Chapter 16, the cerebellum has no direct connections to lower motor neurons, but instead exerts its influence through connections to motor systems of the cortex and brainstem (see Figures 2.17 and 6.6).

The cerebellum is organized into different regions with specialized functions (see Table 15.1). The inferior vermis and flocculonodular lobes regulate balance and eye movements through interactions with the vestibular circuitry. These regions, together with other parts of the vermis, are involved in control of the medial motor systems discussed in Chapter 6 (proximal trunk muscles). More lateral cerebellar regions control the lateral motor systems (distal appendicular muscles). Finally, large regions of the most lateral cerebellar hemispheres are important in motor planning.

Although cerebellar circuitry is complex, the effects of lesions on the cerebellum are relatively easy to understand based on these different cerebellar regions, as we will see. Cerebellar lesions typically result in a characteristic type of irregular uncoordinated movement called **ataxia**. Cerebellar lesions can often be localized on the basis of just a few simple principles (see also KCC 15.2):

1. Ataxia is ipsilateral to the side of a cerebellar lesion.
2. Midline lesions of the cerebellar vermis or flocculonodular lobes mainly cause unsteady gait (truncal ataxia) and eye movement abnormalities, which are often accompanied by intense vertigo, nausea, and vomiting.
3. Lesions lateral to the cerebellar vermis mainly cause ataxia of the limbs (appendicular ataxia).

It is also important to recognize that because of the multiple reciprocal connections between the cerebellum, brainstem, and other regions, ataxia may be seen with lesions in these other locations as well. In addition, cerebellar pathways participate in several other functions, including articulation of speech, respiratory movements, motor learning, and, possibly, certain higher-order cognitive processes.

In this chapter, we will begin our tour of the cerebellum by discussing its overall structure. Next we will discuss the microscopic circuitry and input and output connections of the cerebellum. Finally, we will review the vascular supply to the cerebellum and the clinical effects of lesions on cerebellar networks.

Cerebellar Lobes, Peduncles, and Deep Nuclei

The cerebellum is the largest structure in the posterior fossa (Figure 15.1). It is attached to the dorsal aspect of the pons and rostral medulla by three white matter peduncles (feet) and forms the roof of the fourth ventricle (see Figure 15.1 and Figure 15.3). The cerebellum consists of a midline **vermis**, named for its wormlike appearance, and two large **cerebellar hemispheres** (see Figure 15.3A). There are numerous fissures, the deepest of which is called the **primary fissure** (see Figures 15.1 and 15.3A), separating the cerebellum into an **anterior lobe** and a **posterior lobe**. If the cerebellum is removed from the brainstem by cutting the cerebellar peduncles (Figure 15.2), the ventral surface of the cerebellum becomes visible (see Figure 15.3C). On the ventral inferior surface, the **posterolateral fissure** separates the posterior lobe from the **flocculonodular**

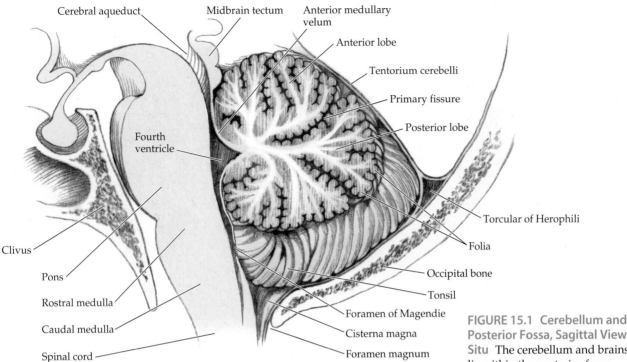

FIGURE 15.1 Cerebellum and Posterior Fossa, Sagittal View in Situ The cerebellum and brainstem lie within the posterior fossa, formed by the occipital bones and clivus inferiorly and by the dura of the tentorium cerebelli superiorly.

lobe, a region with important connections to the vestibular nuclei. The two **flocculi** are connected to the midline structure called the nodulus by thin pedicles (see Figure 15.3C). The **nodulus** is the most inferior portion of the cerebellar vermis. Another important landmark on the inferior surface consists of the **cerebellar tonsils** (see Figures 15.1 and 15.3C). Mass lesions of the cerebrum or cerebellum, or brain swelling with severely elevated intracranial pressure, can cause the tonsils to herniate (see KCC 5.4) through the foramen magnum (see Figure 15.1; see also Figure 5.18), compressing the medulla and causing death because of impingement on medullary respiratory centers.

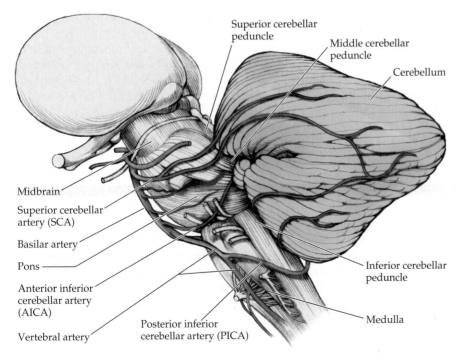

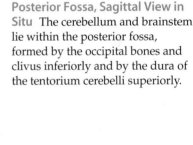

FIGURE 15.2 Lateral View of the Cerebellum Attached to the Brainstem The cerebellar peduncles and the vascular supply to the cerebellum and brainstem are shown.

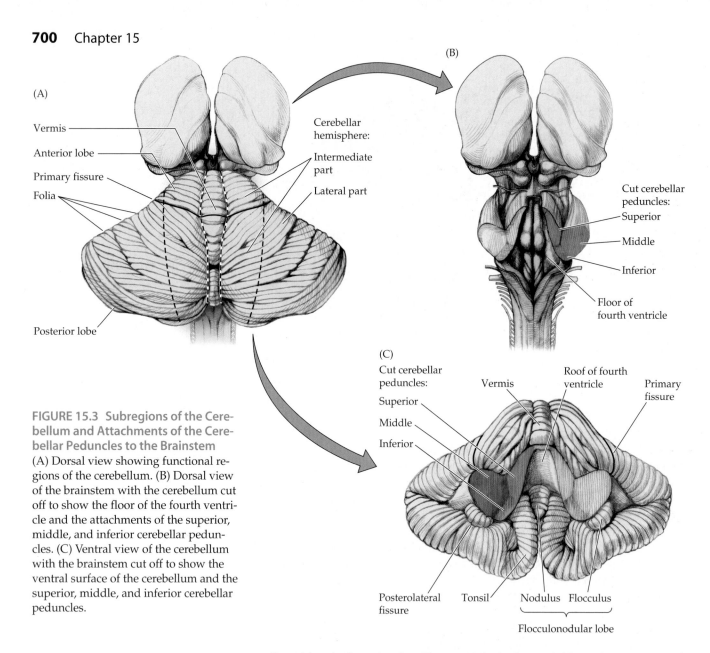

(A)

Vermis

Anterior lobe

Primary fissure

Folia

Posterior lobe

Cerebellar hemisphere:

Intermediate part

Lateral part

(B)

Cut cerebellar peduncles:

Superior

Middle

Inferior

Floor of fourth ventricle

(C)

Cut cerebellar peduncles:

Superior

Middle

Inferior

Vermis

Roof of fourth ventricle

Primary fissure

Posterolateral fissure

Tonsil

Nodulus

Flocculus

Flocculonodular lobe

FIGURE 15.3 Subregions of the Cerebellum and Attachments of the Cerebellar Peduncles to the Brainstem (A) Dorsal view showing functional regions of the cerebellum. (B) Dorsal view of the brainstem with the cerebellum cut off to show the floor of the fourth ventricle and the attachments of the superior, middle, and inferior cerebellar peduncles. (C) Ventral view of the cerebellum with the brainstem cut off to show the ventral surface of the cerebellum and the superior, middle, and inferior cerebellar peduncles.

On midsagittal section (see Figure 15.1), the beautiful branching pattern of the central **cerebellar white matter** and cortical **gray matter** can be appreciated, to which the Latin term "arbor vitae," meaning "tree of life" has been applied. Instead of gyri, the small ridges that run from medial to lateral on the surface of the cerebellum are called **folia**, meaning "leaves" (see Figures 15.1 and 15.3A).

Removing the cerebellum from the brainstem (see Figure 15.3B,C) reveals the three **cerebellar peduncles** (superior, middle, and inferior), which form the walls of the fourth ventricle. The **superior cerebellar peduncle** mainly carries outputs from the cerebellum, while the **middle cerebellar peduncle** and **inferior cerebellar peduncle** mainly carry inputs. The **superior cerebellar peduncle decussates** in the midbrain at the level of the inferior colliculi (see Figure 14.3B). Because of the striking conjunction of fibers in this decussation, another name for the superior cerebellar peduncle is the **brachium conjunctivum**. Because of its massive connections to the pons, an alternative name for the middle cerebellar peduncle is **brachium pontis**. The alternative name for the inferior cerebellar peduncle is **restiform body**, meaning "ropelike body."

The cerebellum can be divided into three functional regions, from medial to lateral, based on their input and output connections (see Figure 15.3A,C; Table 15.1):

TABLE 15.1 Functional Regions of the Cerebellum

REGION	FUNCTIONS	MOTOR PATHWAYS INFLUENCED
Lateral hemispheres	Motor planning for extremities	Lateral corticospinal tract
Intermediate hemispheres	Distal limb coordination	Lateral corticospinal tract, rubrospinal tract
Vermis and flocculonodular lobe	Proximal limb and trunk coordination	Anterior corticospinal tract, reticulospinal tract, vestibulospinal tract, tectospinal tract
	Balance and vestibulo-ocular reflexes	Medial longitudinal fasciculus

1. The **vermis** and **flocculonodular lobes** are important in proximal and trunk muscle control and in vestibulo-ocular control, respectively.
2. The **intermediate part** of the cerebellar hemisphere is mainly involved in control of more distal appendicular muscles in the arms and legs.
3. The largest part of the cerebellum is the **lateral part** of the cerebellar hemisphere, which is involved in planning the motor program for the extremities. Interestingly, a large portion of the lateral cerebellar hemisphere can be removed *unilaterally* without severe deficits.

The deep cerebellar nuclei and vestibular nuclei also fit with this medial to lateral functional organization (**Figure 15.4**). *All outputs from the cerebellum* are relayed by these nuclei (**Figure 15.5**). In addition, these nuclei receive collateral fibers of cerebellar inputs on their way to the cerebellar cortex. The **deep cerebellar nuclei**, or **roof nuclei**, are, from lateral to medial, the dentate nucleus, emboliform nucleus, globose nucleus, and fastigial nucleus (see Figure 15.4). A mnemonic is "**Don't Eat Greasy Foods**" (**D**entate, **E**mboliform, **G**lobose, **F**asti-

MNEMONIC

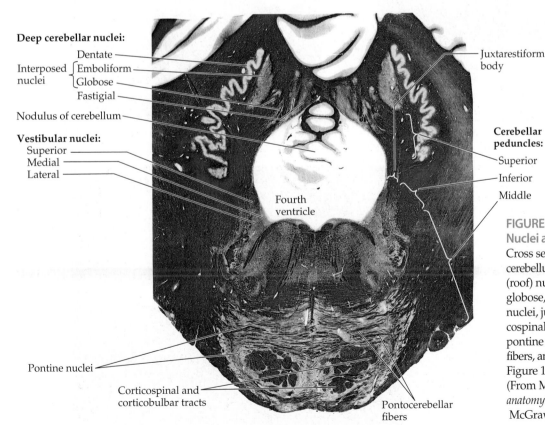

Deep cerebellar nuclei:
Dentate
Interposed { Emboliform
nuclei { Globose
Fastigial
Nodulus of cerebellum

Vestibular nuclei:
Superior
Medial
Lateral

Fourth ventricle

Pontine nuclei

Corticospinal and corticobulbar tracts

Pontocerebellar fibers

Juxtarestiform body

Cerebellar peduncles:
Superior
Inferior
Middle

FIGURE 15.4 Deep Cerebellar Nuclei and Related Structures Cross section through pons and cerebellum showing deep cerebellar (roof) nuclei (dentate, emboliform, globose, and fastigial), vestibular nuclei, juxtarestiform body, corticospinal and corticobulbar fibers, pontine nuclei, pontocerebellar fibers, and cerebellar peduncles (see Figure 14.4C for additional details). (From Martin JH. 1990. *Neuroanatomy: Text and Atlas.* 2nd Ed. McGraw-Hill, New York.)

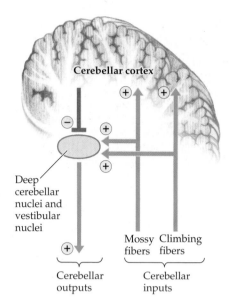

Cerebellar cortex

Deep cerebellar nuclei and vestibular nuclei

Cerebellar outputs

Mossy fibers Climbing fibers

Cerebellar inputs

FIGURE 15.5 Schematic Diagram of Cerebellar Input and Output Pathways
Note that the vestibular nuclei are located in the brainstem rather than cerebellum but function in some ways like additional deep cerebellar nuclei for the inferior vermis and flocculus.

gial). The **dentate nuclei** are the largest of the deep cerebellar nuclei, and they receive projections from the lateral cerebellar hemispheres. The **emboliform** and **globose nuclei** are together called the **interposed nuclei**, and they receive input from the intermediate part of the cerebellar hemispheres. Interestingly, experimental recordings have shown the dentate nucleus to be active just before voluntary movements, while the interposed nuclei are active during and in relation to the movement. The **fastigial nuclei** receive input from the vermis and a small input from the flocculonodular lobe. Most fibers leaving the inferior vermis and flocculi project to the **vestibular nuclei**, (see Figures 12.19 and 15.4) which, though located in the brainstem rather than the cerebellum, function in some ways like additional deep cerebellar nuclei.

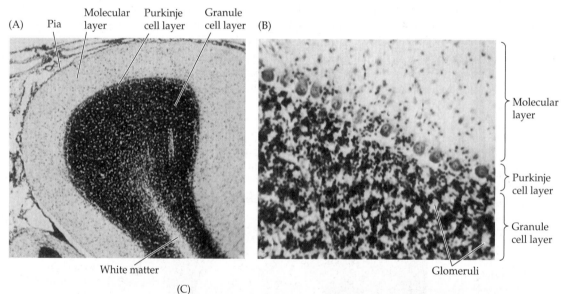

(A) Pia Molecular layer Purkinje cell layer Granule cell layer (B)

White matter

Molecular layer

Purkinje cell layer

Granule cell layer

Glomeruli

FIGURE 15.6 Photomicrographs Showing Cell Layers of Cerebellar Cortex (A,B) Sections through a folium of rhesus monkey cerebellar cortex with Nissl staining showing cell bodies. (A is ×20, B is ×50). (C) Section of rat cerebellar cortex imaged with laser scanning confocal microscopy (×75) following triple immunofluorescent staining. Antibodies to the inositol 1,4,5-trisphosphate receptor (green) are localized mainly to Purkinje cells, while microtubule-associated protein 2 (blue) is localized primarily to Purkinje cell dendrites and to the granule cell layer. Staining for glial fibrillary acidic protein antibodies (red) is seen in the glial astrocytes. (A,B from Parent A. 1996. *Carpenter's Human Neuroanatomy*. 9th Ed. Williams & Wilkins, Baltimore. C courtesy of Tom Deerinck and Mark Ellisman, National Center for Microscopy and Imaging Research.)

(C)

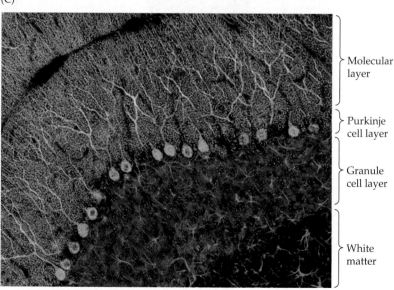

Molecular layer

Purkinje cell layer

Granule cell layer

White matter

Microscopic Circuitry of the Cerebellum

The cerebellar cortex has three layers (Figure 15.6). The **granule cell layer** is tightly packed with small **granule cells** so numerous that they rival the total number of cells in the remainder of the nervous system. The **Purkinje cell layer** contains the cell bodies of large, flask-shaped Purkinje cells. The **molecular layer** consists of the unmyelinated granule cell axons, Purkinje cell dendrites, and several types of interneurons.

There are primarily **two kinds of synaptic inputs** to the cerebellum: mossy fibers and climbing fibers (Figure 15.7; see Figure 15.5). **Mossy fibers**, arise from numerous regions, as we will discuss shortly in the section on cerebellar input pathways. Mossy fibers ascend through the cerebellar white matter to form excitatory synapses onto dendrites of the granule cells. Granule cells, in turn, send axons into the molecular layer, which bifurcate, forming **parallel fibers** that run parallel to the folia (see Figure 15.7). The parallel fibers run perpendicular to the elegant, fanlike dendritic trees of the Purkinje cells. During its course, each parallel fiber forms excitatory synaptic contacts with numerous Purkinje cells. **All output from the cerebellar cortex** is carried by the axons of Purkinje cells into the cerebellar white matter. The Purkinje cells form inhibitory synapses onto the deep cerebellar nuclei and vestibular nuclei, which then convey outputs from the cerebellum to other regions through excitatory synapses (see Figure 15.5).

The other kind of synaptic input to the cerebellar cortex is carried by **climbing fibers**. Climbing fibers arise exclusively from neurons in the contralateral inferior olivary nucleus (see Figure 14.5A). They wrap around the cell body and proximal dendritic tree of Purkinje cells (see Figure 15.7), forming powerful excitatory

REVIEW EXERCISE

1. Cover the right two columns of Table 15.1. For each region of the cerebellum shown in Figure 15.3 that appears in the left column of Table 15.1, list its major functions and the motor pathways influenced.

2. List the deep cerebellar nuclei from lateral to medial. For each cerebellar region in the left column of Table 15.1, state which deep nuclei convey the outputs.

FIGURE 15.7 Summary of Microscopic Circuitry of the Cerebellar Cortex Inputs arrive via mossy and climbing fibers, and outputs leave via Purkinje cell axons. Excitatory neurons include granule cells and inputs from mossy and climbing fibers. Inhibitory neurons include stellate, basket, Golgi, and Purkinje cells.

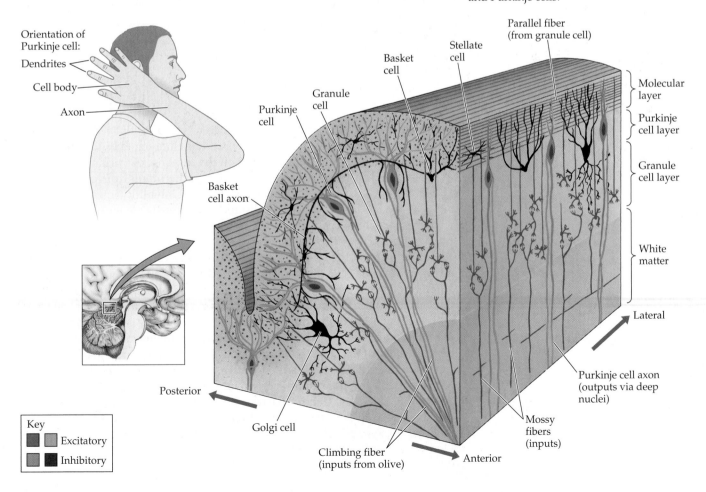

Orientation of Purkinje cell:
Dendrites
Cell body
Axon

Basket cell axon

Purkinje cell

Granule cell

Basket cell

Stellate cell

Parallel fiber (from granule cell)

Molecular layer

Purkinje cell layer

Granule cell layer

White matter

Lateral

Purkinje cell axon (outputs via deep nuclei)

Mossy fibers (inputs)

Anterior

Climbing fiber (inputs from olive)

Golgi cell

Posterior

Key
Excitatory
Inhibitory

FIGURE 15.8 **Enlarged View of Cerebellar Circuit Showing Cerebellar Glomerulus** Synaptic elements of the cerebellar glomerulus include granule cell dendrites that receive excitatory inputs from mossy fiber axonal terminals and inhibitory inputs from Golgi cell axonal terminals.

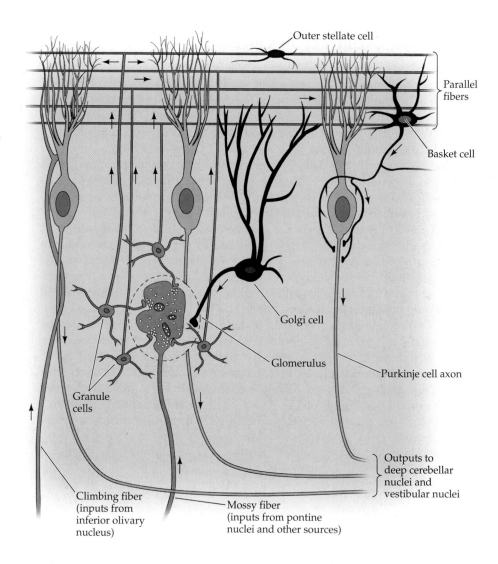

synapses. A single climbing fiber will branch to supply about 10 Purkinje cells; however, each Purkinje cell is excited by just one climbing fiber. Climbing-fiber inputs have a strong modulatory effect on the response of Purkinje cells, causing a sustained decrease in their response to synaptic inputs from parallel fibers.

The cerebellar cortex contains several classes of inhibitory interneurons (Figure 15.8; see Figure 15.7). Basket cells and stellate cells are located in the molecular layer. These cells are excited by synaptic inputs from the granule cell parallel fibers. They then give rise to processes that travel in a rostral–caudal direction, perpendicular to the parallel fibers, to cause lateral inhibition of adjacent Purkinje cells. The stellate cells terminate on Purkinje cell dendrites, while basket cells are named for the strong, inhibitory basketlike connections they form on Purkinje cell bodies. **Golgi cells** are found in the granule cell layer.

The Golgi cells receive excitatory inputs from granule cell parallel fibers in the molecular layer; they then provide **feedback inhibition** onto the granule cell dendrites. This inhibitory feedback tends to shorten the duration of excitatory inputs to the granule cells (enhanced signal resolution in the time domain). Meanwhile, inhibitory lateral connections from stellate and basket cells to adjacent Purkinje cells tend to narrow the spatial extent of excitatory inputs to Purkinje cells (enhanced signal resolution in the spatial domain).

Complex synaptic interactions occur in the granule cell layer in a specialized region called the **cerebellar glomerulus** (see Figure 15.8). Cerebellar glomeruli are visible as small clearings among the granule cells (see Figure 15.6); they contain axons and dendrites encapsulated in a glial sheath. Glomeruli contain two types of inputs (large mossy fiber axon terminals and Golgi cell axon terminals), which form synapses onto one type of postsynaptic cell (granule cell dendrites). To summarize (see Figure 15.7), mossy fibers excite granule cells, which excite the inhibitory Purkinje cells. Climbing fibers excite Purkinje cells directly. Purkinje cells have fanlike dendritic trees, the orientation of which you can imagine by holding up your open hand in a sagittal plane behind your head (see Figure 15.7). Parallel fibers would then pass through your fingers, perpendicular to your palm. The third dimension is provided by basket cells, whose axons would pass from one finger to the next, perpendicular to the parallel fibers.

MNEMONIC

A simple way to remember the excitatory and inhibitory connections of the cerebellar cortex is to recall that all axons projecting upward are excitatory (mossy fibers, climbing fibers, and granule cell parallel fibers), while all axons projecting downward are inhibitory (Purkinje cells, stellate cells, basket cells, and Golgi cells). The outputs of the deep cerebellar nuclei, which are not part of the cerebellar cortex, are excitatory.

MNEMONIC

Cerebellar Output Pathways

As we have discussed already, cerebellar pathways are organized around the three functional regions of the cerebellum: the **lateral** hemispheres, intermediate hemispheres, and vermis plus flocculonodular lobe (see Table 15.1; Figure 15.3A,C). Lesions of the lateral cerebellum therefore affect mainly distal limb coordination, while **medial** lesions affect mainly trunk control, posture, balance, and gait. Another important principle with cerebellar lesions is that **deficits in coordination occur ipsilateral to the lesion**. The reason for this is that the pathways from cerebellum to the lateral motor systems (see Table 15.1) and then to the periphery are "**double crossed**" (see Figure 15.9B). The first crossing occurs as cerebellar outputs exit in the decussation of the superior cerebellar peduncle in the midbrain. The second crossing occurs as the corticospinal and rubrospinal tracts descend to the spinal cord (pyramidal decussation and ventral tegmental decussation, respectively; see Figure 15.9B). Inputs to the cerebellum also follow this organization, so each cerebellar hemisphere receives information about the ipsilateral limbs. In contrast, midline lesions in the cerebellar vermis have effects on the medial motor systems (see Table 15.1; see also Table 6.3). Lesions of the cerebellar vermis do not typically cause unilateral deficits because the medial motor systems influence the proximal trunk muscles bilaterally.

Outputs from the cerebellum are summarized in Figure 15.9 and Table 15.2. Recall that all outputs from the cerebellum are carried by Purkinje cells to the deep cerebellar nuclei or vestibular nuclei (see Figure 15.5). The **lateral cerebellar hemisphere**, involved in motor planning, projects to the dentate nucleus (see Figures 15.4 and 15.9A). The dentate nucleus projects via the superior cerebellar peduncle (see Figure 14.4B), which decussates in the midbrain (see Figure 14.3B) to reach the contralateral **ventral lateral nucleus (VL)** of the thalamus (see Figure 15.9A). The fibers entering this nucleus are called the **thalamic fasciculus**. The more anterior parts of the thalamic fasciculus include outputs from the basal ganglia (as will be described in Chapter 16), which terminate in the anterior VL (VL_a, or VL pars oralis), while cerebellar outputs terminate in the posterior VL (VL_p, or VL pars caudalis) (see Figures 7.7 and 16.6–16.9).* The

---

*Like the basal ganglia, some cerebellar outputs also project to the thalamic ventral anterior (VA) and intralaminar nuclei (see Table 7.3).

FIGURE 15.9 Cerebellar Output Pathways (A) Outputs from the lateral cerebellar hemisphere via the dentate nucleus. (B) Output from the intermediate cerebellar hemisphere via interposed nuclei influencing lateral motor systems. Note the "double crossing" of pathways between cerebellum and spinal cord. (C) Outputs from the cerebellar vermis and flocculonodular lobe via the fastigial nucleus influencing medial motor systems. The flocculonodular lobe and inferior vermis also have direct projections to the vestibular nuclei influencing balance and vestibulo-ocular control.

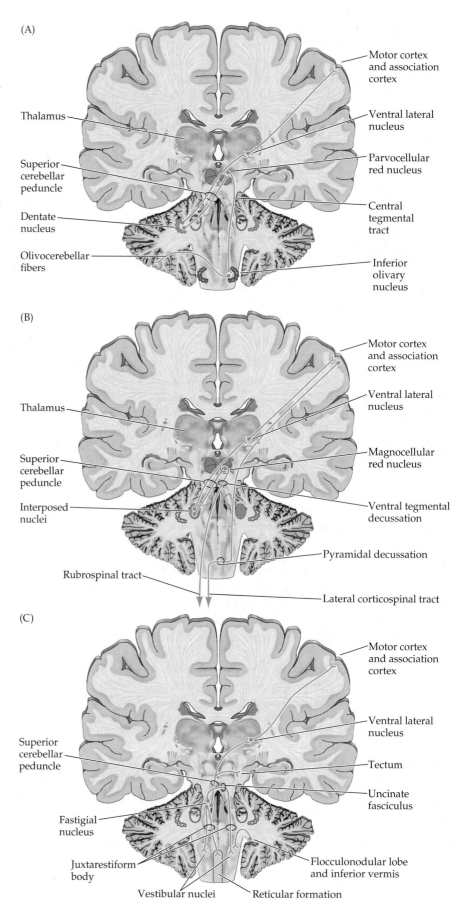

(A)

Motor cortex and association cortex

Thalamus

Ventral lateral nucleus

Superior cerebellar peduncle

Parvocellular red nucleus

Dentate nucleus

Central tegmental tract

Olivocerebellar fibers

Inferior olivary nucleus

(B)

Motor cortex and association cortex

Thalamus

Ventral lateral nucleus

Superior cerebellar peduncle

Magnocellular red nucleus

Interposed nuclei

Ventral tegmental decussation

Rubrospinal tract

Pyramidal decussation

Lateral corticospinal tract

(C)

Motor cortex and association cortex

Superior cerebellar peduncle

Ventral lateral nucleus

Tectum

Uncinate fasciculus

Fastigial nucleus

Juxtarestiform body

Flocculonodular lobe and inferior vermis

Vestibular nuclei

Reticular formation

TABLE 15.2 Main Cerebellar Output Pathways

REGION	DEEP NUCLEI	CEREBELLAR PEDUNCLE	MAIN OUTPUT TARGETS OR EQUIVALENT
Lateral hemispheres	Dentate nucleus	Superior cerebellar peduncle	Ventrolateral nucleus of thalamus (VL), parvocellular red nucleus
Intermediate hemispheres	Interposed nuclei	Superior cerebellar peduncle	VL, magnocellular red nucleus
Vermis	Fastigial nuclei	Superior cerebellar peduncle	VL, tectum
		Uncinate fasciculus,[a] juxtarestiform body[b]	Reticular formation, vestibular nuclei
Inferior vermis and flocculonodular lobe	Vestibular nuclei	Juxtarestiform body[b]	Medial longitudinal fasciculus (eye movement pathways)

[a]The uncinate fasciculus travels with the superior cerebellar peduncle.

[b]The juxtarestiform body travels with the inferior cerebellar peduncle.

VL, in turn, projects to the motor cortex as well as to premotor cortex, supplementary motor area, and parietal lobe to influence motor planning in the **corticospinal** systems (see Figures 6.1 and 6.9). In addition, there is some evidence that outputs from the lateral cerebellum relay in the thalamus to reach the prefrontal association cortex, possibly playing a role in cognitive function. As the output fibers of the dentate nucleus penetrate the red nucleus in the midbrain (see Figure 14.3A), some terminate in the rostral **parvocellular division of the red nucleus**. Recall that the red nucleus has a large rostral parvocellular division that is involved in cerebellar circuitry and a smaller caudal **magnocellular division** that gives rise to the rubrospinal tract. As we will discuss in the next section, the parvocellular division of the red nucleus projects to the inferior olive (see Figure 15.9A).

The **intermediate hemisphere**, involved in the control of ongoing movements of the distal extremities, projects to the emboliform and globose (interposed) nuclei (see Figures 15.4 and 15.9B). Like the dentate nucleus, the interposed nuclei project via the superior cerebellar peduncle to the contralateral thalamic VL, which in turn projects to the motor, supplementary motor, and premotor cortex to influence the **lateral corticospinal tract**. Inputs to the VL from the dentate and interposed nuclei (and from the basal ganglia) do not overlap. The interposed nuclei also project via the superior cerebellar peduncle to the contralateral **magnocellular division of the red nucleus** to influence the rubrospinal tract. The intermediate hemisphere thus influences the lateral motor systems (see Figure 15.9B).

The **cerebellar vermis** and **flocculonodular lobes** influence mainly proximal trunk movements and vestibulo-ocular control, respectively. The vermis influences proximal and trunk muscles through connections to the medial motor pathways (anterior corticospinal, reticulospinal, vestibulospinal, and tectospinal tracts; see Tables 15.1 and 6.3). The vermis projects to the fastigial nucleus (see Figures 15.4 and 15.9C).

Outputs from the fastigial nuclei are carried to some extent by the superior cerebellar peduncle but mainly by fiber pathways called the **uncinate fasciculus** and **juxtarestiform body**, which run along the superior cerebellar peduncle and the inferior cerebellar peduncle, respectively (see Figure 15.9C). The juxtarestiform body (meaning "next to the restiform body") lies on the lateral wall of the fourth ventricle, just medial to the inferior cerebellar peduncle (restiform body) and carries fibers in both directions between the vestibular nuclei and the cerebellum (see Figure 15.4). The uncinate fasciculus (meaning "hook bundle") loops over the superior cerebellar peduncle and then sends fibers to continue caudally via the contralateral juxtarestiform body to reach the contralateral vestibular nuclei (see Figure 15.9C).

Let's briefly review the outputs of the vermis to medial motor systems that influence proximal trunk control (see Table 15.2; Figure 15.9C). Outputs from the vermis reach the fastigial nuclei and are then carried out of the cerebellum via the superior cerebellar peduncle in a crossed pathway. This pathway reaches the thalamic VL and is relayed to the cortex to influence the **anterior corticospinal tract**. Outputs in the same pathway also reach the **tectal area**. Vermis–fastigial outputs also leave the cerebellum via the juxtarestiform body. This pathway reaches the ipsilateral reticular formation and vestibular nuclei to influence the **reticulospinal** and **vestibulospinal tracts**, respectively.

The flocculonodular lobes and inferior vermis are sometimes referred to as the **vestibulocerebellum**, because they project mainly to the vestibular nuclei (via the juxtarestiform body; see Table 15.2). Reciprocal connections between the cerebellum and vestibular nuclei are important for **equilibrium and balance**. In addition, signals from the vestibular nuclei are relayed via the medial longitudinal fasciculus (see Figures 12.19, 14.3, and 14.4) and other **eye movement pathways** to influence vestibulo-ocular reflexes, smooth pursuit, and other eye movements. The inferior vermis and flocculonodular lobes project mainly to the vestibular nuclei but also have a small projection to the fastigial nucleus.

Some fastigial neurons project directly to the upper cervical spinal cord. This small projection appears to be an exception to the principle that the cerebellum and basal ganglia do not project directly to lower motor neurons.

Cerebellar Input Pathways

Inputs to the cerebellum (Table 15.3) arise from widespread areas in the nervous system. These inputs reach the cerebellum from (1) virtually all areas of the cerebral cortex; (2) multiple sensory modalities, including the vestibular, visual, auditory, and somatosensory systems; (3) brainstem nuclei; and (4) the spinal cord. Inputs to the cerebellum have a rough somatotopic organization, with the ipsilateral body represented in both the anterior and posterior lobes, as shown in Figure 15.10. Recall that cerebellar inputs are carried by mossy fibers, except for those from the inferior olivary nucleus, which are carried by climbing fibers. In addition, most inputs to the cerebellar cortex give rise to collaterals that synapse in the deep cerebellar nuclei (see Figure 15.5).

A major source of input consists of **corticopontine fibers** from the frontal, temporal, parietal, and occipital lobes that travel in the internal capsule and

TABLE 15.3 Main Cerebellar Input Pathways

INPUT PATHWAY	MAIN ORIGIN(S) OF INPUT	CELLS PROJECTING TO CEREBELLUM	CEREBELLAR PEDUNCLE OR EQUIVALENT
Pontocerebellar fibers	Cortex	Pontine nuclei	Middle cerebellar peduncle
Spinocerebellar pathways			
Dorsal spinocerebellar tract	Leg proprioceptors	Nucleus dorsalis of Clark	Inferior cerebellar peduncle
Cuneocerebellar tract	Arm proprioceptors	External cuneate nucleus	Inferior cerebellar peduncle
Ventral spinocerebellar tract	Leg interneurons	Spinal cord neurons	Superior cerebellar peduncle
Rostral spinocerebellar tract	Arm interneurons	Spinal cord neurons	Superior and inferior cerebellar peduncles
Climbing fibers	Red nucleus, cortex, brainstem, spinal cord	Inferior olivary nucleus	Inferior cerebellar peduncle
Vestibular inputs	Vestibular system	Vestibular ganglia, vestibular nuclei	Juxtarestiform body

cerebral peduncles (see Figure 14.3A). The primary sensory and motor cortices and part of the visual cortex make the largest contributions to the corticopontine fibers. The corticopontine fibers travel to the ipsilateral pons and synapse in the **pontine nuclei**. These are scattered areas of gray matter in the ventral pons interspersed among the descending corticospinal and corticobulbar fibers (see Figure 15.4; see also Figure 14.4). **Pontocerebellar fibers** then cross the midline to enter the contralateral middle cerebellar peduncle and give rise to mossy fibers that reach almost the entire cerebellar cortex (except for the nodulus).

Another major source of input to the cerebellum consists of **spinocerebellar fibers** (Figure 15.11; see also Table 15.3), which travel in four tracts: the **dorsal and ventral spinocerebellar tracts** for the lower extremities and the **cuneocerebellar and rostral spinocerebellar tracts** for the upper extremities and neck. These spinocerebellar pathways provide feedback information of two different kinds to the cerebellum:

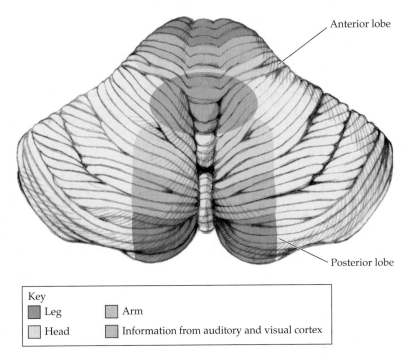

FIGURE 15.10 Somatotopic Organization of Multimodal Inputs to the Cerebellum

Key
- ■ Leg
- ■ Arm
- □ Head
- ■ Information from auditory and visual cortex

1. Afferent information about limb movements is conveyed to the cerebellum by the **dorsal spinocerebellar tract** for the lower extremity and by the **cuneocerebellar tract** for the upper extremity and neck.
2. Information about activity of spinal cord interneurons, thought to reflect the amount of activity in descending pathways, is carried by the **ventral spinocerebellar tract** for the lower extremity and by the **rostral spinocerebellar tract** for the upper extremity.

The **dorsal spinocerebellar tract** ascends in the dorsolateral funiculus, near the surface of the spinal cord, just lateral to the lateral corticospinal tract (see Figure 15.11; see also Figure 7.4). Large, myelinated axons of primary sensory neurons carrying proprioceptive, touch, and pressure sensation from the lower extremities and trunk enter via the dorsal roots and ascend in the gracile fasciculus. Rather than continuing in the posterior columns, some of these fibers form synapses in the **nucleus dorsalis of Clark** (see Figure 15.11; see also Figure 6.4D). This is a long column of cells that runs in the dorsomedial spinal cord gray matter intermediate zone, from C8 to L2 or L3. Fibers arising from the nucleus dorsalis of Clark ascend ipsilaterally in the dorsal spinocerebellar tract (see Figure 15.11). These fibers give rise to mossy fibers that travel to the ipsilateral cerebellar cortex via the inferior cerebellar peduncle. Unlike sensory inputs in the posterior column fibers, the spinocerebellar afferents do not reach conscious perception.

The upper-extremity equivalent of this pathway is the **cuneocerebellar tract**. Large-diameter fibers from the upper extremities enter the cuneate fasciculus and ascend ipsilaterally to synapse in the **external** (or **accessory**, or **lateral**) **cuneate nucleus**, located in the medulla, just lateral to the cuneate nucleus (see Figure 15.11; see also Figure 14.5B). The external cuneate nucleus is the upper-extremity analog of the nucleus dorsalis of Clark. From the external cuneate nucleus, cuneocerebellar fibers ascend in the inferior cerebellar peduncle to the ipsilateral cerebellum. Note that for both the dorsal spinocerebellar tract and the cuneocerebellar tract, unconscious information from the extremities reaches

FIGURE 15.11 Spinocerebellar Pathways The dorsal spinocerebellar, cuneocerebellar, and ventral spinocerebellar tracts. The rostral spinocerebellar tract is not shown.

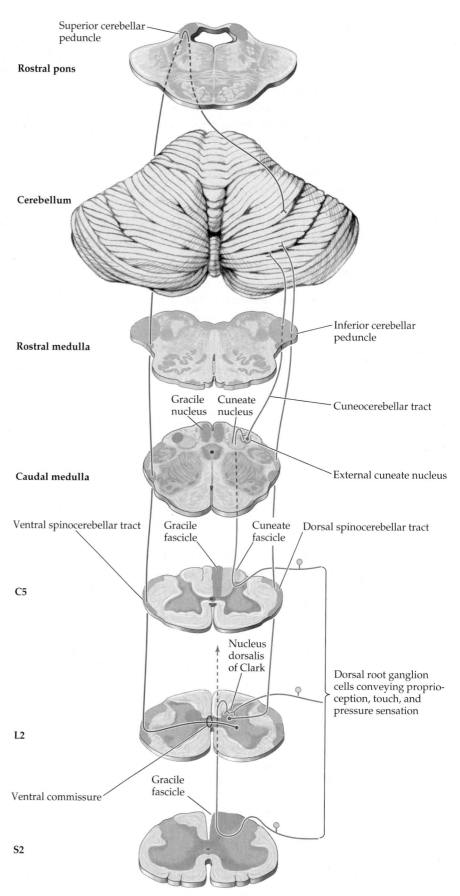

the ipsilateral cerebellum. These pathways provide rapid feedback to the cerebellum about ongoing movements, allowing fine adjustments to be made.

The **ventral spinocerebellar tract** arises from neurons called spinal border cells, located along the outer edge of the central gray matter, and from scattered neurons in the spinal cord intermediate zone (see Figure 15.11). Axons from these cells cross over in the ventral commissure of the spinal cord to ascend in the ventral spinocerebellar tract, just ventral to the dorsal spinocerebellar tract and peripheral to the anterolateral systems (see Figure 7.4). The majority of these fibers then join the superior cerebellar peduncle and cross over a second time, to reach the cerebellum ipsilateral to the side where the pathway began (see Figure 15.11). The **rostral spinocerebellar tract** is the least well characterized of these pathways, but it appears to be the upper-extremity equivalent of the ventral spinocerebellar tract, and it enters the cerebellum through both the inferior and superior cerebellar peduncles. Note that like the output circuits described in the preceding section, the spinocerebellar inputs are either ipsilateral or "double crossed," explaining why cerebellar lesions cause ipsilateral limb ataxia.

The **inferior olivary nuclear complex** gives rise to **olivocerebellar fibers** that cross the medulla to enter the contralateral cerebellum (see Figure 14.5A). These fibers form the major portion of the inferior cerebellar peduncle and terminate as climbing fibers throughout the cerebellum (see Figure 15.7). The parvocellular red nucleus projects to the inferior olive via the central tegmental tract (see Figure 15.9A; see also Figures 14.3–14.5). Note that the parvocellular red nucleus receives inputs from the contralateral dentate nucleus. Thus, a complete loop is formed from lateral cerebellum, to dentate nucleus, to contralateral parvocellular red nucleus, to inferior olive via the central tegmental tract, and then crossing back via the inferior cerebellar peduncle to the original cerebellar hemisphere. The inferior olivary nuclear complex also receives inputs from the cerebral cortex, from other brainstem nuclei, and from the spinal cord. The lateral reticular nucleus is located just dorsal to the inferior olive (see Figure 14.5A,B) and receives similar input connections. The lateral reticular nucleus also projects to the cerebellum via the inferior cerebellar peduncle, but it gives rise to mossy fibers instead of climbing-fiber terminals.

Primary vestibular sensory neurons in Scarpa's **vestibular ganglia** (see Figure 12.15) and secondary vestibular neurons in the **vestibular nuclei** project to the ipsilateral inferior cerebellar vermis and flocculonodular lobe via the juxtarestiform body (see Figure 15.4). Connections between the vestibular system and cerebellum are important in the control of balance and equilibrium, as well as in vestibulo-ocular reflexes. The **flocculus** also receives visual inputs related to retinal slip (disparity of intended and perceived target image) that are important for the control of smooth pursuit eye movements.

Noradrenergic inputs from the locus ceruleus and serotonergic inputs from the raphe nuclei project diffusely throughout the cerebellar cortex (see Figures 14.11 and 14.12). These inputs are not conveyed by mossy-fiber or climbing-fiber terminals, and they are thought to play a neuromodulatory role.

Vascular Supply to the Cerebellum

Blood supply to the cerebellum is provided by three branches of the vertebral and basilar arteries (see Figure 15.2):

1. The **posterior inferior cerebellar artery** (**PICA**)
2. The **anterior inferior cerebellar artery** (**AICA**)
3. The **superior cerebellar artery** (**SCA**)

> ### REVIEW EXERCISE
>
> For each of the four spinocerebellar pathways listed in Table 15.3, name the origin of input, the location of neurons projecting to the cerebellum and the cerebellar peduncle.

Vascular areas

Key
- ■ PICA territory
- ☐ AICA territory
- ▨ SCA territory

FIGURE 15.12 Vascular Territories of Cerebellum Surface view, showing territories supplied by the superior cerebellar artery (SCA), anterior inferior cerebellar artery (AICA), and posterior inferior cerebellar artery (PICA). (A) Dorsal view. (B) View of the cerebellum removed from the brainstem to reveal the ventral surface.

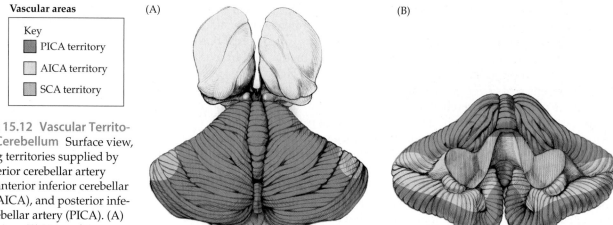

The PICA usually arises from the vertebral artery, the AICA from the lower basilar, and the SCA from the top of the basilar, just below the posterior cerebral artery (see Figure 15.2; see also Figure 14.18).

As these arteries wrap around the brainstem, they supply portions of the lateral medulla and pons in addition to the cerebellum. The **posterior inferior cerebellar artery** supplies the lateral medulla, most of the inferior half of the cerebellum, and the inferior vermis (see Figure 15.2, Figure 15.12, and Figure 15.13A,B). Recall from Chapter 14 that a variable portion of the lateral medulla is also supplied by branches of the vertebral artery (see Figure 14.21D). The **anterior inferior cerebellar artery** supplies the inferior lateral pons, the middle cerebellar peduncle, and a strip of the ventral (anterior) cerebellum between the territories of the PICA and the SCA, including the flocculus (see Figure 15.12 and Figure 15.13B,C; see also Figure 14.21C). The **superior cerebellar artery** supplies the upper lateral pons, the superior cerebellar peduncle, most of the superior half of the cerebellar hemisphere, including the deep cerebellar nuclei, and the superior vermis (see Figure 15.12 and Figure 15.13C,D; see also Figure 14.21B).

Key
- ▨ PICA territory
- ☐ AICA territory
- ▨ SCA territory

FIGURE 15.13 Vascular Territories of the Cerebellar Arteries Reviewed in Axial Sections (A) Caudal cerebellum and mid-medulla. (B) Caudal cerebellum and rostral medulla. (C) Mid-cerebellum and mid-pons. (D) Rostral pons and rostral cerebellum. PICA, posterior inferior cerebellar artery; AICA, anterior inferior cerebellar artery; SCA, superior cerebellar artery.

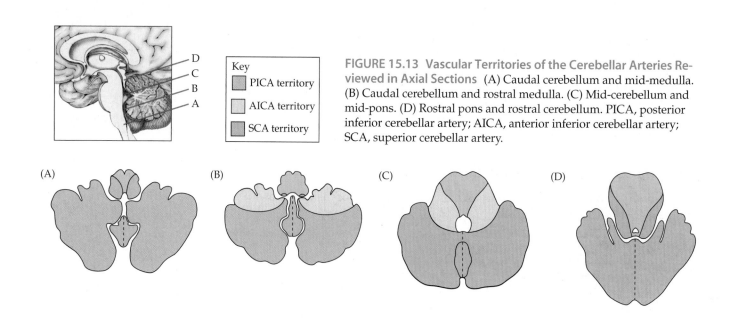

KEY CLINICAL CONCEPT

15.1 CEREBELLAR ARTERY INFARCTS AND CEREBELLAR HEMORRHAGE

Infarcts are more common in the PICA and SCA territories than in the AICA territory. Patients with cerebellar infarcts typically present with vertigo, nausea and vomiting, horizontal nystagmus, limb ataxia (see KCC 15.2), unsteady gait, and headache, which can be occipital, frontal, or in the upper cervical regions. In addition, as we saw in Chapter 14 (see KCC 14.3), many of the important clinical symptoms and signs of cerebellar artery infarcts often result from infarction of the lateral medulla or pons, rather than the cerebellum itself. These include trigeminal and spinothalamic sensory loss, Horner's syndrome, and other findings (see Tables 14.7 and 14.8). Recall that unilateral hearing loss can occur with AICA infarcts because the internal auditory artery often arises from AICA (see Figure 14.18A). Note that infarction of the lateral medulla or pons can cause ataxia because of involvement of the cerebellar peduncles, even if the cerebellum itself is spared.

Infarcts that spare the lateral brainstem and involve mainly the cerebellum itself are more common with SCA infarcts than with PICA or AICA infarcts. Therefore, infarcts causing unilateral (ipsilateral) ataxia with little or no brainstem signs are most commonly in the SCA territory. Infarcts in the PICA and AICA more often involve the lateral brainstem in addition to the cerebellum. In addition, infarcts of the lateral pons or medulla that spare the cerebellum can also sometimes occur, more commonly with the PICA and AICA than with the SCA. Possible mechanisms for infarcts that spare the cerebellum include anastomotic connections sparing the cerebellum or selective occlusion of branches to the lateral brainstem with sparing of cerebellar branches.

Large cerebellar infarcts in the PICA or SCA territories can cause swelling of the cerebellum. The resulting compression of the fourth ventricle can cause hydrocephalus (see KCC 5.7). In addition, compression in the tight space of the posterior fossa (see Figure 15.1) is a life-threatening emergency because the respiratory centers and other vital brainstem structures may be compromised. Large cerebellar infarcts therefore often require surgical decompression of the posterior fossa, including resection of portions of the infarcted cerebellum. Hemorrhage into cerebellar white matter can also occur following an infarct and can cause brainstem compression. Once again, this highlights the importance of carefully evaluating patients who present with vertigo (see KCC 12.6). If vertigo is caused by a large PICA infarct sparing the brainstem, the seriousness of the condition may be overlooked until a few days after onset, when cerebellar swelling and posterior fossa compression develop.

Cerebellar hemorrhage, like spontaneous intraparenchymal hemorrhage in other brain regions, can occur in the setting of chronic hypertension, arteriovenous malformation, hemorrhagic conversion of an ischemic infarct, metastases, or other causes (see KCC 5.6). Patients usually present with headache, nausea, vomiting, ataxia, and nystagmus. If the hemorrhage is large, obstruction of the fourth ventricle can cause hydrocephalus (see KCC 5.7) accompanied by sixth-nerve palsies and impaired consciousness and may eventually cause brainstem compression and death. Sometimes, cerebellar hemorrhage initially presents only with gastrointestinal symptoms of nausea and vomiting, a condition that has been termed "fatal gastroenteritis." Prompt identification and treatment of cerebellar hemorrhage is therefore crucial. Hydrocephalus can be treated with ventriculostomy (see KCC 5.7); however, this carries some risk of upward transtentorial herniation as the posterior fossa hemorrhage and edema expand. For large cerebellar hemor-

REVIEW EXERCISE

Color the regions corresponding to the SCA, AICA, and PICA territories below:

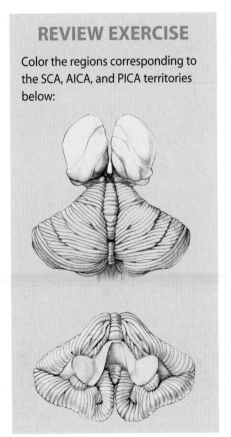

rhages, surgical evacuation of the hemorrhage and decompression of the posterior fossa are often necessary. Patients with cerebellar hemorrhage who are treated promptly usually have a good functional outcome. ■

15.2 CLINICAL FINDINGS AND LOCALIZATION OF CEREBELLAR LESIONS

In this section we will discuss the clinical manifestations and localization of cerebellar lesions. First we will review basic definitions of ataxia and some simple localizing principles. Next we will discuss common symptoms and signs of cerebellar disorders and the expected findings on neurologic examination. Causes of cerebellar disorders are discussed in KCC 15.1 and 15.3. **Ataxia** means, literally, "lack of order." This term refers to the disordered contractions of agonist and antagonist muscles and the lack of normal coordination between movements at different joints, seen in patients with cerebellar dysfunction. As shown in **Figure 15.14A**, even simple movements require coordination of agonists and antagonists around multiple joints in

(A)

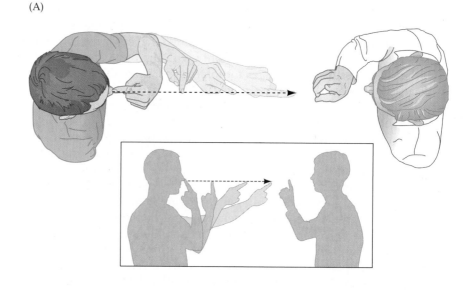

(B)

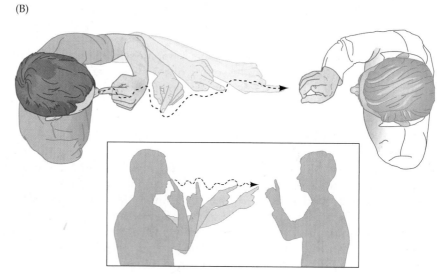

FIGURE 15.14 Finger–Nose–Finger Test (A) Normal individual. (B) An individual with appendicular ataxia. Note that coordinated actions of agonist and antagonist muscles acting on multiple joints, including the shoulder, elbow, and wrist, are required to smoothly perform this movement in a normal fashion.

order to follow a normal smooth path. In ataxia, the movements have an irregular, wavering course that seems to consist of continuous overshooting, overcorrecting, and then overshooting again around the intended trajectory (Figure 15.14B). Ataxic movements have abnormal timing (dysrhythmia) and abnormal trajectories through space (dysmetria). Let's review a few principles of localization with cerebellar lesions.

Truncal Ataxia versus Appendicular Ataxia

Lesions confined to the cerebellar vermis affect primarily the medial motor systems (see Table 15.1). Patients with such lesions therefore often have a **wide-based, unsteady, "drunk-like" gait**, with no other significant abnormalities on exam. This condition is referred to as **truncal ataxia**. In severe truncal ataxia, patients may even have difficulty sitting up without support.

In contrast, lesions of the intermediate and lateral portions of the cerebellar hemisphere affect the lateral motor systems. Therefore, these patients have ataxia on movement of the extremities like that shown in Figure 15.14B, referred to as **appendicular ataxia**. Often, lesions extend to involve both the vermis and the cerebellar hemispheres, and truncal and appendicular ataxia frequently coexist in the same patient. Interestingly, a unilateral lesion in the lateral portion of the cerebellar hemisphere (see Figure 15.3) may produce no appreciable deficit. More severe and lasting deficits are seen with lesions of the intermediate hemisphere, vermis, deep nuclei, or cerebellar peduncles.

Ipsilateral Localization of Ataxia

As we saw in the anatomical review earlier in this chapter, afferent and efferent cerebellar connections involved in the lateral motor systems either are ipsilateral or cross twice between the cerebellum and spinal cord (see Figures 15.9B and 15.11). Therefore, lesions of the cerebellar hemispheres cause ataxia in the extremities **ipsilateral** to the side of the lesion. Similarly, in lesions of the cerebellar peduncles, the ataxia is ipsilateral to the lesion. In contrast, cerebellar lesions affecting the medial motor system cause truncal ataxia, which is a bilateral disorder. Nevertheless, patients with truncal ataxia often tend to fall or sway toward the side of the lesion.

False Localization of Ataxia

Ataxia is often caused by lesions outside the cerebellum that involve the cerebellar input or output pathways. **Lesions of the cerebellar peduncles or pons** can produce severe ataxia, even without involvement of the cerebellar hemispheres. **Hydrocephalus**, which may damage frontopontine pathways, and **lesions of the prefrontal cortex** can both result in gait abnormalities that resemble cerebellar truncal ataxia, as can **disorders of the spinal cord**.

Ataxia-hemiparesis is a syndrome often caused by lacunar infarcts (see KCC 10.4; Table 10.3), in which patients have a combination of unilateral upper motor neuron signs and ataxia, usually affecting the same side. In ataxia-hemiparesis, the ataxia and hemiparesis are both usually contralateral to the side of the lesion. Ataxia-hemiparesis is most often caused by lesions in the corona radiata, internal capsule, or pons that involve both corticospinal and corticopontine fibers. However, it can also be seen in lesions of the frontal lobes, parietal lobes, or sensorimotor cortex, or in midbrain lesions that involve fibers of the superior cerebellar peduncle or red nucleus (see KCC 14.3).

Sensory ataxia occurs when the posterior column–medial lemniscal pathway is disrupted, resulting in loss of joint position sense. Patients with sensory ataxia may have ataxic-appearing overshooting movements of the limbs and a wide-based, unsteady gait resembling that of patients with cerebellar lesions. Unlike cerebellar patients, however, these patients have impaired joint position sense on exam. In addition, sensory ataxia can be improved significantly with visual feedback and is worse with the eyes closed or in the dark. Sensory ataxia is usually caused by lesions of the peripheral nerves or posterior columns, which, if one-sided, result in ipsilateral ataxia. However, it is also occasionally seen in lesions of the thalamus, thalamic radiations, or somatosensory cortex, which cause contralateral ataxia.

Symptoms and Signs of Cerebellar Disorders

Cerebellar disorders affect medial motor systems, lateral motor systems, eye movements, vestibular pathways, and other circuits, resulting in characteristic symptoms as well as specific findings on neurologic examination.

SYMPTOMS OF CEREBELLAR DISORDERS Patients with cerebellar lesions often complain of nausea, vomiting, vertigo, slurred speech, unsteadiness, or uncoordinated limb movements. Headache may occur in the occipital, frontal, or upper cervical areas and is usually on the side of the lesion. Lesions causing incipient tonsillar herniation (see KCC 5.4) can cause depressed consciousness, brainstem findings, hydrocephalus, or head tilt. Head tilt is also seen in cerebellar lesions extending to the anterior medullary velum (see Figures 12.2B and 15.1), which may affect the trochlear nerves.

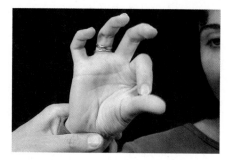

Precision finger tap

OTHER ABNORMALITIES CAN CONFOUND THE CEREBELLAR EXAM When examining a patient with a suspected cerebellar disorder, it is essential to first look carefully for upper or lower motor neuron signs, sensory loss, or basal ganglia dysfunction. Abnormalities in these other systems will significantly affect cerebellar testing. **Upper motor neuron** findings make the exam more difficult, since both corticospinal and cerebellar lesions can cause slow, clumsy movements of the extremities. In addition, if there is severe upper or **lower motor neuron** weakness, cerebellar testing may not be possible. Some tests that require little strength may be helpful, such as repeatedly tapping the tip of the index finger to the crease of the thumb and looking at accuracy (see **neuroexam.com Video 63**). In cerebellar disorders, the tip of the index finger tends to hit a different spot on the thumb each time. **Loss of joint position sense** can cause sensory ataxia. However, the loss of position sense must be severe for significant ataxia to be present, and as mentioned already, sensory ataxia usually improves with visual feedback. **Movement disorders** (see KCC 16.1) such as parkinsonism associated with basal ganglia dysfunction can cause slow, clumsy movements, or gait unsteadiness, which can confound the cerebellar exam. Other movement disorders such as tremor or dyskinesias also make interpretation of cerebellar testing more difficult. We will discuss the typical exam findings in patients with basal ganglia disorders—and how they differ from cerebellar disorders—in Chapter 16.

Finger–nose–finger

TESTING FOR APPENDICULAR ATAXIA Numerous tests can be used to detect appendicular ataxia. These were discussed in Chapter 3 and are demonstrated through video segments on **neuroexam.com**. Most abnormalities can be described as a combination of dysmetria and dysrhythmia. **Dysmetria** is abnormal undershoot or overshoot (also known as past pointing) during movements toward a target. **Dysrhythmia** is abnormal rhythm and timing of movements. The best-known tests for ataxia are the finger–nose–finger test and the heel–shin test.

In the finger–nose–finger test (see **neuroexam.com Video 64**), the patient touches their nose and then the examiner's finger alternately (see Figure

15.14). The examiner can increase the sensitivity of this test by holding the target finger at the limit of the patient's reach, and by moving the target finger to a different position each time the patient touches their nose.

In the heel–shin test (see **neuroexam.com Video 65**), the patient rubs one heel up and down the length of the opposite shin in as straight a line as possible. This test should be done supine so that gravity does not contribute to the downward movement. Variations include tapping the heel repeatedly on the same spot just below the knee or doing a test similar to the finger–nose–finger test, with the patient's foot alternately touching their knee and the examiner's finger.

Heel–shin test

Rapid tapping of the fingers together, of the hand on the thigh (see **neuroexam.com Video 52**) or of the foot on the floor (see **neuroexam.com Video 53**) are good tests for dysrhythmia. In addition, testing the accuracy of tapping the tip of the index finger on the thumb crease, as discussed above (see **neuroexam.com Video 63**) can be useful for identifying both dysmetria and dysrhythmia. Abnormalities of other rapid alternating movements, such as alternately tapping one hand with the palm and dorsum of the other hand (see **neuroexam.com Video 62**) have been called **dysdiadochokinesia** or adiadochokinesia.

The examiner can test for **overshoot**, or **loss of check**, by having the patient raise both arms suddenly from their lap or lower them suddenly to the level of the examiner's hand (see **neuroexam.com Video 66**). Alternatively, the examiner can apply pressure to the patient's outstretched arms and then suddenly release it. Cerebellar lesions can be associated with an irregular large-amplitude **postural tremor** that occurs when the limb muscles are activated to hold a particular position—for example, both arms held outstretched. This characteristic postural tremor seen in disorders of the cerebellar pathways has been called a **rubral tremor** in the past, although more recently the involvement of the red nucleus has been questioned. In addition, the appendicular ataxia during movements toward a target that was described earlier (see Figure 15.14B) is sometimes referred to as an **action tremor** or **intention tremor**. Tremors will be discussed further in Chapter 16. Cerebellar disorders are also often associated with **myoclonus**, a sudden rapid-movement disorder that will be discussed in Chapter 16 as well (see KCC 16.1).

Overshoot

TESTING FOR TRUNCAL ATAXIA Patients with truncal ataxia have a **wide-based, unsteady gait** that resembles the gait of a drunk or of a toddler learning to walk (see also KCC 6.5; Table 6.6). This resemblance is not just coincidental: Alcohol impairs cerebellar function, and cerebellar pathways are not yet fully myelinated in infancy. Truncal ataxia can be demonstrated by having the patient perform a **tandem gait**, in which the heel touches the toe with each step, forcing the patient to assume a narrow stance (see **neuroexam.com Video 68**). During their walking, patients tend to fall or to deviate toward the side of the lesion. The **Romberg test** (see **neuroexam.com Video 67**) can also be helpful in identifying truncal ataxia and may help differentiate cerebellar lesions from lesions of the vestibular or proprioceptive systems (see Chapter 3). A peculiar tremor of the trunk or head called **titubation** can occur with midline cerebellar lesions.

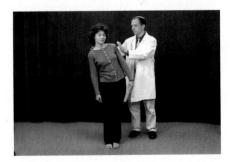

Romberg test

EYE MOVEMENT ABNORMALITIES Patients with cerebellar lesions may exhibit **ocular dysmetria**, in which saccades overshoot or undershoot their target. **Slow saccades** are present in some degenerative conditions of the cerebellum. In addition, during attempted smooth pursuit eye movements there may be jerky saccadic intrusions, particularly when the flocculonodular lobe is involved. **Nystagmus** is often present, usually of the gaze paretic type, meaning that when the patient looks toward a target in the periphery, slow phases occur toward the primary position and fast phases occur back toward the target. Unlike the nystagmus in peripheral vertigo, the nystagmus in cerebellar lesions may change directions, depending on the direction of gaze (see Table 12.7). Vertical nystagmus may also be present in cerebellar disorders.

In normal individuals the vestibulo-ocular reflex (see Chapter 13) is often suppressed by visual inputs. For example, when normal individuals read a newspaper while riding a train as it pulls into the station, their eyes do not jump off the page even though their inner ears detect a large deceleration. This normal **suppression of the vestibulo-ocular reflex (VOR) can be impaired** in cerebellar lesions, particularly if the flocculonodular lobe is involved. There are several ways to test for such impairment, such as asking the patient to fixate on their thumbs, held together at arm's length while they rotate their upper body; asking the patient to fixate on the far end of a straw held in their mouth while they turn their head side-to-side; or doing similar tests using a swivel chair. Younger pediatric patients can be picked up and turned from side to side while they fixate on the examiner's face. Normal patients exhibit no nystagmus during these tests, but nystagmus occurs in patients with impaired suppression of the VOR.

Paraneoplastic cerebellar disorders, encephalitis, and some other conditions can be associated with unusual eye movements called opsoclonus or ocular flutter, in which the eyes show brief bursts of oscillating movements during fixation.

SPEECH ABNORMALITIES Speech can have an ataxic quality in cerebellar disorders, with irregular fluctuations in both rate and volume. This is sometimes called **scanning** or **explosive speech**. Instead of ataxic speech, cerebellar dysfunction can also cause speech to be slurred and difficult to understand. Alcohol intoxication once again provides a good example of this phenomenon.

TABLE 15.4A Differential Diagnosis of Ataxia in Adults	
ACUTE OR RECURRENT ATAXIA	**CHRONIC OR PROGRESSIVE ATAXIA**

Toxin ingestion
    Ethanol; anticonvulsants; other medications
	Cerebellar or other metastasis
	    Lung carcinoma; breast carcinoma; melanoma;
	    other tumors

Ischemic stroke	Multiple sclerosis
Hemorrhagic stroke	Chronic toxin exposure
Basilar migraine	Alcohol/nutritional deprivation; phenytoin; mercury;
Benign paroxysmal vertigo	thallium; toluene (glue, spray paint)
Conversion disorder	Degenerative disorders
Postconcussion syndrome	Olivopontocerebellar atrophy; Machado–Joseph disease
Traumatic hematoma	(SCA-3ª); dentatorubropallidoluysian atrophy (DRPLA);
Multiple sclerosis	other hereditary ataxias including ªSCA-2,5 6, 8, 17, and multiple
Infectious or postinfectious cerebellitis	systems atrophy type-C
Celiac disease	Celiac disease
Brainstem encephalitis	Progressive multifocal leukoencephalopathy (PML)
Miller Fisher Guillain–Barré variant	Toxoplasmosis
Wernicke's encephalopathy	Creutzfeldt–Jakob disease
Toxoplasmosis	Arteriovenous malformation
Brain abscess	Paraneoplastic syndrome (especially in breast or
Brain tumor (usually chronic)	ovarian cancer)
Hereditary episodic ataxias	Wilson's disease
Metabolic disorders	Vitamin E deficiency
Hartnup disease; maple syrup urine disease; pyruvate dehydrogenase deficiency	Adult forms of disorders listed in Table 15.4B
Paraneoplastic syndrome (especially in breast or ovarian cancer)	

OTHER FINDINGS Muscle tone may be somewhat decreased in cerebellar disease, and reflexes can have a "pendular" (swinging) quality. These findings are often not emphasized in the setting of other, more dramatic abnormalities caused by cerebellar lesions. Accumulating evidence suggests that cerebellar damage may contribute to a number of abnormalities in **higher order cognitive function** including impaired attention, processing speed, motor learning, language, and visuospatial processing. However, some argue that the cerebellum does not contribute directly to these functions, and thus this remains a topic of ongoing investigation. ∎

KEY CLINICAL CONCEPT

15.3 DIFFERENTIAL DIAGNOSIS OF ATAXIA

Ataxia can be caused by a wide variety of disorders. The differential diagnosis depends on the age of the patient and on the time course of evolution of the ataxia. In adults (Table 15.4A) the most common causes of acute ataxia are toxin ingestion and ischemic or hemorrhagic stroke. Chronic ataxia in adults is often caused by cerebrovascular disease, brain metastases, chronic toxin exposure (especially to medications or alcohol), multiple sclerosis, or degenerative disorders of the cerebellum and cerebellar pathways. Recent evidence suggests that celiac disease without gastrointestinal symptoms may be an important cause of ataxia in patients with no clear diagnosis. In the pediatric population (Table 15.4B), acute ataxia is most often caused by accidental drug ingestion, varicella-associated cerebellitis, or migraine. Chronic or progressive ataxia in children can be caused by cerebellar astrocytoma, medulloblastoma, Friedreich's ataxia, ataxia-telangiectasia, or a va-

TABLE 15.4B Differential Diagnosis of Ataxia in Infants and Children	
ACUTE OR RECURRENT ATAXIA	**CHRONIC OR PROGRESSIVE ATAXIA**

Toxin ingestion
  Anticonvulsants; other medications; ethanol
	Posterior fossa tumors
	  Medulloblastoma; cerebellar astrocytoma; ependymoma;
	  hemangioblastoma; pontine glioma

Infectious or postinfectious cerebellitis	Congenital malformations
	  Dandy–Walker malformation; cerebellar aplasias;
	  chiari malformation

Brainstem encephalitis	

Basilar migraine	Degenerative disorders
	  Friedreich's ataxia; ataxia-telangiectasia; olivoponto-
	  cerebellar atrophy; Machado–Joseph disease ([a]SCA-3);
	  dentatorubropallidoluysian atrophy (DRPLA); [a]SCA-6 and 17

Benign paroxysmal vertigo	
Conversion disorder	
Postconcussion syndrome	
Traumatic hematoma	Multiple sclerosis

Epileptic pseudoataxia	Metabolic disorders
	  Abetalipoproteinemia; adrenoleukodystrophy;
	  juvenile GM$_2$ gangliosidosis; juvenile sulfatide lipidosis;
	  Hartnup disease; maple syrup urine disease; pyruvate
	  dehydrogenase deficiency; Marinesco–Sjögren syndrome;
	  Ramsay Hunt syndrome; respiratory chain disorders;
	  sea-blue histiocytosis

Brain tumor (usually presents as chronic progressive ataxia)	
Hereditary episodic ataxias	

Metabolic disorders
  Hartnup disease; maple syrup urine disease;
  pyruvate dehydrogenase deficiency	

Neuroblastoma syndrome	
Miller Fisher Guillain–Barré variant	
Multiple sclerosis	
Hemorrhagic stroke	
Ischemic stroke	
Kawasaki disease	

[a]SCA, spinal cerebellar ataxia.
Source: Modified from Fenichel, GM. 2009. *Clinical Pediatric Neurology: A Signs and Symptoms Approach.* 6th Ed. Elsevier: Saunders, Philadelphia.

riety of other conditions (see Table 15.4). There has been an increase in identified genes causing hereditary ataxia syndromes, often called spinocerebellar ataxia (SCA), in both children and adults. Several of the gene defects encode polyglutamine trinucleotide repeats (see KCC 16.3), and inheritance can be autosomal dominant, recessive, or X-linked. Although ataxia is usually a prominent early feature of these disorders, a number of other clinical features can be observed including dementia, parkinsonism, chorea, and corticospinal dysfunction depending on the brain regions involved. ■

CLINICAL CASES

CASE 15.1 SUDDEN ONSET OF UNILATERAL ATAXIA

MINICASE

A 70-year-old semiretired janitor with a history of hypertension went to work one morning and at 7:00 A.M. had sudden onset of **nausea, vomiting, and unsteadiness.** He was taken to the emergency room, where his exam was notable for **mildly slurred speech with slowed tongue movements, dysmetria on finger-to-nose testing on the left, dysmetria on heel-to-shin testing on the left,** and **left dysdiadochokinesia.** Upon **attempting to stand, he fell to the left,** even when he kept his eyes open. The remainder of the exam was unremarkable.

LOCALIZATION AND DIFFERENTIAL DIAGNOSIS

1. On the basis of the symptoms and signs shown in **bold** above, where is the lesion?

2. What is the most likely diagnosis, and what are some other possibilities?

Discussion

The key symptoms and signs in this case are:

- **Left arm and leg ataxia**
- **Unsteadiness, falling to the left**
- **Slurred speech**
- **Nausea and vomiting**

1. This patient had marked appendicular ataxia of the left arm and leg, as well as probable truncal ataxia, causing him to fall to the left side. One possible explanation is an ipsilateral cerebellar lesion involving the left cerebellar hemisphere, extending to the vermis. Another possibility is a lesion of one of the left cerebellar peduncles, which would cause left appendicular ataxia and truncal ataxia. Nausea and vomiting (caused by involvement of cerebellar–vestibular circuits) and slurred speech are also common in cerebellar lesions (see KCC 15.2). Lesions in other locations that cause ataxia are also possible (see KCC 15.2) but would most likely have other associated signs, such as hemiparesis or brainstem findings.

    The most likely *clinical localization* is left cerebellar hemisphere and vermis; or left superior, middle, or inferior cerebellar peduncle.

2. Given the patient's age, history of hypertension, and the sudden onset of symptoms, the most likely diagnosis is an infarct of the left cerebellum. Of the cerebellar arteries, a left superior cerebellar infarct would be most likely to cause pronounced ipsilateral ataxia without other brainstem findings (see KCC 15.1). Another possibility is a cerebellar hemorrhage involving mainly the left cerebellar hemisphere. Less likely possibilities include a left cerebellar abscess or one of the other diagnoses listed in Table 15.4. An acute lesion confined to a

cerebellar peduncle is unlikely, although a lesion extending from the left cerebellar hemisphere to one of the left cerebellar peduncles is possible.

Clinical Course and Neuroimaging

The patient did not arrive at the hospital in time to receive tPA but was admitted for further evaluation and treatment. A **brain MRI** (Image 15.1A,B, page 723) revealed an infarct in the left superior cerebellar artery territory, involving the left superior cerebellar peduncle and the left superior cerebellar hemisphere (compare to Figures 15.12 and 15.13C,D). An embolic workup (see KCC 10.4) was pursued including a transesophageal echocardiogram, 24-hour Holter monitor, transcranial Doppler studies, and MRA, none of which revealed an obvious embolic source. His ataxia gradually improved, and 1 week later he had normal speech with minimal left ataxia and was ambulating with assistance from a physical therapist. He was entered into a randomized trial of aspirin versus Coumadin therapy to prevent recurrent stroke in patients with stroke of unknown cause (this trial ultimately demonstrated that aspirin is as effective in preventing recurrent stroke as Coumadin).

RELATED CASES. Image 15.1C (page 724) shows an MRI from a different patient, who had bilateral superior cerebellar artery infarcts. This coronal image nicely outlines the SCA territories, with sparing of the PICA territories, inferiorly. The precise SCA and PICA boundaries are variable, but it is possible that the right PICA was involved as well (compare to Figure 15.12). This patient had severe vertebrobasilar disease and ultimately died. Note that infarcts in the PCA territory were present as well (see Image 15.1C), suggesting that these strokes were caused by disease of the basilar artery (see KCC 14.3).

Yet another patient is shown in the axial MRI Image 15.1D (page 724), showing bilateral posterior inferior cerebellar artery (PICA) infarcts. Note the sparing of the AICA territories (compare to Figure 15.13B). Recall that the posterior fossa is a small, tight compartment and that large infarcts of the cerebellum present a high risk for brainstem compression and herniation. Therefore, surgical decompression is sometimes required.

CASE 15.2 WALKING LIKE A DRUNKARD

MINICASE

A 76-year-old man with a history of cigarette smoking developed progressive **difficulty walking** over the course of 1 month. He noticed that when he stood up he felt "woozy," and he described his gait as feeling like he was drunk, saying "my legs go one way, and I go the other." His family said he frequently lost his balance, with **staggering and unsteadiness**. He also had frequent **mild headaches** that occurred at any time of the day and night and seemed to be getting worse. Exam was unremarkable except for a **wide-based, unsteady gait, tending to fall to the left, especially with tandem walk-** ing. Of note, there was no ataxia on finger-to-nose or heel-to-shin testing, and rapid alternating movements were normal. There was no history of alcohol intake.

LOCALIZATION AND DIFFERENTIAL DIAGNOSIS

1. On the basis of the symptoms and signs shown in **bold** above, where is the lesion?

2. What is the most likely diagnosis, and what are some other possibilities?

Discussion

The key symptoms and signs in this case are:

- **Unsteady gait, wide-based, falling to the left, especially with tandem walking**
- **Headache**

1. The patient had truncal ataxia, with no significant appendicular ataxia. This symptom may be caused by a lesion of the cerebellar vermis. Other possibilities of this gait disorder include hydrocephalus or a lesion of the frontal lobes or spinal cord (see KCC 15.2; see also KCC 6.5), although additional abnormalities on exam are often (but not always) present with these disorders. The presence of headache suggests that the lesion is intracranial (see KCC 5.1).

    The most likely *clinical localization* is cerebellar vermis.

2. Given the history of cigarette smoking and the gradual onset of symptoms, metastatic lung cancer to the cerebellar vermis should be seriously considered. Other important causes of chronic ataxia in adults are listed in Table 15.4A. In addition, as already mentioned, the patient's gait abnormalities could possibly be caused by hydrocephalus or by lesions of the frontal lobe or spinal cord.

Clinical Course and Neuroimaging

A **head CT** with intravenous contrast (Image 15.2, page 727) revealed an enhancing, cystic lesion in the cerebellar vermis. Although the patient often fell to the left on exam, the lesion did not have any obvious asymmetries. This case demonstrates that lesions of the cerebellar vermis can cause truncal ataxia manifesting mainly as a gait abnormality with little or no appendicular ataxia (see KCC 15.2).

On admission, a chest X-ray revealed a left apical 2 to 3 cm opacity, and prominence of the left pulmonary hilum. A CT-guided needle biopsy of the lung lesion showed adenocarcinoma. An MRI of the brain again demonstrated the vermian lesion, but it also showed a small area of enhancement in the left parietal lobe. The patient was offered surgery but preferred to be treated with radiation therapy and steroids as an outpatient. As a result, the left parietal lesion disappeared, the vermian lesion decreased in size, and the patient's gait improved. Four months later, however, he was readmitted with recurrent gait difficulty, and imaging showed that the vermian lesion had increased in size. He underwent surgery to decompress the posterior fossa, and the vermian lesion was resected, with pathology showing adenocarcinoma. Postoperatively he had some dysmetria, which gradually improved. Unfortunately, metastatic carcinoma is not curable, and although further follow-up information is not available, he presumably eventually died of his illness.

CASE 15.1 SUDDEN ONSET OF UNILATERAL ATAXIA

IMAGE 15.1A,B Left Superior Cerebellar Artery (SCA) Infarct (A) Axial, diffusion-weighted MRI at the level of the superior cerebellum and rostral pons. A bright region of decreased diffusion coefficient, consistent with infarction, is visible in the left superior cerebellar peduncle and left superior cerebellum. (B) Axial, T2-weighted MRI at the level of the mid to upper pons and cerebellum. A T2-bright region, consistent with infarction, is seen in the left superior cerebellar hemisphere.

(A)

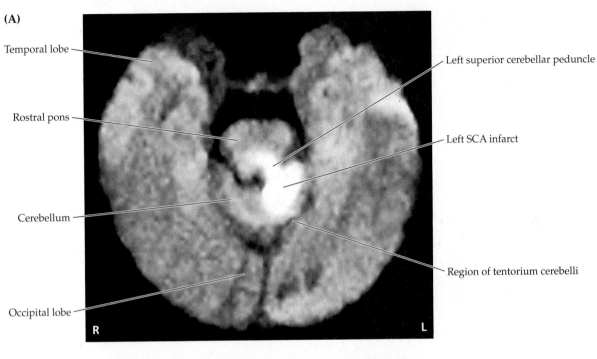

(B)

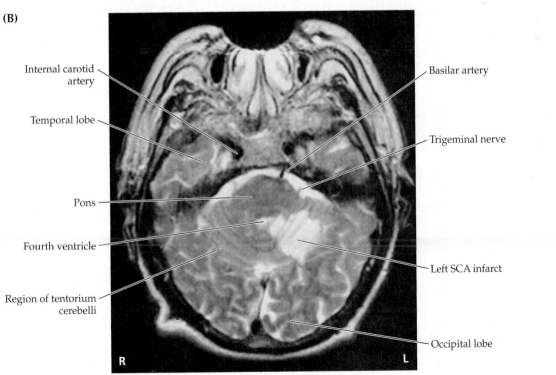

CASE 15.1 *RELATED CASE*

IMAGE 15.1C,D Superior Cerebellar Artery (SCA) and Posterior Inferior Cerebellar Artery (PICA) Infarcts (C) Coronal, T2-weighted MRI in a patient with bilateral SCA infarcts. (D) Axial, T2-weighted MRI in a patient with bilateral PICA infarcts.

(C)

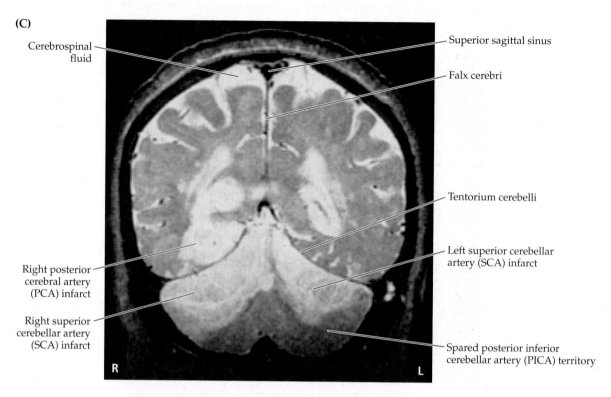

Cerebrospinal fluid

Superior sagittal sinus

Falx cerebri

Tentorium cerebelli

Left superior cerebellar artery (SCA) infarct

Right posterior cerebral artery (PCA) infarct

Right superior cerebellar artery (SCA) infarct

Spared posterior inferior cerebellar artery (PICA) territory

R   L

(D)

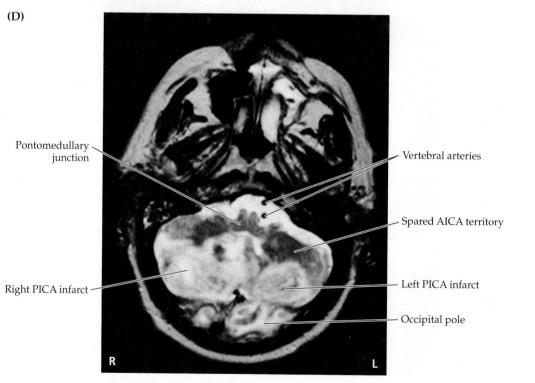

Pontomedullary junction

Vertebral arteries

Spared AICA territory

Right PICA infarct

Left PICA infarct

Occipital pole

R   L

CASE 15.3 A BOY WITH HEADACHES, NAUSEA, SLURRED SPEECH, AND ATAXIA

CHIEF COMPLAINT

A 13-year-old boy was brought to the pediatrician's office because of 2 months of progressive left occipital headaches, nausea, slurred speech, and unsteadiness.

HISTORY

The patient was well until 2 months previously, when he began having headaches, which were initially attributed to a sinus infection. The headaches gradually worsened, with **headache mainly in the left occipital area**, and sometimes accompanied by **nausea and vomiting**, but no visual changes. The headaches were worse at night and in the early morning hours. His teachers noticed that over the past few months he had had some **difficulty concentrating** and learning new material at school. During the week prior to presentation his mother noted **increasing gait instability and mildly slurred speech** and decided to bring him to the pediatrician.

PHYSICAL EXAMINATION

Vital signs: T = 98°F, P = 90, BP = 130/88, R = 16.

Neck: Supple.

Lungs: Clear.

Heart: Regular rate with no murmurs.

Abdomen: Soft, nontender.

Extremities: Normal.

Neurologic exam:

  MENTAL STATUS: Alert and oriented × 3. Speech fluent, with normal repetition and comprehension.

  CRANIAL NERVES: Pupils equal round and reactive to light. Visual fields full. **Fundi had blurred disc margins (mild papilledema) bilaterally.** Extraocular movements were full, but there was **horizontal nystagmus on lateral gaze bilaterally and vertical nystagmus, worse on upgaze than downgaze.** In addition, the **vestibulo-oc-ular reflex was not fully suppressed by visual fixation.** Facial sensation and corneal reflexes intact. Face symmetrical. Hearing intact to whisper bilaterally. **Speech slightly slurred and with an irregular rate.** Normal palate movements and gag reflex. Normal sternomastoid and trapezius strength. Tongue midline.

MOTOR: No drift. Normal tone. 5/5 power throughout.

REFLEXES:

COORDINATION: **Marked dysmetria on finger-to-nose testing, worse on the left, with approximately 2 inches of error.** There was also **dysdiadochokinesia, with inaccurate rapid alternating movements, worse on the left side. Heel-to-shin movements were ataxic on the left** but normal on the right.

GAIT: **Wide-based, with feet approximately 2 feet apart, and unsteady, staggering to the left. Unable to perform tandem gait.** On the Romberg test with feet 4 inches apart there was no worsening of unsteadiness (unable to stand with feet together even with eyes open).

SENSORY: Intact light touch, pinprick, vibration, and joint position sense. Intact graphesthesia and stereognosis.

LOCALIZATION AND DIFFERENTIAL DIAGNOSIS

1. On the basis of the symptoms and signs shown in **bold** above, where is the lesion?

2. What is the most likely diagnosis, and what are some other possibilities?

Discussion

1. The key symptoms and signs in this case are:

   - **Arm and leg ataxia, greater on the left side than the right**
   - **Unsteady gait, wide-based, staggering to the left**
   - **Slurred speech, with an irregular rate**
   - **Bilateral horizontal nystagmus on lateral gaze and vertical nystagmus, worse on upgaze than downgaze**
   - **Vestibulo-ocular reflex not fully suppressed by visual fixation**
   - **Left occipital headache**
   - **Nausea and vomiting**
   - **Difficulty concentrating**
   - **Bilateral papilledema**

This patient had multiple symptoms and signs suggesting diffuse cerebellar dysfunction. The left-sided appendicular ataxia is consistent with a lesion in the left cerebellar hemisphere, while the truncal ataxia suggests involvement of the vermis. The patient also had ataxic speech and typical eye movement abnormalities seen in cerebellar lesions, including impaired suppression of the vestibulo-ocular reflex (see KCC 15.2). The left occipital headache, nausea, and vomiting also fit with a diagnosis of left cerebellar lesion. However, when seen together with papilledema and difficulty concentrating, these symptoms strongly suggest elevated intracranial pressure (see KCC 5.3). Elevated intracranial pressure in cerebellar lesions is commonly due to compression of the fourth ventricle causing hydrocephalus (see KCC 5.7).

The most likely *clinical localization* is a large left cerebellar lesion with associated noncommunicating hydrocephalus.

2. The slowly progressive course in this child suggests a tumor of the posterior fossa, such as cerebellar astrocytoma or medulloblastoma (see KCC 5.8). Given this patient's age, an astrocytoma would be somewhat more likely, since medulloblastoma occurs in the first decade about 90% of the time, while it is not uncommon for cerebellar astrocytoma to occur after age 10. Table 15.4B lists other, less likely possibilities.

Clinical Course and Neuroimaging

A **brain MRI** (Image 15.3A,B, page 729) showed an enhancing mass in the cerebellum with a large, fluid-filled cyst occupying almost the entire left cerebellum. A cyst with a mural nodule of this kind is typical of cerebellar astrocytoma (see KCC 5.8). The fourth ventricle was compressed (see Image 15.3A), and hydrocephalus was present, with dilation of the lateral and third ventricles (see Image 15.3B). The patient was admitted to the hospital and started on steroids to reduce swelling. Surgery was performed 2 days after admission. A midline incision was made over the occipital bone. A bone flap was temporarily removed to gain access to the posterior fossa, and the dura over the cerebellum was opened, revealing tumor and cyst in the vermis extending into the left cerebellum more than the right. All visible tumor was carefully removed, and the cyst was drained. Pathology was consistent with juvenile pilocytic astrocytoma. This is a histologically benign lesion that, unlike medulloblastoma, can often be cured by resection alone (see KCC 5.8). The patient did very well postoperatively, but he did have some residual left greater than right ataxia.

CASE 15.2 WALKING LIKE A DRUNKARD

IMAGE 15.2 **Lung Metastasis in the Cerebellar Vermis** Axial head CT scan with intravenous contrast. An en-hancing, cystic, lung carcinoma metastasis is visible in the midline cerebellar vermis.

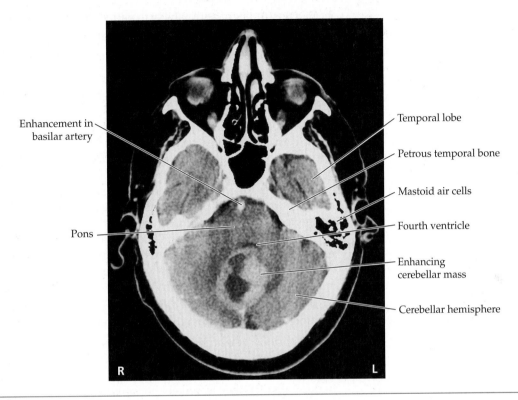

CASE 15.4 NAUSEA, PROGRESSIVE UNILATERAL ATAXIA, AND RIGHT FACE NUMBNESS

MINICASE

A 72-year-old right-handed woman with a history of heavy cigarette smoking came to see an internist because of several months of worsening **nausea and vomiting**. Gastroenterology evaluation had revealed a hiatal hernia, and she was being treated with anti-emetics with little benefit. On discussion of her overall symptoms, it became clear that about 2 months previously she noticed her **handwriting was deteriorating**, becoming less precise. Later, she noted some difficulty opening bottles with her right hand, as if she "didn't have good control," and difficulty putting on her earrings. She also was **slightly unsteady**, slipping more frequently than usual on the ice. Most recently she had noticed that the **right side of her face felt cool**. She had lost 3 pounds over the past 2 months and had one episode of hemoptysis (blood-tinged sputum). She did not complain of headaches. On examination, she had **nystagmus on horizontal and vertical end gaze**, as well as **decreased temperature sensation in the right face**. She also had mild to moderate **ataxia on finger-to-nose and heel-to-shin testing on the right. On attempted tandem gait she fell to the right.** The remainder of the exam was normal.

LOCALIZATION AND DIFFERENTIAL DIAGNOSIS

1. On the basis of the symptoms and signs shown in **bold** above, where is the lesion?
2. Given the history of smoking and recent hemoptysis, what is the most likely diagnosis?

Discussion

The key symptoms and signs in this case are:

- **Nausea and vomiting**
- **Right arm and leg ataxia**
- **Mild unsteadiness, and inability to perform tandem gait, falling to the right**
- **Nystagmus on horizontal and vertical end gaze**
- **Right face cool paresthesias and decreased temperature sensation**

1. This patient has appendicular ataxia affecting the right arm and leg, as well as truncal ataxia with an unsteady tandem gait, falling to the right. These findings strongly suggest a right cerebellar lesion (see KCC 15.2). Nystagmus is also common in cerebellar disorders. Loss of temperature sensation in the right face suggests a lesion affecting the right spinal trigeminal nucleus or tract (see Figure 12.8; Table 12.6). The right middle and inferior cerebellar peduncles also pass by this region (see Figures 14.4C, 14.5A), which could explain the right-sided ataxia. Nausea and vomiting can have many causes, both neurologic and systemic. Lesions involving the cerebellum and its connections to the vestibular system, the vestibular system itself, or the chemotactic trigger zone (see Chapter 14) are particularly likely to cause nausea. Elevated intracranial pressure can also cause nausea and vomiting (see KCC 5.3), although there is no evidence of elevated intracranial pressure on the basis of history or physical examination in this patient.

   The most likely *clinical localization* is right middle or inferior cerebellar peduncle along with the right spinal trigeminal nucleus.

2. The history of weight loss and hemoptysis in a longtime smoker strongly suggests lung cancer. Lung cancer commonly metastasizes to the brain, so a metastasis to the right cerebellar peduncles is the most likely diagnosis in this patient. Other, less likely possibilities in this patient include primary neoplasm, infection, or vascular malformation.

Clinical Course and Neuroimaging

The internist ordered a **brain MRI** scan with gadolinium (Image 15.4A,B, page 730). An enhancing lesion was found in the right middle cerebellar peduncle. This case illustrates that ataxia is often caused by lesions outside the cerebellum proper, in the cerebellar peduncles, brainstem, or other locations in the cerebellar circuitry. This patient's clinical findings resemble, in some ways, that of the patient we discussed in Chapter 14, who had a lateral pontine infarct in the AICA territory (see Case 14.4).

A second, very small (5 mm) lesion was found on the lateral surface of the right cerebellum (not shown). Chest X-ray revealed a right peribronchial lesion. A CT-guided needle biopsy of the peribronchial lesion revealed adenocarcinoma of the lung. Abdominal CT demonstrated another lesion in the adnexal area extending into the cervix. Cervical biopsy confirmed metastatic lung adenocarcinoma in this location as well. Because of the presence of multiple brain metastases and the location of a major lesion extending into the brainstem (see Image 15.4A,B), resective neurosurgery was not performed. The patient was treated with radiation therapy to the brain and chest and with steroids. She initially worsened with the radiation therapy but then improved for a while before ultimately requiring admission to a specialized hospice facility for patients with terminal cancer.

CASE 15.3   A BOY WITH HEADACHES, NAUSEA, SLURRED SPEECH, AND ATAXIA

IMAGE 15.3A,B Cerebellar Astrocytoma in the Vermis and Left Cerebellar Hemisphere Axial T1-weighted MRI images with intravenous gadolinium contrast enhancement, with images A, B progressing from inferior to superior. (A) Cystic lesion in the vermis and left cerebellar hemisphere, with enhancing mural nodule, compatible with cerebellar astrocytoma. Fourth ventricle is compressed. (B) Dilation of lateral and third ventricles with effacement of cortical sulci due to noncommunicating hydrocephalus.

(A)

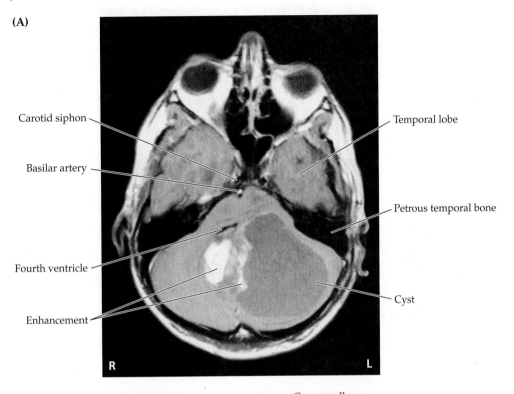

Carotid siphon

Basilar artery

Fourth ventricle

Enhancement

Temporal lobe

Petrous temporal bone

Cyst

R   L

(B)

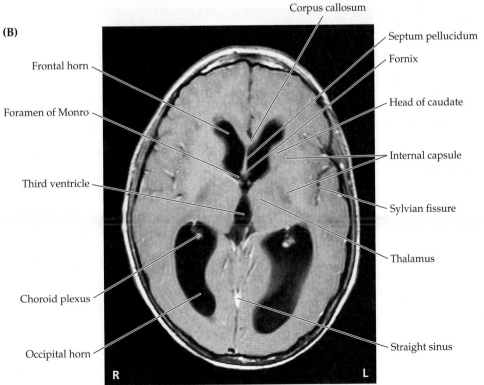

Corpus callosum

Septum pellucidum

Fornix

Frontal horn

Foramen of Monro

Third ventricle

Head of caudate

Internal capsule

Sylvian fissure

Thalamus

Choroid plexus

Occipital horn

Straight sinus

R   L

CASE 15.4 NAUSEA, PROGRESSIVE UNILATERAL ATAXIA, AND RIGHT FACE NUMBNESS

IMAGE 15.4A,B Metastasis of Lung Adenocarcinoma to the Right Middle Cerebellar Peduncle T1-weighted MRI images with intravenous gadolinium contrast. (A) Axial image demonstrating an enhancing lung adenocarcinoma metastasis in the right middle cerebellar peduncle. (B) Coronal view of the same region.

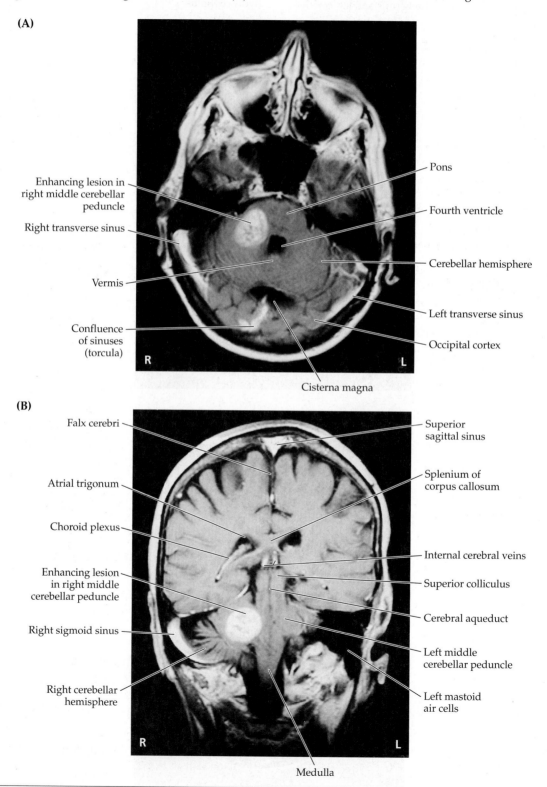

(A)

Enhancing lesion in right middle cerebellar peduncle

Right transverse sinus

Vermis

Confluence of sinuses (torcula)

Pons

Fourth ventricle

Cerebellar hemisphere

Left transverse sinus

Occipital cortex

Cisterna magna

R L

(B)

Falx cerebri

Atrial trigonum

Choroid plexus

Enhancing lesion in right middle cerebellar peduncle

Right sigmoid sinus

Right cerebellar hemisphere

Superior sagittal sinus

Splenium of corpus callosum

Internal cerebral veins

Superior colliculus

Cerebral aqueduct

Left middle cerebellar peduncle

Left mastoid air cells

Medulla

R L

CASE 15.5   A FAMILY WITH SLOWLY PROGRESSIVE ATAXIA AND DEMENTIA

CHIEF COMPLAINT

A 43-year-old man was evaluated at a neurogenetics clinic for progressive ataxia, other abnormal movements, and dementia affecting the patient as well as several other members of his family.

HISTORY

The patient had a normal birth and childhood developmental history. In adulthood, he worked as a bookkeeper for a large plumbing supply company and was an outstanding tennis player. At age 35 he first noticed he "wasn't doing well." He developed **difficulties with his coordination and balance**, leading him to stop riding his bicycle, and he eventually stopped playing tennis. His **gait became progressively unsteady**, causing him to have several falls and minor injuries. In addition, he had **difficulty with calculations** and began having **memory problems** such as difficulty remembering telephone numbers. He was forced to leave his job, and over the subsequent years his symptoms progressed very slowly.

FAMILY HISTORY

The patient was single. He had three siblings (one male age 41, and two females age 38 and 35) and several nieces all with no symptoms. His father developed unsteady gait and mood problems in his 30s, became estranged from the family and died of myocardial infarction in his 50s. His paternal grandmother developed an unsteady, shuffling gait and was forgetful, tending to repeat herself by age 51 years. She went to a nursing home at age 67 and died at 70 years. His paternal grandfather had no gait or cognitive problems when he died of a myocardial infarction at age 57. His mother's family was unaffected. The following pedigree gives further details:

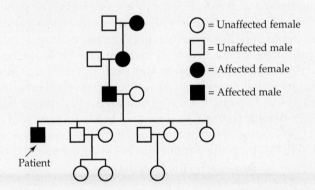

○ = Unaffected female
□ = Unaffected male
● = Affected female
■ = Affected male

Patient

PHYSICAL EXAMINATION

Vital signs: T = 98.8°F, P = 82, BP = 120/80.
General: No dysmorphic features.
Neck: Supple.
Lungs: Clear.

Heart: Regular rate with no murmurs.
Abdomen: Benign.
Extremities: Normal.
Skin: No lesions.
Neurologic exam:

MENTAL STATUS: Alert and oriented × 3. Normal language, including normal reading, writing, and command following. Good insight. Was able to spell "world" forwards and backwards, but **performed poorly on "serial 7s" test**, only able to subtract 7 from 100, but was then unable to continue to subtract 7 repeatedly. **Recalled 2/3 objects after 3 minutes.** Was **unable to copy a picture of two intersecting pentagons**.

CRANIAL NERVES: Normal, except for **saccadic intrusions during smooth pursuit eye movements in all directions**.

MOTOR: **Lower extremities showed moderately increased tone bilaterally. Lower extremities also showed rare, brief chorea-like movements** while legs were dangling from examination table. Strength was 5/5 throughout.

REFLEXES:

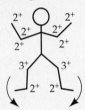

COORDINATION: **Upper extremities showed dysmetria on bilateral finger-nose-finger test, and clumsy rapid alternating movements.** Heel-to-shin testing was not performed.

GAIT: **Stance was wide-based and gait had a slight "staggering" quality. Tandem gait was somewhat unsteady.**

SENSORY: Intact.

LOCALIZATION AND DIFFERENTIAL DIAGNOSIS

1. Which of the symptoms and signs shown in **bold** above are compatible with dysfunction in each of the following systems:

   a. cerebellar dysfunction

   b. cortical dysfunction

   c. basal ganglia dysfunction

2. Which mode of inheritance is suggested by the patient's pedigree, assuming he is affected by the same disorder as other family members?

3. Given the slow progression and family history, what is the most likely type of disorder affecting this patient and his family (see KCC 15.3)?

Discussion

The key symptoms and signs in this case are:

- Moderately impaired memory, attention, calculations, and picture drawing
- Saccadic intrusions on smooth pursuit eye movements
- Increased tone in bilateral lower extremities
- Mild chorea-like movements
- Upper extremity dysmetria, dysdiadochokinesia, and wide-based unsteady gait

1. The patient's symptoms and signs can be localized to dysfunction in the following systems:

   Cerebellar dysfunction Saccadic intrusions on smooth pursuit, dysmetria, dysdiadochokinesia, wide-based unsteady gait (see KCC 15.2).

   Cortical dysfunction Impaired memory, attention, calculations, and picture drawing suggestive of chronic mental status changes (see KCC 19.16). Increased lower extremity tone may also be due to spasticity from corticospinal dysfunction (see KCC 6.1).

   Basal ganglia dysfunction Chorea-like movements. Also, the lower extremity increased tone could be either due to corticospinal or basal ganglia dysfunction (see KCC 16.1).

2. The family history suggests an autosomal dominant inheritance pattern.

3. The patient has slowly progressive ataxia and dementia with possible basal ganglia and corticospinal involvement, occurring with an autosomal dominant inheritance pattern. This is most compatible with a degenerative disorder, such as one of the inherited spinocerebellar ataxias (see KCC 15.3; Table 15.4).

Clinical Course and Neuroimaging

At the time of initial evaluation, genetic testing was not available for most of the hereditary spinocerebellar ataxias. Tests for Huntington's disease and Wilson's disease (see KCC 16.1, 16.3) were negative. Shortly after his initial evaluation, his younger sister also developed gait unsteadiness. The patient was followed clinically. Four years later he had significantly worse gait unsteadiness and was unable to perform tandem gait. On mental status exam he was oriented to his name but not to date or location, and he could recall only 1 of 3 words after 3 minutes. A **brain MRI** performed at that time showed marked atrophy of the cerebellum, as well as the cerebellar peduncles and pons (Image 15.5A,C, page 734–735) compared to normal (Image 15.5B,D, page 734–735). **Genetic testing** had improved in the interim, and revealed an expanded CAG repeat in this patient consistent with spinocerebellar ataxia type 17 (SCA 17). Both the patient and his sister required long-term care in a nursing home shortly after the final diagnosis was made.

Additional Cases

Related cases for the following topics can be found in other chapters: **brainstem lesions associated with ataxia** (Cases 14.1, 14.4, and 14.7); **ataxia in childhood** (Case 5.7); and **ataxic-appearing gait** (Cases 5.9 and 7.6). Other relevant cases can be found using the **Case Index** located at the end of this book.

Brief Anatomical Study Guide

1. The cerebellum is located in the posterior fossa (see Figure 15.1). It consists of the midline **vermis**, the **intermediate part of the cerebellar hemisphere**, and the **lateral part of the cerebellar hemisphere** (see Figure 15.3A). The cerebellum is attached to the brainstem by the **superior, middle, and inferior cerebellar peduncles**, which contain the input and output fibers of the cerebellum (see Figure 15.3B,C).

2. All outputs of the cerebellum are carried by the deep cerebellar nuclei and the vestibular nuclei (see Figures 15.4 and 15.5). The cerebellar cortex and deep nuclei can be divided into three functional zones (see Table 15.1):

   A. The **vermis** (via **fastigial nuclei**) and **flocculonodular lobes** (via **vestibular nuclei**) are important in the control of proximal and trunk muscles and in vestibulo-ocular control, respectively.

   B. The **intermediate part** of the cerebellar hemisphere (via **interposed nuclei**) is involved in the control of more distal appendicular muscles, mainly in the arms and legs.

   C. The largest part of the cerebellum is the **lateral part** of the cerebellar hemisphere (via **dentate nuclei**), which is involved in planning the motor program for the extremities.

3. The microscopic circuitry of the cerebellum involves excitatory inputs carried by **mossy fibers** and **climbing fibers**. These inputs synapse directly or indirectly onto **Purkinje cells**, which carry the outputs to the deep cerebellar and vestibular nuclei (see Figures 15.5 and 15.7). Important local cerebellar neurons include **granule cells** and the inhibitory **Golgi, basket, and stellate cells**.

4. Cerebellar **input and output pathways** are fairly complex; they are summarized in Figures 15.9, 15.10, and 15.11 and in Tables 15.2 and 15.3. The most clinically important points are that these pathways again follow a medial–lateral organization and that all pathways to the lateral motor systems are either ipsilateral or **double crossed**, with the result that cerebellar lesions cause ipsilateral deficits.

5. **Ataxia** is a characteristic irregular movement abnormality seen in cerebellar disorders (see Figure 15.14B). On the basis of the anatomical organization of cerebellar pathways, the following principles of localizing cerebellar lesions emerge:

   A. **Ataxia is ipsilateral** to the side of a cerebellar lesion.

   B. Midline lesions of the cerebellar vermis or flocculonodular lobes cause mainly unsteady gait (**truncal ataxia**), disequilibrium, and eye movement abnormalities.

   C. Lesions of the intermediate part of the cerebellar hemisphere cause mainly ataxia of the limbs (**appendicular ataxia**).

   D. Ataxia is often caused by **lesions of cerebellar circuitry** in the brainstem or other locations rather than in the cerebellum itself, which can lead to false localization.

   E. Because of the strong reciprocal connection between the cerebellum and vestibular system, cerebellar lesions are often associated with **vertigo, nausea, vomiting, and nystagmus**.

CASE 15.5 A FAMILY WITH SLOWLY PROGRESSIVE ATAXIA AND DEMENTIA

Image 15.5A–D Spinocerebellar Ataxia SCA-17
T1-weighted MRI performed 11 years after first onset of symptoms. (A) Axial image showing marked atrophy of cerebellum, pons and middle cerebellar peduncle com-pared to (B) MRI from a normal individual at the same level. (C) Sagittal image also shows marked atrophy compared to (D) a normal MRI.

(A)

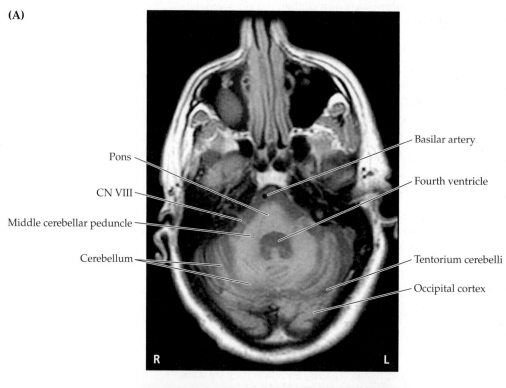

(B)

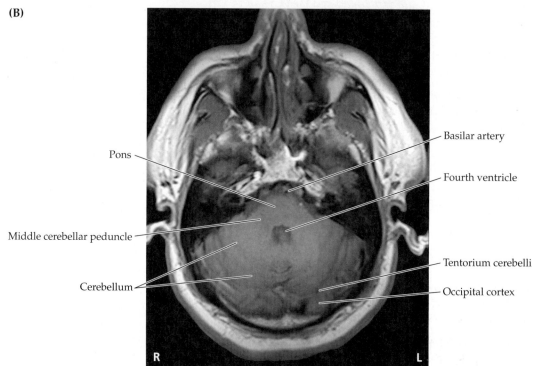

CASE 15.5 (*continued*)

(C)

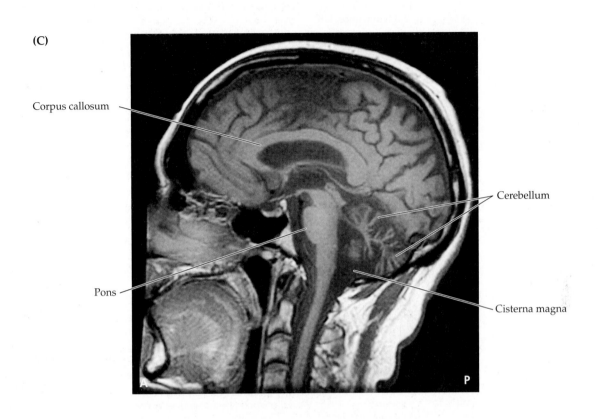

Corpus callosum

Cerebellum

Pons

Cisterna magna

(D)

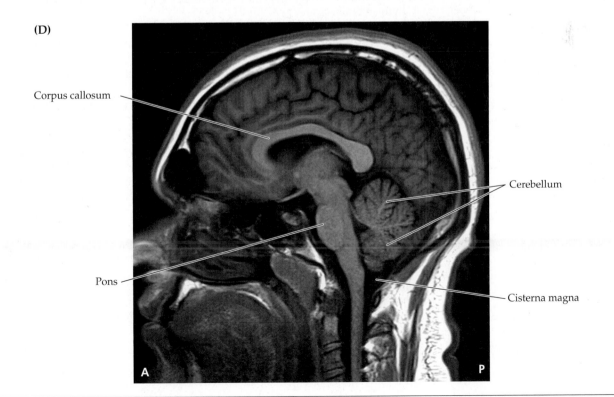

Corpus callosum

Cerebellum

Pons

Cisterna magna

References

General

Barlow JS. 2002. *The Cerebellum and Adaptive Control*. Cambridge University Press, Cambridge, UK.

Blanks RHI. 1988. Cerebellum. *Rev Oculomot Res* 2: 225–272.

de Zeeuw CI, Cicirata F (eds.). 2005. *Creating Coordination in the Cerebellum, Progress in Brain Research*, Vol. 148, 1–114. Elsevier Science, New York.

Manto M-U, and Pandolfo M (eds.). 2002. *The Cerebellum and Its Disorders*. Cambridge University Press, Cambridge, UK.

Schmahmann JD, Jenner P, Harris RA (eds.). 1997. *The Cerebellum and Cognition*, Vol. 41 (International Review of Neurobiology). Academic Press, New York.

Cerebellar Vascular Disorders

Adams RD. 1943. Occlusion of the anterior inferior cerebellar artery. *Arch Neurol Psychiatry* 49: 765–770.

Edlow JA, Newman-Toker DE, Savitz SI. 2008. Diagnosis and initial management of cerebellar infarction. *Lancet Neurol* 7 (10): 951–964.

Hiraga A, Uzawa A, Kamitsukasa I. 2007. Diffusion weighted imaging in ataxic hemiparesis. 78 (11): 1260–1262.

Jensen MB, St Louis EK. 2005. Management of acute cerebellar stroke. *Arch Neurol* 62 (4): 537–544.

Manto M, Marmolino D. 2009. Cerebellar ataxias. *Curr Opin Neurol* 22 (4): 419–429.

Marinkovic S, Kovacevic M, Gibo H, Milisavljevic M, Bumbasirevic L. 1995. The anatomical basis for the cerebellar infarcts. *Surg Neurol* 44 (5): 450–460.

Moh JP, Choi D, Grotta J, Wolf P. 2004. Vertobrobasilar occlusive disease. In *Stroke: Pathophysiology, Diagnosis, and Management*, 4th ed., JM Barnett, JP Mohr, and MS Bennett (eds.), Chapter 10. Churchill Livingstone, New York.

Moulin T, Bogousslavsky J, Chopard JL, Ghika J, Crepin-Leblond T, Martin V, Maeder P. 1995. Vascular ataxic hemiparesis: A re-evaluation. *J Neurol Neurosurg Psychiatry* 58 (4): 422–427.

Tatu L, Moulin T, Bogousslavsky J, Duvernoy H. 1996. Arterial territories of the human brain. *Neurology* 47 (5): 1125–1135.

Other Cerebellar Disorders

Daszkiewicz P, Maryniak A, Roszkowski M, Barszcz S. 2009. Long-term functional outcome of surgical treatment of juvenile pilocytic astrocytoma of the cerebellum in children. *Childs Nerv Syst* 25 (7): 855–860.

Fenichel, GM. 2009. *Clinical Pediatric Neurology: A Signs and Symptoms Approach*. 6th Ed. Elsevier: Saunders, Philadelphia.

Globas C, Tezenas du Montcel S, Baliko L, et al. 2008. Early symptoms in spinocerebellar ataxia type 1, 2, 3, and 6. *Movement Disorders* 23 (15): 2232–2238.

Melo TP, Bogousslavsky J, Moulin T, Nader J, Regli F. 1992. Thalamic ataxia. *J Neurol* 239 (6): 331–337.

Schijman E. History, anatomic forms, and pathogenesis of Chiari I malformations. 2004. *Childs Nerv Syst* 20 (5): 323–328.

Solomon DH, Barohn RJ, Bazan C, Grissom J. 1994. The thalamic ataxia syndrome. *Neurology* 44 (5): 810–814.

Steinlin M. 2008. Cerebellar disorders in childhood: cognitive problems. *Cerebellum* 7 (4): 607–610.

Subramony SH. 2004. Ataxic disorders. In *Neurology in Clinical Practice: The Neurological Disorders*, 4th Ed., Vol. 1, WG Bradley, RB Daroff, GM Fenichel, and CD Marsden (eds.), Chapter 23. Butterworth-Heinemann, Boston.

CONTENTS

Chapter 16

Basal Ganglia

A 35-year-old man and his wife came to see a psychiatrist because of marital problems. The wife reported that during recent months her husband had become increasingly argumentative and had also developed occasional irregular jerking movements of the head, trunk, and limbs. The husband denied having any involuntary movements. His father and several other paternal relatives had developed a similar syndrome, caused by a devastating neurodegenerative disease that destroys the basal ganglia.

In this chapter, we will learn about the anatomy, circuitry, and functional neurochemistry of the basal ganglia and will see cases in which damage to the basal ganglia produces movement disorders and other deficits, including behavioral and cognitive abnormalities.

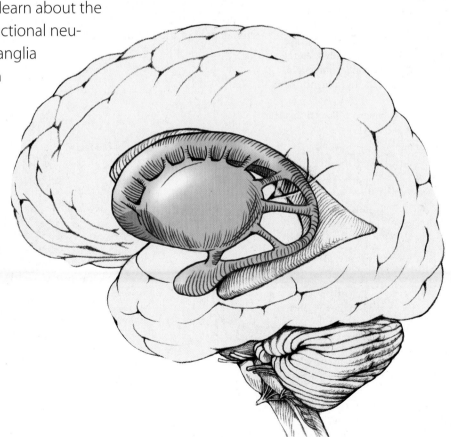

ANATOMICAL AND CLINICAL REVIEW

L IKE THE CEREBELLUM, the basal ganglia participate in complex networks that influence the descending motor systems (see Figures 2.17 and 6.6). Also like the cerebellum, the basal ganglia do not themselves project directly to the periphery. However, the movement abnormalities seen with basal ganglia disorders differ markedly from those seen with cerebellar lesions. Patients with basal ganglia lesions can have either hyperkinetic or hypokinetic movement disorders. **Hyperkinetic movement disorders** are typified by Huntington's disease, in which uncontrolled involuntary movements produce a random pattern of jerks and twists. **Hypokinetic movement disorders** are typified by Parkinson's disease, which is characterized by rigidity, slowness, and marked difficulty initiating movements. Often there will be a mixture of these two kinds of movement disorders in any given patient.

In the sections that follow we will review the basic three-dimensional anatomy of the basal ganglia and then discuss their network connections in an attempt to understand the mechanisms underlying hyperkinetic and hypokinetic movement disorders. We will also discuss some of the other functions of the basal ganglia including emotional control, cognition, and eye movements.

Basic Three-Dimensional Anatomy of the Basal Ganglia

The basal ganglia are a collection of gray matter nuclei located deep within the white matter of the cerebral hemispheres. The main components of the **basal ganglia** are the caudate nucleus, putamen, globus pallidus, subthalamic nucleus, and substantia nigra (Table 16.1; Figure 16.1; see also Figure 16.4). Other nuclei, such as the nucleus accumbens and ventral pallidum, which participate in limbic and basal ganglia circuits, are usually included as well. Some authors also include the amygdala; however, this nucleus functions primarily as part of the limbic system (see Chapter 18).

FIGURE 16.1 Spatial Relationships of Basal Ganglia, Thalamus, and Amygdala The subthalamic nucleus and substantia nigra are not shown (see Figure 16.4D). (A) Lateral view showing basal ganglia, amygdala, and lateral ventricle of the left hemisphere. (B) Anterolateral view showing basal ganglia, amygdala, and the thalamus of the right hemisphere.

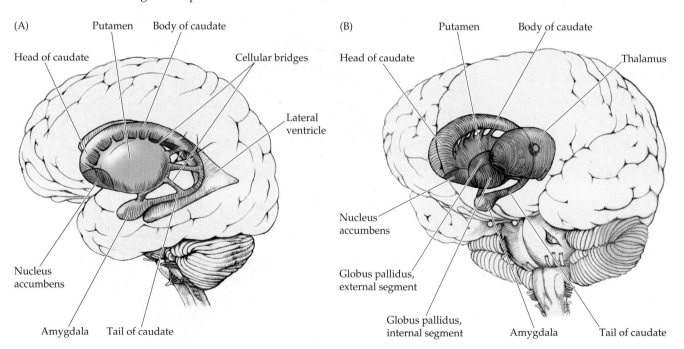

(A)

Putamen Body of caudate

Head of caudate

Cellular bridges

Lateral ventricle

Nucleus accumbens

Amygdala Tail of caudate

(B)

Putamen Body of caudate

Head of caudate

Thalamus

Nucleus accumbens

Globus pallidus, external segment

Globus pallidus, internal segment

Amygdala Tail of caudate

The caudate and putamen are histologically and embryologically closely related and can be thought of as a single large nucleus called the **neostriatum** or simply **striatum**. The striatum receives virtually all inputs to the basal ganglia. The caudate and putamen are separated by penetrating fibers of the internal capsule but remain joined in some places by **cellular bridges** (see Figure 16.1A). The cellular bridges appear as stripes, or striations, connecting the caudate and putamen in histological sections, giving rise to the name "striatum." The **caudate nucleus** is one of the C-shaped structures that we discussed in Chapter 5: Like the corpus callosum and the fornix, it has a constant relationship with the lateral ventricle, as we will discuss shortly. The caudate (meaning "possessing a tail"), is divided into three parts, the **head**, **body**, and **tail**, which do not have distinct boundaries from each other (see Figure 16.1). The amygdala lies just anterior to the tip of the caudate tail, in the temporal lobe.

The **putamen** is a large nucleus forming the lateral portion of the basal ganglia (see Figure 16.1). Anteriorly and ventrally the putamen fuses with the head of the caudate. This region, called the **ventral striatum**, is important in limbic circuitry and is often considered part of the striatum because of its similar embryological development and input and output connections. Most of the ventral striatum consists of the **nucleus accumbens**.

Just medial to the putamen lies the **globus pallidus** (or pallidum), meaning "pale globe," so named because of the many myelinated fibers traversing this region. The globus pallidus has an **internal segment** and an **external segment** (see Figure 16.1B). The putamen and globus pallidus together are called the **lenticular or lentiform** (meaning "lentil- or lens-shaped") **nucleus**. In fact, as we will see in the sections that follow, these nuclei more closely resemble an ice cream cone lying on its side, with the putamen representing the ice cream and the globus pallidus the cone.

To better appreciate the three-dimensional nature of the basal ganglia and related structures, let's review these structures through stained brain sections and artistic renderings (see Figures 16.2, 16.3, and 16.4). Moving from lateral to medial in the horizontal brain sections shown in **Figure 16.2**, we can identify the following structures in sequence:

- Insula
- Extreme capsule
- Claustrum*
- External capsule
- Putamen
- External medullary lamina
- External segment of the globus pallidus
- Internal medullary lamina
- Internal segment of the globus pallidus
- Internal capsule

As we discussed in Chapter 6, the **internal capsule** is a V-shaped collection of fibers going to and from the cortex (see Figure 16.2). The **anterior limb of the internal capsule** passes between the lentiform nucleus and the head of the caudate. The **posterior limb of the internal capsule** passes between the lentiform

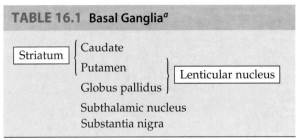

TABLE 16.1 Basal Ganglia[a]

Striatum { Caudate / Putamen / Globus pallidus } Lenticular nucleus

Subthalamic nucleus
Substantia nigra

[a]The nucleus accumbens and ventral pallidum can also be considered part of the basal ganglia.

---

*The exact functions of the claustrum remain unknown, although it may play a role in visual attention.

(A)

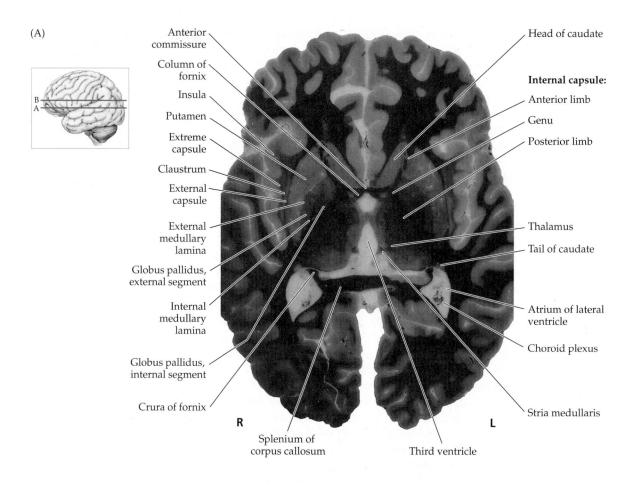

Anterior commissure

Column of fornix

Insula

Putamen

Extreme capsule

Claustrum

External capsule

External medullary lamina

Globus pallidus, external segment

Internal medullary lamina

Globus pallidus, internal segment

Crura of fornix

Splenium of corpus callosum

Third ventricle

Head of caudate

Internal capsule:

Anterior limb

Genu

Posterior limb

Thalamus

Tail of caudate

Atrium of lateral ventricle

Choroid plexus

Stria medullaris

R   L

(B)

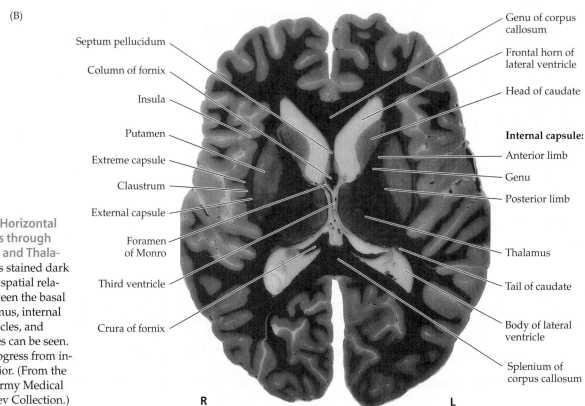

Septum pellucidum

Column of fornix

Insula

Putamen

Extreme capsule

Claustrum

External capsule

Foramen of Monro

Third ventricle

Crura of fornix

Genu of corpus callosum

Frontal horn of lateral ventricle

Head of caudate

Internal capsule:

Anterior limb

Genu

Posterior limb

Thalamus

Tail of caudate

Body of lateral ventricle

Splenium of corpus callosum

R   L

FIGURE 16.2 Horizontal Brain Sections through Basal Ganglia and Thalamus Myelin is stained dark (Weigert). The spatial relationships between the basal ganglia, thalamus, internal capsule, ventricles, and other structures can be seen. (A) and (B) progress from inferior to superior. (From the Walter Reed Army Medical Center, Yakovlev Collection.)

nucleus and the thalamus. Recall that the corticobulbar and corticospinal tracts lie in the posterior limb of the internal capsule (see Figure 6.9). Note that the **caudate and thalamus are always medial** to the internal capsule, while the **lentiform nucleus** (putamen and globus pallidus) **is always lateral** to the internal capsule (see Figure 16.2).

These relationships are again reviewed in Figure 16.3, which shows a left lateral view of these structures. In Figure 16.3A the putamen is visible most laterally, concealing the globus pallidus. The caudate and the thalamus lie behind the internal capsule. In Figure 16.3B, the putamen has been removed to reveal the globus pallidus. The external and internal segments of the globus pallidus have been removed in Figure 16.3C to fully expose the internal capsule. Finally, in Figure 16.3D, the internal capsule has been removed to reveal the caudate and the thalamus. Note the relationship between these structures and the ventricular system. The head and body of the caudate form a bulge on the lateral wall of the lateral ventricle, while the tail of the caudate runs along the roof of the temporal horn (see Figures 16.2, 16.3D, and 16.4). The thalamus forms the lateral walls of the third ventricle (see Figures 16.2, 16.3D, and 16.4D) and lies along the floor of the body of the lateral ventricle.

Coronal sections (Figure 16.4) provide additional perspective. In Figure 16.4A (the most anterior section), the head of the caudate, the putamen, and the nucleus accumbens can be seen. The internal capsule seen at this level is part of the anterior limb, since it is separating caudate from lentiform nucleus, with no thalamus visible. Note that this section includes the putamen but not the globus pallidus. We can understand this if we imagine that the lentiform nucleus is an ice cream cone lying on its side with the cone pointing medially (see Figure 16.4B,C). The most anterior coronal sections would thus cut through ice cream (putamen) without cone (globus pallidus).

The globus pallidus first appears in the next coronal section (see Figure 16.4B) moving toward the back. The external segment of the globus pallidus can be seen at this level. The head of the caudate nucleus is visible as a bulge along the lateral wall of the lateral ventricle. Moving posteriorly, in the next section (see Figure 16.4C) the internal segment of the globus pallidus (tip of the ice cream cone) can be seen, along with all of the structures listed above that were identified in Figure 16.2. The head of the caudate can still be seen bulging into the lateral ventricle. Since the thalamus is not yet visible, we are still in the anterior limb of the internal capsule.

In the most posterior section (see Figure 16.4D) we have begun to lose the globus pallidus again, and in more posterior sections (not shown), only the putamen (ice cream) is visible. In Figure 16.4D the thalamus can be seen, meaning that we are at the level of the posterior limb of the internal capsule, separating the thalamus from the lentiform nucleus. Both the body and the tail of the caudate can be seen in this section, adjacent to the body and temporal horn of the lateral ventricle, respectively. In addition, by following the internal capsule downward, we see the beginnings of the cerebral peduncles of the midbrain. The **substantia nigra** is visible, just dorsal to the cerebral peduncles (see also Figure 14.3A,B). The substantia nigra has a ventral portion called the **substantia nigra pars reticulata**, which contains cells very similar to those of the internal segment of the globus pallidus. The internal segment of the globus pallidus and the substantia nigra pars reticulata are separated from each other by the internal capsule (see Figure 16.4D), in much the same way that it separates the caudate and putamen. The more dorsal **substantia nigra pars compacta** contains the darkly pigmented dopaminergic neurons that give this nucleus its name. Degeneration of these dopaminergic neurons is an important pathogenetic mechanism in Parkinson's disease. Under the thalamus lies the spindle- or cigar-shaped **subthalamic nucleus** (see Figure 16.4D). Unlike the thala-

MNEMONIC

MNEMONIC

REVIEW EXERCISE

1. Beginning with the insula and moving medially, name each gray and white matter structure encountered on the way to the third ventricle. Repeat this while covering the labels in Figure 16.2.

2. Using Figures 16.2–16.4, confirm that the thalamus and caudate head lie medial to the internal capsule in all planes of section, while the lentiform nucleus (putamen and globus pallidus) lies lateral to the internal capsule in all planes of section.

3. Using Figures 16.2 and 16.4, confirm that:

   The anterior limb of the internal capsule is bordered by the head of the caudate medially and the lentiform nucleus laterally.

   The posterior limb of the internal capsule is bordered by the thalamus medially and by the lentiform nucleus laterally.

   The genu of the internal capsule occurs at the level of the foramen of Monro.

4. Cover the labels in Figure 16.4A–D and name as many structures as possible.

(A)

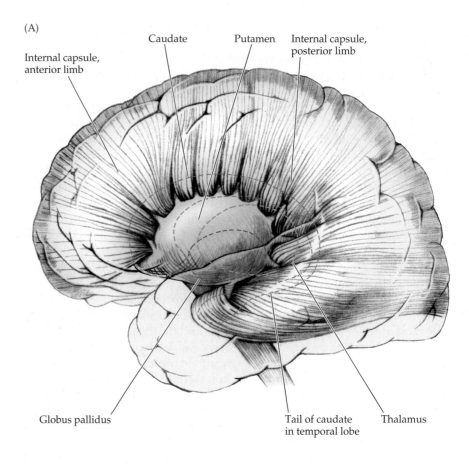

Internal capsule, anterior limb

Caudate

Putamen

Internal capsule, posterior limb

Globus pallidus

Tail of caudate in temporal lobe

Thalamus

FIGURE 16.3 **Basal Ganglia and Thalamus in Relation to the Internal Capsule and Lateral Ventricle** (A) View from the lateral aspect, showing relation of internal capsule to basal ganglia and the thalamus (see also Figure 16.2). The anterior limb of the internal capsule passes between the lentiform nucleus (putamen and globus pallidus) and the caudate, while the posterior limb of the internal capsule passes between the lentiform nucleus and the thalamus. (B) View after removal of the putamen, revealing the globus pallidus more medially. (C) View after removal of the globus pallidus, showing the entire internal capsule. (D) View after removal of the internal capsule, showing relations of the caudate to the lateral ventricle and of the thalamus to both the lateral ventricle and the third ventricle.

(B)

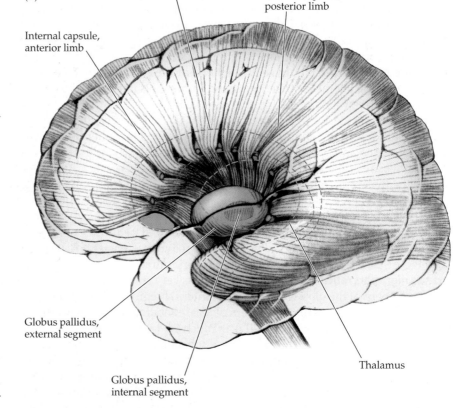

Caudate

Internal capsule, posterior limb

Internal capsule, anterior limb

Globus pallidus, external segment

Globus pallidus, internal segment

Thalamus

(C)

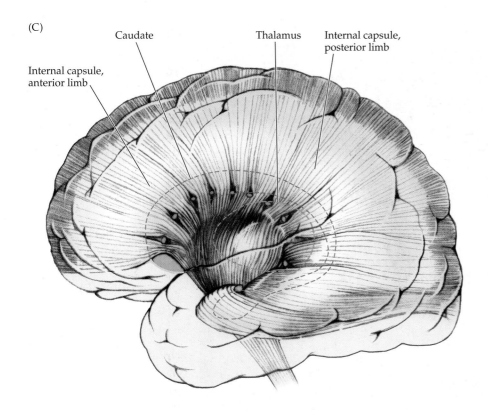

Caudate · Thalamus · Internal capsule, posterior limb · Internal capsule, anterior limb

(D)

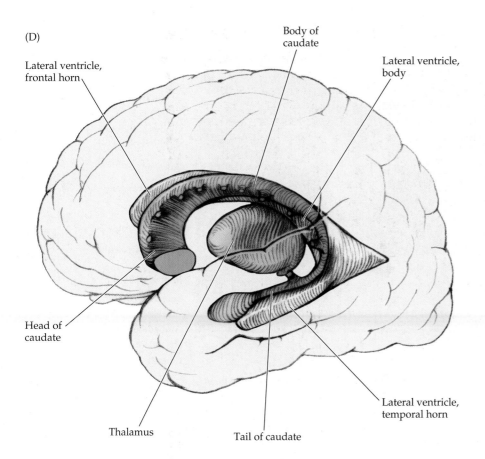

Body of caudate · Lateral ventricle, body · Lateral ventricle, frontal horn · Head of caudate · Thalamus · Tail of caudate · Lateral ventricle, temporal horn

(A)

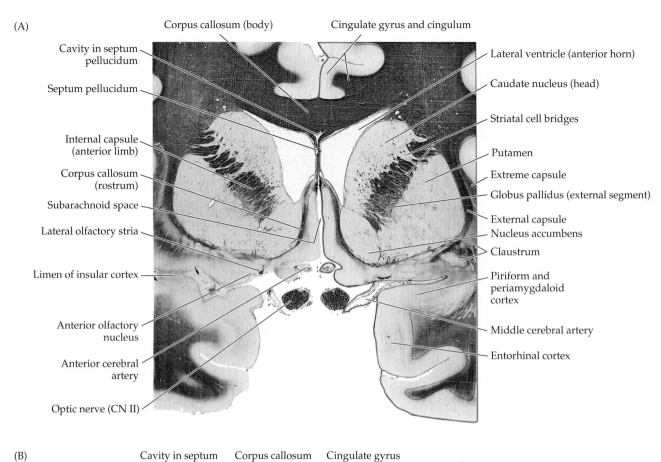

Corpus callosum (body)

Cingulate gyrus and cingulum

Cavity in septum pellucidum

Septum pellucidum

Internal capsule (anterior limb)

Corpus callosum (rostrum)

Subarachnoid space

Lateral olfactory stria

Limen of insular cortex

Anterior olfactory nucleus

Anterior cerebral artery

Optic nerve (CN II)

Lateral ventricle (anterior horn)

Caudate nucleus (head)

Striatal cell bridges

Putamen

Extreme capsule

Globus pallidus (external segment)

External capsule

Nucleus accumbens

Claustrum

Piriform and periamygdaloid cortex

Middle cerebral artery

Entorhinal cortex

(B)

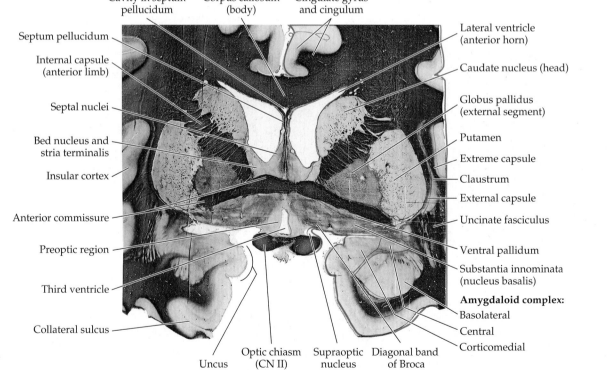

Cavity in septum pellucidum

Corpus callosum (body)

Cingulate gyrus and cingulum

Septum pellucidum

Internal capsule (anterior limb)

Septal nuclei

Bed nucleus and stria terminalis

Insular cortex

Anterior commissure

Preoptic region

Third ventricle

Collateral sulcus

Uncus

Optic chiasm (CN II)

Supraoptic nucleus

Diagonal band of Broca

Lateral ventricle (anterior horn)

Caudate nucleus (head)

Globus pallidus (external segment)

Putamen

Extreme capsule

Claustrum

External capsule

Uncinate fasciculus

Ventral pallidum

Substantia innominata (nucleus basalis)

Amygdaloid complex:
Basolateral
Central
Corticomedial

FIGURE 16.4 Coronal Brain Sections through Basal Ganglia and Thalamus Myelin is stained dark. The spatial relationships between the basal ganglia, thalamus, internal capsule, ventricles, and other structures can be seen. A–D progress from anterior to posterior. (From Martin JH. 1996. *Neuroanatomy: Text and Atlas*. 2nd Ed. McGraw-Hill, New York.)

(C)

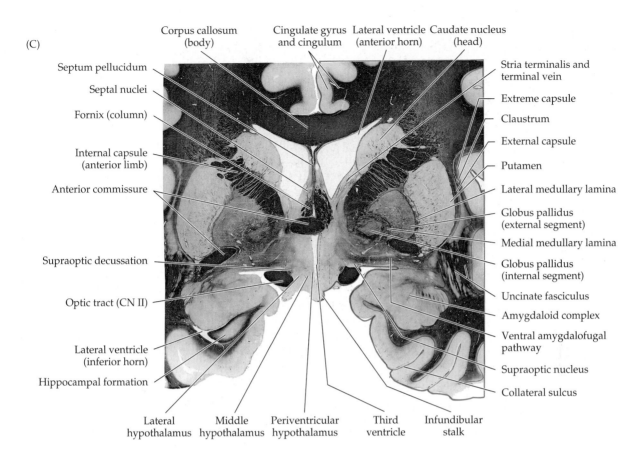

Corpus callosum (body)
Cingulate gyrus and cingulum
Lateral ventricle (anterior horn)
Caudate nucleus (head)
Septum pellucidum
Septal nuclei
Fornix (column)
Internal capsule (anterior limb)
Anterior commissure
Supraoptic decussation
Optic tract (CN II)
Lateral ventricle (inferior horn)
Hippocampal formation
Stria terminalis and terminal vein
Extreme capsule
Claustrum
External capsule
Putamen
Lateral medullary lamina
Globus pallidus (external segment)
Medial medullary lamina
Globus pallidus (internal segment)
Uncinate fasciculus
Amygdaloid complex
Ventral amygdalofugal pathway
Supraoptic nucleus
Collateral sulcus
Lateral hypothalamus
Middle hypothalamus
Periventricular hypothalamus
Third ventricle
Infundibular stalk

(D)

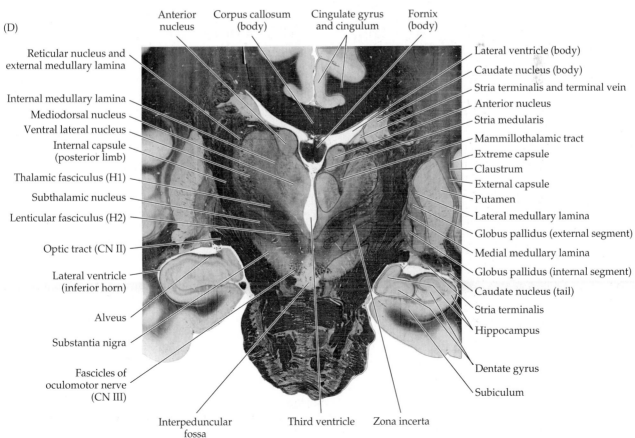

Anterior nucleus
Corpus callosum (body)
Cingulate gyrus and cingulum
Fornix (body)
Reticular nucleus and external medullary lamina
Internal medullary lamina
Mediodorsal nucleus
Ventral lateral nucleus
Internal capsule (posterior limb)
Thalamic fasciculus (H1)
Subthalamic nucleus
Lenticular fasciculus (H2)
Optic tract (CN II)
Lateral ventricle (inferior horn)
Alveus
Substantia nigra
Fascicles of oculomotor nerve (CN III)
Lateral ventricle (body)
Caudate nucleus (body)
Stria terminalis and terminal vein
Anterior nucleus
Stria medularis
Mammillothalamic tract
Extreme capsule
Claustrum
External capsule
Putamen
Lateral medullary lamina
Globus pallidus (external segment)
Medial medullary lamina
Globus pallidus (internal segment)
Caudate nucleus (tail)
Stria terminalis
Hippocampus
Dentate gyrus
Subiculum
Interpeduncular fossa
Third ventricle
Zona incerta

mus, the subthalamic nucleus is derived embryologically from the midbrain rather than the forebrain.

As discussed in Chapter 10, the blood supply to the striatum and globus pallidus is mainly from the lenticulostriate branches of the middle cerebral artery, although the medial globus pallidus is often supplied by the anterior choroidal artery (branch of internal carotid artery), and the caudate head and anterior portions of the lentiform nucleus are often supplied by the recurrent artery of Heubner (branch of anterior cerebral artery) (see Figures 10.7–10.9).

At this point you may find it worthwhile to review the basic anatomy of the basal ganglia one more time, as depicted in Figures 16.1–16.4, before proceeding to the network connections of these structures.

Input, Output, and Intrinsic Connections of the Basal Ganglia

Virtually all inputs to the basal ganglia arrive via the **striatum** (caudate, putamen, and nucleus accumbens). Outputs leave the basal ganglia via the **internal segment of the globus pallidus** and the closely related **substantia nigra pars reticulata**. Inputs and outputs through the basal ganglia are thus easily visualized as a funnel, with the spout pointing medially (see Figure 16.4D). Within the basal ganglia there are a variety of complex excitatory and inhibitory connections utilizing several different neurotransmitters. In addition, there appear to be several parallel pathways in the basal ganglia for different functions, including:

- General motor control
- Eye movements
- Cognitive functions
- Emotional functions

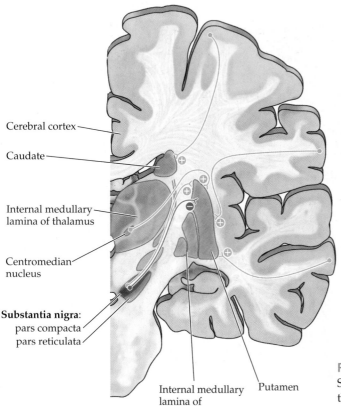

Cerebral cortex

Caudate

Internal medullary lamina of thalamus

Centromedian nucleus

Substantia nigra:
pars compacta
pars reticulata

Internal medullary lamina of globus pallidus

Putamen

In this section we will review the main pathways involved in motor control and discuss a model to explain hyperkinetic and hypokinetic movement disorders. It is important to recognize that the simplified model presented here does not tell the whole story, and these pathways are still under active investigation. In the next section (see also Table 16.2) we will briefly discuss some of the circuitry involved in other basal ganglia functions.

Inputs to the Basal Ganglia

The main input to the basal ganglia comes from massive projections from the entire **cerebral cortex** to the striatum (**Figure 16.5**). The putamen is the most important input nucleus of the striatum for motor control pathways. Most cortical inputs to the striatum are excitatory and use **glutamate** as the neurotransmitter. Another important input to the striatum is the substantia nigra pars compacta. This **dopaminergic nigrostriatal pathway** is excitatory to some cells and inhibitory to others in the striatum (see Figures 16.5 and 16.7). In-

FIGURE 16.5 Basal Ganglia Inputs Arrive at the Striatum Schematic coronal section. Note that, for clarity, inputs other than those from the cortex are shown projecting only to the putamen. In reality, inputs project to the caudate as well.

puts to the pars compacta are still under investigation. However, one important source of input may arise from a subpopulation of striatal neurons located in patches called **striosomes**. The striatum also receives excitatory (glutamatergic) inputs from **intralaminar nuclei** lying within the internal medullary lamina of the thalamus, especially the **centromedian** and **parafascicular nuclei**. The internal medullary lamina of the thalamus should be distinguished from the internal medullary lamina of the globus pallidus (see Figure 16.5). Finally, there are **serotonergic inputs** to the basal ganglia that arise from the raphe nuclei of the brainstem.

Outputs from the Basal Ganglia

Basal ganglia outputs arise from the internal segment of the globus pallidus and from the substantia nigra pars reticulata (**Figure 16.6**). For motor control, the substantia nigra pars reticulata appears to convey information for the head and neck, while the internal segment of the globus pallidus conveys information for the rest of the body. These output pathways are inhibitory and use the neurotransmitter **gamma-aminobutyric acid** (**GABA**). The main output pathways are to the **ventral lateral** (**VL**) and **ventral anterior** (**VA**) nuclei of the thalamus via the **thalamic fasciculus**. The more anterior parts of the thalamic fasciculus carry outputs from the basal ganglia to the anterior portion of VL (VL_A) also called VL pars oralis (VL_O), while the more posterior parts of the thalamic fasciculus carry cerebellar outputs to the posterior VL (VL_P) also called VL pars caudalis (VL_C) (mnemonic: **C**audal part of VL receives inputs from **C**erebellum) (see Chapter 15). Thalamic neurons convey information from the basal ganglia to the entire frontal lobe. However, information for motor control travels mainly to the premotor cortex, supplementary motor area, and primary motor cortex (see Figure 16.8).

MNEMONIC

Basal ganglia outputs also travel to other thalamic nuclei. These include **intralaminar nuclei** (centromedian and parafascicular), which project back to the striatum, and the **mediodorsal nucleus**, which is involved primarily in limbic pathways. In addition, the internal segment of the globus pallidus and the substantia nigra pars reticulata project to the pontomedullary **reticular formation**, thereby influencing the descending reticulospinal tract. The substantia nigra pars reticulata also projects to the **superior colliculus**, to influence tectospinal pathways. In this way, the basal ganglia influence both the lateral motor systems (e.g., the lateral corticospinal tract) and the medial motor systems (e.g., the reticulospinal and tectospinal tracts) (see Table 6.3).

Intrinsic Basal Ganglia Connections

An understanding of the excitatory and inhibitory connections in these pathways provides some insight into the mechanisms of hyperkinetic and hypokinetic movement disorders. There are two predominant pathways from input to output nuclei through the basal ganglia (**Figure 16.7**). The **direct pathway** travels from the striatum directly to the internal segment of the globus pallidus or the substantia nigra pars reticulata. The **indirect pathway** takes

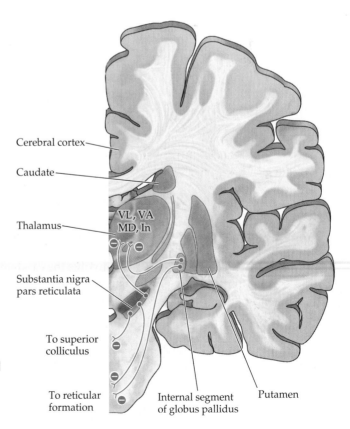

FIGURE 16.6 Basal Ganglia Outputs Arise from the Internal Globus Pallidus and the Substantia Nigra Pars Reticulata Schematic coronal section. Thalamic nuclei: VL, ventral lateral; VA, ventral anterior; MD, mediodorsal; In, intralaminar.

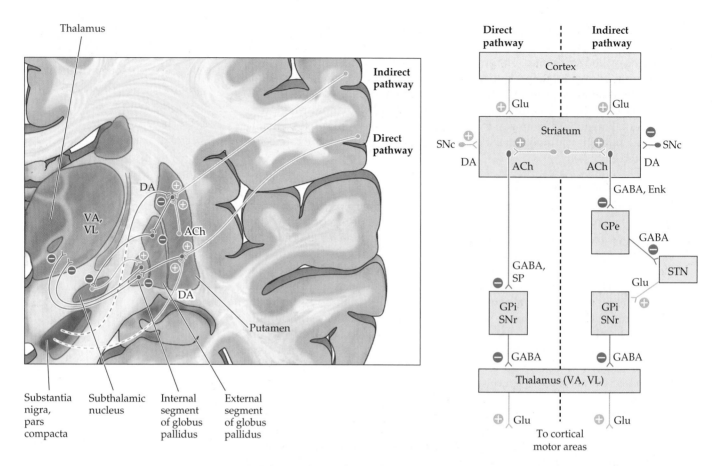

FIGURE 16.7 **Basal Ganglia Internal Connections, Showing the Direct and Indirect Pathways** (A) Schematic coronal section. For simplicity, the striatum is represented by the putamen only, although the caudate has the same connections as well. Similarly, the connections shown for the internal segment of the globus pallidus are the same for the substantia nigra pars reticulata (not shown). The excitatory connections from the thalamus to the cortex are also not shown. (B) Circuit diagram for direct and indirect pathways. Neurotransmitters: Ach, acetylcholine; DA, dopamine; Glu, glutamate; Enk, enkephalin; SP, substance P. Nuclei: SNc, substantia nigra pars compacta; SNr, substantia nigra pars reticulata; GPe, globus pallidus pars externa; GPi, globus pallidus pars interna; STN, subthalamic nucleus; VL, ventral lateral nucleus; VA, ventral anterior nucleus.

REVIEW EXERCISE

List the four major thalamic nuclei to which the basal ganglia project. List two additional basal ganglia outputs. From what structures do the basal ganglia outputs arise? What is the major neurotransmitter for all outputs from the basal ganglia?

MNEMONIC

a detour from the striatum, first to the external segment of the globus pallidus and then to the subthalamic nucleus, before finally reaching the internal segment of the globus pallidus or the substantia nigra pars reticulata. For simplicity, only the pathways through the putamen and internal segment of the globus pallidus are shown in Figure 16.7, although similar pathways also involve the caudate and the substantia nigra pars reticulata.

Figure 16.7B shows that the net effect of excitatory input from the cortex through the direct pathway will be excitation of the thalamus, which in turn will facilitate movements through its connections with the motor and premotor cortex. On the other hand, the net effect of excitation of the indirect pathway will be inhibition of the thalamus, resulting in inhibition of movements through connections back to the cortex (mnemonic: Indirect Inhibits).

Let's look at each of the major synapses in the direct and indirect pathways in series. Striatal projection neurons for both pathways are primarily **inhibitory spiny neurons**, which contain the neurotransmitter **GABA**. In the direct pathway, spiny striatal neurons project to the internal segment of the globus pallidus (and to the substantia nigra pars reticulata) and contain the peptide **substance P** in addition to GABA. Output neurons from the internal globus pallidus and substantia nigra pars reticulata to the thalamus are also inhibitory and contain GABA. In the indirect pathway, striatal neurons project to the external segment of the globus pallidus and contain the inhibitory neurotransmitter GABA, plus the peptide **enkephalin** (mnemonic: **E**nkephalin to **E**xternal pallidum). Neurons of the external globus pallidus, in turn, send inhibitory GABAergic projections to the subthalamic nucleus. Excitatory neurons in the subthalamic nucleus containing **glutamate** then project to the internal segment of the globus pallidus and to the substantia nigra pars reticulata. As in the direct pathway, outputs from these nuclei to the thalamus are inhibitory and are mediated by GABAergic neurotransmission.

A simple way to understand the net effects of these pathways (see Figure 16.7) is to remember from mathematics that $(-1)(-1) = +1$, and similarly, two inhibitory synapses in the direct pathway cause net excitation. By the same logic, three inhibitory synapses in the indirect pathway cause net inhibition since $(-1)(-1)(-1) = -1$ or $(-1)(-1)(+1)(-1) = -1$.

MNEMONIC

Hyperkinetic and Hypokinetic Movement Disorders

Several aspects of hyperkinetic and hypokinetic movement disorders can be understood from the scheme in Figure 16.7. In **Parkinson's disease** (see KCC 16.2), dopamine-containing neurons in the substantia nigra pars compacta degenerate. Dopamine appears to have excitatory effects on striatal neurons of the direct pathway but inhibitory effects on striatal neurons of the indirect pathway (see Figure 16.7). Therefore, dopamine normally has a net excitatory effect on the thalamus. Conversely, loss of dopamine will result in net inhibition of the thalamus, through both the direct and the indirect pathways, which may account for the paucity of movement seen in Parkinson's disease. Drugs that bolster dopaminergic transmission can improve the symptoms of Parkinson's disease.

In addition, anticholinergic drugs can be beneficial. The striatum contains large interneurons called **aspiny neurons**, some of which contain the neurotransmitter **acetylcholine**. Some evidence suggests that these cholinergic interneurons preferentially form excitatory synapses onto striatal neurons of the indirect pathway. Removal of cholinergic excitation of the indirect pathway produces a net decrease in inhibition of the thalamus, which may account for the beneficial effects of anticholinergic agents in parkinsonism (see Figure 16.7). Note that this model does not account for the tremor commonly seen in Parkinson's disease, and the model therefore continues to evolve.

In **hemiballismus** (see KCC 16.1) there are unilateral wild flinging movements of the extremities contralateral to a lesion in the basal ganglia. The lesion in hemiballismus often involves the subthalamic nucleus. Figure 16.7 shows how damage to the subthalamic nucleus could decrease excitation of the internal segment of the globus pallidus, resulting in less inhibition of the thalamus, causing a hyperkinetic movement disorder. In **Huntington's disease**, striatal neurons in the caudate and putamen degenerate. There is histological evidence that, at least initially, the enkephalin-containing striatal neurons of the indirect pathway are more severely affected. This would cause removal of inhibition from the external segment of the globus pallidus, allowing it to inhibit the subthalamic nucleus (see Figure 16.7). Inhibition of the subthalamic nucleus produces a situation similar to a lesion of the subthalamic nucleus and may account for the hyperkinetic movement disorder seen in Huntington's disease. In more advanced stages of Huntington's disease, both the direct and the indirect pathways degenerate, and a rigid hypokinetic parkinsonian state results.

REVIEW EXERCISE

Follow the inhibitory and excitatory pathways in Figure 16.7 to confirm that: (1) Excitatory inputs to the direct pathway produce net excitation of thalamocortical outputs; and (2) Excitatory inputs to the indirect pathway produce net inhibition of thalamocortical outputs.

Parallel Basal Ganglia Pathways for General Movement, Eye Movement, Cognition, and Emotion

The basal ganglia contain multiple parallel channels of information processing for different functions. **Four channels** have been well described (Table 16.2), although others probably exist as well. Each channel passes through slightly different pathways and projects to different regions of the frontal lobes (Figure 16.8). In another classification scheme, the first three channels are lumped together as **dorsal striatal pathways**, while the limbic channel is distinguished because it involves **ventral striatal pathways**. We will see in KCC 16.1–16.3 that basal ganglia disorders can affect all four of these parallel channels, not just the motor system.

The **motor channel** is the best known and forms the basis for most of the discussion in the previous section. Cortical inputs travel mainly to the putamen, and outputs arise from the internal segment of the globus pallidus and the substantia nigra pars reticulata to reach the VL and VA of the thalamus (see Table 16.2). From the thalamus the motor channel continues to the supplementary motor area, premotor cortex, and primary motor cortex (see Figure 16.8).

A separate **oculomotor channel** subserves basal ganglia regulation of eye movements. The input for this pathway is predominantly from the body of the caudate nucleus. Output is to the frontal eye fields and supplementary eye fields of the frontal lobes, areas important for the higher control of eye movements, as discussed in Chapter 13. The **prefrontal channel** is probably important in cognitive processes involving the frontal lobes (see Chapter 19). Input is primarily from the head of the caudate, and output reaches the prefrontal cortex (see Table 16.2; Figure 16.8).

Finally, the **limbic channel** is an important ventral pathway through the basal ganglia, involved in limbic regulation of emotions and motivational drives. Inputs arise from major areas for the limbic system (see Chapter 18), such as the limbic cortex, hippocampus, and amygdala, and project to the **nucleus accumbens** and other regions of the ventral striatum. Outputs arise from

TABLE 16.2 Four Parallel Channels through the Basal Ganglia

SOURCES OF CORTICAL INPUT	BASAL GANGLIA INPUT NUCLEI	BASAL GANGLIA OUTPUT NUCLEI[a]	THALAMIC RELAY NUCLEI[b]	CORTICAL TARGETS OF OUTPUT
MOTOR CHANNEL				
Somatosensory cortex; primary motor cortex; premotor cortex	Putamen	GPi, SNr	VL, VA	Supplementary motor area; premotor cortex; primary motor cortex
OCULOMOTOR CHANNEL				
Posterior parietal cortex; prefrontal cortex	Caudate, body	GPi, SNr	VA, MD	Frontal eye fields; supplementary eye fields
PREFRONTAL CHANNEL				
Posterior parietal cortex; premotor cortex	Caudate, head	GPi, SNr	VA, MD	Prefrontal cortex
LIMBIC CHANNEL				
Temporal cortex; hippocampus; amygdala	Nucleus accumbens; ventral caudate; ventral putamen	Ventral pallidum; GPi; SNr	MD, VA	Anterior cingulate; orbital frontal cortex

Source: Based on Martin, JH. 1996. *Neuroanatomy: Text and Atlas.* McGraw-Hill, New York.

[a]GPi, globus pallidus, internal segment; SNr, substantia nigra pars reticulata.

[b]MD, mediodorsal nucleus; VA, ventral anterior nucleus; VL, ventral lateral nucleus.

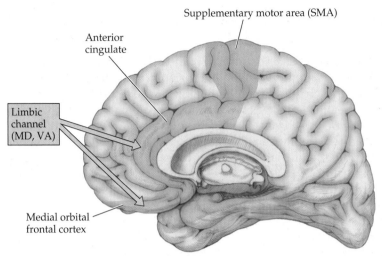

Supplementary motor area (SMA)

Anterior cingulate

Limbic channel (MD, VA)

Medial orbital frontal cortex

FIGURE 16.8 Frontal Lobe Outputs of the Four Parallel Channels through the Basal Ganglia See Table 16.2. Thalamic origins of the outputs are indicated. Thalamic nuclei: VL, ventral lateral; VA, ventral anterior; MD, mediodorsal.

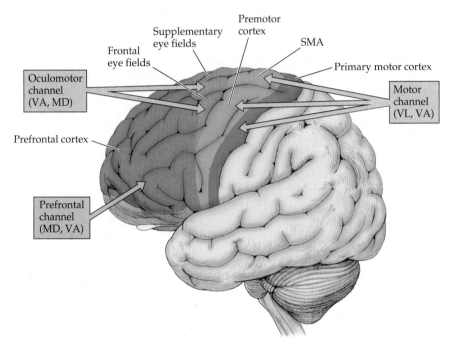

Premotor cortex

Supplementary eye fields

Frontal eye fields

SMA

Primary motor cortex

Oculomotor channel (VA, MD)

Motor channel (VL, VA)

Prefrontal cortex

Prefrontal channel (MD, VA)

the **ventral pallidum**, a region just ventral to the globus pallidus, to reach the thalamic mediodorsal and ventral anterior nuclei. The **mediodorsal nucleus** is particularly important to limbic circuitry. Projections from these thalamic nuclei reach the limbic cortex of the **anterior cingulate gyrus** and **medial orbital frontal gyri** (see Figure 16.8). The limbic channel through the basal ganglia is thought to play a central role in many neurobehavioral and psychiatric disorders (see KCC 18.3). Another component of this pathway is the dopaminergic projection from the **ventral tegmental area**, which lies just medial and dorsal to the substantia nigra of the midbrain, at the base of the interpeduncular fossa (see Figure 14.3A). The ventral tegmental area provides dopaminergic inputs to the nucleus accumbens as well as to other limbic structures and to the frontal lobes (see Figure 14.10). The dopaminergic projections of the ventral tegmental area may be affected in the pathophysiology of schizophrenia and other psychotic disorders, and they appear to play an important role in drug addiction.

REVIEW EXERCISE

Name the four main channels through the basal ganglia. For each channel give the input nuclei in the striatum and the output targets in the cortex.

Ansa Lenticularis, Lenticular Fasciculus, and the Fields of Forel

The output pathways of the globus pallidus and related structures have been dubbed with some very esoteric-sounding names. We will review this nomenclature briefly here, since it occasionally crops up in neuroanatomical discussions. The structures in question are summarized in Figure 16.9.

The internal segment of the globus pallidus sends outputs to the thalamus through two different pathways. One pathway is the **ansa lenticularis** (meaning "lenticular loop"), named for the looping course it takes ventrally under the internal capsule before passing dorsally to reach the thalamus (see Figure 16.9). The ansa lenticularis actually passes slightly rostrally as it loops around the inferior medial edge of the internal capsule, and it then turns back toward the thalamus. Recall that the globus pallidus is lateral to the internal capsule, while the thalamus is medial to the posterior limb of the internal capsule (see Figures 16.2 and 16.4D).

The other pathway is the **lenticular fasciculus** (see Figure 16.9). Instead of taking a looping course, fibers of the lenticular fasciculus penetrate straight through the internal capsule. They then pass dorsal to the subthalamic nucleus and ventral to the zona incerta before turning superiorly and laterally to enter the thalamus. The **zona incerta** (see Figures 16.4D and 16.9) is the inferior extension of the **reticular nucleus of the thalamus** (see Figure 7.6), which should not be confused with the reticular formation of the brainstem. The fibers of the ansa lenticularis and lenticular fasciculus join together to form the **thalamic fasciculus**, which enters the thalamus. The thalamic fasciculus also contains fibers ascending to the thalamus from the deep cerebellar nuclei. At this point it would be helpful to review the course of the ansa lenticularis, lenticular fasciculus, and thalamic fasciculus as summarized in Figure 16.9.

Another nomenclature for these regions was contributed by the Swiss neurologist and psychiatrist Auguste H. Forel. His terminology describes the Haubenfelder ("hauben" is a German term for "hood" or "cap") fields of the

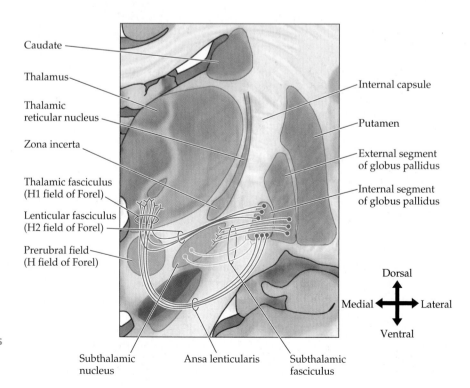

FIGURE 16.9 Terminology of Basal Ganglia Output Pathways, and Fields of Forel Coronal section through the basal ganglia and the thalamus.

subthalamic tegmentum, currently abbreviated as the **H fields of Forel** (see Figure 16.9). The **H1 field of Forel** is the thalamic fasciculus, and the **H2 field of Forel** is the lenticular fasciculus, where it lies dorsal to the subthalamic nucleus. The **prerubral field**, or **H field of Forel**, is the region where the ansa lenticularis and lenticular fasciculus join together.

We will add one more name to the list of tracts discussed so far. In addition to the ansa lenticularis, lenticular fasciculus, and thalamic fasciculus, there is the subthalamic fasciculus (see Figure 16.9). The **subthalamic fasciculus** carries fibers of the indirect pathway from the external segment of the globus pallidus to the subthalamic nucleus, and from the subthalamic nucleus to the internal segment of the globus pallidus.

KEY CLINICAL CONCEPT

16.1 MOVEMENT DISORDERS

In the broadest sense, abnormal movements can be caused by dysfunction anywhere in the complex hierarchical motor network, including upper motor neurons, lower motor neurons, cerebellar circuitry, basal ganglia circuitry, motor association cortex, and even sensory systems. Although movement disorders specialists may be asked to see patients with any motor impairment, when clinicians discuss **movement disorders**, they are very often referring to abnormal movements resulting from basal ganglia pathology. Basal ganglia disorders tend to look different from disorders in other parts of the motor system, and there are several well-recognized syndromes caused by lesions of the basal ganglia.

Lesions in other systems are often referred to by other names, rather than being called movement disorders. For example, slow, clumsy, stiff movements and hyperreflexia resulting from corticospinal, upper motor neuron lesions are called spasticity (see KCC 6.1). Irregular, uncoordinated movements caused by lesions of cerebellar circuitry are called ataxia or a variety of other names (see KCC 15.2). On the other hand, abnormal movements caused by basal ganglia dysfunction may be referred to as **dyskinesia**, meaning simply "abnormal movement."

As in the cerebellar exam (see KCC 15.2), when examining a patient with abnormal movements thought to be of basal ganglia origin, it is essential to first look carefully for abnormalities in other systems that can also cause abnormal movements, including upper or lower motor neuron signs, sensory loss, or ataxia (see **neuroexam.com Videos 48–78**). In addition, abnormal movements are occasionally caused by psychological conditions such as conversion disorder (see Chapter 3).

There is also a historical basis for focusing on the basal ganglia when discussing movement disorders. In the beginning of the twentieth century it was believed that two independent "pyramidal" and "extrapyramidal" motor systems converged on lower motor neurons. The pyramidal system was similar to current corticospinal or upper motor neuron pathways. However, the extrapyramidal system was mistakenly thought to constitute an independent pathway from the striatum descending through polysynaptic connections to the spinal cord. As we have discussed in this chapter, the basal ganglia are in fact part of a network of complex loops that exert their major influence on descending motor systems through projections to the motor and premotor cortex. Nevertheless, movement disorders resulting from basal ganglia dysfunction are still often referred to as extrapyramidal syndromes.

Some of the abnormal movements seen in movement disorders are slow, and some are fast. Some occur at rest, and others are accentuated by move-

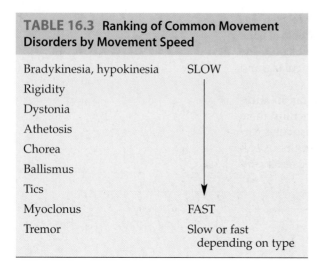

TABLE 16.3 Ranking of Common Movement Disorders by Movement Speed

Bradykinesia, hypokinesia	SLOW
Rigidity	
Dystonia	
Athetosis	
Chorea	
Ballismus	
Tics	
Myoclonus	FAST
Tremor	Slow or fast depending on type

ment or occur only during movement. One common, even if simplified, way of discussing abnormal movements is on a spectrum from slow to fast (Table 16.3). Some movement disorders, such as tremor, can be either slow or fast. Movement disorders can be focal or generalized, unilateral or bilateral. In unilateral movement disorders caused by focal basal ganglia lesions such as, infarct, hemorrhage, abscess, tumor, or degeneration, the **movement disorder is contralateral** to the basal ganglia lesion.

During sleep, most obvious movement abnormalities cease, with a few exceptions, such as palatal myoclonus (palatal tremor), periodic leg movements of sleep, and some tic disorders. Nevertheless, certain aspects of movement disorders often persist in sleep, causing a marked disruption in the normal stages of sleep, and insomnia in some patients. Gait is frequently abnormal in movement disorders, as discussed in KCC 6.5. We will now define several different types of abnormal movements and briefly discuss their differential diagnosis. Localization will also be discussed when possible; however, note that for many movement abnormalities the precise localization is still under investigation. In the sections that follow (see KCC 16.2 and KCC 16.3), we will then review a few specific syndromes in greater detail. In addition, we will see how basal ganglia dysfunction affects not only general body movements, but also eye movements, cognition, and emotional regulation (see Table 16.2).

Bradykinesia, Hypokinesia, and Akinesia

Bradykinesia means "slowed movements"; **hypokinesia** "decreased amount of movements"; and **akinesia** "absence of movement." These terms are traditionally reserved for localizations at levels higher than the upper motor neurons. In other words, these terms are not used for corticospinal, corticobulbar, lower motor neuron, or muscular disorders. Bradykinetic movement disorders can be caused by increased inhibitory basal ganglia outflow to the thalamus. Reviewing the connections in Figure 16.7 should make it clear that lesions in several regions of the basal ganglia can indirectly cause an increase in inhibitory output from the internal globus pallidus and substantia nigra pars reticulata to the thalamus. Examples include loss of function of the dopaminergic nigrostriatal system, loss of inhibitory pathways from the striatum to the substantia nigra and internal pallidum, or loss of inhibitory neurons projecting from the external pallidum to the subthalamic nucleus. Bradykinesia and hypokinesia resulting from basal ganglia dysfunction are an important feature of Parkinson's disease and related disorders (see KCC 16.2). In addition, decreased spontaneous movements, without coma, can be seen in diffuse lesions of the frontal lobes (see KCC 19.11), subcortical white matter, thalami, or brainstem reticular formation (see KCC 14.2). These disorders have been given a variety of names, including abulia and akinetic mutism (see Table 14.3). Depression and advanced schizophrenia can also cause marked psychomotor retardation, which in the extreme is called catatonia.

Rigidity

Increased resistance to passive movement of a limb is called **rigidity**. Rigidity is often present in disorders that cause bradykinesia or dystonia. There are numerous different types of rigidity, seen in different conditions. In spasticity, which results from upper motor neuron lesions, rigidity is velocity dependent. Resistive tone initially increases as the muscles of the limb are stretched, but it may then decrease, giving rise to the term **clasp-knife rigidity** in corticospinal disorders. In contrast, rigidity caused by basal ganglia disorders tends to be more continuous throughout attempts to bend the limb, and it has therefore been called **plastic, waxy,** or **lead pipe rigidity**. A particular kind of plastic rigidity is often seen in parkinsonian disorders (see KCC 16.2); it is called **cogwheel rigidity** because of the ratchet-like interruptions in tone that can be felt as the limb is bent. Cogwheel rigidity is often thought of as rigidity with a superimposed tremor. Patients with frontal lobe dysfunction often actively resist movement of their limbs. This condition is called **paratonia** or **gegenhalten** and can sometimes be distinguished from other causes of rigidity because it has a more active, inconsistent, or almost voluntary quality.

Dystonia

In **dystonia** the patient assumes abnormal, often distorted positions of the limbs, trunk, or face that are more sustained or slower than in athetosis (see Table 16.3). Dystonia can be generalized, unilateral, or focal. **Focal dystonias** include **torticollis**, which involves the neck muscles; **blepharospasm**, which involves the facial muscles around the eyes; **spasmodic dysphonia**, which involves the laryngeal muscles; and **writer's cramp**. These disorders are presumed to be caused by basal ganglia dysfunction, although usually no focal lesion is found.

Treatment with numerous medications has been tried, with modest success. Many cases of dystonia respond well to injection of small amounts of **botulinum toxin** ("Botox") into the affected muscles, which needs to be repeated every few months. Botulinum toxin acts by interfering with presynaptic acetylcholine release at the neuromuscular junction.

Primary idiopathic torsion dystonia, previously called dystonia musculorum deformans, is an uncommon hereditary disorder causing generalized dystonia. Dystonia can also be seen in other disorders known to involve the basal ganglia, such as tumor, abscess, infarct, carbon monoxide poisoning, Wilson's disease, Huntington's disease, and Parkinson's disease. Episodes of dystonic posturing also occur during certain kinds of focal seizures, due to the spread of seizure activity into the basal ganglia.

Dystonias, or faster dyskinesias such as athetosis or chorea, are commonly seen both acutely and after long-term use of dopaminergic antagonists, including many **antipsychotic** and **anti-emetic medications**. Long-term use can produce delayed, or **tardive, dyskinesia**, in which oral and lingual choreic dyskinesias are often prominent. In about one-third of cases, tardive dyskinesia remains severe despite discontinuation of the offending medication. Other dyskinesias induced more acutely by these medications, including dystonia and parkinsonism, are usually reversible, although sometimes up to several months are required for recovery. Parkinsonism (see KCC 16.2) is fairly common with antipsychotic and anti-emetic dopaminergic antagonists, and prophylactic anticholinergic medications are therefore often used with neuroleptic agents, especially in young patients.

An important disorder to recognize early is **Wilson's disease**, an autosomal recessive disorder of biliary copper excretion that causes progressive degeneration of the liver and basal ganglia. Typical neurologic manifestations include the gradual onset of dysarthria, dystonia, rigidity, tremor, choreoathetosis, and prominent psychiatric disturbances. Some patients have a characteristic "wing beating" tremor, in which the arms are abducted with the elbows flexed, and facial dystonia causing a wry smile called risus sardonicus. Liver failure can be the presenting feature, especially in children under the age of 10 years. Almost all patients with neurologic manifestations have brownish outer corneal deposits of copper called Kayser–Fleischer rings that are visible on ophthalmological slit-lamp examination, elevated 24-hour urine copper levels, and serum ceruloplasmin concentrations of less than 20 milligrams per deciliter. In equivocal cases, liver biopsy may be necessary for diagnosis. Treatment with copper-chelating agents such as penicillamine or with zinc to reduce copper absorption can arrest progress of this disorder, thus, early diagnosis is critical for improving outcome. Siblings of affected individuals should also be tested for presymptomatic Wilson's disease.

Athetosis

Athetosis is characterized by writhing, twisting movements of the limbs, face, and trunk that sometimes merge with faster choreic movements, giving rise to the term **choreoathetosis**. Important causes include perinatal hypoxia involving the basal ganglia, kernicterus caused by severe neonatal jaundice, Wilson's disease, ataxia telangiectasia, Huntington's disease, and antipsychotic or anti-emetic medications, among others. In addition, patients with Parkinson's disease (see KCC 16.2) who are being treated with levodopa can experience disabling hyperkinetic dyskinesias, ranging from athetosis to ballismus, at certain times after taking their medications.

Chorea

The word **chorea** means literally "dance" and is applied to movement disorders characterized nearly continuous involuntary movements that have a fluid or jerky, constantly varying quality. In mild cases, low-amplitude chorea may be mistaken for fidgeting or restless movements of the extremities, face, or trunk. Choreic movements are often incorporated into voluntary movements in an attempt to conceal their occurrence. In severe cases, larger-amplitude movements resemble frantic "break dancing," occurring constantly, interrupting voluntary movements, and increasing during distraction or during ambulation. Chorea can involve proximal and distal extremities, the trunk, neck, face, and respiratory muscles.

A major cause of chorea is **Huntington's disease**, an autosomal dominant neurodegenerative disorder that is discussed in KCC 16.3. In addition to chorea, patients with Huntington's disease have severe neuropsychiatric disturbances and ultimately become unable to walk; they die about 15 years after onset, usually from respiratory infections. Most other causes of chorea have a more favorable long-term prognosis. **Benign familial chorea** also usually has an autosomal dominant inheritance pattern, but the chorea is nonprogressive and is not accompanied by cognitive or emotional decline.

Sydenham's chorea, or **rheumatic chorea**, is now rare, except in populations in which group A streptococcal infections are not treated with antibiotics. Onset is usually in adolescence and is more common in females. About 4 months after streptococcal infection, patients insidiously develop increased fidgetiness and emotional lability that may initially be mistaken for normal adolescent behavior. Chorea becomes more apparent over weeks and then gradually subsides, but it recurs later in about one-fifth of patients. Impulsive or obsessive-compulsive behaviors can occasionally persist. The cause of Sydenham's chorea is thought to be antistreptococcal antibodies that cross-react with striatal neurons. However, anti-streptolysin O titers may no longer be elevated by the time chorea occurs. Patients should be treated with antibiotics because rheumatic fever occurs in up to one-third of patients with Sydenham's chorea.

Another important cause of chorea, with increased incidence in young females, is **systemic lupus erythematosus (SLE)**. Chorea can be the first manifestation of SLE. The antinuclear antibody or other rheumatological blood tests are often positive, which can be helpful in distinguishing this condition from Sydenham's chorea. Chorea occurring during pregnancy (**chorea gravidarum**) or while oral contraceptives are being taken can represent an initial episode or recurrence of Sydenham's chorea or SLE.

Chorea is commonly seen as a dyskinetic side effect of **levodopa** in Parkinson's disease or as an early or delayed (tardive) side effect in patients taking **antipsychotic or anti-emetic medications**. There are numerous other causes of chorea, including perinatal anoxia, carbon monoxide poisoning, hyperthyroidism, hypoparathyroidism, electrolyte and glucose abnormalities, phenytoin and other drugs or toxins, neuroacanthocytosis, Wilson's disease, Lesch–Nyhan syndrome, amino acid disorders, and lysosomal storage disorders. **Hemichorea** can occur contralateral to infarct, hemorrhage, tumor, abscess, or other focal lesions of the basal ganglia.

Ballismus

Movements of the proximal limb muscles with a larger-amplitude, more rotatory or flinging quality than chorea are referred to as **ballism** or **ballismus**. The most common type is **hemiballismus**, in which there are unilateral flinging movements of the extremities contralateral to a lesion in the basal ganglia. The classic cause is a lacunar infarct of the subthalamic nucleus, which leads to decreased pallidal inhibition of the thalamus (see Figure 16.7). However, lacunes in other regions of the basal ganglia, especially the striatum, can also cause contralateral hemiballismus. Hemiballismus usually gives way over days or weeks to subtler choreoathetotic movements. However, the movements are often initially quite disabling, and they can be improved with dopaminergic antagonists such as haloperidol. Additional causes of hemiballismus include other unilateral lesions of the basal ganglia, such as hemorrhage, tumor, infection, or inflammation.

Tics

A sudden brief action that is preceded by an urge to perform it and is followed by a sense of relief is called a tic. **Motor tics** usually involve the face or neck and, less often, the extremities. **Vocal tics** can be brief grunts, coughing sounds, howling or barking-like noises, or even more elaborate vocalizations that sometimes include obscene words (coprolalia). Tic disorders make up a spectrum ranging from transient single motor or vocal tics of childhood to **Tourette's syndrome** (also known as Gilles de la Tourette's syndrome), which is characterized by persistent motor and vocal tics.

Tourette's syndrome is four times more common in boys than girls, and it appears to have an autosomal dominant inheritance pattern with incomplete penetrance. Onset is usually in late childhood, and there is often some spontaneous improvement during adolescence. There is an increased incidence of attention-deficit hyperactivity disorder and obsessive-compulsive disorder in patients with Tourette's syndrome, as well as in family members. Diagnosis is based on clinical presentation, since MRI and other tests are generally unrevealing. The most important aspect of treatment is counseling and education of the patient, family, and other contacts about the nature of the disorder, to reduce stigmatization. Symptoms tend to wax and wane, and during severe periods treatment with dopaminergic antagonists such as haloperidol or pimozide may be beneficial. Alternative pharmacological agents, such as clonidine (a central α_2-receptor antagonist) are increasingly being tried first, despite modest efficacy, because of the long-term side effects of anti-dopaminergic agents.

Other resting tremors seen in cerebellar disorders include trunk and head titubation, associated with lesions of the vermis, and palatal tremor. **Palatal tremor** is famous for its notable persistence during sleep, which distinguishes it from most other forms of tremor. Palatal tremor was formerly called **palatal myoclonus** but is now classified as a tremor because the movements are biphasic rather than monophasic. Movements of the palate occur at 1 to 2 hertz and can sometimes extend to the face and even proximal upper extremities. Some patients complain of hearing "clicking sounds" due to movements of the Eustachian tube caused by the contractions of the tensor veli palatine muscles. Treatment with Botox injections to the tensor veli palatini can be effective. Palatal tremor is typically caused by lesions of the central tegmental tract (see Figure 15.9A), most commonly with brainstem infarcts, but also due to multiple sclerosis or trauma.

Some other disorders that cause rhythmic or semirhythmic movements maybe confused with tremor; these include clonus, myoclonus, flapping tremor (asterixis), fasciculations, and focal clonic seizures. These disorders are not considered true tremors. ◼

TABLE 16.5 Differential Diagnosis of Parkinsonism

Parkinson's disease

Drug-induced parkinsonism (dopamine antagonists)

Multisystem atrophy

    Striatonigral degeneration

    Shy–Drager syndrome

    Olivopontocerebellar atrophy

Progressive supranuclear palsy
(Steele–Richardson–Olszewski syndrome)

Dementia with Lewy bodies

Cortical basal ganglionic degeneration

Machado–Joseph disease (SCA-3)[a]

Dentatorubropallidoluysian atrophy

Juvenile-onset Huntington's disease

Wilson's disease

Carbon monoxide poisoning

MPTP toxicity

Von Economo's encephalitis lethargica

Dementia pugilistica

Vascular parkinsonism

Other metabolic and neurodegenerative disorders

Disorders resembling parkinsonism:

    Hydrocephalus

    Frontal lobe dysfunction (abulia, catatonia)

    Diffuse subcortical lesions

    Hypothyroidism

    Depression

[a]SCA-3, spinocerebellar ataxia, type 3.

KEY CLINICAL CONCEPT

16.2 PARKINSON'S DISEASE AND RELATED DISORDERS

Parkinson's disease is a common idiopathic neurodegenerative condition caused by loss of dopaminergic neurons in the substantia nigra pars compacta. It is characterized by asymmetrical resting tremor, bradykinesia, rigidity, and postural instability, which usually respond to therapy with levodopa. **Parkinsonism** and **parkinsonian signs** are more general terms used to describe several other conditions that have some features of Parkinson's disease, especially the bradykinesia and rigidity. In this section we will first discuss idiopathic Parkinson's disease and then briefly review several other conditions associated with parkinsonism (Table 16.5).

Idiopathic Parkinson's Disease

Parkinson's disease is a sporadic disorder of unknown etiology that occurs worldwide. The usual age of onset is between 40 and 70 years. About 1% of individuals over age 65 are affected. There is generally no familial tendency, except in rare cases of familial Parkinson's. Pathologically, there is loss of pigmented dopaminergic neurons in the substantia nigra pars compacta, causing the substantia nigra to appear pale to the eye on cross section (Figure 16.10A). Remaining dopaminergic neurons often contain characteristic cytoplasmic inclusions called **Lewy bodies**, which are eosinophilic, contain ubiquitin and α-synuclein, and have a faint halo (Figure 16.10B). There is also loss of pigmented neurons in other regions of the nervous system.

Diagnosis is based on clinical features. Initially, patients may have only some subtle difficulty using one limb, slowing of movements, or an asymmetrical resting tremor. Eventually, patients with idiopathic Parkinson's disease usually have the classic triad of **resting tremor**, **bradykinesia**, and

(A) Dorsal

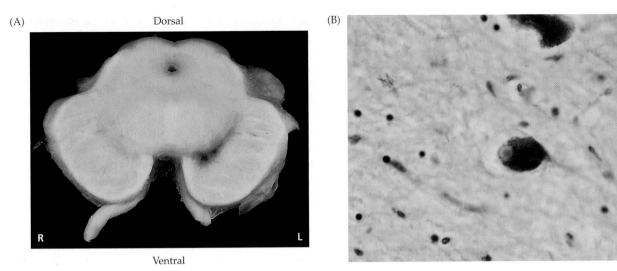

Ventral

(B)

FIGURE 16.10 **Pathologic Changes of Parkinson's Disease** (A) Section through the midbrain from a patient who died with Parkinson's disease. Asymmetrical pallor affecting the right substantia nigra greater than the left can be seen, caused by degeneration of pigmented dopaminergic neurons in the substantia nigra pars compacta. (B) Microscopic section through the substantia nigra from a different patient with Parkinson's disease showing a typical Lewy body. The Lewy body has a characteristic dense (pink) center with a lighter halo and is located in the cytoplasm of a darkly pigmented dopaminergic neuron. (A from Nolte J. 1999. *The Human Brain*. 4th Ed. Mosby, St. Louis; courtesy of Naomi Rance, University of Arizona College of Medicine. B courtesy of Jean Paul G. Vonsattel, Massachusetts General Hospital, Harvard School of Medicine.)

cogwheel **rigidity**, accompanied by postural instability that causes an **unsteady gait**. The disorder is initially unilateral but later becomes bilateral, although the severity often remains **asymmetrical**. It nearly always **improves when treated with levodopa**.

A diagnosis other than idiopathic Parkinson's should be suspected in patients who do not respond to levodopa and in cases of symmetrical symptoms or absence of resting tremor (although up to 30% of Parkinson's disease patients do not develop tremor). Other reasons to suspect alternative diagnoses are the presence of features of atypical parkinsonian disorders (which we will discuss shortly), postural instability very early in the course, or rapid progression of symptoms. In Parkinson's disease, **progression is usually insidious**, over the course of 5 to 15 years, and can eventually cause severe disability and death. Since onset is late in life, many patients die of other causes rather than as a direct result of Parkinson's disease.

Some of the clinical features of Parkinson's disease are worth discussing in greater detail. The resting "pill rolling" tremor has already been discussed in KCC 16.1, as has the rigidity, which is often of the cogwheel type. There are numerous manifestations of bradykinesia and hypokinesia in Parkinson's disease. There is a characteristic decrease in spontaneous blink rate and in facial expression called **masked facies** or **hypomimia**. The voice becomes **hypophonic**, and speech has a hurried, muttering quality. Saccades are slow, and smooth pursuit eye movements are often broken up into a series of catch-up saccades. Writing becomes small, a symptom termed **micrographia**. The posture is stooped, and some patients may have some dystonia. **Postural instability**, the diminished ability to make reflex postural adjustments that maintain balance, can be disabling and results in a characteristic **parkinsonian gait** (see Table 6.6). If pulled backward slightly, patients may exhibit **retropulsion**, in which they take several backward steps to regain balance, or they may fall. They often cannot rise from a chair without using their hands, and they have difficulty initiating gait. Once they have started, they tend to walk with small shuffling steps, termed a **festinating gait**. They sometimes appear to be continually falling and shuffling forward, referred to as **anteropulsion**. Arm swing is diminished, and they often exhibit **en bloc turning**, in which turns are executed without the normal twist of the torso. Inability to suppress blinking when the center of the brow ridge (glabella) is tapped repeatedly (**Myerson's sign**) is a nonspecific finding that can be seen in other neurodegenerative conditions as well.

Dementia is not an early feature of Parkinson's disease, but estimates of dementia later in the course vary from 15% to 40% or higher. Some of these cases are likely due to the coincidence of Alzheimer's disease and Parkinson's disease, and others may represent diffuse Lewy body disease (discussed shortly), but additional cases may not be explained by either of these causes. Patients with advanced Parkinson's disease often have **bradyphrenia**, in which responses to questions are slowed but may be accurate if enough time is allowed. Depression and anxiety are common, especially in advanced Parkinson's disease. Other associated features in Parkinson's disease of uncertain etiology include seborrhea and hypersalivation. Interestingly, one of the earliest symptoms of Parkinson's disease is often impaired smell (anosmia), likely related to degeneration of the olfactory bulb and anterior olfactory nucleus (see Figure 18.5).

The most effective drug for treatment of Parkinson's disease is **levodopa**. Most formulations also contain **carbidopa***, a decarboxylase inhibitor that cannot cross the blood–brain barrier. Carbidopa inhibits the breakdown of levodopa to dopamine in peripheral tissues, making more levodopa available for conversion to dopamine in the central nervous system where it is needed. The most common peripheral side effects of dopamine are gastrointestinal disturbances and orthostatic hypotension, and these are substantially reduced by carbidopa. Higher doses of levodopa can sometimes precipitate psychiatric symptoms such as psychosis.

As Parkinson's disease progresses, patients have other problems with levodopa therapy. Troublesome **wearing off** can occur toward the end of the time between doses, during which the patient may experience **freezing**, becoming almost unable to move. The opposite problem—levodopa-induced **dyskinesias**—also becomes increasingly troublesome. With advanced Parkinson's disease, patients may increasingly experience **on–off phenomena**, in which they fluctuate between dyskinesias and immobility, with very little time during which they are functional. These fluctuations, which occur in some patients who have advanced Parkinson's, are thought to be caused by two factors: abnormal regulation of dopamine levels and an abnormal physiological response to intermittent exogenous dosing, which leads to an unstable network. On–off phenomena may be helped somewhat by sustained-release formulations. In addition, catechol *O*-methyltransferase (COMT) inhibitors and monamine oxidase (MAO) inhibitors have recently shown some promise in maintaining dopamine delivery to the brain and decreasing wearing-off symptoms.

There is currently some controversy about whether to start therapy with levodopa early in the course of Parkinson's disease or to reserve it for later, when other agents are no longer effective. Other treatments used for Parkinson's disease include anticholinergic agents such as benztropine mesylate (Cogentin) and trihexyphenidyl (Artane). The antiviral agent amantadine has an anticholinergic and antiglutamatergic effect and probably also increases the release of dopamine in the striatum. Dopaminergic agonists such as ropinirole and pramipexole have had a growing role in treatment. Selegiline works by inhibiting the breakdown of dopamine. Early claims that selegiline may slow progression of Parkinson's disease have not been substantiated. More recently, the potent irreversible MAO-B inhibitor rasagiline has been shown to be effective for treatment. Surgical treatments for Parkinson's disease are discussed in KCC 16.4.

---

*Outside the United States the peripheral dopa decarboxylase inhibitor benserazide is often used instead of carbidopa.

It should be apparent from reviewing Figure 16.7 how decreased dopaminergic input to the striatum in Parkinson's disease results in increased inhibition of the thalamus by basal ganglia outputs (through both the direct and indirect pathways), causing a hypokinetic movement disorder. In addition, Figure 16.7 should make clear the beneficial effects of agents that enhance dopaminergic actions or inhibit cholinergic actions. However, it is important to recall that this simplified schematic is still incomplete; for example, it fails to account for the tremor seen in Parkinson's disease.

Other Causes of Parkinsonism

Antipsychotic and anti-emetic **dopaminergic antagonists** such as haloperidol (Haldol), and prochlorperazine (Compazine) commonly cause parkinsonian signs such as rigidity, hypokinesia, and even resting tremor. Unlike Parkinson's disease, onset is usually abrupt and symptoms are symmetrical. Occasionally, symptoms can persist for a few months after the offending agent has been discontinued, making a careful review of prior medication history essential when evaluating patients with subacute onset of parkinsonism.

Several neurodegenerative conditions other than Parkinson's disease are associated with parkinsonism (see Table 16.5). These are sometimes referred to as **parkinsonism plus** syndromes. They often produce atypical parkinsonism, which differs from idiopathic Parkinson's disease by having relatively symmetrical symptoms, absence of resting tremor, early appearance of postural instability, and little response to dopaminergic agents. One group of neurodegenerative conditions associated with **atypical parkinsonism** falls under the heading of **multisystem atrophy**. These disorders include striatonigral degeneration, Shy–Drager syndrome, and olivopontocerebellar atrophy. In multisystem atrophy there is loss of dopaminergic neurons of the substantia nigra pars compacta (see Figure 16.7). However, there is also loss of striatal neurons projecting to the globus pallidus and substantia nigra pars reticulata. Therefore, even if dopaminergic transmission is enhanced pharmacologically, there will still be decreased inhibition of the basal ganglia output nuclei, resulting in increased inhibition of the thalamus, and parkinsonism (see Figure 16.7). This may explain the relative insensitivity of multisystem atrophy to levodopa compared to Parkinson's disease. In **striatonigral degeneration**, atypical parkinsonism is often present. **Shy–Drager syndrome** is accompanied by marked atrophy of the intermediolateral cell column of the spinal cord (see Figures 6.4D and 6.12B). Therefore, patients with Shy–Drager syndrome have parkinsonism together with autonomic disturbances such as marked orthostatic hypotension, impotence, and urinary incontinence (see KCC 7.5). **Olivopontocerebellar atrophy** is characterized by parkinsonism together with ataxia. Often these different syndromes of multisystem atrophy overlap significantly.

Another important neurodegenerative condition in which parkinsonism is prominent is **progressive supranuclear palsy** (**PSP**), also known as **Steele–Richardson–Olszewski syndrome**. In this disorder, multiple structures around the midbrain–diencephalic junction degenerate; these include the superior colliculus, red nucleus, dentate nucleus, subthalamic nucleus, and globus pallidus. The range of **vertical eye movement** is usually markedly limited, including both upward and downward saccades, relatively early in the illness (see Chapter 13). This finding should be distinguished from mildly decreased upward eye movements seen in numerous neurodegenerative conditions, and even in normal aging. Patients with PSP also have waxy rigidity, bradykinesia, and a tendency to experience falls early in the course of the illness. There is often a characteristic wide-eyed stare. In contrast to

Parkinson's disease, rigidity early in the course of PSP tends to be more prominent proximally—for example, in the neck rather than in the limbs.

Dementia with Lewy bodies (also called diffuse Lewy body disease) is increasingly being recognized as an important cause of parkinsonism and dementia. Lewy bodies in this disorder are found in the substantia nigra and throughout the cerebral cortex. Patients often have prominent psychiatric symptoms relatively early in the course of the disorder, including visual hallucinations, which tend to have episodic exacerbations. In **cortical basal ganglionic degeneration**, there is parkinsonism that resembles Parkinson's disease by being asymmetrical, together with limb dystonia and marked cortical features such as apraxia (see KCC 19.7), wandering or alien limb syndrome ("my hand/leg has a mind of its own"), and corticospinal abnormalities.

Machado–Joseph disease (also called spinocerebellar ataxia type 3) and **dentatorubropallidoluysian atrophy** (**DRPLA**) are rare neurodegenerative disorders that often include parkinsonian features. Both are transmitted by autosomal dominant inheritance and are caused by expanded trinucleotide repeats (see KCC 16.3). **Huntington's disease**, another trinucleotide repeat disorder, can present with predominantly parkinsonian features in the unusual patients in which onset is in childhood or early adulthood. **Wilson's disease** (see KCC 16.1) can also cause tremor, rigidity, and bradykinesia. Parkinsonism can be seen as a delayed effect, several weeks after **carbon monoxide** poisoning. Illicit drug users who were exposed to the toxin **MPTP** while taking a synthetic heroine-like meperidine analog developed parkinsonism caused by destruction of pars compacta dopaminergic neurons.

In 1914–1930 there was an epidemic of **von Economo's encephalitis lethargica**, a condition that has since disappeared. Many patients were left with severe parkinsonism following this illness. Boxers may develop **dementia pugilistica**, in which there is parkinsonism and cognitive decline. The possible existence in rare cases of **vascular parkinsonism** caused by lacunar strokes in the striatum or substantia nigra has been difficult to confirm because asymptomatic lacunar disease is so common. Pathologically, such cases should have lacunar disease but no Lewy bodies. The numerous other causes of parkinsonism are beyond the scope of this text. (See the References section at the end of this chapter for more details). We should mention, however, that bradykinesia, rigidity or paratonia, hypophonia, and unstable gait can be seen in **hydrocephalus, frontal lobe lesions** (abulia) or advanced schizophrenia (catatonia), and **diffuse subcortical disorders**, and these disorders can sometimes be hard to distinguish from parkinsonism. In addition, severe **hypothyroidism** or **depression** can cause paucity of movement that may be mistaken for parkinsonism. ■

KEY CLINICAL CONCEPT

16.3 HUNTINGTON'S DISEASE

Huntington's disease is an autosomal dominant neurodegenerative condition characterized by a progressive, usually choreiform movement disorder, dementia, and psychiatric disturbances, ultimately leading to death. The pathologic hallmark of Huntington's disease is progressive atrophy of the striatum, especially involving the caudate nucleus. Clinical manifestations include abnormalities in all four domains of basal ganglia function discussed earlier (see Table 16.2; Figure 16.8). Specifically, Huntington's disease results in abnormalities of body movements, eye movements, emotions, and cognition.

The overall prevalence of Huntington's disease is about 4 to 5 cases per 100,000 people, although it is higher in those of northern European ancestry. Usual age of onset is between 30 and 50 years, although early-onset and late-onset cases are occasionally seen. Initial symptoms are usually subtle chorea (see KCC 16.1) and behavioral disturbances. These symptoms may be denied by the patient and brought to medical attention by family members or other contacts. While taking the history, the examiner often can elicit abnormalities extending back several years in retrospect. Interestingly, those experienced with Huntington's disease can often detect subtle eye movement abnormalities before other manifestations become apparent. These abnormalities include slow saccades, impaired smooth pursuit, sluggish optokinetic nystagmus (see Chapter 13), and a characteristic difficulty initiating saccades without moving the head or blinking.

Early movement abnormalities include clumsiness and subtle chorea as described in KCC 16.1, such as mild jerking, fidgety movements. Mild chorea may be voluntarily suppressed, and examiners can make it more obvious by having patients walk or by asking them to hold their arms outstretched with eyes closed. In addition to chorea, other abnormal movements include tics, athetosis, and dystonic posturing. In the rare cases of juvenile onset, a more parkinsonian phenotype is often present.

Common psychiatric manifestations include affective disorders such as depression and anxiety, obsessive-compulsive disorder, impulsive or destructive manic-like behavior, and, occasionally, psychosis. Cognitive impairments are multiple and include decreased attention (see KCC 19.14), a memory disorder that affects both recent and remote memories, anomic aphasia (see KCC 19.6), and impaired executive functions (see KCC 19.11). In advanced Huntington's disease, patients are profoundly demented and lose the ability to make nearly all purposeful movements. They are bedbound, cannot speak, and usually die of respiratory infections. Median survival from the onset of first symptoms is about 15 years.

Pathologically, the most dramatic change in Huntington's disease is progressive atrophy of the caudate nucleus. The putamen is also involved, and to a lesser extent the nucleus accumbens atrophies as well. As already noted, the degeneration initially affects striatal neurons of the indirect pathway (see Figure 16.7), possibly explaining why a hyperkinetic movement disorder usually results. Atrophy of the caudate and putamen can lead the lateral ventricles to appear enlarged on CT and MRI scans. This condition is easily distinguished from hydrocephalus because in Huntington's disease the caudate head no longer forms the bulge normally seen on the walls of the lateral ventricles in coronal sections (see Figure 4.14C). As the disease progresses, milder atrophy of the cerebral cortex also occurs.

In a landmark achievement for human genetics, the abnormal gene causing Huntington's disease was mapped in 1983 and cloned in 1993. The gene is located on chromosome 4, and it includes a region containing multiple repeats of the trinucleotide sequence CAG in tandem. Normal individuals have fewer than 34 CAG repeats in this gene. Individuals with over 40 CAG repeats either have Huntington's disease or will ultimately develop this disorder. The higher the number of CAG repeats, the earlier the onset of symptoms. The gene causing Huntington's disease encodes a protein called huntingtin. Active investigation is under way to clarify how an increase in CAG repeats, encoding multiple copies of the amino acid glutamine in the huntingtin protein, can lead to Huntington's disease. It is hoped that this information will suggest a cure for this devastating condition. In addition to the gene for Huntington's disease, several other genetic disorders, many of which have prominent neurologic manifestations, have also been found to be caused by expanded trinucleotide repeats.

A suspected diagnosis of Huntington's disease can be made on the basis of the appearance of typical clinical features, especially if there is a positive family history. Inheritance is autosomal dominant with complete penetrance. However, in many cases the family history is sketchy or there may be only suggestive leads, such as a parent who died young or was institutionalized. Other causes of chorea should be considered in the differential diagnosis (see KCC 16.1). With the cloning of the Huntington's disease gene, it is now possible to perform a genetic test for Huntington's disease even before symptoms appear. Because the disorder remains incurable, however, such testing raises many ethical and philosophical issues. The test should therefore be performed only on consenting adults in a setting where specialized counseling is available.

Treatment for Huntington's disease is currently directed at alleviating symptoms and does not alter the course of the disease. Chorea can be reduced somewhat by the dopamine depleting agent tetrabenazine or by dopamine receptor blocking neuroleptics. Psychiatric manifestations can be treated with counseling and psychotropic medications. The coming years may bring advances in molecular medicine enabling more definitive treatments for Huntington's disease and other degenerative disorders. ■

KEY CLINICAL CONCEPT

16.4 STEREOTACTIC SURGERY AND DEEP BRAIN STIMULATION

Stereotactic (or **stereotaxic**) **surgery** is a relatively old technique that has had increasing applications in neurosurgery in recent years. This method allows relatively precise localization in three-dimensional space of structures in the brain, based on surface landmarks. There are several variations on this technique, and only the basic concept will be discussed here.

To establish a stereotactic coordinate system, reference points are first applied to the patient's head. Examples of reference systems used in stereotactic surgery include a series of small radiopaque markers placed on the patient's scalp, or a rigid frame affixed to the skull under local anesthesia. The patient, with the reference system in place, is then taken to the CT or MRI scanner, and images are obtained of the brain and reference system. A computer program can then calculate the location of any point in the brain relative to the external reference system.

The patient is then taken to the operating room with the reference system still in place. Using the information provided by the scanned images, the surgeon can guide the tip of a needle or probe to a precise location in the brain through a small hole in the skull. This procedure can be performed under local anesthesia. Applications of this stereotactic technique in neurosurgery have been numerous. For example, an instrument can be inserted through a narrow tube to obtain a **biopsy** of a lesion located deep within the brain. Prior to the development of stereotactic methods, a similar biopsy may have required highly invasive surgery or may have been impossible to perform. Fluid collections such as abscesses can be drained stereotactically, providing therapeutic benefit. In addition to surgery, stereotactic methods are used in radiation therapy. Known as **stereotactic radiosurgery**, this technique can be used to target highly focused beams of radiation such as the gamma knife, or Cyberknife applied externally, to a specific location within the brain.

In the treatment of movement disorders, stereotactic methods are used to carefully place stimulators or lesions at specific points in basal ganglia pathways. **Neurostimulation** has been a growing field in recent years and includes placement of electrical stimulation devices on peripheral nerves, the spinal cord, cerebral cortex, or deep brain structures. Depending on the site of stimulation, these devices have been used in the treatment of chronic pain, movement disorders, epilepsy, and psychiatric disease. **Deep brain stimulation (DBS)** involves placement of electrodes in deep brain structures such as the basal ganglia or the thalamus. DBS is a rapidly growing field in functional neurosurgery. As the use of DBS has increased, the stereotactic placement of lesions has become less common for treatment of movement disorders, since DBS has advantages of being reversible and allowing flexible adjustment of stimulus parameters.

The mechanism of DBS is still under investigation, but is thought to involve **depolarization block**, or reversible dysfunction of neurons near the tip of the stimulating electrodes due to continuous stimulation. In DBS, the stimulus is delivered by a chronically implanted stimulation device and can be externally programmed or turned off at any time. In contrast, with stereotactically placed lesions, sufficient electrical current is passed in the operating room to produce a permanent lesion near the electrode tip. No device is implanted after lesion placement, which may have some advantages in select patients due to reduced infection risk and lower cost. Lesioning procedures are referred to based on location, most commonly **pallidotomy** (internal globus pallidus), **thalamotomy** (VL_p nucleus of the thalamus), or **subthalamotomy** (subthalamic nucleus). Direct comparison trials are underway of DBS vs. lesion placement, however, DBS is currently far more widely used.

The three movement disorders most commonly treated by DBS or lesion placement are medically refractory Parkinson's disease, dystonia, and essential tremor. In advanced **Parkinson's disease**, targeting the subthalamic nucleus or internal segment of the globus pallidus can reduce on-off fluctuations (see KCC 16.2), increasing "on" time and reducing bradykinesia and rigidity. Treatment is initially contralateral to the side of worse deficits but in some cases can be bilateral. The beneficial effects of treatment can be explained by reduced inhibitory output to the thalamus due to depressed function of the internal globus pallidus, either directly or through reduced function of the subthalamic nucleus leading secondarily to reduced pallidal output (see Figure 16.7). For unclear reasons, these procedures also greatly improved drug-induced dyskinesias. It is also unclear why interfering with basal ganglia outputs to the thalamus does not typically produce deficits, a fact that highlights the need for continued research of these circuits.

For refractory **dystonia** (see KCC 16.1), DBS or lesions of the subthalamic nucleus or internal globus pallidus are similarly beneficial. **Essential tremor** (see KCC 16.1) has been shown to be associated with synchronized neuronal discharges in the VL_p, also called the **ventral intermediate nucleus (VIM)** of the thalamus. Recall that this is the main thalamic relay for *cerebellar* outputs en route to the cortex. DBS or lesioning of the thalamic VIM is effective for treating refractory essential tremor, severe parkinsonian tremor, or severe tremor from other causes, such as multiple sclerosis.

Another surgical approach has been transplantation of fetal midbrain neurons or adrenal chromaffin cells into the striatum of patients with Parkinson's disease. The long-term benefits of transplantation remain uncertain, and enthusiasm for these procedures has waned in recent years with the increasing popularity of stereotactically placed lesions and, more recently, DBS. ■

CLINICAL CASES

CASE 16.1 UNILATERAL FLAPPING AND FLINGING*

MINICASE

A 65-year-old HIV-positive man began having **involuntary flinging movements of the right arm and leg**, which became progressively worse over the course of 1 month, making gait and use of the right hand difficult. On exam, he had **continuous wild, uncontrollable flapping and circular movements of the right arm and occasional jerky movements of the right leg**, with an unsteady gait, falling to the right. The remainder of the exam was unremarkable.

LOCALIZATION AND DIFFERENTIAL DIAGNOSIS

1. On the basis of the symptoms and signs shown in **bold** above, where is the lesion?

2. What is the most likely diagnosis, and what are some other possibilities?

*A description of this patient was published previously by Provenzale and Schwarzschild in 1994; see the References section at the end of this chapter.

Discussion

The key symptoms and signs in this case are:

- **Involuntary, wild flinging movements of the right arm and leg**

1. This patient had a unilateral hyperkinetic movement disorder that could be described as hemiballismus or hemichorea (see KCC 16.1). As discussed in the section on hyperkinetic and hypokinetic movement disorders earlier in the chapter, and in KCC 16.1, hyperkinetic movement disorders are often caused by dysfunction of the contralateral subthalamic nucleus, or of indirect-pathway neurons in the striatum (see Figure 16.7). A review of Figure 16.7 should make it clear that a lesion in either location would result in less inhibitory output from the internal segment of the globus pallidus (and the substantia nigra pars reticulata) to the thalamus, resulting in increased excitatory activity traveling from the thalamus to the motor cortex.

   The most likely *clinical localization* is left subthalamic nucleus or left striatum.

2. Given the patient's age, a lacunar infarction in the left subthalamic nucleus or left striatum is the most likely diagnosis. A small hemorrhage in these locations is also possible. Nevertheless, the history of gradual onset over the course of a month would be somewhat unusual for either an infarct or a hemorrhage. Especially because the patient has a history of HIV, other brain lesions should also be considered. The most common intracranial mass lesions in patients with HIV (see KCC 5.9) are toxoplasmosis and primary central nervous system lymphoma, either of which could occur in the subthalamic nucleus or striatum.

Clinical Course and Neuroimaging

A **brain MRI** with gadolinium (Image 16.1, page 773) showed a ring-enhancing lesion in the region of the left subthalamic nucleus (compare to Figures 16.4D and 16.9). Given the clinical setting, the patient was treated empirically with the antitoxoplasmosis medications pyrimethamine and sulfadiazine (see KCC 5.9). Serum and cerebrospinal fluid *Toxoplasma* titers were positive. Four weeks later the patient's right-sided hemiballismus had subsided, but he still had some mild wavering movements on finger-to-nose and heel-to-shin testing on the right side. Repeat MRI showed that the lesion had shrunk markedly in response to therapy. MRI after 4 months showed the lesion to be nearly gone, with only a small homogeneous region of enhancement remaining.

CASE 16.2 IRREGULAR JERKING MOVEMENTS AND MARITAL PROBLEMS

CHIEF COMPLAINT

A 35-year-old man with a recent history of jerking movements went with his wife to see a psychiatrist for marital problems.

HISTORY

The wife reported that during recent months her husband had been having **occasional irregular jerking movements of the head, trunk, and limbs**. At night he sometimes would grind his teeth in his sleep, grip his wife's hand tightly without knowing it, or swallow noisily. The **husband denied having any involuntary movements** but said "they could be there." He did admit to **occasional stumbling**, having recently fallen down a flight of stairs. However, he denied that this represented any significant change in his gait over recent years. He worked as a salesman in a small business and denied any depression or problems due to intellectual function or psychiatric problems. In the army he was a high-70s golfer. At the time of consultation he was shooting 120, but he attributed this decrease in performance to practicing golf less often than previously. His wife felt something was wrong and urged him to seek medical attention, and his refusal led to **bitter arguments**.

FAMILY HISTORY

The wife's family history was unremarkable. The husband had no siblings. His father had died at age 50 years from Huntington's disease (diagnosed at 44 years). His mother's family was unaffected. The following pedigree gives further details:

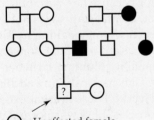

○ = Unaffected female

□ = Unaffected male

● = Female with Huntington's disease

■ = Male with Huntington's disease

?⃞ = The man in this case

PHYSICAL EXAMINATION

At the end of the counseling session, the psychiatrist set aside some time to examine the husband.

Vital signs: T = 98.6°F, P = 76, BP = 140/80, R = 16.

Neck: Supple.

Lungs: Clear.

Heart: Regular rate with no murmurs.

Abdomen: Benign.

Extremities: Normal.

Neurologic exam:

MENTAL STATUS: Alert and fully oriented. Speech fluent. Good attention and calculation skills. Recalled 3/3 objects after 5 minutes. **Affect slightly to moderately blunted**, with some apathy regarding emotional impact of the consultation, stating, for example, "If I've got it, I've got it."

CRANIAL NERVES: Normal, except that **saccadic eye movements were moderately slowed**.

MOTOR: **Rare, brief, irregular, restless-appearing movements in the face, neck, trunk, and upper extremities. Tone was normal to slightly decreased** in all extremities. 5/5 power throughout.

REFLEXES:

2+ 2+
2+ 2+
2+ 2+
2+ 2+

COORDINATION: Normal on finger-to-nose and heel-to-shin testing.

GAIT: Normal. Involuntary movements were not noticeably increased during walking. **Tandem gait was slightly unsteady.**

SENSORY: Intact.

LOCALIZATION AND DIFFERENTIAL DIAGNOSIS

1. On the basis of the symptoms and signs shown in **bold** above, which of the four parallel channels through the basal ganglia (see Table 16.2) are abnormal in this patient?

2. Does this patient have a predominantly hyperkinetic or hypokinetic movement disorder? Referring to Figure 16.7, identify the parts of the basal ganglia in which dysfunction could explain these movements.

3. Which mode of inheritance is suggested by the patient's pedigree, assuming he is affected by the same disorder as other family members? What is the most likely diagnosis? What genetic abnormality causes this disorder? What parts of the brain are predominantly affected?

Discussion

The key symptoms and signs in this case are:

- **Irregular jerking movements, slightly decreased tone, and unsteady gait**
- **Moderately slowed saccadic eye movements**
- **Flat affect, argumentative, and denied having any involuntary movements**

1. The abnormal movements suggest involvement of the motor channel; impaired saccades suggest involvement of the oculomotor channel (see Table 16.1). The patient's emotional changes and disinhibited behavior suggest involvement of the limbic channel and possibly the prefrontal channel as well (see KCC 19.11).

2. This patient has a bilateral hyperkinetic movement disorder, best described as a mild tic or chorea (see KCC 16.1). Hyperkinetic movements of this kind may be caused by bilateral dysfunction in the subthalamic nuclei or in the striatal neurons of the indirect pathway (see Figure 16.7).

3. The pedigree suggests an autosomal dominant inheritance pattern. The most likely diagnosis is Huntington's disease (see KCC 16.3). This disorder is caused by an expanded CAG trinucleotide repeat in the gene encoding the huntingtin protein, located on chromosome 4. In early Huntington's disease, the indirect-pathway neurons of the striatum are preferentially involved (see Figure 16.7). Later in the course of this disorder, there is gross degeneration of the bilateral caudate and putamen, with lesser involvement of the cerebral cortex as well.

Clinical Course and Neuroimaging

The psychiatrist was concerned that the husband in this couple could have early Huntington's disease, and he referred the patient to a neurologist. This occurred in the 1970s, before genetic testing for Huntington's disease was available. Treatment with baclofen had a modest effect in decreasing the patient's involuntary movements.

Three years later his symptoms had progressed substantially, and he was briefly admitted to the hospital for further evaluation. In the interim he had lost his job as a salesman because "business was slow." He was then fired from work as a newspaper distributor because he "was robbed." In addition, he had recently divorced. On exam, mental status was notable for mild irritability and slightly garbled speech when speaking at high speeds. In addition, he made some errors when asked to repeat nonsense syllables. He had developed frequent paroxysmal involuntary twitches of all four limbs, worse distally than proximally. Gait had a slight waddling character, with some dystonic arm carriage. Head CT was normal except for a very slight enlargement of the lateral ventricles. EEG was normal except for low voltage.

Eleven years after initial presentation the patient was readmitted after expressing suicidal thoughts when his driver's license was revoked. He said he was stopped by the police, but he was unable to say why, stating only that he was "trying to make a left turn." He had last worked about 1 year previously, doing part-time paper deliveries. He was still able to live independently at home. On exam, he was alert and oriented × 3. Speech was fluent but had an abnormal rhythm. He was able to follow complex commands. Memory and calculating skills were intact. He had a labile affect, behaving in a frustrated, angry, and impulsive manner. For example, he threatened suicide if he could not drive. He had poor insight into his illness, denying any abnormalities in speech or movements. Motor exam was notable for hypotonia, motor impersistence (see KCC 19.14), and continuous choreiform writhing movements of the tongue, arms, neck, and torso. The involuntary movements were worsened by walking, and he was unable to perform tandem gait.

A **brain MRI** performed at this time showed marked flattening of the lateral walls of the lateral ventricles (Image 16.2A, page 774) when compared to an MRI from a normal individual (Image 16.2B, page 774). This shape suggests bilateral atrophy of the head of the caudate nucleus, which normally bulges into the lateral ventricle in this location (see Image 16.2B). Note that some degree of cortical atrophy was present as well.

The patient was treated with additional medications, including dopamine antagonists such as haloperidol, with little benefit. He was discharged home once it was ascertained that he was no longer suicidal, and he was followed by both a psychiatrist and a neurologist as an outpatient. Within 2 years he could no longer be managed at home and required admission to a chronic inpatient psychiatric facility, where he died soon afterward.

RELATED CASE. Image 16.2C, page 775, shows coronal sections of two half brains. The right half is from the brain of a patient who died with Huntington's disease (this is a different patient from the one presented in Case 16.1). The left half is from the brain of a normal individual who died in an automobile accident. Note that the patient with Huntington's disease had severe atrophy of the caudate nucleus and putamen, as well as some atrophy of the nucleus accumbens and cerebral cortex.

CASE 16.1 UNILATERAL FLAPPING AND FLINGING

IMAGE 16.1 Ring-enhancing Toxoplasmosis Lesion in Left Subthalamic Nucleus Coronal T1-weighted MRI with intravenous gadolinium.

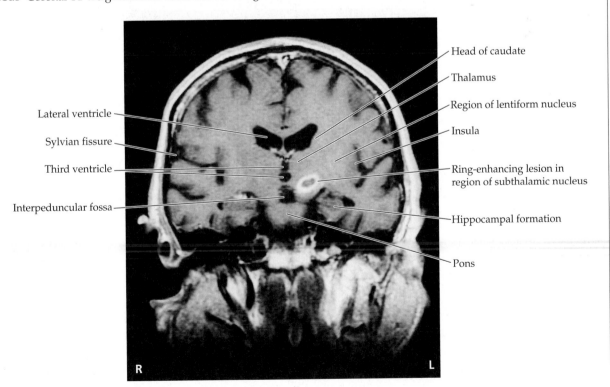

CASE 16.2 IRREGULAR JERKING MOVEMENTS AND MARITAL PROBLEMS

IMAGE 16.2A,B Atrophy of Caudate Head Associated with Huntington's Disease Coronal T1-weighted MRI scans. (A) The lateral ventricular walls have an abnormal concave shape due to severe atrophy of heads of the caudate nuclei as well as the putamen. The cortex is slightly atrophied as well. These findings are typical for Huntington's disease. (B) Normal MRI scan from a different patient at the same plane of section for comparison. Note the normal convex shape of the lateral ventricular walls, formed by the bulges of the heads of the caudate nuclei.

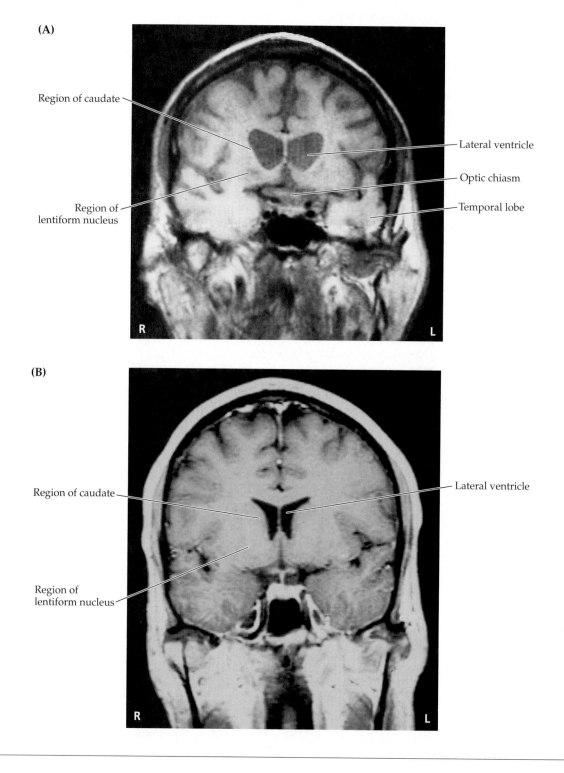

(A)

Region of caudate

Region of lentiform nucleus

Lateral ventricle

Optic chiasm

Temporal lobe

R   L

(B)

Region of caudate

Lateral ventricle

Region of lentiform nucleus

R   L

CASE 16.2 *RELATED CASE*

IMAGE 16.2C Gross Pathologic Changes of Huntington's Disease Coronal brain sections. The right half-section is from a patient who died of Huntington's disease. The caudate and putamen are markedly atrophied. Some cortical atrophy is present as well. For comparison, the left half-section is from a normal brain. (Courtesy of Jean Paul G. Vonsattel, Massachusetts General Hospital, Harvard School of Medicine.)

(C)

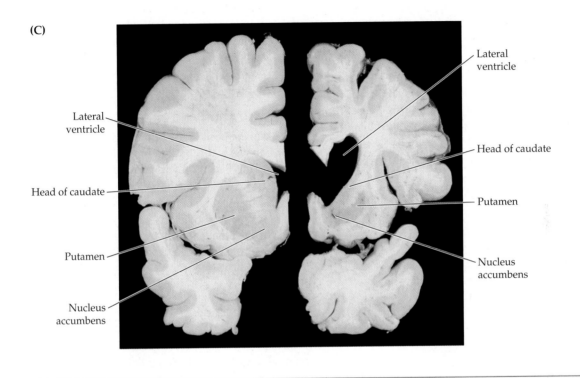

CASE 16.3 ASYMMETRICAL RESTING TREMOR, RIGIDITY, BRADYKINESIA, AND GAIT DIFFICULTIES

CHIEF COMPLAINT

A 53-year-old right-handed man was referred for a second opinion regarding progressive bradykinesia, tremor, rigidity, and unsteady gait.

HISTORY

The patient was well until 10 years previous, when, while working as a fireman, he had noticed some **slowness and difficulty using his right arm.** This symptom gradually progressed, and 2 years later he had to change jobs and began working for the phone company. In the interim he developed occasional **shaking of the right arm and right leg.** He saw a neurologist, who diagnosed him with Parkinson's disease. Treatment with **levodopa plus carbidopa (Sinemet) provided significant benefit.** Bromocriptine (a dopaminergic agonist) was also helpful. He was later enrolled in an experimental trial of deprenyl (selegiline) plus vitamin E, but his symptoms **gradually continued to worsen.** The tremor spread to involve his whole body, and he became progressively slower and stiffer, complaining that he often had **difficulty initiating movements.**

He had no family history of Parkinson's disease and no history of use of dopaminergic antagonist medications, toxin exposure, strokes, or encephalitis. CT and MRI scans were normal, and blood tests for Wilson's disease were negative.

PHYSICAL EXAMINATION

Vital signs: T = 97.1°F, P = 80, BP = 130/80, R = 14.

Skin: Slightly oily and flaky, with ceruminous buildup in ear canals.

Neck: Supple with no bruits.

Lungs: Clear.

Heart: Regular rate with no murmurs, gallops, or rubs.

Abdomen: Normal bowel sounds; nontender.

(*continued on p. 776*)

CASE 16.3 *(continued)*

Extremities: No edema.

Neurologic exam:

MENTAL STATUS: Alert and oriented × 3. Normal language. Recalled 3/3 words after 5 minutes. Spelled "world" forward and backward correctly. Copied shapes correctly. In writing a sentence, exhibited **micrographia**.

CRANIAL NERVES: Normal, except for **masklike decreased facial expression and slightly hypophonic voice.**

MOTOR: **4 Hz tremor of the head and all extremities, worse on the right side and worse at rest. Cogwheel rigidity, especially of the right arm. Finger tapping and rapid alternating movements slow bilaterally.** No pronator drift. 5/5 power throughout.

REFLEXES: **No extinction of the glabellar reflex (positive Myerson's sign).**

COORDINATION: Slow but not ataxic on finger-to-nose and heel-to-shin testing.

GAIT: **Unable to rise from a chair without assistance. Slow, stiff gait, with stooped posture, short steps, decreased arm swing,** and narrow (normal) base. **Turned slowly without twisting body (en bloc turning). Exhibited retropulsion of two steps when pulled gently backward.**

SENSORY: Intact light touch, pinprick, vibration, and joint position sense.

LOCALIZATION AND DIFFERENTIAL DIAGNOSIS

1. On the basis of the symptoms and signs shown in **bold** above, is this patient more likely to have typical idiopathic Parkinson's disease or atypical parkinsonism (KCC 16.1, 16.2)?

2. Degeneration of neurons in which structure is primarily responsible for idiopathic Parkinson's disease, and what is their main neurotransmitter? How does loss of these neurons result in a hypokinetic movement disorder?

Discussion

The key symptoms and signs in this case are:

- **Asymmetrical bradykinesia, cogwheel rigidity, and resting tremor**
- **Stooped gait with short steps, decreased arm swing, en bloc turning, and retropulsion**
- **Significant benefit from levodopa**
- **Gradual progression over a period of years**

1. This patient had all of the typical core features of idiopathic Parkinson's disease listed earlier (see KCC 16.1, 16.2), so this is the most likely diagnosis. No atypical features, such as lack of response to levodopa, early gait instability, symmetrical findings, absence of resting tremor, impaired vertical eye movements, orthostatic hypotension, or early psychiatric features were present. The gradual progression makes an acute cause, such as drug exposure, unlikely. Other associated features seen in this patient, such as the masked facies, hypophonia, micrographia, and Myerson's sign are consistent with parkinsonism but are not specific to idiopathic Parkinson's disease.

2. Parkinson's disease is caused by loss of dopaminergic neurons in the substantia nigra pars compacta. Pathologic changes typical of Parkinson's disease are shown in Figure 16.10. Dopaminergic neurons in the substantia nigra normally project to the striatum. A review of Figure 16.7 shows that loss of dopaminergic excitation of the direct pathway and loss of dopaminergic inhibition of the indirect pathway *both* ultimately result in more inhibitory output from the internal segment of the globus pallidus (and the substantia nigra pars reticulata) to the thalamus. This, in turn, leads to less excitatory activity from the thalamus to the motor and premotor cortices, resulting in a hypokinetic movement disorder.

Problems with On–Off Fluctuations

In taking the history from this patient, the examiner uncovered another problem in addition to the hypokinesia. Escalating doses of Sinemet (carbidopa and levodopa combined), amantadine, and dopaminergic agonists had helped his symptoms somewhat. However, he was having more and more problems with fluctuations, from being "on" after taking a dose of Sinemet to "off" just before a dose. When "off," he could not rise from a chair unassisted, had some trouble walking, occasionally "freezing" in place, had difficulty rolling over in bed, was slow in using utensils and carrying out hygienic activities, and could not button his shirt without assistance. When "on," he still had some difficulties walking and carrying out his daily activities, as well as excessive jerky involuntary movements (dyskinesias) of all limbs. Despite the use of a sustained-release formulation or frequent small doses, the therapeutic window had gradually narrowed, and his on–off symptoms were becoming more severe.

When examined at a different time in relation to his Sinemet dose from the exam described above, the patient had bilateral hyperkinetic dyskinesias and no tremor. Rigidity and bradykinesia were improved but still present, worse on the right than the left.

1. How might excess dopamine cause the hyperkinetic dyskinesias seen in this patient (see Figure 16.7)?

2. Given the unacceptable response to medications in this patient, what neurosurgical procedures might be tried to improve his hypokinetic parkinsonian movements? Why are these procedures expected to benefit hypokinesia?

Discussion

1. We have just seen how too little dopamine in the nigrostriatal projections could cause the hypokinetic features of Parkinson's disease (see Figure 16.7). Conversely, too much dopamine acting on striatal neurons of both the direct and the indirect pathways could inhibit the internal segment of the globus pallidus, thereby reducing the inhibitory output to the thalamus. With less inhibition of the thalamus, thalamocortical projections to motor and premotor cortex would be more active, resulting in hyperkinesia.

2. Deep brain stimulation (DBS) or a stereotactically placed lesion in the subthalamic nucleus or internal segment of the globus pallidus is an effective surgical treatment for medically refractory parkinsonism (see KCC 16.4). DBS or a lesion in the internal segment of the globus pallidus interrupts its inhibitory output to the thalamus (see Figures 16.7, 16.9). This, in turn, leads to increased thalamocortical excitatory activity reaching the motor and premotor cortex and to less hypokinesia. It is not known why this procedure does not produce a hyperkinetic movement disorder. In fact, for unclear reasons, pallidotomy actually causes a marked improvement in medication-related hyperkinetic dyskinesias. Another procedure that is beneficial and has been used more frequently in recent years for treating refractory parkinsonism is implantation of a subthalamic stimulator (see KCC 16.4). By decreasing function of the subthalamic nucleus, this reduces excitatory inputs to the internal globus pallidus, again reducing inhibition of the thalamus, and relieving hypokinesia (see Figures 16.7, 16.9). Another option is subthalamotomy, which, like DBS of the subthalamic nucleus, leads to reduced inhibitory output from the internal globus pallidus.

Clinical Course and Neuroimaging

After reviewing this patient's history and his increasingly narrow therapeutic window in response to medications, a neurosurgical procedure was recommended. Because this took place before the recent increase in popularity of DBS, a stereotactic pallidotomy was performed. The procedure was carried out on the left side because his parkinsonism was worse on the right. The surgery was done with mild sedation and local anesthesia so that the patient could remain awake and be tested neurologically during the procedure. An MRI was done with the stereotactic frame in place, and he was then taken to the operating room with the frame in place. The MRI was used to calculate the coordinates of the left globus pallidus with respect to the frame. A small burr hole was then placed in the left frontal bone, the dura was opened, and an electrode was advanced by 2 mm increments until it reached the internal segment of the left globus pallidus. Before a permanent lesion was created in this area, the location of the electrode tip was tested by passage of a high-frequency electrical stimulus. This caused reversible dysfunction of the cells near the electrode tip, resulting in a dramatic improvement in the patient's right-sided bradykinesia and rigidity. In addition, no visual changes or hemiparesis occurred, indicating that the electrode tip was not too close to the optic tract or internal capsule, respectively (see Figure 16.4D). A permanent lesion was next created by the passage of sufficient current to heat the electrode tip to 70°C.

A postoperative **brain MRI** is shown in Image 16.3A,B, page 781. The stereotactically placed lesion is visible in Image 16.3A, just ventral to the left globus pallidus (see Image 16.3B). One day after surgery, the patient no longer had any tremor on the right side, the tremor on the left was decreased, and he had only a single episode of mild dyskinesia lasting about 10 minutes. He also showed marked improvement in his rigidity and his gait was faster, with larger steps and increased arm swing. Three months after surgery he was seen in follow-up while taking the same medications. He continued to enjoy a marked improvement in his symptoms, saying, "I'm doing a lot, lot better. I can walk. I can walk straight, and people don't look at me like I'm a weirdo anymore." He was "on" for about 14.5 hours per day and "off" for only 2 hours per day. He had dyskinesias for about 3 of his "on" hours, but these were much milder than they had been preoperatively and did not interfere with his activities. On exam, his speech was normal, he had no tremor, he had mild facial bradykinesia, mild dyskinesias of the left leg, minimal rigidity on the right side, moderate rigidity on the left side, and mild slowing of finger tapping. He was able to rise from a chair without difficulty and had a normal gait.

RELATED CASE. As we have already mentioned, the treatment of movement disorders by stereotactic lesions is now less common, since DBS has advantages of being reversible and allowing flexible adjustment of stimulus parameters (see KCC 16.4). More recently than the case just discussed, a 56-year-old woman with advanced Parkinson's disease came to see a neurologist specializing in movement disorders. This patient had a nine-year history of bradykinesia and rigidity with symptoms initially on the left side. She also had a tremor, and later developed some gait difficulties. She was treated with carbidopa plus levodopa (Sinemet) which helped her symptoms, and several other medications were tried over the years, however she developed severe on-off fluctuations—alternating between incapacitating dyskinesias and total body freezing. On examination, she had normal mental status, and just before a Sinemet dose she was bradykinetic and rigid more on the left than the right side, and had a rest tremor worse on the left. After a Sinemet dose she had marked dyskinetic movements, also worse on the left side, with jaw dyskinesias so severe that it was difficult for her to speak.

The patient was referred for neurosurgical placement of a deep brain stimulator targeting the subthalamic nucleus on the right side (see KCC 16.4). She was taken to the operating room where a stereotactic frame was attached to the patient's head under local anesthesia. She then underwent an MRI scan and was returned to the operating room with the frame still in place. A point at the middle of the anterior commissure was identified on MRI (see Figures 4.14C, 4.15A). Using the stereotactic system, a trajectory was planned targeting the subthalamic nucleus, located based on a standard reference atlas at 12 mm lateral, 3 mm posterior, and 5 mm inferior to the mid-commissural point. The red nucleus was also identified on MRI and the location of the target in the subthalamic nucleus was confirmed to be in the appropriate location just rostral to the red nucleus (see Figures 14.3A, 16.4D, and Image 16.1). Again using the stereotactic reference system a scalp incision was made and burr hole drilled in the skull to allow passage of a cannula to a depth of 15 mm above the target. A microelectrode was passed through the guide cannula towards the target, and the typical firing patterns of neurons were identified as the electrode passed progressively from dorsal to ventral through the thalamus, zona incerta, subthalamic nucleus, and substantia nigra (see Figure 16.4D). These firing patterns, and the response to a stimulating electrode were further used to identify the correct final location for the stimulating electrode. Following the procedure, a **brain MRI** was performed (Image 16.3C,D, page 782) demonstrating the tip of the stimulating electrode in the right subthalamic nucleus.

Six days after the electrode placement, a second minor surgical procedure was performed to connect the electrodes as they exited the skull to a set of wires that were tunneled under the skin of the skull and neck to reach the programmable stimulus generator located under the skin in the chest just beneath the clavicle. The patient was seen again by the movement disorders neurologist in the weeks following surgery for programming and adjustment of her stimulus settings. Her bradykinesia and rigidity on the left side were greatly improved, and there was a marked reduction of her on-off fluctuations and dyskinesias.

CASE 16.4 BILATERAL BRADYKINESIA, RIGIDITY, AND GAIT INSTABILITY WITH NO TREMOR

MINICASE

A 48-year-old woman gradually developed difficulty with handwriting and typing, saying that her fingers were stiff and slow. In addition, her **gait became unsteady**, and she had several falls. She was **treated with levodopa plus carbidopa with no significant benefit**. When examined 5 years after symptom onset, she had a normal mental status, **slow saccades, masked facies, slow dysarthric speech, prominent bilateral bradykinesia and rigidity, especially of the axial and neck muscles, no tremor, and a slow, shuffling gait with retropulsion**. There was no evidence of autonomic dysfunction, ataxia, or dementia. Although saccades were slow in the vertical direction, there was no significant limitation of up- or downgaze.

LOCALIZATION AND DIFFERENTIAL DIAGNOSIS

1. On the basis of the symptoms and signs shown in **bold** above, is this patient more likely to have typical idiopathic Parkinson's disease or atypical parkinsonism (see KCC 16.1, 16.2)?

2. What is the most likely diagnosis?

3. Which neurons degenerate in this condition? How does this result in a hypokinetic movement disorder? Referring to Figure 16.7, explain why patients with this disorder usually do not respond very well to levodopa therapy.

Discussion

The key symptoms and signs in this case are:

- **Bilateral bradykinesia and waxy rigidity**
- **No tremor**
- **Early gait unsteadiness; later slow, shuffling gait with retropulsion (parkinsonian gait)**
- **No significant benefit from levodopa**
- **Gradual progression over a period of years**
- **Masked facies, slow saccades, slow dysarthric speech**

1. Bilaterally symmetrical symptoms, absence of tremor, early gait unsteadiness, and lack of response to levodopa make this a case of atypical parkinsonism (see KCC 16.2; Table 16.5).

2. Most of the disorders listed in Table 16.5, are characterized by other significant abnormalities in addition to atypical parkinsonism. Atypical parkinsonism alone, without other significant abnormalities, can be seen in striatonigral degeneration (a form of multisystem atrophy). Drug-induced parkinsonism is another possibility; however, the patient's gradual progression over years does not fit this etiology. The masked facies, slow saccades, and dysarthria are nonspecific findings present in most parkinsonian disorders. In conclusion, the most likely diagnosis is striatonigral degeneration.

3. In striatonigral degeneration, as in idiopathic Parkinson's disease, there is loss of dopaminergic neurons from the substantia nigra pars compacta. The result is more inhibitory activity to the thalamus from the internal globus pallidus and substantia nigra pars reticulata, causing a hypokinetic movement disorder (see Figure 16.7). In idiopathic Parkinson's disease, this dopamine deficiency can be corrected by the administration of levodopa. The medication levodopa is converted to dopamine in the brain, which can then act on striatal neurons (see Figure 16.7). In striatonigral degeneration, however, the striatal neurons degenerate as well. Therefore, administration of levodopa is usually not as beneficial to these patients as it is to patients with Parkinson's disease.

Clinical Course and Postmortem Examination

The patient's dysarthria, dysphagia, bilateral bradykinesia, and rigidity without tremor continued to progress. Six years after symptom onset, she underwent pallidotomy (see KCC 16.4), which provided only some transient benefit. She eventually became bedbound and died of aspiration pneumonia 7 years after initial symptoms.

In accordance with the patient's previously stated wishes, her family consented to an autopsy. On examination of brain sections by eye, the substantia nigra appeared somewhat pale (similar to the left side of Figure 16.10A). In addition, the caudate, putamen, and external segment of the globus pallidus were markedly atrophied (Image 16.4A,B, page 783). Microscopic examination of the substantia nigra revealed marked loss of pigmented neurons and an increase in glial cells when compared to a normal control (Image 16.4C,D, page 784). However, unlike the typical findings in Parkinson's disease, Lewy bodies were not present. Microscopic examination of the striatum also showed marked neuronal loss and gliosis compared to a normal control (Image 16.4E,F, page 784). Some atrophy was also found in the external segment of the globus pallidus, subthalamic nucleus, pons, locus ceruleus, and cerebellum. These findings were compatible with a diagnosis of multisystem atrophy of the striatonigral degeneration type.

CASE 16.3 ASYMMETRICAL RESTING TREMOR, RIGIDITY, BRADYKINESIA, AND GAIT DIFFICULTIES

IMAGE 16.3A,B Left Pallidotomy Performed for Advanced Parkinson's Disease Horizontal T1-weighted MRI scans. A and B are adjacent horizontal sections progressing from inferior to superior. (A) Stereotactically placed lesion is visible. (B) By comparison of this image with A, it can be seen that the lesion was placed along the ventral edge of the globus pallidus.

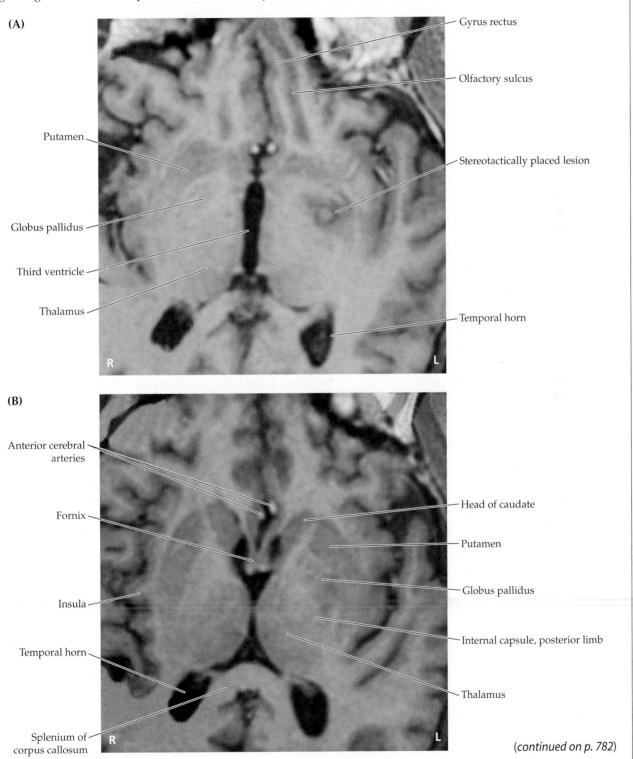

(A)

Gyrus rectus

Olfactory sulcus

Putamen

Stereotactically placed lesion

Globus pallidus

Third ventricle

Thalamus

Temporal horn

R   L

(B)

Anterior cerebral arteries

Fornix

Head of caudate

Putamen

Globus pallidus

Insula

Internal capsule, posterior limb

Temporal horn

Thalamus

Splenium of corpus callosum

R   L

(continued on p. 782)

IMAGE 16.3C,D Subthalamic Stimulator for Advanced Parkinson's Disease Sagittal T1-weighted MRI scans. C and D are adjacent sagittal sections progressing from lateral to medial. (C) Sagittal image showing track of deep brain stimulator (DBS) electrode passing through thalamus towards subthalamic nucleus. (D) Sagittal image showing DBS electrode tip in right subthalamic nucleus.

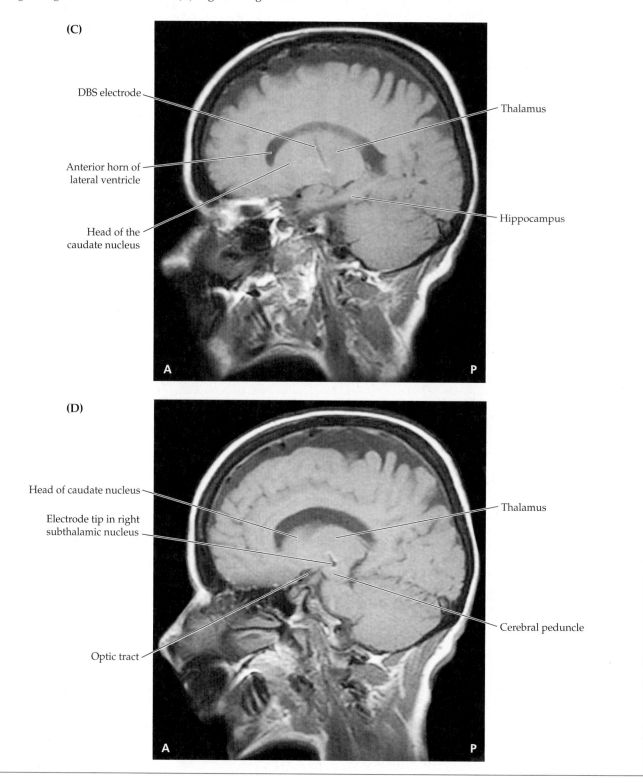

(C)

DBS electrode

Thalamus

Anterior horn of lateral ventricle

Head of the caudate nucleus

Hippocampus

A   P

(D)

Head of caudate nucleus

Thalamus

Electrode tip in right subthalamic nucleus

Optic tract

Cerebral peduncle

A   P

CASE 16.4 BILATERAL BRADYKINESIA, RIGIDITY, AND GAIT INSTABILITY WITH NO TREMOR

IMAGE 16.4A,B **Gross Pathologic Changes in a Patient with Striatonigral Degeneration** Coronal brain sections from patient in Case 16.4; A, B progress from anterior to posterior. Note the severe atrophy of the striatum, including the caudate and putamen. The external segment of the globus pallidus appears atrophied as well.

(A)

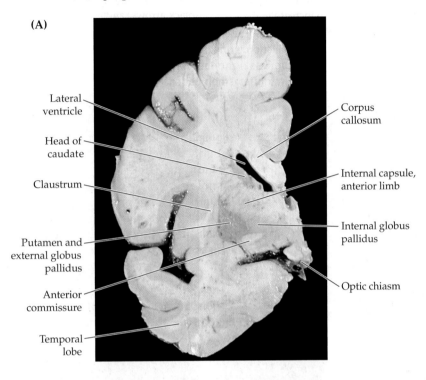

(B)

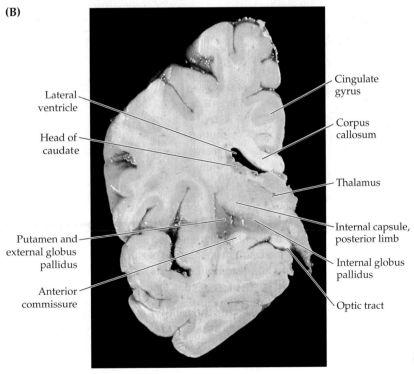

(*continued on p. 784*)

References

Anatomy and Circuit Connections of the Basal Ganglia

Albin RL, Young AB, Penny JB. 1989. The functional anatomy of basal ganglia disorders. *Trends Neurosci* 12 (10): 366–375.

DeLong MR, Wichmann T. 2007. Circuits and circuit disorders of the basal ganglia. *Arch Neurol* 64 (1): 20–24.

Galvan A, Wichmann T. 2007. GABAergic circuits in the basal ganglia and movement disorders. *Prog Brain Res* 160: 287–312.

Haber SN, Calzavara R. 2009. The cortico-basal ganglia integrative network: the role of the thalamus. *Brain Res Bull* 78 (2–3): 69–74.

Kreitzer AC, Malenka RC. 2008. Striatal plasticity and basal ganglia circuit function. *Neuron* 60 (4): 543–554.

Martin JH. 2003. *Neuroanatomy: Text and Atlas.* 3rd Ed., Chapter 14. Appleton & Lange, Stamford, CT.

Obeso JA, Rodríguez-Oroz MC, Benitez-Temino B, Blesa FJ, Guridi J, Marin C, Rodriguez M. 2008. Functional organization of the basal ganglia: therapeutic implications for Parkinson's disease. *Mov Disord* 23 Suppl 3: S548–559.

General Movement Disorders

Riley DE, Lang AE. 2008. Movement disorders. In WG Bradley, RB Daroff, GM Fenichel, CD Marsden (eds.), *Neurology in Clinical Practice: Principles of Diagnosis and Management.* 5th Ed., Chapter 24. Elsevier, Boston.

Hemichorea or Hemiballismus

Hamani C, Saint-Cyr JA, Fraser J, Kaplitt M, Lozano AM. 2004. The subthalamic nucleus in the context of movement disorders. *Brain* 127 (Pt 1): 4–20.

Lee MS, Marsden CD. 1994. Movement disorders following lesions of the thalamus or subthalamic region. *Mov Disorders* 9 (5): 493–507.

Lownie SP, Gilbert JJ. 1990. Hemichorea and hemiballismus: Recent concepts. *Clin Neuropathol* 9 (1): 46–50.

Park SY, Kim HJ, Cho YJ, Cho JY, Hong KS. 2009. Recurrent hemichorea following a single infarction in the contralateral subthalamic nucleus. *Mov Disord* 24 (4): 617–618.

Provenzale JM, Schwarzschild MA. 1994. Hemiballismus. *Am J Neuroradiol* 15 (7): 1377–1382.

Vidailhet M. 2000. Paroxysmal dyskinesias as a paradigm of paroxysmal movement disorders. *Curr Opin Neurol* 13 (4): 457–462.

Vidakovic A, Dragasevic N, Kostic VS. 1994. Hemiballism: Report of 25 cases. *J Neurol Neurosurg Psychiatry* 57 (8): 945–949.

Tourette's Syndrome

Bloch MH. 2008. Emerging treatments for Tourette's disorder. *Curr Psychiatry Rep* 10 (4): 323–330.

Jankovic J. 2001. Tourette's syndrome. *N Engl J Med* 345 (16): 1184–1192.

Lombroso PJ, Scahill L. 2008. Tourette syndrome and obsessive-compulsive disorder. *Brain Dev* 30 (4): 231–237.

Wilson's Disease

Ala A, Walker AP, Ashkan K, Dooley JS, Schilsky ML. 2007. Wilson's disease. *Lancet* 369 (9559): 397–408.

Das SK, Ray K. 2006. Wilson's disease: an update. *Nat Clin Pract Neurol* 2 (9): 482–493.

Demirkiran M, Jankovic J, Lewis RA, Cox DW. 1996. Neurologic presentation of Wilson disease without Kayser–Fleischer rings. *Neurology* 46 (4): 1040–1043.

Tremor

Benito-León J, Louis ED. 2006. Essential tremor: emerging views of a common disorder. *Nat Clin Pract Neurol* 2 (12): 666–678.

Deng H, Le W, Jankovic J. 2007. Genetics of essential tremor. *Brain* 130 (Pt 6): 1456–1464.

Findley LJ, Cleeves L. 1989. Classification of tremor. In NP Quinn, PG Jenner (eds.), *Disorders of Movement: Clinical, Pharmacological and Physiological Aspects*, Chapter 36. Academic Press, London.

Koller WC, Huber SJ. 1989. Tremor disorders of aging: Diagnosis and management. *Geriatrics* 44 (5): 33–37.

Parkinson's Disease and Related Disorders

Berger Y, Salinas JN, Blaivas JG. 1990. Urodynamic differentiation of Parkinson disease and the Shy Drager syndrome. *Neurourol Urodynamics* 9: 117–121.

Dale RC, Webster R, Gill D. 2007. Contemporary encephalitis lethargica presenting with agitated catatonia, stereotypy, and dystonia-parkinsonism. *Mov Disord* 22 (15): 2281–2284.

Gouider-Khouja N, Vidailhet M, Bonnet A, Pichon J, Agid Y. 1995. "Pure" striatonigral degeneration and Parkinson's disease: A comparative clinical study. *Mov Disord* 10 (3): 288–294.

Köllensperger M, Geser F, Seppi K, Stampfer-Kountchev M, Sawires M, Scherfler C, Boesch S, Mueller J, et al.; European MSA Study Group. 2008. Red flags for multiple system atrophy. *Mov Disord* 23 (8): 1093–1099.

Lees AJ, Hardy J, Revesz T. 2009. Parkinson's disease. *Lancet* 373 (9680): 2055–2066.

Lipp A, Sandroni P, Ahlskog JE, Fealey RD, Kimpinski K, Iodice V, Gehrking TL, Weigand SD, et al. 2009. Prospective differentiation of multiple system atrophy from Parkinson disease, with and without autonomic failure. *Arch Neurol* 66 (6): 742–750.

Nutt JG, Wooten GF. 2005. Clinical practice. Diagnosis and initial management of Parkinson's disease. *N Engl J Med* 353 (10): 1021–1027.

Obeso JA, Marin C, Rodriguez-Oroz C, Blesa J, Benitez-Temiño B, Mena-Segovia J, Rodríguez M, Olanow CW. 2008. The basal ganglia in Parkinson's disease: current concepts and unexplained observations. *Ann Neurol* 64 Suppl 2: S30–46.

Poewe W. 2006. The natural history of Parkinson's disease. *J Neurol* 253 Suppl 7: VII 2–6.

Poewe W. 2009. Treatments for Parkinson disease—past achievements and current clinical needs. *Neurology* 72 (7 Suppl): S65–73.

Susatia F, Fernandez HH. 2009. Drug-induced parkinsonism. *Curr Treat Options Neurol* 11 (3): 162–169.

Wenning GK, Shlomo YB, Magalhaes M, Daniel SE, Quin NP. 1994. Clinical features and natural history of multiple system atrophy. An analysis of 100 cases. *Brain* 117 (Pt. 4): 835–845.

Williams DR, Lees AJ. 2009. Progressive supranuclear palsy: clinicopathological concepts and diagnostic challenges. *Lancet Neurol* 8 (3): 270–279.

Yoshida M. 2007. Multiple system atrophy: alpha-synuclein and neuronal degeneration. *Neuropathology* 27 (5): 484–493.

Huntington's Disease

Bossy-Wetzel E, Petrilli A, Knott AB. 2008. Mutant huntingtin and mitochondrial dysfunction. *Trends Neurosci* 31 (12): 609–616.

Cui L, Jeong H, Borovecki F, Parkhurst CN, Tanese N, Krainc D. 2006. Transcriptional repression of PGC-1alpha by mutant huntingtin leads to mitochondrial dysfunction and neurodegeneration. *Cell* 127: 59–69.

Greenamyre JT. 2007. Huntington's disease—making connections. *N Engl J Med* 356 (5): 518–520.

Gusella JF, Wexler NS, Conneally PM, Naylor SL, Anderson MA, Tanzi RE, Watkins PC, Ottina K, et al. 1983. A polymorphic DNA marker genetically linked to Huntington's disease. *Nature* 306 (17): 234–238.

Hodges A, Strand AD, Aragaki AK, Kuhn A, Sengstag T, Hughes G, Elliston LA, Hartog C, et al. 2006. Regional and cellular gene expression changes in human Huntington's disease brain. *Hum Mol Genet* 15 (6): 965–977.

Landles C, Bates GP. 2004. Huntingtin and the molecular pathogenesis of Huntington's disease. *EMBO Rep* 5: 958–963.

Lanska DL. 1995. George Huntington and hereditary chorea. *J Child Neurol* 10 (1): 46–48.

Paulsen JS. 2009. Functional imaging in Huntington's disease. *Exp Neurol* 216 (2): 272–277.

Phillips W, Shannon KM, Barker RA. 2008. The current clinical management of Huntington's disease. *Mov Disord* 23 (11): 1491–1504.

The Huntington's Disease Collaborative Research Group. 1993. A novel gene containing a trinucleotide repeat that is expanded and unstable on Huntington's disease chromosomes. *Cell* 72 (16): 971–983.

van der Burg JM, Björkqvist M, Brundin P. 2009. Beyond the brain: widespread pathology in Huntington's disease. *Lancet Neurol* 8 (8): 765–774.

Stereotactic Surgery and Deep Brain Stimulation

Ackermans L, Temel Y, Visser-Vandewalle V. 2008. Deep brain stimulation in Tourette's Syndrome. *Neurotherapeutics* 5 (2): 339–344.

Benabid AL, Chabardes S, Mitrofanis J, Pollak P. 2009. Deep brain stimulation of the subthalamic nucleus for the treatment of Parkinson's disease. *Lancet Neurol* 8 (1): 67–81.

Charles PD, Gill CE, Davis TL, Konrad PE, Benabid AL. 2008. Is deep brain stimulation neuroprotective if applied early in the course of PD? *Nat Clin Pract Neurol* 4 (8): 424–426.

Diamond A, Jankovic J. 2005. The effect of deep brain stimulation on quality of life in movement disorders. *J Neurol Neurosurg Psychiatry* 76 (9): 1188–1193.

Dogali M, Sterio D, Fazzini E, Kolodny E, Eidelberg D, Berie A. 1996. Effects of posteroventral pallidotomy on Parkinson's disease. *Adv Neurol* 69: 585–590.

Esselink RAJ, de Bie RMA, de Haan RJ, Lenders MWPM, Nijssen PCG, van Laar T, Schuurman PR, Bosch DA, et al. 2009. Long-term superiority of subthalamic nucleus stimulation over pallidotomy in Parkinson disease. *Neurology* 73: 151–153.

Guridi J, Obeso JA, Rodriguez-Oroz MC, Lozano AA, Manrique M. 2008. L-dopa-induced dyskinesia and stereotactic surgery for Parkinson's disease. *Neurosurgery* 62 (2): 311–323; discussion 323–325.

Iacono RP, Lonser RR, Mandybur G, Morenski JD, Yamoda S, Shima F. 1994. Stereotactic pallidotomy results for Parkinson's exceed those for fetal graft. *Am Surg* 60 (10): 777–782.

Kluger B, Klepitskaya O, Okun M. 2009. Surgical Treatment of Movement Disorders. *Neurol Clin* 27: 633–677.

Kopell BH, Rezai AR, Chang JW, Vitek JL. 2006. Anatomy and physiology of the basal ganglia: implications for deep brain stimulation for Parkinson's disease. *Movement Disorders* 21 (Suppl 14): S238–246.

Lang AE, Lozano AM, Montgomery E, Duff J, Tasker R, Hutchinson W. 1997. Posteroventral medial pallidotomy in advanced Parkinson's disease. *N Engl J Med* 337 (15): 1036–1042.

Uc EY, Follett KA. 2007. Deep brain stimulation in movement disorders. *Sem Neurol* 27 (2): 170–182.

Wichmann T, Delong MR. 2006. Deep brain stimulation for neurologic and neuropsychiatric disorders. *Neuron* 52 (1): 197–204.

CONTENTS

Chapter *17*

Pituitary and Hypothalamus

The hypothalamus and pituitary exert complex and fine control over the endocrine system, but because of their anatomical relations to adjacent structures, pituitary or hypothalamic lesions can cause visual deficits as well. A 50-year-old woman developed gradually worsening vision problems over the course of several months that eventually interfered with her driving. She also had a long-standing history of menstrual irregularity and infertility. Her examination was normal except for decreased vision bilaterally, primarily in the temporal portions of her visual fields. Eventually, it was discovered that this patient had a lesion in the pituitary region compressing her optic chiasm.

In this chapter we will learn about the anatomy and neuroendocrine functions of the hypothalamus and pituitary and clinical ramifications of lesions in these structures.

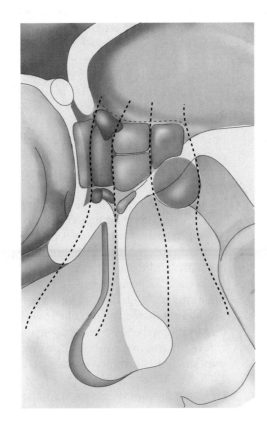

ANATOMICAL AND CLINICAL REVIEW

MNEMONIC

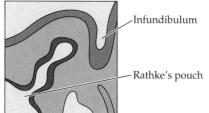

(A) 22 days

Prosencephalon Ventricle

Mesencephalon

Ectodermal thickening

Rhombencephalon

Pharyngeal opening Notochord

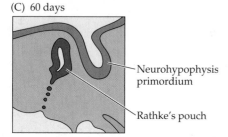

(B) 42 days

Infundibulum

Rathke's pouch

(C) 60 days

Neurohypophysis primordium

Rathke's pouch

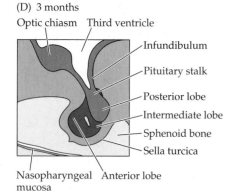

(D) 3 months

Optic chiasm Third ventricle

Infundibulum

Pituitary stalk

Posterior lobe

Intermediate lobe

Sphenoid bone

Sella turcica

Nasopharyngeal Anterior lobe
mucosa

FIGURE 17.1 Embryological Development of the Anterior and Posterior Pituitary

THE PITUITARY AND HYPOTHALAMUS constitute a unique region of the nervous system. In addition to communicating through conventional synaptic transmission, both of these structures utilize soluble humoral factors as a major source of afferent and efferent information. The pituitary and hypothalamus form the link between the **neural and endocrine systems**. In addition, the hypothalamus is the central regulator of **homeostasis** (mnemonic: **H**ypothalamus **H**omeostasis), and has been informally nicknamed the "homeostatic head ganglion." The hypothalamus maintains homeostasis in the body by interacting with and exerting important regulatory influences over four other systems, thereby participating in:

1. **Homeostatic** mechanisms controlling hunger, thirst, sexual desire, sleep–wake cycles, etc.
2. **Endocrine** control, via the pituitary
3. **Autonomic** control
4. **Limbic** mechanisms (see Chapter 18)
   (Mnemonic: **HEAL**)

In the sections that follow, we will first review the overall anatomy of the pituitary and hypothalamus. We will then discuss the major hypothalamic nuclei and their roles in each of the above functions, focusing in most detail on neuroendocrine control of pituitary hormones. Finally, we will review the clinical effects of pituitary and hypothalamic dysfunction.

Overall Anatomy of the Pituitary and Hypothalamus

The pituitary, or hypophysis, is derived from two different embryological pouches (**Figure 17.1**). The **anterior pituitary**, or **adenohypophysis**, is formed by a thickened area of ectodermal cells on the roof of the developing pharynx that invaginate, forming **Rathke's pouch**. The **posterior pituitary**, or **neurohypophysis**, forms from an evagination of the floor of the developing ventricular system. The anterior pituitary contains glandular cells that secrete a variety of hormones into the circulation. Release of hormones from the anterior pituitary is controlled by the hypothalamus through factors carried in a specialized vascular portal system, as we will discuss later (see Figure 17.5). The posterior wall of Rathke's pouch forms a small region called the intermediate part of the anterior lobe (also called the intermediate lobe) of the pituitary (see Figure 17.1D), which has less prominent endocrine functions in humans. The posterior pituitary does not contain glandular cells. Instead, it contains axons and terminals of neurons whose cell bodies are located in the hypothalamus. These terminals in the posterior pituitary secrete the hormones oxytocin and vasopressin into the circulation.

The **hypothalamus** is part of the diencephalon, and it is named for its location underneath the thalamus (**Figure 17.2A**). The hypothalamus forms the walls and floor of the inferior portion of the third ventricle (see Figure 17.2A; see also Figure 16.4C). The hypothalamus is separated from the thalamus by a shallow groove on the wall of the third ventricle called the **hypothalamic sulcus**. On the ventral surface of the brain, the hypothalamus can be seen just posterior to the optic chiasm, forming the tuber cinereum and mammillary bodies (**Figure 17.2B**). However, portions of the hypothalamus are located dorsal to the optic chiasm as well (see Figure 17.3). The **tuber cinereum**, meaning "gray protuberance," is a bulge located between the optic chiasm and the mammillary bodies. The **mammillary**

(A)

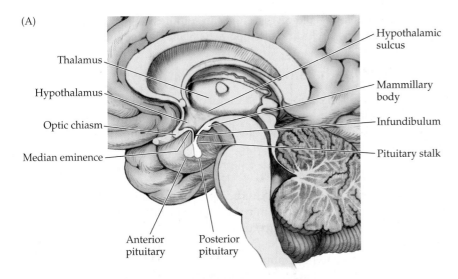

Thalamus

Hypothalamus

Optic chiasm

Median eminence

Hypothalamic sulcus

Mammillary body

Infundibulum

Pituitary stalk

Anterior pituitary

Posterior pituitary

(B)

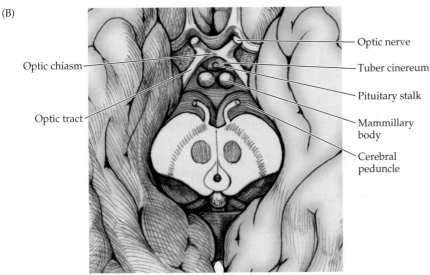

Optic chiasm

Optic tract

Optic nerve

Tuber cinereum

Pituitary stalk

Mammillary body

Cerebral peduncle

FIGURE 17.2 Anatomical Relations of Hypothalamus and Pituitary (A) Medial view. (B) Inferior view with the brainstem and pituitary removed.

bodies are paired structures that form the posterior portion of the hypothalamus. The **infundibulum**, meaning "funnel," arises from the tuber cinereum and continues inferiorly as the **pituitary stalk** (see Figure 17.2A). The anterior portion of the infundibulum is slightly elevated and is called the **median eminence**. The median eminence is the region where hypothalamic neurons release regulating factors that are carried by portal vessels to the anterior pituitary (see Figure 17.5).

The pituitary gland lies within the **pituitary fossa** (see Figures 12.1 and 12.3A). The pituitary fossa is bounded by the **anterior clinoid process** and the **posterior clinoid process**, which, together with the intervening portions of the sphenoid bone, form the fancifully named **sella turcica**, meaning "Turkish saddle." Just beneath the floor of the sella turcica lies the **sphenoid sinus**, allowing the pituitary fossa to be accessed by a transsphenoidal surgical approach (see KCC 17.1). Within the pituitary fossa the pituitary is surrounded by dura. The dura covering the superior portion of the pituitary fossa is called the **diaphragma sella**, and the pituitary stalk communicates with the main cranial cavity through a round hole in the middle of the diaphragma sella (see Figure 10.11B). The pituitary fossa is bounded laterally on both sides by the **cavernous sinus** (see Figure 13.11). Note that the pituitary and other sellar and suprasellar structures lie just behind and inferior to the optic chiasm (see Fig-

REVIEW EXERCISE

1. Which of the following is derived embryologically from Rathke's pouch and which from the prosencephalon?

   A. Anterior pituitary

   B. Posterior pituitary

2. Enlargement of the pituitary by a tumor may compress which structure of the visual pathway?

TABLE 17.1 Some Important Hypothalamic Nuclei
PERIVENTRICULAR AREA
Periventricular nucleus
MEDIAL HYPOTHALAMIC AREA
Preoptic area
Medial preoptic nucleus
Anterior (supraoptic) region
Anterior hypothalamic nucleus
Supraoptic nucleus
Paraventricular nucleus
Suprachiasmatic nucleus
Middle (tuberal) region
Arcuate nucleus
Ventromedial nucleus
Dorsomedial nucleus
Posterior (mammillary) region
Medial mammillary nucleus
Intermediate mammillary nucleus
Lateral mammillary nucleus
Posterior hypothalamic nucleus
LATERAL HYPOTHALAMIC AREA
Lateral preoptic nucleus
Lateral hypothalamic nucleus

ure 17.2). Tumors in this region therefore can compress the optic chiasm, causing visual problems, including **bitemporal hemianopia** (see KCC 11.2).

Important Hypothalamic Nuclei and Pathways

In this section we will discuss the anatomy of the major hypothalamic nuclei and the hypothalamic regions that specialize in homeostatic, autonomic, and limbic functions. The neuroendocrine functions of the hypothalamus will be discussed in detail in the next section.

Major Hypothalamic Nuclei

The hypothalamic nuclei can be divided into four major regions from anterior to posterior (Figure 17.3) and into three areas from medial to lateral (Figure 17.4). Most medially, the **periventricular nucleus** is a thin layer of cells that lies closest to the third ventricle. The fibers of the fornix pass through the hypothalamus on the way to the mammillary body, dividing the major portions of the hypothalamus into a **medial hypothalamic area** and a **lateral hypothalamic area** (see Figure 17.4). The lateral hypothalamic area consists of the **lateral hypothalamic nucleus** and several smaller nuclei. Running through the lateral hypothalamic area in the rostrocaudal direction is a diffuse group of fibers called the **medial forebrain bundle** (**MFB**) (see Figure 17.4), which carries many connections to and from the hypothalamus, and between other regions. Note that the *medial* forebrain bundle runs through the *lateral* hypothalamus.

The **medial hypothalamic area** consists of several different nuclei (Table 17.1; see Figures 17.3 and 17.4), which divide into four regions from anterior to posterior. Most anteriorly, the **preoptic area** is derived embryologically from the telencephalon, while the hypothalamus is derived from the diencephalon. Nevertheless, the preoptic area is functionally part of the hypothalamus. The **lat-**

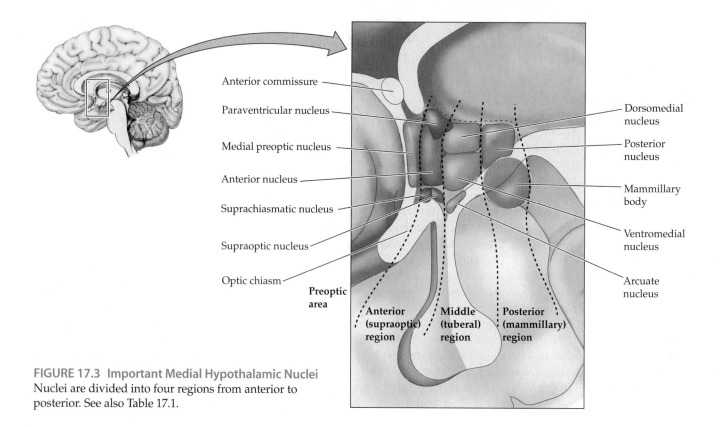

FIGURE 17.3 Important Medial Hypothalamic Nuclei
Nuclei are divided into four regions from anterior to posterior. See also Table 17.1.

(A) Preoptic area

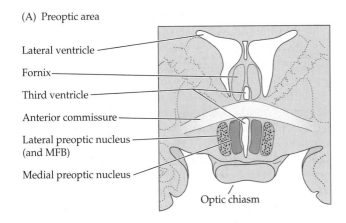

Lateral ventricle

Fornix

Third ventricle

Anterior commissure

Lateral preoptic nucleus (and MFB)

Medial preoptic nucleus

Optic chiasm

(B) Anterior (supraoptic) region

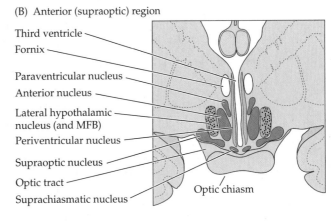

Third ventricle

Fornix

Paraventricular nucleus

Anterior nucleus

Lateral hypothalamic nucleus (and MFB)

Periventricular nucleus

Supraoptic nucleus

Optic tract

Suprachiasmatic nucleus

Optic chiasm

(C) Middle (tuberal) region

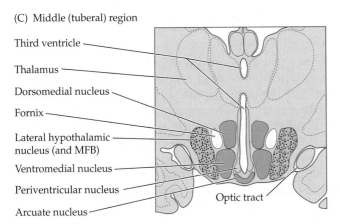

Third ventricle

Thalamus

Dorsomedial nucleus

Fornix

Lateral hypothalamic nucleus (and MFB)

Ventromedial nucleus

Periventricular nucleus

Arcuate nucleus

Optic tract

(D) Posterior (mammillary) region

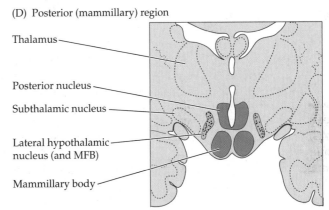

Thalamus

Posterior nucleus

Subthalamic nucleus

Lateral hypothalamic nucleus (and MFB)

Mammillary body

FIGURE 17.4 Coronal Sections through the Hypothalamus Periventricular, medial, and lateral nuclei are shown in four coronal sections proceeding from anterior to posterior. See also Table 17.1. (After Purves D, Augustine GJ, Fitzpatrick D, Katz LC, LaMantia A-S, McNamara JO, Williams SM (eds.). 2001. *Neuroscience.* 2nd ed. Sinauer, Sunderland, MA.)

eral preoptic nucleus** and **medial preoptic nucleus** (see Figure 17.4A) are the rostral continuations of the lateral and medial hypothalamic areas, respectively. The remaining medial hypothalamic area can be divided into three regions from anterior to posterior (see Table 17.1; Figures 17.3 and 17.4). The **anterior hypothalamic region**, or **supraoptic region**, includes the **anterior hypothalamic nucleus**, **supraoptic nucleus**, **paraventricular nucleus**, and **suprachiasmatic nucleus** (see Figures 17.3 and 17.4B). Some neurons located in both the supraoptic and the paraventricular nuclei contain oxytocin or vasopressin and project to the posterior pituitary (see Figure 17.5). The suprachiasmatic nucleus is the "master clock" for circadian rhythms. It receives inputs from specialized retinal ganglion cells containing the photopigment melanopsin, which convey information about day–night cycles directly to the suprachiasmatic nucleus via the retinohypothalamic tract arising from the optic chiasm. The **middle hypothalamic region**, or **tuberal region** (see Table 17.1; Figures 17.3 and 17.4C), includes the **arcuate nucleus**, **ventromedial nucleus**, and **dorsomedial nucleus**. The arcuate nucleus is one of the hypothalamic nuclei projecting to the median eminence to control the anterior pituitary. The **posterior hypothalamic region**, or **mammillary region** (see Table 17.1; Figures 17.3 and 17.4D), includes the **medial mammillary nucleus**, **intermediate mammillary nucleus**, **lateral mammillary nucleus**, and **posterior hypothalamic nucleus**.

Hypothalamic Control of the Autonomic Nervous System

The hypothalamus has important descending projections that influence both the sympathetic and the parasympathetic divisions of the autonomic nervous system. Descending autonomic fibers originate mainly from the **paraventricular nucleus** but also from the **dorsomedial** hypothalamic nucleus and from the

REVIEW EXERCISE

What are the four major types of functions carried out by the hypothalamus (mnemonic: **HEAL**)?

lateral and posterior hypothalamus. The descending autonomic fibers travel initially in the medial forebrain bundle, then in the dorsolateral brainstem, likely via polysynaptic pathways, and in the periaqueductal gray matter. Ultimately they synapse onto preganglionic parasympathetic nuclei in the brainstem and intermediate zone of the sacral spinal cord, and onto preganglionic sympathetic neurons in the intermediolateral cell column of the thoracolumbar spinal cord (see Figures 6.12 and 6.13). Aside from the descending autonomic pathways from the hypothalamus, autonomic pathways also descend from several brainstem nuclei, including the nucleus solitarius, noradrenergic nuclei, raphe nucleus, and pontomedullary reticular formation. Many of these nuclei also receive inputs from the hypothalamus.

Inputs to the hypothalamus that regulate autonomic function come from numerous synaptic and humoral sources. One important source of input is the amygdala and certain regions of the limbic cortex (see Chapter 18), including the orbital frontal, insular, anterior cingulate, and temporal cortices.

Hypothalamic–Limbic Pathways

The limbic system and its connections with the hypothalamus will be discussed in detail in Chapter 18. Here we will simply mention the main input and output connections between the limbic system and the hypothalamus. The subiculum of the **hippocampal formation**, a limbic structure, projects to the mammillary bodies of the hypothalamus via the **fornix**. Meanwhile, the mammillary bodies project via the **mammillothalamic tract** to the **anterior thalamic nucleus**, which in turn projects to limbic cortex in the cingulate gyrus. The **amygdala**, another important limbic structure, has reciprocal connections with the hypothalamus via two pathways: the **stria terminalis** and the **ventral amygdalofugal pathway**. The limbic–hypothalamic interconnections may be an important mechanism for emotional influences on autonomic pathways (explaining why your palms get sweaty and your stomach churns when you are anxious) and on homeostatic pathways, including the immune system (explaining why depressed individuals may be more susceptible to infection). In addition, connections from the hypothalamus to limbic pathways may enable complex motivational and emotional programs to be activated in the service of homeostatic and reproductive functions.

Limbic-hypothalamic interactions are also illustrated by the clinical manifestiations seen in patients with **hypothalamic hamartoma**. This is a rare, histologically benign tumor-like growth that causes unusual seizures consisting of laughing episodes (gelastic epilepsy), usually beginning in early childhood. In most cases, hypothalamic hamartomas are also associated with disturbances in emotional behavior including irritability and aggression, and with cognitive impairment. Endocrinological abnormalities can also occur, and some hypothalamic hamartomas secrete gonadotropin releasing hormone, leading to precocious puberty.

Other Regionalized Functions of the Hypothalamus

In addition to its roles in endocrine, autonomic, and limbic function, the hypothalamus is important in regulating a variety of appetitive, homeostatic, and other behaviors that are often essential to survival of the organism. The regional aspects of these functions have been studied predominantly through lesion and stimulation studies in animals. However, evidence is accumulating for similar localization for many of these functions in humans as well. As we have already mentioned, the **suprachiasmatic nucleus** in the anterior hypothalamus (see Figures 17.3 and 17.4B) is an important regulator of circadian rhythms. Recall also that GABAergic neurons in the ventral lateral preoptic area (VLPO) contribute to nonREM sleep by inhibiting the arousal systems including histaminergic neurons in the tuberomammillary nucleus (TMN) and orexin-containing neurons in the posterior lateral hypothalamus, as well as brainstem serotonergic, noradrenergic, dopaminergic and cholinergic nuclei (see Figures 14.13 and 14.15A). Therefore, **lesions of the anterior hypothalamus** including the VLPO tend to cause **insomnia**. Conversely, **lesions of the posterior hypothalamus**, which destroy the histaminergic neurons in the TMN and orexin-containing neurons, tend to cause **hypersomnia**. The **lateral hypothalamus** is important in **appetite**, and lateral hypothalamic lesions cause a decrease in body weight. Conversely, the **medial hypothalamus**, especially the ventromedial nucleus, appears to be important in **inhibiting appetite**, and medial hypothalamic lesions can cause obesity. Recently, **leptin**, a hormone that is produced by adipose tissue, was discovered. Leptin binds to **Ob receptors** in the hypothalamus and plays an important role in feedback regulation of food intake, reducing appetite and obesity. **Ghrelin**—the opposing hormone—is elaborated by gastric mucosal cells, binds in the hypothalamus, and stimulates appetite. **Thirst** appears to result from the activation of osmoreceptors in the anterior regions of the hypothalamus. Hypovolemia or elevated body temperature can also activate thirst. Lesions of the lateral hypothalamus decrease water intake.

Thermoregulation involves the control of multiple systems, including sweat production; smooth muscles that affect core and surface blood flow; skeletal muscles involved in shivering, panting, and other motor activity; and endocrine systems that control the metabolic rate. The **anterior hypothalamus** appears to detect increased body temperature and activates mechanisms of **heat dissipation**. Anterior hypothalamic lesions can cause hyperthermia. In contrast, the **posterior hypothalamus** functions to **conserve heat**. Bilateral lesions of the posterior hypothalamus usually cause poikilothermia, in which the body temperature varies with the environment because these lesions destroy both heat conservation mechanisms of the posterior hypothalamus and descending pathways for heat dissipation arising from the anterior hypothalamus. The hypothalamus probably also participates in circuitry involved in sexual desire and other complex motivational states. Recently, the hormone oxytocin, produced in the hypothalamus and released in the posterior pituitary has been shown to increase nurturing behaviors. In addition, sexual development and differentiation involve an interplay of neural and endocrine signals, many of which appear to be regulated by the hypothalamus.

> **REVIEW EXERCISE**
>
> What are the effects of lateral hypothalamic lesions vs. medial hypothalamic lesions on body weight? What are the effects of anterior hypothalamic lesions vs. posterior hypothalamic lesions on sleep and on temperature control?

Endocrine Functions of the Pituitary and Hypothalamus

The anterior pituitary produces six important hormones, many of which regulate endocrine systems in other parts of the body, such as the adrenal cortex, thyroid, and gonads. These anterior pituitary hormones are **adrenocorticotropic hormone (ACTH)**, **growth hormone (GH)**, **prolactin**, **thyroid-stimulating hormone (TSH)**, **luteinizing hormone (LH)**, and **follicle-stimulating hormone (FSH)** (Table 17.2). The intermediate lobe is rudimentary in humans, produces pro-opio-melanocortin (POMC) and melanocyte-stimulating hormone (MSH), and has little known clinical significance. Two hormones are released in the posterior pituitary: (1) **oxytocin** and (2) **vasopressin**, which is also called **arginine vasopressin (AVP)** or **antidiuretic hormone (ADH)**.

Release of the **anterior pituitary** hormones by glandular cells is controlled by neurons in the hypothalamus through the **hypophysial portal system** (Figure 17.5). The pituitary receives arterial blood from the **inferior and superior hypophysial arteries**, both of which are branches of the internal carotid artery. The first capillary plexus of the portal system occurs in the **median eminence**. Neurons lying adjacent to the third ventricle in several hypothalamic nuclei project to the median eminence, where they secrete inhibitory and releasing factors (see Table 17.2). Nuclei projecting to the median eminence include the **arcuate nucleus**, **periventricular nucleus**, **medial preoptic nucleus**, and medial parvocellular portions of the **paraventricular nucleus**.

TABLE 17.2 Anterior Pituitary Hormones and Hypothalamic Releasing and Inhibitory Factors

PITUITARY HORMONE	HYPOTHALAMIC RELEASING FACTORS	HYPOTHALAMIC INHIBITORY FACTORS
Adrenocorticotropic hormone (ACTH)	Corticotropin-releasing hormone (CRH), vasopressin, and other peptides	—
Thyroid-stimulating hormone (TSH)	Thyrotropin-releasing hormone (TRH)	Growth hormone–inhibiting hormone (GIH, somatostatin)
Growth hormone (GH)	Growth hormone–releasing hormone (GHRH)	Growth hormone–inhibiting hormone (GIH, somatostatin)
Prolactin	Prolactin-releasing factor (PRF) and thyrotropin-releasing hormone (TRH)	Prolactin release–inhibiting factor (PIF, dopamine)
Luteinizing hormone (LH)	Luteinizing hormone–releasing hormone (LHRH)	—
Follicle-stimulating hormone (FSH)	Luteinizing hormone–releasing hormone (LHRH)	—

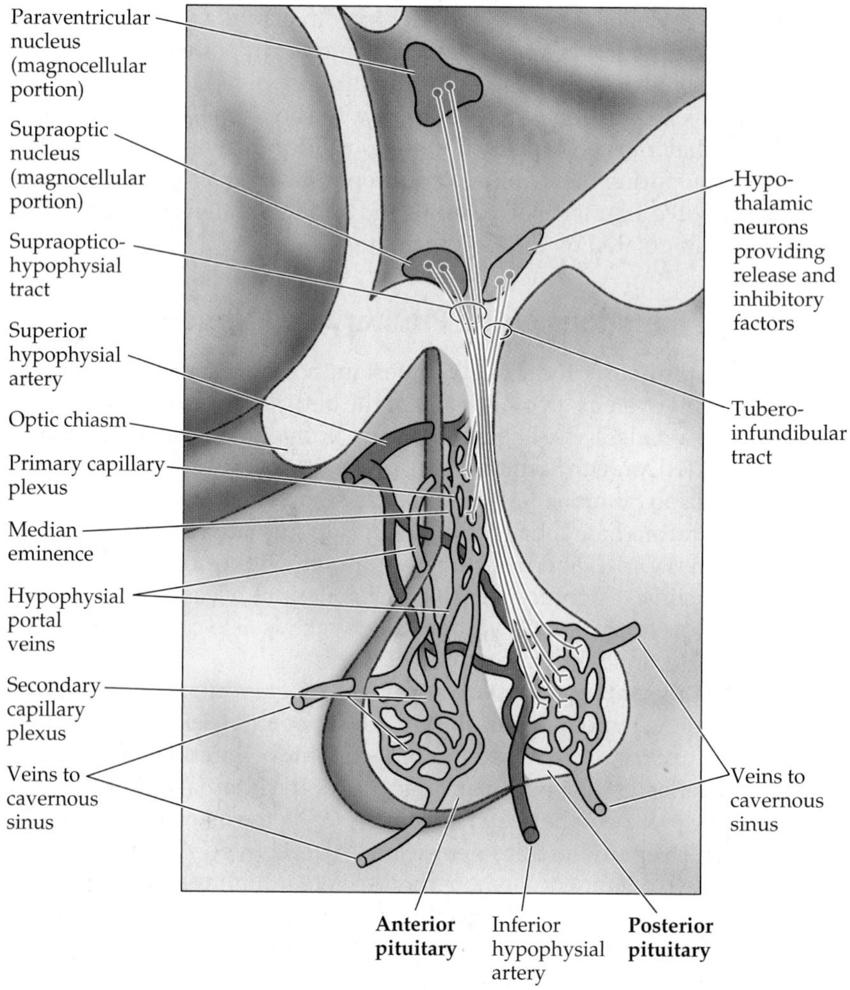

FIGURE 17.5 **Regulation of Anterior and Posterior Pituitary Hormones by the Hypothalamus** Hypothalamic neurons producing inhibitory and releasing factors project to the median eminence, where these factors are carried by the hypophysial portal veins to modulate anterior pituitary hormone release. Hypothalamic neurons in the supraoptic and paraventricular nuclei secrete oxytocin and vasopressin in the posterior pituitary.

Inhibitory and releasing factors enter the capillary plexus of the median eminence (see Figure 17.5; see also Figure 5.15) and are carried by the **hypophysial portal veins** to the anterior pituitary. Most of these factors are peptides, except for prolactin release–inhibiting factor (PIF), which is dopamine (see Table 17.2). Hormones released in the anterior pituitary are picked up by the secondary capillary plexus of the portal system and carried by draining veins to the cavernous sinus. Recall that the cavernous sinus drains primarily via the superior and inferior petrosal sinuses to reach the internal jugular vein (see Figure 10.11A,B).

The **posterior pituitary** also has a capillary plexus (see Figure 17.5), which picks up oxytocin and vasopressin and carries these hormones into the systemic circulation. Oxytocin and vasopressin are secreted in the posterior pituitary by terminals of neurons whose cell bodies lie in the **supraoptic** and **paraventricular nuclei**. Both nuclei contain both hormones, but separate neurons appear to contain either oxytocin or vasopressin, not both.

We will now briefly review the most important functions of each of the pituitary hormones (Figure 17.6). **ACTH** stimulates the **adrenal cortex** to produce

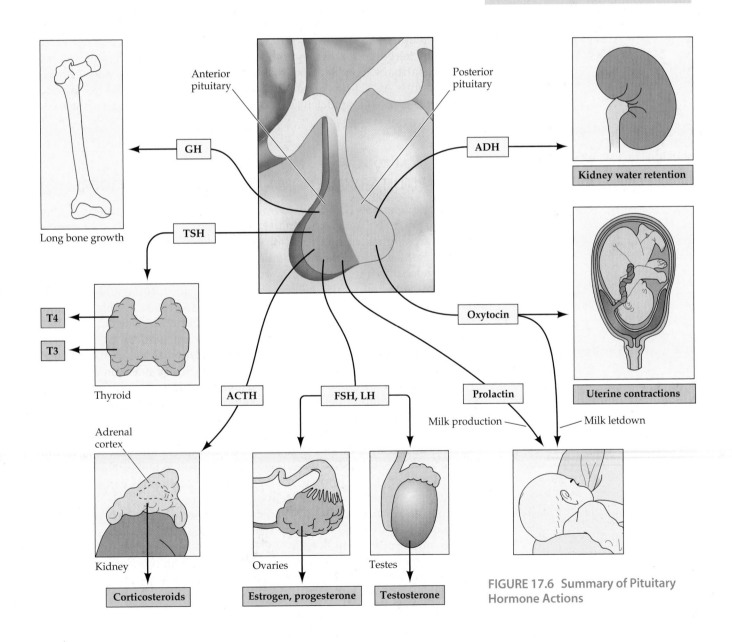

FIGURE 17.6 Summary of Pituitary Hormone Actions

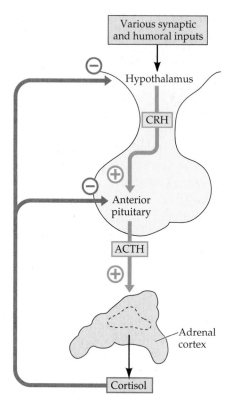

FIGURE 17.7 Feedback Control of CRH and ACTH Production in the Hypothalamic–Pituitary Axis

corticosteroid hormones, especially the glucocorticoid hormone **cortisol**, and to a lesser extent the mineralocorticoid hormone **aldosterone**. These steroid hormones are important for maintaining blood pressure, controlling electrolyte balance, promoting glucose mobilization into the bloodstream, and a variety of other functions. Recall that the adrenal medulla, which is under direct control of the preganglionic sympathetic neurons, releases epinephrine and norepinephrine (see Figure 6.13). **TSH** stimulates the **thyroid gland** to produce **thyroxine (T4)**, and **triiodothyronine (T3)**. These hormones promote cellular metabolism. **Growth hormone** causes the liver, kidneys, and other organs to produce **somatomedins** or **insulin-like growth factors (IGF)**, which promote increased growth of the long bones and other tissues. **Prolactin** causes the mammillary glands to produce milk. **LH** and **FSH** regulate ovarian hormones responsible for the menstrual cycle and oogenesis in females, and they regulate testicular hormones and spermatogenesis in males. **Oxytocin** causes contractions of smooth muscle in the breast for milk letdown and contractions of the uterus during labor. **Vasopressin**, or ADH, participates in osmotic regulation by promoting water retention by the kidneys, allowing concentration of the urine.

Release of hormones in the **hypothalamic–pituitary axis** is regulated by multiple neuroendocrine feedback loops. For example (Figure 17.7), release of corticotropin-releasing hormone (CRH) by the hypothalamus and release of ACTH by the anterior pituitary both receive feedback inhibition from circulating cortisol in the bloodstream. Chronic administration of exogenous steroids can suppress ACTH production to the point that the adrenals atrophy and are unable to provide sufficient cortisol to support life if the exogenous steroids are abruptly discontinued.

KEY CLINICAL CONCEPT

17.1 PITUITARY ADENOMA AND RELATED DISORDERS

Pituitary adenoma is a slow-growing, histologically benign tumor arising from glandular epithelial cells in the anterior pituitary. It is a fairly common tumor, accounting for about 12% of all intracranial neoplasms in adults (see Table 5.5). Mean age at the time of diagnosis is 40 years, although cases occasionally occur in adolescents or in the elderly. Pituitary adenomas can arise from any of the endocrine cell types in the anterior pituitary, and 85% secrete one or more pituitary hormones. Hormone secretion by pituitary adenomas is often in excess of normal levels and is not under normal hypothalamic control, resulting in several endocrinological syndromes, which we will discuss shortly. Even small pituitary **microadenomas** less than 1 millimeter in diameter can cause significant endocrinological abnormalities.

In contrast, nonfunctioning (nonhormone-secreting, or "silent") adenomas often grow larger before causing symptoms. Headache may be present even in small pituitary adenomas because of irritation of pain fibers in the adjacent cavernous region; however, headache is more common in large pituitary tumors. In addition, large tumors can compress the optic chiasm and cause visual disturbances, including a characteristic **bitemporal hemianopia** (see KCC 11.2). If left untreated, large pituitary adenomas can eventually cause hydrocephalus and brainstem compression.

Prolactin is the most commonly secreted hormone in pituitary adenomas, accounting for about 50% of all pituitary adenomas. The next most common is growth hormone, followed by ACTH. TSH-, LH-, and FSH-secreting tumors are more rare, as are tumors secreting more than one hormone. "Nonfunctioning tumors" that secrete no active hormones account for about 15% of pituitary adenomas.

Treatment options for pituitary adenomas include medication, surgery, and radiotherapy. Prolactin-secreting tumors (prolactinomas) often show a good response to treatment with dopaminergic agonists such as bromocriptine or cabergoline, which inhibit prolactin release (see Table 17.2) and shrink tumors. Treatment of non–prolactin-secreting tumors is usually with surgery, since pharmaceutical treatment is less effective. The somatostatin analogue octreotide, which inhibits growth hormone release (see Table 17.2) and shrinks tumors, has shown some promising results in treatment of growth hormone–secreting tumors. Surgical resection offers the advantages of potential immediate cure and relatively low risk. Surgery is also used for prolactin-secreting tumors that do not respond adequately to medical therapy. Usually, a **transsphenoidal approach** is taken, in which, under general anesthesia, the floor of the pituitary fossa is entered through the roof of the sphenoid sinus (see Figure 12.1), with instruments inserted through the nose. With suprasellar pituitary tumors (extending above the sella turcica), an intracranial approach is sometimes necessary to attain adequate tumor removal, although recent advances in **endoscopic neurosurgery** have enabled greater access to skull-base structures even in the suprasellar region using the transsphenoidal approach. Radiotherapy with gamma knife (see KCC 16.4) is used mainly for cases that fail to respond to surgery or in patients who cannot undergo surgery due to operative risk.

Let's discuss the **clinical presentation and diagnosis** of each type of hormone-secreting pituitary adenoma. **Prolactin-secreting adenomas** typically cause amenorrhea in women; hypogonadism in men; and galactorrhea, infertility, hair loss, decreased libido, and weight gain in both sexes. Some of these effects of elevated prolactin are mediated by inhibition of hypothalamic LHRH, which in turn leads to decreased LH and FSH levels (see Table 17.2). In normal women this effect of prolactin on LH and FSH delays the resumption of menses during lactation. As with all other pituitary tumors, headache and visual symptoms can also occur.

Elevated prolactin levels can have many causes, but very high levels (>150 micrograms per liter in nonpregnant patients) are virtually diagnostic of pituitary adenoma. MRI is useful for diagnosis and can now be used to detect microadenomas as small as 0.5 to 1 millimeter in diameter through indirect effects on pituitary shape, although smaller tumors may not be visualized despite significant endocrine abnormalities. Hypothalamic lesions can also sometimes cause elevated prolactin levels due to decreased PIF (dopamine) production, but the increase is not as high as is typically seen in pituitary adenomas.

Growth hormone–secreting adenomas in adults cause **acromegaly**, a slowly progressive overgrowth of bones and soft tissues. Acromegaly is characterized by enlarged hands and feet, coarsened facial features, and a protuberant jaw. Growth hormone excess in children beginning before epiphyseal closure (adolescence) causes **gigantism**. Other common problems in patients with growth hormone excess include carpal tunnel syndrome, arthritis, infertility, hypertension, and diabetes. Diagnosis is by typical clinical features, elevated IGF-1, elevated GH levels of greater than 2 micrograms per liter even after glucose administration, and MRI.

ACTH-secreting adenomas cause Cushing's disease. **Cushing's syndrome** is a general term for the clinical features of glucocorticoid excess of any cause, inlcuding endogenous cortisol excess or exogenous administration of glucocorticoid medications (such as prednisone, methylprednisolone, dexamethasone, or hydrocortisone). **Cushing's disease** is an important cause of Cushing's syndrome and means specifically that the syndrome is caused by an ACTH-secreting pituitary adenoma. In Cushing's syndrome there is a characteristic **cushingoid** appearance, with a round "moon-shaped" facies and deposition of fat on the trunk more than the extremities, resulting in truncal obesity. The body habitus of the cushingoid patient has therefore been described as "spiderlike." Glucocorticoid excess can also cause acne, hirsutism, purplish skin striae, thin-appearing skin, easy bruising, poor wound healing, hypertension, diabetes, edema, immunosuppression, osteoporosis, avascular necrosis of the femoral head, amenorrhea, decreased libido, myopathy, fatigue, and psychiatric disturbances including mania, psychosis, and depression. Endogenous Cushing's syndrome is caused by primary adrenal adenomas or adenocarcinomas in only about 15% of cases. The remaining 85% are caused by ACTH oversecretion by pituitary adenomas (70%) or by nonpituitary tumors that secrete ACTH, such as bronchial carcinoma (15%), referred to as "ectopic" ACTH production.

A series of endocrinological tests is done to localize the cause of endogenous cortisol excess. Very low ACTH levels usually suggest an adrenal source, since adrenal cortisol excess will cause feedback reduction of ACTH production (see Figure 17.7). If an ACTH producing tumor is suspected, the **dexamethasone suppression test** can be useful. This test works on the principle that administration of a dose of dexamethasone at midnight normally acts through negative feedback, like cortisol (see Figure 17.7), to suppress cortisol levels or urine cortisol metabolites measured the next morning. A low-dose (1 to 3 milligram) overnight dexamethasone suppression test is often used as an initial screening test for excess cortisol production. If cortisol production is not suppressed with the low-dose test, the high-dose (8 milligram) dexamethasone suppression test is then helpful because ACTH-secreting pituitary tumors are usually suppressible with this dose, while ectopic ACTH-secreting tumors and adrenal tumors are not. Another strategy is to administer CRH (see Figure 17.7; Table 17.2), which causes an excessive rise in plasma ACTH and cortisol in pituitary adenomas but not in ectopic ACTH or adrenal tumors. MRI is useful in diagnosis as well. Finally, when results of these tests are equivocal, **petrosal sinus sampling** can be helpful to distinguish pituitary from nonpituitary ACTH overproduction. In addition, petrosal sinus sampling can often correctly localize the side of a microadenoma not visible on MRI. In this way, selective surgery on the side of the microadenoma may be possible while function of the remaining pituitary gland is spared.

In petrosal sinus sampling, catheters are inserted through the femoral veins and passed upward under radiological guidance through the internal jugular veins to reach the inferior petrosal sinuses on both sides (see Figure 10.11A,B). Aliquots are first removed to determine baseline ACTH levels. In ACTH-secreting pituitary adenomas, ACTH levels in at least one petrosal sinus should be more than two times the ACTH levels in a peripheral vein. An intravenous dose of CRH (see Figure 17.7) is then given, and ACTH measurements from each inferior petrosal sinus are taken approximately every 5 minutes. A threefold increase in ACTH is diagnostic of a pituitary adenoma. In addition, the ACTH rise is usually 2 to 20 times higher on the side of the tumor than on the contralateral side.

TSH-secreting adenomas are a rare cause of hyperthyroidism. **Hyperthyroidism** is much more commonly caused by primary thyroid disorders such as Graves' disease, thyroiditis, toxic multinodular goiter, and thyroid adenomas. Clinical manifestations of hyperthyroidism include nervousness, insomnia, weight loss, tremor, excessive sweating, heat sensitivity, increased sympathetic output, and frequent bowel movements. Note that thyroid ophthalmopathy can occur in Grave's disease but not in TSH-secreting adenomas. Graves' disease is characterized by inflammatory involvement of the thyroid gland, skin, and orbital tissues leading to proptosis, and, ultimately, extraocular muscle fibrosis. Other important neurologic manifestations of hyperthyroidism include proximal muscle weakness, tremor, dyskinesias, and dementia. Particularly in the elderly, many of the other clinical manifestations of hyperthyroidism may be absent, and hyperthyroidism can mimic dementia (see KCC 19.16) or depression. In hyperthyroidism caused by primary thyroid disorders, TSH levels are completely suppressed, while in TSH-secreting pituitary adenomas, TSH levels may be elevated.

Hypothyroidism is also usually caused by primary thyroid disorders such as autoimmune thyroid disease, iodine deficiency, or previous ablative treatment for hyperthyroidism and rarely is caused by pituitary or hypothalamic insufficiency. However, when lesions of the hypothalamus or pituitary are present, including medium-to-large pituitary adenomas of any type, it is relatively common for TSH production to be impaired, resulting in hypothyroidism. Manifestations of hypothyroidism of any cause include lethargy, weight gain, cold intolerance, smooth, dry skin, hair loss, depression, and constipation. Eventually, **myxedema coma** and cardiac involvement can occur. Other important neurologic manifestations include neuropathy, carpal tunnel syndrome, myalgias, ataxia, and dementia. Like hyperthyroidism, hypothyroidism can present in the elderly with a dementia-like or depression-like picture. Untreated hypothyroidism in utero or in infancy can cause cretinism, which is characterized by mental retardation, short stature, microcephaly, and other abnormalities.

LH- or FSH-secreting adenomas often cause **infertility**, although tumors can reach a relatively large size before being detected clinically. Interestingly, these tumors may produce either high or low testosterone and estradiol levels. Because LH- or FSH-secreting tumors are often large, patients may present with headache and visual changes as the major manifestations.

Other lesions can also occur in the sellar and suprasellar region, causing endocrine disturbances or compressing the optic chiasm. Although pituitary adenoma is the most common, other lesions seen in this region include craniopharyngioma, aneurysms, meningioma, optic glioma, hypothalamic glioma, chordoma, teratoma, epidermoid, dermoid, Rathke's pouch cysts, empty sella syndrome, sarcoidosis, lymphocytic hypophysitis, Langerhans cell histiocytosis, lymphoma, and metastases.

Finally, it should be noted that up to 10% of patients who undergo MRI scans for any reason may have **pituitary incidentalomas**, meaning endocrinologically inert and clinically benign tumors of the pituitary that are discovered as "incidental findings" on MRI scans. The majority pituitary incidentalomas are small and are generally addressed by basic clinical evaluation, endocrine hormone measurements when indicated, and periodic follow-up. ■

KEY CLINICAL CONCEPT

17.2 DIABETES INSIPIDUS AND SIADH

Diabetes insipidus (**DI**) is the production of large amounts of dilute urine. This condition can be caused by deficiency of ADH (**central** or **neurogenic DI**) or by insensitivity of the kidneys to ADH (**nephrogenic DI**). Symptoms of DI include severe thirst, polyuria, and polydipsia. Patients who are able to drink consume large amounts of water to maintain fluid balance. Patients who cannot drink adequately become dehydrated rapidly and die if not treated. The diagnosis of DI is established if a patient with polyuria has relatively low urine osmolality despite increased plasma osmolality. To detect this condition, sometimes the patient must be asked to temporarily stop drinking while in a supervised setting. A dose of subcutaneous vasopressin will cause urine osmolality to rise in neurogenic but not in nephrogenic DI. Common causes of neurogenic DI include neurosurgery, head trauma, and infiltrative or neoplastic lesions in the pituitary–hypothalamic region (see KCC 17.1) or in the third ventricle. Interestingly, lesions of the posterior pituitary do not cause DI unless the lesion is high enough in the pituitary stalk to result in retrograde degeneration of hypothalamic neurons in the supraoptic and paraventricular nuclei (see Figure 17.5). This suggests that neurons in these nuclei are capable of releasing vasopressin in locations other than the posterior pituitary. Treatment of DI is with subcutaneous or intranasal administration of synthetic vasopressin analogs.

In the **syndrome of inappropriate antidiuretic hormone** (**SIADH**), excess ADH production causes a low serum sodium (**hyponatremia**), together with inappropriately elevated urine osmolality. Note that hyponatremia with elevated urine osmolality is not always caused by SIADH, and it can also be seen in hypovolemia or in edematous states such as heart failure or cirrhosis. SIADH can be caused by many neurologic and non-neurologic conditions, including head trauma, meningitis, and numerous other neurologic disorders, pulmonary disorders, medication side effects, and ADH-secreting neoplasms. Severe hyponatremia can cause lethargy, coma, or seizures. When SIADH is the cause of hyponatremia, it should be treated by restriction of daily fluid intake. Treatment can also include vaprisol, which acts as vasopressin blocker. In severe cases, infusions of hypertonic saline are sometimes used, but care must be taken not to correct hyponatremia too rapidly because **central pontine myelinolysis** can result from this approach.

Some conditions can cause the consecutive appearance of SIADH and DI in a single patient. For example, following surgery in the pituitary region there is occasionally a triphasic response, with DI shortly after surgery, followed by SIADH, and finally DI again, which may then gradually improve. Patients with other intracranial disorders, such as catastrophic hemorrhage or infarct, may initially have SIADH. If brain death then ensues, all ADH production ceases, resulting in DI. ■

KEY CLINICAL CONCEPT

17.3 PANHYPOPITUITARISM

Deficiency of multiple pituitary hormones can occur in several conditions of the pituitary and hypothalamic regions. When all pituitary hormones are involved, the condition is called **panhypopituitarism**. **ACTH deficiency** causes hypocortisolism, with fatigue, weakness, decreased appetite, and impaired response to stress resulting in hypotension, fever, hypoglycemia, and a high mortality rate. **TSH deficiency** causes hypothyroidism (see KCC 17.1), and **ADH deficiency** causes diabetes insipidus (see KCC 17.2). **LH and FSH deficiencies** cause hypogonadism, including decreased libido, amenorrhea, and infertility. **GH deficiency** in children causes abnormally short stature. **Prolactin deficiency** in women causes inability to lactate, and **oxytocin deficiency** can cause impaired milk letdown.

There are multiple possible causes of panhypopituitarism, however, primary pituitary tumors and their treatment are the most common. Other lesions in this region include (see KCC 17.1) large, nonfunctioning pituitary adenomas, meningioma, craniopharyngioma, hypothalamic tumors, metastases, and other infiltrative processes, including sarcoidosis, lymphocytic hypophysitis, infections, and autoimmune disorders. On rare occasions, pituitary tumors can undergo spontaneous hemorrhage, resulting in **pituitary**

apoplexy. Patients with pituitary apoplexy often present with sudden headache, meningeal signs, unilateral or bilateral cavernous sinus syndrome (see KCC 13.7), visual loss, hypotension, and depressed level of consciousness. Panhypopituitarism is a common sequela of pituitary apoplexy. Other causes of panhypopituitarism include head trauma, surgery, radiation therapy, pituitary infarct, postpartum pituitary necrosis (Sheehan's syndrome), and congenital abnormalities.

Panhypopituitarism is treated by exogenous replacement of pituitary hormones. ACTH insufficiency is treated by daily administration of steroids such as prednisone or hydrocortisone, with increased doses given in situations of stress such as infection or surgery. Diabetes insipidus is treated with synthetic ADH analogs, and hypothyroidism is treated with synthetic thyroid hormones. Hypogonadism is treated with testosterone or estrogen–progesterone combinations, and fertility can sometimes be achieved with LH and FSH substitution therapy. GH replacement is used in children to improve growth and in adults because of beneficial effects on lipid profile and on other systems. ■

CLINICAL CASES

CASE 17.1 MOON FACIES, ACNE, AMENORRHEA, AND HYPERTENSION

CHIEF COMPLAINT

A 33-year-old woman presented to an endocrinology clinic with multiple complaints, including truncal obesity, acne, amenorrhea, and hypertension.

HISTORY

Symptoms had begun 3 years previously with **increased facial hair, new-onset acne, 45-pound weight gain, especially in the abdomen, easy bruisability, excessive sweating, and striae on her skin**. Two years prior to presentation her **menstrual periods stopped**, and she developed **hypertension** requiring medication. During recent months she had become **irritable and depressed**, and had **decreased energy, with difficulty walking up stairs**.

PHYSICAL EXAMINATION

General appearance: **Round face ("moon facies"), truncal obesity with a "buffalo hump" of fat in the posterior neck, and thin legs.**

Vital signs: T = 98°F, BP = 125/85.

Neck: Supple, obese. Thyroid not enlarged.

Lungs: Clear.

Breasts: No masses.

Heart: Regular rate with no murmurs.

Abdomen: Obese; normal bowel sounds; no masses.

Extremities: No clubbing or edema.

Skin: **Ruddy faced, with facial hair, abdominal striae, ecchymoses, and thin-appearing skin in some areas.**

Genitalia: Normal female.

Neurologic exam: Normal mental status, cranial nerves, motor exam, reflexes, coordination, gait, and sensation.

DIAGNOSIS AND INITIAL LOCALIZATION

1. On the basis of the symptoms and signs shown in **bold** above, what endocrinological syndrome is present in this patient? This syndrome is caused by excess of which hormone?

2. What are the possible localizations for this disorder?

Discussion

The key symptoms and signs in this case are:

- Truncal obesity, increased facial hair, new-onset acne, easy bruisability, excessive sweating, ruddiness and skin striae, amenorrhea, hypertension, irritability and depression, decreased energy, and difficulty walking up stairs

CASE 17.3 A CHILD WITH GIGGLING EPISODES AND AGGRESSIVE BEHAVIOR

IMAGE 17.3A,B Hypothalamic Hamartoma Coronal T2-weighted MRI scan images. A and B are adjacent coronal sections progressing from posterior to anterior.

(A)

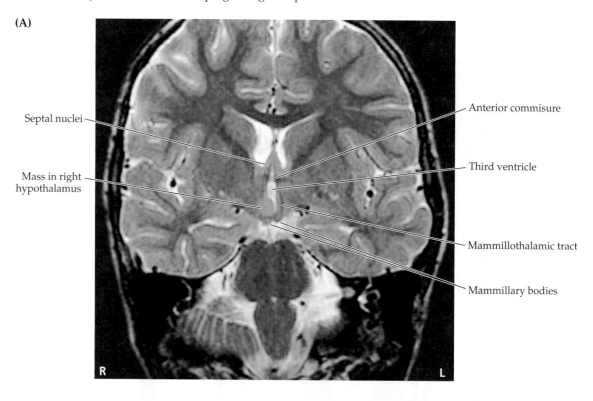

Septal nuclei

Mass in right hypothalamus

Anterior commisure

Third ventricle

Mammillothalamic tract

Mammillary bodies

(B)

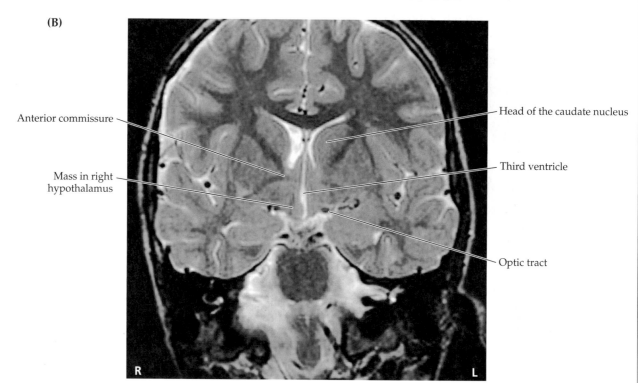

Anterior commissure

Mass in right hypothalamus

Head of the caudate nucleus

Third ventricle

Optic tract

Additional Cases

Related cases for the following topics can be found in other chapters: **lesions of the sellar region** (Cases 11.3 and 13.5); and **disorders of hypothalamic connections to autonomic or limbic circuits** (Cases 14.1 and 18.1–18.5). Other relevant cases can be found by using the **Case Index** located at the end of this book.

Brief Anatomical Study Guide

1. The pituitary and hypothalamus link the nervous system through both synaptic and humoral forms of communication with numerous other systems in the body. The **pituitary gland** is connected to the bottom of the hypothalamus via the pituitary stalk (see Figure 17.2). This location of the pituitary gland just underneath the optic chiasm can result in compression of crossing optic nerve fibers by pituitary region tumors, classically causing **bitemporal hemianopia** or other visual deficits (see Figure 11.15).

2. The **hypothalamus** is located beneath the thalamus and can be divided from anterior to posterior into the **preoptic region, anterior (supraoptic) region, middle (tuberal) region,** and **posterior (mammillary) region.** The hypothalamus can also be divided from medial to lateral into a **periventricular area, medial hypothalamic area,** and **lateral hypothalamic area** (see Figures 17.3, 17.4). Table 17.1 lists the **main nuclei** of the hypothalamus.

3. The hypothalamus participates in a variety of neural and non-neural systems that mainly regulate **homeostasis** through multiple complex feedback loops. The hypothalamus participates in the following functions:

   A. **Homeostatic** control of appetite, thirst, thermoregulation, sleep-wake cycles, and sexual desire

   B. **Endocrine** control

   C. **Autonomic** nervous system

   D. **Limbic** system (see Chapter 18)

4. Important nuclei and connections of the hypothalamus include:

   A. The **paraventricular nucleus, dorsomedial nucleus,** and the **lateral and posterior hypothalamus,** which control autonomic function (see Figures 13.10, 17.3, and 17.4)

   B. Limbic inputs to the **mammillary bodies** from the hippocampal formation via the **fornix** (see Figures 18.9 and 18.13)

   C. Limbic outputs from the mammillary bodies via the **mammillothalamic tract** to the anterior thalamic nucleus (see Figure 18.9)

   D. Reciprocal limbic connections with the amygdala via the **stria terminalis** and **ventral amygdalofugal pathway** (see Figure 18.17)

   E. Control of circadian rhythms by the **suprachiasmatic nucleus** (see Figures 17.3 and 17.4)

   F. Release of oxytocin and vasopressin by the **supraoptic and paraventricular nuclei** into the posterior pituitary (see Figures 17.5 and 17.6)

Brief Anatomical Study Guide (continued)

G. Release of stimulatory and inhibitory factors by the **arcuate nucleus**, **periventricular nucleus**, **medial preoptic nucleus**, and medial parvocellular portions of the **paraventricular nucleus** for control of the anterior pituitary (see Table 17.2; see also Figure 17.5)

5. The **pituitary gland** is composed of an **anterior lobe**, or **adenohypophysis**, derived embryologically from the roof of the pharynx (see Figure 17.1), and a **posterior lobe**, or **neurohypophysis**, derived embryologically from the diencephalon. The anterior pituitary is composed of endocrine tissue and secretes the following six hormones: **adrenocorticotropic hormone (ACTH)**, **growth hormone (GH)**, **prolactin**, **thyroid-stimulating hormone (TSH)**, **luteinizing hormone (LH)**, and **follicle-stimulating hormone (FSH)**. The posterior pituitary is composed of neural tissue extending from the hypothalamus and releases the hormones **oxytocin** and **vasopressin**. Figure 17.6 summarizes the functions of these anterior and posterior pituitary hormones.

6. Secretion of anterior pituitary hormones is controlled by hypothalamic **releasing and inhibitory factors** (see Table 17.2) that travel from the **median eminence** to the anterior pituitary via a **hypophysial portal circulatory system** (see Figure 17.5). Posterior pituitary hormones are released at axon terminals of magnocellular neurons located in the **supraoptic** and **paraventricular** nuclei of the hypothalamus (see Figure 17.5). Multiple feedback loops control the hypothalamic–pituitary axis as shown, for example, with the adrenocortical system in Figure 17.7.

References

Aghi MK. 2008. Management of recurrent and refractory Cushing disease. *Nat Clin Pract Endocrinol Metab* 4 (10): 560–568.

Buchfelder M, Kreutzer J. 2008. Transcranial surgery for pituitary adenomas. *Pituitary* 11(4): 375–384.

Cappabianca P, Cavallo LM, Esposito F, De Divitiis O, Messina A, De Divitiis E. 2008. Extended endoscopic endonasal approach to the midline skull base: the evolving role of transsphenoidal surgery. *Adv Tech Stand Neurosurg* 33: 151–199.

Cooper P. 2004. Neuroendocrinology. In WG Bradley, RB Daroff, GM Fenichel, and CD Marsden (eds.), *Neurology in Clinical Practice: Principles of Diagnosis and Management.* 4th ed., Chapter 47. Butterworth-Heinemann, Boston.

Decaux G, Musch W. 2008. Clinical laboratory evaluation of the syndrome of inappropriate secretion of antidiuretic hormone. *Clin J Am Soc Nephrol* 3 (4): 1175–1184.

Ellison DH, Berl T. 2007. Clinical practice. The syndrome of inappropriate antidiuresis. *N Engl J Med* 356 (20): 2064–2072.

Friedman JM, Halaas JL. 1998. Leptin and the regulation of body weight in mammals. *Nature* 395 (6704): 763–770.

Jagannathan J, Kanter AS, Olson C, Sherman JH, Laws ER Jr, Sheehan JP. 2008. Applications of radiotherapy and radiosurgery in the management of pediatric Cushing's disease: a review of the literature and our experience. *J Neurooncol* 90 (1): 117–124.

Joshi SM, Cudlip S. 2008. Transsphenoidal surgery. *Pituitary* 11 (4): 353–360.

Klok MD, Jakobsdottir S, Drent ML. 2007. The role of leptin and ghrelin in the regulation of food intake and body weight in humans: a review. *Obesity Reviews* 8 (1): 21–34.

Lee MJ, Fried SK. 2009. Integration of hormonal and nutrient signals that regulate leptin synthesis and secretion. *Am J Physiol Endocrinol Metab* 296 (6): E1230–1238.

Maghnie M, Cosi G, Genovese E, Manca-Bitti ML, Cohen A, Zecca S, Tinelli C, Gallucci M, et al. 2000. Central diabetes insipidus in children and young adults. *N Engl J Med* 343 (14): 998–1007.

Murad-Kejbou S, Eggenberger E. 2009. Pituitary apoplexy: evaluation, management, and prognosis. *Curr Opin Ophthalmol* 20 (6): 456–461.

Nawar RN, AbdelMannan D, Selman WR, Arafah BM. 2008. Pituitary tumor apoplexy: a review. *J Intensive Care Med* 23 (2): 75-90.

Powell M. 2009. Microscope and endoscopic pituitary surgery. *Acta Neurochir* (Wien) 151 (7): 723–728.

Rivkees SA. 2008. Differentiating appropriate antidiuretic hormone secretion, inappropriate antidiuretic hormone secretion and cerebral salt wasting: the common, uncommon, and misnamed. *Curr Opin Pediatr* 20 (4): 448–452.

Vance ML. 2008. Pituitary adenoma: a clinician's perspective. *Endocr Pract* 14 (6): 757–763.

CONTENTS

Chapter *18*

Limbic System: Homeostasis, Olfaction, Memory, and Emotion

The structures of the limbic system regulate emotions, olfaction, memory, drives, and homeostasis. A 40-year-old woman awoke from sleep and complained to her husband of an indescribable unpleasant odor, nausea, and a panicky, fearful sensation. During the following week, she had repeated stereotyped episodes of this kind followed by decreased responsiveness and slow, inappropriate speech lasting 2 to 3 minutes.

As we shall see, limbic system abnormalities can cause paroxysmal disorders as seen in this patient. In this chapter we will learn about this important and diverse neural system and the consequences of limbic system damage or dysfunction.

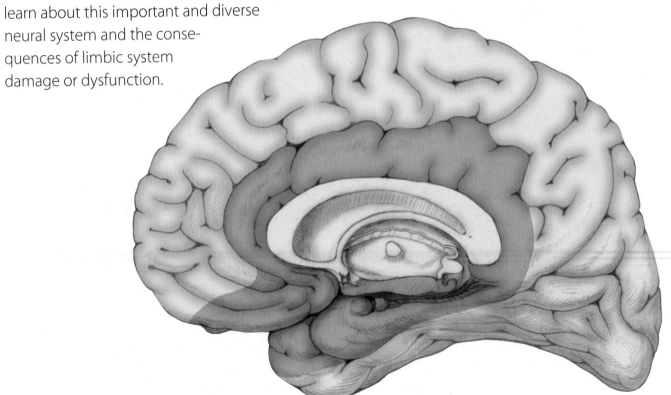

ANATOMICAL AND CLINICAL REVIEW

THE LIMBIC SYSTEM includes diverse cortical and subcortical structures located mainly in the medial and ventral regions of the cerebral hemispheres. These structures are unified by their evolutionarily ancient origins, and they constitute the major portion of the forebrain in many species. Only in higher mammals has the larger neocortical mantle surpassed the limbic system in size.

The functions of the limbic system are also ancient, and they play an important role for survival in the animal kingdom. Limbic functions can be divided into the following four basic categories:

1. **Olfaction**
2. **Memory**
3. **Emotions and drives**
4. **Homeostatic functions, including autonomic and neuroendocrine control**

MNEMONIC

An aid to remembering these functions is **HOME** (**H**omeostasis, **O**lfaction, **M**emory, and **E**motion).* Numerous limbic structures participate in each of these functions, as we will see. The various components of the limbic system (Table 18.1) form a complex network, with multiple reciprocal connections (Figure 18.1). However, simplifying somewhat, one limbic structure can be thought of

*Recall that a similar mnemonic, HEAL, was used to remember hypothalamic functions in Chapter 17. Limbic system and hypothalamic functions are strongly interconnected.

TABLE 18.1 Main Components of the Limbic System

Limbic cortex
  Parahippocampal gyrus
  Cingulate gyrus
  Medial orbitofrontal cortex
  Temporal pole
  Anterior insula
Hippocampal formation
  Dentate gyrus
  Hippocampus
  Subiculum
Amygdala
Olfactory cortex
Diencephalon
  Hypothalamus
  Thalamus
    Anterior nucleus
    Mediodorsal nucleus
  Habenula
Basal ganglia
  Ventral striatum
    Nucleus accumbens
    Ventral caudate and putamen
  Ventral pallidum
Basal forebrain
Septal nuclei
Brainstem

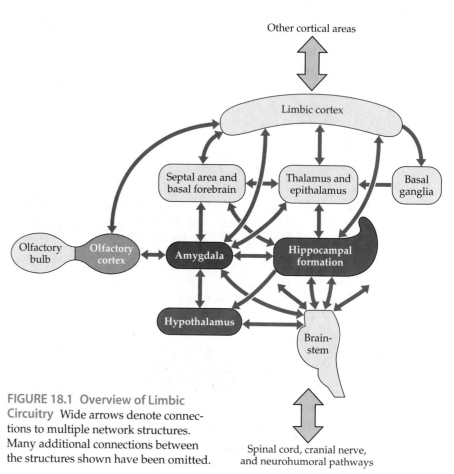

FIGURE 18.1 Overview of Limbic Circuitry Wide arrows denote connections to multiple network structures. Many additional connections between the structures shown have been omitted.

TABLE 18.2 Simplification of Limbic Functions and Corresponding Key Structures

LIMBIC FUNCTION	KEY STRUCTURE
Olfaction	Olfactory cortex
Memory	Hippocampal formation
Emotions and drives	Amygdala
Homeostasis; autonomic and neuroendocrine control	Hypothalamus

as central to each of these four functions (Table 18.2): The **olfactory cortex** is essential to olfaction, the **hippocampal formation** to memory, the **amygdala** to emotions and drives, and the **hypothalamus** to homeostasis (see Chapter 17). It cannot be emphasized enough that in reality, each of these structures participates in a complex network involving numerous limbic components to carry out these functions.

In this chapter we will first review the overall structure of the limbic system and briefly discuss each of its main components. We will then discuss specific subsystems responsible for the four major categories of limbic function (see Table 18.2). Finally, we will review important clinical disorders of the limbic system, including memory loss and epileptic seizures.

Overview of Limbic Structures

The **limbic system** includes a variety of structures extending from the forebrain to the brainstem (see Table 18.1). Most of these structures lie hidden within the medial and ventral regions of the cerebral hemispheres and are not readily visible from the lateral surface. *Limbus* means "border" or "edge" in Latin. The **limbic cortex** (Figure 18.2) forms a ringlike **limbic lobe** around the edge of the cortical mantle, which surrounds the corpus callosum and upper brainstem–diencephalic junction. This "grand lobe limbique" was first described by Broca in 1878. The main components of limbic cortex visible on a medial view are the **cingulate gyrus** (*cingulum* means "girdle" or "belt" in Latin) and the parahippocampal gyrus (see Figure 18.2A). The **parahippocampal gyrus** is separated from the remainder of the temporal lobe by the **collateral sulcus**, which continues anteriorly as the **rhinal sulcus** (see Figure 18.2B). The **uncus** is a bump visible on the anterior medial parahippocampal gyrus. The cingulate gyrus continues anteriorly and inferiorly as the subcallosal and paraterminal gyri. The cingulate gyrus joins the parahippocampapal gyrus posteriorly at the isthmus (see Figure 18.2A). In addition to the cingulate and parahippocampal gyri, other regions of limbic cortex include the **medial orbitofrontal gyri**, the **temporal poles**, and the **anterior insular cortex** (see Figure 18.2A,C; Table 18.1). The limbic cortices share certain surface immunological markers. For example, the herpes simplex virus has a tropism for limbic cortex and can cause severe encephalitis involving predominantly limbic cortical areas (Figure 18.3). In some texts, what we have called limbic cortex is referred to as paralimbic cortex or limbic association cortex.

The **hippocampal formation** (see Table 18.1) is the medial and dorsal continuation of the parahippocampal gyrus. It is buried within the medial temporal lobe, forming the floor of the temporal horn of the lateral ventricle (see Figure 18.8). The hippocampal formation is one of several C-shaped structures in the limbic system. As we will discuss shortly, the hippocampal formation plays an important role in the memory functions of the limbic system.

Unlike the six-layered neocortex, the hippocampal formation has only three layers and is called archicortex. About 95% of the cortex in humans is six-lay-

REVIEW EXERCISE

What are the five regions where limbic cortex is found?

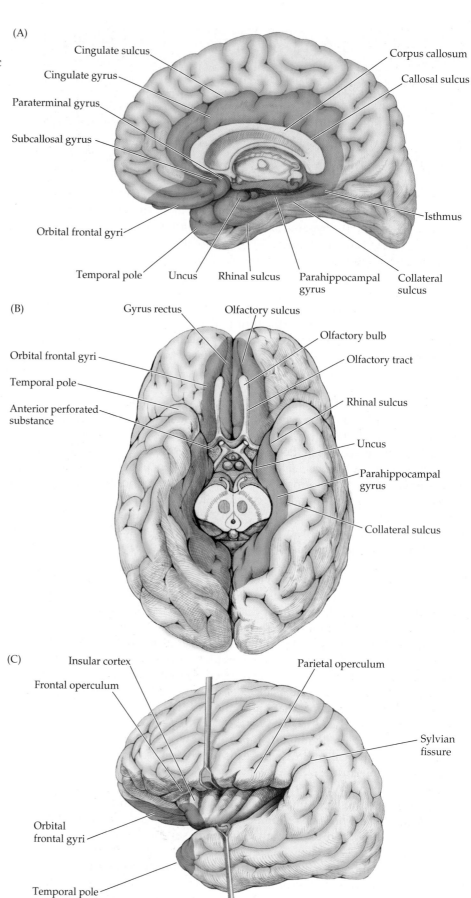

FIGURE 18.2 Limbic Cortex Blue regions represent limbic cortex, also known as paralimbic cortex or limbic association cortex.

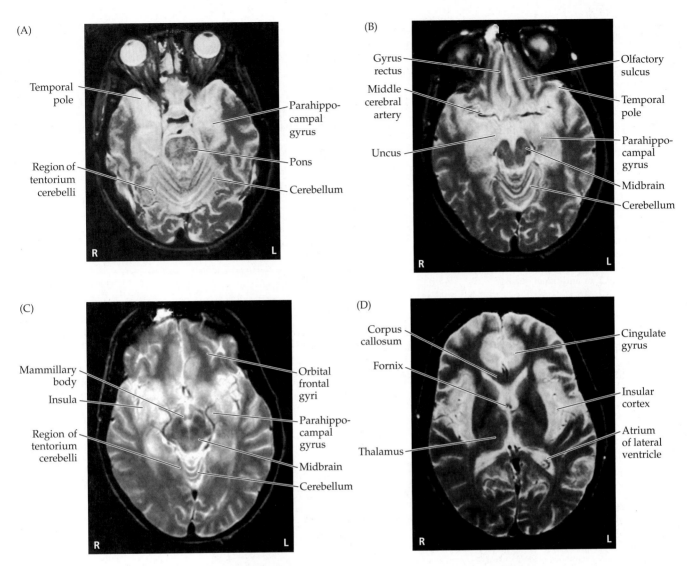

(A)

Temporal pole

Region of tentorium cerebelli

Parahippocampal gyrus

Pons

Cerebellum

R   L

(B)

Gyrus rectus

Middle cerebral artery

Uncus

Olfactory sulcus

Temporal pole

Parahippocampal gyrus

Midbrain

Cerebellum

R   L

(C)

Mammillary body

Insula

Region of tentorium cerebelli

Orbital frontal gyri

Parahippocampal gyrus

Midbrain

Cerebellum

R   L

(D)

Corpus callosum

Fornix

Thalamus

Cingulate gyrus

Insular cortex

Atrium of lateral ventricle

R   L

FIGURE 18.3 Herpes Encephalitis Affecting the Limbic Cortex Bilaterally Axial T2-weighted MRI images from a patient with herpes encephalitis. Slices A–D progress from inferior to superior.

ered **neocortex** (meaning "new cortex"; also called isocortex, meaning "same cortex") (Table 18.3). The different types of neocortex will be discussed in Chapter 19 (see Table 19.2). More phylogenetically ancient forms of cortex, which do not have six distinct layers, are referred to as **allocortex** (meaning "other cortex"). Allocortex includes the three-layered **archicortex** (meaning "first" or "original cortex") of the hippocampal formation, as well as **paleocortex** (meaning "old cortex"), which is found predominantly in the piriform cortex of the **olfactory area** (see Figure 18.6). Olfactory pathways will be discussed in the next section. The regions between three- and six-layered cortex form **transitional cortex**, or **mesocortex** (meaning "middle cortex"), which is found, for example, in the limbic cortex of the parahippocampal gyrus and anterior inferior insula (see Figure 18.2). Some authors classify portions of the transitional cortex as paleocortex rather than as mesocortex. The term **corticoid areas** (see Table 18.3) is used for the simple-structured cortex-like regions that overlie or merge with subcortical nulcei such as the amygdala, substantia innominata, and septal region. Corticoid areas do not contain consistent layers and are considered the most rudimentary form of cortex. In fish and amphibians, the archicortex, paleocortex, and corticoid areas form the major portion of the cerebral hemispheres; only in mammals does the neocortex predominate.

TABLE 18.3 Terminology for Classifying Cerebral Cortex

NAME	EQUIVALENT NAME(S)	DESCRIPTION	EXAMPLE(S)
Neocortex[a]	Isocortex, neopallium	Six-layered cortex	Majority of the cerebral cortex
Mesocortex	Limbic cortex, paralimbic cortex, transitional cortex	Forms gradual transition between three- and six-layered cortex	Parahippocampal gyrus, cingulate gyrus, anterior insula, orbitofrontal cortex, temporal pole
Allocortex	—	Cortex with fewer than six layers	Archicortex, paleocortex
Archicortex	Archipallium	Three-layered hippocampal cortex	Hippocampal formation
Paleocortex	Paleopallium	Three-layered olfactory cortex	Piriform cortex
Corticoid areas	—	Simple cortex that merges with subcortical nuclei	Amygdala, substantia innominata, septal region

[a]See Table 19.2 for additional details.

The **amygdala** is a nuclear complex that lies in the anteromedial temporal lobe. It overlaps the anterior end of the hippocampus and lies dorsal to the tip of the temporal horn of the lateral ventricle (Figure 18.4B,C; see also Figure 18.10). The posterior amygdala and anterior hippocampus lie just underneath the **uncus**, a bump visible on the medial surface of the temporal lobe (see Figures 18.2 and 18.4B). The amygdala is composed of three main nuclei: the

(A)

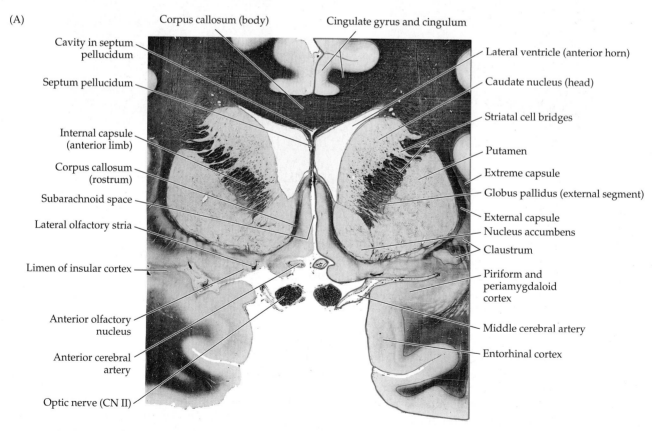

FIGURE 18.4 Coronal Brain Sections through Basal Forebrain and Septal Region Myelin is stained dark. A–C progress from anterior to posterior. (From Martin JH. 1996. *Neuroanatomy: Text and Atlas*. 2nd ed. McGraw-Hill, New York.)

(B)

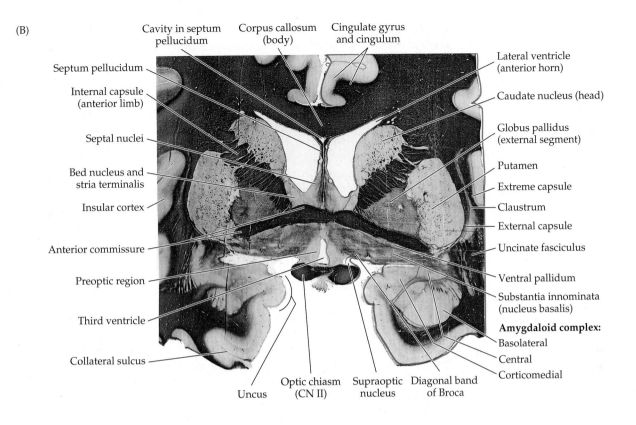

Cavity in septum pellucidum
Corpus callosum (body)
Cingulate gyrus and cingulum
Septum pellucidum
Internal capsule (anterior limb)
Septal nuclei
Bed nucleus and stria terminalis
Insular cortex
Anterior commissure
Preoptic region
Third ventricle
Collateral sulcus
Uncus
Optic chiasm (CN II)
Supraoptic nucleus
Diagonal band of Broca

Lateral ventricle (anterior horn)
Caudate nucleus (head)
Globus pallidus (external segment)
Putamen
Extreme capsule
Claustrum
External capsule
Uncinate fasciculus
Ventral pallidum
Substantia innominata (nucleus basalis)
Amygdaloid complex:
Basolateral
Central
Corticomedial

(C)

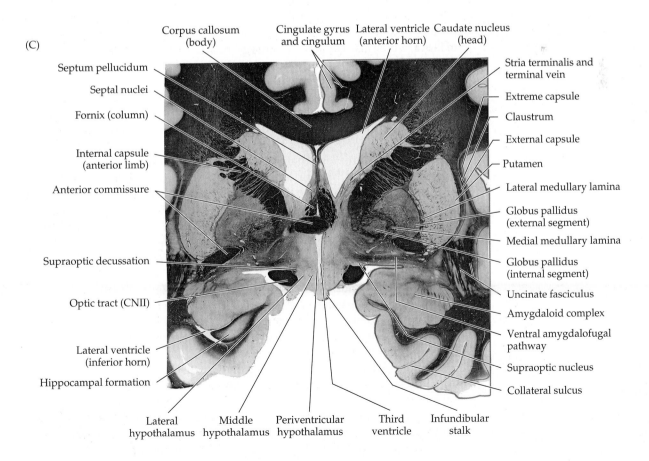

Corpus callosum (body)
Cingulate gyrus and cingulum
Lateral ventricle (anterior horn)
Caudate nucleus (head)
Septum pellucidum
Septal nuclei
Fornix (column)
Internal capsule (anterior limb)
Anterior commissure
Supraoptic decussation
Optic tract (CNII)
Lateral ventricle (inferior horn)
Hippocampal formation
Lateral hypothalamus
Middle hypothalamus
Periventricular hypothalamus
Third ventricle
Infundibular stalk

Stria terminalis and terminal vein
Extreme capsule
Claustrum
External capsule
Putamen
Lateral medullary lamina
Globus pallidus (external segment)
Medial medullary lamina
Globus pallidus (internal segment)
Uncinate fasciculus
Amygdaloid complex
Ventral amygdalofugal pathway
Supraoptic nucleus
Collateral sulcus

corticomedial, **basolateral**, and **central nuclei**. In addition, the *C*-shaped **bed nucleus of the stria terminalis** is considered part of the amygdala. The amygdala serves important functions in emotional, autonomic, and neuroendocrine circuits of the limbic system.

Diencephalic structures (see Table 18.1) participate in all functions of the limbic system. These structures include the **hypothalamus**, the **mediodorsal nucleus** of the thalamus, the **anterior nucleus** of the thalamus, and the **habenula**.

As we discussed in Chapter 16, the ventral portions of the **basal ganglia** process limbic information. Limbic inputs to the basal ganglia arrive at the **ventral striatum** and **nucleus accumbens** (see Figure 18.4) and are then relayed via the **ventral pallidum** to the **mediodorsal nucleus** of the thalamus (see Figure 16.4D). The mediodorsal nucleus of the thalamus projects to the orbitofrontal and anterior cingulate limbic cortices (see Figure 16.8).

The basal forebrain and septal region are contiguous and are sometimes lumped together into one category, but we will discuss them separately here. The **basal forebrain** includes several structures that participate in limbic circuits. These structures lie just anterior and lateral to the hypothalamus, at the base of the frontal lobes near the midline (Figure 18.4A,B). Although these structures are on surface of the forebrain, they are more histologically like gray matter nuclei than like cortex. Therefore, like the amygdala, the nuclei of the basal forebrain and septal region are **corticoid** structures. The term **substantia innominata** is applied variably to the entire basal forebrain or to a nucleus within it called the **nucleus basalis of Meynert**, which lies just ventral to the anterior commissure (see Figure 18.4B).

The nucleus basalis contains cholinergic neurons that provide the major cholinergic innervation for the entire cerebral cortex (see Figure 14.9B). Other nuclei of the basal forebrain include the **olfactory tubercle**, which lies just underneath the anterior perforated substance (see Figure 18.6), and the following nuclei, which can be identified in Figure 18.4B: the **ventral pallidum**, which participates in limbic basal ganglia circuits; the **nucleus of the diagonal band of Broca**, which also contains cholinergic neurons; and the **preoptic area**, which is the rostral extension of the hypothalamus (see Figure 17.3). Portions of the amygdala also lie close to or within the region of the basal forebrain.

The **septal region** lies just dorsal to the basal forebrain, near the septum pellucidum, and it also participates in limbic pathways (see Figure 18.4B,C). The main septal nuclei lie within and just caudal to the subcallosal and paraterminal gyri and are named the **medial septal nucleus** and the **lateral septal nucleus**. The medial septal nucleus contains cholinergic neurons (see Figure 14.9B) that project to the hippocampal formation and may play a role in modulation of memory function. Inputs from the hippocampal formation are mainly to the lateral septal nucleus, and outputs are mainly from the medial septal nucleus. The lateral septal nucleus has a large projection to the medial septal nucleus, completing this circuit. The **nucleus accumbens** (see Figure 18.4A) is sometimes included in the septal region or basal forebrain and is involved in basal ganglia–limbic circuitry, as we have already discussed. Another nearby nucleus with limbic connections is the **bed nucleus of the stria terminalis** (see Figure 18.4B).

Numerous **brainstem nuclei** have reciprocal connections with limbic pathways and are sometimes considered part of the limbic system. Examples include the interpeduncular nucleus, superior central nucleus, dorsal tegmental nucleus, ventral tegmental nucleus, parabrachial nucleus, periaqueductal gray, reticular formation, nucleus solitarius, and dorsal motor nucleus of the vagus. These nuclei may help link limbic pathways to mechanisms for autonomic and behavioral arousal.

The gray matter structures of the limbic system are interconnected by white matter pathways, some of which form prominent tracts. These pathways are summarized in Table 18.4 and will be discussed further in the sections that follow.

REVIEW EXERCISE

Cover the labels in Figures 18.2 and 18.4 and name as many structures as possible.

TABLE 18.4 Summary of Major Limbic Pathways

PATHWAY	FIBERS CARRIED[a]	
	FROM	TO
Fornix	Subiculum	Medial and lateral mammillary nuclei; lateral septal nuclei
	Hippocampus	Lateral septal nuclei
	Hippocampal formation	Anterior thalamic nucleus
	Medial septal nucleus	Hippocampal formation
	Nucleus of the diagonal band	Hippocampal formation
Mammillothalamic tract	Medial mammillary nucleus	Anterior thalamic nucleus
Cingulum	Cingulate gyrus	Parahippocampal gyrus
Anterior commissure, anterior part	Anterior olfactory nucleus	Contralateral anterior olfactory nucleus
Anterior commissure, posterior part	Amygdala	Contralateral amygdala
	Anterior temporal cortex	Contralateral anterior temporal cortex
Medial olfactory stria	Anterior olfactory nucleus	Anterior commissure
	Anterior commissure	Anterior olfactory nucleus
Lateral olfactory stria	Olfactory bulb	Piriform cortex; periamygdaloid cortex; corticomedial amygdala
Stria terminalis	Corticomedial amygdala	Hypothalamus
	Amygdala	Septal nuclei
Uncinate fasciculus (temporal stem)	Piriform and entorhinal cortex	Orbitofrontal olfactory cortex
	Amygdala	Orbitofrontal and cingulate cortex
Inferior thalamic peduncle	Amygdala; anteromedial temporal cortex; insula	Medial diencephalon
Ventral amygdalofugal pathway	Amygdala	Hypothalamus; nucleus basalis; ventral striatum; brainstem nuclei
	Brainstem nuclei	Amygdala
Medial forebrain bundle	Amygdala; other forebrain structures	Brainstem nuclei
	Brainstem nuclei	Amygdala, other forebrain structures
Stria medullaris	Medial septal nuclei	Habenula
Habenulointerpeduncular tract (fasciculus retroflexus)	Habenula	Interpeduncular nucleus
Mammillotegmental tract	Mammillary bodies	Brainstem
Perforant pathway	Entorhinal cortex	Dentate gyrus granule cells
Alvear pathway	Entorhinal cortex	Hippocampal pyramidal cells

[a]For conciseness, many additional reciprocal connections have not been listed here.

Olfactory System

Although olfactory structures dominate the cerebral hemispheres of lower vertebrates, the olfactory system in humans is relatively small, even compared to the remainder of the limbic system. The term **rhinencephalon**, meaning literally "nose brain," was formerly used for many limbic structures but is now more appropriately used for only the few structures that are directly involved in olfaction.

Smell contributes to both sensation of odors in the external world and to the sensation of taste through what is referred to as "retronasal smell." Bipolar **olfactory receptor neurons** in the **olfactory mucosa** express the products of several hundred recently discovered olfactory receptor genes. A single odor mol-

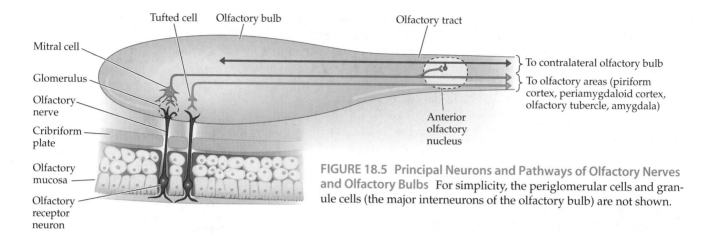

FIGURE 18.5 Principal Neurons and Pathways of Olfactory Nerves and Olfactory Bulbs For simplicity, the periglomerular cells and granule cells (the major interneurons of the olfactory bulb) are not shown.

ecule usually activates several olfactory receptors, enabling a virtually infinite number of different odors to be identified through combinatorial processing. Olfactory receptor neurons send unmyelinated axons in the **olfactory nerves** through the **cribriform plate** to reach the **olfactory bulb** (Figure 18.5). The olfactory bulb is part of the central nervous system that lies in a groove called the **olfactory sulcus**, between the gyrus rectus and orbitofrontal gyri (Figure 18.6). In the **glomeruli** of the olfactory bulb (see Figure 18.5), olfactory receptor neurons synapse onto **mitral cells** and **tufted cells**, both of which have long axons that enter the **olfactory tract** to reach the **olfactory cortex**. Collaterals in the olfactory tract synapse onto scattered neurons, forming the **anterior olfactory nu-**

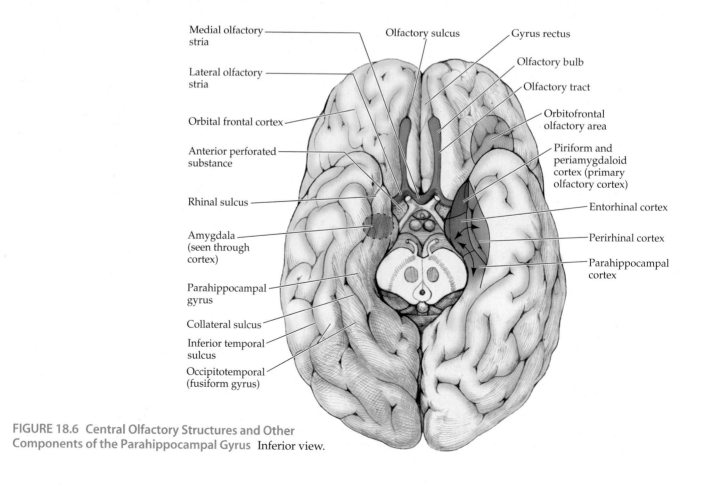

FIGURE 18.6 Central Olfactory Structures and Other Components of the Parahippocampal Gyrus Inferior view.

cleus. Neurons of the anterior olfactory nucleus, in turn, feed back both to the ipsilateral olfactory bulb and to the contralateral olfactory bulb, which is reached via the **medial olfactory stria** and the anterior part of the **anterior commissure** (see Figure 18.6; Table 18.4). Atrophy of the anterior olfactory nucleus may be responsible for impaired olfaction seen in Alzheimer's disease. In addition to mitral cells and tufted cells, the olfactory bulb contains interneurons called **periglomerular cells** and **granule cells**.

The primary olfactory cortex is unique among sensory systems in that it receives direct input from secondary sensory neurons without an intervening thalamic relay. Mitral and tufted cells of the olfactory bulb project directly via the **lateral olfactory stria** of the olfactory tract to the primary olfactory cortex (see Figure 18.6). The **primary olfactory cortex** consists of the **piriform cortex** and the **periamygdaloid cortex**, which are located near the medial anterior tip of the temporal lobe (see Figures 18.4A and 18.6). The piriform cortex is named for its pearlike shape in some nonhuman species (*pirus* means "pear" in Latin). The periamygdaloid cortex is a small region just rostral and dorsal to the amygdala. In addition to the primary olfactory cortex, fibers of the olfactory tract project to the **corticomedial nucleus** of the amygdala (see Figure 18.4B), and a small number of fibers project to the **olfactory tubercle**, located in the **anterior perforated substance** (see Figure 18.6). These projections may be important in emotional and motivational aspects of olfaction.

The primary olfactory cortex projects to several secondary olfactory areas. The **anterior entorhinal cortex** receives projections from the piriform cortex. Given the role of the entorhinal cortex in memory (see the next section), this projection may explain the occasional ability of odors to evoke vivid memories. The piriform cortex projects to the **orbitofrontal olfactory area** both directly and indirectly via relays in the entorhinal cortex, or in the **mediodorsal nucleus of the thalamus**. In monkeys, lesions of the orbitofrontal olfactory areas cause deficits in olfactory discrimination. Other projections of the piriform cortex include the **basolateral amygdala**, the lateral preoptic area, and the nucleus of the diagonal band. Interestingly, there are no direct projections from the piriform cortex to the hippocampal formation, and the hippocampal formation does not appear to have a significant role in olfactory processing.

Hippocampal Formation and Other Memory-Related Structures

One of the most fascinating and important functions of the brain is its remarkable ability to form memories. In this section we will discuss the anatomy of the hippocampal formation and other important structures involved in memory. In the next section (see KCC 18.1), we will discuss memory disorders resulting from lesions of these structures.

Based on the study of lesions in humans and animals, two main regions of the brain appear to be critical to memory formation: the **medial temporal lobe memory areas**, including the hippocampal formation and adjacent cortex of the parahippocampal gyrus, and the **medial diencephalic memory areas**, including the thalamic mediodorsal nucleus, anterior nucleus of the thalamus, mammillary bodies, and other diencephalic nuclei lining the third ventricle. The medial temporal and medial diencephalic memory areas are interconnected both with each other and with widespread regions of cortex by a variety of pathways, crucial for memory consolidation and retrieval, that we will discuss shortly. Therefore, these **white matter network connections** constitute a third component that is necessary for normal memory function. The **basal forebrain** may also play a role in memory, primarily through its widespread cholinergic projections to the cerebral cortex, including the medial temporal lobes.

REVIEW EXERCISE

Refer to Figures 18.5 and 18.6 and review the olfactory pathway, which runs from (1) olfactory receptor neurons to (2) mitral and tufted cells to (3) the primary olfactory cortex (piriform and periamygdaloid cortex) and other olfactory areas (olfactory tubercle, amygdala), and, ultimately, to (4) the orbitofrontal olfactory area.

However, some effects of lesions in the basal forebrain may be explained by damage to nearby fibers of passage from medial temporal or diencephalic memory pathways. In the subsections that follow we will discuss the anatomy of medial temporal and medial diencephalic memory systems in more detail. We will also briefly review the cholinergic pathways of the basal forebrain and septal region.

Hippocampal Formation and Parahippocampal Gyrus

The key structures of the medial temporal lobe memory system are the **hippocampal formation** and the **parahippocampal gyrus**. The hippocampal formation has an elaborate, curving *S* or inverted **S** shape on coronal sections (Figure 18.7 and Figure 18.8). This appearance inspired the names **hippocampus** (meaning "sea horse" in Greek) and **cornu Ammonis** (Latin for "horn of the ancient Egyptian ram-headed god Ammon"). The three components of the hippocampal formation are the **dentate gyrus** (named for the toothlike bumps on its medial surface; Figure 18.9), the **hippocampus**, and the **subiculum** (Latin for "support"). Sometimes the term "hippocampus" is used to refer to all three components. During embryological development, the three-layered archicortex of the medial temporal lobe folds over on itself twice (see Figure 18.7). As a result of this double folding, the pial, or gray matter, surfaces of the dentate gyrus and subiculum fuse, and the ventricular, or white matter, surfaces of the subiculum and parahippocampal gyrus fuse.

The principal neurons of the dentate gyrus are called **granule cells**. The three layers of the dentate gyrus, moving inward from the pia, are the **molecular layer**, **granule cell layer**, and **polymorphic layer** (see Figure18.8A). Note the similarities of these names to the names for the six layers of neocortex (see Table 2.3). The principal neurons of the hippocampus and subiculum are **pyramidal cells**,

REVIEW EXERCISE

Refer to Figures 18.7 and 18.8 and name the three components of the hippocampal formation.

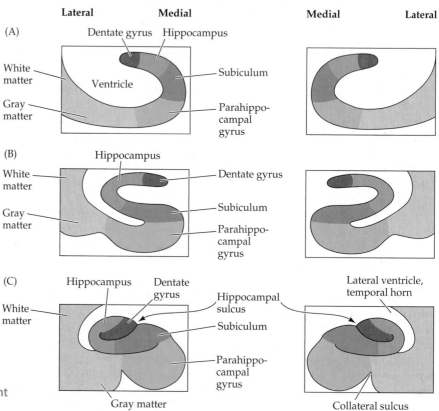

FIGURE 18.7 Embryological Development of the Hippocampal Formation

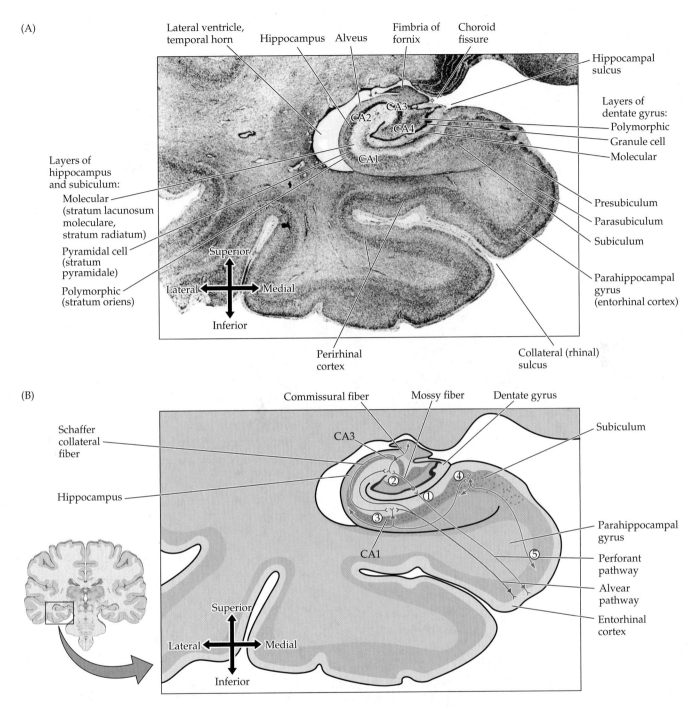

(A)

Lateral ventricle, temporal horn
Hippocampus
Alveus
Fimbria of fornix
Choroid fissure
Hippocampal sulcus

CA3
CA2
CA4
CA1

Layers of dentate gyrus:
Polymorphic
Granule cell
Molecular

Layers of hippocampus and subiculum:
Molecular (stratum lacunosum moleculare, stratum radiatum)
Pyramidal cell (stratum pyramidale)
Polymorphic (stratum oriens)

Superior
Lateral — Medial
Inferior

Presubiculum
Parasubiculum
Subiculum
Parahippocampal gyrus (entorhinal cortex)

Perirhinal cortex
Collateral (rhinal) sulcus

(B)

Commissural fiber
Mossy fiber
Dentate gyrus

Schaffer collateral fiber
CA3
Subiculum

Hippocampus

②
④
①

③
⑤

CA1

Parahippocampal gyrus
Perforant pathway
Alvear pathway
Entorhinal cortex

Superior
Lateral — Medial
Inferior

FIGURE 18.8 Hippocampal Formation
(A) Coronal section through the hippocampus and parahippocampal gyrus. Nissl stain, in which cell bodies appear dark. (B) Schematic circuit diagram showing the perforant and alvear pathway inputs to the hippocampal formation arising from the entorhinal cortex. Intrinsic hippocampal circuitry sending major outputs back to the entorhinal cortex is also shown, as well as outputs leaving via the fornix. (A after Martin JH. 1996. *Neuroanatomy: Text and Atlas.* 2nd ed. McGraw-Hill, New York. Histology courtesy of David Amaral, University of California, Davis.)

and the layers of these structures are the **molecular layer**, **pyramidal cell layer**, and **polymorphic layer** (see Figure 18.8A). The molecular layers of the dentate gyrus and subiculum are apposed, forming the **hippocampal sulcus**. The groove in the medial temporal lobe just dorsal to the hippocampal formation is called the **choroid fissure** (see Figures 18.8A and 18.9). The hippocampal formation is largest anteriorly, where it forms the **pes hippocampi**, also called the **hippocampal head**. The hippocampal formation curves back along the floor of the temporal horn, tapers to a smaller **hippocampal tail**, and finally disappears as it curves under the ventral posterior edge of the splenium of the corpus callosum. A minor vestigial remnant of the hip-

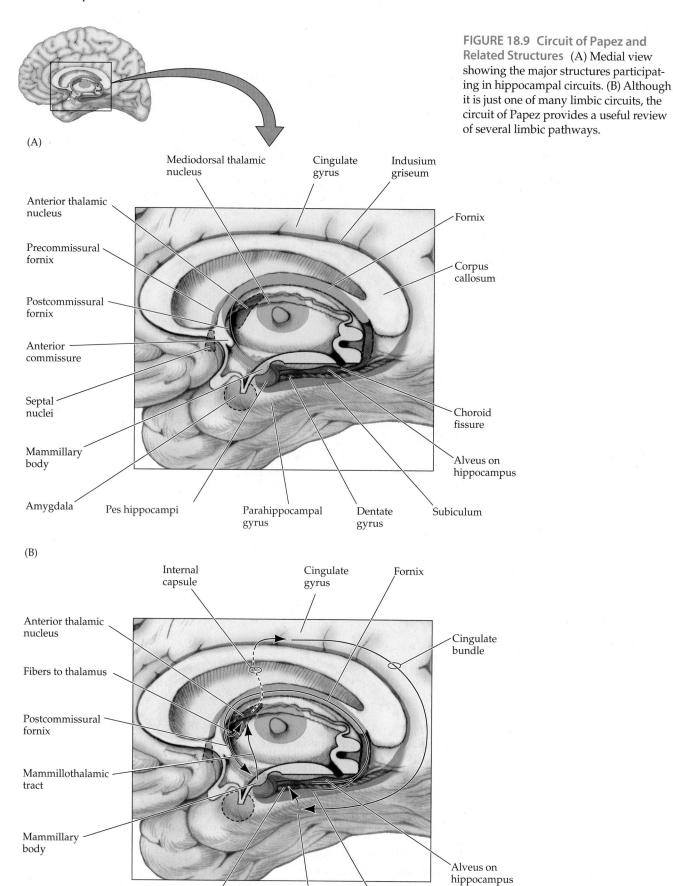

(A)

FIGURE 18.9 Circuit of Papez and Related Structures (A) Medial view showing the major structures participating in hippocampal circuits. (B) Although it is just one of many limbic circuits, the circuit of Papez provides a useful review of several limbic pathways.

Mediodorsal thalamic nucleus

Cingulate gyrus

Indusium griseum

Anterior thalamic nucleus

Precommissural fornix

Postcommissural fornix

Anterior commissure

Septal nuclei

Mammillary body

Amygdala

Pes hippocampi

Parahippocampal gyrus

Dentate gyrus

Fornix

Corpus callosum

Choroid fissure

Alveus on hippocampus

Subiculum

(B)

Internal capsule

Cingulate gyrus

Fornix

Anterior thalamic nucleus

Fibers to thalamus

Postcommissural fornix

Mammillothalamic tract

Mammillary body

Cingulate bundle

Alveus on hippocampus

Dentate gyrus

Entorhinal cortex

Subiculum

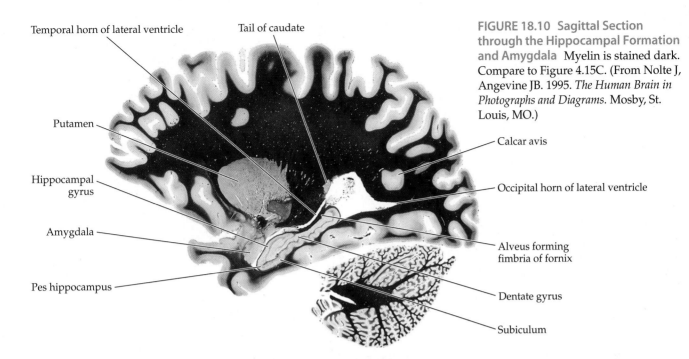

Temporal horn of lateral ventricle

Tail of caudate

Putamen

Hippocampal gyrus

Amygdala

Pes hippocampus

Calcar avis

Occipital horn of lateral ventricle

Alveus forming fimbria of fornix

Dentate gyrus

Subiculum

FIGURE 18.10 Sagittal Section through the Hippocampal Formation and Amygdala Myelin is stained dark. Compare to Figure 4.15C. (From Nolte J, Angevine JB. 1995. *The Human Brain in Photographs and Diagrams.* Mosby, St. Louis, MO.)

pocampal formation, called the indusium griseum, continues along the dorsal surface of the corpus callosum (see Figure 18.9). The layers of the hippocampus can also be appreciated in sagittal sections (Figure 18.10).

The **parahippocampal gyrus** includes several cortical areas with connections to the hippocampal formation, the most important of which is the **entorhinal cortex** (see Figures 18.6, 18.8, 18.9, and 18.11). The entorhinal cortex (Brodmann's area 28; see Figure 2.15) lies in the anterior portions of the parahippocampal gyrus, adjacent to the subiculum, and serves as the major input and output relay between association cortex and the hippocampal formation. The posterior portion of the parahippocampal gyrus is simply called the **parahippocampal cortex** (see Figure 18.6). Laterally, the parahippocampal gyrus is delimited by the **collateral sulcus**, which continues anteriorly as the **rhinal sulcus** (see Figure 18.6). Along both medial and lateral walls of the rhinal sulcus, and continuing laterally onto the adjacent occipitotemporal gyrus, lies the **perirhinal cortex** (Brodmann's areas 35 and 36). About two-thirds of the input from association cortex reaches the entorhinal cortex via relays in the adjacent perirhinal cortex and parahippocampal cortex (Figure 18.11; see also Figure 18.6).

For the sake of completeness, we will now briefly mention the names of the other parts of the parahippocampal gyrus aside from the entorhinal and perirhinal cortex (Table 18.5). (Knowledge of these other terms has no direct clinical relevance.) As we discussed in the earlier section on olfaction, the small regions of olfactory cortex adjacent to the lateral olfactory stria are called piriform and periamygdaloid cortex (see Figure 18.6). Laterally, some authors distinguish a region of cortex called prorhinal cortex lying between the entorhinal and perirhinal cortex. Medially, there are transition zones between the entorhinal mesocortex (parahippocampal gyrus) and the three-layered archicortex of the subiculum (hippocampal formation). Moving from entorhinal cortex toward subiculum, one first encounters the parasubiculum, then the presubiculum, and then the subiculum (see Figure 18.8A). Finally, although this structure is not part of the parahippocampal gyrus, we will mention here that some authors also refer to an additional transition zone, called prosubiculum, lying between the subiculum and the CA1 field of the hippocampus (see the next subsection).

REVIEW EXERCISE

Refer to Figure 18.6 and name the region of the parahippocampal gyrus that is the most important input and output relay between the hippocampal formation and the association cortex.

TABLE 18.5 Components of the Parahippocampal Gyrus

Piriform cortex
Periamygdaloid cortex
Presubicular cortex
Parasubicular cortex
Entorhinal cortex
Prorhinal cortex
Perirhinal cortex
Parahippocampal cortex

FIGURE 18.11 **Summary of Hippocampal Input and Output Connections** D = dentate gyrus; HC = hippocampus; S = subiculum; EC = entorhinal cortex; PRC = perirhinal cortex; PHC = parahippocampal cortex.

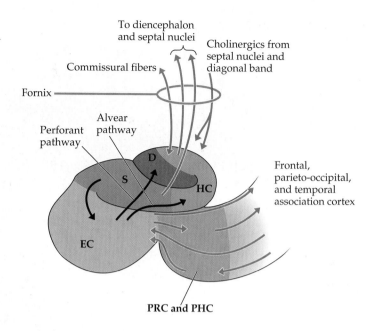

Intrinsic Circuitry of the Hippocampal Formation

The circuitry of the hippocampal formation has been the subject of intensive scientific investigation because of its possible important role in human memory. Figure 18.8B illustrates the major circuits for information flow from the entorhinal cortex, through the hippocampal formation, and back to the entorhinal cortex. Pyramidal cells in layers 2 and 3 of the entorhinal cortex project to the hippocampal formation via the perforant pathway and the alvear pathway. (See Figure 18.8B, synapse 1.)

The **perforant pathway** is named for its course through the subiculum and across the hippocampal sulcus to reach the granule cell layer of the dentate gyrus. The hippocampus has different pyramidal cell sectors named CA (for cornu Ammonis) 1 through 4 (see Figure 18.8A). **CA4** lies within the hilus of the dentate gyrus. **CA3** is adjacent to CA4, **CA2** comes next, and **CA1** lies closest to the subiculum. Granule cells of the dentate gyrus give rise to axons called **mossy fibers**, which synapse on the dendrites of CA3 pyramidal cells (see Figure 18.8B, synapse 2). The axons of CA3 pyramidal cells leave the hippocampal formation via the fornix. However, these axons also give rise to the **Schaffer collaterals**, which synapse on the dendrites of CA1 pyramidal cells (see Figure 18.8B, synapse 3). Axons of CA1 pyramidal cells also leave the hippocampal formation via the fornix. In addition, CA1 pyramidal cells project to the next cellular relay, which lies in the subiculum (see Figure 18.8B, synapse 4). Finally, pyramidal cells of the subiculum project both into the fornix and back to neurons in the deeper layers of the entorhinal cortex, completing the loop (see Figure 18.8B, synapse 5).

In addition to the perforant pathway, neurons of the entorhinal cortex project via the **alvear pathway** directly to CA1 and CA3 (see Figures 18.8B and 18.11). As in the perforant pathway, outputs in the alvear pathway are primarily from CA1 and CA3 to the subiculum. Although earlier work tended to emphasize the importance of the so-called **tri-synaptic** (perforant) pathway from entorhinal cortex to CA1, recent studies have shown that the monosynaptic **direct pathway** from entorhinal cortex to CA1 may dominate (not to be confused with the direct pathway in basal ganglia circuitry; see Figure 16.7).

An interesting form of synaptic plasticity called **long-term potentiation** (**LTP**) is found in the perforant pathway–granule cell, mossy fiber–CA3, and Schaffer

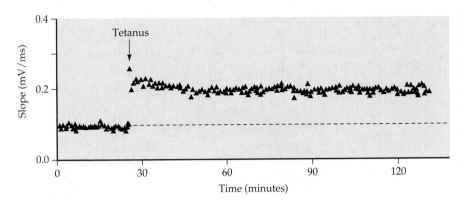

FIGURE 18.12 Long-Term Potentiation (LTP) in the CA1 Sector of the Hippocampus Input fibers to the CA1 pyramidal cells (see Figure 18.8) were stimulated with a test stimulus every 10 seconds, and the CA1 EPSPs were recorded extracellularly. Slope of the EPSPs (an indication of synaptic strength) is indicated on the vertical axis. At the time marked "Tetanus," two 100 Hz stimulus trains lasting 1 second each were delivered at twice the usual stimulus intensity. Following this strong stimulus, a sustained increase in the EPSP slope is evident, lasting several hours. (From Nicoll RA, Kauer JA, and Malenka RC. 1988. The current excitement in long-term potentiation. *Neuron* 1: 97–103.)

collateral–CA1 connections (Figure 18.12). High-frequency activity at any of these synapses causes a long-lasting increase in synaptic strength between the involved neurons. In particular, the perforant pathway–granule cell and Schaffer collateral–CA1 (but not mossy fiber–CA3) synapses require simultaneous pre- and postsynaptic activity for LTP to be elicited. This interesting property may allow these synapses to perform an **associative function**, similar to the learning rule proposed by the psychologist Donald Hebb in the 1940s. The **Hebb rule** states, "When an axon of cell A . . . excite(s) cell B and repeatedly or persistently takes part in firing it, some growth process or metabolic change takes place in one or both cells so that A's efficiency as one of the cells firing B is increased."

Since its initial discovery, LTP has also been demonstrated at synapses in several other areas of the nervous system. In addition, numerous other forms of excitatory and inhibitory, short-term and long-term synaptic modulation have been described. The cellular mechanisms underlying LTP and other types of synaptic plasticity are currently an important and very active area for further investigation and are thought to play an important role in memory formation.

Input and Output Connections of the Medial Temporal Lobe Memory System

The **main input** to the hippocampal formation arrives at the **entorhinal cortex** from association cortex in the frontal, parieto-occipital, and temporal lobes (see Figure 18.11). Much of this information is relayed in the adjacent perirhinal cortex and parahippocampal cortex before reaching the entorhinal cortex. These inputs are thought to contain higher-order information from multiple sensorimotor modalities that is processed further by the medial temporal structures for memory storage. The storage process itself is believed to occur not in the medial temporal structures, but back in the association and primary cortices that allow a particular memory to be reactivated. An **important output** pathway of the hippocampal formation is therefore the projection from the **subiculum to the entorhinal cortex**, and from there back to multimodal association cortex (see Figure 18.11). The mechanisms by which medial temporal structures induce storage, consolidation, and retrieval of memories in the association cortex are not currently known.

An additional major output pathway of the hippocampal formation is the **fornix**, which carries outputs to the diencephalon and septal nuclei (Figure 18.13; see also Figure 18.11). These pathways will be discussed in greater detail in the next subsection. Note that the **subiculum** is the main source of output fibers from the hippocampal formation to the fornix, as well as to the entorhinal cortex. In addition, the subiculum sends monosynaptic connections to the amygdala, orbitofrontal cortex, and ventral striatum. The subiculum is thus an important structure in hippocampal outputs.

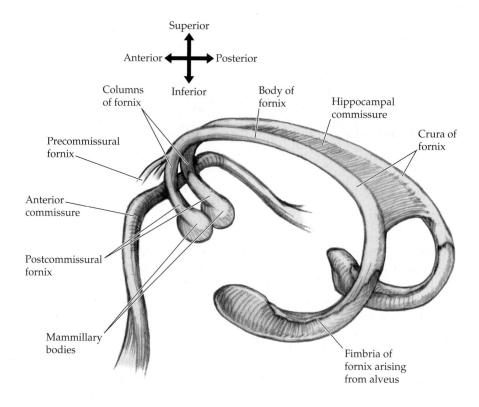

Some inputs reach the hippocampus from the contralateral hippocampus through the **hippocampal commissure** (see Figures 18.11 and 18.13). Finally, the hippocampal formation receives important modulatory inputs via the fornix arising from **cholinergic neurons** in the medial septal nucleus and the nucleus of the diagonal band. The entorhinal cortex and the remainder of the cerebral cortex also receive cholinergic inputs, which arise primarily from the nucleus basalis of Meynert (see Figure 18.4B; see also Figure 14.9). These cholinergic pathways activate muscarinic receptors and may be important in modulating neuronal excitability and synaptic plasticity. Additional modulatory influences also reach the medial temporal lobes from noradrenergic, dopaminergic, and serotonergic nuclei in the brainstem (see Chapter 14).

The Fornix and Medial Diencephalic Memory Pathways

Fornix, meaning "arch" in Latin, is an appropriate name for this white matter structure that curves through the ventricular system from the hippocampal *formation to the diencephalon and septal area* (see Figure 18.13). As we discussed in Chapter 5, several C-shaped structures follow the curve of the lateral ventricles, including the fornix, corpus callosum, and caudate nucleus. Please review the three-dimensional relations of the fornix to these structures, as discussed in the "Brief Anatomical Study Guide" and "A Scuba Expedition through the Brain" in Chapter 5.

Output fibers from the hippocampal formation form a white matter layer on the ventricular surface of the hippocampus called the **alveus** (see Figures 18.8A, 18.9A, 18.10, and 18.13). As these fibers sweep medially, they form a discrete bundle called the **fimbria** of the fornix (see Figures 18.8A and 18.13). The **crura** (meaning "legs") of the fornix leave the hippocampal formation and curve under the corpus callosum (see Figure 18.9; see also Figure 16.2) to run adjacent to each other in the midline, at which point their name changes to the **body** of the fornix (see Figure18.13; see also Figure 16.4D). Between the crura of the fornix, on the undersurface of the corpus callosum, the **hippocampal commissure** provides a route for fibers arising from one hippocampus to reach the contralateral side. The body of the fornix curves anteriorly and downward to form the **columns** of the fornix (see Figures 18.4C, 18.9, and 18.13; see also Figure 16.2).

Axons traveling forward in the fornix have three main targets. The majority of the fibers arise from the **subiculum** and descend behind the anterior commissure in the **postcommissural fornix** to reach the **medial and lateral mammillary nuclei** of the hypothalamus (see Figures 18.9 and 18.13). A smaller contingent of fibers, arising from both the **subiculum and hippocampus**, pass anterior to the anterior commissure in the **precommissural fornix** to reach the **lateral septal nucleus**. Finally, some fibers leave the fornix to terminate in the **anterior thalamic nucleus** (see Figure 18.9).

As mentioned previously, some fibers travel back in the fornix, predominantly from cholinergic neurons in the **medial septal nuclei** and diagonal bard of Broca, to reach the hippocampal formation (see Figure 18.11). This pathway can be influenced by projections from the hippocampal formation forward to the lateral septal nuclei because the lateral septal nuclei project strongly to the medial septal nuclei. These cholinergic projections, along with inhibitory GABAergic projections that also travel in the fornix from the septal nuclei to the hippocampal formation, may play an important modulatory role in memory function.

In 1937 the anatomist James Papez described a circuit involving several limbic structures, thereby stimulating the development of the concept of the limbic system in the 1950s. Although the structures in this circuit have subsequently been shown to have many other important connections as well, the **Papez circuit** remains a useful heuristic device for reviewing some of the major limbic pathways (see Figure 18.9B). The Papez circuit begins with fibers arising from the subiculum of the **hippocampal formation**, which enter the fornix and travel forward to both the medial and the lateral mammillary nuclei. The **medial mammillary nucleus** then projects through the **mammillothalamic tract** to the **anterior thalamic nucleus** (see Figure 16.4D). Recall that the anterior thalamic nucleus also receives a direct projection from the fornix (see Figure 18.9B). The anterior thalamic nucleus next projects through the **internal capsule** to the **cingulate gyrus**. Finally, a prominent white matter pathway underlying the cingulate gyrus called the **cingulate bundle**, or **cingulum**, passes from the cingulate cortex to the **parahippocampal gyrus**. From the parahippocampal gyrus, projections continue to the **entorhinal cortex** and hippocampal formation, completing the loop.

In summary, the **medial temporal lobe memory systems** communicate with the association cortex mainly through bidirectional connections via the entorhinal cortex (see Figure 18.11). The **medial diencephalic memory systems** communicate with the medial temporal memory systems through several pathways. The fornix connects the hippocampal formation with the mammillary bodies and septal nuclei, as well as with the anterior thalamic nuclei, both directly and indirectly, through the mammillothalamic tract. Other medial diencephalic structures implicated in memory function include the **mediodorsal nucleus** of the thalamus and the midline thalamic nuclei (see Figure 7.6). These medial diencephalic nuclei are connected to the limbic structures of the medial temporal lobe and insula via fibers of the inferior thalamic peduncle (see Table 18.4) traveling near the auditory radiation (see Figure 6.9B) and with the amygdala via the ventral amygdalofugal pathway (discussed below). It has also been postulated that damage to the internal medullary lamina of the thalamus (see Figure 7.6) may be important for memory loss with medial diencephalic lesions. The functional roles of individual medial diencephalic nuclei and their relative importance in memory are still under investigation.

REVIEW EXERCISE

Using Figure 18.13 as a guide, study the axial, coronal, and sagittal MRI scan images in Figures 4.13F–H, 4.14A–C, and 4.15A, and review the three-dimensional course of the fornix. Identify the crura (see Figures 4.13G and 4.14A), body (see Figures 4.13H, 4.14B, and 4.15A), and columns (see Figures 4.13G and 4.14C), and review the relationships of the fornix to the hippocampus, lateral ventricle, third ventricle, corpus callosum, septum pellucidum, foramen of Monro, anterior commissure, and mammillary bodies.

REVIEW EXERCISE

Review the major hippocampal input and output connections using Figure 18.11 and referring to Figures 18.6, 18.8, and 18.9.

KEY CLINICAL CONCEPT

18.1 MEMORY DISORDERS

In this section we will discuss memory disorders, beginning with a specific, well known case of amnesia. This case will be used to illustrate the different kinds of memory loss that can occur, before we discuss the differential diagnosis of memory loss.

Patient H.M.: A Landmark Case of Amnesia

In 1953 a 27-year-old man with the initials H.M. underwent an operation in which the *bilateral* medial temporal lobes, including the hippocampal formations and parahippocampal gyri, were resected in an attempt to control his medically refractory epileptic seizures (Figure 18.14). Following the surgery, his seizures improved, but he had severe memory problems, with no other significant deficits. He was unable to learn new facts or recall new experiences. For example, when given a list of three or four words to remember, he was able to correctly recite them back immediately. Within 5 minutes, however, he had no recall of any of the words, even with hints, and he did not even remember being given the list to remember in the first place. In contrast, his memory of remote events from his childhood and up to several years before the surgery was intact; however, he had no recall of events from that point on. Despite his profound memory deficit, H.M.'s personality and general intelligence assessed by IQ testing were normal. In addition, he retained the ability to learn certain tasks that did not require conscious recall. For example, his performance on a mirror drawing task improved on successive days similarly to normal controls, despite his having no recollection of doing the task the previous day. Similarly, when primed by exposure to a word such as "DEFINE" and then asked to complete the stem "DEF-," he

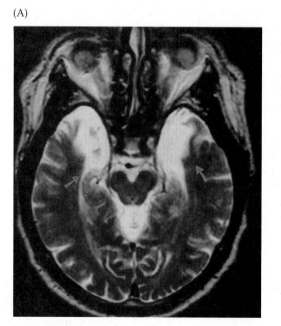
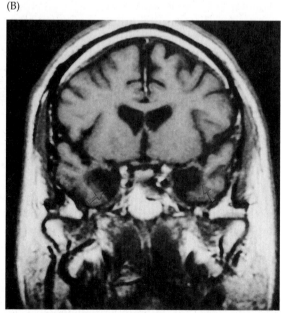

(A)

(B)

FIGURE 18.14 MRI from Patient H.M. This patient underwent bilateral resection of the medial temporal lobe structures. (A) Axial T2-weighted MRI image. (B) Coronal T1-weighted image. Regions of resection are indicated by arrows. (From Corkin S, Amaral DG, Gonzalez RG, Johnson KA, Hyman BT. 1997. H.M.'s medial temporal lesion: findings from magnetic resonance imaging. *J Neurosci* 17 (10): 3964–3979.)

chose the word he had seen previously at higher than chance levels, despite having no conscious recall of having seen the word before.

The selective yet devastating and permanent effects of this operation on H.M.'s memory led to intensive investigations of the medial temporal lobes' role in human memory. H.M. continued to participate in these tests for several decades until shortly before his death in 2008, and his name—Henry Gustav Molaison—was then released. Patients today with medically refractory temporal lobe epilepsy can often be cured with *unilateral* medial temporal lobe resection (see KCC 18.2); after the unfortunate experience with Henry Molaison, bilateral medial temporal lobe resection is no longer performed.

Lessons Learned from H.M.: Classification of Memory and Memory Disorders

On the basis of studies of patient H.M. and numerous other studies of patients and experimental animals in subsequent years, several types of memory and memory disorders have been defined. Although many of these distinctions did not emerge until years after the original studies on H.M., we will review the case of patient H.M. to help illustrate and understand some of these distinctions.

DECLARATIVE VERSUS NONDECLARATIVE MEMORY One major distinction is between **declarative** (or **explicit**) **memory**, which involves conscious recollection of facts or experiences, and **nondeclarative** (or **implicit**) **memory**, which involves nonconscious learning of skills, habits, and other acquired behaviors (**Figure 18.15**). H.M. was severely impaired in his ability to recollect new facts or experiences; however, his behavior could be modified by previous experiences in a nonconscious manner. Thus, H.M. suffered a loss of declarative memory, while his nondeclarative memory remained intact. The term **amnesia** is typically used for declarative memory loss. This form of selective loss of **declarative memory** is typical of **bilateral medial temporal lobe** or **bilateral medial diencephalic** lesions. Unilateral lesions do not usually produce severe memory loss. However, unilateral lesions of the dominant (usually left) medial temporal or diencephalic structures can cause some deficits in **verbal memory**, while unilateral lesions of the nondominant (usually right) hemisphere can cause deficits in **visual-spatial memory**.

Unlike declarative memory, specific lesions do not usually result in clinically significant selective loss of nondeclarative memory (see Figure 18.15). Learning of **skills**, such as mirror drawing by H.M., and of **habits** most likely involves plasticity in several areas, including the basal ganglia, cerebellum, and motor cortex. The caudate nucleus appears to be particularly important

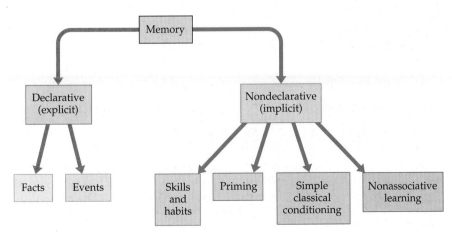

FIGURE 18.15 Classification of Memory (After Squire LR, and Zola-Morgan S. 1991. The medial temporal lobe memory system. *Science* 253: 1380–1385.)

in habit learning, and interestingly, pathology in this region may be linked to obsessive-compulsive disorder (see KCC 18.3). **Priming**, as exhibited by H.M. in the word stem completion task, depends on several cortical areas. Simple associative learning such as **classical conditioning** and nonassociative learning such as **habituation** and **sensitization** have been studied extensively in animals, and appear to involve a variety of structures, including the cerebellum (in classical conditioning), amygdala (in conditioned fear), cerebral cortex, brainstem nuclei, and even spinal cord.

TEMPORAL ASPECTS OF MEMORY AND MEMORY LOSS Although H.M. was able to repeat a short list of words immediately after hearing them, within 5 minutes he could not recall any of the words. How are memories converted from short-term to long-term storage? At least two classes of mechanisms appear to be involved (Table 18.6). First of all, a variety of different cellular mechanisms store information in the nervous system on different time scales (see Table 18.6A). Secondly, different anatomical regions of the brain are important for storing memories at different times. The brain regions thought to be involved in declarative memory at different times are listed in Table 18.6B.

When examining patients at the bedside, practitioners perform tests of **immediate recall**, **attention**, and **working memory** often by asking the patient to repeat back lists of digits or words forward and backward (see **neuroexam.com Video 4**). These functions, which operate on the time scale of less than 1 or 2 minutes, must be intact in order for information to be encoded successfully in declarative memory. However, these functions do not depend on the medial temporal or medial diencephalic memory system (see Table 18.6B). As we will discuss in Chapter 19, alertness and attention are mediated by an interaction of brainstem–diencephalic and frontoparietal networks, acting on the specific regions of cortex involved in portraying a particular concept in conscious awareness. In addition, **working memory** involves holding a particular concept briefly in awareness while a mental operation, such as the carrying function in arithmetic, is performed. Working memory requires the participation of dorsolateral prefrontal association cortex (see Chapter 19). After testing attention and confirming the patient's ability to register new information, **recent memory** should be tested by giving the patient several words to remember and then testing for recall of these words 4 to 5 minutes later (see **neuroexam.com Video 6**). Recent mem-

TABLE 18.6 Memory Mechanisms in the Time Domain and in the Spatial Domain

A. CELLULAR MECHANISMS INVOLVED AT DIFFERENT TIMES IN MEMORY STORAGE

SECONDS TO MINUTES	MINUTES TO HOURS	HOURS TO YEARS
Ongoing electrical activity of neurons; changes in intracellular Ca^{2+} and other ions; changes in second messenger systems	Protein phosphorylation and other covalent modifications; expression of immediate early genes	Additional changes in gene transcription and translation resulting in structural changes of proteins and neurons

B. ANATOMICAL STRUCTURES INVOLVED AT DIFFERENT TIMES IN STORAGE OF EXPLICIT MEMORIES

LESS THAN 1 SECOND ("ATTENTION" OR "REGISTRATION")	SECONDS TO MINUTES ("WORKING MEMORY")	MINUTES TO YEARS ("CONSOLIDATION")	YEARS
Brainstem–diencephalic activating systems; frontoparietal association networks; specific unimodal and heteromodal cortices	Frontal association cortex; specific unimodal and heteromodal cortices	Medial temporal structures; medial diencephalic structures; specific unimodal and heteromodal cortices	Specific unimodal and heteromodal cortices

ory is impaired in dysfunction of the bilateral medial temporal or medial diencephalic regions. **Remote memory** should then also be tested by asking the patient about either verifiable personal information such as previous addresses or schools and about well-known current events (see **neuroexam.com Video 7**). In addition to these simple bedside tests, more precise and quantitative neuropsychology testing can often be useful in evaluating memory dysfunction (see also KCC 19.16).

H.M. and other patients with bilateral medial temporal or medial diencephalic lesions are unable to recall facts and events for more than a few minutes. **Medial temporal and diencephalic structures** appear to mediate a process by which declarative memories are gradually **consolidated** in the neocortex (see Table 18.6B). Ultimately, through this process, declarative memories can be recalled through activity of specific regions of **neocortex** without requiring medial temporal or medial diencephalic involvement.

Anterograde amnesia is the deficit in forming new memories, seen in H.M. and other similar patients from the time of their brain injury onward (**Figure 18.16**). For example, beginning with his surgery and for the rest of his life, H.M. was not able to learn his address, and when asked, he would still cite the address of his childhood; he could not learn the date, and despite watching the news every day, he did not remember most events from after the 1950s. **Retrograde amnesia** is the loss of memories from a period of time before the brain injury. For example, H.M.'s memories of his childhood and adolescence remained relatively normal, but his memories stopped several years before the surgery (see Figure 18.16). This combination of retrograde and anterograde amnesia for declarative memories is typical of lesions of the medial temporal lobe or medial diencephalic memory systems (although it can also be seen in concussion or other diffuse disorders).

The phenomenon of retrograde amnesia suggests that recent memories, for a period of up to several years, are dependent on the normal functioning of medial temporal and diencephalic structures, while more remote memories are not. Retrograde amnesia is often graded, with the memories from the time just before the injury being the most severely impaired (although, like most biological phenomena, the time gradient is typically not perfectly uniform). In patients with reversible causes of amnesia (described in the next subsection), the period of retrograde amnesia often gradually shrinks forward; older memories are recovered before more recent ones. Ultimately, these patients will have a short time period (several hours) of permanently lost memories from before the injury and a period of lost memories from the injury until the time they recover from the anterograde amnesia.

The expressions "short-term" and "long-term" memory are sometimes used for memories enduring less than or more than a few minutes, respectively. This terminology is not ideal because so-called long-term memory includes even recent memories that are disrupted by bilateral medial temporal or diencephalic disorders. Numerous cognitive models of memory function have been developed that are beyond the scope of this book. One common way that memory has been described is as a three-step process involving

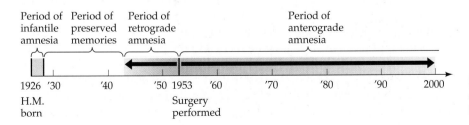

FIGURE 18.16 Diagram of Anterograde and Retrograde Amnesia in Patient H.M.

TABLE 18.7   Causes of Memory Loss

ANATOMICAL LESIONS USUALLY VISIBLE ON IMAGING STUDIES

Bilateral medial temporal lesions
  Surgery
  Cerebral contusions
  Infarct (posterior cerebral arteries)
  Hippocampal sclerosis (usually with chronic epilepsy)
  Herpes encephalitis
  Paraneoplastic limbic encephalitis
  Neoplasm
  Inflammatory process, such as sarcoidosis
Bilateral medial diencephalic lesions
  Wernicke–Korsakoff syndrome
  Infarct (thalamoperforator arteries)
  Whipple's disease
  Neoplasm
Basal forebrain lesions
  Anterior communicating artery aneurysmal rupture
  Neoplasm
Diffuse disorders
  Multiple sclerosis
  Numerous other diffuse cerebral disorders
    (additional deficits common)

NO ANATOMICAL LESIONS VISIBLE ON CONVENTIONAL IMAGING[a]

Seizures, including electroconvulsive therapy
Concussion
Ischemia (bilateral medial temporal or medial diencephalic structures)
Diffuse cerebral anoxia
Transient global amnesia
Early Alzheimer's disease and other degenerative disorders
Diffuse infectious or toxic/metabolic encephalopathies (additional deficits common), including those caused by medications such as benzodiazepines
Psychogenic amnesia
  Dissociative disorders
  Repression
  Conversion disorder
  Malingering

"NORMAL" MEMORY LOSS

Infantile amnesia
During or shortly after awakening from sleep
Passage of time (forgetting)

[a]In some of these disorders, abnormalities may be visible on functional imaging studies such as PET or fMRI (see Chapter 4).

encoding, storage, and retrieval. However, it has proven very difficult to demonstrate a selective deficit in human or animal studies in just one of these processes.

Differential Diagnosis of Memory Loss

Impaired memory can have many causes, as summarized in Table 18.7. These can be conveniently divided into those in which there are usually anatomical abnormalities visible on conventional imaging studies, those in which there are not, and, finally, "normal" forms of memory loss. We will comment on only a few of these conditions in detail.

Cerebral contusions caused by head trauma often involve the anteromedial temporal lobes as well as the basal orbitofrontal cortex (see Figure 5.21), resulting in permanent deficits in memory. In contrast, **concussion** (see KCC 5.5) can be associated with memory loss that is usually reversible, except for the hours around the time of the injury.

Infarcts or **ischemia** (see KCC 10.3 and KCC 10.4) can cause memory deficits, especially when bilateral medial temporal or medial diencephalic structures are affected. Recall that the medial temporal lobes are supplied by distal branches of the posterior cerebral arteries (see Figure 10.5). The medial thalami are supplied by paramedian thalamoperforator arteries, which arise from the initial segments of the posterior cerebral arteries (see Figures 10.8 and 10.9). Thus, arterial lesions at the top of the basilar artery (see Figure 14.18A; KCC 14.3) are well positioned to cause either bilateral medial temporal or medial diencephalic infarcts. In some cases, a single paramedian thalamoperforator artery bifurcates shortly after its origin from the top of the basilar and supplies both medial thalami. Occlusion of this vessel, called an "artery of Percheron," provides another mechanism for bilateral infarctions in this area.

In **global cerebral anoxia**, such as that caused by cardiac arrest, memory loss is often prominent. This may be related to the particular vulnerability of the hippocampus to anoxic injury, especially CA1, in which marked loss of pyramidal cells can be seen. As has already been mentioned, rupture of an **anterior communicating artery aneurysm** can damage the basal forebrain, causing memory loss together with other deficits seen in frontal lobe lesions (see KCC 19.11). It is unclear whether the memory loss in these patients is due to damage to the basal forebrain, medial diencephalon, frontal lobes, or a combination of these locations.

Wernicke–Korsakoff syndrome is caused by thiamine deficiency, seen most often in alcoholics, but also occasionally in patients on chronic parenteral nutrition. Pathologically, these patients have bilateral necrosis of the mammillary bodies and of a variety of medial diencephalic and other periventricular nuclei. Acutely, patients with thiamine deficiency have a triad of ataxia; eye movement abnormalities ranging from horizontal gaze paresis, or nystagmus, to ophthalmoplegia; and a confusional state. Severe cases can result in coma or death. Patients who survive the acute stages are left with anterograde and retrograde amnesia, thought to be caused by the bilateral diencephalic lesions. In addition to amnesia, however, patients with Wernicke–Korsakoff syndrome usually have other neuropsychological deficits that suggest frontal lobe dysfunction (see KCC 19.11). These include impairments in judgment, initiative, impulse control, and sequencing tasks. In contrast to patients with "pure" medial diencephalic or medial temporal lesions, patients with Wernicke–Korsakoff syndrome often lack an awareness of their memory deficit, and in fact they tend to **confabulate**, providing spurious answers to questions rather than saying that they do not remember. Confabulation is also probably related to frontal lobe dysfunction, which causes disinhibition and a loss of self-monitoring capabilities.

Patients with complex partial and generalized tonic-clonic **seizures** (see KCC 18.2) usually have memory loss of events during the seizure and postictal period (the period immediately following the seizure). Memory between seizures may be normal unless the seizures are severe or caused by lesions of the medial temporal lobe, such as **hippocampal sclerosis** (see KCC 18.2). **Electroconvulsive therapy** (**ECT**) is an effective mode of therapy for selected patients with refractory depression. In ECT, seizures are induced while the patient is under anesthesia during multiple sessions, usually over the course of several weeks. During the treatment period, patients develop retrograde and anterograde amnesia similar to that seen in patients with bilateral temporal or diencephalic lesions. The amnesia gradually resolves after the course of treatment is complete, but it typically leaves a gap, including retrograde and anterograde memory loss from around the treatment period.

Transient global amnesia is a somewhat mysterious disorder in which patients abruptly develop retrograde and anterograde amnesia with no obvious cause and no other deficits. Episodes often occur in the setting of physical exertion or emotional stress. During the amnesia, patients characteristically ask the same questions over and over, with no recollection of having asked them a few minutes earlier. The amnesia typically lasts for approximately 4 to 12 hours, after which the patient recovers fully, except for a permanent loss of memories for a period of a few hours before and after onset. In about 85% of patients, a similar episode never happens again.

The cause of this intriguing syndrome is not known. It differs from other common causes of transient neurologic episodes (see KCC 10.3), such as seizures and transient ischemic attacks (TIAs), in several ways. Seizures can cause periods of memory loss; however, other manifestations of seizures, such as abnormal movements or decreased responsiveness, are usually present as well. Complex partial seizures can occasionally cause brief episodes of amnesia with no other obvious behavioral changes. Unlike transient global amnesia, however, the duration of amnesia is only a few minutes, the episodes recur multiple times in a stereotyped fashion, and the EEG is often abnormal. EEG recordings during transient global amnesia do not show epileptic activity. TIAs can cause transient memory loss; however, the typical duration of TIAs is minutes, not hours. In addition, patients who have had an episode of transient global amnesia are not at increased risk for subsequent stroke compared to the general population. A migraine-like (see KCC 5.1) phenomenon has also been proposed as the mechanism for transient global amnesia, and indeed a history of migraine is common in patients with transient global amnesia. Functional imaging studies during transient global amnesia have demonstrated decreased blood flow or decreased glucose metabolism in the medial temporal lobes, as well as in other brain regions. In conclusion, the cause of transient global amnesia remains unknown. It is possible that this syndrome is produced by different etiologies in different patients, but the relative uniformity of this disorder argues for a common mechanism in at least the majority of cases.

In the early stages of several neurodegenerative disorders, especially early **Alzheimer's disease** (see KCC 19.16), memory loss for recent events is often prominent, with no other obvious abnormalities. This phenomenon may occur because early Alzheimer's disease tends to preferentially affect the bilateral hippocampal, temporal, and basal forebrain structures (see Figure 19.15). As Alzheimer's disease progresses, other neurobehavioral abnormalities occur as well, as we will discuss in Chapter 19. Memory loss can also be seen as a part of numerous other diffuse or multifocal disorders of the nervous system stemming from many etiologies. In these disorders—which include multiple sclerosis, brain tumors, intracerebral hemorrhage, infarcts, CNS infections, various toxic or metabolic encephalopathies, CNS vasculitis, hydrocephalus, and many other conditions—a variety of other abnormalities are usually present in addition to memory loss. It is sometimes difficult to distinguish true memory disorders in these conditions from deficits in attention or language processing.

Psychogenic amnesia can occur in several settings, including dissociation, repression, conversion, and malingering. In contrast to medial temporal lobe or diencephalic amnesia, patients with psychogenic amnesia usually do not have a pattern of retrograde and anterograde amnesia affecting mainly recent memories. Instead, patients with psychogenic amnesia often have memory loss for events of particular emotional significance. In psychogenic amnesia there may also be loss of autobiographical memories such as one's name and birthplace—memories that are ordinarily preserved in medial temporal lobe or diencephalic amnesia unless other severe cognitive deficits are present as well.

"Normal" memory loss occurs in several situations. **Infantile amnesia** is the inability for adults to recall events from the first 1 to 3 years of life. A variety of mechanisms have been proposed, but infantile amnesia is most likely the result of ongoing central nervous system maturational processes, such as myelination, that are quite active during infancy and early childhood. At the other extreme of life, **benign senescent forgetfulness** is the presumably normal mild decline in memory function that occurs gradually over the decades. This should be contrasted with Alzheimer's disease and other forms of dementia in which memory loss is more severe and occurs over the course of a few years.

In another normal form of memory loss, dreams can be recalled immediately after awakening from **sleep**, but they often can no longer be remembered a few minutes later. Similarly, a common experience is being awakened from a deep sleep and having a conversation on the telephone that, the next day, one cannot recall. Finally, with the passage of time, **forgetting** normally occurs, in which memories gradually become less distinct and eventually may not be recalled. ▪

The Amygdala: Emotions, Drives, and Other Functions

The **amygdala** (meaning "almond" in Greek), or **amygdaloid nuclear complex**, is a group of nuclei located in the anteromedial temporal lobe, just dorsal to the anterior tip of the hippocampus and temporal horn. It has three main nuclei: the **corticomedial**, **basolateral**, and **central nuclei** (see Figures 18.4B, 18.6, and 18.10). The **bed nucleus of the stria terminalis** (see Figure 18.4B) is also considered part of the amygdala. In humans, the **basolateral nucleus** is largest and is predominantly involved in direct and indirect connections of the amygdala to diverse cortical areas as well as to the basal forebrain and medial thalamus. The smaller **corticomedial nucleus** derives its name from its corticoid structure, located on the medial surface of the temporal lobe, near the basal forebrain and olfactory areas (see Figure 18.4A,B). The main connections of the corticomedial nucleus are involved in olfaction and in interactions with the hypothalamus related to appetitive states. The **central nucleus** is smallest and has connections with the hypothalamus and brainstem that are important in autonomic control.

As discussed in the beginning of the chaper, the amygdala plays a pivotal role in emotions and drives (see Table 18.2). However, through its extensive connections to other structures in the limbic network (see Figure 18.1), the amygdala is an active participant in all four major limbic functions (see Table 18.2). We will first discuss functional roles of the amygdala and then review its main input and output connections.

Emotions and drives appear to be mediated by complex interactions among numerous brain regions, including the heteromodal association cortex, limbic cortex, amygdala, septal area, ventral striatum, hypothalamus, and brainstem autonomic and arousal pathways (Figure 18.17A,B). The amygdala plays a central role, but the other components of the network are essential as well. On the basis of the effects of lesions in humans and experimental animals, the amygdala is important for attaching emotional significance to various stimuli perceived by the association cortex. When both amygdalas have been ablated, behavior tends to be placid. Tame, nonaggressive behavior, together with other behavioral changes, constitutes the **Klüver–Bucy syndrome** studied in monkeys with bilateral lesions of the amygdala and adjacent temporal structures. (Klüver–Bucy syndrome has been observed only rarely in humans.) Seizures (see KCC 18.2) involving the amygdala and adjacent cortex cause powerful emotions of fear and panic.

Interestingly, while activity in the amygdala has been found to be important in states of fear, anxiety, and aggression, activity in the septal area appears to be important in pleasurable states. For example, experimental animals will press a lever repeatedly to obtain electrical stimulation of the septal area, even to the point of neglecting to eat in order to continue the stimulation. Increased activity has been recorded in the septal area during orgasm, and lesions of the septal area in animals cause "sham rage," in which sudden outbursts of aggressive behavior occur. Sham rage behaviors have also been elicited by stimulating certain regions of the midbrain tegmentum.

Reciprocal connections between the amygdala and hypothalamic and brainstem centers for **autonomic control** mediate changes in heart rate, peristalsis, gastric secretion, piloerection, sweating, and other changes commonly seen with strong emotions. The limbic cortex, including the orbitofrontal, insula, anterior cingulate, and temporal cortex, (see Figure 18.2) has important connections with the hypothalamus as well. In addition, connections between the limbic cortex, amygdala, and the hypothalamus are important for **neuroendocrinological** changes seen in different emotional states. For example, patients with severe depression appear to have an increased susceptibility to infection,

(A)

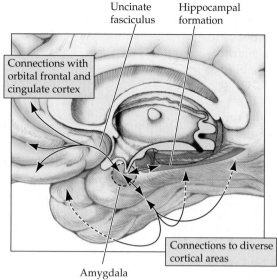

Uncinate fasciculus

Hippocampal formation

Connections with orbital frontal and cingulate cortex

Connections to diverse cortical areas

Amygdala

(B)

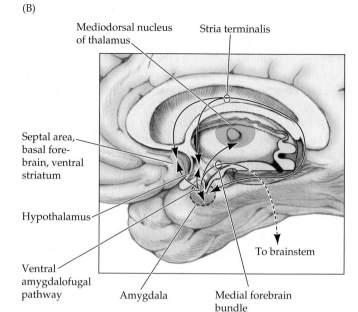

Mediodorsal nucleus of thalamus

Stria terminalis

Septal area, basal fore-brain, ventral striatum

Hypothalamus

To brainstem

Ventral amygdalofugal pathway

Amygdala

Medial forebrain bundle

(C)

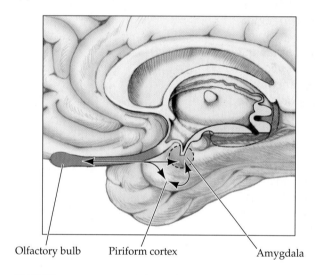

Olfactory bulb

Piriform cortex

Amygdala

FIGURE 18.17 Input and Output Connections of the Amygdala (A) Cortical connections. (B) Subcortical connections. (C) Olfactory connections.

MNEMONIC

possibly resulting from endocrinological effects on the immune system. Although the amygdala was previously thought to be crucial for **memory functions**, more recent studies have instead emphasized the importance of the hippocampal formation. However, the amygdala does appear to play an important role in attaching emotional significance to memories. The role of the amygdala in **olfaction**, particularly in emotional and motivational aspects of olfaction, was mentioned earlier in this chapter.

Let's review the main input and output connections of the amygdala with these functional considerations in mind (see Figure 18.17). Most connections of the amygdala are bidirectional. In analogy to the hippocampal formation, the amygdala both receives and transmits information from **diverse cortical areas**, including heteromodal association cortex and limbic cortex (see Figure 18.17A). These connections occur through two pathways. Fibers pass posteriorly and laterally from the amygdala to reach most cortical areas, often via relays in the anterior temporal and insular cortices. In addition, the **uncinate fasciculus** passes anteriorly to connect the amygdala with the medial orbitofrontal and cingulate cortices (see Figures 18.4B,C, and 18.17A; Table 18.4). The **hippocampal formation** and amygdala have reciprocal connections that may be important for emotional aspects of memory. For example, the amygdala has been demonstrated to be involved in fear memory and in modulating the effects of arousal on memory strength.

Subcortical connections of the amygdala (see Figure 18.17B) are important in the motivational, autonomic, and neuroendocrine aspects of limbic function. Connections that are mainly efferent from the amygdala, but have a smaller afferent component as well, pass through two major pathways: the stria terminalis and the ventral amygdalofugal pathway (see Table 18.4). In simple terms, the stria terminalis can be thought of as "the long way around," while the ventral amygdalofugal pathway can be thought of as "the short-cut" (see Figure 18.17B). The **stria terminalis** is a C-shaped structure that runs from the amygdala along the wall of the lateral ventricle to ultimately reach the hypothalamus and septal area. The stria terminalis can thus be thought of as the "fornix of the amygdala" (compare Figures 18.9A and 18.17B). The stria terminalis can be seen in coronal brain sections to lie just medial to the caudate nucleus (see Figure 18.4B,C) or in the groove between the caudate and thalamus (see Figure 16.4D) and, like the tail of the caudate, can also be seen running along the roof of the temporal horn of the lateral ventricle (see Figure 16.4D). One connection carried by the stria terminalis from the cortico-medial amygdala to the ventromedial hypothalamus may carry olfactory information that either increases or decreases appetite, depending on the odor. The **ventral amygdalofugal pathway** (see Figures 18.4C and 18.17B) passes anteriorly from the amygdala to reach several forebrain and brainstem structures. Projections to the forebrain involved in emotion, motivation, and cognitive functions include connections to the nucleus basalis, septal nuclei,* ventral striatum, and mediodorsal nucleus of the thalamus. Recall that the mediodorsal nucleus serves as a limbic relay, projecting to the frontal lobes (see Figure 7.8). The ventral amygdalofugal pathway also carries bidirectional signals involved in homeostatic (autonomic and neuroendocrine) function and behavioral arousal between the amygdala and hypothalamus (see Chapter 17), as well as between the amygdala and brainstem nuclei such as the nucleus solitarius, parabrachial nucleus, dorsal motor nucleus of the vagus, periaqueductal gray, and reticular formation. These connections between the amygdala, hypothalamus, and brainstem travel in the medial forebrain bundle (see Figure 18.17B; see also Figure 17.4).

As we discussed earlier, **olfactory inputs** are carried from the olfactory bulb through the lateral olfactory stria to the corticomedial nucleus of the amygdala (see Figure 18.17B). In addition, indirect olfactory projections are relayed in the piriform cortex to reach the basolateral amygdala.

Other Limbic Pathways

As should be clear by now, limbic circuits are fairly complex, and there are many other pathways in addition to those already mentioned. We will summarize only a few additional limbic pathways here (see Table 18.4), selected because they form prominent anatomical landmarks. The **stria medullaris** is a band of fibers that runs rostrocaudally along the walls of the third ventricle, on the medial surface of the thalamus (see Figures 16.2A, 16.4D). It carries fibers from the medial septal nuclei to the **habenula**, a small epithalamic structure located just lateral to the pineal gland. The habenula, in turn, projects via the **habenulointerpeduncular tract** (fasciculus retroflexus) to the **interpeduncular nuclei** in the midbrain. The interpeduncular nuclei project to the serotonergic **raphe nuclei** as well as to dopaminergic nuclei, which have diffuse projections throughout the brain (see Figures 14.10 and 14.12).

The anterior portion of the **anterior commissure** (see Figures 18.4B,C and 18.13) connects the anterior olfactory nucleus of the olfactory bulb to the same structure on the contralateral side. The posterior portion of the anterior commissure interconnects the bilateral amygdalas and anterior temporal lobes.

---

*Ventral amygdalofugal fibers travel to the septal nuclei in the **diagonal band of Broca**. The diagonal band of Broca also contains fibers connecting the septum with the hypothalamus and the cortex with the striatum, as well as cholinergic neurons with projections to the medial temporal lobes (see Figure 14.9).

REVIEW EXERCISE

Consult Table 18.4 and Figure 18.17 to answer the following questions:

1. What is the name of the pathway connecting the amygdala and other medial temporal structures with the orbitofrontal cortex?

2. What are the names of two pathways connecting the amygdala with the hypothalamus and the septal region?

3. What is the name of a major pathway connecting the amygdala and the hypothalamus with the brainstem?

KEY CLINICAL CONCEPT

18.2 SEIZURES AND EPILEPSY

Definitions and Incidence

A **seizure** is an episode of abnormally synchronized and high-frequency firing of neurons in the brain that results in abnormal behavior or experience of the individual. As we will discuss in this section, different types of seizures can occur depending on the location in the brain and the duration and type of abnormal electrical activity. **Epilepsy** is a disorder in which there is a tendency to have recurrent unprovoked seizures. Thus, a seizure is a *symptom* of abnormal brain function, which can occur in patients with or without epilepsy. For example, seizures can occur in patients with epilepsy but can also be provoked in normal individuals by conditions such as electrolyte abnormalities, alcohol withdrawal, electroshock therapy, toxins, etc.

Epilepsy in a given individual can be caused by genetic, structural, metabolic or other abnormalities, or the cause may be unknown. Epilepsy is a fairly common disorder, affecting close to 1% of the population. The lifetime risk of having a single seizure is even higher, with estimates ranging from 10% to 15% of the population.

A few more definitions will be helpful at this point to understand the material that follows. **Ictal** means during a seizure, **post-ictal** means immediately after a seizure, and **interictal** means between seizures.

Classification

No classification is perfect, and new classification schemes of seizures and epilepsy are currently being developed. The classification discussed in this section is still the most widely used and is based on that developed by the International League Against Epilepsy in the 1980s. Seizures are classified as either partial or generalized (Table 18.8). In **partial (focal, local) seizures**, abnormal paroxysmal electrical activity occurs in a localized region of the brain; in **generalized seizures**, the abnormal electrical discharge involves the entire brain. Note that a seizure can begin as a partial seizure and then spread to become **secondarily generalized**.

A similar classification scheme is used for epilepsy syndromes. Specific epilepsy syndromes are defined on the basis of the types of seizures that occur, together with other clinical features, such as age of onset, family history, and associated medical abnormalities other than seizures. In broad terms, however, epilepsy syndromes are divided into **localization-related (partial, focal, or local) epilepsy** and **generalized, or primary generalized, epilepsy**. Some seizures or epilepsy syndromes cannot easily be assigned to the local or generalized categories and remain unclassified. Ultimately, as our genetic and pathophysiological understand-

---

TABLE 18.8 International Classification of Epileptic Seizures

I. Partial (focal, local) seizures
  A. **Simple partial**[a] seizures
    1. With motor signs
    2. With somatosensory or special sensory symptoms
    3. With autonomic symptoms or signs
    4. With psychic symptoms
  B. **Complex partial** seizures
    1. Simple partial onset followed by impairment of consciousness
    2. With impairment of consciousness at onset
  C. Partial seizures evolving to secondarily generalized seizures
    1. Simple partial seizures evolving to generalized seizures
    2. Complex partial seizures evolving to generalized seizures
    3. Simple partial seizures evolving to complex partial seizures evolving to generalized seizures
II. Generalized seizures
  A. Absence seizures
    1. Typical **absence (petit mal)**
    2. Atypical absence
  B. Myoclonic seizures
  C. Clonic seizures
  D. Tonic seizures
  E. **Tonic-clonic (grand mal)** seizures
  F. Atonic seizures
III. Unclassified epileptic seizures

Source: Epilepsia 1981, 22: 489–501.

[a]The most common and clinically important seizure types are shown in **boldface**.

ing of epilepsy improves, epilepsy syndromes will be defined on the basis of specific gene defects or cellular abnormalities.

Partial seizures can be further subdivided into simple partial and complex partial seizures (see Table 18.8). In **simple partial seizures**, consciousness is spared. For example, if a patient has rhythmic twitching movements of the left hand caused by a simple partial seizure in the right motor cortex hand area, they will remain alert, talk normally during the seizure, and clearly recall the entire episode afterward. Partial seizures can have positive symptoms, such as hand twitching, or negative symptoms, such as impaired language abilities. The manifestations of partial seizures depend on the anatomical regions of the brain in which the seizure activity occurs (Table 18.9). For example, as we discussed in Chapter 11 (see KCC 11.1), seizures in the primary visual cortex can cause simple geometric shapes and flashes of light to appear in the contralateral visual field, while seizures in visual association cortex can produce more elaborately formed visual hallucinations such as people's faces or complex scenes. Patients with seizures in auditory cortex report simple sounds like a roaring engine or horn, most often coming from the direction opposite the involved cortex, or they may report having difficulty hearing, as if they were submerged underwater. Seizures in auditory association cortex can cause the patient to hear voices or music. Musical hallucinations are more common in nondominant hemisphere seizures (see KCC 19.13). Contralateral somatosensory phenomena occur during seizures in the somatosensory cortex.

The term **aura** means "breeze" and was originally used in ancient Greece by Galen's teacher Pelops to describe the sensation felt by one of his patients in his leg prior to having a large seizure. Auras are brief, simple partial seizures of any type that are experienced by a patient with no outward behavioral manifestations. They can occur in isolation, or they may serve as a warning for a larger seizure, which begins as an aura and then spreads. Patients with seizures arising from medial temporal limbic structures (see Table 18.9) often report auras of a rising visceral sensation in the epigastric area, a feeling of deja vu, strange unpleasant odors, or feelings of extreme fear and anxiety. **Odors** and **panic** are thought to originate from the amygdala and nearby cortex rather than from the hippocampus. Some reports suggest that olfactory phenomena during seizures can also arise from the orbitofrontal olfactory cortex (see Figure 18.6).

As mentioned earlier, patients with seizures in primary motor cortex usually have simple rhythmic jerking **clonic** movements or sustained **tonic** contractions in the contralateral extremities. However, seizures in frontal motor association cortex, such as the supplementary motor area, can produce more elaborate movements, including a characteristic "fencing posture" with extension of the contralateral arm; bilateral leg cycling movements; turning of the eyes, head, or entire body; and the production of unusual sounds. Typical duration for simple partial seizures is 10 to 30 seconds, although longer or shorter seizures are not uncommon. Recall that the time during a seizure is called the ictal period, and the time immediately after a seizure is called the post-ictal period. With brief, simple partial seizures there are often no new deficits post-ictally. If the seizures are prolonged or recurrent, there may be post-ictal depressed function of the local region of cortex involved, producing focal weakness (**Todd's paresis**) or other deficits.

Unlike simple partial seizures, in **complex partial seizures** (see Table 18.8) there is **impairment of consciousness**. The impaired consciousness with complex partial seizures is presumably due to seizure activity affecting wider regions of cortex, or deep brainstem and diencephalic regions. Impairment of consciousness in complex partial seizures may be complete or mild, some-

TABLE 18.9 Clinical Manifestations of Partial Seizures in Different Brain Regions

TEMPORAL LOBE

Medial: An indescribable sensation, rising epigastrium ("butterflies" in the stomach), nausea, déjà vu, fear, panic, unpleasant odor, autonomic phenomena (tachycardia, pupillary dilation, piloerection, borborygmi, belching, pallor, flushing), bland staring with unresponsiveness, oroalimentary automatisms (lipsmacking, chewing, swallowing), bilateral or unilateral gestural automatisms, contralateral dystonia with ipsilateral automatisms.

Lateral: Vertigo (temporoparietal operculum), inability to hear, simple auditory hallucinations (buzzing, roaring engine, tones), elaborate auditory hallucinations (voices, music). Aphasia, including inability to understand what people are saying, is more common with dominant temporal seizures. Saying words or phrases repeatedly and having musical hallucinations are more common with nondominant temporal seizures.

Comments: Usual duration 1 to 2 minutes, often with post-ictal amnesia, tiredness, headache, emotional changes, or other focal deficits. Most common cause of complex partial seizures. Head or eye deviation probably results from spread to frontal or parietal lobes (see below). Medial temporal lobe seizures associated with hippocampal sclerosis usually do not generalize once treated with anticonvulsant medications; however, the complex partial seizures in this condition are often medically refractory and frequently can be cured by surgery.

FRONTAL LOBE

Dorsolateral convexity: Contralateral tonic or clonic activity (primary motor cortex); strong version (turning) of eyes, head, and body away from side of the seizure (prefrontal cortex and frontal eye fields). Aphasia (dominant hemisphere).

Supplementary motor area: Fencing posture with extension of contralateral upper extremity, other tonic postures, speech arrest, unusual sounds.

Orbitofrontal and cingulate: Elaborate motor automatisms, making unusual sounds, autonomic changes, olfactory hallucinations (orbitofrontal), incontinence (cingulate).

Comments: Seizures are often brief, occur multiple times per day, and may have no post-ictal deficits. Nocturnal exacerbation is common. Elaborate motor automatisms without loss of consciousness or postictal deficits often lead to misdiagnosis as psychogenic episodes.

PARIETAL LOBE

Vertigo, contralateral numbness, tingling, burning, sensation of movement or need to move, aphasia (dominant hemisphere), contralateral hemineglect (nondominant hemisphere). Eyes and head may deviate toward or away from the side of the seizure.

OCCIPITAL LOBE

Sparkles, flashes, pulsating colored lights, scotoma, or hemianopia in contralateral visual field (primary visual cortex), formed visual hallucinations (inferior temporo-occipital association cortex), nystagmoid or oculogyric jerks, palpebral jerks, eye blinking, sensations of eye oscillation.

Comments: May be precipitated by changes in lighting conditions. Often associated with migrainelike symptoms. Variable spread to other lobes is common and can lead to mislocalization.

Note: Features of both simple partial and complex partial seizures are included here. Seizures may begin in one region and spread to another, producing signs and symptoms corresponding to multiple neuroanatomical areas.

times making the distinction from simple partial seizures difficult. The most common localization for complex partial seizures is the temporal lobes. Patients with complex partial seizures arising from the temporal lobes are said to have temporal lobe epilepsy (formerly known as psychomotor epilepsy). **Medial temporal lobe–onset complex partial seizures** (see Table 18.9) often begin with an aura, as described earlier, of an unusual, often indescribable sensation, or with epigastric, emotional, or olfactory phenomena, or with deja vu. Sometimes no aura can be recalled. The initial symptoms are followed by unresponsiveness and loss of awareness, during which the patient may have **automatisms**, which are usually repetitive behaviors such as lip smacking, swallowing, or stereotyped hand or leg movements such as stroking, wringing, or patting.

Interestingly, in temporal lobe seizures the ipsilateral basal ganglia are often involved, causing contralateral dystonia or immobility (see KCC 16.1), while the ipsilateral extremities remain free to exhibit automatisms. These behaviors can be falsely localizing to the unpracticed eye, since the side ipsilateral to the temporal lobe seizure shows movements (automatisms), while the contralateral side is relatively still (dystonia). Sometimes, instead, there is simply bland staring, immobility, and unresponsiveness. Autonomic phenomena such as tachycardia, pupillary dilation, and piloerection can occur. Typical duration is 30 seconds to 1 to 2 minutes. Post-ictal deficits may last from minutes to hours and can include unresponsiveness, confusion, amnesia, tiredness, agitation, aggression, and depression. Headache is common. Patients with left temporal–onset seizures may have post-ictal language deficits, although care must be taken to distinguish such deficits from overall decreased responsiveness, which is nonlocalizing. There may also be some mild post-ictal weakness or hyperreflexia contralateral to the side of onset. This is usually subtle, unless the seizure has spread to the motor cortex, in which case there would be unilateral tonic or clonic activity during the seizure. Table 18.9 describes the clinical features of temporal lobe seizures and other partial seizures including frontal, parietal, and occipital lobe seizures, in more detail.

The most common type of **generalized seizure** is a **generalized tonic-clonic seizure**, or **grand mal seizure** (see Table 18.8). A generalized tonic-clonic seizure can be generalized from the onset, or it may begin focally and then secondarily generalize. It typically begins with a **tonic phase** characterized by loss of consciousness and generalized contraction of all muscles lasting for 10 to 15 seconds. This often results in stiff extension of the extremities, during which the patient may fall "like a tree," and injure themselves, and a characteristic expiratory gasp or moan may occur as air is forced past the closed glottis. Next is a **clonic phase**, characterized by rhythmic bilateral jerking contractions of the extremities, usually in flexion, at a frequency of about 1 hertz, which gradually slow down and then stop. Incontinence or tongue biting is common. There is usually a massive ictal autonomic outpouring with tachycardia, hypertension, hypersalivation, and pupillary dilation. Typical duration is 30 seconds to 2 minutes. Immediately post-ictally, patients lie immobile, flaccid, and unresponsive, with eyes closed, breathing deeply to compensate for the mixed metabolic and respiratory acidosis produced by the seizure. Within a few minutes they usually begin to move and respond. Post-ictal deficits last from minutes to hours and include profound tiredness, confusion, amnesia, headache, and other deficits related to the location of seizure onset.

In addition to generalized tonic-clonic, or grand mal, seizures, there are a variety of other generalized seizure types (see Table 18.8). Most of these are relatively uncommon and will not be discussed further here, with the exception of absence seizures. Typical **absence seizures (petit mal seizures)** are brief episodes of staring and unresponsiveness lasting for about 10 seconds or less. There are no post-ictal deficits, except for a lack of awareness of what occurred during the brief time of the seizure. These seizures are accompanied by a characteristic **generalized 3–4 hertz spike and wave** discharge on electroencephalogram (EEG) recordings (see Chapter 4).

TABLE 18.10 Staring Spells: Complex Partial Seizures versus Absence Seizures

TYPE OF SEIZURE	AURA	DURATION	AUTOMATISMS	POST-ICTAL DEFICITS	FREQUENCY	INTERICTAL BRAIN FUNCTION	ICTAL EEG
Typical medial temporal complex partial seizures	May be present	30–120 s	May be present	May be present	3–4/month	Focal abnormalities	Unilateral or asymmetrical 5–8 Hz rhythmic activity over the temporal lobe
Typical absence seizures	None	<10 s	None[a]	None	Multiple daily	Normal	Generalized 3–4 Hz spike–wave

[a]Minor clonic movements of the eyelids or corners of the mouth at 3–4 Hz sometimes occur.

Although absence seizures are generalized, they differ from grand mal seizures, which begin with a prolonged high-frequency electrical discharge that causes a more severe disruption of brain function. Absence seizures are most common in childhood and can occur multiple times per day, causing impaired school performance. They often can be provoked by hyperventilation, strobe lights, or sleep deprivation. About 70 to 80% of children with typical absence seizures remit spontaneously. It should be noted that both absence seizures and complex partial seizures can cause episodes of staring and unresponsiveness. Typical cases of medial temporal lobe complex partial seizures can usually easily be distinguished from typical absence seizures on clinical grounds (Table 18.10). However, in atypical absence seizures or brief complex partial seizures, the distinction may be more difficult.

When seizures of any type occur either continuously or repeatedly in rapid succession, the condition is referred to as **status epilepticus**. Generalized tonic-clonic status epilepticus is a medical emergency that requires immediate and aggressive treatment. When first-line agents such as benzodiazepines and phenytoin are ineffective, intubation and general anesthesia may be necessary. EEG recording should always be performed urgently in this situation to ensure that electrographic seizure activity has stopped. Blood tests (see the next subsection), head CT, and lumbar puncture (when appropriate) should be performed without delay so that other specific treatments can be initiated. Prognosis depends mainly on the promptness of treatment and on the underlying etiology. Other forms of status epilepticus aside from tonic-clonic seizures should also be treated promptly, although a balance must be struck between the risks of aggressive treatment and the risks of ongoing seizures. The behavioral features of status epilepticus can sometimes be subtle, so the prompt performance of an EEG is often needed for the diagnosis to be made, and for treatment to be initiated.

Diagnosis and Etiology

The diagnostic evaluation of patients with epilepsy is essential because the correct diagnosis has a major impact on treatment. The first step is to ascertain whether the episodes are epileptic seizures or another type of transient event (see Tables 10.2 and 16.3). Epileptic seizures are usually brief events that are **stereotyped** from one episode to the next in a given patient and often fit one of the typical seizure patterns already described. If the events are epileptic seizures, the next step is to determine the type of seizure (see Tables 18.8 and 18.10), as well as the localization if the seizures are focal in onset (see Table 18.9). Finally, a cause for the seizures should be sought (Table 18.11).

The fundamental tools used for diagnosis in epilepsy include a detailed clinical history; physical exam; basic blood tests; MRI scan with special thin coronal cuts and pulse sequences used to view the medial temporal, cortical, and subcortical structures in detail; and an interictal EEG (see Chapter 4). When the diagnosis remains uncertain, additional tests can be helpful in diagnosis and localization. These include admission for continuous video and EEG monitoring in an attempt to obtain an ictal recording; ictal and interictal nuclear medicine tests, such as SPECT (single photon emission computed tomography) and PET (positron emission tomography) scans; and neuropsychology testing. In patients who are being considered for epilepsy surgery, additional tests, including a Wada test (see the next subsection), are often performed.

Let's briefly discuss some of the individual causes of seizures listed in Table 18.11. The most common causes vary with age, resulting in a bimodal age distribution for the risk of developing new-onset seizures. The risk of new-onset seizures is high in infancy and childhood, declines in adulthood, and then rises again in the elderly population. The most common causes of seizures in infancy and childhood are **febrile seizures**, **congenital disorders**, and **perinatal injury**. In contrast, the most common cause in patients over age 60 years is **cerebrovascular disease**, followed by **brain tumors** and **neurodegenerative conditions**.

The risk of seizures after **head trauma** increases with the severity of the injury. Minor head injuries with no clear structural damage and only brief confusion or loss of consciousness (for less than 30 minutes) do not pose a significant risk for subsequent seizures. **Hypoglycemia**, **electrolyte abnormalities** such as hyponatremia, hypernatremia, hypocalcemia, or hypomagnesemia; **metabolic abnormalities**, or exposure to a variety of endogenous or exogenous **toxic substances** can provoke seizures. It is therefore essential to check blood chemistries, including glucose, sodium, calcium, magnesium, liver function tests, creatinine, and a toxicology screen when assessing a patient with new-onset seizures, especially in the acute setting, so that these abnormalities can be corrected.

Febrile seizures are fairly common, occurring in 3 to 4% of all children, usually between the ages of 6 months and 5 years. These are usually brief, generalized tonic-clonic seizures called **simple febrile seizures**, which are not associated with increased risk of epilepsy. However, there is an increased risk of subsequent epilepsy in children with **complex febrile seizures**, defined as seizures lasting more than 15 minutes, or occurring more than once in 24 hours, or having focal features. Some of these children may have an underlying cause for epilepsy that is first unmasked in the setting of fever. It has also been hypothesized that prolonged febrile seizures cause subsequent temporal lobe epilepsy in some patients, through a pathologic process called **mesial temporal sclerosis** or **hippocampal sclerosis**, in which there is marked neuronal loss and gliosis, particularly in the CA1 sector of the hippocampus, as well as in other medial temporal structures. In addition to febrile seizures, mesial temporal sclerosis may also be triggered by other initial precipitating injuries in infancy and childhood, such as head trauma or CNS infections. Once established, there is often a latent period of up to several years between the precipitant and the onset of complex partial seizures. The seizures usually consist of a hard-to-describe aura of fear or epigastric sensation and other features typical of medial temporal lobe seizures (see Table 18.9). Once treated with anticonvulsant medications, seizures in patients with mesial temporal sclerosis rarely generalize. However, the complex partial seizures in these patients can be quite incapacitating and are often associated with memory decline. In addition, unlike the tonic-clonic seizures, the complex

TABLE 18.11 Causes of Seizures[a]
Head trauma
Cerebral infarct
Intracerebral hemorrhage
Vascular malformation
Cerebral venous thrombosis
Anoxia
Mesial temporal (hippocampal) sclerosis
Electrolyte abnormalities
Hypoglycemia
High fever
Toxin exposure
Alcohol, benzodiazepine, or barbiturate withdrawal
Meningitis
Encephalitis
Brain abscess
Vasculitis
Neoplasm
Inborn error of metabolism
Neuronal migrational abnormality
Hereditary epilepsy syndrome
Neurodegenerative disorder
Nonepileptic seizures[b]

[a]Following format of Figure 1.1.

[b]Formerly called "pseudoseizures," these include psychogenic episodes, syncopse, arrhythmias, and other causes of nonepileptic transient episodes listed in Table 10.2.

partial seizures in these patients are often medically refractory. Based on the diagnostic methods discussed earlier, unilateral surgical resection of the medial temporal lobe structures has a cure rate of over 90% in patients with seizures localized to one temporal lobe.

Numerous other causes of **focal brain lesions**, some of which are listed in Table 18.11, can result in seizures, and many can be detected by a good-quality MRI scan. **Family history** is also essential in assessing patients with seizures. Many epilepsy syndromes, especially primary generalized epilepsies, but also some focal disorders, have a genetic component. **Rolandic epilepsy** is a common cause of focal, mostly nocturnal seizures in children that probably has autosomal dominant inheritance with incomplete penetrance. The EEG shows characteristic **centrotemporal spikes**. Onset is usually between ages 3 and 13 years, and seizures are often mild, not always requiring medications. Remission is nearly always complete by age 15 years.

There are several familial **primary generalized epilepsy** syndromes, including **childhood absence epilepsy** (pyknolepsy), characterized by typical absence seizures (discussed earlier), **juvenile myoclonic epilepsy**, and other disorders. For a more detailed discussion of the many other causes of epilepsy, see the References at the end of this chapter.

Treatment

The basic goals of epilepsy treatment are to reduce the risk of seizures while minimizing side effects, in order to achieve the best possible overall quality of life. Major considerations include impact on driving (regulations vary from state to state), ability to work, the unfortunate public stigma associated with seizures, and effects on pregnancy and lactation. **Medications** can be used to achieve satisfactory control of seizures in over 70% of cases. The first-line agents for treatment of localization-related epilepsy and primary generalized epilepsy differ somewhat. From the 1970s to early 1990s no major new anticonvulsants were introduced, and most cases of localization-related epilepsy were initially treated with carbamazepine (Tegretol) or phenytoin (Dilantin), childhood absence epilepsy without convulsions was treated with ethosuximide (Zarontin), and childhood absence epilepsy with convulsions was treated with valproate (Depakote). However, since 1993 the pharmaceutical industry has released an average of one or two new anticonvulsants per year, greatly increasing the available treatment options. Some of these new anticonvulsants have advantages over the older medications, with greater efficacy and fewer side effects. Choice of anticonvulsant medications should be individualized for each patient based on a number of considerations including type of epilepsy, undesirable side effects (for example, in patients with headaches, obesity, or depression it is preferred to avoid medications that may worsen these conditions), potential teratogenicity, hepatic metabolism, drug–drug interactions, and other factors. (see the References at the end of this chapter for additional details on epilepsy medications.)

About 20 to 30% of patients with epilepsy have seizures that cannot be adequately controlled with medications and are considered **medically refractory**. Some children with refractory epilepsy may improve on a high-fat, low-carbohydrate **ketogenic diet**, but the effect is often temporary, and the diet is hard to maintain in the long term.

Many patients with medically refractory epilepsy are candidates for **epilepsy surgery**. In the ideal candidate for epilepsy surgery, seizures always start in the same location, and this location can be safely resected without creating deficits. The best surgical outcome statistics are seen for patients with unilateral medial temporal lobe epilepsy, in which surgery controls seizures in 90% or more of cases, depending on the study. Prior to epilepsy surgery, patients undergo a detailed evaluation and education process, and recommendations to the patient for or against surgery are usually based on a multidisciplinary team approach.

In attempting to localize the region of onset, no single test is definitive. Therefore, combined information is used from each of the diagnostic tests listed in the previous section. In addition, an **angiogram Wada test** is often helpful in planning surgery. In this test the sedative agent sodium amytal is injected through an angiographic catheter (see Figure 4.9) directly into each common carotid artery, causing transient inhibition of the injected hemisphere for approximately 10 minutes. After each injection, **language** is tested, and the presence or absence of aphasia is used to localize which hemisphere is dominant for language so that language areas can be avoided during surgery.

In addition, **memory** is tested after each injection. In patients with normal bilateral medial temporal memory function, injection of one hemisphere does not eliminate memory, since the other hemisphere can compensate. However, when a medial temporal lobe is not functioning properly, as is often the case in patients with medial temporal epilepsy, injection of the contralateral hemisphere causes severe memory difficulties. Preserved memory with injection of the hemisphere ipsilateral to seizure onset is reassuring because it suggests that the contralateral hemisphere will be able to support memory function after the ipsilateral medial temporal structures are resected. Note that in the Wada test, amytal is injected into the common carotid artery, which in most individuals perfuses the anterior (ACA) and middle (MCA) cerebral artery territories, but not the posterior cerebral artery (PCA) territory (see Figures 4.16 and 4.17). Since the medial temporal structures are perfused by the PCA, it is not immediately obvious why the Wada test should inhibit medial temporal function. The most likely explanation is that the large ACA/MCA-territory injection inhibits most of the hemispheric cortex, white matter, and corpus callosum, thereby indirectly inhibiting the medial temporal lobe by cutting off its major sources of input.

When these tests are all concordant, suggesting seizure onset in a single area, surgical resection can be contemplated. However, when there are discrepancies in localization between different tests, **intracranial EEG monitoring** can be employed to record seizure activity directly from the brain. Multicontact subdural electrodes are used to record from the cortical surface, and depth electrodes are used to record from deep structures such as the hippocampus.

In appropriately selected candidates, several different surgical approaches can be used to treat medically refractory epilepsy. As we have already mentioned, the best candidates for **focal surgical resection** have epilepsy localized to a single temporal lobe; these patients have a surgical success rate of over 90%. Patients with confirmed localization to a single location other than the temporal lobe can also often be treated successfully with surgical resection. For these patients with extratemporal epilepsy, the best outcomes are usually seen when all diagnostic studies point to a single location; however, certain other features, including a focal lesion visible on MRI scan, may also predict a favorable prognosis. It is also crucial in these cases to carefully map critical areas of cortical function, including motor and language cortex, using techniques such as direct brain stimulation and fMRI and to avoid these areas during surgery.

Surgery is considered successful if patients no longer have seizures and have no adverse effects from the operation. Although many patients must continue anticonvulsant medications following surgery to maintain freedom from seizures, this is still a major lifestyle improvement for patients who, prior to surgery, had frequent seizures even with medications. Following successful surgery, for example, many previously disabled patients are able to drive and to pursue productive employment.

In some patients, if the region of seizure onset lies in a functionally critical area such as the motor or language cortex, surgical resection cannot safely be performed. In these patients, **multiple subpial transection** may be helpful. In this procedure, a special sharpened probe is inserted under the pia and is used to sever the cortical–cortical connections, thereby making multiple parallel tracts that functionally disconnect the epileptogenic cortex. Patients with severe epilepsy arising from multiple locations in the brain may benefit from **callosotomy**. In this procedure, the corpus callosum is cut, preventing seizures from propagating from one hemisphere to the other. Callosotomy does not cure seizures. The procedure is reserved mainly for patients who have frequent falls and injuries when their seizures generalize. These patients may be helped if generalization is prevented, by allowing them to avoid such falls during seizures. Deficits associated with callosotomy are discussed in KCC 19.8.

In some patients, the seizure onset is not localized to a specific region but rather to multiple regions within a single hemisphere. In patients younger than 2 to 3 years of age, hemispheric specialization is still under development. Therefore, in some of these patients, **hemispherectomy** (surgical removal of an entire hemisphere) can be considered. Remarkably, many patients do quite well following hemispherectomy and are able to lead functional lives. Seizures are often cured, allowing language and motor development (commonly arrested by severe epilepsy) to proceed, with language and motor representations for both sides of the body forming in the single remaining hemisphere.

Neurostimulation is a growing area of treatment for patients with medically refractory epilepsy, particularly those who are not candidates for resective surgery or disconnection procedures described above. Chronically implanted **vagus nerve stimulators** reduce seizure frequency in some patients; however, only rarely do the seizures stop completely with this device. The device chronically stimulates the vagus nerve with current pulses at a set rate, and patients or family members can also trigger the stimulus externally, using a magnet, in an effort to abort individual seizures. As we have discussed in this chapter and in Chapter 14, vagal afferents reach the nucleus solitarius (see Figures 12.12 and 14.5A,B) and are relayed to the limbic system and other forebrain structures via the para-brachial nucleus of the pons (see Figure 14.4A). The mechanisms by which vagal stimulation raises seizure threshold are still under active investigation. **Deep brain stimulation** (see KCC 16.4) has shown some promising results in refractory epilepsy, with investigational trials targeting thalamic nuclei. **Responsive neurostimulation** is another area of active research, with devices designed to automatically detect seizure onset and then deliver an electrical stimulus to interrupt focal seizure initiation. Other promising treatment avenues are continually being developed for those few patients who do not respond to conventional therapy.

In summary, the majority of patients with epilepsy can be successfully treated with either medications or other therapies discussed here and are able to resume their normal lives. ■

KEY CLINICAL CONCEPT

18.3 ANATOMICAL AND NEUROPHARMACOLOGICAL BASIS OF PSYCHIATRIC DISORDERS

The neurobiological basis of many psychiatric disorders is beginning to be better understood through a convergence of information derived from pathologic and anatomical imaging studies, functional imaging studies, neuropharmacological analysis, and other methods of investigation, such as molecular genetics. In this section we will briefly discuss on a basic level some of the major pathophysiological findings for a few common psychiatric disorders.

Schizophrenia

Patients with schizophrenia exhibit a variety of abnormalities of thought including delusions, hallucinations, disorganized, tangential speech; flat affect, and occasionally, a profound decrease in spontaneous activity called catatonia. Cognition, particularly working memory, is also often affected. Studies of the pathophysiology of schizophrenia have suggested abnormalities in the limbic system, frontal lobes, and basal ganglia. Both pathologic studies and high-resolution MRI have demonstrated bilateral subtle decreases in the volume of the amygdala, hippocampal formation, and parahippocampal gyrus in patients with schizophrenia. More variable anatomical changes have also been reported in the basal ganglia and other regions. Functional imaging studies such as fMRI and PET have shown decreased activation of the dorsolateral prefrontal cortex in patients with schizophrenia during tasks such as the Wisconsin Card Sorting Test. Much evidence suggests that an abnormality in dopamine is important in schizophrenia; for example, psychotic symptoms improve with antidopaminergic agents.

Recall from Chapter 16 that dopaminergic neurons in the ventral tegmental area project to the nucleus accumbens and ventral striatum, as well as to the prefrontal cortex and limbic cortex (see also Figure 14.10). Some other neurotransmitters may be important as well in the pathogenesis of schizophrenia, including glutamate, gamma-aminobutyric acid (GABA), serotonin, and norepinephrine. A complete pathophysiological understanding of schizophrenia is complicated by the fact that this disorder can include both positive symptoms such as psychotic delusions and hallucinations and negative symptoms such as flat affect and impaired executive function. The most likely cause of schizophrenia is a combination of abnormalities in several anatomical areas and neurotransmitter systems.

Obsessive-Compulsive Disorder

In obsessive-compulsive disorder, recurrent, intrusive obsessive thoughts cause the patient much anxiety, while the performance of repetitive, compulsive behaviors such as hand washing or checking the door lock provide temporary relief. The improvement of obsessive-compulsive symptoms with serotonin-enhancing medications suggests a role for serotonin in this disorder; however, other neurotransmitters may be important as well. Functional imaging studies have shown abnormally increased activity in the basal ganglia, especially the head of the caudate, as well as in the anterior cingulate gyrus and orbitofrontal cortex. These changes improve with pharmacological or behavioral treatment.

MRI studies have shown a subtle bilateral decreased volume in the head of the caudate, but these findings are not conclusive. Thus, obsessive-compulsive disorder appears to result from dysfunction in a network consisting of the caudate, cingulate gyrus, and orbitofrontal cortex. This may be analogous to hyperkinetic movement disorders (see Chapter 16), but with unwanted thoughts or compulsions instead of movements. Indeed, there may be some overlap, since obsessive-compulsive disorder is present in about half of patients with Tourette's syndrome and can also occur in Huntington's disease, Sydenham's chorea, and other basal ganglia disorders.

Anxiety

Anxiety disorders encompass a variety of conditions, including panic disorder, phobias, posttraumatic stress disorder, and generalized anxiety disorder. Obsessive-compulsive disorder is also currently classified as an anxiety disorder. Anxiety is thought to be associated with an increase in noradrenergic and serotonergic transmitter systems in the central nervous system. In addition, anxiety symptoms can be controlled with benzodiazepines, which act on $GABA_A$ receptors. Symptoms of anxiety, panic, and fear are also accompanied by increased peripheral sympathetic tone and increased release of epinephrine by the adrenal glands. Functional imaging studies during episodes of panic have shown inconsistent results, but they may demonstrate increased activation in the anterior cingulate and temporal cortices. Evidence from patients with panic as an epileptic phenomenon suggests involvement of the amygdalar region of the medial temporal lobes.

Depression and Mania

Patients with depression have a sad mood and lack of enjoyment together with other findings such as, impaired concentration, increased or decreased sleep, appetite, or level of activity; feelings of worthlessness or guilt; or suicidal thoughts or actions. In contrast, patients with mania have an abnormally elevated, irritable mood, with other features including grandiosity, decreased sleep, pressured speech, racing thoughts, distractibility, increased activity, and impulsive behavior. A variety of evidence suggests that depression is marked by deficits in both the noradrenergic and serotonergic neurotransmitter systems. Other transmitters may be important as well, including dopamine, amino acid neurotransmitters, and neuropeptides.

Structural and functional neuroimaging studies in depression are still an area of active investigation, but there is some evidence of a global decrease in activity of the cerebral cortex, with a more prominent decrease in the frontal lobes. In addition, several studies have revealed decreased hippocampal volume in depression, particularly on the left side. Neuroendocrine changes occur in depression as well, including an increased release of cortisol in about 40% of patients with severe depression, resulting from excessive release of corticotropin-releasing hormone by the hypothalamus. Recent work suggests that stress and other factors may decrease the expression of neurotrophic factors such as brain-derived neurotrophic factor (BDNF) or vascular endothelial growth factor (VEGF) in critical brain regions, while most effective treatments increase the expression of these factors. Slow excitotoxic-like effects and inflammatory cytokines may also play a role. Histopathological studies of the prefrontal cortex have shown decreased number and density of glial cells and decreases in some populations of interneurons. The effects of brain lesions on mood have also shed some light on possible mechanisms for mood disorders. Some studies suggest that left frontal lesions are more likely to produce a depressed mood, while right frontal lesions tend to produce an abnormally elevated mood, although these associations are not absolutely consistent. Similarly, bilateral lesions of the dorsolateral prefrontal cortex tend to produce a flat affect resembling depression (see KCC 19.11), while bilateral lesions of the medial orbitofrontal cortex may produce an abnormally elevated affect. ■

CLINICAL CASES

CASE 18.1 SUDDEN MEMORY LOSS AFTER A MILD HEAD INJURY

MINICASE

A 33-year-old neurology resident with a history of migraine fell backward while attempting a ski jump, struck his occiput on the snow, and suddenly developed amnesia. His wife witnessed the incident and reported that he had not lost consciousness from the fall. He stood himself up, skied part way down the slope, and then stopped skiing to tell his wife that he must have amnesia because he did not know the date, where he was, or how he had gotten there. He also reported seeing a scintillating scotoma in the left upper part of his visual field, like that accompanying his typical migraines.

At first his wife thought he was joking around, but when he began **asking the same questions repeatedly**, it soon became clear that he had a serious problem. He was **unable to retain any new information for more than 1 to 2 minutes**. In addition, **he did not recall any events that had occurred during the previous approximately 1 year**, including the fact that his wife was pregnant. Aside from the memory loss, he had no other significant deficits. He was taken to a local hospital, where a head CT was normal, and he was discharged in the care of his wife and friends. Being a neurologist, he constantly urged those around him to test his memory during the long car ride home.

Meanwhile, he was able to enjoy the exciting "news" that his wife was pregnant, over and over again. Approximately 5 hours after onset, he was finally able to recall three words after a 3 to 4 minute delay, but this ability was still inconsistent. At about the same time, memories from the previous year began to return gradually, with the **most remote memories generally returning first**. By the next morning he was able to consistently recall new facts about as well as at baseline. In addition, he remembered everything from the previous year. He recalled nothing, except for the period extending from 2 to 3 hours before the injury to about 5 hours after the injury. An MRI scan performed a few days later was normal.

LOCALIZATION AND DIFFERENTIAL DIAGNOSIS

1. On the basis of the symptoms and signs shown in **bold** above, what type of memory was impaired in this patient? How would you describe the chronological features of his amnesia? Dysfunction in which locations in the brain could produce these findings (see KCC 18.1)?

2. What is the most likely diagnosis, and what are some other possibilities?

Discussion

The key symptoms and signs in this case are:

- **Inability to retain any new information for more than 1 to 2 minutes and asking the same questions repeatedly**
- **Inability to recall any events that had occurred during the previous approximately 1 year**
- **During recovery, return of the most remote memories generally first**

1. The patient was unable to remember facts and events; therefore, he had a deficit in declarative (explicit) memory (see Figure 18.15). He had anterograde amnesia, since he was unable to learn new information, and retrograde amnesia for a period of about 1 year prior to the injury (see Figure 18.16). This pattern of selective declarative memory loss with anterograde and retrograde amnesia is characteristic of bilateral medial temporal or bilateral medial diencephalic dysfunction (see KCC 18.1). Asking the same questions repeatedly is also typical of acute dysfunction in these areas, and recovery of remote memories prior to more recent ones is often seen when the memory loss is reversible.

   The most likely *clinical localization* is bilateral medial temporal or bilateral medial diencephalic structures.

2. Given the onset after a head injury, concussion must be considered as the cause of amnesia in this patient (see KCC 18.1; see also KCC 5.5). No loss of consciousness was observed; however, it may have been too brief to be no-

ticed. Interestingly, it is unclear whether concussion causes amnesia due to direct impact to the medial temporal structures, or whether the mechanism is diffuse dysfunction of the white matter pathways necessary for normal function of the medial temporal and medial diencephalic memory systems.

Alternatively, the clinical features of the memory loss and recovery in this patient could represent transient global amnesia (see KCC 18.1) that occurred coincidentally following a minor head injury. The onset in the setting of stress and history of migraine are also suggestive of this diagnosis, as are the typical migraine symptoms of a scintillating scotoma (see KCC 5.1) reported by the patient during the episode. Other causes of transient amnesia, which are less likely, given the history in this patient, include transient ischemic attack, seizure, Wernicke's encephalopathy, psychogenic amnesia, or administration of a benzodiazepine or other medication (see Table 18.7).

Clinical Course

The patient resumed work a few days later and had no further episodes of amnesia. On follow-up over 5 years, he had no deficits except for continued lack of memories for the period of a few hours before and after onset of his amnesia.

CASE 18.2 PROGRESSIVE SEVERE MEMORY LOSS, WITH MILD CONFABULATION

MINICASE

A 75-year-old semiretired businessman was brought to the emergency room by his friends because of several weeks of **severe progressive memory problems**. At baseline, the patient had normal cognition, exercised avidly, and maintained an active schedule, driving himself to appointments with friends and business associates. Ten days prior to admission he met a friend for lunch and had a normal, clear, precise conversation, except that he did not remember the name of the hostess, whom he had known for several years. Four days later, the same friend spoke with the patient on the phone and discovered that the patient had no recollection of having lunch together or of any of their conversation. He seemed normal otherwise. The next day the patient missed an important business meeting. When the patient's son contacted him over subsequent days, his conversation seemed appropriate except that he was totally unaware of current events, including a recent highly publicized plane crash. There was no known history of alcoholism.

Exam was normal except for **profound problems with recent memories and milder problems with remote memories**. He said the year was 1964 (it was 1994) and realized he was in a hospital but did not know which one. Attention and immediate recall were normal, with a digit span forward of 7 and backward of 5. He was able to repeat three words immediately when asked to memorize them. After 3 minutes, however, he did not even recall the task, and he got 0/3 words correct, even with multiple choice. When the examiner left the room and came back a few minutes later, the patient had no recollection of having met him before.

The patient had no knowledge of recent current events and was completely unaware of the highly publicized O. J. Simpson trial going on at the time. More remote memory was better but still not perfect. For example, he was able to describe his hometown, childhood, family members, marriage, and the fact that he had fought in World War II. However, he could not recall any battles he fought in, and he was surprised to hear that John F. Kennedy had been shot. He did not know that Johnson and Nixon were presidents during the Vietnam War or that his wife had died a few years back. With some prompting, however, he was able to generate the name of the current president: "Clinton." There was also a mild **tendency to confabulate**. For example, when asked why he was in the hospital, he said he "came here to pick up some things and leave," and when asked if anyone had visited him, he mentioned several names despite having had no visitors. The remainder of the mental status exam was normal, including normal attention span (see above), pleasant affect and behavior, normal language, good calculations, normal reading and writing, normal drawing of a clock face and cube, and good interpretation of similarities and proverbs. The rest of the neurologic exam was likewise normal.

LOCALIZATION AND DIFFERENTIAL DIAGNOSIS

1. On the basis of the symptoms and signs shown in **bold** above, where is the lesion?

2. What is the most likely diagnosis, and what are some other possibilities?

Discussion

The key symptoms and signs in this case are:

- **Severe progressive problems with recent memory and milder problems with remote memory**
- **Tendency to confabulate**

1. As in Case 18.1, this patient had anterograde and retrograde amnesia affecting declarative memory, which could be caused by dysfunction of the bilateral medial temporal or bilateral medial diencephalic structures (see KCC 18.1). Unlike the patient in Case 18.1, however, this patient also had a mild tendency to confabulate. This symptom suggests additional dysfunction affecting the frontal lobes (see KCC 19.11).

2. The time course of the memory decline in this patient was too rapid to be a neurodegenerative disorder such as Alzheimer's disease, which progresses over months to years, rather than over weeks as in this case (see KCC 19.16). Wernicke–Korsakoff syndrome resulting from thiamine deficiency should be considered, since this disorder causes declarative memory loss, often with confabulation (see KCC 18.1). However, there was no known alcoholism or nutritional deficiency in this case, and onset of Wernicke–Korsakoff syndrome is often more abrupt. Given the insidious onset, other important possibilities to consider, among those listed in Table 18.7, include tumor, paraneoplastic limbic encephalitis, or other inflammatory or infiltrative disorders affecting the bilateral medial temporal or medial diencephalic structures and extending to the frontal lobes. Other, less likely possibilities include an anterior communicating artery aneurysm that had a small hemorrhage followed by a larger hemorrhage, or several transient ischemic attacks or small infarcts followed by a larger infarct.

Clinical Course and Neuroimaging

The patient was given thiamine with no benefit and was admitted to the hospital, where a **brain MRI** with gadolinium (Image 18.2A–C, pages 862–863) revealed markedly abnormal enhancement in the bilateral medial temporal lobes, including the anterior hippocampal formations and amygdalae. In addition, there was less dramatic enhancement in the fornix, bilateral periventricular region of the third ventricle, and foramina of Monro (see Image 18.2B) and extending to the bilateral basal forebrain and medial orbitofrontal cortex (see Image 18.2C). Several lumbar punctures (see KCC 5.10) and blood tests were done but did not yield a diagnosis. A follow-up examination revealed that the patient had **lost his sense of smell**, suggesting that the lesion had spread to involve adjacent olfactory structures such as the bilateral piriform cortex or olfactory bulbs (see Figures 18.5, 18.6). Over the course of several days he became more disoriented, stating that his location was "Israel," and he responded more slowly to questions. This change raised concerns about more diffuse spread, and a stereotactic biopsy (see KCC 16.4) of the right orbitofrontal cortex was therefore done. Pathologic examination revealed an atypical-appearing B cell lymphoma (see KCC 5.8).

The patient was treated with steroids and multiple cycles of intravenous chemotherapy with methotrexate and had a dramatic improvement in his memory. One month after diagnosis he was able to recall 3/4 words at 5 minutes and 4/4 with hints. Formal neuropsychology testing revealed that he still had some subtle residual memory deficits in both verbal and visual-spatial domains. Repeat MRI scan 3 months after diagnosis showed complete disappearance of the enhancing lesions, and the patient had resumed most of his previous activities. He continued chemotherapy and at last follow-up, 15 months after diagnosis, he was doing well, recalling 3/3 words after 10 minutes.

CASE 18.2 PROGRESSIVE SEVERE MEMORY LOSS, WITH MILD CONFABULATION

IMAGE 18.2A–C **Bilateral Medial Temporal Lymphoma** T1-weighted MRI
scan with intravenous gadolinium. (A) Axial image. (B,C) Coronal images progressing from posterior to anterior.

(A)

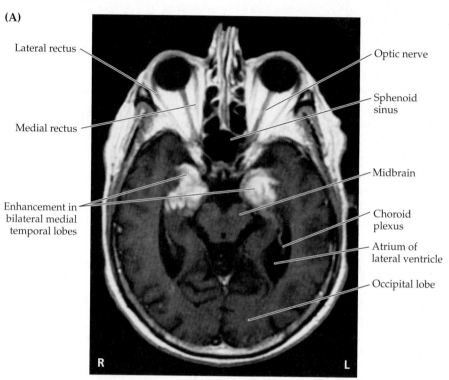

Lateral rectus

Medial rectus

Enhancement in
bilateral medial
temporal lobes

Optic nerve

Sphenoid
sinus

Midbrain

Choroid
plexus

Atrium of
lateral ventricle

Occipital lobe

R   L

(B)

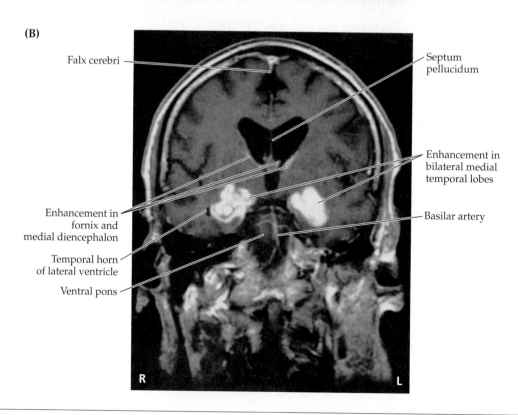

Falx cerebri

Enhancement in
fornix and
medial diencephalon

Temporal horn
of lateral ventricle

Ventral pons

Septum
pellucidum

Enhancement in
bilateral medial
temporal lobes

Basilar artery

R   L

CASE 18.2 (*continued*)

(C)

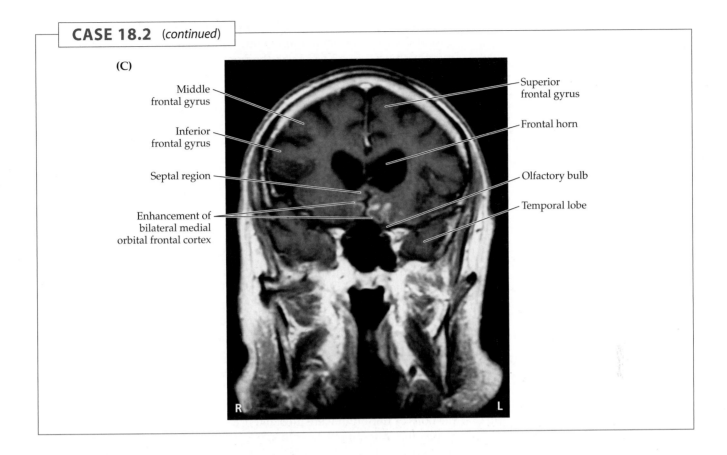

Middle frontal gyrus

Inferior frontal gyrus

Septal region

Enhancement of bilateral medial orbital frontal cortex

Superior frontal gyrus

Frontal horn

Olfactory bulb

Temporal lobe

R   L

CASE 18.3 TRANSIENT DIPLOPIA, LETHARGY, AND HEMIPARESIS, FOLLOWED BY A SUSTAINED MEMORY DEFICIT

MINICASE

A 45-year-old right-handed research technician woke up late for work one morning and noticed that he had **horizontal diplopia**, which disappeared if he covered either eye. His wife found him in the bathroom with his head resting on the sink. She helped him back to bed and thought he seemed **unusually somnolent, with slurred speech**, so she called 911. In the emergency room the patient was initially back to baseline, but then he had **a transient episode of right-sided weakness** lasting approximately 30 minutes. Neurologic exam, after he recovered from this episode, was entirely normal except for a **deficit in recent memory**. He was alert and oriented × 3, with good attention, spelling "world" forward and backward cor-rectly, and he knew the names of the past three presidents. However, he **recalled 0/3 words after 3 minutes**.

LOCALIZATION AND DIFFERENTIAL DIAGNOSIS

1. Dysfunction in which part of the brain could account for all of the transient symptoms listed in **bold** above (excluding the memory loss)?

2. In what locations could a lesion produce memory loss as described in this case?

3. What is the most likely diagnosis, and what are some other possibilities?

Discussion

The key symptoms and signs in this case are:

- **Horizontal diplopia**
- **Somnolence**
- **Dysarthria**
- **Transient right-sided weakness**
- **Deficit in recent memory**

1. Horizontal diplopia, together with somnolence and dysarthria, strongly suggest brainstem dysfunction involving, respectively, horizontal eye movement pathways (see KCC 13.8), the pontomesencephalic reticular activating systems (see KCC 14.2), and corticobulbar or cerebellar pathways (see KCC 14.3). The episode of transient right-sided weakness could also be explained by involvement of brainstem corticospinal fibers (see KCC 6.3, 14.3). These abnormalities together suggest dysfunction of the medial pons (abducens fascicles or nucleus, paramedian pontine reticular formation, or medial longitudinal fasciculus; reticular activating systems; corticobulbar or pontocerebellar fibers; and corticospinal tract), or of the midbrain (oculomotor fascicles or nucleus, convergence center, or medial longitudinal fasciculus; reticular activating systems; corticobulbar, corticopontine, or rubropontine fibers; and corticospinal tract). Note that horizontal diplopia with no vertical component, and hemiparesis, are more common with pontine than with mesencephalic lesions (see KCC 14.3).

2. After recovering from the initial transient episode, this patient had anterograde amnesia for declarative memories. He was more mildly affected than the patients in Cases 18.1 and 18.2, since he was able to remember some recent information, such as his location and the correct date. Some retrograde amnesia may have been present as well, although this was not specifically described. This patient's deficit in declarative (explicit) memory, with preserved attention and other cognitive skills, suggests bilateral medial temporal or bilateral medial diencephalic dysfunction (see KCC 18.1).

3. The transient neurologic symptoms in this patient (see KCC 10.3) affecting the pons or midbrain were most likely caused by transient ischemic attack, or possibly migraine. Given this, the sudden onset of a permanent memory deficit suggests that the transient episodes heralded a subsequent infarct in the medial temporal lobes or diencephalon. Together these findings could indicate a "basilar scrape" syndrome (see KCC 14.3), with an embolus migrating up the basilar artery causing transient symptoms as various penetrators were temporarily occluded, followed by an infarct when the embolus lodged at the top of the basilar, obstructing bilateral thalamoperforator arteries or branches of the bilateral PCAs supplying both medial temporal lobes.

Clinical Course and Neuroimaging

A head CT revealed no hemorrhage or infarct, and the patient was admitted and treated with intravenous heparin while an embolic workup was pursued. A **brain MRI** done the day after admission (Image 18.3A,B, page 866) showed bilateral T2-bright areas consistent with infarcts in both medial thalami, larger on the left than on the right side. A magnetic resonance angiogram, echocardiogram, and 24-hour cardiac Holter monitor were negative. However, a hypercoagulation profile showed activated protein C resistance, a disorder that can predispose affected individuals to blood clot formation. Heparin was therefore changed over to oral anticoagulation with Coumadin by the time of discharge. By hospital day 2 the patient was able to recall 3/5 words after 5 minutes.

Neuropsychology testing was done on hospital day 3 and repeated on hospital day 5. Subtests of the Wechsler Memory Scale—Revised were used to test verbal and visual memory. To test verbal memory, two paragraphs were read to the patient, and he was scored on his ability to recall items from the paragraphs immediately after they were read and after a 20-minute delay. To test visual memory, the patient was shown three cards with printed geometric designs, and he was scored on his ability to draw the designs from memory immediately and after a 20-minute delay. The results are shown in the table below.

Results of the Wechsler Memory Scale Test[a]

	Hospital day 2	Hospital day 5
Verbal memory		
Immediate	**1%**	21%
Delayed	**2%**	**8%**
Visual memory		
Immediate	50%	72%
Delayed	**16%**	45%

[a]Scores are expressed as the percentile of the normal population. Scores that were significantly lower than the normal range are shown in bold.

On hospital day 2 the patient had severe deficits in verbal memory and a barely significant deficit in delayed visual memory. On hospital day 5, he continued to have a significant deficit in delayed verbal memory. The discrepancy between verbal and visual memory in this patient was likely due to the larger lesion on the left side (see Image 18.3A,B). Similar discrepancies are seen in patients with asymmetrical lesions in medial temporal memory structures.

Repeat neuropsychology testing 2 months later showed further improvements in his verbal memory. At a follow-up appointment 6 months after the stroke he was still able to recall only 3/5 words after 5 minutes, but he was back to his previous level of activity at work, and he was not aware of any language or memory difficulties.

CASE 18.4 EPISODES OF PANIC, OLFACTORY HALLUCINATIONS, AND LOSS OF AWARENESS

MINICASE

A 40-year-old, right-handed woman came to the emergency room because of unusual episodes. She had been healthy until 2 weeks previously, when she awoke one night from sleep and complained to her husband of an **indescribable, unpleasant odor, nausea, and a panicky, fearful sensation.** This lasted for 2 to 3 minutes, after which she felt very tired and went back to sleep. During the following week, while traveling with a Girl Scout troup, she began having stereotyped episodes about three times per day. These always began with panic, nausea, and a strong unpleasant odor "like smelling salts," followed by **decreased responsiveness and slow, inappropriate speech** lasting 2 to 3 minutes. After each episode she had a bifrontal headache and felt very tired. On the day of

presentation, she had six episodes of this kind, so her husband brought her to the emergency room. General exam and a detailed neurologic exam were normal.

LOCALIZATION AND DIFFERENTIAL DIAGNOSIS

1. On the basis of the symptoms shown in **bold** above, where in the brain is dysfunction occurring during the episodes?

2. What type of transient neurological episodes was this patient most likely experiencing (see KCC 10.3)?

3. What is the most likely diagnosis, and what are some other possibilities?

CASE 18.3 TRANSIENT DIPLOPIA, LETHARGY, AND HEMIPARESIS, FOLLOWED BY A SUSTAINED MEMORY DEFICIT

IMAGE 18.3A,B Bilateral Medial Thalamic Infarcts Axial T2-weighted MRI images. Images A,B progress from inferior to superior.

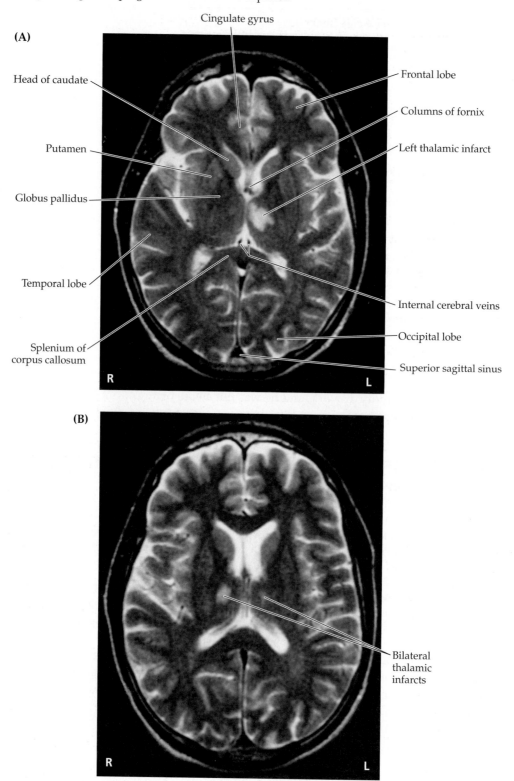

Discussion

The key symptoms and signs in this case are:

- **Episodes of an indescribable, unpleasant odor, nausea, and a panicky, fearful sensation**
- **Decreased responsiveness and slow, inappropriate speech**

1. Episodes of olfactory phenomena, nausea, and panic could arise from dysfunction in the amygdala and adjacent cortex in the medial temporal lobe. Decreased responsiveness can be caused by dysfunction of the brainstem–diencephalic activating systems or of bilateral diffuse regions of the cerebral hemispheres (see KCC 14.2). The slow, inappropriate speech could either signify involvement of language areas in the left hemisphere (see KCC 19.6), or it could be secondary to the impaired responsiveness, as already mentioned.

   The overall *clinical localization* in this patient is focal dysfunction of the amygdala and adjacent cortex (most likely of the left hemisphere), as well as milder, more diffuse dysfunction of both hemispheres or of the activating systems.

2. The transient episodes in this patient (see KCC 10.3) could be caused by seizures arising from the medial temporal lobe. Because there was decreased responsiveness, these events would be classified as complex partial seizures (see KCC 18.2). Additional support for this diagnosis comes from the fact that the episodes were stereotyped, had the appropriate duration for complex partial seizures, and were followed by post-ictal lethargy and headache. The slow, inappropriate speech may have been caused by language dysfunction resulting from spread of left mesial temporal–onset seizures to the left lateral temporal neocortex. However, since the patient was not examined during an actual seizure, one cannot exclude speech difficulties simply caused by decreased responsiveness.

   Psychogenic episodes, such as panic attacks, could also be considered in this case. However, panic attacks are not accompanied by olfactory phenomena, usually last longer, and do not have post-ictal deficits. Other causes of transient neurologic episodes listed in Table 10.2, such as migraine or transient ischemic attack, are also unlikely to have this time course or to be accompanied by positive olfactory phenomena.

3. Numerous focal lesions in the medial temporal lobe could produce seizures of the type seen in this patient (see Table 18.11). The most likely possibilities in a 40-year-old woman are a tumor such as glioblastoma, metastatic breast or lung cancer, other neoplasms (see KCC 5.8), a bacterial brain abscess, toxoplasmosis, cysticercosis (see KCC 5.9), vascular malformation, previous head trauma, sarcoidosis, or other infectious or inflammatory disorders. For unknown reasons, neuronal developmental disorders can sometimes present with seizure onset in adulthood. Onset at age 40 years would be older than usual, but not impossible. In mesial temporal (hippocampal) sclerosis, seizure onset is typically at a younger age than in this patient, there is often an initial precipitating injury such as prolonged febrile seizures in childhood, and olfactory auras are not commonly present.

Clinical Course and Neuroimaging

A head CT with contrast was performed in the emergency room, and a subsequent **brain MRI** with gadolinium confirmed the presence of a large enhancing mass in the left anterior temporal lobe, involving the amygdala (Image 18.4A, page 871) and extending to the left orbitofrontal region (Image 18.4B, page 871). Note that the mass had a heterogeneous, hypointense center, suggesting central necrosis. There was also surrounding hypointensity on T1-weighted images, with effacement of the Sylvian fissure and temporal horn, consistent with edema and mass effect (see Chapter 4). This appearance was felt to be most compatible with a glioblastoma multiforme or, less likely, a metastasis or other tumor. The patient was started on the anticonvulsant medication phenytoin, which stopped her seizures, and she underwent a surgical resection of the left anterior temporal lobe. Pathology revealed a glioblastoma multiforme (see KCC 5.8). She was treated with radiation therapy and multiple cycles of chemotherapy. On this regimen she remained stable for a while, but then she developed progressive deficits as the tumor grew. At last follow-up, 9 months after presentation, she was lethargic and disoriented, with a right visual field cut and right-sided hemiparesis (2/5 to 3/5 strength). Glioblastoma multiforme unfortunately remains a relatively common, yet incurable brain tumor.

CASE 18.5 EPISODES OF STARING, LIP SMACKING, AND UNILATERAL SEMIPURPOSEFUL MOVEMENTS

MINICASE

A 26-year-old right-handed woman was evaluated at an epilepsy referral center for medically refractory seizures. She was the product of a normal birth and delivery, but at age 9 months she had two prolonged febrile seizures in one day. These consisted of generalized tonic-clonic activity, with the first seizure lasting nearly 2.5 hours and the second seizure, after arrival in the hospital, lasting about 1 hour. She was treated with anticonvulsant medications, which initially stopped the seizures. Within a few years, however, she began having stereotyped episodes that continued into adulthood and were refractory to multiple medications. Her episodes typically began with an **aura of a "vague" and "scary feeling,"** followed **by loss of awareness.** Her mother has witnessed many of her seizures and described her as next having **staring, unresponsiveness, lip smacking, rigid posturing of the left arm, and groping, purposeless movements of the right arm lasting for about 1 minute.** Sometimes **she spoke during the episodes without grammatical errors**, but what she said was inappropriate for the questions being asked. Post-ictally she **was fatigued and somewhat confused but had no other deficits.** These episodes occurred two to three times per week, occasionally up to three times per day. She also had occasional isolated auras, without progression to full-blown episodes.

Her longest seizure-free interval in adulthood was about 3 months. Rarely (at intervals of greater than about 3 to 5 years),

mostly in the setting of missing her medications, she had an episode that progressed to a **generalized convulsion.** Numerous medications, including phenobarbital, phenytoin, carbamazepine, mephobarbital, valproate, gabapentin, and felbamate were tried in various combinations, without effective seizure control. Aside from the febrile seizures, she had no other seizure risk factors, such as head injury, CNS infections, or family history of seizures. She graduated from high school, worked briefly as a cashier, was married, and had three children. She was unable to drive because of her seizure risk. General physical exam and neurologic exam were normal.

LOCALIZATION AND DIFFERENTIAL DIAGNOSIS

1. On the basis of the features shown in **bold** above, this patient had simple partial, complex partial, and generalized seizures at different times. How would you classify the different seizures in this patient according to this scheme (see Table 18.8)?

2. How can these different seizure types all be explained by onset in one region of the brain and spread to different areas?

3. In what part of the brain and on which side is the most likely region of onset?

4. What is the most likely cause of the seizures, and what are some other possibilities?

Discussion

The key symptoms and signs in this case are:

- Aura of a "vague" and "scary feeling"
- Loss of awareness, staring, and unresponsiveness
- Lip smacking, rigid posturing of the left arm, and groping, purposeless movements of the right arm

- Ability to produce speech during the episodes that was inappropriate, but without grammatical errors
- Post-ictal fatigue and confusion without other deficits
- Rare generalized convulsions

1. According to the classification in Table 18.8, the auras in this patient were simple partial seizures, since consciousness was spared. The episodes with loss of consciousness but no generalized convulsion were complex partial seizures. Finally, the febrile convulsions in infancy and the rare generalized convulsions in adulthood were generalized tonic-clonic seizures.

2. A simple partial seizure beginning in one part of the brain can spread to become a complex partial seizure, and then spread further to become a generalized tonic-clonic seizure.

3. A vague feeling and fear are features typically seen in seizures arising from the medial temporal lobe (see Table 18.9). In addition, complex partial seizures with staring, unresponsiveness, oral automatisms, unilateral gestural automatisms, and contralateral dystonia are commonly seen in temporal lobe seizures. The fact that dystonic posturing occurred on the left and automatisms on the right suggests that the seizures arose from the right side (see KCC 18.2). The fact that the patient had nonaphasic speech during the seizures and no post-ictal aphasia also suggests that the seizures arose from the nondominant (usually right) hemisphere. The most likely *clinical localization* is right medial temporal lobe.

4. The history of prolonged febrile seizures in infancy, followed by the development of medial temporal–onset complex partial seizures, is the classic presentation of hippocampal (mesial temporal) sclerosis (see KCC 18.2). Although only a small number of children with febrile seizures go on to develop epilepsy, children in the subgroup with complex febrile seizures are at increased risk for epilepsy. In addition, about 30 to 40% of patients with epilepsy and hippocampal sclerosis have a known history of febrile convulsions. Other features in this patient that are typical of hippocampal sclerosis are complex partial seizures that are notoriously refractory to medication therapy, and only rare secondary generalization while on medications. Other possible lesions in the medial temporal lobe that could be the cause of epilepsy in this patient are listed in Table 18.11.

Clinical Course and Neuroimaging

Given this patient's frequent, medically refractory seizures, a comprehensive evaluation was pursued in an effort to localize the region of onset and to determine whether she was a suitable candidate for epilepsy surgery. She was admitted to the hospital for continuous video and simultaneous EEG monitoring. Ten stereotyped seizures were recorded, with loss of consciousness, lip smacking, left-sided dystonia, and right-sided automatisms. EEG during the seizures showed rhythmic 8 Hz activity over the right temporal lobe. Interictal EEG showed occasional right temporal spikes and right temporal slow waves. A **brain MRI** was performed, with special thin coronal sections used to evaluate epilepsy patients. This technique is capable of demonstrating detailed hippocampal anatomy (Image 18.5A, page 872). The MRI revealed marked atrophy of the right hippocampal formation compared to the left (Image 18.5C–F, page 873). In addition, signal intensity in the right hippocampal formation was increased, suggesting gliosis (Image 18.5B, page 872). The volumes of the right and left hippocampal formations were 562 and 983 (arbitrary units) respectively, based on measurements performed on the entire sequence of coronal MRI images. This discrepancy between the two sides was over five standard deviations greater than normal. A **fluorodeoxyglucose PET scan** (see Chapter 4) performed interictally showed markedly reduced glucose metabolism in the right temporal lobe, especially medially and anteriorly (Image 18.5G–J, page 874), a finding commonly seen in medial temporal lobe epilepsy.

Neuropsychology testing revealed decreased visual-spatial memory on the Wechsler Memory Scale, with preserved verbal memory. In addition, on selective reminding tests (another measure of recent memory), the patient had visual-spatial memory that was two standard deviations below normal and verbal memory that was normal. A Wada test (see KCC 18.2) was performed as well. Following injection of the right carotid with amytal, the patient developed left hemiplegia and was shown 10 test items. Her speech was not aphasic. Ten minutes later the hemiplegia had worn off, and the patient was able to remember 5/10 items spontaneously and 10/10 with multiple choice. These results demonstrated good memory in the left hemisphere. Injection of the left carotid produced right hemiplegia and global aphasia. The patient was shown 10 test items. During recovery she had paraphasic errors and comprehension difficulties that resolved, as did the hemiplegia, within 10 minutes. She was then able to recall 0/10 test items and got only 2/10 correct with multiple choice. These results demonstrated poor memory in the right hemisphere. In addition, the Wada test results demonstrated that language dominance was in the left hemisphere.

These findings were discussed by the multidisciplinary team at the epilepsy referral center, and all results were felt to be concordant, indicating onset of her seizures in the right medial temporal lobe. Given this location and the results of the patient's Wada test and neuropsychology testing suggesting that the left temporal lobe could support adequate memory functions, it was felt that surgical treatment of her right mesial temporal sclerosis could be offered with a low risk of producing deficits and a good chance of curing her seizures. She decided to pursue this option and underwent a right anteromedial temporal resection.

She had no deficits as a result of the surgery, and her seizures stopped completely. Pathology revealed right hippocampal cell loss and gliosis, consistent with hippocampal sclerosis. About 1 month after surgery and again 6 months after surgery, she tried briefly to discontinue her antiepileptic medications, but on both occasions she had recurrent seizures. As long as she took her medications, she remained without seizures or auras at last follow-up, a year and a half after surgery.

Additional Cases

Related cases for the following topics can be found in other chapters: **impaired olfaction** (Case 12.1); **abnormal emotions** (Cases 12.8, 16.2, and 19.7); **memory decline** (Cases 5.9, 14.8, 19.7, and 19.11); and **seizures** (Cases 10.13, 12.3, and 19.10). Other relevant cases can be found using the **Case Index** located at the end of this book.

CASE 18.4 EPISODES OF PANIC, OLFACTORY HALLUCINATIONS, AND LOSS OF AWARENESS

IMAGE 18.4A,B Left Anterior Temporal Glioblastoma Multiforme
Coronal T1-weighted MRI images, with intravenous gadolinium. Images A,B progress from posterior to anterior.

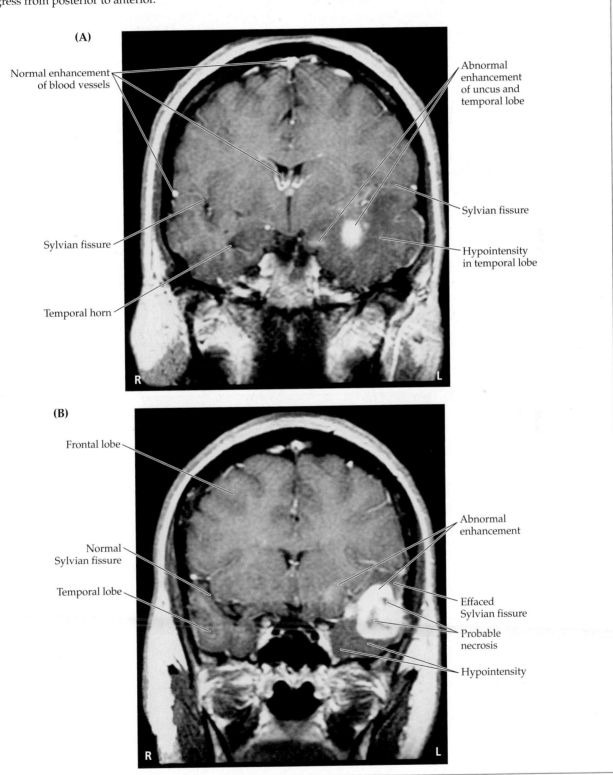

CASE 18.5 EPISODES OF STARING, LIP SMACKING, AND UNILATERAL SEMIPURPOSEFUL MOVEMENTS

IMAGE 18.5A–F Right Hippocampal Sclerosis (A) Enlarged view of coronal thin-slice, T1-weighted MRI done with a special epilepsy protocol revealing normal structures of the left medial temporal lobe (compare to Figure 18.8A). (B) Coronal T2-weighted image. Increased signal intensity is evident in the right hippocampal formation, compatible with gliosis. (C–F) Coronal thin-slice, T1-weighted images. Atrophy is seen in the right hippocampal formation compared to the left. Images C–F progress from posterior to anterior.

(A)

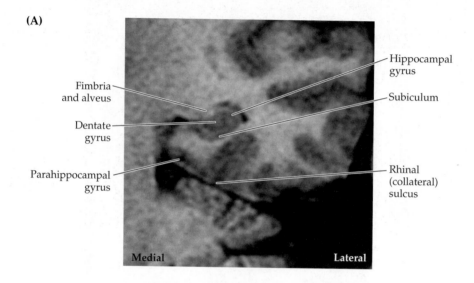

(B)

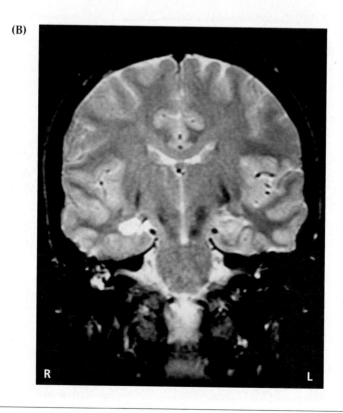

CASE 18.5 *(continued)*

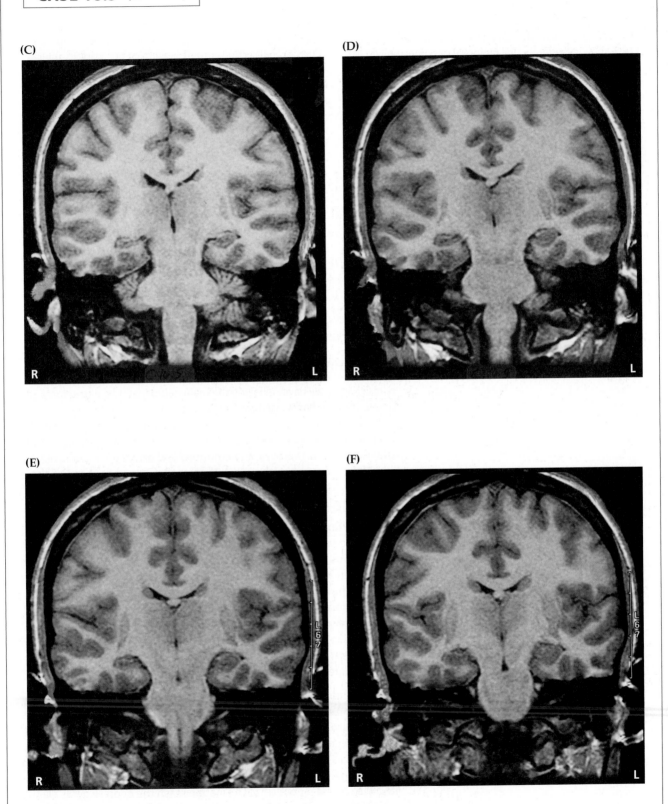

(C)

(D)

(E)

(F)

CASE 18.5 (continued)

IMAGE 18.5G–J **Right Temporal Hypometabolism**
Fluorodeoxyglucose (FDG) PET scan for patient in Case
18.5, showing right medial temporal hypometabolism.

Darker colors indicate regions of decreased metabolism.
(G,H) Axial sections progressing from inferior to superior.
(I,J) Coronal sections progressing from posterior to anterior.

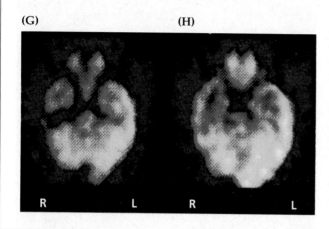

(G)   **(H)**

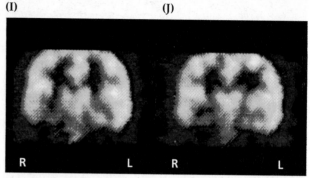

(I)   **(J)**

Brief Anatomical Study Guide

1. In this chapter we have reviewed the main neuroanatomical structures and functions of the **limbic system**. The limbic system can be defined as a network of brain structures lying on the medial and inferior aspects of the brain (see Figures 18.1 and 18.2), involved in four general functions: **homeostasis**, **olfaction**, **memory**, and **emotions and drives** (mnemonic HOME). Simplifying somewhat, the most important structure for each of these functions is listed in Table 18.2, although in reality these functions are mediated by a widely distributed network (see Figure 18.1).

2. **Limbic cortex**, also called paralimbic cortex or limbic association cortex, forms a ring on the medial aspect of the brain consisting mainly of the **cingulate gyrus** and the **parahippocampal gyrus** (see Figure 18.2). Within this ring and in the temporal lobe lies the simple three-layered **archicortex** of the **hippocampal formation** (see Table 18.3; Figure 18.9). The hippocampal formation consists of three gyri named, from medial to lateral, the **dentate gyrus**, **hippocampal gyrus**, and **subiculum** (see Figures 18.7 and 18.8). Unlike the six-layered neocortex making up most of the brain surface, these gyri of the hippocampal formation each consist of three layers (see Figure 18.8).

3. The hippocampal formation is crucial to memory function and has many input and output network connections (see Figure 18.11). The connections with the **association cortex** may be particularly important for memory, and they occur mainly via the **perirhinal** and **parahippocampal cortex**, connected to the adjacent **entorhinal cortex** (see Figure 18.6), which in turn is connected to the hippocampal formation. Inputs from the entorhinal cortex to the hippocampal formation reach the dentate gyrus via the **perforant pathway** and reach the CA1 and CA3 fields of the hippocampal gyrus via the **alvear pathway** (see Figure 18.8B). Outputs from the

hippocampal formation travel from the subiculum back to the entorhinal cortex. The hippocampal formation also has connections with subcortical structures, especially the **mammillary bodies**, **mediodorsal nuclei**, **other medial diencephalic nuclei** and the **septal nuclei** via the **fornix** (see Figures 18.9, 18.11, and 18.13) and other connections.

4. Lesions of the **medial temporal lobe memory system** (hippocampal formation and parahippocampal gyrus) or of the **medial diencephalic memory system** (thalamic mediodorsal nucleus, anterior nucleus, and other periventricular diencephalic nuclei) cause a characteristic **anterograde amnesia**, in which new information cannot be learned, together with **retrograde amnesia**, in which material from a period of time prior to the lesion cannot be recalled (see Figure 18.16).

5. The **amygdala** is a nuclear complex located just anterior to the hippocampal formation in the anterior temporal lobe (see Figure 18.10). It has three main nuclei: the **corticomedial**, **basolateral**, and **central nuclei** (see Figure 18.4B). Like the hippocampal formation, the amygdala has connections with association cortex and subcortical structures (see Figure 18.17A,B), but unlike the hippocampal formation, it also has direct connections with olfactory structures (see Figure 18.17C). The amygdala participates in networks mediating all of the major limbic functions, including autonomic and neuroendocrine control, olfaction, emotional aspects of memory (e.g., fear memory), and emotions.

6. Olfaction is relatively poorly developed in humans, but it is a major function of limbic structures in other species. Olfactory inputs travel via the **olfactory nerves** to reach the **olfactory bulbs**, where **mitral cells** and **tufted cells** relay information to olfactory areas, including the piriform cortex, periamygdaloid cortex, olfactory tubercle, and amygdala (see Figures 18.5 and 18.6). Unlike other sensory modalities, olfactory inputs reach the **primary olfactory cortex** (**piriform cortex** and **periamygdaloid cortex**) directly, with no obligatory relay in the thalamus. Secondary olfactory areas include the **orbitofrontal olfactory cortex**, which receives inputs from the piriform cortex both directly and via the mediodorsal nucleus of the thalamus.

References

General References

Aggleton JP (ed.). 2000. *The Amygdala: A Functional Analysis.* 2nd ed. Oxford University Press.

Andersen P, Morris R, Amaral D, Bliss T, O'Keefe J. 2006. *The Hippocampus Book.* Oxford University Press.

Carpenter MB. 1991. *Core Text of Neuroanatomy.* 4th ed., Chapter 12. Williams & Wilkins, Baltimore, MD.

Ehrlich I, Humeau Y, Grenier F, Ciocchi S, Herry C, Lüthi A. 2009. Amygdala inhibitory circuits and the control of fear memory. *Neuron* 62 (6): 757–771.

Gartner A, Frantz D. 2010. *Hippocampus: Anatomy, Functions and Neurobiology.* Nova Science Pub. Inc.

Lautin A. 2001. *The Limbic Brain.* Springer.

Martin JH. 2003. *Neuroanatomy: Text and Atlas.* 3rd ed., Chapter 16. Appleton & Lange, Stamford, CT.

Shepherd GM. 2006. Smell images and the flavour system in the human brain. *Nature.* 444 (7117): 316–321.

Memory and Memory Disorders

Arasaki K, Kwee IL, Nakada T. 1987. Limbic lymphoma. *Neuroradiology* 29: 389–392.

Bailey CH, Kandel ER. 2008. Synaptic remodeling, synaptic growth and the storage of long-term memory in Aplysia. *Prog Brain Res* 169: 179–198.

Bailey CH, Kandel ER, Si K. 2004. The persistence of long-term memory: a molecular approach to self-sustaining changes in learning-induced synaptic growth. *Neuron* 44 (1): 49–57.

Bauer RM, Tobias B, Valenstein E. 2003. Amnesic disorders. In *Clinical Neuropsychology,* 4th ed., KM Heilman and E Valenstein (eds.), Chapter 18. Oxford University Press, New York.

Baxter MG. 2009. Involvement of medial temporal lobe structures in memory and perception. *Neuron* 61 (5): 667–677.

Bliss TVP, Collingridge GL. 1993. A synaptic model of memory: Long-term potentiation in the hippocampus. *Nature* 361: 31–39.

Budson AE. 2009. Understanding memory dysfunction. *Neurologist* 15 (2): 71–79.

Corkin S. 1984. Lasting consequences of bilateral medial temporal lobectomy: Clinical course and experimental findings in H.M. *Sem Neurol* 4: 249–259.

Gentilini M, Renzi E, Crisi G. 1987. Bilateral paramedian thalamic artery infarcts: Report of eight cases. *J Neurol Neurosurg Psychiatry* 50: 900–999.

Kandel ER. 2001. The molecular biology of memory storage: a dialogue between genes and synapses. *Science* 294 (5544): 1030–1038.

Kopelman MD, Thomson AD, Guerrini I, Marshall EJ. 2009. The Korsakoff syndrome: clinical aspects, psychology and treatment. *Alcohol Alcohol* 44 (2): 148–154.

Levin HS, High W, Meyers CA, Von Laufen A, Hayden ME, Eisenberg HM. 1985. Impairment of remote memory after closed head injury. *J Neurol Neurosurg Psychiatry* 49: 556–563.

Lim C, Alexander MP, LaFleche G, Schnyer DM, Verfaellie M. 2004. The neurological and cognitive sequelae of cardiac arrest. *Neurology* 63 (10): 1774–1778.

McGaugh JL. 2004. The amygdala modulates the consolidation of memories of emotionally arousing experiences. *Annu Rev Neurosci* 27: 1–28.

Rosazza C, Minati L, Ghielmetti F, Maccagnano E, Erbetta A, Villani F, Epifani F, Spreafico R, Bruzzone MG. 2009. Engagement of the medial temporal lobe in verbal and nonverbal memory: assessment with functional MR imaging in healthy subjects. *Am J Neuroradiol* 30 (6): 1134–1141.

Rosenbaum RS, Winocur G, Moscovitch M. 2001. New views on old memories: re-evaluating the role of the hippocampal complex. *Behav Brain Res* 127 (1–2): 183–197.

Scoville WB, Milner B. 1996. Loss of recent memory after bilateral hippocampal lesions. *J NIH Res* 8: 42–51.

Sechi G, Serra A. 2007. Wernicke's encephalopathy: new clinical settings and recent advances in diagnosis and management. *Lancet Neurol* 6 (5): 442–455.

Shaw NA. 2002. The neurophysiology of concussion. *Prog Neurobiol* 67 (4): 281–344.

Shekhar R. 2008 Transient global amnesia—a review. *Int J Clin Pract* 62 (6): 939–942. Epub 2008 Jan 30.

Squire LR, Kandel ER. 1999. *Memory: From Mind to Molecules.* Freeman, New York.

Squire LR, Zola-Morgan S. 1991. The medial temporal lobe memory system. *Science* 253: 1380–1386.

Sullivan EV, Pfefferbaum A. 2009. Neuroimaging of the Wernicke–Korsakoff syndrome. *Alcohol Alcohol* 44 (2): 155–165.

Tulving E, Schacter DL. 1990. Priming and human memory systems. *Science* 247: 301–306.

Zola-Morgan S, Squire LR. 1993. Neuroanatomy of memory. *Annu Rev Neurosci* 16: 547–563.

Seizures and Epilepsy

Acharya V, Acharya J, Luders H. 1998. Olfactory epileptic auras. *Neurology* 51: 56–61.

Engel J. 1989. *Seizures and Epilepsy*. FA Davis, Philadelphia.

Engel J, Pedley TA, Aicardi J, Dichter MA, Moshé S (eds.). 2007. *Epilepsy: A Comprehensive Textbook*, 3 vols. Lippincott-Williams & Wilkins, Baltimore.

French JA, Pedley TA. 2008. Clinical practice. Initial management of epilepsy. *N Engl J Med* 359 (2): 166–176.

Panayiotopoulos CP. 2005. *The Epilepsies: Seizures, Syndromes and Management*. Blandon, Oxfordshire, England.

Wyllie E, Gupta A, Lachhwani DK (eds.). 2005. *The Treatment of Epilepsy*. 4th ed. Lippincott Williams & Wilkins, Baltimore, MD.

Psychiatric Disorders

Buchanan RW, Freedman R, Javitt DC, Abi-Dargham A, Lieberman JA. 2007. Recent advances in the development of novel pharmacological agents for the treatment of cognitive impairments in schizophrenia. *Schizophr Bull* 33 (5): 1120–1130.

Freedman R. 2003. Schizophrenia. *N Engl J Med* 349 (18): 1738–1749.

Insel TR. 1992. Toward a neuroanatomy of obsessive-compulsive disorder. *Arch Gen Psychiatry* 49: 739–744.

Jenike MA. 2004. Clinical practice. Obsessive-compulsive disorder. *N Engl J Med* 350 3): 259–265.

Lilly R, Cummings JL, Benson DF, Frankel M.1983. The human Klüver–Bucy syndrome. *Neurology* 33: 1141–1145.

Ravindran AV, da Silva TL, Ravindran LN, Richter MA, Rector NA. 2009. Obsessive-compulsive spectrum disorders: a review of the evidence-based treatments. *Can J Psychiatry* 54 (5): 331–343.

Sadock VA, Sadock BJ. 2008. *Kaplan and Sadock's Concise Textbook of Clinical Psychiatry*. 3rd ed. Lippincott Williams & Wilkins, Baltimore, MD.

CONTENTS

Chapter *19*

Higher-Order Cerebral Function

A 64-year-old woman progressively developed difficulties with reading, along with a right visual field defect. When she presented at the clinic, she was completely unable to read but was able to write normally. For example, she wrote "Today is a nice day" and "It is a sunny day in Boston" but was unable to read her own writing a few minutes later.

As we shall see, higher-order cerebral functions (such as reading) depend on both local cortical functions (vision) together with more distributed cortical network functions (language). In this chapter, we will learn about the local higher-order functions of the cerebrum and about the network connections that are essential for distributed functions such as language and cognition.

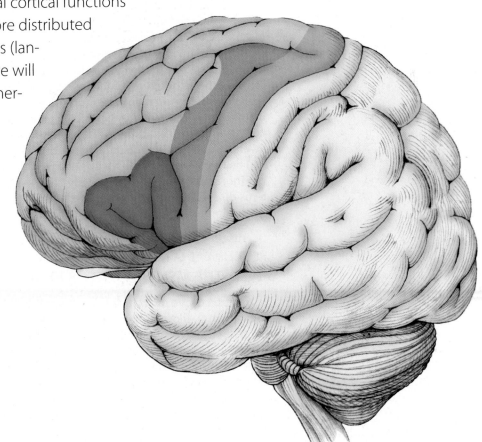

ANATOMICAL AND CLINICAL REVIEW

IN HUMANS, the majority of the brain surface is composed of association cortex. The functions of this vast cortical mantle are perhaps more difficult to understand than those of any other brain area, yet they are also what make us uniquely human. Functions of the association cortex include sophisticated operations such as higher-order sensory processing, motor planning, language processing and production, visual-spatial orientation, determining socially appropriate human behavior, and perhaps even what we would call "abstract thought."

Although the emphasis of this chapter is association cortex, it should be noted that numerous subcortical structures participate in these functions as well. We saw in Chapter 18 that lesions of the medial diencephalon can mimic the memory deficits seen in medial temporal lesions. Similarly, lesions of the basal ganglia, thalamus, subcortical white matter, and other structures can produce deficits such as aphasia or unilateral neglect that resemble lesions of the association cortex. Although cerebral cortical localization will be emphasized in this chapter, in reality, networks of both cortical and subcortical structures mediate virtually all brain functions. For example, we have learned that the thalamus and basal ganglia participate in association cortex networks (see Figures 7.8 and 16.8), that brainstem activating systems are crucial for behavioral arousal (see Figure 14.7), and that the amygdala has widespread connections participating in emotions and drives (see Figure 18.17). Thus, in addition to deficits arising from focal cortical lesions, specific neurobehavioral deficits can be caused by lesions that involve only subcortical structures or by lesions that disrupt cortical–cortical or cortical–subcortical network connections.

Historically there has been a dichotomy in theories of brain function. Some investigators have attributed brain functions to distributed networks, while others have assigned functions to specific localizations. In reality, both network and localized mechanisms participate in brain function, and we will therefore discuss both in this chapter. We will begin by reviewing the overall structure of the mental status examination introduced in Chapter 3. Next, we will discuss localized aspects of cerebral function and focal clinical disorders in four general regions of the association cortex: dominant (usually left) hemisphere, nondominant (usually right) hemisphere, frontal lobes, and visual association cortex. Finally, we will discuss more widely distributed functions such as attention and awareness and clinical disorders resulting from more global brain dysfunction.

The epilogue that follows this chapter will briefly present an overview of the integrated functions of the nervous system, including the various systems discussed throughout this book. An attempt will be made to unify these into a simple working model of the mind.

KEY CLINICAL CONCEPT

19.1 THE MENTAL STATUS EXAM

In Chapter 3, we introduced the **mental status exam**. This part of the neurological exam, like all other parts, should be performed and interpreted in the context of a more general clinical assessment that includes the patient history, general physical examination, and appropriately selected diagnostic tests. The mental status exam provides a useful bedside evaluation of mental performance. Formal **neuropsychology testing** can provide more detailed, accurate, and quantitative information when needed. There are many variations in the mental status exam, depending on the clinician's style and on the specific clinical situation. However, the mental status exam usually includes the basic elements listed in Table 19.1.

The patient's level of alertness, attention, and cooperation will influence virtually every other part of the exam. Therefore, the exam usually begins with an evaluation of these functions, which tend to depend on more widely distributed networks. Any abnormalities should be carefully documented, since they will affect the interpretation of all other parts of the exam and could lead to false diagnosis of a focal condition. For example, patients who are in a global confusional state because of a toxic or metabolic disorder (see KCC 19.15) are often inattentive and may perform poorly on writing tests. Unless the level of alertness and attention is carefully tested, this poor performance could be misinterpreted as a focal deficit in written language.

Next, the patient's orientation and memory should be tested. Memory is a crucial element of mental status that was covered in detail in Chapter 18. The remaining parts of the mental status exam (see Table 19.1) mainly test more localized brain regions such as the dominant (usually left) hemisphere (language and related functions), right hemisphere (neglect and constructions), and frontal lobes (sequencing tasks and frontal release signs). In the sections that follow, we will discuss the anatomy and testing of each of these brain regions in greater detail. In addition, we will discuss testing and disorders of visual processing that are not listed explicitly in Table 19.1. Finally, the examiner should evaluate the patient for several additional, more global functions (logic and abstractions) and for psychiatric disorders (see KCC 18.3). In addition to the formal exam test items, it is crucial to carefully observe the behavioral interactions and comportment of the patient during the interview to obtain additional useful information about mood, affect, speech, thought, judgment, and insight.

To avoid confusion, note that the general organization of this chapter does not follow the order of Table 19.1, but instead begins with localized functions and then later continues with more global functions such as attention. The mental status exam is described in greater detail in Chapter 3 (and in **neuroexam.com Videos 3–23**) and should be reviewed carefully at this point before reading the rest of this chapter, since it will make the material presented here much easier to understand. ■

TABLE 19.1 Overview of the Mental Status Exam

1. Level of alertness, attention, and cooperation
2. Orientation
3. Memory
   Recent memory
   Remote memory
4. Language
   Spontaneous speech
   Comprehension
   Naming
   Repetition
   Reading
   Writing
5. Calculations, right–left confusion, finger agnosia, agraphia
6. Apraxia
7. Neglect and constructions
8. Sequencing tasks and frontal release signs
9. Logic and abstraction
10. Delusions and hallucinations
11. Mood

Unimodal and Heteromodal Association Cortex

Association cortex can be divided into **unimodal (modality-specific) association cortex** and **heteromodal (higher-order) association cortex** (Figure 19.1; Table 19.2).

TABLE 19.2 Terminology Used for Classifying Neocortex[a]

NAME	EQUIVALENT NAME(S)	EXAMPLES
Primary sensory and motor cortex	Idiotypic cortex, heterotypic cortex[b]	
Primary sensory cortex	Koniocortex, hypergranular cortex, granular cortex	Primary somatosensory cortex, primary visual cortex, primary auditory cortex
Primary motor cortex	Macropyramidal cortex, agranular cortex	Primary motor cortex
Association cortex	Homotypic cortex[c]	
Unimodal association cortex	Modality-specific association cortex	Somatosensory, visual, or auditory association cortex, premotor cortex, supplementary motor area
Heteromodal association cortex	Higher-order association cortex	Prefrontal cortex, parietal and temporal heteromodal association cortex

[a]Six-layered cortex, sometimes called isocortex or neopallium. See Table 18.3 for classification of other types of cerebral cortex.
[b]Cortex in which the layers are unequal.
[c]Cortex in which the layers are relatively equal.

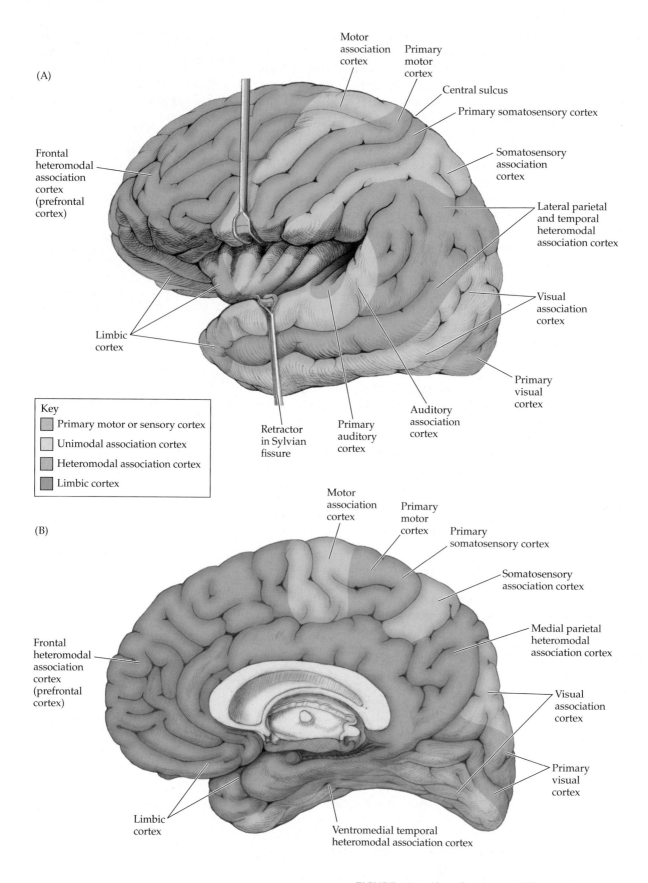

(A)

Motor association cortex

Primary motor cortex

Central sulcus

Primary somatosensory cortex

Somatosensory association cortex

Frontal heteromodal association cortex (prefrontal cortex)

Lateral parietal and temporal heteromodal association cortex

Visual association cortex

Limbic cortex

Primary visual cortex

Key

- Primary motor or sensory cortex
- Unimodal association cortex
- Heteromodal association cortex
- Limbic cortex

Retractor in Sylvian fissure

Primary auditory cortex

Auditory association cortex

(B)

Motor association cortex

Primary motor cortex

Primary somatosensory cortex

Somatosensory association cortex

Frontal heteromodal association cortex (prefrontal cortex)

Medial parietal heteromodal association cortex

Visual association cortex

Primary visual cortex

Limbic cortex

Ventromedial temporal heteromodal association cortex

FIGURE 19.1 Classification of Different Types of Cortex See also Tables 18.3 and 19.2.

Examples of unimodal association cortex include somatosensory association cortex, visual association cortex, auditory association cortex, and motor association cortex (premotor cortex and supplementary motor area)(see Figure 19.1). Unimodal sensory association cortex receives its predominant input from primary sensory cortex of a specific sensory modality and performs higher-order sensory processing for that modality. Unimodal motor association cortex projects predominantly to primary motor cortex, and it is important in formulating the motor program for complex actions involving multiple joints.

In contrast, heteromodal association cortex has bidirectional connections with both motor and sensory association cortex of all modalities. In addition, heteromodal association cortex has bidirectional connections with limbic cortex. This anatomical arrangement enables heteromodal association cortex to perform the highest-order mental functions. These functions apparently require integration of abstract sensory and motor information from unimodal association cortex, together with emotional and motivational influences provided by limbic cortex. Heteromodal association cortex is found in the frontal lobes and at the parieto-occipitotemporal junctions (see Figure 19.1).

In the sections that follow, we will survey some of the main functions of unimodal and heteromodal association cortex in different brain regions and study disorders that result from lesions in these areas.

REVIEW EXERCISE

Cover the labels in Figure 19.1. Name each region and state whether it is primary motor or sensory cortex, unimodal association cortex, heteromodal association cortex, or limbic cortex.

Principles of Cerebral Localization and Lateralization

As we have already mentioned, in the 1800s and early 1900s there was a controversy between those who described the brain as a network and those who described it as a collection of specialized areas. In reality, the brain is both. Localized regions of the brain do carry out specific functions, but they do so through network interactions with many other regions of the nervous system. **Focal brain lesions can cause specific deficits**, as we will see in the sections that follow. By understanding the main functions of the different cortical regions, deficits such as aphasia, unilateral neglect, impaired executive function, or inability to process visual information normally, can often be localized based on the neurological exam. However, because brain functions are mediated by networks involving not single, but multiple areas, **false localization** can sometimes occur. For example, so-called frontal lobe functions involve networks that encompass diverse regions including the frontal, parietal, and limbic cortices, the thalamus, basal ganglia, cerebellum, and brainstem. Therefore, lesions in these other structures, or in their white matter connections, can sometimes produce deficits that mimic frontal lobe lesions. **Disconnection syndromes** (see KCC 19.8) are another result of the network properties of brain function. For example, when a lesion in the white matter disconnects the network connections between visual cortex and the language processing areas, a patient may lose the ability to read.

Another important principle that can help localize deficits on clinical grounds is the tendency of some functions to be lateralized to the left or right hemisphere, resulting in **hemispheric specialization**. The human brain appears fairly symmetrical anatomically between the left and right hemispheres, and many basic sensory and motor functions in the brain are distributed symmetrically. Homologous regions of cerebral cortex on either side of the brain are connected to each other via long association fibers carried by the corpus callosum. For unknown reasons, however, there are marked asymmetries in several brain functions. It has been postulated that these asymmetries allow certain functions to be processed mainly within one hemisphere, eliminating delays caused by long callosal transmission times.

The most obvious asymmetry in cerebral function is **handedness**. Approximately 90% of the population is right-handed. The degree of asymmetry in manual dexterity varies, but most individuals are remarkably clumsy in performing tasks such as writing or closing buttons with the nondominant (usually left) hand. Functional neuroimaging and the results of lesions have suggested that although each hemisphere controls simple movements of the contralateral limbs, skilled complex motor tasks for both right and left limbs are programmed mainly by the dominant, usually left, hemisphere. Lesions of the dominant hemisphere therefore are more commonly associated with apraxia, a disorder of formulating skilled movements (KCC 19.7).

Language is another well-known example of hemispheric specialization. In most individuals, language function depends predominantly on the left hemisphere. **The left hemisphere is dominant for language in over 95% of right-handers, and in over 60 to 70% of left-handers.** Thus, lesions of the left hemisphere language areas usually cause language dysfunction, even in left-handed individuals. However, many left-handed individuals have significant bilateral representation of language, especially if there is a family history of left-handedness or ambidexterity. Thus, after a left hemisphere lesion, it is believed that left-handed individuals recover language more quickly than right-handers (although this has not yet been rigorously tested).

The **nondominant hemisphere is specialized for certain nonverbal functions** and appears to be generally more important for complex visual-spatial skills, for imparting emotional significance to events and language, and for music perception. Although the right and left hemispheres are each involved in attention to the contralateral environment, only the right hemisphere is significantly involved in attending to both sides. Lesions of the right hemisphere usually cause marked inattention to the contralateral (left) side, even in individuals who are right hemisphere dominant for language. Right hemisphere specialization for spatial attention may therefore be even more highly conserved than left hemisphere dominance for language.

To summarize the "flavor" of lesions of each hemisphere, dominant (usually left) hemisphere lesions cause impairments of language, detailed analytical abilities, and complex motor planning (praxis), while nondominant (usually right) hemisphere lesions cause impairments of spatial attention and complex visual-spatial abilities, especially those involving spatial orientation and perception of the overall gestalt, or big picture. As Table 19.3 shows, many important activities are carried out by combinations of different specialized skills of the left and right hemispheres. These functions of the dominant and

TABLE 19.3 Functions of the Dominant and Nondominant Hemispheres

DOMINANT (USUALLY LEFT) HEMISPHERE FUNCTIONS	NONDOMINANT (USUALLY RIGHT) HEMISPHERE FUNCTIONS
Language	Prosody (emotion conveyed by tone of voice)
Skilled motor formulation (praxis)	Visual-spatial analysis and spatial attention
Arithmetic: sequential and analytical calculating skills	Arithmetic: ability to estimate quantity and to correctly line up columns of numbers on the page
Musical ability: sequential and analytical skills in trained musicians	Musical ability: in untrained musicians, and for complex musical pieces in trained musicians
Sense of direction: following a set of written directions in sequence	Sense of direction: finding one's way by overall sense of spatial orientation

nondominant hemispheres are mediated by distributed networks involving frontoparietal connections, connections with limbic memory structures, and reciprocal connections with subcortical nuclei. Lesions that disconnect these networks, either within one hemisphere or between hemispheres at the corpus callosum, can cause specific disconnection syndromes (see KCC 19.8).

The anatomical basis for hemispheric specialization is only just beginning to be investigated. Although it has been known since the 1960s that the superior temporal plane (planum temporale) is larger in the left hemisphere in most individuals, the functional significance of this finding is still debated. Some preliminary evidence suggests that portions of the parietal lobe are larger in the right hemisphere. Functional and structural asymmetries in the brain are not unique to humans and have been demonstrated in other primates, and even in amphibians. Handedness and other lateralized aspects of cerebral function are not apparent in humans until they are 3 or 4 years of age, suggesting that developmental processes play a crucial role in hemispheric specialization. When lesions of the dominant hemisphere occur early in life, language and other functions often move to the nondominant hemisphere, resulting in a remarkable preservation of function.

In addition to left versus right, brain functions are also organized along the anterior to posterior axis. As in the spinal cord, where more posterior regions are sensory and more anterior regions are motor, the posterior parietal and temporal association cortex are more involved in interpreting perceptual data and assigning meaning to sensory information, while the anterior frontal association cortex is more important for planning, control, and execution of actions.

In the sections that follow, we will move through the brain from left to right and then from front to back. Thus, we will first discuss functions generally associated with the dominant hemisphere, followed by functions associated with the nondominant hemisphere. We will then turn to the frontal lobes and, finally, to visual association cortex.

The Dominant Hemisphere: Language Processing and Related Functions

In this section, we will describe the components of the language network and then discuss disorders of the dominant hemisphere affecting language and other typical functions of the dominant hemisphere (see Table 19.3).

Anatomy of Language Processing

A Historical Note:
The anatomical study of language disorders has had considerable historical importance over the past two centuries in the development of theories of cerebral localization in general. In the 1860s and 1870s, when Pierre Paul Broca and Karl Wernicke described language disorders caused by focal lesions in the brain, these descriptions were taken as important support for the notion that localized regions of the brain are specialized to perform specific cognitive functions. This concept fell into disfavor around the turn of the twentieth century, in the face of more holistic schemes of brain function proposed by Sigmund Freud and Pierre Marie, only to be revitalized eventually by Norman Geschwind and others in the 1960s and subsequent decades. Currently, the most widely accepted model is based on localized regions of cortex that have specialized functions but also participate in networks employing multiple brain regions to perform cognitive tasks.

As mentioned in the previous section, the left hemisphere is **dominant** for language in over 95% of right-handers and in over 60 to 70% of left-handers. Important areas for language processing and production are shown in Figure

FIGURE 19.2 Anatomy of Language Areas (A) Core language circuit composed of Broca's area, Wernicke's area, and the arcuate fasciculus. (B) Network of areas involved in language, including interactions with adjacent anterior and posterior association cortex, subcortical structures, and callosal connections to the contralateral hemisphere.

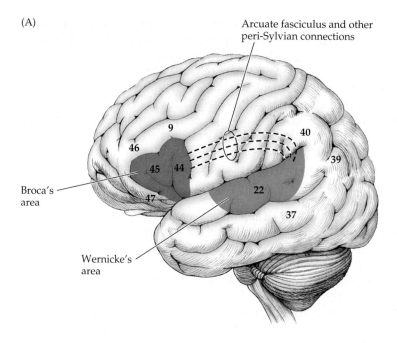

(A)

Arcuate fasciculus and other peri-Sylvian connections

Broca's area

Wernicke's area

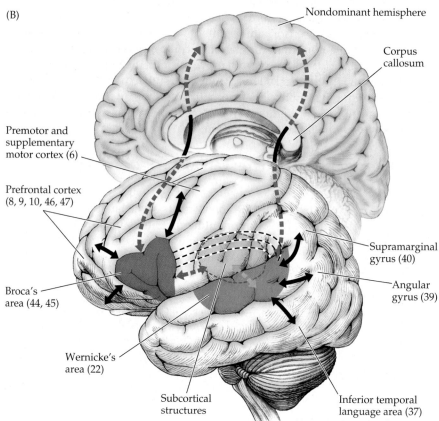

(B)

Nondominant hemisphere

Corpus callosum

Premotor and supplementary motor cortex (6)

Prefrontal cortex (8, 9, 10, 46, 47)

Broca's area (44, 45)

Wernicke's area (22)

Subcortical structures

Supramarginal gyrus (40)

Angular gyrus (39)

Inferior temporal language area (37)

19.2. Figure 19.2A shows the core structures associated with basic linguistic processes, such as hearing a word and then repeating it aloud. Auditory information reaches primary auditory cortex on the superior bank of the Sylvian fissure in the temporal lobe (see Figure 19.1A). The initial steps of language processing that enable particular sequences of sounds to be identified and comprehended as meaningful words are performed in the adjacent association cortex, which has been named **Wernicke's area** (see Figure 19.2A). Wernicke's area corresponds to Brodmann's area 22 (see Chapter 2), which encompasses the posterior two-thirds of the superior temporal gyrus in the dominant hemisphere. Many authors also include in Wernicke's area a rim of adjacent association cortex from Brodmann's areas 37, 39, and 40, because lesions extending to these areas produce Wernicke's aphasia (see KCC 19.5).

Articulation of the sounds that constitute speech depends on the face area of the primary motor cortex, located in the inferior portion of the precentral gyrus (see Figure 6.2). The motor program that activates particular sequences of sounds to produce words and sentences is formulated in the adjacent association cortex, which is called **Broca's area** (see Figure 19.2B). Broca's area corresponds to Brodmann's areas 44 and 45 (easy to remember because Broca's—44, is twice Wernicke's—22). Broca's area lies in the opercular and triangular portions of the inferior frontal gyrus (see Figure 2.11A) in the dominant hemisphere (see Figure 19.2A). Many authors include a rim of the adjacent areas 9, 46, and 47 in Broca's area, or more widely including 6, 8, and 10 (see Figure 2.15), because lesions extending to these areas produce Broca's aphasia (see KCC 19.4).

Continuing with our example, the ability to hear a word and then repeat it aloud requires transfer of information across the Sylvian fissure from Wernicke's area to Broca's area (see Figure 19.2A). Simplifying somewhat, neural representations for sounds are converted into words in Wernicke's area, and neural representations for words are converted back into sounds in Broca's area. Wernicke's and Broca's areas communicate with each other through several different connections, the best known of which is a subcortical white matter pathway called the **arcuate fasciculus** (see Figure 19.2A). In addition, numerous polysynaptic connections convey information from Wernicke's area along the intervening peri-Sylvian cortex to reach Broca's area.

Although Broca's area and Wernicke's area are critical way stations, they do not perform the operations involved in the production and comprehension of language alone. In fact, Broca's and Wernicke's areas have reciprocal connection with a large network of cortical areas engaged in language processing (see Figure 19.2B). Broca's area connects with other regions of the frontal lobes, including the prefrontal cortex, premotor cortex, and supplementary motor area. These areas function together with Broca's area, primarily in higher-order motor aspects of speech formulation and planning. In addition, the correct **syntax**, or grammatical structure, of both language production and comprehension appears to depend critically on these anterior (frontal) structures. Wernicke's area has reciprocal connections with the supramarginal gyrus and angular gyrus of the parietal lobe, as well as with regions of the temporal lobe such as Brodmann's area 37 (see Figure 19.2B). These posterior (temporoparietal) language areas function together with Wernicke's area primarily in language comprehension. In addition, the posterior areas appear to contain the **lexicon**, which is important in mapping sounds to meaning for both the comprehension and the production of meaningful language. The language areas of the dominant parietal lobe, especially the angular gyrus, are also important for written language. When one is **reading**, visual information first reaches primary visual cortex in the occipital lobes, is processed in visual association cortex, and then travels anteriorly via the angular gyrus to reach the language areas.

Connections through the corpus callosum (see Figure 19.2B) allow the **nondominant hemisphere** to participate in the language-processing network. The nondominant hemisphere appears to be important in both the recognition and the production of the affective elements of speech. Thus, patients with lesions of the nondominant hemisphere usually have no obvious language disturbance. However, they may have difficulty judging the intended expres-

MNEMONIC

REVIEW EXERCISE

For each of the following, state whether it is more closely associated with Broca's or Wernicke's area:

1. Superior temporal gyrus

2. Areas 44 and 45

3. Language formulation and planning

4. Inferior frontal gyrus

5. Area 22

6. Lexicon

7. Syntax

8. Language comprehension

Answers: (B = Broca's area, W = Wernicke's area) 1. W; 2. B; 3. B; 4. B; 5. W; 6. W; 7. B; 8. W.

TABLE 19.4 Disorders Commonly Mistaken for Aphasia

Disorders of speech production
    Dysarthria
    Aphemia (verbal apraxia)
    Mutism
Auditory disorders
    Peripheral hearing loss
    Pure word deafness
    Cortical deafness
Defects in arousal and attention
    Global confusional state
    Narcolepsy
Psychiatric disorders
    Schizophrenia
    Conversion disorder and other somatoform disorders
Uncooperative patient

sion imparted by a particular tone of voice, or they may have difficulty producing emotionally appropriate expression in their own voice. Perhaps more importantly, in lesions of the dominant hemisphere, callosal connections may allow the nondominant hemisphere to take over some functions of the damaged areas and to participate in at least partial recovery.

The language network also has important reciprocal connections with **subcortical structures** such as the thalamus and basal ganglia. Lesions of the thalamus, basal ganglia, or subcortical white matter in the dominant hemisphere can produce aphasia that can sometimes be mistaken for a cortical lesion.

KEY CLINICAL CONCEPT

19.2 DIFFERENTIAL DIAGNOSIS OF LANGUAGE DISORDERS

Aphasia, or dysphasia, is a defect in language processing caused by dysfunction of the dominant cerebral hemisphere. Because aphasia is a disorder of language and not a simple sensory or motor deficit, both spoken language and written language are affected. Aphasia is not caused by impaired audition or articulation, although deficits in these modalities may coexist with aphasia. The definition of aphasia will become more intuitively clear as we discuss specific examples in the sections that follow.

It is important to distinguish aphasia from several other disorders that affect language but are not specific disorders of language itself (Table 19.4). In motor disorders such as dysarthria (see KCC 12.8) and aphemia (verbal apraxia) (see KCC 19.7), speech may be difficult to comprehend; however, it has normal content and grammar, and written language is often normal. Mutism may result from severe aphasia or from motor disorders, but these can sometimes be distinguished by testing writing, which may be spared in speech motor disorders but is impaired in aphasia. Similarly, in peripheral (see KCC 12.5) or central (see KCC 19.7) auditory disorders, perception of spoken language is impaired, but reading and other aspects of language are normal.

Disorders of arousal and attention from a variety of causes (see Chapter 14), including toxic or metabolic disorders, post-ictal state, brainstem ischemia, and sleep disorders, are occasionally mistaken for aphasia because of the impaired comprehension and incoherent speech seen in these conditions. Finally, psychiatric disorders are sometimes confused with aphasia. In particular, schizophrenic patients may have very disordered, nonsensical, clanging speech, full of neologisms, which may resemble aphasia. It is important to recognize that the opposite situation can also be hazardous, in which a patient with aphasia is incorrectly given one of the diagnoses in Table 19.4. Through careful examination of the patient, as discussed in the following sections, it is usually possible to make these critical distinctions.

The most common cause of acute onset of aphasia is cerebral infarct, as we will discuss through multiple examples in this chapter. However, aphasia can also be caused by a wide variety of other disorders of the dominant hemisphere (Table 19.5). ■

TABLE 19.5 Causes of Aphasia[a]

Cerebral contusion; subdural or epidural hematoma

Ischemic or hemorrhagic vascular events

Ictal or post-ictal deficit with focal seizures in dominant hemisphere

Mass lesions such as brain tumor, abscess, or toxoplasmosis

Inflammatory or autoimmune disorders such as multiple sclerosis or vasculitis

Developmental disorders such as language delay or autism

Degenerative disorders such as progressive nonfluent aphasia, semantic dementia, moderately advanced Alzheimer's disease, and Huntington's disease

[a]Following the format of Figure 1.1.

19.3 BEDSIDE LANGUAGE EXAM

Examination of language is an integral part of the mental status exam (see KCC 19.1). The language exam is often very helpful in alerting the examiner to the presence of a focal brain lesion and in identifying its localization. In some patients, the presence of subtle language dysfunction goes undetected until it is formally tested for.

Benson and Geschwind popularized a six-step bedside examination for language that is still widely used today (Table 19.6). This language exam is demonstrated on **neuroexam.com Videos 8–12**. We will elaborate on some of the test items as we discuss specific language disorders in the sections that follow. In some situations, referral for more extensive formal neuropsychological testing and speech therapy evaluation is appropriate. Note that language testing should be done in the patient's primary language whenever possible. ■

19.4 BROCA'S APHASIA

Broca's aphasia is usually caused by lesions affecting Broca's area and adjacent structures in the dominant frontal lobe (see Figure 19.2). The most common etiology is infarct in the territory of the left middle cerebral artery (MCA) superior division (**Figure 19.3A**; see also KCC 10.1), although other lesions in this location can produce Broca's aphasia as well. Clinically, the most salient feature of Broca's aphasia is **decreased fluency** of spontaneous speech (see Table 19.6). The impaired fluency in Broca's (in contrast to Wernicke's) aphasia can be remembered by the mnemonic *Broca's broken boca* ("boca" means "mouth" in Spanish). **Fluency** can be surprisingly difficult to define and assess in an objective manner. Some helpful guidelines are that patients with decreased fluency tend to have a **phrase length** of fewer than five words, and the number of **content words** (e.g., nouns) exceeds the number of **function words** (e.g., prepositions, articles, and other syntactic modifiers). Word generation tasks, such as the FAS test (see KCC 19.11) can be useful for detecting subtle decreases in verbal fluency. In addition, **prosody** (the normal melodious intonation of speech that conveys the meaning of sentence structure) is lacking in patients with Broca's aphasia. The resulting speech in Broca's aphasia has an effortful, telegraphic quality, with a lack of grammatical structure and a monotonous sound.* Speech output is often better for certain overlearned, semiautomatic tasks, such as naming the days of the week or singing familiar songs like "Happy Birthday," and performance is often improved by cuing—that is, providing the first sound of a word during naming tests. Paraphasic errors (see KCC 19.5) occasionally occur, although these are less common than in Wernicke's aphasia.

The decreased fluency in Broca's aphasia is associated with marked **naming difficulties**. In addition, lesions in Broca's area cause a disconnection of

---

*In the extreme, patients in the early stages of severe Broca's aphasia may be virtually mute, and then they may gradually develop the more typical pattern of decreased fluency.

TABLE 19.6 Bedside Language Exam
1. Spontaneous speech
Fluency
Prosody
Grammar and meaning
Paraphasias
Articulation
2. Naming
Visual confrontation naming
Responsive naming
Objects and parts
Nouns, verbs, proper nouns, colors, etc.
3. Comprehension
Commands, simple to complex
Yes/no questions and multiple choice
Point to objects
Syntax-dependent meaning
4. Repetition
Single words
Simple sentences
Complex sentences
5. Reading
Aloud
Comprehension
6. Writing
Patient's name
Copy sentence
Spontaneous sentence

MNEMONIC

(A)

(B)

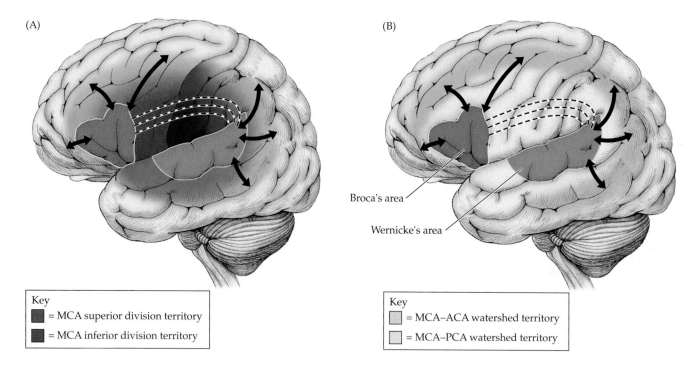

Broca's area

Wernicke's area

Key
■ = MCA superior division territory
■ = MCA inferior division territory

Key
□ = MCA–ACA watershed territory
□ = MCA–PCA watershed territory

FIGURE 19.3 Vascular Territories of Language Areas (A) Middle cerebral artery (MCA) superior and inferior division territories. (B) MCA–ACA and MCA–PCA watershed territories. (ACA, anterior cerebral artery; PCA, posterior cerebral artery.)

this structure from Wernicke's area (see Figure 19.3A). Therefore, in Broca's aphasia **repetition is impaired** as well. Patients tend to have the most difficulty repeating phrases with a high content of function words, such as "No ifs, ands, or buts" or "If I were here, she would be there." In contrast, because the posterior language structures are spared, **comprehension is relatively intact** in Broca's aphasia. The one notable exception is impaired comprehension of **syntactically dependent structures**. For example, when hearing a passive sentence such as "The lion was killed by the tiger," a patient with Broca's aphasia often incorrectly chooses the tiger as the animal that is dead. **Writing** and **reading** aloud in Broca's aphasia have a slow, effortful, agrammatical quality that is similar to the deficits in spoken language. Reading comprehension is often relatively spared, except for syntactically dependent structures.

Commonly **associated features** in Broca's aphasia include **dysarthria**, and **right hemiparesis** affecting the face and arm more than the leg, especially when left MCA superior division infarct is the cause. Visual fields are usually normal. Other common features are **frustration** and **depression**. **Apraxia** (see KCC 19.7) may also be present, often affecting the nonparetic left side of the body and oral–buccal–lingual structures.

A distinction is often made between so-called **little Broca's** aphasia and **big Broca's** aphasia. Big Broca's aphasia is caused by large lesions, such as MCA superior division infarcts (see Figure 19.3A), which often involve much of the dominant frontal lobe, as well as subcortical structures. In big Broca's there is initially a global aphasia (see KCC 19.6), which improves during recovery to settle into a Broca's aphasia. Little Broca's is caused by smaller lesions, confined to the region of the frontal operculum, including Broca's area. In little Broca's, there is initially a Broca's aphasia, which improves during recovery to only mildly decreased fluency and some naming difficulties. ■

KEY CLINICAL CONCEPT

19.5 WERNICKE'S APHASIA

Wernicke's aphasia is usually caused by a lesion of Wernicke's area and adjacent structures in the dominant temporoparietal lobes (see Figure 19.2). The most common etiology is infarct in the left MCA inferior division territory (see Figure 19.3A; see also KCC 10.1), although other lesions can also produce Wernicke's aphasia. Clinically, patients with Wernicke's aphasia have markedly **impaired comprehension** (see Table 19.6). Patients with severe Wernicke's aphasia do not respond appropriately to questions and follow virtually no commands. Interestingly, a few commands relating to axial muscles, especially "close your eyes" and sometimes "stick out your tongue," may elicit a correct response despite severe Wernicke's aphasia (and even in global aphasia, discussed in KCC 19.6). Spontaneous speech in Wenicke's aphasia has normal fluency, prosody, and grammatical structure. However, impaired lexical function results in speech that is **empty, meaningless, and full of nonsensical paraphasic errors. Paraphasic errors** can be either inappropriate substitutions of a word for one of similar meaning (verbal or semantic paraphasias), or of a part of a word for one with a similar sound (literal or phonemic paraphasias). For example, a semantic paraphasia is when a patient says "ink" instead of "pen" or "bus" instead of "taxi." A phonemic paraphasia is when a patient says "pish" instead of "fish" or "rot" instead of "rock." Occasional **neologisms**, or nonwords, are used. **Naming** is similarly impaired in Wernicke's aphasia, with frequent paraphasic errors or other irrelevant responses. Lesions of Wernicke's area also result in disconnection from Broca's area (see Figure 19.3A), causing **impaired repetition. Reading and writing** show similar impairments to the speech deficits in Wernicke's aphasia, and consist of fluent, but meaningless, paraphasic renditions.

Commonly **associated features** in Wernicke's aphasia include a **contralateral visual field cut**, especially of the right upper quadrant due to involvement of the optic radiation (see Figure 11.15E). **Apraxia** may be present (see KCC 19.7), but it can be difficult to demonstrate because of impaired comprehension. Dysarthria and right hemiparesis are usually absent or very mild. In addition, in marked contrast to Broca's aphasia, patients often appear **unaware of their deficit** (anosognosia), behaving as if carrying on a normal conversation despite their markedly abnormal speech. **Angry or paranoid behavior** may occur, causing Wernicke's aphasia occasionally to be misdiagnosed as a psychotic disorder (recall that patients with schizophrenia may also have abnormal speech). Examining a patient with severe Wernicke's aphasia can be difficult because the patient may respond to all questions or commands with nothing but incomprehensible paraphasic jargon. When examining the patient with Broca's aphasia, the patient often feels frustrated, while when examining the patient with Wernicke's aphasia, the examiner may feel frustrated.

Other names are sometimes used to refer to Broca's aphasia and Wernicke's aphasia, respectively, including expressive and receptive aphasias, motor and sensory aphasias, anterior and posterior aphasias, and nonfluent and fluent aphasias. However, these terms each have drawbacks. For example, Broca's aphasia is not simply an expressive deficit, since comprehension of syntactically dependent structures is impaired. Conversely, Wernicke's aphasia is not simply a receptive deficit, since speech expression is highly paraphasic and largely uninterpretable. Similarly, although Broca's and Wernicke's aphasias are usually caused by anterior and posterior lesions, respectively, this is not always the case. The simple syndromic names Broca's aphasia and Wernicke's aphasia are preferable. ■

MNEMONIC

KEY CLINICAL CONCEPT

19.6 SIMPLIFIED APHASIA CLASSIFICATION SCHEME

Broca's aphasia and Wernicke's aphasia are prototypical aphasia syndromes that, once understood, allow for easier classification of other aphasia syndromes. In this section we will present a simplified scheme for classifying aphasias that is useful because it is easy to apply and makes anatomical sense (Figure 19.4). Note, however, that aphasias do not always fit neatly into these categories. Left-handed patients, in particular, have more variable distribution of their language areas between the two hemispheres, and they may have aphasia syndromes that do not fit the classification presented here. In addition, many aphasia researchers consider this classification to be oversimplified. Nevertheless, this scheme is useful for most basic clinical purposes.

When examining patients with aphasia, recall that neurologic deficits are not all-or-none phenomena. In addition to deciding on the absence or presence of a deficit such as decreased fluency or impaired comprehension, it is important to assess the deficit's severity. This assessment can help both to clarify the diagnosis and to track the clinical progression of the disorder. For example, consider a patient with normal fluency who can comprehend and repeat simple phrases but who has difficulty comprehending and repeating more complex phrases and has occasional paraphasic errors. Despite the fact that some comprehension is present, this patient would be considered to have Wernicke's aphasia (see KCC 19.5), but in a relatively mild form.

The classification scheme is based on three parts of the language exam: fluency, comprehension, and repetition (see Figure 19.4). The other portions of the exam are important as well, to complete the clinical picture and to distinguish aphasias from other disorders. For example, naming difficulty and paraphasias do not appear explicitly in Figure 19.4 because these two disorders of language can occur in virtually any aphasia syndrome and have little localizing value. Nevertheless, the presence of naming difficulty or paraphasic errors may help suggest the presence of a language disturbance, as opposed to another disorder.

FIGURE 19.4 Classification of Language Disorders Aphasias can be classified based on fluency, comprehension, and repetition. All patients are assumed to have impaired naming and some paraphasic errors. The usual lesion locations for different forms of aphasia are indicated on the brain inset. Anomic aphasia can occur with lesions in many locations in the language network.

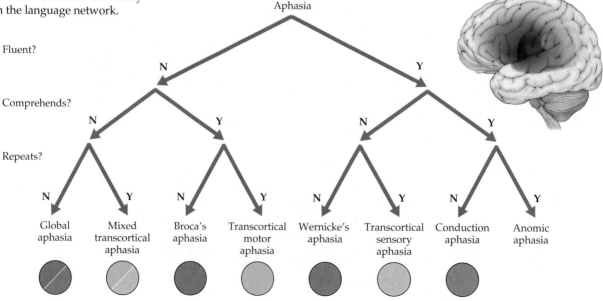

Let's begin tracing the different alternative pathways in the decision tree of Figure 19.4. A patient with impaired fluency, normal comprehension, and impaired repetition has Broca's aphasia (see KCC 19.4). This is often caused by a left MCA superior division infarct (see Figure 19.3A).

Normal fluency with impaired comprehension and impaired repetition is Wernicke's aphasia (see KCC 19.5). Wernicke's aphasia is often caused by a left MCA inferior division infarct (see Figure 19.3A).

A patient with impaired fluency, impaired comprehension, and impaired repetition has **global aphasia**. Global aphasia can be seen in large left MCA infarcts that include both the superior and the inferior divisions (see Figure 19.3A). It can also be seen in the initial stages of large left MCA superior division infarcts that eventually improve to become Broca's aphasia (big Broca's) and in large subcortical infarcts, hemorrhages, or other lesions.

A patient with normal fluency and normal comprehension, but impaired repetition, has **conduction aphasia**. This condition is caused by an infarct or other lesions in the peri-Sylvian area that interrupt the arcuate fasciculus or other pathways in the vicinity of the supramarginal gyrus that connect Wernicke's area to Broca's area (see Figure 19.2A). Speech is fluent, paraphasic errors are common, and naming is often impaired, which can lead to a misdiagnosis of Wernicke's aphasia. Unlike the case with Wernicke's aphasia, however, comprehension is spared.

Transcortical aphasias resemble Broca's, Wernicke's, and global aphasias, except that **repetition is spared** (see Figure 19.4). The classic cause of transcortical aphasia is watershed infarcts (Figure 19.3B; see also KCC 10.2), which spare Broca's area, Wernicke's area, and their interconnections but damage other language areas in the frontal or temporoparietal cortices (see Figure 19.2B). Transcortical aphasias are also common in subcortical lesions, such as those involving the basal ganglia or thalamus in the dominant hemisphere. In addition, transcortical aphasia is a common pattern seen during recovery from other aphasia syndromes.

A patient with impaired fluency and normal comprehension, as in Broca's aphasia, but with spared repetition, has **transcortical motor aphasia**. One possible cause is ACA–MCA watershed infarct (see Figure 19.3B). This lesion destroys connections to other regions of the frontal lobe that are needed for Broca's area to function in language formulation (see Figure 19.2B). However, peri-Sylvian connections from posterior to anterior language areas are left intact, enabling repetition.

A patient with normal fluency but impaired comprehension, as in Wernicke's aphasia, but with intact repetition, has **transcortical sensory aphasia**. MCA–PCA watershed infarcts (see Figure 19.3B) are one possible cause of this disorder. Connections to structures in the parietal lobe and temporal lobe that are needed for Wernicke's area to function are destroyed (see Figure 19.2B), while the peri-Sylvian area is left intact. The result is a condition that resembles Wernicke's aphasia, except that repetition is spared.

A patient with impaired fluency and impaired comprehension, as in global aphasia, but with intact repetition, has **mixed transcortical aphasia**, also called isolation of the language areas. One possible cause is combined MCA–ACA and MCA–PCA watershed infarcts (see Figure 19.3B), although this form of aphasia is often seen in subcortical lesions as well.

Finally, let's consider patients who show normal fluency, normal comprehension, and normal repetition but, like the other patients discussed here, have some naming difficulties and occasional paraphasias. This disorder is called **anomia** or **dysnomia** (see Figure 19.4). Naming difficulties can be severe or relatively mild. Careful testing for subtle dysnomia can be a sensitive indicator of language dysfunction because naming is often the first function

REVIEW EXERCISE

1. On a blank sheet of paper, draw the three-tiered classification tree shown in Figure 19.4 until you are able to reproduce it without referring to the figure.

2. Referring to Figure 19.4, draw a rough sketch showing a lateral view of the left hemisphere. For each of the disorders listed in Figure 19.4 (except anomia), shade in the usual area of infarction and explain how involvement of Broca's area, Wernicke's area, other related association cortex, or white matter connections are responsible for the observed deficits. For each disorder, list the vascular territory or watershed territory most commonly involved.

Naming

to be impaired and the last to recover in language disorders. A careful test of naming is thus an excellent **screening test** for aphasia. Patients with subtle dysnomia often have particular difficulty naming lower-frequency words or parts of objects (see **neuroexam.com Video 10**). For example, asking the patient to identify the parts of a watch (face, band, clasp) or a shirt (collar, pocket, sleeve, cuff) is a useful bedside test. Sometimes a similar but incorrect word will be used, such as "clock" for watch or "pencil" for pen (semantic paraphasias). Causes of anomic aphasia are numerous and include subcortical or cortical lesions in the dominant hemisphere and recovery from more severe forms of aphasia.

Recovery from Aphasia

Although the extent of recovery from aphasia can vary, the nature of recovery tends to follow certain common patterns. Global aphasia seen in big Broca's usually recovers to a Broca's aphasia. Broca's aphasia may recover to a transcortical motor aphasia and, eventually, to a subtle dysnomia. Similarly, Wernicke's aphasia may recover to a transcortical sensory aphasia and then to a dysnomia. Other patients have primarily naming and repetition difficulties following recovery, resembling a conduction aphasia. Dysnomia is the most common long-term deficit, although some patients have other more severe residual deficits as well. Subtle decreases in fluency can be tested for with word generation tasks, as discussed in KCC 19.11. ■

KEY CLINICAL CONCEPT

19.7 OTHER SYNDROMES RELATED TO APHASIA

Several important syndromes are related to the aphasic disorders of the dominant hemisphere. These disorders can occur either together with aphasia or in isolation.

Alexia and Agraphia

Alexia and **agraphia** are impairments in reading or writing ability, respectively, that are caused by deficits in central language processing and not by simple sensory or motor deficits. Alexia and agraphia can each occur in isolation, or they can occur together. In patients with aphasia, agraphia is invariably present. This co-occurrence may be due to the fact that normal writing requires intact functioning of the entire language apparatus. When alexia or agraphia occurs as part of an aphasic disorder, the reading and writing abnormalities tend to parallel those of the aphasic syndrome for spoken language. For example, reading aloud is nonfluent and agrammatical in patients with Broca's aphasia, but comprehension is relatively spared, except for syntactically dependent structures. Reading comprehension is impaired in Wernicke's aphasia, and reading aloud is fluent but full of paraphasic errors. Writing in Broca's aphasia is usually performed with the ipsilateral (usually left) nonparetic hand and is labored, agrammatical, and sparse. Writing in Wernicke's aphasia is paraphasic and largely incomprehensible. Lesions that cause aphasia are the most common cause of alexia and agraphia.

Alexia or agraphia can occasionally occur without significant aphasia. **Agraphia without aphasia** can be seen in lesions of the inferior parietal lobule of the language-dominant hemisphere. This may or may not be accompanied by the other features of Gerstmann's syndrome (see the next subsection). Writing requires focused attention and is therefore usually severely abnormal in patients with global confusional disorders (see KCC 19.15). In addition, agraphia in the hand ipsilateral to the language-dominant hemisphere can occasionally be seen in lesions of the corpus callosum because of disconnection of language (usually left hemisphere) from motor function (right hemisphere for left hand) (see KCC 19.8).

The classic syndrome of **alexia without agraphia** was first described by Dejerine in 1892, and it is caused by a lesion in the dominant occipital cortex extending to the posterior corpus callosum, often a PCA infarct (Figure 19.5). The lesion in the dominant (usually left) occipital cortex prevents the processing of visual information from the right hemifield, including written material. A right hemianopia is therefore usually present (see Figure 11.15). Meanwhile, information about the left hemifield that has reached the right occipital cortex is prevented from crossing to the language areas by the lesion in the posterior corpus callosum (see Figure 19.5), another example of a disconnection syndrome (see KCC 19.8).

Patients who have alexia without agraphia are characteristically able to write normally, but they cannot read even their own writing. Interestingly, they can name words that are spelled out loud to them. Other mild associated deficits are sometimes present, including some degree of anomia, especially **color anomia** (see KCC 19.12). However, significant aphasia is not usually present. It is not known why the lesion of alexia without agraphia impairs reading so much more than other visual naming tasks, but it is possible that the information for these other tasks crosses the callosum at a different point.

Alexia with agraphia can occur with lesions of the dominant inferior parietal lobule, in the region of the angular gyrus (see the next section). In these patients, aphasia is sometimes absent or may consist of only mild dysnomia and paraphasias. Alexia, which is an acquired deficit in reading, should be distinguished from **dyslexia**, a developmental reading disorder.

Before testing reading and writing abilities, it is important to obtain a history of the patient's educational level and previous literacy skills because these factors can greatly influence the results of assessment.

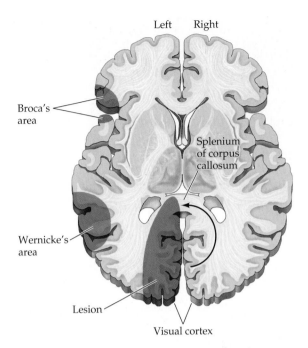

FIGURE 19.5 Schematic Circuit Diagram of Alexia without Agraphia

Gerstmann's Syndrome

Gerstmann's syndrome consists of the following somewhat odd tetrad of clinical findings: (1) **agraphia**, (2) **acalculia** (impaired arithmetic calculating abilities), (3) **right–left disorientation** (difficulty identifying the right versus left side of the body), and (4) **finger agnosia** (inability to name or identify individual fingers). **Agnosia** has been defined by Teuber as "a normal percept stripped of its meanings." Specific agnosias have been described for a variety of visual, auditory, and other percepts, as we will see in this chapter. Each of the single components of Gerstmann's syndrome, when found in isolation, has little specific localizing value and can be seen in a variety of brain disorders. However, when all four components are present in the absence of a global confusional state or other diffuse disorder, this syndrome is strongly localizing to the **dominant inferior parietal lobule**, in the region of the angular gyrus. Gerstmann's syndrome can occur as a relatively pure syndrome, but it is more often accompanied by other deficits localizing to the dominant inferior parietal lobule, such as a contralateral visual field cut (see Figure 11.15), alexia, anomia, or more severe aphasia.

Apraxia

Apraxia, or, more specifically, **ideomotor apraxia**, is the inability to carry out an action in response to verbal command, in the absence of any comprehension deficit, motor weakness, or incoordination. It is caused by an inability

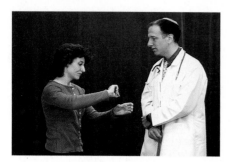

Praxis

REVIEW EXERCISE

Using Figure 19.5 as a guide, draw a rough sketch of the circuit involved in reading, and explain how a lesion of the left occipital cortex and posterior corpus callosum can cause alexia without agraphia.

to formulate the correct movement sequence. In testing for apraxia, the patient is usually asked to carry out imaginary actions, such as saluting the flag, brushing their hair, lighting a match and blowing it out, and so on (see **neuroexam.com Video 15**). Patients with apraxia make awkward-looking attempts and perform tasks ineffectually. In mild apraxia, patients may exhibit **body part substitution** (or "body part as tool")—for example, using their index finger like a toothbrush instead of holding an imaginary toothbrush between their fingers in the normal fashion. Intact comprehension of the command should be confirmed by multiple choice, with different actions demonstrated by the examiner. In addition, the patient should have intact motor skills that would allow performance of the task, which can be demonstrated if the patient spontaneously performs the same task or a similar task involving the same muscles at another time. These strict criteria are hard to meet, and sometimes a patient's inability to perform a task is best described as "probably" resulting from apraxia.

Apraxia is not a very well localized disorder, and it can be caused by lesions in many locations. However, there is an association between apraxia and aphasia: At least one-third of patients with aphasia have some apraxia, and apraxia of the oral and buccal muscles is particularly common in Broca's aphasia. Apraxia can affect orofacial, proximal, or distal limb movements differentially. Thus, some patients may have particular difficulty puckering their lips or sticking out their tongues on command, while others may have more difficulty with other body movements.

In addition to ideomotor apraxia (usually known simply as apraxia), the term "apraxia" has been applied to a variety of other, seemingly unrelated disorders. For discussion of conditions such as dressing apraxia, ocular apraxia, constructional apraxia, gait apraxia, ideational apraxia, and so on, consult the references at the end of this chapter.

Aphemia (Verbal Apraxia)

In **aphemia**, patients have severe apraxia of the speech articulatory apparatus, without a language disturbance. Aphemia is usually caused by a small lesion of the dominant frontal operculum restricted to Broca's area. In contrast to patients with Broca's aphasia, these patients have normal written language. Patients with aphemia have effortful, poorly articulated speech sometimes referred to as **foreign accent syndrome**. Severe aphemia can cause muteness, with preserved writing ability. Aphemia also occurs as a developmental disorder in children, often without a visible lesion on imaging studies. It is referred to as **verbal apraxia** in this context.

Cortical Deafness, Pure Word Deafness, and Nonverbal Auditory Agnosia

Patients with **cortical deafness** have bilateral lesions of the primary auditory cortex in Heschl's gyrus (see Figure 19.1). These patients are often aware that a sound has occurred but are unable to interpret verbal stimuli and cannot identify nonverbal stimuli such as a telephone ringing or a dog barking. In contrast, patients with **pure word deafness**, or verbal auditory agnosia, can identify nonverbal sounds but cannot understand any spoken words. Unlike Wernicke's aphasia patients, these patients can read and write normally. Although a few paraphasic errors may occur early on, speech is usually normal within a few days of onset.

The lesion in pure word deafness is usually an infarct in the auditory area of the dominant hemisphere that extends to the subcortical white matter, cutting off auditory input from the contralateral hemisphere as well. Like alexia without agraphia, there is a lesion in one hemisphere and disconnec-

REVIEW EXERCISE

Fill in the blank:

Pure word deafness is to Wernicke's aphasia as aphemia is to

_____.

(Answer: The syndrome described in KCC 19.4.)

tion with the other hemisphere. Patients with pure word deafness can usually speak normally but cannot understand speech, even their own speech if it is recorded and played back to them. Disconnections syndromes are discussed further in the next section (see KCC 19.8). Some patients with pure word deafness have also been reported with bilateral lesions of the superior temporal gyrus.

In **nonverbal auditory agnosia**, patients understand speech but cannot identify nonverbal sounds. The lesion in nonverbal auditory agnosia is usually located in the nondominant hemisphere. ■

KEY CLINICAL CONCEPT

19.8 DISCONNECTION SYNDROMES

In **disconnection syndromes**, cognitive dysfunction is caused by damage to pathways that connect one cortical area to another. We have already discussed several disconnection syndromes in this chapter, including conduction aphasia, impaired repetition in Broca's and Wernicke's aphasia (see Figures 19.3 and 19.4), alexia without agraphia (see Figure 19.5), and pure word deafness (see KCC 19.7). We also mentioned, when discussing Broca'a aphasia (see KCC 19.4), that left MCA superior division infarcts can cause apraxia of the left (nonweak) hand. Figure 19.6 demonstrates how a lesion in this location can disconnect the language areas and premotor cortex of the dominant left hemisphere from the right hemisphere premotor cortex, resulting in left hand apraxia.

Lesions of the corpus callosum can produce several additional classic disconnection syndromes. Naturally occurring lesions **involving primarily the corpus callosum** are relatively uncommon, but they can be seen in multiple sclerosis, gliomas, metastases, lymphoma, lipoma, and infarcts (ACA or PCA territory). **Corpus callosotomy** is occasionally performed in cases of medically refractory, poorly localized epilepsy, in which falls are a major problem (see KCC 18.2). The goal in these cases is to prevent secondary generalization—not to cure the seizures. After corpus callosotomy, the right hemisphere is unable to access language functions in the left hemisphere. Therefore, there may be agraphia of the left hand, inability to name objects placed in the left hand with the eyes closed, and inability to read in the left hemifield (usually detectable only with special testing apparatus). Other information also cannot be transferred between the hemispheres. Therefore, for example, if the patients are blindfolded and given an object to feel with one hand, they may then be unable to select that object from others using the contralateral hand. Some patients may have difficulty with tasks that require bimanual coordination, and in extreme cases the two hands may even work against each other. The classic example is a patient who buttons their shirt with one hand, while the other hand follows unbuttoning right behind. In most cases of callosotomy, however, what is surprising is not the deficits, but rather how functional these patients are in daily life despite the separation of the two hemispheres.

A Wada test (see KCC 18.2) can be helpful to predict some of the more severe deficits that occasionally occur following callosotomy. For example, the Wada test could show that a patient is left hemisphere dominant for language but has significant memory function only in the right hemisphere. In this case, callosotomy would disconnect language (left hemisphere) from memory functions (right hemisphere) and could produce a se-

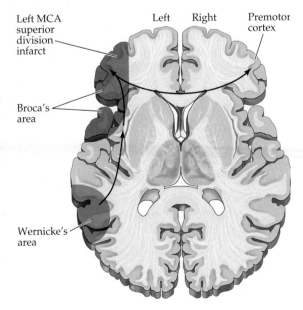

FIGURE 19.6 Disconnection Syndrome in which a Left Middle-Cerebral Artery (MCA) Superior Division Infarct Causes Left-Hand Apraxia

Left MCA superior division infarct

Broca's area

Wernicke's area

Left   Right   Premotor cortex

vere verbal memory deficit. Similarly, a patient who writes with the left hand could be found on Wada testing to be left hemisphere dominant for language. In this case, callosotomy would disconnect left hand motor control (right hemisphere) from language (left hemisphere) and could produce agraphia. ■

The Nondominant Hemisphere: Spatial Processing and Lateralized Attention

Exploring the functions of the **nondominant (usually right) hemisphere** brings us tantalizingly close to the fundamental mechanisms of consciousness. While the dominant hemisphere is specialized for language and step-by-step formulation and execution of motor tasks, the nondominant hemisphere is more important for **attention** and for generating an **integrated visual-spatial gestalt**. The more global mechanisms of attention (such as generalized vigilance, concentration, and behavioral arousal) and the **consciousness system** introduced in Chapter 14 will be discussed in greater detail later in this chapter. In this section we will touch on these concepts only briefly, emphasizing instead the more localized functions of the nondominant hemisphere in directed attention to the contralateral hemispace and in visual-spatial processing.

Lateralized Aspects of Attention

As we discussed in Chapter 14, the **consciousness system** includes brain networks that participate in alertness, attention, and awareness. Thus, attention (like alertness and awareness) depends on the medial and intralaminar thalamic nuclei (see Table 7.3); widespread projecting neuromodulatory systems in the upper brainstem, hypothalamus and basal forebrain (see Figures 14.7–14.13); the cingulate gyrus, medial and lateral fronto-parietal association cortex (see Figure 19.1); and possibly other structures such as the basal ganglia and cerebellum.

Although both hemispheres are involved in attention, there is a marked asymmetry in the relative importance of the two hemispheres, and **the right hemisphere is more important for attentional mechanisms in most individuals**. As we will discuss in KCC 19.9, lesions of the right hemisphere often lead to prominent and long-lasting deficits in attention to the contralateral side, while in left hemisphere lesions, contralateral neglect is relatively mild or undetectable. In studies of normal individuals using functional neuroimaging or electrophysiological investigations, the left hemisphere responds to stimuli on the right side, while the right hemisphere responds to both left- and right-sided stimuli but more strongly to stimuli on the left. (This is analogous to the involvement of the right premotor cortex in movement of the left hand and of the left premotor cortex in movement of both the right and left hands, discussed earlier.)

The hemispheric asymmetry of attentional mechanisms, and the effects of lesions, are shown schematically with "attention rays" in Figure 19.7. Under normal conditions (see Figure 19.7A), the right hemisphere attends strongly to the left side and less strongly to the right side, while the left hemisphere attends mainly to the right side. The result is a very slight net attentional bias toward the left in most individuals, which may explain why many languages are written from left to right. With right hemisphere lesions (see Figure 19.7B), the left hemisphere is still able to attend to the right side, but there is a profound deficit in attention to the left. In addition, there is a milder deficit in attention on the right side (ipsilateral to the lesion) as well. With left hemisphere lesions

(A) Normal

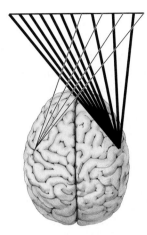

(B) Right hemisphere lesion (severe left neglect)

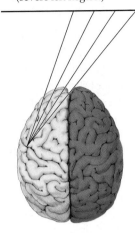

(C) Left hemisphere lesion (minimal right neglect)

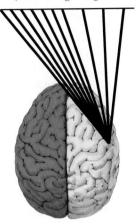

(D) Bilateral lesion (severe right neglect)

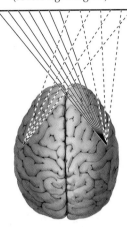

FIGURE 19.7 Hemispheric Asymmetry in Attention, Demonstrated through Attentional Rays

(see Figure 19.7C), the right hemisphere is still able to attend to the right side, so only mild right-sided deficits or no deficits in attention are seen. Finally, with bilateral partial lesions (see Figure 19.7D), there is some residual ability of the right hemisphere to attend to the left side only, resulting in a marked deficit in attention to the right. As we discussed in the section on cerebral lateralization earlier in this chapter, the reasons for left hemisphere specialization for language and right hemisphere specialization for attention and spatial analysis are not known.

Spatial Analysis and Integration

Spatial analysis depends on integration of information from multiple sensory modalities; however, since vision plays such an important role in human perception, the term "visual-spatial analysis" is often used. Like other mental functions, visual-spatial analysis is performed by a distributed network and depends on bilateral regions of the frontal and parietal association cortex. However, the **parietal association cortex** at the junction of the parietal, temporal, and occipital lobes is especially important for spatial analysis, and the **nondominant (usually right) hemisphere** is more important than the left.

As we discussed in Chapter 11, and will further discuss later in this chapter, visual information is analyzed by two streams of higher-order information processing: a *"What?"* stream in the *ventral* occipital, temporal, and prefrontal cortex, and a *"Where?"* stream in the *dorsal* occipital, parietal, and prefrontal cortex (see Figure 19.12). The parietal association cortex at the junction of the parietal, temporal, and occipital lobes (see Figure 19.1) lies directly in the dorsal stream, analyzing **location and movement of visual objects in space**. The posterior parietal cortex is also ideally situated to integrate other sources of spatial information from adjacent cortical areas. Spatial analysis thus encompasses both the surrounding environment and the relative position of the individual's body in space, using **visual**, **proprioceptive**, **vestibular**, **auditory**, and other information from adjacent cortical areas (see Figure 19.1). As we will discuss in KCC 19.10, disorders of spatial analysis such as impaired visual-spatial judgment or spatial constructional abilities are most commonly seen in lesions of the right parietal cortex, but they can occur with lesions on other areas as well.

REVIEW EXERCISE

What deficits in attention are expected with:

Right hemisphere lesions?

Left hemisphere lesions?

Bilateral partial lesions?

19.9 HEMINEGLECT SYNDROME

One of the most dramatic syndromes in clinical neurology is hemineglect syndrome, seen most often with infarcts or other acute lesions of the right parietal or right frontal lobes. Patients with this syndrome often exhibit profound neglect for the contralateral half of the external world, as well as for the contralateral half of their own bodies. Most strikingly, despite their profound deficits, these patients are often unaware that anything is wrong, and they sometimes even fail to recognize that the left sides of their bodies belong to them.

Contralateral hemineglect occurs most often with lesions of the **right parietal** or **frontal cortex** (Figure 19.8). Cases of contralateral neglect also occasionally occur with lesions of the cingulate gyrus, thalamus, basal ganglia, or midbrain reticular formation. As we have already discussed, neglect is usually much more pronounced and lasts longer with right hemisphere lesions, but milder forms of neglect can occur with left hemisphere lesions as well (see Figure 19.7).

Neglect is most severe in lesions of sudden onset such as infarct, hemorrhage, seizures, or head trauma, but it can also be seen in more slowly developing lesions, such as brain tumors or other space-occupying lesions. In large strokes, recovery from hemineglect can take weeks to months, and some patients remain with a permanent deficit in contralateral attention. During the recovery period, patients with hemineglect are more prone to injury and falls, and they may inadvertently bump or injure their contralateral side. Driving should be avoided until patients are able to demonstrate normal attention to both sides.

Testing for Hemineglect on Patient Examination

Clues to the presence of hemineglect include a history of the patient's bumping into objects on one side, ignoring food on one side of the plate, or being

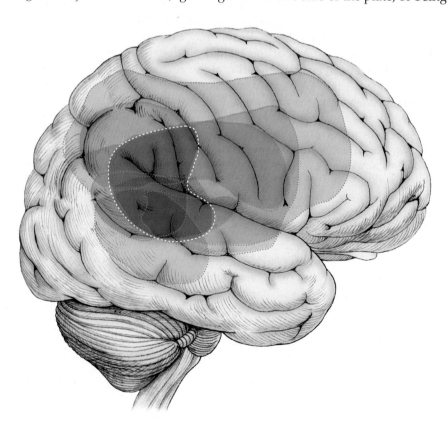

FIGURE 19.8 Lesions Causing Left Hemineglect Lateral view of right hemisphere with superimposed lesions, as determined by CT scan, for eight patients with left hemineglect. (After Heilman H, Valenstein E. 1985. *Clinical Neuropsychology.* 2nd ed. Oxford, New York.)

unaware of deficits. In addition, careful observation of the patient's behavior, movements, and grooming (some patients may comb or shave only their right side!) can be helpful. Other associated features of nondominant hemisphere lesions are discussed in KCC 19.10.

Four main types of testing can be performed on patient examination to evaluate different aspects of the hemineglect syndrome (Table 19.7). Tests usually evaluate **sensory neglect**, in which patients ignore visual, tactile, or auditory stimuli in the contralateral hemispace; **motor-intentional neglect**, in which patients perform fewer movements in the contralateral hemispace; **combined sensory and motor neglect**; and **conceptual neglect**, in which the patients' internal representations of their own bodies or of the external world exhibit contralateral hemineglect. Many tasks used to test for hemineglect depend on more than one component of the neglect syndrome. For example, drawing the face of a clock from memory depends on all of the above components. Although it has been postulated that more posterior or anterior lesions may cause more sensory or motor neglect, respectively, this has not been consistently demonstrated, and the sublocalization of different aspects of the hemineglect syndrome is still under active investigation.

In the subsections that follow, we will describe several useful tests for evaluating different aspects of the hemineglect syndrome.

Testing for Sensory Neglect

Hemineglect can be present in just one or in more than one sensory modality. **Tactile hemineglect** is most common, but **visual hemineglect** is fairly common as well, with **auditory hemineglect** detected less frequently. As described in Chapter 3, **visual**, **tactile**, or **auditory extinction on double simultaneous stimulation** are useful tests of sensory hemineglect (see **neuroexam.com Videos 27, 42, and 77**). For the results of extinction testing to be valid, it is important first to establish normal primary sensation by testing each side alone. Next, unilateral and bilateral presentations of stimuli should be randomly intermixed and the patient asked to report whether the stimulus was on the right, left, or both sides (eyes should be closed for tactile and auditory testing). With subtle neglect, extinction may be inconsistent. In addition, to bring out evidence of subtle tactile neglect, a proximal stimulus on the normal side (touching the left cheek) may produce extinction of a distal stimulus on the neglected side (touching the right hand), whereas when the sides are reversed, both stimuli may be reported (see **neuroexam.com Video 77**).

Patients with hemineglect commonly exhibit **allesthesia**, in which they erroneously report the location of a stimulus given to the left side of the body as being on the right. The extent of sensory neglect may vary depending on the relative position of the stimulus to the patient's eyes, head, or body. The frame of reference that is most important varies from patient to patient. For example, some patients may completely ignore both sides of a visual stimulus when it is placed to the left of their bodies, while ignoring only the left half of the stimulus when it is placed directly in front of them. Therefore, it is usually best to examine patients with their eyes, head, and body aligned straight ahead, and to present the stimuli symmetrically. This may be difficult with patients who have a marked unilateral gaze preference (see the next subsection).

Visual extinction

Tactile extinction

TABLE 19.7 Types of Neglect Tested on Patient Examination

1. Sensory neglect (visual, tactile, or auditory)

2. Motor-intentional neglect

3. Combination of sensory and motor neglect

4. Conceptual neglect

It is also important to distinguish neglect from a primary visual field defect (see KCC 11.2). For example, patients with a left homonymous hemianopia due to a right occipital lesion usually do not neglect stimuli on the left side if allowed to move their eyes, while patients with a right parietal lesion often neglect the left hemifield even when eye movements are not constrained.

Testing for Motor-Intentional Neglect

Patients should be observed for akinesia or decreased spontaneous movements of unilateral limbs or trunk, or decreased eye movements towards the neglected side. A marked ipsilateral gaze preference (toward the lesion) is common, especially in acute frontal or parietal lesions (see Figure 13.15A). Patients may exhibit motor impersistence (see KCC 19.11), especially of the contralateral limbs. They may have apparently decreased motor power on the neglected side, yet normal power may be demonstrated with increased effort, increased motivation, and active redirection of the patient's attention to the neglected side. The examiner can demonstrate **motor extinction** with the patient's eyes closed by randomly intermixing commands to raise the right arm, left arm, or both. **Allokinesia** may also be present, in which the patient inappropriately moves the normal limb when asked to move the neglected limb.

A useful test of hemineglect in patients who are encephalopathic (see KCC 19.14 and KCC 19.15) or have difficulty following commands is the **tactile response test**. Patients are instructed to raise whichever limb is touched, obviating the need for them to attend to and interpret the commands "right," "left," or "both." Once they understand the task, more subtle deficits can be detected during this test if they are asked to close their eyes. Note that the tactile response test is sensitive to both sensory and motor neglect (see the next subsection).

To test motor neglect in isolation, a variant of the tactile response test, called the **crossed response test**, can be used. In this test the patient is asked to move the limb *opposite* the one touched. Some patients may have difficulty understanding this task. Other tests of motor neglect, or of **directional motor bias**, include asking the patient to close their eyes and then point to a spot directly in front of their sternum, or asking them to collect coins from a table while blindfolded. Some patients exhibit **spatial akinesia**, in which limb movements are impaired when the limbs are located in the neglected hemispace. The examiner can demonstrate this deficit by asking patients to cross their arms during testing.

Combined Testing for Sensory and Motor Neglect

Many tests of neglect combine sensory and motor modalities, as well as conceptual or representational functions described in the next section. One simple example of combined sensory and motor neglect testing is the tactile response test that was described in the previous section. Other useful tests that combine sensory and motor function often involve the use of pen and paper. Several of these tests are demonstrated on **neuroexam.com Video 16**. In administering **pen-and-paper tests**, it is essential to ensure that the patient is centered at the workspace and that the test objects are centered in front of the patient and immobile if possible, to prevent patients from moving the entire stimulus into their non-neglected field. In addition, the stimuli should be large enough that they extend into both fields, again to prevent patients from easily capturing the entire stimulus within their normal hemifield.

In the **line bisection** task, the patient is instructed to cross out a horizontal line by making a mark right in the middle. The horizontal line should appear on an otherwise blank piece of paper without other cues and should be about 10 inches long. Normal individuals bisect the line right in the middle, or up to about 1 cm to the left of center, whereas patients with hemineglect often bisect the line far to the right of midline (**Figure 19.9A** and Figure

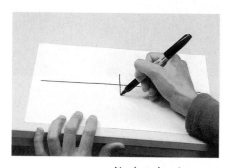

Neglect drawing tests

FIGURE 19.9 Cancellation Tasks in a Patient with Left Hemineglect Cancellation tasks from a 50-year-old right-handed man, 2 days after he sustained a right MCA inferior division infarct involving the right temporoparietal cortex. (A) Line bisection task, showing left neglect. (B) A more difficult line cancellation task, making deficit more obvious. (C) A still more difficult task involving letter cancellation among distractors, showing even worse performance.

19.10A). Other, more difficult **cancellation tasks** are useful in detecting and quantifying more subtle neglect. Such tasks include presenting patients with a page filled with numerous small lines for them to cross out, or harder still, a page full of mixed letters or other objects among which they must cancel only certain targets (e.g., the letter A, or all star shapes) while ignoring the distractors (Figure 19.9B,C). Patients with neglect tend to miss the targets

Copy drawing

mainly on the left side of the page, but may also show some evidence of decreased attention to the right side of the page.

Drawing is a useful test, and there are famous examples of artists who drew only half of objects or faces after developing hemineglect. A standard test is to ask the patient to draw a clock face, taking up as much of the page as possible, and to then fill in the numbers (Figure 19.10B). Patients can also be asked to draw other objects, such as a flower or a house, and to copy simple or complex figures (see **neuroexam.com Video 17**). In addition to testing for neglect, drawing tests **constructional abilities** (see KCC 19.10). These visual-spatial functions are often impaired in patients with lesions of the nondominant hemisphere, even when significant hemineglect is not present.

Reading a newspaper or magazine can be helpful in testing for hemineglect. Patients with hemineglect often read only the rightmost few letters or the rightmost few words of headlines. They may also ignore the left half of a picture when asked to **describe a visual scene**. In addition, while **writing**, patients with hemineglect tend to crowd their words onto one side of the page.

Conceptual Neglect

One of the more striking features of hemineglect syndrome is the common occurrence of **anosognosia**, meaning lack of awareness of the illness. Patients with hemiplegia, hemianopia, and hemisensory loss caused by right hemisphere lesions are often perplexed as to why they are in the hospital and may ask to be discharged. Anosognosia is not unique to right hemisphere lesions; it can be seen in other disorders as well. For example, anosognosia is also seen in patients with Wernicke's aphasia (see KCC 19.5), frontal lobe disorders (see KCC 19.11), amnesia with confabulation such as that seen in Wernicke–Korsakoff syndrome (see KCC 18.1), and cortical blindness (see KCC 19.12). In addition, patients with some psychiatric conditions such as schizophrenia or bipolar disorder may show profound lack of insight regarding their dysfunction.

Aside from anosognosia, some patients with right hemisphere lesions may exhibit **anosodiaphoria**, in which they are aware that they have severe deficits yet show no emotional concern or distress about it. An even more bizarre manifestation of hemineglect is **hemiasomatognosia**, in which patients deny that the left half of their body belongs to them. A patient may become distressed because "someone left an arm in my bed." When shown their left extremities, the patient may claim that they belong to someone else, or that they are not even limbs at all.

FIGURE 19.10 Pen-and-Paper Tasks from Another Patient with Left Hemineglect Test results from an 88-year-old right-handed woman, 1 day after sustaining a right MCA inferior division infarct, again involving the right temporoparietal cortex. (A) Line bisection task. On her first attempt, the patient missed the line entirely and made a mark in the far right margin of the page. When the examiner actively directed her attention to the line, she bisected it, but to the right of midline. Of note, a heavy line was used for this patient because on initial testing with a normal thickness line (as in Figure 19.9A) the patient ignored the line no matter how much her attention was directed to it. (B) Clock-drawing task, demonstrating left hemineglect. Note that this patient demonstrated some perseveration as well (see KCC 19.11) in the tests shown in both (A) and (B).

(A)

(B)

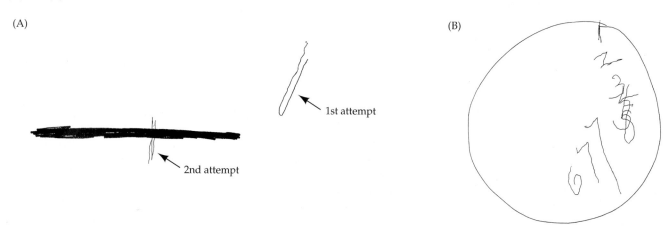

1st attempt

2nd attempt

Other aspects of what we have called conceptual neglect include abnormal internal representations or memories of imagined scenes or experiences. For example, Bisiach and Luzzatti described patients who were asked to recall various landmarks from a square in Milan while facing the cathedral. Patients with left hemineglect could recall details on only the right side of the square. When they were then asked to imagine themselves facing away from the cathedral, the patients were able to describe details on the opposite side of the square. ■

KEY CLINICAL CONCEPT

19.10 OTHER CLINICAL FEATURES OF NONDOMINANT HEMISPHERE LESIONS

Not all patients with nondominant (right) hemisphere lesions have significant hemineglect, and in some patients other deficits may be more troublesome. The most common difficulty encountered by patients with nondominant-hemisphere lesions is in tasks requiring **visual-spatial analysis**, or **constructional abilities**, such as drawing pictures or arranging blocks in specific geometric patterns. Patients may also have marked difficulty judging or matching the orientation of lines displayed at different angles. Deficits in these tasks are usually most severe with right parietal lesions, but they can also be seen with lesions in other locations in the right hemisphere, as well as with lesions of the left parietal lobe. Patients with nondominant-hemisphere lesions often have difficulty appreciating the **gestalt**, or overall spatial arrangement, in such tasks. These patients tend to make errors in the overall organization of the picture or block design, often rotating elements inappropriately in space. In contrast, patients with dominant-hemisphere lesions tend to understand the overall concept yet omit certain important details.

Patients with nondominant-hemisphere lesions often have relatively severe personality and emotional changes. They may appear bland or apathetic and, in addition to hemineglect, they often display an overall decrease in the level of alertness and attention, especially if they have acute lesions. On the other hand, irritability is common. Patients with acute right parietal infarcts sometimes exhibit bilateral ptosis, lying in bed with both eyes forcibly closed and becoming very nasty when one attempts to examine them. Some patients are overtly psychotic and have delusions and hallucinations. Patients with nondominant hemisphere lesions also often have difficulty comprehending the emotional content of others' speech (receptive aprosody) and conveying appropriate emotional expression in their own speech (expressive aprosody).

Patients with lesions of the right hippocampal formation, such as mesial temporal sclerosis (see KCC 18.2), may have deficits in visual-spatial memory. For example, these patients may be able to copy a picture, but they cannot draw it from memory a few minutes later. Impaired **geographic orientation**, or sense of direction, was thought to be associated with right temporal lesions, although recent studies have emphasized the importance of the right parietal or right occipitotemporal cortex. Many patients with right temporal seizures or lesions have an increased tendency to experience **déjà vu** and other mystical or religious phenomena. Right hemisphere lesions may also be associated with rare but intriguing disorders such as **Capgras syndrome**, in which patients insist that their friends or family members have all been replaced by identical-looking imposters; **Fregoli syndrome**, in which patients believe that different people are actually the same person who is in disguise; and **reduplicative paramnesia**, in which patients believe that a person, place, or object exists as two identical copies. ■

TABLE 19.8   Some Functions of the Frontal Lobes

RESTRAINT	INITIATIVE	ORDER
Judgment	Curiosity	Abstract reasoning
Foresight	Spontaneity	Working memory
Perseverance	Motivation	Perspective taking
Delaying gratification	Drive	Planning
Inhibiting socially inappropriate responses	Creativity	Insight
	Shifting cognitive set	Organization
Self-governance	Mental flexibility	Sequencing
Concentration	Personality	Temporal order

The Frontal Lobes: Anatomy and Functions of an Enigmatic Brain Region

More than any other part of the brain, the frontal lobes enable us to function as effective and socially appropriate human beings. It should perhaps be no surprise, therefore, that the frontal lobes are also among the most enigmatic, contradictory, and difficult to study brain regions. The importance of the frontal lobes has been debated over the years, with some earlier researchers believing that the frontal lobes are generally superfluous, and others feeling that they are the most important part of the brain. These different opinions arose because patients with frontal lobe lesions often have no deficits that can be detected on routine testing, yet they are completely unable to function normally in the "laboratory" of the real world. Additional complexity is evident in the wide variety of functions ascribed to the frontal lobes, some of which are listed in Table 19.8. Similarly, lesions of the frontal lobes produce highly variable behavioral syndromes, many of which seem contradictory even within a single patient (Table 19.9).

It should be clear from this discussion that there is no single frontal lobe syndrome. Frontal lobe function is probably best viewed as comprising several different realms, which we will present in this section, in somewhat simplified fashion, as clustering around the following three domains: restraint, initiative, and order (see Table 19.8). First, however, we will review the regional anatomy of the frontal lobes and some of the more important connections of frontal lobe circuitry.

Regional Anatomy of the Frontal Lobes

The frontal lobes are the largest region of the brain, comprising nearly one-third of the cerebral cortex. Recall that the frontal lobes are separated from the parietal lobes by the central sulcus, and from the temporal lobes by the Sylvian fissure (Figure 19.11A). The frontal lobes have three surfaces: lateral, medial, and orbitofrontal (see Figure 19.11 A, B, and C, respectively). Let's study these three surfaces and briefly review the regions of frontal cortex discussed in previous chapters. On the lateral surface, the primary motor cortex lies in the precentral gyrus (see Figure 19.11A). The primary motor cortex continues medially in the anterior portion of the paracentral lobule. Just in front of the primary motor cortex, on the lateral surface, lie the premotor cortex and, in the dominant hemisphere, Broca's area. In front of the primary motor cortex on the medial surface is the supplementary motor area (Figure 19.11B). The motor, premotor, and supplementary motor cortex were discussed in Chapters 6, 15, and 16. The frontal eye fields are located in the premotor cortex on the lateral convexity and extend anteriorly, as discussed in Chapter 13 (see Figure 13.14). The medial frontal lobes contain a micturition inhibitory area (see Figure 19.11B; KCC 7.5). In addition, as we discussed in Chapter 18, the anterior cingulate gyrus and posteromedial orbitofrontal cortex are important limbic areas that lie in the frontal lobes (see Figure 19.11B), and the ventral frontal lobes contain the orbitofrontal olfactory area (Figure 19.11C).

TABLE 19.9   Apparently Contradictory Behavior Seen in Frontal Lobe Syndromes

Apathetic indifference	vs.	Explosive emotional lability
Abulia	vs.	Environmental dependency
Akinesia	vs.	Distractibility
Perseveration	vs.	Impersistence
Mutism	vs.	Confabulation
Depression	vs.	Mania
Hyposexuality	vs.	Hypersexuality

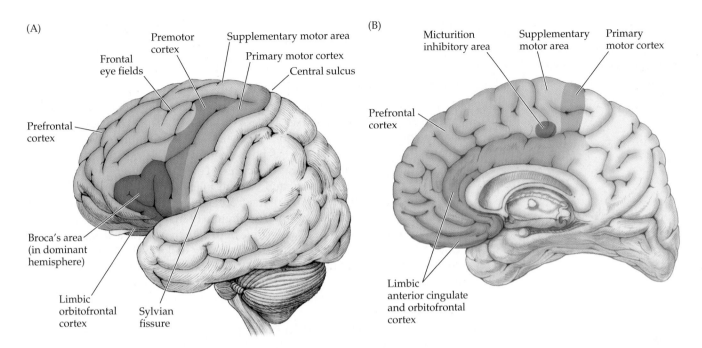

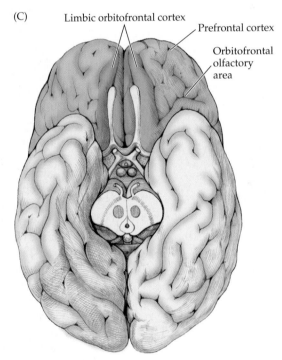

FIGURE 19.11 Main Functional Areas of Frontal Cortex Prefrontal cortex, motor cortex, and limbic cortex are indicated. (A) Lateral surface. (B) Medial surface. (C) Orbitofrontal surface.

The discussion in this section will focus on the frontal cortex lying anterior to the motor, premotor, and limbic areas, which is called the **prefrontal cortex** (see Figure 19.11). The prefrontal cortex is the largest part of the frontal lobes and consists of higher-order heteromodal association cortex. It is the prefrontal cortex that is usually being referred to when disorders of the frontal lobe are discussed.

Connections of the Prefrontal Cortex

The prefrontal cortex has numerous cortical and subcortical connections, which have been studied most extensively in nonhuman primates. Most of these connections are bidirectional. The anatomical configuration of frontal lobe connections is compatible with the role of this lobe in higher-order processes that require integration of multimodal sensory, motor, and limbic information. **Cortical connections** are mainly with the **association cortex** of the parietal, occipital, and temporal lobes, including unimodal sensory association cortex and heteromodal association cortex. In addition, there are connections with motor association cortex in the frontal lobes. Important connections of the prefrontal cortex with **limbic cortex** exist as well, especially with the anterior cingulate gyrus and posteromedial orbitofrontal cortex. **Subcortical connections** are numerous as well. The **amygdala** is connected with the orbital and medial portions of the frontal lobes by the uncinate fasciculus (see Figure 18.4B,C). The frontal lobes are connected to anteromedial temporal cortex by the uncinate fasciculus and to the **hippocampal formation** via the cingulate gyrus and parahippocampal gyrus (see Figure 18.9).

The most important **thalamic nucleus** that relays information to and receives projections back from the prefrontal cortex is the **mediodorsal nucleus**, although connections with the medial pulvinar and intralaminar nuclei are present as well (see Figure 7.8). The prefrontal cortex projects to the **basal ganglia** mainly via the head of the caudate nucleus (see Figure 16.8; Table 16.2). Other important subcortical connections also exist with the hypothalamus, septal region, subthalamic region, cerebellum, and midbrain. Finally, like all other cortical

MNEMONIC

areas, the frontal lobes receive projections from multiple subcortical and brainstem modulatory neurotransmitter systems, including dopamine, acetylcholine, serotonin, norepinephrine, histamine, and orexin.

Functions of the Frontal Lobes

As mentioned already, frontal lobe functions are quite diverse and apparently contradictory at times (see Tables 19.8 and 19.9). The frontal lobes are crucial for the sophisticated decisions we make, and for the subtle social interactions we continually engage in as normal humans. The vast range of human proclivities have been classified into three categories by philosophers and scientists from Plato to Freud, and we will follow this traditional simplification in describing the functions of the frontal lobes. These can be classified as functions important for (1) **restraint** (inhibition of inappropriate behaviors), (2) **initiative** (motivation to pursue positive or productive activities), and (3) **order** (the capacity to correctly perform sequencing tasks and a variety of other cognitive operations) (mnemonic RIO). The functions listed in Table 19.8 are arranged according to this scheme. Note, however, that this is only a simplification, and that some frontal lobe functions do not fit easily into any one of these categories.

You will get a better sense for frontal lobe functions when we review frontal lobe disorders in KCC 19.11. In this section we will discuss only briefly a few frontal lobe functions that have been studied extensively in the research setting. **Working memory** (see Table 18.6) is the ability to hold a limited amount of information in an immediately available store while a variety of cognitive operations are performed. An example is the carrying function in arithmetic. Both studies in animals and functional imaging in humans have shown the importance of the dorsolateral prefrontal cortex in working memory.

Recent functional imaging studies have also shown that the dorsolateral prefrontal cortex may function together with the medial temporal lobes in **learning new material**. In these studies, the left frontal lobe and medial temporal lobes showed activation during learning of new verbal information that was later successfully recalled. Similarly, the right frontal lobe and medial temporal lobes showed activation during learning of new nonverbal information that was later successfully recalled.

Activation of the dorsolateral frontal cortices has also been shown during tasks that require **shifting cognitive set**. An example is the Wisconsin Card Sorting Test, in which subjects must have the mental flexibility to infer that the rules for the sorting task they are performing change repeatedly. Interestingly, the frontal lobes are also activated during tasks that require **selective attention**—for example, listening to words while visual or tactile stimuli are presented simultaneously. Another important area of research has been the role of the frontal lobes in **integrating information from limbic and heteromodal association cortex** in decision making. This emotional weighting of abstract decision making is thought to enable subtle emotional and motivational factors to participate in human judgment so that more efficient, or "intuitive," decisions can be made when limited information and time are available.

KEY CLINICAL CONCEPT

19.11 FRONTAL LOBE DISORDERS

As shown in Table 19.9, the effects of frontal lobe lesions are often perplexing, with apparently contradictory features seen in different patients, or even within a single patient. Several explanations have been offered for this observation. First, the frontal lobes are large, encompassing many different functional areas. Frontal lobe deficits are often subtle at first, and lesions may reach a large size and involve several different functional areas before becoming clinically apparent. Similarly, bilateral lesions usually produce more clinically obvious deficits than unilateral lesions, and behaviorally complex disorders result from bilateral lesions. Finally, the functions of the frontal lobes are themselves complex and therefore difficult to study, especially in the formal examination setting.

On the basis of studies of animals and humans with frontal lobe lesions, a distinction is sometimes made between lesions of the dorsolateral convexity and lesions of the ventromedial orbitofrontal cortex. According to this scheme, **dorsolateral convexity lesions** tend to produce an apathetic, lifeless, **abulic** (opposite of ebullient) state, while ventromedial **orbitofrontal lesions** lead to impulsive, disinhibited behavior and poor judgment. In clinical practice, numerous exceptions to this dichotomy exist. In addition, many frontal lesions affect both dorsolateral and orbitofrontal regions, making the usefulness of this classification somewhat limited. Another more tenuous distinction has been made between **left frontal lesions**, which may be more associated with depression-like symptoms, and **right frontal lesions**, which are more associated with behavioral disturbances resembling mania, however numerous exceptions exist.

Despite these contradictions and uncertainties, certain characteristic features revealed during evaluation of a patient can suggest frontal lobe dysfunction. Familiarity with these features is important for recognizing patients with probable frontal lobe dysfunction, so we will now discuss them in detail.

Evaluating Patients with Suspected Frontal Lobe Dysfunction

The basic steps of clinical assessment for patients with suspected frontal lobe dysfunction are summarized in Table 19.10. The most important information in evaluating patients with suspected frontal dysfunction is often not obtained during the formal neurologic exam. Critical evidence of frontal lobe dysfunction may be obtained from the patient's **history** and through discussions with family and other contacts who have witnessed the patient's abnormal functioning in the real world. In addition, careful observations should be made for certain **behavioral abnormalities** that may be seen in patients with frontal lobe dysfunction (see Table 19.10).

Patients with **abulia** are passive and apathetic, exhibiting little spontaneous activity, markedly delayed responses, and a tendency to speak briefly or softly. In the extreme, abulic patients may be totally immobile, akinetic, and mute (see KCC 14.2) but will continue to appear awake, sitting with their eyes open. In contrast, **disinhibition** may also be seen, including silly behavior, crass jokes, and aggressive outbursts. Some patients exhibit **inappropriate jocularity**, or **witzelsucht**, seeming unconcerned about potentially serious matters. Patients may have **limited insight** into their condition and may **confabulate**. Patients who display **utilization behavior** or **environmental dependency** tend to respond to whatever stimuli are at hand, even when not appropriate. For example, they may put on glasses that are not theirs. In a remarkable combination of abulia and utilization behavior, some patients described by C.M. Fisher would sit profoundly unresponsive and indifferent until the phone rang, upon which they would pick up the phone and speak normally for a brief time (the "telephone effect"). Another example of extreme utilization behavior is the "next bed over syndrome," which occurs when the patient in the next bed shouts out answers from behind the curtain in response to every question that you ask your patient and will not stop even when

TABLE 19.10 Evaluating Frontal Lobe Function

I. History and behavioral observations

    A. Best test is real world; history from family or other contacts may be more revealing than patient exam

    B. Behavioral observations, looking especially for:

        1. Abulia

        2. Inappropriate jocularity (witzelsucht)

        3. Other abnormalities of comportment or insight

        4. Confabulation

        5. Utilization behavior and environmental dependency

        6. Perseveration, impersistence, and spontaneous frontal release signs

        7. Incontinence

II. Mental status exam

    A. Attention

        1. Digit span forward and backward

        2. Months forward and backward

    B. Memory

    C. Perseveration and set-shifting ability

        1. Bedside: Luria alternating sequencing tasks (written, manual)

        2. Formal testing: Trails B, Wisconsin Card Sorting Test

    D. Ability to suppress inappropriate response

        1. Bedside: auditory or visual go-no-go tasks

        2. Formal testing: Stroop test

    E. Word generation, figure generation

        1. FAS test or other word generation tasks

        2. Figure generation

    F. Abstract reasoning

        1. Similarities

        2. Proverb interpretation

        3. Logic problems

    G. Judgment; influence of future consequences on current behavior

        1. Difficult to test; questions about situations (e.g., fire in theater) are artificial and test mainly general knowledge and reasoning ability

        2. Gambling task[a]

    H. Language testing

    I. Testing for hemineglect

III. Other exam findings

    A. Skull shape (hyperostosis may signify a frontal meningioma)

    B. Olfaction (anosmia may signify an orbitofrontal tumor)

    C. Optokinetic nystagmus testing (impaired saccades occur away from side of lesion)

    D. Hemiparesis or upper motor neuron signs

    E. Motor impersistence (stick out tongue or hold up arms for 20 seconds)

    F. Gegenhalten (paratonia)

    G. Primitive reflexes, or "frontal release signs" (grasp, suck, snout, root)

    H. Frontal "magnetic" gait disturbance

[a]See Bechara A, Damasio AR, Damasio H, Anderson SW. 1994. Insensitivity to future consequences following damage to human prefrontal cortex. *Cognition* 50: 7–15.

kindly reminded that they are not being addressed. **Perseveration, impersistence**, and **frontal release signs** are tested for during the neurologic exam, but they may also be observed spontaneously, especially in more severely affected patients. Patients with severe perseveration may repeatedly answer the same question, even when the examiner is trying to move on. **Incontinence** is sometimes seen in frontal lobe disorders, especially those affecting the medial frontal regions. Patients are characteristically unconcerned about their incontinence.

During the **mental status** portion of the neurologic exam, important clues can be obtained about possible frontal lobe dysfunction. Note that many of these tests do not test frontal lobe function alone but depend on intact functioning of numerous other systems, which should be assessed during the remainder of the exam. The tests listed here were selected for their clinical usefulness in detecting subtle frontal lobe abnormalities.

In patients with frontal lobe dysfunction, **attention** may be impaired; such impairment can be evaluated with digit span and other simple tests (see KCC 19.14; Table 19.10). Deficits on **memory testing** (see KCC 18.1) can be seen due to impaired attention and problems with memory retrieval caused by frontal lobe lesions. Subtle **perseveration** may be detected by use of one of the **Luria sequencing tasks** (see **neuroexam.com Videos 19 and 20**). For example, when a patient was asked to copy the sequence shown in Figure 3.1 and continue it to the end of the page, the patient clearly perseverated. Figure 3.1 also demonstrates the **closing-in** phenomenon with this test, in which some patients' drawing gradually approaches that of the examiner, possibly exhibiting a form of environmental dependency. Another useful test of perseveration is the Luria manual sequencing task, in which the patient is asked to tap their thigh with a closed fist, then with the palm of their hand, and then with the side of their open hand, repeatedly. A verbal form of the "Trails B" test can be administered at bedside by asking the patient to continue the sequence A, 1, B, 2, C, 3 and so on. Some formal neuropsychological tests of perseveration and set-shifting ability are also listed in Table 19.10.

A useful bedside test of a patient's inability to suppress inappropriate responses is the **go–no-go** task (see **neuroexam.com Video 21**). This task is similar to (but easier than) the children's game "Simon Says." In the auditory go-no-go task, the patient is instructed to raise a finger in response to one tap and to keep the finger still in response to two taps. The examiner then produces a random sequence of one or two tapping sounds. In the visual form of the task, the examiner randomly shows one or two fingers. More formal testing can be accomplished with the Stroop test. In the **Stroop test**, patients are given a list of color names such as "red, yellow, green" and so on, which are printed in colored ink, but the colors do not match the meanings of the words (e.g., the word "yellow" might be printed in green ink). Patients are then instructed to list the ink colors of the words without reading the words.

The ability to spontaneously **generate lists** of related words or drawings is often impaired in patients with frontal lobe lesions. **Word generation tasks** can be used to detect subtle decreases in verbal fluency and are a sensitive measure of dominant frontal dysfunction. A useful standardized task of this kind is the FAS test, in which the patient is given 60 seconds to produce as many words as possible starting with the letter F, then 60 seconds to produce words starting with the letter A, and so on. Proper nouns, such as names of persons or places, are not allowed. Normal individuals produce 12 or more words for each letter. Similarly, normal individuals can name at least 15 animals in 60 seconds. For detecting nondominant frontal lobe dysfunction, similar tasks based on the generation of simple drawings can be used (referred to as tests of figural fluency).

Frontal dysfunction often interferes with **abstract reasoning** ability. Two common tools for testing this are **proverbs** and **similarities**. (see **neuroexam.com Video 22**). Responses are rated as either normal or concrete. It is important to use proverbs and similarities ranging from easy to difficult (Table 19.11) to gauge the

Manual alternating sequencing test

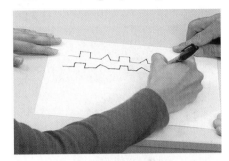

Written alternating sequencing test

Auditory go–no-go

TABLE 19.11 Examples of Proverbs and Similarities Used for Testing Abstraction		
Proverbs: "What does it mean if I say . . ."		
Don't cry over spilled milk.		Easiest
People who live in glass houses should not throw stones.		
The tongue is the enemy of the neck.		
One swallow does not a summer make.		
You never cross the same stream twice.		Most difficult
Similarities: "How are a _____ and a _____ alike?"		
Hammer	Screwdriver	Easiest
Orange	Apple	
Table	Couch	
Train	Airplane	
Poem	Sculpture	
Moth	Tree	Most difficult

patient's level of functioning. A variety of logic problems can also be used to test abstract reasoning. **Judgment** is very difficult to test in the artificial setting of the neurologic exam and is best evaluated on the basis of reports of the patient's performance in the real world. Questions can be asked, such as, "What would you do if you saw a fire in a crowded theatre?" or "What would you do if you found a stamped and addressed letter on the street?" However, patients usually respond to this kind of question on the basis of their general knowledge and reasoning ability, and they may behave quite differently in real-life situations. Bechara, Damasio, and colleagues devised a task resembling gambling that may be useful for detecting impaired judgment or overemphasis on short-term versus long-term consequences in frontal lobe patients (see Table 19.10).

Two other parts of the mental status exam may provide clues about frontal dysfunction. In language testing, a nonfluent aphasia suggests possible left frontal lobe dysfunction (see KCC 19.4). In addition, although left-sided neglect is more common with right parietal lesions, it can occasionally be seen with right frontal lesions as well (see KCC 19.9). Left-sided neglect, together with other signs of frontal lobe dysfunction, suggests a lesion in the right frontal lobe.

Finally, certain findings on other parts of the general and neurologic exam can be helpful for identifying frontal lobe lesions (see Table 19.10). On general exam, **skull shape** should be noted. For example, meningiomas can occasionally cause hyperostosis of the overlying skull, resulting in a large palpable bump on the head. **Olfaction** should be checked (see **neuroexam.com Video 24**) because tumors in the orbitofrontal area often cause anosmia (see Case 12.1). **Saccades** should be carefully checked because involvement of the frontal eye field can cause impaired saccades away from the lesion (see Figure 13.14). **Optokinetic nystagmus (OKN) testing** is a sensitive way of evaluating saccades (see Chapter 13 and **neuroexam.com Videos 33 and 34**). Subtle asymmetries in OKN, with a decreased fast phase in one direction, may indicate a lesion in the contralateral frontal lobe.

Motor findings, such as hemiparesis or upper motor neuron signs, may be present, with lesions affecting the primary motor cortex of the precentral gyrus or underlying white matter pathways in the frontal lobe. In contrast, lesions of the prefrontal cortex are often associated with other, more elaborate motor findings, reflex abnormalities, or gait abnormalities. The examiner can test for **motor impersistence** by having the patient sustain an action such as holding out both arms or sticking out the tongue for 20 seconds. Impersistence can also be seen in right parietal lesions, Huntington's disease, and other conditions that cause inattention (see KCC 19.14). Patients with frontal lobe lesions may have **paratonia**, or **gegenhalten**, in which tone is increased, but in a manner in which the patient appears to resist the movements of the examiner in an almost willful fashion (see also KCC 16.1).

In addition, so-called **frontal release signs**, or **primitive reflexes**, normally seen in infants, may return in adults with frontal lesions. The most useful of these is the **grasp reflex**, which the examiner can elicit by stroking the patient's palm (see **neuroexam.com Video 18**). When the grasp reflex is severe, the patient cannot release their grasp even when explicitly told not to grab the examiner's hand. A more subtle grasp reflex may be elicitable only if the examiner distracts the patient—for example, by engaging the patient in conversation. If asked why they are grabbing the examiner's hand, patients usually respond that they do not know.

Other, less specific reflexes are the **suck**, elicited by touching the patient's lips with a cotton swab; **snout**, elicited by tapping the patient's lips; and **root**, elicited by stroking the patient's cheek or holding an object near the patient's mouth. The palmomental reflex, in which stroking the thenar eminence causes the ipsilateral chin muscles to contract, is not generally useful because it is seen

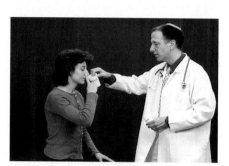

Olfaction

OKN

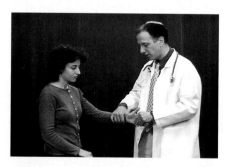

Grasp reflex

in many normal individuals. Myerson's glabellar sign is associated with movement disorders such as Parkinson's disease (see KCC 16.2). Frontal lobe lesions can cause characteristic **frontal gait abnormalities** (see KCC 6.5; Table 6.6), with a shuffling, unsteady, **magnetic gait**, in which the patient's feet barely leave the floor.

Differential Diagnosis of Frontal Lobe Disorders

Disorders that commonly affect the frontal lobes are listed in part I of Table 19.12. Often, patients with clinical syndromes that closely resemble frontal lobe lesions turn out to have disorders that affect the brain diffusely (see Table 19.12, part II). One reason for this may be that the frontal lobes make up a large proportion of the brain, and therefore multifocal disorders have a greater chance of affecting the frontal lobes before other brain regions. In the case of hydrocephalus, it has been proposed that the dilated ventricles compress frontal subcortical white matter pathways, or possibly the anterior cerebral arteries.

Other disorders, some of which have lesions located remotely from the frontal lobes, may also mimic frontal lobe pathology (see Table 19.12, part III). This is likely because, as already discussed, the frontal lobes form part of a network of widespread cortical and subcortical connections. In addition, abnormalities of certain neurotransmitter systems, especially dopamine, can lead to frontal lobe dysfunction. These numerous other conditions listed in Table 19.12, which can mimic frontal lobe disorders, are discussed in more detail elsewhere (see KCC 5.7, 6.5, 6.6, 14.2, 15.2, 16.2, 18.3, 19.15 and 19.16). ■

TABLE 19.12 Differential Diagnosis of Frontal Lobe Disorders

I. Disorders commonly affecting the frontal lobes
  Head trauma
  ACA or MCA infarcts
  Hemorrhage, from hypertension, tumor, or AComm aneurysm
  Gliomas, metastases, meningiomas, or other tumors
  Brain abscess, toxoplasmosis, or herpes simplex encephalitis (limbic cortex)
  Frontotemporal lobar degeneration (FTLD)
  Developmental abnormalities
  Frontal lobe seizures
II. Diffuse processes causing frontal-like syndromes
  Toxic or metabolic disorders
  Hydrocephalus
  Binswanger's encephalopathy
  Diffuse anoxic injury
  Demyelination and other subcortical degenerative disorders
  Advanced Alzheimer's disease
III. Other disorders causing frontal-like syndromes
  Schizophrenia (negative symptomatology)
  Depression
  Parkinson's disease
  Huntington's disease
  Cerebellar lesions
  Lesions of brainstem ascending reticular activating system or thalamus

Visual Association Cortex: Higher-Order Visual Processing

As we discussed in Chapter 11, after arrival at the primary visual cortex, visual information is processed in two streams of association cortex (Figure 19.12). The **dorsal pathways** project to parieto-occipital association cortex. These pathways answer the question **"Where?"** by analyzing motion and spatial relationships between objects, and between the body and visual stimuli. The **ventral pathways** project to occipitotemporal association cortex. These pathways answer the question **"What?"** by analyzing form, with specific regions identifying colors, faces, letters, and other visual stimuli (Figure 19.13). In reality, this processing is very complex and involves the cooperation of multiple specialized cortical modules (see Figure 11.11A). The functions of the dorsal and ventral higher-order visual

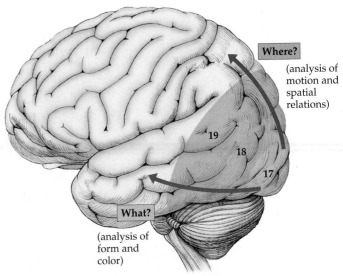

FIGURE 19.12 "What?" and "Where?" Streams of Visual Processing

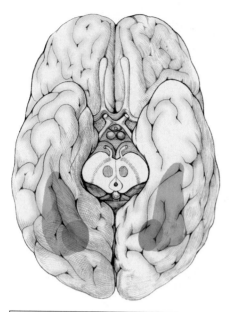

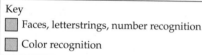

Key
▨ Faces, letterstrings, number recognition
▨ Color recognition

FIGURE 19.13 Color, Face, Letter, and Number Recognition Areas of the Inferior Occipitotemporal Cortex Regions of the inferior occipitotemporal cortex (fusiform gyrus) activated on human functional imaging and evoked potential studies by colors, faces, letterstrings, and numbers. (After Allison T, McCarthy G, Nobre A, Puce A, and Belger A. 1994. Human extrastriate visual cortex and the perception of faces, words, numbers, and colors. *Cerebral Cortex* 4 (5): 544–554.)

pathways in humans are well illustrated by the clinical syndromes that will be discussed in KCC 19.12. Here we will briefly mention a few areas of active research in higher-order visual processing.

One important area of research is the question of how information from different cortical areas, both visual and nonvisual, is combined to form the unified perceptions that we experience. For example, our experience of someone speaking to us is not fragmented into a perception of sounds, verbal comprehension, facial recognition, facial color, location of the face in space, and emotional impact of the words spoken, even though these functions each occur in a different cortical area. The question of how unified percepts are formed in the brain is called the **binding problem**. Much research and hypothesis development is taking place to investigate the binding problem, especially in the visual system; however, no conclusive mechanism has yet been found.

A second area of research is **visual mental imagery**—that is, our ability to imagine a scene even when it is not present. This process appears to utilize the same visual areas that perceive exogenous visual stimuli; however, there is some controversy over the extent to which primary visual cortex is involved.

A third research area is the phenomenon of **blindsight**, in which some individuals with lesions of the primary visual cortex are able to perform tasks such as inserting an envelope correctly in a slot, despite having no conscious visual perception of the slot. Blindsight apparently depends on information transmitted to association cortex by extrageniculate visual pathways (see Figure 11.6), bypassing the lateral geniculate nucleus and primary visual cortex. Some studies have also demonstrated small islands of preserved vision in the blind hemifield measuring only a few minutes of degree that are not able to support conscious vision but may influence behavior.

KEY CLINICAL CONCEPT

19.12 DISORDERS OF HIGHER-ORDER VISUAL PROCESSING

Visual hallucinations, illusions, and other visual phenomena were introduced in Chapter 11; review this material before reading further (see KCC 11.1). In this section we will build on this background and describe several syndromes of abnormal visual processing associated with specific localization to the primary visual cortex, inferior occipitotemporal cortex ("What?" pathways), or dorsolateral parieto-occipital cortex ("Where?" pathways).

Syndromes of Primary Visual Cortex

Visual hallucinations, seizures, and migraine-related phenomena associated with the primary visual cortex were discussed in KCC 11.1, and alexia without agraphia associated with color anomia was discussed in KCC 19.7. Another well-known syndrome is **cortical blindness**, or **Anton's syndrome**, caused by bilateral lesions of the primary visual cortex. In Anton's syndrome, patients have complete visual loss on confrontation testing, yet they have anosognosia and are completely unaware of the deficit. Other exam findings include loss of blink to threat, loss of eye closure in response to bright lights, and loss of optokinetic nystagmus (OKN). Some patients may have blindsight (see the previous section). Anosognosia for visual loss can also be seen in other situations that are not, strictly speaking, Anton's syndrome, such as combined occipital and frontal lesions (resulting in confabulation) or combined occipital and right parietal lesions (resulting in neglect).

Syndromes of the Inferior Occipitotemporal Cortex

The inferior occipitotemporal cortex lies in the "What?" stream of visual analyis and processes color and visual form involved in object identification. Specialized regions that are involved in recognizing colors, faces, and letter strings (see Figure 19.13) have been described. Lesions of the inferior occipitotemporal cortex can therefore cause deficits in the recognition of colors, faces, and other objects, as well as other visual phenomena related to color and form. Complex formed visual hallucinations resulting from seizures of the inferior occipitotemporal visual association cortex were described in KCC 11.1 and KCC 18.2.

In **prosopagnosia**, patients are unable to recognize people by looking at their faces. The usual lesion location in prosopagnosia is the bilateral inferior **occipitotemporal cortex**, also known as the **fusiform gyrus** (see Figure 19.13; see also Figure 2.11C). Some evidence suggests that the right hemisphere is more important in face recognition, but in most reported cases of prosopagnosia with an enduring deficit, the lesions are bilateral. Recall that **agnosia** is defined as normal perception stripped of its meaning. In prosopagnosia, for example, patients can describe or even name the different parts of a face and can identify a face as being a face, and they may even be able to match faces on the basis of similar features, but they cannot recognize a face as belonging to a particular individual, even if it is someone they know very well. This "pure" form of agnosia without perceptual deficits is also known as associative agnosia, which should be distinguished from perceptual (apperceptive) agnosia, in which impairments of the primary sensory modality may contribute to difficulties with recognition. One way to remember the definition of agnosia is to think of it as a higher-order deficit, in contrast to simple impairments of perception.

Patients with prosopagnosia cannot recognize people by their faces, but they can identify people by their clothes, voices, or other cues. Interestingly, the recognition deficit in prosopagnosia is not restricted to human faces. Farmers with prosopagnosia have been reported to have difficulty recognizing their cows, and bird-watchers have been reported to have difficulty recognizing specific birds. One formulation is that prosopagnosia results in intact "generic recognition"—for example, the ability to recognize faces as faces, cars as cars, buildings as buildings—but impaired "specific recognition" of individuals within a class—for example, the inability to recognize specific faces, cars, buildings, and so on. Prosopagnosia is often associated with achromatopsia (see the next paragraph). In addition, it is sometimes associated with alexia and with upper-quadrant or bilateral upper visual field defects (see Figures 11.15J and 11.17B).

Achromatopsia is a central disorder of color *perception*. It can be thought of as cortical color blindness and should be contrasted with color agnosia, in which color perception is intact (see the next paragraph). Patients with achromatopsia cannot name, point to, or match colors presented visually. They can, however, name the appropriate color for an object described verbally. Usually patients with achromatopsia are aware of the deficit and describe the affected vision as appearing in shades of gray. Achromatopsia may occur in a quadrant, a hemifield, or the entire visual field. When the whole field is involved, the deficit is usually associated with prosopagnosia and is caused by lesions in bilateral inferior occipitotemporal cortex (see Figure 19.13). Hemiachromatopsia is caused by lesions in the contralateral inferior occipitotemporal cortex. Other deficits that are sometimes associated with achromatopsia include alexia and upper-quadrant or bilateral upper visual field defects.

Although **color anomia**, more correctly referred to as **color agnosia**, is not caused by lesions of the inferior occipitotemporal cortex, we discuss it briefly here to contrast it with achromatopsia. Color agnosia is caused by lesions of the primary visual cortex of the dominant hemisphere extending into the corpus callosum, and it is associated with alexia without agraphia and right hemianopia (see KCC 19.7; Figure 19.5). Color agnosia has also been reported with lesions of the medial occipitotemporal association cortex, adjacent to the splenium of the corpus callosum. Patients cannot name or point to colors presented visually. However, perception of colors is preserved, as demonstrated by patients' ability to match colors presented visually. This is not a true anomia or language disorder because patients can name the appropriate color for an object described verbally, so it should be distinguished from anomic aphasia.

Other visual agnosias are less widely known. For example, investigators have described several category-specific visual agnosias for living things, human-made tools, or other specific categories. Category-specific visual agnosias of this kind are sometimes seen with atrophy of the temporal neocortex in semantic dementia (see KCC 19.16). Large bilateral lesions of the inferior occipitotemporal region may cause a generalized visual-object agnosia applying to both generic and specific recognition of all visual objects, including those just described. This may be considered a perceptual agnosia because vision is often described as appearing hazy. Interestingly, some patients with lesions of this kind have a so-called visual static agnosia, in which they are able to recognize an object only when it moves.

Other illusory phenomena can occur with lesions of the inferior occipitotemporal cortex that are less specifically localized to this region. In **micropsia**, objects appear unusually small; in **macropsia**, objects appear unusually big; and either disorder can sometimes occur in only part of the visual field. **Metamorphopsia** is a more general term describing a condition in which objects have distorted shape and size. These disorders are sometimes referred to as "Alice in Wonderland" syndrome and can occur in migraine, infarct, hemorrhage, tumors, or other disorders of the inferior or lateral visual association cortex. They can also occasionally be seen in retinal pathology or toxic or metabolic disturbances.

In **visual reorientation**, the environment appears tilted or inverted to the patient. This condition has been associated with vestibular or lateral medullary dysfunction. In **palinopsia**, lesions of the visual association cortex cause a previously seen object to reappear periodically. For example, one patient looked at a plant, and then a few minutes later the plant reappeared and seemed to be growing out of her omelet. Another patient saw an aide entering her hospital room, and then later that evening she saw the image of the aide entering her room over and over again. Palinopsia can occasionally be caused by medications such as trazodone.

In **cerebral diplopia** or **polyopia**, patients see two or more images, respectively, of an object. Diplopia caused by dysconjugate gaze was discussed in KCC 13.1. The appearance of more than two images, as well as monocular diplopia, is sometimes psychiatric in origin. However, monocular or binocular double vision, triple vision, and so on can also occasionally be seen with occipital lesions, corneal lesions, or cataracts. Another visual illusion occasionally seen with cortical lesions is **erythropsia**, which is characterized by gold, red, purple, or other unnatural coloring of the visual field. Disturbances of color vision can also be seen with certain drugs, such as in digitalis toxicity, in which objects may appear to have a yellowish halo.

Syndromes of the Dorsolateral Parieto-Occipital Cortex

The dorsolateral parieto-occipital cortex lies in the "Where?" stream of visual analyis and processes motion as well as spatial localization and integration (see Figure 19.12). Lesions of the dorsolateral parieto-occipital cortex can therefore cause deficits in these aspects of visual processing. Constructional impairments and other deficits of visual-spatial analysis occurring especially in lesions of the parietal lobes and more commonly in the nondominant hemisphere were mentioned earlier in this chapter (see KCC 19.10).

In **Balint's syndrome**, caused by bilateral lesions of the dorsolateral parieto-occipital association cortex, there is a clinical triad consisting of (1) simultanagnosia, (2) optic ataxia, and (3) ocular apraxia. **Simultanagnosia** is the core abnormality of Balint's syndrome and consists of impaired ability to perceive parts of a visual scene as a whole. Patients with simultanagnosia can perceive only one small region of the visual field at a time. This region shifts around unpredictably, often causing patients to lose track of what they were looking at. Patients have particular difficulty scanning a complex visual scene or identifying moving objects (this is the opposite of visual static agnosia, described earlier). When confronted with a large complex visual stimulus, patients tend to describe small, isolated parts seemingly at random and have no awareness of the overall unified object or scene. Simultanagnosia can be thought of as a deficit in visual-spatial binding.

Optic ataxia is the impaired ability to reach for or point to objects in space under visual guidance. This condition can be distinguished from cerebellar ataxia because in optic ataxia the ability to point using proprioceptive or auditory cues is intact, and once an object has been touched, a patient with optic ataxia can perform smooth movements back and forth to it even with the eyes closed. **Ocular apraxia** is difficulty voluntarily directing one's gaze toward objects in the peripheral vision through saccades. Some patients need to move their heads to initiate a voluntary redirection of gaze. Again, this condition may be related to the defect in visual perception of stimuli other than in a small region of the visual field.

Patients with Balint's syndrome may be diagnosed incorrectly with visual agnosias or alexia, which can be shown not to be present when care is taken to ensure that the visual stimulus is in a region the patient can see. Because Balint's syndrome is caused by lesions of the dorsolateral parieto-occipital association cortex, associated deficits may include inferior-quadrant visual field cuts (see Figure 11.15), aphasia, or hemineglect. Most often these bilateral lesions of dorsolateral parieto-occipital cortex are caused by MCA–PCA watershed infarcts (see Figure 19.3B; see also Figure 10.10B), although bilateral hemorrhage, tumors, dementia with posterior cortical atrophy, or other lesions can also produce this syndrome.

Isolated features of Balint's syndrome may occur in some patients with bilateral parieto-occipital lesions who do not exhibit the full syndrome. For example, some patients may exhibit optic allesthesia (a false localization of objects in visual space) or cerebral akinetopsia (an inability to perceive moving objects). ■

REVIEW EXERCISE

For each of the following, state whether it is usually associated with lesions of the dorsal parieto-occipital ("Where?") stream, the ventral occipital temporal ("What?") stream (see Figure 19.12), or the primary visual cortex:

Prosopagnosia

Simultanagnosia

Achromatopsia

Blindness with anosognosia

KEY CLINICAL CONCEPT

19.13 AUDITORY HALLUCINATIONS

Disturbances of higher-order auditory processing involving the auditory cortex and adjacent association cortex (see Figure 19.1) are just one cause of auditory hallucinations and other positive auditory phenomena. **Tinnitus** is a common disorder consisting of a persistent ringing tone or buzzing in one or both ears, usually caused by peripheral auditory disorders affecting the tympanic membrane, middle ear ossicles, cochlea, or eighth cranial nerve (see KCC 12.5). **Self-audible bruits** are pulsatile "whooshing" sounds that can be associated with turbulent flow in arteriovenous malformations, carotid dissection, or the extracranial-to-intracranial pressure gradient that is produced by elevated intracranial pressure. Some positive auditory phenomena are analogous to similar disturbances of the visual system (see KCC 11.1 and KCC 19.12). For example, elderly patients with sensorineural deafness can develop elaborate auditory hallucinations (music, voices, etc.), which may be a **release phenomenon** analogous to Bonnet syndrome (visual hallucinations caused by visual loss). Lesions or ischemia of the **pontine tegmentum** involving the trapezoid body, superior olivary nucleus, and other auditory circuits (see Figure 12.17) can, rarely, cause auditory hallucinations, such as rain on a roof, buzzing, or musical tones like an orchestra tuning up, which are analogous to the visual phenomena of peduncular hallucinosis. In **paracusis**, another rare disorder, a sound that is heard once is then heard repeatedly, analogous to palinopsia. **Psychotic disorders** are a relatively common cause of either simple or elaborate auditory hallucinations. Functional neuroimaging studies during auditory hallucinations in schizophrenic patients have recently demonstrated activation of diverse brain regions. As we discussed in KCC 18.2, **seizures** in primary auditory cortex can cause simple auditory phenomena, such as the sound of a train approaching or of an airplane taking off, often perceived as coming from the side opposite to the involved cortex. Seizures in auditory cortex can also cause transiently decreased hearing. Involvement of the auditory association cortex gives rise to more elaborate auditory phenomena such as voices or music. Musical hallucinations are more often caused by seizures in the nondominant hemisphere than in the dominant hemisphere, although musical hallucinations can also occur with peripheral or pontine lesions, as we have just discussed. Other disturbances of higher-order auditory perception, such as cortical deafness and nonverbal auditory agnosia, were discussed in KCC 19.7. ■

The Consciousness System Revisited: Anatomy of Attention and Awareness

Consciousness is one of the most fascinating topics in neuroscience. The quest to explain conscious thought in terms of brain activity has become somewhat of a Holy Grail for neuroscience, and has attracted many individuals into this field, especially in recent years. No brain system comes as close to explaining consciousness as do the networks underlying mechanisms of attention and awareness.

Consciousness was described by Plum and Posner as having two components:

1. Content of consciousness
2. Level of consciousness

The **content of consciousness** is the substrate upon which consciousness acts and includes sensory, motor, emotional, and mnemonic systems acting at multiple levels. It comprises many of the systems described throughout this book, including memory, emotion and drives, language, executive function, visual processing, motor planning, and so on (see the figure in the Epilogue at the end of this book). Without normal operation of these systems, completely normal consciousness is not possible. The **level of consciousness** is regulated by several brain networks that act on this substrate and constitute the **consciousness system** (Figure 19.14). Like other major functional systems in the brain (such as the motor system, somatosensory system, and so on), the consciousness system consists of cortical and subcortical networks that carry out specialized functions—in this case, regulating the level of consciousness. As we

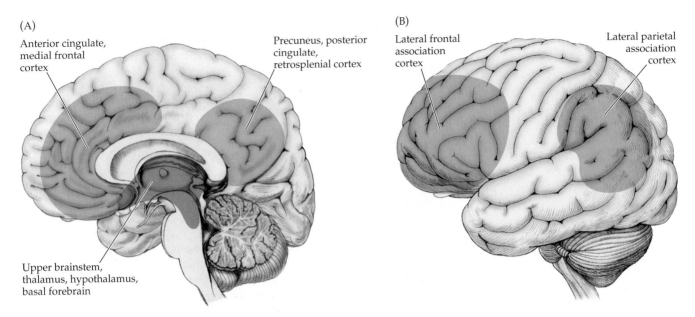

(A)

Anterior cingulate, medial frontal cortex

Precuneus, posterior cingulate, retrosplenial cortex

Upper brainstem, thalamus, hypothalamus, basal forebrain

(B)

Lateral frontal association cortex

Lateral parietal association cortex

FIGURE 19.14 The Consciousness System Anatomical structures involved in regulating the level of alertness, attention and awareness. (A) Medial view showing cortical (blue) and subcortical (red) components of the consciousness system. (B) Lateral cortical components of the consciousness system. Note that other circuits not pictured here, such as the basal ganglia and cerebellum, may also play a role in attention and other aspects of consciousness.

discussed in Chapter 14, the level of consciousness can be described in terms of three distinct but related processes that maintain (1) **alertness**; (2) **attention**; and (3) **awareness** of self and environment (mnemonic **AAA**). Recall that the consciousness system networks that control these functions include the upper brainstem, thalamic, hypothalamic, and basal forebrain activating systems (see Figures 14.7–14.15), along with the medial and lateral frontoparietal association cortex and cingulate gyrus (see Figure 19.14). Other structures such as the basal ganglia and cerebellum may also play a role through their participation in attention mechanisms. In Chapter 14 we covered the role of the consciousness system in regulating basic alertness, and the awake state. We will now continue our exploration of consciousness by moving from functions that prevent coma (and promote alertness), first to a discussion of attention, and later to the more controversial topic of awareness.

MNEMONIC

General Mechanisms of Attention

Earlier in this chapter, we first introduced the lateralized aspects of attention and emphasized that the nondominant (usually right) hemisphere is more important for attention in most individuals (see Figure 19.7; KCC 19.9). We will now describe in greater detail the networks operating in both hemispheres that contribute to more general aspects of attention. **Attention** can be described in many different ways and has a variety of different functional components. A coherent, physiologically or anatomically based taxonomy of attention has not yet emerged. However, attention includes **at least two major functions**: (1) Selective, or directed, attention involves focusing attention on a particular domain above others. (2) Sustained attention includes functions such as vigilance, concentration, and nondistractability. The circuits involved in these two aspects of attention share many common components and mechanisms.

Selective attention, or directed attention, implies attention to certain objects, stimuli, or concepts to the exclusion of others. Selectivity is an essential component of attention and has been studied extensively. In selective attention, attention is directed at specific components of the "substrate," or content of consciousness, described in the previous paragraph. Examples of selective attention are quite diverse; they include:

- Attention to a visual, tactile, or auditory stimulus in a particular location in space (discussed earlier in this chapter)
- Attention to inputs of a specific sensory modality
- Attention to a specific higher-order aspect of a stimulus (e.g., color versus shape)
- Attention to a particular object, including inputs from various modalities
- Attention to an object, emotion, plan, or concept that is not physically present but is either remembered or imagined

Functional MRI and evoked-potential studies in humans, and recordings from animal models, suggest that these examples of directed attention are reflected in the activation of specific brain regions, following known anatomical principles. For example, attention to a somatosensory stimulus on the body activates the corresponding somatotopic region of somatosensory cortex, attention to a visual stimulus in a particular location activates the corresponding retinotopic region of visual cortex, and so on. In addition, specific areas of limbic and association cortex are involved in aspects of higher-order processing.

Sustained attention (concentration, vigilance, nondistractibility) is a second major set of attentional functions. Sustained attention can be directed at a specific task, object, or modality, or it can involve a more generalized increased level of vigilance—for example, while one is awaiting an anticipated stimulus.

As an illustration of different types of attentional mechanisms, imagine a student trying to study for a final exam while sitting in the middle of a rowdy fraternity party. The student demonstrates directed (selective) attention by looking at her book and reading, attending to the visual, linguistic, and conceptual aspects of its content while ignoring the loud noises, flashing lights, and scents of the room around her. She demonstrates sustained attention and concentration by continuing to read even if she finds the subject slightly boring, and she exhibits nondistractibility by ignoring other students' pleas for her to join in the dancing. Finally, she demonstrates vigilance by being prepared to duck quickly out of harm's way when flying objects occasionally enter her peripheral vision.

It should be evident from this discussion that both selective and sustained attention may involve enhanced activity in stimulus-relevant regions of the brain ("signal") and/or decreased activity in stimulus-irrelevant brain regions ("noise"). The relative importance of **signal enhancement versus noise suppression** in attention is still under active investigation. Similarly, attention must be able to shift from one target to the next, which may involve mechanisms that both **engage** the relevant stimulus and **disengage** the irrelevant (or no longer relevant) stimulus. Some authors include an **executive** or **control** function in describing the components of attention. Other, harder questions still under active investigation include the mechanisms through which directed attention has a **limited capacity** in the spatial, temporal, modality-specific, and other domains, and the mechanisms through which attended targets from disparate brain regions are unified into a single coherent concept (the **binding problem**).

Anatomy of Attention

Networks involved in attention are distributed through many cortical and subcortical structures within the consciousness system (see Figure 19.14). Brain systems involved in attention include (1) widespread projection systems (discussed in Chapter 14); (2) frontal and parietal association cortex; (3) anterior cingulate cortex and limbic pathways; (4) tectum, pretectal area, and pulvinar; and (5) other structures, such as the cerebellum and basal ganglia. Interestingly, consciousness system networks involved in generalized alertness also participate in mechanisms of selective and sustained attention. Recall that attentional networks are asymmetrical and that the nondominant (usually right) hemisphere plays a more important role (see KCC 19.9). Let's review each of the anatomical systems involved in selective and sustained attention in more detail.

WIDESPREAD PROJECTION SYSTEMS A prerequisite for attention is an awake, alert state. Mechanisms of arousal were discussed in Chapter 14; they include widespread projection systems of the upper brainstem, thalamus, hypothalamus, and basal forebrain. In addition to aiding in generalized arousal, many of the same systems may contribute to directed and sustained attention. We will therefore briefly review these systems again here (see Table 14.2; Figures 14.7–14.13). **Upper brainstem projection systems** include cholinergic (pedunculopontine and laterodorsal tegmental nuclei) and noncholinergic (pontomesencephalic reticular formation, possibly glutamatergic) projections to thalamus, hypothalamus, and basal forebrain systems, which in turn have widespread cortical projections. In addition, noradrenergic (locus ceruleus and lateral tegmental area) and serotonergic (dorsal and medial raphe) systems project widely to the cortex and other structures, while dopaminergic (substantia nigra pars reticulata, ventral tegmental area) systems project to striatum, limbic cortex, and prefrontal cortex.

Thalamic systems involved in arousal include the intralaminar, midline, ventral medial, and possibly other thalamic nuclei that transfer inputs from the upper brainstem reticular formation and cholinergic nuclei to widespread areas of cerebral cortex. In addition, the thalamic reticular nucleus has been postulated to play a role in gating information transfer through the thalamus because it receives inputs from cortex, thalamus, and brainstem systems and sends inhibitory (GABAergic) projections to the thalamus (and back to the brainstem as well).

Hypothalamic systems important to arousal include the posterior lateral hypothalamic histaminergic neurons (tuberomammillary nucleus) and orexinergic neurons, which receive inputs from anterior hypothalamus and brainstem and project widely to cortex and thalamus.

Finally, **basal forebrain systems** involved in arousal include the nucleus basalis, diagonal band, and medial septal cholinergic and GABAergic neurons, which receive inputs from the brainstem and project to the entire cortex and thalamus.

FRONTAL AND PARIETAL ASSOCIATION CORTEX The frontal and parietal association cortices (see Figure 19.1) communicate with each other via strong reciprocal connections and play an important role in attentional mechanisms. The lateral **parietal cortex** and adjacent temporal and occipital association cortex lie at the nexus of auditory, visual, and somatosensory unimodal association cortex (see Figure 19.1). This region is thus ideally situated for heteromodal integration in attention. As we discussed earlier in this chapter, the parietal association cortex plays an important role in heteromodal spatial representations, encoding the location of attended objects in space. Lesions of the parietal cortex, especially in the nondominant hemisphere, are the best-known cause of deficits in contralateral directed attention, or hemineglect (see KCC 19.9).

The **frontal heteromodal association cortex** (prefrontal cortex; see Figure 19.1), also discussed earlier, plays important roles in both directed and sustained attentional mechanisms. In particular, the region of the frontal eye fields is important in directed attention to the contralateral side and in the initiation of eye movements toward attended targets. In addition, the prefrontal cortex may play an important role in motor-intentional aspects of attention toward the contralateral side. The ability to initiate spontaneous movements of the contralateral limbs or toward the contralateral hemispace may depend on prefrontal cortex and on ascending dopaminergic modulation of the prefrontal cortex and striatum. Prefrontal cortex also participates more generally in reorienting attention to relevant stimuli. Finally, on the basis of both functional neuroimaging studies and the effects of lesions, the prefrontal cortex is crucial to sustaining attention and reducing distractibility.

ANTERIOR CINGULATE CORTEX AND LIMBIC PATHWAYS   The **anterior cingulate cortex** (see Figure 19.11B) is important in motivational aspects of attention. It forms a network together with the amygdala, medial orbitofrontal cortex, thalamic mediodorsal nucleus, and other **limbic structures** (see Chapter 18) that may play an important role in motivating directed and sustained attention toward a relevant or interesting stimulus. The importance of motivational factors in attention is illustrated by the example in which a patient performs poorly when asked to remove objects such as pieces of paper from a table but then shows a dramatic improvement when the objects are replaced with dollar bills.

TECTUM, PRETECTAL AREA, AND PULVINAR   The **superior colliculi**, **pretectal area**, and **pulvinar** participate together with the parietotemporo-occipital cortex and frontal eye fields in directing visual attention toward relevant visual stimuli for saccadic eye movements (see Figures 11.6 and 13.14). Directed attention for other modalities, such as audition, may also be processed by these pathways.

OTHER STRUCTURES   Accumulating evidence suggests that certain parts of the **basal ganglia** and **cerebellum** also participate in mechanisms of directed attention.

Awareness of Self and Environment

One of the great remaining mysteries of modern science is the mechanism for our subjective and personal experience of awareness. As we have discussed in this chapter, consciousness certainly includes a "substrate," or **content**, represented by sensorimotor systems, memory systems, and limbic systems, as well as "consciousness system" mechanisms for controlling the **level** of alertness, attention, and awareness. So far, we have discussed alertness and attention, but what are the mechanisms of awareness? Philosophers debate whether a biological explanation for the awareness aspects of consciousness, sometimes referred to as **qualia**, is even possible. As this debate continues, and although the final answers are not in yet, exciting new developments in neuroscience have begun to shed light on systems that may participate in generating our subjective personal experience of awareness. To facilitate the progress in this field, we can define awareness in simple functional terms as follows: **Conscious awareness** is our ability to combine various forms of sensory, motor, emotional, and mnemonic information into *an efficient summary of mental activity that can potentially be remembered at a later time.*

Like other functions of the nervous system, awareness most likely is mediated by a network involving both specialized regions of local processing and widespread regions of distributed processing. We saw in Chapter 18 that memory functions are segregated anatomically into **declarative memory**, which involves conscious awareness and is processed by medial temporal and diencephalic regions, and **nondeclarative memory**, which does not involve consciousness and is processed by other brain regions (see KCC 18.1). Further investigation of the neural circuits underlying declarative memory may ultimately help us understand what is "special" about conscious versus nonconscious memories.

We have discussed the **hemineglect syndrome**, in which inattention causes circumscribed loss of awareness of both the self and the environment (see KCC 19.9). This syndrome suggests that the same mechanisms that are involved in attention play an important role in awareness as well, and some investigators even debate whether a distinction between attention and awareness is valid. One aspect of awareness that is difficult to explain in terms of current theories of attention is the **binding** of sensory, motor, emotional, and mnemonic information from disparate brain regions into what we perceive as a single unified experience. Where or how is this synthesis of multiple forms of information perceived? Some investigators have contended that binding is a distributed process occurring over widespread networks. Theories for binding on a cellular level have included widespread horizontal connections between certain cortical layers, and synchronized or coherent gamma-frequency (about 40 hertz) oscillations in neuronal activity occurring between regions involved in binding. Others have proposed that specific regions of higher-order association cortex, such as the frontal or parietal lobes, may be critical for binding. From a clinical perspective, **Balint's syndrome** (see KCC 19.12) provides an interesting example of how focal lesions of the parieto-occipital cortex can cause a profound deficit in the ability to bind various individual parts of a visual scene into a single integrated whole.

The role of the prefrontal cortex in **working memory**, or the ability to hold a certain amount of information in an active short-term store, is also likely to play an important role in any process that mediates awareness, as are the senses of **chronological sequence** and **self-monitoring** that are mediated by the frontal lobes. Similarly, studies of **mental imagery** have begun to demonstrate the involvement of certain regions of primary and association cortex in generating internal representations of both sensory and motor phenomena—important ingredients for any internal representation or engram of awareness.

Recent work in functional neuroimaging has shown that when people are awake but not engaged in a specific task, brain networks show slow (duration greater than 10 s) spontaneous fluctuations in activity levels, referred to as **resting state activity**. This resting activity is correlated within distinct functional brain systems such as the motor, visual and other brain networks. Interestingly, resting activity is also seen in the cortical components of the consciousness system (see Figure 19.14) when awake individuals are not engaged in a specific task and are simply allowed to "daydream" or to "introspect," and it is referred to in this context as **default mode network activity**. In particular, studies have suggested a potentially important role for the **medial parietal region** including the precuneus, along with the adjacent posterior cingulate, and retrosplenial cortex (cortex located just behind splenium of the corpus callosum) in **self-reflection, introspection, and self-awareness**.

Finally, although significant advances are being made in understanding **limbic networks**, the question of how neural activity gives rise to emotions remains as difficult for us to answer as the question of how neural activity gives rise to conscious thought. In summary, conscious awareness may yet be explained on the basis of an understanding of both distributed and specialized local network processing in the brain. However, this important and interesting quest for understanding is still very much under way and will undoubtedly continue to provide a rich vein for researchers.

KEY CLINICAL CONCEPT

19.14 ATTENTIONAL DISORDERS

Because attention depends on so many different systems, as described in the preceding sections, it is not surprising that deficits in attention can be produced by focal lesions in many locations, as well as by diffuse disorders affecting large regions of the nervous system. In this section we will discuss the assessment and diagnosis of patients with general disorders of **sustained attention** (see KCC 19.9 for the special case of unilateral deficits in directed attention). In the next two sections we will discuss other disorders that tend to affect the nervous system in a widespread manner and that may also cause prominent deficits in global attention (see KCC 19.15 and KCC 19.16).

Like most other disorders of the nervous system, attentional disorders can range in severity from mild to severe. Mild inattention may cause patients to have difficulty registering new information, and they may occasionally fail to complete tasks. Severe inattention, on the other hand, may cause patients to be completely unresponsive to outside stimuli. Of note, patients may be fully awake and yet profoundly inattentive.

Testing Sustained Attention

Because attention is so important for performance on the mental status examination and can affect nearly all other tests, it is essential to evaluate and document the patient's level of attention toward the beginning of the examination. Clues that an attentional disorder is present can often be obtained from the history and by observation of the patient's behavior during all parts of the examination. A very useful test and well-standardized measure of attention (and working memory) is **digit span**. In this test, a random series of numbers is recited to the patient, and the patient is asked to repeat them back immediately to the examiner. Normal digit span is five to seven or more digits. Next, patients can again be given a series of numbers but then asked to repeat them backward. This task is slightly more difficult, and **backward digit span** is normally four or more digits, or two less than the patient's forward digit span.

A similar test that is easier to administer is to ask the patient to recite the **months of the year forward and then backward** (see **neuroexam.com Video 4**). Although this test is not standardized, with some experience the examiner gets a sense of what is normal and what is abnormal. Normally, reciting the months backward should take no longer than twice as long as reciting them forward, and this task should be performed without errors. Other similar but perhaps less useful tests include asking the patient to spell "world" backward or to count backward by threes from 30, or by sevens from 100.

Motor impersistence is another useful bedside indication of impaired attention, which we discussed earlier in the evaluation of frontal lobe disorders (see KCC 19.11; Table 19.10). The patient is asked to stick out their tongue, or hold up their arms for 20 seconds, without subsequent prompting. If they fail to do so, they have motor impersistence, a form of inattention. **Vigilance** can also be tested at the bedside with the **"A" random letter test**, in which the examiner recites a random sequence of letters at a rate of about one per second, and the patient is instructed to tap the desk each time they hear the letter A.

Formal neuropsychological tests of attention exist as well and can be useful for obtaining a more quantitative assessment. In addition, some patients may sustain their attention for the brief time required for bedside tests such as digits forward and backward, but have difficulty on more extended neuropsychological measures. As already noted, impaired attention may also cause patients to perform poorly on many other parts of the mental status examination. This poor overall performance can lead to false localization unless the attentional deficit is recognized. Finally, note that there is some overlap between the definitions of sustained attention and working memory. Attention, however, is involved in instantaneous stimulus selection, while working memory serves as a temporary storage depot without necessarily being directly involved in stimulus selection. The frontal lobes play an important role in both working memory and in attention mechanisms, and tests of frontal lobe function (discussed in KCC 19.11) are often quite sensitive to deficits in attention.

Differential Diagnosis of Disorders of Sustained Attention

Some common causes of impaired general attention are listed in Table 19.13. As already noted, impaired attention can be mild or quite severe. **Encephalopathy** is a nonspecific term that means simply diffuse brain dysfunction, which we will discuss further in KCC 19.15 and KCC 19.16. Various forms of diffuse encephalopathy are the most common causes of impaired attention. Encephalopathy, especially when acute in onset, can also be associated with an impaired level of alertness ranging from mild lethargy to coma. **Focal lesions** can also cause impaired attention. This is particularly true of lesions in the frontal lobes, parietal lobes, or brainstem-diencephalic activating systems, although focal lesions in numerous other brain regions can also result in impaired attention.

Attention-deficit hyperactivity disorder (**ADHD**) is a fairly common condition affecting 1 to 5% of elementary school children. Onset is typically by age 3 years, although problems do not usually emerge until school is started. In some children the disorder of attention is predominant, while in others problems with impulsive and hyperkinetic behaviors are predominant. Most children with ADHD have normal neuroimaging studies and essentially normal neurologic evaluations, except for markedly impaired attention, impulsivity, and perhaps some "soft" findings on exam. The impaired attention in ADHD is somewhat different from the other disorders listed in Table 19.13. ADHD is more likely, for example, to cause problems with high-level executive functions, organizational skills, and time management abilities rather than simpler impairments of digit span. ADHD is occasionally seen in patients with other neurologic conditions, but in most cases the cause of ADHD is unknown. ADHD is three to five times more common in boys than girls, and siblings are at increased risk for having the disorder. The condition is treated with CNS stimulants such as methylphenidate (Ritalin), selective norepinephrine reuptake inhibitors such as atomoxetine (Strattera) or other medications, combined with individual and family behavioral therapy. It is interesting that stimulants that enhance dopaminergic and noradrenergic neurotransmission are beneficial in this disorder. Long-term outcome is variable: Many individuals show remission upon reaching adolescence, but about 30% of patients continue to have ADHD into adulthood.

Psychiatric disorders (see KCC 18.3) are a very important cause of impaired attention. Patients with depression, anxiety, mania, schizophrenia, or other milder conditions are often severely inattentive on examination. The remainder of the mental status examination in these patients should therefore be carefully interpreted in this context to avoid misdiagnosis of the patient as having a memory disorder or other focal dysfunction. Formal neuropsychological testing can often be helpful in distinguishing "pseudodementia" seen in psychiatric conditions from true dementia (see KCC 19.16). ◼

TABLE 19.13 Common Causes of Impaired General Attention
Diffuse encephalopathy (see Tables 19.14, 19.15)
Focal lesions, especially of the frontal or parietal lobes or of the brainstem-diencephalic activating systems, but also of many other brain regions
Attention-deficit hyperactivity disorder (ADHD)
Psychiatric disorders—e.g., depression, mania, schizophrenia

19.15 DELIRIUM AND OTHER ACUTE MENTAL STATUS DISORDERS

One of the most common reasons for neurologic consultation is the nonspecific condition referred to as "mental status changes." Although focal lesions can cause altered mental status, usually the consultation is for a patient with no significant focal findings on examination, other than impaired cognitive abilities. The impairments of mental status are often relatively nonlocalizing as well, consisting of prominent inattention, confusion, and memory deficits, together with variable degrees of impaired alertness and attention. These patients may be referred to as **encephalopathic**, meaning simply that they exhibit diffuse brain dysfunction.

In evaluating patients of this kind, the first important piece of information is whether the change in mental status is **acute** (or subacute), meaning that it is of recent onset (hours to weeks or a few months), or **chronic**, meaning that it is part of a long-standing or more gradual deterioration (months to years). Acute encephalopathy is most often toxic or metabolic in origin and is often reversible (Table 19.14); chronic encephalopathy has a poorer prognosis, and in elderly patients it most often represents dementia of the Alzheimer's type (see KCC 19.16). It is often surprisingly difficult to determine whether mental status changes are acute or chronic, even after consultation with family members, because family members might not provide the most objective information about the mental status of loved ones and might minimize changes and attribute them to "aging." When it is available, the most useful information is a prior mental status examination, or at least historical information about the patient's level of functioning, including specific things they previously could or could not do on their own, ranging from complex activities (e.g., managing finances, driving) to more simple activities (e.g., dressing, grooming). Of note, impaired attention is usually prominent in acute mental status changes, while it may be minimal in chronic encephalopathies, especially early on (see KCC 19.16).

Acute mental status changes are also referred to as **acute confusional states**, acute global confusional states, organic psychosis, acute organic brain syndrome, or delirium. All of these terms have a similar meaning and apply to patients who have recent onset of encephalopathy, with prominent inattention, confusion, waxing

---

TABLE 19.14 Causes of Acute or Subacute Mental Status Changes

Toxic or metabolic encephalopathies
    Drug or alcohol toxicity
    Withdrawal from alcohol or other sedatives
    Electrolyte abnormality (especially elevated sodium, calcium, or magnesium)
    Hypoglycemia
    Diffuse anoxia
    Hypothyroidism, hyperthyroidism
    Hypoadrenalism
    Thiamine deficiency (Wernicke–Korsakoff encephalopathy)
    Hepatic failure
    Renal failure
    Pulmonary failure
    Sepsis
    Inborn errors of metabolism
    Paraneoplastic syndrome
    Hereditary endogenous benzodiazepine production
Head trauma
Diffuse or focal cerebral ischemia or infarct
Intracranial hemorrhage
Migraine
Seizures or post-ictal state
Hydrocephalus
Elevated intracranial pressure
Diffuse cerebral edema
Meningitis, encephalitis, brain abscess
Vasculitis, diffuse subcortical demyelination (e.g., multiple sclerosis)
Intracranial neoplasm
Paraneoplastic syndrome
Mild insult (e.g., urinary infection or change in environment) in setting of underlying impaired mental status
Psychiatric disorders (e.g., depression, mania, schizophrenia, etc.)
Sleep deprivation
Visual deprivation or more generalized sensory deprivation
Hypotension
Hypertensive crisis
Posterior reversible encephalopathy syndrome

and waning level of alertness, and difficulty registering new memories due to both inattention and diffuse dysfunction of the memory network (see KCC 18.1). Patients often have marked difficulty with writing, calculations, and constructional abilities, any of which could be misconstrued as focal deficits if the underlying deficit in attention is not recognized. Acute confusional states tend to wax and wane in severity over the course of hours, and they are often exacerbated in the evening (a phenomenon referred to as "sundowning"). Although all patients with acute confusional states are inattentive, the level of alertness may be normal in mild cases and can range from agitation to near coma in more severe cases. The term **delirium** is commonly used for acute confusional state in which agitation and hallucinations (auditory, visual, and tactile) are prominent. The classic example is **delirium tremens**, which occurs in the setting of alcohol withdrawal. However, it is important to recognize that delirium and acute confusional state can also occur without agitation, particularly in elderly hospitalized patients, making it easy to miss the diagnosis.

The most common causes of acute confusional states are **toxic or metabolic disorders**, followed by infection, head trauma, and seizures (see Table 19.14). Patients should therefore be evaluated promptly by assessment of vital signs, respiratory status (including arterial blood gas), and other appropriate blood tests, including blood glucose, electrolytes, blood urea nitrogen, creatinine, liver function tests, ammonia level, complete blood count, thyroid function tests, and toxicology screen. The patient's medications should be reviewed for those with known central nervous system side effects (e.g., anticholinergic, sedative–hypnotic, narcotic). When the diagnosis remains uncertain, the patient should immediately undergo neuroimaging and a lumbar puncture. If these are negative, an EEG should be performed promptly to rule out subtle ongoing seizure activity. Aside from seizure activity, the EEG in many forms of encephalopathy listed in Table 19.14 may show diffuse slowing or other abnormalities. In elderly patients or patients with previous neurologic disorders, acute confusional states are often provoked by seemingly minor causes, such as a urinary tract infection, or even by a change from home to the hospital setting. Patients in intensive care units are prone to acute confusional states from the combination of sedative use, immobilization, and sleep and sensory deprivation, although they should be evaluated for treatable causes as described above.

As we will discuss in the next section, **chronic mental status changes** can also have numerous causes and are more typically gradually progressive, as in Alzheimer's disease, or are static, as in anoxic brain damage (see KCC 19.16). **Dementia** is a broad term, meaning literally "decline in mental function;" however, it is usually applied more specifically to gradually progressive disorders such as Alzheimer's disease. Although both acute and chronic mental status changes can be treatable, acute mental status changes usually carry a somewhat better prognosis. Therefore, an important goal is often to disinguish **delirium versus dementia**.

To summarize, **acute confusional states** such as delirium typically develop over the course of hours to weeks, have prominent attentional disturbances, tend to wax and wane over the course of hours, often have marked slowing on the EEG, and are most often caused by toxic or metabolic disorders, alcohol withdrawl, head trauma, infection, and seizures. **Chronic mental status changes**, typified by Alzheimer's disease and other gradually progressive dementias (see KCC 19.16), usually develop over months to years, do not tend to fluctuate as rapidly (although exacerbations of function can occur in certain settings), and early in their course tend to have less prominent disturbances in attention and a relatively normal EEG. Note that there are many exceptions to these general principles. For example, patients with acute mental status changes do not always show waxing and waning function, and many patients with chronic mental status changes (Huntington's disease, for example) show prominent deficits in attention early in the course of the disorder.

The vast majority of cases of encephalopathy and global confusional states are caused by conditions that affect the brain in a diffuse manner *bilaterally*. However, unilateral lesions, particularly in the right parietal, right frontal, or right medial temporo-occipital regions, can occasionally produce global states of impaired alertness and attention that mimic diffuse encephalopathy. ■

KEY CLINICAL CONCEPT

19.16 DEMENTIA AND OTHER CHRONIC MENTAL STATUS DISORDERS

As we approach the end of this book, it is appropriate to discuss a group of disorders that may await many of us in the future; hopefully this discussion will motivate us to pursue greater understanding of these disorders and ultimately to discover more effective treatments for them. **Dementia** is defined as a decline in memory and other cognitive abilities from a previously higher level of function to impaired functional status. Although this definition includes patients in which the decline is sudden or nonprogressive (e.g., the result of a single head injury or other insult), the term "dementia" is more typically used when there is gradually progressive deterioration over the course of months to years.

Patients with dementia typically show a decline in memory and other more generalized cognitive abilities, although patients with focal decline, such as in progressive nonfluent aphasia, are often included in the definition of dementia as well. Dementia is not restricted to the elderly; however, disorders causing dementia are much more common in the later years of life. Dementia should be distinguished from the apparently "normal" mild deterioration of processing speed and memory that occurs with age and does not significantly interfere with daily function. In contrast, a prodromal phase of more prominent cognitive failures called **mild cognitive impairment (MCI)** is seen at the onset of many forms of progressive dementia. Patients with early dementia exhibit deterioration well beyond statistical norms for age-matched controls.

Aside from dementia, several other terms for **chronic mental status changes** are sometimes used. **Static encephalopathy** is a term that refers to permanent nonprogressive brain damage as a result of head injury, anoxia, or congenital abnormalities of brain development, for example. **Mental retardation** is defined as impaired general intellectual and social adaptive function originating during development that is approximately two or more standard deviations below average. Although chronic mental status changes can occur in the pediatric population, they are more common in adults, and we will focus on adult-onset dementias in this section. Please see the References section at the end of the chapter for additional reading about dementia in pediatric patients.

Cortical dementias, with prominent disturbances in language, praxis, visual-spatial functions, and other typically cortical functions, are sometimes distinguished from **subcortical dementias** (such as Huntington's disease or progressive supranuclear palsy), in which these features are not present, although the usefulness of this distinction has been questioned. Dementia can also be subdivided into **primary dementia**, typically associated with neurodegenerative conditions for which definitive treatments are usually unavailable, and **secondary dementia**, caused by other conditions that may, in some cases, be reversible. In the sections that follow we will discuss the different causes of dementia and other chronic mental status changes and the general approach to evaluating patients with these disorders before focusing on Alzheimer's disease in greater detail.

Causes and Evaluation of Dementia

Fifty years ago, senile dementia was thought to be caused mainly by cerebrovascular disease, known colloquially as "hardening of the arteries," and Alzheimer's disease was considered relatively rare. With increased knowledge and understanding of Alzheimer's disease, it is now recognized that over 50% of the cases of senile dementia are caused by Alzheimer's disease, making it the most common cause by far. We will discuss Alzheimer's disease in more detail in the sections that follow; here we focus on the other causes of chronic mental status changes (Table 19.15). Many of the causes of chronic mental status changes listed in Table 19.15 are the same as for acute mental status changes (see Table 19.14) but are the result of long-standing treatable disorders, such as chronic hypothyroidism or chronic hydrocephalus, or are the outcome of permanent brain injury, such as previous head trauma or encephalitis. Note that because Alzheimer's disease is relatively common, many elderly patients will have Alzheimer's disease coexisting with one of the other disorders listed in Table 19.15.

Because there are so many possible causes of dementia, **evaluation of patients with dementia** should focus on and target treatable disorders, with the recognition that a treatable cause is found in only about 10% of cases. Patients should be evaluated with a **thorough history**, including an assessment of activities of daily living, family history, and description of any possible precipitants. As in the evaluation of patients with acute mental status changes (see KCC 19.15), it is crucial to review the patient's medications for any that may cause central nervous system side effects. A careful **general and neurologic exam** should then be performed.

Formal neuropsychological testing can be useful in distinguishing dementia from

TABLE 19.15 Causes of Chronic Mental Status Changes
Primary neurodegenerative disorders
Alzheimer's disease
Dementia with Lewy bodies
Frontotemporal lobar degeneration (FTLD)
Frontotemporal dementia
Progressive nonfluent aphasia
Semantic dementia
Other related disorders
Parkinson's disease with dementia
Huntington's disease
Progressive supranuclear palsy
Cerebellar atrophies
Cortical–basal ganglionic degeneration
Dentatorubropallidoluysian atrophy
Other disorders
Vascular dementia
Multi-infarct dementia
Binswanger's disease
Intracranial hemorrhage
Cerebral amyloid angiopathy
Psychiatric pseudodementia (especially depression, schizophrenia, and conversion disorder)
Thiamine deficiency (Wernicke–Korsakoff encephalopathy) and other alcohol-related causes
Intracranial neoplasms and paraneoplastic syndromes
Normal-pressure hydrocephalus and noncommunicating hydrocephalus
Head trauma, including chronic subdural hematoma and dementia pugilistica
HIV–associated neurocognitive disorder (HAND) and other infections including meningitis, encephalitis, neurosyphilis, Lyme disease, Creutzfeldt–Jakob disease, other prion diseases, and CNS Whipple's disease
Diffuse anoxic injury
Prolonged status epilepticus
Chronically elevated intracranial pressure
Multiple sclerosis and other demyelinating disorders
Vasculitis
Chronic electrolyte abnormalities or hepatic, renal, or pulmonary failure
Heavy-metal toxicity (lead, arsenic, gold, bismuth, manganese, or mercury)
Hypothyroidism, hyperthyroidism
Vitamin B_{12} deficiency, pellagra, folate deficiency
Inborn errors of metabolism
Wilson's disease
Congenital or developmental disorders

"pseudodementia" caused by depression or other treatable psychiatric disorders. In addition, complete neuropsychology testing can be useful to detect **mild cognitive impairment** (**MCI**), which occurs early in the course of Alzheimer's and other progressive dementias, since testing results can help guide decisions for treatment. More abbreviated testing batteries that can be administered easily at the bedside and that are especially useful in following patients during serial evaluations over time include the Blessed Dementia Scale (Table 19.16), the Folstein Mini-Mental State Examination, and the Activities of Daily Living Rating Scale. **Blood tests** should include routine chemistries, as well as thyroid function tests, B_{12} and folate levels, serum syphilis test (in endemic areas), and other blood tests (see Table 19.15), depending on the patient's age and the clinical situation. **Neuroimaging** should ideally include an MRI scan of the brain. If the patient has any atypical features such as rapid progression, headaches, or young onset (i.e., dementia before age 60) then a lumbar puncture should be performed to test for chronic meningitis (see KCC 5.9, 5.10). In selected cases, an EEG can be helpful (for example, to detect periodic sharp waves in Creutzfeldt–Jakob disease, or triphasic waves in hepatic encephalopathy), and occasionally a brain biopsy is indicated, when the results may guide treatment (e.g., in suspected CNS vasculitis).

Often the distinction between acute and chronic mental status changes is not obvious, as discussed in KCC 19.15. In addition, patients may have a combination of acute mental status changes superimposed on an underlying chronic mental status disorder, such as the "sundowning" seen in demented patients, especially when they are in unfamiliar surroundings, or the delirium caused by a minor infection in a patient with underlying mild dementia (resulting in a "beclouded dementia"). These cases can be challenging at times, especially in attempts to find treatable causes of impaired mental status, and they require careful evaluation.

Aside from Alzheimer's disease, numerous other primary neurodegenerative disorders can lead to dementia, some of which are listed in Table 19.15. Many (such as dementia with Lewy bodies, Parkinson's disease, and others) are associated with movement disorders and are described in Chapters 15 and 16 (see KCC 15.3 and KCC 16.1–16.3). Briefly, recall that **dementia with Lewy bodies** typically begins with fluctuating dementia, parkinsonism, and visual hallucinations. Pathologically, there are intraneuronal inclusions called Lewy bodies in the substantia nigra (as in Parkinson's disease) as well as in more widespread cortical and subcortical structures, often coexisting with the pathologic changes of Alzheimer's disease.

After Alzheimer's disease and dementia with Lewy bodies, the next most common group of primary degenerative dementias is **frontotemporal lobar degeneration** (**FTLD**) (see Table 19.15). Patients with FTLD dementia develop **lobar atrophy**, usually involving the anterior frontal and temporal lobes out of proportion to other structures. The atrophy may be asymmetrical in some patients. Onset is often accompanied by behavioral changes suggestive of frontal lobe dysfunction, such as personality changes, abulia, and disinhibition (see KCC 19.11) or temporal lobe dysfunction including impaired understanding of language and visual information. Classification of FTLD is continually evolving with improved molecular understanding of these disorders. Pathologically, the majority of FTLD cases show neuronal inclusions containing ubiquinated **TDP-43**, which stands for transactive response (TAR) DNA-binding protein of 43 kDa. Other FTLD cases have neuronal inclusion bodies containing the microtubule-associated protein **tau**. The most common form of FTLD is **frontotemporal dementia** (**FTD**), in which there is prominent atrophy of the frontal and anterior temporal lobes, leading to personality

TABLE 19.16 Blessed Dementia Scale[a]

A. CHANGES IN PERFORMANCE OF EVERYDAY ACTIVITIES

Inability to perform household tasks . . . . 0 .5 1

Inability to cope with small sums of money . . . . . 0 .5 1

Inability to remember short lists of items
(e.g., in shopping) . . . . 0 .5 1

Inability to find way about indoors. . . . . 0 .5 1

Inability to find way about familiar streets . . . . . . 0 .5 1

Inability to interpret surroundings (e.g., to
recognize whether in hospital or at home,
to discriminate between patients, doctors,
nurses, relatives, hospital staff, etc.). . . . . 0 .5 1

Inability to recall recent events (e.g., recent
outings, relatives' or friends' visits, etc.). . . . . . . 0 .5 1

Tendency to dwell in the past . . . . 0 .5 1

B. CHANGES IN HABIT

Eating:	Cleanly with proper utensils .	0
	Messily with spoon	1
	Simple solids (no utensils) . .	2
	Has to be fed	3
Dressing:	Unaided.	0
	Occasionally misplaced buttons, etc.	1
	Wrong sequence, often forgets items	2
	Unable to dress	3
Sphincter control:	Complete.	0
	Occasionally wets bed	1
	Frequently wets bed	2
	Doubly incontinent	3

C. CHANGES IN PERSONALITY

Increased rigidity   0 1

Increased egocentricity 0 1

Impairment of regard for feelings of others. 0 1

Coarsening affect   0 1

Impairment of emotional control
(e.g., increased petulance and irritabilty) 0 1

Hilarity in inappropriate situations. 0 1

Diminished emotional responsiveness
(e.g., depression) 0 1

Sexual misdemeanor (de novo in old age). 0 1

D. CHANGES IN INTERESTS AND DRIVES

Hobbies relinquished. 0 1

Diminished initiative or growing apathy. 0 1

Purposeless hyperactivity 0 1

LEFT SCORE _____

E. INFORMATION

Name 0 1

Age 0 1

Time (hour) 0 1

Time of day 0 1

Day of week. 0 1

Date. 0 1

Month. 0 1

Season. 0 1

Year. 0 1

Place: Name. 0 1

Street 0 1

Town 0 1

Type of place (home, hospital, etc.). 0 1

Recognition of persons (any 2 available) 0 1 2

F. PERSONAL MEMORY

Date of birth 0 1

Place of birth 0 1

School attended 0 1

Occupation 0 1

Name of sibling or spouse 0 1

Name of any town where patient worked 0 1

Name of employer 0 1

G. NONPERSONAL MEMORY

Date of WWI (1914–1918) 0 1

Date of WWII (1939–1945) 0 1

President 0 1

Vice president 0 1

H. 5-MINUTE RECALL

(Mr.) John Brown 0 1 2

42 West (Street) 0 1 2

Cambridge, (MA) 0 1

I. CONCENTRATION

Months backwards. 0 1 2

Counting 1–20. . . . . 0 1 2

Counting 20–1. . . . . 0 1 2

RIGHT SCORE _____

TOTAL SCORE _____

[a]From Blessed G, Tomlinson BE, Roth M. 1968. *Br J Psychiatry* 114: 797–811.

changes, and inappropriate disinhibited behavior as early features. Originally described by Pick in the late 1800s and previously called Pick's disease, only about half of FTD patients have the characteristic "Pick bodies" consisting of tau-containing neuronal inclusions, while the other half have inclusions containing TDP-43. In some patients with FTD there is concurrent motor neuron disease resembling ALS (amyotrophic lateral sclerosis; see KCC 6.7). FTD can be familial, with a growing list of identified genes including FTD with parkinsonism localized to chromosome 17, several mutations in the genes encoding the proteins tau and progranulin on chromosome 17, and other familial FTDs localized to chromosome 3 or 9.

Frontotemporal dementia with prominent language dysfunction at the outset was previously called **primary progressive aphasia**, but has now been further subdivided into progressive nonfluent aphasia and semantic dementia. **Progressive nonfluent aphasia** is characterized by left hemisphere perisylvian atrophy and nonfluent aphasia with relative sparing of language comprehension (analogous in some ways to Broca's aphasia; see KCC 19.4). **Semantic dementia** can affect the temporal neocortex of both hemispheres, but when left hemisphere involvement is more prominent, patients have more impaired comprehension of words, while bilateral involvement causes visual agnosias for faces and other objects. In another variant of FTD, patients have **progressive apraxia** and exhibit difficulty carrying out skilled motor tasks or mouth movements (see KCC 19.7) associated with mainly frontoparietal rather than frontotemporal atrophy. In the related disorder **cortical-basal ganglionic degeneration** (cortico-basal degeneration), there is asymmetrical onset of a movement disorder such as dystonia, accompanied by cortical features most often consisting of a marked apraxia.

Although Alzheimer's disease is now recognized as being the most common cause, **vascular dementia** remains the second most common single cause, accounting for 10 to 15% of all cases of dementia. In **multi-infarct dementia**, several cortical or subcortical infarcts cause a stepwise decline in cognitive function. Diffuse subcortical infarcts, often associated with chronic hypertension, cause a form of subcortical dementia termed **Binswanger's disease**. With the advent of CT and MRI scans, however, nonspecific diffuse white matter changes termed leukoaraiosis are often seen in older patients but are not always associated with dementia. When diffuse and severe white matter degeneration of any cause are present, a clinical picture emerges consisting of dementia, pseudobulbar affect, and frontal lobelike features such as shuffling, magnetic gait, and gegenhalten (see KCC 19.11). **Cerebral amyloid angiopathy**, often familial, can cause dementia through multifocal recurrent hemorrhages, as well as through white matter ischemic disease. Patients who recover from severe intracranial hemorrhage of any cause may remain with a static encephalopathy.

Psychiatric pseudodementia resulting from depression or conversion disorder can sometimes be mistaken for dementia. The diagnosis of depression, especially in the elderly, is often overlooked. It is critical to recognize the presence of so-called masked depression in these cases because depression is usually much more treatable than dementia, can have at least as significant an impact on function, and may be life-threatening because of suicide risk. In cases of uncertainty, neuropsychology testing can be helpful in making the distinction. A clinical rule of thumb with many exceptions is that depressed patients often complain about having memory problems, while patients with dementia usually do not. Schizophrenia can also occasionally mimic dementia but, in addition to typically younger age of onset, delusions and hallucinations are more prominent in schizophrenia than in most forms of dementia. In schizophrenia, attention is often profoundly impaired, making examination of cognition difficult. When patients are able to attend, however, they usually are oriented and have intact recent memory, unlike patients with dementia.

Alcoholism is another major cause of dementia. Dementia in such patients is likely multifactorial, with possible causes including thiamine deficiency, other nutritional deficiencies, multiple head injuries, and seizures. Whether alcohol itself causes permanent cortical neuronal injury remains controversial, although it is likely to cause cerebellar degeneration.

Intracranial neoplasms, which were discussed in KCC 5.8, can cause cognitive decline through local cerebral injury or through elevated intracranial pressure. It is important to perform a neuroimaging study as part of the evaluation of dementia because sometimes treatable tumors such as meningiomas can present with no significant neurologic findings other than slowly progressive cognitive decline over months to years. Similarly, imaging studies can elucidate other treatable causes, such as **normal-pressure hydrocephalus** or **chronic subdural hematoma** (see KCC 5.6 and KCC 5.7). In the elderly, chronic subdural hematomas can occur with little or no significant history of head injury.

Diffuse anoxic brain injury such as that occurring after cardiac arrest, complicated delivery, or carbon monoxide poisoning is another important and common cause of static encephalopathy. Similarly, **traumatic brain injury** is an unfortunately common cause of permanent brain damage and chronic mental status changes. In addition, some epidemiological studies suggest that patients who have suffered head injuries with loss of consciousness may have increased risk of later developing Alzheimer's disease. **Dementia pugilistica** is another form of delayed head trauma–related dementia, seen mainly in boxers who suffer recurrent head injuries.

HIV-associated neurocognitive disorder (**HAND**) is common in advanced AIDS, but it can also occur early in the course of the illness, and it responds to treatment with antiretroviral therapy. Dementia can also occur as a result of other HIV-associated illnesses, such as PML (progressive multifocal leukoencephalopathy), and other CNS infections (see KCC 5.9). **Prion diseases**, such as Creutzfeldt–Jakob disease, are unfortunately untreatable at present, and they lead to a relatively rapid cognitive decline often associated with an exaggerated startle response, myoclonus, visual distortions or hallucinations, and ataxia (see KCC 5.9).

We will now briefly touch on a few of the other secondary causes of chronic mental status changes listed in Table 19.15, particularly those that are treatable. **Electrolyte abnormalities**, especially of calcium, magnesium, or sodium, or **hepatic, renal, or pulmonary failure** can cause reversible cognitive impairment. Either **hypothyroidism** or **hyperthyroidism** can impair cognition. Particularly in the elderly, cognitive impairments may occur without other obvious manifestations of thyroid dysfunction. **Vitamin B$_{12}$** deficiency causes megaloblastic anemia, along with **subacute combined degeneration** of the spinal cord (posterior columns more than corticospinal tracts). Subacute combined degeneration can also involve the cerebral white matter, resulting in dementia. The degree of reversibility depends on how quickly vitamin B$_{12}$ deficiency is treated. **Folic acid** deficiency may also be associated with an increased risk of dementia. In **pellagra**, niacin deficiency can cause **d**ementia, **d**ermatitis, and **d**iarrhea (mnemonic: three **D**s).

MNEMONIC

Treatable infectious disorders (see KCC 5.9) such as **chronic meningitis**, particularly caused by *Cryptococcus*, can occasionally present as dementia (most often seen in elderly patients or patients with HIV). **Neurosyphilis** was formerly common and has returned somewhat in recent decades, possibly in association with the rise of HIV. **Lyme disease** is a spirochetal illness that sometimes causes impaired cognition (see KCC 5.9).

Wilson's disease is an important treatable cause of dementia, often presenting in adolescence with hepatic dysfunction, dysarthria, movement disorders, or psychotic manifestations (see KCC 16.1). **Heavy-metal toxicity** can result in cognitive impairment, often with other neurologic signs, such as peripheral neuropathy. **Dialysis dementia** has become uncommon since aluminum was removed from dialysate solutions.

Pathophysiology of Alzheimer's Disease

In Alzheimer's disease, the distribution of pathologic changes parallels the typical clinical features of this disorder (Figure 19.15). The major pathologic changes are **cerebral atrophy**, **neuronal loss**, **amyloid plaques**, and **neurofibrillary tangles**. These changes typically occur initially and are most severe in the following locations (see Figure 19.15), in decreasing order: (1) medial temporal lobes, including the amygdala, hippocampal formation (especially CA1), and entorhinal cortex; (2) basal temporal cortex extending over the lateral posterior temporal cortex, parieto-occipital cortex, and posterior cingulate gyrus; and (3) frontal lobes. Of note, the primary motor, somatosensory, visual, and auditory cortices are relatively spared, at least initially (see Figure 19.15). Cell loss and neurofibrillary tangles are also prominent in the nucleus basalis, septal nuclei, and nucleus of the diagonal band, where cholinergic projections arise (see Figure 14.9B), and to a lesser extent in the locus ceruleus (norepinephrine), and raphe nuclei (serotonin).

The cause of neuronal loss and the other pathologic changes of Alzheimer's disease is still under investigation, although much progress has been made in recent years. **Senile plaques** are composed of an insoluble protein core containing β-amyloid, along with apolipoprotein E, surrounded by a rim of abnormal axons and dendrites called dystrophic neurites. **Neurofibrillary tangles** are intracellular accumulations of hyperphosphorylated microtubule-associated proteins or paired helical filaments known as tau proteins. Amyloid is a general term for insoluble protein deposits that can occur in various organ systems in different forms of amyloidosis. β-**Amyloid** is a specific protein associated with Alzheimer's disease, derived from proteolytic cleavage of a transmembrane protein of unknown function called **amyloid precursor protein** (**APP**). Cleavage of APP can occur at several different sites, but cleavage at an intracellular location by a protein called γ-secretase is thought to promote the formation of toxic soluble β-amyloid oligomers, which later aggregate into insoluble extracellular β-amyloid plaques.

FIGURE 19.15 Typical Distribution of Pathologic Changes in Alzheimer's Disease Severity of composite gray matter changes (including plaques, tangles, gliosis, microvacuolation, neuronal loss, laminar blurring, and shrinkage of gray matter) is indicated by the density of colored dots. (After figure provided by Arne Brun, Lund University Hospital, Lund, Sweden.)

(A)

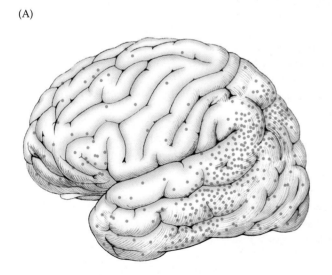

(B)

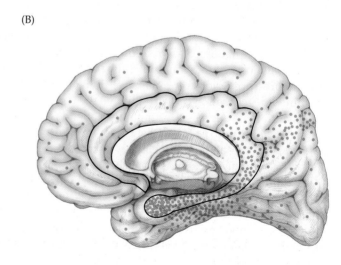

Investigations of the genetic basis of Alzheimer's disease have begun to shed some light on pathogenesis. **Most cases of Alzheimer's disease are sporadic** and occur after the age of 60 years. However, the *ε4* **allele of apolipoprotein E (ApoE4)** carries a 3-fold increased risk of developing late-onset Alzheimer's disease in heterozygotes, and a 15-fold increase in homozygotes. Because ApoE participates in lipid transport, it has been postulated to play a role in modulating plaque formation and clearance, but it may also contribute to disease susceptibility through other mechanisms. Several other potential susceptibility genes are currently under investigation for increased risk of late-onset Alzheimer's disease.

In addition, Alzheimer's disease can rarely be inherited as an **autosomal dominant disorder in some families with early onset**, as early as the third or fourth decade of life. In these families, mutations have been found in three different locations that can cause early-onset disease: (1) the **APP** gene located on chromosome 21, (2) the **presenilin 1 gene** on chromosome 14, and (3) the **presenilin 2 gene** on chromosome 1. These findings are very interesting because APP is the precursor for β-amyloid, while the presenilins appear to be involved in APP cleavage. Additional genetic evidence suggesting the importance of abnormal APP processing in the pathogenesis of Alzheimer's disease comes from Down's syndrome. In Down's syndrome there is an extra copy of chromosome 21, which contains the *APP* gene. Patients with Down's syndrome develop early pathologic and clinical features of Alzheimer's disease after the age of 30 years.

Clinical Features of Alzheimer's Disease

Alzheimer's disease is a common disorder affecting a large proportion of the elderly population. The prevalence of Alzheimer's disease increases rapidly with age, from 1% below the age of 65 to 40% over the age of 85 years. Aside from age, the strongest risk factor is presence of the ApoE4 allele, as already discussed. In rare familial forms of Alzheimer's disease, multiple family members may be affected by early onset. Another possible risk factor is a history of traumatic brain injury.

The clinical course of Alzheimer's disease is variable; however, a description of the typical evolution is instructive and parallels the anatomical distribution of pathologic changes (see Figure 19.15). Obtaining the history from patients with Alzheimer's disease may be difficult because patients often, but not always, have anosognosia and are unaware of their deficits. Family members may be helpful, although some will also deny deficits in loved ones, so the best data comes from either previous mental status assessments (when available) or objective descriptions of functional status at specific times, such as ability to balance a checkbook, pay the bills, shop unaccompanied, and so on.

The earliest clinical feature, often appreciated only in retrospect, is typically a subtle loss of interest in or withdrawal from some previous activities. However, **the dominant early feature of Alzheimer's disease is usually memory loss**, particularly for recent memories, with a relative sparing of remote memories. Patients may have difficulty remembering recent events, where they left their keys, or what they were planning on buying in a store. Although remote memories are less severely impaired, there is often some memory loss for less recent events as well. These early changes in memory function parallel the prominent pathologic involvement of the medial temporal lobe structures in Alzheimer's disease (see Figure 19.15; see also KCC 18.1).

Next, patients often develop word-finding difficulty, or an **anomic aphasia** (see KCC 19.6), along with other features of posterior temporoparieto-occipital dysfunction, including **apraxia** and **visual-spatial deficits** (see KCC 19.7 and KCC 19.10). At some point in the course of the illness, patients develop various behavioral abnormalities that become very difficult for caregivers. Typically, behavioral abnormalities occur later in this illness than in frontotemporal dementia, but eventually they occur in Alzheimer's disease as well. Patients may wander and become lost in the neighborhood; leave the house unclothed; become paranoid or accusatory, sexually inappropriate, agitated, or aggressive; fail to recognize family members; or perform unusual activities, such as placing food in the oven without turning it on or, even worse, turning on all the stove's burners and then leaving the house. One woman cut holes in her pillows, poured orange juice into them, and then explained that she was "feeding the babies." Later in the course of the disease, patients develop more severe **frontal lobe dysfunction**, with gait impairment, abulia, and incontinence.

Of note, motor disturbances are not usually present early in the course, and if abnormal gait is present early on, other diagnoses should be considered. Similarly, while hallucinations may occur in Alzheimer's disease, they are not usually an early feature, and they are more common in dementia with Lewy bodies. In late Alzheimer's disease, patients eventually become akinetic, mute, unresponsive, and bedridden, ultimately succumbing to infection or other illnesses. The median time of survival from onset is approximately 8 years but is age-dependent.

The evaluation of patients with suspected Alzheimer's disease is similar to that of other patients with dementia, as described earlier.

Treatment of Alzheimer's Disease

Although a definitive treatment for Alzheimer's disease has not yet been developed, a combination of symptomatic treatment and counseling of the patient and family members can substantially improve quality of life. The NMDA glutamate receptor antagonist memantine and several cholinesterase inhibitors, including donepezil, rivastigmine, and galantamine, have been shown to produce a modest improvement in cognitive function in patients with Alzheimer's disease. The occurrence of depression, psychosis, or agitation should be treated carefully with medications, with care being taken to avoid worsening functional status with medication side effects.

Investigations are ongoing for a number of agents that may slow the progression of Alzheimer's disease. These include γ-secretase inhibitors, specific β-amyloid antibodies, gamma globulin (which includes anti-amyloid antibodies), inhibition of amyloid-related inflammatory responses by blocking receptors for advanced glycated endproducts (RAGE), and other strategies. It is hoped that with further research, a cure for this common and debilitating disorder will be developed in the near future. ■

CLINICAL CASES

CASE 19.1 ACUTE SEVERE APHASIA, WITH IMPROVEMENT

CHIEF COMPLAINT

A 74-year-old right-handed woman was brought to the emergency room because of **sudden inability to speak and right-sided weakness**.

HISTORY

One month prior to admission, the patient was diagnosed with atrial fibrillation associated with hyperthyroidism. Her medications included Coumadin anticoagulation, but she may have been noncompliant. During dinner on the evening of admission, she suddenly stopped speaking and was noted by her family to have right-sided weakness. By the time they brought her to the emergency room, the right-sided weakness had mostly resolved, but her speech remained very abnormal.

PHYSICAL EXAMINATION (HOSPITAL DAY 1)

Vital signs: T = 99.1°F, P = 100, BP = 138/84, R = 20.

Neck: Supple with no bruits; thyroid slightly enlarged.

Lungs: A few crackles at the bases bilaterally.

Heart: Irregular rate; tachycardic; 2/6 systolic murmur heard at apex.

Abdomen: Normal bowel sounds; soft, nontender.

Extremities: No edema.

Neurologic exam:

MENTAL STATUS: Alert.

1. SPONTANEOUS SPEECH: **Marked paucity of spontaneous speech**; said only single words and rare rote phrases, such as "No" (inappropriately), "I can't," or "I don't know." Cleared throat as if about to speak, but then didn't say anything or said very little. Uttered occasional meaningless **paraphasic** phonemes. Voice **dysarthric** and hypophonic. Did better with rote tasks and singing. For example, once the examiner got her started, the patient was able to continue counting up to 7 and to sing a few phrases of "Happy Birthday." She appeared frustrated by her difficulties.

2. COMPREHENSION: **Followed no commands** correctly except "close your eyes." Did not answer yes/no questions or point appropriately at named objects. Did imitate or obey gestures occasionally and caught on to tasks on the basis of context.

3. REPETITION: **Could repeat "no," but not "cat,"** "Alice," or other words.

4. NAMING: **Named no objects.**

5. READING: **Unable to read.**

6. WRITING: **Scribbled** with pen, producing no letters.

CRANIAL NERVES: Pupils 3 mm bilaterally. Preserved blink to threat bilaterally. Extraocular movements intact. Corneal reflex intact bilaterally. **Marked weakness of right face, sparing forehead.** Normal gag.

MOTOR: Normal tone. **Finger taps slightly slowed on the right.** Normal strength bilaterally, but testing limited by motor impersistence. Able to use both hands well when putting on socks.

REFLEXES:

COORDINATION: No ataxia of arm movements when reaching for objects bilaterally.

GAIT: Somewhat tentative, but narrow based.

SENSORY: Flinched equally in response to pinprick in all extremities.

FOLLOW-UP EXAM (HOSPITAL DAY 3)

On motor exam, the patient had **mild clumsiness and pronator drift of the right arm**. Spontaneous speech consisted of **only a few single words**. Comprehension was much improved, and she could answer most yes/no questions appropriately and followed one-step commands. **Repetition, naming, reading, and writing were still very poor or nonexistent.**

FOLLOW-UP EXAM (HOSPITAL DAY 5)

Spontaneous speech was markedly nonfluent, with brief agrammatical phrases and occasional paraphasias. For example, when asked to describe what she saw in a complex scene that included water overflowing a sink, she said, "Walur over here." She continued to seem frustrated and discouraged by her limitations. Comprehension continued to improve, and she was able to answer simple questions related to herself with 70% accuracy, and to point to objects with 75% accuracy, with occasional perseveration. When asked to name a pen, she said "pencil," and she could not name a watch. She wrote "Thank you" instead of her name, and she could not read it back.

LOCALIZATION AND DIFFERENTIAL DIAGNOSIS

1. Summarize the findings on Day 1. What kind of aphasia did the patient have that day, and what is the localization? What is the most likely cause?

2. Summarize the findings on Days 3 and 5. What kind of aphasia did the patient have, and what is the localization?

Discussion

1. The key symptoms and signs in this case on Day 1 are:

   - **Poor fluency**
   - **Poor comprehension**
   - **Poor repetition**
   - **Paraphasias, anomia, alexia, agraphia, and frustration**
   - **Dysarthria, with right face and hand weakness**

   Severely impaired fluency, comprehension, and repetition together are classified as global aphasia (see Figure 19.4; KCC 19.6), a diagnosis that is compatible with the other language abnormalities seen in this patient. Persistent global aphasia is usually caused by large lesions involving the anterior and posterior language areas, or by large subcortical lesions (see KCC 19.6). However, global aphasia can also occur acutely in new-onset lesions confined to the anterior or posterior language areas. The fact that her visual fields were not involved based on her preserved blink to threat suggests that the more posterior parietotemporal regions were likely spared. In addition, combined anterior–posterior language area lesions or large subcortical lesions usually result in contralateral hemiplegia, whereas this patient had relatively mild arm weakness and no leg weakness. The pattern of weakness in this patient is compatible with a lesion encroaching on the left lateral precentral gyrus face and hand areas (see Figures 6.2, 6.14D), suggesting that she has acute global aphasia resulting from a lesion in the left lateral frontal cortex, including the anterior language areas. Given her vascular risk factors of atrial fibrillation and possible noncompliance with anticoagulation therapy, as well as the sudden onset of her deficits, the most likely diagnosis is embolic infarction in the territory of the left middle cerebral artery superior division (see Figure 19.3A). Another possibility is a hemorrhage in the left frontal lobe, especially because she is taking Coumadin. Focal seizures or migraine are also possible but much less likely (see Table 19.5).

2. The key symptoms and signs in this case on Days 3 and 5 are:

   - **Poor fluency**
   - **Poor repetition**
   - **Comprehension relatively spared**
   - **Poor naming, reading, and writing, with paraphasias and frustration**
   - **Mild right arm weakness**

   By days 3–5 the patient's global aphasia had improved to a typical Broca's aphasia (see KCC 19.4). This pattern of evolution is often seen in acute lesions of the left lateral frontal lobe. The most common cause would be left MCA superior division infarct.

Clinical Course and Neuroimaging

Initial head CT was normal. The patient's admission INR (International Normalized Ratio), a measure of the amount of anticoagulation, was subtherapeutic at 1.1 (target 2.0–3.0), and levels of her heart medication were also low, supporting medication noncompliance as the cause of a presumed embolic infarction in the territory of the left MCA superior division. Because of the time of her arrival at the emergency room, she was not a candidate for tPA (see KCC 10.4), however she was admitted to the hospital for further evaluation and treatment. Two days after admission, a **brain MRI** (Image 19.1A,B, page 941) revealed a medium-sized infarct in the left frontal operculum involving Broca's area, just in front of the precentral gyrus face and arm areas.

As already noted, the patient's global aphasia rapidly evolved into a Broca's aphasia, and this condition continued to improve, as did her trace right-sided weakness. She was seen in follow-up 2 months later, at which time she was able to speak in sentences with occasional pauses to find words, and occasional paraphasias. Comprehension, repetition, reading, and writing were relatively good (although not perfect), and her main difficulties were with naming and finding words. Thus, as is commonly seen in cases with good recovery, her residual deficit was predominantly a dysnomic aphasia.

CASE 19.2 NONSENSICAL SPEECH

MINICASE

An 81-year-old right-handed woman with a history of hypertension was brought to the emergency room by her family one morning when she was suddenly **"unable to communicate properly,"** speaking with words and sentences that did not make any sense. On exam, she had an irregular pulse, and spontaneous speech was fluent with normal prosody and grammatical constructs; however, most of what she said was **meaningless and did not fit the context, showing frequent paraphasic errors and repetition. She followed no commands, except to close her eyes. When asked to raise her arms, she said, "What do you want?" She could not repeat even single words. On testing of naming, she called a pen "red rains," a watch "round thing," a tie "po, do, bi, fisdo," and a pen (on second presentation) "like when you want** to write down something." Writing sample: **"the wond youg wweng whaweta."** She could not read what she wrote. When asked to read, **"The dog ran down the road,"** she read, **"The roth ran ra a goth."** She had a pleasant, unconcerned affect and seemed oblivious to any deficits. Blink to threat was decreased on the right side. The remainder of her exam was normal.

LOCALIZATION AND DIFFERENTIAL DIAGNOSIS

1. What kind of aphasia does this patient have? Where is her lesion?

2. What is the most likely cause, and what are some other possibilities?

Discussion

The key symptoms and signs in this case are:

- Fluent, meaningless speech
- Poor comprehension
- Poor repetition
- Poor naming, reading, and writing; frequent paraphasias
- Lack of concern about and obliviousness to deficits
- Right visual field defect

1. This patient has typical findings of Wernicke's aphasia (see KCC 19.5). The localization therefore is most likely in the left posterior superior temporal and left inferior parietal region. A lesion in this location would also explain the patient's right visual field cut because it could interrupt the left optic radiations (see Figure 11.8).

2. The left temporoparietal region lies in the territory of the left MCA inferior division. The most likely diagnosis, especially given the patient's age, history of hypertension, and acute onset, is an infarct in this territory (see Figure 19.3A). Hemorrhage in this location is also possible, with focal seizure or migraine again possible but unlikely (see also Table 19.5).

Clinical Course and Neuroimaging

The patient did not arrive at the emergency room in time for tPA therapy (see KCC 10.4). A head CT on the day of admission showed an infarct in the left temporoparietal region, in the territory of the left MCA inferior division. This was confirmed 4 days later by an **MRI scan** (Image 19.2A,B, page 943). Note that the infarct included Wernicke's area along with the adjacent temporoparietal cortex, as well as the left optic radiation where it passes just lateral to the atrium of the lateral ventricle. The MRI also demonstrated moderate cortical atrophy as an incidental finding in this patient. Admission electrocardiogram showed new onset of atrial fibrillation. Other embolic workup (see KCC 10.4) was negative. The patient was treated with warfarin (Coumadin) anticoagulation and discharged home in the care of her family after 1 week. Further follow-up was not available.

Additional Basic Aphasia Cases

For additional cases illustrating **basic types of aphasia**, see Chapter 10 (Cases 10.5, 10.6, and 10.8).

CASE 19.3 APHASIA WITH PRESERVED REPETITION

MINICASE

A 63-year-old right-handed woman with a history of breast cancer in remission was talking on the phone with her sister one evening and suddenly began to have **difficulty getting the words out, and she could not answer simple questions**. She was taken to the hospital, and on exam she was alert and able to state her name, the month, and the year, but not the date, and she correctly chose her location by multiple choice. Her language exam was as follows:

1. *SPONTANEOUS SPEECH:* **Halting, labored, telegraphic, with decreased use of function words (verbs and prepositions). She frequently used fillers such as "um," "aaah," and "you know" and had many paraphasic errors and neologisms. She did better once started by the examiner on automatic tasks such as naming the days of the week. On word generation tasks, she could name only six animals in 1 minute, and she came up with no words starting with the letter "s" in 1 minute.**

2. *COMPREHENSION:* Followed three-step verbal commands. Was able to rapidly identify body parts, objects, shapes, and letters by pointing. Correctly answered 6/8 questions on verbally presented paragraphs. Had difficulty following only lengthy, complex commands.

3. *REPETITION:* Repeated words and sentences with 100% accuracy (much better than spontaneous speech). With low-probability sentences, she made occasional paraphasic errors.

4. *NAMING:* **Could name only 1/6 objects and no shapes.** Naming was improved when she was given the first sound of the word as a hint.

5. *READING:* Reading aloud was not tested.

6. *WRITING:* **Wrote the following sentence spontaneously: "I this a mighty foyin am y."**

The remainder of her exam was normal.

LOCALIZATION AND DIFFERENTIAL DIAGNOSIS

1. What kind of aphasia was present in this patient? Where was her lesion?

2. What are some possible causes in this patient?

Discussion

The key symptoms and signs in this case are:

- **Poor fluency**
- **Repetition and comprehension relatively spared**
- **Poor naming and writing; paraphasias**

1. This patient had a transcortical motor aphasia (see Figure 19.4). This can be produced by lesions located in the lateral frontal cortex in the dominant hemisphere that spare the arcuate fasciculus, the other peri-Sylvian conduct-

CASE 19.1 ACUTE SEVERE APHASIA, WITH IMPROVEMENT

IMAGE 19.1A,B Left MCA Superior Division Infarct
MRI scans (A) Axial T2-weighted image showing bright region compatible with infarct in the left frontal opercu-lum including Broca's area. (B) Sagittal T1-weighted image showing hypointense region compatible with infarct.

(A)

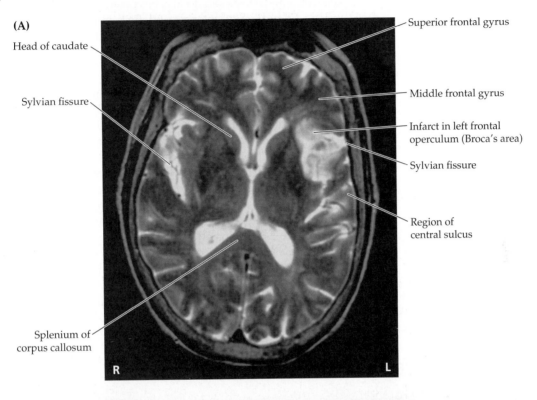

Head of caudate

Sylvian fissure

Superior frontal gyrus

Middle frontal gyrus

Infarct in left frontal operculum (Broca's area)

Sylvian fissure

Region of central sulcus

Splenium of corpus callosum

R   L

(B)

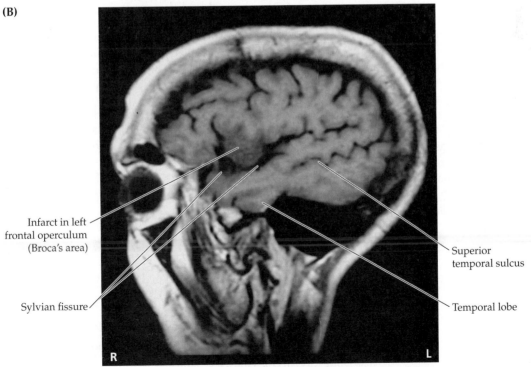

Infarct in left frontal operculum (Broca's area)

Sylvian fissure

Superior temporal sulcus

Temporal lobe

R   L

ing pathways, and Broca's area (see Figure 19.3B), or occasionally by sub-cortical lesions of the dominant hemisphere sparing the same regions.

2. Possible causes of a lesion in these locations resulting in sudden onset of deficits include watershed infarct in the ACA–MCA territory (see Figure 19.3B) or hemorrhage (see KCC 5.6). Given the patient's history of breast cancer, a metastasis should also be considered, possibly resulting in hemorrhage. Of note, sudden onset of deficits in brain tumors can sometimes occur even without hemorrhage, although this is relatively uncommon.

Clinical Course and Neuroimaging

Head CT (Image 19.3A, page 945) showed a hemorrhage in the left frontal lobe, sparing the peri-Sylvian cortex and sparing Broca's area, but lying just dorsal to it. This hemorrhage would be expected to disconnect Broca's area from other structures in the left frontal lobe needed for language formulation, producing the transcortical motor aphasia (impaired fluency, preserved repetition) seen in this patient. An **MRI scan** (Image 19.3B, page 945) done with gadolinium was negative for brain metastases but again showed a hemorrhage in this location. The exam and neuroimaging findings in this patient should be contrasted against those for the patient with Broca's aphasia in Case 19.1.

Follow-up MRI scans at 1, 4, and 9 months after the hemorrhage showed no evidence of metastases or vascular malformation. The patient did not undergo an angiogram, but MRAs were unremarkable. Her language gradually improved, and 1 year later her fluency was nearly normal. She remained in active speech therapy, and her main residual problems were some word-finding difficulties and circumlocution.

CASE 19.4 IMPAIRED REPETITION

MINICASE

A 67-year-old right-handed woman was in a motor vehicle accident with **brief loss of consciousness and mild confusion at the scene**. Past history was significant for a mechanical mitral valve and coronary artery disease, treated with aspirin and Coumadin anticoagulation. After the accident she complained of **left frontal headache, neck pain, nausea, and vomiting**. She was brought to the emergency room, where cervical spine X-rays were normal and initial neurologic exam was described as normal except for **amnesia for the accident and uncertainty about the exact date**, although she knew the correct month and the year. Two days later her headache worsened, and she was noted to have some new speech difficulties. A neurology consult was called. Exam showed some **nuchal rigidity**, a systolic murmur, fluent speech, intact comprehension following three-step commands, but **difficulties with repetition, being unable to repeat "no ifs, ands, or buts" or other similar short phrases**. In addition, she had a **slight right pronator drift and an upgoing toe on the right side**.

LOCALIZATION AND DIFFERENTIAL DIAGNOSIS

1. What kind of aphasia did this patient have on exam, and what is the localization of her aphasia and associated mild motor abnormalities?

2. What diagnosis is suggested by the patient's history and her initial symptoms?

Discussion

The key symptoms and signs in this case are:

- Poor repetition
- Relatively preserved fluency and comprehension
- Mild right pronator drift and right Babinski's sign
- Brief loss of consciousness, amnesia, left frontal headache, nausea, vomiting, and nuchal rigidity

CASE 19.2 NONSENSICAL SPEECH

IMAGE 19.2A,B Left MCA Inferior Division Infarct Axial T2-weighted images.
A, B progress from inferior to superior. The infarct includes Wernicke's area.

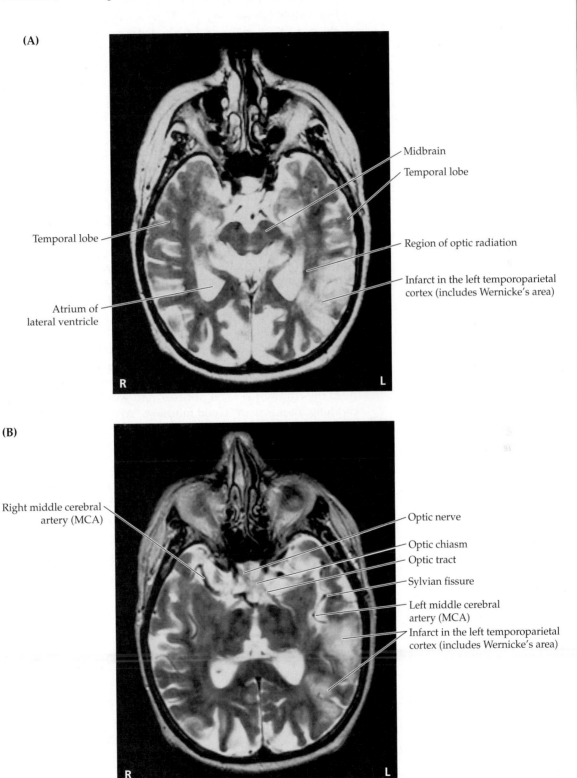

(A)

Midbrain

Temporal lobe

Temporal lobe

Region of optic radiation

Infarct in the left temporoparietal
cortex (includes Wernicke's area)

Atrium of
lateral ventricle

R   L

(B)

Right middle cerebral
artery (MCA)

Optic nerve

Optic chiasm

Optic tract

Sylvian fissure

Left middle cerebral
artery (MCA)

Infarct in the left temporoparietal
cortex (includes Wernicke's area)

R   L

1. This patient had conduction aphasia (see Figure 19.4), which can be caused by subcortical lesions involving the arcuate fasciculus or by cortical lesions of the peri-Sylvian region in the dominant hemisphere (see Figure 19.2A). The mild right corticospinal findings also suggest a lesion in the left hemisphere causing mild impingement on the nearby corticospinal tract (see Figure 6.10A).

2. The brief loss of consciousness and confusion suggest a possible head injury during the motor vehicle accident. The headache, nausea, vomiting, and nuchal rigidity are signs of meningeal irritation (see Table 5.6), which, together with the patient's use of oral anticoagulation and aspirin, suggest intracranial hemorrhage (see KCC 5.6) located in the peri-Sylvian region of the dominant (left) hemisphere. Other, less likely possibilities include infarct, neoplasm, or infection in this location.

Clinical Course and Neuroimaging

Head CT (Image 19.4A–D, pages 946–947) demonstrated a traumatic hemorrhage (see KCC 5.5, 5.6) in the left Sylvian fissure. This included layering of subarachnoid blood in the sulci, as well as a more confluent parenchymal hematoma. The hemorrhage involved the peri-Sylvian region, including the region of the arcuate fasciculus, and could thus disconnect the anterior and posterior language areas (see Figure 19.2A). A small subdural hematoma was present as well, adjacent to the right falx cerebri (not shown on Image 19.4A–D), likely unrelated to this patient's findings.

The patient's aspirin was discontinued and her Coumadin temporarily stopped. She was also given a few units of fresh frozen plasma to partly reverse her anticoagulation. She remained stable over the next few days and was carefully treated with low-dose heparin before resuming her Coumadin. Nine days after admission, her repetition had improved, but she also practiced the commonly used phrases, and would greet the neurology team on morning rounds with "Good morning. No ifs, ands, or buts. If I were here, she would be there....How was that?" She continued to have some mild difficulties, especially when repeating new sentences that she had not heard before.

CASE 19.3 APHASIA WITH PRESERVED REPETITION

IMAGE 19.3A,B Left Frontal Hemorrhage The hemorrhage lies just dorsal to Broca's area and spares the peri-Sylvian region. (A) Axial CT scan image. (B) Coronal T1-weighted MRI with intravenous gadolinium.

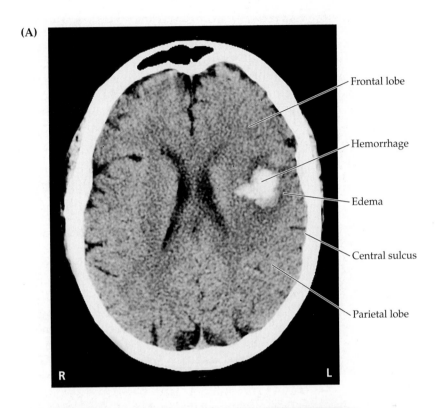

(A)

- Frontal lobe
- Hemorrhage
- Edema
- Central sulcus
- Parietal lobe

R   L

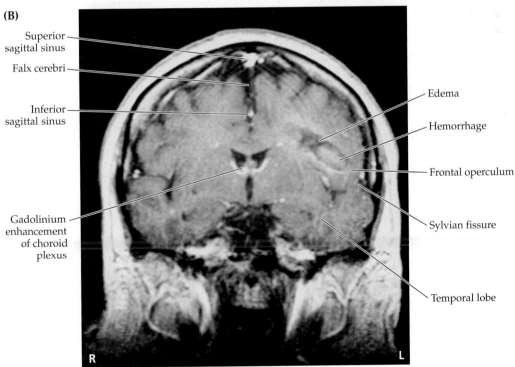

(B)

- Superior sagittal sinus
- Falx cerebri
- Inferior sagittal sinus
- Gadolinium enhancement of choroid plexus
- Edema
- Hemorrhage
- Frontal operculum
- Sylvian fissure
- Temporal lobe

R   L

CASE 19.4 IMPAIRED REPETITION

IMAGE 19.4A–D **Left Peri-Sylvian Intraparenchymal Hemorrhage in Region
of Arcuate Fasciculus** Axial CT images. A–D progress from inferior to superior.

(A)

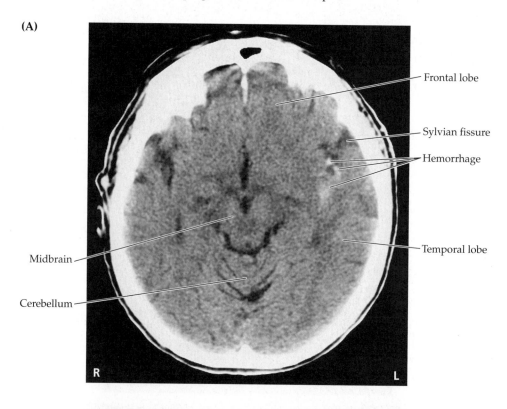

(B)

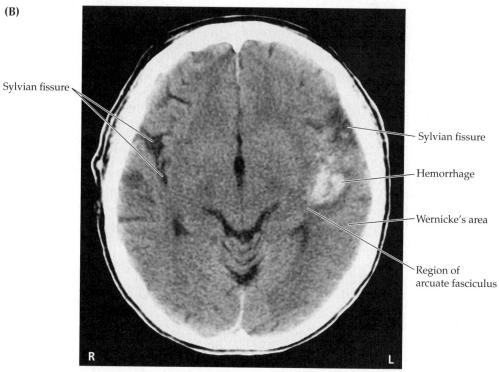

CASE 19.4 *(continued)*

(C)

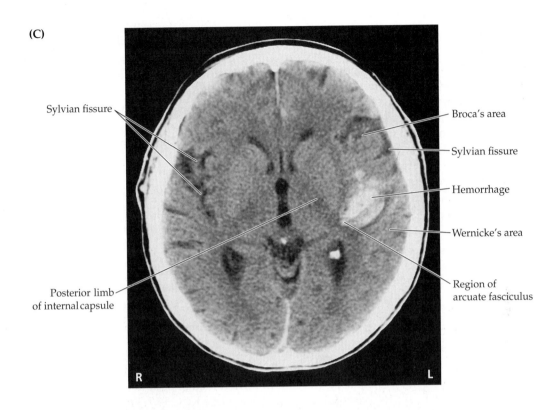

Sylvian fissure

Broca's area

Sylvian fissure

Hemorrhage

Wernicke's area

Posterior limb
of internal capsule

Region of
arcuate fasciculus

R   L

(D)

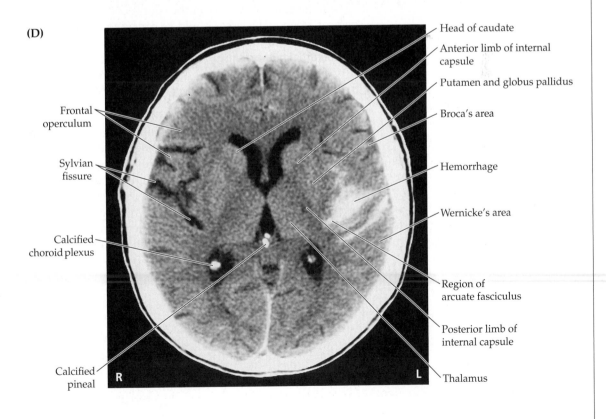

Head of caudate

Anterior limb of internal
capsule

Putamen and globus pallidus

Frontal
operculum

Broca's area

Sylvian
fissure

Hemorrhage

Wernicke's area

Calcified
choroid plexus

Region of
arcuate fasciculus

Posterior limb of
internal capsule

Calcified
pineal

Thalamus

R   L

CASE 19.5 INABILITY TO READ, WITH PRESERVED WRITING SKILLS

MINICASE

A 64-year-old right-handed woman had 1 week of progressive **troubles with her vision and reading**. Of note, she had a 3-year history of colon cancer with metastases to the liver. Exam was normal, **except for inability to read written words**, **some difficulties with short-term verbal memory**, **and a right homonymous hemianopia**. She was able to write normally, writing "Today is a nice day" and "It is a sunny day in Boston," but when shown her own writing a few minutes later, she was unable to read it. She also had a **subtle dysnomia**, naming a watch, ring, finger, elbow, and lips, but unable to name knuckle, nail, veins, or hand. Color naming was not tested.

LOCALIZATION AND DIFFERENTIAL DIAGNOSIS

1. What is the name for the syndrome that includes the inability to read and preserved writing seen in this patient, and where is it localized?
2. Given the gradual onset of deficits and the patient's history, what are some possible causes of a lesion in this location?

Discussion

The key symptoms and signs in this case are:

- **Ability to write normally, but inability to read**
- **Right homonymous hemianopia**
- **Some difficulties with short-term verbal memory**
- **Mild dysnomia**

1. This patient has the classic findings of alexia without agraphia, which is caused by lesions of the left medial occipital visual cortex extending to the posterior corpus callosum (see KCC 19.7; Figure 19.5).

2. Given the gradual onset over the course of a week and the history of colon cancer, brain metastases should be considered in this patient. Other possibilities include a slowly evolving infarct in the left PCA territory, evolving intracranial hemorrhage, other intracranial neoplasm, or abscess.

Clinical Course and Neuroimaging

A **head CT** with intravenous contrast (Image 19.5A,B, page 951) revealed a large, cystic enhancing lesion in the left medial occipital lobe with edema and mass effect extending into the posterior corpus callosum. The patient was admitted and treated with steroids to reduce edema. By means of a stereotactic approach (see KCC 16.4), fluid was drained from the cyst with a needle, leading to a partial improvement in her hemianopia and reading difficulties. In addition, a stereotactic biopsy revealed metastatic adenocarcinoma. She was then treated with stereotactic proton beam radiosurgery (see KCC 16.4) directed at the lesion, and she remained stable at last available follow-up, 3 months after presentation.

CASE 19.6 LEFT HEMINEGLECT*

MINICASE

A 61-year-old left-handed security guard with a history of cigarette smoking had an **episode of left hand tingling lasting less than an hour** that was reported to medical staff by a friend. The next day he was at the grocery store buying a lottery ticket and reportedly slumped briefly to the floor. He **denied that anything was wrong** but said, "They called an ambulance because they said I had a stroke." On examination, he was **unaware of having any deficits** and was quite **impatient and grouchy**, wanting to go home. He had **profound left visual neglect**, describing only the curtains to the far right in a picture of a complex visual scene (Figure 19.16) and reading only the right two words on each line of a magazine article. When asked to write or draw a clock, he **moved the pen in the air off to the right of the page.** He then handed the pen back, saying, "I'm finished," apparently thinking he had completed the task. He had **no blink to threat on the left**, a **marked right gaze preference**, and **mildly decreased left nasolabial fold**. Spontaneous **movements were decreased on the left side**, but with provocation he was able to achieve **4/5 strength in the left arm and leg**. He was able to feel touch on the left side but had **extinction on the left to double simultaneous tactile stimulation. Reflexes were slightly brisker on the left.**

LOCALIZATION AND DIFFERENTIAL DIAGNOSIS

1. Summarize the different types of neglect demonstrated by this patient (see Table 19.7).
2. On the basis of the symptoms and signs shown in **bold** above, what is the most likely location of the lesion in this patient?
3. What is the most likely diagnosis?

*Note: This patient was also presented in Case 10.10.

Copyright © 1972 by Lea and Febiger

FIGURE 19.16 Complex Picture Used for Visual Neglect Testing Asked to describe what he saw, the patient mentioned only the curtains on the far right edge of the page. (Reproduced with permission of Lea and Febiger.)

Discussion

The key symptoms and signs in this case are:

- Anosognosia
- Left visual neglect
- Extinction on the left to double simultaneous tactile stimulation
- Movement of the hand off to the right of the page when asked to draw, and decreased spontaneous movements on the left side, but ability to achieve 4/5 strength when pressed
- Impatience and irritability
- Right gaze preference
- No blink to threat on the left
- Mildly decreased left nasolabial fold, 4/5 strength in the left arm and leg, and slightly brisker reflexes on the left side
- Transient episode of left hand tingling

scribed and in Case 19.7. In other cases, more subtle perseveration can be detected by tests such as manual or written alternating sequencing tasks (see **neuroexam.com Videos 19 and 20**). Image 19.7C (page 958) shows an example of a different patient who exhibited perseveration on a written alternating sequencing task. Interestingly, this patient had benzodiazepine toxicity but did not have a frontal lobe lesion, demonstrating that diffuse disorders can sometimes mimic lesions of the frontal lobes (see KCC 19.11).

CASE 19.6 LEFT HEMINEGLECT

IMAGE 19.6A Right MCA Inferior Division Infarct Axial head CT image 2 months after onset of symptoms, demonstrating a right temporoparietal hypodensity compatible with infarct.

(A)

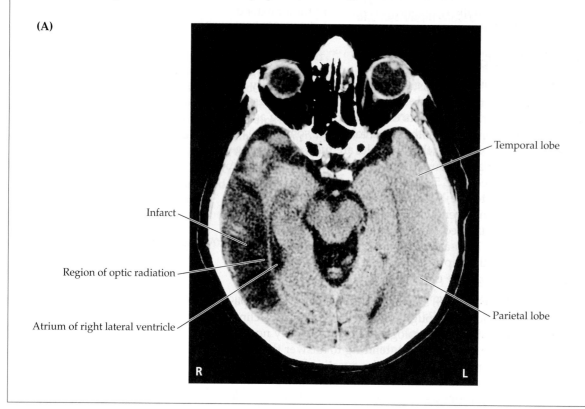

CASE 19.6 *RELATED CASE*

IMAGE 19.6B–E Head CT Scans from a Patient with Left Hemispherectomy and Hydrocephalus Axial CT scan images. (B,C) Before shunting. (D,E) After ventriculoperitoneal shunt placement. (From Kalkanis SN, Blumenfeld H, Sherman JC, Krebs DE, Irizarry MC, Parker SW, Cosgrove GR. 1996. Delayed complications 36 years after hemispherectomy: a case report. *Epilepsia* 37 (8): 758–762.)

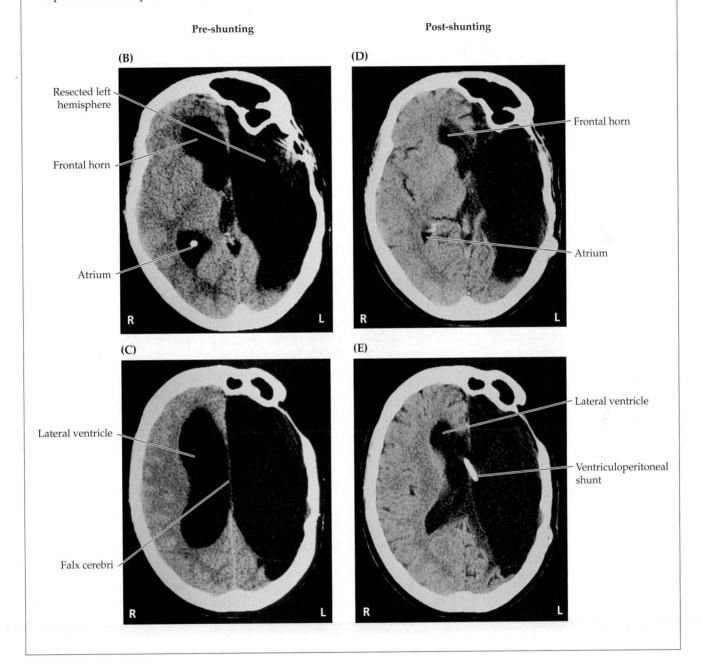

Pre-shunting

Post-shunting

(B)

Resected left hemisphere

Frontal horn

Atrium

R   L

(D)

Frontal horn

Atrium

R   L

(C)

Lateral ventricle

Falx cerebri

R   L

(E)

Lateral ventricle

Ventriculoperitoneal shunt

R   L

CASE 19.6 *RELATED CASE*

IMAGE 19.6F,G Clock Drawings by a Patient with Left Hemispherectomy and Hydrocephalus Same patient as in Image 19.6B–E. (F) Before shunting, the patient exhibited right hemineglect. (G) Neglect recovered after ventriculoperitoneal shunt placement. (From Kalkanis SN, Blumenfeld H, Sherman JC, Krebs DE, Irizarry MC, Parker SW, Cosgrove GR. 1996. Delayed complications 36 years after hemispherectomy: a case report. *Epilepsia* 37 (8): 758–762.)

(F)

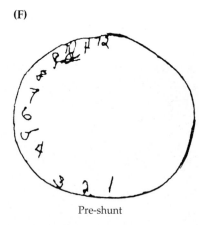

Pre-shunt

(G)

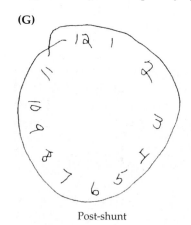

Post-shunt

CASE 19.8 BLINDNESS WITHOUT AWARENESS OF DEFICIT

MINICASE

An 82-year-old woman was brought to the emergency room by her family because she seemed to have been having vision problems for 3 days. She **needed to hold on to the walls to find her way around her apartment**, but she seemed otherwise normal, except for a **bifrontal headache. The patient claimed that nothing was wrong and that she could see normally.** On exam, **she was unable to recognize large ob-jects even when they were held right in front of her face, and she had no blink to threat.**

LOCALIZATION AND DIFFERENTIAL DIAGNOSIS

What syndrome does this patient have, and what is its localization? What is the most likely diagnosis?

Discussion

The key symptoms and signs in this case are:

- **Complete blindness**
- **Unawareness of blindness**
- **Bifrontal headache**

This patient had complete visual loss, together with anosognosia. This syndrome can be seen with bilateral lesions of the visual cortex and is called cortical blindness, or Anton's syndrome (see KCC 19.12). Given the patient's age and the sudden onset, the most likely diagnosis is infarction of the bilateral medial occipital lobes. This can be caused by an embolus or emboli that reach the top of the basilar artery and then break up to occlude the PCAs bilaterally (see Figures 10.3, 10.5). Another possibility is bilateral hemorrhage in the visual cortex. Multiple hemorrhages, often in a posterior distribution, can be seen in amyloid angiopathy (see KCC 5.6). The patient could also have a tumor or infection involving both occipital lobes. The bifrontal headache is consistent with any of these causes of occipital pathology (see Table 14.6). Finally, the patient may have a lesion in the frontal lobes or right parietal lobe causing anosognosia, along with separate lesions somewhere in the visual pathway causing bilateral visual loss.

Clinical Course and Neuroimaging

A **head CT** (Image 19.8A, page 959) revealed bilateral infarcts of the calcarine (primary visual) cortex. In addition, there were bilateral infarcts higher up in the white matter underlying the posterior parietal lobes bilaterally (Image 19.8B, page 959). The patient did not arrive in time for treatment with tPA, but was admitted while an embolic workup was done, revealing no clear cause for her infarcts. Within a few days of admission, she regained some vision in a small sector of the right superior quadrant in each eye. She was able to consistently detect finger movements and to identify the pattern on the examiner's tie in this area of her visual field. With this regained vision, however, some elements of Balint's syndrome emerged (see KCC 19.12), including difficulty reaching out and grasping objects in front of her, making inaccurate groping motions (optic dysmetria). These features may have been caused by the presence of bilateral infarcts underlying the dorsolateral parietal visual association cortex (see Image 19.8B; see also Figure 19.12). She was discharged to a rehabilitation hospital to work on strategies to compensate for her new deficits.

CASE 19.7 ABULIA

IMAGE 19.7A "Butterfly" Astrocytoma of the Bilateral Frontal Lobes Axial CT scan demonstrating tumor extending through the corpus callosum, leading to extensive involvement of both frontal lobes.

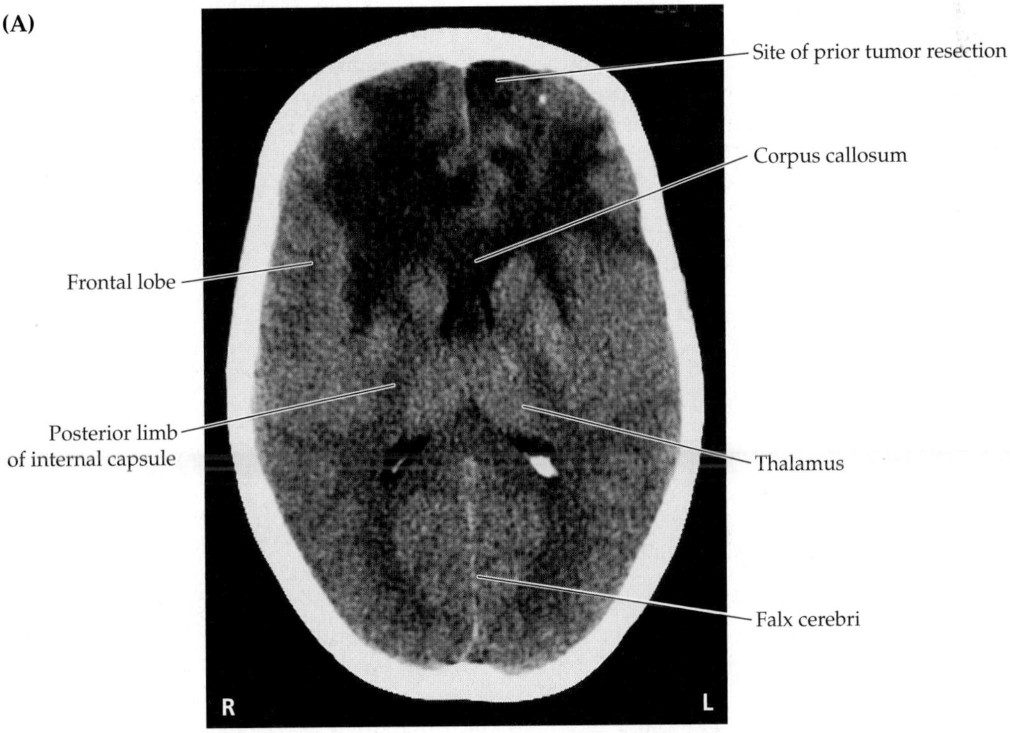

CASE 19.7 *RELATED CASE*

IMAGE 19.7B Large Falcine Meningioma Compressing the Frontal Lobes
Axial T2-weighted MRI.

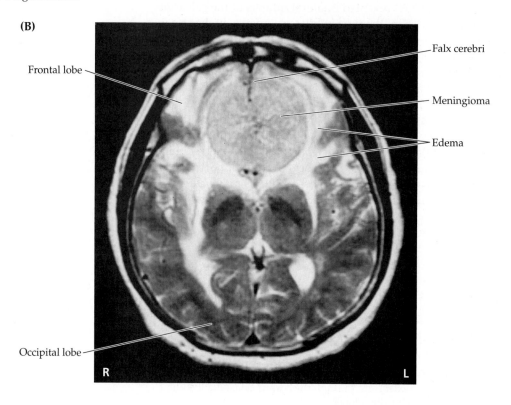

(B)

Frontal lobe

Falx cerebri

Meningioma

Edema

Occipital lobe

R   L

IMAGE 19.7C Written Alternating Sequencing Task
Patient was instructed to copy a pattern drawn by the examiner and continue it to the end of the page. Perseveration and "closing in" are evident in this patient with ben-zodiazepine toxicity. This is an example of apparently focal deficits that can be seen in patients with diffuse encephalopathy (see KCC 19.15).

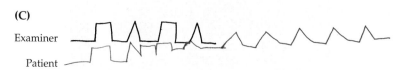

(C)

Examiner

Patient

CASE 19.8 BLINDNESS WITHOUT AWARENESS OF DEFICIT

IMAGE 19.8A,B Bilateral Infarcts of the Visual Cortex Axial CT scan images. A, B progress from inferior to superior.

(A)

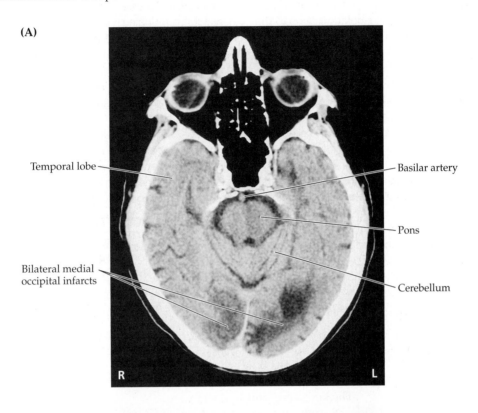

Temporal lobe

Basilar artery

Pons

Bilateral medial occipital infarcts

Cerebellum

R   L

(B)

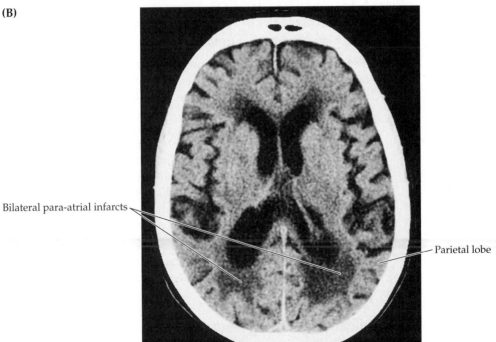

Bilateral para-atrial infarcts

Parietal lobe

R   L

CASE 19.9 SUDDEN INABILITY TO RECOGNIZE FACES

MINICASE

A 78-year-old right-handed man with a history of atrial fibrillation treated with Coumadin anticoagulation came to the emergency room because of unusual visual disturbances. He had a history of a cerebral infarct 6 years previously, resulting in a left homonymous hemianopia, which subsequently improved. On the day prior to admission, while driving his car, he suddenly saw **"a translucent white curtain" obscuring the upper part of his vision** and making it hard to see traffic. He felt that his **motion perception was "off"** because he nearly hit a truck that was stopped in front of him. By the morning of admission, he no longer saw a curtain but instead had the perception of **"shimmering glass balls or water drops" dotting the upper areas of his vision with both eyes**. He then encountered a close friend on the street, who began walking next to him, but his friend looked unfamiliar. "His facial features were moving all around his face. I couldn't put it together. I know this guy like I know my brother." The patient said to the friend, "Who are you?" and his friend answered, "What are you talking about? You've known me for years!" The patient then recognized him, but by his voice only.

On exam in the emergency room, the patient had normal mental status, including naming presidents back to Wilson and 3/3 memory at 5 minutes. He had normal visual fields, except for a **homonymous scotoma in the left inferior quadrant**, and he was able to see through the teardrop-like distortion in his upper fields. In addition, he had **difficulty recognizing faces** in a magazine. For example, when shown a picture of George Bush, he said, "I think I should know him." When shown a picture of Bill Clinton, he said, "Get that thing away from me!" (the patient was a staunch Republican). He was able to name objects other than faces well. Reading and writing abilities were normal, and he did not have achromatopsia. The remainder of his exam was normal.

LOCALIZATION AND DIFFERENTIAL DIAGNOSIS

1. What is the name for disorders in which perception and language are normal but recognition is impaired? Given the findings shown in **bold** above, what is the most likely localization for this patient's lesion(s)?
2. What is the most likely diagnosis?

Discussion

The key symptoms and signs are

- Difficulty recognizing faces
- Visual obscuration of upper fields bilaterally
- Left inferior-quadrant scotoma

1. Impaired recognition in the absence of defects in primary perception or in naming ability is called agnosia (see KCC 19.12). This patient demonstrates impaired ability to recognize faces, with preserved ability to see, recognize, and name other objects. This syndrome is known as prosopagnosia and is usually caused by bilateral lesions of the face recognition areas in the visual association cortex of the inferior occipitotemporal (fusiform) gyri (see Figure 19.13). Associated features can include upper visual field obscuration of the kind seen in this patient because of the proximity to the inferior calcarine regions (see Figure 11.15). This patient's left inferior-quadrant scotoma is most likely related to his prior history of a cerebral infarct 6 years earlier that had caused a left homonymous hemianopia at the time, which subsequently improved.

2. Given his history of atrial fibrillation and treatment with Coumadin, the most likely diagnoses are bilateral inferior occipitotemporal infarcts or hemorrhages.

Clinical Course and Neuroimaging

A **head CT** (Image 19.9, page 962) demonstrated an old right occipital infarct, along with a recent infarct of the left inferior occipitotemporal (fusiform) gyrus. The patient did not arrive in the emergency room within the time window for tPA (see KCC 10.4) but was admitted for further evaluation. On admission, his anticoagulation INR (International Normalized Ratio) was in the therapeutic range. An MRA showed no significant stenoses. His ability to recognize faces improved markedly within 2 days of admission, and the visual distortion in his upper fields resolved more gradually. He was discharged home on continued Coumadin anticoagulation.

CASE 19.10 MUSICAL HALLUCINATIONS

MINICASE

A 21-year-old right-handed man from El Salvador was brought to the emergency room after having a seizure. Beginning 3 years previously, he had developed frequent brief episodes in which he heard music playing and voices. These episodes were often followed by a generalized convulsion. He was living in a rural area at the time, and the episodes stopped on their own after about 3 months without medical treatment. On the day of admission, while at work, the patient again suddenly **heard music playing and voices**. He then **felt dizzy, saw spots, and lost consciousness**. Witnesses reported that he had had a **generalized tonic-clonic seizure** lasting 2 to 3 minutes, followed by **post-ictal** lethargy and generalized headache, but **no clear language deficit**. He had a second identical episode upon arrival at the emergency room. Exam performed a few hours after the last seizure was entirely normal.

LOCALIZATION AND DIFFERENTIAL DIAGNOSIS

1. What anatomical structures could be involved sequentially during this patient's seizures to produce the sequence of reported ictal phenomena? Which side of the brain and what location were likely involved in seizure onset (see KCC 18.2)?

2. What are some possible causes of seizures in this patient?

Discussion

The key symptoms and signs in this case are:

- **Hearing music playing and voices**
- **Dizziness**
- **Seeing spots**
- **Loss of consciousness and generalized convulsion**
- **No clear language deficit post-ictally**

1. Although auditory hallucinations can be caused by lesions in many possible locations along the auditory pathways (see KCC 19.13), only lesions in the cortex would be expected to produce seizures.* Therefore, this patient's seizures probably began in the auditory cortex or auditory association cortex; spread to the vestibular cortex of the parietal operculum, producing dizziness; then to parieto-occipital cortex, producing visual spots; and finally to widespread bilateral brain regions, producing generalized tonic-clonic activity (see KCC 18.2). Musical abilities depend more on the right hemisphere than on the left in most individuals who are not trained musicians (see Table 19.3), and musical hallucinations are more often seen with right rather than left temporal seizures (see KCC 19.13). In addition, left hemisphere seizures are often associated with post-ictal aphasia, which was not present in this patient. Therefore, the most likely site of seizure onset is the right superior temporal gyrus, in the vicinity of Heschl's gyri (primary auditory cortex) or auditory association cortex (see Figure 19.1).

---

*Certain rat strains exhibit seizures triggered by brainstem auditory pathways, but a similar condition has not been reported in humans.

2. The most common cause of nontraumatic early-adult–onset seizures in patients from Central America is CNS cysticercosis (see KCC 5.9; see also Image 5.7D–H). Other possibilities include other CNS infections, a low-grade brain tumor, or a cortical developmental abnormality.

Clinical Course and Neuroimaging

A **brain MRI** revealed an enhancing cyst near the right superior gyrus and Heschl's gyri (Image 19.10A,B, page 965). The patient underwent lumbar puncture and had blood serology sent to confirm the diagnosis of cysticercosis. He was treated with anticonvulsant medications and given a course of the antiparasitic agent praziquantel, together with a brief course of steroids to prevent swelling that can sometimes occur in response to treatment. He did well and was discharged home without further problems.

CASE 19.9 SUDDEN INABILITY TO RECOGNIZE FACES

IMAGE 19.9 Bilateral Inferior Occipitotemporal Infarcts Axial CT image.

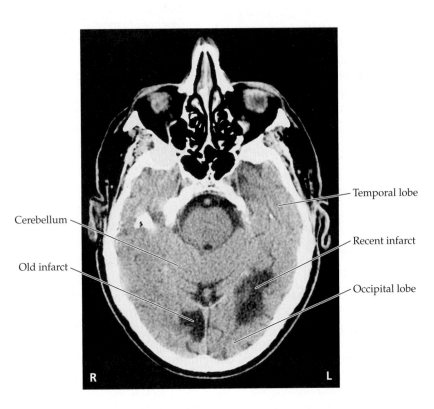

CASE 19.11 PROGRESSIVE DEMENTIA, BEGINNING WITH MEMORY PROBLEMS

CHIEF COMPLAINT

A 76-year-old right-handed woman was referred to a memory disorders clinic because of 4 years of worsening memory problems.

HISTORY

The patient had previously been in good health. Her **memory problems began when she retired as a secretary at age 72 and had progressed slowly ever since.** Her husband reported that she initially had difficulty making certain recipes at home, as well as trouble recalling things she had been told in recent days or weeks. She seemed to recall events from the distant past somewhat better. Her memory gradually grew worse, and she began losing her keys and her pocketbook. She often repeated questions that she had asked a few minutes before. She occasionally had **difficulty distinguishing the letter "O" from the number zero on the touch-tone phone.** Most recently, her husband had been concerned that if she were left four or five blocks from home, she **would get lost.** She also became **somewhat irritable and became angry over little things that would not have bothered her in the past.** For example, one morning she was very insistent on making her bed before going to church. She saw her physician, who ordered a head CT and routine blood tests, including thyroid function tests, VDRL (test for syphilis), B_{12} level, and erythrocyte sedimentation rate, all of which were normal.

PHYSICAL EXAMINATION

Vital signs: T = 98°F, BP = 140/82.

Neck: No bruits.

Lungs: Clear.

Heart: No murmurs.

Abdomen: Soft, nontender.

Extremities: No edema.

Neurologic exam:

MENTAL STATUS: Alert. Named five objects easily, and had no apraxia. No delusions or hallucinations. The **Blessed Dementia Scale (BDS) showed significant memory impairment** (see Table 19.16). She scored 3 on the left column and **11 on the right column, losing points mostly for questions on 5-minute recall, nonpersonal memory, and orientation. Her Activities of Daily Living (ADL) score was 22%.** (Higher scores on the BDS and ADL indicate more severe impairment.)

CRANIAL NERVES: Normal.

MOTOR: Normal tone. No pronator drift. Normal power.

REFLEXES: No grasp, root, suck, or snout reflex.

COORDINATION: Normal on finger-to-nose testing.

GAIT: Normal. Normal tandem gait and hopping on either foot. No Romberg sign.

SENSORY: Normal.

LOCALIZATION AND DIFFERENTIAL DIAGNOSIS

Dysfunction in what anatomical structures could explain the abnormalities shown in **bold** above? What is the most likely diagnosis, and what are some other possibilities?

Discussion

The key symptoms and signs in this case are:

- **Slowly progressive memory impairment, worse for recent memories**
- **Difficulty distinguishing the letter "O" from zero on a touch-tone phone**
- **Tendency to get lost**
- **Irritability**

Impairment of recent memory can be caused by bilateral dysfunction of the medial temporal lobe or medial diencephalic memory systems (see KCC 18.1). This patient's geographic disorientation and her spatial disorientation with regard to O and zero on the telephone (or possibly reading difficulties) suggest possible parietotemporal dysfunction (see KCC 19.7, 19.10). Irritability and personality changes are nonspecific, but they can also be caused by dysfunction of limbic or association cortex. Overall, these findings and the slowly progressive time course fit both anatomically and clinically with early Alzheimer's disease (see Figure 19.15; KCC 19.16). In a patient over the age of 65 with progressive deficits in recent memory, mild temporoparietal dysfunction, no motor abnormalities on neurologic exam, and normal basic blood tests and neuroimaging, as in this patient, Alzheimer's disease is by far the most likely diagnosis. Other possible causes of chronic mental status changes are listed in Table 19.15.

Clinical Course

The patient had an MRI scan of the brain that was unremarkable except for mild atrophy. Follow-up in the months after her first visit was as follows:

- **7 months: BDS left = 2.5, right = 11; ADL = 23%. The patient was still carrying out activities around the house, such as making the beds and doing the cooking, and she was doing volunteer work with her husband at a hospital.**
- **11 months: BDS right = 10; ADL = 36%. Still doing some volunteer work.**
- **17 months: BDS right = 16; ADL = 32%.**
- **25 months: BDS right = 15; ADL = 47%.**
- **31 months: BDS right = 21; ADL = 37%. No longer cooking. Could not be left home alone, and was enrolled in a day care program. No longer wanted to take regular baths, and instead took sponge baths. Neurologic exam and general health still normal, other than her dementia.**
- **37 months: BDS right = 25; ADL = 52%. Increased agitation, with daily episodes of inconsolable screaming and swearing at her husband.**
- **43 months: BDS right = 26; ADL = 59%. Needed assistance in getting dressed. Continued occasional agitation. Husband attending support groups.**
- **47 months: BDS right = 23; ADL = 60%. Having occasional hallucinations. For the first time, her tandem gait and hopping were mildly unsteady on exam. General health remained good, and she was on no medications.**
- **52 months: She became angry and aggressive with increasing frequency and was placed in a nursing home.**
- **70 months: The patient died at age 82, about 10 years after her first symptoms.**

Pathology

The patient's family consented to a postmortem examination, which revealed typical changes of Alzheimer's disease, including cortical atrophy, neuronal loss, and plaques and tangles in a widespread distribution but most prominent in the medial temporal lobes. A typical amyloid plaque and intracytoplasmic neurofibrillary tangle seen on silver staining from the frontal polar cortex of this patient are shown in Image 19.11 (page 966).

Additional Cases

Related cases can be found in other chapters for the following topics: **acute mental status changes** (Cases 5.3, 5.10, 7.1, and 14.8); **chronic mental status changes** (Cases 5.1, 5.9, 15.3, and 16.2); **aphasia** (Cases 7.1, 10.5, 10.6, and 10.8); **hemineglect** (Cases 5.1, 10.2, and 10.11); and **limbic system disorders** (Cases 18.1–18.5). Other relevant cases can be found using the **Case Index** located at the end of the book.

CASE 19.10 MUSICAL HALLUCINATIONS

IMAGE 19.10A,B Cysticercosis Lesion in the Right Superior Temporal Gyrus T1-weighted MRI. (A) Sagittal image. (B) Coronal image with intravenous gadolinium enhancement.

(A)

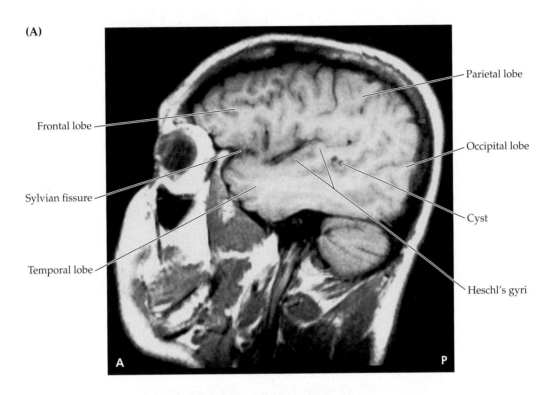

Frontal lobe

Parietal lobe

Sylvian fissure

Occipital lobe

Temporal lobe

Cyst

Heschl's gyri

A   P

(B)

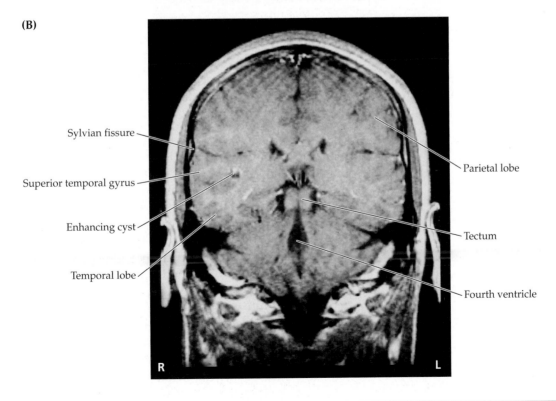

Sylvian fissure

Superior temporal gyrus

Parietal lobe

Enhancing cyst

Temporal lobe

Tectum

Fourth ventricle

R   L

CASE 19.11 PROGRESSIVE DEMENTIA, BEGINNING WITH MEMORY PROBLEMS

IMAGE 19.11 **Plaque and Tangles** Pathology slide from the frontal cortex prepared with a silver stain demonstrating an amyloid plaque and neurofibrillary tangles typical of Alzheimer's disease.

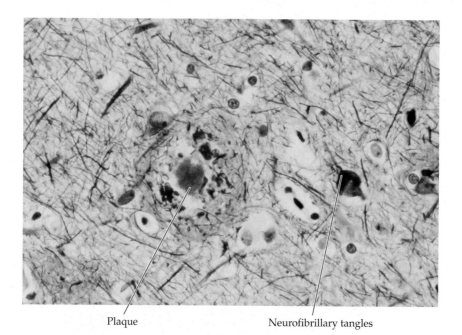

Plaque   Neurofibrillary tangles

Brief Anatomical Study Guide

1. In this chapter we discussed networks involved in higher-order cognitive functions, emphasizing the cerebral cortex. A majority of the cerebral cortical surface is composed of association cortex. Association cortex can be divided into **unimodal (modality-specific) association cortex** and **heteromodal (higher-order) association cortex** (see Figure 19.1; Table 19.2).

2. Although most structures and sensorimotor functions are distributed symmetrically in the brain, there are also marked asymmetries in brain function. The **left hemisphere** is typically dominant for skilled motor functions (handedness) and language, and the **right hemisphere** plays a greater role in attentional mechanisms and spatial analysis (see Table 19.3). The **left hemisphere** is dominant for language in over 95% of right-handers and in over 60% to 70% of left-handers. Language is mediated by a network centered on structures in the dominant (usually left) hemisphere but extending to the right hemisphere as well (see Figure 19.2).

3. **Broca's area**, located in the frontal lobe adjacent to the speech articulatory motor cortex, is important in language formulation. **Wernicke's area**, located in the temporal lobe adjacent to the primary auditory cortex, is important in associating meaning with words. Both Broca's and Wernicke's areas function through interactions between the posterior and anterior language areas via the **arcuate fasciculus** and other connections in the peri-Sylvian region (see Figure 19.2). Figure 19.4 summarizes a simplified

classification scheme for language disorders, or **aphasias**, resulting from different lesions.

4. In Chapter 14 we described the activating systems of the brainstem pontomesencephalic region—the basal forebrain, thalamus, and cortex—essential for maintaining the normal awake state (see Figures 14.7 and 14.8; Table 14.2). In this chapter we discussed mechanisms of attention, touching briefly on the more controversial topic of awareness. In discussing **networks mediating attention**, we reviewed the roles of widespread projection systems, frontal and parietal association cortex, anterior cingulate cortex and limbic pathways, tectal circuits, and other structures, such as the basal ganglia and cerebellum.

5. Although both hemispheres are involved, **the right hemisphere is more important for attentional mechanisms in most individuals** (see Figure 19.7). Lesions of the right hemisphere often cause prominent neglect of the contralateral side. In addition, the right hemishere, particularly the **right parietal region**, is most important in **visual-spatial analysis**.

6. The **frontal lobes** are large in humans, constituting nearly one-third of the cerebral hemispheres. They have three surfaces: lateral, medial, and orbitofrontal (see Figure 19.11). In this chapter we focused on the frontal cortex lying anterior to the motor, premotor, and limbic areas, which is called the **prefrontal cortex** (see Figure 19.1) and consists of higher-order heteromodal association cortex. The prefrontal cortex has **cortical connections** with **heteromodal association cortex** (parietal, occipital, and temporal lobes), **motor association cortex** (frontal lobes), and **limbic cortex** (anterior cingulate and posteromedial orbitofrontal cortex). **Subcortical connections** include the **amygdala** (connected with the orbital and medial portions of the frontal lobes by the uncinate fasciculus; see Figure 18.4B,C), the **hippocampal formation** (via the cingulate gyrus and parahippocampal gyrus; see Figure 18.9), **thalamus** (mediodorsal nucleus, medial pulvinar, and intralaminar nuclei; see Figure 7.8), and the **basal ganglia** (mainly via the head of the caudate nucleus; see Table 16.2; Figure 16.8).

7. Frontal lobe functions are quite diverse and apparently contradictory at times (see Tables 19.8 and 19.9). The frontal lobes are crucial for the sophisticated decisions we make and for the subtle social interactions we continually engage in as normal humans. The functions of the frontal lobes can be classified as important for (1) **restraint**, or inhibition of inappropriate behaviors; (2) **initiative**, or motivation to pursue positive or productive activities; and (3) **order**, or the capacity to correctly perform sequencing tasks and a variety of other cognitive operations.

8. As we discussed in Chapter 11, after arrival at the primary **visual cortex,** visual information is processed in two streams of association cortex (see Figure 19.12). The **dorsal pathways** project to parieto-occipital association cortex. These pathways answer the question "**Where?**" by analyzing motion as well as spatial relationships between objects and between the body and visual stimuli. The **ventral pathways** project to occipitotemporal association cortex. These pathways answer the question "**What?**" by analyzing form, with specific regions identifying colors, faces, letters, and other visual stimuli (see Figure 19.13). The functions of these two streams of higher-order visual information processing are well illustrated by clinical syndromes of the dorsal and ventral visual association pathways (see KCC 19.12).

> **Brief Anatomical Study Guide** (continued)
>
> **9.** We concluded this chapter with a discussion of dementia, because it demonstrates the functional importance of multiple neuroanatomical systems (see Figure 19.15; KCC 19.16) and because discovering new treatments for dementia and other presently incurable neurologic disorders may lie just over the horizon if we continue, through study and investigation, to increase our understanding of the brain.

References

General References

Feinberg T, Farah M. 2003. *Behavioral Neurology and Neuropsychology.* McGraw-Hill Professional, New York.

Heilman KM, Valenstein E (eds.). 2003. *Clinical Neuropsychology.* 4th ed. Oxford University Press, New York.

Mesulam MM. 2000. *Principles of Behavioral and Cognitive Neurology.* 2nd ed. Oxford University Press, New York.

Miller BL, Boeve BF. 2009. *The Behavioral Neurology of Dementia.* Cambridge University Press, New York.

Pincus JH, Tucker GJ. 2002. *Behavioral Neurology.* Oxford University Press, New York.

Reviews of Aphasia

Benson DF, Geschwind N. 1985. Aphasia and related disorders: A clinical approach. In *Principles of Behavioral Neurology*, MM Mesulam (ed.). FA Davis, Philadelphia.

Critchley M. 1964. The Broca-Dax controversy. *Rev Neurol (Paris)* 110: 73.

Damasio AR. 1992. Aphasia. *N Engl J Med* 326 (8): 531–539.

Jordan LC, Hillis AE. 2006. Disorders of speech and language: aphasia, apraxia and dysarthria. *Curr Opin Neurol* 19 (6): 580–585.

Lazar RM, Antoniello D. 2008. Variability in recovery from aphasia. *Curr Neurol Neurosci Rep*: 497–502.

Broca's Aphasia

Alexander MP, Naesser MA, Palumbo C. 1990. Broca's area aphasia: Aphasia after lesions including the frontal operculum. *Neurology* 40 (2): 353–362.

Keller SS, Crow T, Foundas A, Amunts K, Roberts N. 2009. Broca's area: nomenclature, anatomy, typology and asymmetry. *Brain Lang* 09 (1): 29–48.

Transcortical Aphasia

Freedman M, Alexander MP, Naeser MA. 1984. Anatomic basis of transcortical motor aphasia. *Neurology* 34 (4): 409–417.

Grossi D, Trojano L, Chiacchio L, Soricelli A, Mansi L, Postiglione A, Salvatore M. 1991. Mixed transcortical aphasia: Clinical features and neuroanatomical correlates. *Eur Neurol* 31 (4): 204–211.

Conduction Aphasia

Bernal B, Ardila A. 2009. The role of the arcuate fasciculus in conduction aphasia. *Brain* 132 (Pt. 9): 2309–2316.

Damasio H, Damasio AR. 1980. The anatomical basis of conduction aphasia. *Brain* 103 (2): 337–350.

Kempler D, Metter EJ, Jackson CA, Hanson WR, Riege WH, Mazziotta JC, Phelps ME. 1998. Disconnection and cerebral metabolism. *Arch Neurol* 45 (3): 275–279.

Subcortical Aphasia

Crosson B, Parker JC, Kim AK, Warren RL, Kepes JJ, Tully R. 1986. A case of thalamic aphasia with postmortem verification. *Brain Lang* 29 (2): 301–314.

Hillis AE, Barker PB, Wityk RJ, Aldrich EM, et al. 2004. Variability in subcortical aphasia is due to variable sites of cortical hypoperfusion. *Brain and Lang* 89 (3): 524–530.

Mega MS, Alexander MP. 1994. Subcortical aphasia: The core profile of capsulostriatal infarction. *Neurology* 44 (10): 1824–1829.

Gerstmann's Syndrome

Benton AL. 1992. Gerstmann's syndrome. *Arch Neurol* 49 (5): 445–447.

Levine, DN, Mani, RB, Calvanio, R. 1988. Pure agraphia and Gerstmann's syndrome as a visuospatial-language dissociation: an experimental case study. *Brain and Language* 35: 172–196.

Pearce JMS. 1996. Gerstmann's syndrome. *J Neurol Neurosurg Psychiatry* 61 (1): 56.

Alexia without Agraphia

Quint DJ, Gilmore JL. 1992. Alexia without agraphia. *Neuroradiology* 34 (3): 210–214.

Sheldon CA, Malcolm GL, Barton JJ. 2008. Alexia with and without agraphia: an assessment of two classical syndromes. *Can J Neurol Sci* 35 (5): 616–624.

Aphasia and Apraxia

Benson DF, Geschwind N. 1985. Aphasia and related disorders: A clinical approach. In *Principles of Behavioral Neurology*, MM Mesulam (ed.). FA Davis, Philadelphia.

Jordan LC, Hillis AE. 2006. Disorders of speech and language: aphasia, apraxia and dysarthria. *Curr Opin Neurol* 19 (6): 580–585.

Petreska B, Adriani M, Blanke O, Billard AG. 2007. Apraxia: a review. *Prog Brain Res* 164: 61–83.

Callosal Disconnection Syndromes

Bogen, JE. 2003. The callosal syndromes. In *Clinical Neuropsychology,* 4th ed., KM Heilman and E Valenstein (eds.), Chapter 14. Oxford University Press, New York.

Devinsky O, Laff R. 2003. Callosal lesions and behavior: history and modern concepts. *Epilepsy Behav* 4 (6): 607–617.

Jea A, Vachhrajani S, Widjaja E, et al. 2008. Corpus callosotomy in children and the disconnection syndromes: a review. *Childs Nerv Syst* 24 (6): 685–692.

Right Parietal Lobe and Attention Mechanisms

Heilman KH, Watson RT, Valenstein E. 2003. Neglect and related disorders. In *Clinical Neuropsychology,* 4th ed., KM Heilman and E Valenstein (eds.), Chapter 13. Oxford University Press, New York.

Hier DB, Mondlock J, Caplan LR. 1983. Behaviorial abnormalities after right hemisphere stroke. *Neurology* 33 (3): 337–344.

Hillis AE. 2006. Neurobiology of unilateral spatial neglect. *Neuroscientist* 12 (2): 153–163.

Luauté J, Halligan P, Rode G, Rossetti Y, Boisson D. 2006. Visuo-spatial neglect: a systematic review of current interventions and their effectiveness. *Neurosci Biobehav Rev* 30 (7): 961–982.

Mesulam MM. 1985. Attention, confusional states, and neglect. In *Principles of Behavioral Neurology*, MM Mesulam (ed.). FA Davis, Philadelphia.

Frontal Lobes

Bechara A, Damasio AR, Damasio H, Anderson SW. 1994. Insensitivity to future consequences following damage to human prefrontal cortex. *Cognition* 50 (1–3): 7–15.

Bechara A, Van Der Linden M. 2005. Decision-making and impulse control after frontal lobe injuries. *Curr Opin Neurol* 18 (6): 734–739.

Bogousslavsky J. 1994. Frontal stroke syndromes. *Eur Neurol* 34 (6): 206–215.

Damasio AR. 1994. Unpleasantness in Vermont—Phineas P. Gage. In *Descartes' Error: Emotion, Reason, and the Human Brain*, Part 1. GP Putnam, New York.

Damasio AR, Anderson SW. 2003. The frontal lobes. In *Clinical Neuropsychology,* 4th ed., KM Heilman and E Valenstein (eds.), Chapter 15. Oxford University Press, New York.

Duncan J. 2005. Frontal lobe function and general intelligence: why it matters. *Cortex* 41 (2): 215–217.

Fisher CM. 1983. Honored Guest Presentation: abulia minor vs. agitated behavior. *Clin. Neurosurg* 31: 9–31.

Goldberg E, Bougakov D. 2005. Neuropsychologic assessment of frontal lobe dysfunction. *Psychiatr Clin North Am* 28 (3): 567–580.

Kövari E. 2009. Neuropathological spectrum of frontal lobe dementias. *Front Neurol Neurosci* 24: 149–159.

Lhermitte F. 1983. Utilization behavior and its relation to lesions of the frontal lobes. *Brain* 106 (Pt. 2): 237–255.

Rossi AF, Pessoa L, Desimone R, Ungerleider LG. 2009. The prefrontal cortex and the executive control of attention. *Exp Brain Res* 192 (3): 489–497.

Schott, J.M., Rossor, M.N., 2003. The grasp and other primitive reflexes. *Journal of Neurology, Neurosurgery and Psychiatry* 74: 558–560.

Working Memory

Cowan N. 2008. What are the differences between long-term, short-term, and working memory? *Prog Brain Res* 169: 323–338.

Dash PK, Moore AN, Kobori N, Runyan JD. 2007. Molecular activity underlying working memory. *Learn Mem* 14 (8): 554–563.

Goldman-Rakic PS. 1992. Working memory and the mind. *Sci Am* 267 (3): 110–117.

Linden DE. 2007. The working memory networks of the human brain. *Neuroscientist* 13 (3): 257–267.

Soto D, Hodsoll J, Rotshtein P, Humphreys GW. 2008. Automatic guidance of attention from working memory. *Trends Cogn Sci* 12 (9): 342–348.

Mental Imagery and Blindsight

Danckert J, Rossetti Y. 2005. Blindsight in action: what can the different sub-types of blindsight tell us about the control of visually guided actions? *Neurosci Biobehav Rev* 29 (7): 1035–1046.

Georgopoulos AP, Lurito JT, Petrides M, Schwartz AB, Massey JT. 1989. Mental rotation of the neuronal population vector. *Science* 243 (4888): 234–236.

Goodale MA, Milner AD, Jakobson LS, Carey DP. 1991. A neurological dissociation between perceiving objects and grasping them. *Nature* 349 (6305): 154–156.

Kosslyn SM, Thompson WL, Kim IJ, Alpert NM. 1995. Topographical representations of mental images in primary visual cortex. *Nature* 378 (6556): 496–498.

Marshall JC, Halligan PW. 1988. Blindsight and insight in visuo-spatial neglect. *Nature* 336 (6201): 766–767.

Naccache L. 2005. Visual phenomenal consciousness: a neurological guided tour. *Prog Brain Res* 150: 185–195.

Stoerig P. 2006. Blindsight, conscious vision, and the role of primary visual cortex. *Prog Brain Res* 155: 217–234.

Stoerig P, Cowey A. 2007. Blindsight. *Curr Biol* 17 (19): R822–R824.

Weiskrantz L. 2009. Is blindsight just degraded normal vision? *Exp Brain Res* 192 (3): 413–416.

Visual Hallucinations and Related Phenomena

Tekin S, Cummings J. 2003. Hallucinations and Related Conditions. In *Clinical Neuropsychology,* 4th ed., KM Heilman and E Valenstein (eds.), Chapter 17. Oxford University Press, New York.

Wilkinson F. 2004. Auras and other hallucinations: windows on the visual brain. *Prog Brain Res* 144: 305–320.

Syndromes of Visual Association Cortex

Allison T, McCarthy G, Nobre A, Puce A, Belger A. 1994. Human extrastriate visual cortex and the perception of faces, words, numbers, and colors. *Cerebral Cortex* 4 (5): 544–554.

Bauer RM. 2003. Agnosia. In *Clinical Neuropsychology*, 4th ed., KM Heilman and E Valenstein (eds.), Chapter 12. Oxford University Press, New York.

Bornstein B, Kidron DP. 1959. Prosopagnosia. *J Neuro Neurosurg Psychiatry* 22: 124.

Damasio AR. 1985. Disorders of complex visual processing: Agnosias, achromatopsia, Balint's syndrome, and related difficulties of orientation and construction. In *Principles of Behavioral Neurology*, MM Mesulam (ed.). FA Davis, Philadelphia.

Damasio AR, Damasio H, Van Hoesen GW. 1982. Prosopagnosia: Anatomic basis and behavioral mechanisms. *Neurology* 32 (4): 331–341.

Grüter T, Grüter M, Carbon CC. 2008. Neural and genetic foundations of face recognition and prosopagnosia. *J Neuropsychol* 2 (Pt. 1): 79–97.

Hecaen H, Angelergues R. 1962. Agnosia for faces (prosopagnosia). *Arch Neurol* 7: 24.

Miller NR. 1985. *Clinical Neuro-Ophthalmology*, Vol. 2. Williams & Wilkins, Baltimore, MD.

Puce A, Allison T, Asgari M, Gore JC, McCarthy G. 1996. Differential sensitivity of human visual cortex to faces, letterstrings, and textures: A functional magnetic resonance imaging study. *J Neurosci* 16 (16): 5205–5215.

Zeki S. 1990. A century of cerebral achromatopsia. *Brain* 113 (Pt. 6): 1721–1777.

Musical Hallucinations

Evers S, Ellger T. 2004. The clinical spectrum of musical hallucinations. *J Neurol Sci* 227 (1): 55–65.

Patel AD. 2003. Language, music, syntax and the brain. *Nat Neurosci* 6 (7): 674–681.

Williams VG, Tremont G, Blum AS. 2008. Musical hallucinations after left temporal lobectomy. *Cogn Behav Neurol* 21 (1): 38–40.

Disorders of Consciousness

Fort P, Bassetti CL, Luppi PH. 2009. Alternating vigilance states: new insights regarding neuronal networks and mechanisms. *Eur J Neurosci* 29 (9): 1741–1753.

Gregoriou GG, Gotts SJ, Zhou H, Desimone R. 2009. Long-range neural coupling through synchronization with attention. *Prog Brain Res* 176: 35–45.

Jones BE. 2008. Modulation of cortical activation and behavioral arousal by cholinergic and orexinergic systems. *Ann N Y Acad Sci* 1129: 26–34.

Kinomura S, Larsson J, Gulyas B, Roland PE. 1996. Activation by attention of the human reticular formation and thalamic intralaminar nuclei. *Science* 271 (5248): 512–515.

Laureys S, Tononi G (eds.). 2009. *The Neurology of Consciousness: Cognitive Neuroscience and Neuropathology*. Elsevier, Ltd.

Mesulam MM. 1985. Biology of the attentional matrix. In *Principles of Behavioral Neurology*, MM Mesulam (ed.). FA Davis, Philadelphia.

Parasuraman R (ed.). 1998. *The Attentive Brain*. The MIT Press, Cambridge, MA.

Steriade M, Curro Dossi R, Contreras D. 1993. Electrophysiological properties of intralaminar thalamocortical cells discharging rhythmic (~40 Hz) spike-bursts at ~1000 Hz during waking and rapid eye movement sleep. *Neuroscience* 56 (1): 1–9.

Uhlhaas PJ, Singer W. 2006. Neural synchrony in brain disorders: relevance for cognitive dysfunctions and pathophysiology. *Neuron* 52 (1): 155–168.

Dementia

Cummings JL. 2004. Alzheimer's Disease. *N Engl J Med* 351: 56.

Josephs KA. 2008. Frontotemporal dementia and related disorders: deciphering the enigma. *Ann Neurol* 64 (1): 4–14.

Rademakers R, Rovelet-Lecrux A. 2009. Recent insights into the molecular genetics of dementia. *Trends in Neurosciences* 32 (8): 451–561.

Snowden, J, Neary D, et al. 2007. Frontotemporal lobar degeneration: clinical and pathological relationships. *Acta Neuropathol* 114 (1): 31–38.

Wadia PM, Lang AE. 2007. The many faces of corticobasal degeneration. *Parkinsonism Relat Disord* 13 Suppl 3: S336–S340.

Epilogue

A Simple Working Model of the Mind

Where is the mind, and what is the mind? These questions have taunted scientists and philosophers throughout human history. Although we cannot yet answer these questions with certainty, investigation of the nervous system allows at least tentative conjectures in this realm.

Basic Assumptions

Although some would argue otherwise, the burden of evidence currently available suggests that the mind is manifested through ordinary physical processes located within the body. Note that these first two fundamental conjectures about where the mind is (in the body) and what the mind is (normal physical processes) remain hypotheses, perhaps with growing evidence in their favor, yet remaining unproven nonetheless. Within the body, one can conjecture further, the mind appears to be manifested in the nervous system. Interactions with the rest of the body and with the external environment are clearly important, yet most evidence suggests that the main functions of the mind are carried out within the nervous system. Finally, there appears to be a gradation in the importance of different parts of the nervous system in what we consider to be the mind. For example, although the peripheral nervous system and spinal cord play a significant role in channeling (and even in modulating) inputs and outputs to the remainder of the nervous system, the brain is likely more important to the mind. It should be emphasized, however, that there is no sharp division between "mind" and "nonmind" parts of the nervous system. A gradient of relative importance to mind may exist within the nervous system, yet reciprocal interactions ensure that the tapering of the gradient extends to all parts and possibly even to structures outside the nervous system.

We will now briefly summarize some of the abundant evidence supporting these conjectures before moving on to a more specific model of the mind. As for the location and nature of the mind, we know that physical injuries to the brain cause changes in mind function. Furthermore, changes in mind function vary from mild disturbances in cognition, to profound abnormalities in thought, to brain death, and they are related to both the anatomical sites involved and the mechanisms of the injury. We have seen numerous examples of this kind in the patients discussed throughout this book, and an overwhelming number of other examples exist in the literature.

In addition to negative effects of lesions, positive evidence for the participation of specific regions of the nervous system in mind functions comes from a multitude of studies employing electrophysiological recordings, functional neuroimaging, brain stimulation, and other methods. Lesion studies and recordings of neural function come not just from humans, but from other animals as well. In this regard, another line of evidence supporting the dependence of mind on brain comes from the parallel evolutionary trends in the complexity of brain and mind. However, despite these and a mountain of other studies, some lingering doubts remain as to whether the mind is truly an ordinary physical process manifested within the nervous system.

Summary and Model of Mind Functions

Perhaps the reason for these lingering doubts is that many processes of the mind, particularly certain aspects of consciousness and emotions, remain difficult to explain fully in neurophysiological terms. Plausible and testable hypotheses for the relationship between brain activity and conscious thought are still under development. However, as understanding of the brain increases, functions of the mind once thought beyond the ken of scientific investigation come more and more firmly into the arena of neuroscience. Examples include memory, language, planning, and attention, to name a few. With continued investigation, my opinion is that consciousness will eventually migrate as well into the domain of accepted neurophysiological phenomena. In the interim, and as a conclusion to this neuroanatomy text, it may be useful to consider an overall framework for summarizing the functions of the nervous system and the mind, encompassing both those mechanisms that are relatively well understood and those still under intense investigation.

The first task in this summary is to discuss inputs and outputs (see figure). In the earlier chapters of this book, we discussed numerous sensory and motor systems. Different sensory inputs include vision, somatic sensation, hearing, smell, taste, and vestibular sense, as well as various chemical, mechanical, and other signals arising from the body's internal milieu. Similarly, numerous different effector pathways leave the nervous system, including motor outputs to skeletal muscle, autonomic outputs to smooth muscle and glands, and neuroendocrine outputs. These sensory and motor pathways are organized in multiple parallel channels entering and leaving the nervous system and subserving different sensory and motor functions.

Information processing in sensory and motor systems is hierarchically organized (see figure). For example, primary somatosensory information is processed at the most rudimentary level by the receptor cell and primary somatosensory neuron entering the spinal cord. This information is refined further and integrated with inputs and influences from other neurons at successively higher-order levels of processing in the nervous system, including the brainstem, thalamus, primary somatosensory cortex, unimodal association cortex, and heteromodal association cortex. In motor systems, similar hierarchical processing occurs, but in reverse. For example, higher-order signals for motor planning that arise from heteromodal association cortex are conveyed to successively lower levels of processing in premotor cortex, primary motor cortex, and lower motor neurons in the spinal cord before traveling to the periphery.

Information flow in the hierarchically organized sensory and motor pathways is not strictly linear, however. Direct connections exist between sensory and motor systems beginning at the spinal cord and continuing throughout successively higher levels of processing. In addition, from the spinal cord and upward, chains of interneurons of varying complexity further process sensory and motor information at various levels and carry information between these systems. Numerous feed-forward and feedback loops occur between higher-level and lower-level sensory and motor systems, involving both local and long-range network interactions. The integrated action of these hierarchically organized sensory and motor systems is capable of fantastic feats of information processing, including visual recognition of abstract shapes, programming of complex motor tasks in response to the environment, understand-

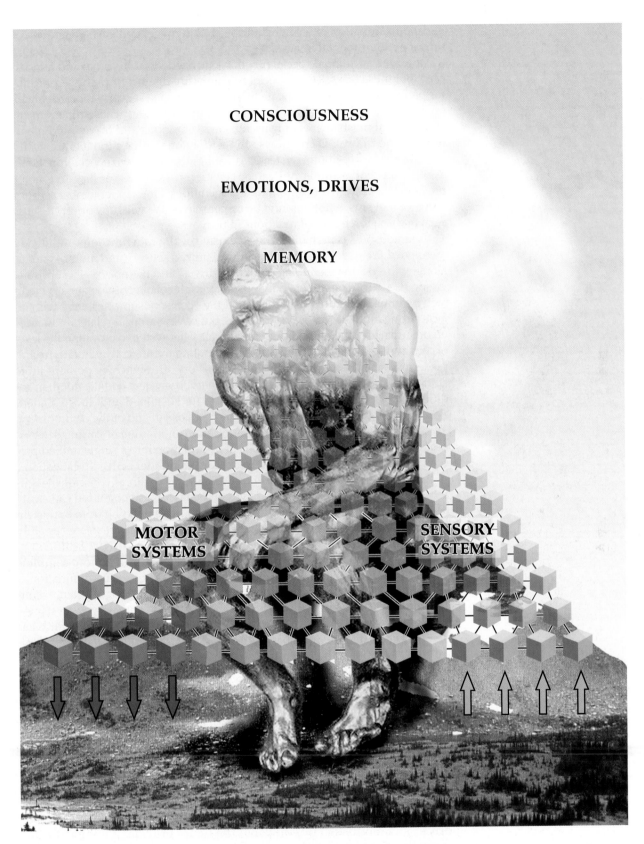

A Working Model of the Mind Parallel interconnected and hierarchically organized sensory and motor systems receive inputs, generate outputs, and perform internal processing on multiple levels, from relatively simple to highly abstract. Three additional special functions—consciousness, emotions and drives, and memory—act on these systems in a widely distributed manner, especially at the highest levels of processing.

ing and formulation of written and spoken language, and even generation of sensory–motor mental images.

However, additional special functions are needed for the brain and mind to carry out these tasks. These special functions are distinct from the sensorimotor pathways, yet they permeate them and influence sensory–motor systems on many levels (represented by the cloud in the figure). These functions include neural mechanisms of consciousness, emotions and drives, and memory (see the figure). Each of these three functions utilizes the hierarchically organized sensory–motor systems, as well as the other two special functions, as substrate. For example, one can be conscious of various sensory and motor phenomena, as well as of emotions and memories. One can remember sensory and motor events, emotions, and being conscious. We will now briefly discuss each of these three special functions in turn.

Consciousness is perhaps the most elusive of all brain functions. As already mentioned, it utilizes the hierarchically organized sensory and motor systems, along with systems mediating emotions, drives, and memory as its substrate, giving rise to the content of consciousness. The level of consciousness is mediated by brain structures we have called the "consciousness system," with effects permeating the forebrain structures, but having arousal-related influences on other parts of the nervous system as well. This system controlling the level of consciousness includes mechanisms promoting the alert, awake state. In simple terms, alertness could be considered the "on–off" switch, or perhaps more appropriately the "dimmer" switch, of the nervous system and the mind. In addition, consciousness involves mechanisms for sustained and directed attention. Finally, consciousness involves awareness of self and of the environment. This awareness requires integration, association, or binding of various sensory, motor, mnemonic, and emotional bits of information into a single unified percept, and the mechanisms for this process are still under active investigation.

Emotions and drives are almost as difficult to explain as consciousness itself. The effects of emotions and drives on behavior and cognition can easily be observed, and so-called limbic system structures extending from brainstem to diencephalon, to basal ganglia, amygdala, and limbic cortex, are certainly involved. Yet the experience of emotions and drives is still very difficult to explain in neuroscientific terms, and it may need to await a more complete mechanistic understanding of conscious awareness for its full illumination.

Memory, defined most broadly as a change in behavior of a system resulting from its past experience, affects the function of the entire nervous system—indeed, of the whole body. Within the nervous system, abundant evidence has demonstrated that memory occurs as a result of physiological and anatomical changes influencing communication between neurons. A specific kind of memory, called explicit or declarative memory, involves the conscious recollection of experiences. Persistence of explicit memories for more than a few minutes depends on normal functioning of the medial temporal lobes and medial diencephalon. This dependence suggests a critical role for these structures in conscious memory and may provide further clues to understanding the mechanisms of consciousness.

It seems plausible that the mind arises from the integrated actions of all the components we have just discussed. Thus, the mind involves many parallel pathways of sensory–motor information processing occurring on both a conscious and an unconscious level, invoking memories, motivated by drives, and modulated by emotions. The figure provides a framework and summary of the integrated actions of these systems. By continuing to investigate the nervous system, perhaps someday soon we can fill in some of the gaps in our understanding and reach a more complete picture of the relationship between brain activity and conscious and unconscious thought.

Case Index
by signs, symptoms, and diagnoses